Developments in Sedimentology
Volume 64

Trace Fossils as Indicators of Sedimentary Environments

Developments in Sedimentology

Volume 64

Trace Fossils as Indicators of Sedimentary Environments

Editors

Dirk Knaust
Statoil ASA
Stavanger
Norway

Richard G. Bromley
Geological Museum—SNM
Copenhagen
Denmark

ELSEVIER

AMSTERDAM • BOSTON • HEIDELBERG • LONDON • NEW YORK • OXFORD
PARIS • SAN DIEGO • SAN FRANCISCO • SINGAPORE • SYDNEY • TOKYO

Elsevier
Radarweg 29, PO Box 211, 1000 AE Amsterdam, The Netherlands
The Boulevard, Langford Lane, Kidlington, Oxford OX5 1GB, UK
30 Corporate Drive, Suite 400, Burlington, MA 01803, USA
525 B Street, Suite 1900, San Diego, CA 92101-4495, USA

First edition **2012**

British Library Cataloguing in Publication Data
A catalogue record for this book is available from the British Library

Library of Congress Cataloging-in-Publication Data
A catalog record for this book is available from the Library of Congress

ISBN: 978-0-444-53813-0
ISSN: 0070-4571

Printed and bound by CPI Group (UK) Ltd, Croydon, CR0 4YY

Transferred to digital print 2013

Dedication

To my parents, my spouse Eva-Maria and my children
Per Ragnar and Solveig Alida, without their effort
this book would not exist.
Dirk Knaust.

Contents

Part II
Continental and Glacial Systems

Part III
Shallow-Marine Siliciclastic Systems

Part V
Marine Carbonate Systems

Part VI
Ichnology in Hydrocarbon-Reservoir and Aquifer Characterization

Contributors

Baniak, Greg (837), Department of Earth & Atmospheric Sciences, University of Alberta, Edmonton, Alberta, Canada

Bann, Kerrie L. (103, 157, 563, 607), Ichnofacies Analysis Inc., Calgary, Alberta, Canada

Baucon, Andrea (3), UNESCO Geopark Naturtejo Meseta Meridional, Geology and Palaeontology Office, Centro Cultural Raiano, Idanha-a-Nova, Portugal; and Università degli Studi di Milano, Dipartimento di Scienze della Terra, Via Mangiagalli, Milano, Italy

Beatty, Tyler W. (807), Department of Geology and Geophysics, University of Calgary, Calgary, Alberta, Canada

Benner, Jacob S. (299), Department of Earth and Ocean Sciences, Tufts University, Medford, Massachusetts, USA

Bordy, Emese (3), Geology Department, Rhodes University, Grahamstown, South Africa

Bottjer, David J. (807), Department of Earth Sciences, University of Southern California, Los Angeles, California, USA

Bromley, Richard G. (139, 419), Geological Museum—SNM, Copenhagen, Denmark

Brustur, Titus (3), National Institute of Marine Geology and Geo-ecology (GEOECOMAR), Bucharest, Romania

Buatois, Luis A. (3, 195, 299, 329, 379, 529), Department of Geological Sciences, University of Saskatchewan, Saskatoon, Saskatchewan, Canada

Catuneanu, Octavian (157), Department of Earth and Atmospheric Sciences, University of Alberta, Edmonton, Alberta, Canada

Chaplin, James R. (807), Oklahoma Geological Survey, University of Oklahoma, Norman, Oklahoma, USA

Cunningham, Kevin J. (869), U.S. Geological Survey, Davie, Florida, USA

Cunningham, Tyron (3), International Society of Professional Trackers, Missouri, Cameron, USA

Curran, H. Allen (705, 869), Department of Geosciences, Smith College, Northampton, Massachusetts, USA

Dashtgard, Shahin E. (103, 157, 273, 463, 563), Applied Research in Ichnology and Sedimentology (ARISE), Department of Earth Sciences, Simon Fraser University, Burnaby, British Columbia, Canada

de Gibert, Jordi M. (441), Departament d'Estratigrafia, Paleontologia I Geociències Marines, Universitat de Barcelona, Barcelona, Spain

De, Chirananda (3), Palaeontology Division-1, CHQ, Geological Survey of India, Kolkata, India

Desjardins, Patricio R. (529), Shell International Exploration and Production Inc., Houston, Texas, USA

Domènech, Rosa (441), Departament d'Estratigrafia, Paleontologia I Geociències Marines, Universitat de Barcelona, Barcelona, Spain

Dronov, Andrei V. (705), Russian Academy of Sciences, Geological Institute, Moscow, Russia

Duffin, Christoffer (3), 146 Church Hill Road, Sutton, Surrey, United Kingdom

Ekdale, Allan A. (139, 419), Department of Geology and Geophysics, University of Utah, Salt Lake City, Utah, USA

Felletti, Fabrizio (3), Università degli Studi di Milano, Dipartimento di Scienze della Terra, Via Mangiagalli, Milano, Italy

Gaigalas, Algirdas (299), Department of Geology and Mineralogy, Vilnius University, Čiurlionio, Vilnius, Lithuania

Gaillard, Christian (3), Paléoenvironments et Paléobiosphère, UFR, Sciences de la Terre, Université claud Bernard Lyon 1, Villeurbanne cedex, France

Genise, Jorge F. (329), CONICET, División Icnología, Museo Argentino de Ciencias Naturales, Buenos Aires, Argentina

Gingras, Murray K. (103, 273, 463, 563, 807, 837), Ichnology Research Group, Department of Earth and Atmospheric Sciences, University of Alberta, Edmonton, Alberta, Canada

Gordon, John (837), Husky Energy, Calgary, Alberta, Canada

Greene, Sarah E. (807), Department of Earth Sciences, University of Southern California, Los Angeles, California, USA

Hovikoski, Jussi (837), Geological Survey of Denmark and Greenland (GEUS), Copenhagen, Denmark

Hu, Bin (3), Key Laboratory of Biogenic Traces and Sedimentary Minerals of Henan Province, Henan Polytechnic University, Jiaozuo City, Henan Province, China

Hu, Lei (3), Key Laboratory of Biogenic Traces and Sedimentary Minerals of Henan Province, Henan Polytechnic University, Jiaozuo City, Henan Province, China

Hubbard, Stephen M. (607), Department of Geoscience, University of Calgary, Calgary, Alberta, Canada

Hutchings, Pat (751), Australian Museum, Sydney, New South Wales, Australia

Jensen, Sören (3), Área de Paleontología, Facultad de Ciencias, Universidad de Extremadura, Badajoz, Spain

Kazakauskas, Vaidotas (299), Institute of Geology and Geography, Vilnius, Lithuania

Knaust, Dirk (3, 79, 139, 157, 245, 705), Statoil ASA, Stavanger, Norway

Konhauser, Kurt O. (837), Department of Earth & Atmospheric Sciences, University of Alberta, Edmonton, Alberta, Canada

La Croix, Andrew (837), Department of Earth Sciences, Simon Fraser University, Burnaby, British Columbia, Canada

Lemiski, Ryan (837), Talisman Energy Inc., Calgary, Alberta, Canada

Lockley, Martin (3), Dinosaur Tracks Museum, University of Colorado Denver, Denver, Colorado, USA

Lowe, Pat (3), Backroom press, Broome, Western Australia, Australia

MacEachern, James A. (103, 157, 463, 563, 607), Applied Research in Ichnology and Sedimentology (ARISE), Department of Earth Sciences, Simon Fraser University, Burnaby, British Columbia, Canada

MacNaughton, Robert B. (195), Geological Survey of Canada, Natural Resources Canada, Calgary, Alberta, Canada

Martindale, Rowan C. (807), Department of Earth Sciences, University of Southern California, Los Angeles, California, USA

Martinell, Jordi (441), Departament d'Estratigrafia, Paleontologia I Geociències Marines, Universitat de Barcelona, Barcelona, Spain

Mata, Scott A. (807), Department of Earth Sciences, University of Southern California, Los Angeles, California, USA

Mayor, Adrienne (3), Classics Department, Stanford University, Stanford, California, USA

Mayoral, Eduardo (3), Departamento de Geodinámica y Paleontología, Facultad de Ciencias Experimentales, Campus de El Carmen, Universidad de Huelva, Huelva, Spain

McHugh, Luke P. (807), Department of Earth and Atmospheric Sciences, University of Alberta, Edmonton, Alberta, Canada

Melchor, Ricardo N. (329), INCITAP (CONICET and Universidad Nacional de La Pampa), Santa Rosa, La Pampa, Argentina

Mendoza, Carl (837), Department of Earth & Atmospheric Sciences, University of Alberta, Edmonton, Alberta, Canada

Mikuláš, Radek (3), Institute of Geology, Academy of Sciences of the Czech Republic, Praha, Czech Republic

Mángano, M. Gabriela (195, 299, 379, 529), Department of Geological Sciences, University of Saskatchewan, Saskatoon, Saskatchewan, Canada

Muttoni, Giovanni (3), Università degli Studi di Milano, Dipartimento di Scienze della Terra, Via Mangiagalli, Milano, Italy

Neto de Carvalho, Carlos (3), UNESCO Geopark Naturtejo Meseta Meridional, Geology and Palaeontology Office, Centro Cultural Raiano, Idanha-a-Nova, Portugal

Netto, Renata G. (299), Geology Graduate Program, Universidade do Vale do Rio dos Sinos, São Leopoldo, Rio Grande do Sul, Brazil

Pemberton, S. George (3, 103, 157, 463, 563, 807, 837), Ichnology Research Group, Department of Earth and Atmospheric Sciences, University of Alberta, Edmonton, Alberta, Canada

Pollard, John (3), School of Earth, Atmospheric and Environmental Sciences, University of Manchester, United Kingdom

Polo, Camilo (837), Department of Earth & Atmospheric Sciences, University of Alberta, Edmonton, Alberta, Canada

Ranger, Michael J. (463), 808 West Chestermere Drive, Chestermere, Alberta, Canada

Ridge, John C. (299), Department of Earth and Ocean Sciences, Tufts University, Medford, Massachusetts, USA

Rindsberg, Andrew K. (3, 45), University of West Alabama, Livingston, USA; and Department of Biological & Environmental Sciences, University of West Alabama, Livingston, Alabama, USA

Santos, Ana (3), Departamento de Geodinámica y Paleontología, Facultad de Ciencias Experimentales, Campus de El Carmen, Universidad de Huelva, Huelva, Spain

Savrda, Charles E. (777), Department of Geology and Geography, Auburn University, Auburn, Alabama, USA

Schoengut, Jesse A. (807), Department of Earth and Atmospheric Sciences, University of Alberta, Edmonton, Alberta, Canada

Schoengut, Jesse (463), Canadian Natural Resources Limited Calgary, Alberta, Canada

Scott, Jennifer J. (379), Department of Earth & Atmospheric Sciences, University of Alberta, Edmonton, Alberta, Canada

Seike, Koji (3), Department of Earth and Planetary Science, The University of Tokyo, Bunkyo-ku, Tokyo, Japan

Song, Hui-bo (3), Key Laboratory of Biogenic Traces and Sedimentary Minerals of Henan Province, Henan Polytechnic University, Jiaozuo City, Henan Province, China

Sukop, Michael C. (869), Department of Earth and Environment, Florida International University, Miami, Florida, USA

Tapanila, Leif (751), Department of Geosciences, Idaho State University, Pocatello, Idaho, USA; and Earth Sciences Division, Idaho Museum of Natural History, Pocatello, Idaho, USA

Tonkin, Nicola S. (507), Department of Earth Sciences, Memorial University, St. John's, Newfoundland and Labrador, Canada

Turner, Susan (3), Geology & Palaeontology, Queensland Museum and Monash University School of Geosciences, Hendra, Queensland, Australia

Uchman, Alfred (3, 299, 643, 673), Institute of Geological Sciences, Jagiellonian University, Kraków, Poland

Umazano, Aldo M. (329), Universidad Nacional de La Pampa, Santa Rosa, La Pampa, Argentina

Wang, Yuan-yuan (3), Key Laboratory of Biogenic Traces and Sedimentary Minerals of Henan Province, Henan Polytechnic University, Jiaozuo City, Henan Province, China

Wetzel, Andreas (643, 673), Geologisch-Paläontologisches Institut der Universität, Basel, Switzerland

Wisshak, Max (213), Senckenberg am Meer, Marine Research Department, Wilhelmshaven, Germany

Yi-ming, Gong (3), Key Laboratory of Biogenic Traces and Sedimentary Minerals of Henan Province, Henan Polytechnic University, Jiaozuo City, Henan Province, China

Zhang, Lu (3), Key Laboratory of Biogenic Traces and Sedimentary Minerals of Henan Province, Henan Polytechnic University, Jiaozuo City, Henan Province, China

Zhang, Wen-tao (3), Key Laboratory of Biogenic Traces and Sedimentary Minerals of Henan Province, Henan Polytechnic University, Jiaozuo City, Henan Province, China

Zonneveld, John-Paul (103, 463, 563, 807, 837), Ichnology Research Group, Department of Earth & Atmospheric Sciences, University of Alberta, Edmonton, Alberta, Canada

Preface

Utilization of trace fossils (ichnofossils) in the reconstruction of sedimentary environments in the geological record is nearly as old as the proper reconstruction of trace fossils as such. Consequently, many early studies attempted to recognize the tracemaker in order to apply information learned from modern tracemakers to the fossil record. This, of course, has led to misinterpretations as one and the same trace can result from producers belonging to different taxonomic groups of organisms and, on the contrary, one producer may be responsible for the production of traces having contrasting morphologies. However, in a different approach, reaching back to the 1950s, the behavior of an organism has served as the basis for a classification of trace fossils. Building on this smart method, the concept of ichnofacies was first presented by Adolf Seilacher in 1967, in which general trace-fossil patterns are used in the delineation of the bathymetry of broad facies zones, such as abyssal-bathyal, slope and shelf. Since then, the ichnofacies concept has been modified and applied in numerous studies, and its limits have been debated (Seilacher, 2007). Simultaneously, much effort was used in neoichnological investigations in all major environments and the results, together with the outcomes from ichnological experiments, have boosted the whole research field to a higher level. The understanding of the close interaction of organisms with sediment led to the recognition of bioturbate textures and has guided the development of the ichnofabric concept. Finally, ichnology is a powerful tool in stratigraphy and the reconstruction of evolutionary trends and mass extinctions, and bioturbation is increasingly recognized to influence the quality of hydrocarbon reservoirs and aquifers.

The idea of this book originated in September 2009 at an IAS (International Association of Sedimentologists) meeting in Alghero, Italy. The meeting was very successful, comprising several hundreds of participants from all over the world. However, none of the given oral presentations had ichnology as its topic and only very few posters covered this subject. This situation seems not to be unique, as we have learned from several other international conferences on geology and sedimentology in particular. On the other hand, specific meetings such as Ichnia 2004 (Trelew, Argentina) and 2008 (Cracow, Poland) enjoyed a great and active participation, and the same is true for numerous workshops and special sessions during various meetings of geological societies and associations (IGC, AAPG, EAGE). In a postconference field trip, it became suddenly clear that there is a strong interest in the utilization of ichnological methods in facies reconstruction, but often the methods and concepts are rather obscure to the non-specialist. Thus, the world of geologists seems to be divided into a small group of specialized ichnologists and a wide range of sedimentologists mostly disregarding trace-fossil studies. To bridge this gap is the main purpose of this book.

The first request of the series editor of *Developments in Sedimentology* to publish a book about trace fossils for sedimentologists (either single author or multiauthor approach) had to be denied because of the huge amount of data that became available during the past years, which would certainly be overwhelming to an individual researcher in ichnology and sedimentology. However, we have also recognized the desperate need to bring all that ichnological information to a level where it can be applied by the sedimentologist. The last such attempt to cover all major depositional systems in terms of their trace-fossil content and their use in interpreting depositional environments was made almost 30 years ago when Ekdale et al. (1984) and Curran (1985) published their compilations, although Pemberton et al. (2001), McIlroy (2004), Bromley et al. (2007), MacEachern et al. (2007), Miller (2007), Avanzini and Petti (2008), Gerard and Bromley (2008), and Buatois and Mángano (2011) made very good progress in the right direction.

The present book tries to reduce ichnological principles and ichnotaxonomy to a common denominator that can be shared between ichnologists and sedimentologists. After an extensive introduction to the subject and various methods for studying it, the selection of chapters follows major sedimentary environments as they are recognized from a continental to deep-marine realm. Unfortunately, our ichnological knowledge of these sedimentary environments is not equally distributed. Siliciclastic deposits are generally better studied and understood than carbonates, and relatively little information exists for intermediate systems (e.g., mixed siliciclastic/carbonates, chalk deposits). Ichnology in the marine sphere has a longer tradition and a more extensive record than do the continental systems, and a similar situation seems to be true for shallow-marine *versus* deep-marine ichnology. Finally, some important sedimentary environments are only poorly characterized in terms of their trace-fossil content and its relationship to facies zones or depositional units, including alluvial fans and shelves with all their specific situations. This unbalanced source was the starting point for the establishment of the concept of this book. These circumstances have been mastered by a crew of highly qualified and motivated ichnologists and sedimentologists from different schools and having a broad background.

The editors and authors gratefully acknowledge the reviewers for their thorough and time-consuming effort. We are indebted to Tom van Loon who initiated our activities on this volume and Elsevier for cooperation in publishing our work.

CHAPTER REVIEWERS:

Brett, Carlton E.
Buatois, Luis A.
Dalrymple, Robert W.
Edinger, Evan
Embry, Ashton
Flemming, Burghard W.
Gérard, Jean
Gierlowski-Kordesch, Elizabeth H.
Glennie, Kenneth W.
Hampson, Gary J.
Helland-Hansen, William
Hertweck, Günther
Isbell, John
Jensen, Sören
Kane, Ian
Löwemark, Ludvig
Lukasik, Jeffrey
Martin, Anthony J.
Martinius, Allard W.
McCann, Tom
McIlroy, Duncan
Miller, William
Mountney, Nigel
Radtke, Gudrun
Rindsberg, Andrew K.
Rodríguez-Tovar, Francisco J.
Schneider, Jörg
Shinn, Eugene
Suhr, Peter
Surlyk, Finn
Uchman, Alfred
Vallon, Lothar H.
van den Berg, Janrik
Walter, Harald
Weidlich, Oliver
Wilson, Mark A.

REFERENCES

Avanzini, M., Petti, F., (Eds.) 2008. Italian Ichnology. Studi Trent. Sci. Nat. Acta Geol. 83, 347 pp.

Bromley, R.G., Buatois, L.A., Mángano, G., Genise, J.F., Melchor, R.N. (Eds.), 2007. Sediment-Organism Interactions: A Multifaceted Ichnology. SEPM Spec. Publ., vol. 88, 393 pp.

Buatois, L.A., Mángano, M.G., 2011. Ichnology. Organism–Substrate Interactions in Space and Time. Cambridge University Press, Cambridge, 358 pp.

Curran, H.A. (Ed.), 1985. Biogenic Structures: Their Use in Interpreting Depositional Environments. SEPM Spec. Publ., vol. 35, 364 pp.

Ekdale, A.A., Bromley, R.G., Pemberton, S.G., 1984. Ichnology. The Use of Trace Fossils in Sedimentology and Stratigraphy. SEPM Short Course 15, 317 pp.

Gerard, J.R.F., Bromley, R.G., 2008. Ichnofabrics in Clastic Sediments: Applications to Sedimentological Core Studies. Jean R.F. Gerard, Madrid, 97 pp.

MacEachern, J.A., Bann, K.L., Gingras, M.K., Pemberton, S.G. (Eds.), 2007. Applied Ichnology. SEPM Short Course Notes 52, 380 pp.

McIlroy, D. (Ed.), 2004. The Application of Ichnology to Palaeoenvironmental and Stratigraphic Analysis. Geol. Soc. London, Spec. Publ., vol. 228, 493 pp.

Miller III, W. (Ed.), 2007. Trace Fossils. Concepts, Problems, Prospects. Elsevier, Amsterdam, 611 pp.

Pemberton, S.G., Spila, M., Pulham, A.J., Saunders, T., MacEachern, J.A., Robbins, D., Sinclair, I.K., 2001. Ichnology & Sedimentology of Shallow to Marginal Marine Systems: Ben Nevis & Avalon Reservoirs, Jeanne d'Arc Basin. Geol. Ass. Can., Short Course Notes 15, 343 pp.

Seilacher, A., 1967. Bathymetry of trace fossils. Mar. Geol. 5, 413–428.

Seilacher, A., 2007. Trace Fossil Analysis. Springer, Berlin, 226 pp.

Dirk Knaust (Stavanger) and Richard G. Bromley (Copenhagen)
February 2012

Part I

History, Concepts, and Methods

Chapter 1

A History of Ideas in Ichnology

Andrea Baucon,[*,†,1] Emese Bordy,[‡] Titus Brustur,[§] Luis A. Buatois,[¶] Tyron Cunningham,[||] Chirananda De,[#] Christoffer Duffin,[**] Fabrizio Felletti,[†] Christian Gaillard,[††] Bin Hu,[‡‡] Lei Hu,[‡‡] Sören Jensen,[§§] Dirk Knaust,[¶¶] Martin Lockley,[||||] Pat Lowe,[##] Adrienne Mayor,[***] Eduardo Mayoral,[†††] Radek Mikuláš,[‡‡‡] Giovanni Muttoni,[†] Carlos Neto de Carvalho,[*] S. George Pemberton,[§§§] John Pollard,[¶¶¶] Andrew K. Rindsberg,[||||||] Ana Santos,[###] Koji Seike,[****] Hui-bo Song,[‡‡] Susan Turner,[††††] Alfred Uchman,[‡‡‡‡] Yuan-yuan Wang,[‡‡] Gong Yi-ming,[‡‡] Lu Zhang[‡‡] and Wen-tao Zhang[‡‡]

**UNESCO Geopark Naturtejo Meseta Meridional, Geology and Palaeontology Office, Centro Cultural Raiano, Idanha-a-Nova, Portugal, [†]Università degli Studi di Milano, Dipartimento di Scienze della Terra, Via Mangiagalli, Milano, Italy, [‡]Geology Department, Rhodes University, Grahamstown, South Africa, [§]National Institute of Marine Geology and Geo-ecology (GEOECOMAR), Bucharest, Romania, [¶]Department of Geological Sciences, University of Saskatchewan, Saskatoon, Canada, [||]International Society of Professional Trackers, Missouri, Cameron, USA, [#]Palaeontology Division-1, CHQ, Geological Survey of India, Kolkata, India, **146 Church Hill Road, Sutton, Surrey, United Kingdom, [††]Paléoenvironments et Paléobiosphère, UFR, Sciences de la Terre, Université claud Bernard Lyon 1, Villeurbanne cedex, France, [‡‡]Key Laboratory of Biogenic Traces and Sedimentary Minerals of Henan Province, Henan Polytechnic University, Jiaozuo City, Henan Province, China, [§§]Área de Paleontología, Facultad de Ciencias, Universidad de Extremadura, Badajoz, Spain, [¶¶]Statoil ASA, Stavanger, Norway, [||||]Dinosaur Tracks Museum, University of Colorado Denver, Denver, Colorado, USA, [##]Backroom press, Broome, Western Australia, ***Classics Department, Stanford University, Stanford, California, USA, [†††]Departamento de Geodinámica y Paleontología, Facultad de Ciencias Experimentales, Campus de El Carmen, Universidad de Huelva, Huelva, Spain, [‡‡‡]Institute of Geology, Academy of Sciences of the Czech Republic, Praha, Czech Republic, [§§§]Ichnology Research Group, University of Alberta, Edmonton, Canada, [¶¶¶]School of Earth, Atmospheric and Environmental Sciences, University of Manchester, United Kingdom, [||||||]University of West Alabama, Livingston, USA, [###]Departamento de Geodinámica y Paleontología, Facultad de Ciencias Experimentales, Campus de El Carmen, Universidad de Huelva, Huelva, Spain, ****Department of Earth and Planetary Science, The University of Tokyo, Bunkyo-ku, Tokyo, Japan, [††††]Geology & Palaeontology, Queensland Museum and Monash University School of Geosciences, Hendra, Queensland, Australia, [‡‡‡‡]Institute of Geological Sciences, Jagiellonian University, Kraków, Poland*

[1]Corresponding author: e-mail: andrea@tracemaker.com

Developments in Sedimentology, Vol. 64. http://dx.doi.org/10.1016/B978-0-444-53813-0.00001-0

1. INTRODUCTION

Among one and another rock layer, there are the traces of the worms that crawled in them when they were not yet dry.
Leonardo da Vinci, Leicester Codex, folio 10 v

Since the beginnings of ichnology, trace fossils have been recognized for their twofold nature as biological and sedimentological objects. As a discipline studying biogenic sedimentary structures, ichnology proved to be very important for paleontology and sedimentary geology as well.

The tendency of applying trace fossils in the characterization of past depositional environments manifested itself very early. Leonardo da Vinci used trace fossils to prove the marine origin of the sedimentary successions of the Apennines (Baucon, 2010), but it took ichnology four centuries to develop comprehensive and precise scientific tools for the needs of paleoenvironmental analysis. Nowadays, ichnology is a matter of great interest due to the huge spectrum of potential applications such as facies interpretation, paleoenvironmental reconstruction and recognition of discontinuities, prospecting and exploration of hydrocarbon resources.

This chapter aims to delineate the progression from empirical knowledge toward the modern concepts of ichnology, with particular regard to the application of ichnology to facies analysis. This has guided our areas of emphasis so that, for example, we give invertebrate traces a greater allocation of space than vertebrate traces because they find more sedimentological applications.

The purpose of this chapter is to provide information not only about ichnologists and their theories but also why they developed an idea as they did. For this reason, particular attention has been given to the specific social and historical circumstances surrounding the individual line of thought. Similarly, we followed a global perspective, based on the belief in the international character of ichnology.

Tracing the global history of ichnology is possible through the texts which have survived, hence the necessity of compiling a comprehensive bibliographic database on the history of ichnology (see the Supplementary Material: http://booksite.elsevier.com/9780444538130). The purpose is to show semiquantitatively the chronological relationships among various branches of ichnology based upon similarities and differences in the interpretation of traces.

2. THE AGES OF ICHNOLOGY

Although the study of trace fossils is an important field for the solution of fundamental and applied problems of geology, there is a lack of general historical overviews. With the exception of studies on specific episodes and geographical areas, the only general historical accounts are those of Osgood (1970, 1975; invertebrate ichnology) and Sarjeant (1987; vertebrate ichnology).

Osgood (1975) attempted to subdivide the history of ichnology on the basis of periods of time with relatively stable characteristics:

1. *Age of Fucoids* (*1823–1881*): This stage started with Brongniart (1823), who considered invertebrate trace fossils as *fucoids*, or seaweed. During this historical stage, the botanical interpretation dominated the scientific view of trace fossils.
2. *Period of Reaction* (*or Age of Controversy*) (*1881–1925*): Based on analogies with modern traces, Nathorst (1881) argued that many *fucoids* were trace fossils. This aroused a consistent debate, with prominent scientists like Lebesconte and de Saporta supporting the botanical interpretation.
3. *Development of the Modern Approach* (*1925–1953*): This stage started with the establishment of the Senckenberg Laboratory, a marine institute devoted to neoichnology (Cadée and Goldring, 2007). The geologists of the period agreed about the ichnological nature of trace fossils, opening the avenues to the decisive steps toward modern ichnology.

In more recent times, two historical stages were added to Osgood's classical periods: Pemberton et al. (2007) recognized a *Modern Era of Ichnology*, extending from 1953 to the present day. This period saw the foundation of the central concepts of modern ichnology, starting with Seilacher's (1953) seminal publication on the methods of ichnology. Recently, Baucon (2010) established the *Age of Naturalists*, spanning roughly the fifteenth to the seventeenth centuries. During this stage, several Renaissance intellectuals depicted and studied trace fossils, although ichnology existed as disconnected ideas about traces. Holding these stages as a chronological reference, this chapter will explore the evolution in the study of trace fossils, from Paleolithic times to the present.

3. FROM PALEOLITHIC TIMES TO GRECO-ROMAN ANTIQUITY

Archeological evidence indicates that humans have recognized trace fossils since Paleolithic times. Bioeroded Miocene mollusks are commonly found within the cultural layers of Pavlov and Dolní Věstonice (Czech Republic, Late Paleolithic, 29,000–24,000 years ago; Fig. 1A). Statistical data from the primary collection sites indicate that humans selectively collected mollusks with bioerosional trace fossils (*Oichnus*) in order to use them as items of personal adornment (i.e., segments of collars; Jarošová et al., 2004 in Supplementary Material: http://booksite.elsevier.com/9780444538130).

Such archeological evidence shows that trace fossils were a subject of ancient interest for humans, although it gives no information about their interpretation in past hunter/gatherer societies. In this regard, anthropological analogy with modern indigenous populations represents a valuable tool of analysis. As one of the oldest continuous cultures in the world, Australian Aborigines provide an exceptional insight in the anthropology of biogenic traces (Lowe, 2002). Native

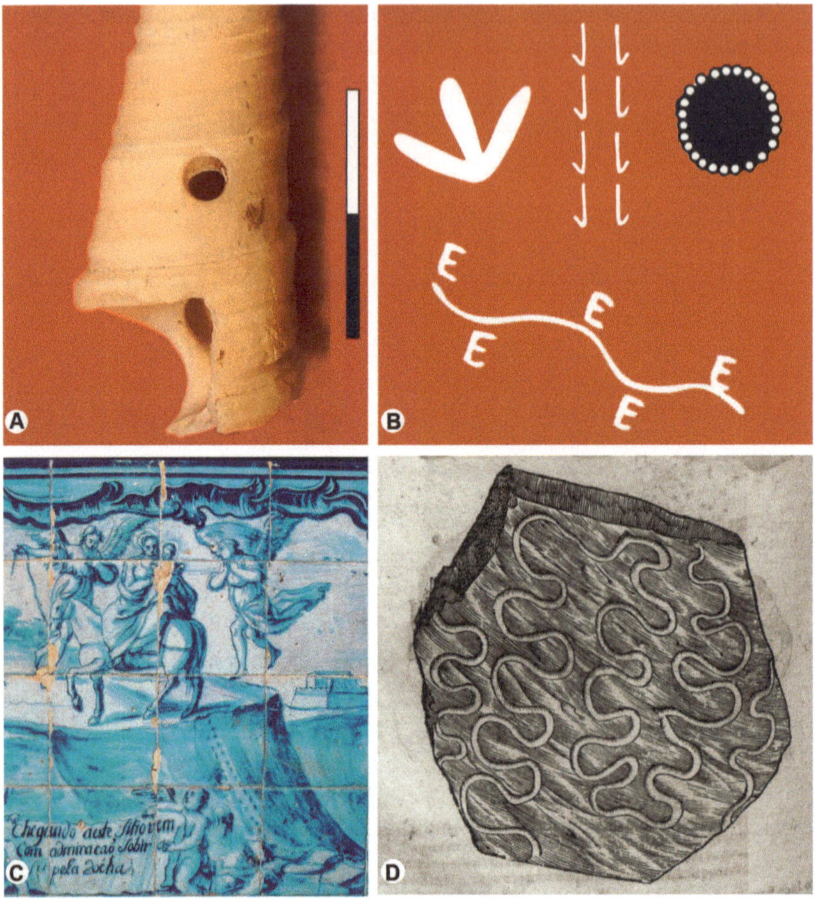

FIGURE 1 Beginnings of ichnology. (A) Miocene bioeroded mollusk found within Paleolithic cultural layers. Scale bar: 1 cm (Dolní Věstonice II, Czech Republic; Jarošová et al., 2004 in Supplementary Material: http://booksite.elsevier.com/9780444538130). (B) Ichnological symbols commonly used in Australian Aboriginal art. Clockwise from top left: emu footprint, kangaroo track, burrow, goanna (monitor lizard) track. (C) Eighteenth century *azulejo* (painted, glazed ceramic tilework) depicting *Nossa Senhora da Pedra Mua* (Our Lady of the Mule Stone). A set of footprints is clearly visible on the sea cliff. Memory Chapel, Cabo Espichel (Portugal). (D) Trace fossils in the Italian Renaissance: *Cosmorhaphe*, as represented in Aldrovandi's *Musaeum Metallicum*.

Australian people developed remarkable neoichnological abilities, gaining a detailed understanding of animal behavior through the interpretation of tracks and burrows (Ellena and Escalante, 2007). The crucial role of tracking is mirrored in the rich vocabulary for traces and their conditions, often without equivalents in European languages. In Walmajarri language, tracks left after rain are called *murrmarti*, and a goanna burrow in soft sand that is too deep for the animal to be reached is called *purruj*. Similarly, Aboriginal art has a visual vocabulary for traces, consisting of standard symbols for each kind of track (Fig. 1B).

Trace fossils are also an object of interest for Australian Aborigines. Dinosaur tracks are part of the *Ngarrangkarni* (or "Dreamtime"), a mythical past when giant ancestor beings left traces of their exploits, shaping the landscape to its present form. Native people attribute theropod tracks to an enormous feathered "Emu-man", *Marrala* (also spelled *Marella*; Mayor and Sarjeant, 2001). Marrala's footprints represent ichnohierophanies (*sensu* Baucon et al., 2008), which are supposed traces of supernatural entities in objects that are an integral part of our natural world (i.e., rocks). Ichnohierophanical accounts are reported from cultures all over the world and include both vertebrate and invertebrate trace fossils (Baucon et al., 2008).

One of the best examples are the sauropod trackways and the trace fossil *Rhizocorallium* of Cape Espichel (Portugal), object of curiosity and devotion since the thirteenth century. Fishermen from the region of Sesimbra interpreted such trace fossils as the footprints of a giant mule that carried the Virgin Mary (Lockley et al., 1994). During the eighteenth century, a sanctuary was erected and "Nossa Senhora da Pedra Mua" (Our Lady of the Mule Stone) became an object of national worship (Fig. 1C).

Ichnohierophanic interpretations were common in the Greco-Roman world, as exemplified by Lucianus of Samostata (approx. AD 125–after AD 180), who satirized the frequent claims of "footprints in rock" (Mayor and Sarjeant, 2001). The figure of Heracles often provided a mythological explanation for fossil footprints. For instance, Pseudo-Aristotle reported that "near Pandosia in Iapygia (present-day Heraclea, Italy), the footprints of Heracles are shown and no one is allowed to step on them". First attributed to Pleistocene mammal tracks (Mayor and Sarjeant, 2001), Heracles' footprints are most probably dinosaur footprints, as evidenced by the abundant dinosaur tracksites in the same area of Pandosia.

Although only a minor part of the texts of natural philosophers survived to us, neoichnological observations are found in Aristotle's (384–322 BC) *History of Animals* and Theophrastus' (approx. 371–287 BC) *On Fish* (Sharples, 1995). Trace fossils were apparently ignored by early natural philosophers, with the exception of Pliny the Elder (23–79 AD). In his *Naturalis Historia*, the Roman author described *phycites* as an "alga-like stone", possibly some kind of branched trace fossil. In fact, the term "phycites" or "ficite" was successively used to indicate *Chondrites* (i.e., Targioni-Tozzetti, 1777).

4. THE AGE OF NATURALISTS

Despite these early interpretations, the move toward a rational understanding of trace fossils began in the European Renaissance. This cultural movement flourished in Italy during the fourteenth century and rapidly spread across Europe, bringing the start of a revolution in the way to investigate nature.

In Renaissance times, the inquiry into the natural world took place both in science and art, which were connected by a coherent line of continuity. This phenomenon can be clearly seen in the work of Leonardo da Vinci, the founding father of ichnology (Baucon, 2010). Indeed Leonardo not only sketched

Paleodictyon, but he also examined the subject of trace fossils when dealing with the origin of body fossils or, with Leonardo's words, *nichi petrificati* (petrified seashells). While his contemporaries supported an inorganic origin for body fossils, Leonardo referred to bioeroded specimens to disprove this idea. Similarly, da Vinci used bioturbational traces (*andamenti delli lombrici* or "traces of worms") as a paleoenvironmental tool to demonstrate the marine origin of the sedimentary layers. Da Vinci's accurate analysis shows that, from the beginnings of ichnology, trace fossils have been an indispensable tool for facies analysis and paleobiology. However, Leonardo did not have influence on his contemporaries because he wrote in mirror-image Italian, at a time when Latin was the idiom of erudition. Despite his revolutionary conclusions, Leonardo remained an isolated voice, not participating in the academic discussion. In contrast, Ulisse Aldrovandi, the first professor of natural sciences at the University of Bologna, left a considerable legacy, including the word *geology* (Vai and Cavazza, 2003). The *Musaeum Metallicum*, his most extensive work in geology and paleontology, contains several trace fossils, frequently described in detail and with magnificent illustrations. Aldrovandi described *Cosmorhaphe* as a snake-like structure, possibly suggesting an inorganic origin, but he accurately interpreted bioerosional structures as produced by bioeroding mollusks (Baucon, 2009).

Known for his vast cabinet of curiosities, Aldrovandi exchanged specimens and ideas with another prominent naturalist of the Renaissance, the Swiss Conrad Gesner. *De omni rerum fossilium genere*, Gesner's (1565) geological treatise, exemplifies a common trait of the Renaissance pioneers of ichnology: a non-exclusive interest in traces. In fact, Gesner, as many other contemporaries, investigated any aspect of the natural world that aroused his interest, from trace fossils to gemstones. Though younger than Gesner, Johann Bauhin was the main pupil of the influential naturalist. He published a tourist guide in which he described a branched trace fossil (*Phymatoderma*). The accompanying illustration shows intricate angelic figures, but Bauhin commented, "The human figures have been mistakenly added either by the painter or the wood cutter" (Seilacher, 2007).

A chondritid was also reported by Legati (1677) when describing Ferdinando Cospi's *Wunderkammer*, placed next to the Aldrovandi's Museum. Legati followed the interpretation of Aldrovandi, who previously studied the same specimen, and indicated it as "Wooded-stone" ("Alberina", "Dendrite", "Pietra Imboscata"; Legati, 1677; Fig. 2A). The "Alberina stone" was commonly excavated in the surroundings of Bologna and used as a building material after calcination. For this reason, Ovidio Montalbani proposed *Chondrites* as the symbol of stone masonry, accompanied by the Latin motto, *Silex medio cremium sibi viscere sculpsit, an cupit in calcem versa parare domos?* ("Hard stone has carved out its bowels into brushwood, perhaps does it want to build houses after becoming lime?"; Legati, 1677: 174). Bioerosional traces were also present in the Cospi collection and compared to petrified ant burrows (Legati, 1677: 175).

FIGURE 2 Interpretation of trace fossils through time. (A) Legati's *Pietra Alberina* (Tree stone) exemplifies the fact that chondritids have been compared to plants since the seventeenth century or earlier. (B) William Buckland, pioneer of the study of fossil feces, owned a table made with sectioned coprolites from Carboniferous deposits (Duffin, 2009). (C) Fanciful representation of a funnel-weaving spider, placed by Eberhard Werner Happel among *vermes lapidum*, or rock-boring organisms (Happel, 1707, published posthumously).

These naturalists pioneered the study of ichnology and added many descriptions of trace fossils to their natural histories. Their work demonstrates that ichnology has its roots firmly anchored in Renaissance times, although it existed as disconnected ideas about traces.

5. SEVENTEENTH TO EIGHTEENTH CENTURY: A PERIOD OF TRANSITION

For many years after Osgood's (1970, 1975) work, it was argued that ichnology was born in the nineteenth century, Age of Fucoids—a not implausible suggestion, given the difficulty to find ichnological studies from prior to the 1800s. However, recent works not only highlight the Renaissance roots of ichnology (Section 4; Fig. 2A) but also evidence some ichnological observations dating back to the 1700s (i.e., Duffin, 2009). As a consequence, one question might arise: what historical phenomena connected the Age of Naturalists to the Age of Fucoids?

By the late 1600s, the emphasis on reason in the intellectual life of Europe started to become pervasive. The early Enlightment in Europe went hand in hand with an expansion of scientific societies which, by the start of the eighteenth century, had transformed the organization of scientific research (Gohau, 1991). This aspect affected also the study of trace fossils, as evidenced by the central role of the Royal Society of London in the study of coprolites. Indeed, John Woodward, fellow of the Royal Society, provided one of the first explanations for their origin by comparing Mesozoic specimens to "Iuli" or cones of larch trees (Woodward, 1729). However, he was not the only researcher to describe coprolites. Indeed another fellow of the Royal Society, Lhwyd (1760), figured a spiral coprolite, and Mylius (1709) previously illustrated some Permian specimens. The botanical interpretation of coprolites has been predominant since the time when Buckland (1822), an influential fellow of the Royal Society, recognized them as fossil feces (Duffin, 2009; Fig. 2B). Bioerosional structures also received attention during the eighteenth century. Intellectuals such as Lorentz Spengler and Gottfried Sellius focused on modern bioeroding bivalves (Supplementary Material: http://booksite.elsevier.com/9780444538130). Bioeroding organisms, collectively named as *vermes lapidum*, were attributed by some authors to funnel-weaving spiders (Happel, 1707; Lémery, 1716; Fig. 2C). By the late 1700s, the attention given to bioturbational structures gradually rose. This phenomenon emerged from the same cultural areas—the Italian and the German-speaking regions—which dominated the Age of Naturalists.

From the eighteenth to the early nineteenth century, the term *fucite* was widely adopted by Italian scholars to indicate bioturbated rocks; the term may refer to Pliny's *phycites*, or alga-like stone. In his travel reports from Tuscany, Targioni-Tozzetti (1777) accurately described *fuciti*: "they are similar to the Worm-Stones (Pietre Lombricarie), and when they are split-off ... they reveal impressions of algae (Fuci)". Despite the alga-like appearance, Tozzetti questioned the vegetal nature of *fuciti*, of which the origin "botanical or animal, is not known" (Targioni-Tozzetti, 1777). To solve this issue, he collaborated with the abbot Alberto Fortis, who had previously described *elmintoliti* (*Helminthoida*) from Istria and Dalmatia (Fortis, 1774; Surić et al., 2007). Adolphe Brongniart, the initiator of the Age of Fucoids, studied *Fucoides* (=*Chondrites*) *targionii* from the collections of the Italian scientist, proving Targioni-Tozzetti's far-reaching influence (Brongniart, 1828: 56–57). After the 1830s, the term *fucite* fell into disuse, being quickly replaced by the etymologically analogous *fucoid*.

In Germany, interest for natural history became popular in Thuringia, Saxony, and Bavaria, giving particular attention to Permian and Triassic units, among which the trace-fossil-rich Muschelkalk. In naming rocks, researchers came up with specific descriptors, often based on those aspects of the texture resulting from bioturbation. Among these early contributors, Schütte (1761) used a Renaissance term, *osteocolla*, for describing bone-like rocks characterized by *Protovirgularia* and *Planolites*. However, the best example of this tendency is Batsch (1802), who meticulously described *Zungenkalkstein* ("tongue limestone", for its *Rhizocorallium*-dominated ichnofabric), *Loecherkalkstein* ("holey limestone", for the abundant *Balanoglossites*), and *Wurmkalkstein* ("worm limestone", for the prominent presence of *Planolites* and *Protovirgularia*).

The botanical interpretation, successively prominent in the Age of Fucoids, was followed by a German student of Couvier and friend of Alexander von Humboldt, namely, Johann Gotthelf Fischer von Waldheim, who reported *Umbellularia logimna* (now *Zoophycos*) from the Ural Mountains (Supplementary Material: http://booksite.elsevier.com/9780444538130). Invertebrate trace fossils were also noticed by the explorer Hinrich Lichtenstein during his travel in South Africa (Master, 2010 in Supplementary Material: http://booksite.elsevier.com/9780444538130).

While in continental Europe, the major part of the ichnological observations appeared in treatises; in Britain, they regularly appeared in periodic journals published by scientific societies. Together with the Philosophical Transactions of the Royal Society, the Transactions of the Geological Society of London were a major channel for disseminating ichnological research. For instance, McCulloch (1814) published a description of the pipe rock of north-western Scotland, suggesting similarity to sabellid marine worm-burrows. Other influential contributors include Buckland, who described "paramoudras" from the Chalk of Northern Ireland, and Webster, who figured *Thalassinoides* and *Ophiomorpha* as zoophytes (plant-like animals, e.g., alcyonarian cnidarians; Supplementary Material: http://booksite.elsevier.com/9780444538130).

In conclusion, the voyage from the Age of Naturalists to the Age of Fucoids was tortuous (Fig. 3A and B), and the corresponding transitional period incorporated concepts and media from both cultural stages.

6. THE AGE OF FUCOIDS

6.1 Emergence of the Paleobotanical Interpretation

The end of the French Revolution, the expansion of colonial empires, and the Industrial Revolution were conditions that made the nineteenth century a period of profound social and economic change (Frey and Frey, 2004). This milieu served as the background to the French botanist Adolphe Brongniart, defined as the "typical child of the best of the French revolution" (Stafleu, 1966; Fig. 3B). Adolphe held a name with status: he was the grandson of Napoleon's architect and the son of the eminent geologist Alexandre Brongniart. In 1817, Adolphe joined his father in a geological *grand tour* through the Swiss and

FIGURE 3 Pioneers in the study of trace fossils. (A) Leonardo da Vinci. (B) Adolphe Brongniart. (C) Alfred Nathorst.

Italian Alps, where he observed abundant trace fossils from the Apennines foredeep basin (Brongniart, 1828: 45).

A new stage in the history of ichnology began in the year 1823, when Brongniart published his *Observations sur les fucoides et sur quelques autres plantes marines fossiles* (Brongniart, 1823; Osgood, 1970, 1975). By his attitude toward comparative anatomy, Brongniart associated some branching forms of trace fossils (e.g., *Chondrites*) to modern algae, and according to the resemblance to the brown alga *Fucus*, he used the term *fucoid* to indicate such fossils. Although other researchers previously supported the botanical origin of trace fossils (i.e., von Schlotheim, 1822), none of these reached the influence of the French botanist. One reason may be found in the scientific authority of Brongniart, who was recognized as the founding father of paleobotany even during his lifetime (Stafleu, 1966). Second, Brongniart provided the scientific community with relevant tools: the first classification of all known fossil plants (including fucoids) and its biostratigraphical application (Brongniart, 1828). Finally, French was one of the most used international languages in Europe, being the language of diplomacy from the seventeenth to mid-twentieth centuries (Chew, 2009).

The French School and the rising discipline of paleobotany offered convincing arguments in support to the fucoid hypothesis (e.g., de Saporta, 1873), which was readily accepted by the international scientific community. For instance, the German paleontologist Bronn (1837) described and figured numerous *Fucoiditen*, an interpretation that was also provided for *Asterosoma* (von Otto, 1854). Much of the fucoid research originated from the flysch in the Swiss Alps, where Fischer-Ooster followed Brongniart in using "fucoids" for stratigraphical correlation (Supplementary Material: http://booksite.elsevier.com/9780444538130). Among the various works that have dealt with fucoids, *Flora Fossilis Helvetiae* (Heer, 1876/1877) stands preeminent for the detail of the descriptions.

The champions of the fucoid hypothesis came from the most disparate parts of Europe. Specifically, the Portuguese paleontologist Joaquim Nery Delgado had a central role in supporting the botanical origin of trace fossils, although he successively interpreted *Nereites* and *Skolithos* as trace fossils (e.g., Delgado, 1903). Similarly, the Bohemian geologist Sternberg (1833–1838), one of the founding fathers of paleobotany, was one of the most prominent authors of the Age of Fucoids for his numerous descriptions of *Chondrites*.

The term *fucoid* soon became part of the international geological lexicon, from Spain (De Prado, 1864) to Poland (Pusch, 1837), and from Britain (Buckland, 1836) to Italy (Savi and Meneghini, 1851). The important legacy of the fucoid approach included not only substantial advances in descriptive knowledge but also the establishment of still valid ichnogenera such as *Rhizocorallium*, *Scolicia*, *Daedalus*, *Paleodictyon*, *Spirophyton*, *Diplocraterion*, and *Zoophycos* (Häntzschel, 1975). Particularly noteworthy is the case of *Cruziana*, described as a body fossil by the French naturalist Alcide d'Orbigny during a mission for the Paris Museum. He established the (ichno)-genus *Cruziana* after the Bolivian president Mariscal Santa Cruz, who had sponsored his visit to South America (Seilacher, 2007). The significance of his work resides in the definition of a taxon that was subsequently reinterpreted as a trace fossil, gaining immense popularity worldwide.

6.2 Zoophytes and other Popular Interpretations

Together with the dominant botanical hypothesis, the interpretation of trace fossils followed three other themes, namely, as zoophytes (plant-like animals); as worm-like body fossils; and as true invertebrate burrows, tracks, and trails.

The plant-like aspect of branched burrows suggested the zoophyte interpretation, which appeared even before Brongniart's (1823) seminal paper. Indeed, Webster (1814) appears to have been the first to interpret *Thalassinoides paradoxica* and *Ophiomorpha* as alcyonarian cnidarians. The "zoophyte" interpretation was usually applied to such burrows as *Oldhamia* (Forbes, 1848). The zoophyte interpretation had a particular success in eastern Europe. For instance, Zaręczny (1878) described *Spongia sudolica* from the Cretaceous marls near Cracow. As the etymology may suggest, the trace fossil was considered a sponge, being later included in *Spongeliomorpha* (Raciborski, 1890). Alongside to the description of several fucoids from Russia, Estonia, and Ukraine, Eichwald (1860–1968) presented *Paleodictyon* (his *Cephalites maximus*) as a sponge. A less conspicuous number of trace fossils were described as body fossils of annelids. This is the case for *Nereites* MacLeay (1839) and "*Nemertilites*" (now *Scolicia*) *strozzii* (Savi and Meneghini, 1851), both interpreted as marine worms.

In the same years, a minor part of scholars suggested the ichnological nature of some trace fossils, but their morphological diversity was a particularly challenging subject. It is therefore not surprising that ichnological interpretations often coexisted with other explanations. For instance, Savi and Meneghini (1851) interpreted "*Nemertilites*" (now *Scolicia*) *strozzii* as a "giant marine worm" and *Chondrites* as seaweed, and in the same work, they admitted the ichnological origin of "*Nemertilites meandrites*" (possibly *Scolicia*). Specifically, they argued, "as it is impossible to see any animal remains, at least it is necessary to recognize the action of an animal, that is, a physiological imprint (*impronta fisiologica*)". This term is strongly reminiscent of d'Orbigny's "*emprintes physiologiques*" (d'Orbigny, 1849: 27–29). Another example is given by the paleobotanist Heer, who not only recognized several species of *Zoophycos*, *Chondrites*, and *Gyrophyllites* as fucoids, but also described *Wurmsteine* (*Helminthoiden*) as sediment-filled burrows of marine worms (Heer, 1876/1877; Supplementary Material: http://booksite.elsevier.com/9780444538130).

Britain was a very fertile ground for the fucoid hypothesis (i.e., Buckland, 1836), but Victorian geologists were also familiar with animal traces on modern tidal flats. For instance, Charles Lyell devoted some space in his *Principles of Geology* to explain burrowing mollusks (Lyell, 1833: 288). It should also be noted that bioerosional traces played a central role for the development of uniformitarianism, as Lyell recognized recent relative sea-level fluctuations through the observations of bioeroded Roman columns (Baucon et al., 2008; de Gibert et al., 2012). Animal architects received significant attention also in popular science books (Wood, 1866).

Victorian geologists applied their neoichnological attitude to the rock record since the 1850s. They recognized the true nature of U-burrows and bivalve traces, annelid trails, and especially arthropod trackways (e.g., Binney, Hancock, Salter, Roberts in Supplementary Material: http://booksite.elsevier.com/9780444538130). On the question of bioturbational structures, Nicholson (1873) clearly established that many *fucoids* of earlier British workers were annelid burrows or trails. Such studies reflect, but clearly predate, Nathorst's (1881) classic work to refute the nature of fucoids.

By the same time, geologist Henry Thomas de la Beche drew one of the earliest paleoecological reconstitutions (*Duria Antiquior*), based on coprolites and body fossils found by Mary Anning (Duffin, 2009). De la Beche also used fossil borings for recognizing unconformities (de la Beche, 1846: 290). In the same period, the study of microborings was made possible by the use of light microscopy, which played a dominant role in microbioerosional research until the advent of scanning electron microscopy ("embedding-casting technique"; Golubic et al., 1975; Tapanila, 2008).

In some cases, the study of trace fossils benefited from the expansion of the British Empire, which represented the leading superpower of the nineteenth century. A clear example is given by Edward John Dunn, who left Bedminster (England) for New South Wales (Australia), where he trained as a geologist. Successively, he traveled to southern Africa, accounting for "trails of worms and tracks of crustaceans" in the Permian Ecca Group (Dunn, 1872). Similarly, the missionary Stephen Hislop (1860) pioneered the study of coprolites in India.

Among British researchers, Charles Darwin provided an important contribution to neoichnology through the study of earthworms and the production of vegetable mold (Meysman et al., 2006; Pemberton and Frey, 1990). Darwin acknowledged the importance of fossil tracks in a letter to the American pioneer of ichnology Edward Hitchcock: "In my opinion these footsteps [. . .] make one of the most curious discoveries of the present century and highly important in its several bearings" (Burkhardt and Smith, 1987).

6.3 An Independent Ichnological Center: North America

Charles Darwin referred to the first inspiration of ichnology in North America, the discovery of Triassic–Jurassic vertebrate trackways in New England in the 1830s (Burkhardt and Smith, 1987). Hitchcock's (1858) *magnum opus*, *Ichnology of New England*, was an instant classic. Although Hitchcock was originally inspired by vertebrate trace fossils, he also studied the invertebrate ichnotaxa, even making neoichnological observations for comparison. However, these terrestrial trace fossils were difficult to compare with the Paleozoic examples found by later generations of geologists and, until continental ichnology emerged as a field of research, Hitchcock's work on invertebrate traces lay fallow.

Other early workers on invertebrate trace fossils, for example, James Hall, Elkanah Billings, and Léo Lesquereux (Supplementary Material: http://booksite.elsevier.com/9780444538130) usually ascribed them to fucoids. The first trace fossil to be named in North America was *Fucoides* (*Cladorytes*, now *Arthrophycus*) *alleghaniensis* (Rindsberg and Martin, 2003) from the Silurian of Pennsylvania, but it was set within the realm of botany, not ichnology. James Hall, who had firsthand experience of the New England coast, did figure a few Paleozoic "molluscan" trails. Hall taught many North American geologists, and they followed his interpretation of *fucoids*.

Accordingly, the development of ichnology in North America radiated from two independent centers: a Canadian School of professional geologists centered at the Geological Survey of Canada, and a partly New York-inspired Cincinnati School consisting of self-taught paleontologists (Pemberton et al., 2007). Both schools had impact and worked somewhat in isolation from active European researchers. This isolation not only had its negative aspects but also allowed American workers to explain these obscure structures in novel ways.

The Canadian geoscientists William Logan and J. William Dawson, who were well acquainted with the seashore, were quick to dismiss the fucoid origin and ready to consider other options. This resulted in insightful observations such as Dawson's interpretation of *Rusophycus*, *Arthrophycus*, and *Nereites* as products of burrowing trilobites. The interpretation of *Rusophycus* as a trilobite trace predated the first European to come to the same conclusion by almost 15 years (Pemberton et al., 2007). Although these views were widely disseminated, for the most part Europeans did not agree, having already set their minds. However, when Logan similarly interpreted traces unknown in Europe, *Climactichnites* and *Protichnites* as locomotion traces, there was no disagreement that they were made by animals.

The Cincinnati School started work in a landscape as rich in fossils as to demand attention from amateurs, some of whom became authors. Samuel Almond Miller, C. B. Dyer, Uriah P. James, and others contributed to the roster of fucoids before Joseph F. James, a nephew of U. P. James, ushered in the Period of Reaction (Supplementary Material: http://booksite.elsevier.com/9780444538130).

6.4 The Rise of Vertebrate Ichnology

For its influence, Edward Hitchcock subsequently became known as the father of vertebrate ichnology. Nevertheless, he was not the first to study vertebrate tracks. The first scientific studies on vertebrate traces, which appeared in a series of short newspapers, magazines, and journal articles between 1828 and 1831, referred to tracks found in Permian sandstones in Scotland. First interpreted as turtle tracks, and therefore named *Chelichnus*, they attracted much attention and stimulated William Buckland, the first professor of Geology at Oxford, to conduct experiments with modern turtles walking in pastry dough. We now know these tracks to be those of mammal-like reptiles (synapsids) that inhabited ancient dune fields (Supplementary Material: http://booksite.elsevier.com/9780444538130). According to Häntzschel (1975: W2), Buckland's legacy includes the term "ichnology" itself.

In 1835, the famous "hand-shaped" track *Chirotherium* was reported from the Triassic of Germany before any equivalent skeletal remains were known. This gave rise to many fanciful interpretations, and it was not until the 1930s that it was convincingly attributed to an archosaur (Seilacher, 2007). Likewise, in 1836, Edward Hitchcock described the first vertebrate tracks known from North America, before the concept of dinosaurs was established in 1842. Famously, he named large, emu- to moa-sized Jurassic tracks from the Connecticut valley region as *Ornithoichnites*, implying that they were made by giant birds (Hitchcock, 1836). In naming tracks, he established the tradition of vertebrate ichnotaxonomy (naming tracks rather than body fossils) and his seminal work is frequently cited to this day.

During his scientific career, Hitchcock assembled a vast ichnological collection, housed at the Appleton Cabinet at Amherst. Hitchock's ichnological cabinet served as a reference collection and attracted scientists from all over the world. Among others, these include the Italian geologist Capellini (1867) and the Austrian paleontologist Abel (1926), who studied the famous Upper Triassic vertebrate track sites of Connecticut and Massachusetts.

7. PERIOD OF REACTION

7.1 Fucoids *versus* Traces

What did begin to emerge in the late Age of Fucoids was an increased interest in invertebrate trace fossils as biogenic sedimentary structures. However, yet even where the fucoid hypothesis was rejected or modified, the ichnological interpretation was still not persuasive. Of crucial importance to the history of ichnology are the publications by Nathorst (Supplementary Material: http://booksite.elsevier.com/9780444538130; Fig. 3C). Indeed, his 1881 work is seen as a major watershed in the history of ichnology in that it generated a broad acceptance for a trace-fossil origin of various structures that at the time were considered remains of plants or animal body fossils (Osgood, 1970).

Nathorst conducted systematic neoichnological experiments by introducing various animals into dishes with plaster of paris and observing their traces. Nathorst pointed out the correspondence between modern invertebrate traces and fucoids, challenging traditional ideas about how trace fossils formed. Among Nathorst's arguments was also the common preservation along bedding planes and in such pronounced relief that he found an algal origin impossible. He also remarked on the absence of any organic material.

Thanks to the extended summary in French that accompanied his work, Nathorst provided the impetus for the development of paleoichnology. Throughout this historical stage, discussion and argument over the nature of fucoids took on increasing prominence among scientists. In fact, by the late nineteenth century, French scholars began to mount a sustained attack on the rising paleoichnology and published voluminous descriptive and taxonomic works (e.g., de Saporta, 1882; Lebesconte, 1883). Only in 1886 did Nathorst take part in the discussion (Cadée and Goldring, 2007). In holding his opinion against such authorities as Lebesconte and de Saporta, Nathorst showed his sturdy independence of judgment.

7.2 The Period of Reaction: a Worldwide Phenomenon

The Period of Reaction was a worldwide phenomenon, although not all of the elements of debate appeared everywhere in the same order and with the same strength. For instance, the reaction against the fucoid interpretation had much less impact in Britain than on the continent where the extensive fucoid monographs of Brongniart, Heer, and de Saporta had been published. Nathorst's invertebrate interpretation was readily accepted by several workers, such as Keeping (flysch traces), Taylor, Smith, Bather (worm trails and burrows), and Beasley, Smith (arthropod tracks; Supplementary Material: booksite.elsevier.com/9780444538130).

Conversely, the Iberian Peninsula saw one of the strongest advocates of the fucoid interpretation, the Portuguese geologist Joaquim Nery Delgado. However, by the early 1900s, Delgado interpreted *Cruziana* as "crustacean trails"; *Nereites* as "annelid trails"; and *Skolithos* as "worm burrows" (Supplementary Material: http://booksite.elsevier.com/9780444538130). Spanish workers maintained vague interpretations, due mostly to the controversial explanations that characterized the Period of Reaction (e.g., Palacios, 1918). Within a recently unified Italy, Federico Sacco published his *Notes on Italian Palaeoichnology* (Sacco, 1888), but geological literature often referred to fucoids. Other influential studies were performed by Peruzzi, Gabelli, Stoppani, and Gortani (Supplementary Material: http://booksite.elsevier.com/9780444538130).

During most of the Period of Reaction, the geopolitical scenario of central-eastern Europe was dominated by the Austrian-Hungarian Empire. Despite its multinational nature and the contentious questions concerning language, the Austrian-Hungarian Empire contributed to spread the use of German, which rapidly became one of the most important languages of science and scholarship (Ammon, 1998). As Chew (2009) argued, in many disciplines, knowledge of German became a basic requirement up to 1945. In this context, Austrian geologist Theodor Fuchs added valid arguments in deciphering the true nature of *fucoids* as burrows, although this view was controversial for a long time (Supplementary Material: http://booksite.elsevier.com/9780444538130). Fuchs (1895) introduced also the first classification of trace fossils, based on flysch trace fossils. He distinguished three family groups:

1. *Vermiglyphen*: threadlike, straight or winding reliefs occurring mostly on bed soles;
2. *Rhabdoglyphen*: straight bulges on lower bedding surfaces;
3. *Graphoglypten*: reliefs resembling ornaments or letters. The term partly corresponds to the similar term *Hieroglyphen* (Fuchs, 1895: 394; Häntzschel, 1975).

The term *Graphoglypten* had a considerable success as its English analog (graphoglyptid) is still used for indicating a group of ornamental trace fossils occurring on the sole of flysch sandstones (Seilacher, 2007; Wetzel, 2006).

By the early 1900s, the fucoid debate resolved in favor of Nathorst's ideas as emblematically shown by Othenio Abel, the founding father of paleobiology. Among his contributions, there is the introduction of the term *Lebensspur* (Häntzschel, 1975). In the same years, scholars from the German-speaking area recognized the applied value of ichnology, integrating trace fossils with facies reconstruction (e.g., Reis, Schindewolf, and Soergel in Supplementary Material: http://booksite.elsevier.com/9780444538130). The Austrian-Hungarian Empire comprised also the present-day Czech Republic, which saw the ichnological studies of Anton Fritsch (also spelled as Antonín Frič; Häntzschel, 1975).

During the Period of Reaction, the eastern part of Europe attracted the gaze of geologists and paleontologists, who discovered more than enough sites to inspire further exploration and research. The mentioned area encompassed the vast borderland between Austria-Hungary and the Russian Empire, roughly corresponding to the modern-day Poland, Ukraine, western Russia, Baltic, and Caucasian countries. By the early Period of Reaction, Polish geologist Łomnicki (1886) described "*Glossifungites saxicava*" (now *Rhizocorallium jenense*) as a sponge from the Miocene deposits of Lviv (capital city of Galicia, under the influence of Austria-Hungary; now Ukraine). This successively became the area of the nomenclatural archetype of the *Glossifungites* Ichnofacies (Uchman et al., 2000).

However, the Carpathian flysch was to dominate ichnological research in eastern Europe. The extraordinary morphological diversity of traces was to arouse a number of different interpretations: branched structures were mainly explained as fucoids, radial structures as medusae, winding structures as fossil worms (i.e., the studies of Niedźwiecki, Maas, and Keller in Supplementary Material: http://booksite.elsevier.com/9780444538130). Zuber (1918), in addition to *fucoids* and *medusae*, used already such expressions as "traces of crawling" or "traces of worms" in figure captions. Thanks to its ichnological richness and geographical position, the Carpathian flysch became the cradle of the raising Polish school of ichnology. Likewise, the Carpathian Mountains played an important role in the development of Romanian ichnology. Italian and French scholars began the studies of the Romanian ichnological heritage, but it was Protescu who inaugurated the study of the eastern Carpathian flysch. Intriguingly, he compared trace fossils with modern traces (Brustur, 1997).

Russian scholars laid the groundwork for later scientific exploration by beginning the process of documenting the ichnoheritage of Crimea, Ukraine, and Caucasus. Among these early contributors, Bogatschev (1908) presented "*Taonurus*" (now *Spongeliomorpha*) from the Tertiary of Russia and eastern Ukraine and questioned its algal origin. This hypothesis was supported by Kryschtofowitsch (1911), who interpreted *Zoophycos* from eastern Siberia as a trace fossil. Despite these progressive explanations, fucoid and poriferan interpretations were also common (e.g., Bogdanovich, Karakasch; Supplementary Material: http://booksite.elsevier.com/9780444538130).

In North America, James, working independently and in ignorance of Nathorst, arrived at many of the same criteria that Nathorst used to criticize the fucoid origins of many problematic fossils (Osgood, 1975). With his restudy of the systematics of *Fucoides*, *Skolithos*, and *Arthrophycus*, James can rightfully be considered as the first ichnotaxonomist (Pemberton et al., 2007). Still, James did not prevail; his publications, though often prescient, were backward in some respects and not widely available (Osgood, 1970), and his neoichnological comparisons, like Hitchcock's, were to freshwater and terrestrial animals such as insect larvae. James' biographer wrote that it was a pity that such a promising scientist had wasted so much time on the taxonomy of useless material. Although momentum gathered for the ichnological interpretation of problematic structures (notably, with Matthew, Barbour, and Sarle), the fucoid interpretation remained standard in North America until the 1920s and was common even into the 1950s (e.g., Newberry, White; Supplementary Material: http://booksite.elsevier.com/9780444538130).

During the Period of Reaction, the ichnological heritage of southern Africa began to be extensively explored. In these early days, *Spirophyton* was attributed to fucoids, inorganic processes, and impressions of seaweeds of screw-like form. Its first ichnological interpretation was given almost half of a century later (Du Toit, 1954).

8. DEVELOPMENT OF THE MODERN APPROACH

8.1 Decline of Ichnology

By challenging the long-standing belief in fucoids, Nathorst profoundly altered the conceptual fabric that underlay the understanding of trace fossils. This transformation operated at two levels: (1) Nathorst showed that neoichnology was the key to understanding trace fossils, and (2) he dismissed the idea of fucoids. The former level led to the creation of the first organized study of ichnology in the Wadden Sea, which marked the start of the development of the Modern Approach (Osgood, 1970, 1975). The second fact caused a widespread crisis in European ichnology, which represents the predominant context of this historical stage. For these reasons, we will start the discussion from the latter aspect.

Contrary to what one might expect, the acceptance of Nathorst's ideas did not result in a new ichnological revolution. Paradoxically, the disprovement of fucoids brought a decline in interest for trace fossils and a consequent stagnation in ichnological research. The factors accounting for this phenomenon include the nomenclatural problems associated to traces, which consequently fell into a no-man's land (see Knaust, 2012a). Additionally, *fucoids* apparently lost their biostratigraphical value and their application as indicators of shallow-marine, euphotic environments (Cadée and Goldring, 2007; Osgood, 1975). Possibly, this crisis was exacerbated by the aftermath of World War I, corresponding to dramatic social, economic, and geopolitical effects. Notably, four empires disappeared: the German, Austrian-Hungarian, Ottoman, and the Russian.

The French School was the first to be struck by the ichnological crisis, and its once dominant contribution stopped after the previous polemic. Only diverse short and obscure works were published; the nature of *Zoophycos* (=*Cancellophycus*) created some debate, but French scholars interpreted it as an alcyonarian or a sponge (see the studies of Lucas, Dangeard in Supplementary Material: http://booksite.elsevier.com/9780444538130). The same happened in the other major center of the Period of Reaction, Scandinavia, where trace fossils received sporadic attention. Nathorst himself returned to the study of fossil plants (Cadée and Goldring, 2007).

This was a critical moment for ichnology at a global scale. For instance, Italy's ichnological school waned and trace fossils were only reported as a support of stratigraphical works. In Britain, the pioneer work on modern traces in the Wadden Sea by the German school at Frankfurt and Wilhelmshaven in the 1920s and 1930s (Cadée and Goldring, 2007), which established modern scientific ichnology, passed virtually unnoticed. This resulted from the language barrier, lack of European cooperation after World War I, and very little field paleontology or sedimentology in this period. Spain and Portugal saw numerous but regional works, driven by the first extensive geological mapping of the Iberian Peninsula.

A more complex scenario involved eastern Europe. Poland envisaged the same decline as the rest of Europe, with only marginal mentions of trace fossils. At the same time, Czechoslovak geology gained a couple of internationally recognized experts in ichnology, Bedřich Bouček and Ferdinand Prantl. The first one focused on *Skolithos*; the second suggested a facies division of regional units based on trace fossils, to some extent looking forward to the future ichnofacies concept (Supplementary Material: http://booksite.elsevier.com/9780444538130).

In many Eurasian countries, bedding-plane structures were referred to as *hieroglyphs* (Fuchs, 1895), composed of physical and biogenic sedimentary structures; in their strict sense, *hieroglyphs* corresponded to graphoglyptids (Häntzschel, 1975: W17). The area of influence of the Soviet Union produced a vast number of studies on this subject. Specifically, much of the work on *hieroglyphs* is associated with the names of Grossheim, Vassoievitch, and Bogatschev, who studied material from the Caucasus, Carpathians, and

Dagestan. Such studies started in the 1930s and significantly deepened the knowledge of trace fossils, although their exact nature was not always clear. For instance, Grossheim (1946) noted traces of crawling worms but considered *Paleodictyon* as a cast of mudcracks. Thanks to their abundance within the Carpathian flysch, *hieroglyphs* were also commonly reported in the Romanian geological literature. For instance, Mrazec compared *hieroglyphs* to modern insect and gastropod traces, pioneering neoichnological studies in Romania. Interestingly, Filipescu described *Paleodictyon* as a "burrow system" (Brustur, 1997). Despite these insightful interpretations, throughout the first half of the twentieth century, the major part of Romanian geologists used the term *fucoid* to indicate trace fossils.

A period of stagnation was experienced also on the other side of the world. In North America, ichnology was viewed as a minor branch of paleontology, and minor progress was made. Carroll Lane Fenton and his wife Mildred Adams Fenton were paleontologists who turned their careers to popular science writing—and who, like Dawson and Logan, had seashore experience that enabled them to propose plausible makers of Paleozoic trace fossils by use of actualistic reasoning. For example, they showed that phoronids could make *Skolithos linearis* (Supplementary Material: http://booksite.elsevier.com/9780444538130). Other influential studies included those of the Cincinnati-based Kenneth Caster (Devonian limuloid trackways), Lionel Brady (Permian arthropod trackways), and Benjamin F. Howell (*Skolithos* and other early Paleozoic burrows) (Supplementary Material: http://booksite.elsevier.com/9780444538130). The piecemeal nature of this work made progress slow; few attempts were made outside Europe to systematize ichnology at this time. However, the stage was set as the value of actualistic marine research, especially in Germany, France, and England, became more broadly appreciated.

In the same period, ichnology started blooming in South America and Asia. It should be emphasized, however, that these ichnological schools had a different development under different circumstances than European and North American ichnology. For this reason, Osgood's (1970, 1975) periodization serves here only as a chronological reference. South American ichnology went through a slow start. Trace fossils were only occasionally mentioned, almost invariably within the context of regional studies. The Uruguayan school teacher Lucas Rosselli represents an exception, as he accurately documented insect trace fossils from Paleogene paleosols in Uruguay (Roselli, 1938). These deposits host the world's most diverse paleosol ichnofauna, and Rosselli's studies are at the foundation of the present-day school of insect paleoichnology in Argentina and Uruguay. Likewise, Asian ichnological schools followed a gradual development. In India, few studies characterized the development of Modern Approach and the earliest years of the Modern Era. In these early years, Davitashili (1945) coined the term ichnocoenosis to indicate the traces of a biological community (biocoenosis); the term was also independently proposed by Lessertisseur (1956).

In the early stages of ichnological study in Japan, trace fossils were simply referred to as sand pipes or problematica. *Archaeozostera*, a trace fossil occurring in Cretaceous turbidites, has a place of relevance in the ichnological history of Japan. In Japanese folklore, this structure is referred to as *Shobu-ishi*, which relates to plants. Koriba and Miki (1931) assigned the name *Archaeozostera* and considered it a plant fossil of an ancestor of marine seagrass. Nevertheless, the most accepted hypothesis interprets *Archaeozostera* as a trace fossil (e.g., Seilacher, 2008). Ichnology in China started with the report of vertebrate footprints (Young, 1929) and *Cruziana* (Yin, 1932). During this stage, Chinese scholars had only a rudimentary knowledge of trace fossils and provided prevalently morphological descriptions.

8.2 The Senckenberg Marine Institute

The pictured scenario reveals a widespread crisis in almost all Europe and a slow development in the rest of the world. Nevertheless, this historical period is synchronously marked by major advances related to the Senckenberg Laboratory, the first marine institute devoted to the comparison of modern and fossil depositional environments. Founded by Rudolf Richter in 1928 at Wilhelmshaven (Wadden Sea), it is regarded as the birthplace of modern ichnology (Cadée and Goldring, 2007).

What prompted this change in the way of studying traces was a combination of factors. Translations and reviews of Lyell's *Principles of Geology* were published, especially in France and Germany, from the late 1830s onward (Vaccari, 1998). German geoscientists were particularly receptive to modern environments, as shown by the influential work of Johannes Walther (Middleton, 2003). Of further significance for neoichnology, Nathorst's work promptly penetrated the German-speaking area, as recognized by Fuchs (1895).

In this scientific context, Rudolf Richter developed a strong interest for tidal areas as open-air laboratories where one may see geology at work. For instance, he spent several weeks on a *Wohnbake* (a tidal palafitte-like observation post) with his wife Emma (Cadée and Goldring, 2007). It was probably during that and similar fieldwork that he sought a permanent facility for the study of modern environments. Richter fulfilled this idea in 1928, when he founded the "Senckenberg am Meer" Institute in Wilhelmshaven (Germany). From the very beginning of the institute, ichnology occupied a central place. Soon, the Senckenberg Marine Institute became renowned as a center for ichnology and a magnet for biologists and paleontologists alike.

As a result of his actualistic research, Richter coined the terms *Aktuopaläontologie* (Richter, 1928) and *bioturbate Textur* (Richter, 1952). The Senckenberg Marine Institute had profound and far-reaching effects on the history of ichnology, being a source of inspiration for generations of ichnologists. Richter's articles on modern analogs of *Skolithos* were influential for his contemporaries, such as the Fentons (Cadée and Goldring, 2007). In this regard, numerous

English reviews contributed to disseminate Richter's work outside from Germany (Cadée and Goldring, 2007).

When Richter was designated director of the Senckenberg Museum, Walter Häntzschel became his successor at Wilhelmshaven. Walter Häntzschel, who first worked on trace fossils from the Createceous of Saxony, led the Senckenberg Institution from 1934–1938 before he continued his ichnological work in Dresden (see Cadée and Goldring, 2007). In the same period, an important contribution originated far from Senckenberg: it was Othenio Abel's *Vorzeitliche Lebenspuren*, which became the standard reference book over the next 20 years (Abel, 1935; Osgood, 1975).

In 1938, Wilhelm Schäfer succeeded Häntzschel at Wilhelmshaven. Schäfer continued to develop the *Aktuopaläontologie* as an important branch of geosciences, with special focus on facies relationships. Unfortunately, the world was entering a tumultuous period, which culminated in the dramatic events of World War II. Ichnology suffered heavily from these years: Richter tried to reduce Nazi influence in Senckenberg, Häntzschel was drafted in a Russian prison camp, Schäfer was in military service, and *Senckenberg am Meer* was almost destroyed.

After the war, Häntzschel worked as a librarian in Hamburg, becoming an expert in ichnological literature (Cadée and Goldring, 2007). This opened the doors for his most influential contribution, the ichnological section of the Treatise of Invertebrate Paleontology (Häntzschel, 1975). In 1947, Schäfer started to rebuild the Marine Institute and its research facilities. Though never part of the Wilhelmshaven group, Adolf Seilacher had proficient contacts with Schäfer and Häntzschel (Seilacher, 2007). This cultural exchange, together with the studies on the Wadden Sea, formed the base for Seilacher's seminal papers, which initiated the Modern Era of Ichnology (Cadée and Goldring, 2007).

9. MODERN ERA

9.1 The Ethological Revolution

The Seilacherian approach of the early 1950s reshaped ichnology by opening up new pursuits unexpected in the classical approaches of the previous stages. Adolf Seilacher (Fig. 4A) significantly contributed to the conceptual framework of modern ichnology, establishing trace fossils as facies indicators. A considerable part of the conceptual innovations introduced by Seilacher evolved around one concept, *viz.* that trace fossils are a manifestation of behavior. On this side, Seilacher (1953) developed an ethological classification, consisting of basic categories named for the prevailing behavioral pattern. For instance, cubichnia correspond to resting traces. As Cadée and Goldring (2007) argued, Seilacher's ethological classification was possibly influenced

FIGURE 4 Important personalities of the Modern Era. (A) Adolf Seilacher. (B) James Howard. (C) Marian Książkiewicz.

by earlier taxonomical categorizations such as those by Richter and Krejci-Graf (see also Häntzschel, 1975: W17).

During the same years, ichnologists started to pay attention to the spatial arrangement of trace fossils. Seilacher (1953, 1964) and Martinsson (1965) developed two different toponomical classifications, both based on the spatial relationship between a trace fossil and its casting medium (see Bromley, 1996 and Rindsberg, 2012, for a detailed treatment of both classifications). The set of terms introduced by Seilacher and Martinsson were rapidly adapted as a powerful and easy-to-use descriptive tool.

In his forward-looking article, Häntzschel (1955) recognized the value of the *Spuren-Vergesellschaftung* (trace-fossil association) for the characterization of sedimentary environments and *Ichno-Fazies*. These principles were emphasized by Seilacher's ethological approach, stating that trace fossils reveal fossil behavior and, as such, can be modified by the environment. Seilacher discerned the divergence between shallow-marine and deep-sea trace-fossil assemblages at a global scale and throughout the Phanerozoic (Pemberton et al., 2001; Seilacher, 2007). These propositions were the cornerstone for the ichnofacies concept, originally consisting of six sets of traces named for a characteristic ichnotaxon and related to a bathymetric gradient (Seilacher, 1967). Ichnofacies are distinctive, recurring ethological groupings of traces with specific environmental significance. Initially, paleobathymetry was the principal controlling factor, but more environmental implications underlie the modern concept (i.e., substrate consistency, hydrodynamics, turbidity, food resources; MacEachern et al., 2007, 2012).

Seilacher built the archetypal ichnofacies on a great deal of empirical observations (Seilacher, 2007: 205). Indeed, the ichnofacies concept was chronologically preceded by contacts with Schäfer and Häntzschel at the Senckenberg

Institute in Wilhelmshaven, examination of flysch ichnocoenoses in Italy, and an expedition to the Salt Range in Pakistan (Seilacher, 2007: VII). Later, Seilacher has put much effort in the use of trace fossils for paleoecological reconstructions; recognition of depositional environments; trace-fossil evolution; and ichnostratigraphy, ideas, and results which are manifested in his book on *Trace Fossil Analysis* (Seilacher, 2007).

It is important to point out that, in the same period, the school of researchers of the Senckenberg Marine Institute continued to achieve significant results, among which are the works of Häntzschel, Schäfer, Reineck, Dörjes, and Hertweck (Supplementary Material). In particular, Reineck (1963) established a classification of bioturbation from the North Sea as the basis of the modern bioturbation index scheme (Knaust, 2012b). Based on neoichnological studies in the North Sea and on Sapelo Island (Georgia, USA), Günther Hertweck developed methods for the analysis of ichnocoenoses in the light of their sedimentological context, environmental significance, and facies zonation (Hertweck, 1970, 1972). This approach was later developed into the study of ichnofabrics, a concept widely used today (Ekdale et al., 2012). An important milestone in the history of ichnology is the *Trace Fossils and Problematica* volume by Häntzschel. Its second edition (Häntzschel, 1975), chronologically corresponding with other seminal contributions about trace fossils (Basan, 1978; Crimes and Harper, 1970; Frey, 1975) and the history of ichnology (Osgood, 1975; Sarjeant, 1974), broadly defines the limit of the early Modern Era.

9.2 Early Modern Era: A New Impetus for the Study of Traces

Ichnology owes an immense debt to Seilacher, not only for the development of clear goals and methods (Osgood, 1975), but also for giving new impetus to the study of trace fossils. Indeed, in many regions, the resurgence of interest in trace fossils was largely due to the work of Seilacher. Seilacher had frequent contacts with Jacques Lessertisseur in France; Roland Goldring in Britain; and Richard Osgood, among others, in the United States. Osgood (1975) pointed out that, as early as 1956, Lessertisseur authored *Traces fossiles d'activité animale et leur signification paléobiologique*, which gave an excellent panorama of the discipline and reviewed the German work of Seilacher in French. This important volume initiated the Modern Era in France although, despite the recognized importance of trace fossils, ichnology was not considered as a major discipline of the earth sciences.

Goldring et al. (2000) indicated that the development of modern ichnology in Britain emerged from three main phases:

1. The pioneer work of Scott Simpson, who was familiar with the pre-World War II work of Richter's Frankfurt School. From 1955 to 1956, Simpson was the Ph.D. supervisor of Roland Goldring, who met many of the leaders

of the Senckenberg School of researchers, including Richter, Häntzschel, Schäfer, Seilacher, and Reineck (Pollard, 2007).

2. Seilacher's visit in 1962 and subsequent publication of his work in English. During the 1960s, Roland Goldring, George Farrow, Peter Crimes, James Harper, Richard Bromley, and J. I. Chisholm gave a new impulse to ichnology, applying trace fossils to sedimentology, facies analysis, stratigraphical correlation, study of hardgrounds, and core analysis (Supplementary Material: http://booksite.elsevier.com/9780444538130).
3. The First International Trace Fossil Meeting (Liverpool, 1970), which resulted in the volume *Trace Fossils* (Crimes and Harper, 1970), provided a great stimulus to British ichnologists.

The seminal publications of Adolf Seilacher in Germany and Jacques Lessertisseur in France during the 1950s did not have as immediate an impact in North America as they did in Europe, though Ager, Goldring, Simpson, and others were inspired by it in England. When the new ideas were presented in English, geologists in North America began to pay attention. It was Seilacher's demonstration that trace fossils could be used as bathymetric indicators, presented at the Annual Meeting of the Geological Society of America in Cincinnati in 1962, that seized the attention of American researchers. This was a seminal talk indeed, for Seilacher inspired Richard Osgood to begin his innovative dissertation, published in 1970 as *Trace Fossils of the Cincinnati Area*. This work emphasized the relationship of formal ichnotaxa to behavior as opposed to morphology and still ranks as one of the most influential ichnological studies ever performed.

Geologists of the Humble Petroleum Company (later Exxon) invited Seilacher to return to the United States for a summer to explore trace fossils in critical outcrops and core. Graduate student James D. Howard accompanied him and was inspired to conduct his own research in ichnology; and he inspired C. Kent Chamberlain. Seilacher's report was proprietary, but similar work was published (e.g., Seilacher, 1964: fig. 7). To place the significance of the bathymetric use of trace fossils in context, one must recall that until the work of Kuenen and Bouma, flysch was thought by many geologists to be of intertidal origin. Seilacher's research, along with that of micropaleontologists, sedimentologists, and structural geologists, allowed the discovery of vast resources of petroleum.

9.3 The Golden Age of North American Ichnology

The 1970s were a golden age for ichnology, and Robert W. Frey stands out as the most important ichnologist in North America, for he inspired a generation of researchers (Pemberton, 1992). Frey ascribed his initial inspiration in trace fossils to reading Derek Ager's *Principles of Paleoecology* (Rindsberg, 1999). Frey began his research in the Cretaceous chalk of the Western Interior Basin. Realizing that he needed experience with modern processes, he took a course in 1965 at the marine station at Beaufort, North Carolina, where he heard of James Howard

(Fig. 4B), a chance that eventually led to Frey's being hired by the University of Georgia. Several happy years of ichnological research followed at the University of Georgia Marine Institute on Sapelo Island, where the two joined forces and began to publish works that attracted a long series of graduate students. The petroleum-savvy James Howard invited Hans-Erich Reineck and other researchers from the Senckenberg Institute, resulting in two thematic volumes of *Senckenbergiana Maritima* and establishing Sapelo Island as a touchstone for neoichnology. Marginal-marine environments were extensively studied, leading to the discovery of stratigraphic traps in the Western Interior Basin and elsewhere, notably by Robert Weimer and his students at the Colorado School of Mines.

Howard established the *Ichnology Newsletter* in 1968 as a means to share recent findings, and this "gray" periodical, the first devoted to ichnology, has continued episodically to the present day. Frey corresponded with ichnologists throughout the world to put the basics of the science, especially its terminology, on a firm basis. He organized and edited the first summation of ichnology, *The Study of Trace Fossils* (Frey, 1975), the publication of which galvanized many students to enter the field. About the same time, the *Society of Economic Paleontologists and Mineralogists (now: Society for Sedimentary Research, SEPM)* established its *Trace Fossil Research Group*, which was influential during the crucial 1970s, when surges in the price of oil encouraged new methods of exploration. The recognized importance of ichnology in the petroleum industry fostered the study of trace fossils in core. Among the first to bring this approach in practice, C. Kent Chamberlain studied the core appearance of several ichnogenera (Bromley, 1996: 261) and made important achievements in the fields of ichnotaxonomy and paleoethology (Seilacher, 2007: 105).

Although this section deals with the early Modern Era, it is important to note that this golden age laid the basis of present-day North American ichnology, acting by direct transmission of knowledge. An intricate set of connections departs from John E. Warme, who focused on deep-sea ichnology and on bioerosion during the 1970s and 1980s. He taught Allan "Tony" A. Ekdale and others at Rice University before joining the faculty of the Colorado School of Mines, where he taught Andrew K. Rindsberg before moving onto sedimentological topics. Ekdale went on to the University of Utah and developed the ichnofabric approach together with long-term collaborator Richard Bromley. Together with J. F. Bockelie, the pair founded the long-lasting series of *International Ichnofabric Workshops* (see Ekdale et al., 2012); Bromley (1990) successively authored the influential textbook *Trace Fossils: Biology and Taphonomy*. Notably, the book was translated into German, Japanese, and Chinese (Bromley, personal communication). Another example is represented by Frey and Pemberton, who established a fruitful collaboration in the 1980s. Among other accomplishments, they taught one of the first courses in ichnology in North America and founded the first standard journal in the field, *Ichnos* (1990). Pemberton went on to the University of Alberta, teaching many of the most active researchers in petroleum ichnology and leading numerous workshops for petroleum companies around the world.

David Bottjer taught the first ichnological course in North America at the University of Southern California. He is probably best known for developing the tiering concept along with William Ausich, among many other accomplishments (Supplementary Material: http://booksite.elsevier.com/9780444538130). His students include many of today's most active ichnologists. The loose association of researchers connected with the Gerace Research Station on San Salvador (Bahamas) should be mentioned as well. Centered on H. Allen Curran of the Smith College, they include many accomplished researchers. Notably, Curran (1985) edited the first ichnological book treating all major depositional environments. Space does not allow a complete listing of every North American ichnologist here; our intent is to make clear that many of the most influential were inspired and trained by other ichnologists, and many indeed at a few centers.

9.4 The Eastern Bloc During the Early Modern Era

The social and political consequences of World War II influenced the development of ichnology. The rise of English and the decline of German as the international language of science created new channels for disseminating research at a global scale. On the geopolitical side, the Soviet Union and the United States emerged as rival superpowers after the end of the war. The so-called Eastern Bloc, under direct influence of the Soviet Union, developed an independent but complementary school of ichnology.

East German paleontologist Arno Herman Müller worked, based on his foundations of biostratinomy, on many ichnological issues from the Carboniferous, Permian, Triassic, and Cretaceous of Germany. In his classic textbook *Lehrbuch der Paläozoologie*, he dedicated a long appendix on the ichnology of invertebrates with focus on facies reconstruction (Müller, 1989, and references therein).

The Polish geologist Marian Książkiewicz (Fig. 4C) occupies a prominent place in the fields of ichnology and sedimentary geology. At the early beginnings of the Modern Era, Książkiewicz (1954) distinguished between pre- and post-depositional trace fossils; this fundamental idea was developed and popularized in the successive decade by Seilacher (see Seilacher, 2007: 206). Książkiewicz had also a profound impact in ichnotaxonomy; his numerous contributions were summarized in two classical works (Książkiewicz, 1970, 1977), in which trace fossils are classified in nine morphological groups. During his career, Książkiewicz formed one of the largest ichnological collections in the world, which has been studied by later workers, such as Kern and Uchman (Supplementary Material: http://booksite.elsevier.com/9780444538130). Książkiewicz also worked on paleoecology, stratigraphy, taphonomy, and paleoenvironmental significance of flysch trace fossils, trying to calibrate bathymetry of trace fossils on the basis of benthic Foraminifera (Supplementary Material: http://booksite.elsevier.com/9780444538130).

In the same period, an active group of researchers assembled in Warsaw. Centered on Radwański, Karaszewski, Orłowski, and Roniewicz, they focused particularly on early Paleozoic and Jurassic units from the Holy Cross Mountains. Notably, Radwański and Roniewicz developed the ichnocoenosis concept (Häntzschel, 1975: W2). Radwański pioneered the application of borings for defining paleoshorelines, although he did not follow the parataxonomical system in naming trace fossils (Häntzschel, 1975). Among other Polish contributions may be noted those by Nowak, Birkenmajer, Ślączka on ichnotaxonomy and ethological analysis of flysch trace fossils (Häntzschel, 1975).

As concerns the Russian academic scenario, the Modern Era saw many of the main characters of the previous stage. For instance, Vassoievich (1953) supported the ichnological nature of "fucoids" by analyzing several traces from the Caucasus and the Carpathians. Grossheim gave some correct interpretations but withdrew his opinion (Grossheim, 1946) about a mudcrack origin of *Paleodictyon* and placed it among biogenics (Imprints of snail eggs or imprints of algae). Similarly, he regarded *Ubinia wassoevitschi* as seaweed and interpreted some star-shaped trace fossils (*Lorenzinia*, *Atollites*) as imprints of ammonite shells (Supplementary Material: http://booksite.elsevier.com/9780444538130).

Within the middle to late twentieth century, ichnologists came to see Russia and central Asia as a place of interest for field investigation. This period of intense ichnological traveling was inaugurated between 1930 and 1960 by Hecker and Vialov, who described many invertebrate and vertebrate ichnotaxa from Russia, Ukraine, and central Asia (Häntzschel, 1975). Vialov also proposed a new classification of trace fossils (Häntzschel, 1975: W23–W24). However, his definition of trace fossils was broad, unfitting to the mainstream opinion (i.e., Bertling et al., 2006; Knaust, 2012a). Nevertheless, Vialov's ideas are still vivid in many Russian scientific institutions, partly because of his influential textbook (Mikuláš and Dronov, 2006; Vialov, 1966).

Between the late 1950s and 1970s, Romanian paleontology saw significant achievements in the field of vertebrate ichnology, for which reason Brustur (1997) distinguished a Stage of Vertebrate Footprints in Romanian Ichnology. An important figure of this stage is the academician Gheorghe Murgeanu, who promoted several sedimentological investigations, while Nicolae Panin gave a great impulse to paleoichnological studies. In the same period, Miroslav Plička, Ilja Pek, and Ivo Chlupáč contributed to Czech ichnology by ichnotaxonomical proposals, descriptions of trace fossils in unusual substrates and settings, and ethological interpretations (Supplementary Material: http://booksite.elsevier.com/9780444538130).

9.5 Global Ichnology

The fruitful advancements made at the European and North American schools were accompanied by a global momentum guiding the development of ichnology at a large scale. Nevertheless, mode and tempo of this expansion were slightly different in Asia, South America, and Australia with respect to the other areas.

Between 1950 and 1965, Rodolfo Casamiquela and the Italian-Argentinean Joaquin Frenguelli opened the avenues for the development of modern South American ichnology. While previous studies mostly mentioned trace fossils as accessory components, their works were entirely devoted to ichnology (Supplementary Material). The birth of South American Ichnology took place between the 1960s and the 1980s and is illustrated by the work of researchers in Argentina, Venezuela, and Brazil. Although the scope of some of these studies is still mostly regional, trace fossils moved from the margins to the core. This tendency undoubtedly reflected global developments in the discipline, most likely the influential work of Adolf Seilacher, who was personally connected with some of the classic figures of this age during his frequent trips to South America. At the same time that Borrello undertook his research in Argentina, Macsotay (1967) dealt with Cenozoic turbidite trace fossils from Venezuela. Macsotay's work gave a strong impulse to the use of trace fossils in paleoenvironmental reconstructions, particularly oriented to the flourishing oil industry in Venezuela, an applied side of ichnology that was subsequently explored by another Venezuelan ichnologist, Nicolas Muñoz. In the same period, Rodolfo Casamiquela and the Italian-Brazilian priest Giuseppe Leonardi investigated vertebrate ichnology (see Fernandes et al., 2002, for a comprehensive bibliographic review).

The next phase of studies within this age took place in the 1970s and 1980s, particularly with the work by Florencio Aceñolaza and his research associates in the Precambrian and early Paleozoic units of north-western Argentina. In the same years, there was a rapid development of Brazilian invertebrate ichnology, as shown by multiple contributions of Gerardo Muñiz and Antonio Fernandes (Fernandes et al., 2002). In Brazil, new studies were undertaken during the 1980s by Henrique Godoy Ciguel in Paleozoic units and Ismar de Souza Carvalho, mostly on vertebrate ichnology (Fernandes et al., 2002). Among other Argentinean contributions may be noted those by Ricardo Alonso on vertebrate ichnology, Jose Laza on ant trace fossils, Poiré on the sedimentological application of trace fossils, and Luis Buatois and Gabriela Mángano whose multifaceted work is summarized in a textbook (Supplementary Material: http://booksite.elsevier.com/9780444538130).

On the other side of the ocean, Asian ichnology followed a composite development, to some extent comparable to the recent history of South American ichnology. In Japan, trace fossils have received vague interpretations ("sand pipes", "problematica") since the 1960s, when the description and taxonomy of some Japanese trace fossils were carried out by Katto, Shuto, and Shiraishi (Supplementary Material: http://booksite.elsevier.com/9780444538130). During the early 1970s and 1980s, there was an increased interest in trace fossils as environmental indicators. For example, Kikuchi (1972) suggested the use of "white vermiform trace fossils" (successively identified as *Macaronichnus segregatis*) for recognizing beach environments. Studies on paleoethology and tracemaking mechanisms began in the 1980s with Kotake, famed for his model of *Zoophycos* (Kotake, 1989). In the following years, scientists such as Nara, Naruse, and Nifuku continued to actively develop paleoichnology, greatly improving the knowledge of the Japanese ichnological heritage (Supplementary Material: http://booksite.elsevier.com/9780444538130).

At the end of the 1970s, science and technology in China became an integral part of the socio-economic development of the country. At the same time, the concepts and methods of trace-fossil analysis were introduced from abroad, being formalized in textbooks such as *Introduction to Ichnology* (Wu, 1986). This led to a period of intense ichnological research, characterized by systematic studies. Chinese sedimentary units yield great abundance of both shallow-marine and deep-sea trace fossils, including excellent trace-fossil assemblages at the Precambrian/Cambrian boundary. Marine trace fossils were studied by, among others, Yang Shi-pu, Yang-Zun-yi, Gong Yi-ming, Wang Yue, Hu Bin, Jin Hui-juan, and Li Yu-ci (Yang et al., 2004). Since the late 1980s, scientists such as Wu Xian-tao, Hu Bin, and Shi Zhen-sheng studied continental trace fossils. At this stage, ichnological analysis aimed at both paleoecological reconstruction and applied investigations (i.e., energy exploration; Hu et al., 1997) while ichnofabric analysis started by the end of the 1990s (Supplementary Material: http://booksite.elsevier.com/9780444538130).

Over the course of the Modern Era, ichnological investigations rapidly diversified in India. The study of Phanerozoic trace fossils started at the end of the 1960s, stimulated by the rich ichnological heritage of the Cretaceous units of central-western India. In particular, Chiplonkar and Badve initiated a series of influential studies on the Bagh Beds, which successively attracted a vast number of researchers (i.e., Verma, Ghare, Sanganwar, Kundal, Kumar; Supplementary Material: http://booksite.elsevier.com/9780444538130). By the same years, the first reports of Neoproterozoic burrows (Among others Misra and Awasthi, 1962) raised a “trace-fossil rush” in the Vindhya Mountains. As a consequence, Proterozoic trace fossils were abundantly explored by ichnologists such as Sarkar, Maithy, Singh, and Chakraborti, just to cite some (see Maithy, 1990, for a comprehensive reference list). Despite the bewildering diversity of tracemaking invertebrate communities and depositional environments, neoichnology is the most recent step in Indian ichnology. Since 1980, the Sundarban Delta Complex, hosting the world’s largest mangrove forest, has been studied by Bakshi, Chakrabarti, Chattopadhyay, and De, among others.

Following a slow start, the development of ichnology into an established subdiscipline of geology in southern Africa only occurred in the early 1970s and 1980s (Modern Era of Ichnology). This period is marked by the publication of the first detailed trace-fossil descriptions, ichnotaxonomic treatments, and ichnologically based biostratigraphic and paleoecological reconstructions. This golden era of southern African ichnology is most of all featured in contributions of international importance by Ann Anderson (Paleozoic invertebrate trace fossils, South Africa), Paul Ellenberger (Mesozoic vertebrate tracks, Lesotho), and Gerald Germs (Precambrian and Cambrian invertebrate trace fossils, Namibia and South Africa; Supplementary Material). These active years also led to pioneering application of trace fossils to sedimentary facies interpretation and incorporation of neoichnological studies in trace-fossil interpretations (see works by Hobday, Mason, Shone, Smith, Stanistreet, Rust, and Turner; Supplementary Material: http://booksite.elsevier.com/9780444538130).

10. CONCLUSIONS AND DISCUSSION

10.1 Evolution of the Interpretation of Trace Fossils

The collection of numerous bibliographic resources allows building a database (Supplementary Material: http://booksite.elsevier.com/9780444538130), graphically resulting in a semiquantitative evolutionary tree of ichnology (Fig. 5). This "tree of ichnology" supports the reliability of the current periodization of the history of ichnology and shows very well the phases of ferment and stasis in

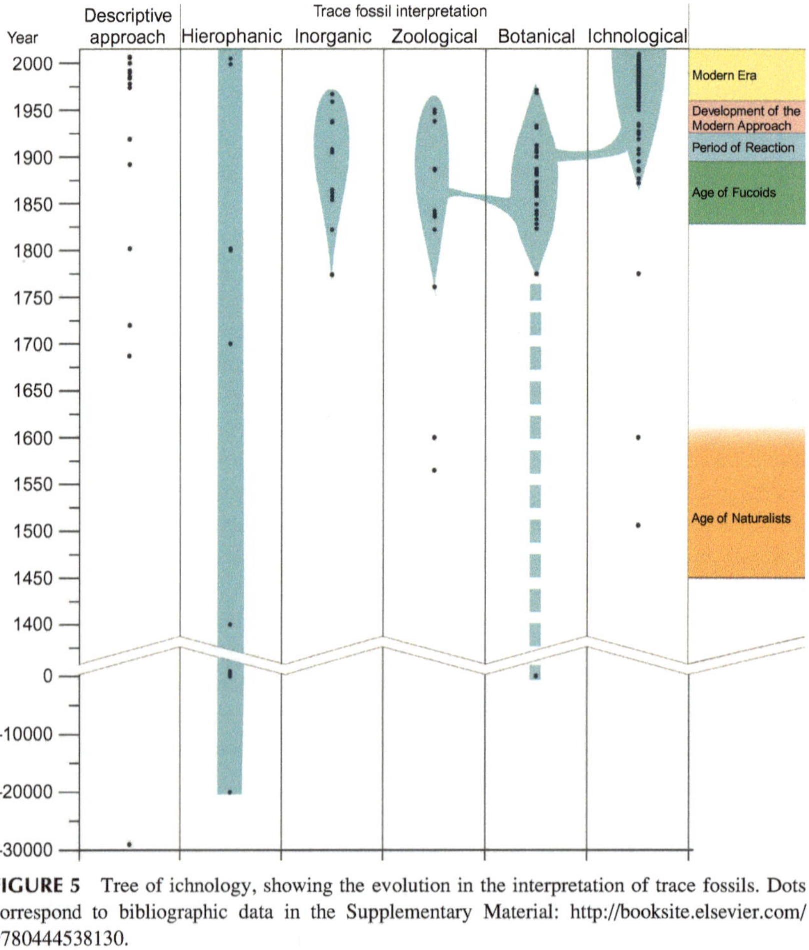

FIGURE 5 Tree of ichnology, showing the evolution in the interpretation of trace fossils. Dots correspond to bibliographic data in the Supplementary Material: http://booksite.elsevier.com/9780444538130.

ichnological research, that is, the crisis that followed the dismissal of fucoids (Fig. 5). Accordingly, the history of ichnology is paced by discontinuities and dominant interpretations; although at a narrower temporal scale, the development of ichnological concepts has an incremental (conservative) evolution.

10.2 Modern Centers of Ichnological Research

A great part of modern-day ichnological schools is best seen in the light of past historical events. The recent history of ichnology pictures a decline in working invertebrate ichnologists in many cultural areas, such as Italy, France, and Britain. While this critical situation is a recent phenomenon in Britain, the Italian and French decline is a direct by-product of the Age of Reaction. However, fields such as the ethology of trace fossils, facies analysis, and the study of arthropod trackways still remain active in these cultural areas. In contrast, new centers have arisen in Europe, South America, and Asia (e.g., Spain, Brazil, and China) as a direct consequence of the global impetus of the early modern era. Likewise, direct transmission of knowledge played an important role in the development of many ichnological schools, that is, the major part of active North American researchers was trained by other ichnologists during the golden age of the 1970s.

10.3 Modern Trends in Ichnology as a Legacy from the Past

Isaac Newton's famous admonition about standing on the shoulders of giants fits well with present-day ichnology. In fact, modern ichnology expanded substantially on the achievements of prior stages, which were influencing and guiding its contemporary trends. For instance, the ethological revolution of the 1950s established trace fossils as representing fossil behavior, and for this reason, they are nowadays crucial ingredients of both evolutionary paleoecology and paleobiology. The mentioned assumptions are valid even for those concepts explained by the impact of individual researchers, such as ichnofacies. Obviously, the ichnofacies concept primarily derived from the brilliant intuitions of Adolf Seilacher, although it would not exist in the present form without a set of antecedents such as Nathorst and the Wilhelmshaven school.

However, the ichnofacies approach is based on the interpretation of individual morphologies, and for this reason, it is difficult to apply to well cores (Seilacher, 2007: 201). Consequently, since the 1980s, some workers have encouraged consideration of those aspects of the texture that result from bioturbation and bioerosion: ichnofabrics (Bromley and Ekdale, 1986; Ekdale et al., 2012). The ichnofabric paradigm appeared to be an excellent methodology for paleoenvironmental reconstruction, complementary to the ichnofacies approach. It found particular application in the study of coalfields (i.e., Pollard, 1988) and hydrocarbon reservoirs (Knaust, 2012b; Taylor and Goldring, 1993).

Not only does the importance of historical awareness regard the question of memory, but also analogs from history can serve as a guide or inspiration for future insights. For instance, history reminds us of the cultural value of trace fossils. Already in the seventeenth century, Bauhin included invertebrate trace fossils in his tourist guide of Bad Boll, Germany (Seilacher, 2007) and Jacinto Pedro Gomes pioneered geoconservation of dinosaur tracksites in the nineteenth century (Supplementary Material: http://booksite.elsevier.com/9780444538130). Vertebrate tracksites play a major role in ichnological geotourism, although invertebrate geosites are also under development, providing new economic activities and additional sources of income, especially in rural regions (i.e., the Ichnological Park of Penha Garcia, Portugal; Neto de Carvalho et al., 2009).

As concerns scientific research, history shows us that almost every major idea in ichnology has depended on neoichnological observations; it is a pervasive influence. As Gingras et al. (2011) argued, the models we have for animal/sediment relations are largely based on neoichnological studies of the 1950s, 1960s, and 1970s. History warns us that for higher-resolution models, new studies on modern environments are required.

10.4 Disciplinary and Interdisciplinary Aspects in Ichnology

Owing to its nature, ichnology is a system of knowledge with a very well-defined nucleus (traditional ichnology) but with poorly defined borders. "Traditional ichnology" is often based on a geological background, but it blurs in a vast gray area shared with other disciplines. A typical example is given by recent traces, the study of which is often referred to biology, archeology, or forensic science, avoiding mention to ichnology (Baucon et al., 2008; Bromley, 1996: xi). This tendency can be traced back to the nineteenth century (i.e., Wood, 1866), and it is still influencing the study of traces. For instance, significant overlap exists between neoichnology and tracking, a discipline practiced by hunting guides, biologists, search-and-rescue teams, soldiers in war, and forensic investigators (Cunningham, 2004; Liebenberg et al., 2010). According to Liebenberg et al. (2010), tracking involves each and every indication of an animal's presence. It typically includes not only traces (i.e., footprints, feces, burrows) but also other structures (i.e., eggs, auditory signs, etc.).

The present organization of scientific knowledge is a product of historical phenomena, and the study of traces makes no exception. While earth sciences had a prevalent role in the evolution of traditional ichnology, the development of modern tracking has been guided by military, zoological, and forensic specialties since the early nineteenth century. Despite of the commonality of subject matters and internal logic, ichnology is poorly connected to tracking, although recent cases of mutual recognition are recorded (i.e., Eiseman et al., 2010: ix; Lockley and Meyer, 2000: 1).

This scenario fits a long-standing trend in science: the uncommunicative piling-up of similar fields of research (Campbell, 1969). Such clustering of specialties results in lack of communication (i.e., unshared knowledge between ichnologists and biologists) and duplication of effort (i.e., ichnology and tracking embody their own separate terminology, nomenclature, and community). Additionally, this phenomenon produces knowledge gaps between or at the edges of disciplinary clusters. This is the case of root-related structures, which fall in a land of convergence between body and trace fossils, therefore posing a semantic problem in their categorization (Gregory et al., 2004). Although Sarjeant (1975) clearly recognized root systems as trace fossils, they still remain an understudied field. Yet, the assessment of plant trace fossils is typically limited to identifying such structures as "root traces" (Gregory et al., 2004).

Besides the aforementioned semantic issues, the interplay of interpretative, cultural, and historical factors explains the present state of plant ichnology. In fact, plant trace fossils offer significant interpretative challenges, while most of the active ichnologists have a prevailing zoological training (Gregory et al., 2004). Additionally, two distinct historical phenomena played a crucial role. First, Nathorst's neoichnological experiments involved crustaceans, annelids, and bivalves (Cadée and Goldring, 2007), thus suggesting animals as the main actors in ichnology. Second, and of equal importance, the path between the Period of Reaction and the Modern Era was mediated by the Senckenberg Institute. Being a marine research center, it focused on marine invertebrates, consolidating the leaning toward animal tracemakers.

10.5 Ichnology as a Historical Product

An important point, for our understanding of ichnological innovation, is that the historical background plays a parallel role in respect to the scientific one. Paraphrasing Spencer (1896), ichnologists are the products of their societies, and their actions would be impossible without the conditions built before their lifetime. This question does not only involve catastrophic geopolitical events (e.g., the French Revolution), but also relies on the whole social and cultural scenario. For instance, success of an ichnological concept depends as much on the idea itself as on its recognition by the scientific community. A clear example comes from the roots of ichnology: the innovative ideas of Leonardo da Vinci did not influence the course of ichnology because he compiled handwritten manuscripts in mirror-image Italian, at a time when scientific communication rested on Latin treatises. It is therefore evident that the question of language and scientific communication plays a crucial role. Dominant schools of thought often coincide with the *lingua franca* of the moment: just coincidence or factual interrelation?

Technology has been a driving force both in the field of scientific communication and in the design of analytic tools. For instance, the invention of movable-type printing spreads the results of the Scientific Revolution and allowed the establishment of periodical reports such as the *Philosophical Transactions of the Royal Society*, one of the oldest scientific journals disseminating ichnological ideas. This historical aspect is clearly seen in the radical changes introduced by the advent of digital media. The impact of digital technologies such as internet, email, and PDF, brought the ability to easily move ichnological information between media and to access or distribute it remotely. Similarly, they implied a greater interconnectedness between researchers and encouraged social research. Nowadays, the *Skolithos* forum and the *Ichnology Newsletter* are digital-based media widely used for informal communications. However, in contrast with other scientific disciplines, ichnology is lacking a database of ichnological data. With increased accessibility to and elaboration upon advanced analytical techniques, new methods and perspectives in the study of trace fossils are recently emerging. These include molecular paleontology methods, 3-D visualization, and computer modeling of trace fossils, topoichnology, theoretical foraging (i.e., the studies by Gong, Hu, Si, Plotnick in Supplementary Material: http://booksite.elsevier.com/9780444538130 see Knaust, 2012b). It is desirable that these innovations should bring the same stimuli of the techniques designed and improved by the Senckenberg am Meer scientists (e.g., box cores, sediment peels, resin casts; Cadée and Goldring, 2007).

In conclusion, this historical synthesis shows that innovation in ichnology is fueled by a complex interplay of factors: not only scientific but also geopolitical, social, and technological. Historical knowledge of such factors will inspire new directions in further investigations and define the place of ichnology in our culture.

ACKNOWLEDGMENTS

Andrea Baucon sincerely thanks Richard G. Bromley and Dirk Knaust for the invitation to compile this contribution. The chapter greatly benefited from the constructive reviews of Günther Hertweck and Richard G. Bromley. This chapter directly benefited from helpful correspondence with Leonardo Ambasciano, Matteo Belvedere, David Bressan, Mathias Heckroth, Scott Linklater, Duncan McIlroy, Fabio Massimo Petti, and Maša Surić. The work of Hu, Hu, Song, Wang, Yi-ming, Zhang, and Zhang was supported by the National Natural Science Foundation of China (No. 40872001) and by the Foundation from the Higher Education Innovative Teams of Henan Province, Developing Projects of Science and Technology of Henan Province (No. 092300410167), and the Doctoral Program Fund of High education of China (No.20094116110002), as well as Open Projects of Key Laboratory of Biogenic Traces & Sedimentary Minerals of Henan Province (No. OTMP0904).

REFERENCES

Abel, O., 1926. Amerikafahrt. Fischer, Jena, 462 pp.

Abel, O., 1935. Vorzeitliche Lebensspuren. Fischer, Jena, 644 pp.

Ammon, U., 1998. Ist Deutsch noch internationale Wissenschaftssprache? Englisch auch für die Lehre an den deutschsprachigen Hochschulen. De Gruyter, Berlin, New York, 330 pp.

Basan, P.B. (Ed.), 1978. Trace Fossil Concepts. SEPM Short Course 5, 181 pp.

Batsch, A.J.G.C., 1802. Taschenbuch für mineralogische Excursionen in die umliegende Gegend um Jena. Industrie–Comptoirs, Weimar, 361 pp.

Baucon, A., 2009. Ulisse Aldrovandi (1522–1605): the study of trace fossils during the Renaissance. Ichnos 16, 245–256.

Baucon, A., 2010. Leonardo da Vinci, the founding father of ichnology. Palaios 25, 361–367.

Baucon, A., Privitera, S., Morandi Bonacossi, D., Canci, A., Neto de Carvalho, C., Kyriazi, E., Laborel-Deguen, F., Morhange, C., Marriner, N., 2008. Principles of ichnoarchaeology: new frontiers for studying past times. In: Avanzini, M., Petti, F. (Eds.), Italian Ichnology. Studi Trent. Sci. Nat. Acta Geol., vol. 83, pp. 43–72.

Bertling, M., Braddy, S.J., Bromley, R.G., Demathieu, G.R., Genise, J., Mikuláš, R., Nielsen, J.K., Nielsen, K.S.S., Rindsberg, A.K., Schlirf, M., Uchman, A., 2006. Names for trace fossils: a uniform approach. Lethaia 39, 265–286.

Bogatschev (=Bogachev), V.V., 1908. Die problematische Alge *Taonurus* im russischen Paläogen. Ezhegodnik po Geologii y Mineralogii Rossi 10, 221–226.

Bromley, R.G., 1990. Trace Fossils: Biology and Taphonomy. Unwin Hyman, London, 280 pp.

Bromley, R.G., 1996. Trace Fossils: Biology, Taphonomy and Applications. Chapman & Hall, London, 361 pp.

Bromley, R.G., Ekdale, A.A., 1986. Composite ichnofabrics and tiering of burrows. Geol. Mag. 231, 59–65.

Brongniart, A., 1823. Observations sur les fucoides et sur quelques autres plantes marines fossiles. Mém. Soc. d'Hist. Nat. Paris 1, 301–321.

Brongniart, A., 1828. Histoire des végétaux fossiles ou recherches botaniques et géologiques sur les végétaux renfermés dans les diverses couches du globe. Dufour et d'Ocagne, Paris 1, 1–488.

Bronn, H.G., 1837. Lethaea Geognostica oder Abbildungen und Beschreibungen der für die Gebirgs-Formationen bezeichnendsten Versteinerungen, vol. 1. Schweizerbart, Stuttgart, 544 pp.

Brustur, T., 1997. The stages of paleoichnological studies in Romania. Geo-Eco-Marina 2, 205–216.

Buckland, W., 1822. Account of an assemblage of fossil teeth and bones of elephant, rhinoceros, hippopotamus, bear, tiger and hyaena, and sixteen other animals; discovered in a cave at Kirkdale, Yorkshire, in the year 1821. Phil. Trans. R. Soc. Lond. 112, 171–236.

Buckland, W., 1836. Geology and Mineralogy Considered with Reference to Natural Theology, vol. 2. William Pickering, London, 599 pp., 128 pp.

Burkhardt, F., Smith, S., 1987. The Correspondence of Charles Darwin: 1844–1846. Cambridge University Press, Cambridge, 523 pp.

Cadée, G.C., Goldring, R., 2007. The Wadden Sea, cradle of invertebrate ichnology. In: Miller III, W. (Ed.), Trace Fossils. Concepts, Problems, Prospects. Elsevier, Amsterdam, pp. 3–13.

Campbell, D.T., 1969. Ethnocentrism of disciplines and the fish-scale model of omniscience. In: Sherif, M., Sherif, C.W. (Eds.), Interdisciplinary Relationships in the Social Sciences. Aldine, Chicago, pp. 328–348.

Capellini, G., 1867. Ricordi di un viaggio scientifico nell' America Settentrionale nel MDCCCLXIII. G. Vitali, Bologna, 279 pp.

Chew, P.G.L., 2009. Emergent Lingua Francas and World Orders: The Politics and Place of English as a World Language. Taylor & Francis, New York, 287 pp.

Crimes, T.P., Harper, J.C. (Eds.), 1970. Trace Fossils. Seel House Press, Liverpool. Geol. J., Spec. Iss. 3, 547 pp.

Cunningham, T.J., 2004. Scout Tracking Operations: Basic Trailing and Surveillance Concepts for Operational Trackers. Lost Trail Ranch-Scout Tracking School, Cheyenne, 33 pp.

Curran, H.A., 1985. Biogenic Structures: Their Use in Interpreting Depositional Environments. SEPM Spec. Publ., vol. 35, 347 pp.

Davitashili, L.S., 1945. Tsenozy zhivykh organizmovI organicheskikh ostatkov [Coenoses of living organisms and organic remains]. Akademiya Nauk Gruzinskoy SSR 6, 527–534 (in Russian).
de Gibert, J.M., Domènech, R., Martinell, J., 2012. Rocky shorelines. In: Knaust, D., Bromley, R.G. (Eds.), Trace Fossils as Indicators of Sedimentary Environments. Developments in Sedimentology, vol. 64. Elsevier, Amsterdam, pp. 441–462.
De La Beche, H.T., 1846. On the formation of the rocks of south Wales and south western England. Mem. Geol. Surv. Great Britain 1, 1–296.
d'Orbigny, A., 1849. Cours Élémentaire de Paléontologie et de Géologie Stratigraphique, vol. 1. Masson, Paris, 299 pp.
Delgado, J.F.N., 1903. Note sur *Scolithus* Dufrenoyi ROUAULT. Com. Serv. Geol. Portugal 5, 251–253.
De Prado, C., 1864. Descripción Física y Geológica de la Provincia de Madrid. Junta General de Estadística, Madrid 219 pp.
de Saporta, G., 1873. Plantes Jurassiques, Tome 1. Algues, Equisétacées, Characées, Fougères. Paléontologie Française ou Description des Fossiles de la France. 2 série, Végétaux. Masson, Paris, 506 pp., Atlas 70 pl.
de Saporta, G., 1882. A Propos des Algues Fossiles. Masson, Paris, 82 pp., 9 pl.
Duffin, C.J., 2009. "Records of warfare...embalmed in the everlasting hills": a history of early coprolite research. Mercian Geol. 17, 101–111.
Dunn, E.J., 1872. Through Bushmanland, Part 1. Cape Monthly Magazine 5, 375–376.
Du Toit, A., 1954. The Geology of South Africa. Oliver and Boys, Edinburgh, 611 pp.
Eiseman, C., Charney, N., Carlson, J., 2010. Tracks and Sign of Insects and Other Invertebrates: A Guide to North American Species. Stackpole Books, Mechanicsburg, 582 pp.
Ekdale, A.A., Bromley, R.G., Knaust, D., 2012. The ichnofabric concept. In: Knaust, D., Bromley, R.G. (Eds.), Trace Fossils as Indicators of Sedimentary Environments. Developments in Sedimentology, vol. 64. Elsevier, Amsterdam, pp. 139–155.
Ellena, E., Escalante, V., 2007. The Last Trackers of the Outback. Amun Ra Films, 51 min.
Fernandes, A.C.S., Borghi, L., Carvalho, I.S., Abreu, C.G., 2002. Guia dos Icnofósseis de Invertebrados do Brasil. Interciencia, Rio de Janeiro, 260 pp.
Forbes, E., 1848. On *Oldhamia*, a new genus of Silurian fossils. J. Geol. Soc. Dublin 4, 20.
Fortis, A., 1774. Alvise Milocco. Venice, vol. 2. 204 pp.
Frey, R.W. (Ed.), 1975. The Study of Trace Fossils. Springer, Berlin, 562 pp.
Frey, L., Frey, M., 2004. The French Revolution. Greenwood Press, Westport, 195 pp.
Fuchs, T., 1895. Studien über Fucoiden und Hieroglyphen. Denkschr. Kais. Akad. Wiss. Wien, Math.-Nat. Kl) 62, 369–448.
Gesner, K. (Ed.), 1565. De Omni Rerum Fossilium Genere. Iacobus Gesnerus, Zürich, 169 pp.
Gingras, M.K., MacEachern, J.A., Dashtgard, S.E., 2011. Process ichnology and the elucidation of physico-chemical stress. Sediment. Geol. 237, 115–134.
Gohau, G., 1991. A History of Geology. Rutgers University Press, New Brunswick, 259 pp.
Goldring, R., Pollard, J.E., Rindsberg, A.K., 2000. Development of invertebrate ichnology in UK: Roland Goldring and John Pollard in interview. Ichnol. Newslett. 22, 40–45.
Golubic, S., Perkins, R.D., Lukas, K.J., 1975. Boring microorganisms and microborings in carbonate substrates. In: Frey, R.W. (Ed.), The Study of Trace Fossils. Springer, Berlin, pp. 229–259.
Gregory, M.R., Martin, A.J., Campbell, K.A., 2004. Compound trace fossils formed by plant and animal interactions: quaternary of northern New Zealand and Sapelo Island, Georgia (USA). Fossils Strata 51, 88–105.
Grossheim, V.A., 1946. O znachienii i metodike izucheniya ieroglifov (na materiale Kavkazkogo flisha) [About the meaning and methods of the study of hieroglyphs (based on material from the Khazak flysch)]. Izvestiya Akad. Nauk SSR, Ser. Geol. 2, 111–120 (in Russian).

Häntzschel, W., 1955. Lebensspuren als Kennzeichen des Sedimentationsraumes. Geol. Rundsch. 43, 551–662.

Häntzschel, W., 1975. Trace fossils and problematica. In: Teichert, C. (Ed.), Treatise on Invertebrate Paleontology, Part W, Miscellanea, Supplement 1. Geological Society of America and University of Kansas Press, Boulder, Colorado and Lawrence, Kansas, pp. W1–W269.

Happel, E.G., 1707. Relationes Curiosae oder Denkwürdigkeiten der Welt. Remannischer Verlag, Hamburg, 675 pp.

Heer, O., 1876/1877. Flora Fossilis Helvetiae. Die vorweltliche Flora der Schweiz. J. Wurste, Zürich, 182 pp.

Hertweck, G., 1970. The animal community of a muddy environment and the development of biofacies as effected by the life cycle of the characteristic species. In: Crimes, T.P., Harper, J.C. (Eds.), Trace Fossils. Geol. J., Spec. Iss. 3, pp. 235–242.

Hertweck, G., 1972. Georgia coastal region, Sapelo Island, U.S.A.: sedimentology and biology. V. Distribution and environmental significance of lebensspuren and in-situ skeletal remains. Senckenberg. Marit. 4, 125–167.

Hislop, S., 1860. On the Tertiary deposits associated with Trap rocks in the East Indies. Quat. J. Geol. Soc. Lond. 16, 154–166.

Hitchcock, E., 1836. Ornithichnology. Description of the footmarks of birds (*Ornithichnites*) on New Red Sandstone in Massachusetts. Am. J. Sci. 29, 307–340.

Hitchcock, E., 1858. Ichnology of New England. William White, Boston, 348 pp.

Hu, B., Wang, G., Qi, Y., 1997. Hen Ji Xue Li Lun Yu Ying Yong [Theory and Application of Ichnology]. University of Mining Press, Xuzhou (Jiangsu), 209 pp. (in Chinese).

Kikuchi, T., 1972. Narita-so san hakuhanjo seikon-kaseki to sono kochirigakuteki igi [A characteristic trace fossil in the Narita Formation and its paleogeographical significance]. J. Geol. Soc. Japan 78, 137–144 (in Japanese, with English summary).

Knaust, D., 2012a. Trace-fossil systematics. In: Knaust, D., Bromley, R.G. (Eds.), Trace Fossils as Indicators of Sedimentary Environments. Developments in Sedimentology, vol. 64. Elsevier, Amsterdam, pp. 79–101.

Knaust, D., 2012b. Methodology and techniques. In: Knaust, D., Bromley, R.G. (Eds.), Trace Fossils as Indicators of Sedimentary Environments. Developments in Sedimentology, vol. 64. Elsevier, Amsterdam, pp. 245–271.

Koriba, K., Miki, S., 1931. Hakua-ki izumi-sagan no kaseki kodaiamamo (shinshu) ni kansuru kousatsu [Consideration on *Archaeozostera* (new name) from the Cretaceous Izumi Sandstone]. Chikyu 15, 165–204 (in Japanese).

Kotake, K., 1989. Paleoecology of the *Zoophycos* producers. Lethaia 22, 327–341.

Kryschtofowitsch, A.N., 1911. Ueber problematische Algenreste *Taonurus-Spirophyton* aus Juraablagerungen des Ussuri–Golfes. Izvestia Geologicheskogo Komiteta (Bulletins du Comité Géologique) 191, 477–486.

Książkiewicz, M., 1954. Uwarstwienie frakcjonalne i laminowane we fliszu karpackim [Graded and laminated bedding in the Carpathian Flysch]. Rocznik Polskiego Towarzystwa Geologicznego 22, 399–471 (in Polish, with English summary).

Książkiewicz, M., 1970. Observations on the ichnofauna of the Polish Carpathians. In: Crimes, T.P., Harper, C. (Eds.), Trace Fossils. Geol. J., Spec. Iss. 3, pp. 283–322.

Książkiewicz, M., 1977. Trace fossils in the flysch of the Polish Carpathians. Palaeont. Polon. 36, 1–208.

Lebesconte, P., 1883. Œuvres posthumes de Marie Rouault; suivies de les *Cruziana* et *Rysophycus*, connus sous le nom général de *Bilobites*, sont–ils des végétaux ou des traces d'animaux? Oberthur, Rennes-Paris, pp. 1–73.

Lessertisseur, J., 1956. Traces fossiles d'activité animale et leur signification paléobiologique. Mémoires de la Société Géologique de France, Nouvelle série 74, 150 pp.

Legati, L., 1677. Museo Cospiano Annesso a Quello del Famoso Ulisse Aldrovandi. Giacomo Monti, Bologna, 532 pp.

Lémery, N., 1716. Dictionnaire ou traité universel des drogues simples. Aux Dépens de la Compagnie, Amsterdam, 695 pp.

Lhwyd, E., 1760. Lithophylacii Britannici ichnographia. Edition Alterata. Clarendon Press, Oxford, 156 pp.

Liebenberg, L., Louw, A., Elbroch, M., 2010. Practical Tracking: A Guide to Following Footprints and Finding Animals. Stackpole Books, Mechanicsburg, 344 pp.

Lockley, M.G., Meyer, C., 2000. Dinosaur Tracks and Other Fossil Footprints of Europe. Columbia University Press, New York, 323 pp.

Lockley, M.G., Novikov, V., Santos, V.F., Nessov, L.A., Forney, G., 1994. "Pegadas de Mula": an explanation for the occurrence of Mesozoic traces that resemble mule tracks. Ichnos 3, 125–133.

Łomnicki, A.M., 1886. Słodkowodny utwór trzeciorzędny na Podolu galicyjskim [Tertiary freshwater deposits in the Galician Podolia]. Sprawozdanie Komisji Fizjograficznej Akademii Umiejętności w Krakowie 20, 48–119 (in Polish).

Lowe, P., 2002. Hunters and Trackers of the Australian Desert. Rosenberg Publishing, Dural, 128 pp.

Lyell, C., 1833. Principles of Geology, vol. 2. J. Murray, London, 556 pp.

MacEachern, J.A., Bann, K.L., Gingras, M.K., Zonneveld, J.-P., Dashtgard, S.E., Pemberton, S.G., 2012. The ichnofacies paradigm. In: Knaust, D., Bromley, R.G. (Eds.), Trace Fossils as Indicators of Sedimentary Environments. Developments in Sedimentology, vol. 64. Elsevier, Amsterdam, pp. 103–138.

MacEachern, J.A., Pemberton, S.G., Gingras, M.K., Bann, K.L., 2007. The ichnofacies paradigm: a fifty-year retrospective. In: Miller III, W. (Ed.), Trace Fossils. Concepts, Problems, Prospects. Elsevier, Amsterdam, pp. 52–77.

MacLeay, W.S., 1839. Note on the Annelida. In: Murchison, R.I. (Ed.),The Silurian System. Part 2. J. Murray, London, pp. 699–701.

Macsotay, O., 1967. Huellas problemáticas y su valor paleoecológico en Venezuela. Geos 16, 7–79.

Maithy, P.K., 1990. Metaphyte and metazoan remains from the Indian Proterozoic successions. Palaeobotany 38, 20–28.

Martinsson, A., 1965. Aspects of a Middle Cambrian thanatotope on Öland. Geol. Fören. Stockholm, Förhandl. 87, 181–230.

Mayor, A., Sarjeant, W.A.S., 2001. The folklore of footprints in stone: from classical antiquity to the present. Ichnos 8, 143–163.

McCulloch, J., 1814. Remarks on several parts of Scotland which exhibit quartz rocks, and on the nature and connections of these rocks in general. Trans. Geol. Soc. Lond. 1, 450–487.

Meysman, F.J.R., Middelburg, J.J., Heip, C.H.R., 2006. Bioturbation: a fresh look at Darwin's last idea. Trends Ecol. Evol. 21, 688–695.

Middleton, G.V., 2003. Encyclopedia of Sediments & Sedimentary Rocks. Kluwer Academic Press, Dordrecht, 928 pp.

Mikuláš, R., Dronov, A., 2006. Paleoihnologija—Vvedenie v Izuchenie Iskopaemyh Sledov Zhiznedejatel'nosti (Palaeoichnology—Introduction to the Study of Trace Fossils). Academy of Sciences of Czech Republic, Prague, 122 pp. (in Russian).

Misra, R.C., Awasthi, N., 1962. Sedimentary markings and other structures in the rocks of the Vindhyan Formations of the Son Valley and Maihar-Rewa area, India. J. Sediment. Petrol. 32, 764–775.

Müller, A.H., 1989. Lehrbuch der Paläozoologie. Band II, Invertebraten, Teil 3, Arthropoda 2—Hemichordata. VEB Gustav Fischer, Jena, 775 pp.

Mylius, G.F., 1709. Memorabilium Saxoniæ Subterraneæ, Pars Prima. Friedrich Groschuffen, Leipzig, 89 pp.
Nathorst, A.G., 1881. Om spår av nagra evertebrede djur och deras paleontologiska betydelse. Kongl. Svenska Vetenskaps-Akademien, Handlingar 18, 1–104.
Neto de Carvalho, C., Baucon, A., Catana, M.M., Rodrigues, J., 2009. The Ichnological Park of Penha Garcia: bringing it back to nature and sustainable enjoyment (what's next?). In: Neto de Carvalho, C., Rodrigues, J., Jacinto, A. (Eds.), Geotourism & Local Development Printmor - Rio Maior. Câmara Municipal de Idanha-a-Nova, pp. 161–164.
Nicholson, H.A., 1873. Contributions to the study of the errant annelids of the older Palaeozoic rocks. Proc. R. Soc. Lond. 21, 288–290.
Osgood, R.G., 1970. Trace fossils of the Cincinnati area. Paleont. Am. 6, 281–444.
Osgood, R.G., 1975. The history of invertebrate ichnology. In: Frey, R.W. (Ed.), The Study of Trace Fossils. Springer, Berlin, pp. 3–12.
Palacios, P., 1918. La constitución estratigráfica del Moncayo. Bol. Real Soc. Esp. Hist. Nat. 18, 101–103.
Pemberton, S.G., 1992. Memorial to Robert Wayne Frey (1938–1992). Ichnos 2, 1–6.
Pemberton, S.G., Spila, M.V., Pulham, A.J., Saunders, T., MacEachern, J.A., Robbins, D., Sinclair, I.K., 2001. Ichnology & Sedimentology of Shallow to Marginal Marine Systems. Ben Nevis & Avalon reservoirs, Jeanne d'Arc Basin. Geol. Ass. Canada, Short Course Notes 15, 343 pp.
Pemberton, S.G., MacEachern, J.A., Gingras, M.K., 2007. The antecedents of invertebrate ichnology in North America: the Canadian and Cincinnati schools. In: Miller III, W. (Ed.), Trace Fossils. Concepts, Problems, Prospects. Elsevier, Amsterdam, pp. 32–51.
Pemberton, S.G., Frey, R.W., 1990. Darwin on worms: the advent of experimental neoichnology. Ichnos 1, 65–71.
Pollard, J.E., 1988. Trace fossils in coal–bearing sequences. J. Geol. Soc. London 145, 339–350.
Pollard, J.E., 2007. Memorial to Roland Goldring (1928–2005). In: Miller III, W. (Ed.), Trace Fossils. Concepts, Problems, Prospects. Elsevier, Amsterdam, pp. xiii–xiv.
Pusch, G.G., 1837. Polens Paläontologie oder Abbildung und Beschreibung der vorzüglichsten und der noch unbeschriebenen Petrefacten aus den Gebirgsformationen in Polen, Volhynien und den Karpathen. Schweizerbart, Stuttgart, 218 pp.
Raciborski, M., 1890. *Taonurus ultimus* Sap. et Mar. in Galizien. Verhandlungen der kaiserlich-königlichen Geologischen Reichsanstalt 1890, 265–266 (in German).
Reineck, H.-E., 1963. Sedimentgefüge im Bereich der südlichen Nordsee. Abh. Senckenberg. Naturf. Ges. 505, 1–138.
Richter, R., 1928. Aktuopaläontologie und Paläobiologie, eine Abgrenzung. Senckenbergiana 10, 285–292.
Richter, R., 1952. Fluidal-Textur in Sediment-Gesteinen und über Sedifluktion überhaupt. Notizbl. Hess. Landesamtes Bodenforsch. Wiesbaden 6, 67–81.
Rindsberg, A.K., 1999. Interviews in 1987 with the late Robert W. Frey. Ichnol. Newslett. 21, 29–32.
Rindsberg, A.K., 2012. Ichnotaxonomy: finding patterns in a welter of information. In: Knaust, D., Bromley, R.G. (Eds.), Trace Fossils as Indicators of Sedimentary Environments. Developments in Sedimentology, vol. 64. Elsevier, Amsterdam, pp. 45–78.
Rindsberg, A.K., Martin, A.J., 2003. *Arthrophycus* in the Silurian of Alabama (USA) and the problem of compound trace fossils. Palaeogeogr. Palaeoclimat. Palaeoecol. 192, 187–219.
Roselli, F.L., 1938. Apuntes de geología y paleontología uruguaya. Sobre insectos del Cretácico del Uruguay o descubrimiento de admirables instintos constructivos de esa época. Bol. Soc. Amigos Ciencias Naturales "Kraglievich-Fontana" 1, 72–102.
Sacco, F., 1888. Note di Paleoicnologia Italiana. Atti Soc. Ital. Sci. Nat. 31, 151–192.

Sarjeant, W.A.S., 1974. History and bibliography of the study of vertebrate footprints in the British Isles. Palaeogeogr. Palaeoclimat. Palaeoecol. 16, 265–379.
Sarjeant, W.A.S., 1975. Plant trace fossils. In: Frey, R.W. (Ed.), The Study of Trace Fossils. Springer, Berlin, pp. 163–179.
Sarjeant, W.A.S., 1987. The study of fossil vertebrate footprints: a short history and selective bibliography. In: Leonardi, G. (Ed.), Glossary and Manual of Tetrapod Footprint Palaeoichnology, Republica Federativa do Brasil. Ministério das Minas e Energia, Brasilia, Brazil, pp. 1–19.
Savi, P., Meneghini, G., 1851. Osservazioni Stratigrafiche e Paleontologiche Concernenti la Geologia della Toscana e Paesi Limitrofi. Stamperia Granducale, Firenze, 259 pp.
Schütte, J.H., 1761. Oryctographia Jenensis, sive fossilium et mineralium in agro Jenensi. Güthuius, Jena, 141 pp.
Seilacher, A., 1953. Studien zur Palichnologie. I. Über die Methoden der Palichnologie. N. Jb. Geol. Paläont. Abh. 96, 421–452.
Seilacher, A., 1962. Biogenic sedimentary structures. GSA Spec. Pap. 68, 266–267.
Seilacher, A., 1964. Biogenic sedimentary structures. In: Imbrie, J., Newell, N.D. (Eds.), Approaches to Paleoecology. John Wiley & Sons, Hoboken, pp. 296–313.
Seilacher, A., 1967. Bathymetry of trace fossils. Mar. Geol. 5, 413–428.
Seilacher, A., 2007. Trace Fossil Analysis. Springer, Berlin, 226 pp.
Seilacher, A., 2008. Fossil Art. An Exhibition of the Geologisches Institut Tuebingen University Germany. CBM-Publishing, Laasby, 97 pp.
Sharples, R.W., 1995. Theophrastus of Eresus. Sources for His Life, Thoughts and Influence. Commentary 5, Sources on Biology. E.J. Brill, Leiden, 279 pp.
Spencer, H., 1896. The Study of Sociology. Appleton, New York, 438 pp.
Stafleu, F.A., 1966. Brongniart's Histoire des végétaux fossiles. Taxon 15, 320–324.
Sternberg, K.M. Graf. von, 1833–1838. Versuch einer geognostisch–botanischen Darstellung der Flora der Vorwelt. Fr. Fleischer, Leipzig, Prague, vols. 5–6, vols. 7–8.
Surić, M., Lončarić, R., Čuka, A., Faričić, J., 2007. Geological issues in Alberto Fortis' Viaggio in Dalmazia (1774). Compt. Rend. Geosci. 339, 640–650.
Tapanila, L., 2008. The medium is the message: imaging a complex microboring (*Pyrodendrina cupra* n. igen. and isp.) from the early Paleozoic of Anticosti Island, Canada. In: Wisshak, M., Tapanila, L. (Eds.), Current Developments in Bioerosion. Erlangen Earth Conference Series. Springer, Berlin, pp. 123–146.
Targioni-Tozzetti, G., 1777. Relazioni d'Alcuni Viaggi Fatti in Diverse Parti della Toscana. Stamperia Granducale, Firenze, Tomo decimo, 466 pp.
Taylor, A.M., Goldring, R., 1993. Description and analysis of bioturbation and ichnofabric. J. Geol. Soc. London 150, 141–148.
Uchman, A., Bubniak, I., Bubniak, A., 2000. *Glossifungites* Ichnofacies in the area of its nomenclatural archetype, Lviv, Ukraine. Ichnos 7, 183–193.
Vaccari, E., 1998. Lyell's reception on the continent of Europe: a contribution to an open historiographical problem. In: Blundell, D.J., Scott, A.C. (Eds.), Lyell: The Past is the Key to the Present. Geol. Soc. London, Spec. Publ., vol. 143, pp. 39–52.
Vai, G.B., Cavazza, W. (Eds.), 2003. Four Centuries of the Word Geology. Ulisse Aldrovandi 1603 in Bologna. Minerva Edizioni, Bologna, p. 315.
Vassoievich, N.B., 1953. O niektorih flishovih teksturah [About some flysch textures]. Trudy Lvovskoho Geologicheshoho Obschestva pri Gosudarstvennom Universitetie imeni Ivana Franko. Geologitcheskaya seriya 3, 17–85 (in Russian).

Vialov, O.S., 1966. Sledy Zhiznedeyatel'nosti Organizmov i Ikh Paleontologicheskoe Znachenie (Trace Fossils and their Palaeontological Significance). Naukova Dumka, Kiev, 219 pp. (in Russian).
von Otto, E., 1854. Additamente zur Flora des Quadergebirges in Sachsen, II. Heft. Gustav Mayer, Leipzig, 53 pp.
von Schlotheim, E.F., 1822. Nachträge zur Petrefactenkunde. Becker, Gotha, 114 pp.
Webster, T., 1814. On some new varieties of fossil Alcyonia. Trans. Geol. Soc. London 1, 377–387.
Wetzel, A., 2006. Theodor Fuchs´ experiments on the formation of solemarks in flysch. Ichnol. Newslett. 27, 16–19.
Wood, J.G., 1866. Homes without Hands. Longmans, Green and Co, London, 632 pp.
Woodward, J., 1729. An Attempt towards a Natural History of the Fossils of England; in a Catalogue of the English Fossils in the Collection of J. Woodward, Tome II. Fayram Senex Osborn and Longman, London, 110 pp.
Wu, X.-T., 1986. Hen Ji Xue Ru Men [Introduction to Ichnology]. Coal Industry Press, Beijing, 155 pp. (in Chinese).
Yang, S.-P., Zhang, J.-P., Yang, M.-F., 2004. Zhong Guo Yi Ji Hua Shi [Trace Fossils of China]. Science Press, Beijing, 353 pp. (in Chinese, English summary).
Yin, Z.-X., 1932. Zai Yunnan He Sichuan De Fa Xian [On the occurrence of *Cruziana* (bilobites) in Yunnan and Szechuan]. Bull. Geol. Soc. China 12, 75–82 (in Chinese).
Young, C.-C., 1929. On some traces of vertebrate life in the Jurassic and Triassic beds of Shansi and Shensi. Bull. Geol. Soc. China 8, 131–133.
Zaręczny, S., 1878. O warstwach kredowych w krakowskim okręgu [About Cretaceous beds in the Kraków region]. Sprawozdanie Komisyi Fizyjograficznej, Akademija Umiejętności w Krakowie 12, 176–246 (in Polish).
Zuber, R., 1918. Flisz i Nafta [Flysch and Oil]. Prace naukowe, dział II, 2. Towarzystwo dla Popierania Nauki Polskiej, Lwów, 381 pp. (in Polish).

Chapter 2

Ichnotaxonomy: Finding Patterns in a Welter of Information

Andrew K. Rindsberg[1]
Department of Biological & Environmental Sciences, University of West Alabama, Livingston, Alabama, USA
[1]*Corresponding author: e-mail: arindsberg@uwa.edu*

1. INTRODUCTION

The chief purpose of ichnotaxonomy is to communicate information about biogenic structures, which is best done by naming them. As Frey (1973) and others have pointed out, in the absence of names, it is difficult to discuss objects with other people or to recognize previous descriptions as referring to the same object. Names allow data retrieval in indexed and catalogued form.

Names also allow the classification of the objects that they describe. A secondary purpose of ichnotaxonomy is to classify biogenic structures in ways that encourage clarity of thought about them. After many false starts and proposals (described in detail by Lessertisseur, 1956 and Häntzschel, 1975: W16–W24), the several classifications of Seilacher (1953) provided the paradigm within which ichnologists work today. In brief, he proposed that any biogenic structure could be classified simultaneously in several ways: (1) toponomically, according to its relationship with contrasting materials within the substrate; (2) biologically, according to its relationship to its maker; (3) ethologically, according to its biological function; and (4) systematically, according to its morphology. Although Seilacher drew on the previous work of others, no one had presented these ideas so clearly before. These classifications have been modified and extended since 1953 (e.g., by Frey and Seilacher, 1980; Müller, 1962). All four systems are used by ichnologists simultaneously, but only the systematic system is called ichnotaxonomy in the strict sense of providing names within the standard Linnaean classification.

Good classifications reflect real relationships and therefore have predictive power—which is generally a good sign that science is taking place rather than mere fiddling. Many classifications have been proposed that failed this crucial test, and most of them have not gained acceptance among scientists. For example, Faul (1951) proposed a rigidly geometric classification of trackways

Developments in Sedimentology, Vol. 64. http://dx.doi.org/10.1016/B978-0-444-53813-0.00002-2

according to the number of foot imprints per set and the number of digits per foot imprint. This had the advantage of grouping the trace fossils of some clades together, for example, most footprints of birds (two tracks per set, three forward and one backward digit imprint per track). However, it also had the effect of grouping trace fossils of distant clades together (e.g., trails of diverse invertebrates) while separating others (e.g., trackways of horses, cattle, and dogs). Morphologically based classifications that ignore biological relationships are unlikely to have predictive value.

If so, the most obvious way to base a classification of trace fossils would be one based directly on biological relationships. Biologists and sedimentologists alike ask, why can we not name trace fossils for the organisms that made them? But in most cases we do not know, nor can we obtain, this information. The trace fossils must be understood on their own terms.

2. OBSERVING TRACE FOSSILS

To classify trace fossils, we must first observe them. Many studies have gone astray because the researcher did not know what to look for, or even how to look. Like any other fossils, trace fossils have discernible features, some of which are shared among populations and others of which are unique to individuals. In the classification of trace fossils, we are usually more interested in shared traits than in unique ones, most of which represent slight variations of behavior. We try to observe as many specimens as possible to determine the full range of behavior represented by these fossil objects. The traits or features that observation shows are shared by many objects can be called morphological *characters*.

Generally, there are many more characters to observe in body fossils than in trace fossils, which makes the classification of trace fossils simpler but also more ambiguous. Even the best footprint does not preserve all the features of the body that made it, and a burrow or boring usually shows even less. Fortunately, these trace fossils do show aspects of tracemaker behavior that may not be inferable from body fossils.

How does the geologist differentiate between sedimentary structures made by organisms from those made by inorganic processes? This is not always simple, and particular cases have been debated for decades. However, some guidelines apply. Bearing in mind that animals of any one species generally have much the same shape, usually have only a limited size range, and often belong to cohorts of the same age and size, the burrows, footprints, and borings made by animals are likely to have a limited range in size and shape as well. Sediment riddled by burrows may at first resemble a conglomerate, but a section of the burrowed sediment is likely to present only a few consistent sizes and shapes of cross-sections, while the conglomerate may have pebbles of more diverse size and form. In particular, burrows in the third dimension will generally show that they are more or less cylindrical, in contrast to pebbles, which should be relatively spheroidal. The angle of intersection between burrow and outcrop (or section) determines the burrow's visible outline, which is ideally an ellipse, and is only circular in the case of a burrow at right angles with the outcrop.

The beginner who wishes to pick up pointers on how to observe trace fossils should read Howard (1978) as well as Bromley (1996). Chamberlain's (1978) well-illustrated primer on the recognition of burrows in cores is still valuable although outdated by the ichnofabric atlas of Gerard and Bromley (2008). The best teacher is practice: the best ichnotaxonomist is the one who has seen the most traces!

3. DESCRIBING TRACE FOSSILS

The terminology of trace-fossil description was standardized by Frey (1973), who obtained an international consensus on terms. For further development of usage, see Häntzschel (1975: W4), Ekdale et al. (1984), and especially Bromley (1996). This discussion will focus on the terminology of trace fossils that are most commonly given Linnaean names: tracks, trails, burrows, and borings. A general term for any sign of organismal activity without regard to time is *trace*.

Traces are physical objects consisting of previously existing material (a *substrate*) that has been altered by biological activity. They extend in three dimensions, have an orientation, are made up of sediment or other material, and have boundaries with regard to the surrounding substrate. In short, they can be described in a rigorous and repeatable manner. All classifications of traces ultimately depend on morphological description.

In general, a *track* is a single imprint made by the appendages of an animal on the surface of the substrate, while a *trackway* is a series of tracks. A *trail* is a continuous trace made on the surface of a substrate, while a *burrow* is a continuous trace made within a substrate. The *orientation* of a trace refers to the angle of its longest dimension with regard to the surface of the substrate; this is usually simplified as *horizontal, subhorizontal, oblique, subvertical,* and *vertical.* It is worth knowing that, even if an observer has no other information on the trace fossils of a bed, their general orientation can yield information on surface conditions such as wave energy, because animals dig vertical burrows to escape from difficult surface conditions (Seilacher, 1953). Most nameable burrows have a more or less defined contact with the surrounding sediment; this is the burrow's *wall* or *boundary*. Material added to this boundary may be called a *lining,* as discussed further by Bromley (1996) and Zorn et al. (2010).

4. STRATINOMIC CLASSIFICATIONS

4.1 Substrate Consistency

Stratinomy is the study of sedimentary processes that lead to the preservation of a sedimentary record, with biostratinomy referring to sedimentary processes that affect the remains, including traces, of organisms. Processes that lead to burial and preservation of fossils (taphonomy) are of particular interest in stratinomy. For a recent work on the taphonomy of trace fossils, see Savrda (2007).

The physical properties of the substrate are of paramount importance in the classification of trace fossils for biological reasons (Bertling et al., 2006). Organisms are generally restricted to living in either soft substrates or hard substrates, but not both; the distinction is so sharp that burrows and borings are automatically classified separately, with very little overlap. Further, burrowers in soft substrates usually are restricted to relatively narrow ranges of grain size (mostly mud or sand). Animals that live in relatively pure sand are commonly unable to burrow or even respire in muddy substrates. However, because a complete gradation exists between muddy and sandy substrates and their inhabitants, it is best to consider them all together when classifying their traces.

Borers are mostly restricted by the chemical composition of the substrate—carbonate, silicate, bone, or wood—and most specialists focus on only one of these substrates for study. For example, an extensive literature deals with borings in wooden ships, pilings, and other marine constructions. It has been shown that some borers remove the substrate chemically and others mechanically; a chemical process that works in limestone may be effective in loosening the grains of a calcareous sandstone, but not siliceous sandstone. Accordingly, there are sound biological reasons to classify borings made in these four kinds of substrates separately, but always using common sense to recognize larger patterns (Bertling et al., 2006).

4.2 Toponomic Classification

Toponomy is the study of form. In this classification, initiated by Seilacher (1953) and modified by Martinsson (1965, 1970), trace fossils are classified according to their preservation within sedimentary beds, particularly with regard to a more resistant casting medium such as sandstone (Fig. 1). Traces are made at the surface of a substrate, within a homogeneous substrate, or at the contact between substrates of different lithology. In the simplest version of toponomic classification (Martinsson, 1970), the observer merely notes the current relationship of lithologies without considering how they originated. A trace fossil preserved at the top surface of a bed is called an *epichnion* (plural, *epichnia*); one preserved at the base is called a *hypichnion* (plural, *hypichnia*); a trace fossil preserved within the bed is an *endichnion* (plural, *endichnia*). Epichnia and hypichnia together are referred to as *semireliefs,* while trace fossils preserved "in the round" are called *full reliefs*. Semireliefs that protrude from the bed are said to have positive relief, or alternatively may be called convex, in opposition to concave trace fossils with negative relief. For example, a convex hypichnion is a trace fossil preserved by casting on the base of a bed and which protrudes downward from it. Needless to say, if a sandstone slab is found as talus, then directional indicators must be used to determine the original orientation (Reineck and Singh, 1980; Shrock, 1948).

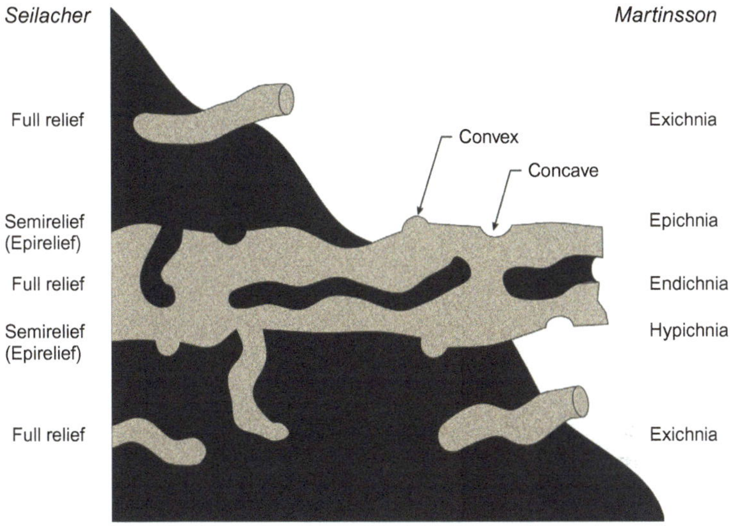

FIGURE 1 Toponomy of trace fossils, after Seilacher (1953) and Martinsson (1970), with inspiration from Bromley (1996).

The above classification is simple, easy to use, and informative, but more advanced observers can glean much more from the trace fossils. For example, a convex hypichnion may be produced by casting a surface trace, by collapsing an open burrow, by passive fill of a previously open burrow, by active fill of a burrow by its maker, or by scouring out a previously buried burrow (Fig. 2).

Distinction of *passive fill* and *active fill* (sometimes called *infill*) is particularly important, as a clue to deposition and to the trace fossil's ethological function. Passive fill is defined as fill resulting from physical sedimentation without the active involvement of the tracemaker; it occurs in abandoned burrows and in burrows that are filled too rapidly for the maker to resist the process. Accordingly, passive fill is generally laminated more or less horizontally. Active fill is defined as fill made by the tracemaker during its lifetime. This fill is more usually obliquely or vertically laminated than horizontally disposed, may show signs of manipulation or pressure (scratches, tension fractures, laminae of alternating lithology), and may include fecal pellets. Actively filled burrows include burrows that were never open, that is, burrows made by animals that moved within the substrate while feeding on sediment or on other creatures.

Recognition of scoured hypichnia is also necessary, especially in the study of storm beds and turbidites. The original work by Seilacher (1959) is still well worth consulting, with its discussion of *pre-depositional* trace fossils made before a turbidity current (including scoured hypichnia) and *post-depositional* trace fossils made after it (Uchman and Wetzel, 2012). Many of the most delicate forms, originally excavated in mud, would not be preserved, had they not been partially scoured by currents and then filled with a more resistant casting

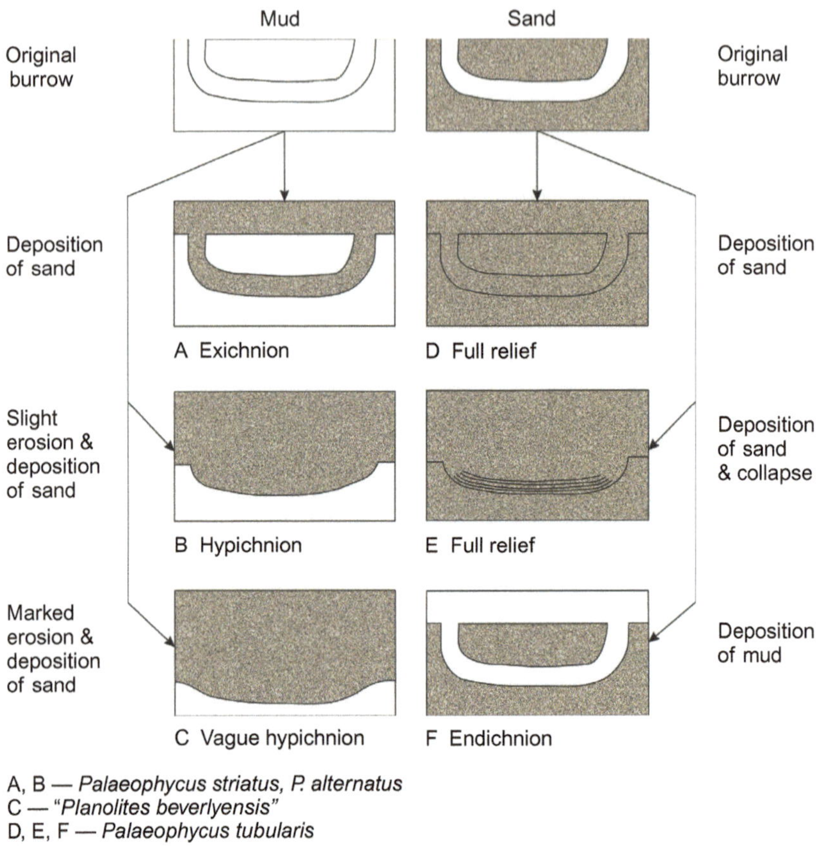

FIGURE 2 Taphonomy of *Palaeophycus alternatus* Pemberton and Frey, 1982 at its type locality (Ringgold Gap) in the Silurian Red Mountain Formation of Georgia. Without recognition of taphonomic changes, burrows C-F would be given these misleading names. Instead, they should be called "probable eroded *Palaeophycus alternatus*", "cf. *Palaeophycus* isp.", or the like.

medium such as sand. For storm beds, the work of Aigner (1985) is a good place to start; Seilacher (1977, Fig. 1) showed how shallow burrows are preserved in turbidites. For additional terms, see Frey (1973) and Bromley (1996). Because of erosion, shafts usually are preserved without their apertures or other near-surface features such as thickened linings or chimneys. It is unusual to see *Monocraterion* preserved with the crater-shaped aperture or *Ophiomorpha* with the characteristically thickened lining and narrow lumen at the top. In cases where such features are preserved, additional description and study often result in valuable information on environmental conditions at the time of deposition (e.g., Howard, 1978; Schlirf, 2003).

Trace fossils can be further characterized as *endogenic* (made within the substrate) or *exogenic* (made at the sediment/water interface) (Chamberlain, 1971), although these terms are sparsely used. It is now realized that some animals do not make a hard-and-fast distinction between moving at the surface of the substrate (which can be soupy) and the sediment immediately beneath it. There is thus a continuum between surface trails and subsurface locomotion burrows that makes hash of fine theoretical distinctions between epifaunal and infaunal behavior. The terms exogenic and endogenic are useful because they do neatly distinguish between what the modern observer can and cannot see on a substrate, but the taxonomist should classify an animal's behavior from the animal's point of view.

In short, the ichnotaxonomist should classify a trace fossil based on its original form, as deduced rigorously from its appearance in a large number of examples, and not rely solely on the current appearance of each individual following erosion (Fig. 2). This advice should not be interpreted as a license to name trace fossils based on imaginary characters!

5. BIOLOGICAL CLASSIFICATIONS

5.1 Phylogenetic Classification

In the phylogenetic classification, traces are called after their makers, for example, a *Phoronis* burrow, the coprolite of *Tyrannosaurus rex,* or more informally, a bird footprint or buffalo wallow. This classification is the most intuitive and is also potentially the most informative, and it is used in virtually all treatments of modern traces (e.g., Eiseman et al., 2010; Elbroch, 2003; Elbroch and Marks, 2001). The aspects of a trace that are useful in identifying its maker may be called its *bioprint* (Rindsberg and Kopaska-Merkel, 2005). However, in the absence of the organisms and distinctive bioprint, traces cannot always be identified correctly as to their maker even in modern environments, especially those that are most unfamiliar to us, such as the deep sea (Ekdale, 1977; Wetzel, 1981; Wetzel and Uchman, 2012). In ancient examples, it is even more difficult to link a trace fossil with its maker. The two are only rarely preserved together, and even where a body fossil is found within a trace fossil, a logical case must be constructed to test whether the organism made the trace instead of simply harboring or being deposited there.

Another problem is that trace fossils commonly do not contain the right kind of data to allow them to be classified in a manner even roughly paralleling that of their biological makers. At the present state of the art, it is not possible to say that all *Diplocraterion* belong to any one clade, on the contrary. Neither is it possible to classify all the trace fossils made by a single organism together as one taxon, because they can be very diverse, a point that is discussed further under Section 6.

The trace fossils of some organisms are more apt for biological classification than others, especially in cases where the trace fossils are complex and are well preserved. The effort is worthwhile for an increasingly long list of trace fossils, notably, trilobite trails and resting traces (*Cruziana* and *Rusophycus;* Seilacher, 1970), the footprints of fossil vertebrates, the communal burrows of social insects (Genise, 2000), arthropod trackways (Minter et al., 2007), and many kinds of borings, including those of sponges, bryozoans, and bivalves (Bromley, 1970). In some cases, specialists have gone so far as to classify trace fossils in parallel with modern biological species, for example, boring bryozoans (Pohowsky, 1978) and vertebrate trackways. In the case of boring bryozoans, the borings accurately represent the shapes of several internal organs; moreover, the animals die and decay within minutes so that the standard technique for studying them is to remove the tissues as a matter of course before study of the borings (J.D. Soule, Allan Hancock Foundation, University of Southern California, Los Angeles, oral communication, 1982). This is a more ambiguous case than that of vertebrate trackways, which at best can represent only a small part of an animal's body and behavior, but which nevertheless often have been harnessed as representatives of the missing animal itself.

It is always worthwhile to ask what made a trace fossil, but at present the makers of most trace fossils cannot be identified unambiguously. Fossil footprints represent only a small part of the morphology of the animal. Even in the most convincing case, that of boring bryozoans, this is an error. It is true that, a few decades ago, boring bryozoans could be best classified by the morphology of the boring; comparison between living and fossil specimens was thus facilitated. But as molecular analysis replaces morphology in the study of living organisms, the classification of modern and fossil borings will become disconnected again. Fossil borings lack the necessary DNA.

5.2 Ecological Classifications

Trace fossils can also be classified according to their ecological functions. The most used ecological classification is Seilacher's ethological classification, which will be discussed in detail. It is only one of many possible ecological classifications, however. The tiering and ichnoguild concepts also merit brief discussion here.

Ichnologically, a *tier* is a vertical zone within the substrate that has a characteristic assemblage or association of traces (Ausich and Bottjer, 1982). Each burrowing or boring organism penetrates only so far into the substrate and may even determine the texture of that tier. New substrates may be characterized by a pioneer community; substrates reaching a climax ecosystem may have more fully developed tiers. The preservation of tiers depends on rates of deposition, and particularly on the frequency of sudden burial, which preserves well-defined tiers that might otherwise be erased during more continuous deposition (Howard, 1978).

Traces can also be classified ecologically into *ichnoguilds* representing similar behavior without regard to their biological affinities (Bromley, 1996). Traces may be grouped according to functional morphology, food source, or use of space (tiering).

While these concepts are useful in grouping trace fossils found in particular studies, they have not yet yielded general classifications of trace fossils. Seilacher's ethological classification is another matter.

6. ETHOLOGICAL CLASSIFICATION

Ethology is the study of behavior. One of Seilacher's (1953) crowning achievements was to recognize that trace fossils could be classified according to their ethological function because similar functions can result in similar morphology. A U-shaped burrow made by a polychaete might look much like one made by a crustacean. One might say that this is a classification made from the tracemaker's point of view. Seilacher further recognized that these functions depend on the environment—that is, on the use of available resources and avoidance of danger. By taking the focus of attention away from identifying tracemakers and instead analyzing the functions of trace fossils in whole communities, Seilacher made it possible to study *ichnofacies*—archetypal facies based on their trace-fossil content.

In Seilacher's original ethological classification, traces are categorized into five main groups: resting traces (cubichnia), dwelling traces (domichnia), crawling or locomotion traces (repichnia), grazing traces (pascichnia), and feeding traces (fodinichnia). The singular of each of these terms ends in *-ichnion*. These are still the most commonly used terms, but additional terms have been added over the years to incorporate other behavior (Table 1). The morphology of trace fossils and comparison with modern traces are used to deduce ancient behavior as follows. For a fuller treatment of the underlying concept, see Frey and Seilacher (1980).

6.1 Resting Traces (Cubichnia)

Resting traces, or cubichnia, include traces made by organisms that have dug into the substrate in order to rest, hide, or perform other relatively stationary actions such as respiration or rehydration for a relatively brief period of time (Martin, 2006). For instance, a ghost crab may dig into the sediment, leaving only its eyes and antennae showing (Martin, 2006). Sea stars, brittle stars, crustaceans, and flatfishes commonly bury themselves in this way (Schäfer, 1972). In hard substrates, limpets and sea urchins are examples of animals that bore shallow pits that might be considered as resting traces (Bromley, 1970), because they meet the morphological criteria. Predators such as stonefish may hide under sediment to ambush their prey. Reflecting this diversity of functions, Lessertisseur (1956) distinguished informally between resting and hiding traces; these may, however, prove difficult to distinguish in the fossil record.

In most cases, a thin veneer of sediment suffices to conceal the animal; so, cubichnia tend to be much shallower than they are broad (Frey and Seilacher, 1980). If the animal is startled, then it can emerge rapidly from the sediment. No lining or other reinforcement is usually provided; none is needed because the sediment is supported by the tracemaker's body, except perhaps around respiratory organs to maintain an unimpeded water flow. Because the animals burrow into the sediment once and then remain nearly motionless, the trace may retain the sculpture made by the digging appendages—feet, fins, claws, or podia—which are clues not only to the function of the trace but also to

TABLE 1 Generalized Morphological Characteristics of Some Ethological Categories of Trace Fossils. *Modified in Part after Frey and Seilacher (1980)*

Ethological category	Characteristic morphology
Resting traces (cubichnia): traces made as temporary hiding or resting places, or for related purposes.	Shallow, trough-like depressions, recording to some extent the maker's ventral morphology; traces usually isolated, but in some cases overlapping or sequential, or even intergradational with locomotion or escape traces.
Dwelling traces (domichnia): traces made as permanent or long-term homes.	Burrows or borings consisting of originally open hollows, or of a permanent cavity filled by the tracemaker, reflecting to some degree the outline of the maker and in many cases with sculpture reflecting its appendages or its method of excavation; traces simple to U-shaped, Y-shaped, or complexly branched, typically vertical to oblique near the surface but commonly incorporating horizontal elements at depth; fill usually passive; walls commonly lined.
Locomotion traces (repichnia): traces made while moving from one place to another.	Traces generally simple and shallow, either continuous or discontinuous; may incorporate resting traces. Trackways made up of repeated sets of discontinuous impressions reflecting the maker's appendages and their motion in the substrate; trails consisting of continuous grooves or ridges of sediment with more or less parallel sides, in some cases meniscate or annulated.
Grazing traces (pascichnia): traces made by surface-feeding.	Shallow grooves and pits, either continuous or discontinuous, generally reflecting maximum utilization of a substrate surface, as in meandering, coiled, or otherwise patterned course.
Feeding traces (fodinichnia): traces of deposit-feeding within the substrate.	Burrows or borings originally held open only temporarily; traces simple to complex; orientation variable; fill commonly active; complex burrows with repeatedly branched probes or with gradually shifting burrows (spreite); pattern commonly showing avoidance of reburrowing previously used sediment; wall usually with minimal or no lining.

TABLE 1 **Generalized Morphological Characteristics of Some Ethological Categories of Trace Fossils.** ***Modified in Part after Frey and Seilacher (1980)*—Cont'd**

Ethological category	Characteristic morphology
Farming traces (agrichnia): traces made to farm microbes or to trap meiobenthos.	Burrows originally open, parallel to substrate surface, usually narrow; typically complexly patterned (coiled, meandering, networked), yet with unburrowed space left between galleries.
Escape traces (fugichnia): traces made in response to rapid burial.	Burrows showing upward adjustment of the tracemaker, for example, vertically repetitive resting traces, biogenic laminae in vertical series.
Brooding traces (calichnia): traces made to protect eggs and juveniles.	Burrows simple to complex, commonly consisting of one or more chambers that may be connected to adult-sized tunnels as access but much smaller exit apertures or tunnels for juveniles.

the identity of the tracemaker. And the overall form of the trace may outline the body of the tracemaker itself. Trace fossils commonly classified as cubichnia include *Asteriacites, Rusophycus,* and *Limulicubichnus.*

6.2 Dwelling Traces (Domichnia)

Dwelling traces, or domichnia, are open structures of which the primary function is to give the organism a permanent home: a place that affords protection not only from competitors and predators, but also from temporary changes in the local environment. Dwelling traces include burrows in soft and firm substrates and borings in hard substrates. However, not all sediment dwellers live in domichnia; many live permanently surrounded by the substrate. To excavate and maintain an open burrow requires special adaptations on the part of the tracemaker. An extensive literature may be consulted on burrowing methods, morphological adaptations to burrowing, and burrow irrigation (maintenance of water circulation) (e.g., Bromley, 1996; Schäfer, 1972; Trueman and Ansell, 1969; Vogel, 1978).

Domichnia are typically reinforced, lined, or at least smoothed off internally. Open borings in hard substrates and open burrows in firmgrounds typically need little reinforcement, but may be smoothly finished, and they may be lined to limit influx of pore water from the substrate. Open burrows in soft substrates tend to be lined to prevent collapse and to fend off other burrowers as well as to control water flow.

The dwellers of domichnia can be among the longest lived members of a community, with burrows changing through time. The burrow morphology of a juvenile may be differently shaped from that of an adult; for example, the burrow of the fiddler crab *Uca pugnax* is initially a simple shaft and is only later extended as a U-shaped burrow (Basan and Frey, 1977). The diameter of a burrow or boring may increase through time (Bromley, 1970, 1996); the size of sculptural elements also increases. Branches may be added to a burrow system, and others filled up or closed off (Schäfer, 1972). Walls and linings may be reinforced, opened for expansion, or repaired. Burrows may also be shifted by removing material from one side of the inner wall while adding it to the other, eventually resulting in a planar structure called a *spreite,* which can deepen or enlarge a dwelling burrow. A spreite may also be present in feeding burrows.

Large domichnia may be inhabited by different organisms at the same time (e.g., callianassid shrimp and the innkeeper worm *Urechis caupo;* Bromley, 1996; Fischer and MacGinitie, 1928), and they may be inherited by later generations of individuals. After the abandonment of the dwelling trace, it typically collapses or is filled passively. The trace fossil represents the end product of all these processes. The modifications made through time by the maker of a domichnion should be included in its description.

The water circulation of open burrows with more than one aperture, including U-shaped burrows and some network burrows, has been extensively studied by biologists. Most researchers have focused on animals that irrigate their burrows by active motions of gills or appendages, such as the polychaete *Chaetopterus variopedatus* (Bromley, 1996). Irrigation of the burrow increases the flux of oxygen and nutrients, allowing burrowers such as the echiuran *Urechis caupo* to be filter feeders (Bromley, 1996). Where one aperture is raised higher into the water column than the other, the difference in current velocity generates a current through the burrow without active irrigation by the trace-maker (Vogel, 1978).

The uppermost parts of dwelling burrows are usually not preserved because they are removed by physical erosion, or because other animals rework the substrate. Where preserved, the openings (*apertures*) can be highly distinctive and useful in ichnotaxonomy. They are often narrowed to prevent entry by sedimentary particles or unwanted visitors and to enhance the circulation of water (Howard, 1978; Vogel, 1978). Others with a feeding function open out into funnels, as in *Monocraterion* (Jensen, 1997). Many are reinforced more strongly than the deeper parts of the burrow, in some cases with chimneys protruding above the substrate as in *Diopatrichnus* (Kern, 1978).

In the 1970s, the ichnological consensus was to avoid utilizing easily eroded parts of trace fossils in ichnotaxonomy (Fürsich, 1974) and to emphasize features that could be seen in a small piece of the fossil (e.g., for *Ophiomorpha*, Frey et al., 1978). This made trace-fossil identification easier, particularly with incomplete material, and gave a great boost to ichnology among sedimentologists, but a more recent approach is to use all parts of a trace fossil holistically. A trilobite specialist, who ignored important characteristics of trilobites because it made fragments of trilobites easier to identify, would not be taken seriously; many ichnologists are now making the extra effort needed to add detail and make finer distinctions where necessary. Where this hampers exact identification, open nomenclature is used. Examples of domichnia include *Skolithos*, *Arenicolites*, *Diplocraterion*, *Bergaueria*, *Lockeia*, and *Thalassinoides*.

6.3 Locomotion Traces (Repichnia)

Traces, the chief function of which is movement from one place to another, are called crawling traces or repichnia (Seilacher, 1953). A better term is *locomotion traces* because this category includes walking and running traces as well as those made by crawling. Swimming traces (natichnia; Müller, 1962) can be included here but are often so distinctive morphologically that they are given their own category.

Locomotion from one place to another generally produces a straight or curved trace of which the sides are more or less parallel. Excess movement of sediment is inefficient for this purpose, and locomotion traces tend to be shallow. Movement may be effected by appendages, as in arthropods, echinoderms, and most vertebrates, or by the whole body, as in earthworms, often creating a characteristic bioprint of the tracemaker. Where the animal's body is supported above the substrate by appendages, their contact with the surface leaves a series of imprints (*tracks*) that together make up a *trackway*. Where the animal's body touches the surface, the resulting continuous trace is called a *trail*. *Locomotion burrows* represent the subsurface passage of an animal through the sediment without feeding on it. Historically, there has been some confusion over burrows that are now thought to represent combined feeding and locomotion, such as *Cruziana* (Goldring, 1985). If the tracemaker moved a considerable amount of sediment while moving on or just below the substrate, then feeding may have occurred as well as locomotion. There is no firm distinction between traces at the substrate surface and those just below it, which has caused some confusion as well.

Locomotion traces commonly cross over themselves or one another (Rindsberg and Martin, 2003), in some cases making simple traces appear as if they were branched. One animal may retrace its own path or even follow another's, and predators commonly follow the path of their prey. This can lead to compound traces requiring skilful interpretation.

The bioprint of trackways is more distinctive than that of most trace fossils and so is used extensively in the diagnosis of ichnogenera. The repetition of imprints in characteristic series allows them to be identified as the work of animals with two, four, six, eight, or more appendages. Care must be taken to avoid mistakes; for example, decapod crustaceans such as crabs have ten legs, but walk on only eight, holding the pincers above the substrate. Also, because some pairs of appendages may penetrate more deeply into the sediment than others, not all of the appendages may be registered in the *underprint*, that is, the deeper parts of the trackway. This can lead to a gross overestimate of ichnotaxa in an assemblage.

The imprints of vertebrate trackways are large enough to show great detail where favorably preserved. Even individual scales may be represented, and vertebrate ichnologists have become accomplished at identifying footprints with particular clades of makers. However, footprints are not simple molds of feet, but the active product of feet that change shape and direction while moving through the substrate. The analysis of the motions that produce footprints—tracking—has occupied humans since before history (Brown, 1999) and has lately entered the computer era with regard to ancient footprints (Gatesy et al., 1999). To give an example of a commonly encountered pitfall, consider the motion of a foot of which five digits penetrate the substrate to different depths. If the rock bearing this footprint is broken open at a shallow level, all five digits will be represented in the trace. However, if broken at a deeper level, only three or even two of the digits may be represented. Historically, ichnogenera were erected on the basis of the number of digits per footprint, and it has taken time to recognize that some footprint taxa represent incomplete versions of others. Repichnia include all trackways and most trails, for example, *Chirotherium, Petalichnus, Climactichnites,* and *Cochlichnus* (in its strict sense).

6.4 Grazing Traces (Pascichnia)

Ecologists differentiate surface and subsurface deposit-feeding because they require different life strategies and adaptations. In Seilacher's (1953) original conception, the traces made by surface and subsurface deposit-feeding were called grazing traces and feeding traces, respectively. Unfortunately, most of his examples, such as *Nereites* and *Paleodictyon,* are now recognized as having been made at depth within the substrate. The concept remains valid, but at present, there are few named examples.

Grazing traces are horizontal or subhorizontal and were originally made on or just below the sediment/water interface. Accordingly, their preservation potential is low and depends on rapid burial without much erosion. They may be accompanied by other surface traces such as locomotion traces and the apertures of dwelling burrows, which are usually not preserved (see Section 6.2).

Grazing traces are mostly rather shallow in comparison to their width. In most cases, they consist of continuous trails, but discontinuous grazing traces can be made by animals that support their bodies (or at least their eating organs) off the substrate. After processing the surface sediment, the makers of most grazing traces avoid crossing their own and other individuals' paths where possible. Segments of the traces commonly parallel one another. Efficient grazing may result in back-and-forth or spiral patterns of trails (Seilacher, 1977).

Examples of grazing traces include the dabblings of shorebirds as they feed on animals and microscopic plants in the sediment and the raspings of land snails on bark and rocks as they feed on algae.

6.5 Feeding Traces (Fodinichnia)

Feeding traces are deposit-feeding traces, that is, the traces made by animals ingesting subsurface sediment. Most are burrows rather than borings, as rock is not very nutritious, but feeding traces in wood exist, such as bivalve-made *Teredolites.* Usually, fodinichnia incorporate a dwelling structure that may be open during the life of the animal and which may be filled passively. The feeding part of the structure may be actively filled or left open during the life of the animal. Feeding traces may have horizontal, oblique, or vertical orientation, and they are extensive. In the simplest case, they consist of simple burrows of indefinite length, such as *Planolites* or *Macaronichnus;* but feeding traces may branch or shift by excavating one side of the lumen (the open part of the burrow) while depositing material on the other. This shifting results in a spreading planar structure called a *spreite,* which is also seen in dwelling traces. Feeding traces include some of the morphologically diverse and distinctive trace fossils, such as linear *Nereites,* branched *Chondrites,* and helical *Zoophycos.*

6.6 Farming Traces (Agrichnia)

Farming traces are a category based on the discovery that some marine invertebrates use mucus-lined burrows as nets to capture meiobenthos (e.g., echiuran *Urechis caupo,* Fischer and MacGinitie, 1928; polychaete *Paraonis fulgens,* Röder, 1971). This inspired the conjecture that some rather complex burrows might even be microbial farms (Ekdale et al., 1984), similar to the galleries in which leafcutter ants (e.g., *Atta colombica*) cultivate fungi. The tracemakers irrigate the burrows, which are constructed in such a way as to maximize the inner surface area. Some of these burrows even pump sulfide-rich pore water through the burrow to encourage the growth of sulfide-metabolizing bacteria; these burrows may include downward extensions to tap this pore water (e.g., burrows of the bivalve *Solemya*). Apertures may be few or numerous. Much is still unknown about these systems; their biology has been studied in shallow-marine environments, but most of the ichnotaxonomic work has been directed to deep-sea environments, where these traces are well preserved in turbidites. Meandering *Cosmorhaphe,* coiled *Spirorhaphe,* zigzag *Belorhaphe,* and netlike *Paleodictyon* are possible examples.

A related category, chemichnia, has been proposed for farming traces that apparently make use of redox differences between oxic and sulfide-rich tiers in order to cultivate microbes (Bromley, 1996). It would go far to explain the oddities of *Chondrites* and *Zoophycos,* which are currently classified as feeding burrows. In this case, the scientific value of a good classification is clear. The morphology and sedimentology of agrichnia and chemichnia raise questions that cannot be answered within the existing classification, which suggests new categories and new directions for research.

6.7 Escape Traces (Fugichnia)

Escape traces (Frey, 1973) are those made by animals that have been rapidly buried by a layer of sediment. This requires that the animal be reasonably muscular and able to respire while temporarily buried. Not all animals can escape rapid burial; some succumb to a light dusting of sediment while others can dig upward for many centimeters. Accordingly, the bodies of some hard-shelled animals, such as gastropods, can be found in escape position where they died within a storm bed (tempestite) or turbidite, and some fugichnia likewise do not penetrate the full thickness of a bed.

Fugichnia are temporary structures and accordingly are not open or reinforced. They chiefly consist of disturbed laminae (Ekdale et al., 1984; Schäfer, 1972). Collapsed shafts may resemble fugichnia, but the laminae of fugichnia tend to have a relatively constant angle with relation to the burrow wall.

6.8 Brooding Traces (Calichnia)

Genise and Bown (1994) proposed calichnia as a name for biogenic structures used for raising young, such as wasp nests and the balls made by dung beetles. Despite a series of articles by Genise, Laza, Hasiotis, and others, only a few researchers were familiar with these structures until Ichnia 2004, when Genise and his colleagues led a field trip to the magnificent Cenozoic succession exposed in Patagonia. Since then, it has been clear that calichnia constitute one of the most important categories of trace fossils in paleosols.

The most complex calichnia are those made by the social activity of colonial organisms, especially social insects such as bees, wasps, termites, and ants. These burrow systems are commonly hybrid, incorporating parts devoted to specialized functions such as dwelling and farming in addition to reproduction, all with different morphologies. In general, these systems consist of densely branched networks of open burrows in all orientations. Branching may be regularly patterned, or as irregular as the streets of a medieval city. Some are very deep, in the case of termites extending as much as tens of meters to the water table, and they may be very extensive. The hybrid nature of these social traces, used for other functions in addition to reproduction, makes them different from structures that are used exclusively for brooding. An alternative suggestion is to use the name polychresichnia for this meaning (Hasiotis, 2003).

The taxonomic study of calichnia is a specialized one that requires training in entomology as well as ichnology. Details of the wall structure must be observed in addition to the architecture of the whole system. Calichnia are among the most challenging trace fossils to study, yet also some of the most rewarding, because they are potentially so informative about the history of terrestrial environments (Hasiotis, 2003). Because of their complexity, calichnia are more easily linked to tracemakers than most other trace fossils, adding to the history of life as well.

6.9 Other Ethological Categories

Many other categories have been proposed, some of which are used more or less frequently by researchers. Among them are swimming traces (natichnia; Müller, 1962), flying traces (volichnia; Müller, 1962), predation traces (praedichnia; Ekdale, 1985), equilibrium traces (equilibrichnia; Bromley, 1996: 197; Frey and Pemberton, 1985), aboveground traces (aedifichnia; Bown and Ratcliffe, 1988), and the list shows no sign of ending (Buatois and Mángano, 2011). Some of these categories represent behaviors that are uncommon in the geological record, while others represent special cases of broader categories. At this point, one has to wonder whether additional terminology is worth the effort of memorization of two new terms, formal and informal, for each concept.

6.10 Overlapping Functions

Most tracemakers produce traces that fit readily into Seilacher's ethological classification. However, many trace fossils represent more than one function and thus cannot be shoehorned into a single ethological category. Many feeding burrows include an open dwelling burrow, for instance, and the distinction between surface-feeding and simple locomotion is not always unambiguous, as in trilobite trace fossils (Goldring, 1985). The burrows of bivalves are commonly considered as dwelling burrows by those who emphasize the traces' relative permanence, but they are not open burrows and their makers are generally capable of moving from place to place at need; so, they are classified also as resting traces by other researchers.

In an ideal taxonomy, each ichnogenus should fit into only one ethological category, and most researchers have followed this guideline. Where single specimens incorporate complex behavior, as in *Hillichnus* (Bromley et al., 2003), the observer has the choice of giving more than one name to the trace fossil along its path or subsuming all the behaviors under one name. In practice, the first choice is the most common, because most organisms apparently perform only one behavior at a time, and the different behaviors typically result in traces of different morphology that are not usually found connected. This is the case for the grazing trace *Cruziana* and the resting trace *Rusophycus,* which are sometimes found connected; Seilacher (1970) emphasized the identity of the trilobite tracemakers while others emphasized the difference in ethology. In the case of *Hillichnus,* a bivalve tracemaker incorporated behaviors as different as locomotion, feeding, and respiration so intimately into one complex burrow structure that the authors somewhat reluctantly gave the trace a single name.

7. SYSTEMATIC CLASSIFICATION

Trace fossils can also be classified with formal Linnaean names. Classification (the grouping of items into categories) and nomenclature (the naming of items) are not the same processes, but the two are merged in the system of binominal nomenclature.

Binominal (alternatively, *binomial*) names consist of two parts, one of which represents a general category, and the other a more particular category within it. Any object can be classified binominally, even minerals or clouds (e.g., *altocumulus castellanus, altocumulus lenticularis*). The method is highly mnemonic and serves well for information retrieval. Trace fossils do not need DNA in order to be classified binominally.

7.1 International Code of Zoological Nomenclature

Ichnotaxonomy in the strict sense is stabilized by the various codes of nomenclature that govern animals, plants (including algae and fungi), and bacteria. Only the *International Code of Zoological Nomenclature* (ICZN) recognizes biogenic structures (the work of animals) as worthy of formal Linnaean names, and only trace fossils are so recognized, not modern structures. The *International Code of Botanical Nomenclature* and the *International Code of Bacterial Nomenclature* do not recognize trace fossils made by plants or bacteria. This means that the names of trace fossils interpreted by researchers as the work of plants or bacteria are not automatically protected under these codes; however, researchers who are interested in these objects are free to name these objects as if they were protected, and this seems the only practical way to proceed, especially because it is not always clear what kind of organism made a trace fossil. This is particularly so in the case of microborings, whose makers include algae and fungi.

The *ICZN* governs the formal names of trace fossils: Very well then, what does it say? Unfortunately, although the first few editions of the *Code* were written in fairly plain language, the fourth and fifth were not (ICZN, 1985, 1999). Even a taxonomist is obliged to read interpretive works such as those of Schenck et al. (1948) and Blackwelder (1967) to understand not only what the *Code* says in words but also the historical intent behind its rules. And not one of these interpretive texts includes a complete and accurate summary of the rules concerning trace fossils. In particular, the otherwise excellent textbook *Describing Species* (Winston, 1999) suffers from having been written just before the fifth edition of the *Code* took effect in 2000, with several changes in the rules concerning ichnotaxa. Instead, ichnologists must rely on works such as those by Bromley and Fürsich (1980), Rindsberg (1990), Pickerill (1994), and Bertling et al. (2006) to understand how the rules have evolved in recent decades and how they apply to trace fossils today.

One aspect of the *Code* that is especially troubling to newcomers is the fact that many rules incorporate cut-off dates. Taxa named before January 1, 1931 could be based on an illustration without a diagnosis; taxa named on or after that date cannot. Before 2000, ichnogenera could be named without designating a type ichnospecies; thenceforward, they must have a type ichnospecies. And so on. Anyone can learn the complexities of the *Code* with practice, though for most non-specialists the time could be better spent in cultivating a systematist as coauthor. Take a taxonomist out to dinner!

7.2 A Brief History of Ichnotaxonomy

A little history will go a long way toward understanding the current status of ichnotaxonomy. In the early years, trace fossils were either mostly misidentified as plants and named as such (e.g., modern *Fucus,* ancient *Fucoides* Brongniart, 1823), or correctly identified as the burrows, trackways, and so on of animals. Where a maker could be inferred, their trace fossils tended to be named as ancient examples of them (e.g., *Arenicolites* Salter, 1857 for supposed maker *Arenicola*). Where the maker was unknown, the trace was often treated as evidence of the animal, and the unknown animal could be named on this basis (e.g., trackway *Chirotherium* Kaup, 1835, the name meaning "hand animal"; burrow *Histioderma* Kinahan, 1858, not merely attributed to worms but regarded as one in its own right). When an enterprising entomologist extended this line of reasoning to modern gall wasps, using the structure of plant galls as evidence of otherwise unknown species of wasps, the taxonomic community, predominated by entomologists, reacted in 1964 by banning all names for "the work of animals" that were established after a cut-off date of December 31, 1930.

Paleontologists who worked with trace fossils were inconvenienced by this turn of events, because names of trace fossils that were established in 1930 or earlier were protected by the ICZN and names erected afterward were not. The result was a broad range of proposals on how to classify trace fossils inside or outside the *Code*, which we need not examine here, as most of these proposals did not take hold (Häntzschel and Kraus, 1972). Häntzschel and Kraus, and later Bromley and Fürsich (1980), argued forcefully in favor of acting as if trace fossil names were covered by the *Code* while petitioning the International Commission for Zoological Nomenclature to amend the *Code* to include ichnotaxa. Most ichnologists followed this advice, and in its fourth edition (1985), the *Code* was revised to include names for "the work of animals" again, with the proviso that only fossil examples be included. Rindsberg (1990) reported on this progress and pointed out some remaining problems, which were discussed by the taxonomic community and largely resolved in the fifth edition of the *Code* (1999).

A few problems remained. What constitutes a trace? What is a fossil? What about the activity of plants? These areas are still open to interpretation by any taxonomist, but the consensus of the first two Workshops on Ichnotaxonomy (Bertling et al., 2006) is the new starting point for any further discussion. Bertling et al. (2006) recognized that many kinds of organismal behavior are reflected in the fossil record, but not all should be named. For example, soils are made by the collective behavior of many species of plants, animals, and microbes, but it seems counterproductive to give these behavioral complexes Linnaean names. Likewise, human artifacts are undoubtedly traces, but anthropologists and archeologists already have their own systems of classification that would only be hampered by the bestowal of Linnaean names to such items as stone tools. The consensus of the two workshops on "What is a trace?" is summarized in Table 2.

The status of trace fossils that are interpreted to have been made by plants, such as root traces, is unsatisfactory as the ICZN deals only with the "work of animals", not of members of other kingdoms. The status of reaction tissues, such as galls grown by plants in reaction to the activity of wasps, is also unsettled. For the time being, I would encourage researchers to name these fossils as if they were unchallenged trace fossils. They can be sorted out later.

The question of what is a *fossil* constitutes another gray area, and one that is ultimately unanswerable. Bertling et al. (2006) wrestled with this problem and concluded that, in general, pre-Holocene material should be considered as fossil while Holocene material should not. Even this distinction proved to be too sharp for some researchers. Nielsen et al. (2003) pointed out that their deep-sea

TABLE 2 Categories of Fossils that are Considered as Traces for the Purpose of Binominal Nomenclature. *Modified from Bertling et al. (2006)*

Classified as traces under the ICZN	Not classified as traces under the ICZN
Footprints, trackways	Skeletons, other hard parts
Trails	Soft tissues
Burrows	Hair, feathers
Borings	Eggs
Coprolites	Calculi
Gastroliths	Pearls
Regurgitaliths	Embedment structures
Nests	Secreted cocoons
Spider webs	Plant reaction tissues (e.g., to insects)
Woven cocoons	Soils
Caddisfly cases	Stromatolites
"Sand reefs"	Pathological structures (signs of disease)
Bite and gnaw structures, other signs of predation	Signs of human technological activity
Signs of human biological activity	

samples of microborings were of assuredly mixed Pleistocene and Holocene age; should they delay the progress of science until more precise samples could be collected? Accordingly, Bertling et al. (2006) left the ultimate decision to the researchers most concerned with their material.

Another problem, pointed out chiefly by R. G. Bromley, was that there is no material difference between an empty boring and a fossil one or between an abandoned burrow and a fossil burrow in an unconsolidated substrate. In this case, the workshop ichnologists recommended for the sake of clarity that the Linnaean names given to trace fossils be extended to modern examples but that the ban on erecting new Linnaean names based on modern material be continued. In the end, the ban exists to demand that scientists make every possible effort to identify modern tracemakers rather than simply name the trace.

Which trace fossils should be named, and which should not? Not every jot and tittle of life activity deserves a formal name. Recall that names are erected for the purpose of communication with others. If the information represented by a fossil is not worth communicating, then there is no reason to name it. For example, a short, poorly preserved fragment of a horizontal burrow with no distinctive features need not be called a *Planolites* merely because this ichnogenus is described as including simple horizontal burrows, and a non-descript vertical burrow need not be categorized as *Skolithos*. Instead, the non-descript burrow can be called simply an indeterminate horizontal burrow. Bear in mind that an incomplete, apparently simple burrow may be part of a branched system!

Open nomenclature can be used effectively to indicate uncertainty. Thus, *Planolites? tubularis* indicates that the ichnogeneric assignment is uncertain, while the ichnospecific assignment is clear; *Planolites tubularis?* indicates that only the ichnospecific assignment is uncertain; and *?Planolites tubularis* indicates that both are uncertain. Other nuances can be expressed by the use of abbreviations and quotation marks. "Cf." (Latin *confer*) means "compare"; hence, cf. *Planolites* means that the assignment to *Planolites* is tentative because the material is inadequate for a definitive determination. "Aff." (Latin *affinis*, "related to") is used to indicate that the material is adequate but that it does not closely resemble the taxon in question, for example, *Planolites* aff. *P. tubularis* refers to a well-preserved specimen that is definitely *Planolites* and similar to *P. tubularis* but may belong to another ichnospecies, possibly an undescribed one. "*Planolites*" indicates that the author is aware that the name is inappropriate but that no better name can serve under the circumstances. The added terms (?, cf., aff., "") are not considered to be parts of a binominal name, but modifiers of it. The use of open nomenclature is encouraged.

7.3 The Mechanics of Naming New Ichnospecies and Ichnogenera

If the researcher has decided that the trace fossil material at hand is distinctive enough for a name, but no previous name for the material is available, he or she may give it a new specific name. The new ichnospecies must be proposed within an ichnogenus, which may also be new. It must be proposed in a scientific publication that is available to the general community, either for freely or at a price. This may be a journal article, a book, or even a website as long as it follows current ICZN guidelines (Winston, 1999). A thesis distributed only among the members of a university is considered to be an unpublished manuscript, as are labels in a museum collection. Reports available only within a petroleum company are also considered to be unpublished with regard to the scientific community at large. However, new species can be, and have been, named in the journals of obscure local academies of science, and in any language, testing the limits of ichnotaxonomists' patience.

The name must be definitely proposed as new. It may not be proposed conditionally, that is, one may not name a fossil that has never been seen, but is thought to exist, say, as a theoretical transitional form. The author may not be anonymous.

A new name must be spelt using the Latin alphabet, which may include the letters *j, k, w,* or *y*, even though they were not used in Classical Latin. It may not include diacritical marks. Originally, virtually all names were proposed in the Latin language, or as latinized versions of surnames or Greek words, but today the net has been cast more broadly. However, a genus name is still considered as being a noun in the nominative singular case, and if it is proposed as a Latin or latinized word, then it must obey Latin grammar. Latin nouns are of masculine, feminine, or neuter gender, most commonly ending in *-us, -a,* and *-um,* respectively. However, there are many exceptions and it is always a good idea to consult *Botanical Latin* (Stearn, 1992) or a Latin dictionary.

Accordingly, a specific name must harmonize grammatically with its generic name. If the name is given a Latin form, as is usual, then it is treated as (1) an adjective or participle in the nominative singular; (2) a noun in the nominative singular, treated as an appositive of the generic name; (3) a noun in the genitive case, either singular or plural; or (4) an adjective that is used as a substantive. Most specific names are intended to describe the species, but some honor the discoverer of the specimens (not the author!), or indicate the location, age, or formation of discovery.

Ichnospecies and ichnogenera are erected in similar but not identical ways. Ichnospecies are sets of material that a researcher considers to be similar. For each new ichnospecies, a *type specimen* (or in some cases more than one type specimen) is designated, which can be consulted in case of any later questions about identification. If many specimens of the same kind are preserved on a slab, then it is a good idea to designate one as the holotype. If a later researcher decides to split the original set of material into two ichnospecies, then the subset including the type specimen retains the original ichnospecies name. As a rule, this type specimen is deposited in a museum collection rather than retained by the researcher, because material in personal collections has a high probability of being lost. Historically, a great many type specimens were lost before this lesson was learned.

A single type specimen is called a *holotype;* multiple type specimens are called series of *paratypes*. Other names are available for special kinds of type specimens.

Ichnogenera are not based on type specimens, but on *type ichnospecies*. An ichnogenus is therefore more of an abstraction than an ichnospecies: a set of sets. An ichnogenus may contain any number of additional ichnospecies, but if some of the ichnospecies are later split off, then the type ichnospecies remains with the ichnogenus for which it is the name bearer, regardless of how inconvenient this may be for the researcher. In practice, many researchers use only ichnogeneric names for trace fossils, but this is a short cut; new ichnogenera must be based on type ichnospecies to be available.

A new name must be available, that is, it must not have been used before within the areas covered by any one code. However, because plants and animals are covered by different codes, and ichnologists currently act as though all trace fossils were covered by the zoological *Code* (though inconveniently, they are not!), it is possible for a trace fossil to have the same name as a plant. For instance, the trace fossil *Hartsellea* Rindsberg, 1994 and the fossil plant *Hartsellea* Gastaldo et al., 2006 were both named for the Hartselle Sandstone. Determining whether a name has been published before can often be accomplished by computer searches; the *Zoological Record, Biological Abstracts, GeoRef,* and similar databases such as that of Knaust (2012) should be consulted before publication. A genus or species can have only one valid name. If a name is determined to be a junior synonym of an older name, the senior synonym is used in favor of the junior one. (If the priority is ambiguous, for instance, if two synonyms were erected in the same issue of one journal, the "first reviser" of the work is allowed the privilege of deciding which name to use henceforth, so long as the reviser announces that this is a definite choice.)

A new name can be in any language and indeed may consist of a random combination of Latin letters, if it is pronounceable. But ideally, a new ichnogeneric name should be in Latin or latinized Greek and follow the grammatical rules of Botanical Latin, which is a late dialect of Latin with a specialized vocabulary developed by naturalists, in which whole books were formerly written (Stearn, 1992). As Latin is no longer routinely learned by zoologists, a distinct falling off has occurred in new Latin names, but Latin remains important in botany and botanists may be consulted on the matter.

The two languages should not be mixed within a single word, though latinizing the ending is acceptable. Compounds are usual, because it is almost impossible to find a simple Latin or Greek word that can conceivably refer to plants or animals that has not already been published as a generic name. It is no longer easy to find a reasonable compound based on two roots, and compounds of three roots are generally too long. Many ichnologists have, therefore, followed Seilacher's (1953) and Frey's (1973) advice by constructing new ichnogenera based on a root followed by the latinized Greek *-ichnus,* meaning "trace": *Hillichnus* to honor Gary Hill, *Oichnus* to describe the O-shape of a drill hole, *Asterichnus* to describe the stellate shape of a trace, and so on. Unfortunately, as Frey recognized in private, this has led to stylistically awkward but unavoidable phrases like "ichnogenus *Teichichnus*", and perhaps it is time to become more creative.

One cannot simply put two Latin words together to make a new Latin word. In general, the first word should be reduced to its root, that is, the part without the ending. If the first root ends in a consonant and the second begins in a consonant, usually a connecting vowel *-i-* is inserted, for example, *Arenicolites,* because few consonant clusters are allowed in Latin, unlike German, for instance. The usual connecting vowel for Greek roots is *-o-*, for example, *Asterosoma, Diplocraterion,* and *Ophiomorpha.* It is déclassé to mix unlatinized Greek and Latin in the same word, though it is permissible to mix latinized Greek with Latin, for example, *Stelloglyphus, Ancorichnus.*

7.4 Diagnosis and Description

The naming of new trace fossils, to paraphrase T. S. Eliot, is a difficult matter. It may seem trivial to mention this, but the first step is to observe the trace fossils. The initial description may be followed by a tentative identification, which may suggest further morphological details to seek out. In practice, description is partly informed by the process of identification, and it is often necessary to return to an outcrop or a museum drawer to gain new information.

How can one find out if the new trace fossil is really new? There is no one source that includes descriptions of all ichnotaxa; identification may require access to a large library. The most extensive key for invertebrate ichnogenera is that of Knaust (2012). Brief descriptions and illustrations of many ichnogenera are available in the second edition of the "Trace Fossils and Problematica" part of the *Treatise on Invertebrate Paleontology,* a posthumous work by Häntzschel (1975). This is the closest we have to a compendium of trace fossils thought to be of invertebrate origin; every such ichnogenus known at that time was listed alphabetically. However, it was never intended as an identification manual. Other, more specialized works that may be of use in identification are summarized briefly in Table 3.

The *Treatise* is a very useful work, but the researcher should be aware of some additional problems with using it as a source. Häntzschel's research was incomplete at the time of his death (Teichert, 1975: vi), and some well-defined ichnogenera were relegated to a waste pile entitled "Unrecognized and Unrecognizable Genera" from which a few good names (e.g., *Ptychoplasma*) have since been salvaged, usually after investigation of type specimens. Another problem with this work is the fact that many trace fossils were already described in other parts of the *Treatise* and the editor evidently did not want to duplicate them under "Trace Fossils". Many of the genera listed as medusae (Harrington and Moore, 1956) or worms (Howell, 1962) are trace fossils. Moreover, some genera of fossil "boring animals" are now considered as ichnogenera of fossil borings, a

TABLE 3 Some Reference Works for the Identification of Invertebrate Trace Fossils, Including Examples of Systematic Monographs

Source	Description
Kuhn (1963)	List of available and unavailable vertebrate ichnogenera with extensive bibliography
Häntzschel (1965)	List of available and unavailable invertebrate ichnogenera with extensive bibliography
Bromley (1970)	Borings made by modern marine animals
Osgood (1970)	Systematics of Ordovician trace fossils of the Cincinnati area, USA
Chamberlain (1971)	Systematics of Carboniferous trace fossils of Oklahoma, USA
Häntzschel (1975)	Descriptions and illustrations of available ichnogenera with extensive bibliography
Chamberlain (1977)	Systematics of Ordovician and Devonian trace fossils of Nevada, USA
Seilacher (1977)	Pattern analysis of selected graphoglyptids (agrichnia)
Książkiewicz (1977)	Systematics of Cretaceous–Tertiary flysch trace fossils of Poland
Chamberlain (1978)	Guide to identification of common burrows in core
Noda (1982)	List and bibliography of trace fossils in Japan and vicinity
Haubold (1984)	Overview of fossil vertebrate trackways
Fillion and Pickerill (1990)	Systematics of Late Cambrian(?) to Early Ordovician trace fossils of eastern Newfoundland, Canada
Leonardi (1994)	Atlas of vertebrate trace fossils of South America
Rindsberg (1994)	Systematics of Mississippian trace fossils of Alabama, USA
Jensen (1997)	Systematics of Early Cambrian trace fossils of Sweden
Uchman (1998)	Systematics of Cretaceous–Tertiary flysch traces of Poland
Genise (2000)	Calichnia of social insects, including *Celliforma*
Fernandes et al. (2002)	Manual of invertebrate trace fossils of Brazil
Schlirf (2003)	Systematics of Jurassic trace fossils of France
Yang et al. (2004)	Systematics of trace fossils of China
Wilson (2012)	Bibliography of marine bioerosion
Knaust (2012)	Key to all currently known invertebrate trace fossils

These works contain extensive bibliographies.

distinction that was clarified by Bromley (1970). Still another problem is that the figures emphasize type material which is not always representative of the ichnogenus as it is understood today, while other figures taken from significant articles are neither type material nor representative; some indeed are no longer thought to belong to the ichnotaxon in question. However, the main problem with Häntzschel's compendium is simply that it is out of date: Many ichnogenera have been named, emended, or clarified since 1975 (Knaust, 2012). The last problem can be addressed, if tediously, by reference to the extensive bibliographies and lists of new ichnotaxa in issues of the *Ichnology Newsletter*, *Biological Abstracts*, *GeoRef*, *Zoological Record*, and other serials. A more rapid method may be to consult an expert.

As pointed out previously, identification of ichnotaxa may require better material than the researcher has at hand, and very often better material than exists. Only well-preserved, distinctive material should be used as the basis for new ichnospecies (Bertling et al., 2006).

The new name is commonly accompanied by a *diagnosis:* an account of the characters that distinguish the new ichnospecies from all others. This is not a full description but may consist of a selection of descriptive features. Alternatively, a diagnosis may contrast the features found in the new material with those of previously known ichnotaxa. This is, however, less satisfying than a diagnosis expressed in more positive terms. A diagnosis should be succinct and cover only the essential characters, saving the remainder for a separate description.

An effective diagnosis uses the same kinds of characters as have been used in the most similar previously described ichnotaxa. These kinds of characters are called *ichnotaxobases*. When ichnotaxonomists meet, the proper selection of ichnotaxobases is likely to take up a good deal of their time. At present, the best summaries of ichnotaxobases for invertebrate burrows are those of Bromley (1996) and Bertling (2007); for invertebrate borings, consult Bromley (1970) and Kelly and Bromley (1984).

Not all of the description of a trace fossil is used in the diagnosis. In an ichnotaxonomic work, the diagnosis is ordinarily followed by a separate *description* section that includes features that are not considered to be crucial to the naming of the trace fossil, but nonetheless have their own significance. Good descriptions last forever: The names and interpretations that Brongniart (1828–1838) gave to trace fossils are no longer current, but his illustrations and descriptions are still useful after nearly 180 years. Poor descriptions are usually forgotten except when they are attached to new names; so, it is particularly important to get it right the first time.

7.5 Ichnofamilies and higher Ichnotaxa

Biologists classify living and fossil organisms into a hierarchy of many additional ranks, extending beyond species and genera: families, orders, classes, phyla (in botany, *divisions*), and kingdoms. But all life has a shared ancestry as shown by its DNA. In the absence of an objectively determined classification, can such a hierarchy be useful for traces and trace fossils?

Nothing in the ICZN prevents such classification (Rindsberg, 1990). The ICZN governs only the formal ranks of species, genus, and family, together with subdivisions such as subspecies, superspecies, subgenus, supergenus, subfamily, and superfamily, as well as others that we need not consider here. (Incidentally, because the *Code* does not apply to higher taxa, it is possible for different researchers to use alternate names for the same taxon, for example, Bivalvia, Pelecypoda, and Lamellibranchiata. The fact that it is possible, however, does not make it a good idea.)

The use of ichnofamilies and higher ichnotaxa has, in fact, already been applied to traces of vertebrate and invertebrate origin. In the case of vertebrate ichnology, a relatively stable classification is possible because of the complexity of characters, which commonly allow recognition of a clade of tracemakers, and it is far preferable to maintain a higher classification of vertebrate trace fossils than to shoehorn them into the classification of tracemakers.

In the case of invertebrate trace fossils, no one higher classification has taken hold. Though it can be argued that no higher classification of such traces is theoretically possible, the fact is that researchers have repeatedly attempted such classifications and that some of the categories they have proposed are useful. One of the best approaches toward an eventual consensus is that of Książkiewicz (1977) and Uchman (1995, 1998), whose papers on flysch trace fossils have gradually refined morphological categories over a period of time without formal designation as ichnofamilies. By grouping similar forms rather than listing them alphabetically, Uchman (1995) was able to show comparisons among trace fossils that represent related behaviors. These morphological groups include an *Ophiomorpha* group (*Ophiomorpha, Thalassinoides, Psilonichnus*), a *Scolicia* group (*Scolicia, Bichordites*), a *Glockerichnus* group (*Glockerichnus, Capodistria*), and others. By using informal but *named* groups, he allowed discussion to continue while avoiding unnecessary taxonomic debates.

7.6 Ichnosubspecies and other Subdivisions

In another direction, it is occasionally helpful in ichnotaxonomy to utilize finer subdivisions, such as subspecies. As in the case of ichnofamilies, the ICZN allows the use of any established biological rank as an ichnological rank as well. The prefix *ichno-* is added to make the distinction clear, for example, *ichnosubspecies* (not "subichnospecies"). According to the ICZN, the naming of a species, genus, or family automatically generates a series of subdivisions (subspecies, superspecies, subgenus, and the rest), all of which share the same type, are attributed to the same author, and are assigned the same date. Very few such ichnotaxa have been named, and only within relatively well-studied ichnogenera. Examples include ichnosubspecies named in *Arthrophycus* and

Paleodictyon by Seilacher (1977, 2000). There is *no need* to name ichnosubspecies to reflect every slight difference found in material.

7.7 Exceptional Cases

In some cases, a strict adherence to rules is intolerable. For example, imagine a researcher who rediscovers the type specimen of the type ichnospecies of a common ichnogenus in a museum drawer, where it has lain unexamined for a century and a half, and finds that it not only does not match the current consensus of the diagnosis of that ichnogenus but is actually a plant. Following the rules would lead to a situation where every specimen bearing a common name would have to be renamed, while the single specimen to retain the name would have no meaningful connection to the vast majority of the literature. The International Code of Zoological Nomenclature (1999) allows for the resolution of such exceptional cases by petitioning the Commission to overturn one or more of its rules. The process is fairly lengthy and should not be considered as a method of solving one researcher's distress, but of resolving a problem of interest to the whole ichnological community.

7.8 Beyond Systematics

Part of the skill involved in taxonomy is knowing what aspect of it to use and which to set aside. The strict systematic approach is most useful when one wants to answer questions relating to particular trace fossils and the organisms that made them. This is not restricted to questions such as "What made this sedimentary structure?" Systematics is necessary, for instance, when one wishes to understand the biodiversity represented by an assemblage of trace fossils. This approach is also needed, generally speaking, when exact answers are wanted.

The other approaches promoted by Seilacher (1953) can be more appropriate in cases where large datasets based on other criteria are used to determine facies. The two facies approaches that are most in use are the studies of archetypical ichnofacies (Frey and Seilacher, 1980; MacEachern et al., 2007, 2012; Seilacher, 1963, 1964) and ichnofabrics (Ekdale and Bromley, 1983). The ichnofacies approach depends on the proportion of different ethological categories, the *Skolithos* Ichnofacies being dominated by vertical dwelling traces, the *Nereites* Ichnofacies by feeding traces, and so on. Although most ichnofacies work does utilize ichnotaxa, ichnofacies can be used in the absence of identifiable ichnotaxa; for example, "dominance by vertical dwelling burrows" is sufficient to determine that a stratum belongs to the *Skolithos* Ichnofacies without identification of *Skolithos, Diplocraterion,* or other ichnotaxa.

The more flexible ichnofabric approach (Ekdale and Bromley, 1983; Ekdale et al., 2012; Gerard and Bromley, 2008; Taylor et al., 2002) uses repetitive, observable characters of biogenic structures, without necessarily identifying them systematically, to categorize the textures of rocks into different *ichnofabrics*. In cores, where systematic identification of trace fossils can be severely limited by poor contrast and restriction to cross-sections, the ichnofabric approach is powerful indeed. It is also highly useful when time is limited. Nevertheless, in this case as well, systematic identification of trace fossils will yield additional insight—if time and proper material are available for such study.

8. CONCLUSIONS

Names are necessary for discussing traces and trace fossils. Modern traces are named according to their biological makers and classified chiefly in terms of behavior and ecology. Trace fossils, in contrast, are named according to use of binominal systematics and may additionally be classified according to their toponomy, behavior (ethology), ecology, and other characteristics. This is the paradigm established for ichnology by Seilacher (1953). It is the basis for the ichnofacies and ichnofabric concepts that have proved so useful in sedimentology.

No one classification of trace fossils exists, but several, and each, including binominal systematics, is partially arbitrary. Rather than bemoan the lack of order, ichnologists revel in the freedom to use several classifications at once in order to glean as much information as possible from these interesting fossils. The best classifications are those that allow the generation of new hypotheses and predictions, reflecting patterns in the data rather than in the mind of the observer. To think about worm burrows, think like a worm!

ACKNOWLEDGMENTS

This contribution would not have been possible without the many hours of discussion at the first two Workshops on Ichnotaxonomy, hosted by Richard G. Bromley and Alfred Uchman and directed largely by Markus Bertling. A similar round of discussions took place in one of the first courses on ichnology, taught by Robert W. Frey and S. George Pemberton at the University of Georgia in 1979, and this can be said to have shaped my ideas on the subject. I also thank the reviewers, Lothar Vallon, Dirk Knaust, and Richard G. Bromley, for their insightful comments and encouragement. Larry A. Herr drafted the figures.

REFERENCES

Aigner, T., 1985. Storm Depositional Systems: Dynamic Stratigraphy in Modern and Ancient Shallow-Marine Sequences. Lecture Notes in Earth Sciences 3. Springer, Berlin, 50 pp.

Ausich, W.I., Bottjer, D.J., 1982. Tiering in suspension feeding communities on soft substrata throughout the Phanerozoic. Science 216, 173–174.

Basan, P.B., Frey, R.W., 1977. Actual-palaeontology and neoichnology of salt marshes near Sapelo Island, Georgia. In: Crimes, T.P., Harper, J.C. (Eds.), Trace Fossils 2. Geol. J., Spec. Iss. 9, pp. 41–70.

Bertling, M., 2007. What's in a name? Nomenclature, systematics, ichnotaxonomy. In: Miller III, W. (Ed.), Trace Fossils. Concepts, Problems, Prospects. Elsevier, Amsterdam, pp. 92–109.

Bertling, M., Braddy, S.J., Bromley, R.G., Demathieu, G.R., Genise, J., Mikuláš, R., Nielsen, J.K., Nielsen, K.S.S., Rindsberg, A.K., Schlirf, M., Uchman, A., 2006. Names for trace fossils: a uniform approach. Lethaia 39, 265–286.

Blackwelder, R.E., 1967. Taxonomy: A Text and Reference Book. John Wiley & Sons, New York, 698 pp.

Bown, T.R., Ratcliffe, B.C., 1988. The origin of *Chubutolithes* Ihering, ichnofossils from the Eocene and Oligocene of Chubut Province, Argentina. J. Paleontol. 62, 163–167.

Bromley, R.G., 1970. Borings as trace fossils and *Entobia cretacea* Portlock, as an example. In: Crimes, T.P., Harper, J.C. (Eds.), Trace Fossils. Geol. J., Spec. Iss. 3, pp. 49–90.

Bromley, R.G., 1996. Trace Fossils: Biology, Taphonomy and Applications. Chapman and Hall, London, 361 pp.

Bromley, R.G., Fürsich, F.T., 1980. Comments on the proposed amendments to the International Code of Zoological Nomenclature regarding ichnotaxa. Bull. Zool. Nomencl. 37, 6–10 Z.N. (S.) 1973.

Bromley, R.G., Uchman, A., Gregory, M.R., Martin, A.J., 2003. *Hillichnus lobosensis* igen. et isp. nov., a complex trace fossil produced by tellinacean bivalves, Paleocene, Monterey, California, USA. Palaeogeogr. Palaeoclimatol. Palaeoecol. 192, 157–186.

Brongniart, A.T., 1823. Observations sur les *Fucoïdes*. Mémoires de la Société d'Histoire Naturelle de Paris 1, pp. 301−320, pls. 19−21.

Brongniart, A.T., 1828–1838. Histoire des végétaux fossiles ou recherches botaniques et géologiques sur les végétaux renfermés dans les diverses couches du globe. G. Dufour & E. d'Ocagne, Paris, 560 pp., 195 pls.

Brown Jr., T., 1999. The Science and Art of Tracking: Nature's Path to Spiritual Discovery. Berkley Books, New York, 219 pp.

Buatois, L.A., Mángano, M.G., 2011. Ichnology. Organism–Substrate Interactions in Space and Time. Cambridge University Press, Cambridge, 358 pp.

Chamberlain, C.K., 1971. Morphology and ethology of trace fossils from the Ouachita Mountains, southeast Oklahoma. J. Paleontol. 45, 212–246.

Chamberlain, C.K., 1977. Ordovician and Devonian trace fossils from Nevada. Nevada Bureau of Mines and Geology, Bull. 90, 24 pp.

Chamberlain, C.K., 1978. Recognition of trace fossils in cores. In: Basan, P.B. (Ed.), Trace Fossil Concepts. SEPM Short Course 5, pp. 119–166.

Eiseman, C., Charney, N., Carson, J., 2010. Tracks and Sign of Insects and Other Invertebrates. Stackpole Books, Mechanicsburg, 582 pp.

Ekdale, A.A., 1977. Abyssal trace fossils in worldwide Deep Sea Drilling Project cores. In: Crimes, T.P., Harper, J.C. (Eds.), Trace Fossils 2. Geol. J., Spec. Iss. 9, pp. 163–182.

Ekdale, A.A., 1985. Paleoecology of the marine endobenthos. Palaeogeogr. Palaeoclimatol. Palaeoecol. 50, 63–81.

Ekdale, A.A., Bromley, R.G., 1983. Trace fossils and ichnofabric in the Kjølby Gaard Marl, uppermost Cretaceous, Denmark. Bull. Geol. Soc. Denmark 31, 107–119.

Ekdale, A.A., Bromley, R.G., Knaust, D., 2012. The ichnofabric concept. In: Knaust, D., Bromley, R.G. (Eds.), Trace Fossils as Indicators of Sedimentary Environments. Developments in Sedimentology, vol. 64. Elsevier, Amsterdam, pp. 139–155.

Ekdale, A.A., Bromley, R.G., Pemberton, S.G., 1984. Ichnology—The use of Trace Fossils in Sedimentology and Stratigraphy. SEPM Short Course 15, 317 pp.

Elbroch, M., 2003. Mammal Tracks & Sign: A Guide to North American Species. Stackpole Books, Mechanicsburg, 779 pp.

Elbroch, M., Marks, E., 2001. Bird Tracks & Sign: A Guide to North American Species. Stackpole Books, Mechanicsburg, 456 pp.

Faul, H., 1951. The naming of fossil footprint "species". J. Paleontol. 25, 409.

Fernandes, A.C.S., Borghi, L., Carvalho, I.de.S., de Abreu, C.J.de, 2002. Guia dos Icnofósseis de Invertebrados do Brasil. Interciência, Rio de Janeiro, 260 pp.

Fillion, D., Pickerill, R.K., 1990. Ichnology of the Upper Cambrian? to Lower Ordovician Bell Island and Wabana groups of eastern Newfoundland, Canada. Palaeontogr. Can. 7, 1–119.

Fischer, W.K., MacGinitie, G.E., 1928. The natural history of an echiuroid worm. Ann. Mag. Nat. Hist. 10, 204–213.

Frey, R.W., 1973. Concepts in the study of biogenic sedimentary structures. J. Sedimt. Petrol. 43, 6–19.

Frey, R.W., Pemberton, S.G., 1985. Biogenic structures in outcrops and cores. I. Approaches to ichnology. Bull. Can. Petrol. Geol. 35, 333–357.

Frey, R.W., Seilacher, A., 1980. Uniformity in marine invertebrate ichnology. Lethaia 13, 183–207.

Frey, R.W., Howard, J.D., Pryor, W.A., 1978. *Ophiomorpha:* its morphologic, taxonomic, and environmental significance. Palaeogeogr. Palaeoclimatol. Palaeoecol. 23, 199–229.

Fürsich, F.T., 1974. On *Diplocraterion* Torell 1870 and the significance of morphological features in vertical, spreiten-bearing, U-shaped trace fossils. J. Paleontol. 48, 952–962.

Gastaldo, R.A., Gibson, M.A., Blanton-Hooks, A., 2006. A Late Mississippian back-barrier marsh ecosystem in the Black Warrior and Appalachian basins. In: Greb, S.F., DiMichele, W.A. (Eds.), Wetlands Through Time. GSA Spec. Pap. 399, pp. 139–154.

Gatesy, S.M., Middleton, K.M., Jenkins Jr., F.A., Shubin, N.H., 1999. Three-dimensional preservation of foot movements in Triassic theropod dinosaurs. Nature 399, 141–144.

Genise, J.F., 2000. The ichnofamily Celliformidae for *Celliforma* and allied ichnogenera. Ichnos 7, 267–282.

Genise, J.F., Bown, T.M., 1994. New Miocene scarabeid and hymenopterous nests and early Miocene (Sanctacrucian) paleoenvironments, Patagonian Argentina. Ichnos 3, 107–117.

Gerard, J.R.F., Bromley, R.G., 2008. Ichnofabrics in Clastic Sediments: Applications to Sedimentological Core Studies. Jean R.F. Gerard, Madrid, 97 pp.

Goldring, R., 1985. The formation of the trace fossil *Cruziana*. Geol. Mag. 122, 65–72.

Häntzschel, W., 1965. Vestigia Invertebratorum et Problematica. Fossilium Catalogus, Ser. I, Animalia 108, 142 pp.

Häntzschel, W., 1975. Trace fossils and problematica. In: Teichert, C. (Ed.), Treatise on Invertebrate Paleontology, Part W, Miscellanea, Supplement 1. Geological Society of America and University of Kansas Press, Boulder, Colorado and Lawrence, Kansas, pp. W1–W269.

Häntzschel, W., Kraus, O., 1972. Names based on trace fossils (ichnotaxa): request for a recommendation. Bull. Zool. Nomencl. 29, 137–141 Z. N. (S.) 1973.

Harrington, H.J., Moore, R.C., 1956. Medusae incertae sedis and unrecognizable forms. In: Moore, R.C. (Ed.), Treatise on Invertebrate Paleontology F, Coelenterata. Geological Society of America/University of Kansas Press, New York/Lawrence, Kansas, pp. F153–F161.

Hasiotis, S., 2003. Complex ichnofossils of solitary and social soil organisms: understanding their evolution and roles in terrestrial paleoecosystems. Palaeogeogr. Palaeocolimatol. Palaeoecol. 192, 259–320.

Haubold, H., 1984. Saurierfährten. A. Ziemsen, Lutherstadt Wittenberg, 231 pp.

Howard, J.D., 1978. Sedimentology and trace fossils. In: Basan, P.B. (Ed.), Trace Fossil Concepts. SEPM Short Course 5, pp. 11–42.

Howell, B.F., 1962. Worms. In: Moore, R.C. (Ed.), Treatise on Invertebrate Paleontology W, Miscellanea: Conodonts, Conoidal Shells of Uncertain Affinities, Worms, Trace Fossils and Problematica. Geological Society of America/University of Kansas Press, New York/Lawrence, Kansas, pp. W144–W177.

International Commission on Zoological Nomenclature (ICZN), 1985. International Code of Zoological Nomenclature, 3rd ed. International Trust for Zoological Nomenclature and British Museum (Natural History)/University of California Press, London/Berkeley and Los Angeles Adopted by the XX General Assembly of the International Union of Biological Sciences, 338 pp.

International Commission on Zoological Nomenclature (ICZN), 1999. International Code of Zoological Nomenclature, 4th ed. International Trust for Zoological Nomenclature. Adopted by the International Union of Biological Sciences. http://www.nhm.ac.uk/hosted-sites/iczn/code.

Jensen, S., 1997. Trace fossils from the Lower Cambrian Mickwitzia sandstone, south-central Sweden. Fossils Strata 42, 1–110.

Kaup, J.F., 1835. Mitteilungen über Tierfährten von Hildburghausen. N. Jb. Geogr. Geol. Petrol. Abh. 1835, 327–328.

Kelly, S.R.A., Bromley, R.G., 1984. Ichnological nomenclature of clavate borings. Palaeontology 27, 793–807.

Kern, J.P., 1978. Paleoenvironment of new trace fossils from the Eocene Mission Valley Formation, California. J. Paleontol. 52, 186–194.

Kinahan, J.R., 1858. On the organic relations of the Cambrian rocks of Bray (County of Wicklow) and Howth County (County of Dublin), with notices of the most remarkable fossils. J. Geol. Soc. Dublin 8, 68–72.

Knaust, D., 2012. Trace-fossil systematics. In: Knaust, D., Bromley, R.G. (Eds.), Trace Fossils as Indicators of Sedimentary Environments. Developments in Sedimentology, vol. 64. Elsevier, Amsterdam, pp. 79–101.

Książkiewicz, M., 1977. Trace fossils in the flysch of the Polish Carpathians. Palaeont. Polon. 36, 1–208.

Kuhn, O., 1963. Ichnia Tetrapodorum. Fossilium Catalogus, Ser. I Animalia 101, 176 pp.

Leonardi, G., 1994. Annotated Atlas of South America Tetrapod Footprints (Devonian to Holocene). Companhia de Pesquisa de Recursos Minerais, Brasília, 247 pp.

Lessertisseur, J., 1956. Traces fossiles d'activité animale et leur signification paléobiologique. Mém. Soc. Géol. France 74, 150 pp.

MacEachern, J.A., Bann, K.L., Gingras, M.K., Zonneveld, J.-P., Dashtgard, S.E., Pemberton, S.G., 2012. The ichnofacies paradigm. In: Knaust, D., Bromley, R.G. (Eds.), Trace Fossils as Indicators of Sedimentary Environments. Developments in Sedimentology, vol. 64. Elsevier, Amsterdam, pp. 103–138.

MacEachern, J.A., Pemberton, S.G., Gingras, M.K., Bann, K.L., 2007. The ichnofacies paradigm: a fifty-year retrospective. In: Miller III, W. (Ed.), Trace Fossils. Concepts, Problems, Prospects. Elsevier, Amsterdam, pp. 52–77.

Martin, A.J., 2006. Resting traces of *Ocypode quadrata* associated with hydration and respiration: Sapelo Island, Georgia, USA. Ichnos 13, 57–67.

Martinsson, A., 1965. Aspects of a Middle Cambrian thanatotope on Öland. GFF 87, 181–230.

Martinsson, A., 1970. Toponomy of trace fossils. In: Crimes, T.P., Harper, J.C. (Eds.), Trace Fossils. Geol. J., Spec. Iss. 3, pp. 323–330.

Minter, N.J., Braddy, S.J., Davis, R.B., 2007. Between a rock and a hard place: arthropod trackways and ichnotaxonomy. Lethaia 40, 365–375.

Müller, A.H., 1962. Zur Ichnologie, Taxiologie und Ökologie fossiler Tiere. Freiberger Forschungshefte C151, 5–49.

Nielsen, K.S.S., Nielsen, J.K., Bromley, R.G., 2003. Palaeoecological and ichnological significance of microborings in Quaternary foraminifera. Palaeontologica. Electronica 6, 13. http://palaeo-electronica.org/2003_1/ichno/ichno.pdf.

Noda, H., 1982. Check list and bibliography of trace fossils and related forms in Japan (1889–1980) and neighbourhood (1928–1980). Introduction to study of trace fossils, Part 2. Institute of Geoscience, University of Tsukuba, Ibaraki, Japan, 80 pp.

Osgood Jr., R.G., 1970. Trace fossils of the Cincinnati area. Palaeontogr. Am. 6, 277–444.

Pickerill, R.K., 1994. Nomenclature and taxonomy of invertebrate trace fossils. In: Donovan, S.K. (Ed.), The Palaeobiology of Trace Fossils. Johns Hopkins University Press, Baltimore, pp. 3–42.

Pohowsky, R.A., 1978. The boring ctenostomate Bryozoa: taxonomy and paleobiology based on cavities in calcareous substrata. Bull. Am. Paleontol. 73, 1–192.

Reineck, H.-E., Singh, I.B., 1980. Depositional Sedimentary Environments. Springer, Berlin, 549 pp.

Rindsberg, A.K., 1990. Ichnological consequences of the 1985 Code of Zoological Nomenclature. Ichnos 1, 59–63.

Rindsberg, A.K., 1994. Ichnology of the Upper Mississippian Hartselle Sandstone of Alabama, with notes on other Carboniferous formations. Geol. Surv. Alabama, Bull. 158, 1–107.

Rindsberg, A.K., Kopaska-Merkel, D.C., 2005. *Treptichnus* and *Arenicolites* from the Steven C. Minkin Paleozoic footprint site (Langsettian, Alabama, USA). In: Buta, R.J., Rindsberg, A.K., Kopaska-Merkel, D.C. (Eds.), Pennsylvanian Footprints in the Black Warrior Basin of Alabama. Alabama Paleontol. Soc., Monogr. 1, 121–141.

Rindsberg, A.K., Martin, A.J., 2003. *Arthrophycus* in the Silurian of Alabama (USA) and the problem of compound trace fossils. Palaeogeogr. Palaeoclimatol. Palaeoecol. 192, 187–219.

Röder, H., 1971. Gangsysteme von *Paraonis fulgens* Levinsen 1883 (Polychaeta) in ökologischer, ethologischer und aktuo-paläontologischer Sicht. Senckenberg. Marit. 3, 3–49.

Salter, J.W., 1857. On annelide-burrows and surface-markings from the Cambrian rocks of the Longmynd. Q. J. Geol. Soc. Lond. 13, 199–206.

Savrda, C.E., 2007. Taphonomy. In: Miller III, W. (Ed.), Trace Fossils. Concepts, Problems, Prospects. Elsevier, Amsterdam, pp. 92–109.

Schäfer, W., 1972. Ecology and Paleoecology of Marine Environments. Oliver and Boyd, Edinburgh, 568 pp.

Schenck, E.T., McMasters, J.H., Keen, A.M., Muller, S.W., 1948. Procedure in Taxonomy. Stanford University Press and Geoffrey Cumberledge/Oxford University Press, Stanford/London, 93 pp.

Schlirf, M., 2003. Palaeoecologic significance of Late Jurassic trace fossils from the Boulonnais, N France. Acta Geol. Polon. 53, 123–142.

Seilacher, A., 1953. Studien zur Palichnologie. I. Über die Methoden der Palichnologie. N. Jb. Geol. Paläont. Abh. 96, 421–452.

Seilacher, A., 1959. Zur ökologischen Charakteristik von Flysch und Molasse. Eclogae Geol. Helv. 51, 1062–1078.

Seilacher, A., 1963. Lebensspuren und Salinitätsfazies. Fortschr. Geol. Rheinland Westfalen 10, 81–94.

Seilacher, A., 1964. Biogenic sedimentary structures. In: Imbrie, J., Newell, N. (Eds.), Approaches to Paleoecology. Wiley, New York, pp. 296–316.

Seilacher, A., 1970. *Cruziana* stratigraphy of "nonfossiliferous" Palaeozoic sandstones. In: Crimes, T.P., Harper, J.C. (Eds.), Trace Fossils. Geol. J., Spec. Iss. 3, pp. 447–476.

Seilacher, A., 1977. Pattern analysis of *Paleodictyon* and related trace fossils. In: Crimes, T.P., Harper, J.C. (Eds.), Trace Fossils 2. Geol. J., Spec. Iss. 9, pp. 289–334.

Seilacher, A., 2000. Ordovician and Silurian arthrophycid ichnostratigraphy. In: Sola, M.A., Worsley, D. (Eds.), Geological Exploration in Murzuk Basin. Elsevier, Amsterdam, pp. 237–258.

Shrock, R.R., 1948. Sequence in Layered Rocks. McGraw-Hill, New York, 507 pp.

Stearn, W.T., 1992. Botanical Latin: History, Grammar, Syntax, Terminology and Vocabulary. David & Charles, Newton Abbot, 546 pp.

Taylor, A., Goldring, R., Gowland, S., 2002. Analysis and application of ichnofabrics. Earth-Sci. Rev. 60, 227–259.

Teichert, C., 1975. Editorial preface. In: Teichert, C. (Ed.), Treatise on Invertebrate Paleontology, Part W, Miscellanea, Supplement 1. Geological Society of America/University of Kansas Press, Boulder/Lawrence, pp. v–viii.

Trueman, E.R., Ansell, A.D., 1969. The mechanisms of burrowing into soft substrata by marine animals. Oceanogr. Mar. Biol. Ann. Rev. 7, 315–366.

Uchman, A. 1995. Taxonomy and palaeoecology of flysh trace fossils: the Marnoso-arenacea Formation and associated facies (Miocene, Northern Appennines, Italy). Beringeria 15, 115 pp.

Uchman, A., 1998. Taxonomy and ethology of flysch trace fossils: a revision of the Marian Książkiewicz collection and studies of complementary material. Ann. Soc. Geol. Polon. 68, 105–218.

Uchman, A., Wetzel, A., 2012. Deep-sea fans. In: Knaust, D., Bromley, R.G. (Eds.), Trace Fossils as Indicators of Sedimentary Environments. Developments in Sedimentology, vol. 64. Elsevier, Amsterdam, pp. 643–672.

Vogel, S., 1978. Organisms that capture currents. Sci. Am. 239 (2), 108–117.

Wetzel, A., 1981. Ökologische und stratigraphische Bedeutung biogener Gefüge in quartären Sedimenten am NW-afrikanischen Kontinentalrand. "Meteor" Forsch.-Ergebnisse Reihe C 34, 1–47.

Wetzel, A., Uchman, A., 2012. Hemipelagic and pelagic basin plains. In: Knaust, D., Bromley, R.G. (Eds.), Trace Fossils as Indicators of Sedimentary Environments. Developments in Sedimentology, vol. 64. Elsevier, Amsterdam, pp. 673–702.

Wilson, M.A., 2012. Marine Bioerosion Bibliography. http://www3.wooster.edu/geology/bioerosion/BioerosionBiblio.pdf, 196 pp.

Winston, J.E., 1999. Describing Species: Practical Taxonomic Procedure for Biologists. Columbia University Press, New York, 518 pp.

Yang, S., Zhang, J., Yang, M., 2004. Trace Fossils of China. Science Press, Beijing, 353 pp., 64 pl.

Zorn, M.E., Gingras, M.K., Pemberton, S.G., 2010. Variation in burrow-wall micromorphologies of select intertidal invertebrates along the Pacific Northwest Coast, U.S.A. behavioral and diagenetic implications. Palaios 25, 59–72.

Chapter 3

Trace-Fossil Systematics

Dirk Knaust[1]
Statoil ASA, Stavanger, Norway
[1]*Corresponding author: e-mail: dkna@statoil.com*

1. INTRODUCTION

The application of trace-fossil studies to sedimentological investigations, particularly the reconstruction of depositional environments, is largely dependent on the exact recognition of ichnotaxa. The heterogeneous composition of many ichnogenera, with ichnospecies of similar morphology and size, but contrasting in behavior and paleoecological requirements of their producers, in many cases calls for an exact determination of the ichnotaxa at the ichnospecies level. Unfortunately, the taxonomical treatment of trace fossils is not straightforward and is far away from being resolved (Bertling, 2007). Ichnotaxonomy, the manner in which to name and classify trace fossils, has been the subject of many proposals and discussions (see Miller, 2012; Rindsberg, 2012), and regularly arranged workshops on special issues of ichnotaxonomy have resulted in further guidelines toward a uniform approach in the procedure of naming trace fossils (Bertling et al., 2006; Rindsberg, 1990).

Moreover, the lack of a consistent classification scheme for trace fossils often hinders integrated studies in which ichnological information is fully employed in sedimentological analysis. Our current basis for taxonomical and systematic work of invertebrate trace fossils is the *Treatise on Invertebrate Paleontology, Part W*, compiled in its second edition by Häntzschel (1975), and listing ichnogenera in alphabetical order. Before and after that milestone, there were several attempts to classify trace fossils by using various criteria, such as their ethology (Chamberlain, 1971; de Gibert et al., 2004; Krejci-Graf, 1932), supposed producers (Dzik, 2005, 2007; Lessertisseur, 1956; Vialov, 1968), preservation (Martinsson, 1965, Seilacher, 1964), overall morphology (Książkiewicz, 1970, 1977; Richter, 1927; Simpson, 1975), or a combination (Seilacher, 1953; Zherikhin, 2003). Gong and Si (2002) attempted to distinguish and describe morphological characteristics of trace fossils by topological parameters as the basis for the classification and evolution of metazoan traces. Furthermore, Ekdale and Lamond (2003) employed trace-fossil phenograms to depict evolutionary relationships of behavioral components reflected in some common trace fossils.

Developments in Sedimentology, Vol. 64. http://dx.doi.org/10.1016/B978-0-444-53813-0.00003-4

Today, many workers have adopted an ethological classification for trace fossils, and a "*purely morphological classification, without references to any ethological model and without taphonomic interpretation, seems to be one of the dead-ends of ichnotaxonomy*" (Uchman, 1998: 107). This trend is only confirmed by Buatois and Mángano (2011: 25) who write in their textbook, "*Although there should be a conscious effort to focus on significant morphological details, there is no such a thing as a purely morphological classification. The ichnotaxonomic classification is permeated by our understanding of the ethology*".

Since Häntzschel (1975), the number of valid ichnogenera has been more than doubled (increased by 58%) and it becomes increasingly difficult to keep track of the continual growing number of new ichnotaxa. An overload of ichnotaxonomical data without a proper classification hinders a transformation of this information into knowledge. This problem not only complicates the introduction and description of new ichnotaxa and the evaluation of their relationships to existing ones but also hinders the simple application of trace fossils in facies reconstructions.

Despite the preferential application of an ethological classification scheme, many workers have attempted to group trace fossils morphologically in particular studies (e.g., Haubold, 1996; Keighley and Pickerill, 2003; Książkiewicz, 1977; Mägdefrau, 1937; Schlirf, 2000; Seilacher, 2007; Trewin, 1994; Uchman, 1995, 1998). Compared with the widely used ethological classification, a key based on morphological characteristics has the advantage of being descriptive and independent from behavioral interpretation and promotes morphological grouping toward higher ichnotaxonomic levels (e.g., ichnofamilies).

In view of an optimal facies reconstruction and interpretation of sedimentary environments, trace fossils with assignable producers would be the perfect tool to perform this task. In the past, sporadic attempts have been made in this direction (e.g., Dzik, 2005, 2007; Mayoral, 2001; Vialov, 1968), but these were subsequently rejected because of a too broad approach or ichnotaxonomical issues. Nevertheless, given the accelerated progress in the assignment of tracemakers to their trace fossils made in the past few years (e.g., by direct evidence, functional interpretations, or analogy with modern traces), this approach may be revitalized in the near future to support a stronger link between ichnology and sedimentology. This procedure is already a common practice in the analysis of bioerosional trace fossils (particularly microborings; see Bromley, 2004; Wilson, 2007; Wisshak, 2006) and the achievements in terms of environmental reconstructions are obvious (see Wisshak, 2012). However, until then the way out is probably precise applications of well-defined ichnotaxa at the ichnospecific level, with diagnostic signatures that result from the activity of distinctive tracemakers.

2. NEED OF A ROBUST ICHNOTAXONOMY AND TRACE-FOSSIL CLASSIFICATION

Several examples demonstrate the importance of the separation of trace fossils produced by contrasting groups of organisms, a procedure that favors the approach of "splitters" rather than "lumpers" (Bertling et al., 2006). For instance, burrows or surface trails of the well-known ichnogenus *Cruziana* d'Orbigny, 1842 were commonly produced by trilobites in Paleozoic, shallow-marine shelf deposits, whereas morphologically similar but overall smaller trace fossils from late Paleozoic and Mesozoic continental deposits result from the activity of notostracan branchiopods and thus were originally assigned to *Isopodichnus* Bornemann, 1889. Moreover, large and robust burrows with analogous scratches were introduced as *Cruziana seilacheri* Zonneveld et al., 2002, a potential junior synonym of the Paleocene *Carpatichnis zuberi* Vialov in Vialov and Fedonkin (1993), but which may be part of a burrow system that resulted from the activity of lobsters (e.g., *Pholeus elongatus*; Knaust, 2007). All three categories of *Cruziana* differ from each other by their distinctive size classes, although overlaps can occur.

Another example is the recently introduced ichnogenus *Parmaichnus* Pervesler and Uchman, 2009, which resembles *Psilonichnus* Fürsich, 1981, but differs from it by the presence of turning chambers in the upper part of the burrow (Fig. 1). *Parmaichnus* is a feeding burrow attributed to upogebiid decapod crustaceans, while *Psilonichnus* results from the activity of sheltering ocypodid crabs. The *Psilonichnus* Ichnofacies is characteristic of backshore environments, but a much broader distribution within supra-, inter-, and subtidal environments was inferred (e.g., Nesbitt and Campbell, 2006) before the strict separation of *Parmaichnus* and *Psilonichnus*. The distinction of the two ichnogenera is thus beneficial to sedimentological studies.

Similarly, only the intensely scratched and passively filled *Rhizocorallium jenense* serves for the reconstruction of the *Glossifungites* Ichnofacies in connection with firm bedding surfaces and thus may have a greater importance for sequence-stratigraphic interpretations (see MacEachern et al., 2012). In contrast, *Rhizocorallium commune* with its active spreite fill and incorporation of fecal pellets *Coprulus oblongus* characterizes predominantly soft substrates of the *Cruziana* Ichnofacies (Knaust et al., 2012).

A further example includes the similar ichnogenera *Thalassinoides* and *Balanoglossites*, which are easy to confuse. However, *Balanoglossites* is a worm burrow and differs from arthropod-produced *Thalassinoides* by its U- and Y-shaped morphology, high irregularity and bluntly terminated side branches (Knaust, 2008; Knaust and Costamagna, 2012).

Finally, the ichnogenus *Skolithos* Haldeman, 1840 has degenerated into a basket for simple vertical burrows with a quite heterogeneous appearance and produced by a wide range of organisms (different kinds of worms, crustaceans, and plants) which often hinders the proper ichnospecific assignment of such burrows. However, due to their poor diagnostic features, a subtle distinction based on the length/width ratio, departure from a strict vertical orientation, annulations or constrictions, presence or absence of lining and funnel, ornamentation, burrow termination, etc., would certainly help to increase the value of *Skolithos* for paleoenvironmental reconstructions. For instance, *Cylindricum* Linck, 1949 (like the similar *Capayanichnus vinchinensis* Melchor et al., 2010) is known from fluvial deposits and was likely produced by freshwater crustaceans. This ichnotaxon differs from *Skolithos* by a lower length/width ratio (Hasiotis, 2008) and thus justifies its own status.

FIGURE 1 *Psilonichnus versus Skolithos* Ichnofacies. (A) *Psilonichnus upsilon* Frey, Curran and Pemberton, 1984 in cross-bedded eolian calcarenite. Pleistocene, San Salvador Island, Bahamas. Knife is 15 cm long. (B) *Parmaichnus stironensis* Pervesler and Uchman, 2009 in low-angle cross-stratified grainstone, where these burrows occur together with *Ophiomorpha borneensis* Keij, 1965 in an assumed shallow subtidal to lower intertidal facies. Holocene, Enjefa Beach, Salwa District, Kuwait. The diagnostic subspherical swellings in the upper part of the Y-shaped burrows distinguish the ichnogenus *Parmaichnus* from similar *Psilonichnus* ichnospecies, which lack such swellings. Moreover, this delicate discrimination at the ichnogenus level leads to a better separation of the *Skolithos* Ichnofacies in foreshore to shoreface environments from the *Psilonichnus* Ichnofacies characteristic of a backshore setting.

The ranking of various ichnotaxobases is crucial in the success of a robust trace-fossil classification. In many examples, different ranking levels have been applied over time to ichnotaxobases in the definition of ichnogenera. The ichnogenus *Ophiomorpha* Lundgren, 1891 demonstrates this inconsistency very well, as it originally was separated from other ichnogenera by its overall morphology, way of branching, and peloidal burrow lining. By consensus, *Ophiomorpha* consisted of passively filled burrows (Ekdale, 1992), but later on, burrows with actively filled meniscate segments were added (*O. annulata*, *O. rudis*;

see Uchman, 1995). This procedure is not very favorable in order to delineate contrasting behavior of the tracemaker, which in this case would probably be dwelling *versus* feeding. In contrast, a number of unbranched meniscate burrows are now differentiated at the ichnogenus level basically by the kind and organization of their backfill (Smith et al., 2008), corresponding to a smaller rank of ichnotaxobases compared with the current concept of *Ophiomorpha*.

3. A NEWLY PROPOSED NOMENCLATURE KEY

3.1 Purpose and Advantages

In this chapter, a nomenclature key to the main ichnogenera is proposed, based on a hierarchy of diagnostic features (ichnotaxobases). The purpose of this ichnosystematic classification is the organization of key ichnotaxa with respect to their ichnotaxobases. The nomenclature key allows for the determination of major groups of trace fossils down to the ichnogenus level. Although the proposed classification certainly has weaknesses such as the subjectively selected ichnotaxobases and their hierarchical order, partly inaccurate assignments, and missing ichnogenera, it may serve as a basis for further evaluation and has the following potential:

1. It is the first compilation of valid invertebrate trace fossils at the ichnogenus level since the publication of the *Trace Fossils* part of the *Treatise of Invertebrate Paleontology* by Häntzschel (1975). This overview of available ichnogenera could be the basis for an updated version of the *Treatise*.
2. It guides the identification of synonymous or established ichnotaxa when planning new descriptions or revisions. Examples of potential subjective synonyms include *Akinereites* and *Nereites*, *Catenarichnus* and *Catenichnus*, *Cornulatichnus* and *Lithoplaision*, *Digitichnus* and *Cylindricum*, *Fimbrichnium* and *Psammichnites*, *Furculosus* and *Hoplichnus*, *Katbergia* and *Katarrhedrites*, *Ophioichnus* and *Zhadaichnus*, *Ovatiichnum* and *Ptychoplasma*, *Ramosichnus* and *Arborichnus*, *Sphaerichnus* and *Sabiuracolites*, and *Tursia* and *Paradictyodora*.
3. The nomenclature key facilitates the assessment of the deployed ichnotaxobases for certain trace-fossil groups (e.g., ichnofamilies, ichnogenera).
4. The classification scheme is intended to be an open document, which in the future may be transformed into a digital database. In this way, it can be continuously updated, improved, and extended in terms of ichnospecies, geological ages, key references, figures, locality information, etc.
5. Finally, this approach hopefully stimulates further reviews of remaining groups of trace fossils.

3.2 Structure of the Key and Challenges

A major uncertainty exists in the definition of "trace fossils", and workers diverge concerning which groups are to be included. The key deals with the main groups of trace fossils, for example, burrows, bioerosional trace fossils, and surface traces (trackways, imprints, and trails), while other trace-fossil groups such as coprolites, root traces, plant–insect traces, etc., are excluded. Bioclaustration, bioimmuration, and embedment structures remain more controversial in respect to their (trace-fossil) origin.

The past comprehensive review of burrows was compiled by Häntzschel (1975), whereas bioerosional trace fossils were subsequently reviewed by Taylor and Wilson (2003) and Bromley (2004). Information about invertebrate trails, tracks, and trackways is scattered in many case studies, although Trewin (1994) revised a morphological system to identify arthropod trackways. Furthermore, Haubold (2010) provided an update of fossil tetrapod footprints, tracks, and trackways. Ediacaran trace fossils require special attention, as they are easily confused with body fossils and sedimentary structures (see review by Jensen et al., 2006). For bryozoan-produced trace fossils, the reader is referred to Rosso (2008). The group of coprolites and fecal pellets (microcoprolites), from both invertebrates and vertebrates, is very heterogeneous and in need of revision. The interactive database FAVRIS contains data on structured crustacean coprolites (Blau, 1993–2000), whereas Schmitz and Binda (1991) and Eriksson et al. (2011) offer classification schemes for vertebrate coprolites, and Hunt et al. (2007) review Triassic vertebrate coprolites. Plant–insect trace fossils are assessed by Labandeira (2007), whereas bioclaustration is reviewed by Tapanila (2005). Root traces (rhizoliths, rhizoconcretions, etc.) are important for paleoenvironmental reconstruction and are usually described in relation to their host sediment or paleosol. Root-trace classification is approached by Klappa (1980), Bockelie (1994), Mikuláš (1999), and Kraus and Hasiotis (2006).

The hierarchy of the selected ichnotaxobases may be most controversial, as it is strongly subject to the investigator and varies considerably from one ichnotaxon to another. With respect to burrows, most authors agree on the following ichnotaxobases with decreasing importance (see Bertling, 2007): (1) orientation, (2) presence or absence of branching, (3) shape, and (4) presence or absence of lining or mantle. Furthermore, ornamentation is regarded as an accessory ichnotaxobase, having minor importance at the ichnogenus level. Based on the highest ichnotaxobase, orientation, burrows are classified into (1) subhorizontal, (2) subvertical, and (3) complex (Fig. 2). Several transition fields hinder a proper discrimination of different trace-fossil groups. Numerous ichnotaxa can occur as both burrows and surface traces and thus are highlighted in the key. Transitions between burrows and borings also exist (e.g., *Gastrochaenolites*, *Balanoglossites*). Nests (e.g., those produced by insects) are commonly regarded as a separate trace-fossil group (calichnia), although some were created below the surface and partly comprise burrow elements. These are included in the key, but for a comprehensive classification, the reader is referred to Genise (2004). Other ichnogenera comprise branched and unbranched ichnospecies such as *Palaeophycus*, *Planolites*, *Rhizocorallium*, and *Trichichnus*. These ichnogenera are indicated accordingly but only listed once, in the category that is regarded as most significant.

Although the vast majority of invertebrate trace fossils with an ichnotaxonomically valid status is considered in this classification, it cannot be excluded that a minor number of ichnogenera escaped the critical review, as they were

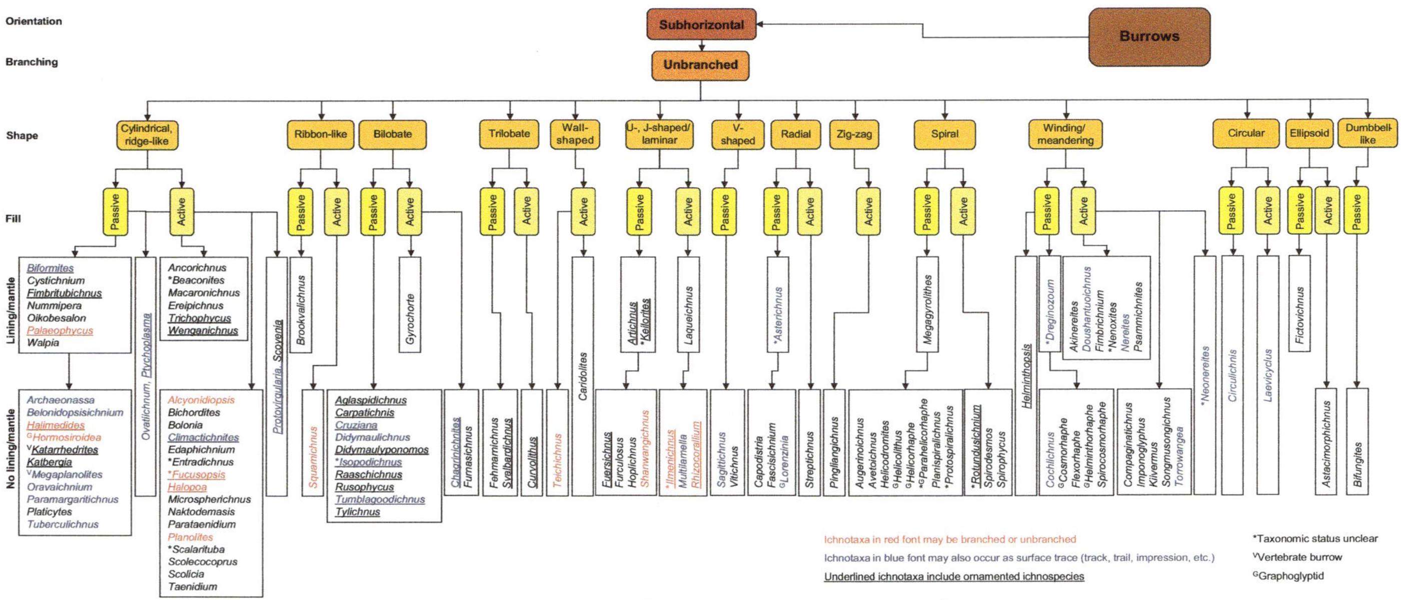

FIGURE 2 Morphologic classification of burrows, as the major group of trace fossils, following a hierarchy of ichnotaxobases. See text for details and Lists 1 and 2 for complete lists of valid and invalid ichnogenera, including references (Supplementary Material: http://booksite.elsevier.com/9780444538130). (A) Subhorizontal, unbranched burrows.

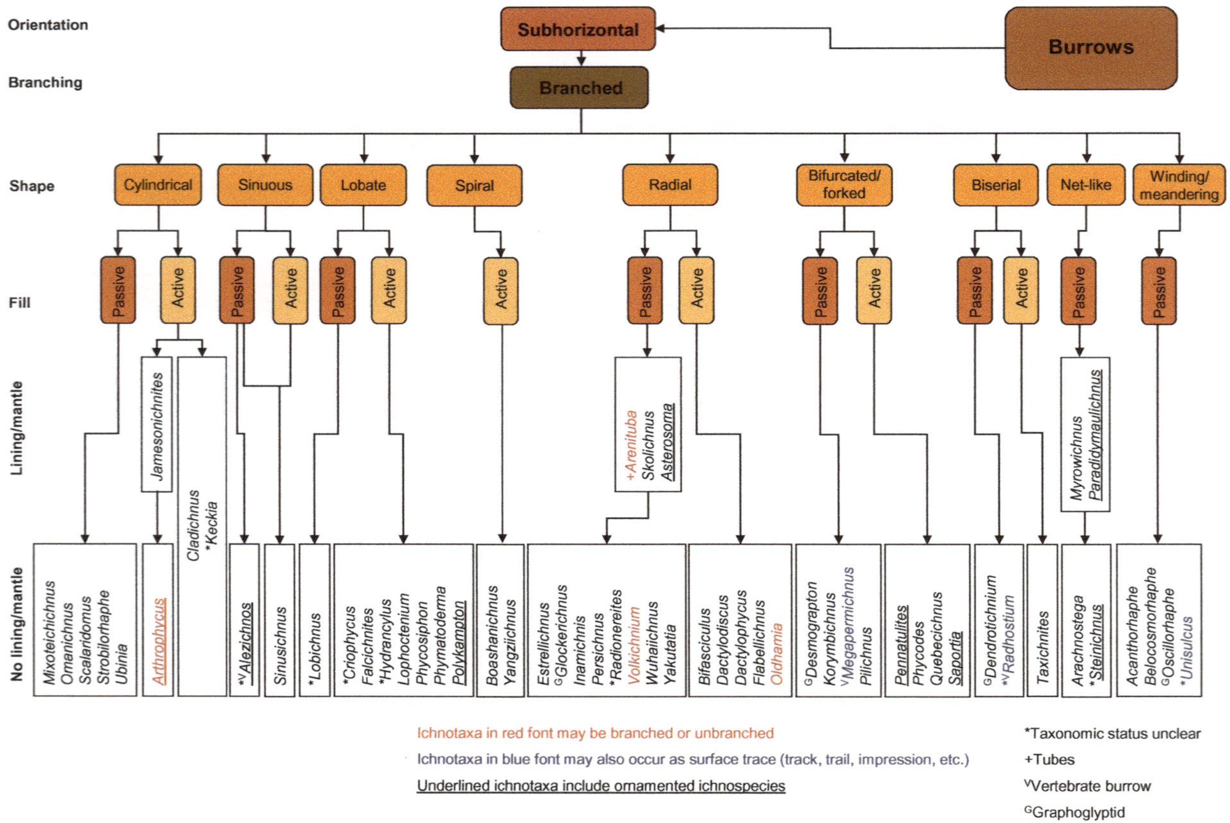

FIGURE 2—Cont'd. (B) Subhorizontal, branched burrows.

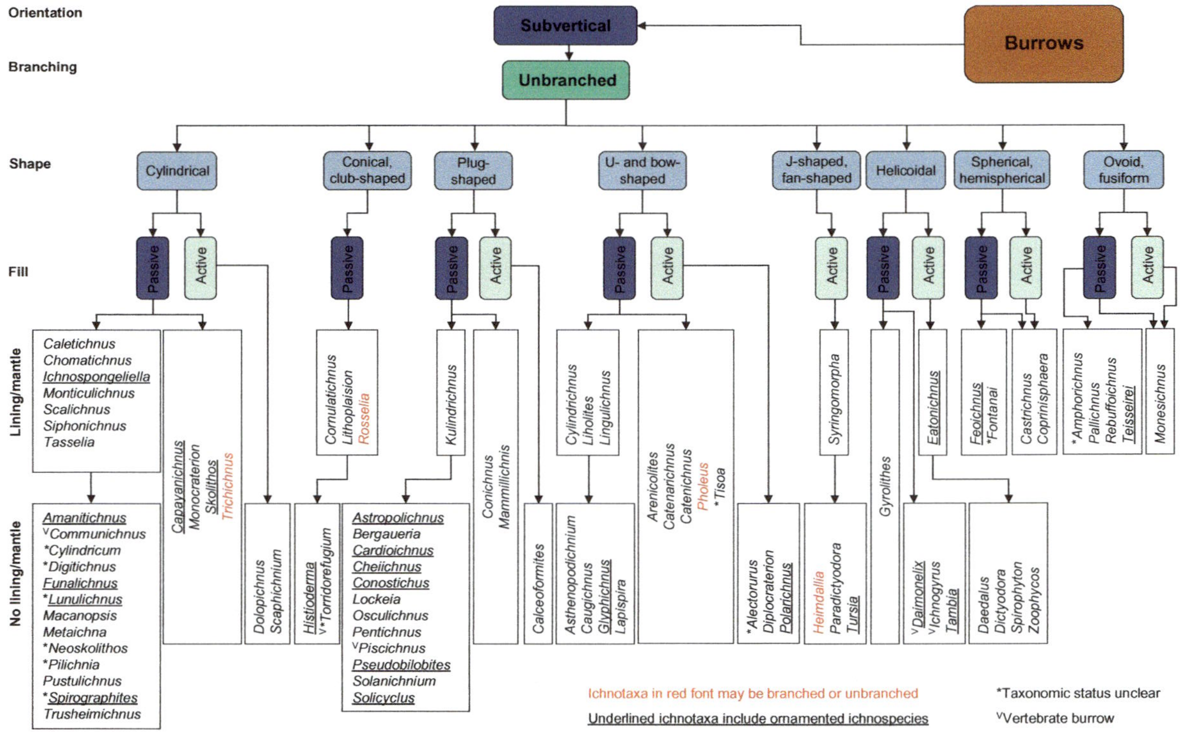

FIGURE 2—Cont'd. (C) Subvertical, unbranched burrows.

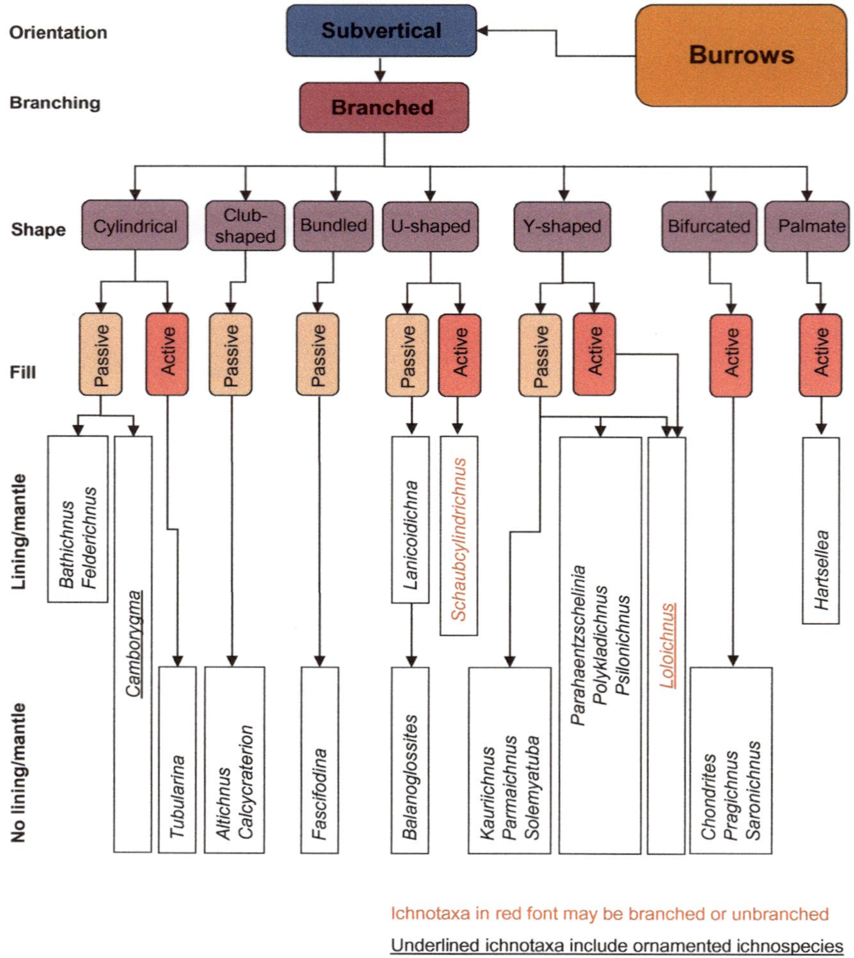

FIGURE 2—Cont'd. (D) Subvertical, branched burrows.

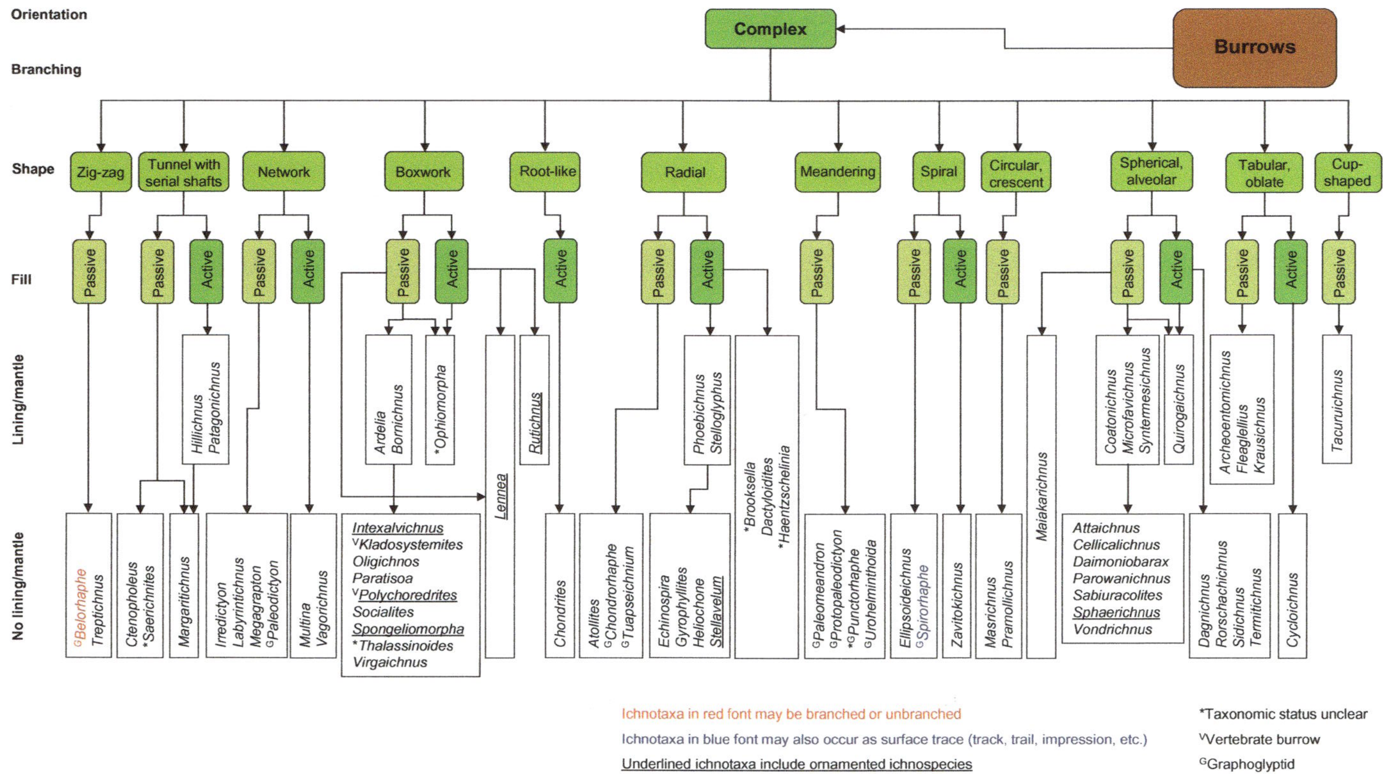

FIGURE 2—Cont'd. (E) Complex burrows.

introduced in publications that are either difficult to access, published in a non-English language, or simply overlooked.

It seems to be natural to separate vertebrate from invertebrate burrows as practiced by Häntzschel (1975) and subsequent workers. However, in some instances, burrows occur in a transitional field and are not unequivocally assignable to one category or another. For instance, *Daimonelix* was regarded as an invertebrate trace fossil by Häntzschel (1975) but later revealed as the burrow of vertebrates (see Hembree and Hasiotis, 2008). Large burrows (e.g., *Katarrhedrites, Megaplanolites*) may be produced by invertebrates (e.g., crustaceans) or vertebrates (e.g., mammals) or both. The fact that only about 3% of all burrows are attributed to the activity of vertebrates, and no clear boundary can be drawn between invertebrate and vertebrate burrows, justifies the procedure to include all burrows within this key. Relevant characteristics for recognizing and describing vertebrate burrows are discussed by Miller et al. (2001).

All burrow ichnogenera regarded as valid ichnotaxa and included in the key (Fig. 2) are alphabetically listed (Supplementary Material, List 1: http://booksite.elsevier.com/9780444538130) together with their author, year of publication, and one or more recent references with more ichnotaxonomic information (such as diagnoses, descriptions, and interpretations). Furthermore, burrow ichnogenera that were introduced since Häntzschel (1975), or which status was invalidated after that date, are listed in the same way (Supplementary Material, List 2: http://booksite.elsevier.com/9780444538130) in order to provide a record.

The key for bioerosional trace fossils is kept rather simple and only differentiates branched and unbranched forms, followed by groupings of morphotypes (Fig. 3). Transitional forms, microbioerosion, and superficial etching traces are highlighted. Similar to burrows, all valid (Supplementary Material, List 3: http://booksite.elsevier.com/9780444538130) and invalid ichnogenera (Supplementary Material, List 4: http://booksite.elsevier.com/9780444538130) are listed with references.

The approach to classify invertebrate trackways (Fig. 4) is based on the distinguishing of uniserial and biserial trackways and, on a lower hierarchical level, the maximum number of tracks within a trackway. This latter ichnotaxobase includes a relatively high uncertainty which is related to the varying number of tracks, preservation style, and interpretation biases. For instance, uniform tracks may belong to the group of one or even numerous tracks. All valid (Supplementary Material, List 5: http://booksite.elsevier.com/9780444538130) and invalid ichnogenera (Supplementary Material, List 8: http://booksite.elsevier.com/9780444538130) are listed with references.

Imprints are simply classified according to their overall morphology (Fig. 5), and all valid and invalid ichnogenera are listed with references (Supplementary Material, Lists 6 and 8: http://booksite.elsevier.com/9780444538130). Trails are distinguished on the basis of their number of lateral elements (e.g., lobes, grooves, etc.) and furthermore their overall morphology (e.g., straight, winding, meandering) (Fig. 6). All valid (Supplementary Material, List 7: http://booksite.elsevier.com/9780444538130) and invalid ichnogenera (Supplementary Material, List 8: http://booksite.elsevier.com/9780444538130) are listed with references.

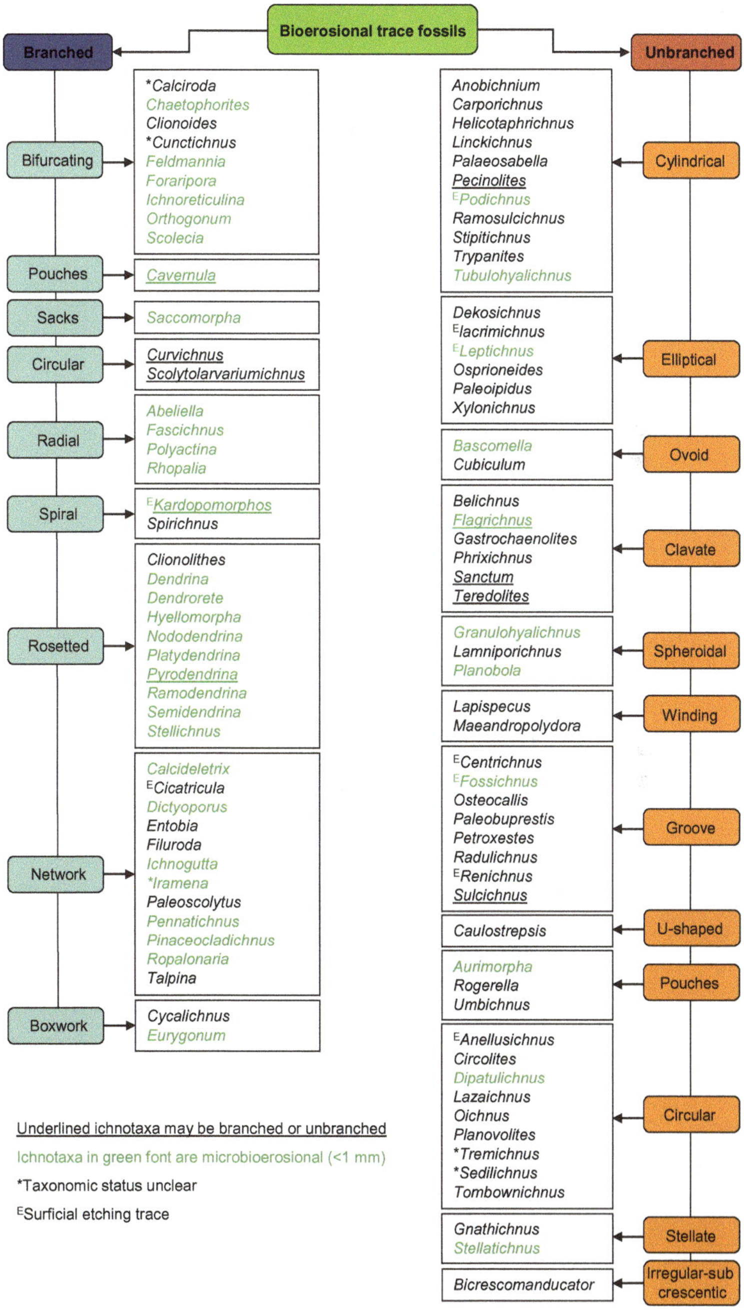

FIGURE 3 Morphological classification of bioerosional trace fossils made by invertebrates. See text for details and Lists 3 and 4 for complete lists of valid and invalid ichnogenera, including references (Supplementary Material: http://booksite.elsevier.com/9780444538130).

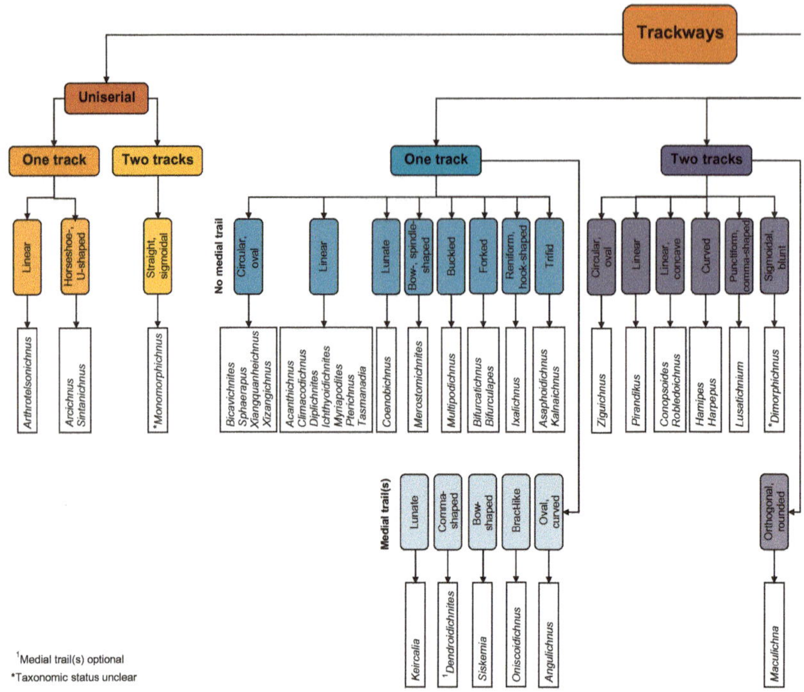

FIGURE 4 Morphological classification of trackways made by invertebrates. See text for details and Lists 5 and 8 for complete lists of valid and invalid ichnogenera, including references (Supplementary Material: http://booksite.elsevier.com/9780444538130).

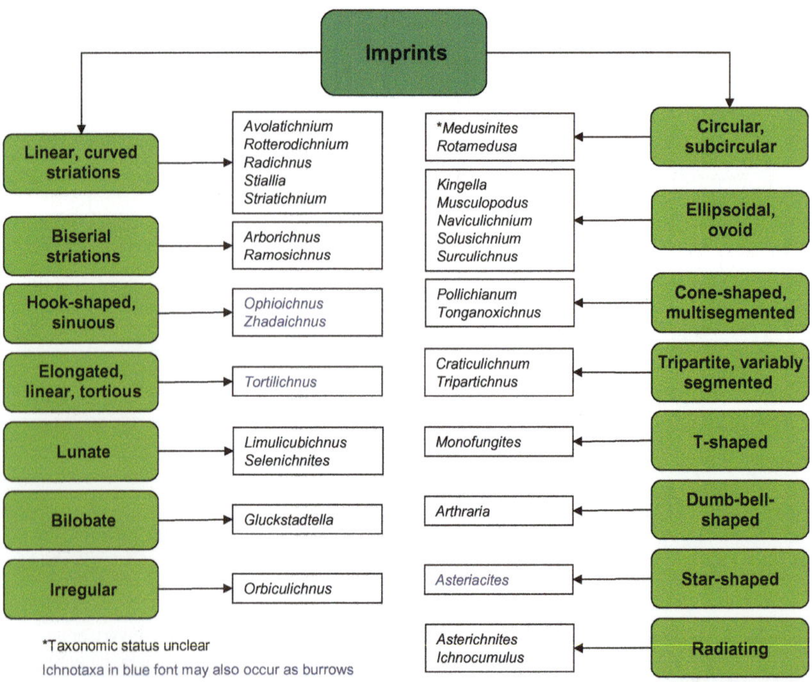

FIGURE 5 Morphological classification of imprints made by invertebrates. See text for details and Lists 6 and 8 for complete lists of valid and invalid ichnogenera, including references (Supplementary Material: http://booksite.elsevier.com/9780444538130).

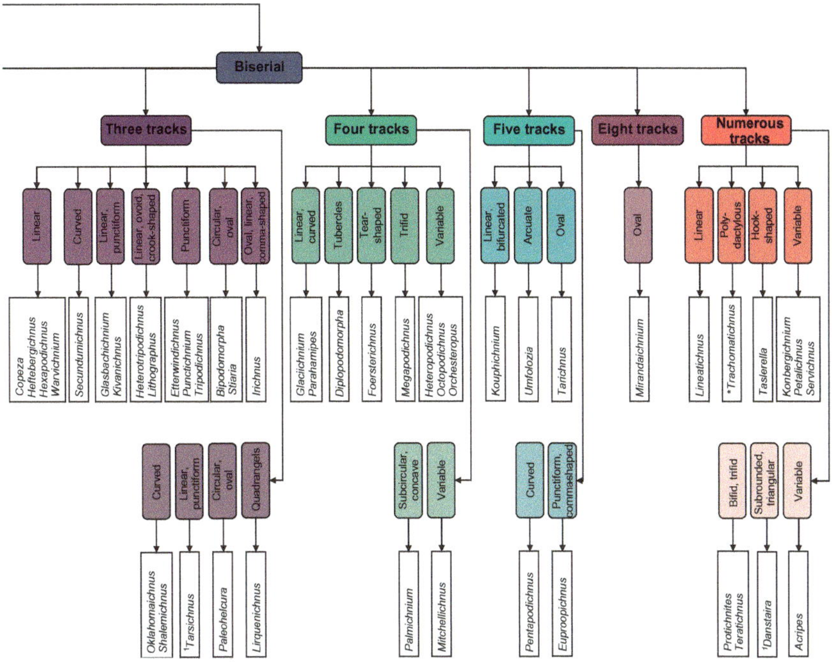

FIGURE 4—Cont'd

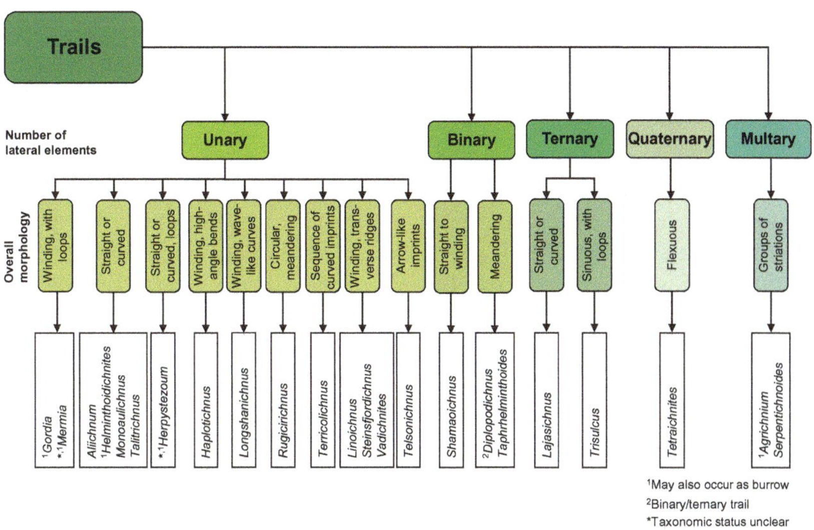

FIGURE 6 Morphological classification of trails made by invertebrates. See text for details and Lists 7 and 8 for complete lists of valid and invalid ichnogenera, including references (Supplementary Material: http://booksite.elsevier.com/9780444538130).

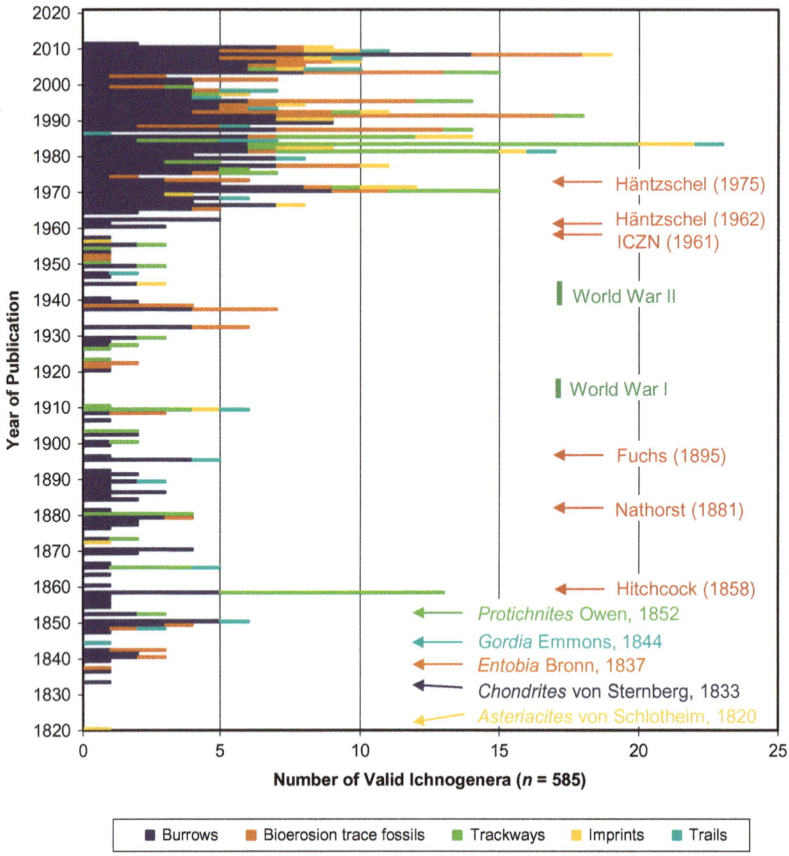

FIGURE 7 Cross plot of the number of valid invertebrate ichnogenera erected over time ($n = 585$) as included in this database and published until 2011 (inclusive), subdivided into different trace-fossil groups. The introduction of the first trace fossil in each category is highlighted, and major events and publications with an impact on ichnotaxonomy are indicated.

3.3 Data Analysis

Figure 7 displays a graph indicating the number of invertebrate ichnogenera erected each year ($n = 585$), as included in this database and published until end of 2011, with some interesting trends worthy of note. Since the erection of *Asteriacites* von Schlotheim, 1820 as the first valid ichnogenus (but see the discussion by Schlirf, 2012), a considerable number of ichnotaxa was introduced as presumed body fossils (mainly seaweeds, e.g., fucoids) during the nineteenth century. Their interpretation as trace fossils by Nathorst (1881) was slowly accepted by contemporaneous researchers but became more accepted after the comprehensive work done by Fuchs (1895). This change in paradigm seems to be the reason for a considerably reduced activity in the introduction of new ichnotaxa until the 1960s. Furthermore, this trend was influenced by the two world wars and the post-World War II situation, with periods of very restricted ichnotaxonomical work.

The rules of the third edition of the *International Code of Zoological Nomenclature* (ICZN), published in 1961, regarded trace-fossil taxa established after 1930 as to be not available. This fact seems to contradict the significant upturn in the establishment of new ichnogenera around the same time, which may be related to the first edition of the *Treatise* section on *Invertebrate Trace Fossils* (Häntzschel, 1962). New proposals for the ichnotaxonomical treatment of ichnotaxa and their coverage under the ICZN during the 1970s led to a revitalization of ichnotaxonomical investigations (Basan, 1979; Pemberton and Frey, 1982). The second edition of the *Treatise* (Häntzschel, 1975) provided an excellent basis for further ichnotaxonomical work. In the 1970s, new ichnotaxa were introduced from a mainly paleontological point of view, whereas more recent descriptions are often related to more complex situations and a sedimentological perspective.

An analysis of the established database reveals some interesting facts worth mentioning. First of all, many new ichnogenera were introduced in the past few decades following the rules of the ICZN. This is very welcome, as it helps to provide a stable platform for communication and to make ichnotaxa available for paleontological and sedimentological studies. Seventy percent of all invertebrate trace fossils (at the ichnogenus level) included in this database was introduced after the publication of the first edition of Häntzschel's *Treatise* in 1962. As a result, many new forms were properly described, although the rapidly increasing number of ichnotaxa would hamper their handling if no robust classification scheme were provided.

A lot of good work was done to review different groups of trace fossils, which in turn led to an increase of stability in ichnotaxonomy. This effort has resulted in the relatively high status of many (mainly common) ichnogenera. Conversely, the ichnotaxonomical status of many less common ichnotaxa is poor and diminishes their value for further use. This is often the case where new ichnotaxa are introduced without an appropriate study of their morphological features, relationship with other ichnotaxa, and potential synonyms.

Figure 8 displays a graph with the frequency of major trace-fossil groups as regarded in this classification. Most of the analyzed ichnogenera are burrows (60%), followed by bioerosional trace fossils (17%), trackways (14%), imprints (5%), and trails (4%) (transitional forms with multiple behavior styles were counted in those groups where they are most common). It has long been a custom to name ichnogenera with the suffix *-ichnus*, which applies to about 44% of them.

4. WAY FORWARD

The proposed nomenclatural key is intended to test applied ichnotaxobases, existing trace-fossil groups, ichnofamilies, and ichnogenera for the classification of trace fossils. If it succeeds, it may be the basis for the establishment of further keys for other trace-fossil groups. In the short term, it may assist in the review of existing ichnogenera and ichnofamilies and aid the proper description of ichnotaxa, leading toward a unified application of ichnotaxobases for

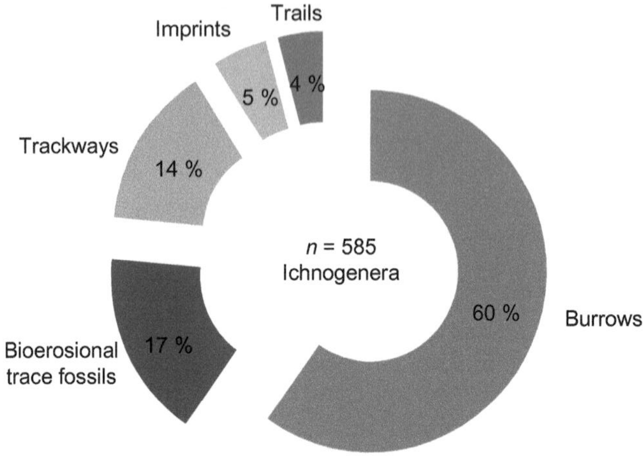

FIGURE 8 Frequency of major trace-fossil groups as included in this classification.

particular trace-fossil groups. Of course, the existing key is premature in that it contains numerous inconsistencies and gaps, which certainly will need improvement. One of the foremost tasks will be the maintenance of the key with a continuous update of missing and newly introduced ichnogenera (suggestions for new entries are welcome).

In the longer term, it would be a clear advantage to transform the provided information into a structured online database, containing all ichnotaxa as a platform for further ichnological work. As known from other, similar databases, a wealth of useful functionality is available, including filter-based searchability and linking with additional data (such as images, age distribution, ethological interpretation, interpreted producer, references). The *Zoological Record Online*® would probably best suit the needs of invertebrate and vertebrate ichnology.

Finally, after having this basic work done, there is still the possibility for a producer-related interpretation and classification of trace fossils, as many ichnotaxa already have a producer interpreted at a higher taxonomic level. This would clearly have the advantage of better linking ecological relationships between traces and producers and thus would directly contribute to advanced reconstructions of paleoenvironments based on trace fossils.

5. CONCLUSIONS

The classification of trace fossils is an important requirement for their systematic usage and for their employment within the reconstruction of sedimentary environments. Widely applied is the grouping of trace fossils according to the behavioral interpretation of their producers, although this may mislead environmental analysis in cases where incorrect assumptions on the ethology of trace fossils were made.

A classification scheme based on significant morphological features (ichnotaxobases) is proposed in this chapter for burrows, bioerosional trace fossils, trackways, imprints, and trails. All treated ichnogenera are listed together with recent references for more detailed data. Even those ichnogenera regarded as invalid and not included in Häntzschel (1975) are listed with references.

A critical review of all ichnogenera currently regarded as valid shows that the number of ichnogenera erected after the last compilation of invertebrate ichnogenera by Häntzschel (1975) has more than doubled and is now approaching 600. Burrows are by far the largest subdivision of them (60%), followed by bioerosional trace fossils (17%), trackways (14%), imprints (5%), and trails (4%).

This classification scheme is intended to serve as a solid basis for further ichnotaxonomical work toward a more stable trace-fossil systematics. As a next step, it could be the basis of a searchable online database in which further criteria of ichnogenera may be regarded and filtered.

ACKNOWLEDGMENTS

I am indebted to Alfred Uchman (Kraków) and Andrew K. Rindsberg (Livingston) for their thorough review, and Ricardo N. Melchor (Santa Rosa) and Max Wisshak (Wilhelmshaven) for valuable comments. Richard G. Bromley (Copenhagen) provided insightful suggestions and comments that improved the chapter.

REFERENCES

Basan, P.B., 1979. Trace fossil nomenclature: the developing picture. Palaeogeogr. Palaeoclimatol. Palaeoecol. 28, 143–167.

Bertling, M., 2007. What's in a name? Nomenclature, systematics, ichnotaxonomy. In: Miller III, W. (Ed.), Trace Fossils. Concepts, Problems, Prospects. Elsevier, Amsterdam, pp. 92–109.

Bertling, M., Braddy, S.J., Bromley, R.G., Demathieu, G.R., Genise, J., Mikuláš, R., Nielsen, J.K., Nielsen, K.S.S., Rindsberg, A.K., Schlirf, M., Uchman, A., 2006. Names for trace fossils: a uniform approach. Lethaia 39, 265–286.

Blau, J., 1993–2000. FAVRIS—the interactive database on structured crustacean coprolites. http://www.uni-giessen.de/~gg13/FAVRIS/FAVRIS.HTM

Bockelie, J.F., 1994. Plant roots in core. In: Donovan, S.K. (Ed.), The Palaeobiology of Trace Fossils. John Wiley & Sons, Chichester, pp. 177–199.

Bornemann, J.G., 1889. Über den Buntsandstein in Deutschland und seine Bedeutung für die Trias. Beitr. Geol. Paläont. 1, 1–61.

Bromley, R.G., 2004. A stratigraphy of marine bioerosion. In: McIlroy, D. (Ed.), The Application of Ichnology to Palaeoenvironmental and Stratigraphic Analysis. Geol. Soc. London, Spec. Publ., vol. 228, pp. 455–479.

Buatois, L.A., Mángano, M.G., 2011. Ichnology. Organism-Substrate Interactions in Space and Time. Cambridge University Press, Cambridge, 358 pp.

Chamberlain, C.K., 1971. Morphology and ethology of trace fossils from the Ouachita Mountains, southeast Oklahoma. J. Paleontol. 45, 212–246.

de Gibert, J.M., Domènech, R., Martinell, J., 2004. An ethological framework for animal bioerosion trace fossils upon mineral substrates with proposal of a new class, fixichnia. Lethaia 37, 429–437.

d'Orbigny, A., 1842. Voyage dans l'Amérique méridionale (le Brésil, la République orientale de l'Uruguay, la République d'Argentine, la Patagonie, la République du Chili, la République de Bolivia, la République du Pérou) éxécuté pendants les années 1826, 1827, 1829, 1830, 1831, 1832 et 1833. v. 3, pt. 4 (Paléontologie), Pitois-Levrault édit., Paris, 188 pp.

Dzik, J., 2005. Behavioral and anatomical unity of the earliest burrowing animals and the cause of the "Cambrian explosion". Paleobiology 31, 503–521.

Dzik, J., 2007. The Verdun syndrome: simultaneous origin of protective armour and infaunal shelters at the Precambrian–Cambrian transition. In: Vickers-Rich, P., Komarower, P. (Eds.), The Rise and Fall of the Ediacaran Biota. Geol. Soc. London, Spec. Publ., vol. 286, pp. 405–414.

Ekdale, A.A., 1992. Muckraking and mudslinging: the joys of deposit-feeding. In: Maples, C.G., West, R.R. (Eds.), Trace Fossils. Short Courses in Paleontology, vol. 5. Paleontological Society, Knoxville, pp. 145–171.

Ekdale, A.A., Lamond, R.E., 2003. Behavioral cladistics of trace fossils: evolution of derived tracemaking skills. Palaeogeogr. Palaeoclimatol. Palaeoecol. 192, 335–343.

Eriksson, M.E., Lindgren, J., Chin, K., Månsby, U., 2011. Coprolite morphotypes from the Upper Cretaceous of Sweden: novel views on an ancient ecosystem and implications for coprolite taphonomy. Lethaia 44, 455–468.

Frey, R.W., Curran, H.A., Pemberton, S.G., 1984. Tracemaking activities of crabs and their environmental significance: the ichnogenus *Psilonichnus*. J. Paleontol. 58, 333–350.

Fuchs, T., 1895. Studien über Fucoiden und Hieroglyphen. Denkschr. Kais. Akad. Wiss. Wien, Math.-Nat. Kl. 62, 369–448.

Fürsich, F.T., 1981. Invertebrate trace fossils from the Upper Jurassic of Portugal. Comm. Serv. Geol. Portugal 67, 53–168.

Genise, J.F., 2004. Ichnotaxonomy and ichnostratigraphy of chambered trace fossils in palaeosols attributed to coleopterans, termites and ants. In: McIlroy, D. (Ed.), The Application of Ichnology to Palaeoenvironmental and Stratigraphic Analysis. Geol. Soc. London, Spec. Publ., vol. 228, pp. 419–453.

Gong, Y.-M., Si, Y.-L., 2002. Classification and evolution of metazoan traces at a topological level. Lethaia 35, 263–274.

Haldeman, S.S., 1840. Supplement to number one of "A monograph of the Limniades, or freshwater univalve shells of North America", containing descriptions of apparently new animals in different classes, and the names and characters of the subgenera in Paludina and Anculosa. Philadelphia, 3 pp.

Häntzschel, W., 1962. Trace fossils and problematica. In: Moore, R.C. (Ed.), Treatise on Invertebrate Paleontology, Part W, Miscellanea. Geological Society of America/University of Kansas Press, New York/Lawrence, pp. W177–W245.

Häntzschel, W., 1975. Trace fossils and problematica. In: Teichert, C. (Ed.), Treatise on Invertebrate Paleontology, Part W, Miscellanea Supplement 1. Geological Society of America/University of Kansas Press, Boulder/Lawrence, pp. W1–W269.

Hasiotis, S.T., 2008. Reply to the comments by Bromley et al. of the paper "Reconnaissance of the Upper Jurassic Morrison Formation ichnofossils, Rocky Mountain Region, USA: paleoenvironmental, stratigraphic, and paleoclimatic significance of terrestrial and freshwater ichnocoenoses" by Stephen T. Hasiotis. Sediment. Geol. 208, 61–68.

Haubold, H., 1996. Ichnotaxonomie und Klassifikation von Tetrapodenfährten aus dem Perm. Hallesches Jb. Geowiss. B 18, 23–88.

Haubold, H., 2010. Literaturbericht: Tetrapodenfährten 2005-2009—Karbon bis Känozoikum (fossil footprints—tracks—trackways—tetrapod resp. vertebrate ichnology). Zbl. Geol. Paläont. II 2010, 1–462.

Hembree, D.I., Hasiotis, S.T., 2008. Miocene vertebrate and invertebrate burrows defining compound paleosols in the Pawnee Creek Formation, Colorado, U.S.A. Palaeogeogr. Palaeoclimatol. Palaeoecol. 270, 349–365.

Hitchcock, E., 1858. Ichnology of New England. A Report on the Sandstone of the Connecticut Valley, Especially its Fossil Footmarks. Commonwealth of Massachusetts. William White, Boston, 220 pp.

Hunt, A.P., Lucas, S.G., Spielmann, J.A., Lerner, A.J., 2007. A review of vertebrate coprolites of the Triassic with descriptions of new Mesozoic ichnotaxa. In: Lucas, S.G., Spielmann, J.A. (Eds.), The Global Triassic. New Mexico Mus. Nat. Hist. Sci., Bull., vol. 41, pp. 88–107.

ICZN (International Commission for Zoological Nomenclature), 1961. International Code of Zoological Nomenclature, 1st ed. International Trust for Zoological Nomenclature, London, 176 pp.

Jensen, S., Droser, M.L., Gehling, J.G., 2006. A critical look at the Ediacaran trace fossil record. In: Xiao, S., Kaufman, A.J. (Eds.), Neoproterozoic Geobiology and Paleobiology. Springer, Berlin, pp. 115–157.

Keighley, D.G., Pickerill, R.K., 2003. Ichnocoenoses from the Carboniferous of eastern Canada and their implications for the recognition of ichnofacies in nonmarine strata. Atlantic Geol. 39, 1–22.

Keij, A.J., 1965. Miocene trace fossils from Borneo. Paläont. Z. 39, 220–228.

Klappa, C.F., 1980. Rhizoliths in terrestrial carbonates: classification, recognition, genesis and significance. Sedimentology 27, 613–629.

Knaust, D., 2007. Invertebrate trace fossils and ichnodiversity in shallow-marine carbonates of the German Middle Triassic (Muschelkalk). In: Bromley, R.G., Buatois, L.A., Mángano, G., Genise, J.F., Melchor, R.N. (Eds.), Sediment-Organism Interactions: A Multifaceted Ichnology. SEPM Spec. Publ., vol. 88, pp. 221–238.

Knaust, D., 2008. *Balanoglossites* Mägdefrau, 1932 from the Middle Triassic of Germany: part of a complex trace fossil probably produced by burrowing and boring polychaetes. Paläont. Z. 82, 347–372.

Knaust, D., Costamagna, L.G., 2012. Ichnology and sedimentology of the Triassic carbonates of north-west Sardinia, Italy. Sedimentology 59, 1190–1207.

Knaust, D., Curran, H.A., Dronov, A., 2012. Shallow-marine carbonates. In: Knaust, D., Bromley, R.G. (Eds.), Trace Fossils as Indicators of Sedimentary Environments. Developments in Sedimentology, vol. 64. Elsevier, Amsterdam, pp. 703–750.

Kraus, M.J., Hasiotis, S.T., 2006. Significance of different modes of rhizolith preservation to interpreting paleoenvironmental and paleohydrologic settings: examples from Paleogene paleosols, Bighorn Basin, Wyoming, U.S.A. J. Sediment. Res. 76, 633–646.

Krejci-Graf, K., 1932. Definition der Begriffe Marken, Spuren, Fährten, Bauten, Hieroglyphen und Fucoiden. Senckenbergiana 14, 19–39.

Książkiewicz, M., 1970. Observations on the ichnofauna of the Polish Carpathians. In: Crimes, T.P., Harper, J.C. (Eds.), Trace Fossils. Geol. J., Spec. Iss. 3, pp. 283–322.

Książkiewicz, M., 1977. Trace fossils in the flysch of the Polish Carpathians. Palaeont. Polon. 36, 1–208.

Labandeira, C.C., 2007. Assessing the fossil record of plant-insect associations: ichnodata *versus* body-fossil data. In: Bromley, R.G., Buatois, L.A., Mángano, G., Genise, J.F., Melchor, R.N. (Eds.), Sediment-Organism Interactions: A Multifaceted Ichnology. SEPM Spec. Publ., vol. 88, pp. 9–26.

Lessertisseur, J., 1956. Traces fossiles d'activité animale et leur signification paléobiologique. Soc. Géol. France, Mém. 74, 1–150.

Linck, O., 1949. Lebens-Spuren aus dem Schilfsandstein (Mittl. Keuper km 2) NW-Württembergs und ihre Bedeutung für die Bildungs-Geschichte der Stufe. Jh. Ver. Vaterl. Naturkd. Württemberg, Jahresh. 97–101, pp. 1–100.

Lundgren, S.A.B., 1891. Studier öfver fossilförande lösa block. Geol. Fören. Stockholm, Förhandl. 13, 111–121.

MacEachern, J.A., Dashtgard, S.E., Knaust, D., Catuneanu, O., Bann, K.L., Pemberton, S.G., 2012. Sequence stratigraphy. In: Knaust, D., Bromley, R.G. (Eds.), Trace Fossils as Indicators of Sedimentary Environments. Developments in Sedimentology, vol. 64. Elsevier, Amsterdam, pp. 157–194.

Mägdefrau, K., 1937. Lebensspuren fossiler 'Bohr'-Organismen. Beitr. Naturkundl. Forsch. Südwestdeutschl. 2, 54–67.

Martinsson, A., 1965. Aspects of a Middle Cambrian thanatotope on Öland. Geol. Fören. Stockholm, Förhandl. 87, 181–230.

Mayoral, E., 2001. Pistas fósiles de artrópodos. Bol. S.E.A. 28, 15–33.

Melchor, R.N., Genise, J.F., Farina, J.L., Sánchez, M.V., Sarzetti, L., Visconti, G., 2010. Large striated burrows from fluvial deposits of the Neogene Vinchina Formation, La Rioja, Argentina: a crab origin suggested by neoichnology and sedimentology. Palaeogeogr. Palaeoclimatol. Palaeoecol. 291, 400–418.

Mikuláš, R., 1999. Notes to the concept of plant trace fossils related to plant-generated sedimentary structures. Věstník Českého geol. úst. 74, 39–42.

Miller, M.F., Hasiotis, S.T., Babcock, L.E., Isbell, J.L., Collinson, J.W., 2001. Tetrapod and large burrows of uncertain origin in Triassic high paleolatitude floodplain deposits, Antarctica. Palaios 16, 218–232.

Miller III, W., 2012. On the doctrine of ichnotaxonomic conservatism: the differences between ichnotaxa and biotaxa. N. Jb. Geol. Paläont. Abh. 265, 295–304.

Nathorst, A.G., 1881. Om spår av några evertebrerade djur och deras paleontologiska betydelse. Kongl. Svenska Vetenskaps-Akademiens Handlingar 18, 1–104.

Nesbitt, E.A., Campbell, K.A., 2006. The paleoenvironmental significance of *Psilonichnus*. Palaios 21, 187–196.

Pemberton, S.G., Frey, R.W., 1982. Trace fossil nomenclature and the *Planolites-Palaeophycus* dilemma. J. Paleontol. 56, 843–881.

Pervesler, P., Uchman, A., 2009. A new Y−shaped trace fossil attributed to upogebiid crustaceans from Early Pleistocene of Italy. Acta Palaeont. Polon. 54, 135–142.

Richter, R., 1927. Die fossilen Fährten und Bauten der Würmer, ein Überblick über ihre biologischen Grundformen und deren geologische Bedeutung. Paläont. Z. 9, 193–240.

Rindsberg, A.K., 1990. Ichnological consequences of the 1985 International Code of Zoological Nomenclature. Ichnos 1, 59–63.

Rindsberg, A.K., 2012. Ichnotaxonomy: finding patterns in a welter of information. In: Knaust, D., Bromley, R.G. (Eds.), Trace Fossils as Indicators of Sedimentary Environments. Developments in Sedimentology, vol. 64. Elsevier, Amsterdam, pp. 45–78.

Rosso, A., 2008. *Leptichnus tortus* isp. nov., a new cheilostome etching and comments on other bryozoan-produced trace fossils. In: Avanzini, M., Petti, F. (Eds.), Italian Ichnology. Studi Trent. Sci. Nat. Acta Geol., vol. 83, pp. 75–85.

Schlirf, M., 2000. Upper Jurassic trace fossils from the Boulonnais (northern France). Geol. et Paleont. 34, 145–213.

Schlirf, M., 2012. *Heliophycus seilacherin.* isp. and *Biformites insolitus* Linck, 1949 (trace fossils) from the Late Triassic of the Germanic Basin: their taxonomy and palaeoecological relevance. N. Jb. Geol. Paläont. Abh. 263, 185–198.

Schmitz, M., Binda, P.L., 1991. Coprolites from the Maastrichtian Whitemud Formation of southern Saskatchewan: morphological classification and interpretation on diagenesis. Paläont. Z. 65, 199–211.

Seilacher, A., 1953. Studien zur Palichnologie. I. Über die Methoden der Palichnologie. N. Jb. Geol. Paläont. Abh. 96, 421–452.

Seilacher, A., 1964. Sedimentological classification and nomenclature of trace fossils. Sedimentology 3, 253–256.

Seilacher, A., 2007. Trace Fossil Analysis. Springer, Berlin, 226 pp.

Simpson, S., 1975. Classification of trace fossils. In: Frey, R.W. (Ed.), The Study of Trace Fossils. Springer, New York, pp. 39–54.

Smith, J.J., Hasiotis, S.T., Kraus, M.J., Woody, D.T., 2008. *Naktodemasis bowni*: new ichnogenus and ichnospecies for adhesive meniscate burrows (AMB), and paleoenvironmental implications, Paleogene Willwood Formation, Bighorn Basin, Wyoming. J. Paleontol. 82, 267–278.

Tapanila, L., 2005. Palaeoecology and diversity of endosymbionts in Palaeozoic marine invertebrates: trace fossil evidence. Lethaia 38, 89–99.

Taylor, P.D., Wilson, M.A., 2003. Palaeoecology and evolution of marine hard substrate communities. Earth-Sci. Rev. 62, 1–103.

Trewin, N.H., 1994. A draft system for the identification and description of arthropod trackways. Palaeontology 37, 811–823.

Uchman, A., 1995. Taxonomy and palaeoecology of flysch trace fossils: the Marnoso-arenacea Formation and associated facies (Miocene, Northern Apennines, Italy). Beringeria 15, 3–115.

Uchman, A., 1998. Taxonomy and ethology of flysch trace fossils: a revision of the Marian Książkiewicz collection and studies of complementary material. Ann. Soc. Geol. Pol. 68, 105–218.

Vialov, O.S., 1968. Materialy k klassifikatsii iskopaemykh sledov i sledov zhiznedeyatelnosti organizmov. [Materials and classification of traces of fossils and vital activity of organisms]. Paleont. Sbornik 1, 125–129 (In Russian).

Vialov, O.C., Fedonkin, M.A., 1993. Trace Fossils of Extinct Organisms. Nauka, Moscow, 126 pp (In Russian).

Wilson, M.A., 2007. Macroborings and the evolution of marine bioerosion. In: Miller III, W. (Ed.), Trace Fossils. Concepts, Problems, Prospects. Elsevier, Amsterdam, pp. 356–367.

Wisshak, M., 2006. High-latitude bioerosion: the Kosterfjord experiment. Lect. Notes Earth Sci. 109, 1–202.

Wisshak, M., 2012. Microbioerosion. In: Knaust, D., Bromley, R.G. (Eds.), Trace Fossils as Indicators of Sedimentary Environments. Developments in Sedimentology, vol. 64. Elsevier, Amsterdam, pp. 213–243.

Zherikhin, V.V., 2003. Insect trace fossils: their diversity, classification, and scientific importance. Acta Zool. Cracoviensia 46 (suppl.—Fossil Insects), 59–66.

Zonneveld, J.-P., Pemberton, S.G., Saunders, T.D.A., Pickerill, R.K., 2002. Large, robust *Cruziana* from the Middle Triassic of northeastern British Columbia: ethologic, biostratigraphic, and paleobiologic significance. Palaios 17, 435–448.

Chapter 4

The Ichnofacies Paradigm

James A. MacEachern,*[,1] Kerrie L. Bann,[†] Murray K. Gingras,[‡] John-Paul Zonneveld,[‡] Shahin E. Dashtgard* and S. George Pemberton[‡]
*ARISE, Department of Earth Sciences, Simon Fraser University, Burnaby, British Columbia, Canada, [†]Ichnofacies Analysis Inc., Calgary, Alberta, Canada, [‡]IRG, Department of Earth and Atmospheric Sciences, University of Alberta, Edmonton, Alberta, Canada
[1]Corresponding author: e-mail: jmaceach@sfu.ca

1. INTRODUCTION

The ichnofacies paradigm has evolved across a nearly six-decade period since its original inception (Seilacher, 1953a,b). Today, the ichnofacies paradigm stands as a multidimensional framework that is underpinned by recurring, facies-controlled (i.e., environmentally related) groupings of biogenic structures that reflect animal responses (ethology) to paleoenvironmental conditions (see Buatois and Mángano, 2011; MacEachern et al., 2007a,b for summaries). As such, ichnofacies provide a logical first step in the interpretation of depositional settings and constitute a crucial component in the analysis of infaunal behavioral diversity in different environmental settings.

1.1 Rise of the Ichnofacies Concept

The concept of "ichnofacies" was developed by Dolf Seilacher in the 1950s and 1960s (Seilacher, 1953a,b, 1964), primarily focussed on marine successions. The concept was originally based on empirical observation of recurring associations of trace fossils coupled with the recognition that the parameters controlling the distribution of marine tracemakers tend to change with increasing distance from the paleocoast. In response to a highly influential paper in a theme volume dedicated to discerning paleobathymetry (Seilacher, 1967), trace fossils within the "Seilacherian" ichnofacies framework were soon (albeit erroneously) regarded as paleobathymeters (e.g., Farrow, 1966; Pickett et al., 1971; Weimer and Hoyt, 1964). While ichnofacies remain essential to paleoenvironmental reconstructions, paleobathymetry constitutes only one aspect of the modern concept and is largely confined to marine associations. Organism behaviors (ethology) and their resulting biogenic structures in subaqueous regimes are principally controlled by factors including substrate consistency, sediment grain size, energy

Developments in Sedimentology, Vol. 64. http://dx.doi.org/10.1016/B978-0-444-53813-0.00004-6

conditions at the bed, food resource types and availability, water turbidity, water salinity, depositional rates, oxygenation, and temperature. These factors tend to change in a proximal to distal trend, such that marine Seilacherian ichnofacies are properly recognized to display only a passive relationship to bathymetry (e.g., Ekdale, 1988; Ekdale et al., 1984; Frey et al., 1990; Fürsich, 1975). Much of the ethological focus in marine ichnofacies is related to substrate consistency and trophic style (e.g., the food-resource paradigm), although a wide range of behavioral and trophic groupings can be identified. Continental ichnofacies and their distributions, by contrast, are largely independent of such proximal–distal bathymetric trends (with the exception of some lacustrine associations) but instead reflect organism responses to different environmental factors (see Buatois and Mángano, 2011; Melchor et al., 2012 for summaries). For the continental realm, climate is an overriding control, closely linked to the availability of water, either as surface water or as pore water (Buatois and Mángano, 1995; Dubois et al., 2012; Genise et al., 2000; Hasiotis et al., 2007). In freshwater settings, energy, oxygenation, and substrate consistency remain critical factors as well. There is also a close association between vegetation patterns and terrestrial ichnofacies (Buatois and Mángano, 2011; Genise et al., 2010), which is largely dictated by climate (temperature and precipitation).

Many ichnofacies practitioners argue that the archetypal Seilacherian ichnofacies are proper names. As the namesake is derived from an ichnogenus, it must be capitalized and italicized. There is only one *Skolithos* Ichnofacies (that is, not "a" *Skolithos* Ichnofacies), and so it is also a proper name, and the word "ichnofacies" must be capitalized when following the namesake identifier. Suites and ichnofabrics are not formally defined and do not have broad temporal or spatial recurrence—hence, they are not proper names. While the ichnogenus or ichnogenera identifiers are capitalized and italicized, the words "suite" or "ichnofabric" are not capitalized.

Seilacher (1967) defined the six original ichnofacies, each named for a characteristic ichnotaxon. These correspond to four of the five softground marine types (*Skolithos*, *Cruziana*, *Zoophycos*, and *Nereites*) (Table 1), one of three substrate-controlled types (*Glossifungites*), and one of the six softground continental types (*Scoyenia*) (Table 2). Frey and Seilacher (1980) added the *Trypanites* Ichnofacies to characterize borings in hardgrounds, and Bromley et al. (1984) introduced the *Teredolites* Ichnofacies to encompass borings into woodgrounds. Frey and Pemberton (1987) proposed the *Psilonichnus* Ichnofacies for permanent vertical to inclined dwellings, continental biogenic structures, and surface locomotion traces in sand-prone substrates of high intertidal and supratidal coastal settings. Historically, the dataset for this ichnofacies has been somewhat sparse, largely owing to the setting's low preservation potential. This has been alleviated more recently through the addition of well-described case studies (e.g., Curran, 2007; Curran and White, 1991; Marshall, 2003; Nesbitt and Campbell, 2006; Netto and Grangeiro, 2009). Buatois and Mángano (2011) have indicated that the *Psilonichnus* Ichnofacies can be subdivided into landward and seaward expressions using the proportions of terrestrial *versus* marine elements (Curran, 2007).

The greatest expansion in ichnofacies archetypes has been seen in the continental realm (see Buatois and Mángano, 2011 for a thorough summary). It was always clear that the *Scoyenia* Ichnofacies could never hope to serve the range of variability in animal/sediment relationships that occur in the continental realm (Bromley and Asgaard, 1979), but rather, it reflected the early focus of ichnologists on marine and marginal-marine successions. Ichnofacies analysis of continental settings led first to three ichnofacies, with the revision of the *Scoyenia* Ichnofacies (Buatois and Mángano, 1995), followed by the addition of the *Mermia* Ichnofacies and the *Coprinisphaera* Ichnofacies (Buatois and Mángano, 2004; Genise et al., 2000). More recently, the range of continental ichnofacies has been expanded to six, with the inclusion of the *Termitichnus* Ichnofacies (cf. Genise et al., 2000, 2010; Smith et al., 1993), *Celliforma* Ichnofacies (Genise et al., 2010), and *Octopodichnus– Entradichnus* Ichnofacies (Ekdale et al., 2007; Hunt and Lucas, 2007; see Buatois and Mángano, 2011; Melchor et al., 2012; Tables 1 and 2). The latter three ichnofacies are based on very few case studies and therefore are regarded as preliminary at this time.

1.2 Variation in the Use of Ichnofacies

The use of ichnofacies has been uneven in the literature, in some ways paralleling the variations in the use of "facies". Lockley et al. (1994) and Lockley (2007) summarize the different scales of usage of the ichnofacies concept. McIlroy (2004b) also provides a succinct table outlining some of the different ichnofacies. The broadest usage of "ichnofacies" encompasses those erected by Seilacher (1953a,b, 1964, 1967) as well as the few others that have global and temporal recurrence, and have come to be known as "Seilacherian ichnofacies" (Bromley, 1990, 1996). These are discussed in greater detail in Section 3.

A number of distinctive, spatially recurring assemblages of trace fossils have been given status as ichnofacies but do not adhere to the Seilacherian ichnofacies paradigm. Lockley et al. (1987) regard these to be "medium-scale ichnofacies" which operate as subsets of the Seilacherian ichnofacies. Some workers consider these as "ichnosubfacies" (e.g., Seilacher, 1974; Uchman, 2001, 2009). Such subsets have been suggested to show widespread distributions in space and time, but insufficient case studies have been presented to date to support this contention.

These include the *Arenicolites* Ichnofacies, the *Curvolithus* Ichnofacies, and the *Entobia* Ichnofacies, among others. These non-Seilacherian ichnofacies do not represent theoretical constructs that adhere to the rigors of the ichnofacies concept. Additionally, they do not record distinctive ethological responses and therefore do not aid in facies analysis. Such occurrences are not mappable and

TABLE 1 General Characteristics and Environmental Aspects of the Marine and Marginal-Marine Seilacherian Ichnofacies

	Ichnofacies	Environmental constraints	Settings
Marine softground ichnofacies	*Psilonichnus*	Variably sorted, periodically exposed sandy substrates, deposition rates low. Food present as detritus. Softgrounds to stiffgrounds.	Supratidal to upper intertidal positions, subject to moderate- to low-energy marine and/or eolian settings such as beach backshore and dunes, washover fans and supratidal flats.
	Skolithos	Moderately to well-sorted, shifting sandy substrates with variable sedimentation and erosion rates. Food mainly held in suspension, sieved organics present in shallow water. Softgrounds to stiffgrounds.	Lower intertidal to shallow subtidal settings, in moderate to high-energy conditions such as foreshore, shoreface (above fair-weather wave base), tidal inlets and channels, sandy shoals and bars, and sandy deep-sea fans.
	Cruziana	Relatively low-energy, semicohesive muddy substrates with intercalated silt and sand. Sedimentation and erosion rates low. Food deposited on or buried in the sediment. Softground near surface, deep tiers encounter stiffgrounds.	Permanently subtidal settings lying below fair-weather wave base, but above storm wave base. Common to fully marine quiet-water conditions such as offshore, shelf and open epeiric embayments.
	Zoophycos	Fine-grained, cohesive muddy to sandy substrates, locally showing oxygen deficiency. Sedimentation rates very low. Food deposited on substrate. Shallow tiers in softgrounds, deep tiers in stiffgrounds and firmgrounds.	Low-energy subtidal settings near to or below storm wave base. Settings are generally free of turbidites. Common environments include silled basins, restricted lagoons, continental shelves and upper continental slopes.
	Nereites	Pelagic and hemipelagic cohesive muddy substrates reflecting very slow sedimentation. Commonly interstratified with sandy turbidites. Patchy distribution of food confined mainly at surface and shallow tiers. Predominantly softgrounds.	Basinal subaqueous settings ranging from bathyal to abyssal, typically low energy but oxygenated. Settings include lower slope, submarine canyon and basin-floor (abyssal plain) environments.
Substrate-controlled ichnofacies	*Glossifungites*	Firm, compacted, semi-lithified substrates characterized by very slow sedimentation. Food dominantly held in suspension.	Marine- or marginal-marine-generated omission surfaces on firm but unlithified substrates. Reflects recolonization of dewatered, compacted and/or incipiently cemented media. Commonly associated with erosional exhumation.
	Teredolites	Xylic [woody] substrates characterized by slow sedimentation. Food mainly in suspension, with some organisms adapted to consuming wood.	Marine omission surfaces on peat/coal deposits and driftwood pavements. Not applied to single wood clasts. Typical of erosion in peat-forming areas (e.g., estuarine margin, delta plain or back-barrier settings).
	Trypanites	Hard, lithified substrates characterized by erosion and non-deposition. Food predominantly held in suspension.	Marine-influenced omission surfaces on lithified media such as rocky coasts, beachrock, hardgrounds, reefs or bone beds and coquinas. Not applied to individual bones, shells or clasts.

TABLE 1 General Characteristics and Environmental Aspects of the Marine and Marginal-Marine Seilacherian Ichnofacies—Cont'd

Ethologies and trophic styles	Common traces	References
Vertical J-, Y-, and U-shaped dwelling, crawling and foraging traces, vertebrate tracks. Formed by mixture of terrestrial and marine predators, scavengers and herbivores.	*Psilonichnus, Psammichnites, Polykladichnus, Planolites, Macanopsis, Lockeia,* roots.	Frey and Pemberton (1987), Marshall (2003), and Nesbitt and Campbell (2006)
Vertical, cylindrical, branching and/or U-shaped dwellings, equilibrium and escape traces. Formed by suspension feeders, surface-detritus feeders and/or passive (tubicolous) carnivores.	*Ophiomorpha, Conichnus, Arenicolites, Skolithos, Bergaueria, Palaeophycus, Diplocraterion, Gyrolithes, Schaubcylindrichnus coronus, Siphonichnus.*	Seilacher (1964), Howard and Frey (1984), and Hiscott et al. (1984)
Predominantly horizontal and vertical dwelling burrows as well as mobile deposit-feeding traces. Dominant trophic behaviors are deposit-feeding, with minor suspension-feeding, detritus-feeding, grazing and foraging traces. Bedding planes commonly show abundant locomotion traces.	*Planolites, Teichichnus, Thalassinoides, Cylindrichnus, Asterosoma, Chondrites, Rhizocorallium, Phycosiphon, Zoophycos, Rosselia, Scolicia, Palaeophycus, Ophiomorpha, Diplocraterion, Siphonichnus, Phoebichnus, Cruziana.*	Seilacher (1964), Pemberton and Frey (1984), and Frey (1990)
Predominantly mobile foraging, sessile deposit-feeding, surface-grazing and locomotion traces. Organisms display epifaunal and infaunal feeding strategies. Chemosymbiosis locally important.	*Zoophycos, Phycosiphon, Planolites, Scolicia, Chondrites, Helminthopsis, Thalassinoides, Cosmorhaphe, Spirophyton, Schaubcylindrichnus freyi, Nereites irregularis, Asterosoma.*	Seilacher (1964), Olivero and Gaillard (1996), and Uchman and Demírcan (1999)
Complex grazing and patterned feeding/dwelling structures below the sediment/water interface (graphoglyptids), reflecting organized, efficient feeding strategies to accommodate slow accumulation of food. Some ichnogenera reflect composite traces and chemosymbiotic relationships with bacteria.	*Nereites, Paleodictyon, Chondrites, Cosmorhaphe, Helminthopsis, Zoophycos, Spirophyton, Scolicia, Planolites, Hormosiroidea, Phycosiphon, Spirorhaphe, Lorenzinia, Spirodesmos, Urohelminthoida, Spirophycus, Thalassinoides.*	Seilacher (1964), Leszczyński and Seilacher (1991), and Wetzel, (2008)
Vertical, cylindrical, U- or tear-shape pseudo-borings or sparsely to densely ramified dwelling burrows. Mainly constructed as dwellings for suspension feeders or as refugia for scavengers and predators.	*Diplocraterion, Skolithos, Thalassinoides, Arenicolites, Rhizocorallium jenense, Zoophycos, Psilonichnus, Palaeophycus, Bergaueria, Balanoglossites, Gastrochaenolites.*	Seilacher (1964), Pemberton and Frey (1985), MacEachern et al. (1992), MacEachern and Burton (2000), and Uchman et al. (2000)
Stumpy to elongate, clavate to subcylindrical, locally subparallel borings. Most structures are dwellings or refugia made by suspension feeders, passive carnivores or scavengers. Wood-boring bivalves also consume the wood itself.	*Teredolites, Thalassinoides, Diplocraterion, Rhizocorallium, Rogerella, Psilonichnus, Maeandropolydora, Caulostrepsis, Entobia.*	Bromley et al. (1984), Savrda et al. (1993), and Gingras et al. (2004)
Vase-, tear- and/or U-shaped to irregular borings serving as dwellings. Some shallow, anastomose borings. Excavated by suspension feeders and/or passive carnivores. Surface raspings and gnawings of fauna such as algal grazers.	*Gastrochaenolites, Trypanites, Rogerella, Entobia,* ploychaete burrows, echiniod grooves.	Bromley (1975), Frey and Seilacher (1980), and Pemberton et al. (1980)

The ichnofacies are separated into those that are contemporaneous with deposition and represent softground conditions, and those encompassing omission suites and post-depositional (substrate-controlled) conditions.

TABLE 2 General Characteristics and Environmental Aspects of the Continental Seilacherian Ichnofacies

	Ichnofacies	Environmental constraints	Settings
Continental ichnofacies	*Scoyenia*	Moist to wet, argillaceous to sandy substrates; sporadically to persistently exposed. Pore-moisture contents and temperatures variable. Food present as surface detritus. Mainly stiffgrounds in low-energy sites.	Low-energy continental settings with intermittent subaerial conditions. Settings are inundated with freshwater, particularly during river floods. Environments include upper point bars, channel banks, crevasse splays, lake margins and floodplains adjacent to channels.
	Mermia	Low-energy, permanently subaqueous substrates. Sedimentation rates low but locally variable. Dissolved oxygen and total dissolved solids vary locally. Food deposited on or buried in the sediment. Softgrounds and soupgrounds typical.	Freshwater lakes, shallow lake margins, glacial lakes, carbonate lakes and freshwater delta complexes.
	Coprinisphaera	Permanently subaerially exposed continental substrates characterized by minimal deposition or erosion. Variable pore moisture contents and temperatures. Food deposited as surface detritus or buried in the sediment. Stiffgrounds to firmgrounds.	Paleosols of herbaceous (grass) communities. Common to alluvial floodplains, desiccated floodplains, crevasse splays, levees, abandoned point bars and vegetated eolian environments.
	Termitichnus	Warm, humid conditions with high water tables. Abundant plant growth including flank-buttressed trees.	Paleosols associated with closed forest ecosystems.
	Celliforma	Warm to cold, arid conditions.	Carbonate-rich, well-aerated paleosols; palustrine settings and calcretes.
	Octopodichnus-Entradichnus	Arid conditions, looseground mobile substrates.	Eolian dune settings.

represent, instead, local variations in organism/sediment interactions that can be accommodated by the established Seilacherian ichnofacies.

The post-depositional "*Arenicolites* Ichnofacies" (cf. Bromley and Asgaard, 1991), for example, is well preserved in sandstone beds deposited under storm-related high-energy environments of the lower to middle shoreface. The bioturbation is restricted to vertical burrows and is characterized by a low-diversity assemblage of traces constructed by opportunistic organisms responding to an open niche. Bromley and Asgaard (1991) concluded that such occurrences in these storm deposits do not fall within the *Skolithos* Ichnofacies *sensu stricto*

TABLE 2 General Characteristics and Environmental Aspects of the Continental Seilacherian Ichnofacies—Cont'd

Ethologies and trophic styles	Common traces	References
Mobile deposit-feeding, locomotion tracks including tracks and trails, fish-fin traces, invertebrate and vertebrate dwelling structures, footprints and roots.	*Camborygma, Macanopsis, Naktodemasis, Planolites, Palaeophycus, Skolithos, Scoyenia, Beaconites, Fuersichnus, Umfolozia, Cochlichnus, Daimonelyx,* roots, footprints.	Seilacher (1964, 1967), Smith (1993), Buatois and Mángano (1995), and Buatois and Mángano (2004)
Grazing and feeding structures produced by mobile and sessile deposit- and detritus-feeding organisms. Some permanent to temporary dwelling structures of surface detritus feeders. Locomotion and resting structures of various freshwater organisms.	*Cochlichnus, Gordia, Mermia, Helminthoidichnites, Planolites, Treptichnus, Undichna, Palaeophycus, Diplichnites, Circulichnus, Tuberculichnus.*	Buatois and Mángano (1995, 2004), Melchor (2004), and Genise et al. (2010)
Insect nests (ants, bees, beetles, wasps, termites), permanent to temporary vertebrate and invertebrate dwellings, mobile deposit-feeding structures and roots.	*Attaichnus, Coprinisphaera, Celliforma, Daimonelyx, Eatonichnus, Naktodemasis, Pallichnus, Skolithos, Teisseirei, Uruguay,* roots and/or rhizoliths, footprints.	Genise et al. (2000, 2004, 2010), and Duringer et al. (2007)
Dominance of termite nests, vertebrate and invertebrate dwellings, low diversity, high abundance.	*Termitichnus, Vondrichnus, Fleaglellius, Krausichnus, Masrichnus, Naktodemasis,* abundant roots.	Smith et al. (1993), Genise et al. (2000, 2010), and Buatois and Mángano (2011)
Dominance of calichnia (particularly halictid bee nests), pupichnia, vertebrate and invertebrate dwellings, association with hackberry endocarps, moderate diversity, high abundance.	*Celliforma, Rosselichnus, Pallichnus, Rebuffoichnus, Palmiraichnus, Teisseirei, Naktodemasis,* root structures.	Genise et al. (2000, 2010), and Buatois and Mángano (2011)
Vertebrate and invertebrate (particularly arthropod) trackways, vertebrate and invertebrate dwellings, invertebrate backfilled traces, moderate diversity, low abundance.	*Octopodichnus, Entradichnus, Palaeophycus, Macanopsis, Skolithos, Planolites, Digitichnus, Arenicolites, Diplocraterion, Naktodemasis.*	Ekdale et al. (2007), Hunt and Lucas (2007), and Buatois and Mángano (2011)

Recent ichnotaxonomic work suggests that structures identified as *Naktodemasis* are better regarded as *Taenidium bowni* (e.g., Krapovickas et al., 2009).

and proposed its occurrence to be assigned to a new ichnofacies—the "*Arenicolites* Ichnofacies". However, according to Buatois and Mángano (2004), assemblages attributed to the *Arenicolites* Ichnofacies can be difficult to distinguish from those of the *Skolithos* Ichnofacies. In fact, the constituent trace fossils of the "*Arenicolites* Ichnofacies" can be important elements of the *Skolithos* as well as the *Cruziana* Ichnofacies. Ultimately, the "*Arenicolites* Ichnofacies" is difficult to map and does not adhere to Walther's Law; that is, tempestites do not represent a depositional environment *per se*. Rather, such suites represent *in loco* fluctuations of energy levels brought on by episodic tempestite deposition *within* an environment covered by the Seilacherian ichnofacies. Where such suites occur in tempestites emplaced in an upper offshore setting, they represent temporal fluctuations in depositional energy within the *Cruziana* Ichnofacies. Where comparable suites occur in tempestites of the shoreface, they are encompassed by the *Skolithos* Ichnofacies.

In a similar manner, Lockley et al. (1987) proposed the "*Curvolithus* Ichnofacies" as a subset of the *Cruziana* Ichnofacies. Attributable suites of the *Curvolithus* Ichnofacies are dominated by infaunal locomotion/feeding traces in environments characterized by the relatively rapid deposition of sand at rates exceeding physical reworking. Such conditions are common to deltaic systems. The authors stress that the recurring assemblage is, in fact, part of the *Cruziana* Ichnofacies and represents local variations in sedimentation. Such ichnosubfacies have not, as yet, been demonstrated to possess spatial or temporal recurrence and are not recognized in the Seilacherian ichnofacies paradigm.

The *Trypanites* Ichnofacies, as erected by Frey and Seilacher (1980), accommodates hard-substrate trace-fossil assemblages but otherwise lacks paleoenvironmental implications. Bromley and Asgaard (1993) erected the "*Entobia* Ichnofacies" to encompass long-duration boring assemblages in paleocliffs and the "*Gnathichnus* Ichnofacies" to characterize bored shells and boulders deposited in slightly deeper environments prone to rapid burial. They went on to propose that these two ichnofacies potentially could serve as archetypal hardground associations that would replace the *Trypanites* Ichnofacies, or act as subsets of it. The *Entobia* and *Gnathichnus* ichnofacies were established from a single occurrence in the Pliocene of Rhodes. MacEachern et al. (2007a) argued that these are better regarded as suites that correspond to tiers within the *Trypanites* Ichnofacies rather than true ichnofacies. Detailed work by de Gibert et al. (1998) and Cachão et al. (2009) established a broader range of occurrence for the *Entobia* Ichnofacies, but the association nevertheless demonstrates both ethological and morphological conformity with the existing *Trypanites* Ichnofacies and so can be regarded as equivalent (e.g., Buatois and Mángano, 2011; MacEachern et al., 2007a; see also de Gibert et al., 2012).

The case for incorporating the *Gnathichnus* association with the *Trypanites* Ichnofacies, however, is not so clear-cut. MacEachern et al. (2007a) argued that the *Gnathichnus* Ichnofacies is restricted to clasts that can be easily transported out of the depositional environment. As such, occurrences of the *Gnathichnus* Ichnofacies do not necessarily reflect the environment in which the borings occurred, and consequently, this "ichnofacies" does not meet the criteria for inclusion in the Seilacherian paradigm. More recently, however, intervals representing rapid burial of bioeroded surfaces and shells have been identified in the rock record, and it was found that they form mappable constructs (e.g., de Gibert et al., 2007; Mayoral and Muñiz, 1996), allowing discrimination of settings prone to long-term bioerosion with little or no sedimentation (*Trypanites* Ichnofacies) from settings prone to short-term colonization and moderate energy (*Gnathichnus* Ichnofacies). Buatois and Mángano (2011) point out that differentiating shallow-tier associations from mid- and deep-tier associations as discrete ichnofacies is not really very different from the distinction between the *Zoophycos* Ichnofacies (dominated by deep-tier associations) and the *Nereites* Ichnofacies (largely defined by shallow-tier suites). We believe, however, that this distinction can be captured within the details of the hardground suites themselves and do not require a discrete ichnofacies. In defense of the *Zoophycos* and *Nereites* ichnofacies distinction, it is the presence of graphoglyptids (organized farming and grazing/foraging structures) that is the diagnostic indicator of the *Nereites* Ichnofacies (e.g., Frey and Seilacher, 1980; Seilacher, 1967), not merely shallow-tier associations. The graphoglyptids tend to be cast on the soles of low-energy and/or low-density turbidites that were emplaced into the deep basin. In no instances known to the authors have thin, low-energy tempestites been found to cast shallow-tier associations containing graphoglyptids within the *Zoophycos* Ichnofacies. That is, settings residing above storm wave base on the shelf do not appear to possess graphoglyptid associations, indicating that the two archetypal ichnofacies are geographically discrete. Suites attributable to continental slope associations of the *Zoophycos* Ichnofacies do have the potential, however, for some overlap with the *Nereites* Ichnofacies (cf. Hubbard et al., 2012).

The foregoing demonstrates that medium-scale subdivisions of the Seilacherian ichnofacies ultimately may be warranted. However, we believe that, (1) it is necessary to demonstrate their recurrence in time and space in a manner similar to the original Seilacherian ichnofacies, and (2) ichnosubfacies must demonstrate their application to the characterization of subenvironments, *before* being formally designated and applied. Until so shown, such ichnosubfacies cannot be regarded as formal ichnofacies and are better regarded as discrete ichnofabrics or trace-fossil suites. Indeed, MacEachern et al. (2007c) and MacEachern and Bann (2008) have characterized trace-fossil suites in the context of "proximal", "archetypal", and "distal" expressions of some Seilacherian ichnofacies, in order to define some subenvironments.

Finally, "ichnofacies" have also been employed to characterize specific recurring assemblages in particular rock units (e.g., D'Alessandro et al., 1986; Hayward, 1976; Hunt et al., 1994; Lockley, 2007; Lockley et al., 1994). This approach has become particularly common for workers focused on the widespread and extensive tetrapod trackways of the continental realm (e.g., Hunt and Lucas, 2007; Hunt et al., 1994; Lockley et al., 1994). Several workers established that specific associations of track types and coprolites represent particular depositional environments and subenvironments (e.g., Hunt et al., 1994, 1995, 1998; Lockley et al., 1994). Between the establishment of the first tetrapod ichnofacies (Lockley et al., 1994) and 2007, approximately 15 tetrapod and vertebrate coprolite ichnofacies were named.

Although commonly characteristic of specific environments, most named vertebrate ichnofacies are limited in temporal distribution (Hunt and Lucas, 2007; Santi and Nicosia, 2008). The *Chelichnus* and *Brasilichnium* ichnofacies, characteristic of eolian successions, are limited to the Permian and Late Triassic/Early Jurassic, respectively (Hunt and Lucas, 2006, 2007; Lockley et al., 1994). The *Brontopodus*, *Carichnium*, and *Pteraichnus* ichnofacies are restricted to Early Cretaceous clastic successions, Early Cretaceous carbonate intervals, and Late Jurassic marine shoreline successions, respectively (Hunt and Lucas, 2007; Lockley and Meyer, 2000; Lockley et al., 1994, 2001). Other Seilacherian ichnofacies have been established to cover a wide variety of continental habitats (see Buatois and Mángano, 2011, and the discussion in Melchor et al., 2012). However, all of the tetrapod and vertebrate coprolite ichnofacies named prior to 2007 are limited in their temporal distribution, and in some cases, in geographic distribution as well. Indeed, the *Laoporus* Ichnofacies and *Brasilichnium* Ichnofacies are virtually indistinguishable from one another and differentiated mainly on stratigraphic grounds owing to the age limitations of the ichnofacies' namesakes (cf. Lockley et al., 1994; McIlroy, 2004b). Hunt and Lucas (2007) referred to this as the biotaxonomic tradition of ichnofacies. In this regard, the tetrapod ichnofacies framework has more in common with the Seilacherian *Cruziana* biostratigraphy than it does with the archetypal ichnofacies concept.

The *Cruziana* biostratigraphic paradigm, established by Seilacher (1970, 1991, 1994), has been shown to be useful, in certain parts of the world, for biostratigraphic zonation of otherwise non-fossiliferous successions. Similarly, the original tetrapod "ichnofacies" (Hunt et al., 1994, 1995, 1998; Lockley, 2007; Lockley et al., 1994), while not comprising true ichnofacies, most certainly exhibit biostratigraphic utility for continental successions that commonly contain few, if any, body fossils (Mángano et al., 2012).

Most of the early tetrapod ichnofacies were downgraded to ichnocoenoses by Hunt and Lucas (2007), with the definitions of the remainder broadened to include several former groupings. The remaining five ichnofacies include the *Batrachichnus* Ichnofacies (characteristic of intertidal flats and alluvial plains), the *Brontopodus* Ichnofacies (carbonate and clastic coastal plains and shorelines), the *Characichnos* Ichnofacies (shallow lacustrine), the *Chelichnus* Ichnofacies (eolian successions), and the *Grallator* Ichnofacies (marginal-lacustrine settings). Although the Hunt and Lucas (2007) revision represents a substantial improvement over the previous chaos of temporally and geographically restricted tetrapod ichnofacies, their amended ichnofacies still exhibit significant temporal restrictions, albeit much less severe than previous exemplars. Unfortunately, these "new" tetrapod ichnofacies continue to differ from true ichnofacies in terms of their taxonomic restriction. Ichnofacies, in the sense they were intended (cf. Seilacher, 1953a,b), are theoretical constructs encompassing ethologically distinct ichnocoenoses that are widely distributed, both geographically and through geological time (Bromley and Asgaard, 1991; Buatois and Mángano, 2011; Frey et al., 1990; Hunt and Lucas, 2007; MacEachern et al., 2007a,b; Pemberton et al., 1992). Likewise, these ichnofacies are not restricted to particular taxonomic groupings. The archetypal Seilacherian ichnofacies represent the activities of a wide variety of organisms, including (but not restricted to) arthropods, echinoderms, polychaetes, echiurans, mollusks, nematodes, and tetrapods. Erecting a parallel ichnofacies framework for the tracks and trackways of some (but not all) of the representatives of a single one of the ~35 phyla in Kingdom Animalia is neither necessary nor particularly helpful from a depositional- or environmental-interpretation standpoint. Consequently, we do not consider any of the tetrapod "ichnofacies" to be valid. Rather, these ichnocoenoses are taxonomically restricted components of globally occurring, taxonomically inclusive continental or marginal-marine ichnofacies such as the *Psilonichnus* Ichnofacies, *Scoyenia* Ichnofacies, and *Octopodichnus–Entradichnus* Ichnofacies.

The sedimentary facies concept traveled a similar path in the mid-1900s, with some advocating the limiting of facies to "areally restricted parts of a designated stratigraphic unit" (e.g., Moore, 1949). This departed markedly from the original intended use of the concept (Gressley, 1838) and ultimately was abandoned. Middleton (1973, 1978) and Reading (1996) provide excellent and instructive summaries of the rationales for the stratigraphically unrestricted usage of facies. Ultimately, the identification of temporally and spatially restricted "ichnofacies" undermines the paradigm's broad application to environmental interpretation. As Bromley (1996) points out, temporally and spatially restricted but nevertheless recurring groupings of trace fossils have a use, but should be designated by a different term. We would argue that the Seilacherian approach yields the only true ichnofacies, and that the term should be employed in the manner that it was originally intended. This preserves the usefulness of the paradigm for facies analysis and depositional environment reconstruction.

2. NEOICHNOLOGICAL UNDERPINNING OF SEILACHERIAN ICHNOFACIES

Neoichnological research—and a range of biological studies that are unwittingly neoichnological in character—ratifies contemporary interpretations of ichnofacies (e.g., Dashtgard, 2011a; Genise et al., 2000, 2010; Gingras et al., 2004; Hertweck, 1972; Lawfield and Pickerill, 2006; Scott et al., 2009; Virtasalo et al., 2011; Wetzel, 2008). That is to say, neoichnological observations support the assertion that ichnofacies reflect broad animal behavioral responses to environmental conditions, wherein animal ethologies collectively are associated with strategies of survival and procreation. Specifically, neoichnological observations confirm the relationship between food-resource distribution, sediment caliber and consistency, and feeding behavior, which forms the premise for the ichnofacies paradigm (e.g., Curran and Martin, 2001; Dafoe et al., 2011; Dashtgard et al., 2008; Genise et al., 2000, 2010; Gingras et al., 2007a,b, 2008a,b, 2011; Howard et al., 1974; Reineck, 1968). Departures from established archetypes are apparent where factors such as salinity and oxygenation play an increasingly important role (e.g., Bann et al., 2004, 2008; Dashtgard, 2011b; Gingras et al., 2001; Howard and Frey, 1975; MacEachern and Gingras, 2007; MacEachern et al., 2005, 2007b; Rhoads and Morse, 1971). Buatois and Mángano (2011) provide an excellent up-to-date synopsis of each ichnofacies, including the ethological and trophic implications of the dominant tracemakers.

To illustrate the foregoing, neoichnological considerations of two of the most common ichnofacies in shallow-marine settings (*Skolithos* and *Cruziana*) are compared with one another. Ichnocoenoses attributable to the *Skolithos* and *Cruziana* ichnofacies are readily observed in modern settings and bear out the idea that the same ichnofacies, as expressed by the trace-fossil suites in the rock record, encompass collective and recurring behavioral and tropic responses (see Tables 1 and 2).

Trace fossils that are commonly associated with the *Skolithos* Ichnofacies include *Skolithos*, *Ophiomorpha*, and *Diplocraterion*, and to a lesser extent, *Rosselia* and *Conichnus*. Modern analogs are known for all of these trace fossils, and the ethologic purposes of the traces have been well documented (e.g., Dashtgard and Gingras, 2012; Dashtgard et al., 2008; Gingras et al., 1999, 2008a,b; Häntzschel, 1939; Reineck, 1965; Seilacher, 1964). For the constituents of the *Skolithos* Ichnofacies, at least one of two important survival strategies is consistently recognized (see Table 3): use of the biogenic structure for (1) filter feeding; and/or (2) for coping with shifting substrates and/or episodic deposition (see Hiscott et al., 1984; Howard and Frey, 1984). The use of some burrows for multiple purposes ensures that those that are suitable for mobile substrates and energetic settings (typically characterized by the *Skolithos* Ichnofacies) can also be employed in some lower-energy settings that are characterized by the *Cruziana* Ichnofacies. For example, the domichnion *Ophiomorpha* is produced in low-cohesive substrates and is used for filter-feeding, farming (fermentation), and subsurface deposit-feeding; therefore, *Ophiomorpha* is common to both high- and low-energy environments (e.g., Bromley, 1996; Heinberg and Birkelund, 1984).

By way of comparison, the *Cruziana* Ichnofacies is typified by trace fossils such as *Asterosoma*, *Scolicia*, *Planolites*, *Palaeophycus*, *Thalassinoides*, and *Teichichnus*. All of these trace fossils have modern analogs that strongly support the interpretation of these biogenic structures as deposit-feeding traces in resource-rich sediments (Dworschak and Ott, 1993; Gingras et al., 1999; Howard et al., 1974; Wetzel, 2008). Importantly, each trace can also serve other purposes, including acting as a domicile (*Teichichnus*, *Thalassinoides*, *Palaeophycus*, and *Asterosoma*), for motility (*Scolicia* and *Planolites*), or to reequilibrate to the sediment/water interface (equilibrichnia, *Lingulichnus*, some *Siphonichnus* and *Teichichnus*). Some neoichnological studies suggest that *Scolicia*, in various occurrences, may be produced by chemosymbionts (e.g., Bromley et al., 1995). Other trace fossils, such as *Arenicolites*, *Skolithos*, and *Rosselia*, commonly occur within suites of the *Cruziana* Ichnofacies, but are not characteristic of deposit-feeding behaviors. Nevertheless, vertical traces are used for a variety of feeding behaviors. For example, *Skolithos* are used by some infauna for head-down deposit-feeding (e.g., *Maldanus maldanus*; Gingras et al., 1999). Similarly, J-shaped *Arenicolites* is an optimal configuration for mining sediment (e.g., *Abarenicola pacifica*), whereas conventional U-shapes are well suited for interface deposit-feeding (e.g.,

TABLE 3 Ethological Characteristics of some common Biogenic Structures. *After Gingras et al. (2007a)*

	Ichnogenera	Domicile	Sediment stabilization	Filter-feeding	Carnivory or scavenging	Interface deposit-feeding	Stoping or mining	Shallow-tier deposit-feeding	Deep-tier deposit-feeding	Spatial-optimized deposit-feeding	Intraburrow waste stowage	Readjustment	Agrichnia
Suspension-feeding	*Cylindrichnus*	commonly	rarely	very strongly		very strongly					rarely		
	Diplocraterion	commonly	rarely	very strongly		rarely			rarely			very strongly	
	Entobia			very strongly									
	fugichnia											very strongly	
	Gastrochaenolites	very strongly		very strongly		rarely							
	Gyrolithes	rarely		commonly		rarely			commonly				
	Lingulichnus	very strongly		very strongly								commonly	
	Lockeia	commonly		very strongly		commonly							
Suspension-feeding and predation	*Ophiomorpha*	very strongly	very strongly	rarely			rarely		very strongly		rarely		
	Rosselia		commonly	very strongly		commonly					commonly	rarely	
	Arenicolites	commonly		very strongly	very strongly	commonly	commonly	commonly					
	Bergaueria	very strongly		very strongly	very strongly								
	Conichnus	very strongly		commonly	very strongly								
	Schaubcylindrichnus	very strongly		commonly	very strongly	rarely							
	Schaubcylindrichnus freyi	commonly		commonly	very strongly								
	Skolithos	commonly		very strongly	commonly	rarely	rarely						
	Palaeophycus	commonly			very strongly	commonly							
	Psilonichnus	very strongly			very strongly								
Deposit-feeding increasingly systematic	cryptic bioturbation							commonly					
	Planolites							very strongly					
	Thalassinoides	commonly		commonly					very strongly				
	Beaconites							very strongly	rarely				
	Scolicia							very strongly	rarely				
	Teichichnus	commonly						commonly	commonly			very strongly	
	Macaronichnus							very strongly	commonly	rarely			
	Cosmorhaphe							very strongly		very strongly			
	Phycodes							very strongly		commonly			
	Helminthopsis									commonly			
	Asterosoma	commonly	rarely						very strongly	commonly	commonly	rarely	
	Chondrites	rarely							very strongly	very strongly	rarely		
	Rhizocorallium	commonly		commonly		commonly			commonly	commonly			
	Zoophycos	commonly				rarely			very strongly	very strongly	commonly		rarely
	Paleodictyon												very strongly
	Phycosiphon								very strongly	commonly	commonly		

very strongly associated behavior — commonly associated behavior — rarely associated behavior

Corophium volutator). Moreover, both *Rosselia* and *Skolithos* can be used for surface deposit-feeding at the sediment/water interface (e.g., *Cirriformia luxuriosa*; Gingras et al., 1999; cf. Nara, 1995, 2002). As with the *Skolithos* Ichnofacies, it is evident that a range of behaviors and trophic styles can be employed in order to survive in *Cruziana* Ichnofacies-associated settings, both to exploit the available food resources and to contend with other aspects of the local environment (e.g., mixed grain-size substrates, high population densities, or lower levels of dissolved oxygen).

One of the most common misinterpretations of ichnofacies is that any one ichnofacies is characterized by a key behavior that fundamentally defines the ethology expressed by *each* trace fossil within it (e.g., that all elements of the *Skolithos* Ichnofacies equate to domiciles for the purpose of filter-feeding). As alluded to above, the adaptive landscape in any depositional setting is complex, and it is naive to ascribe a single behavior or trophic style as a panacea for survival to a specific set of environmental conditions. As is evident from neoichnological distributions, modern trace assemblages represent a community response to a broad range of environmental parameters, rather than simply the nature of the food resource. In the marine realm, other factors aside from food resources affect burrow distributions, including the sedimentation rate, shifting of sediment, grain sizes of sediment, and consistency of the substrate, oxygenation, and biomass (see "Environmental Constraints" in Tables 1 and 2; Dashtgard, 2011a,b; Dashtgard and Gingras, 2012; Dashtgard et al., 2008; Gingras et al., 2008b). In terrestrial settings, food distribution is commonly secondary to pore-water concentrations, climate, and precipitation (e.g., Buatois and Mángano, 2011; Genise et al., 2000; Hasiotis, 2002; Hasiotis et al., 2007; Hembree and Hasiotis, 2007).

Although in a trace-fossil assemblage it is the range of ethologies as well as the dominant ethological response(s) that are important to identify, there are clear associations between the dominant ethology expressed in a trace-fossil suite and the environmental conditions to which the tracemakers respond. A key control on animal behavior is the availability and delivery of food, wherein suspended food tends to favor filter-feeding behaviors, and suspension settling of food favors surface- and subsurface-deposit-feeding behaviors. As such, the fundamental basis of the marine ichnofacies paradigm resides in the food-delivery mechanism and food distribution (i.e., the food-resource paradigm) and its association with substrate consistency. However, as ichnofacies evolve to address less-static marine settings, other environmental factors will be seen to play an increasing role in character of the allied ichnological suites.

A fundamental basis for the interpretation of ichnofacies is the worker's ability to recognize the range of behaviors that a trace fossil may represent (see Table 3). For many trace fossils, this is broadly established in neoichnological studies, and for others, careful analysis of their morphology has led to refined interpretations of the behavior they represent. To this end, most of the ichnofacies defining ethological groups (i.e., dwelling traces, deposit-feeding traces, grazing traces, and farming systems) can be reliably identified (cf. Rindsberg, 2012).

Of these ethological categories, dwelling traces are the most problematic in that most domiciles are used as a platform for feeding as well. For example, in the marine realm, domiciles can be used for filter-feeding (*Lingulichnus*, *Siphonichnus*, *Skolithos*), interface deposit-feeding (*Siphonichnus*, *Skolithos*, *Arenicolites*), carnivory and predation (*Skolithos*, *Palaeophycus*), waste stowage (*Asterosoma*, *Zoophycos*; Bromley, 1991; Kotake, 1991, 1992), or scavenging (*Psilonichnus*). In these instances, ichnological analyses depend, in part, on sedimentological analyses and in establishing the knowledge of how food was likely distributed in the environment. Deposit-feeding traces generally are intrastratal in nature, focused within certain sedimentary levels, and show evidence of sediment ingestion or reworking (e.g., *Macaronichnus*, *Phycosiphon*, *Planolites*, *Scolicia*). As a result, many deposit-feeding trace fossils are easily identified and interpreted. Grazing traces are associated with the systematic processing of shallow-sediment tiers (e.g., *Nereites*, *Helminthopsis*), an interpretation supported by modern observations (e.g., Wetzel, 1991, 2002) and computer-generated behavioral evolution models (Papentin, 1973). Finally, farming traces dominantly comprise complex, permanent architectures that show no evidence of sediment processing (e.g., *Spirorhaphe*; Seilacher, 1977).

3. THE SEILACHERIAN ICHNOFACIES

The Seilacherian ichnofacies (Bromley, 1990, 1996) comprise temporally and spatially recurring, ethologically distinct groupings of trace-fossil suites (e.g., Buatois and Mángano, 2011; Frey et al., 1990; Hunt and Lucas, 2007; MacEachern et al., 2007a,b,c, 2010; Pemberton et al., 1992). We regard there to be 14 formally defined ichnofacies that adhere to Seilacher's (1953a,b) original approach (see Tables 1 and 2). Five ichnofacies encompass the marine to marginal-marine softground realm: the *Psilonichnus* Ichnofacies for high intertidal and supratidal, predominantly sandy settings (Fig. 1); the *Skolithos* Ichnofacies for intertidal to shallow subtidal, high-energy, sandy marine environments (Fig. 2); the *Cruziana* Ichnofacies for shallow subtidal, moderate energy silty to muddy marine settings (Fig. 3); the *Zoophycos* Ichnofacies for subtidal, low-energy, silty, and muddy marine conditions (Fig. 4); and the *Nereites* Ichnofacies for subtidal, very low-energy silty to muddy marine environments characterized by extremely slow deposition (Fig. 5). Three ichnofacies encompass post-depositional colonization of substrates, generally regarded as substrate-controlled (omission) ichnofacies: the *Glossifungites* Ichnofacies, reflecting colonization of firmground (dewatered and compacted but unlithified) substrates in marginal-marine and marine settings (Fig. 6A); the *Trypanites* Ichnofacies, associated with lithified (hard) substrates in marine settings (Fig. 6B); and the *Teredolites* Ichnofacies for colonization of xylic (woodground) substrates in marine and marginal-marine environments (Fig. 6C and D). Finally, six ichnofacies encompass the continental realm: the *Scoyenia* Ichnofacies associated with intermittently subaqueous, softground substrates (Fig. 7A and B); the *Mermia* Ichnofacies for permanently subaqueous freshwater settings (Fig. 7C); the

FIGURE 1 Biogenic structures common to the *Psilonichnus* Ichnofacies. (A) and (B) Bedding-plane view of modern crab-generated dwelling burrows (arrows and cb), variably representing *Psilonichnus* and *Polykladichnus*. (C) Cross section through a modern crab burrow, attributable to *Psilonichnus* (*Ps*). (D) Sandstone with *Psilonichnus* (*Ps*). Lens cap is approximately 4.5 cm in diameter.

FIGURE 2 Trace-fossil suites attributable to the *Skolithos* Ichnofacies. (A) Well-sorted Late Cretaceous sandstone (BI = 4) containing *Ophiomorpha nodosa*. (B) Sandstone (BI = 5) with a monospecific suite of *Diplocraterion habichi*. Scale is 3 cm long. (C) Sandstone with laminated-to-burrowed bedding (BI = 1–4), displaying *Ophiomorpha* (*O*) and *Skolithos* (*S*). Scale is 3 cm long. (D) Sandstone with well-developed trough cross-stratification (BI = 1) displaying an isolated *Conichnus* to the left. (E) Sandstone (BI = 3) with robust *Diplocraterion parallelum*.

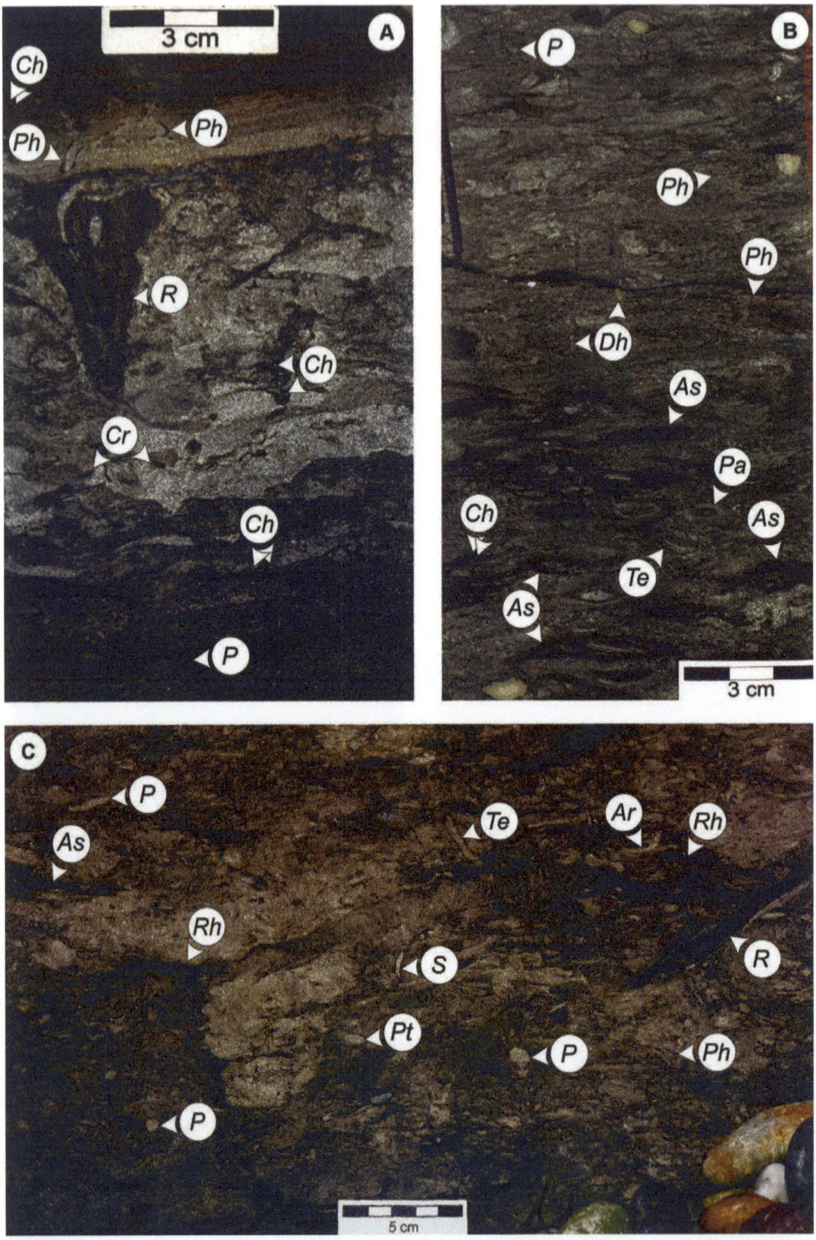

FIGURE 3 Trace-fossil suites attributable to the *Cruziana* Ichnofacies. (A) Sandy mudstones (BI=4) with a suite consisting of *Chondrites* (*Ch*), *Phycosiphon* (*Ph*), *Rosselia* (*R*), *Cosmorhaphe* (*Cr*), and *Planolites* (*P*). (B) Bioturbated silty to sandy mudstones (BI= 5) showing a suite consisting of *Planolites* (*P*), *Phycosiphon* (*Ph*), *Diplocraterion habichi* (*Dh*), *Chondrites* (*Ch*), *Asterosoma* (*As*), *Palaeophycus* (*Pa*), and *Teichichnus* (*Te*). (C) Bioturbated sandy mudstone (BI=5) with a suite consisting of *Arenicolites* (*Ar*), *Planolites* (*P*), *Asterosoma* (*As*), *Rhizocorallium* (*Rh*), *Palaeophycus tubularis* (*Pt*), *Phycosiphon* (*Ph*), *Rosselia* (*R*), *Skolithos* (*Sk*), and *Teichichnus* (*Te*).

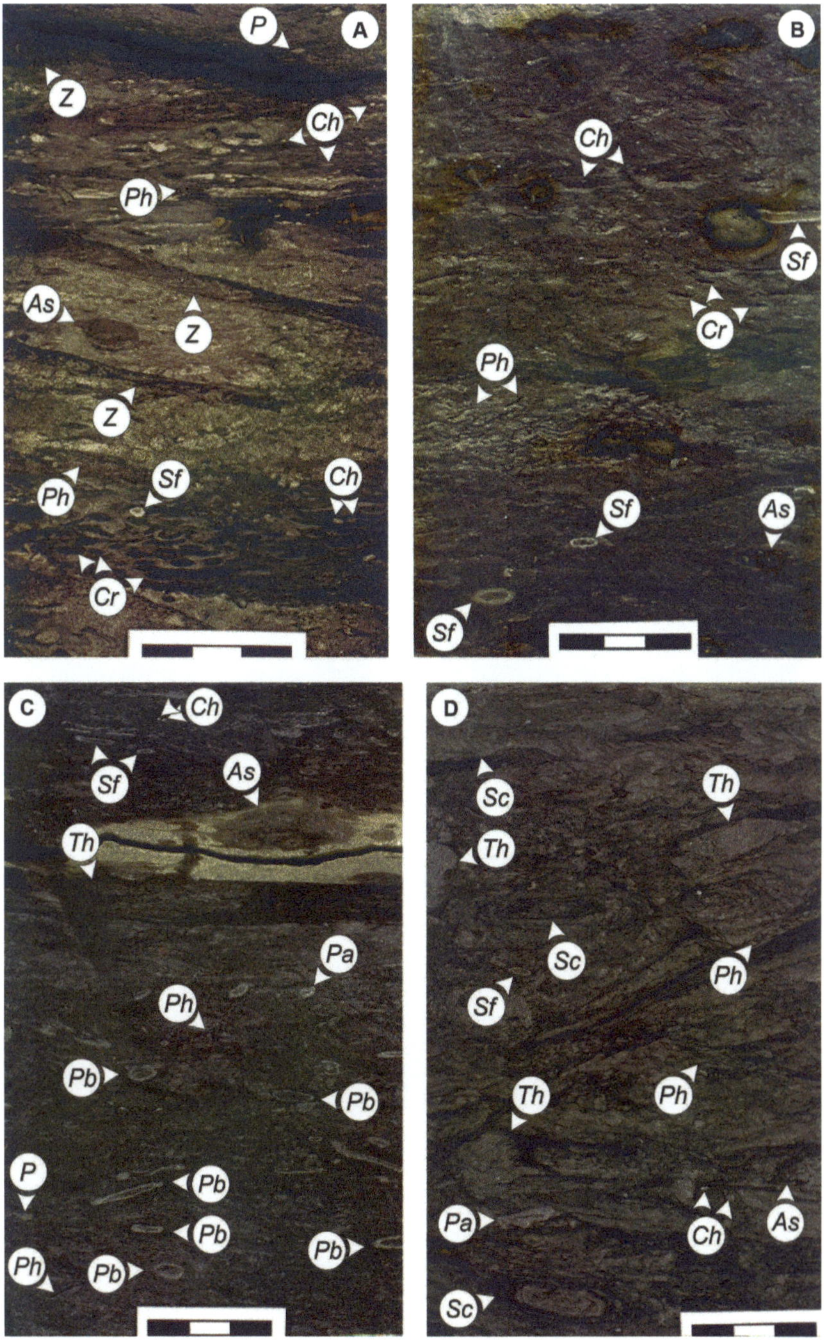

FIGURE 4 Trace-fossil suites attributable to the *Zoophycos* Ichnofacies. (A) Sandy mudstone (BI = 5) with a suite consisting of *Planolites* (*P*), *Zoophycos* (*Z*), *Chondrites* (*Ch*), *Phycosiphon* (*Ph*), *Asterosoma* (*As*), *Cosmorhaphe* (*Cr*), and *Schaubcylindrichnus freyi* (*Sf*). (B) Bioturbated silty mudstone (BI = 5) with a suite consisting of *Chondrites* (*Ch*), *Cosmorhaphe* (*Cr*), *Phycosiphon* (*Ph*), *Asterosoma* (*As*), and *Schaubcylindrichnus freyi* (*Sf*). (C) Bioturbated silty mudstone (BI = 5) with a suite consisting of *Chondrites* (*Ch*), *Schaubcylindrichnus freyi* (*Sf*), *Phycosiphon* (*Ph*), *Asterosoma* (*As*), *Phoebichnus* (*Pb*), *Planolites* (*P*), *Thalassinoides* (*Th*), and *Palaeophycus heberti* (*Pa*).

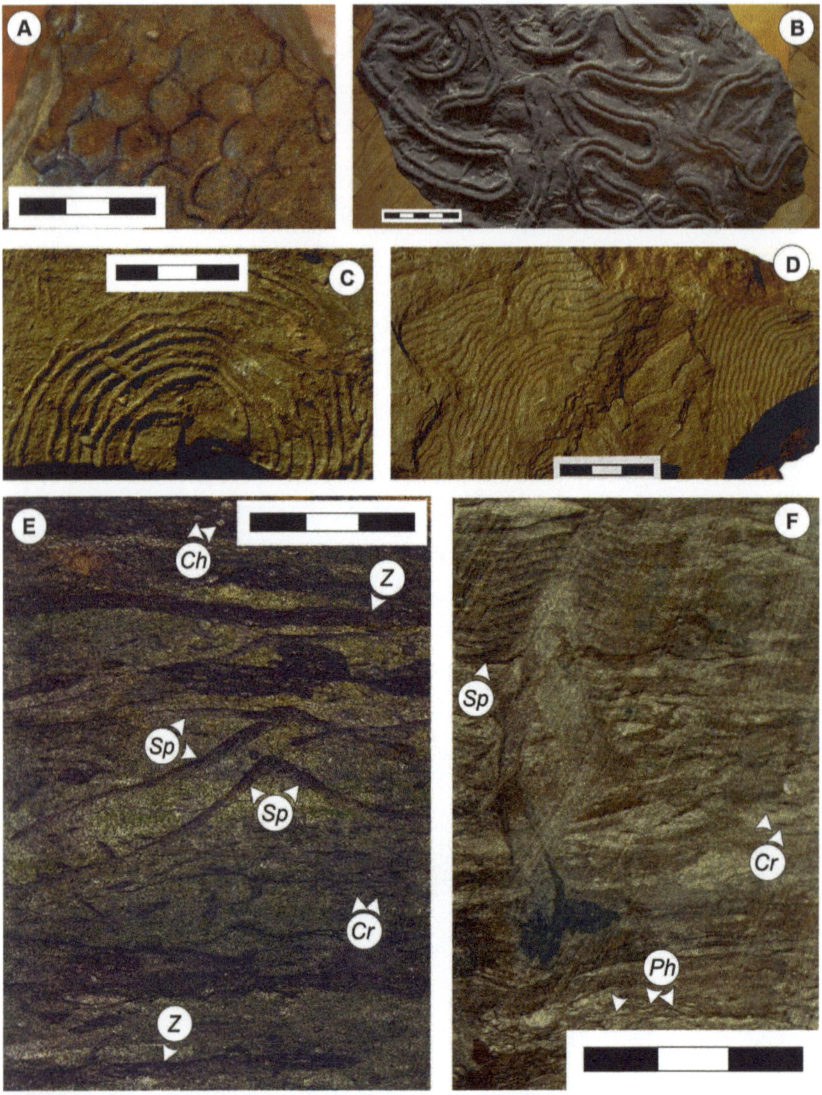

FIGURE 5 Trace fossils and trace-fossil suites attributable to the *Nereites* Ichnofacies. (A) Bedding-plane view with *Paleodictyon*. (B) Bedding plane with *Taphrhelminthopsis (=Scolicia strozzii)*. (C) Bedding plane with *Spirorhaphe*. (D) Bedding plane with *Nereites*. (E) Muddy sandstone (BI = 5) with a suite consisting of *Chondrites* (*Ch*), *Zoophycos* (*Z*), *Spirophyton* (*Sp*), and *Cosmorhaphe* (*Cr*). (F) Sandy mudstone (BI = 5) with a suite consisting of *Spirophyton* (*Sp*), *Cosmorhaphe* (*Cr*), and *Phycosiphon* (*Ph*). Scale in B is 5 cm long; all other scales are 3 cm long.

Figure 4—cont'd (D) Bioturbated muddy sandstone (BI = 5) with a suite consisting of *Scolicia* (*Sc*), *Thalassinoides* (*Th*), *Schaubcylindrichnus freyi* (*Sf*), *Phycosiphon* (*Ph*), *Palaeophycus heberti* (*Pa*), *Chondrites* (*Ch*), and *Asterosoma* (*As*). Scales are 3 cm long. NB: Recent ichnotaxonomic work suggests that structures identified as *Cosmorhaphe* in this chapter are better regarded as *Nereites missouriensis* (e.g., Angulo and Buatois, 2012).

FIGURE 6 Trace fossils and trace-fossil suites attributable to the substrate-controlled ichnofacies. (A) Firmground *Diplocraterion* (*D*) of the *Glossifungites* Ichnofacies. (B) The present-day boring *Gastrochaenolites* (*Ga*) of the *Trypanites* Ichnofacies. (C) and (D) Woodground borings *Teredolites* (*Te*), *Rhizocorallium* (*Rh*), and *Thalassinoides* (*Th*) of the *Teredolites* Ichnofacies. Scale in D is 3 cm long.

Coprinisphaera Ichnofacies for permanently subaerially exposed semiarid paleosols in terrestrial settings associated with savannahs, grasslands, prairies, and steppes (Fig. 7D); the *Termitichnus* Ichnofacies for warm, humid paleosols associated with closed forests; the *Celliforma* Ichnofacies for arid, carbonate-rich paleosols in scrubs, woodlands, and palustrine environments; and the *Octopodichnus–Entradichnus* Ichnofacies for eolian settings. For more detailed characterization of the individual Seilacherian ichnofacies, the reader is referred to MacEachern et al. (2007a,b, 2010), MacEachern and Bann (2008), and Buatois and Mángano (2011).

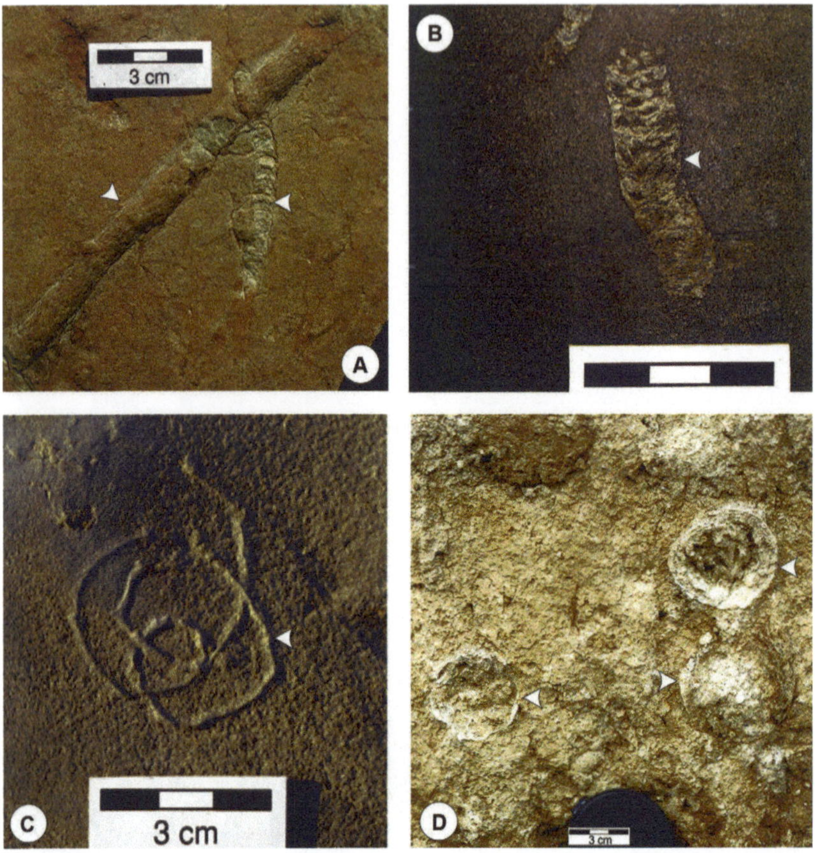

FIGURE 7 Trace fossils and trace-fossil suites attributable to the continental ichnofacies. (A) Sandstone bedding plane containing *Scoyenia* (*Sc*) of the *Scoyenia* Ichnofacies. (B) Vertical section through *Naktodemasis* of the *Scoyenia* Ichnofacies. Scale is 3 cm long. NB: Recent ichnotaxonomic work suggests that structures identified as *Naktodemasis* in this chapter are better regarded as *Taenidium bowni* (e.g., Krapovickas et al., 2009). (C) Bedding plane with *Mermia* of the *Mermia* Ichnofacies. (D) Sandstone with *Coprinisphaera ecuadoriensis* of the *Coprinisphaera* Ichnofacies (photo courtesy of Luis Buatois).

In the Seilacherian framework, ichnofacies are *not* recurring trace-fossil suites, nor are they ichnocoenoses (Bromley et al., 2007; MacEachern et al., 2007a, 2010; Pemberton et al., 1992). Rather, they reside at a hierarchal level above trace-fossil suites and constitute the *theoretical models* that encompass distinct ethological and trophic groupings of trace fossils for the purpose of assisting in the paleoenvironmental interpretation of the suites. Individual assemblages, ichnocoenoses, or suites may be *attributable* to an ichnofacies, but the ichnofacies cannot be said to be equivalent to them (Bromley et al., 2007; MacEachern et al., 2007a). For the purpose of facies analysis, trace fossils are regarded more as biogenic sedimentary structures (akin to physical sedimentary structures in lithofacies) than as ichnotaxonomic entities. Although various ichnogenera may come and go throughout the geological record and trace-fossil suites as well as "ichnofabrics" may differ in time and space, the "essence" of the behavioral and trophic groupings that define the ichnofacies remains consistent. In this way, the temporal limitations of individual ichnogenera and ichnospecies have less importance, and ichnological suites and ichnofabrics consisting of different ichnotaxa can be compared and contrasted ethologically with one another. This truth lies at the heart of the ichnofacies paradigm and explains the enduring strength of the concept.

Ichnological suites constitute part of the total aspect of the rock interval and consist of the primary biogenic structures created by organisms that inhabited the depositional environment. Insights into the depositional environment are derived from the fact that organisms respond in predictable ways to variations in environmental conditions (see Section 2). Although in the marine realm many of these conditions change progressively with increasing water depth, ichnogenera display only a passive relationship to bathymetry (Frey et al., 1990). This passive relationship *does* tend to lead to predictable proximal–distal trends in many successions, but there are numerous exceptions, particularly those associated with dynamic depositional regimes (e.g., submarine fans, fan deltas, sandy tide-dominated shelves, etc.). The resulting ichnological suites and ichnofabrics tend to vary spatially, so much so that they are not readily mappable at the scale of depositional facies (for an exception, see Bockelie, 1991). Ichnofacies serve as the conceptual basis for grouping these spatially variable suites and ichnofabrics, allowing them to be employed to interpret the depositional environment. In this way, ichnofacies, like lithofacies, are subject to Walther's Law. The utility of ichnofacies to paleoenvironmental reconstruction lies in their lateral continuity and predictable vertical succession, which leads to mappable units. Accurate interpretations of depositional environments favor reliable predictions of laterally adjacent settings and their associated ichnofacies. Like all facies analyses, interpretations of ichnofaunas are improved substantially when they are evaluated in the context of their sedimentological (i.e., lithofacies) and stratigraphic context (e.g., Bann et al., 2004, 2008; Buatois et al., 2012; Crimes et al., 1981; Dam, 1990; Ekdale et al., 1984; Frey, 1990; Genise et al., 2004; Gingras et al., 2001; Mángano and Buatois, 2004; MacEachern et al., 2007a,c, 2010; McIlroy, 2004a, 2008; Melchor, 2004; Pemberton and Frey, 1984; Pickerill, 1992; Savrda et al., 2001a,b; Smith, 1993; Smith et al., 1993; Taylor and Goldring, 1993).

Ichnological suites that show departures from the archetypal Seilacherian ichnofacies are also common, but their recognition and interpretation are possible only by comparison with these established norms (e.g., Bann et al., 2004; Buatois and Mángano, 2011; Gingras et al., 2012; MacEachern et al., 2007b, 2010; cf. Section 4). By their very nature, anomalous ichnological suites yield important insights into the specific characteristics of the depositional setting, highlighting animal/sediment interactions in response to imposed environmental stresses. In this way, brackish-water environments, anoxic to dysaerobic settings, and areas of fluvio-deltaic deposition can be readily recognized (e.g., Bann et al., 2004, 2008; Buatois et al., 2012;

Dashtgard, 2011a; Gingras et al., 1999, 2001; MacEachern and Bann, 2008; MacEachern and Gingras, 2007; MacEachern et al., 2005, 2007b; McIlroy, 2004a, 2007; Savrda, 2007; Savrda et al., 2001a,b).

4. THE ROLE OF SEILACHERIAN ICHNOFACIES

The Seilacherian ichnofacies operate as *facies models*, addressing animal/sediment responses in the depositional environment (MacEachern et al., 2010). They have been built through the distillation of recurring characteristics of trace-fossil suites from numerous case studies (Fig. 8). The distillation of observations from these case studies allows the ichnofacies to retain the recurring characteristics of the attributable suites (e.g., diversities, ethological groupings, responses to substrate consistency), independent of their case-specific characteristics and minimizes the impact of changes in suites owing to evolution. Ichnogenera may differ, but their ethological and trophic affinities are equivalent, and hence, their significance to the interpretation of the depositional environment is comparable. The unique characteristics of the suites are retained in order to evaluate the details of the specific case study. Ichnofacies have also been validated and refined *via* neoichnological studies (see Section 2). This "process ichnology" (e.g., Dashtgard, 2011a,b; Gingras et al., 2007a,b, 2011) underpins the ichnofacies concept just as process sedimentology underpins lithofacies-based facies models.

Ichnofacies models serve the same five functions (Fig. 8) as do lithofacies models (e.g., Walker, 1992; see also Dalrymple, 2010; MacEachern et al., 2010). First, an archetypal ichnofacies acts as a norm for the purposes of comparison. Archetypal ichnofacies, by acting as norms for comparison, allow one to determine whether a new example meets the criteria for inclusion within a specific ichnofacies. Without a norm, it cannot be determined whether a new example contains unusual or anomalous characteristics. If it conforms to an archetypal ichnofacies, it strengthens the concept, and interpretation of the new example is simplified. If it differs from the established norm, then the worker can specify precisely *how* it differs and refine its interpretation.

The second role of the ichnofacies concept is to act as a *framework or guide for future observations*. When one is working in particular bioturbated units, mental search criteria are erected. Certain ichnofacies have commonly associated trace-fossil elements. Knowing these affinities helps to guide the worker and focuses the worker's eye to search out outcrop or core orientations that display the entire suite.

The third role of the ichnofacies concept is that it *serves as a predictor in new situations*. The Seilacherian ichnofacies paradigm demonstrates that the various archetypal ichnofacies correspond to predictable combinations of environmental conditions. Seilacherian ichnofacies show a close relationship with

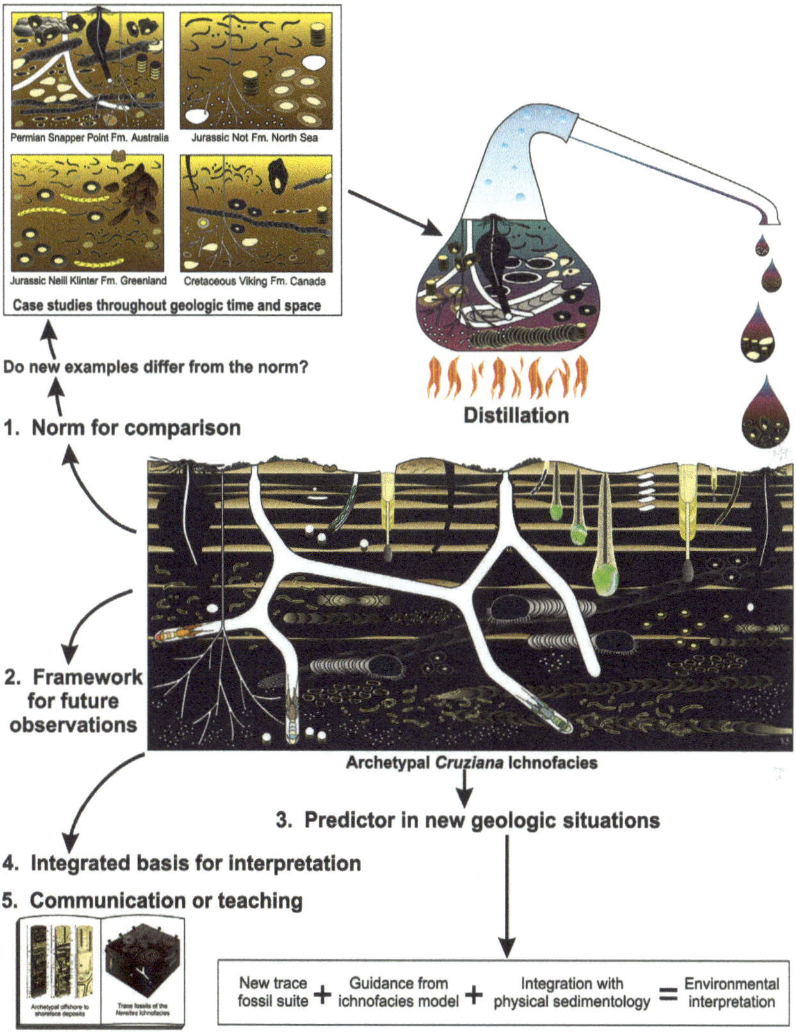

FIGURE 8 Stylized model showing the development of a Seilacherian ichnofacies (e.g., *Cruziana* Ichnofacies) and the five main uses to which it is put. *Diagram modeled after Walker (1992) and MacEachern et al. (2010).*

lithofacies, facies models, and obey Walther's Law, making their mappable distributions and occurrences in vertical successions predictable. Perturbations in their predictable occurrence commonly indicate the presence of an omission surface.

A fourth use of ichnofacies, as with physically based facies models, is that it may act as an *integrated basis for interpretation* of the system that it represents

(Walker, 1992). The linkage of neoichnology to the ichnofacies concept forms this integrated basis for interpretation, through "process ichnology" (Dashtgard and Gingras, 2012; Gingras et al., 2007a, 2011). The environmental parameters that constrain the faunal behaviors and the types and diversities of the resulting ichnogenera allow evaluations of depositional energy, substrate consistency, deposition rates, oxygenation, salinities, etc. As a result, no matter what specific paleoenvironment is ultimately determined for a sedimentary unit, it *must conform to the physico-chemical parameters* that are diagnostic of the ichnofacies to which the trace-fossil suites belong (Tables 1 and 2).

Finally, ichnofacies serve a fifth role: they *facilitate teaching and communication* (cf. Dalrymple, 2010; MacEachern et al., 2010). The ichnofacies concept provides a basis for understanding the ways in which organisms respond to their environment and forms the theoretical basis for evaluating the trace-fossil suites. Using ichnofacies, these relationships can be effectively presented to students of the science.

5. CONCLUSIONS

The ichnofacies paradigm is based on temporally and spatially recurring ethological groupings as indicated by ichnological suites. Although the Seilacherian ichnofacies are especially linked to rock-record occurrences, the interpretations and understanding of most ichnofacies have been and will continue to be refined through studies of modern animals and their burrowing behaviors (e.g., Buatois and Mángano, 2004; Dafoe et al., 2011; Dashtgard and Gingras, 2005, 2012; Dworschak and Ott, 1993; Frey and Pemberton, 1987; Frey et al., 1984; Genise et al., 2000, 2004; Gingras et al., 1999, 2008a, 2011; Hauck et al., 2008, 2009; Hertweck, 1972; Netto and Grangeiro, 2009; Virtasalo et al., 2011; Wetzel, 2008). When coupled with specific ichnological and sedimentological data, ichnofacies can be used not merely to discern the general character of the depositional environment, but also to derive specific information regarding salinity, oxygenation, sedimentation rates, rhythmicity of sedimentation, and the character of event deposition. Ichnofacies analysis can be applied to sequence-stratigraphic concepts, including the identification of key surfaces and the interpretation of general transgressive–regressive cycles (MacEachern et al., 1992, 2007d, 2012). Spatially, ichnofacies aid in delineating genetically related strata and help to identify facies boundaries.

In hindsight, naming Seilacherian ichnofacies for a characteristic ichnogenus may not have been the wisest course, as it has led to considerable confusion. The namesake element may not appear in a given suite, and indeed, owing to biological evolution and extinction events, the namesake element may not be common outside of particular time periods. For example, *Cruziana* are exceedingly uncommon after the Triassic, although the *Cruziana* Ichnofacies is a common expression of shallow-marine environments throughout the entire Phanerozoic. Additionally, the ichnogenus *Coprinisphaera* has not been identified in rocks older than the Paleocene, although equivalent suites lacking the namesake could extend the utility of the ichnofacies further back in time. In an even more confusing manner, the *Glossifungites* Ichnofacies continues to exist, while the namesake ichnogenus "*Glossifungites*" has been deemed invalid, and is considered a scratch-marked expression of *Rhizocorallium* (e.g., Fürsich, 1974; Uchman et al., 2000).

Future work should see the increased integration of process-based neoichnological data with observations from the rock record. Additionally, development of numerical analysis and statistical evaluation schemes is desirable as a means of expressing ichnological relationships. Refinements in the application of the ichnofacies concept to continental strata have seen great strides over the past decade (see Buatois and Mángano, 2011; Melchor et al., 2012). Nevertheless, the application of ichnofacies to these complex settings has proved difficult, owing to taphonomic considerations and the extreme lateral variability in many of the environmental parameters. For those that enjoy a challenge, this is an exciting area of research that is likely to continue to expand. Finally, most Seilacherian ichnofacies have been defined to accommodate ambient (largely unstressed) settings. The development of additional ichnofacies is likely but will need to be approached carefully. Defining new Seilacherian ichnofacies requires multiple case studies (both modern and ancient) and clear-cut ethological groupings by which the appropriate trace-fossil suites can be recognized. Analysis of case studies from all over the world and from all life-bearing periods of geological time is essential in order to ensure the validity of any newly defined Seilacherian ichnofacies.

Critics of the ichnofacies approach question its utility by implying that trace-fossil suites and ichnofabrics are "lumped" into broad entities that are then assigned to a limited number of environments (e.g., Goldring, 1993; McIlroy, 2004b; Taylor and Goldring, 1993; Taylor et al., 2003). Their concern is that the details imparted by the observed ichnological suites and ichnofabrics become downplayed in the interpretation of the facies. A similar criticism has been leveled against sedimentological facies models (Dalrymple, 2010). However, to turn the phrase of Walker (1992), it is the generality embodied by the ichnofacies, as opposed to the summary of one particular case study, which enables the ichnofacies paradigm to serve its most valuable functions.

One does not evaluate a sandstone interval and assign a shoreface interpretation simply because it hosts suites attributable to the *Skolithos* Ichnofacies, for example. The *Skolithos* Ichnofacies is *not* diagnostic of middle and upper shoreface environments. Rather, the middle shoreface and upper shoreface environments commonly host suites of the *Skolithos* Ichnofacies, precisely because they possess the depositional conditions that favor those ethological and trophic groupings. Other environments host suites attributable to the *Skolithos* Ichnofacies as well, because they also possess the physico-chemical parameters upon which the ichnofacies is founded. The details of the trace-fossil suites, integrated with physical sedimentological characteristics, allow the specific paleoenvironment to be identified.

The ichnofacies paradigm's strength lies in the environmental validity of its behavioral and trophic groupings. Assigning a trace-fossil suite to a particular ichnofacies does not lead to identification of a particular depositional environment. Rather, the paradigm specifies the depositional parameters necessary to produce such suites. Whether a worker employs trace-fossil suites integrated with lithofacies analysis and genetic stratigraphic frameworks, or ichnofabric analysis, the ultimate paleoenvironmental interpretation of the ichnological data requires an understanding of animal/sediment interactions as constrained by the environment. The ability to understand and therefore apply ichnology in sedimentology and stratigraphy is contingent upon the unifying paradigm of the Seilacherian ichnofacies.

ACKNOWLEDGMENTS

The authors would like to thank the Natural Science and Engineering Research Council of Canada (NSERC) for research funding; Discovery Grant 184293 to J. A. M., Discovery Grant 238530 to M. K. G., Discovery Grant (PRG) 341789 to S. E. D., Discovery Grant 371662 to J.-P. Z., and Operating Grant A0816 to S. G. P. Pemberton would like to thank the Canada Research Chairs program for their support of his research. This chapter benefited greatly from the excellent review of Luis Buatois and insights provided by Dirk Knaust and Richard Bromley. Gabriela Mángano and Luis Buatois kindly responded to our numerous questions during the revision process.

REFERENCES

Angulo, S., Buatois, L.A., 2012. Ichnology of an Upper Devonian—Early Carboniferous low-energy seaway: the Bakken Formation of subsurface Saskatchewan, Canada: assessing paleoenvironmental controls and biotic responses. Palaeogeogr. Palaeoclimatol. Palaeoecol. 315–316, 46–60.

Bann, K.L., Fielding, C.R., MacEachern, J.A., Tye, S.C., 2004. Differentiation of estuarine and offshore marine deposits using integrated ichnology and sedimentology: Permian Pebbley Beach Formation, Sydney Basin, Australia. In: McIlroy, D. (Ed.), The Application of Ichnology to Palaeoenvironmental and Stratigraphic Analysis. Geol. Soc. London, Spec. Publ., vol. 228, pp. 179–211.

Bann, K.L., Tye, S.C., MacEachern, J.A., Fielding, C.R., Jones, B.G., 2008. Ichnological and sedimentologic signatures of mixed wave- and storm-dominated deltaic deposits: examples from the Early Permian Sydney Basin, Australia. In: Hampson, G.J., Steel, R.J., Burgess, P.B., Dalrymple, R.W. (Eds.), Recent Advances in Models of Siliciclastic Shallow-Marine Stratigraphy. SEPM Spec. Publ., vol. 90, pp. 293–332.

Bockelie, J.F., 1991. Ichnofabric mapping and interpretation of Jurassic reservoir rocks of the Norwegian North Sea. Palaios 6, 206–215.

Bromley, R.G., 1975. Trace fossils at omission surfaces. In: Frey, R.W. (Ed.), The Study of Trace Fossils. Springer, Berlin, pp. 399–428.

Bromley, R.G., 1990. Trace Fossils: Biology and Taphonomy. Unwin Hyman, London, 280 pp.

Bromley, R.G., 1991. *Zoophycos*: strip mine, refust dump, cache or sewage farm? Lethaia 24, 460–462.

Bromley, R.G., 1996. Trace Fossils: Biology, Taphonomy and Applications. Chapman and Hall, London, 361 pp.

Bromley, R.G., Asgaard, U., 1979. Triassic freshwater ichnocoenoses from Carlsberg Fjord, East Greenland. Palaeogeogr. Palaeoclimatol. Palaeoecol. 28, 39–80.

Bromley, R.G., Asgaard, U., 1991. Ichnofacies: a mixture of taphofacies and biofacies. Lethaia 24, 153–163.

Bromley, R.G., Asgaard, U., 1993. Two bioerosion ichnofacies produced by early and later burial associated with sea-level change. Geol. Rundsch. 82, 872–874.

Bromley, R.G., Pemberton, S.G., Rahmani, R.A., 1984. A Cretaceous woodground: the *Teredolites* Ichnofacies. J. Paleontol. 58, 488–498.

Bromley, R.G., Jensen, M., Asgaard, U., 1995. Spatangoid echinoids: deep-tier trace fossils and chemosymbiosis. N. Jb. Geol. Paläont. Abh. 195, 25–35.

Bromley, R.G., Buatois, L.A., Genise, J.F., Labandeira, C.C., Mángano, M.G., Melchor, R.N., Schlirf, M., Uchman, A., 2007. Comments on the paper "Reconnaissance of Upper Jurassic Morrison Formation ichnofossils, Rocky Mountain Region, USA: paleoenvironmental, stratigraphic, and paleoclimatic significance of terrestrial and freshwater ichnocoenoses" by Stephen T. Hasiotis. Sediment. Geol. 200, 141–150.

Buatois, L.A., Mángano, M.G., 1995. The paleoenvironmental and paleoecological significance of the lacustrine *Mermia* ichnofacies: an archetypical subaqueous nonmarine trace fossil assemblage. Ichnos 4, 151–161.

Buatois, L.A., Mángano, M.G., 2004. Animal-substrate interactions in freshwater environments: applications of ichnology in facies and sequence stratigraphic analysis of fluvio-lacustrine successions. In: McIlroy, D. (Ed.), The Application of Ichnology to Palaeoenvironmental and Stratigraphic Analysis. Geol. Soc. London, Spec. Publ., vol. 228, pp. 311–333.

Buatois, L.A., Mángano, M.G., 2011. Ichnology: Organism-Substrate Interactions in Space and Time. Cambridge University Press, New York, 358 pp.

Buatois, L.A., Santiago, N., Herrera, M., Plink-Björklund, P., Steel, R., Espin, M., Parra, K., 2012. Sedimentological and ichnological signatures of changes in wave, river and tidal influence along a Neogene tropical deltaic shoreline. Sedimentology 59, 1568–1612.

Cachão, M., Da Silva, C.M., Santos, A., Domèmech, R., Martinell, J., Mayoral, E., 2009. The bioeroded megasurface of Oura (Algarve, south Portugal): implications for the Neogene stratigraphy and tectonic evolution of southwest Iberia. Facies 55, 213–225.

Crimes, T.P., Goldring, R., Homewood, P., Van Stuijvenberg, J., Winkler, W., 1981. Trace fossil assemblages of deep-sea fan deposits, Grunigel and Schlieren flysch (Cretaceous-Eocene, Switzerland). Eclogae Geol. Helv. 74, 953–995.

Curran, H.A., 2007. Ichnofacies, ichnocoenoses and ichnofabrics of Quaternary shallow-marine to dunal tropical carbonates: a model and implications. In: Miller III, W. (Ed.), Trace Fossils. Concepts, Problems, Prospects. Elsevier, Amsterdam, pp. 232–247.

Curran, H.A., Martin, A.J., 2001. Intertidal mounds of tropical callianassids provide substrates from complex upogebiid shrimp burrows: modern and Pleistocene examples from the Bahamas. PaleoBios 21, 44 Program and abstracts, NAPC 2001: paleontology in the new millennium.

Curran, H.A., White, B., 1991. Trace fossils of shallow subtidal to dunal ichnofacies in Bahamian Quaternary carbonates. Palaios 6, 498–510.

Dafoe, L.T., Rygh, A.L., Yang, B., Gingras, M.K., Pemberton, S.G., 2011. A new technique for assessing tubificid burrowing activities, and recognition of biogenic grading formed by these oligochaetes. Palaios 26, 66–80.

D'Alessandro, A., Ekdale, A.A., Sonnino, M., 1986. Sedimentologic significance of turbidite ichnofacies in the Saraceno Formation (Eocene), southern Italy. J. Sediment. Petrol. 56, 294–306.

Dalrymple, R., 2010. Interpreting sedimentary successions: facies, facies analysis and facies models. In: James, N.P., Dalrymple, R.W. (Eds.), Facies Models 4. Geol. Ass. Can., St. John's, Geotext 6, pp. 3–18.

Dam, G., 1990. Palaeoenvironmental significance of trace fossils from the shallow marine Lower Jurassic Neill Klinter Formation, East Greenland. Palaeogeogr. Palaeoclimatol. Palaeoecol. 79, 221–248.

Dashtgard, S.E., 2011a. Linking invertebrate burrow distributions (neoichnology) to physicochemical stresses on a sandy tidal flat: implications for the rock record. Sedimentology 58, 1303–1325.

Dashtgard, S.E., 2011b. Neoichnology of the lower delta plain: Fraser River Delta, British Columbia, Canada: implications for the ichnology of deltas. Palaeogeogr. Palaeoclimatol. Palaeoecol. 307, 98–108.

Dashtgard, S.E., Gingras, M.K., 2005. Facies architecture and ichnology of recent salt-marsh deposits: Waterside Marsh, New Brunswick, Canada. J. Sediment. Res. 75, 596–607.

Dashtgard, S.E., Gingras, M.K., 2012. Marine invertebrate neoichnology. In: Knaust, D., Bromley, R.G. (Eds.), Trace Fossils as Indicators of Sedimentary Environments. Developments in Sedimentology, vol. 64. Elsevier, Amsterdam, pp. 273–295.

Dashtgard, S.E., Gingras, M.K., Pemberton, S.G., 2008. Grain-size controls on the occurrence of bioturbation. Palaeogeogr. Palaeoclimatol. Palaeoecol. 257, 224–243.

de Gibert, J.M., Martinell, J., Domènech, R., 1998. *Entobia* ichnofacies in fossil rocky shores, Lower Pliocene, northwestern Mediterranean. Palaios 13, 476–487.

de Gibert, J.M., Domèneche, R., Martinell, J., 2007. Bioerosion in shell beds from the Pliocene Roussillon Basin, France: implications for the (macro) bioerosion ichnofacies model. Acta Palaeontol. Polon. 52, 783–798.

de Gibert, J.M., Domèneche, R., Martinell, J., 2012. Rocky shorelines. In: Knaust, D., Bromley, R.G. (Eds.), Trace Fossils as Indicators of Sedimentary Environments. Developments in Sedimentology, vol. 64. Elsevier, Amsterdam, pp. 441–462.

Dubois, M.K., Goldstein, R.H., Hasiotis, S.T., 2012. Climate-controlled aggradation and cyclicity of continental loessic siliciclastic sediments in Asselian-Sakmarian cyclothems, Permian, Hugoton embayment, USA. Sedimentology 59, 1782–1816.

Duringer, P., Schuster, M., Genise, J.F., Mackaye, H.T., Vignaud, P., Brunet, M., 2007. New termite trace fossils: galleries, nests and fungus combs from the Chad basin of Africa (upper Miocene-Lower Pliocene). Palaeogeogr. Palaeoclimatol. Palaeoecol. 251, 323–353.

Dworschak, P.C., Ott, J.A., 1993. Decapod burrows in mangrove-channel and back-reef environments of the Atlantic Barrier Reef, Belize. Ichnos 2, 277–290.

Ekdale, A.A., 1988. Pitfalls of paleobathymetric interpretations based on trace fossil assemblages. Palaios 3, 464–472.

Ekdale, A.A., Bromley, R.G., Pemberton, S.G., 1984. Ichnology. The Use of Trace Fossils in Sedimentology and Stratigraphy. SEPM Short Course 15, 317 pp.

Ekdale, A.A., Bromley, R.G., Loope, D.B., 2007. Ichnofacies of an ancient erg: a climatically influenced trace fossil association in the Jurassic Navajo Sandstone, southern Utah, USA. In: Miller III, W. (Ed.), Trace Fossils. Concepts, Problems, Prospects. Elsevier, Amsterdam, pp. 562–564.

Farrow, G.E., 1966. Bathymetric zonation of Jurassic trace fossils from the coast of Yorkshire, England. Palaeogeogr. Palaeoclimatol. Palaeoecol. 2, 103–151.

Frey, R.W., 1990. Trace fossils and hummocky cross-stratification, Upper Cretaceous of Utah. Palaios 5, 203–218.

Frey, R.W., Pemberton, S.G., 1987. The *Psilonichnus* ichnocoenose and its relationship to adjacent marine and nonmarine ichnocoenoses along the Georgia coast. Bull. Can. Petrol. Geol. 35, 333–357.

Frey, R.W., Seilacher, A., 1980. Uniformity in marine invertebrate ichnology. Lethaia 13, 183–207.

Frey, R.W., Curran, H.A., Pemberton, S.G., 1984. Tracemaking activities of crabs and their environmental significance: the ichnogenus *Psilonichnus*. J. Paleontol. 58, 333–350.

Frey, R.W., Pemberton, S.G., Saunders, T.D.A., 1990. Ichnofacies and bathymetry: a passive relationship. J. Paleontol. 64, 155–158.

Fürsich, F.T., 1974. Ichnogenus *Rhizocorallium*. Paläont. Z. 48, 16–28.

Fürsich, F.T., 1975. Trace fossils as environmental indicators in the Corallian of England and Normandy. Lethaia 8, 151–172.

Genise, J.F., Mángano, M.G., Buatois, L.A., Laza, J.H., Verde, M., 2000. Insect trace fossil associations in paleosols: the *Coprinisphaera* Ichnofacies. Palaios 15, 49–64.

Genise, J.F., Bellosi, E.S., Gonzalez, M.G., 2004. An approach to the description and interpretation of ichnofabrics in palaeosols. In: McIlroy, D. (Ed.), The Application of Ichnology to Palaeoenvironmental and Stratigraphic Analysis. Geol. Soc. London, Spec. Publ., vol. 228, pp. 355–382.

Genise, J.F., Melchor, R.N., Bellosi, E.S., Verde, M., 2010. Invertebrate and vertebrate trace fossils from continental carbonates. In: Alonso-Zarza, A.M., Tanner, L. (Eds.), Carbonates in Continental Settings: Facies, Environments, and Processes. Developments in Sedimentology, vol. 61. Elsevier, Amsterdam, pp. 319–369.

Gingras, M.K., Pemberton, S.G., Saunders, T., Clifton, H.E., 1999. The ichnology of modern and Pleistocene brackish-water deposits at Willapa Bay, Washington: variability in estuarine settings. Palaios 14, 352–374.

Gingras, M.K., Pemberton, S.G., Saunders, T.D.A., 2001. Bathymetry, sediment texture, and substrate cohesiveness: their impact on *Glossifungites* trace assemblages at Willapa Bay, Washington. Palaeogeogr. Palaeoclimatol. Palaeoecol. 169, 1–21.

Gingras, M.K., MacEachern, J.A., Pickerill, R.K., 2004. Modern perspectives on the *Teredolites* Ichnofacies: observations from Willapa Bay, Washington. Palaios 19, 79–88.

Gingras, M.K., Bann, K.L., MacEachern, J.A., Pemberton, S.G., 2007a. A conceptual framework for the application of trace fossils. In: MacEachern, J.A., Bann, K.L., Gingras, M.K., Pemberton, S.G. (Eds.), Applied Ichnology. SEPM Short Course Notes 52, pp. 1–26.

Gingras, M.K., Lalonde, S.V., Amskold, L., Konhauser, K.O., 2007b. Wintering chironomids mine oxygen. Palaios 22, 433–438.

Gingras, M.K., Dashtgard, S.E., MacEachern, J.A., Pemberton, S.G., 2008a. Biology of shallow marine ichnology: a modern perspective. Aquatic Biol. 2, 255–268.

Gingras, M.K., Pemberton, S.G., Dashtgard, S., Dafoe, L., 2008b. How fast do marine invertebrates burrow? Palaeogeogr. Palaeoclimatol. Palaeoecol. 270, 280–286.

Gingras, M.K., MacEachern, J.A., Dashtgard, S.E., 2011. Process ichnology and the elucidation of physico-chemical stress. Sediment. Geol. 237, 115–134.

Gingras, M.K., Baniak, G., Gordon, J., Hovikoski, J., Konhauser, K.O., La Croix, A., Lemiski, R., Mendoza, C., Pemberton, G.S., Polo, C., Zonneveld, J.-P., 2012. Porosity and permeability in bioturbated sediments. In: Knaust, D., Bromley, R.G. (Eds.), Trace Fossils as Indicators of Sedimentary Environments. Developments in Sedimentology, vol. 64. Elsevier, Amsterdam, pp. 835–868.

Goldring, R., 1993. Ichnofacies and facies interpretation. Palaios 8, 403–405.

Gressley, A., 1838. Observations géologiques sur le Jura Soleurois. N. Denkschr. allg. schweiz. Ges. ges. Naturwiss. 2, 1–112.

Häntzschel, W., 1939. Die Lebens-Spuren von *Corophium volutator* (Pallas) und ihre paläontologische Bedeutung. Senckenbergiana 21, 215–227.

Hasiotis, S.T., 2002. Continental Trace Fossils. SEPM Short Course Notes 51, pp. 1–132.

Hasiotis, S.T., Kraus, M.J., Demko, D.M., 2007. Climatic controls on continental trace fossils. In: Miller III, W. (Ed.), Trace Fossils. Concepts, Problems, Prospects. Elsevier, Amsterdam, pp. 172–195.

Hauck, T.E., Dashtgard, S.E., Gingras, M.K., 2008. Relationships between organic carbon and fodinichnia morphology. Palaios 23, 336–343.

Hauck, T.E., Dashtgard, S.E., Pemberton, S.G., Gingras, M.K., 2009. Brackish-water ichnological trends in a microtidal barrier island/embayment system, Kouchibouguac National Park, New Brunswick, Canada. Palaios 24, 478–496.

Hayward, B.W., 1976. Lower Miocene bathyal and submarine canyon ichnocoenoses from Northland, New Zealand. Lethaia 9, 149–162.

Heinberg, C., Birkelund, T., 1984. Trace-fossil assemblages and basin evolution of the Vardekløft Formation (Middle Jurassic, Central East Greenland). J. Paleontol. 58, 362–397.

Hembree, D.I., Hasiotis, S.T., 2007. Paleosols and ichnofossils of the White River Formation of Colorado: insight into soil ecosystems of the North American Midcontinent during the Eocene-Oligocene transition. Palaios 22, 123–142.

Hertweck, G., 1972. Georgia coastal region, Sapelo Island, U.S.A.: sedimentology and biology. V. Distribution and environmental significance of lebensspuren and in-situ skeletal remains. Senckenberg. Marit. 4, 125–167.

Hiscott, R.N., James, N.P., Pemberton, S.G., 1984. Sedimentology and ichnology of the lower Cambrian Bradore Formation, coastal Labrador: fluvial to shallow-marine transgressive sequence. Bull. Can. Petrol. Geol. 32, 11–26.

Howard, J.D., Frey, R.W., 1975. Regional animal-sediment characteristics of Georgia estuaries. Senckenberg. Marit. 7, 33–107.

Howard, J.D., Frey, R.W., 1984. Characteristic trace fossils in nearshore to offshore sequences, Upper Cretaceous of east central Utah. Can. J. Earth Sci. 21, 200–219.

Howard, J.D., Reineck, H.-E., Rietschel, S., 1974. Biogenic sedimentary structures formed by heart urchins. Senckenberg. Marit. 6, 185–201.

Hunt, A.P., Lucas, S.G., 2006. Tetrapod ichnofacies and their utility in the Paleozoic. In: Buta, R.J., Rindsberg, A.K., Kopaska-Merkel, D.C. (Eds.), Pennsylvanian Footprints in the Black Warrior Basin of Alabama. Alabama Paleontol. Soc., Monogr. 1, pp. 113–119.

Hunt, A.P., Lucas, S.G., 2007. Tetrapod ichnofacies: a new paradigm. Ichnos 14, 59–68.

Hunt, A.P., Chin, K., Lockley, M.G., 1994. The palaeobiology of vertebrate coprolites. In: Donovan, S.K. (Ed.), The Palaeobiology of Trace Fossils. Wiley and Sons, New York, pp. 221–240.

Hunt, A.P., Lucas, S.G., Lockley, M.G., Haubold, H., Braddy, S., 1995. Tetrapod ichnofacies in Early Permian red beds of the American Southwest. New Mexico Mus. Nat. Hist. Sci., Bull. 6, 295–301.

Hunt, A.P., Lucas, S.G., Lockley, M.G., 1998. Taxonomy and stratigraphic and facies significance of vertebrate coprolites of the Upper Triassic Chinle Group, western United States. Ichnos 5, 225–234.

Kotake, N., 1991. Non-selective surface deposit feeding by the *Zoophycos* producers. Lethaia 24, 379–385.

Kotake, N., 1992. Deep-sea echiurans: possible producers of *Zoophycos*. Lethaia 25, 311–316.

Krapovickas, V., Ciccioli, P.L., Mángano, M.G., Marsicano, C.A., Limarino, C.O., 2009. Paleobiology and paleoecology of an arid–semiarid Miocene South American ichnofauna in anastomosed fluvial deposits. Palaeogeogr. Palaeoclimatol. Palaeoecol. 284, 129–152.

Lawfield, A.M.W., Pickerill, R.K., 2006. A novel contemporary fluvial ichnocoenose: unionid bivalves and the *Scoyenia-Mermia* Ichnofacies transition. Palaios 21, 391–396.

Leszczyński, S., Seilacher, A., 1991. Ichnocoenoses of a turbidite sole. Ichnos 1, 293–303.

Lockley, M.G., 2007. A tale of two ichnologies: the different goals and potentials of invertebrate and vertebrate (tetrapod) ichnotaxonomy and how they relate to ichnofacies analysis. Ichnos 14, 39–57.

Lockley, M.G., Meyer, C.A., 2000. Dinosaur Tracks and Other Fossil Footprints of Europe. Columbia University Press, New York, 323 pp.

Lockley, M.G., Rindsberg, A.K., Zeiler, R.M., 1987. The paleoenvironmental significance of the nearshore *Curvolithus* Ichnofacies. Palaios 2, 255–262.

Lockley, M.G., Hunt, A.P., Meyer, C.A., 1994. Vertebrate tracks and the ichnofacies concept: implications for palaeoecology and palichno-stratigraphy. In: Donovan, S.K. (Ed.), The Palaeobiology of Trace Fossils. Wiley and Sons, New York, pp. 241–268.

Lockley, M.G., Wright, J.L., Matsukawa, M., 2001. A new look at *Magnoavipes* and so-called big bird tracks from Dinosaur Ridge (Cretaceous Colorado). Mountain Geol. 38, 137–146.

MacEachern, J.A., Bann, K.L., 2008. The role of ichnology in refining shallow marine facies models. In: Hampson, G.J., Steel, R.J., Burgess, P.B., Dalrymple, R.W. (Eds.), Recent Advances in Models of Siliciclastic Shallow-Marine Stratigraphy. SEPM Spec. Publ., vol. 90, pp. 73–116.

MacEachern, J.A., Burton, J.A., 2000. Firmground *Zoophycos* in the Lower Cretaceous Viking Formation, Alberta: a distal expression of the *Glossifungites* Ichnofacies. Palaios 15, 387–398.

MacEachern, J.A., Gingras, M.K., 2007. Recognition of brackish-water trace-fossil suites in the Cretaceous Western Interior Seaway of Alberta, Canada. In: Bromley, R., Buatois, L.A., Mángano, M.G., Genise, J., Melchor, R. (Eds.), Sediment-Organism Interactions: A Multifaceted Ichnology. SEPM Spec. Publ., vol. 88, pp. 149–193.

MacEachern, J.A., Raychaudhuri, I., Pemberton, S.G., 1992. Stratigraphic applications of the *Glossifungites* Ichnofacies: delineating discontinuities in the rock record. In: Pemberton, S.G. (Ed.), Applications of Ichnology to Petroleum Exploration. SEPM Core Workshop Notes 17, pp. 169–198.

MacEachern, J.A., Bann, K.L., Bhattacharya, J.P., Howell, C.D., 2005. Ichnology of deltas: organism responses to the dynamic interplay of rivers, waves, storms and tides. In: Giosan, L., Bhattacharya, J.P. (Eds.), River Deltas: Concepts, Models and Examples. SEPM Spec. Publ., vol. 83, pp. 49–85.

MacEachern, J.A., Bann, K.L., Pemberton, S.G., Gingras, M.K., 2007a. The ichnofacies paradigm: high resolution paleoenvironmental interpretation of the rock record. In: MacEachern, J.A., Bann, K.L., Gingras, M.K., Pemberton, S.G. (Eds.), Applied Ichnology. SEPM Short Course Notes 52, pp. 27–64.

MacEachern, J.A., Pemberton, S.G., Bann, K.L., Gingras, M.K., 2007b. Departures from the archetypal ichnofacies: effective recognition of environmental stress in the rock record. In: MacEachern, J.A., Bann, K.L., Gingras, M.K., Pemberton, S.G. (Eds.), Applied Ichnology. SEPM Short Course Notes 52, pp. 65–93.

MacEachern, J.A., Pemberton, S.G., Gingras, M.K., Bann, K.L., 2007c. The ichnofacies concept: a fifty-year retrospective. In: Miller III, W. (Ed.), Trace Fossils. Concepts, Problems, Prospects. Elsevier, Amsterdam, pp. 50–75.

MacEachern, J.A., Pemberton, S.G., Gingras, M.K., Bann, K.L., Dafoe, L.T., 2007d. Use of trace fossils in genetic stratigraphy. In: Miller III, W. (Ed.), Trace Fossils. Concepts, Problems, Prospects. Elsevier, Amsterdam, pp. 105–128.

MacEachern, J.A., Pemberton, S.G., Gingras, M.K., Bann, K.L., 2010. Ichnology and facies models. In: James, N.P., Dalrymple, R.W. (Eds.), Facies Models 4. Geol. Ass. Can., St. John's, Geotext 6, pp. 19–58.

MacEachern, J.A., Dashtgard, S.E., Knaust, D., Catuneanu, O., Bann, K.L., Pemberton, S.G., 2012. Sequence stratigraphy. In: Knaust, D., Bromley, R.G. (Eds.), Trace Fossils as Indicators of Sedimentary Environments. Developments in Sedimentology, vol. 64. Elsevier, Amsterdam, pp. 157–194.

Mángano, M.G., Buatois, L.A., MacNaughton, R.B., 2012. Ichnostratigraphy. In: Knaust, D., Bromley, R.G. (Eds.), Trace Fossils as Indicators of Sedimentary Environments. Developments in Sedimentology, vol. 64. Elsevier, Amsterdam, pp. 195–212.

Marshall, P., 2003. Ichnofossils of the *Psilonichnus* Ichnofacies and their paleoecological and paleoenvironmental significance in the Scottish Middle Jurassic. Ichnos 9, 95–108.

Mayoral, E., Muñiz, F., 1996. La ichnofacies de *Gnathichnus* en el sector suroccidental de la Cuenca del Guadalquivir (Lepe, Huelva, España). Coloquios de Paleontologia 48, 71–82.

McIlroy, D., 2004a. Ichnofabrics and sedimentary facies of a tide-dominated delta: Jurassic Ile Formation of the Kristin Field, Haltenbanken, Offshore Mid-Norway. In: McIlroy, D. (Ed.), The Application of Ichnology to Palaeoenvironmental and Stratigraphic Analysis. Geol. Soc. London, Spec. Publ., vol. 228, pp. 3–27.

McIlroy, D., 2004b. Some ichnological concepts, methodologies, applications and frontiers. In: McIlroy, D. (Ed.), The Application of Ichnology to Palaeoenvironmental and Stratigraphic Analysis. Geol. Soc. London, Spec. Publ., vol. 228, pp. 3–27.

McIlroy, D., 2007. Ichnology of a macrotidal tide-dominated deltaic depositional system: Lajas Formation, Neuquén Province, Argentina. In: Bromley, R.G., Buatois, L.A., Mángano, M.G., Genise, J.F., Melchor, R.N. (Eds.), Sediment-Organism Interactions: A Multifaceted Ichnology. SEPM Spec. Publ., vol. 88, pp. 195–211.

McIlroy, D., 2008. Ichnological analysis: the common ground between ichnofacies workers and ichnofabric analysts. Palaeogeogr. Palaeoclimatol. Palaeoecol. 270, 332–338.

Melchor, R.N., 2004. Trace fossil distribution in lacustrine deltas: examples from the Triassic rift lakes of the Ischigualasto—Villa Unión basin, Argentina. In: McIlroy, D. (Ed.), The Application of Ichnology to Palaeoenvironmental and Stratigraphic Analysis. Geol. Soc. London, Spec. Publ., vol. 228, pp. 335–354.

Melchor, R.N., Genise, J.F., Buatois, L.A., Umazano, A.M., 2012. Fluvial environments. In: Knaust, D., Bromley, R.G. (Eds.), Trace Fossils as Indicators of Sedimentary Environments. Developments in Sedimentology, vol. 64. Elsevier, Amsterdam, pp. 329–378.

Middleton, G.V., 1973. Johannes Walther's Law of the correlation of facies. GSA Bull. 84, 979–988.

Middleton, G.V., 1978. Facies. In: Fairbridge, R.W., Bourgeois, J. (Eds.), Encyclopedia of Sedimentology. Dowden, Hutchinson and Ross, Stroudsburg, pp. 323–325.

Moore, R.C., 1949. Meaning of facies. In: Longwell, C.R. (Ed.), Sedimentary Facies in Geological History. GSA Mem. 39, pp. 1–34.

Nara, M., 1995. *Rosselia socialis* Dahmer 1937: a dwelling structure of a probable terebellid polychaete. Lethaia 28, 171–178.

Nara, M., 2002. Crowded *Rosselia socialis* in Pleistocene inner shelf deposits: benthic paleoecology during rapid sea-level rise. Palaios 17, 268–276.

Nesbitt, E.A., Campbell, K.A., 2006. The paleoenvironmental significance of *Psilonichnus*. Palaios 21, 187–196.

Netto, R., Grangeiro, M., 2009. Neoichnology of the seaward side of Peixe Lagoon in Mostardas, southernmost Brazil: the *Psilonichnus* ichnocoenosis revisited. Rev. Brasileira de Paleontol. 12, 211–224.

Olivero, D., Gaillard, C., 1996. Paleoecology of Jurassic *Zoophycos* from south-eastern France. Ichnos 4, 249–260.

Papentin, F., 1973. A Darwinian evolutionary system III: experiments on the evolution of feeding patterns. J. Theor. Biol. 39, 431–445.

Pemberton, S.G., Frey, R.W., 1984. Ichnology of storm-influenced shallow marine sequence: Cardium Formation (Upper Cretaceous) at Seebe, Alberta. In: Stott, D.F., Glass, D.L. (Eds.), The Mesozoic of Middle North America. Can. Soc. Petrol. Geol., Mem. 9, pp. 281–304.

Pemberton, S.G., Frey, R.W., 1985. The *Glossifungites* Ichnofacies: modern examples from the Georgia coast, U.S.A. In: Curran, H.A. (Ed.), Biogenic Structures: Their Use in Interpreting Depositional Environments. SEPM Spec. Publ., vol. 35, pp. 237–259.

Pemberton, S.G., Kobluk, D.R., Yeo, R.K., Risk, M.J., 1980. The boring *Trypanites* at the Silurian-Devonian disconformity in southern Ontario. J. Paleontol. 54, 1258–1266.

Pemberton, S.G., MacEachern, J.A., Frey, R.W., 1992. Trace fossil facies models: environmental and allostratigraphic significance. In: Walker, R.G., James, N. (Eds.), Facies Models: Response to Sea Level Change. Geol. Ass. Can., St. John's, Geotext 1, pp. 47–72.

Pickerill, R.K., 1992. Carboniferous nonmarine invertebrate ichnocoenoses from southern New Brunswick, eastern Canada. Ichnos 2, 21–35.

Pickett, T.E., Kraft, J.C., Smith, K., 1971. Cretaceous burrows: Chesapeake and Delaware Canal, Delaware. J. Paleontol. 45, 209–211.

Reading, H.G., 1996. Sedimentary Environments and Facies. Blackwell Publications, Oxford, 688 pp.

Reineck, H.-E., 1965. Der Knechtsand. 6. Die Materialumlagerung auf Grund von Oberflächen- und Gefügestudien. Jahrb. 1964, Foch.-Stelle Norderney. pp. 169–181 (as cited in Reineck, H.-E., Singh, I.B., 1980. Depositional Sedimentary Environments. Springer-Verlag, Berlin, 551 pp.).

Reineck, H.-E., 1968. Lebensspuren von Herzigeln. Senck. Leth. 49, 311–319.

Rhoads, D.C., Morse, J.W., 1971. Evolution and ecologic significance of oxygen-deficient marine basins. Lethaia 4, 413–428.

Rindsberg, A.K., 2012. Ichnotaxonomy: finding patterns in a welter of information. In: Knaust, D., Bromley, R.G. (Eds.), Trace Fossils as Indicators of Sedimentary Environments. Developments in Sedimentology, vol. 64. Elsevier, Amsterdam, pp. 45–78.

Santi, G., Nicosia, U., 2008. The ichnofacies concept in vertebrate ichnology. In: Avanzini, M., Petti, F.M. (Eds.), Italian Ichnology. Studi Trent. Sci. Nat. Acta Geol., vol. 83, pp. 223–229.

Savrda, C.E., 2007. Trace fossils and marine benthic oxygenation. In: Miller III, W. (Ed.), Trace Fossils. Concepts, Problems, Prospects. Elsevier, Amsterdam, pp. 149–158.

Savrda, C.E., Ozalas, K., Demko, T.H., Hutchinson, R.A., Scheiwe, T.D., 1993. Log grounds and the ichnofossil *Teredolites* in transgressive deposits of the Clayton Formation (Lower Paleocene), western Alabama. Palaios 8, 311–324.

Savrda, C.E., Browning, J.V., Krawinkel, H., Hesselbo, S.P., 2001a. Firmground ichnofabrics in deep-water sequence stratigraphy, Tertiary clinoform-toe deposits, New Jersey slope. Palaios 16, 294–305.

Savrda, C.E., Krawinkel, H., McCarthy, F.M.G., McHugh, C.M.G., Olson, H.C., Mountain, G., 2001b. Ichnofabrics of a Pleistocene slope succession, New Jersey margin: relations to climate and sea-level dynamics. Palaeogeogr. Palaeoclimatol. Palaeoecol. 171, 41–61.

Scott, J.J., Renault, R.W., Owen, R.B., Buatois, L.A., 2009. Trace fossils in exhumed surfaces around saline lakes: an example from Lake Bogoria, Kenya Rift Valley. Palaeogeogr. Palaeoclimatol. Palaeoecol. 272, 176–198.

Seilacher, A., 1953a. Studien zur Palichnologie. I. Über die Methoden der Palichnologie. N. Jb. Geol. Paläont. Abh. 96, 421–452.

Seilacher, A., 1953b. Studien zur Palichnologie. II. Die fossilen Ruhespuren (Cubichnia). N. Jb. Geol. Paläont. Abh. 98, 87–124.

Seilacher, A., 1964. Biogenic sedimentary structures. In: Imbrie, J., Newell, N. (Eds.), Approaches to Paleoecology. Wiley, New York, pp. 296–316.
Seilacher, A., 1967. Bathymetry of trace fossils. Mar. Geol. 5, 413–428.
Seilacher, A., 1970. *Cruziana* stratigraphy of "nonfossiliferous" Palaeozoic sandstones. In: Crimes, T.P., Harper, J.C. (Eds.), Trace Fossils. Geol. J., Spec. Iss. 3, pp. 447–476.
Seilacher, A., 1974. Flysch trace fossils: evolution of behavioural diversity in the deep-sea. N. Jb. Geol. Paläont. Mh. 1974, 233–245.
Seilacher, A., 1977. Pattern analysis of *Paleodictyon* and related trace fossils. In: Crimes, T.P., Harper, J.C. (Eds.), Trace Fossils 2. Geol. J., Spec. Iss. 9, pp. 289–334.
Seilacher, A., 1991. An updated *Cruziana* stratigraphy of Gondwanan Palaeozoic sandstones. In: Salem, M.J., Hammuda, O.S., Eliagoubi, B.A. (Eds.), The Geology of Libya, vol. 4. Elsevier, Amsterdam, pp. 1565–1581.
Seilacher, A., 1994. How valid is *Cruziana* stratigraphy? Geol. Rundsch. 83, 752–758.
Smith, R.M.H., 1993. Sedimentology and ichnology of floodplain paleosurfaces in the Beaufort Group (Late Permian), Karoo sequence, South Africa. Palaios 8, 339–357.
Smith, R.M.H., Mason, T.R., Ward, J.D., 1993. Flash-flood sediments and ichnofacies of the Later Pleistocene Homeb Silts, Kuiseb River, Namibia. Sediment. Geol. 85, 579–599.
Taylor, A.M., Goldring, R., 1993. Description and analysis of bioturbation and ichnofabric. J. Geol. Soc. London 150, 141–148.
Taylor, A.M., Goldring, R., Gowland, S., 2003. Analysis and application of ichnofabric. Earth-Sci. Rev. 60, 227–259.
Uchman, A., 2001. Eocene flysch trace fossils from the Hecho Group of the Pyrenees, northern Spain. Beringeria 28, 3–41.
Uchman, A., 2009. The *Ophiomorpha rudis* ichnosubfacies of the *Nereites* ichnofacies: characteristics and constraints. Palaeogeogr. Palaeoclimatol. Palaeoecol. 276, 107–119.
Uchman, A., Demírcan, H., 1999. A *Zoophycos* group trace fossil from Miocene flysch deposits of the Istria peninsula, Croatia and Slovenia. Ichnos 5, 1–22.
Uchman, A., Bubniak, I., Bubniak, A., 2000. *Glossifungites* Ichnofacies in the area of its nomenclatural archetype, Lviv, Ukraine. Ichnos 7, 183–193.
Virtasalo, J.J., Bonsdorff, E., Moros, M., Kabel, K., Kotilainen, A.T., Ryabchuk, D., Kallonen, A., Haemaelaeinen, K., 2011. Ichnological trends along an open-water transect across a large marginal-marine epicontinental basin, the modern Baltic Sea. Sediment. Geol. 241, 40–51.
Walker, R.G., 1992. Facies, facies models and modern stratigraphic concepts. In: Walker, R.G., James, N.P. (Eds.), Facies Models: Response to Sea Level Change. Geol. Ass. Can., St. John's, Geotext 1, pp. 1–14.
Weimer, R.J., Hoyt, J.H., 1964. Burrows of *Callianassa major* Say, geologic indicators of littoral and shallow neritic environments. J. Paleontol. 38, 761–767.
Wetzel, A., 1991. Ecologic interpretation of deep-sea trace fossil communities. Palaeogeogr. Palaeoclimatol. Palaeoecol. 85, 47–69.
Wetzel, A., 2002. Modern *Nereites* in the South China Sea—ecological association with redox conditions in the sediment. Palaios 17, 507–515.
Wetzel, A., 2008. Recent bioturbation in the deep South China Sea: a uniformitarian ichnologic approach. Palaios 23, 601–615.

Chapter 5

The Ichnofabric Concept

Allan A. Ekdale,*[,1] Richard G. Bromley[†] and Dirk Knaust[‡]
*Department of Geology and Geophysics, University of Utah, Salt Lake City, Utah, USA,
[†]Geological Museum—SNM, Copenhagen, Denmark, [‡]Statoil ASA, Stavanger, Norway
[1]Corresponding author: e-mail: a.ekdale@utah.edu

1. INTRODUCTION

Most people would agree that ichnology is the study of traces and trace fossils, which of course is quite true, at least in part. But ichnology includes much more than just identifying and interpreting trace fossils. Ichnology broadly encompasses the preservable and preserved effects that active organisms (animals, plants, bacteria, etc.) produce in the substrate (both unconsolidated *and* lithified). It includes analytical approaches to understanding the processes of bioturbation and bioerosion, as well as the products of those processes.

While recognition and description of identifiable trace fossils are of paramount importance in ichnological studies, the taphonomic overprint and preservation mode of trace fossils also hold great value for interpreting the post-depositional history of sedimentary units. Even more broadly, recognition of characteristic ways in which the texture and fabric of a sediment has been affected by organism activity offers significant sedimentological, paleoecological and even stratigraphical information. This is the arena of ichnofabric.

2. EARLY DEVELOPMENT OF THE CONCEPT

The *fabric* of a bolt of fine cloth, or an ornate oriental carpet, or an intricately constructed tapestry is composed of many multihued, multitextured, intertwined threads that create an exquisite whole. A single thread does not a tapestry make. Yet, the omission of a single thread may alter the appearance and possibly even the overall character of the fabric, and removal of one thread may cause the entire fabric to unravel. The fabric of a sedimentary deposit is not only composed of many different threads (involving the mineralogy, size, shape, orientation, distribution, etc. of grains and matrix) but also constructed by the complex interplay of numerous physical and biological processes. Fabrics are not woven instantaneously, and so the fabric of a sedimentary deposit (especially

Developments in Sedimentology, Vol. 64. http://dx.doi.org/10.1016/B978-0-444-53813-0.00005-8

one developed on a sedimentary surface) may reflect substantial changes with the passage of time. Ignoring any of the many threads that are woven together in a sedimentary deposit will yield an incomplete, or possibly misleading, view of the geological history and paleoenvironmental significance of that deposit.

Beginning only about three decades ago, the ichnofabric approach is a relatively young direction in ichnology, although its roots can be traced back a few decades further to the neoichnological studies of the German scientists Rudolf Richter, Wilhelm Schäfer, and Hans-Erich Reineck at the Senckenberg Institute in Wilhelmshaven (see Baucon et al., 2012). Based on experiments and observations in the North Sea, Schäfer (1956) noticed that benthic organisms modify the sediment in many different ways, which can lead to total sediment homogenization (*Fossitexturae deformativae*) with subsequent overprinting and preservation of discrete burrows (*Fossitexturae figurativae*). Reineck (1958) applied these features to delineate the amount of reworking in tidal deposits, and subsequently documented biologically generated sediment fabrics in combination with other primary sedimentary features based on box-core samples from the North Sea shelf Reineck (1963). He also developed a semiquantitative scheme for estimating relative amounts of bioturbation in such samples, a method that has subsequently been adapted in ichnofabric analysis (see discussion of the bioturbation index in Knaust, 2012). Refinement of the box-core sampling procedure allowed for the collection and detailed description of biogenic sedimentary fabrics in modern deposits (Reineck et al., 1967). In the 1970s, this methodology was expanded and refined by the use of X-ray radiography to carry out neoichnological studies, especially along the Georgia coast of North America (Howard and Elders, 1970; Howard and Frey, 1975).

Also in the 1970s, trace fossils and complex sedimentary fabrics resulting from burrowing were beginning to be recognized and documented in thoroughly bioturbated sediment in deep-sea cores (Berger et al., 1979; Chamberlain, 1975; Ekdale, 1977, 1978; Ekdale and Berger, 1978; Van der Lingen, 1973; Warme et al., 1973). Accompanying those descriptive investigations of ichnological features in cores were analytical attempts to quantify the nature and extent of vertical mixing of pelagic sediment by burrowing organisms in the deep sea (Berger and Heath, 1968; Guinasso and Schink, 1975), because that aspect of ichnology has direct implications for the sharpness (or fuzziness) of biostratigraphic horizons based on microfossils. Observations in deep-sea box cores led to the interpretation of a three-tiered vertical stratification (mixed layer, transition layer, and historical layer) of burrow emplacement in deep-sea pelagic deposits (Berger et al., 1979; Ekdale et al., 1984b). Wetzel (1981, 1983, 1984, 1985) interpreted the ecological and stratigraphical significance of biogenic sedimentary structures in box cores of modern deep-sea sediments, and he outlined their characteristic position in vertical tiers within the sediment.

Following the aforementioned seminal investigations, the concept of *ichnofabric per se* was born in the 1980s and grew out of a need to decipher a meaning from the results of bioturbation and/or bioerosion in situations where individually distinct and identifiable trace fossils cannot be observed. The parallel concept of *ichnofacies*, introduced by Seilacher (1964, 1967), is an extension of the biofacies approach by recognizing recurrent associations of ichnotaxa that represent particular paleoenvironments or specific sets of environmental conditions, such as bathymetry, salinity, substrate consistency, etc. (MacEachern et al., 2012). In contrast to ichnofacies, ichnofabric extends beyond a simple listing of common associations of ichnotaxa by highlighting the broader effects of organism behavior on the substrate itself.

In the first formal use of the term "ichnofabric" in a refereed publication, Ekdale and Bromley (1983) exemplified the concept by illustrating in great detail the ichnofabric of the 15-cm-thick Kjølby Gård Marl, a thoroughly bioturbated marly chalk layer in the uppermost Cretaceous of western Denmark. They wrote that ichnofabric includes "those aspects of the texture and internal structure of the bed resulting from all phases of bioturbation" (Ekdale and Bromley, 1983: 110). In the glossary of an SEPM short course text on ichnology, they further defined ichnofabric as "all aspects of the texture and internal structure of a sediment that result from bioturbation and bioerosion at all scales; includes both bioturbation fabric and bioerosion fabric" (Ekdale et al., 1984a: 308).

From that point, the practical application of the ichnofabric concept ramified in several different, complementary directions. Some ichnofabrics may be thought of as *simple ichnofabrics* in cases where they contain just one type of trace fossil. In certain cases, the ichnofabric consists of a single ichnotaxon superimposed on primary stratification with portions of the original sedimentary laminae still discernible behind the trace fossils (Fig. 1). In other cases, the entire bed is totally bioturbated with only a single ichnotaxon evident in

FIGURE 1 Monoichnospecific ichnofabrics in incompletely bioturbated sediment, as depicted in these two examples, often reflect high-energy depositional environments, such as intertidal settings (as in A) or storm-related deposits (as in B). In such situations, the interplay of sedimentation, bioturbation, and erosional processes is directly reflected in the resultant ichnofabrics. (A) *Skolithos* ichnofabric in the Watson Ranch Quartzite, Lower Ordovician, Confusion Range, Millard County, Utah. (B) *Ophiomorpha* ichnofabric alternating with low-angle, cross-stratified beds in a "lam-scram" succession (after Ekdale, 1985a) in the Blackhawk Formation, Upper Cretaceous, Wasatch Plateau, Carbon County, Utah.

the resultant ichnofabric (Fig. 2). Monoichnospecific ichnofabrics can yield important insights concerning the response of benthic organisms to sedimentary dynamics in high-energy depositional settings (de Gibert and Goldring, 2007; Droser and Bottjer, 1989; Nara, 1997, 2002).

Often, however, the situations are not so simple, as in the creation of *composite ichnofabrics* by the superimposition of different (successive) suites of biogenic structures. Commonly this occurs when a vertically tiered arrangement of different types of burrows occurs in unconsolidated sediment (Bromley and Ekdale, 1986a). That idea was expanded upon by Wetzel and Aigner (1986), who likened a composite ichnofabric to a measuring stick for understanding the magnitude and sequence of depositional and erosional events reflected in a sedimentary deposit. Ekdale and Bromley (1991) further exemplified the concept of composite ichnofabrics by illustrating the tiered structure of ichnocoenoses in Late Cretaceous pelagic chalks of Denmark, where "detailed ichnofabric analysis reveals over a dozen successive burrowing episodes in the Danish chalk, testifying to the fact that the original ooze passed through guts of perhaps hundreds of organisms before it lithified" (Ekdale and Bromley, 1991: 232).

Another, slightly different application of the ichnofabric concept involved the development of ichnofabric indices to describe the overall intensity of bioturbation that had affected a sedimentary deposit (Droser and Bottjer, 1986, 1989; see Knaust, 2012). This approach followed earlier (pre-"ichnofabric") attempts to categorize bioturbated sediment on the basis of how much sediment had been disturbed by organisms (Reineck, 1967), as well as the simple recognition that sediment had been burrowed without resulting in preservation of

FIGURE 2 *Thalassinoides* ichnofabrics. In both cases, the intense burrowing by infaunal shrimps has completely churned the sediment, yielding a totally bioturbated—but monoichnospecific—ichnofabric (*ii* = 5; BI = 6). (A) Modern *Thalassinoides* ichnofabric, produced by burrowing decapod crustaceans (*Upogebia* sp.) in high-intertidal sediment at Estero Morua, a macrotidal flat on the coast of the northern Gulf of California near Puerto Peñasco, Sonora, Mexico. (B) *Thalassinoides* ichnofabric in Pleistocene intertidal rock in a coastal outcrop in the same area as (A).

individually distinguishable burrows, sometimes referred to as "bioturbate texture" (Frey, 1973).

The role of ichnofabric analysis in understanding paleo-oxygen conditions within the sediment has been explored and outlined by a large number of workers (Bromley and Ekdale, 1984a; Ekdale, 1985b; Ekdale and Mason, 1988; Leszczyński, 1991; Savrda and Bottjer, 1986, 1989a,b, 1994; Savrda and Ozalas, 1993; Wignall, 1991, 1993). This application of ichnofabric recognition is very important for several reasons, including the enhancement of our understanding of paleoecological adaptations of certain benthic organisms to low-oxygen environments, tectonic development of silled marine basins, and generation and accumulation of hydrocarbons in organic-rich sedimentary units. Currently, there is a strong economic emphasis on exploration for unconventional hydrocarbon resources in black shale strata, which frequently exhibit characteristic, *Chondrites*-dominated, oxygen-influenced ichnofabrics (Bromley and Ekdale, 1984a; Savrda and Bottjer, 1989a; Schieber, 2003).

Trace-fossil taphonomy contributes to the development of ichnofabrics. Bromley and Ekdale (1984b, 1986a,b) recognized the role of ichnofabric in influencing the early diagenetic preservation of trace fossils, particularly in the case of burrow flints in pelagic chalk deposits (Bromley and Ekdale, 1984b, 1986a) and also in early stage pressure-solution seams in fine-grained carbonates (Ekdale and Bromley, 1988).

In 1990, Pollard and Ekdale convened the first international symposium with ichnofabric as the focal point at the International Sedimentological Congress in Nottingham, England (Ekdale and Pollard, 1991). Participants presented a diverse array of seminal ideas of how ichnofabric approaches can aid in describing the biogenic aspects of various sedimentological situations and solving particular sedimentological problems. A dedicated theme issue dealing with ichnofabrics in the journal *Palaios* (1991, volume 6, number 3) stemmed directly from this symposium. Several papers in this issue expanded upon the description of local ichnofabrics to demonstrate their use in regional studies, such as in stratigraphic correlation (Mortimore and Pomerol, 1991), basin dynamics (Droser and Bottjer, 1991), and petroleum exploration (Bockelie, 1991).

Also as a result of the symposium in Nottingham, an ongoing series of International Ichnofabric Workshops (IIW) was initiated in order to bring together diverse views for advancement of ichnofabric investigations. The 1st IIW was held in Norway in 1991, the 2nd in the United States (Utah) in 1993, the 3rd in Denmark in 1995, the 4th in the Bahamas in 1997, the 5th in England in 1999, the 6th in Venezuela in 2001, the 7th in Switzerland in 2003, the 8th in New Zealand in 2005, the 9th in Canada in 2007, the 10th in China in 2009, and the 11th in Spain in 2011. These workshops have grown steadily in size and diversity over time, and all indications are that they will continue on a regular basis for many years to come.

3. A CONTROVERSIAL CONCEPT?

The ichnofabric concept has not been without controversy. Just as there are varying approaches to the application of the concept, there have been differing opinions about the validity and/or appropriateness of the concept. For example, Frey and Pemberton (1990, 1991) even objected to the word itself, arguing that "ichnofabric" is an inappropriate term in an etymological sense and that it is simply synonymous with the previously proposed term "bioturbate texture". Ekdale et al. (1991) responded in defense, arguing that there is no etymological problem with the word and that the "ichnofabric" concept encompasses far more ichnological aspects than can be described as "bioturbate texture". In the long run, "ichnofabric" has survived and is currently in widespread use in the literature, whereas "bioturbate texture" is employed infrequently.

Another controversy has been the employment of three competing index schemes to describe the intensity of bioturbation that has occurred in a sediment (Knaust, 2012). Droser and Bottjer (1986) introduced the rank scale of "ichnofabric indices" (*ii*), which spans from $ii=1$ (no bioturbation) to $ii=5$ (completely bioturbated). They created a visual scale for various facies, including shallow-marine carbonates, *Skolithos*-rich sandstones, *Ophiomorpha*-rich sandstones, and pelagic deposits (Droser, 1991). This method has proven useful in field observations because simple *ii* flash cards can be carried easily and used for visual assessment of the *ii* of successive beds in outcrops and cores.

As an alternative, some other workers (Taylor and Goldring, 1993; Taylor et al., 2003) have championed the use of "bioturbation indices" (BI), which span a scale of $\mathrm{BI}=0$ (no bioturbation) to $\mathrm{BI}=6$ (complete bioturbation). Although the BI method is very similar to the *ii* method, confusion arises, because the index numbers are slightly different in the amount of bioturbation that they represent. Some workers prefer the BI approach because it seems intuitive that zero bioturbation should have an index value of zero. But other workers prefer the *ii* approach because assigning the first level (no bioturbation) with the integer one allows for numerical manipulations of the data that avoid the problem of a zero in a denominator. Thus, *ii* data can be used to summarize intensity of bioturbation in vertical successions via statistical techniques.

Since both the *ii* and BI approaches are applied in vertical faces in outcrops and cores, a third index was proposed by Miller and Smail (1997) to assess the amount of burrowing on a horizontal plane. They developed the "bedding plane bioturbation index" (BPBI) with a scale extending from $\mathrm{BPBI}=0$ (no horizontal burrows) to $\mathrm{BPBI}=5$ (full coverage of a bedding plane by horizontal burrows). Like the *ii* method of Droser and Bottjer (1986, 1989), Miller and Smail (1997) created BPBI flash cards for easy use in the field. In order to ensure objectivity, the categories of successive index numbers were constructed using an image analysis program that calculates coverage (in %) of a surface by burrows. Some workers employ BPBI with success, but other workers have not embraced the scheme, because it offers little of sedimentological significance, since it addresses situations with no appreciable vertical (stratigraphic) mixing.

A third controversial topic in the realm of ichnofabric involves differing ideas of the role of biogenic activity in influencing, or even controlling, early diagenetic processes in the sediment. Virtually, all ichnologists agree that diagenetic features should be shunned in ichnotaxonomic and paleoethological studies of trace fossils. However, sedimentary rocks displaying an ichnofabric also commonly display a diagenetic fabric, which may have developed along a pathway outlined by burrows and other ichnological attributes of the original sediment. A diagenetically induced ichnofabric may include differential cementation and/or mineral replacement of burrow walls or burrow-fill sediment. The geological record is replete with nodular fabrics and mineralized burrows, where the resultant diagenetic fabric reflects and enhances the ichnofabric, particularly in carbonate sediments (Bromley, 1967; Bromley and Ekdale, 1986b; Kennedy and Garrison, 1975; Knaust et al., 2012). Caution is warranted, of course, because some types of non-biogenic features resulting from diagenetic processes (including compaction and dissolution phenomena) may be mistaken for biologically generated ichnofabrics.

4. IMPORTANCE OF ICHNOFABRIC

Ichnofabric, at its center, pertains to the potential interpretations one can make from the mode and style of preservation of trace fossils and other evidence of bioturbation and bioerosion. In recent years, ichnofabric has been compared and contrasted with ichnofacies and ichnocoenoses by some workers, sometimes confusing their definitions and applications. In a way, they (ichnofabrics, ichnofacies, and ichnocoenoses) simply are three different approaches to interpreting the same thing. In a strict sense, ichnofacies analysis is centered on interpreting physical aspects of the depositional environment (e.g., bathymetry, salinity, oxygenation, substrate character, etc.) based on the association of the identifiable trace fossils that occur. Some workers apply the ichnofacies concept more broadly to include ichnological aspects that go beyond the recognition of trace fossils (MacEachern et al., 2007; Pemberton et al., 2004), but the principal objective of ichnofacies analysis nevertheless has a distinctly paleoenvironmental focus, whereas the scope of ichnofabric analysis is much broader (Buatois and Mángano, 2011; Gerard and Bromley, 2008; Gingras et al., 2011; McIlroy, 2007, 2008).

In a strict sense, ichnocoenosis studies are centered on interpreting paleoecological aspects of the benthic community based on the wide range of results of their burrowing and bioeroding activities. The ichnoguild approach, championed by Bromley (1996), contributes significantly to such studies. Ichnoguild offers a combined view of the microhabitat within the substrate, especially in terms of infaunal tiering (Bromley and Ekdale, 1986a) or ecological stratification (Seilacher, 1978), and the ecological niche (i.e., functional role within the community). In a sense, an ichnoguild is an amalgam of the ichnological clues to what type of food a group of organisms ate in the context of where and how those organisms obtained their food. Thus, the ichnoguild concept lies at the heart of analyzing ichnocoenoses.

Although there are obvious overlaps of ichnofabrics with ichnofacies and ichnocoenoses, there is one important area of ichnology that is solely the domain of ichnofabric, namely, trace-fossil diagenesis. Diagenetic processes have nothing to do with ichnocoenoses and are not appropriate concerns of ichnofacies, but diagenetic processes are influenced by—and in some cases controlled by—the sediment fabric, including those aspects of the sediment fabric that are biogenic in origin (Knaust et al., 2012).

A very important application of ichnofabric studies involves the effects of reworking on sediment porosity and permeability. As burrowers move through the sediment, grain packing typically is disturbed, cohesiveness is reduced, and porosity may be increased. Of course, a change in permeability naturally results, especially if clay redistribution inside the burrow reduces the continuity that would be provided in cleaner sediment. In some cases, the mean (and/or the maximum) grain size is decreased by means of the ingestion, digestion, and egestion activities of sediment feeders. In situations where the sediment consistency is firm, burrow-fill material generally is more loosely packed and more porous than the surrounding host material; so, the burrows serve as permeability conduits (Cunningham et al., 2012; Gingras et al., 1999, 2004, 2012). This effect is clearly manifested where burrows are preferentially cemented (or alternatively uncemented) and/or mineralized (dolomitized or chertified) relative to the host rock (Fig. 3).

FIGURE 3 These two Paleozoic examples display elite burrows that have been accentuated by preferential mineralization of the latest (and presumably) deepest burrows. (A) "*Thalassinoides*" (or probably *Balanoglossites*) ichnofabric illustrating preferentially dolomitized burrows in the Marjum Formation, Middle Cambrian, House Range, Millard County, Utah. (B) *Thalassinoides* ichnofabric illustrating preferentially silicified (chertified), large *Thalassinoides* burrows in the Great Blue Limestone, Mississippian (Lower Carboniferous), South Lakeside Mountains, Tooele County, Utah.

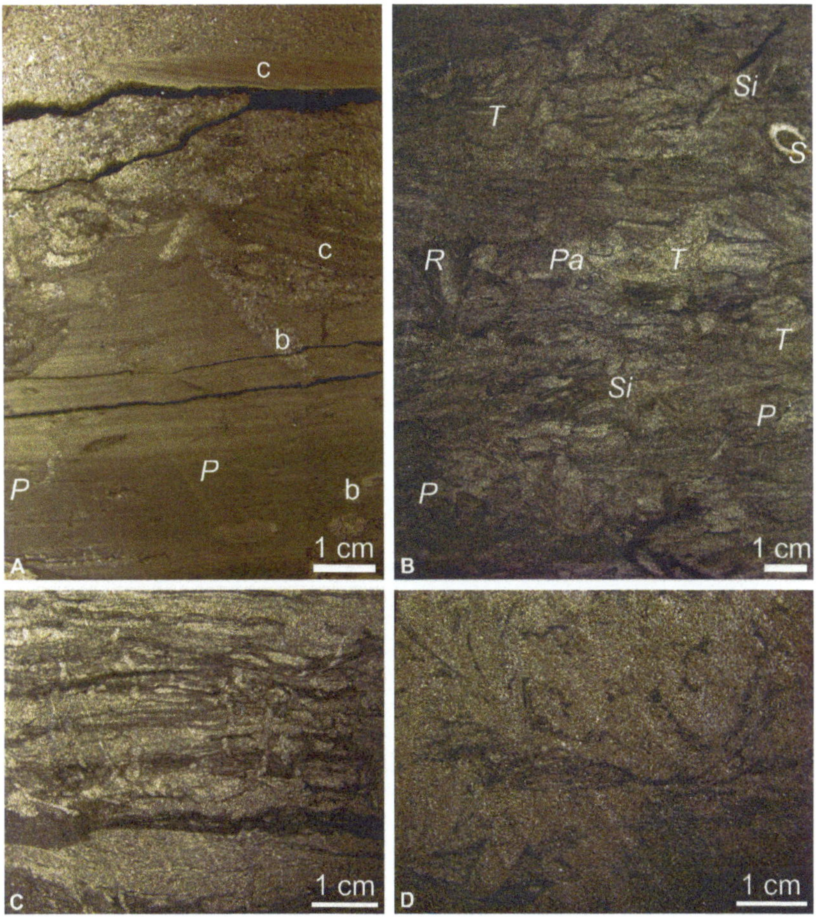

FIGURE 4 Ichnofabrics in vertically slabbed cores from wells in the marginal-marine Cook Formation (Lower Jurassic), offshore Norway (Block 34/2 and 35/10). (A) Laminated siltstone with sparse *Phycosiphon* (*P*), sharply eroded at its top and overlain by coarse-grained sandstone with rip-up clasts (c). This erosion surface occurs regionally and is associated with deeply penetrating, sand-filled, uncompacted burrows (b) indicative of a firmground omission surface. (B) Complex, intensely bioturbated ichnofabric, including *Siphonichnus* (*Si*), *Palaeophycus* (*Pa*), *Teichichnus* (*T*), *Rosselia* (*R*), *Phycosiphon* (*P*), and *Schaubcylindrichnus* (*S*). This interval is interpreted as a lower shoreface deposit. (C) Heterolithic, ripple-laminated sandstone with monoichnospecific occurrence of minute *Siphonichnus* as indicative of a restricted environment (e.g., decreased salinity) on a tidal flat. (D) Sandstone with the complex trace fossil *Hillichnus*, probably resulting from the deep-infaunal activity of a tellinacean bivalve.

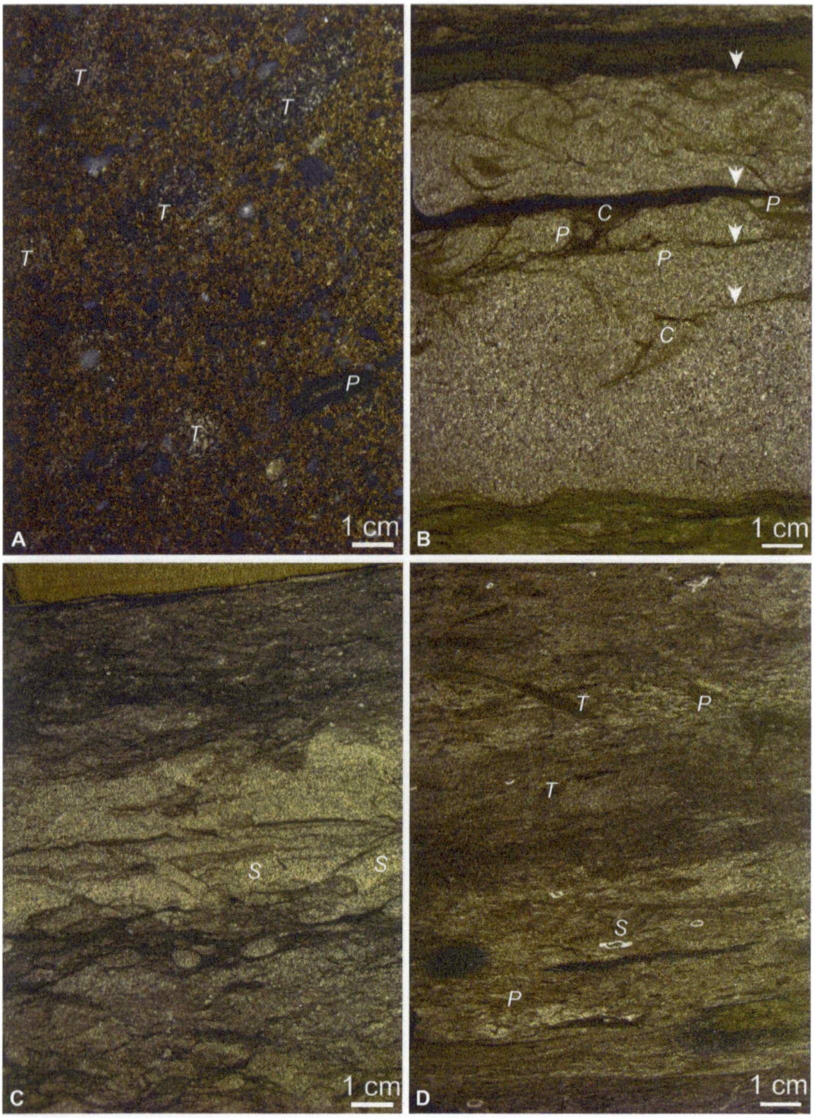

FIGURE 5 Ichnofabrics in vertically slabbed cores from wells in the Fram Field area, offshore Norway (Block 35/11), containing Upper Jurassic (Oxfordian) shelf turbidite deposits. (A) Gravel, deposited on channelized fans, with abundant sand-filled *Thalassinoides* (*T*) and an thickly lined *Palaeophycus* ichnoclast (*P*) in a deep-tier colonization. (B) Proximal turbidite deposit, with internal subdivision that can be recognized by multiple colonization surfaces (arrows). *Cylindrichnus* (*C*) occurring as a discrete trace fossil, while the upper sand layer shows a mottled fabric (mixed zone). The muddy burrow fills and drapes include *Palaeophycus* (*P*). (C) Thin-bedded turbidite deposit, dominated by *Scolicia* (*S*), interbedded with intensely bioturbated heterolithic shelf sandstone containing minute *Planolites*, *Palaeophycus*, and *Phycosiphon*. (D) Completely bioturbated sandy claystone with *Phycosiphon* (*P*), *Schaubcylindrichnus* (*S*), and *Teichichnus* (*T*).

FIGURE 6 Ichnofabrics in slabbed cores from wells in the Vøring Basin, offshore Norway (Block 6706/12 and 6707/10), containing deep-marine fan deposits of the Santonian Kvitnos Formation (B, C) and the Campanian Nise Formation (A, D). (A) Top of a sandy turbidite unit with complete (upper part) to moderate (lower part) bioturbation, consisting of monoichnospecific, irregularly branched, walled, and meniscate burrows indicative of cf. *Ophiomorpha* (or *Keckia*) (45° deviated well). (B) Highly bioturbated (mottled) heterolithic sandstone with relict microbial mats (laminated chips, l), discrete spreiten burrows (s, some of which preserve the cast of their producer as a bright spot, p), *Palaeophycus* (*Pa*), and *Planolites* (*Pl*). (C) Highly bioturbated sandstone with fragmented microbial mats (laminated chips), fractured carbonate concretion (lower right), and calcitic bioclasts (bright spots). (D) Thinly bedded sandstone to siltstone with low bioturbation consisting of *Taenidium* (thick backfilled burrow) and *Zoophycos* (thin spreiten) (45° deviated well).

FIGURE 7 Ichnofabrics in slabbed well cores from the Oseberg Field, offshore Norway (Block 30/9), with deep-marine, redeposited Shetland Chalk (Upper Maastrichtian–Danian). The wavy to nodular fabric is enhanced by the abundant occurrence of dissolution seams. UV light (A, B) and white light (C, D) (~60° deviated well). (A) The overall degree of bioturbation is high (approx. 85%), and the ichnofabric comprises a monoichnospecific suite of branched and partly lined networks with a meniscate or passive fill assigned to *Thalassinoides*/*Ophiomorpha*, probably produced by burrowing shrimp. Two colonization phases can be recognized: (1) predeformation ichnofabric consisting of relatively homogeneous, oil-stained chalk (yellow color) of the background sediment, and (2) post-deformation ichnofabric resulting from subsequent massive colonization, which led to the destruction and *in-situ* brecciation of the primary sedimentary fabric and the incorporation of gray mud. (B) Higher up in the succession, sediment destruction during the second colonization phase is increased and contributes to subsequent sediment reworking, which again leads to considerably reduced reservoir properties. (C) Mass-transport complex consisting of debrite units (D) with well-rounded chalk intraclasts (some with bioerosion traces, arrows) and larger fragments of semiconsolidated rocks. The rock fragment in the upper portion of the core shows a gradual transition from original soft (S) to firm (F) sediment and preserves a diffuse ichnofabric with *Planolites*, *Palaeophycus*, and *Chondrites* from the pelagic background. The debris-flow deposit is weakly bioturbated (~15%) and only comprises discrete *Thalassinoides*/*Ophiomorpha* filled with gray mud and partly incorporated chalk clasts. (D) Detail of (C), showing a *Thalassinoides* burrow with an active (meniscate) fill in the debrite and softground (left and middle portion of the image), gradually changing to a passive fill in the firmground (right portion of the image).

This post-depositional aspect of ichnofabric obviously is of considerable interest to exploration geologists and hydrogeologists because it demonstrates the direct influence that biogenic processes have on the migration and pooling of fluids in porous media. Since most Phanerozoic sedimentary rocks exhibit an ichnofabric of one sort or another, it follows that ichnofabric data must be a component in the characterization and evaluation of oil, gas, and water reservoirs. This is especially true when examining drill cores for hydrocarbon exploration. Ichnofabrics are commonly described in core (Bockelie, 1991; Gerard and Bromley, 2008; Knaust, 2009, 2010; Martin and Pollard, 1996; McIlroy, 2007; Taylor and Goldring, 1993), whereas specific ichnotaxa are not always recognized with ease and identified with certainty in core.

Given the expense of recovering drill cores from the subsurface, and in the light of the invaluable information they can provide to the exploration or reservoir geologist, ichnological analysis is a crucial part in the description and interpretation of well cores (see Knaust, 2012). A few selected examples of common ichnofabrics in typical marine reservoirs from offshore Norway illustrate common ichnofabrics and their appearance in slabbed cores and demonstrate their value for reservoir characterization (Figs. 4–7).

5. CONCLUSIONS

The ichnofabric concept and application of the techniques of ichnofabric analysis that derive from the concept have proven to be robust and resilient. Sedimentologically, ichnofabric can be used as a way to assess the degree or intensity of bioturbation. It is also a direct reflection of sediment consistency and sedimentary dynamics within the depositional environment, and as such it may be a crucial tool in interpreting event sedimentation. Geochemically, the dynamics of interstitial oxygen concentrations will be reflected in different ichnofabrics. Paleoecologically, ichnofabric provides a direct reflection of the endobenthic community, including the vertical tiering structure of the ichnocoenosis. Taphonomically, ichnofabric ostensibly contributes to the diagenetic fabric that results from lithification of a burrowed sediment. Stratigraphically, recurrent ichnofabrics can be used effectively for correlation in certain situations within sedimentary basins. Practically, the analysis of ichnofabrics can be used as a powerful describer of reservoirs.

ACKNOWLEDGMENTS

The authors' ideas about ichnofabrics have benefited considerably from the shared thoughts and insights of many astute colleagues over the past three decades. Much of the early work on ichnofabrics by A. A. E. was supported by research grants from the National Science Foundation. D. K. is grateful to Statoil for permission to publish the data presented in Figs. 4–7. The review of Jean Gerard greatly improved this chapter.

REFERENCES

Baucon, A., Bordy, E., Brustur, T., Buatois, L.A., Cunningham, T., De, C., Duffin, C., Felletti, F., Gaillard, C., Hu, B., Hu, L., Jensen, S., Knaust, D., Lockley, M., Lowe, P., Mayor, A., Mayoral, E., Mikuláš, R., Muttoni, G., Neto de Carvalho, C., Pemberton, S.G., Pollard, J., Rindsberg, A.K., Santos, A., Seike, K., Song, H., Turner, S., Uchman, A., Wang, Y., Yi-ming, G., Zhang, L., Zhang, W., 2012. A history of ideas in ichnology. In: Knaust, D., Bromley, R.G. (Eds.), Trace Fossils as Indicators of Sedimentary Environments. Developments in Sedimentology, vol. 64. Elsevier, Amsterdam, pp. 3–43.

Berger, W.H., Heath, G.R., 1968. Vertical mixing in pelagic sediments. J. Mar. Res. 26, 134–143.

Berger, W.H., Ekdale, A.A., Bryant, P.F., 1979. Selective preservation of burrows in deep-sea carbonates. Mar. Geol. 32, 205–230.

Bockelie, J.F., 1991. Ichnofabric mapping and interpretation of Jurassic reservoir rocks of the Norwegian North Sea. Palaios 6, 206–215.

Bromley, R.G., 1967. Some observations on burrows of thalassinidean Crustacea in chalk hardgrounds. Q. J. Geol. Soc. Lond. 123, 157–182.

Bromley, R.G., 1996. Trace Fossils: Biology, Taphonomy and Applications. Chapman and Hall, London, 361 pp.

Bromley, R.G., Ekdale, A.A., 1984a. *Chondrites*: a trace fossil indicator of anoxia in sediments. Science 224, 872–874.

Bromley, R.G., Ekdale, A.A., 1984b. Trace fossil preservation in flint in the European chalk. J. Paleontol. 58, 298–311.

Bromley, R.G., Ekdale, A.A., 1986a. Composite ichnofabric and tiering of burrows. Geol. Mag. 123, 59–65.

Bromley, R.G., Ekdale, A.A., 1986b. Flint and fabric in the European chalk. In: de Sieveking, G., Hart, M.B. (Eds.), The Scientific Study of Flint and Chert. Cambridge University Press, Cambridge, pp. 71–82.

Buatois, L.A., Mángano, M.G., 2011. Ichnology. Organism-Substrate Interactions in Space and Time. Cambridge University Press, Cambridge, 358 pp.

Chamberlain, C.K., 1975. Trace fossils in DSDP cores of the Pacific. J. Paleontol. 49, 1074–1096.

Cunningham, K.J., Sukop, M.C., Curran, H.A., 2012. Carbonate aquifers. In: Knaust, D., Bromley, R.G. (Eds.), Trace Fossils as Indicators of Sedimentary Environments. Developments in Sedimentology, vol. 64. Elsevier, Amsterdam, pp. 869–896.

de Gibert, J.M., Goldring, R., 2007. An ichnofabric approach to the depositional interpretation of the intensely burrowed Bateig Limestone, Miocene, SE Spain. Sediment. Geol. 194, 1–16.

Droser, M.L., 1991. Ichnofabric of the Paleozoic *Skolithos* Ichnofacies and the nature and distribution of *Skolithos* piperock. Palaios 6, 316–325.

Droser, M.L., Bottjer, D.J., 1986. A semiquantitative field classification of ichnofabric. J. Sediment. Petrol. 56, 558–559.

Droser, M.L., Bottjer, D.J., 1989. Ichnofabric of sandstones deposited in high-energy nearshore environments: measurement and utilization. Palaios 4, 598–604.

Droser, M.L., Bottjer, D.J., 1991. Ichnofabric and basin analysis. Palaios 6, 199–205.

Ekdale, A.A., 1977. Trace fossils in worldwide Deep Sea Drilling Project cores. In: Crimes, T.P., Harper, J.C. (Eds.), Trace Fossils 2. Geol. J., Spec. Iss. 9, pp. 163–182.

Ekdale, A.A., 1978. Trace fossils in Leg 42A cores. In: Hsü, K.J., Montadert, L. et al., (Eds.), Initial Reports of the Deep Sea Drilling Project 42, pp. 821–827.

Ekdale, A.A., 1985a. Paleoecology of the marine endobenthos. Palaeogeogr. Palaeoclimatol. Palaeoecol. 50, 63–81.

Ekdale, A.A., 1985b. Trace fossils and Mid-Cretaceous anoxic events in the Atlantic Ocean. In: Curran, H.A. (Ed.), Biogenic Structures: Their Use in Interpreting Depositional Environments. SEPM Spec. Publ., vol. 35, pp. 334–342.

Ekdale, A.A., Berger, W.H., 1978. Deep-sea ichnofacies: modern organism traces on and in pelagic carbonates of the western equatorial Pacific. Palaeogeogr. Palaeoclimatol. Palaeoecol. 23, 263–278.

Ekdale, A.A., Bromley, R.G., 1983. Trace fossils and ichnofabric in the Kjølby Gård Marl, uppermost Cretaceous, Denmark. Bull. Geol. Soc. Denmark 31, 107–119.

Ekdale, A.A., Bromley, R.G., 1988. Diagenetic microlamination in chalk. J. Sediment. Petrol. 58, 857–861.

Ekdale, A.A., Bromley, R.G., 1991. Analysis of composite ichnofabrics: an example in uppermost Cretaceous chalk of Denmark. Palaios 6, 232–249.

Ekdale, A.A., Mason, T.R., 1988. Characteristic trace fossil assemblages in oxygen-poor sedimentary environments. Geology 16, 720–723.

Ekdale, A.A., Pollard, J.E., 1991. Ichno-this and ichno-that. Palaios 6, 197–198.

Ekdale, A.A., Bromley, R.G., Pemberton, S.G., 1984a. Ichnology. The Use of Trace Fossils in Sedimentology and Stratigraphy. SEPM Short Course 15, 317 pp.

Ekdale, A.A., Muller, L.N., Novak, M.T., 1984b. Quantitative ichnology of modern pelagic deposits in the abyssal Atlantic. Palaeogeogr. Palaeoclimatol. Palaeoecol. 45, 189–223.

Ekdale, A.A., Bromley, R.G., Bockelie, J.F., Droser, M.L., Bottjer, D.J., 1991. Ichnofabric it is! Palaios 6, 100–101.

Frey, R.W., 1973. Concepts in the study of biogenic sedimentary structures. J. Sediment. Petrol. 43, 6–19.

Frey, R.W., Pemberton, S.G., 1990. Bioturbate texture or ichnofabric? Palaios 5, 385–386.

Frey, R.W., Pemberton, S.G., 1991. Or is it bioturbate texture? Ichnos 1, 327–329.

Gerard, J.R.F., Bromley, R.G., 2008. Ichnofabrics in Clastic Sediments: Applications to Sedimentological Core Studies. Jean R.F. Gerard, Madrid, 97 pp.

Gingras, M.K., Pemberton, S.G., Mendoza, C.A., Henk, F., 1999. Assessing the anisotropic permeability of *Glossifungites* surfaces. Petrol. Geosci. 5, 349–357.

Gingras, M.K., Pemberton, S.G., Muehlenbachs, K., Machel, H.G., 2004. Conceptual models for burrow-related, selective dolomitization with textural and isotopic evidence from the Tyndall Limestone. Geobiology 2, 21–30.

Gingras, M.K., MacEachern, J.A., Dashtgard, S.E., 2011. Process ichnology and the elucidation of physico-chemical stress. Sediment. Geol. 237, 115–134.

Gingras, M.K., Baniak, G., Gordon, J., Hovikoski, J., Konhauser, K.O., La Croix, A., Lemiski, R., Mendoza, C., Pemberton, S.G., Polo, C., Zonneveld, J.-P., 2012. Porosity and permeability in bioturbated sediments. In: Knaust, D., Bromley, R.G. (Eds.), Trace Fossils as Indicators of Sedimentary Environments. Developments in Sedimentology, vol. 64. Elsevier, Amsterdam, pp. 835–868.

Guinasso, N.L., Schink, D.R., 1975. Quantitative estimates of biological mixing rates in abyssal sediments. J. Geophys. Res. 80, 3032–3043.

Howard, J.D., Elders, C.A., 1970. Burrowing patterns of haustoriid amphipods from Sapelo Island, Georgia. In: Crimes, T.P., Harper, J.C. (Eds.), Trace Fossils. Geol. J., Spec. Iss. 3, pp. 243–260.

Howard, J.D., Frey, R.W., 1975. Estuaries of the Georgia Coast, U.S.A.: sedimentology and biology. II. Regional animal-sediment characteristics of Georgia Estuaries. Senckenberg. Marit. 7, 33–103.

Kennedy, W.J., Garrison, R.E., 1975. Morphology and genesis of nodular chalks and hardgrounds in the Upper Cretaceous of southern England. Sedimentology 22, 311–386.

Knaust, D., 2009. Ichnology as a tool in carbonate reservoir characterization: a case study from the Permian–Triassic Khuff Formation in the Middle East. GeoArabia 14, 17–38.

Knaust, D., 2010. The end-Permian mass extinction and its aftermath on an equatorial carbonate platform: insights from ichnology. Terra Nova 22, 195–202.

Knaust, D., 2012. Methodology and techniques. In: Knaust, D., Bromley, R.G. (Eds.), Trace Fossils as Indicators of Sedimentary Environments. Developments in Sedimentology, vol. 64. Elsevier, Amsterdam, pp. 245–271.

Knaust, D., Curran, H.A., Dronov, A., 2012. Shallow-marine carbonates. In: Knaust, D., Bromley, R.G. (Eds.), Trace Fossils as Indicators of Sedimentary Environments. Developments in Sedimentology, vol. 64. Elsevier, Amsterdam, pp. 703–750.

Leszczyński, S., 1991. Oxygen-related controls on predepositonal ichnofacies in turbidites and hemipelagites, Guipúzoan flysch (Albian-Lower Eocene), northern Spain. Palaios 6, 271–280.

MacEachern, J.A., Bann, K.L., Pemberton, S.G., Gingras, M.K., 2007. The ichnofacies paradigm: high-resolution paleoenvironmental interpretation of the rock record. In: MacEachern, J.A., Bann, K.L., Gingras, M.K., Pemberton, S.G. (Eds.), Applied Ichnology. SEPM Short Course 52, pp. 27–64.

MacEachern, J.A., Bann, K.L., Gingras, M.K., Zonneveld, J.-P., Dashtgard, S.E., Pemberton, S.G., 2012. The ichnofacies paradigm. In: Knaust, D., Bromley, R.G. (Eds.), Trace Fossils as Indicators of Sedimentary Environments. Developments in Sedimentology, vol. 64. Elsevier, Amsterdam, pp. 103–138.

Martin, M.A., Pollard, J.E., 1996. The role of trace fossil (ichnofabric) analysis in the development of depositional models for the Upper Jurassic Fulmar Formation of the Kittiwake Field (Quadrant 21 UKCS). In: Hurst, A. (Ed.), Geology of the Humber Group, Central Graben and Moray Firth, UKCS. Geol. Soc. London, Spec. Publ., vol. 114, pp. 163–183.

McIlroy, D., 2007. Lateral variability in shallow marine ichnofabrics: implications for the ichnofabric analysis method. J. Geol. Soc. Lond. 164, 359–369.

McIlroy, D., 2008. Ichnological analysis: the common ground between ichnofacies workers and ichnofabric analysts. Palaeogeogr. Palaeoclimatol. Palaeoecol. 270, 332–338.

Miller, M.F., Smail, S., 1997. A semiquantitative field method for evaluating bioturbation on bedding planes. Palaios 12, 391–396.

Mortimore, R.N., Pomerol, B., 1991. Stratigraphy and eustatic implications of trace fossil events in the Upper Cretaceous chalk of northern Europe. Palaios 6, 216–231.

Nara, M., 1997. High-resolution analytical method for event sedimentation using *Rosselia socialis*. Palaios 12, 489–494.

Nara, M., 2002. Crowded *Rosselia socialis* in Pleistocene inner shelf deposits: benthic paleoecology during rapid sea-level rise. Palaios 17, 268–276.

Pemberton, S.G., MacEachern, J.A., Saunders, T., 2004. Stratigraphic applications of substrate-specific ichnofacies: delineating discontinuities in the fossil record. In: McIlroy, D. (Ed.), The Application of Ichnology to Palaeoenvironmental and Stratigraphic Analysis. Geol. Soc. London, Spec. Publ., vol. 228, pp. 29–62.

Reineck, H.-E., 1958. Wühlbau-Gefüge in Abhängigkeit von Sediment-Umlagerungen. Senck. Leth. 39, 1–23; 54–56.

Reineck, H.-E., 1963. Sedimentgefüge im Bereich der südlichen Nordsee. Abh. Senckenberg. Naturf. Ges. 505, 1–138.

Reineck, H.-E., 1967. Parameter von Schichtung und Bioturbation. Geol. Rundsch. 56, 420–438.

Reineck, H.-E., Gutmann, W., Hertweck, G., 1967. Das Schlickgebiet südlich Helgoland als Beispiel rezenter Schelfablagerungen. Senck. Leth. 48, 219–275.

Savrda, C.E., Bottjer, D.J., 1986. Trace-fossil model for reconstruction of paleo-oxygenation in bottom waters. Geology 14, 3–6.

Savrda, C.E., Bottjer, D.J., 1989a. Anatomy and implications of bioturbated beds in "black shale" sequences: examples from the Jurassic Posidonienschiefer (southern Germany). Palaios 4, 330–342.

Savrda, C.E., Bottjer, D.J., 1989b. Trace-fossil model for reconstructing oxygenation histories of ancient marine bottom waters: application to Upper Cretaceous Niobrara Formation, Colorado. Palaeogeogr. Palaeoclimatol. Palaeoecol. 74, 49–74.

Savrda, C.E., Bottjer, D.J., 1994. Ichnofossils and ichnofabrics in rhythmically bedded pelagic/hemipelagic carbonates: recognition and evaluation of benthic redox and scour cycles. IAS Spec. Publ. 19, 195–210.

Savrda, C.E., Ozalas, K., 1993. Preservation of mixed-layer ichnofabrics in oxygenation-event beds. Palaios 8, 609–613.

Schäfer, W., 1956. Wirkungen der Benthos-Organismen auf den jungen Schichtverband. Senck. Leth. 37, 183–263.

Schieber, J., 2003. Simple gifts and buried treasures—implications of finding bioturbation and erosion surfaces in black shales. Sediment. Rec. 1, 4–9.

Seilacher, A., 1964. Biogenic sedimentary structures. In: Imbrie, J., Newell, N. (Eds.), Approaches to Paleoecology. Wiley, New York, pp. 296–316.

Seilacher, A., 1967. Bathymetry of trace fossils. Mar. Geol. 5, 413–428.

Seilacher, A., 1978. Use of trace fossils for recognizing depositional environments. In: Basan, P.B. (Ed.), Trace Fossil Concepts. SEPM Short Course 5, pp. 175–201.

Taylor, A.M., Goldring, R., 1993. Description and analysis of bioturbation and ichnofabric. J. Geol. Soc. Lond. 150, 141–148.

Taylor, A.M., Goldring, R., Gowland, S., 2003. Analysis and application of ichnofabrics. Earth-Sci. Rev. 60, 227–259.

Van der Lingen, G.J., 1973. Ichnofossils in deep-sea cores from the southwest Pacific. In: Burns, R.E., Andrews, J.E. et al., (Eds.), Initial Reports of the Deep Sea Drilling Project 21, pp. 693–700.

Warme, J.E., Kennedy, W.J., Scheidermann, N., 1973. Biogenic sedimentary structures (trace fossils) in Leg 15 cores. In: Edgar, N.T., Saunders, J.B. et al., (Eds.), Initial Reports of the Deep Sea Drilling Project 15, pp. 813–831.

Wetzel, A., 1981. Ökologische und stratigraphische Bedeutung biogener Gefüge in quartären Sedimenten am NW-afrikanischen Kontinentalrand. "Meteor" Forsch.-Ergebn. Reihe C 34, 1–47.

Wetzel, A., 1983. Biogenic sedimentary structures in a modern upwelling region: northwest African continental margin. In: Thiede, J., Suess, E. (Eds.), Coastal Upwelling and Its Sediment Record. Part B, Sedimentary Records of Ancient Coastal Upwelling. Plenum Press, New York, pp. 123–144.

Wetzel, A., 1984. Bioturbation in fine-grained sediments: influence of sediment texture, turbidite frequency and rates of environmental change. Geol. Soc. London, Spec. Publ. 15, pp. 595–608.

Wetzel, A., 1985. Biogenic structures in modern slope to deep-sea sediments in the Sulu Sea Basin (Philippines). Palaeogeogr. Palaeoclimatol. Palaeoecol. 42, 285–304.

Wetzel, A., Aigner, T., 1986. Stratigraphic completeness: tiered trace fossils provide a measuring stick. Geology 14, 234–237.

Wignall, P.B., 1991. Dysaerobic trace fossils and ichnofabrics in the Upper Jurassic Kimmeridge Clay of southern England. Palaios 6, 264–270.

Wignall, P.B., 1993. Distinguishing between oxygen and substrate control in fossil benthic assemblages. J. Geol. Soc. Lond. 150, 193–196.

Chapter 6

Sequence Stratigraphy

James A. MacEachern,*[,1] Shahin E. Dashtgard,* Dirk Knaust,[†] Octavian Catuneanu,[‡] Kerrie L. Bann[§] and S. George Pemberton[‡]
*ARISE, Department of Earth Sciences, Simon Fraser University, Burnaby, British Columbia, Canada, [†]Statoil ASA, Stavanger, Norway, [‡]Department of Earth and Atmospheric Sciences, University of Alberta, Edmonton, Alberta, Canada, [§]Ichnofacies Analysis Inc., Calgary, Alberta, Canada
[1]Corresponding author: e-mail: jmaceach@sfu.ca

1. INTRODUCTION

Sequence stratigraphy can be regarded as a methodology that employs stratal stacking patterns and key bounding surfaces to erect a framework in which the depositional facies of sedimentary environments can be mapped and interpreted in the context of paleogeography (cf. Bhattacharya, 2011; Catuneanu et al., 2011). Implicit in the methodology is that changes in stratal stacking patterns are a response to changes in accommodation and sediment supply and that bounding surfaces represent depositional breaks or changes in depositional architecture. The concept of accommodation is central to sequence stratigraphy and indicates the amount of space available for sediments to fill (Jervey, 1988). Changes in accommodation are controlled by shifts in the relative sea level (RSL), which account for the interplay of eustasy and tectonism (Posamentier and Allen, 1999).

The focus of sequence stratigraphy, historically, has been on allogenically induced changes (e.g., eustasy, tectonics, climate), but more recent work has included autogenically produced changes as well (e.g., delta lobe switching). Additionally, unlike the analytical, objectively based stratigraphic methods (e.g., lithostratigraphy and biostratigraphy), sequence stratigraphy is recognized as "a genetic, process-based analytical approach to stratigraphic interpretation that of necessity involves conceptual depositional models" (Catuneanu et al., 2011: 177). Today, our understanding of stratigraphic architecture largely rests upon sequence-stratigraphic frameworks.

Sequence-stratigraphic methodology is designed to be employed at a number of different scales, and as a result, the datasets utilized vary from outcrop, core, geophysical well-logs, to seismic data. Ichnological data are largely restricted to core- and outcrop-scale observations and can be applied to sequence-stratigraphic problems at the same scale of observation as physical

Developments in Sedimentology, Vol. 64. http://dx.doi.org/10.1016/B978-0-444-53813-0.00006-X

sedimentology. The strong facies control on the character and distribution of trace-fossil suites allows this data to be integrated with physical sedimentology to yield more precise depositional interpretations of the sedimentary units. Like sedimentology, ichnology has been employed, for the most part, to assist in the identification and interpretation of bounding surfaces separating different stratal units (e.g., Ainsworth, 1994; Bann et al., 2004; Dam, 1990; Hillgärtner, 1998; MacEachern et al., 1992a,b, 1999a, 2007a,b; Pemberton et al., 1992a; Savrda, 1991a,b; Savrda et al., 2001a,b; Taylor and Gawthorpe, 1993).

The ichnological identification and interpretation of key bounding surfaces has rested on the recognition of (1) ichnologic omission suites, and (2) the juxtaposition of softground trace-fossil suites (cf. MacEachern et al., 1992b, 2007a,b; Pemberton and MacEachern, 1995; Pemberton et al., 2004). Ichnological omission focuses on substrate-controlled ichnofacies and palimpsest suites and has a long history of application to sedimentary successions (e.g., Bromley, 1975; Bromley and Goldring, 1992; Bromley et al., 1984; Dam, 1990; de Gibert and Robles, 2005; Fürsich et al., 1981; Hayward, 1976; Jones and Pemberton, 1989; Kobluk et al., 1977; Miller and Rehmer, 1982; Pemberton and Frey, 1985; Pemberton et al., 1980; Savrda, 1991a,b; Vossler and Pemberton, 1988). The latter approach addresses vertically and laterally juxtaposed trace-fossil associations that contravene Walther's Law and is best achieved through the integration with physical sedimentological facies analysis (e.g., Bann et al., 2004; Bhattacharya, 1993; Fielding et al., 2006, 2008; Ghibaudo et al., 1996; Hampson and Howell, 2005; Hampson and Storms, 2003; Hillgärtner, 1998; MacEachern and Pemberton, 1994; MacEachern et al., 1998, 1999a,b, 2007a,b; McIlroy, 2004, 2008; Mellere and Steel, 1995; Pattison and Walker, 1994, 1998; Pemberton and MacEachern, 1995, 2005; Pemberton et al., 1992a, 2001, 2004; Rodríguez-Tovar et al., 2007; Savrda, 1995; Savrda et al., 2001b; Taylor and Gawthorpe, 1993; Taylor et al., 2003).

Research continues to expand and refine the role of trace-fossil analysis in the sequence-stratigraphic evaluation of siliciclastic and carbonate successions. This chapter summarizes the ichnological signal of the key bounding surfaces of sequence-stratigraphic relevance. Three case studies are presented that showcase some of the ways in which ichnology can assist in the identification of these bounding surfaces and the interpretation of the successions.

1.1 Nomenclature, Key Bounding Surfaces, and Systems Tracts

The working group on sequence stratigraphy, appointed in 2008 by the International Subcommission on Stratigraphic Classification, provides guidelines concerning sequence-stratigraphic nomenclature and key bounding surfaces (Catuneanu et al., 2011).

The subaerial unconformity (SU; Sloss et al., 1949) is an unconformity that forms under subaerial conditions as a result of fluvial erosion or bypass, pedogenesis, wind degradation, or dissolution and karstification. This is considered the sequence boundary within the context of the "depositional sequence" model.

The regressive surface of marine erosion (RSME) *sensu* Plint (1988) constitutes an erosional surface that generally forms by wave scouring during forced regression in wave-dominated shallow-water settings as a consequence of lowered wave base.

The correlative conformity (CC) *sensu* Posamentier et al. (1988) is a marine stratigraphic surface that marks the change in stratal stacking patterns from highstand normal regression (HNR) to forced regression. As such, it corresponds to the sea floor at the *onset* of forced regression. Hunt and Tucker (1992) refer to this surface as the "basal surface of forced regression" (BSFR). By contrast, the CC *sensu* Hunt and Tucker (1992) is a marine stratigraphic surface marking the change in stratal stacking patterns from forced regression to lowstand normal regression and records the sea floor at the *end* of forced regression.

The maximum regressive surface (MRS; Helland-Hansen and Martinsen, 1996) is a stratigraphic surface marking a change in stratal stacking pattern from lowstand normal regression to transgression. The term "maximum regressive surface" is recommended where emphasis is placed on the *end* of regression, whereas the term "transgressive surface (TS)" is recommended where emphasis is placed on the *onset* of transgression.

Transgressive surface of erosion (TSE; Posamentier and Vail, 1988) is a general term referring to transgressive ravinement surfaces (Nummedal and Swift, 1987). Those that form by means of wave scouring are wave ravinement surfaces (WRSs; Swift, 1975). Those formed by tidal scouring are tidal ravinement surfaces (TRSs; Allen and Posamentier, 1993). Such surfaces are generated during transgression in coastal to shallow-water environments.

The maximum flooding surface (MFS; Galloway, 1989; Posamentier et al., 1988; Van Wagoner et al., 1988) marks the change in stratal stacking pattern from transgression to highstand normal regression. The surface is also commonly expressed as downlap by the overlying highstand systems tract (HST).

There are four systems tracts. The falling-stage systems tract (FSST; e.g., Ainsworth, 1994; Catuneanu, 2006; Plint and Nummedal, 2000) is generated by forced regression. The fall in RSL is indicated by erosion of the subaerially exposed sediment surface updip of the coastline at the end of forced regression and by the progradation of shallow-water deposits with a downstepping (offlapping) stacking pattern. The SU may be onlapped by fluvial deposits belonging to the lowstand or the transgressive systems tracts (TSTs). The SU may also be reworked locally by a diachronous TSE overlain by a sediment lag.

The lowstand systems tract (LST) corresponds to deposits that accumulate after the onset of RSL rise but during normal regression, capping the FSST and the corresponding SU lying updip. Stacking patterns exhibit forestepping, aggrading clinoforms that thicken downdip (in the case of siliciclastic systems), with a topset of fluvial, coastal plain and/or delta plain deposits.

The TST comprises the deposits that accumulate from the onset of transgression until maximum transgression of the coast, immediately prior to renewed normal regression during the HST. The TST directly overlies the MRS formed at the end of regression and is bounded at the top by the "MFS" formed when marine sediments reach their most landward position. Stacking patterns exhibit backstepping and onlapping, as well as retrogradational clinoforms that thicken landward in siliciclastic systems.

The HST encompasses progradational deposits formed when sediment accumulation rates exceed the rate of accommodation increase during the later stages of RSL rise. The HST directly overlies the MFS and is capped by the SU and its CC *sensu* Posamentier and Allen (1999). Stacking patterns display prograding and aggrading clinoforms that typically show downdip thinning, capped by a topset of fluvial, coastal plain and/or delta plain deposits.

1.2 Recognition of Trace-Fossil Omission Suites

Ichnological omission highlights a discontinuous record of sediment accumulation and/or of its preservation. Omission generally corresponds to sediment bypass, erosional exhumation, sediment starvation, and/or subaerial exposure, typically leading to changes in substrate consistency. Substrate consistency constitutes one of the most important factors controlling the distribution of organisms in sedimentary environments (see Dashtgard and Gingras, 2012), and therefore, changes in the substrate owing to omission lead to fundamentally different ichnocoenoses. Consequently, omission suites can help to recognize some of the key bounding surfaces in sequence-stratigraphic frameworks, because such surfaces are commonly associated with erosion, prolonged non-deposition, or subaerial exposure (e.g., RSME, TRS, WRS, and MFS). Much of the existing literature has focused on the identification of RSME, TRS, and WRS, as these show erosional exhumation or significant sediment bypass within a marine or marginal-marine environment (cf. MacEachern et al., 2007a,b). The potential exists, however, for the SU to be identified through the presence of continental omission suites (e.g., suites of the *Coprinisphaera* or *Scoyenia* ichnofacies) overprinting marine or marginal-marine facies of the earlier HST (e.g., Buatois and Mángano, 2011; Buatois et al., 1998; Pemberton et al., 1992c).

Omission suites are recognized on the basis of some combination of the following features: (1) trace fossils cross-cut both the primary sedimentary fabric and the palimpsest trace-fossil suites; (2) trace fossils are passively filled with sediment derived from above the discontinuity; (3) trace fossils are commonly unlined; (4) trace fossils are commonly sharp walled and may contain bioglyphs (e.g., scratches); (5) trace fossils may occur in lithologies that would not or could not host such ethological groupings in their original softground consistencies (e.g., dwelling/suspension-feeding structures in shale); (6) trace fossils are borings; (7) trace fossils show little or no post-depositional compaction; and (8) trace fossils are associated with mineralized crusts and/or encrusting biota (e.g., Bromley, 1975; MacEachern et al., 1992a; McIlroy, 2004; Pemberton and MacEachern, 1995, 2005; Pemberton et al., 1980; Taylor et al., 2003).

Omission suites are generally most apparent when developed in hardgrounds (*Trypanites* Ichnofacies) and woodgrounds (*Teredolites* Ichnofacies). Such suites become progressively more challenging to discern where they form in firmgrounds and stiffgrounds (*Glossifungites* Ichnofacies). Palimpsest softground suites are commonly the most difficult to identify. Considerable care must be taken in differentiating palimpsest softground suites from cross-cutting tiers in an association of contemporaneous traces; detailed appraisal of the ichnofabric may facilitate this distinction (cf. Buatois and Mángano, 2011; McIlroy, 2004; Taylor and Goldring, 1993; Taylor et al., 2003).

1.3 Autogenic *versus* Allogenic Discontinuities

In sequence-stratigraphic analysis, differentiation of breaks of autogenic derivation from those of allogenic origin is paramount. Autogenically driven changes in deposition are part of the dynamic nature of the sedimentary environment and generally affect only that setting and those to which it supplies sediment. The scale can be quite variable, such as distributary avulsion and autogenic lobe switching in large deltas; nevertheless, such surfaces are not regional and do not correlate along strike into adjacent settings. Hiatal surfaces or localized discontinuities may be generated and colonized by omission suites, but are of limited temporal significance and more spatially limited (e.g., Gingras et al., 2000, 2001).

Changes induced allogenically (e.g., via tectonics, eustasy, or climate), by contrast, affect RSL, and therefore are imposed upon multiple depositional settings both along strike and down depositional dip by external forces. The resulting discontinuities tend to record fundamental changes in depositional conditions that affect a number of contemporaneous environments. While such surfaces are *produced* by autogenic processes characteristic of various depositional settings (e.g., wave erosion in a shoreface setting, fluvial incision in a river), the changes that occur are a *consequence* of allogenic mechanisms. The resulting discontinuities tend to be of greater temporal significance and are more regionally extensive, allowing correlation through stratal units of a number of depositional environments. Surfaces corresponding to allogenic changes are the most relevant to sequence stratigraphy.

Differentiation of autogenically and allogenically induced surfaces is, however, challenging at the scale of outcrop and core studies, and omission suites may be occur at both. Surfaces observed at these local scales must be demonstrated to possess the geometries and regional extents consistent with allogenic discontinuities in order to have relevance to sequence-stratigraphic architecture. One exception to this is the tendency to employ autogenically induced marine flooding surfaces associated with delta-lobe switching to bound parasequences (e.g., Bhattacharya, 2010; Catuneanu et al., 2011). This arises from the recognition that deltas fail to achieve equilibrium profiles during stable sea level and normal regression, and that autogenic controls continue to dominate the architecture, referred to as "autostratigraphy" (Muto et al., 2007). For large-scale and long-lived deltaic successions this is practical, but practitioners should be aware that such autogenic parasequences will not correlate along strike into contemporaneous shoreline successions such as shoreface settings. Allogenically induced FS, however, will bound parasequences and parasequence sets that will extend along strike into non-deltaic portions of the shoreline.

Autogenic production of a highly firm substrate in the marine and marginal-marine realm is generally difficult to accomplish, because the eroding event must be capable of exhuming a compacted sediment layer without burying the exposed surface during the same event. That is, the firmground must be colonized during a hiatus between the eroding event and subsequent deposition in order to create an ichnologically demarcated omission surface (Fig. 1). In siliciclastic settings, autogenic shallow-marine firmgrounds may be associated with the margins and bases of channels (e.g., tidal channels, tidal inlets, distributary channels, and estuaries); with mud beds that are temporarily buried beneath large, periodically moribund bedforms (e.g., tidal ridges, tidal shoals, and storm-mobilized bedforms; Dashtgard et al., 2008; Yang et al., 2009); and with high intertidal mud flats, which may endure long periods of subaerial exposure and desiccation around neap cycles. Another source of firm, shallowly buried substrates may be found in the foreshore of fine-grained sand beaches, where wave pounding contributes to the repacking of sand grains and dewatering (e.g., Dalrymple, 1979; Kindle, 1936). This results in surprisingly firm media that may favor palimpsest softground omission suites. However, as the sands are still water-saturated, they are unlikely to produce firmgrounds unless incipiently cemented during early diagenesis prior to exhumation. In arid settings, some intertidal siliciclastics may be prone to incipient calcite cementation, forming firmgrounds or even hardgrounds (Gryszczynski, 1986).

Carbonate settings are more susceptible to cementation during autogenic exposure (e.g., beachrock), or to some occurrences of subaqueous cementation due to sediment starvation (e.g., some submarine hardgrounds; Bromley, 1975). In contrast to siliciclastic deposits, carbonates mainly contain biogenic sediments that originate due to the growth of organisms and build extensive platforms with low gradients (Knaust et al., 2012). Consequently, frequent subaerial exposure within the peritidal facies belt is common at the top of shallowing-upward successions. Moreover, microbial mats and biofilms thrive in such environments and contribute to sediment binding and preferred trace-fossil preservation. Calichification and early diagenetic cementation significantly contribute to the lithification of carbonate sediment and promote the development of discontinuity surfaces.

Deep-marine settings are also susceptible to autogenic omission suites, owing to oceanic currents (e.g., contour currents) as well as proximal turbidity currents, which may strip away superficial soupy muds (cf. Savrda et al., 2001a,b; Stow and Holbrook, 1984). Oceanic currents tend to provide a persistent source of energy that continuously removes sediment from the area (sediment stripping), whereas the latter process mainly leads to sediment bypass in the vicinity of the truncated sediment, leaving only a thin veneer of sediment in its wake.

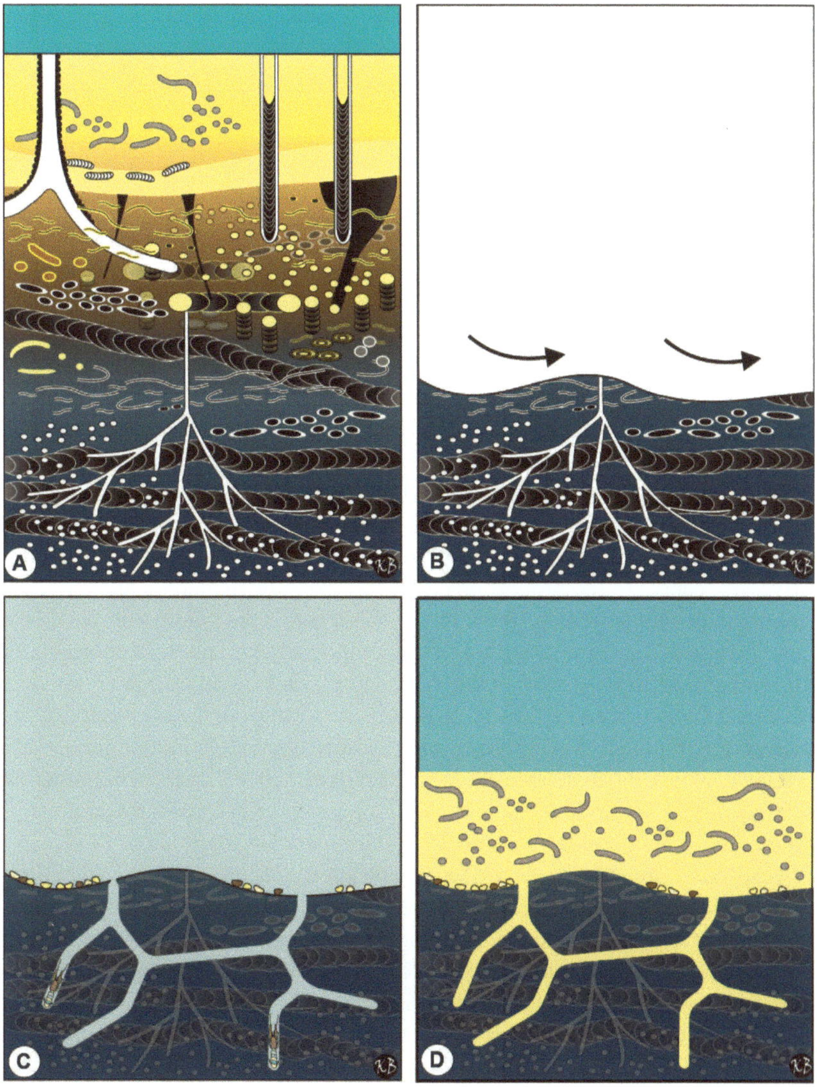

FIGURE 1 Schematic development of an erosionally exhumed omission suite. (A) Muddy substrate is buried, dewatered, and compacted. (B) Erosional exhumation exposes a firm substrate. (C) Colonization of the discontinuity surface by infauna occurs during a depositional hiatus. (D) Burial of the firmground and filling of omission structures. Omission suite in this model is attributable to the *Glossifungites* Ichnofacies. *After MacEachern et al. (2007b).*

The ultimate test for differentiating autogenic erosion surfaces from allogenic discontinuities is *regional correlatability* beyond the spatial limits of any single depositional environment. Autogenic surfaces are limited to the areal extent of the sedimentary environment that produced them and separate genetically related facies successions. Allogenic discontinuities, on the other hand, are marked by regionally mappable extents and separate genetically unrelated successions possessing discrete stratal geometries. It remains the responsibility of the geoscientist to determine the significance of any substrate-controlled omission suites encountered in the rock record, and *not* to blindly assign sequence-stratigraphic importance to them.

1.4 Facies and Ichnofacies Juxtaposition

Sedimentary facies are well positioned to characterize depositional trends and to recognize changes in those trends. As such, changes in facies lend insight into the presence and possible origin of the stratigraphic surfaces that bound the facies. Ichnological assessments can be critical in evaluating and interpreting such sedimentary facies. As well, changes in trace-fossil suites can be useful in discerning shifts in RSL, because they record organism responses to changes in a number of physico-chemical conditions, such as sedimentation rate, depositional energy, salinity, and subaerial exposure. Such ichnological assessments, however, can be challenging as they rely heavily on the interpretation of tracemaker behaviors, determination of the ichnological diversity (particularly with respect to the proportions of facies-crossing elements), and paleoenvironmental implications of the trace-fossil suites in the context of Walther's Law. Juxtaposed softground suites or ichnofabrics that do not reflect originally adjacent depositional environments (i.e., successions of suites that contravene Walther's Law) can be critical in the sequence-stratigraphic assessment of the ancient record (e.g., MacEachern and Pemberton, 1994; Pemberton et al., 1992a; Taylor and Gawthorpe, 1993; Taylor et al., 2003). Used in conjunction with physical sedimentology, they constitute a powerful tool in the characterization and interpretation of facies and facies associations.

RSL fall commonly leads to the juxtaposition of more proximal conditions over more distal, because it records the onset of negative accommodation. Shoreline trajectories change from an aggradational or progradational stacking pattern to a degradational one (Bhattacharya, 2011). Ichnologically, this negative accommodation leads to continental trace-fossil suites or marginal-marine suites to overlie fully marine associations (e.g., Bann et al., 2004), but is also commonly accompanied by localized higher sedimentation rates (leading to lower bioturbation index values), increased sediment calibers (leading to changes in animal/sediment interaction as reflected by ethology), and/or widespread subaerial exposure (favoring development of continental ichnocoenoses).

RSL rise, by contrast, is accompanied by positive accommodation and commonly leads to distal settings overlying more proximal environments, provided that the rate of positive accommodation exceeds sediment supply. This favors aggradational and backstepping shoreline trajectories. Erosion associated with TSE also commonly leads to top-truncated shoreline successions (e.g., Bhattacharya, 2010, 2011), which are commonly demarcated by ichnological omission suites. Creation of accommodation at rates higher than sediment supply leads to the return of more marine or more basinal conditions into a given area (e.g., marine overlying brackish-water deposits or brackish and marine deposits overlying continental successions). Sediment starvation in more basinal positions, because of preferential aggradation in continental and nearshore positions, leads to reduced deposition rates, reflected by higher bioturbation index values. Slower deposition in fully marine environments also generally leads to increased ichnological diversity, and a tendency to shift toward deposit-feeding- and grazing-dominated ichnological suites, as most food arrives to the setting via suspension settling. Where the creation of accommodation is slower than sediment supply, progradation occurs, but shoreline trajectories tend to be somewhat aggradational. Resulting successions typically show suites reflecting such progradation, but the elevated sedimentation rates commonly lead to reductions in bioturbation intensities, particularly in inshore, higher energy settings. The specific characteristics of facies and ichnofacies juxtaposition must, nevertheless, be evaluated on a case-by-case basis and will vary spatially both down depositional dip and along depositional strike.

2. OVERVIEW OF SEQUENCE-STRATIGRAPHIC FRAMEWORKS

2.1 Siliciclastic Responses to RSL Changes

The sequence stratigraphy of siliciclastic systems has been discussed at length in the literature (e.g., Bhattacharya, 2011; Catuneanu, 2003, 2006; Catuneanu et al., 2011; Helland-Hansen and Gjelberg, 1994; Hunt and Tucker, 1992; Posamentier et al., 1988, 1992; Van Wagoner et al., 1990). The response of siliciclastic-dominated coastlines and deep-marine systems to changes in RSL varies significantly as a function of a number of factors including, but not limited to, sediment supply, tectonic regime, and relative hydraulic energy. Sediment supply is a critical factor in determining shifts of siliciclastic shorelines. On coastlines where the sediment supply is high, sediment is delivered to the coastline faster than it can be reworked by basinal processes and, hence, shorelines prograde as deltas and delta-fed strandplains even during periods of RSL rise. Correspondingly, a decline in sediment delivery may lead to local coastal retreat, even when the rest of the system is characterized by normal regression.

The tectonic regime may affect deep-marine systems more profoundly than it does shallow-marine systems. For example, shelf widths on active margins are typically narrower than on passive margins (e.g., Eisma, 1988), and tectonically active margins tend to have higher rates of hinterland uplift and correspondingly, increased erosion and sedimentation. On tectonically active coastlines (as well as passive-margin shelves where available accommodation is largely filled), significant volumes of sediment bypass the shelf even during periods of sea-level rise and relative highstand, such that basin-floor submarine fans commonly consist of coarse-grained sediment (e.g., sand and gravel; Covault et al., 2007).

Finally, the relative hydraulic energy of the coastline affects sedimentation distribution and sedimentation rates. Fluvially dominated systems may maintain a regressive (progradational) character throughout periods of RSL rise and fall, whereas tidal- and wave-dominated coasts more closely track changes in RSL with corresponding cycles of transgression and regression. This difference is mainly a consequence of the ability of basin processes to redistribute sediment along the coastline.

The above-mentioned relationships notwithstanding, siliciclastic systems do respond to changes in RSL in predictable ways. During falling RSL (falling stage), the shoreline is forced to regress. Previously deposited shallow-marine deposits (5–15 m deep; Roy et al., 1994) are eroded by wave action, producing sharp-based "incised" shorefaces (e.g., Downing and Walker, 1988; Plint, 1988; Posamentier et al., 1992; Walker and Eyles, 1988). Shelf and coastal sediments deposited during previous highstand or transgression are eroded and transported across the shelf by rivers. Rivers may incise into the shelf, and deltas are more prevalent than estuaries. In the offshore, sediment is more easily transported across the shelf, and turbidite fans receive increasingly more abundant and generally coarser-grained sediment (Catuneanu, 2003, 2006). Lowstand deposits are similar to falling-stage deposits, although these deposits lie the most seaward, commonly extending to the shelf-slope break and typically display normal regressive or aggradational shoreline trajectories. Fluvial topsets form during lowstand normal regression, starting the process of aggradation within incised valleys.

During transgression, the shoreline shifts landward. Coastal successions that were deposited during RSL fall are cannibalized by wave and tidal processes, and incised valleys continue to be infilled with fluvial and estuarine successions (e.g., Dalrymple et al., 1992; Zaitlin et al., 1994). Small deltas are flooded and evolve into estuaries, whereas large deltas may exhibit a punctuated backstepping pattern, but may prograde across the shelf during intermittent periods of stillstand. Incremental transgression favors erosive shoreface retreat, and pauses in RSL rise permit short periods of shoreline and delta progradation across the transgressively modified SU (e.g., MacEachern et al., 1999a,b; Walker and Wiseman, 1995). In the offshore, increased accommodation on the shelf results in sediment being trapped near the coastline and limits sediment transport to the slope and abyssal plain (Catuneanu, 2003, 2006). Consequently, submarine fans that were sand dominated during RSL fall commonly revert to a mud-dominated condition during RSL rise.

During highstand, shorelines aggrade and/or prograde. Normal progradation occurs when sedimentation rates are high, typical of deltaic or delta-influenced settings. Submarine fans tend to be cut off from sediment supply, particularly in the case of narrow shelves, as most sediment is trapped in landward positions on the shelf (Catuneanu, 2003).

2.2 Carbonate Responses to RSL Changes

Carbonates differ from siliciclastics in a number of characteristics (see Knaust et al., 2012). Most carbonate deposits result from the growth and decay of organisms instead of through the redeposition of clastic debris. Accordingly, carbonate depositional systems typically show very low depositional gradients and comprise broad facies zones (Schlager, 2005), although locally they may produce higher gradients than can be achieved by siliciclastic settings, owing to biolithic development (e.g., reef margins). This correspondingly implies greater sensitivity to small changes in RSL, but with less lithological contrast between facies than observed in siliciclastic systems (Miall, 2010). During RSL fall, the main "carbonate factory" of a tropical carbonate system is readily exposed on the shelf and is, therefore, shut down. As a result, the shelf is subject to widespread karstification (more pronounced under humid climatic conditions) or pedogenesis (more dominant under arid climatic conditions), while the basin center is starved of sediment and/or prone to pronounced evaporite precipitation. During transgression, the "carbonate factory" resumes production with the submergence of the shelf. Slow transgressions may allow the carbonates to catch up with the rise in RSL, whereas rapid transgressions may result in the drowning of the carbonate platform and the formation of drowning unconformities, which are essentially flooding surfaces in the carbonate realm (Catuneanu et al., 2011). Following drowning, sediment starvation may result in the development of hardgrounds or in the accumulation of condensed sections.

All types of sequence-stratigraphic surfaces and systems tracts may develop within carbonate successions (Bover-Arnal et al., 2009; Catuneanu et al., 2011; MacNeil and Jones, 2008). Among the array of sequence-stratigraphic surfaces, SUs and MRSs are arguably surfaces with most prominent physical and/or ichnological expression within carbonate systems. Subaerial exposure surfaces originate where marine deposition is temporarily interrupted by terrestrial conditions, with at least incipient soil formation. Deeply penetrating rhizoliths are a common delineating phenomenon, typically associated with karstification, caliche, and vadose diagenesis (Fig. 2A and B). Such exposure surfaces are potential candidates for SUs.

MRSs occur at the tops of shoaling-upward trends maximized in the shallow-marine, intertidal, or supratidal environments. These indicate a change to deeper-water environments or more distal, open-marine conditions. MRSs are identified by the transition from upward-shallowing intervals to upward-deepening facies. Ichnology is well suited, at the facies level, to identify such transitions (Giannetti and Monaco, 2004; Knaust, 1998; Laporte, 1969; Lukasik and James, 2003; Olivero, 1996). Suites in the underlying regressive interval are characterized by classical transitions of Seilacherian ichnofacies from distal low-energy archetypes (e.g., *Zoophycos* or *Cruziana* ichnofacies) to higher-energy suites

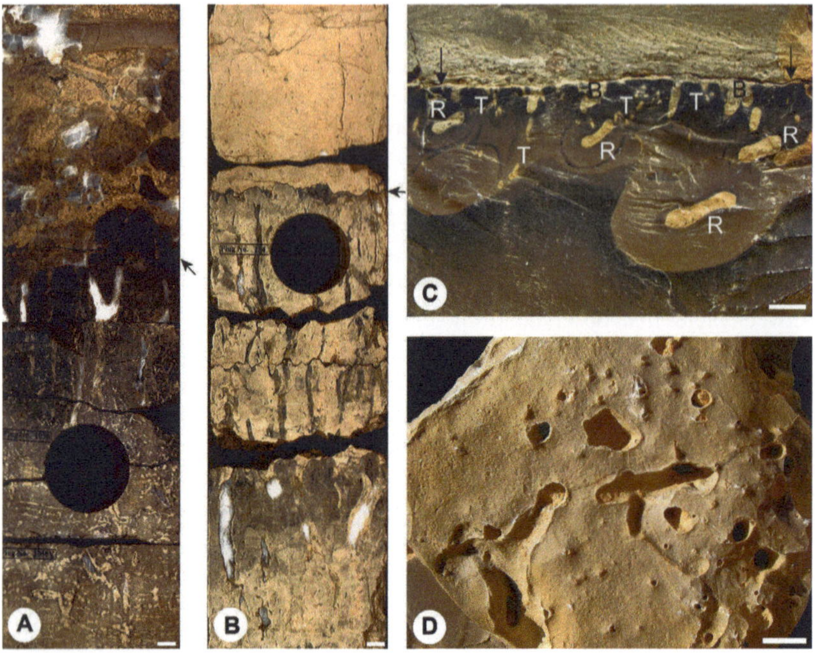

FIGURE 2 Maximum regressive surfaces (MRS) and transgressive surfaces of erosion (TSE; arrows) in vertical subsurface cores from a Late Permian carbonate platform of the Khuff Formation, South Pars gas field, offshore Iran (A and B; see Knaust, 2009). A dense root system (probably resulting from *Cordaitales*) deeply penetrates from an erosive, blackened surface and is filled with organic-rich sediment, dolomite, and gypsum. MRS in vertical section (C, arrows) and in plan view (D) from the Middle Triassic Lower Muschelkalk Group, Thuringia, Germany (Upper Oolithe Bed; see Knaust, 1998; Knaust et al., 2012). The top of the lime mudstone represents an omission surface with *Balanoglossites triadicus* (B), *Rhizocorallium jenense* (R), and *Trypanites weisei* (T), indicative of a gradual transition from firmground to hardground (sub- to intertidal) conditions. It is overlain by a rudstone/packstone (shoal deposit). Scale bars = 1 cm.

typical of landward positions (e.g., *Skolithos* Ichnofacies). Suites in the upward-deepening interval associated with transgression would show the opposite trend. The MRS itself may reflect sediment consolidation and early lithification and be demarcated by omission suites reflecting firmground palimpsest burrows and/or hardground borings belonging to the *Glossifungites* and *Trypanites* ichnofacies (Fig. 2C and D), respectively (e.g., Bromley, 1975). Successions reflecting upward deepening are widespread in the transgressive portions of sequences and are most distinctive during initial flooding across the MRS and at the MFS capping the platform (e.g., Rodríguez-Tovar et al., 2007). Nevertheless, due to their susceptibility to early diagenetic cementation, carbonate successions also host minor firmgrounds and hardgrounds associated with autogenically induced subaqueous erosion and omission (Hillgärtner, 1998). These discontinuities differ from major bounding surfaces by their rather limited lateral extent and lack of change in the overlying stratal stacking pattern.

3. ICHNOLOGICAL APPLICATIONS TO SEQUENCE STRATIGRAPHY (CASE STUDIES)

The applications of ichnology to sequence stratigraphy are best illustrated through case studies. The first two case studies come from the extensively studied Early Cretaceous Viking Formation of the Western Canada Sedimentary Basin of Alberta, Canada. Discontinuities in this interval are widespread and abundant. They are interpreted to have resulted from a combination of tectonic and eustatic processes operating in the foreland basin, as well as deposition associated with a period of low accommodation. The third case study comes from the Middle Permian–Early Triassic Khuff Formation, offshore Iran. This carbonate succession accumulated on a carbonate platform subject of repeated regression and transgression.

3.1 Incised Shorefaces, Viking Formation, Alberta, Canada

The Viking Formation of Alberta contains a number of sharp-based sand bodies of shallow-marine origin, which overlie widespread erosional discontinuities (e.g., Downing and Walker, 1988; MacEachern et al., 1998, 1999a,b; Posamentier et al., 1992; Walker and Wiseman, 1995). These sand bodies are interpreted as incised shorefaces. Incised shorefaces can be generated during the FSST or the TST. The FSST occurrences overlie the RSME and are genetically linked to the Viking valley incisions lying further landward (e.g., Pattison and Walker, 1994; Posamentier et al., 1992; Walker and Wiseman, 1995; see Section 3.2). The LST example directly overlies the FSST at Judy Creek, resting on a CC. The LST displays progradation as well as aggradation. The TST examples are associated with the WRS excavated during erosive shoreface retreat across the SU, to produce a composite surface (WRS/SU; Fig. 3). The incised shorefaces of the FSST and TST are excavated into shallow-marine parasequences that reflect normal regression associated with the HST. These parasequences consist of facies broadly similar to those of the incised shorefaces, making recognition of the FSST, LST, and retrogradationally stacked TST shoreline deposits challenging. Recognition of the *presence* of a stratigraphic discontinuity, therefore, is critical. Ichnologically, these discontinuities are commonly demarcated by firm-ground and palimpsest softground omission suites.

3.1.1 *Shorefaces of the FSST and LST*

The FSST shoreface overlies the RSME (cf. Helland-Hansen and Gjelberg, 1994; Hunt and Tucker, 1992, 1995; Mellere and Steel, 1995; Figs. 4 and 5). The RSME is ideally suited to omission suite colonization, as the discontinuity is cut in a subaqueous marine setting, allowing the dewatered, firm, or cemented substrate to become colonized by infauna prior to its burial during forced regressive shoreface progradation. The RSME is considered to be excavated and available for omission colonization only in positions above fair-weather wave base. Certainly, erosion may persist as far seaward as the storm-wave base, but

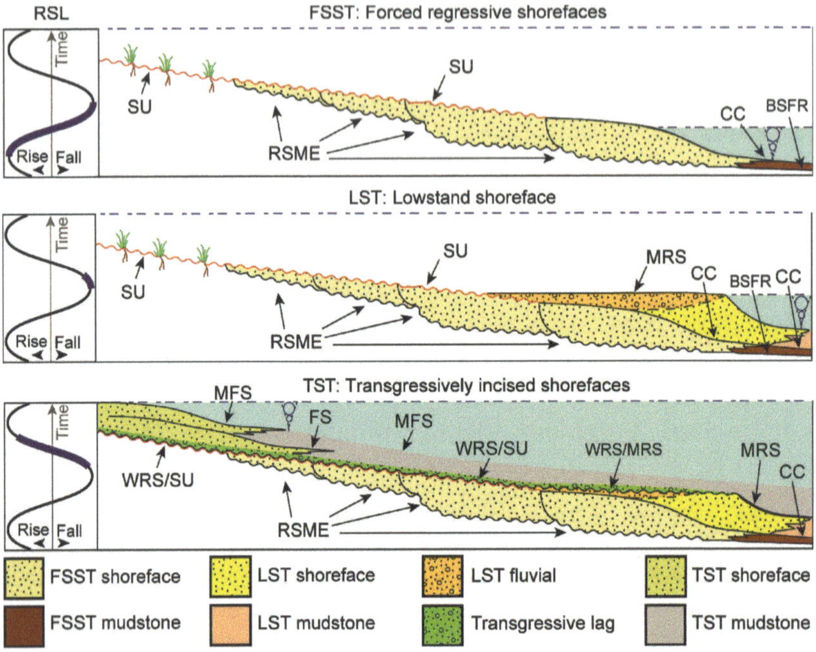

FIGURE 3 Schematic model differentiating incised shorefaces of the Falling Stage Systems Tract (FSST), Lowstand Systems Tract (LST), and Transgressive Systems Tract (TST). SU, Subaerial Unconformity; RSME, Regressive Surface of Marine Erosion; BSFR, Basal Surface of Forced Regression; CC, Correlative Conformity; MRS, Maximum Regressive Surface; WRS, Wave Ravinement Surface; FS, Flooding Surface; MFS, Maximum Flooding Surface. See text for explanation. *Modified from MacEachern et al. (1999a).*

these autogenically induced scour surfaces are mantled by storm beds, shielding them from omission colonization. Seaward of fair-weather wave base, the RSME passes into the non-erosional BSFR (Fig. 5C and D). The RSME typically reworks the BSFR, which is the lower bounding surface of the FSST (i.e., the sea floor at the onset of forced regression). The BSFR is preserved seaward of the downdip termination of the RSME as a conformable surface across which an abrupt increase in sediment caliber is typically recorded (Fig. 5C and D; MacEachern et al., 1999a). The upper bounding surface of the forced regressive shoreface can be quite variable; in proximal positions, it is marked by the SU, whereas in distal locations it is capped by the CC.

The Viking Formation in the Kaybob-Fox Creek Field provides an excellent example of a forced regressive shoreface generated during the early stages of RSL fall (Pemberton and MacEachern, 1995; Fig. 4). This shoreface shows marked incision into the underlying marine parasequences of the previous HST. The truncated highstand parasequences form well-developed, upward-coarsening successions that consist of intensely bioturbated (BI = 5) silty mudstones of the lower offshore, grading through pervasively burrowed (BI = 5)

FIGURE 4 Incised shoreface deposits of the falling stage systems tract (FSST). Core box photo (A) and close-up photo (B) of a forced regressive incised shoreface produced during a relative sea-level fall. The locality corresponds to a more proximal position in the forced regressive shoreface, wherein proximal upper offshore sandy mudstones of the underlying highstand systems tract (HST) are erosionally overlain by bioturbated proximal lower shoreface sandstones of the FSST. The discontinuity (arrow) is a regressive surface of marine erosion (RSME). The contact shows palimpsest softground *Skolithos* (*Sk*) marking the omission suite. Core box photo (C) and close-up photo (D) of the same forced regressive incised shoreface shown in photos A and B, but lying in a more seaward position. Here, the RSME is cut into lower offshore mudstones of the underlying HST normal regressive parasequences and is overlain by bioturbated muddy sandstones of the incised lower shoreface. The discontinuity is demarcated by firmground *Thalassinoides* (*Th*) of the *Glossifungites* Ichnofacies. Early Cretaceous Viking Formation, Kaybob-Fox Creek Field, Alberta. Scale bars = 15 cm (A and C) and 3 cm (D).

sandy mudstones of the upper offshore and into thoroughly burrowed (BI = 4–5) muddy sandstones of the distal lower shoreface (cf. Pemberton et al., 2012).

In proximal positions (Fig. 4A and B), the RSME tends to be incised into lower shoreface sandstones, and omission suites are generally represented by palimpsest softground associations dominated by *Skolithos* (Fig. 4B) and *Diplocraterion*. In slightly more distal positions, the RSME is incised into upper offshore and lower offshore deposits of the earlier highstand. Omission suites are generally characterized by firmground *Thalassinoides* (Fig. 4C and D), *Spongeliomorpha*, and rare *Diplocraterion*. The BSFR is not preserved within the Kaybob-Fox Creek area. Additional RSL falls led to removal of the BSFR by overlying RSME associated with younger, downstepping forced regressive shorefaces that terminate at the Judy Creek Field lying to the northeast.

The forced regressive shoreface deposits in the area are coarser-grained than the underlying highstand succession, form generally thinner coarsening-upward intervals, contain abundant cemented (typically sideritic) horizons and nodules, and commonly display cemented upper surfaces and erosion at the top, attributed to development of the SU. The forced regressive shoreface consists of well-developed, laminated to burrowed sandstones (BI = 0–5) (Fig. 4A). The laminated sandstones contain HCS and rare SCS consistent with storm deposition, with classical trace-fossil evidence of opportunistic recolonization and their replacement by resident communities (Pemberton and MacEachern, 1997). The forced regressive shoreface at Kaybob-Fox Creek reflects a moderately storm-influenced shoreface succession (MacEachern and Pemberton, 1992; Pemberton et al., 2012).

The Judy Creek Field, lying well to the northeast, contains the last (youngest) forced regressive shoreface of the FSST, which is overlain by the lowstand shoreface of the LST (Figs. 5 and 6). The forced regressive shoreface is incised into the marine mudstones of the underlying highstand parasequence, and like the Kaybob-Fox Creek example, the RSME is demarcated by firmground omission suites attributable to the *Glossifungites* Ichnofacies (e.g., *Spongeliomorpha* and firmground *Thalassinoides*; Fig. 5A and B). The incised shoreface is

FIGURE 5 Incised shoreface deposits of the FSST. Core box photo (A) and close-up (B) from a forced regressive incised shoreface in a seaward position. The forced regressive shoreface here represents the distal equivalent of the same falling stage systems tract (FSST) shown in Fig. 4 (see Fig. 3 for context). In this position, bioturbated silty mudstones interpreted as lower offshore deposits are incised by the regressive surface of marine erosion (RSME), and overlain by moderately to intensely bioturbated distal lower shoreface silty sandstones of the incised shoreface. The RSME is demarcated by firmground *Spongeliomorpha* (*S*) of the *Glossifungites* Ichnofacies. Core box photo (C) and close-up (D) of the same lowstand shoreface, but lying in a more distal position. Here, the RSME has passed into the basal surface of forced regression (BSFR), marked by upper offshore sandy mudstones of the regional parasequences sharply (but *not* erosionally) overlain by upper offshore grit-bearing sandy mudstones of the forced regressive shoreface. Early Cretaceous Viking Formation, Judy Creek Field. See text for explanation. Scale bars = 15 cm (A and C) and 3 cm (B).

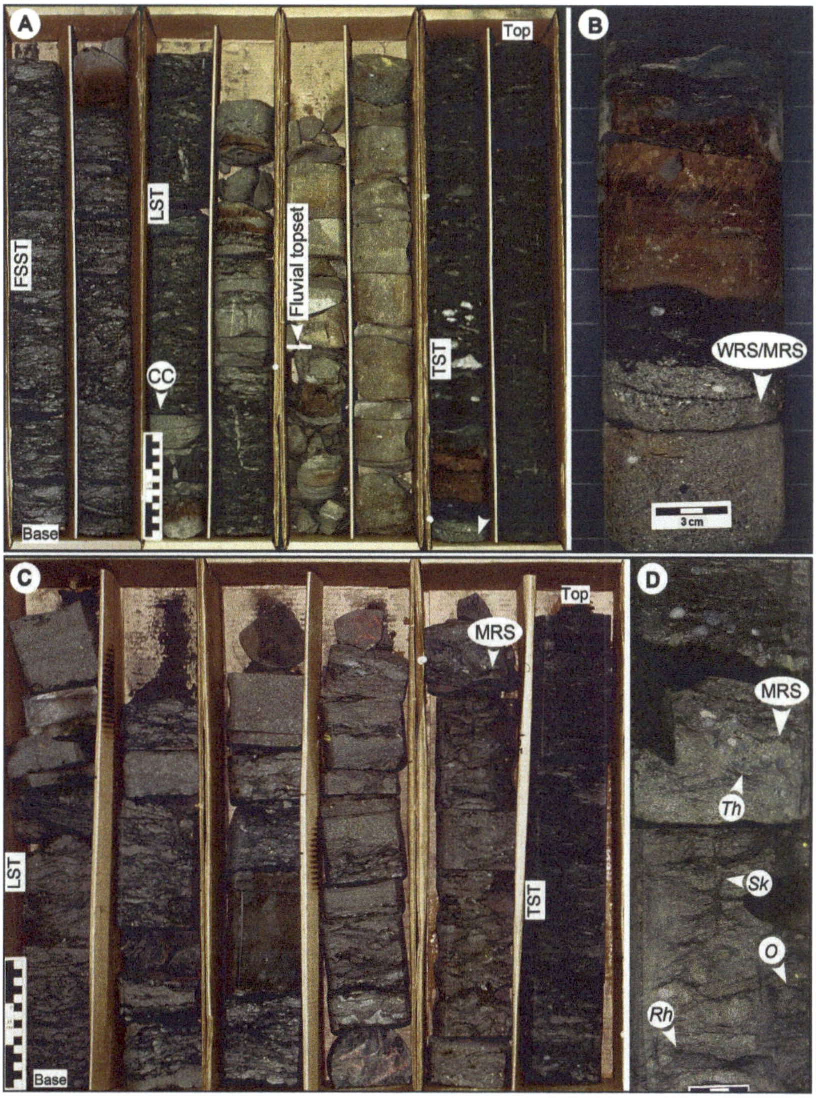

FIGURE 6 Shoreface deposits of the LST. Core box photo (A) and close-up (B) from a LST incised shoreface (see Fig. 3 for context). In the landward position, bioturbated muddy lower shoreface sandstones of the FSST coarsen upward and are overlain by the correlative conformity (CC), which marks the base of the LST shoreface. The LST shoreface consists of finer-grained lower shoreface muddy sandstones that coarsen upward and are sharply overlain by medium-grained cross-stratified sandstones interpreted as fluvial topset beds. These are capped by the maximum regressive surface (MRS) (white arrow in A), which has been erosionally modified by the wave ravinement surface (WRS). The composite surface WRS/MRS is overlain by transgressive mudstones of the TST. Core box photo (C) and close-up (D) of the same lowstand shoreface, but lying in a seaward position. Fluvial topset beds are not present and the LST shoreface is entirely marine. Common trace fossils are *Thalassinoides* (*Th*), *Skolithos* (*Sk*), *Ophiomorpha* (*O*), and *Rhizocorallium* (*Rh*). The shoreface is terminated by marine transgression across the MRS and capped by transgressive mudstones of the TST. Early Cretaceous Viking Formation, Judy Creek Field. See text for explanation. Scale bar = 15 cm (A and C) and 3 cm (D).

characterized by a coarsening-upward succession of bioturbated muddy to silty sandstones with rare tempestites. Sediment grain sizes are significantly larger than in the underlying highstand succession. Further seaward, the RSME passes into the conformable BSFR, identified as the CC by MacEachern et al. (1999a) following the three-systems tract model of Posamentier et al. (1988). Figure 5C and D shows the BSFR at Judy Creek, where underlying upper offshore mudstones of the HST are overlain by similar upper offshore mudstones of the FSST, with the exception that the upper mudstones contain coarser-grained sand and dispersed granules of chert supplied during forced regression. Ichnologically, the two mudstones are identical, showing BI=5 and fully marine suites attributable to a proximal expression of the *Cruziana* Ichnofacies. The BSFR is preferentially preserved in this location, as it resides beyond the seaward termination of the RSME.

The LST shoreface rests on the CC and progrades over the forced regressive deposits (Figs. 3 and 6). The lowstand shoreface corresponds to the seaward limit of shoreface progradation in the area. It lies along the depositional strike of the Viking Joarcam sand body, which has also been interpreted as a lowstand shoreface (Posamentier and Chamberlain, 1993). The CC at the base of the LST has a subtle physical expression, marked by a decrease in the caliber of riverborne sediment delivered to the sea following the onset of RSL rise. At Judy Creek, it is expressed sedimentologically as the boundary separating a coarsening-upward lowstand shoreface succession that is finer-grained than the underlying forced regressive shoreface (Fig. 6A). In landward positions, fluvial topset deposits cap the lowstand shoreface succession (Fig. 6A). The upper surface of the LST shoreface represents the MRS and is recognized by the transition from a coarsening-upward succession recording progradation and aggradation of the shoreline to fining-upward successions recording transgression (Fig. 6). When preserved from erosion by wave ravinement, the MRS is typically a conformable surface or a surface of non-deposition, depending on the sediment supply conditions during the overlying TST (Fig. 6C and D). In landward positions, the MRS may be reworked during transgressive wave ravinement to form a composite surface designated as WRS/MRS (Fig. 6A and B). Omission suites may be present at either expression of the MRS, but are generally palimpsest softground suites. Where wave ravinement is pronounced and where it cuts across compacted facies, firmground omission suites attributed to the *Glossifungites* Ichnofacies may occur. Potentially, suites of the *Teredolites* Ichnofacies may also occur where coastal swamp deposits were exhumed. Hardground suites of the *Trypanites* Ichnofacies are not common in this scenario.

3.1.2 Incised Shorefaces of the TST

The Viking Formation of Alberta contains a number of incised shoreface deposits that directly overlie WRSs (e.g., MacEachern et al., 1999a,b; Pemberton et al., 2004; Walker and Wiseman, 1995). An excellent example of this occurs in the Joffre Field. These WRS are cut through an underlying SU to form the composite surface WRS/SU (Figs. 3 and 7A, B). These shoreface successions record short-lived periods of shoreline progradation across the WRS/SU during an overall transgression. This surface is particularly susceptible to the generation of ichnological omission suites, because units subaerially exposed during formation of the SU are dewatered, erosionally exhumed, and/or cemented. During transgression and wave ravinement, these units are exposed to marginal-marine or marine conditions, favoring colonization of the erosional discontinuity. Such omission suites are most commonly attributable to the *Glossifungites* Ichnofacies, but hardground suites of the *Trypanites* Ichnofacies can

FIGURE 7 Incised shoreface deposits of the transgressive systems tract. Core box photo (A) and close-up (B) from a transgressively incised shoreface. Lower offshore mudstones of the underlying regional parasequences of the highstand systems tract are cross-cut by bioturbated lower shoreface muddy sandstones containing a gravel lag. The discontinuity is a composite surface representing a wave ravinement surface cutting across an earlier subaerial unconformity (WRS/SU). The discontinuity is demarcated by firmground *Thalassinoides* (*Th*) of the *Glossifungites* Ichnofacies. Core box photo (C) and close-up (D) of the same transgressively incised shoreface, but lying in a more distal position. The underlying lower offshore bioturbated mudstones of the highstand systems tract are overlain by pervasively bioturbated, grit-, and gravel-bearing upper offshore mudstones associated with the incised shoreface. Unlike the BSFR of the forced regressive shoreface shown in Fig. 5D, however, these mudstones are separated by a firmground suite of *Thalassinoides* (*Th*) attributable to the *Glossifungites* Ichnofacies. As such, the discontinuity represents erosional conditions prior to deposition, corresponding to the WRS/SU, despite the fact that the overlying facies reflect deposition *below* fair-weather wave base. This shows that the surface was cut while RSL was lower, and was subsequently reworked by waves during transgression, indicating that the incised shoreface comprises part of the transgressive systems tract. Early Cretaceous Viking Formation, Joffre Field. See text for explanation. Scale bar = 15 cm (A and C) and 3 cm (B and D).

also be expected, as well as palimpsest softground suites. In the Joffre area, the WRS/SU of the Viking Formation is most commonly demarcated by firmground *Thalassinoides*, *Spongeliomorpha*, and *Diplocraterion*, or palimpsest softground suites consisting of *Skolithos* and *Diplocraterion*, depending upon the consistency of the exhumed substrate (Fig. 7A and B).

The facies that comprise the transgressively incised shoreface successions in the Viking Formation at Joffre are similar to those of the forced regressive complexes and lowstand shorefaces. They form coarsening-upward successions that contain fully marine trace-fossil suites attributable to the *Zoophycos*, *Cruziana*, and *Skolithos* ichnofacies. Facies record pervasively bioturbated (BI=4–5) silty mudstones, sandy mudstones, muddy sandstones, silty sandstones of the lower offshore, upper offshore, lower shoreface, and middle shoreface, respectively (Fig. 7). Tempestites are locally intercalated, but constitute a minor component, indicating a weakly storm-affected shoreface setting (MacEachern and Pemberton, 1992; Pemberton et al., 2012). The transgressively incised shorefaces most closely resemble the forced regressive shorefaces, particularly in displaying incision into underlying units. In more landward positions, where the overlying facies record deposition above fair-weather wave base, the WRS/SU appears similar to the RSME. Both discontinuities possess omission suites dominated by the *Glossifungites* Ichnofacies. Commonly, however, the basal discontinuity of the transgressively incised shoreface is mantled with a granule to fine pebble lag (e.g., Fig. 7A and B), indicating that sediment eroded during transgression by the WRS is typically coarser than the sediment eroded during forced regression by the RSME.

In contrast to the coarsening-upward forced regressive shoreface deposits, transgressive shorefaces display overall retrogradational (fining-upward) stacking trends that are easier to observe at the systems-tract scale. At the facies scale, however, differentiating incised shorefaces of the TST from those of the FSST can be challenging. TST shoreface successions tend to be thicker than those of the FSST, because they display aggradation reflecting increasing accommodation space.

In the FSST shorefaces, the omission suites related to the RSME are always overlain by facies that record deposition *above* fair-weather wave base (i.e., lower shoreface deposits or shallower). By contrast, in the transgressive scenario, the erosional WRS/SU may be overlain by facies that were deposited *below* fair-weather wave base, because the surface was colonized as RSL continued to rise. Offshore silty and sandy mudstone deposits commonly overlie thin transgressive lags and firmground omission suites associated with the WRS (Fig. 7C and D; MacEachern et al., 1999a). Additionally, omission suites in the TST can record colonization under much lower energy conditions (e.g., upper or lower offshore) than in the FSST. Omission suites in offshore positions are marked by a predominance of firmground *Thalassinoides*, *Rhizocorallium*, and *Zoophycos* (e.g., MacEachern and Burton, 2000; MacEachern et al., 1999a). Careful analysis of the sedimentological and ichnological expressions of the basal discontinuity, coupled with a study of the overlying facies and the overall stacking patterns, is therefore key to determining whether the shoreface or deltaic succession occupies the FSST, LST, or TST.

3.2 Estuarine Incised-Valley Fill, Viking Formation, Alberta, Canada

The Early Cretaceous Viking Formation of Alberta, Canada, has served as a hydrocarbon exploration target since the 1930s. The 1978 discovery of the Crystal Field, however, represents the first Viking interval to be recognized as an incised-valley complex (Reinson et al., 1988). Pattison (1991) and Pattison and Walker (1994) placed the Viking deposits at Crystal into an allostratigraphic/sequence-stratigraphic framework, employing the wave-dominated estuary model of Dalrymple et al. (1992). Pattison (1992), Pemberton et al. (1992b), and MacEachern and Pemberton (1994) addressed the ichnological signature of the various depositional subenvironments of the incised valley, integrating the trace-fossil data with the sedimentological and sequence-stratigraphic models. The Viking Formation of the Willesden Green Field was also assigned an estuarine valley interpretation by Boreen and Walker (1991).

Although the resulting model is relatively robust, detailed resolution of the internal facies architecture has remained elusive, despite the presence of 395 wells in the Crystal Field (149 of them cored). This architectural complexity is a consequence of the valley having served as the locus of more than one phase of valley reincision, leading to the juxtaposition of non-contemporaneous fluvial and estuarine erosional remnants. The wave-dominated estuary model shows spatially discrete facies complexes associated with the bay-head delta, central basin, and estuary mouth (Fig. 8). These deposits are locally cross-cut by channel

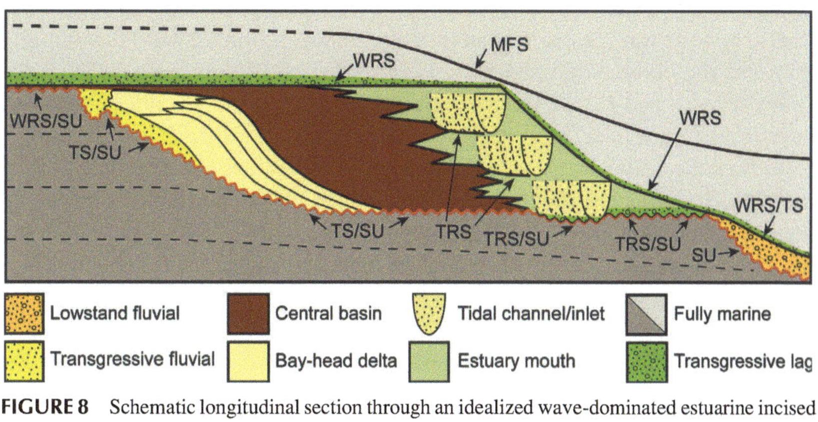

FIGURE 8 Schematic longitudinal section through an idealized wave-dominated estuarine incised valley, showing the distribution of facies associations and internal discontinuities. SU, subaerial unconformity; WRS, wave ravinement surface; TRS, tidal ravinement surface; MFS, maximum flooding surface. See text for explanation. *Model modified from Zaitlin et al. (1994).*

complexes, reflecting fluvial channels, distributary channels of the bay-head delta, tidal channels, and/or tidal inlets. This distribution of facies has led to the designation of the classical tripartite zonation of facies (Dalrymple et al., 1992). Each valley complex reflects a phase of incision during RSL fall, accompanied by sediment bypass to the lowstand shoreline and delta complexes lying further seaward (Pattison and Walker, 1994). Valley fill may have been initiated during the LST with the aggradation of a fluvial topset (Fig. 7), but occurred mainly during early transgression. Successions are characterized by predominantly estuarine conditions (see Gingras et al., 2012) and show onlap of brackish-water bay deposits along valley margins as well as aggradation of bay-head delta lobes. Pattison (1991) and Pattison and Walker (1994) describe bay-head delta successions that locally reach 27.4 m in thickness.

3.2.1 Stratigraphic Discontinuities of the Viking Incised Valleys

The bases and margins of the incised valleys are commonly demarcated by firmground suites attributable to the *Glossifungites* Ichnofacies (MacEachern and Pemberton, 1994; Pemberton and MacEachern, 1995; Pemberton et al., 1992b). The valley mouth and margins correspond to the SU, which may be eroded or draped by different TSs (Figs. 8–11). Tidal scour generates TRSs near the mouth of the estuarine valley (Fig. 9A and B) and are associated with tidal inlets and tidal channels (Fig. 9C and D). TRS tend to show erosional relief and may erode through the SU to form a composite surface (TRS/SU). TRSs are generally not widespread and may form spatially discontinuous surfaces, particularly when associated with tidal inlets that backstep across older central-basin mudstones during erosional retreat of the barrier complex at the mouth of the system (Fig. 8).

WRSs are more widespread and are formed along the seaward face of the backstepping barrier complex fronting the valley (Fig. 8). As RSL continues to rise, wave erosion bevels a relatively flat, seaward-dipping WRS that initially truncates the earlier lowstand fluvial and transgressive estuarine infill of the valley (Figs. 8 and 11A, C), but eventually cuts across the SU along the valley interfluves to generate the WRS/SU once the valley is filled (Fig. 11D and E).

Within the valley, the TS may flood across older lowstand deposits, or may onlap the SU along the valley margins to produce TS/SU, particularly below the bay-head delta and the central-basin mudstones (Fig. 8). In such positions, the TS shows minimal scour, with bay-head delta deposits showing downlap and aggradational trajectories and the central-basin mudstones showing onlap of the valley margins (Figs. 10 and 11 A, C).

TRS/SU, TS/SU, and WRS/SU are readily colonized by substrate-controlled tracemakers and are widespread in many stratigraphic successions (Figs. 9–11). The SU cut during RSL fall typically produces widespread firmground, hardground, and woodground surfaces, which can then be colonized by marine and marginal-marine tracemakers during RSL rise, provided the SU is not

FIGURE 9 TRS/SU of wave-dominated estuarine valleys. Core box photo (A) and close-up photo (B) of the TRS/SU developed near the mouth of the estuarine valley. Shelf mudstones of the regional Viking Formation parasequences are truncated by the SU, which has been subsequently eroded by the TRS representing initial transgression. The discontinuity hosts firmground *Rhizocorallium* (*Rh*) of the *Glossifungites* Ichnofacies. The overlying facies comprise part of the estuary mouth complex. Willesden Green Field, Alberta. Core box photo (C) showing offshore mudstones of the regional Viking Formation parasequences, erosionally overlain by tidal channel trough cross-stratified sandstone. The discontinuity (arrow) represents a tidal ravinement surface cutting the subaerial unconformity (TRS/SU). The surface is demarcated by firmground *Diplocraterion* and *Skolithos* of the *Glossifungites* Ichnofacies, but in this example, they are visible only on the bedding plane. Crystal Field, Alberta. (D) Close-up photo of tidal-inlet trough cross-stratified sandstone containing *Ophiomorpha* (*O*), erosionally overlying offshore bioturbated mudstones. The discontinuity represents a composite TRS/SU, akin to that of photo C. In this example, firmground *Thalassinoides* (*Th*) and *Rhizocorallium* (*Rh*) of the *Glossifungites* Ichnofacies represent the omission suite. Willesden Green Field, Alberta. See Fig. 3 for sequence-stratigraphic surface names. Scale bar = 15 cm (A and C).

FIGURE 10 TS/SU of a wave-dominated estuarine valley. Core box photo (A) and close-up photo (B) showing fully marine offshore mudstones erosionally overlain by brackish-water heterolithic facies of the estuary central-basin complex. The stratigraphic discontinuity (arrow) represents the transgressive surface associated with bay-line flooding, overlying the subaerial unconformity (TS/SU). The surface is demarcated by granules of chert and firmground *Gastrochaenolites* (*Ga*) of the *Glossifungites* Ichnofacies. Crystal Field, Alberta. See Fig. 3 for sequence-stratigraphic surface names. Scale bar = 15 cm (A).

buried beneath fluvial deposits. Omission suites associated with ravinement generally record ethologies consistent with high-energy conditions, typified by permanent vertical dwellings consistent with filter-feeding organisms (e.g., firmground *Skolithos* and *Diplocraterion*; cf. MacEachern et al., 1992a, b, 1999b). Omission suites do not generally accompany the TS unless it directly overlies the SU (e.g., TS/SU). These occurrences typically record ethological groupings more consistent with low-energy conditions, as colonization of the TS/SU occurs in sheltered and/or distal positions during transgression (cf. MacEachern and Burton, 2000; Savrda et al., 2001a).

3.2.2 *Juxtaposed Facies of the Viking Incised Estuarine Valleys*

Recognition of estuarine incised-valley fills is also facilitated by the contrast between the facies into which the valley is incised and the brackish-water deposits of the valley itself. In the case of the Viking Formation at Crystal and Willesden Green, the facies of the estuarine valley complexes are readily identified ichnologically, as the SU is incised into open-marine parasequences that record shelf to distal lower shoreface settings (MacEachern and Pemberton, 1994; Pemberton et al., 1992b). Regionally, the SU is identified by truncation of the underlying parasequences that record normal regression. The underlying parasequences are typified by high bioturbation intensities (BI = 4–5), uniformly distributed bioturbation, a high ichnological diversity, and the common

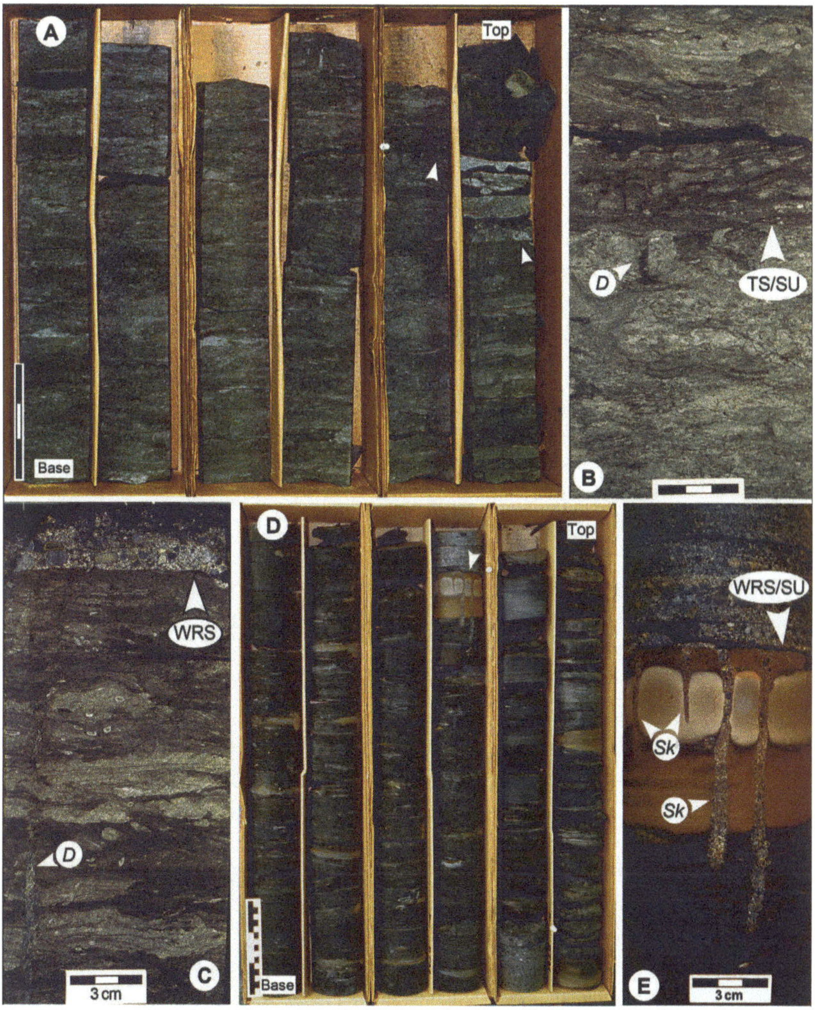

FIGURE 11 WRS, TS/SU, and WRS/SU associated with wave-dominated estuarine valleys. Core box photo (A) and close-ups (B and C) show bioturbated upper offshore sandy mudstones of the underlying marine parasequences cross-cut by late-stage central-basin heterolithic facies. The discontinuity (white arrow and photo B) represents a TS/SU, reflecting a bay-line flooding surface draped onto the subaerial unconformity. The surface is demarcated by palimpsest softground *Diplocraterion* (*D*). The uppermost discontinuity (outlined arrow and photo C) corresponds to a WRS cut across central-basin deposits near the valley margin. The surface is marked by firmground *Diplocraterion habichi* (*D*) of the *Glossifungites* Ichnofacies, capped by a transgressive lag and shelf mudstones. Crystal Field, Alberta. Core box photo (D) and close-up (E) from an off-valley (interfluve) position. The discontinuity, marked by firmground *Skolithos* (*Sk*) of the *Glossifungites* Ichnofacies, corresponds to the WRS/SU and separates underlying lower offshore mudstones from overlying upper offshore mudstones. Gilby Field, adjacent to the Willesden Green valley, Alberta. See Fig. 3 for sequence-stratigraphic surface names. Scale bar = 15 cm (A and D) and 3 cm (B).

presence of ichnogenera characteristic of unstressed, normal-marine conditions (e.g., *Helminthorhaphe*, *Phycosiphon*, *Zoophycos*, *Asterosoma*, and *Scolicia*; MacEachern and Gingras, 2007; MacEachern and Pemberton, 1994). Sedimentologically, these parasequences show distinct coarsening-upward successions of consistent thickness. The parasequences are stacked into a progradational parasequence set, typical of HNR (e.g., Pattison, 1991).

By contrast, the estuarine valley fill successions are characterized by highly complex facies architectures that record incremental transgressive fill, variable estuarine subenvironments, periods of valley reincision, and the presence of internal erosional discontinuities. Ichnological suites within the estuary conform well to the brackish-water (i.e., salinity reduced) ichnological model (e.g., Beynon et al., 1988; Gingras et al., 1999, 2012; MacEachern and Gingras, 2007; MacEachern and Pemberton, 1994; Pemberton et al., 1982).

Bay-head delta lobes occur in the landward zones of the valley and form intervals dominated by sandy successions that coarsen upward in the short term (at the scale of individual lobes), but fine upward in the long term as a result of backstepping. Intervals consist of both wave- and current-generated sedimentary structures, are markedly heterolithic, and display low bioturbation intensities (BI=0–3) that vary spatially at the bed and bedset scale (Fig. 10; MacEachern and Pemberton, 1994; Pattison and Walker, 1994). Suites are of low diversity and dominated by facies-crossing elements that commonly occur in sandy intervals (e.g., *Ophiomorpha irregulaire*, *Cylindrichnus*, *Rosselia*, *Planolites*, and *Teichichnus*; MacEachern and Gingras, 2007).

Central-basin deposits are generally more mudstone dominated, but also coarsen upward, commonly due to the backstepping of the estuary mouth complex (Fig. 11A–C). Most central-basin successions form composite bedsets attributable to wavy and lenticular bedding, but are dominated by wave-generated sedimentary structures (e.g., Boreen and Walker, 1991; MacEachern and Pemberton, 1994; Pattison and Walker, 1994). These bays vary widely in character depending upon the salinity gradient within the bay, the position of the deposits with respect to the mouth of the estuary, and the overall sedimentation rate. Most central-basin successions show moderate intensities of bioturbation (BI=1–3) with localized bands of BI=0 (typically associated with storm beds) and BI=4. The resulting facies shows the highest ichnological diversities of the valley, locally displaying 9–11 ichnogenera. Nevertheless, elements typical of more normal-marine settings, which are common in the underlying highstand parasequences, are notably absent. Common trace fossils in the central-basin deposits include *Planolites*, *Thalassinoides*, *Schaubcylindrichnus freyi*, *Chondrites*, *Teichichnus*, *Rosselia*, *Cylindrichnus*, *Palaeophycus*, *Diplocraterion*, and *Ophiomorpha irregulaire* (MacEachern and Gingras, 2007; MacEachern and Pemberton, 1994).

The estuary mouth complex is sandstone-dominated and consists mainly of bioturbated muddy sandstones (Fig. 9A and B) or current-generated cross-stratified sandstones. Tidal-inlet and tidal-channel deposits are dominated by trough cross-stratified and current-rippled sandstones and pebbly sandstones (Fig. 9C and D), with lesser horizontal planar parallel-laminated sandstone beds (Boreen and Walker, 1991; MacEachern and Pemberton, 1994; Pattison, 1991; Pattison and Walker, 1994). Ichnological suites are sporadically distributed and are commonly associated with pauses in deposition. Ichnogenera are restricted mainly to dwelling structures such as *Ophiomorpha irregulaire*, *Ophiomorpha nodosa*, *Rosselia*, *Cylindrichnus*, *Diplocraterion*, and *Skolithos*. Mudstone interbeds, where present, may contain isolated *Planolites*. Most beds show BI = 0, with zones of BI = 1–2 (cf. MacEachern and Gingras, 2007).

The flood-tidal delta deposits interfinger with bay deposits and are characterized by stacked current and combined flow ripple-lamination, commonly capping horizontal planar parallel-lamination (Pattison, 1991; Pattison and Walker, 1994). These flood-tidal deposits form composite bedsets attributable to wavy bedding. Trace fossils are sporadically distributed and bioturbation intensities are variable (BI = 0–4), but ichnological suites show higher diversities than apparent in the sandy bay-head deltas, and contain higher proportions of more fully marine elements (MacEachern and Gingras, 2007; MacEachern and Pemberton, 1994). Constituents include *Ophiomorpha irregulaire*, *Rosselia*, *Diplocraterion*, *Palaeophycus*, *Skolithos*, *Chondrites*, *Planolites*, and very rare *Phycosiphon*.

Much of the delay in recognizing that the Viking Formation at Crystal and Willesden Green constituted parts of estuarine incised-valley complexes lay in the focus, at the time, on purely sedimentological criteria. Physical sedimentological characteristics are most pronounced in channel deposits (fluvial, distributary, and tidal), the flood-tidal delta deposits, and the sandy bay-head deltas. Discriminating the central-basin deposits from the open-marine parasequences largely relies on ichnological characterization, which was only coming to be appreciated in the late 1970s. Many bay fills were routinely regarded to be part of the underlying marine parasequences and the presence of softground and firmground omission suites went unrecognized.

3.3 Carbonate Platform, Khuff Formation, Middle East

The Middle Permian to Early Triassic Khuff Formation and their correlative formations in the Middle East comprise giant gas accumulations, and thus have been of increasing economical and reservoir technical interest. Good subsurface data coverage combined with superb outcrops has guided the regional correlation and has led to a reasonable understanding of the depositional environments and paleogeographic situation. In addition to the biostratigraphic delineation, a sequence-stratigraphic framework for the Khuff Formation was established early and has been subject of some refinement (see review in Al-Husseini and Matthews, 2010).

Accordingly, six long-period sequences have been identified, subdivided into the Lower Khuff Formation (sequences KS6-5) and the Upper Khuff Formation (sequences KS4-1), and separated by a widespread evaporation event (Middle Anhydrite) on the top of sequence KS5. Each sequence comprises a TST and a HST, separated from one another by a maximum accommodation zone including the MFS. The six sequences consist of stacked, smaller-scale sequences of higher frequency with a pronounced shallowing-upward character.

Based on wells from the Iranian South Pars Field in the Persian Gulf, Knaust (2009) employed core data to supplement the sequence-stratigraphic subdivision by means of ichnology. A first step of that study was the interpretation of facies by using sedimentological and ichnological criteria (e.g., trace-fossil suites, bioturbation index, and ichnodiversity), and their distribution on a proximal-to-distal transect on the Khuff carbonate platform (Fig. 12):

1. *Sabkha and salina*: Mudstone, alternating with anhydrite and gypsum, with breccia, laminated caliche crusts and nodules (pedogenic carbonates, supratidal). Plant roots are common and probably result from the colonization of cordaitalean trees in a coastal environment.
2. *Tidal flat*: Dolomitic mudstones with sporadically distributed root traces and biolaminites dominate in the upper zone. In the lower zone, rooted mudstone layers are intercalated with oolitic and intraclastic grainstone and packstone containing *Asterosoma* and *Rosselia* (beach sands, sand waves, tidal channels, tidal deltas, and storm deposits). The degree of bioturbation is high but the ichnodiversity is moderate.
3. *Restricted lagoon*: Laminated and homogeneous dolomitic mudstone and wackestone with scattered pseudomorphs of evaporite minerals and anhydrite nodules. The ichnodiversity is low to moderate, including *Taenidium*, *Planolites,* and *Arenicolites,* and the degree of bioturbation is moderate.
4. *Open lagoon*: Mud-dominated lithofacies with wackestone, grainstone, and packstone interlayers due to washover and storm deposition. Grain-filled *Zoophycos* dominate, accompanied by *Asterosoma*, *Siphonichnus*, and *Chondrites*. The degree of bioturbation and ichnodiversity are moderate to high, indicating a suitable environment with adequate water circulation.
5. *Platform margin*: An extensive rim of grainstone and packstone with thin mudstone interlayers separates the interior platform from the open sea. The leeward margin of the rim is weakly bioturbated and has a moderate ichnodiversity. It contains *Asterosoma*, while toward its central part, root traces are more abundant. The high-energy foreshore environment on the seaward side of the margin is dominated by a high to intense degree of bioturbation and by a low ichnodiversity with *Macaronichnus*. *Arenicolites* results from opportunistic colonization of sand shoals and sand waves, whereas *Balanoglossites* occurs in firm- and hardgrounds.
6. *Shoreface/inner ramp*: Packstone, grainstone, and mudstone with a low degree of bioturbation and a moderate ichnodiversity, as well as intensely mottled event beds. *Skolithos* dominates, accompanied by *Asterosoma* and *Cylindrichnus*.

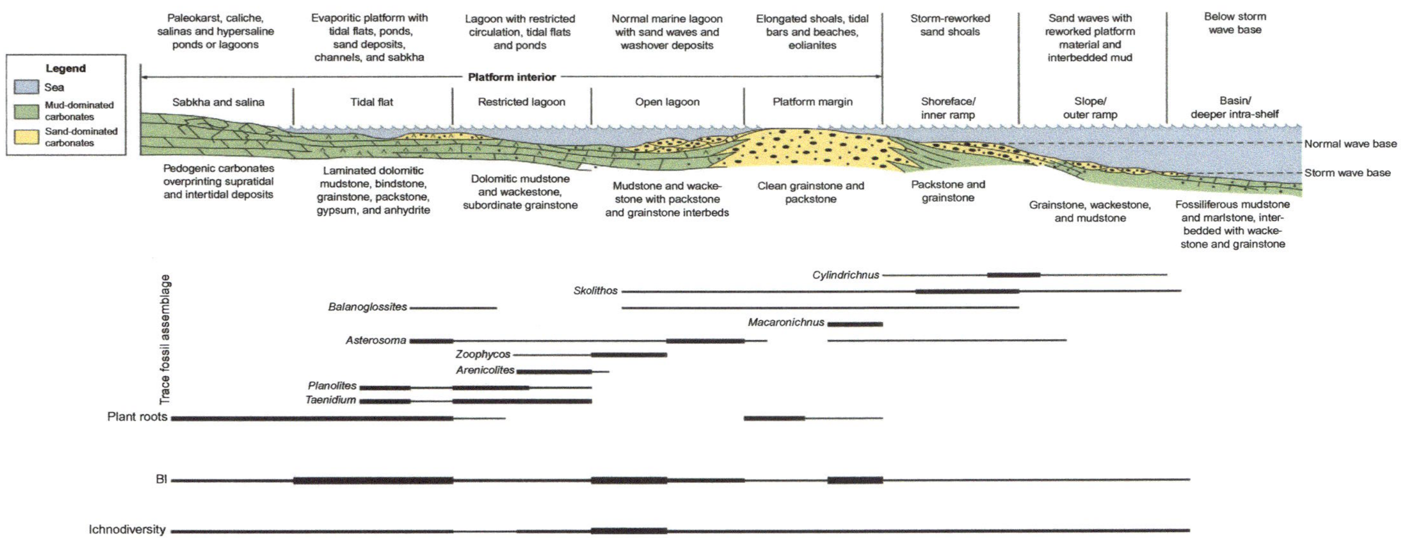

FIGURE 12 Generalized cross section of the Khuff carbonate platform in the Persian Gulf area from south (proximal to the left) to north (distal to the right), based on Wilson (1975). The simplified facies zones are based on sedimentological and ichnological core observations and are the building blocks of small-scale shallowing-upward sequences. The bars in the lower part refer to recognized trace-fossil assemblages, bioturbation index (BI), and ichnodiversity with increasing abundance (rare, moderate, and abundant). *From Knaust (2009).*

7. *Slope/outer ramp*: Grainstone, wackestone, and mudstone with low bioturbation intensities and moderate ichnodiversity. *Cylindrichnus* dominates, accompanied by *Skolithos* and *Asterosoma*.
8. *Basin/deeper intrashelf*: Laminated mudstone/marlstone with thin wackestone and grainstone layers, weakly bioturbated, and displaying a moderate ichnodiversity.

Small-scale sequences of higher frequency, serving as building blocks of the long-period sequences, were identified, based on changing lithofacies, bioturbation characteristics, trace-fossil assemblages, and tiering patterns. These peritidal carbonate sequences (parasequences) can be hard to differentiate based purely on sedimentological data, but are evident when ichnological criteria are employed. The meter-scale, shallowing-upward sequences are composed of subtidal to supratidal sedimentary units and are bounded at their tops by MRSs and TSEs. Finally, major stratigraphic surfaces, such as SUs, have the potential to be mapped and traced regionally, based on their trace-fossil contents (Fig. 2A and B).

A special aspect of the Khuff Formation is the influence of the end-Permian mass extinction on the evolution of the carbonate platform. This not only had an impact on the carbonate-building organisms but also significantly controlled the abundance and distribution of ichnotaxa and the resulting bioturbation (Knaust, 2010).

4. CONCLUSIONS

The strong facies control implicit in ethologically based trace-fossil analysis allows ichnology to be applied to a wide range of sequence-stratigraphic problems. The applications of ichnology to genetic stratigraphy are twofold. The most obvious use lies in the identification of omission suites (e.g., substrate-controlled suites attributable to the *Glossifungites*, *Teredolites*, and *Trypanites* ichnofacies, as well as palimpsest softground assemblages), which demarcate erosional discontinuities that correspond to stratigraphically significant surfaces. Unfortunately, insufficient numbers of case studies exist to indicate whether there are suites unique to particular discontinuity types. At present, ichnological omission, on its own, only helps to identify the presence of a stratigraphic break and to help identify the surface's position in a stratal succession.

The second use is facies based. It relies on interpreting organism behaviors, as preserved in trace fossils, to help identify the depositional environments above and below a stratigraphic break, and in using Walther's Law to map their paleogeographic distributions. Juxtaposed softground suites that do not reflect originally adjacent depositional environments (i.e., successions of suites that contravene Walther's Law) are highly effective in assisting in the sequence-stratigraphic assessment of the ancient record. Indeed, the integration of ichnological and sedimentological datasets can be crucial for the reliable delineation and interpretation of facies and facies associations at the local scale.

Given the spatial and temporal complexities in the genesis of stratigraphic discontinuities, continued refinement in the usage of ichnology is essential. The infaunal colonization of discontinuities yields critical information about the origin of the discontinuity and the paleoenvironmental conditions that existed during its faunal colonization. The integration of ichnological and physical sedimentological datasets with sequence-stratigraphic models is essential in order to achieve high-resolution reconstructions of the ancient record.

ACKNOWLEDGMENTS

The authors would like to thank the Natural Science and Engineering Research Council of Canada (NSERC) for research funding; Discovery Grant No. 184293 to J. A. M., Discovery Grant (PRG) No. 341789 to S. E. D., Discovery Grant No. 227176 to O. C., and Operating Grant No. A0816 to S. G. P. S. G. P. would like to thank the Canada Research Chairs program for their support of his research. D. K. thanks Statoil ASA for support and permission to publish on the Khuff Formation, and Stephen N. Ehrenberg, Tore A. Svånå, and Joanna Garland who kindly provided information. Murray Gingras provided valuable insights into the autogenic compaction of sand in nearshore positions. The thorough reviews of William Helland-Hansen and Ashton Embry were instrumental in the wholesale revision of this manuscript, and they are warmly thanked for their assistance.

REFERENCES

Ainsworth, R.B., 1994. Marginal marine sedimentology and high resolution sequence analysis; Bearpaw-Horseshoe Canyon transition, Drumheller, Alberta. Bull. Can. Petrol. Geol. 42, 26–54.

Al-Husseini, M.I., Matthews, R.K., 2010. Calibrating Mid-Permian to Early Triassic Khuff sequences with orbital clocks. GeoArabia 15, 171–206.

Allen, G.P., Posamentier, H.W., 1993. Sequence stratigraphy and facies model of an incised valley fill: the Gironde Estuary, France. J. Sediment. Petrol. 63, 378–391.

Bann, K.L., Fielding, C.R., MacEachern, J.A., Tye, S.C., 2004. Differentiation of estuarine and offshore marine deposits using integrated ichnology and sedimentology: Permian Pebbley Beach Formation, Sydney Basin, Australia. In: McIlroy, D. (Ed.), The Application of Ichnology to Palaeoenvironmental and Stratigraphic Analysis. Geol. Soc. London, Spec. Publ., vol. 228, pp. 179–211.

Beynon, B.M., Pemberton, S.G., Bell, D.A., Logan, C.A., 1988. Environmental implications of ichnofossils from the Lower Cretaceous Grand Rapids Formation, Cold Lake Oil Sands Deposit. In: James, D.P., Leckie, D.A. (Eds.), Sequences, Stratigraphy, Sedimentology: Surface and Subsurface. Can. Soc. Petrol. Geol., Mem. 15, pp. 275–290.

Bhattacharya, J.P., 1993. The expression and interpretation of marine flooding surfaces and erosional surfaces in core: examples from the Upper Cretaceous Dunvegan Formation in the Alberta foreland basin. In: Summerhayes, C.P., Posamentier, H.W. (Eds.), Sequence Stratigraphy and Facies Associations. IAS Spec. Publ., vol. 18, pp. 125–160.

Bhattacharya, J.P., 2010. Deltas. In: James, N.P., Dalrymple, R.W. (Eds.), Facies Models 4. Geol. Ass. Can., St. John's, Geotext 6, pp. 233–264.

Bhattacharya, J.P., 2011. Practical problems in the application of the sequence stratigraphic method and key surfaces: integrating observations from ancient fluvial–deltaic wedges with Quaternary and modelling studies. Sedimentology 58, 120–169.

Boreen, T., Walker, R.G., 1991. Definition of allomembers and their facies assemblages in the Viking Formation, Willesden Green area, Alberta. Bull. Can. Petrol. Geol. 39, 123–144.

Bover-Arnal, T., Salas, R., Moreno-Bedmar, J.A., Bitzer, K., 2009. Sequence stratigraphy and architecture of a late Early–Middle Aptian carbonate platform succession from the western Maestrat Basin (Iberian Chain, Spain). Sediment. Geol. 219, 280–301.

Bromley, R.G., 1975. Trace fossils at omission surfaces. In: Frey, R.W. (Ed.), The Study of Trace Fossils. Springer, New York, pp. 399–428.

Bromley, R.G., Goldring, R., 1992. The palaeoburrows at the Cretaceous to Palaeocene firmground unconformity in southern England. Tertiary Res. 13, 95–102.

Bromley, R.G., Pemberton, S.G., Rahmani, R.A., 1984. A Cretaceous woodground: the *Teredolites* Ichnofacies. J. Paleontol. 58, 488–498.

Buatois, L.A., Mángano, M.G., 2011. Ichnology: Organism-Substrate Interactions in Space and Time. Cambridge University Press, Cambridge, 358 pp.

Buatois, L.A., Mángano, M.G., Maples, C.G., Lanier, W.P., 1998. Allostratigraphic and sedimentologic applications of trace fossils to the study of incised estuarine valleys: an example from the Virgilian Tonganoxie Sandstone Member of eastern Kansas. Curr. Res. Earth Sci. 241, 1–27.

Catuneanu, O., 2003. Sequence Stratigraphy of Clastic Systems. Geological Association of Canada, St. John's, Newfoundland, 248 pp.

Catuneanu, O., 2006. Principles of Sequence Stratigraphy. Elsevier, Amsterdam, 336 pp.

Catuneanu, O., Galloway, W.E., Kendall, C.G.St.C, Miall, A.D., Posamentier, H.W., Strasser, A., Tucker, M.E., 2011. Sequence stratigraphy: methodology and nomenclature. Newsl. Strat. 44, 173–245.

Covault, J.A., Normark, W.R., Romans, B.W., Graham, S.A., 2007. Highstand fans in the California borderland: the overlooked deep-water depositional systems. Geology 35, 783–786.

Dalrymple, R.W., 1979. Wave-induced liquefaction: a modern example from the Bay of Fundy. Sedimentology 26, 835–844.

Dalrymple, R.W., Zaitlin, B.A., Boyd, R., 1992. Estuarine facies models: conceptual basis and stratigraphic implications. J. Sediment. Petrol. 62, 1130–1146.

Dam, G., 1990. Palaeoenvironmental significance of trace fossils from the shallow marine Lower Jurassic Neill Klinter Formation, East Greenland. Palaeogeogr. Palaeoclimatol. Palaeoecol. 79, 221–248.

Dashtgard, S.E., Gingras, M.K., 2012. Marine invertebrate neoichnology. In: Knaust, D., Bromley, R.G. (Eds.), Trace Fossils as Indicators of Sedimentary Environments. Developments in Sedimentology, vol. 64. Elsevier, Amsterdam, pp. 273–295.

Dashtgard, S.E., Gingras, M.K., Pemberton, S.G., 2008. Grain-size controls on the occurrence of bioturbation. Palaeogeogr. Palaeoclimatol. Palaeoecol. 257, 224–243.

de Gibert, J.M., Robles, J.M., 2005. Firmground ichnofacies recording high-frequency marine flooding events (Langhian transgression, Vallès Basin, Spain). Geol. Acta 3, 295–305.

Downing, K.P., Walker, R.G., 1988. Viking Formation, Joffre Field, Alberta: shoreface origin of long, narrow sand body encased in marine mudstones. AAPG Bull. 72, 1212–1228.

Eisma, D., 1988. An introduction to the geology of continental shelves. In: Postma, H., Zijlstra, J.J. (Eds.), Continental Shelves. Ecosystems of the World, vol. 27. Elsevier, Amsterdam, pp. 39–91.

Fielding, C.R., Bann, K.L., MacEachern, J.A., Tye, S.C., Jones, B.G., 2006. Cyclicity in the nearshore marine to coastal, Lower Permian, Pebbley Beach Formation, southern Sydney Basin, Australia: a record of relative sea-level fluctuations at the close of the Late Palaeozoic Gondwanan ice age. Sedimentology 53, 435–463.

Fielding, C.R., Frank, T.D., Birgenheier, L.P., Rygel, M.C., Jones, A.T., Roberts, J., 2008. Stratigraphic record and facies associations of the late Paleozoic ice age in eastern Australia (New

South Wales and Queensland). In: Fielding, C.R., Frank, T.D., Isbell, J.L. (Eds.), Resolving the Late Paleozoic Ice Age in Time and Space. GSA Spec. Pap. 441, pp. 41–57.

Fürsich, F.T., Kennedy, W.J., Palmer, T.J., 1981. Trace fossils at a regional discontinuity surface: the Austin/Taylor (Upper Cretaceous) contact in central Texas. J. Paleontol. 55, 537–551.

Galloway, W.E., 1989. Genetic stratigraphic sequences in basin analysis I: architecture and genesis of flooding surface bounded depositional units. AAPG Bull. 73, 125–142.

Ghibaudo, G., Grandesso, P., Massari, F., Uchman, A., 1996. Use of trace fossils in delineating sequence stratigraphic surfaces (Tertiary Venetian Basin, northeastern Italy). Palaeogeogr. Palaeoclimatol. Palaeoecol. 120, 261–279.

Giannetti, A., Monaco, P., 2004. Burrow decreasing-upward parasequence (BDUP): a case study from the Lower Jurassic of the Trento Carbonate Platform (Southern Alps), Italy. Riv. Ital. Paleontol. Strat. 110, 77–86.

Gingras, M.K., Pemberton, S.G., Saunders, T., Clifton, H.E., 1999. The ichnology of brackish water Pleistocene deposits at Willapa Bay, Washington: variability in estuarine settings. Palaios 14, 352–374.

Gingras, M.K., Pemberton, S.G., Saunders, T.D.A., 2000. Firmness profiles associated with tidal creek deposits: the temporal significance of *Glossifungites* assemblages. J. Sediment. Res. 70, 1017–1025.

Gingras, M.K., Pemberton, S.G., Saunders, T.D.A., 2001. Bathymetry, sediment texture, and substrate cohesiveness: their impact on *Glossifungites* trace assemblages at Willapa Bay, Washington. Palaeogeogr. Palaeoclimatol. Palaeoecol. 169, 1–21.

Gingras, M.K., MacEachern, J.A., Dashtgard, S.E., Zonneveld, J.-P., Schoengut, J., Ranger, M.J., Pemberton, S.G., 2012. Estuaries. In: Knaust, D., Bromley, R.G. (Eds.), Trace Fossils as Indicators of Sedimentary Environments. Developments in Sedimentology, vol. 64. Elsevier, Amsterdam, pp. 463–505.

Gryszczynski, M., 1986. Hardground and ecological succession in the light of early diagenesis (Jurassic, Holy Cross Mountains, Poland). Acta Palaeontol. Polon. 31, 163–212.

Hampson, G.J., Howell, J.A., 2005. Sedimentologic and geomorphic characterization of ancient wave-dominated deltaic shorelines, Upper Cretaceous Blackhawk Formation, Book Cliffs, Utah, U.S.A. In: Giosan, L., Bhattacharya, J.P. (Eds.), River Deltas: Concepts, Models and Examples. SEPM Spec. Publ., vol. 83, pp. 133–154.

Hampson, G.J., Storms, J.E.A., 2003. Geomorphological and sequence stratigraphic variability in wave-dominated, shoreface-shelf parasequences. Sedimentology 50, 667–701.

Hayward, B.W., 1976. Lower Miocene bathyal and submarine canyon ichnocoenoses from Northland, New Zealand. Lethaia 9, 149–162.

Helland-Hansen, W., Gjelberg, J.G., 1994. Conceptual basis and variability in sequence stratigraphy: a different perspective. Sediment. Geol. 92, 31–52.

Helland-Hansen, W., Martinsen, O.J., 1996. Shoreline trajectories and sequences: description of variable depositional-dip scenarios. J. Sediment. Res. 66, 670–688.

Hillgärtner, H., 1998. Discontinuity surfaces on a shallow-marine carbonate platform (Berriasian, Valanginian, France and Switzerland). J. Sediment. Res. 68, 1093–1108.

Hunt, D., Tucker, M.E., 1992. Stranded parasequences and the forced regressive wedge systems tract: deposition during base level fall. Sediment. Geol. 81, 1–9.

Hunt, D., Tucker, M.E., 1995. Stranded parasequences and the forced regressive wedge systems tract: deposition during base level fall—reply. Sediment. Geol. 95, 147–160.

Jervey, M.T., 1988. Quantitative geological modelling of siliciclastic rock sequences and their seismic expression. In: Wilgus, C.K., Hastings, B.S., Kendall, C.G.St.C., Posamentier, H.W., Ross, C.A., Van Wagoner, J.C. (Eds.), Sea-Level Changes: An Integrated Approach. SEPM Spec. Publ., vol. 42, pp. 47–69.

Jones, B., Pemberton, S.G., 1989. Sedimentology and ichnology of a Pleistocene unconformity-bounded, shallowing-upward carbonate sequence: the Ironshore Formation, Salt Creek, Grand Cayman. Palaios 4, 343–355.

Kindle, E.M., 1936. Dominant factors in the formation of firm and soft sand beaches. J. Sediment. Petrol. 6, 16–22.

Knaust, D., 1998. Trace fossils and ichnofabrics on the Lower Muschelkalk carbonate ramp (Triassic) of Germany: tool for high-resolution sequence stratigraphy. Geol. Rundsch. 87, 21–31.

Knaust, D., 2009. Ichnology as a tool in carbonate reservoir characterization: a case study from the Permian–Triassic Khuff Formation in the Middle East. GeoArabia 14, 17–38.

Knaust, D., 2010. The end-Permian mass extinction and its aftermath on an equatorial carbonate platform: insights from ichnology. Terra Nova 22, 195–202.

Knaust, D., Curran, H.A., Dronov, A.V., 2012. Shallow-marine carbonates. In: Knaust, D., Bromley, R.G. (Eds.), Trace Fossils as Indicators of Sedimentary Environments. Developments in Sedimentology, vol. 64. Elsevier, Amsterdam, pp. 703–750.

Kobluk, D.R., Pemberton, S.G., Karolyi, M., Risk, M.J., 1977. The Silurian-Devonian disconformity in southern Ontario. Bull. Can. Petrol. Geol. 25, 1157–1186.

Laporte, L.F., 1969. Recognition of a Transgressive Carbonate Sequence within an Epeiric Sea: Helderberg Group (Lower Devonian) of New York State. SEPM Spec. Publ. 14, pp. 98–119.

Lukasik, J.J., James, N.P., 2003. Deepening-upward subtidal cycles, Murray Basin, South Australia. J. Sediment. Res. 73, 653–671.

MacEachern, J.A., Burton, J.A., 2000. Firmground *Zoophycos* in the Lower Cretaceous Viking Formation, Alberta: a distal expression of the *Glossifungites* Ichnofacies. Palaios 15, 387–398.

MacEachern, J.A., Gingras, M.K., 2007. Recognition of brackish-water trace fossil assemblages in the Cretaceous Western Interior Seaway of Alberta. In: Bromley, R., Buatois, L.A., Mángano, M.G., Genise, J., Melchor, R. (Eds.), Sediment-Organism Interactions: A Multifaceted Ichnology. SEPM Spec. Publ., vol. 88, pp. 149–193.

MacEachern, J.A., Pemberton, S.G., 1992. Ichnological aspects of Cretaceous shoreface successions and shoreface variability in the Western Interior Seaway of North America. In: Pemberton, S.G. (Ed.), Applications of Ichnology to Petroleum Exploration. SEPM Core Workshop Notes 17, pp. 57–84.

MacEachern, J.A., Pemberton, S.G., 1994. Ichnological character of incised valley fill systems from the Viking Formation of the Western Canada Sedimentary Basin, Alberta, Canada. In: Dalrymple, R., Boyd, R., Zaitlin, B. (Eds.), Incised-Valley Systems: Origin and Sedimentary Sequences. SEPM Spec. Publ., vol. 51, pp. 129–157.

MacEachern, J.A., Bechtel, D.J., Pemberton, S.G., 1992a. Ichnology and sedimentology of transgressive deposits, transgressively related deposits and transgressive systems tracts in the Viking Formation of Alberta. In: Pemberton, S.G. (Ed.), Applications of Ichnology to Petroleum Exploration: A Core Workshop. SEPM Core Workshop Notes 17, pp. 251–290.

MacEachern, J.A., Raychaudhuri, I., Pemberton, S.G., 1992b. Stratigraphic applications of the *Glossifungites* Ichnofacies: delineating discontinuities in the rock record. In: Pemberton, S.G. (Ed.), Applications of Ichnology to Petroleum Exploration: A Core Workshop. SEPM Core Workshop Notes 17, pp. 169–198.

MacEachern, J.A., Zaitlin, B.A., Pemberton, S.G., 1998. High-resolution sequence stratigraphy of early transgressive incised shoreface and early transgressive valley/embayment deposits of the Viking Formation, Joffre Field, Alberta, Canada. AAPG Bull. 82, 729–756.

MacEachern, J.A., Zaitlin, B.A., Pemberton, S.G., 1999a. A sharp-based sandstone succession of the Viking Formation, Joffre Field, Alberta, Canada: criteria for recognition of transgressively incised shoreface complexes. J. Sediment. Res. 69, 876–892.

MacEachern, J.A., Zaitlin, B.A., Pemberton, S.G., 1999b. Coarse-grained, shoreline-attached, marginal marine parasequences of the Viking Formation, Joffre Field, Alberta Canada. In: Bergman, K.M., Snedden, J.W. (Eds.), Isolated Shallow Marine Sand Bodies: Sequence Stratigraphic and Sedimentologic Interpretation. SEPM Spec. Publ., vol. 64, pp. 273–296.

MacEachern, J.A., Gingras, M.K., Bann, K.L., Pemberton, S.G., Dafoe, L.T., 2007a. Applications of ichnology to high-resolution genetic stratigraphic paradigms. In: MacEachern, J.A., Bann, K.L., Gingras, M.K., Pemberton, S.G. (Eds.), Applied Ichnology. SEPM Short Course Notes 52, pp. 95–129.

MacEachern, J.A., Pemberton, S.G., Gingras, M.K., Bann, K.L., Dafoe, L.T., 2007b. Use of trace fossils in genetic stratigraphy. In: Miller III, W. (Ed.), Trace Fossils. Concepts, Problems, Prospects. Elsevier, Amsterdam, pp. 105–128.

MacNeil, A.J., Jones, B., 2008. Nutrient-gradient controls on Devonian reefs: insight from the ramp-situated Alexandra Reef System (Frasnian), Northwest Territories, Canada. In: Lukasik, J., Simo, J.A. (Eds.), Controls on Carbonate Platform and Reef Development. SEPM Spec. Publ., vol. 89, pp. 271–289.

McIlroy, D., 2004. Some ichnological concepts, methodologies, applications and frontiers. In: McIlroy, D. (Ed.), The Application of Ichnology to Palaeoenvironmental and Stratigraphic Analysis. Geol. Soc. London, Spec. Publ., vol. 228, pp. 3–27.

McIlroy, D., 2008. Ichnological analysis: the common ground between ichnofacies workers and ichnofabric analysts. Palaeogeogr. Palaeoclimatol. Palaeoecol. 270, 332–338.

Mellere, D., Steel, R., 1995. Facies architecture and sequentiality of nearshore and "shelf" sandbodies; Haystack Mountains Formation, Wyoming, USA. Sedimentology 42, 551–574.

Miall, A.D., 2010. The Geology of Stratigraphic Sequences. Springer, Berlin, 522 pp.

Miller, M.F., Rehmer, J., 1982. Using biogenic structures to interpret sharp lithologic boundaries: an example from the Lower Devonian of New York. J. Sediment. Petrol. 52, 887–895.

Muto, T., Steel, R.J., Swenson, J.B., 2007. Autostratigraphy: a framework norm for genetic stratigraphy. J. Sediment. Res. 77, 2–12.

Nummedal, D., Swift, D.J.P., 1987. Transgressive stratigraphy at sequence-bounding unconformities: some principles derived from Holocene and Cretaceous examples. In: Nummedal, D., Pilkey, O.H., Howard, J.D. (Eds.), Sea-Level Fluctuation and Coastal Evolution. SEPM Spec. Publ., vol. 41, pp. 241–260.

Olivero, D., 1996. *Zoophycos* distribution and sequence stratigraphy. Examples from the Jurassic and Cretaceous deposits of southeastern France. Palaeogeogr. Palaeoclimatol. Palaeoecol. 123, 273–287.

Pattison, S.A.J., 1991. Sedimentology and Allostratigraphy of Regional, Valley-Fill, Shoreface and Transgressive Deposits of the Viking Formation (Lower Cretaceous). Central Alberta. Ph.D. Thesis, McMaster University, Ontario, Canada, 380 pp.

Pattison, S.A.J., 1992. Recognition and interpretation of estuarine mudstones (central basin mudstones) in the tripartite valley-fill deposits of the Viking Formation, Central Alberta. In: Pemberton, S.G. (Ed.), Applications of Ichnology to Petroleum Exploration: A Core Workshop. SEPM Core Workshop Notes 17, pp. 223–249.

Pattison, S.A.J., Walker, R.G., 1994. Incision and filling of a lowstand valley; late Albian Viking Formation at Crystal, Alberta, Canada. J. Sediment. Res. 64, 365–379.

Pattison, S.A.J., Walker, R.G., 1998. Multiphase transgressive filling of an incised valley and shoreface complex, Viking Formation, Sundance-Edson area, Alberta. Bull. Can. Petrol. Geol. 46, 89–105.

Pemberton, S.G., Frey, R.W., 1985. The *Glossifungites* Ichnofacies: modern examples from the Georgia coast, U.S.A. In: Curran, H.A. (Ed.), Biogenic Structures: Their Use in Interpreting Depositional Environments. SEPM Spec. Publ., vol. 35, pp. 237–259.

Pemberton, S.G., MacEachern, J.A., 1995. The sequence stratigraphic significance of trace fossils: examples from the Cretaceous foreland basin of Alberta, Canada. In: Van Wagoner, J.C., Bertram, G. (Eds.), Sequence Stratigraphy of Foreland Basin Deposits: Outcrop and Subsurface Examples from the Cretaceous of North America, AAPG Mem. 64, pp. 429–475.

Pemberton, S.G., MacEachern, J.A., 1997. The ichnological signature of storm deposits: the use of trace fossils in event stratigraphy. In: Brett, C.E. (Ed.), Paleontological Event Horizons: Ecological and Evolutionary Implications. Columbia University Press, New York, pp. 73–109.

Pemberton, S.G., MacEachern, J.A., 2005. Significance of ichnofossils in applied stratigraphy. In: Koutsoukos, E.A. (Ed.), Applied Stratigraphy. Springer, Dordrecht, pp. 281–302.

Pemberton, S.G., Kobluk, D.R., Yeo, R.K., Risk, M.J., 1980. The boring *Trypanites* at the Silurian-Devonian disconformity in southern Ontario. J. Paleontol. 54, 1258–1266.

Pemberton, S.G., Flach, P.D., Mossop, G.D., 1982. Trace fossils from the Athabasca oil sands, Alberta, Canada. Science 217, 825–827.

Pemberton, S.G., MacEachern, J.A., Frey, R.W., 1992a. Trace fossil facies models: environmental and allostratigraphic significance. In: Walker, R.G., James, N. (Eds.), Facies Models: Response to Sea Level Change. Geol. Ass. Can., St. John's, Geotext 1, pp. 47–72.

Pemberton, S.G., Reinson, G.E., MacEachern, J.A., 1992b. Comparative ichnological analysis of Late Albian estuarine valley fill and shelf shoreface deposits, Crystal Viking Field, Alberta. In: Pemberton, S.G. (Ed.), Applications of Ichnology to Petroleum Exploration: A Core Workshop. SEPM Core Workshop Notes 17, pp. 291–317.

Pemberton, S.G., Van Wagoner, J.C., Wach, G.D., 1992c. Ichnofacies of a wave-dominated shoreline. In: Pemberton, S.G. (Ed.), Applications of Ichnology to Petroleum Exploration: A Core Workshop. SEPM Core Workshop Notes 17, pp. 339–382.

Pemberton, S.G., Spila, M., Pulham, A.J., Saunders, T., MacEachern, J.A., Robbins, D., Sinclair, I.K., 2001. Ichnology & Sedimentology of Shallow to Marginal Marine Systems: Ben Nevis & Avalon Reservoirs, Jeanne d'Arc Basin. Geol. Ass. Canada, Short Course Notes 15, 343 pp.

Pemberton, S.G., MacEachern, J.A., Saunders, T.D.A., 2004. Stratigraphic applications of substrate-specific ichnofacies: delineating discontinuities in the rock record. In: McIlroy, D. (Ed.), The Application of Ichnology to Palaeoenvironmental and Stratigraphic Analysis. Geol. Soc. London, Spec. Publ., vol. 228, pp. 29–62.

Pemberton, S.G., MacEachern, J.A., Dashtgard, S.E., Bann, K.L., Gingras, M.K., Zonneveld, J.-P., 2012. Shorefaces. In: Knaust, D., Bromley, R.G. (Eds.), Trace Fossils as Indicators of Sedimentary Environments. Developments in Sedimentology, vol. 64. Elsevier, Amsterdam, pp. 563–604.

Plint, A.G., 1988. Sharp-based shoreface sequences and "offshore bars" in the Cardium Formation of Alberta: their relationship to relative changes in sea level. In: Wilgus, C.K., Hastings, B.S., Kendall, C.G.St.C., Posamentier, H.W., Ross, C.A., Van Wagoner, J.C. (Eds.), Sea-Level Changes: An Integrated Approach. Soc. Paleont. Min. SEPM Spec. Publ., vol. 42, pp. 357–370.

Plint, A.G., Nummedal, D., 2000. The falling stage systems tract: recognition and importance in sequence stratigraphic analysis. In: Hunt, D., Gawthorpe, R.L. (Eds.), Sedimentary Response to Forced Regression. Geol. Soc. London, Spec. Publ., vol. 172, pp. 1–17.

Posamentier, H.W., Allen, G.P., 1999. Siliciclastic Sequence Stratigraphy: Concepts and Applications. SEPM Concepts in Sedimentology and Paleontology, vol. 7, 210 pp.

Posamentier, H.W., Chamberlain, C.J., 1993. Sequence stratigraphic analysis of Viking Formation lowstand beach deposits at Joarcam field, Alberta, Canada. In: Posamentier, H.W.,

Summerhayes, C.P., Haq, B.U., Allen, G.P. (Eds.), Stratigraphy and Facies Associations in a Sequence Stratigraphic Framework. IAS Spec. Publ., vol. 18, pp. 469–485.

Posamentier, H.W., Vail, P.R., 1988. Eustatic controls on clastic deposition. II. Sequence and systems tract models. In: Wilgus, C.K., Hastings, B.S., Kendall, C.G.St.C., Posamentier, H.W., Ross, C.A., Van Wagoner, J.C. (Eds.), Sea-Level Changes—An Integrated Approach. SEPM Spec. Publ., vol. 42, pp. 125–154.

Posamentier, H.W., Jervey, M.T., Vail, P.R., 1988. Eustatic controls on clastic deposition I. Conceptual framework. In: Wilgus, C.K., Hastings, B.S., Kendall, C.G.St.C., Posamentier, H.W., Ross, C.A., Van Wagoner, J.C. (Eds.), Sea-Level Changes—An Integrated Approach. SEPM Spec. Publ., vol. 42, pp. 110–124.

Posamentier, H.W., Allen, G.P., James, D.P., Tesson, M., 1992. Forced regressions in a sequence stratigraphic framework: concepts, examples, and exploration significance. AAPG Bull. 76, 1687–1709.

Reinson, G.E., Clark, J.E., Foscolos, A.E., 1988. Reservoir geology of Crystal Viking Field, Lower Cretaceous estuarine tidal channel-bay complex, south-central Alberta. AAPG Bull. 72, 1270–1294.

Rodríguez-Tovar, F.J., Pérez-Valera, F., Pérez-López, A., 2007. Ichnological analysis in high-resolution sequence stratigraphy: the *Glossifungites* Ichnofacies in Triassic successions from the Betic Cordillera (southern Spain). Sediment. Geol. 198, 293–307.

Roy, P.S., Cowell, P.J., Ferland, M.A., Thom, B.G., 1994. Wave-dominated coasts. In: Carter, R.W.G., Woodroffe, C.D. (Eds.), Coastal Evolution: Late Quaternary Shoreline Morphodynamics. Cambridge University Press, Cambridge, pp. 121–186.

Savrda, C.E., 1991a. *Teredolites,* wood substrates, and sea-level dynamics. Geology 19, 905–908.

Savrda, C.E., 1991b. Ichnology in sequence stratigraphic studies: an example from the lower Paleocene of Alabama. Palaios 6, 39–53.

Savrda, C.E., 1995. Ichnologic applications in paleoceanographic, paleoclimatic, and sea-level studies. Palaios 10, 565–577.

Savrda, C.E., Browning, J.V., Krawinkel, H., Hesselbo, S.P., 2001a. Firmground ichnofabrics in deep-water sequence stratigraphy, Tertiary clinoform-toe deposits, New Jersey slope. Palaios 16, 294–305.

Savrda, C.E., Krawinkel, H., McCarthy, F.M.G., McHugh, C.M.G., Olson, H.C., Mountain, G., 2001b. Ichnofabrics of a Pleistocene slope succession, New Jersey margin: relations to climate and sea-level dynamics. Palaeogeogr. Palaeoclimatol. Palaeoecol. 171, 41–61.

Schlager, W., 2005. Carbonate Sedimentology and Sequence Stratigraphy. SEPM Concepts in Sedimentology and Paleontology, vol. 8, 200 pp.

Sloss, L.L., Krumbein, W.C., Dapples, E.C., 1949. Integrated facies analysis. In: Longwell, C.R. (Ed.), Sedimentary Facies in Geologic History. GSA Mem. 39, pp. 91–124.

Stow, D.A.V., Holbrook, J.A., 1984. North Atlantic contourites: an overview. In: Stow, D.A.V., Piper, D.J.W. (Eds.), Fine-Grained Sediments: Deep-Water Processes and Facies. Geol. Soc. London, Spec. Publ., vol. 15, pp. 245–256.

Swift, D.J.P., 1975. Barrier-island genesis: evidence from the central Atlantic shelf, eastern U.S.A. Sediment. Geol. 14, 1–43.

Taylor, A.M., Gawthorpe, R.L., 1993. Application of sequence stratigraphy and trace fossil analysis to reservoir description: examples from the Jurassic of the North Sea. In: Parker, J.R. (Ed.), Petroleum Geology of Northwest Europe, Proceedings of the 4th Conference. Geol. Soc. London, pp. 317–335.

Taylor, A.M., Goldring, R., 1993. Description and analysis of bioturbation and ichnofabric. J. Geol. Soc. Lond. 150, 141–148.

Taylor, A.M., Goldring, R., Gowland, S., 2003. Analysis and application of ichnofabric. Earth-Sci. Rev. 60, 227–259.

Van Wagoner, J.C., Posamentier, H.W., Mitchum, R.M., Vail, P.R., Sarg, J.F., Loutit, T.S., Hardenbol, J., 1988. An overview of sequence stratigraphy and key definitions. In: Wilgus, C.K., Hastings, B.S., Kendall, C.G.St.C., Posamentier, H.W., Ross, C.A., Van Wagoner, J.C. (Eds.), Sea-Level Changes—An Integrated Approach. SEPM Spec. Publ., vol. 42, pp. 39–45.

Van Wagoner, J.C., Mitchum Jr., R.M., Campion, K.M., Rahmanian, V.D., 1990. Siliciclastic Sequences, Stratigraphy in Well Logs, Cores, and Outcrops. AAPG Methods in Exploration 7, 55.

Vossler, S.M., Pemberton, S.G., 1988. Ichnology of the Cardium Formation (Pembina oilfield): implications for depositional and sequence stratigraphic interpretations. In: James, D.P., Leckie, D.A. (Eds.), Sequences, Stratigraphy, Sedimentology: Surface and Subsurface. Can. Soc. Petrol. Geol., Mem. 15, pp. 237–253.

Walker, R.G., Eyles, C.H., 1988. Geometry and facies of stacked shallow-marine sandier upward sequences dissected by erosion surface, Cardium Formation, Willesden Green, Alberta. AAPG Bull. 72, 1469–1494.

Walker, R.G., Wiseman, T.R., 1995. Lowstand shorefaces, transgressive incised shorefaces, and forced regressions: examples from the Viking Formation, Joarcam area, Alberta. J. Sediment. Res. 65, 132–141.

Wilson, J.L., 1975. Carbonate Facies in Geologic History. Springer, Berlin, 471 pp.

Yang, B.C., Dalrymple, R.W., Gingras, M.K., Pemberton, S.G., 2009. Autogenic occurrence of *Glossifungites* Ichnofacies: examples from wave-dominated, macrotidal flats, southwestern coast of Korea. Mar. Geol. 260, 1–5.

Zaitlin, B.A., Dalrymple, R.W., Boyd, R., 1994. The stratigraphic organization of incised-valley systems associated with relative sea level change. In: Dalrymple, R.W., Boyd, R., Zaitlin, B.A. (Eds.), Incised Valley Systems: Origin and Sedimentary Sequences. SEPM Spec. Publ., vol. 51, pp. 45–60.

Chapter 7

Ichnostratigraphy

M. Gabriela Mángano,*[,1] Luis A. Buatois* and Robert B. MacNaughton[†]
*Department of Geological Sciences, University of Saskatchewan, Saskatoon, Saskatchewan, Canada, [†]Geological Survey of Canada, Natural Resources Canada, Calgary, Alberta, Canada
[1]Corresponding author: e-mail: gabriela.mangano@usask.ca

1. INTRODUCTION

Many ichnotaxa have long temporal ranges and occur in a narrow range of facies. For these reasons, trace fossils have not been used extensively in biostratigraphic studies. Trace fossils can provide invaluable biostratigraphic data, however, in situations where other fossil age control is limited or lacking. Various approaches to ichnostratigraphy (trace-fossil biostratigraphy) have been documented. Some trace fossils preserve specific fingerprints of their producers that yield biostratigraphic information, as in the long tradition of using tetrapod footprints to date continental successions (Lockley, 1998) or in the use of *Cruziana* and *Rusophycus* ichnospecies to date non-fossiliferous Paleozoic sandstones (Seilacher, 1970, 1990, 1992). Complex trace fossils can have biostratigraphic significance due to a behavioral pattern that has a limited stratigraphic range (e.g., ichnospecies of *Oldhamia* or arthrophycids) or due to a consistent first appearance datum (e.g., first occurrence of *Treptichnus pedum*). Beyond these applications, even long-ranging ichnotaxa can provide useful age controls based on their documented first appearances, as is the case with Ediacaran to early Cambrian trace-fossil biozones.

In this chapter, we briefly review the utility of trace fossils to establish relative ages in both marine and continental settings. Readers desiring a more in-depth discussion of the principles and applications of ichnostratigraphy can consult a number of published reviews (Buatois and Mángano, 2011; Crimes, 1975; MacNaughton, 2007). Ichnostratigraphy has been applied primarily in siliciclastic strata, where the preservation potential of stratigraphically useful ichnotaxa is relatively high, and this is reflected in our discussion. In carbonates, by contrast, late diagenetic processes tend to reduce the diversity of ichnocoenoses and hamper preservation of fine morphological details required for detailed ichnostratigraphy (e.g., Narbonne, 1984).

Developments in Sedimentology, Vol. 64. http://dx.doi.org/10.1016/B978-0-444-53813-0.00007-1

2. CONTINENTAL ENVIRONMENTS

Tetrapod trackways have been extensively used for biostratigraphy in continental environments (e.g., Haubold and Katzung, 1978) (Fig. 1A–D). In a recent review, Lucas (2007) recognized several global time intervals based on the footprint record (Table 1). Lucas (2007) noted that global biochronology based on tetrapod trackways resolves geological time approximately 20–50% as well as the body-fossil record. Notably, this resolution is better than that of invertebrate trace fossils currently used in biostratigraphy (see *Cruziana* and Arthrophycidae stratigraphy). Also, given the uneven resolution of continental body fossils through time, the temporal resolution of tetrapod footprints may be as good as that of body fossils for certain time intervals having a poor skeletal record. Unfortunately, the reliability of vertebrate footprints in stratigraphy is limited by the uneven quality of footprint ichnotaxonomy (Lockley, 2000; Rainforth, 2003). As noted by Lucas (2007), biostratigraphic zonations using ichnotaxa defined on the basis of extramorphological features (i.e., morphological features resulting from foot/substrate interactions) give the false appearance of stratigraphic resolution, but this is an artifact resulting from a splitter approach to ichnotaxonomy. Zonations based on better defined ichnotaxa (e.g., Haubold, 1996, 2000) are sound, albeit with limited resolution (Lucas, 2007).

In addition to the extensive use of footprints, vertebrate coprolites may have some utility for stratigraphic correlation. For example, Hunt et al. (1993, 1998) found that vertebrate coprolites were useful for correlation in the Upper Triassic of the western USA.

Invertebrate trace fossils in continental deposits offer relatively little scope for ichnostratigraphy. The most common invertebrate ichnotaxa in continental settings tend to have long stratigraphic ranges, and as a result, similarities in ichnocoenoses between coeval successions are more a reflection of similar environmental limiting factors than of a common age (e.g., MacNaughton and Pickerill, 2010). One group of continental trace fossils with at least somewhat limited stratigraphic distribution is insect trace fossils in paleosols (see review by Genise, 2004). Most of these ichnotaxa are not known in strata older than Paleogene. For example, *Eatonichnus* is only known from the Paleocene–Eocene and *Teisseirei* from the Eocene–Oligocene. Although the temporal resolution of these ichnotaxa is quite limited, their presence may be used to differentiate Cenozoic paleosols from older continental deposits (Genise et al., 2000). As the ichnotaxonomy of continental trace fossils improves, it is possible that additional ichnotaxa will prove to have at least broad age implications. For example, the burrow network *Spongeliomorpha carlsbergi*, which ranges from the Late Triassic to the Miocene (Melchor et al., 2010), may be useful for distinguishing between Mesozoic and upper Paleozoic continental deposits.

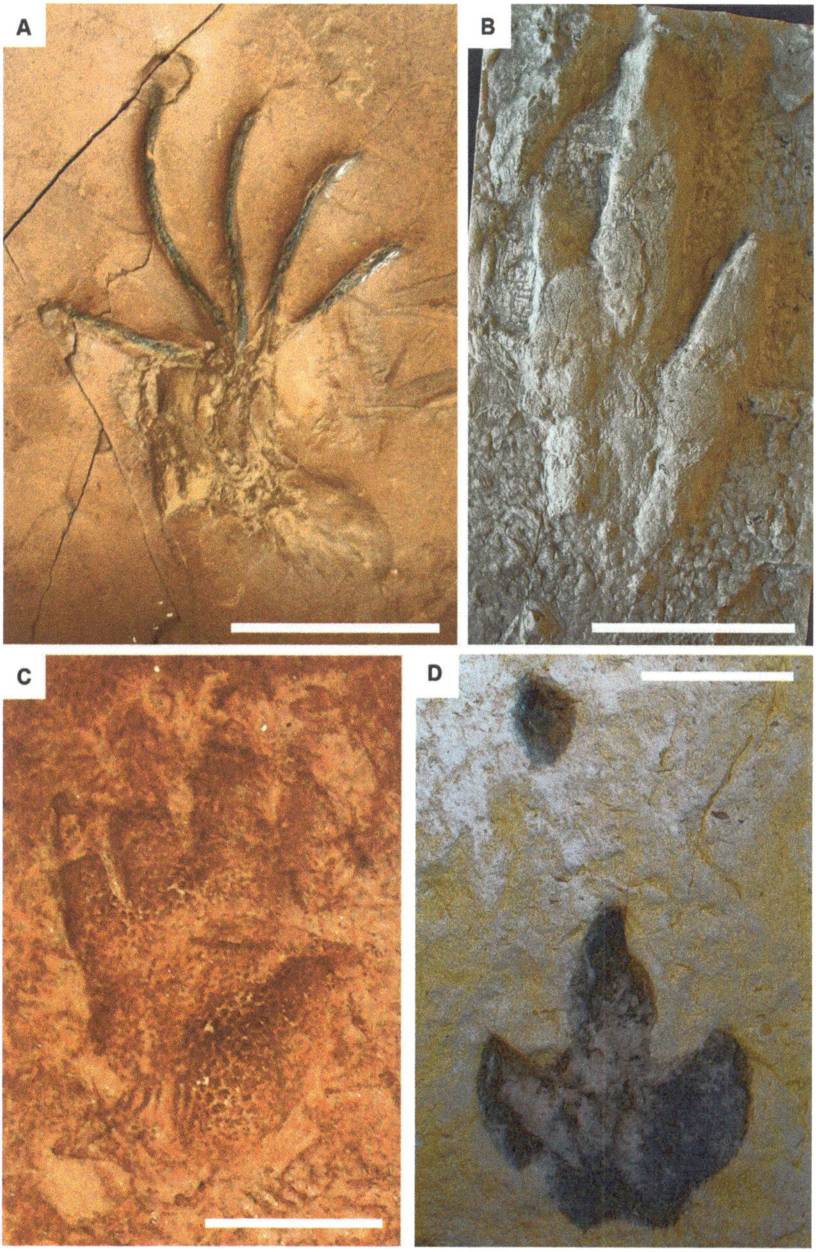

FIGURE 1 Examples of vertebrate ichnotaxa with biostratigraphic significance. (A) *Dimetropus leisnerianus*. Lower Permian, Hueco Formation, Robledo Mountains, New Mexico, USA. Scale bar = 5 cm. (B) *Grallator sulcatus*. Upper Triassic, Brunswick Formation, Clark Quarry, near Milford, New Jersey, USA. Scale bar = 5 cm. Photograph courtesy of Verónica Krapovickas. (C) *Isochirotherium* isp. Lower Triassic, Wióry Formation, Wióry site, Holy Cross Mountains, Poland. Scale bar = 5 cm. (D) *Caririchnium leonardii*. Lower Cretaceous, Dakota Group, Dinosaur Ridge, Colorado, USA. Scale bar = 10 cm. Photograph courtesy of Verónica Krapovickas.

TABLE 1 **Summary of Tetrapod-Footprint Stratigraphy (*Based on Lucas, 2007*)**

Period/epoch	Diagnostic characteristics	Representative ichnotaxa
Neogene	– Dominance of trackways produced by derived ungulates	*Macrauchenichnus*, *Venatoripes*, *Megatherichnum*, *Neomegatherichnum*, *Eumacrauchenichnus*, and *Odocoileinichnum*
Paleogene	– Abundance of trackways produced by basal ungulates	*Sarjeantipes*, *Quirtipes*, and *Albertasuchipes*
Late Cretaceous	– Appearance of ceratopsian, tyrannosaurid, and hadrosaurid footprints	*Haenamichnus* and *Ceratopsipes*
Early Cretaceous	– Dominance of sauropod trackways – Presence of ornithopod and bird tracks	*Parabrontopodus*, *Brontopodus*, *Iguanodontipus*, *Caririchnium*, *Ignotormis*, and *Aquatilavipes*
Middle–Late Jurassic	– Increase in the size of tracks – Dominance of theropod, sauropod, and ornithopod trackways	*Carmelopodus*, *Megalosauripus*, *Gigantosauropus*, *Parabrontopodus*, *Breviparopus*, and *Dinehichnus*
Early Jurassic	– Dominance of non-avian theropod footprints	*Eubrontes*, *Anomoepus*, and *Ameghinichnus*
Late Triassic	– Higher diversity of dinosaur-like and dinosaur footprints – Dinosaur assemblage	*Tetrasauropus*, *Pseudotetrasuropus*, *Grallator*, and *Atreipus*
Late Middle Triassic	– Appearance of tridactyl bipedal trackways that have been attributed to dinosaur or dinosaur-like organisms – Dinosauromorph assemblage	*Parachirotherium*, *Atreipus*, and *Grallator*
"Early" Triassic to early Middle Triassic	– Dominance of archosaur trackways – Chirothere assemblage	*Chirotherium*, *Isochirotherium*, *Rotodactylus*, *Brachychirotherium*, and *Synaptichnium*
Earliest Triassic (Induan)	– Dicynodont tracks	None

TABLE 1 Summary of Tetrapod-Footprint Stratigraphy (*Based on Lucas, 2007*)—Cont'd

Period/epoch	Diagnostic characteristics	Representative ichnotaxa
Late Permian	– Trackways produced by pareiasaurs and dicynodonts – *Rhychosauroides* biochron	*Pachypes, Dicynodontipus,* and *Rhynchosauroides*
Early Permian	– Trackways produced by temnospondyls, "diadectomorphs", seymouriamorphs, procolophonids, and basal synapsids (pelycosaurs) – Continuation of the *Dromopus* biochron	*Batrachichnus, Limnopus, Amphisauropus, Dromopus, Varanopus, Hyloidichnus, Ichniotherium, Dimetropus,* and *Gilmoreichnius*
Late Pennsylvanian	– Beginning of the *Dromopus* biochron	*Batrachichnus, Ichniotherium, Dromopus, Gilmoreichnus,* and *Dimetropus*
Early–Middle Pennsylvanian	– Dominance of stem amniote (captorhinomorphs) trackways – Temnospondyl footprints present also – *Pseudobradypus* biochron	*Pseudobradypus, Limnopus, Paleosauropus, Cursipes,* and *Asperipes*
Mississippian	– Trackways produced by temnospondyls and stem amniotes (captorhinomorphs)	*Peratodactylopus, Megapezia, Baropezia, Hylopus,* and *Paleosauropus*
Late Devonian	– Trackways attributed to ichthyostegalians	None

3. MARGINAL-MARINE ENVIRONMENTS

Marginal-marine brackish-water ichnofaunas display evolutionary changes through the Phanerozoic, providing a macroevolutionary signature (Buatois et al., 2005). However, their potential in high-resolution biostratigraphy seems to be limited. A few ichnotaxa characteristic of marginal-marine deposits may display stratigraphically restricted distributions. For example, *Climactichnites* (Fig. 2A) and its associated resting trace *Musculopodus* are only known from the Upper Cambrian (Furongian) (Getty and

FIGURE 2 Examples of invertebrate ichnotaxa with biostratigraphic significance in marginal-marine rocks. (A) *Climactichnites wilsoni*. Upper Cambrian, Postdam Group, slab exhibited at the Smithsonian Institution. Scale bar = 10 cm. (B) *Heimdallia chatwini*. Ordovician, Peninsula Formation, Table Mountain Group, Matjiesgloof Farm, South Africa. Pencil = 14 cm long.

Hagadorn, 2008; Yochelson and Fedonkin, 1993), though with a highly restricted paleogeographic distribution (Jensen et al., in press). The wall-like trace *Heimdallia* (Fig. 2B) records sophisticated sediment handling and is apparently restricted to Late Ordovician–Early Carboniferous strata, with various ichnospecies displaying even more restricted ranges (e.g., *Heimdallia zigzag* in the Lower Silurian) (Seilacher, 2007). The ichnospecies *Treptichnus pedum*, indicative of the Proterozoic/Cambrian boundary (see below), is typical of shallow to fully marine deposits but also has been recorded in marginal-marine environments (Geyer and Uchman, 1995; MacNaughton and Narbonne, 1999). Similarly, some ichnospecies of *Cruziana* and *Rusophycus* may occur in brackish-water settings, expanding the environmental range of application of *Cruziana* stratigraphy (Mángano and Buatois, 2003) beyond its more typical shallow-marine application (see Section 4.2).

4. SHALLOW-MARINE ENVIRONMENTS

The application of trace fossils to the dating of shallow-marine deposits has focused mainly on Ediacaran and Paleozoic shallow-marine siliciclastic strata. In such strata, ichnostratigraphy has been applied to three main problems: (1) positioning of the Proterozoic/Cambrian boundary and dating of adjacent strata based on the establishment of a number of trace-fossil zones (e.g., Buatois and Mángano, 2011; Jensen, 2003; MacNaughton and Narbonne, 1999); (2) establishment of relative ages in Paleozoic siliciclastic successions based on ichnospecies of *Cruziana* and *Rusophycus* (e.g., Seilacher, 1970, 1992, 1994); and (3) establishment of relative ages in early Paleozoic siliciclastic units based on ornamented burrows, such as *Arthrophycus* and related trace fossils (e.g., Mángano et al., 2005; Seilacher, 2000).

4.1 Proterozoic/Cambrian Boundary

The Proterozoic/Cambrian boundary Global Standard Stratotype-section and Point (GSSP) is the only stratigraphic boundary established on ichnological grounds. It was placed at the base of the *Treptichnus pedum* Zone at Fortune Head, Newfoundland, Canada (Fig. 3; Brasier et al., 1994; Landing, 1994; Narbonne et al., 1987), although more recent work has moved the base of the zone slightly below the GSSP (Gehling et al., 2001).

FIGURE 3 The Proterozoic/Cambrian boundary Global Standard Stratotype-section and Point (GSSP) at Fortune Head, Newfoundland, Canada. Sören Jensen's right hand is at the boundary. Subsequent research demonstrated the presence of *Treptichnus pedum* some 4 m below the GSSP (Gehling et al., 2001). Photograph courtesy of Jim Gehling.

The biostratigraphic significance of Precambrian–Cambrian ichnofaunas has been intensely explored since the 1970s (e.g., Alpert, 1977; Crimes, 1987, 1994; Jensen, 2003; MacNaughton and Narbonne, 1999; Narbonne et al., 1987; Walter et al., 1989). MacNaughton and Narbonne (1999) and Jensen (2003) presented a biostratigraphic scheme that has been slightly modified more recently (Fig. 4; Buatois and Mángano, 2011; Jensen et al., 2006). This scheme comprises two Ediacaran zones and three Early Cambrian zones. The lower Ediacaran zone includes simple grazing trails (e.g., *Helminthoidichnites*, *Helminthopsis*, *Archaeonassa*) as well as dickinsonid resting traces and *Kimberella* radular scratches (*Radulichnus*). The age of this interval is approximately 560–550 Ma (Jensen et al., 2006; Martin et al., 2000). The upper Ediacaran zone includes the oldest branching burrow systems, represented by *Treptichnus* (non *T. pedum*) and *Streptichnus*, as well as trilobate trace fossils similar to *Curvolithus*. The age of this zone is 550–542 Ma (Grotzinger et al., 1995; Jensen et al., 2006). The lowermost Early Cambrian zone is known as the *T. pedum* zone. It is of Fortunian age and its base is marked by the first appearance of *T. pedum* (Fig. 5A). Other first appearances in this zone are *Gyrolithes polonicus* and *Bergaueria*. The overlying zone is referred to as the *Rusophycus avalonensis* zone and contains the oldest bilobate, trilobite-type resting traces (*R. avalonensis*) together with the bilobate epichnial trail "*Taphrelminthopsis*" *circularis* (Jensen et al., 2006). The age of this zone ranges from the Fortunian to Cambrian Stage 2. Highest is the *Cruziana tenella* (=*problematica*) zone, which contains the oldest bilobate, trilobite-like furrowing trails (*Cruziana problematica*), commonly associated with large backfilled traces (*Psammichnites gigas*). The age of this zone is Cambrian Stage 2.

Even simple ichnotaxa with long stratigraphic ranges may be useful to differentiate Precambrian and Phanerozoic strata. For example, the presence of such long-ranging trace fossils as *Planolites* and *Skolithos* has proven to be valuable in distinguishing between Cambrian and early Neoproterozoic siliciclastic formations during bedrock mapping in northern Canada (Aitken et al., 1973; Fallas and MacNaughton, in press).

4.2 *Cruziana* Stratigraphy

Cruziana stratigraphy actually is based on ichnospecies both of ribbon-like bilobate traces, referred to as *Cruziana sensu stricto* (Fig. 5B), and short bilobate traces, referred to as *Rusophycus* (Seilacher, 1970, 1992, 1994). There is a considerable functional and biostratigraphic information (e.g., Fortey and Seilacher, 1997; Seilacher, 1985) to indicate that these structures were produced primarily by trilobites, particularly in early Paleozoic marine deposits, but some ichnotaxa could have been produced by other arthropods. Trilobites were well suited for benthic life and inhabited a wide range of environmental settings during the Early Paleozoic, potentially enhancing the utility of their traces for biostratigraphy (Crimes, 1975; Seilacher, 1970, 1992, 1994).

Cruziana and *Rusophycus* ichnospecies are defined based on fine morphological features. The so-called claw formula (i.e., the distinctive "fingerprint"

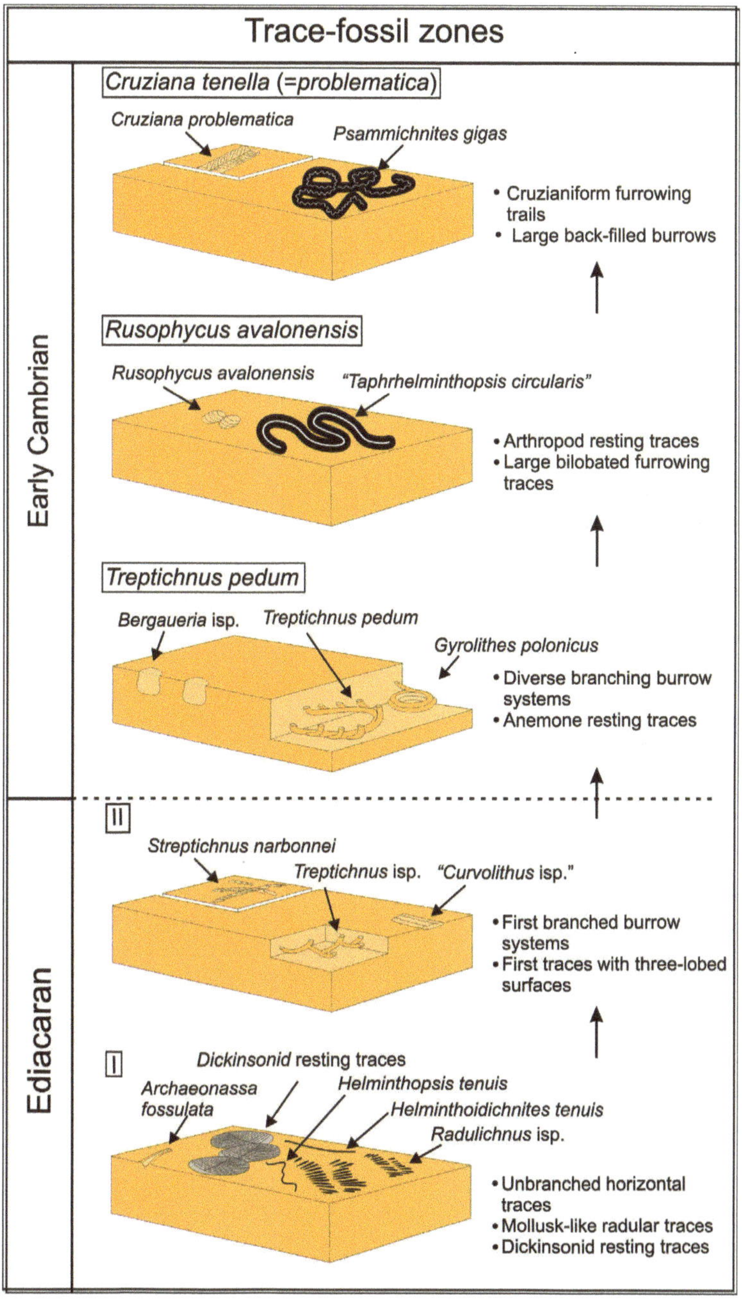

FIGURE 4 Trace-fossil zones above and below the Ediacaran–Cambrian boundary, with listings of trace-fossil first appearances characteristic of each zone (after Buatois and Mángano, 2011).

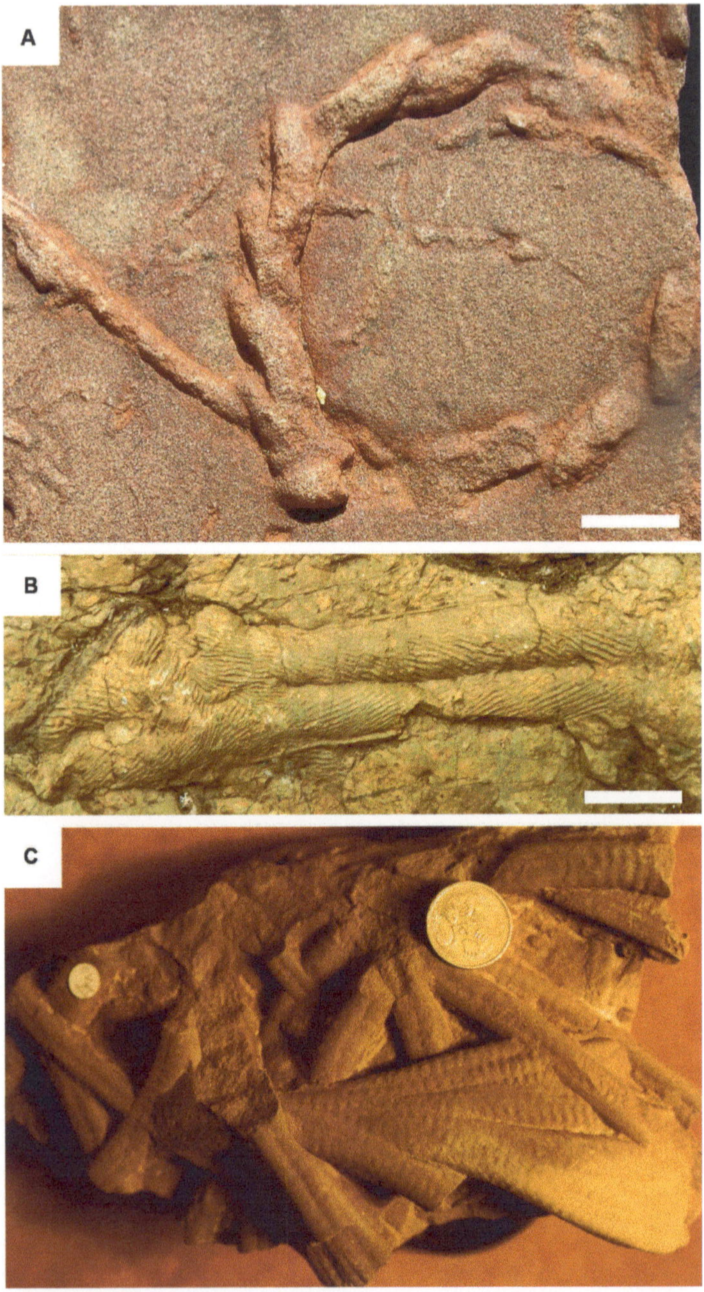

FIGURE 5 Examples of invertebrate ichnotaxa with biostratigraphic significance in shallow-marine rocks. (A) *Treptichnus pedum*. Lower Cambrian, Klipbak Formation, Brandkop Subgroup, near Brandkop, South Africa. Scale bar = 1 cm. (B) *Cruziana rugosa rugosa*. Arenigian–Llanvirnian, Mojotoro Formation, Quebrada del Gallinato, Argentina. Scale bar = 5 cm. (C) *Arthrophycus alleghaniensis*. Lower Silurian, Medina Sandstone, Rochester, USA. Coin = 1.9 cm.

left by claws or setae present in the distal part of the walking appendages) is particularly important, but other features reflecting the producers' gross morphology or burrowing behavior (e.g., presence of cephalic impressions, coxal marks, exopodal brushings, pleural or genal spine impressions) are also significant traits to be taken into account. The underlying assumption is that, if leg morphology displays high rates of evolutionary change, coupled with significant changes in burrowing behavior, it is possible to establish relatively narrow stratigraphic ranges for the different ichnospecies of *Cruziana* and *Rusophycus* (Buatois and Mángano, 2011; Seilacher, 1970, 1992).

Ichnospecific classification of *Cruziana* and *Rusophycus* is challenging and may not always be achievable. The claw formula is not always easy to determine as there are common taphonomic constraints to overcome. Even if the claw formula can be determined from a single, well-preserved specimen, this may not be sufficient for reliable ichnospecific assignment. Accurate reconstruction of the burrowing technique also is essential and typically requires a representative sample of several specimens. The possibility of convergency also needs to be carefully considered (compare, e.g., Magwood and Pemberton, 1990, with Seilacher, 1994). Nonetheless, if both morphological and behavioral traits are taken into consideration, a relatively solid biostratigraphic scheme emerges (e.g., Mángano and Buatois, 2003; Seilacher, 1990, 1992).

Cruziana stratigraphy has been developed and applied almost entirely with reference to Gondwana, where more than 30 ichnospecies of *Cruziana* and *Rusophycus* with biostratigraphic significance have been identified (Seilacher, 1970, 1992). Subsequent efforts have expanded the model into Laurentia (Seilacher, 1994) and Baltica (Jensen et al., 2011; Knaust, 2004). The stratigraphic ranges of these ichnospecies comprise between one and four series, but the majority of these ichnotaxa are restricted to only one or two series. However, the available volume of information is uneven. The most extensive dataset is for the Late Cambrian to Middle Ordovician series (e.g., Baldwin, 1977; Crimes, 1970, 1975; Fillion and Pickerill, 1990; Knaust, 2004; Mángano and Buatois, 2003; Mángano et al., 1996; Pickerill and Fillion, 1983), but some significant proposals have been made more recently for the Early Silurian (Llandovery) (Seilacher, 1996). In the stratigraphic scheme proposed, *Cruziana* and *Rusophycus* ichnospecies are in turn clustered into groups (e.g., *dispar* group, *semiplicata* group, *rugosa* group) (Seilacher, 1970, 1992). The problems involved in *Cruziana* stratigraphy have been recently explored by Buatois and Mángano (2011), who underscored the need for critical analysis based on adequate sampling and the lack of independent biostratigraphic evidence in the case of many ichnospecies (see also MacNaughton, 2007).

4.3 Arthrophycid Stratigraphy

Still more controversial is the so-called Arthrophycid stratigraphy, based on various ichnotaxa (*Arthrophycus*, *Daedalus*, *Phycodes*) included in the ichnofamily Arthrophycidae, which were proposed to have biostratigraphic significance (Seilacher, 2000). Of these, the most distinctive ichnotaxon is *Arthrophycus* (Fig. 5C), the ichnospecies of which can be used for biostratigraphy in Ordovician–Silurian rocks. Five ichnospecies are thought to have biostratigraphic significance: *A. minimus* (Upper Cambrian to Lower Ordovician), *A. brongniartii* (Lower Ordovician to Lower Silurian), *A. alleghaniensis* (Lower Silurian), *A. lateralis* (Lower Silurian), and *A. parallelus* (Carboniferous) (Buatois and Mángano, 2011; Seilacher, 2000). Unfortunately, the evidence supporting the scheme is poor. A few ichnospecies are widespread (e.g., *A. alleghaniensis*), but others are only known from one or two localities, and accurate, independent biostratigraphic control is lacking in most cases. As with *Cruziana* stratigraphy, Arthrophycid stratigraphy is a potential means to date and correlate otherwise unfossiliferous early Paleozoic sandstone-dominated strata, but much more work is needed to solidify these ichnostratigraphic schemes and test their validity.

5. DEEP-MARINE ENVIRONMENTS

Ichnostratigraphy in deep-marine strata has not been explored to the same degree as in shallow-marine counterparts. Interestingly, the paucity or lack of body fossils in many early Paleozoic deep-marine successions hampers high-resolution relative dating of these deposits, and so trace-fossil data can be highly useful in this regard. In these strata, even long-ranging ichnotaxa can provide useful age control (e.g., Pickerill and Fyffe, 1999). The ichnogenus *Oldhamia* (Fig. 6A), widespread in Early Cambrian deep-marine deposits, has long been considered a characteristic fossil in the Cambrian of Ireland (e.g., Baily, 1865) and may have broader potential in biostratigraphy as pointed out by several authors (e.g., Lindholm and Casey, 1990; MacNaughton, 2007; Seilacher, 1974, 2007; Seilacher et al., 2005). *Oldhamia* typically occurs in intensely tectonized and slightly metamorphosed rocks that are devoid of body fossils, which has hampered the calibration of this ichnotaxon against other biostratigraphic data. *Oldhamia curvata*, *O. radiata*, and *O. flabellata* are known from Early Cambrian rocks, and *O. antiqua* has been recorded in Early Cambrian to, more rarely, early Middle Cambrian rocks (Seilacher et al., 2005). An extensive review by Herbosch and Verniers (2011) suggests that the best-constrained occurrences of *Oldhamia* are within Cambrian Stages 3–5. Some authors have proposed that evolutionary lineages can be recognized within this ichnogenus (Lindholm and Casey, 1990; Seilacher, 1974). Such lineages could lend themselves to biostratigraphy, but convincing documentation remains elusive. In any case, *Oldhamia* is a Cambrian indicator and has allowed distinction from Precambrian rocks worldwide (e.g., Mirré and Aceñolaza, 1972).

Another Paleozoic deep-marine ichnogenus with restricted stratigraphic distribution is *Dictyodora* (Fig. 6B) (Benton and Trewin, 1980; Seilacher 1967; Uchman, 2004). *D. simplex* is recorded from the Cambrian and Ordovician; *D. zimmermanni* seems to be restricted to the Ordovician; *D. scotica* and *D. tenuis* are Ordovician–Silurian, and *D. liebeana* is only known from the Early Carboniferous.

FIGURE 6 Examples of invertebrate ichnotaxa with biostratigraphic significance in deep-marine rocks. (A) *Oldhamia radiata*. Lower Cambrian, Bray Group, Bray Head, Ireland. Scale bar = 5 cm. (B) *Dictyodora liebeana*. Lower Carboniferous, Cabo de Favaritz Beds, Menorca, Spain. (C) *Glockerichnus alata*. Lower to Middle Eocene, Tarcau Sandstone, Teherau Valley, Romania. Scale bar = 10 cm.

Uchman (2004) noted that a few graphoglyptid ichnotaxa have rather narrow stratigraphic ranges. Examples include *Glockerichnus alata* (Fig. 6C) (only known from the Eocene) and *Rotundusichnium zumayense* (Maastrichtian-Eocene). Nevertheless, most graphoglyptid ichnotaxa have longer stratigraphic ranges, spanning the Cretaceous to the Neogene (Uchman, 2003, 2004).

6. CONCLUSIONS

The applications of trace fossils in biostratigraphy are more limited than in fields such as facies analysis and sequence stratigraphy. However, in the right circumstances, trace fossils can provide crucial biostratigraphic data. It is worthwhile to consider that the most important boundary in the stratigraphic record, the Proterozoic/Cambrian boundary, is recognized based on the first appearance of *Treptichnus pedum*. Although ichnostratigraphy seldom can compete with high-resolution biostratigraphy based on other fossil groups, many complex trace fossils still do present a remarkable macroevolutionary pattern in the stratigraphic record. Also, where body fossils are absent, scarce, or unevenly distributed (e.g., Early Paleozoic shallow-marine sandstones, Cambrian deep-marine deposits, continental red beds), ichnological data may provide a crucial age constraint.

In short, the application of ichnology in biostratigraphy, the so-called ichnostratigraphy, remains a poorly explored field that deserves further analysis. At the very least, documentation and discussion of stratigraphic ranges should be a standard part of trace-fossil studies. Given the biostratigraphic insights that have arisen from databases compiled for the Proterozoic/Cambrian boundary (Crimes, 1987, 1992; Jensen, 2003) or for deep-marine ichnotaxa through time (Uchman, 2003, 2004), there probably is much scope for compilations focused on other aspects of the trace-fossil record. Such studies may provide a refreshing, new understanding of the ichnological record.

ACKNOWLEDGMENTS

Jim Gehling provided the Fortune Head photograph and Verónica Krapovickas provided some of the vertebrate footprint photographs. We thank Sören Jensen for his valuable review and Dirk Knaust and Richard Bromley for inviting us to write this chapter and for editing the chapter. Funding was provided by Natural Sciences and Engineering Research Council of Canada (NSERC) grants to M. G. M. (Discovery Grant 311727-05 and -08) and L. A. B. (Discovery Grant # 311726-05 and -08). Geological Survey of Canada internal review of the chapter was provided by James White. This is Earth Sciences Sector contribution 20110303.

REFERENCES

Aitken, J.D., Macqueen, R.W., Usher, J.L., 1973. Reconnaissance Studies of Proterozoic and Cambrian Stratigraphy, Lower Mackenzie River Area (Operation Norman), District of Mackenzie. Geol. Surv. Can. Paper 73–79, pp. 1–178.

Alpert, S.P., 1977. Trace fossils and the basal Cambrian boundary. In: Crimes, T.P., Harper, J.C. (Eds.), Trace Fossils 2, Geol. J., Spec. Iss. 9, pp. 1–8.

Baily, W.H., 1865. The Cambrian rocks of the British Islands, with especial reference to the occurrence of this formation and its fossils in Ireland. Geol. Mag. 15, 385–400.

Baldwin, C.T., 1977. The stratigraphy and facies associations of trace fossils in some Cambrian and Ordovician rocks of northwestern Spain. In: Crimes, T.P., Harper, J.C. (Eds.), Trace Fossils 2, Geol. J., Spec. Iss. 9, pp. 9–40.

Benton, M.J., Trewin, N.J., 1980. *Dictyodora* from Silurian of Peeblesshire, Scotland. Palaeontology 23, 501–513.

Brasier, M.D., Cowie, J.W., Taylor, M.E., 1994. Decision on the Precambrian-Cambrian boundary. Episodes 17, 3–8.

Buatois, L.A., Mángano, M.G., 2011. Ichnology: Organism-substrate interactions in space and time. Cambridge University Press, Cambridge, 358 pp.

Buatois, L.A., Gingras, M.K., MacEachern, J., Mángano, M.G., Zonneveld, J.-P., Pemberton, S.G., Netto, R.G., Martin, A.J., 2005. Colonization of brackish-water systems through time: evidence from the trace-fossil record. Palaios 20, 321–347.

Crimes, T.P., 1970. Trilobite tracks and other trace fossils from the Upper Cambrian of North Wales. Geol. J. 7, 47–68.

Crimes, T.P., 1975. The stratigraphical significance of trace fossils. In: Frey, R.W. (Ed.), The Study of Trace Fossils. Springer, New York, pp. 109–130.

Crimes, T.P., 1987. Trace fossils and correlation of late Precambrian and early Cambrian strata. Geol. Mag. 124, 97–119.

Crimes, T.P., 1992. Changes in the trace fossil biota across the Proterozoic-Phanerozoic boundary. J. Geol. Soc. London 149, 637–646.

Crimes, T.P., 1994. The period of early evolutionary failure and the dawn of evolutionary success. In: Donovan, S.K. (Ed.), The Palaeobiology of Trace Fossils. John Wiley & Sons, Chichester, pp. 105–133.

Fallas, K.M., MacNaughton, R.B., 2012. Distribution of Cambrian formations in the eastern Mackenzie Mountains, Northwest Territories. Geological Survey of Canada, Current Research (Online) no. 2012-2, 16 pages, doi:10.4095/289498.

Fillion, D., Pickerill, R.K., 1990. Ichnology of the Upper Cambrian? to Lower Ordovician Bell Island and Wabana groups of eastern Newfoundland, Canada. Palaeontogr. Can. 7, 1–119.

Fortey, R.A., Seilacher, A., 1997. The trace fossil *Cruziana semiplicata* and the trilobite that made it. Lethaia 30, 105–112.

Gehling, J.G., Jensen, S., Droser, M.L., Myrow, P.M., Narbonne, G.M., 2001. Burrowing below the basal Cambrian GSSP, Fortune Head, Newfoundland. Geol. Mag. 138, 213–218.

Genise, J.F., 2004. Ichnotaxonomy and ichnostratigraphy of chambered trace fossils in palaeosols attributed to coleopterans, ants and termites. In: McIlroy, D. (Ed.), The Application of Ichnology to Palaeoenvironmental and Stratigraphic Analysis. Geol. Soc. London, Spec. Publ., vol. 228, pp. 419–453.

Genise, J.F., Mángano, M.G., Buatois, L.A., Laza, J., Verde, M., 2000. Insect trace fossil associations in paleosols: the *Coprinisphaera* Ichnofacies. Palaios 15, 33–48.

Getty, P.R., Hagadorn, J.W., 2008. Reinterpretation of *Climactichnites* Logan 1860 to include subsurface burrows, and erection of *Musculopodus* for resting traces of the trailmaker. J. Paleontol. 82, 1161–1172.

Geyer, G., Uchman, A., 1995. Ichnofossil assemblages from the Nama Group (Neoproterozoic-Lower Cambrian) in Namibia and the Proterozoic-Cambrian boundary problem revisited. Beringeria, Spec. Iss. 2, pp. 175–202.

Grotzinger, J., Bowring, S.A., Saylor, B.Z., Kaufman, A.J., 1995. Biostratigraphic and geochronologic constraints on early animal evolution. Science 270, 598–604.

Haubold, H., 1996. Ichnotaxonomie und Klassifikation von Tetrapodenfährten aus dem Perm. Hallesches Jb. Geowiss. B 18, 23–88.

Haubold, H., 2000. Tetrapodenfährten aus dem Perm—Kenntnisstand und Progress 2000. Hallesches Jb. Geowiss. B 22, 1–16.

Haubold, H., Katzung, G., 1978. Paleoecology and paleoenvironments of tetrapod footprints from the Rotliegend (Lower Permian) of Central Europe. Palaeogeogr. Palaeoclimatol. Palaeoecol. 23, 307–323.

Herbosch, A., Verniers, J., 2011. What is the biostratigraphic value of the ichnofossil *Oldhamia* for the Cambrian?: a review. Geol. Belgica 14, 229–248.

Hunt, A.P., Lockley, M.G., Conrad, K.L., Paquette, M., Chure, D.J., 1993. Late Triassic vertebrates from the Dinosaur National Monument area (Utah, USA) with an example of the utility of coprolites for correlation. In: Lucas, S.G., Morales, M. (Eds.), The Nonmarine Triassic. New Mex. Mus. Nat. Hist. Sci., Bull. 3, pp. 197–198.

Hunt, A.P., Lucas, S.G., Lockley, M.G., 1998. Taxonomy and stratigraphic and facies significance of vertebrate coprolites of the Upper Triassic Chinle Group, western United States. Ichnos 5, 225–234.

Jensen, S., 2003. The Proterozoic and earliest trace fossil record: patterns, problems and perspectives. Integr. Compar. Biol. 43, 219–228.

Jensen, S., Droser, M.L., Gehling, J.G., 2006. A critical look at the Ediacaran trace fossil record. In: Kaufman, J., Xiao, S. (Eds.), Neoproterozoic Geobiology and Paleobiology. Topics in Geobiol., vol. 27, pp. 115–157.

Jensen, S., Bogolepova, O.K., Gubanov, A.P., 2011. *Cruziana semiplicata* from the Furongian (Late Cambrian) of Severnaya, Zemlya Archipelago, Arctic Russia, with a review of the spatial and temporal distribution of this ichnospecies. Geol. J. 46, 26–33.

Jensen, S., Buatois, L.A., Mángano, M.G., in press. Palaeogeographical patterns in the distribution of Cambrian trace fossils. In: Harper, D., Servais, T. (Eds.), Early Palaeozoic Palaeobiogeography and Palaeogeography. Geol. Soc. London, Mem.

Knaust, D., 2004. Cambro-Ordovician trace fossils from the SW-Norwegian Caledonides. Geol. J. 39, 1–24.

Landing, E., 1994. Precambrian-Cambrian boundary global stratotype ratified and a new perspective of Cambrian time. Geology 22, 179–182.

Lindholm, R.M., Casey, J.F., 1990. The distribution and possible biostratigraphic significance of the ichnogenus *Oldhamia* in the shales of the Blow Me Down Brook Formation, western Newfoundland. Can. J. Earth Sci. 27, 1270–1287.

Lockley, M.G., 1998. The vertebrate track record. Nature 396, 429–432.

Lockley, M.G., 2000. Philosophical perspectives on theropod track morphology: blending qualities and quantities in the science of ichnology. Gaia: Rev. Geociencias. Museu Nacional Hist. Nat. Lisbon 15, 279–300.

Lucas, S.G., 2007. Tetrapod footprint biostratigraphy and biochronology. Ichnos 14, 5–38.

MacNaughton, R.B., 2007. The application of trace fossils to biostratigraphy. In: Miller III, W. (Ed.), Trace Fossils. Concepts, Problems, Prospects. Elsevier, Amsterdam, pp. 135–158.

MacNaughton, R.B., Narbonne, G.M., 1999. Evolution and ecology of Neoproterozoic-Lower Cambrian trace fossils, NW Canada. Palaios 14, 97–115.

MacNaughton, R.B., Pickerill, R.K., 2010. A cautionary note on the use of invertebrate trace fossils for correlation in the Triassic-Jurassic Fundy Group. Atlant. Geol. 46, 90–94.

Magwood, J.P., Pemberton, S.G., 1990. Stratigraphic significance of *Cruziana*: new data concerning the Cambrian-Ordovician ichnostratigraphic paradigm. Geology 18, 729–732.

Mángano, M.G., Buatois, L.A., 2003. Trace fossils. In: Benedetto, J.L. (Ed.), Ordovician Fossils of Argentina. Secretaría de Ciencia y Tecnología Universidad Nacional de Córdoba, Argentina, pp. 507–553.

Mángano, M.G., Buatois, L.A., Aceñolaza, G.F., 1996. Trace fossils and sedimentary facies from an Early Ordovician tide-dominated shelf (Santa Rosita Formation, northwest Argentina): implications for ichnofacies models of shallow marine successions. Ichnos 5, 53–88.

Mángano, M.G., Carmona, N.B., Buatois, L.A., Muñiz Guinea, F., 2005. A new ichnospecies of *Arthrophycus* from the Upper Cambrian-lower Tremadocian of northwest Argentina: implications for the Arthrophycid lineage and potential in ichnostratigraphy. Ichnos 12, 179–190.

Martin, M.W., Grazhdankin, D.V., Bowring, S.A., Evans, D.A.D., Fedonkin, M.A., Kirschvink, J.L., 2000. Age of Neoproterozoic bilaterian body and trace fossils, White Sea, Russia: implications for metazoan evolution. Science 288, 841–845.

Melchor, R.N., Bromley, R.G., Bedatou, E., 2010. *Spongeliomorpha* in nonmarine settings: an ichnotaxonomic approach. Trans. R. Soc. Edinburgh, Earth Environ. Sci. 100, 429–436.

Mirré, J.C., Aceñolaza, F.G., 1972. El hallazgo de *Oldhamia* sp. (traza fósil) y su valor como evidencia de edad cámbrica para el supuesto Precámbrico del borde occidental del Aconquija, Prov. de Catamarca. Ameghiniana 9, 72–78.

Narbonne, G.M., 1984. Trace fossils in Upper Silurian tidal flat to basin slope carbonates of Arctic Canada. J. Paleontol. 58, 398–415.

Narbonne, G.M., Myrow, P.M., Anderson, M.M., 1987. A candidate stratotype for the Precambrian-Cambrian boundary, Fortune Head, Burin Peninsula, southeastern Newfoundland. Can. J. Earth Sci. 24, 1277–1293.

Pickerill, R.K., Fillion, D., 1983. On the Tremadoc-Arenig and Lower-Upper Tremadoc boundaries in the Bell Island Group, Conception Bay, eastern Newfoundland. Marit. Sediment. Atlant. Geol. 19, 21–30.

Pickerill, R.K., Fyffe, L.R., 1999. The stratigraphic significance of trace fossils from the Lower Paleozoic Baskahegan Lake Formation near Woodstock, west-central New Brunswick. Atlant. Geol. 35, 205–214.

Rainforth, E.C., 2003. Revision and re-evaluation of the Early Jurassic dinosaurian ichnogenus *Otozoum*. Palaeontology 46, 803–838.

Seilacher, A., 1967. Fossil behavior. Sci. Am. 217, 72–80.

Seilacher, A., 1970. *Cruziana* stratigraphy of "non-fossiliferous" Palaeozoic sandstones. In: Crimes, T.P., Harper, J.C. (Eds.), Trace Fossils, Geol. J., Spec. Iss. 3, pp. 447–476.

Seilacher, A., 1974. Flysch trace fossils: evolution of behavioural diversity in the deep-sea. N. Jb. Geol. Paläont. Mh. 1974, 233–245.

Seilacher, A., 1985. Trilobite palaeobiology and substrate relationships. Trans. R. Soc. Edinburgh Earth Sci. 76, 231–237.

Seilacher, A., 1990. Paleozoic trace fossils. In: Said, R. (Ed.), The Geology of Egypt. A.A. Balkema, Rotterdam, pp. 649–670.

Seilacher, A., 1992. An updated *Cruziana* stratigraphy of Gondwanan Palaeozoic sandstones. In: Salem, M.J. (Ed.), The Geology of Lybia, vol. 5. Elsevier, Amsterdam, pp. 1565–1580.

Seilacher, A., 1994. How valid is *Cruziana* stratigraphy? Geol. Rundsch. 83, 752–758.

Seilacher, A., 1996. Evolution of burrowing behaviour in Silurian trilobites: Ichnosubspecies of *Cruziana acacensis*. In: Salem, M.J. (Ed.), The Geology of the Sirt Basin, vol. 1–3. Elsevier, Amsterdam, pp. 523–530.

Seilacher, A., 2000. Ordovician and Silurian arthrophycid Ichnostratigraphy. In: Sola, M.A., Worsley, D. (Eds.), Geological Exploration in Murzuk Basin. Elsevier, Amsterdam, pp. 237–258.

Seilacher, A., 2007. Trace Fossil Analysis. Springer, Berlin, 226 pp.

Seilacher, A., Buatois, L.A., Mangano, M.G., 2005. Trace fossils in the Ediacaran-Cambrian transition: behavioural diversification, ecological turnover and environmental shift. Palaeogeogr. Palaeoclimatol. Palaeoecol. 227, 323–356.

Uchman, A., 2003. Trends in diversity, frequency and complexity of graphoglyptid trace fossils: evolutionary and palaeoenvironmental aspects. Palaeogeogr. Palaeoclimatol. Palaeoecol. 192, 123–142.

Uchman, A., 2004. Phanerozoic history of deep-sea trace fossils. In: McIlroy, D. (Ed.), The Application of Ichnology to Palaeoenvironmental and Stratigraphic Analysis. Geol. Soc. London, Spec. Publ., vol. 228, pp. 125–139.

Walter, M.R., Elphinstone, R., Heys, G.R., 1989. Proterozoic and Early Cambrian trace fossils from the Amadeus and Georgina Basins, central Australia. Alcheringa 13, 209–256.

Yochelson, E.L., Fedonkin, M.A., 1993. Paleobiology of *Climactichnites*, an enigmatic Late Cambrian fossil. Smithsonian Contr. Paleobiol. 74, 1–74.

Chapter 8

Microbioerosion

Max Wisshak[1]
Senckenberg am Meer, Marine Research Department, Wilhelmshaven, Germany
[1]*Corresponding author: e-mail: max.wisshak@senckenberg.de*

1. INTRODUCTION

Barely visible to the naked eye, a diverse assemblage of microborings—alongside the paleoenvironmental information encoded by them—may be encountered on the spatial scale of a thumbnail-sized shell fragment. These microbioerosion features are less than a millimeter in size with tunnel diameters of commonly less than a hundred microns and are being produced by euendolithic microorganisms, including bacteria, cyanobacteria, chlorophytes, rhodophytes, and fungi, which either find shelter from grazers and physical disturbances or seek nutrition within the mineralized substrate they dissolve (Golubic et al., 1975, 1981). Most microborings perfectly match the outline of the organisms producing them and can thus be assigned to their living tracemaker with a fair degree of confidence (Radtke, 1991). A favorable concert of evolutionary longevity and environmental dependency make them superb paleoenvironmental indicators for all Phanerozoic strata that contain fossil skeletal material in a proper quality of preservation (Glaub and Vogel, 2004; Glaub et al., 2007; Vogel and Glaub, 2004).

Since 90% of, for instance, the known Carboniferous microboring ichnotaxa are still found in today's oceans (Wisshak et al., 2008), neoichnological observations regarding their physiological and ecological demands of their producers allow a proper characterization of these paleoenvironmental indicators. Keeping the obligatory reservations concerning an actualistic approach in mind, such as possible evolutionary changes in ecological demands of certain biota, is of course mandatory. It is worth mentioning that microbioerosion is not limited to calcareous substrates, which have received by far the most attention, but microbial boring activity has been reported also from wood (Genise, 2004), bone (Jans, 2008), conodonts (Glaub and Königshof, 1997), and even magmatic rocks (McLoughlin et al., 2008).

The aim of this chapter is to provide the sedimentologist, ichnologist, or paleoecologist with a comprehensive overview on microboring biota and the trace fossils they produce. This is complemented by a detailed account on the

Developments in Sedimentology, Vol. 64. http://dx.doi.org/10.1016/B978-0-444-53813-0.00008-3

methodological tools at hand, and a review and outlook of their applicability for paleoenvironmental studies with respect to relative bathymetry, temperature, salinity, and the trophic regime. For all of these applications, the current state of knowledge on relevant ichnotaxa, their known or inferred producers, and their geological record is compiled in tables that can serve as a guideline during paleoenvironmental assessments.

For further information on the various aspects of microbioerosion, the reader is referred to the review papers on bioerosion in general (Warme, 1975), microbioerosion in general (Golubic et al., 1975, 1981; Tribollet et al., 2010), cyanobacterial microbioerosion (Schneider and Le Campion-Alsumard, 1999), fungal microbioerosion (Golubic et al., 2005), high-latitude bioerosion (Wisshak, 2006), tropical reef microbioerosion (Tribollet, 2008), the fossil record (Glaub and Vogel, 2004), and geological implications (Glaub et al., 2007). An extensive bibliographic source is the continuously growing online bibliography maintained by Wilson (2008, 2012).

2. THE METHODOLOGICAL TOOLKIT

At present, three methods are at our disposal for visualizing trace fossils left by microbioerosion agents in lithic substrates. These are microscopy of petrographic thin sections, scanning electron microscopy (SEM) of casts obtained by the epoxy cast-embedding technique, and sophisticated microcomputed tomography (micro-CT) that is about to conquer a resolution in the nanometer scale, enabling non-destructive investigations of even minute bioerosion features.

2.1 Thin Sectioning

Being a standard tool in microfacies analyses, transmission microscopy of petrographic thin sections may also be applied for microbioerosion analyses, but bears the limitations adherent to any two-dimensional (2D) visualization of three-dimensional (3D) objects (Fig. 1). On the other hand, this method does comprise a number of advantages such as that (1) microbioerosion patterns can be visualized together with the surrounding matrix, (2) the true penetration depth of microborings can easily be determined since they are not prone to collapse such as in the cast-embedding technique (see below), and (3) even those fossil microborings can be studied that are filled with spar cement or other secondary mineralizations prohibiting the standard cast-embedding technique.

When preparing thin sections of semiconsolidated and/or relatively porous and permeable samples for microbioerosion analyses, an impregnation with a low-viscosity epoxy resin such as Ciba-Geigy Araldite BY158 + Aradur 21 is advised. It is advantageous to stain the resin with 2% (by weight) of Keystone Oil Blue powder for an easy visualization of the pore space within the microbioerosion cavities. The degree of impregnation can be enhanced when placing

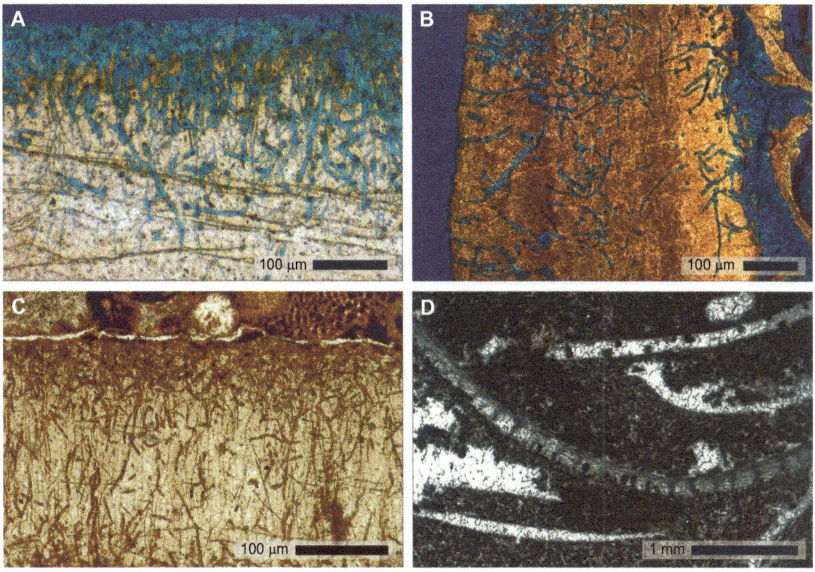

FIGURE 1 Microbioerosion in thin sections. (A) The ichnotaxa *Eurygonum nodosum* and *Scolecia filosa* in a recent bivalve shell from the Azores. The shell was vacuum-impregnated with blue-stained epoxy resin prior to sectioning. (B) The fungal trace *Orthogonum lineare* in a Pleistocene cold-water coral from Rhodes, Greece. The microborings were still empty and accessible for stained vacuum impregnation. (C) Same assemblage as in A, but from the Carboniferous Buckhorn Asphalt Lagerstätte, USA. The microborings are naturally impregnated with asphalt. (D) The fungal microboring *Saccomorpha clava*, well preserved in a brachiopod shell (bottom), and a completely recrystallized bivalve shell (top) from the Silurian of Gotland, Sweden (courtesy of Axel Munnecke).

the sample in a vacuum chamber at about 10 kPa for a couple of minutes and impregnation of the sample surface after cutting.

2.2 Vacuum Cast-Embedding

The most approved method for visualizing the 3D architecture of tunnel systems left by microbioeroders is the vacuum cast-embedding technique (e.g., Nielsen and Maiboe, 2000; Wisshak, 2006), which produces polymer resin casts (Fig. 2). These casts can be studied under the SEM in order to display even the most delicate morphological features. A prerequisite for this technique is, however, that the borings are still empty or filled only with little-lithified matrix. Immersion in a tenside for several hours (e.g., Goldschmidt Rewo GmbH: Rewoquat W3690), directly followed by a treatment with hydrogen peroxide, shatters and removes clayey matrix within the borings. Final cleaning in an ultrasonic bath, rinsing in distilled water, and drying complete the preembedment procedure. The samples are then placed in a small vacuum chamber specifically designed for cast-embedding (Struers CitoVac or its precursor Epovac). The chamber allows the infiltration of several samples with a

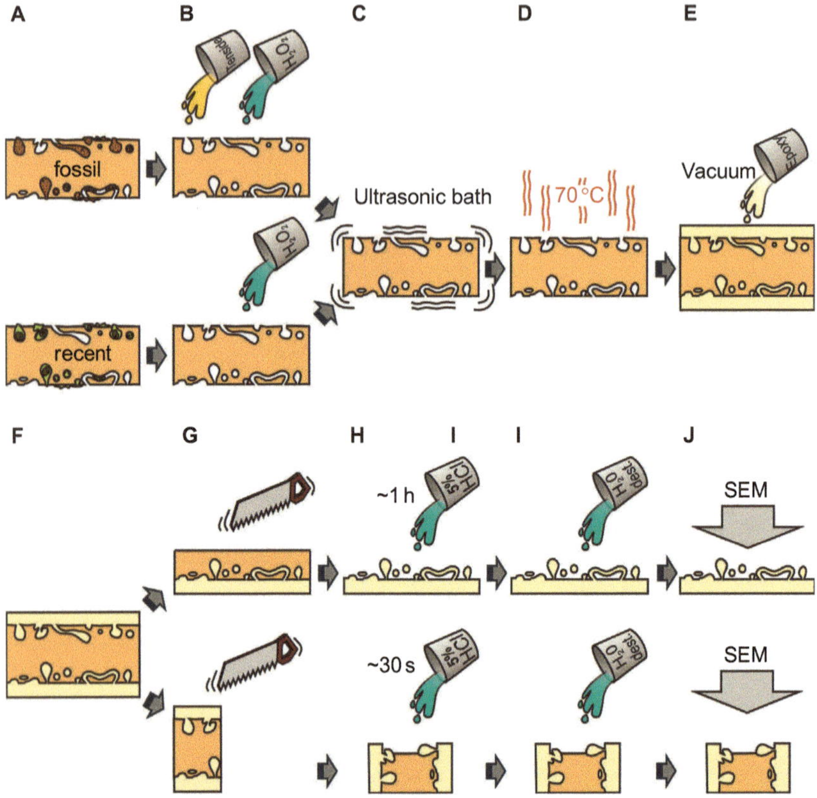

FIGURE 2 Schematic sketch illustrating the vacuum cast-embedding technique. (A) The bored calcareous substrate, either fossil with partial sediment fill or recent with microendoliths and biofilms. (B) Removal of sediment or organic material by treatment with tensides and/or hydrogen peroxide. (C) Cleaning in an ultrasonic bath. (D) Drying at 70 °C. (E) Infill with epoxy resin under vacuum conditions. (F) Curing for 24 h. (G) Formatting of blocks for surface SEM and cross-section SEM with a rock saw. (H) Removal of calcareous substrate via treatment with diluted hydrochloric acid. (I) Rinsing with distilled water. (J) SEM visualization after drying and sputter-coating with gold.

low-viscosity epoxy resin (e.g., Ciba-Geigy: Araldite BY158 + Aradur 21) *while* the samples are held under vacuum. In this way, a complete infiltration of the borings is assured while trapping of air is avoided. After curing, the samples are formatted with a rock saw, decalcified by a treatment with diluted HCl (~5%), rinsed in distilled and filtered water, and dried. The samples—now exhibiting the positive casts of the borings—are then mounted on aluminum stubs before gold sputter-coating and SEM analysis.

In order to prevent delicate casts from collapsing and to aid the recognition of the actual penetration depth of the traces, a complemental procedure is to prepare sample cross sections that are only partially etched (~10–30 s with ~5% HCl), allowing a visualization of the traces in context of the shell or rock matrix.

In case a suitable vacuum chamber is not available, one can alternatively apply the original cast-embedding technique as established by Golubic et al. (1970) where a sequence of different resin/acetone mixtures is used to impregnate and dehydrate the samples. The latter method is still important in neoichnological studies as the first step of the double-embedding technique (Golubic et al., 1970, 1975) allows the study of the embedded euendoliths within their borings. Ultra-thin sectioning of these reembedded biological samples can further be investigated with transmission electron microscopy for study of the cellular structure.

2.3 Micro-CT

X-ray computed tomography (CT), and particularly micro-CT, is becoming an increasingly popular 3D visualization tool for paleontological investigations of various kinds (e.g., Dierick et al., 2007; Sutton, 2008). The highest resolution is currently achieved with synchrotron microtomography that utilizes monochromatic X-ray with very high beam intensity that allows a voxel size well below 1 μm—the spatial magnitude of microborings (e.g., Tafforeau et al., 2006). But even conventional, more affordable, polychromatic X-ray micro-CT is rapidly evolving and is about to enable resolutions at a submicron scale. Related technologies worth mentioning are magnetic resonance imaging that detects subtle differences in water or organic molecule content in fossils (e.g., Mietchen et al., 2008) and neutron tomography that utilizes a neutron beam instead of X-ray and has advantages when visualizing organic fossil material (e.g., Winkler, 2006).

The potential of all these technologies grounds on the fact that they are non-destructive and have the ability to characterize sediment-filled fossil borings otherwise inaccessible for the vacuum cast-embedding technique. Modern CT scanners with sophisticated analysis software are able to detect very subtle density differences that are necessary to visualize lithified structures. For bioerosion analyses, this has been exemplified by Beuck et al. (2008) with the establishment of the macroboring ichnotaxon *Osprioneides kampto* within a Silurian stromatoporoid, and by Tapanila (2008) who described the new microboring *Pyrodendrina cupra* partly based on micro-CT scans of pyrite-filled fossil material. The third major advantage of CT technology is to present results in printable and/or screenable 3D formats such as colored semitransparent views (Fig. 3A), stereoscopic anaglyph techniques (Fig. 3B), or online supplementary material as conventional or anaglyph 3D animations (see, e.g., Beuck et al., 2007, 2008, and http://app.pan.pl/acta53/app53-Beuck_etal_SOM.pdf).

As yet, CT application in bioerosion analysis is still anecdotal but the potential for microbioerosion analysis with respect to 2D and 3D morphometry, drill-core scanning, and the non-destructive study of well-lithified deep-time ichnocoenoses is obvious.

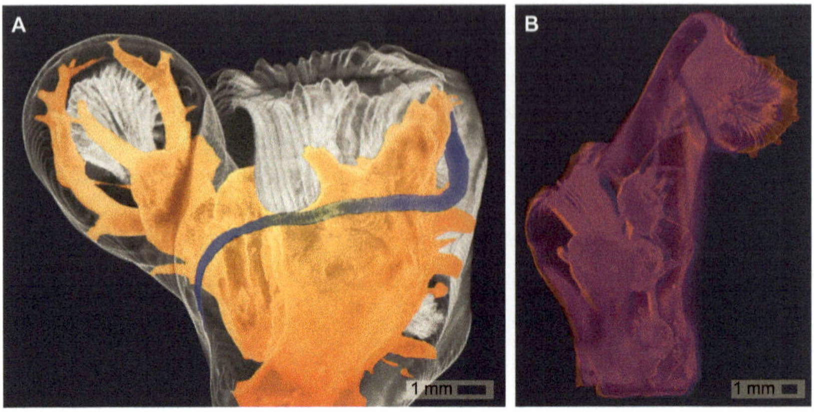

FIGURE 3 Micro-CT scans of the cold-water coral *Lophelia pertusa* attacked *syn vivo* by a bioeroding sponge (courtesy of Lydia Beuck). (A) Colored semitransparent visualization of corallite with the sponge cavity (orange) and an embedded epibiontic worm (blue). (B) Anaglyph 3D visualization of another corallite with heavy sponge infestation (requires anaglyph glasses with left eye red and right eye blue).

3. MICROBIOERODING BIOTA AND THEIR TRACES

In the following account, a brief overview on the physiology, ecology, systematics, and fossil record of the main agents of microbioerosion is given and relevant key ichnotaxa for each group are illustrated by SEM images of vacuum epoxy resin casts.

3.1 Cyanobacteria

The majority of cyanobacteria are obligate photoautotrophs but facultative photoheterotrophy and even chemoheterotrophy have been reported for certain species (e.g., May and Perkins, 1979). Facultative heterotrophy is discussed, for instance, for the endolithic *Plectonema terebrans* which, as a particularly low-light specialist, copes with dysphotic conditions (deepest record in 370 m water depth; Lukas, 1978). Endolithic cyanobacteria reproduce by means of varying vegetative strategies and are most active in the supratidal to upper subtidal range (Radtke et al., 1996; Schneider and Le Campion-Alsumard, 1999). They seek shelter from grazers in calcareous substrates, which they penetrate by chemical means via carbonic acid formed from metabolic CO_2 (Schneider, 1976) or the support of endosymbiotic bacteria (Lukas and Golubic, 1981). Further, biogeochemically more favorable modes of boring have been proposed in the form of a temporal or spatial separation of photosynthesis and boring activity or a "calcium pump" in the sense of an extrusion of calcium ions through an active cellular uptake and transport process (Garcia-Pichel, 2006).

An endolithic mode of life is common in the cyanobacteria with the most characteristic representatives being found in the genera *Hyella*, *Solentia*, *Mastigocoleus,* and *Plectonema* and corresponding ichnospecies are found within the ichnogenera *Fascichnus* (Fig. 4A–D), *Planobola*, *Eurygonum* (Fig. 4E), and *Scolecia* (Fig. 4F).

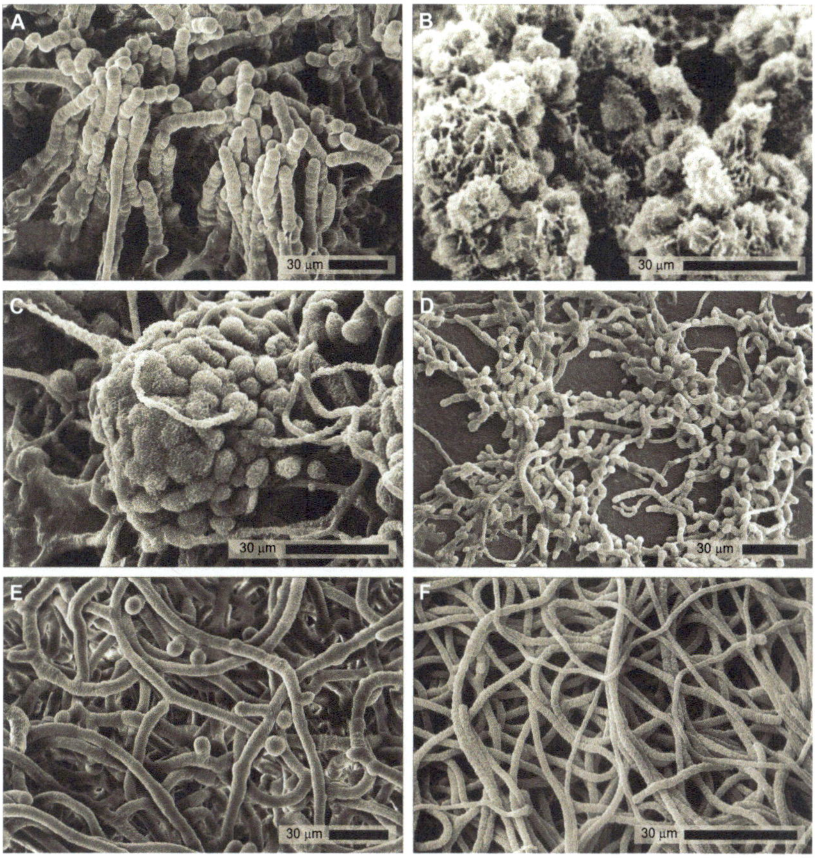

FIGURE 4 SEM images of casted microborings produced by cyanobacteria. (A) *Fascichnus dactylus*, recent, Faial, Azores, intertidal. (B) *Fascichnus acinosus*, Upper Jurassic, Consolacao, Portugal (courtesy of Ingrid Glaub). (C) *Fascichnus rogus*, recent, Kosterfjord, Sweden, intertidal. (D) *Fascichnus parvus*, recent, Faial, Azores, 15 m water depth. (E) *Eurygonum nodosum*, recent, Faial, Azores, intertidal. (F) *Scolecia filosa*, recent, Faial, Azores, 60 m water depth.

The endolithic life habit of cyanobacteria dates at least as far back as the Early Proterozoic with the oldest record of *Euhyella campbellii* reported from 1.7 Ga old stromatolites discovered in China (Zhang and Golubic, 1987). Some representatives of this genus are still around today and illustrate the evolutionary longevity of microendoliths.

3.2 Chlorophytes and Rhodophytes

Bioeroders are found within the Chlorophyta (green algae) as well as Rhodophyta (red algae). They are eukaryotes which reproduce by vegetative or sexual means. Chlorophytes are photoautotrophs and only for the low-light extremist *Ostreobium quekettii* a facultative chemoheterotrophy was suggested (e.g.,

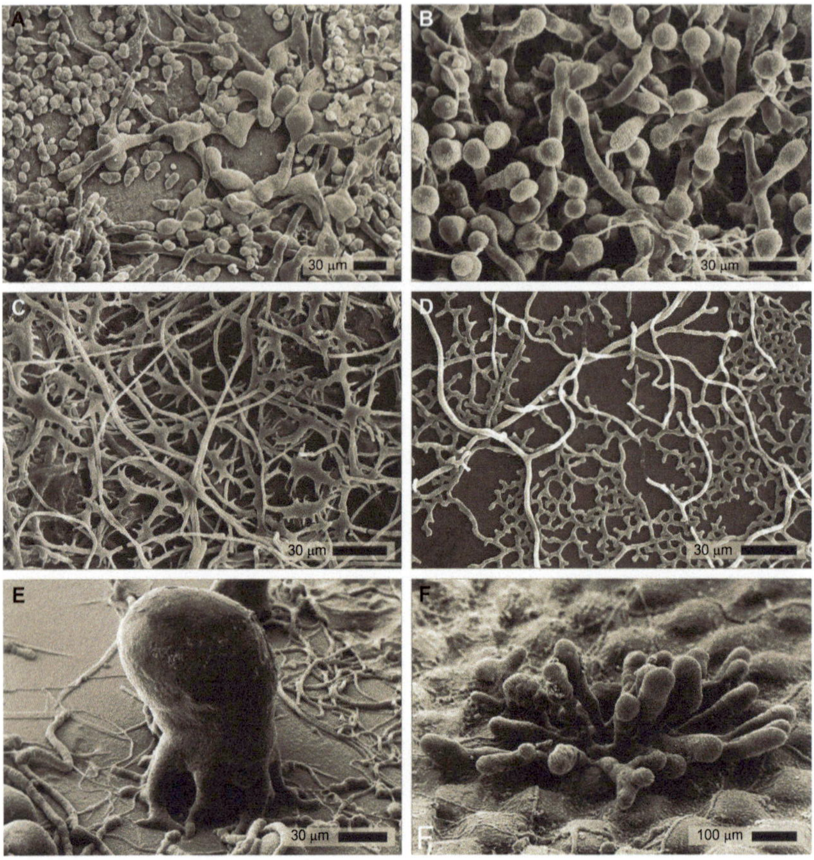

FIGURE 5 SEM images of casted microborings produced by chlorophytes. (A) *Rhopalia catenata*, recent, Sapienza Island, Greece, 15 m water depth (courtesy of Ines Pyko). (B) *Rhopalia clavigera*, recent, Faial, Azores, 15 m water depth. (C) *Rhopalia spinosa*, recent, Faial, Azores, 60 m water depth. (D) *Ichnoreticulina elegans*, recent, Sapienza Island, Greece, 100 m water depth (courtesy of Ines Pyko). (E) *Cavernula pediculata*, recent, Kosterfjord, Sweden, intertidal. (F) *Fascichnus grandis*, Eocene, Paris Basin, Fercourt, France (courtesy of Gudrun Radtke).

Schroeder, 1972) and disputed (Glaub, 1994). Chlorophytes bore their domiciles by chemical means, possibly by utilizing the CO_2 produced during nighttime respiration formed into carbonic acid (Schneider, 1976) or the photosynthesis/dissolution separation or "calcium pump" mechanism proposed by Garcia-Pichel (2006), as noted above.

Important endolithic chlorophytes are represented by the genera *Phaeophila*, *Eugomontia,* and *Ostreobium*. Others are endolithic only during the Codiolum stage of their life cycle, such as *Gomontia*, or they are epibiliths that form endolithic holdfasts such as in *Acetabularia*. Ichnogenera affiliated to chlorophyte tracemakers are *Rhopalia* (Fig. 5A–C), *Ichnoreticulina* (Fig. 5D), *Cavernula* (Fig. 5E), and one ichnospecies of *Fascichnus* (Fig. 5F). Endolithic rhodophytes show little diversity, even though they are understudied, and are found within the

Bangiaceae genera *Porphyra* and *Bangia*. Their complex heteromorphic life cycle includes an endolithic Conchocelis stage that has received semiichnological treatment in the form of *Paleoconchocelis* by Campbell et al. (1979; described as body fossil but including image of cast).

Chlorophytes first appear in the Cambrian and the oldest endolithic forms are known from the Ordovician, as recorded by the ichnospecies *Ichnoreticulina elegans* and *Cavernula pediculata* (Vogel and Brett, 2009). Fossil rhodophytes and particularly the endolithic genus *Bangia* have been reported from strata as old as the Proterozoic (Butterfield et al., 1990), whereas the ichnological record commences in the Ordovician (Vogel and Brett, 2009).

3.3 Fungi

Endolithic marine fungi are chemoheterotrophic and reproduce by vegetative and sexual means. They biodegrade calcareous shells in order to exploit mineralized organic matter such as organic lamellae, conchiolin, or chitin, as well as other endolithic organisms (Golubic et al., 2005; Jones and Pemberton, 1987). As a consequence, they are not restricted to the photic zone but are found down to abyssal depths, except for where they engage in parasitic, saprophytic, and symbiotic relationships, which may result in an indirect light dependency (Golubic et al., 2005). This is, for instance, the case where fungi and cyanobacteria and/or algae consort as lichens, which show a specific zonation pattern in the inter- and supratidal zones (Schneider, 1976). Euendolithic and cryptoendolithic fungi are, together with chlorophytes, often found thriving in a complex interaction within living tropical corals (Priess et al., 2000). Little is known about the mode of penetration in marine fungi. Jones and Pemberton (1987) found organic acids such as citric acid, and Golubic et al. (2005) suggest the presence of specific enzymes for digestion of the organic matter facilitating penetration of shells rich in organic matrices.

At least 16 genera among the Eumycetes (higher fungi) and Phycomycetes (lower fungi) are known to follow an endolithic mode of life (Zeff and Perkins, 1979). Important genera are *Dodgella*, *Conchyliastrum*, *Ostracoblabe*, *Lithopythium*, and *Schizochytrium*. Corresponding ichnospecies are found within the ichnogenera *Saccomorpha* (Fig. 6A–C), *Polyactina* (Fig. 6D), *Orthogonum* (Fig. 6E), and *Flagrichnus* (Fig. 6F).

The oldest marine endolithic fungi, the Chytridiomycetes, were reported from Cambrian shell material (Müller and Löffler, 1992) and the first ichnotaxa *Orthogonum fusiferum* and *Saccomorpha clava* are known from the Ordovician (Vogel and Brett, 2009).

3.4 Other Chemotrophs

A number of further microendoliths or their traces are commonly encountered down to aphotic depths and are thus identified as chemotrophic in nature. The biological identity of two very common traces, *Orthogonum lineare* (Fig. 7A)

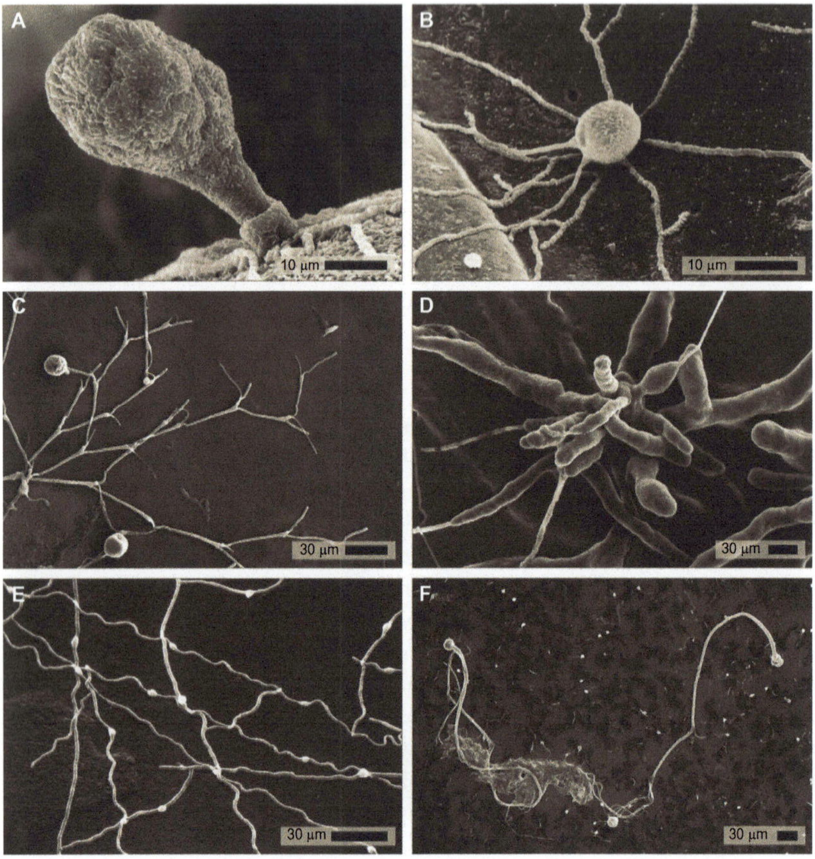

FIGURE 6 SEM images of casted microborings produced by fungi. (A) *Saccomorpha clava*, Pleistocene, Rhodes, Greece. (B) *Saccomorpha sphaerula*, Eocene, Paris Basin, Villiers-St.-Fréderic, France (courtesy of Gudrun Radtke). (C) *Saccomorpha* cf. *terminalis*, recent, Kosterfjord, Sweden, 85 m water depth. (D) *Polyactina araneola*, recent, Kosterfjord, Sweden, 50 m water depth. (E) *Orthogonum fusiferum*, recent, Kosterfjord, Sweden, 1 m water depth. (F) *Flagrichnus profundus*, recent, Faial, Azores, 150 m water depth.

and *Flagrichnus baiulus* (Fig. 7B), is not known to date but is likely to be found among the fungi (Glaub, 1994; Wisshak and Porter, 2006).

Apart from the ubiquitous cyanobacteria, also other (Eu)bacteria and Archaea are likely involved directly in the degradation of calcareous substrates (Friedman et al., 1971) and indirectly through the related process of maceration (Freiwald, 1995, 1998). Bacteria, for instance, are responsible for the karstification of limestone bedrock (Golubic and Schneider, 1979) and are an important catalyst in speleogenesis (Barton and Jurado, 2007). However, they are difficult to address, owing to their minute size, but advances are made via molecular methods. To date, only two ichnotaxa have been tentatively related to the activity of bioeroding bacteria (Budd and Perkins, 1980; Radtke, 1991), *Scolecia serrata* (Fig. 7C), and *Scolecia maeandria*, the former being already known from Carboniferous strata (Wisshak et al., 2008).

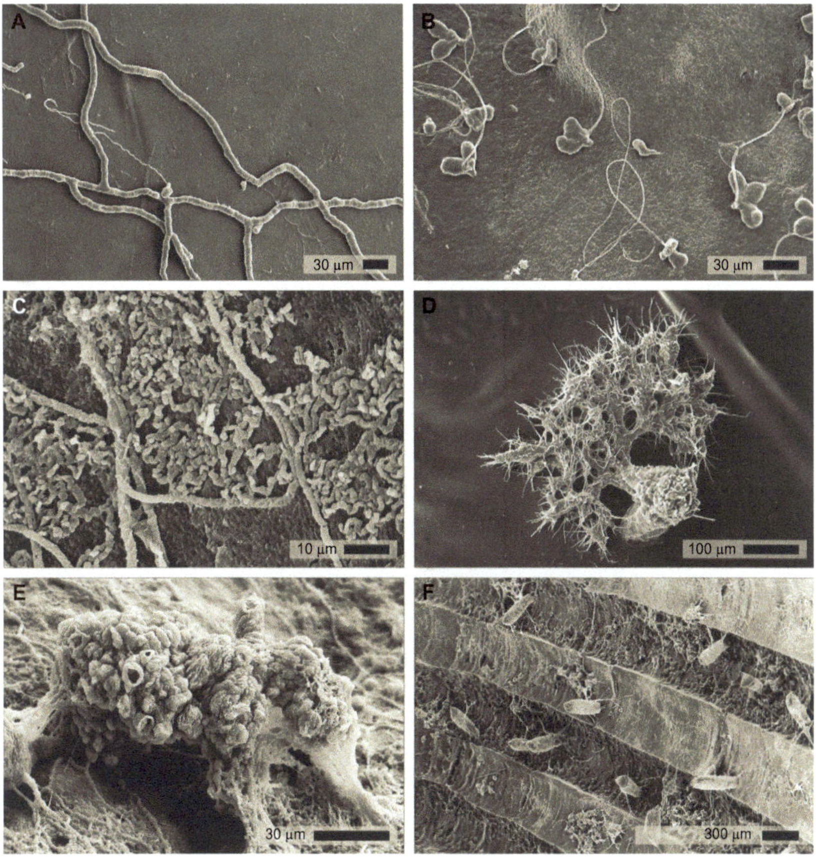

FIGURE 7 SEM images of casted microborings produced by various chemotrophs. (A) *Orthogonum lineare*, recent, Faial, Azores, 500 m water depth. (B) *Flagrichnus baiulus*, Svalbard, Norway. (C) *Scolecia serrata*, Carboniferous, Buckhorn Asphalt Lagerstätte, USA. (D) *Semidendrina pulchra*, recent, Faial, Azores, 60 m water depth. (E) *Entobia mikra*, Pleistocene, Rhodes, Greece. (F) *Imergentia* isp., Pleistocene, Rhodes, Greece.

The ichnofamily Dendrinidae bears a number of distinct fan-shaped traces, the identity of which is not entirely solved either, even though the most likely candidates are naked foraminiferans as discussed for *Semidendrina pulchra* (Bromley et al., 2007; Fig. 7D). Fossil representatives of this ichnofamily comprise the ichnogenera *Dendrina*, *Clionolithes*, *Dendrorete*, *Nododendrina*, *Platydendrina*, *Ramodendrina*, *Pyrodendrina*, and *Hyellomorpha*, which are most diverse in the Devonian and Cretaceous (Hofmann, 1996; Vogel et al., 1987).

With respect to the important group of mainly macroboring sponges, two dwarf forms have been described, as there are the ichnospecies *Entobia mikra* (Fig. 7E) and *Entobia nana* (Wisshak, 2008), the biotaxonomical identities of which have not been revealed as yet.

Finally, there is a diverse set of bioeroding ctenostome bryozoans that have produced characteristic networks of large microborings since the Early

Ordovician (Pohowsky, 1978). Most of these have been established as zootaxa but often mainly based on the morphology of their traces and thus ichnological characters (e.g., Pohowsky, 1978), leading to taxonomical confusion. This has partly been resolved by the establishment of respective ichnotaxa such as *Pennatichnus*, *Iramena* (Fig. 7F), and *Pinaceocladichnus* by Boekschoten (1970) and Mayoral (1988). Their possibly considerable potential as paleoenvironmental indicators has, however, not been explored to date.

4. PALEOENVIRONMENTAL SIGNATURES

4.1 Paleobathymetry

The investigation of microendolithic ichnocoenoses has been shown to be a strong tool for judging light availability and hence relative bathymetry in both modern and ancient environmental settings. This is based on the photoautotrophic character of many euendoliths and their specific low-light tolerance limit and light optimum, resulting in a distinct vertical zonation pattern. Cyanobacteria dominate the shallow euphotic zone in the supratidal to shallow subtidal range, chlorophytes are most abundant in the deep euphotic to dysphotic zone, and exclusively chemotrophs occur in aphotic depths (Akpan and Farrow, 1984; Budd and Perkins, 1980; Golubic et al., 1975; Günther, 1990; Perkins and Tsentas, 1976). From the intertidal toward the supratidal zone, this distribution pattern is complemented by a gradual decrease in endolithic biodiversity governed by the varying ability of the euendoliths to tolerate strong UV fluxes, partial desiccation, and rainfall (Hoffman, 1985; Le Campion-Alsumard, 1978; Radtke et al., 1996). This vertical distribution pattern was translated into a set of bathymetric index ichnocoenoses (Table 1) defined by the co-occurrence of specific key ichnotaxa and supported by general characteristics of the ichnocoenoses (Glaub, 1994; Glaub et al., 2001; Vogel et al., 1995, 2008). For high-latitude settings, *Cavernula pediculata* has been suggested as substitute for the rare *Fascichnus acinosus* (Wisshak, 2006). In addition, it is desirable to nominate *Rhopalia catenata* as a replacement for the problematic "*Paleoconchocelis starmachii*" based on the main occurrence of its producing chlorophyte, *Phaeophila dendroides*, in the shallow euphotic III to deep euphotic zone and its reasonably wide stratigraphic and biogeographic range.

Case studies in various modern tropical to high-latitude settings confirmed a remarkable consistency in the zonation pattern of the ichnocoenoses despite considerable latitudinal telescoping leading to a condensed photic zonation in high latitudes (e.g., Gektidis, 1997; Gektidis et al., 2007; Glaub, 2004; Glaub et al., 2002; Vogel et al., 2000; Wisshak, 2006; Wisshak et al., 2005, 2011). In parallel, the ichnocoenosis scheme has successfully been applied to a variety of fossil carbonate factories throughout the Phanerozoic, allowing, for instance, a comparison of the relative bathymetry of prominent reef occurrences and corresponding faunal associations in the Silurian, Permian, Triassic, and Jurassic (summarized in Vogel et al., 1995, 1999). An extensive microbioerosion

TABLE 1 The Photic Zonation as Reflected by Specific Index Ichnocoenoses and General Microborer Distribution Characteristics (*Modified from Glaub, 1994; Glaub et al., 2001; and Vogel et al., 2008*)

Photic zonation		Index ichnocoenoses	General characteristics
Euphotic zone (>1% surface illumination)	Shallow I (supratidal)	Not yet defined	Dominance of cyanobacteria with sheath pigmentation
	Shallow II (intertidal)	*Fascichnus acinosus*/ *Fascichnus dactylus*	Dominance of cyanobacteria, vertical orientation of borings
	Shallow III (subtidal)	*Fascichnus dactylus*/ *Rhopalia catenata* (replacing "*Paleoconchocelis starmachii*")	Cyanobacteria abundant and eukaryotes, change from vertical to horizontal orientation
	Deep	*Rhopalia catenata* (replacing "*Paleoconchocelis starmachii*")/ *Ichnoreticulina elegans*	Dominance of eukaryotes, mainly rhodophytes and chlorophytes, horizontal orientation, chemotrophs increasing, maximum diversity
Dysphotic zone (0.01–1% surface illumination)		*Saccomorpha clava*/ *Ichnoreticulina elegans*	Dominance of chemotrophs, additionally *Ichnoreticulina elegans* and/or *Scolecia filosa*
Aphotic zone (<0.01% surface illumination)		*Saccomorpha clava*/ *Orthogonum lineare*	Only chemotrophs

analysis of Ordovician strata showed that the early emplacement of the first index ichnocoenoses took place during the Late Ordovician (Vogel and Brett, 2009) and that the scheme can readily be applied to strata as old as the Silurian (Vogel et al., 1999).

Since microbioerosion analyses can be based on even few very small shell fragments, it is a particularly suited approach for paleoenvironmental assessments where only limited sample material is available—such as is the case in drill-core material. Despite this promising potential, corresponding studies so far remain scarce. Radtke (1998) was the first to identify and interpret microboring trace fossils from a drill-core in an Oligocene limnic-brackish lake setting of Sieblos/Rhön, Germany, and Radtke (2007) revealed salinity trends in a marine-limnic succession recorded in a core of the Oligocene Pechelbronn

Formation from the Mainz Basin, Germany. Only recently have two further case studies tackled drill-core material, both targeting Holocene reefal development in the Pacific. In the first study, Heindel et al. (2009) were able to trace the deepening-upward trend during the last postglacial sea-level rise and to link it to a possible onset of eutrophication based on a detailed ichnocoenosis analysis carried out on drill-core material from Tahitian reefs. In the second study, Chazottes et al. (2009) inversely investigated a series of Holocene shallowing-up sequences during reef growth recorded in coral cores from New Caledonia.

Caution for the application of microborings as a paleobathymetrical indicator concerns the following issues: (1) This method always provides only relative bathymetrical evidence in the form of an assignation to a specific photic zone, since other factors controlling the absolute depth (mainly the turbidity) are not known and may vary on a small scale (Perry and Macdonald, 2002). (2) A significant local variance in ichnocoenosis composition may further be encountered in spatially highly structured paleoenvironments such as a reef setting comprising illuminated, shaded, and even cryptic subhabitats ("cryptophotic" *sensu* Heindel et al., 2009) potentially supporting the whole range of index ichnocoenoses even at very shallow depth (Gektidis et al., 2007; Perry and Macdonald, 2002). (3) The investigated substrate needs to be evaluated with respect to potential transport from a shallower photic zone to its final site of deposition, making sessile substrates such as corals more reliable and calling for a reasonable amount of samples to be investigated in order to achieve a solid interpretation.

Table 2 summarizes the current state of knowledge about the bathymetrical distribution of relevant ichnotaxa and their known or inferred producers. While a wealth of studies has been carried out in intertidal to aphotic waters, the supratidal zone is understudied with respect to ichnotaxonomical treatment of the inhabiting euendolithic cyanobacteria, and the respective index ichnocoenosis is yet to be defined.

4.2 Paleothermometry

The evaluation of microborings as a paleotemperature and/or paleolatitude indicator is still in its infancy, even though a respective potential has long been pointed out (e.g., Golubic et al., 1975). In modern seas, many euendoliths are eurythermal and have a cosmopolitan distribution (e.g., *O. quekettii*, *P. terebrans*, *Dodgella priscus*, *Conchyliastrum enderi*), but certain taxa are stenotherm and occur within specific temperature limits resulting in a restricted biogeographic distribution. This applies, for instance, to many chlorophytes and cyanobacteria, which share a similar upper temperature limit of about 35 °C but show considerable differences with respect to their lower tolerance limit (Lüning, 1985). This factor acts in concert with the general trend of depletion in the phototrophic borer spectrum with higher latitude as a result of the seasonal fluctuation of light availability (Schmidt and Freiwald, 1993).

TABLE 2 **Ichnotaxa with Paleobathymetrical Relevance, Given with their Known or Inferred Producers, Known Stratigraphic Range, and Today's Bathymetrical Range (XX = common; X = rare) as Reported by Golubic et al. (1975), Perkins and Tsentas (1976), Budd and Perkins (1980), Akpan and Farrow (1984), Hoffmann (1985), Günther (1990), Glaub (1994, 2004), Gektidis (1997), Bundschuh and Balog (2000), Vogel et al. (2000), Vogel and Glaub (2004), Radtke and Golubic (2005), Golubic and Radtke (2008), Wisshak (2006), Radtke et al. (2010), Wisshak et al. (2011), and Complemented by own Observations**

Ichnotaxon Biotaxon (type)	Stratigraphic range	Shallow euphotic I	Shallow euphotic II	Shallow euphotic III	Deep euphotic	Dysphotic	Aphotic
Fascichnus dactylus (Radtke, 1991) *Hyella caespitosa* Bornet and Flahault, 1889 (cyanobacterium)	Ordovician–recent		XX[a]	XX[a]	X		
Fascichnus acinosus (Glaub, 1994) *Hyella balani* Lehmann, 1903 (cyanobacterium)	Permian–recent	XX	XX[(a)]	X			
Fascichnus frutex (Radtke, 1991) *Hyella gigas* Lukas and Golubic, 1983 (cyanobacterium)	Proterozoic–recent		XX	XX	X		
Fascichnus rogus (Bundschuh and Balog, 2000) *Hyella racemus* Al-Thukair et al., 1994 (cyanobacterium)	Ordovician–recent		X	XX			
Fascichnus parvus (Radtke, 1991) *Hyella pyxis* Lukas and Hoffman, 1984 (cyanobacterium)	Eocene–recent		XX	XX	X		

Continued

TABLE 2 Ichnotaxa with Paleobathymetrical Relevance, Given with their Known or Inferred Producers, Known Stratigraphic Range, and Today's Bathymetrical Range (XX = common; X = rare) as Reported by Golubic et al. (1975), Perkins and Tsentas (1976), Budd and Perkins (1980), Akpan and Farrow (1984), Hoffmann (1985), Günther (1990), Glaub (1994, 2004), Gektidis (1997), Bundschuh and Balog (2000), Vogel et al. (2000), Vogel and Glaub (2004), Radtke and Golubic (2005), Golubic and Radtke (2008), Wisshak (2006), Radtke et al. (2010), Wisshak et al. (2011), and Complemented by own Observations—Cont'd

Ichnotaxon Biotaxon (type)	Stratigraphic range	Shallow euphotic I	Shallow euphotic II	Shallow euphotic III	Deep euphotic	Dysphotic	Aphotic
Abeliella bellafurca Radtke et al., 2010 *Hyella stella* Al-Thukair and Golubic, 1991 (cyanobacterium)	Oligocene–recent		X	XX			
Eurygonum nodosum Schmidt, 1992 *Mastigocoleus testarum* Lagerheim, 1886 (cyanobacterium)	Ordovician–recent	?X	XX	XX	X		
Scolecia filosa Radtke, 1991 *Plectonema terebrans* Bornet and Flahault, 1889 (cyanobacterium)	Ordovician–recent	X	XX	XX	XX	X	?X
Fascichnus grandis (Radtke, 1991) *Acetabularia* sp. (chlorophyte)	Eocene–recent		XX	XX			
Cavernula pediculata Radtke, 1991 *Gomontia polyrhiza* (Lagerheim) Bornet and Flahault, 1889 (chlorophyte)	Ordovician–recent		X	XX			
Rhopalia catenata Radtke, 1991 *Phaeophila dendroides* (Crouan) Batters, 1902 (chlorophyte)	Carboniferous–recent		X	XX[a]	XX[a]		

Rhopalia clavigera Golubic and Radtke, 2008 *Eugomontia sacculata* Kornmann, 1960 (chlorophyte)	Eocene–recent	X	XX	XX		
Rhopalia spinosa Radtke and Golubic, 2005 ? (chlorophyte)	Eocene–recent		XX	X		
Eurygonum pennaforme Wisshak et al., 2005 ? *Epicladia testarum* (Kylin) Nielsen, 1980 (chlorophyte)	Triassic–recent		XX			
Ichnoreticulina elegans (Radtke, 1991) *Ostreobium quekettii* Bornet and Flahault, 1889 (chlorophyte)	Ordovician–recent	X	XX	XX[a]	X[a]	
"*Paleoconchocelis starmachii*" Campbell et al., 1979 *Porphyra* and *Bangia* Conchocelis stages (rhodophytes)	Ordovician–recent	X	X[(a)]	XX[(a)]	X	
Saccomorpha sphaerula Radtke, 1991 *Lithopythium gangliiforme* Bornet and Flahault, 1889 (fungus)	Eocene–recent	X	XX	X		
Saccomorpha clava Radtke, 1991 *Dodgella priscus* Zebrowski, 1937 (fungus)	Ordovician–recent	X	X	X	XX[a]	XX[a]
Orthogonum lineare Glaub, 1994 ? (?fungus)	Ordovician–recent	X	X	X	XX	XX[a]

[a]*Key ichnospecies for this zone.*

While such euendoliths indicate cool- or cold-water conditions by their absence, only few ichnotaxa have been identified to date that indicate low paleotemperatures by their presence. These are the fungal trace *Saccomorpha* cf. *terminalis* (Radtke, 1991; Wisshak, 2006; new ichnotaxon currently under description), the possible thraustochytrid trace *F. baiulus* (Wisshak and Porter, 2006), and the presumed foraminiferal boring *S. pulchra* as discussed by Bromley et al. (2007).

There are three possible pitfalls linked to the evaluation of microborings with respect to (paleo)temperature. First, some traces might have been produced during a relatively short time interval during a warm summer spell by euendoliths that would not survive winter temperatures in the given area or *vice versa*. Second, it is paramount to acknowledge the fact that paleotemperature does not necessarily allow deductions on the (paleo)latitude despite a certain degree of correlation, since "cold-water species" may well occur in tropical latitudes in cold bottom waters or shallow upwelling areas. The proposed cold-water indicator *F. baiulus*, for instance, was found as far south as the warm-temperate to tropical Florida Escarpment but only in cold bottom waters in 3000 m depth (Wisshak and Porter, 2006) and *S. pulchra* was reported from cold upwelling waters off Mauretania (Glaub, 2004). Third, temperature and latitude bear a certain degree of correlation with the trophic regime that shows an overall trend of eutrophication toward higher latitudes. Hence, an apparent temperature dependency might actually reflect a certain nutrient demand instead. However, there are a number of taxa (Table 3) that can yield us valuable information on the paleotemperature of a depositional system, provided that the whole ichnocoenosis is considered and other biogeographic indicators are taken into account. Our present knowledge on the biogeographic distribution, particularly of the named cold-water indicators, is still limited, calling for more research in this direction.

4.3 Paleosalinity

Microendoliths occur over a wide range of salinities from hypersaline lagoons to pure freshwater environments, implying the feasibility of applying microboring trace fossils as paleosalinity indicators, even though they have to date not been widely appreciated as such. Whereas many euendoliths are obligate open marine and thus strictly stenohaline, a number of euryhaline representatives have been identified (Table 4). The endolithic bio- and ichnodiversity strongly decreases when salinity differs from open marine conditions—a trend that is only partially an artifact of the highly understudied freshwater and hypersaline settings.

Radtke (1991, 1992) describes a number of ichnotaxa from brackish-water paleofacies of parts of the Paleogene Paris Basin, France, and concludes that boring assemblages from brackish environments consist mainly of heterotrophic endoliths. From a drill-core in an Oligocene limnic-brackish lake setting

TABLE 3 Ichnotaxa with Paleothermometrical Relevance, Given with their Known or Inferred Producers, Known Stratigraphic Range, and Today's Biogeographic Distribution (XX = common; X = rare) as Reported by Radtke et al. (1997), Lukas and Golubic (1981), Akpan and Farrow (1984), Schmidt and Freiwald (1993), Vogel et al. (2000), Bundschuh and Balog (2000), Radtke and Golubic (2005), Wisshak (2006, 2008), Wisshak and Porter (2006), Golubic and Radtke (2008) and Bromley et al. (2007), Wisshak et al. (2011), and Complemented by own Observations

Ichnotaxon Biotaxon (type)	Stratigraphic range	Tropical	Warm-temperate	Cold-temperate	Arctic
Fascichnus rogus (Bundschuh and Balog, 2000) *Hyella racemus* Al-Thukair et al., 1994 (cyanobacterium)	Ordovician–recent	XX	XX	X	
Fascichnus acinosus (Glaub, 1994) *Hyella balani* Lehmann, 1903 (cyanobacterium)	Permian–recent	XX	XX	?X	
Eurygonum nodosum Schmidt, 1992 *Mastigocoleus testarum* Lagerheim, 1886 (cyanobacterium)	Ordovician–recent	X	XX	X	
Planobola macrogota Schmidt, 1992[a] *Cyanosaccus piriformis* Lukas and Golubic, 1981 (cyanobacterium)	Ordovician–recent	XX	XX		
Fascichnus grandis (Radtke, 1991) *Acetabularia* sp. (chlorophyte)	Eocene–recent	X	XX		
Rhopalia catenata Radtke, 1991 *Phaeophila dendroides* (Crouan) Batters, 1902 (chlorophyte)	Carboniferous–recent	XX	XX	X	
Rhopalia clavigera Golubic and Radtke, 2008 *Eugomontia sacculata* Kornmann, 1960 (chlorophyte)	Eocene–recent	X	XX	X	

Continued

TABLE 3 Ichnotaxa with Paleothermometrical Relevance, Given with their Known or Inferred Producers, Known Stratigraphic Range, and Today's Biogeographic Distribution (XX = common; X = rare) as Reported by Radtke et al. (1997), Lukas and Golubic (1981), Akpan and Farrow (1984), Schmidt and Freiwald (1993), Vogel et al. (2000), Bundschuh and Balog (2000), Radtke and Golubic (2005), Wisshak (2006, 2008), Wisshak and Porter (2006), Golubic and Radtke (2008) and Bromley et al. (2007), Wisshak et al. (2011), and Complemented by own Observations—Cont'd

Ichnotaxon Biotaxon (type)	Stratigraphic range	Tropical	Warm-temperate	Cold-temperate	Arctic
Rhopalia spinosa Radtke and Golubic, 2005 ? (chlorophyte)	Eocene–recent	XX	X		
Saccomorpha sphaerula Radtke, 1991 *Lithopythium gangliiforme* Bornet and Flahault, 1889 (fungus)	Eocene–recent	XX	X	X	
Flagrichnus baiulus Wisshak and Porter, 2006 ? (fungus?)	Oligocene–recent		X[b]	XX	X
Saccomorpha cf. *terminalis* Radtke, 1991[c] *Arborella kohli* Zebrowski, 1937 (fungus)	Paleocene–recent			XX	X
Semidendrina pulchra Bromley et al., 2007 *Globodendrina monile* Plewes et al., 1996 (foraminiferan?)	Carboniferous–recent		X[b]	XX	X
Entobia mikra Wisshak, 2008 ? (sponge)	Pleistocene–recent		X	XX	X

[a]*Only those* Planobola macrogota *that resemble* Cyanosaccus piriformis.
[b]*Cold, deep, or upwelling water only.*
[c]*Alternative ichnospecies description in progress.*

TABLE 4 **Ichnotaxa with Paleosalinity Relevance, Given with their Known or Inferred Euryhaline Producers, Known Stratigraphic Range, and Salinity Range (XX = common; X = rare) as Reported by Wilkinson (1974), Radtke (1991, 1992, 1998, 2007), Glaub (1994), Radtke and Golubic (2005), Wisshak et al. (2005), Wisshak (2006), Pantazidou et al. (2006), Golubic and Radtke (2008), and Complemented by own Observations**

Ichnotaxon Biotaxon (type)	Stratigraphic range	Freshwater	Brackish	Open marine	Hypersaline
Rhopalia catenata Radtke, 1991 *Phaeophila dendroides* (Crouan) Batters, 1902 (chlorophyte)	Carboniferous–recent		XX	XX	X
Rhopalia clavigera Golubic and Radtke, 2008 *Eugomontia sacculata* Kornmann, 1960 (chlorophyte)	Eocene–recent		X	XX	
Rhopalia spinosa Radtke and Golubic, 2005 *Phaeophila dendroides* (Crouan) Batters, 1902 (chlorophyte)	Eocene–recent		X	XX	
Eurygonum pennaforme Wisshak et al., 2005 ?*Epicladia testarum* (Kylin) Nielsen, 1980 (chlorophyte)	Triassic–recent		X	XX	
Cavernula pediculata Radtke, 1991 *Gomontia polyrhiza* (Lagerheim) Bornet and Flahault, 1889 (chlorophyte)	Ordovician–recent		X	XX	X
Ichnoreticulina elegans (Radtke, 1991) *Ostreobium quekettii* Bornet and Flahault, 1889 (chlorophyte)	Ordovician–recent		X	XX	X
Fascichnus dactylus (Radtke, 1991) *Hyella caespitosa* Bornet and Flahault, 1889 (cyanobacterium)	Ordovician–recent	X	XX	XX	X
Fascichnus frutex (Radtke, 1991) *Hyella gigas* Lukas and Golubic, 1983 (cyanobacterium)	Proterozoic–recent		X	XX	

Continued

TABLE 4 Ichnotaxa with Paleosalinity Relevance, Given with their Known or Inferred Euryhaline Producers, Known Stratigraphic Range, and Salinity Range (XX = common; X = rare) as Reported by Wilkinson (1974), Radtke (1991, 1992, 1998, 2007), Glaub (1994), Radtke and Golubic (2005), Wisshak et al. (2005), Wisshak (2006), Pantazidou et al. (2006), Golubic and Radtke (2008), and Complemented by own Observations—Cont'd

Ichnotaxon Biotaxon (type)	Stratigraphic range	Freshwater	Brackish	Open marine	Hypersaline
Fascichnus parvus (Radtke, 1991) *Hyella pyxis* Lukas and Hoffman, 1984 (cyanobacterium)	Eocene–recent		X	X	
Eurygonum nodosum Schmidt, 1992 *Mastigocoleus testarum* Lagerheim, 1886 (cyanobacterium)	Ordovician–recent	?X	X	XX	
Scolecia filosa Radtke, 1991 *Plectonema terebrans* Bornet and Flahault, 1889 (cyanobacterium)	Ordovician–recent		X	XX	X
Polyactina araneola Radtke, 1991 *Conchyliastrum* (fungus)	Silurian–recent		X	XX	
Polyactina fastigata Radtke, 1991 ? (?fungus)	Carboniferous–Eocene		X	XX	
Orthogonum lineare Glaub, 1994 ? (?fungus)	Ordovician–recent		X	XX	
Orthogonum fusiferum Radtke, 1991 *Ostracoblabe implexa* Bornet and Flahault, 1889 (fungus)	Ordovician–recent		X	XX	
Saccomorpha clava Radtke, 1991 *Dodgella priscus* Zebrowski, 1937 (fungus)	Ordovician–recent		XX	XX	
Scolecia serrata Radtke, 1991 ? (?bacterium)	Carboniferous–recent		X	XX	

of Sieblos/Rhön, Germany, Radtke (1998) reports further ichnotaxa. Again touching upon a fossil marine-brackish paleoenvironment, Radtke (2007) identifies 10 ichnotaxa in a drill-core from the Oligocene Pechelbronn Formation from the Mainz Basin, Germany.

Only very limited information on microendolith occurrences is at hand from hypersaline settings as, for instance, the hypersaline lagoon Ahivadolimni on Milos Island, Greece (Pantazidou et al., 2006).

Bioerosion by microendoliths is not limited to marine waters but also takes place in pure freshwater environments such as lakes or below terrestrial soils on limestone bedrock (see Schneider and Le Campion-Alsumard, 1999 for a review). A number of freshwater bioeroding cyanobacteria have been described over the years (e.g., Anagnostidis and Pantazidou, 1988; Seeler and Golubic, 1991), but an ichnotaxonomical assignment of their traces is still lacking. Further research in this direction would be highly promising and could yield key ichnotaxa for identifying fossil lacustrine paleoenvironments. Likewise, by means of intensified research on hypersaline microbioerosion, stenohaline euendoliths and their traces might be identified that could pinpoint paleoenvironments with unusually high salinities.

4.4 Paleotrophodynamics

The interplay of dissolved inorganic nutrients and particulate organic matter in water with an abundance and diversity of bioerosion agents is complex and only partially understood. This is because these environmental factors have different effects on the various types of bioeroding biota and the epibiont cover which, in turn, may inversely affect each other. Moreover, nutrients and organic matter influence other controlling factors which are then difficult to separate in experimental studies and the geological record alike—such as the negative effect of eutrophication promoting plankton blooms on the availability of light for photoautotrophic microendoliths. At the same time, high plankton abundance promotes the activity of planktivorous macroborers such as recorded by, for instance, the global correlation of plankton primary production and the abundance of boring bivalves (Highsmith, 1980).

Hallock and Schlager (1986) and Hallock (1988) compiled the knowledge on the role of nutrient availability on bioerosion and thereby highlighted the important consequences on the balance between carbonate accumulation and degradation in recent carbonate buildups, and the implications of this interplay on the fossil record.

With respect to microbioerosion, Chazottes (1996) and Chazottes et al. (2002) experimentally studied eutrophicated *versus* pristine reef transects at Réunion Island and concluded that the influence of reef-water quality on bioerosion is chiefly governed by the composition of the epibiont cover. Grazing was lowest and microbioerosion highest in the nutrient-enriched transects where macroalgae and encrusting algae flourished, whereas pristine reef areas

TABLE 5 Ichnotaxa with Paleotrophodynamic Relevance, Given with Their Known or Inferred Producers, Known Stratigraphic Range, and Inorganic Nutrient or Particulate Organic Matter (OM) Preference (XX = strong; X = weak) as Experimentally Determined by Carreiro-Silva et al. (2009)

Ichnotaxon Biotaxon (type)	Stratigraphic range	High levels of OM only	High levels of OM + nutrients	High levels of nutrients
Rhopalia catenata Radtke, 1991 *Phaeophila dendroides* (Crouan) Batters, 1902 (chlorophyte)	Carboniferous–recent		XX	XX
Fascichnus dactylus (Radtke, 1991) *Hyella caespitosa* Bornet and Flahault, 1889 (cyanobacterium)	Ordovician–recent		X	X
Scolecia filosa Radtke, 1991 *Plectonema terebrans* Bornet and Flahault, 1889 (cyanobacterium)	Ordovician–recent		XX	X
Eurygonum nodosum Schmidt, 1992 *Mastigocoleus testarum* Lagerheim, 1886 (cyanobacterium)	Ordovician–recent		X	XX
Orthogonum fusiferum Radtke, 1991 *Ostracoblabe implexa* Bornet and Flahault, 1889 (fungus)	Ordovician–recent	X	X	
Saccomorpha sphaerula Radtke, 1991 *Lithopythium gangliiforme* Bornet and Flahault, 1889 (fungus)	Eocene–recent	X		
Saccomorpha clava Radtke, 1991 *Dodgella priscus* Zebrowski, 1937 (fungus)	Ordovician–recent	XX		

supported the development of algal turfs and heavy grazing, consequently limiting microbioerosion. This pattern was confirmed by Zubia and Peyrot-Clausade (2001) in another Réunion study where *Acropora* microbioerosion was found highest in eutrophicated areas. An experimental study in the Great Barrier Reef indicated only minute or insignificant impact of dissolved inorganic nutrients on microendolithic ichnocoenoses and bioerosion rates (Kiene, 1997). Campbell (1982) showed that the diversity of microendoliths is reduced by petrochemical pollution, whereas eutrophication by municipal sewage does not have a corresponding effect.

In a recent elaborate experiment at Glovers Reef in Belize, Carreiro-Silva et al. (2009) were able to differentiate between dissolved inorganic nutrients and particulate organic matter and showed that nitrogen and phosphorous fertilization significantly increases both the dominance of euendolithic photoautotrophs over chemotrophic fungi and the microbioerosion rates. Particulate organic matter, in turn, increased the abundance of chemotrophic fungi but only in the treatments without inorganic nutrient fertilization. In addition, Carreiro-Silva et al. (2005) showed that herbivory by macrograzers lead to a significant decrease in microbioerosion rates. Both studies included a detailed assessment of the ichnocoenoses, providing first insight on the effect of the trophic regime on both the endolithic community structure and the abundance of specific ichnotaxa (Table 5), and thus can be regarded as an important step toward an application as indicator for ancient trophodynamics. However, more detailed studies on the microbioerosion ichnocoenoses in (particularly long-term) fertilization experiments are required in order to foster the identification of key ichnospecies and specific ichnocoenosis compositions diagnostic for either a high content in particulate organic matter and/or high nutrient levels.

5. CONCLUSIONS

The (neo)ichnology and particularly the ichnotaxonomy of microbioerosion are still young disciplines at the interface of bio- and geosciences and thus offer a wide-open field for future discoveries. At present, about three dozen microboring ichnotaxa have been established, chiefly based upon epoxy resin casts taken from well-preserved calcareous fossils. Many of these ichnotaxa can be closely matched with their extant producing euendoliths among the cyanobacteria, chlorophytes, rhodophytes, as well as fungi or other chemotrophs. Thereby, they provide circumstantial evidence for a pronounced evolutionary longevity reflected in a fossil record with a high percentage of ichnotaxa reaching back to Mid-Paleozoic times. A systematic evaluation of the biogeographic distribution and linked ecological demands of the various modern microendoliths, complemented by experimental studies in the field and laboratory, currently pushes forward our knowledge on the ecophysiological tolerances of these organisms and the applicability of their corresponding traces as paleoenvironmental indicators with respect to paleobathymetry, paleothermometry, paleosalinity, and paleotrophodynamics. Ultimately, further progress in the identification of key ichnotaxa and index ichnocoenoses indicative for specific environmental properties and their application to fossil material of any age will tell whether microbioerosion analysis meets its potential of becoming a more regularly applied tool in the characterization of paleoenvironments and sedimentary settings.

ACKNOWLEDGMENTS

Ingrid Glaub (Frankfurt, Germany) and Katrin Heindel (Erlangen, Germany) are acknowledged for fruitful discussions and their critical perusal of an earlier draft of this chapter. Helpful reviews and comments were provided by Gudrun Radtke (Wiesbaden, Germany) and the editors.

REFERENCES

Akpan, E.B., Farrow, G.E., 1984. Shell-boring algae on the Scottish continental shelf: identification, distribution, bathymetric zonation. Transact. R. Soc. Edinburgh 75, 1–12.

Anagnostidis, K., Pantazidou, A., 1988. Endolithic cyanophytes from the saline thermal springs of Aedipsos, Hellas (Greece). Algolog. Stud./Arch. Hydrobiol. 50–53 (Suppl.), 555–559.

Barton, H.A., Jurado, V., 2007. What's up down there? Microbial diversity in caves. Microbe 2, 132–138.

Beuck, L., Vertino, A., Stepina, E., Karolczak, M., Pfannkuche, O., 2007. Skeletal response of *Lophelia pertusa* (Scleractinia) to bioeroding sponge infestation visualised with micro-computed tomography. Facies 53, 157–176.

Beuck, L., Wisshak, M., Munnecke, A., Freiwald, A., 2008. A giant boring in a Silurian stromatoporoid analysed by computer tomography. Acta Palaeont. Polon. 53, 149–160.

Boekschoten, G.J., 1970. On bryozoan borings from the Danian at Fakse, Denmark. In: Crimes, T.P., Harper, J.C. (Eds.), Trace Fossils. Geol. J., Spec. Iss. 3, pp. 43–48.

Bromley, R.G., Wisshak, M., Glaub, I., Botquelen, A., 2007. Ichnotaxonomic review of dendriniform borings attributed to foraminiferans: *Semidendrina* igen. nov. In: Miller III, W. (Ed.), Trace Fossils. Concepts, Problems, Prospects. Elsevier, Amsterdam, pp. 518–530.

Budd, D.A., Perkins, R.D., 1980. Bathymetric zonation and paleoecological significance of microborings in Puerto Rican shelf and slope sediments. J. Sediment. Petrol. 50, 881–904.

Bundschuh, M., Balog, S.-J., 2000. *Fasciculus rogus* nov. isp., an endolithic trace fossil. Ichnos 7, 149–152.

Butterfield, N.J., Knoll, A.H., Swett, K., 1990. A bangiophyte red alga from the Proterozoic of Arctic Canada. Science 250, 104–107.

Campbell, S.E., 1982. Petrochemical pollution: endolithic response. In: Vies Journées ètudiants Pollutions, Cannes, C.I.E.S.M, pp. 183–187.

Campbell, S.E., Kazmierczak, J., Golubic, S., 1979. *Palaeoconchocelis starmachii* gen. n., sp. n., an endolithic rhodophyte (Bangiaceae) from the Silurian of Poland. Acta Palaeont. Polon. 24, 403–408.

Carreiro-Silva, M., McClanahan, T.R., Kiene, W.E., 2005. The role of inorganic nutrients and herbivory in controlling microbioerosion of carbonate substratum. Coral Reefs 24, 214–221.

Carreiro-Silva, M., McClanahan, T.R., Kiene, W.E., 2009. Effects of inorganic nutrients and organic matter on microbial euendolithic community composition and microbioerosion rates. Marine Ecol. Prog. Ser. 392, 1–15.

Chazottes, V., 1996. Étude expérimentale de la bioérosion et de la sédimentogénèse en milieu récifal, effets de l'eutrophisation (Île de la Réunion, Océan Indien). Océanographie (Géologie Marine) 323, 787–794.
Chazottes, V., Le Campion-Alsumard, T., Peyrot-Clausade, M., Cuet, P., 2002. The effects of eutrophication-related alterations to coral reef communities on agents and rates of bioerosion (Réunion Island, Indian Ocean). Coral Reefs 21, 375–390.
Chazottes, V., Cabioch, G., Golubic, S., Radtke, G., 2009. Bathymetric zonation of modern microborers in dead coral substrates from New Caledonia—implications for paleodepth reconstructions in Holocene corals. Palaeogeogr. Palaeoclimatol. Palaeoecol. 280, 456–468.
Dierick, M., Cnudde, V., Masschaele, B., Vlassenbroeck, J., Van Hoorebeke, L., Jacobs, P., 2007. Micro-CT of fossils preserved in amber. Nucl. Instr. Meth. Phys. Res. Sect. A 580, 641–643.
Freiwald, A., 1995. Bacteria-induced carbonate degradation: a taphonomic case study of *Cibicides lobatulus* from a high-boreal carbonate setting. Palaios 10, 337–346.
Freiwald, A., 1998. Microbial maceration and carbonate dissolution on cold-temperate shelves. Hist. Biol. 13, 27–35.
Friedman, G.M., Gebelein, C.D., Sanders, J.E., 1971. Micritic envelopes of carbonate grains are not exclusively of photosynthetic algal origin. Sedimentology 16, 89–96.
Garcia-Pichel, F., 2006. Plausible mechanisms for the boring on carbonates by microbial phototrophs. Sediment. Geol. 185, 205–213.
Gektidis, M., 1997. Vorkommen, Ökologie und Taxonomie von Mikrobohrorganismen in ausgewählten Riffbereichen um die Insel Lee Stocking Island (Bahamas) und One Tree Island (Australien). Ph.D. Thesis, University Frankfurt, 276 pp.
Gektidis, M., Dubinsky, Z., Goffredo, S., 2007. Microendoliths of the shallow euphotic zone in open and shaded habitats at 30°N—Eilat, Israel—paleoecological implications. Facies 53, 43–55.
Genise, J.F., 2004. Fungus trace in wood: a rare bioerosional item. In: Buatois, L.A., Mángano, M.G. (Eds.), Ichnia 2004, First International Congress on Ichnol. Trelew, Argentina, Abstr. p. 37.
Glaub, I., 1994. Mikrobohrspuren in ausgewählten Ablagerungsräumen des europäischen Jura und der Unterkreide (Klassifikation und Palökologie). Cour. Forschungsinst. Senckenberg 174, 1–324.
Glaub, I., 2004. Recent and sub-recent microborings from the upwelling area off Mauritania (West Africa) and their implications for palaeoecology. In: McIlroy, D. (Ed.), The Application of Ichnology to Palaeoenvironmental and Stratigraphic Analysis. Geol. Soc. London, Spec. Publ., vol. 228, pp. 63–77.
Glaub, I., Königshof, P., 1997. Microborings in conodonts. Cour. Forschungsinst. Senckenberg 201, 137–143.
Glaub, I., Vogel, K., 2004. The stratigraphic record of microborings. Fossils Strata 51, 126–135.
Glaub, I., Vogel, K., Gektidis, M., 2001. The role of modern and fossil cyanobacterial borings in bioerosion and bathymetry. Ichnos 8, 185–195.
Glaub, I., Gektidis, M., Vogel, K., 2002. Microborings from different North Atlantic shelf areas—variability of the euphotic zone extension and implications for paleodepth reconstructions. Cour. Forschungsinst. Senckenberg 237, 25–37.
Glaub, I., Golubic, S., Gektidis, M., Radtke, G., Vogel, K., 2007. Microborings and microbial endoliths: geological implications. In: Miller III, W. (Ed.), Trace Fossils. Concepts, Problems, Prospects. Elsevier, Amsterdam, pp. 368–381.
Golubic, S., Radtke, G., 2008. The trace *Rhopalia clavigera* isp n. reflects the development of its maker *Eugomontia sacculata* Kornmann, 1960. In: Wisshak, M., Tapanila, L. (Eds.), Current Developments in Bioerosion. Erlangen Earth Conference Series. Springer, Berlin, pp. 95–108.
Golubic, S., Schneider, J., 1979. Carbonate dissolution. In: Trudinger, P.A., Swaine, D.J. (Eds.), Biogeochemical Cycling of Mineral-forming Elements. Elsevier, Amsterdam, pp. 107–129.

Golubic, S., Brent, G., LeCampion, T., 1970. Scanning electron microscopy of endolithic algae and fungi using a multipurpose casting-embedding technique. Lethaia 3, 203–209.
Golubic, S., Perkins, R.D., Lukas, K.J., 1975. Boring microorganisms and microborings in carbonate substrates. In: Frey, R.W. (Ed.), The Study of Trace Fossils. Springer, Berlin, pp. 229–259.
Golubic, S., Friedmann, I., Schneider, J., 1981. The lithobiontic ecological niche, with special reference to microorganisms. J. Sediment. Petrol. 51, 475–478.
Golubic, S., Radtke, G., Le Campion-Alsumard, T., 2005. Endolithic fungi in marine ecosystems. Trends Microbiol. 13, 229–235.
Günther, A., 1990. Distribution and bathymetric zonation of shell-boring endoliths in recent reef and shelf environments: Cozumel, Yucatan (Mexico). Facies 22, 233–262.
Hallock, P., 1988. The role of nutrient availability in bioerosion: consequences to carbonate buildups. Palaeogeogr. Palaeoclimatol. Palaeoecol. 63, 275–291.
Hallock, P., Schlager, W., 1986. Nutrient excess and the demise of coral reefs and carbonate platforms. Palaios 1, 389–398.
Heindel, K., Wisshak, M., Westphal, H., 2009. Microbioerosion in Tahitian reefs: a record of environmental change during the last deglacial sea-level rise (IODP #310). Lethaia 42, 322–340.
Highsmith, R.C., 1980. Geographic patterns of coral bioerosion: a productivity hypothesis. J. Exp. Biol. Ecol. 146, 177–196.
Hoffman, E.J., 1985. Distribution patterns of recent microbial endoliths in the intertidal and supratidal zones, Bermuda. In: Curran, H.A. (Ed.), Biogenic Structures: Their Use in Interpreting Depositional Environments. SEPM, Spec. Publ., vol. 35, pp. 179–194.
Hofmann, K., 1996. Die mikro-endolithischen Spurenfossilien der borealen Oberkreide Nordwest-Europas und ihre Faziesbeziehungen. Geol. Jahrb. A 136, 1–151.
Jans, M.E., 2008. Microbial bioerosion of bone—a review. In: Wisshak, M., Tapanila, L. (Eds.), Current Developments in Bioerosion. Erlangen Earth Conference Series. Springer, Berlin, pp. 397–413.
Jones, B., Pemberton, S.G., 1987. The role of fungi in the diagenetic alteration of spar calcite. Can. J. Earth Sci. 24, 903–914.
Kiene, W.E., 1997. Enriched nutrients and their impact on bioerosion: results from ENCORE. In: Proceedings of the 8th International Coral Reef Symposium, Panama City, pp. 897–902.
Le Campion-Alsumard, T., 1978. Les Cyanophycées endolithes marines. Systématique, ultrastructure, écologie et biodestruction. Ph.D. Thesis, University Aix-Marseille II, vol., 1, 198 pp.
Lukas, K.J., 1978. Depth distribution and form among common microboring algae from the Florida continental shelf. GSA 10, 448 Abstracts with Program.
Lukas, K.J., Golubic, S., 1981. New endolithic cyanophytes from the North Atlantic Ocean: I. *Cyanosaccus piriformis* gen. et sp. nov. J. Phycol. 17, 224–229.
Lüning, K., 1985. Meeresbotanik: Verbreitung, Ökophysiologie und Nutzung der marinen Makroalgen. Thieme, Stuttgart, 375 pp.
May, J.A., Perkins, R.D., 1979. Endolithic infestation of carbonate substrates below the sediment-water interface. J. Sediment. Petrol. 49, 357–378.
Mayoral, E., 1988. *Pennatichnus* nov. icnogen.; *Pinaceocladichnus* nov. icnogen. e *Iramena.* huellas de bioerosion debidas a Bryozoa perforantes (Ctenostomata, Plioceno Inferior) en la Cuenca del Bajo Guadalquivir. Rev. Españ. Paleont. 3, 13–22.
McLoughlin, N., Furnes, H., Banerjee, N.R., Staudigel, H., Muehlenbachs, K., de Wit, M., Van Kranendonk, M.J., 2008. Micro-bioerosion in volcanic glass: extending the ichnofossil record to Archaean basaltic crust. In: Wisshak, M., Tapanila, L. (Eds.), Current Developments in Bioerosion. Erlangen Earth Conference Series. Springer, Berlin, pp. 371–396.

Mietchen, D., Aberhan, M., Manz, B., Hampe, O., Mohr, B., Neumann, C., Volke, F., 2008. Three-dimensional magnetic resonance imaging of fossils across taxa. Biogeoscience 5, 25–41.

Müller, E., Löffler, W., 1992. Mykologie. Thieme, Stuttgart, 367 pp.

Nielsen, J.K., Maiboe, J., 2000. Epofix and vacuum: an easy method to make casts of hard substrates. Palaeontologica Electronica 3, 10 pp. http://palaeo-electronica.org/2000_1/epofix/issue1_00.htm.

Pantazidou, A., Louvrou, I., Economou-Amilli, A., 2006. Euendolithic shell-boring cyanobacteria and chlorophytes from the saline lagoon Ahivadolimni on Milos Island, Greece. Eur. J. Phycol. 41, 189–200.

Perkins, R.D., Tsentas, C.I., 1976. Microbial infestations of carbonate substrates planted on the St. Croix shelf. Bull. GSA 87, 1615–1628.

Perry, C.T., Macdonald, I.A., 2002. Impacts of light penetration on the bathymetry of reef microboring communities: implications for the development of microendolithic trace assemblages. Palaeogeogr. Palaeoclimatol. Palaeoecol. 189, 101–113.

Pohowsky, R.A., 1978. The boring ctenostomate Bryozoa: taxonomy and paleobiology based on cavities in calcareous substrata. Bull. Am. Paleontol. 73, 1–192.

Priess, K., Le Campion-Alsumard, T., Golubic, S., Gadel, F., Thomassin, B.A., 2000. Fungi in corals: black bands and density-banding of *Porites lutea* and *P. lobata* skeleton. Mar. Biol. 136, 19–27.

Radtke, G., 1991. Die mikroendolithischen Spurenfossilien im Alt-Tertiär West-Europas und ihre palökologische Bedeutung. Cour. Forschungsinst. Senckenberg 138, 1–185.

Radtke, G., 1992. Microendolithic trace fossils of Paris Basin as facies indicators. In: Proceedings of the 7th International Coral Reef Symposium, Guam, pp. 419–426.

Radtke, G., 1998. Mikroendolithische Bohrspuren in Gastropoden-Schalen von Sieblos/Rhön (Unter-Oligozän). Geol. Abh. Hessen 104, 143–155.

Radtke, G., 2007. Mikroendolithen in Molluskenschalen aus den Mittleren Pechelbronn-Schichten (Mainzer Becken). Geol. Abh. Hessen 116, 141–155.

Radtke, G., Golubic, S., 2005. Microborings in mollusc shells, Bay of Safaga, Egypt: morphometry and ichnology. Facies 51, 118–134.

Radtke, G., Le Campion-Alsumard, T., Golubic, S., 1996. Microbial assemblages of the bioerosional "notch" along tropical limestone coasts. Algolog. Stud. 83, 469–482.

Radtke, G., Gektidis, M., Golubic, S., Hofmann, K., Kiene, W.E., Le Campion-Alsumard, T., 1997. The identity of an endolithic alga: *Ostreobium brabantium* Weber-van Bosse is recognized as carbonate-penetrating rhizoids of *Acetabularia* (Chlorophyta, Dasycladales). Cour. Forschungsinst. Senckenberg 201, 341–347.

Radtke, G., Glaub, I., Vogel, K., Golubic, S., 2010. A new dichotomous microboring: *Abeliella bellafurca* isp. nov., distribution, variability and biological origin. Ichnos 17, 25–33.

Schmidt, H., Freiwald, A., 1993. Rezente gesteinsbohrende Kleinorganismen des norwegischen Schelfs. Nat. Mus. 123, 149–155.

Schneider, J., 1976. Biological and inorganic factors in the destruction of limestone coasts. Contrib. Sedimentol. 6, 1–123.

Schneider, J., Le Campion-Alsumard, T., 1999. Construction and destruction of carbonates by marine and freshwater cyanobacteria. Eur. J. Phycol. 34, 417–426.

Schroeder, J.H., 1972. Calcified filaments of an endolithic algae in Recent Bermuda reefs. N. Jb. Geol. Paläont. Mh. 1, 16–33.

Seeler, J.S., Golubic, S., 1991. *Iyengariella endolithica* sp. nova, a carbonate boring stigonematalean cyanobacterium from a warm spring-fed lake: nature or culture. Algolog. Stud. 64, 399–410.

Sutton, M.D., 2008. Tomographic techniques for the study of exceptionally preserved fossils. Proc. Royal Soc. Lond. B 275, 1587–1593.

Tafforeau, P., Boistel, R., Boller, E., Bravin, A., Brunet, M., Chaimanee, Y., Cloetens, P., Feist, M., Hoszowska, J., Jaeger, J.-J., Kay, R.F., Lazzari, V., Marivaux, L., Nel, A., Nemoz, C., Thibault, X., Vignaud, P., Zabler, S., 2006. Applications of x-ray synchrotron microtomography for non-destructive 3D studies of paleontological specimens. Appl. Phys. A 83, 195–202.

Tapanila, L., 2008. The medium is the message: imaging a complex microboring (*Pyrodendrina cupra* igen. n., isp. n.) from the early Paleozoic of Anticosti Island, Canada. In: Wisshak, M., Tapanila, L. (Eds.), Current Developments in Bioerosion. Erlangen Earth Conference Series. Springer, Berlin, pp. 123–146.

Tribollet, A., 2008. The boring microflora in modern coral reef ecosystems: a review of its roles. In: Wisshak, M., Tapanila, L. (Eds.), Current Developments in Bioerosion. Erlangen Earth Conference Series. Springer, Berlin, pp. 67–94.

Tribollet, A., Golubic, S., Radtke, G., Reitner, J., 2010. On microbiocorrosion. Lect. Notes Earth Sci. 131, 243–253.

Vogel, J., Brett, C.E., 2009. Record of microendoliths in different facies of the Upper Ordovician in the Cincinnati Arch region USA: the early history of light-related microendolithic zonation. Palaeogeogr. Palaeoclimatol. Palaeoecol. 281, 1–24.

Vogel, K., Glaub, I., 2004. 450 Millionen Jahre Beständigkeit in der Evolution endolithischer Mikroorganismen? Sitzungsber. Wiss. Ges. J.W. Goethe-Univ. Frankf. a. M. 42, 1–42.

Vogel, K., Golubic, S., Brett, C.E., 1987. Endolith associations and their relation to facies distribution in the Middle Devonian of New York State, U.S.A. Lethaia 20, 263–290.

Vogel, K., Bundschuh, M., Glaub, I., Hofmann, K., Radtke, G., Schmidt, H., 1995. Hard substrate ichnocoenoses and their relations to light intensity and marine bathymetry. N. Jb. Geol. Paläont. Abh. 195, 49–61.

Vogel, K., Balog, S.-J., Bundschuh, M., Gektidis, M., Glaub, I., Krutschinna, J., Radtke, G., 1999. Bathymetrical studies in fossil reefs, with microendoliths as paleoecological indicators. Profil 16, 181–191.

Vogel, K., Gektidis, M., Golubic, S., Kiene, W.E., Radtke, G., 2000. Experimental studies on microbial bioerosion at Lee Stocking Island, Bahamas and One Tree Island, Great Barrier Reef, Australia: implications for paleoecological reconstructions. Lethaia 33, 190–204.

Vogel, K., Schäfer, H., Glaub, I., 2008. Spurengemeinschaften bohrender Mikroorganismen. Nat. Mus. 138, 184–189.

Warme, J.E., 1975. Borings as trace fossils, and the process of marine bioerosion. In: Frey, R.W. (Ed.), The Study of Trace Fossils. Springer, Berlin, pp. 181–227.

Wilkinson, M., 1974. Investigations on the autecology of *Eugomontia sacculata* Kornmann, a shell-boring alga. J. Exp. Mar. Biol. Ecol. 16, 19–27.

Wilson, M.A., 2008. An online bibliography of bioerosion references. In: Wisshak, M., Tapanila, L. (Eds.), Current Developments in Bioerosion. Erlangen Earth Conference Series. Springer, Berlin, pp. 473–478.

Wilson, M.A, 2012. http://www.wooster.edu/geology/bioerosion/bioerosion.html [online bibliography of bioerosion references].

Winkler, B., 2006. Applications of neutron radiography and neutron tomography. Rev. Min. Geochem. 63, 459–471.

Wisshak, M., 2006. High-latitude bioerosion: the Kosterfjord experiment. Lect. Notes Earth Sci. 109, 1–202.

Wisshak, M., 2008. Two new dwarf *Entobia* ichnospecies in a diverse aphotic ichnocoenosis (Pleistocene/Rhodes, Greece). In: Wisshak, M., Tapanila, L. (Eds.), Current Developments in Bioerosion. Erlangen Earth Conference Series. Springer, Berlin, pp. 213–234.

Wisshak, M., Porter, D., 2006. The new ichnogenus *Flagrichnus*—a palaeoenvironmental indicator for cold-water settings? Ichnos 13, 135–145.

Wisshak, M., Gektidis, M., Freiwald, A., Lundälv, T., 2005. Bioerosion along a bathymetrical gradient in a cold-temperate setting (Kosterfjord/SW Sweden): an experimental study. Facies 51, 93–117.

Wisshak, M., Seuß, B., Nützel, A., 2008. Evolutionary implications of an exceptionally preserved Carboniferous microboring assemblage in the Buckhorn Asphalt Lagerstätte (Oklahoma, USA). In: Wisshak, M., Tapanila, L. (Eds.), Current Developments in Bioerosion. Erlangen Earth Conference Series. Springer, Berlin, pp. 21–54.

Wisshak, M., Tribollet, A., Golubic, S., Jakobsen, J., Freiwald, A., 2011. Temperate bioerosion: Ichno- and biodiversity from intertidal to bathyal depths (Azores). Geobiology 9, 492–520.

Zeff, M.L., Perkins, R.D., 1979. Microbial alteration of Bahamian deep-sea carbonates. Sedimentology 26, 175–201.

Zhang, Y., Golubic, S., 1987. Endolithic microfossils (Cyanophyta) from early Proterozoic stromatolites, Hebei, China. Acta Micropalaeontol. Sin. 4, 1–12.

Zubia, M., Peyrot-Clausade, M., 2001. Internal bioerosion of *Acropora formosa* in Réunion (Indian Ocean): microborer and macroborer activities. Oceanol. Acta 24, 251–262.

Chapter 9

Methodology and Techniques

Dirk Knaust[1]
Statoil ASA, Stavanger, Norway
[1]*Corresponding author: e-mail: dkna@statoil.com*

1. INTRODUCTION

Bioturbation, discrete trace fossils, and ichnofabrics appear with a wide range of preservation styles in the sedimentary record. The taphonomic and diagenetic history, together with subsequent weathering, is able to produce different expressions of the same structure, which may have unlike appearances in contrasting lithologies such as sandstone, mudstone, and limestone. The utilization of all ichnological aspects in facies analysis and in the interpretation of sedimentary environments is only possible when the visibility of such biogenic sedimentary structures can be optimized. This, in turn, implies that the methodology for their study varies quite a lot from case to case and the continuously growing number of studies dealing with the methodology in ichnological investigations documents the solutions achieved so far.

Previous overviews about techniques applicable to reveal biogenic structures were provided by Farrow (1975) and Ekdale et al. (1984). For a broader outline, the reader is referred to standard texts on sedimentary techniques by Bouma (1969), Tucker (1988), and Coe (2010). With respect to neoichnological methods, a wealth of information is available in the biological literature but only main aspects can be dealt with here (see Dashtgard and Gingras, 2012). Stockdale et al. (2009) review available technology to investigate microscale biogeochemical heterogeneity in sediments, whereas St-Onge et al. (2007) and Löwemark (2007) reconsider several techniques applied in the examination of cored marine sediments. Finally, image analysis is an important process in the documentation and investigation of biogenic sedimentary structures and is covered in great detail by Francus (2004).

This chapter aims to touch on and to present some of the most relevant methods in ichnology in order to allow the reader to become familiar with them, to provide key references for accessing further information, and finally to utilize these methods in the wide field of facies analysis. It will not repeat details of standard techniques and methodologies, which can be found in the cited references, although some effort is given to the revitalization of significant

Developments in Sedimentology, Vol. 64. http://dx.doi.org/10.1016/B978-0-444-53813-0.00009-5

procedures. It rather focuses on the review of the most appropriate techniques and methodologies applicable in ichnology, with emphasis on those introduced during the past two decades.

2. FIELD TECHNIQUES

2.1 Outcrop Observations

Accurate field observations are crucial in integrated sedimentological–ichnological studies as well as precise ichnotaxonomical assignments. The introduction given by Stow (2005) may help to organize the field work in an optimal way, although ichnological investigations in outcrops may require specialized techniques (see Ekdale et al., 1984). Beside conventional preparation of logs, notes, sketches, and photographs (see Hodges, 2003; Savazzi, 2011), many advanced methods and techniques for investigation of complex ichnological–sedimentological situations in the field have evolved during the past decades.

The applied technique strongly depends on the nature of the rock (composition, degree of lithification, weathering, etc.) and its exposure (e.g., natural outcrop, excavation, quarry, pit, sawn surfaces, etc.). Based on the unique composition and diagenetic history of trace fossils and bioturbate textures, freshly excavated quarry walls may reveal only limited ichnological information, whereas weathered cliffs can exhibit the full range of features. Sawn walls of consolidated rock typically expose detailed characteristics which were otherwise invisible (Fig. 1A; Knaust et al., 2012), whereas ichnofabrics in poorly lithified sediments such as sand and clay can be made visible by sectioning with spades and knives (Bromley and Uchman, 2003; Netto et al., 2012). Much emphasis has been placed on the analysis of ichnofabrics in order to capture different aspects of the biogenic sedimentary structures in a systematic way (Ekdale et al., 2012; McIlroy, 2007; Taylor et al., 2003).

A classical method is the collection of slabs with trace fossils from outcrops, and consequently, numerous collections around the world contain significant and valuable material. Important is a proper collecting procedure, documentation, packing and transportation, cleaning, preparation, and storage similar to the collection of other fossils (cf. Green, 2001). Trace fossils can be utilized to estimate the stratigraphical completeness based on the tiering pattern (Wetzel and Aigner, 1986), to quantify mechanical compaction (Löwemark et al., 2006) and synsedimentary deformation (Gaillard and Jautee, 1987; Naruse and Nifuku, 2008; Ziółkowski, 2005), and to determine finite strain in bedding surfaces (Waldron, 1988).

2.2 Peels, Molds, and Casts

The preparation of peels, molds, and casts is an effective way of capturing biogenic and sedimentary structures (including trace fossils, ichnofabrics, and modern traces), and different materials and methods have been applied successfully in numerous studies.

FIGURE 1 Field techniques in ichnological analysis. (A) Sawn wall in the Babino Quarry south of the Ladoga Lake, Russia, displaying a Middle Ordovician limestone succession with ichnofabrics in stained firm- and hardgrounds. (B) Prof. Dr. Adolf Seilacher explaining a casted bedding plane with *Haentzschelinia ottoi* as part of his Fossil Art collection exhibited during the 33rd International Geological Congress 2008 in Oslo. (C) Koji Seike with a plaster cast of a *Psilonichnus*-like Y-shaped burrow of the ghost crab *Ocypode quadrata* from the East Beach, San Salvador Island, Bahamas. Cast and photo courtesy of Koji Seike and Al Curran. (D) Resin cast from a box core from the southern North Sea with *Arenicolites*-like burrows (upside down, up to 3 cm long) produced by the amphipod *Corophium volutator*. Original from Hans-Erich Reineck, Senckenberg in Wilhemshaven, Germany. (E) Panels made from pelagic nodular limestone with *Thalassinoides*-like burrow networks of the Lower Jurassic Rosso Ammonitico facies, northern Italy. (F) Plaza in a Spanish Pyrenean village paved with slabs exhibiting Paleogene flysch trace fossils.

Unconsolidated, wet or dry sediment sections can be treated with various materials such as lacquer, clues, latex, plaster of Paris, resins, etc., in order to replicate them in the form of a peel (Kidwell et al., 1985; Skipper et al., 1998). A similar approach allows also for collecting modern surface traces in sand (Stidham and Mason, 2009). Molding and casting of trace fossils by using plaster or latex is a sufficient technique in both vertebrate ichnology (Goodwin and Chaney, 1994) and invertebrate ichnology (Seilacher, 2008; Fig. 1B). It can be used not only for preserving vertebrate footprints, burrows, or bioerosion structures from the field but also to prepare duplicates from original specimens in museums and collections.

Neoichnological studies can be carried out based on plaster, polyester, or polyurethane resin casts obtained from open burrows, a technique commonly applied in terrestrial, littoral, and sublittoral substrates (Atkinson and Chapman, 1984; Bouma, 1969; Farrow, 1975; Gingras et al., 2002b; Knaust et al., 2012; Reineck, 1963; Shinn, 1968; Tschinkel, 2003). In this way, the morphology, size, and details of burrows (e.g., from crabs, shrimps, fishes, insects, or others) can be captured *in situ* for later investigations (including measuring, weighting, identification of commensalism, etc.; Fig. 1C). In the future, this technique will be more refined and remotely operated vehicles may allow burrow casting in the deep sea as recently demonstrated by Seike et al. (2012).

2.3 Core Sampling of Surface Sediment

Besides conventional coring of sedimentary rocks, special equipment has been developed to take samples mainly from marine but also lacustrine and fluvial surface sediments that are unlithified. Different kinds of core samplers such as hand-held, gravity, vibra-core, box-core, or piston-core samplers are employed to force a hollow tube into the substrate and to retrieve a vertical profile of the sediment (Ekdale et al., 1984; Farrow, 1975), nowadays sometimes under video guidance. The sediment column including its ichnological content can then be sectioned, X-rayed, or cast before examination (Fig. 1D). Sediment box cores as introduced by Reineck (1963) were subsequently extensively used for ichnological investigation of shallow-marine to deep-sea deposits (Dashtgard and Gingras, 2005; Wetzel, 1983).

2.4 Imaging and Analysis of Dinosaur, Bird, and Mammal Tracks

Tracks and trackways of dinosaur and other vertebrates have received much attention during the past decades and a number of different techniques are applied for their analysis. Lockley (1991) provides an overview of conventional surveying and mapping techniques of dinosaur tracks, and Breithaupt and Matthews (2001) present innovative techniques for a three-dimensional documentation and analysis of track sites. Close-range photogrammetry (Breithaupt et al., 2004; Petti et al., 2008), anaglyph stereo imaging (Gatesy et al., 2005), high-resolution LiDAR (Bates et al., 2008), and laser scanning (Petti et al., 2008) combined with geographic information systems were successfully applied for three-dimensional collection, documentation and analysis of tracks and trackways, contours, and site visualization.

2.5 Building Stones

Building stones are a valuable source for ichnological studies, but as yet, their potential seems not fully utilized. Of special interest are cases where plenty of material was cut from a known outcrop, ground, and polished. The stratigraphical position of the material can often be recognized on the basis of information known from the outcrop, and the wealth of cut surfaces offers ample details for investigation. Examples of ichnologically interesting material include slabs of Mesozoic limestone in Italy (Fig. 1E), Cenozoic sandstone slabs with flysch trace fossils in provincial towns of the Spanish Pyrenees (Fig. 1F), and Turonian sandstone with crustacean burrows from Saxony in Germany used in many historical buildings in Dresden and Berlin (Müller, 1971). Outstanding is the Ordovician limestone from Scandinavia, Baltica, and NW Russia (Knaust et al., 2012), which was quarried for hundreds of years as an excellent building stone for St. Petersburg, Stockholm, Copenhagen and many Hanseatic towns to where it was shipped in the Middle Ages as ship ballast. Building stones of that Ordovician limestone revealed many secrets and assisted Ekdale and Bromley (2001, 2003) in the establishment of new ichnotaxa.

2.6 Quantification of Bioturbation

The quantification of bioturbation provides a clue to the bottom conditions for the infauna and therefore is essential in ichnological studies. Widely accepted is the bioturbation scale from Reineck (1963), which later was revitalized by Taylor and Goldring (1993). A similar semiquantitative classification scheme and visual charts were subsequently introduced by Droser and Bottjer (1986), modified by Miller and Smail (1997) to include bioturbation on bedding planes, by Lockley (1991) for trampling on soil or sedimentary substrates by dinosaurs (dinoturbation), and by Montague et al. (2010) to estimate the extent of microbioerosion. An overview of semiquantitative and quantitative methods was recently given by Marenco and Bottjer (2011).

The Reineck scale was originally developed for the description of bioturbation observed in box cores from the southern North Sea and ranges from grade 0 (no bioturbation) to 6 (bedding completely destroyed). The degree of bioturbation indicates the percentage of bioturbation within a vertical section, for example, the area that is occupied by bioturbation, and is commonly expressed as the bioturbation index (BI). This scheme has the great advantage that it has been widely applied in ichnological studies and thus allows for comparison of the bioturbation degree in different settings. However, two issues are worth discussing that influence the applicability of any existing scheme. These are the scaling issue and the fact of categories having unequal proportions.

Initially, Reineck (1963) implemented his scale for a semiquantification of biogenically disturbed (bioturbated) sediment, while later on, this method has been extended in order to include discrete traces and trace fossils. In the latter respect, there is a scaling relationship between the size of the trace fossils observed in a section and the total size available for investigation. For instance, it makes a difference whether a core slab restricted by a size of 100 × 10 cm comprises tiny burrows such as *Chondrites* or *Trichichnus* with a burrow diameter in the millimeter range or whether it contains large burrows like *Thalassinoides* or *Phoebichnus* in the range of centimeters. The same number of burrows in each example would result in completely different amounts of bioturbation (e.g., the area they occupy within the section), low *versus* high, respectively. Therefore, it is essential to select a reasonably sized section for investigation with a representative statistical distribution.

Another issue is related to the seven categories (0–6) with a disproportionate weight. While the current scheme works fine in semiquantitative analysis, it influences a statistical analysis of the amount of bioturbation, a method which is relevant for the comparison of depositional settings or stratigraphical changes (e.g., Knaust, 2010a; Martin, 1993; Pemberton and Frey, 1984). Moreover, ichnofabric constituent diagrams are typically constructed on the basis of percentage of bioturbation (Taylor and Goldring, 1993). For such analysis, categories with the same proportions (increments) and a linear increase would enhance their comparability in a statistical way (Martin, 1993). In contrast to the semi-quantification approach with varying proportions (percentage) of the categories, statistical analysis should be based on categories with constant proportions between the two extremes 0% and 100%, such as 0%, 20%, 40%, 60%, 80%, and 100% of bioturbation as maxima (Fig. 2). The content of each category can be selected on the basis of the quality and quantity of the available material and the purpose of the analysis. This procedure has been used occasionally (e.g., Rebata-H et al., 2006) but so far is not in common use.

From a technical aspect, point counting can be employed for collecting data on bedding planes and vertical (e.g., core) sections (for details, see Flügel, 2004; Marenco and Bottjer, 2010). Image analysis techniques for quantifying bioturbation were applied by Francus (2001), Löwemark (2003), and Honeycutt and Plotnick (2008). Challenges regarding the logging of the BI in core, particularly in carbonates, are addressed by Knaust (2009b).

3. LABORATORY TECHNIQUES

3.1 Sectioning and Enhancement of Visibility

An important method to enhance the visibility of biogenic structures is the sectioning of rocks and staining the smooth surfaces with oil or other chemicals. The employed method is dependent on the kind of lithology, the degree of consolidation, and the composition of the rock. Cutting and polishing of rock faces may already give some reasonable results (Miller, 1988) but can still be improved by photography under alcohol (Bockelie, 1973), by applying polarizing filters to camera and light sources (Bengtson, 2000), or subsequent staining techniques (Farrow, 1975). Less-consolidated siliciclastic rocks with relicts

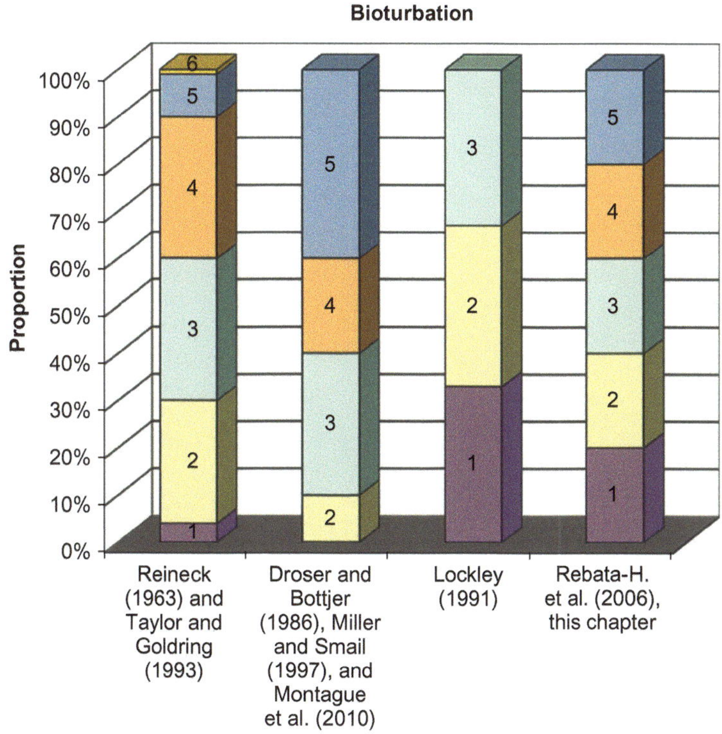

FIGURE 2 Comparison of various classification systems to quantify the degree or index of bioturbation. Established systems use unequal proportions in each category, whereas an equal distribution of categories should be preferred for statistical analysis.

of organic compounds may reveal their internal fabric by staining using the periodic acid-Schiff reaction (Risk and Szczuczko, 1977), marly chalk or limestone samples are best studied after staining with oil (Bromley, 1981; Olóriz and Rodríguez-Tovar, 2002), whereas cherts and siliceous rocks can be etched using hydrofluoric acid (Kakuwa, 2004).

Peels can be retained from polished rock surfaces of different chemical composition, and continuous grinding and peeling may produce serial sections (Miller, 1988). Carbonates, for instance, can be etched with hydrochloric or formic acid, dried, and subsequently treated with acetone before being covered with a transparent film for replication (Fig. 3). Further, serial sectioning is a powerful technique to reveal the three-dimensional nature of trace fossils and has been successfully applied in many studies (Bednarz and McIlroy, 2009; Michalík and Šimo, 2010; Milán and Bromley, 2006; Naruse and Nifuku, 2008; Retallack, 2001). It is an advanced technique commonly applied in biology and medicine, and specific software applications help to process the data in order to reconstruct the three dimensions of the object under investigation (e.g., Fiala, 2005).

FIGURE 3 Part of an acetate peel of a bio- and intraclastic limestone slab from the Middle Triassic of Weimar, Germany, with a *Balanoglossites triadicus* burrow system filled with sucrose dolomite.

3.2 Thin Sectioning

The preparation of thin sections for petrographical and microfacies studies is an old technique (see Miller, 1988) with many possible applications also in the field of ichnology, which can be grouped into the following categories: (1) identification of minute trace fossils (Banerjee et al., 2007; Das and Rao, 1992; Knaust, 1998; Wisshak, 2012), (2) determination of microcoprolites based on their internal structure (Flügel, 2004), (3) examination of the internal structure of burrows and rhizoliths for morphological, behavioral, biostratinomic, and diagenetic interpretations (Knaust, 2009b; Retallack, 2001), (4) enhancement of visibility of indistinct biogenic structures with low color and/or grain-size contrast (Olóriz and Rodríguez-Tovar, 2002; Schieber, 2003), (5) identification of remains of the tracemaker (Fig. 4), (6) investigation of the composition and tiering patterns of ichnofabrics and the related fluid-flow behavior (Gingras et al., 2004), and (7) quantification of bioturbation via image analysis (Francus, 2001).

Large-scale thin sections up to a postcard format are invaluable for the study of trace fossils and ichnofabrics in transmitted light and may reveal continuous structures and greater parts of ichnofabrics. This method has been successfully

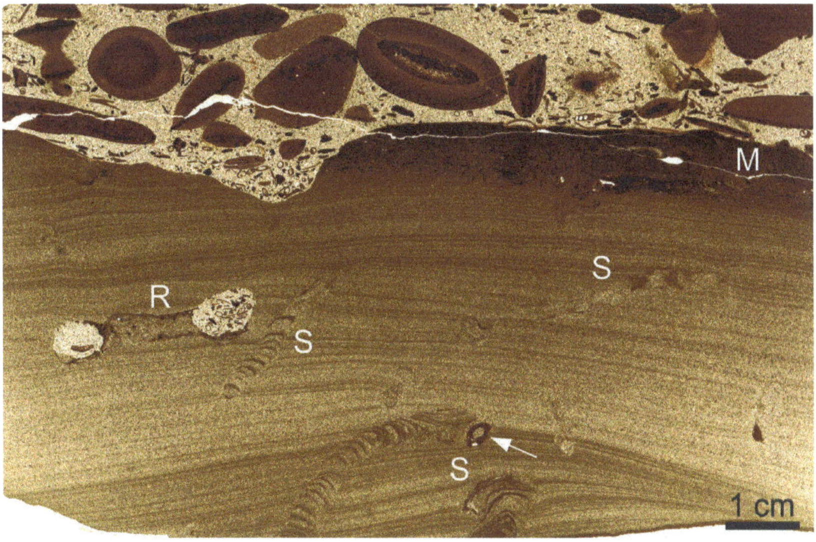

FIGURE 4 Part of a large-scale thin section displaying a vertical section through a Lower Jurassic calcite concretion from the clay pit Grimmen, northern Germany. The visualization in transmitted light via a high-resolution flatbed scanner reveals an insightful ichnofabric that otherwise would be invisible. In its lower part, siltstone with low-angle cross-lamination contains two generations of burrows: (1) actively backfilled meniscate burrows (S), produced by scaphopods (arrow), and (2) a horizontal spreite burrow (*Rhizocorallium*, R) with a passively filled marginal tube around an actively produced spreite. A mixed layer (M) with complete bioturbation and enriched organic matter is developed at the top of the siltstone bed, but it is partly or completely cut through by an erosion surface with a bioclastic conglomerate above it.

applied in homogeneous carbonates (Knaust, 1998) and siliciclastics (Garton and McIlroy, 2006) to enhance the visibility of trace fossils and bioturbate structures (Fig. 4). Semiconsolidated samples are commonly impregnated with low-viscosity and stained epoxy resin before the preparation of thin sections. In order to achieve a better contrast and visualization, it might be worthwhile to diverge from the commonly applied standard thickness of 0.03 mm but prepare it slightly thicker. For imaging, a high-resolution flatbed scanner is suitable. Chemical staining can be applied to emphasize certain structures in carbonates (see Miller, 1988).

3.3 Optical Microscopy

Microscopical analysis of trace fossils in reflected or transmitted light is an important method for ichnological investigations. Given a size range from bacteria up to giants, a wealth of organism groups does exist at the lower size limit with potential tracemakers, which may produce an exceeding number of morphologically different traces compared to their macroscopic counterparts. Thus, microscopy becomes essential in the analysis of traces left by bacteria, fungi, and meiobenthic organisms.

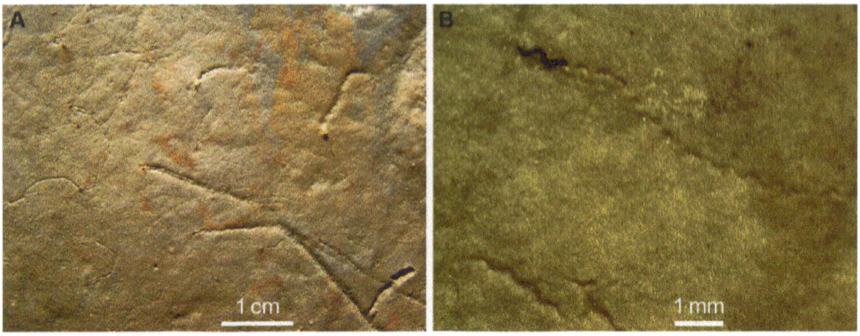

FIGURE 5 Minute trace fossils and their meiobenthic producers on a micritic bedding plane, Middle Triassic of Weimar, Germany (cf. Knaust, 2010b). (A) Overview image made with the macro-function of a digital camera. Diffuse sunlight illumination from the upper left augments the contrast. (B) Microphotograph of a nematode (upper left) and its trail (*Cochlichnus* isp.) taken with a digital camera attached to the microscope.

Microorganisms (such as foraminifers and amoebas) are known to produce traces that require a good microscope for their study. Bengtson and Rasmussen (2009) used modern traces formed by large unicellular organisms such as amoebas as an analog to explain Ediacaran trace-like fossils. Microscopy is commonly applied in the study of microborings, mainly in combination with epoxy casting (Ekdale et al., 1984; Wisshak, 2012).

A problem for unbiased ichnological studies of all size classes are those trace fossils that fall between the microscopic and macroscopic domains. They are typically in the range of their meiobenthic producers, for example, between 0.06 and 1 mm in shortest body size (e.g., platyhelminthes, nematodes, nemerteans, polychaetes, sipunculans, crustaceans, molluscs, etc.), and are often overlooked because of their intermediary size. For instance, the Triassic shallow-marine carbonates of Germany are rich in well-studied macrofossils, but the comprehensive meiobenthic fauna including their trace fossils (partly concentrated in the fills of ammonoid shells) was only recently discovered (Knaust, 2007). For the study of meiobenthic trace fossils, it often becomes necessary to combine conventional microscopy with digital macrophotography in order to capture a wider field of the object of investigation (Knaust, 2010b; Fig. 5).

Another application of microscopy is the study of burrow-internal details such as grain alignment and orientation for the recognition of diagnostic features. The morphological analysis of fecal pellets (microcoprolites) and the internal structure of coprolites commonly requires a microscope (Flügel, 2004) as does the study of trace fossils in amber (Poinar, 1998).

3.4 Scanning Electron Microscopy

Scanning electron microscopy (SEM) provides a useful means in ichnology and is typically combined with other techniques, such as resin casting or thin sectioning. Because of its high level of resolution, microscopical trace fossils or trace-fossil parts can be studied in detail, for instance, microbial fossils within trace fossils (Gong et al., 2007), microborings (Wisshak, 2012), drill holes in prey skeletons (Schiffbauer et al., 2008), subtle bioturbation fabrics and microburrows in fine-grained sediment (O'Brien and Pietraszek-Mattner, 1998), grain-size differences in burrow walls (Zorn et al., 2010) or other microstructural features, and chemical changes related to trace fossils (Driese and Foreman, 1991). Thomsen and Vorren (1984) used SEM, energy dispersive X-ray, microprobe, and X-ray diffractometer to study the composition of pyritized *Trichichnus*-like burrows. An introduction to SEM and related techniques for geologists is given by Reed (2010).

3.5 X-ray Radiography

X-ray radiography is an effective technique to reveal internal structures in various kinds of sediments and rock slabs, especially those taken from core. The gray-scale images acquired from the core slab reflect primarily sediment density as a function of lithology, grain size, and cementation, and reveal bioturbation and trace fossils in high resolution, which otherwise would be invisible (Bouma, 1969; Löwemark, 2007). Digital X-ray systems gradually replace conventional systems working with film and permit a convenient application both in the laboratory and in the field (St-Onge et al., 2007).

There are many examples of applications that mainly cover unlithified and apparently homogeneous Quaternary deposits from shallow-marine to deep-marine environments (Bouma, 1964; Howard, 1968; Löwemark, 2003; Löwemark and Schäfer, 2003; Werner and Wetzel, 1982; Winn, 2006). In contrast, X-ray radiography is not fully utilized in the study of bioturbation and trace fossils in industrial well cores (Gerard and Bromley, 2008; Pemberton and Gingras, 2005). Löwemark et al. (2006) used the deformation of pyritized trace fossils as recognized in X-ray radiographs to estimate core shortening in gravity cores.

3.6 Computer-Aided Tomography

Axial tomography (CAT scan) and high-resolution X-ray microcomputed tomography are relatively young techniques that allow for non-destructive visualization and quantification of sedimentary structures, trace fossils, and organisms, preferably captured in cored samples (Dufour et al., 2005; Fu et al., 1994; St-Onge et al., 2007; Sutton, 2008). During the scan process, a series of X-ray radiographs is captured at constant increments as the sample continuously rotates between the emitter and detector. The different attenuation at each scanned point (voxel) provides the basis for the reconstruction of two-dimensional images, which again allow for generating a three-dimensional image of the burrow or boring inside the sample.

Computer-aided tomography has been applied to quantify the amount of bioturbation in the scope of hydrocarbon and aquifer characterization (Cunningham et al., 2009, 2012; Pemberton and Gingras, 2005), to study marine and terrestrial benthic communities (Dufour et al., 2005; Perez et al., 1999), to

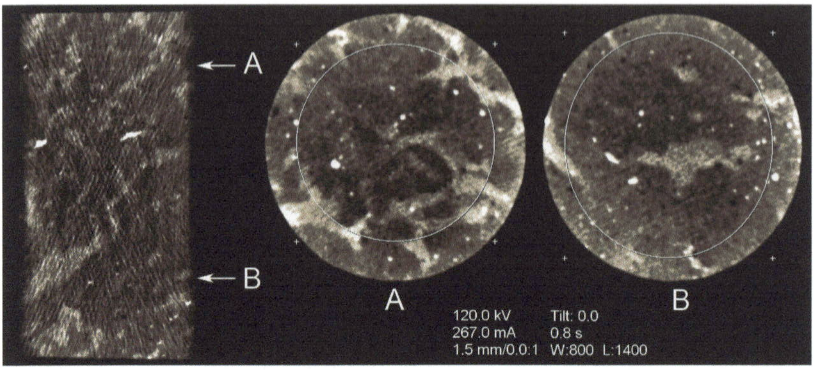

FIGURE 6 CAT-scan of a borehole core (~11 cm in diameter) from a Campanian deep-sea fan system of the Vøring Basin, Norwegian Sea (cf. Knaust, 2009a). The axial and accompanying cross sections reveal a complex burrow system with meniscate backfill cf. *Ophiomorpha* isp., which is difficult to recognize in the two-dimensional expression of the slabbed core (see Fig. 8).

examine trace fossils in core (Fu et al., 1994; Gerard and Bromley, 2008; Fig. 6), and in the investigation of morphological features of burrows and macro- and microborings (Capowiez et al., 2003; Schönberg and Shields, 2008). Gingras et al. (2002a) and Pemberton and Gingras (2005) used magnetic resonance imaging for the visualization of complex burrow systems and related porosity.

3.7 Polymer-Resin Casting

Casting of microborings in carbonate substrate is a standard procedure in ichnology in order to perform SEM studies on the morphology of these tiny traces (Golubic et al., 1975; Wisshak, 2012). After cleaning the open boring systems with hydrogen peroxide and in an ultrasonic bath, the samples are infiltrated with low-viscosity epoxy resin in a vacuum chamber (Nielsen and Maiboe, 2000). The solid resin block can then be cut and the carbonate substrate etched using a dilute acid. The nicely exposed casts are then ready for coating with metal and examination with SEM.

4. BOREHOLE CORES AND IMAGES

4.1 Well Cores

Well cores are an invaluable source for ichnological and sedimentological information from the subsurface and are usually obtained during the exploration of hydrocarbon, aquifers, and deep-sea drilling programs. The limited size, restricted surface, two-dimensional expression on (often vertical) core slabs, etc., require a special treatment of ichnological data in well cores. The acquired core can either be studied as full core or, more commonly, as individual slabbed sections such as the half-cut or a slabbed and resinated slice from the middle (Blackbourn, 2009; Fig. 7A). Because of the economic value, the procedure of studying trace fossils in core has received much attention in numerous

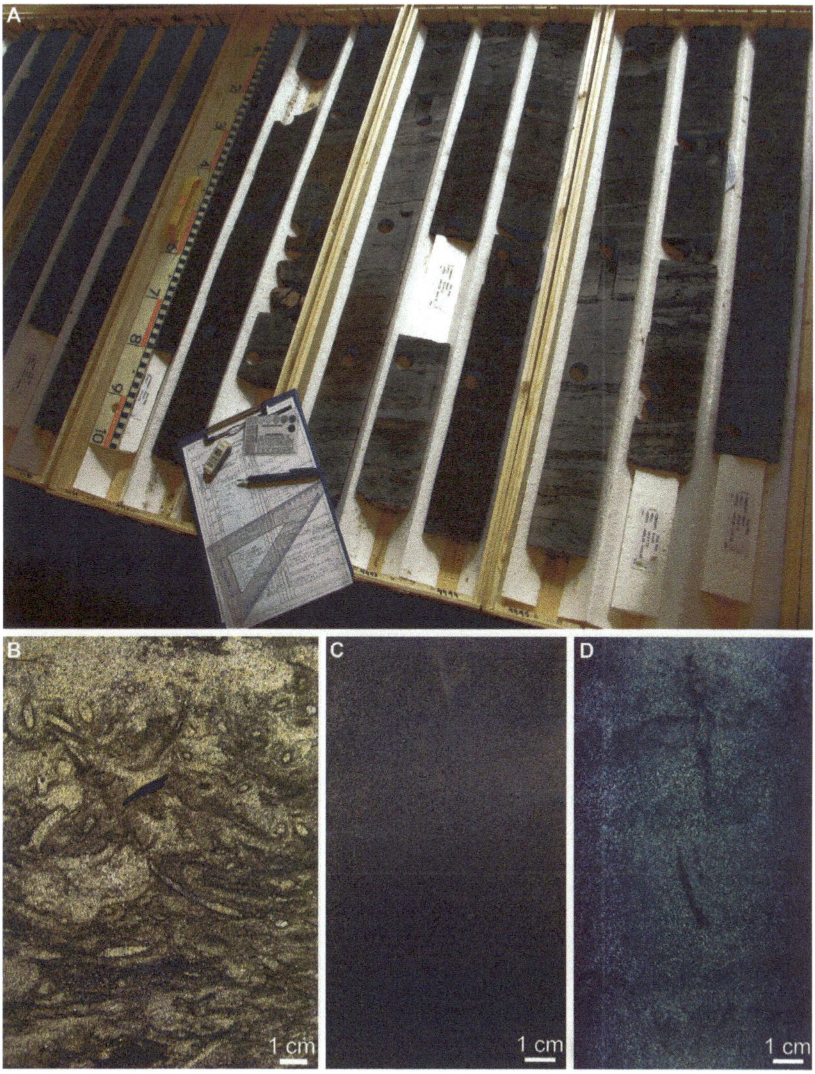

FIGURE 7 Well-core slabs from the Norwegian North Sea. (A) Slabbed core (half cut) on display for sedimentological and ichnological logging. (B) Highly bioturbated sandstone comprising a *Lingulichnus* ichnofabric, Early Jurassic Cook Formation. (C) Oil-saturated sandstone with an apparently homogeneous fabric, Middle Jurassic Hugin Formation. (D) Same sample as in (C) but under UV light, revealing internal bedding, colonization surfaces, and discrete trace fossils.

approaches (Bockelie, 1991; Bromley, 1996; Chamberlain, 1978; Gerard and Bromley, 2008; MacEachern et al., 2007; McIlroy, 2004; Pemberton, 1992).

Trace-fossil identification is not straightforward in two-dimensional core slabs, and often only limited information about the whole morphology, particularly of complex trace fossils, can be achieved. In contrast, different planes of sections through the trace fossils may reveal very precisely the morphological details such as wall structure, fill, cross section, pellets, etc. These circumstances call for an adapted nomenclature for ichnotaxa identified in core, and in many cases, the identification is limited to the ichnogenus level, which again influences the paleoenvironmental interpretation. It therefore becomes obvious that ichnological observations achieved from core are not fully comparable with those obtained from outcrop. For instance, subenvironments of deep-sea fan systems are well characterized from various outcrop studies, whereas vertical core sections from such systems display a very different suite of trace fossils (e.g., Knaust, 2009a). This is also related to the fact that fine-grained deep-sea deposits preferably occur in the form of so-called graphoglyptids along bedding planes, which are hard to study in core.

The ichnological study of cores is crucial in the interpretation of depositional environments, key stratigraphical surfaces, and sequence stratigraphy, and different kinds of data can be collected from core. Foremost, ichnofabric analysis plays an important role and offers a lot of information regarding colonization surfaces, tiering patterns, cross-cut relationships of trace fossils, and the sedimentological context (Ekdale et al., 2012; McIlroy, 2004; Taylor and Goldring, 1993; Taylor et al., 2003; Fig. 7B). Further, the amount of bioturbation (BI), ichnodiversity (number of ichnotaxa per interval), and morphometric data (e.g., burrow diameters) are commonly logged as useful parameters for interpretation.

Besides high-resolution digital imaging, drawing the discrete bioturbation patterns in relation to the primary bedding with permanent color pens on a transparent sheet is a good way to study and document the ichnofabrics and their different components and relationships (Knaust, 2010a). Core slabs of homogeneous carbonates, cemented or oil-stained intervals, may reveal their internal structures under UV light (Fig. 7C and D).

4.2 Borehole Images

Borehole image logs complement to a great extent our information on the subsurface geology, including the ichnological content (Gerard and Bromley, 2008; Lovell et al., 1999). This is particularly important in hydrocarbon and aquifer reservoirs where no core material can be obtained, whereas an overlap of well core and borehole image log allows for a direct calibration of the borehole images based on cores and adds valuable information to characterize the deposits.

Based on the measurement technique, electrical (Schlumberger tools FMS, FMI, OBMI), acoustical (UBI, Schlumberger; CBIL, Baker Hughes), and optical borehole images can be obtained with varying resolution. After processing and visualization of the data in a special software tool, ichnological features can be investigated in addition to the structural and sedimentological interpretation. Particularly, the FMI (Formation Micro Imager) tool allows imaging with a resolution down to about 1 cm or below and thus is suitable for the interpretation of bioturbation and larger trace fossils (such as *Diplocraterion*, *Ophiomorpha*, *Thalassinoides*, and *Skolithos*, Fig. 8) which, in turn, can reveal key surfaces for sequence-stratigraphical interpretations.

FIGURE 8 Example of a dynamic FMI Image (Schlumberger tool) from a Campanian deep-sea fan system of the Vøring Basin, Norwegian Sea (cf. Knaust, 2009a). Bright color (resistive) corresponds to sandstone and dark color (conductive) to mudstone. The 2-m thick interval is accompanied by four representative core image examples (about 10 cm wide) taken from the corresponding slabbed core to illustrate the lithological changes, the bioturbation index (BI), and recognizable trace fossils. Note the muddy, meniscate backfill of cf. *Ophiomorpha* isp. that is also detectable in the sandy portions of the borehole image (e.g., at 3115.7 and 3115.1 m). The nodules at 3116.1 m consist of siderite.

There are not many published examples where ichnological information has been incorporated into the conventional interpretation of borehole image logs (e.g., Cunningham et al., 2009, 2012; Salimullah and Stow, 1995), probably because of the image resolution, which is just about the size of typical burrows. Prosser et al. (1999) provide an overview of applications and methodologies of borehole images with special reference to integrated sedimentological–ichnological studies.

5. STATISTICAL ANALYSIS

An introduction to statistical methods for earth scientists is given by McKillup and Dyar (2010), whereas Harper (1999) reviews computer-based techniques available for the analysis and modeling of paleontological data. Statistical methods can be applied in various ways to analyze ichnological data, and various text books cover different aspects of this topic: (1) Trace fossils can be described and named on the basis of their morphological features (morphometrics) which may be captured by a statistical analysis if sufficient material is available (Orr, 1999; Platt et al., 2010). (2) Distribution analysis and modeling of trace fossils in time and space can provide important clues for paleoecological and taphonomic reconstructions (Kowalewski, 2002). (3) Benthic communities can be numerically analyzed to delineate population densities, subenvironments, changes in substrate, or stress factors. This is a common procedure in neoichnological investigations, but can also be applied to trace-fossil assemblages with some restrictions (Martin, 1993; Pemberton and Frey, 1984). (4) Changes in trace-fossil composition and amount of bioturbation can be detected through the stratigraphic record, which may help in the identification of ecological and evolutionary trends (Uchman, 2004). This is particularly crucial in the recognition of significant changes (e.g., ichnodiversity) across mass extinctions (Knaust, 2010a). (5) Statistical analysis can contribute to quantifying bioturbation, mapping trace-fossil distributions, or to the testing of lateral persistence of ichnofabrics (McIlroy, 2007).

A common technique for statistical data analysis is a multivariate analysis (cluster analysis), which can be used in ichnology, for instance, for finding similarities in trace-fossil assemblages and for defining environmental gradients expressed by trace fossils (Bjerstedt, 1988; Minter and Braddy, 2009). Diagrams containing the extent of bioturbation (ichnograms) can be used in basin analysis (Bottjer and Droser, 1991) but also have a great potential in the characterization and interwell correlation of hydrocarbon reservoirs (Fig. 9).

6. COMPUTER MODELING

Computer simulation of trace fossils and bioturbation has a great potential in different fields of neo- and paleoichnology. Artificial life methods and neural networks are employed to create virtual trace fossils, whereas the computer-based study of the morphogenesis of the spatial distribution of complex traces

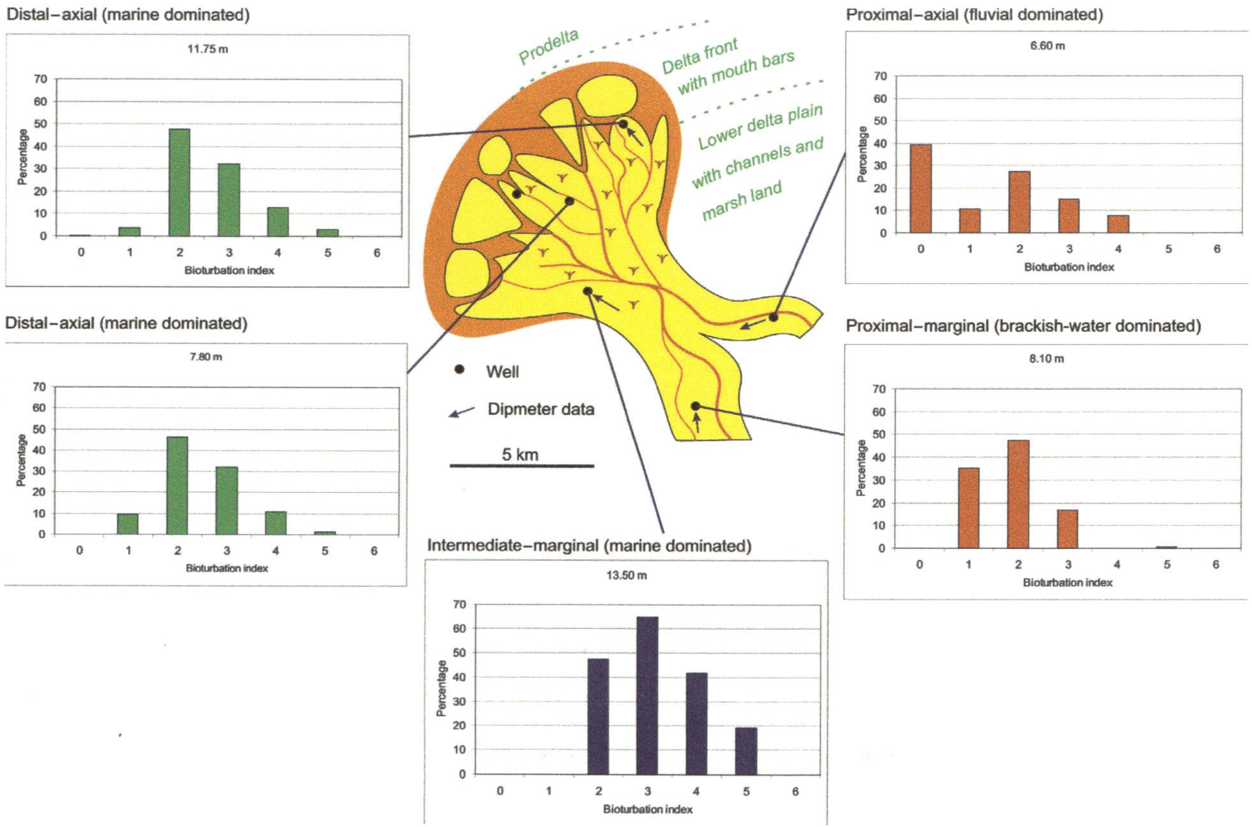

FIGURE 9 Reconstruction of a bay-head delta based on limited well data from a Middle Jurassic oil and gas field under development in the Norwegian North Sea. The performance of an ichnological study helped in the characterization, correlation, and zonation of the reservoir, and in the establishment of a conceptual model. Ichnograms (diagrams containing the extent of bioturbation), taken from a particular reservoir zone dominated by tidal-flat and bar facies, clearly show proximal-to-distal and axial-to-lateral trends as the intensity of bioturbation decreases with increasing input of freshwater from the SE. In combination with other ichnological (e.g., kinds and size of trace fossils, ichnodiversity) and sedimentological data, ichnograms guided the prediction of sand distribution within the area.

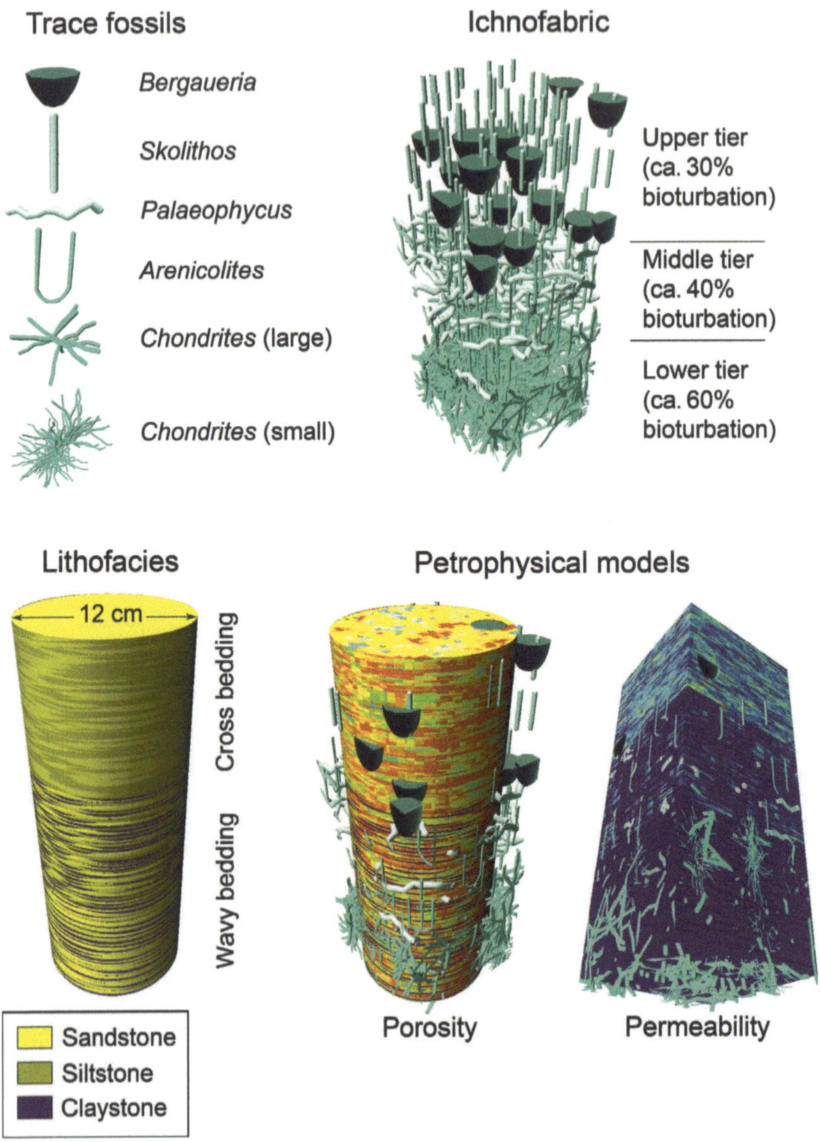

FIGURE 10 Characterization of the impact of bioturbation on reservoir quality, connectivity, and flow behavior by performing scenario modeling in SBED (Geomodeling). Various kinds of trace fossils (burrows) are given a specific morphology, size, and variability before they are arranged as an ichnofabric with different tiers, amount of bioturbation, colonization surfaces, and cross-cut relationships. Based on process modeling, the heterolithic tidal-flat lithofacies are simulated in the form of a digital well core and combined with the ichnofabric. Each trace-fossil group has assigned porosity and permeability values. Numerous realizations, property modeling, and different upscaling methods allow for comparison of non-bioturbated with bioturbated rock and to identify sensitivities in bioturbation-related rock properties.

assists in behavioral and evolutionary studies (Koy and Plotnick, 2007) and the reconstruction of burrow geometry as constituent of porous systems (Bastardie et al., 2002). Hofmann (1990) simulated horizontal trace fossils using algorithms for random patterns to achieve quantitative morphometric data. The fractal geometry of some trace fossils is suitable for distinguishing taxa and thus has implications for ichnotaxonomy. Further, numerous biological and ecological studies focus on the simulation of bioturbation for understanding bioirrigation mechanisms and its environmental impact (Meysman et al., 2006).

Bioturbation may alter sediments with resulting increase or decrease in porosity and permeability, which again has implications for the characterization of hydrocarbon reservoirs and aquifers (Cunningham, 2009; Cunningham et al., 2012; Gingras et al., 2012; Knaust, 2009b; Pemberton and Gingras, 2005). Previous approaches include stochastical simulations of three-dimensional burrow networks for quantification of bioirrigation (Koretsky et al., 2002), assessing the anisotropic permeability of burrowed surfaces (Gingras et al., 1999) and modeling of fluid migration through a two-dimensional bioturbation model using the software program MPath (Permedia; Spila et al., 2007).

The impact of bioturbation on reservoir quality can be simulated together with high-resolution, process-based models of sedimentary bedding using the software application SBED (Geomodeling; Ringrose et al., 2005), where common burrow morphologies (e.g., simple shaft, plug-shaped, U-shaped, irregularly branched) are integrated in a range of bedding structures and lithologies in varying density (Dabek and Knepp, 2011; Fig. 10). After assigning porosity and permeability values (probe permeameter data) to different host-rock lithologies (e.g., sandstone, siltstone, and mudstone) and burrow elements (e.g., core and wall), the upscaled results of the digital rock sample are comparable with those of the non-bioturbated model.

7. CONCLUSIONS

Numerous techniques and methods are available for supporting ichnological analysis and help to make paleoenvironmental reconstructions successful. In the field, the lithology-dependent visibility of bioturbation, trace fossils, and ichnofabrics can be enhanced by brushing, sectioning, weathering, sawing, etc. Further, the preparation of peels, molds, and casts can be applied to preserve a wide range of trace fossil and modern traces. Different kinds of core samplers (e.g., box-core sampler) are available to obtain subaquatic surface sediment. Tracks and trackways are imaged and analyzed by utilizing photogrammetry, anaglyph stereo imaging, high-resolution LiDAR, or laser scanning. Building stones, as excavated from ichnological sites, offer the opportunity to closely study various aspects of ichnology. Semiquantification of bioturbation is widely done with the help of the scale developed by Reineck (1963), whereas categories with equal proportions are more applicable for statistical analysis.

In the laboratory, various procedures can help to increase the outcome of integrated ichnological–sedimentological studies. They include sectioning and enhancement of visibility by staining, peeling, or advanced photography techniques. Thin sections (preferably large-scale) are suitable to solve various problems, and optical microscopy is essential in the study of minute and meiobenthic trace fossils. These techniques may be complemented with SEM in order to reveal still more details. Conventional X-ray radiography is a standing technique to study internal structures particularly in core, whereas computer-aided tomography becomes increasingly important in the three-dimensional reconstruction of trace fossils. Finally, polymer-resin casting is a prerequisite for the proper study of microbioerosion.

Industrial application of ichnology is closely related to borehole cores and images. The study of well cores is challenging due to their limited exposure surfaces and thus requires a special procedure. High-resolution borehole images provide an invaluable source of ichnological information, particularly when calibrated against cored intervals.

Statistical analysis can be widely applied in ichnology, including in ichnotaxonomy, in paleoecology, or to study benthic communities, evolutionary trends, and trace-fossil distributions. Similarly, computer modeling has been increasingly employed to solve several problems, such as quantitative morphometry, behavior of tracemakers, bioirrigation, environmental and substrate changes, and reservoir quality.

ACKNOWLEDGMENT

I thank the reviewer, Ludvig Löwemark (Stockholm), for his helpful comments. I am grateful to Statoil for permission to publish the data presented in Fig. 6 and Section 4–6.

REFERENCES

Atkinson, R.J.A., Chapman, C.J., 1984. Resin casting: a technique for investigating burrows in sublittoral sediments. Prog. Underwat. Sci. 9, 15–25.

Banerjee, N.R., Simonetti, A., Furnes, H., Muehlenbachs, K., Staudigel, H., Heaman, L., Van Kranendonk, M.J., 2007. Direct dating of Archean microbial ichnofossils. Geology 35, 487–490.

Bastardie, F., Cannavacciuolo, M., Capowiez, Y., de Dreuzy, J.-R., Bellido, A., Cluzeau, D., 2002. A new simulation for modelling the topology of earthworm burrow systems and their effects on macropore flow in experimental soils. Biol. Fert. Soils 36, 161–169.

Bates, K.T., Manning, P.L., Vila, B., Hodgetts, D., 2008. Three-dimensional modelling and analysis of dinosaur trackways. Palaeontology 51, 999–1010.

Bednarz, M., McIlroy, D., 2009. Three-dimensional reconstruction of "phycosiphoniform" burrows: implications for identification of trace fossils in core. Palaeontologica Electronica 12, 15 pp. http://palaeo-electronica.org/2009_3/195/index.html.

Bengtson, S., 2000. Teasing fossils out of shales with cameras and computers. Palaeontologica Electronica 3, 14 pp. http://palaeo-electronica.org/2000_1/fossils/issue1_00.htm.

Bengtson, S., Rasmussen, B., 2009. New and ancient trace makers. Science 323, 346–347.

Bjerstedt, T.W., 1988. Multivariate analyses of trace fossil distribution from an Early Mississippian oxygen-deficient basin, Central Appalachians. Palaios 3, 53–68.

Blackbourn, G.A., 2009. Cores and Core Logging for Geoscientists. Whittles Publishing, Dunbeath, 152 pp.

Bockelie, J.F., 1991. Ichnofabric mapping and interpretation of Jurassic reservoir rocks of the Norwegian North Sea. Palaios 6, 206–215.

Bockelie, T.G., 1973. A method of displaying sedimentary structures in micritic limestones. J. Sediment. Petrol. 43, 537–539.

Bottjer, D.J., Droser, M.L., 1991. Ichnofabric and basin analysis. Palaios 6, 199–205.

Bouma, A.H., 1964. Sampling and treatment of unconsolidated sediments for study of internal structures. J. Sediment. Petrol. 34, 349–354.

Bouma, A.H., 1969. Methods for the Study of Sedimentary Structures. Wiley-Interscience, New York, 458 pp.

Breithaupt, B.H., Matthews, N.A., 2001. Preserving paleontological resources using photogrammetry and geographic information systems. In: Harmon, D. (Ed.), Proceedings of the 11th Conference on Research and Resource Management in Parks and on Public Lands "From Crossing Boundaries in Park Management". The George Wright Society Hancock, Michigan, pp. 62–70.

Breithaupt, B.H., Matthews, N.A., Noble, T.A., 2004. An integrated approach to three-dimensional data collection at dinosaur tracksites in the Rocky Mountain West. Ichnos 11, 11–26.

Bromley, R.G., 1981. Enhancement of visibility of structures in marly chalk: modification of the Bushinsky oil technique. Bull. Geol. Soc. Denmark 29, 111–118.

Bromley, R.G., 1996. Trace Fossils: Biology, Taphonomy and Applications. Chapman and Hall, London, 361 pp.

Bromley, R.G., Uchman, A., 2003. Trace fossils from the Lower and Middle Jurassic marginal marine deposits of the Sorthat Formation, Bornholm, Denmark. Bull. Geol. Soc. Denmark 50, 185–208.

Capowiez, Y., Pierret, A., Moran, C.J., 2003. Characterization of the three-dimensional structure of earthworm burrow systems using image analysis and mathematical morphology. Biol. Fert. Soils 38, 301–310.

Chamberlain, C.K., 1978. Recognition of trace fossils in cores. In: Basan, P.B. (Ed.), Trace Fossil Concepts. SEPM Short Course 5, pp. 119–166.

Coe, A.L. (Ed.), 2010. Geological Field Techniques. Wiley-Blackwell, Chichester, 336 pp.

Cunningham, K.J., Sukop, M.C., Huang, H., Alvarez, P.F., Curran, H.A., Renken, R.A., Dixon, J.F., 2009. Prominence of ichnologically influenced macroporosity in the karst Biscayne aquifer: stratiform "super-K" zones. GSA Bull. 121, 164–186.

Cunningham, K.J., Sukop, M.C., Curran, H.A., 2012. Carbonate aquifers. In: Knaust, D., Bromley, R.G. (Eds.), Trace Fossils as Indicators of Sedimentary Environments. Developments in Sedimentology, vol. 64. Elsevier, Amsterdam, pp. 871–898.

Dabek, L.B., Knepp, R., 2011. Bioturbation and its effects on permeability in wave-dominated shoreface rocks of the Spring Canyon Member, Blackhawk Formation, Utah, USA. Search and Discovery Article #50425, 10 pp.

Das, S.S., Rao, C.N., 1992. Micro-burrows from the Charmuria Formation, Madhya Pradesh, India. Palaios 7, 548–552.

Dashtgard, S.E., Gingras, M.K., 2005. Facies architecture and ichnology of Recent salt-marsh deposits: Waterside Marsh, New Brunswick, Canada. J. Sediment. Res. 75, 596–607.

Dashtgard, S.E., Gingras, M.K., 2012. Marine invertebrate neoichnology. In: Knaust, D., Bromley, R.G. (Eds.), Trace Fossils as Indicators of Sedimentary Environments. Developments in Sedimentology, vol. 64. Elsevier, Amsterdam, pp. 273–295.

Driese, S.G., Foreman, J.L., 1991. Traces and related chemical changes in a Late Ordovician paleosol, *Glossifungites* Ichnofacies, southern Appalachians, USA. Ichnos 1, 207–219.

Droser, M.L., Bottjer, D.J., 1986. A semiquantitative classification of ichnofabric. J. Sediment. Petrol. 56, 558–559.

Dufour, S.C., Desrosiers, G., Long, B., Lajeunesse, P., Gagnoud, M., Labrie, J., Archambault, P., Stora, G., 2005. A new method for three-dimensional visualization and quantification of biogenic structures in aquatic sediments using axial tomodensitometry. Limnol. Oceanogr. Meth. 3, 372–380.

Ekdale, A.A., Bromley, R.G., 2001. Bioerosional innovation for living in carbonate hardgrounds in the Early Ordovician of Sweden. Lethaia 34, 1–12.

Ekdale, A.A., Bromley, R.G., 2003. Paleoethologic interpretation of complex *Thalassinoides* in shallow-marine limestones, Lower Ordovician, southern Sweden. Palaeogeogr. Palaeoclimatol. Palaeoecol. 192, 221–227.

Ekdale, A.A., Bromley, R.G., Pemberton, S.G., 1984. Ichnology. The Use of Trace Fossils in Sedimentology and Stratigraphy. SEPM Short Course Notes 15, 317 pp.

Ekdale, A.A., Bromley, R.G., Knaust, D., 2012. The ichnofabric concept. In: Knaust, D., Bromley, R.G. (Eds.), Trace Fossils as Indicators of Sedimentary Environments. Developments in Sedimentology, vol. 64. Elsevier, Amsterdam, pp. 139–155.

Farrow, G.E., 1975. Techniques for the study of fossil and recent traces. In: Frey, R.W. (Ed.), Trace Fossils. Springer, New York, pp. 537–554.

Fiala, J.C., 2005. *Reconstruct*: a free editor for serial section microscopy. J. Microsc. 218, 52–61.

Flügel, E., 2004. Microfacies of Carbonate Rocks: Analysis, Interpretation and Application. Springer, Berlin, 976 pp.

Francus, P., 2001. Quantification of bioturbation in hemipelagic sediments via thin-section image analysis. J. Sediment. Res. 71, 501–507.

Francus, P. (Ed.), 2004. Image Analysis, Sediments and Paleoenvironments. In: Developments in Paleoenvironmental Research, vol. 7. Springer, Dordrecht, 330 pp.

Fu, S., Werner, F., Brossmann, J., 1994. Computed tomography: application in studying biogenic structures in sediment cores. Palaios 9, 116–119.

Gaillard, C., Jautee, E., 1987. The use of burrows to detect compaction and sliding in fine-grained sediments: an example from the Cretaceous of S.E. France. Sedimentology 34, 585–593.

Garton, M., McIlroy, D., 2006. Large thin slicing: a new method for the study of fabrics in lithified sediments. J. Sediment. Res. 76, 1252–1256.

Gatesy, S.M., Shubin, N.H., Jenkins Jr., F.A., 2005. Anaglyph stereo imaging of dinosaur track morphology and microtopography. Palaeontologica Electronica 8, 10 pp. http://palaeo-electronica.org/2005_1/gatesy10/issue1_05.htm.

Gerard, J.R.F., Bromley, R.G., 2008. Ichnofabrics in Clastic Sediments: Applications to Sedimentological Core Studies. Jean R.F. Gerard, Madrid, 97 pp.

Gingras, M.K., Pemberton, S.G., Mendoza, C.A., Floyd, H., 1999. Assessing the anisotropic permeability of *Glossifungites* surfaces. Petrol. Geosci. 5, 349–357.

Gingras, M.K., MacMillan, B., Balcom, B.J., Saunders, T., Pemberton, S.G., 2002a. Using magnetic resonance imaging and petrographic techniques to understand the textural attributes and porosity distribution in *Macaronichnus*-burrowed sandstone. J. Sediment. Res. 72, 552–558.

Gingras, M.K., Pickerill, R., Pemberton, G.S., 2002b. Resin cast of modern burrows provides analogs for composite trace fossils. Palaios 17, 206–211.

Gingras, M.K., Mendoza, C.A., Pemberton, S.G., 2004. Fossilized worm burrows influence the resource quality of porous media. AAPG Bull. 88, 875–883.

Gingras, M.K., Baniak, G., Gordon, J., Hovikoski, J., Konhauser, K.O., La Croix, A., Lemiski, R., Mendoza, C., Pemberton, S.G., Polo, C., Zonneveld, J.-P., 2012. Porosity and permeability in bioturbated sediments. In: Knaust, D., Bromley, R.G. (Eds.), Trace Fossils as Indicators of Sedimentary Environments. Developments in Sedimentology, vol. 64. Elsevier, Amsterdam, pp. 835–868.

Golubic, S., Perkins, R.D., Lukas, K.J., 1975. Boring microorganisms and microborings in carbonate substrates. In: Frey, R.W. (Ed.), The Study of Trace Fossils. Springer, Berlin, pp. 229–259.

Gong, Y.M., Xu, R., Xie, S.C., Huang, X.Y., Hu, B., Qi, Y.A., Zhang, G.C., 2007. Microbial and molecular fossils from the Permian *Zoophycos* in South China. Sci. China Ser. D: Earth Sci. 50, 1121–1127.

Goodwin, M.B., Chaney, D.S., 1994. Molding and casting: techniques and materials. In: Leiggi, P., May, P. (Eds.), Vertebrate Paleontological Techniques, vol. 1. Cambridge University Press, New York, pp. 235–271.

Green, O.R., 2001. A Manual of Practical Laboratory and Field Techniques in Palaeobiology. Kluwer Academic Publishers, Dordrecht, 538 pp.

Harper, D.A.T. (Ed.), 1999. Numerical Palaeobiology: Computer-Based Modelling and Analysis of Fossils and their Distributions. John Wiley & Sons, Chichester, 468 pp.

Hodges, E.R.S. (Ed.), 2003. The Guild Handbook of Scientific Illustration. John Wiley & Sons, Hoboken, 656 pp.

Hofmann, H.J., 1990. Computer simulation of trace fossils with random patterns, and the use of goniograms. Ichnos 1, 15–22.

Honeycutt, C.E., Plotnick, R., 2008. Image analysis techniques and gray-level co-occurrence matrices (GLCM) for calculating bioturbation indices and characterizing biogenic sedimentary structures. Comput. Geosci. 34, 1461–1472.

Howard, J.D., 1968. X-ray radiography for examination of burrowing in sediments by marine invertebrate organisms. Sedimentology 11, 249–258.

Kakuwa, Y., 2004. Trace fossils from the Triassic-Jurassic deep water, oceanic radiolarian chert successions of Japan. Fossils Strata 51, 58–67.

Kidwell, S.M., Moore, J.A., Moore, J.R., 1985. Inexpensive field technique for polyester resin peels of structures in unconsolidated sediments. Mar. Geol. 64, 351–359.

Knaust, D., 1998. Trace fossils and ichnofabrics on the Lower Muschelkalk carbonate ramp (Triassic) of Germany: tool for high-resolution sequence stratigraphy. Geol. Rundsch. 87, 21–31.

Knaust, D., 2007. Meiobenthic trace fossils as keys to the taphonomic history of shallow-marine epicontinental carbonates. In: Miller III, W. (Ed.), Trace Fossils. Concepts, Problems, Prospects. Elsevier, Amsterdam, pp. 502–517.

Knaust, D., 2009a. Characterisation of a Campanian deep-sea fan system in the Norwegian Sea by means of ichnofabrics. Mar. Petrol. Geol. 26, 1199–1211.

Knaust, D., 2009b. Ichnology as a tool in carbonate reservoir characterization: a case study from the Permian–Triassic Khuff Formation in the Middle East. GeoArabia 14, 17–38.

Knaust, D., 2010a. The end-Permian mass extinction and its aftermath on an equatorial carbonate platform: insights from ichnology. Terra Nova 22, 195–202.

Knaust, D., 2010b. Remarkably preserved benthic organisms and their traces from a Middle Triassic (Muschelkalk) mud flat. Lethaia 43, 344–356.

Knaust, D., Curran, H.A., Dronov, A., 2012. Shallow-marine carbonates. In: Knaust, D., Bromley, R.G. (Eds.), Trace Fossils as Indicators of Sedimentary Environments. Developments in Sedimentology, vol. 64. Elsevier, Amsterdam, pp. 703–750.

Koretsky, C.M., Meileb, C., Van Cappellen, P., 2002. Quantifying bioirrigation using ecological parameters: a stochastic approach. Geochem. Trans. 2002, 17–30.

Kowalewski, M., 2002. The fossil record of predation: an overview of analytical methods. In: Kowalewski, M., Kelley, P.H. (Eds.), The Fossil Record of Predation. Paleontol. Soc., Spec. Pap. 8, pp. 3–42.

Koy, K.A., Plotnick, R.E., 2007. Theoretical and experimental ichnology of mobile foraging. In: Miller III, W. (Ed.), Trace Fossils. Concepts, Problems, Prospects. Elsevier, Amsterdam, pp. 419–432.

Lockley, M.G., 1991. Tracking Dinosaurs: A New Look at an Ancient World. Cambridge University Press, Cambridge, 238 pp.

Lovell, M., Williamson, G., Harvey, P. (Eds.), 1999. Borehole Imaging; Applications and Case Histories. Geol. Soc., Spec. Publ., vol. 159, 294 pp.

Löwemark, L., 2003. Automatic image analysis of X-ray radiographs: a new method for ichnofabric evaluation. Deep-Sea Res. I 50, 815–827.

Löwemark, L., 2007. Importance and usefulness of trace fossils and bioturbation in paleoceanography. In: Miller III, W. (Ed.), Trace Fossils. Concepts, Problems, Prospects. Elsevier, Amsterdam, pp. 413–427.

Löwemark, L., Schäfer, P., 2003. Ethological implications from a detailed X-ray radiograph and ^{14}C study of the modern deep-sea *Zoophycos*. Palaeogeogr. Palaeoclimatol. Palaeoecol. 192, 101–121.

Löwemark, L., Schönfeld, J., Schäfer, P., 2006. Deformation of pyritized burrows: a novel technique for the detection and quantification of core shortening in gravity cores. Mar. Geol. 233, 37–48.

MacEachern, J.A., Pemberton, S.G., Gingras, M.K., Bann, K., 2007. Applied Ichnology. SEPM Short Course Notes 52, 380 pp.

Marenco, K.N., Bottjer, D.J., 2010. The intersection grid technique for quantifying the extent of bioturbation on bedding planes. Palaios 25, 457–462.

Marenco, K.N., Bottjer, D.J., 2011. Quantifying bioturbation in Ediacaran and Cambrian rocks. In: Laflamme, M., Schiffbauer, J.D., Dornbos, S.Q. (Eds.), Quantifying the Evolution of Early Life. Numerical Approaches to the Evaluation of Fossils and Ancient Ecosystems. Topics in Geobiology, vol. 36. Springer, Berlin, pp. 135–160.

Martin, A.J., 1993. Semiquantitative and statistical analysis of bioturbate textures, Sequatchie Formation (Upper Ordovician), Georgia and Tennessee, USA. Ichnos 2, 117–136.

McIlroy, D., 2004. Some ichnological concepts, methodologies, applications and frontiers. In: McIlroy, D. (Ed.), The Application of Ichnology to Palaeoenvironmental and Stratigraphic Analysis. Geol. Soc. London, Spec. Publ., vol. 228, pp. 3–27.

McIlroy, D., 2007. Lateral variability in shallow marine ichnofabrics: implications for the ichnofabric analysis method. J. Geol. Soc. 164, 359–369.

McKillup, S., Dyar, M.D. (Eds.), 2010. Geostatistics Explained. An Introductory Guide for Earth Scientists. Cambridge University Press, Cambridge, 412 pp.

Meysman, F.J.R., Galaktionov, O.S., Gribsholt, B., Middelburg, J.J., 2006. Bioirrigation in permeable sediments: advective pore-water transport induced by burrow ventilation. Limnol. Oceanogr. 51, 142–156.

Michalík, J., Šimo, V., 2010. A new spreite trace fossil from Lower Cretaceous limestone (Western Carpathians, Slovakia). Trans. R. Soc. Edinburgh, Earth Sci. 100, 417–427.

Milán, J., Bromley, R.G., 2006. True tracks, undertracks and eroded tracks, experimental work with tetrapod tracks in laboratory and field. Palaeogeogr. Palaeoclimatol. Palaeoecol. 231, 253–264.

Miller, J., 1988. Microscopical techniques: I. Slices, slides, stains and peels. In: Tucker, M.E. (Ed.), Techniques in Sedimentology. Blackwell Science, Oxford, pp. 86–107.

Miller, M.F., Smail, S.E., 1997. A semiquantitative field method for evaluating bioturbation on bedding planes. Palaios 12, 391–396.

Minter, N.J., Braddy, S.J., 2009. Ichnology of an Early Permian Intertidal Flat: The Robledo Mountains formation of Southern New Mexico, USA. Spec. Pap. Palaeont. 82, 1–107.
Montague, K.E., Walton, A.W., Hasiotis, S.T., 2010. Euendolithic microborings in basalt glass fragments in hyaloclastites: extending the ichnofabric index to microbioerosion. Palaios 25, 393–399.
Müller, A.H., 1971. Bioturbation durch Decapoda (Crustacea) in Sandsteinen der sächsischen Oberkreide. Dt. Akad. Wiss. Berlin, Monatsber. 13, 696–707.
Naruse, H., Nifuku, K., 2008. Three-dimensional morphology of the ichnofossil *Phycosiphon incertum* and its implication for paleoslope inclination. Palaios 23, 270–279.
Netto, R.G., Benner, J.S., Buatois, L.A., Uchman, A., Mángano, M.G., Ridge, J.C., Kazakauskas, V., Gaigalas, A., 2012. Glacial environments. In: Knaust, D., Bromley, R.G. (Eds.), Trace Fossils as Indicators of Sedimentary Environments. Developments in Sedimentology, vol. 64. Elsevier, Amsterdam, pp. 299–327.
Nielsen, J.K., Maiboe, J., 2000. Epofix and vacuum: an easy method to make casts of hard substrates. Palaeontologica Electronica 3, 10. http://palaeo-electronica.org/2000_1/epofix/issue1_00.htm.
O'Brien, N.R., Pietraszek-Mattner, S., 1998. Origin of the fabric of laminated fine-grained glaciolacustrine deposits. J. Sediment. Res. 68, 832–840.
Olóriz, F., Rodríguez-Tovar, F.J., 2002. Trace-fossils and minor discontinuities in a marl limestone rhythmite, Lower–Middle Kimmeridgian, southern Spain. Geobios 35, 581–593.
Orr, P.J., 1999. Quantitative approaches to the resolution of taxonomic problems in invertebrate ichnology. In: Harper, D.A.T. (Ed.), Numerical Palaeobiology: Computer-based Modelling and Analysis of Fossils and their Distributions. John Wiley & Sons, Chichester, pp. 395–431.
Pemberton, S.G., 1992. Applications of Ichnology to Petroleum Exploration: A Core Workshop. SEPM Core Workshop 17, 429 pp.
Pemberton, S.G., Frey, R.W., 1984. Quantitative methods in ichnology: spatial distribution among populations. Lethaia 77, 33–49.
Pemberton, S.G., Gingras, M.K., 2005. Classification and characterizations of biogenically enhanced permeability. AAPG Bull. 89, 1493–1517.
Perez, K.T., Davey, E.W., Moore, R.H., Burn, P.R., Rosol, M.S., Cardin, J.A., Johnson, R.L., Kopans, D.N., 1999. Application of computer-aided tomography (CT) to the study of estuarine benthic communities. Ecol. Appl. 9, 1050–1058.
Petti, F.M., Avanzini, M., Belvedere, M., deGasperi, M., Ferretti, P., Girardi, S., Remondino, F., Tomasoni, R., 2008. Digital 3D modelling of dinosaur footprints by photogrammetry and laser scanning techniques: integrated approach at the Coste dell'Anglone tracksite (Lower Jurassic, Southern Alps, Northern Italy). In: Avanzini, M., Petti, F.M. (Eds.), Italian Ichnology. Studi Trentini di Scienze Naturali, Acta Geol., vol. 83. pp. 303–315.
Platt, B.F., Hasiotis, S.T., Hirmas, D.R., 2010. Use of low-cost multistripe laser triangulation (MLT) scanning technology for three-dimensional, quantitative paleoichnological and neoichnological studies. J. Sediment. Res. 80, 590–610.
Poinar Jr., G.O., 1998. Trace fossils in amber: a new dimension for the ichnologist. Ichnos 6, 47–52.
Prosser, J., Buck, S., Saddler, S., Hilton, V., 1999. Methodologies for multi-well sequence analysis using borehole image and dipmeter data. In: Lovell, M., Williamson, G., Harvey, P. (Eds.), Borehole Imaging: Applications and Case Histories. Geol. Soc., Spec. Publ., vol. 159, pp. 91–121.
Rebata-H, L.A., Gingras, M.K., Räsänen, M.E., Barberi, M., 2006. Tidal-channel deposits on a delta plain from the Upper Miocene Nauta Formation, Marañón foreland Sub-basin, Peru. Sedimentology 53, 971–1013.

Reed, S.J.B., 2010. Electron Microprobe Analysis and Scanning Electron Microscopy in Geology. Cambridge University Press, Cambridge, 212 pp.

Reineck, H.-E., 1963. Sedimentgefüge im Bereich der südlichen Nordsee. Abh. Senckenb. Naturf. Ges. 505, 1–138.

Retallack, G.J., 2001. *Scoyenia* burrows from Ordovician palaeosols of the Juniata Formation in Pennsylvania. Palaeontology 44, 209–235.

Ringrose, P., Nordahl, K., Wen, R., 2005. Vertical permeability estimation in heterolithic tidal deltaic sandstones. Petrol. Geosci. 11, 29–36.

Risk, M.J., Szczuczko, R.B., 1977. A method for staining trace fossils. J. Sediment. Petrol. 47, 855–859.

Salimullah, A.R.M., Stow, D.A.V., 1995. Ichnofacies recognition in turbidites/hemiturbidites using enhanced FMS images: examples from ODP Leg 129. The Log Analyst 36, 38–49.

Savazzi, E., 2011. Digital Photography for Science. Close-up Photography, Macrophotography and Photomacrophotography. Enrico Savazzi, 698 pp.

Schieber, J., 2003. Simple gifts and buried treasures: implications of finding bioturbation and erosion surfaces in black shales. Sediment. Rec. 1, 4–8.

Schiffbauer, J.D., Yanes, Y., Tyler, C.L., Kowalewski, M., Leighton, L.R., 2008. The microstructural record of predation: a new approach for identifying predatory drill holes. Palaios 23, 810–820.

Schönberg, C.H.L., Shields, G., 2008. Micro-computed tomography for studies on *Entobia*: transparent substrate *versus* modern technology. In: Wisshak, M., Tapanila, L. (Eds.), Current Developments in Bioerosion. Springer, Berlin, pp. 147–164.

Seike, K., Jenkins, R.G., Watanabe, H., Nomaki, H., Sato, K., 2012. Novel use of burrow casting as a research tool in deep-sea ecology. Biol. Lett. 8, 648–651.

Seilacher, A., 2008. Fossil Art. An Exhibition of the Geologisches Institut Tuebingen University Germany. CBM-Publishing, Laasby, 97 pp.

Shinn, E.A., 1968. Burrowing in Recent lime sediments of Florida and the Bahamas. J. Paleontol. 42, 879–894.

Skipper, J.A., Ward, D.J., Johnson, R., 1998. A rapid, lightweight sediment peel technique using polyurethane foam. J. Sediment. Res. 68, 516–517.

Spila, M.V., Pemberton, S.G., Rostron, B., Gingras, M.K., 2007. Biogenic textural heterogeneity, fluid flow and hydrocarbon production: bioturbated facies Ben Nevis Formation, Hibernia Field, Offshore Newfoundland. In: MacEachern, J.A., Bann, K.L., Gingras, M.K., Pemberton, S.G. (Eds.), Applied Ichnology. SEPM Short Course Notes 52, pp. 363–380.

Stidham, T.A., Mason, J., 2009. A quick method for collecting modern small-scale ichnological and sedimentological structures. J. Paleontol. Tech. 4, 1–4.

Stockdale, A., Davison, W., Zhang, H., 2009. Micro-scale biogeochemical heterogeneity in sediments: a review of available technology and observed evidence. Earth-Sci. Rev. 92, 81–97.

St-Onge, G., Mulder, T., Francus, P., Long, B., 2007. Continuous physical properties of cored marine sediments. Dev. Mar. Geol. 1, 63–98.

Stow, D.A.V., 2005. Sedimentary Rocks in the Field. A Colour Guide. Manson Publishing, London, 320 pp.

Sutton, M.D., 2008. Tomographic techniques for the study of exceptionally preserved fossils. Proc. R. Soc. B 275, 1587–1593.

Taylor, A.M., Goldring, R., 1993. Description and analysis of bioturbation and ichnofabric. J. Geol. Soc. 150, 141–148.

Taylor, A., Goldring, R., Gowland, S., 2003. Analysis and application of ichnofabrics. Earth-Sci. Rev. 60, 227–259.

Thomsen, E., Vorren, T.O., 1984. Pyritization of tubes and burrows from Late Pleistocene continental shelf sediments off North Norway. Sedimentology 31, 481–492.

Tschinkel, W.R., 2003. Subterranean ant nests: trace fossils past and future? Palaeogeogr. Palaeoclimatol. Palaeoecol. 192, 321–333.

Tucker, M.E. (Ed.), 1988. Techniques in Sedimentology. Blackwell Science, Oxford, 394 pp.

Uchman, A., 2004. Phanerozoic history of deep-sea trace fossils. In: McIlroy, D. (Ed.), The Application of Ichnology to Palaeoenvironmental and Stratigraphic Analysis. Geol. Soc. London, Spec. Publ., vol. 228, pp. 125–139.

Waldron, J.W.F., 1988. Determination of finite strain in bedding surfaces using sedimentary structures and trace fossils: a comparison of techniques. J. Struct. Geol. 10, 273–281.

Werner, F., Wetzel, A., 1982. Interpretation of biogenic structures in oceanic sediments. Bull. l'Inst. Géol. Bass. d'Aquit. 31, 275–288.

Wetzel, A., 1983. Biogenic structures in modern slope to deep-sea sediments in the Sulu Sea Basin (Philippines). Palaeogeogr. Palaeoclimatol. Palaeoecol. 42, 285–304.

Wetzel, A., Aigner, T., 1986. Stratigraphic completeness: tiered trace fossils provide a measuring stick. Geology 14, 234–237.

Winn, K., 2006. Bioturbation structures in marine Holocene sediments of the Great Belt (Western Baltic). Meyniana 58, 157–178.

Wisshak, M., 2012. Microbioerosion. In: Knaust, D., Bromley, R.G. (Eds.), Trace Fossils as Indicators of Sedimentary Environments. Developments in Sedimentology, vol. 64. Elsevier, Amsterdam, pp. 213–243.

Ziółkowski, P., 2005. Deformations of trace fossils *versus* compaction in micritic limestones, an example from Korzkiew, Upper Jurassic (Kraków Upland). Tomy Jurajskie 3, 55–61.

Zorn, M.E., Gingras, M.K., Pemberton, S.G., 2010. Variation in burrow-wall micromorphologies of select intertidal invertebrates along the Pacific Northwest Coast, USA: behavioural and diagenetic implications. Palaios 25, 59–72.

Chapter 10

Marine Invertebrate Neoichnology

Shahin E. Dashtgard*[,1] and Murray K. Gingras†
*Applied Research in Ichnology and Sedimentology (ARISE) Group, Department of Earth Sciences, Simon Fraser University, Burnaby, British Columbia, Canada, †Ichnology Research Group, Department of Earth and Atmospheric Sciences, University of Alberta, Edmonton, Alberta, Canada
[1]Corresponding author: e-mail: shahin_dashtgard@sfu.ca

1. INTRODUCTION

Ichnology draws heavily on the linkage between the burrowing behaviors and distributions of modern animals, and the trace fossils that are commonly observed in sedimentary strata. In contrast with sedimentary structures, which record the hydrodynamic conditions under which sediments are deposited, ichnology is a proxy for the physical and chemical (physico-chemical) stresses at the time of sediment deposition. Neoichnology (studies of modern animal burrows, trackways, and trails) details the linkage between physico-chemical stress and animal response, where animal response to various environmental factors can be assessed and the resultant trace assemblage determined. Taken further, quantification of environmental factors and burrow distributions in modern settings may enable semiquantification of paleoenvironmental conditions. Through semiquantitative analysis, it may be possible to constrain subaerial exposure times, sedimentation rates, substrate consistencies, and salinity at the time of sediment colonization.

A fundamental principle of ichnology is the interpretation of burrow morphology as a function of animal behavior. With few exceptions, these behaviors are consistent through the Phanerozoic (Buatois et al., 2005), reflecting the necessity for animals to employ specific burrowing strategies in response to similar environmental conditions (Ginsburg et al., 1966; MacEachern et al., 2007; Pemberton et al., 2001; Seilacher, 1953, 1964). By studying animal response, and specifically animal burrowing behavior under varying environmental conditions, neoichnologists can associate depositional parameters and organism response to the environment (Clifton and Thompson, 1978; Dafoe et al., 2008; Dashtgard, 2011a,b; Dashtgard and Gingras, 2005; Dashtgard et al., 2008; Gingras et al., 1999, 2008b; Howard and Frey, 1975; Howard et al., 1975).

Developments in Sedimentology, Vol. 64. http://dx.doi.org/10.1016/B978-0-444-53813-0.00010-1

The focus of this chapter is on the neoichnology of siliciclastic, marginal-marine to marine depositional environments. The chapter is divided into two parts. Part one is a review of the major tracemaking groups, and the similarities and diversity in their burrow morphologies. Part two is a discussion of the impact of various environmental stresses on modern trace distributions and draws extensively on published examples.

2. MAJOR INFAUNAL GROUPS AND THEIR COMMON TRACES

A majority of burrowing marine macrofauna are invertebrates. Of the major invertebrate groups, vermiform animals (e.g., annelids, nemerteans, and hemichordates) are the dominant burrowers, followed by, in no particular order, bivalves, crustaceans, echinoderms, and cnidarians.

2.1 Vermiform Animals: Annelids, Hemichordates, and Nemerteans

Vermiform animals produce smooth-walled burrows that are usually mucous lined (Fig. 1). Preservation of the lining is dependent upon the incorporation of sediment with the mucous, and the geochemical microenvironment in and near the burrow. Sessile vermiform organisms construct mainly unbranching, vertical (Figs. 1 and 2G; e.g., *Clymenella* spp., *Pectinaria* spp.) to U-shaped (Figs. 1 and 2A) burrows. Motile vermiform animals (e.g., *Nereis* sp., *Nephtys* sp., *Cerebratulus* sp.) commonly produce three-dimensional burrow networks that are either branching (if the burrow network acts as both a dwelling and a feeding structure; Figs. 1 and 2H, I) or unbranching where the burrow is a temporary structure mainly used for locomotion (i.e., sediment intrusion or swimming). In the case of both sessile and motile polychaetes, the diameter of the tube is typically uniform along the length of the burrow, enabling identification of these structures as vermiform-generated traces.

Variations from the morphological generalities described above are common in both sessile and motile vermiform burrows. For sessile vermiform burrows, variations include vertical readjustments of burrows, thin and thick burrow linings, and spiral morphologies (Gingras et al., 2008a). Vertical readjustment of U-shaped burrows produces spreiten either below or above the occupied burrow (e.g., *Polydora ligni*, Fig. 1; Gingras et al., 2000b, 2001; Ginsburg et al., 1966; Howard and Frey, 1975; Pemberton and Frey, 1985; Rijken, 1979; Schäfer, 1972). Burrow linings are commonly preserved when sediment is incorporated into the lining. For example, pectinariids (e.g., *Pectinaria gouldii*; Fig. 1) construct brick-layer style sand tubes that surround the animal as it moves through the sediment (Hauck et al., 2009; Rhoads, 1967). Thin, sandy burrow linings are employed by vertical deposit feeders, in both sand- and gravel-dominated sediments (Fig. 2D, E, and G; Bromley, 1990; Dashtgard et al., 2008; Rhoads, 1967; Schäfer, 1972). Thick burrow linings, composed mainly of mud, are common for

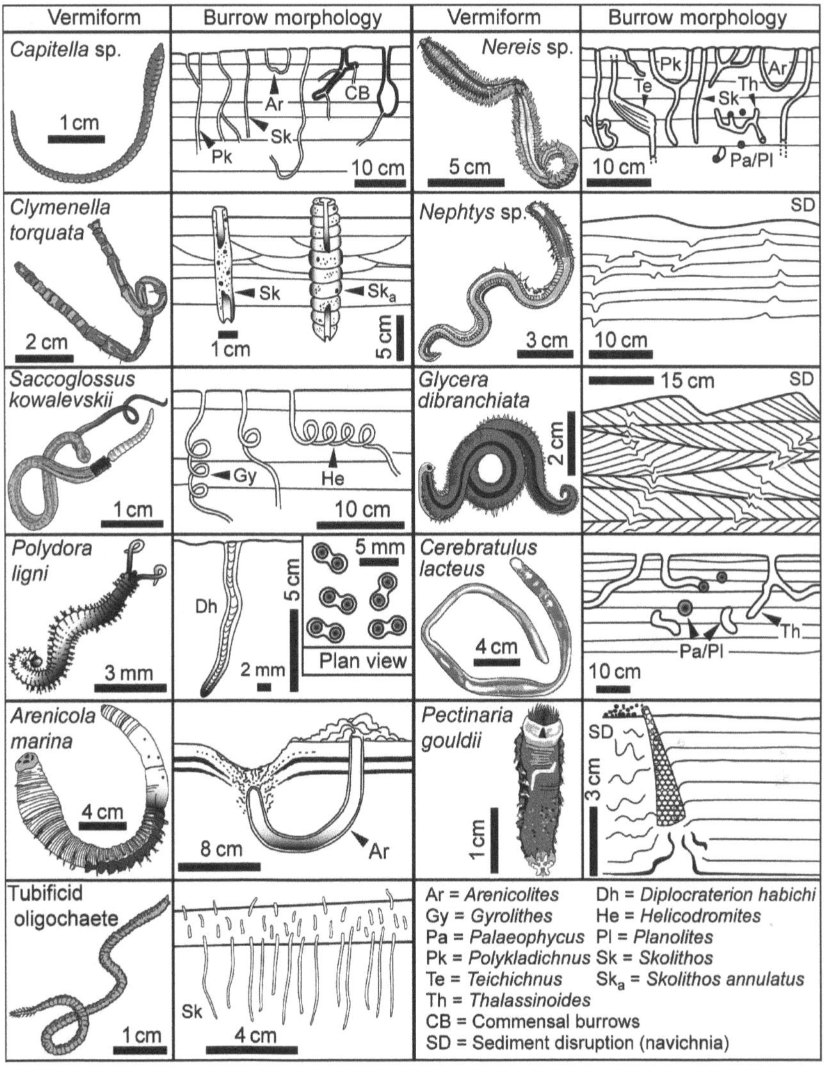

FIGURE 1 Common burrowing vermiform animals and their burrow morphologies. Most of the animals are polychaetes with the exception of *Saccoglossus kowalevskii* (hemichordate), *Cerebratulus lacteus* (nemertean), and the tubificid oligochaetes (closely related to polychaetes). Trace names are assigned based on morphological similarities between the burrows and common trace fossils. All animals burrow in intertidal and shallow subtidal settings. *Modified after Hauck et al. (2009).*

terebellid polychaetes in all substrate types (Aller and Yingst, 1978; Bromley, 1990; Dashtgard et al., 2008; Gingras et al., 1999). The exact purpose of the thick lining is unknown, although it may be used to (1) maintain water movement through the tube, (2) stabilize the substrate, (3) protect the animal, and/or (4) grow

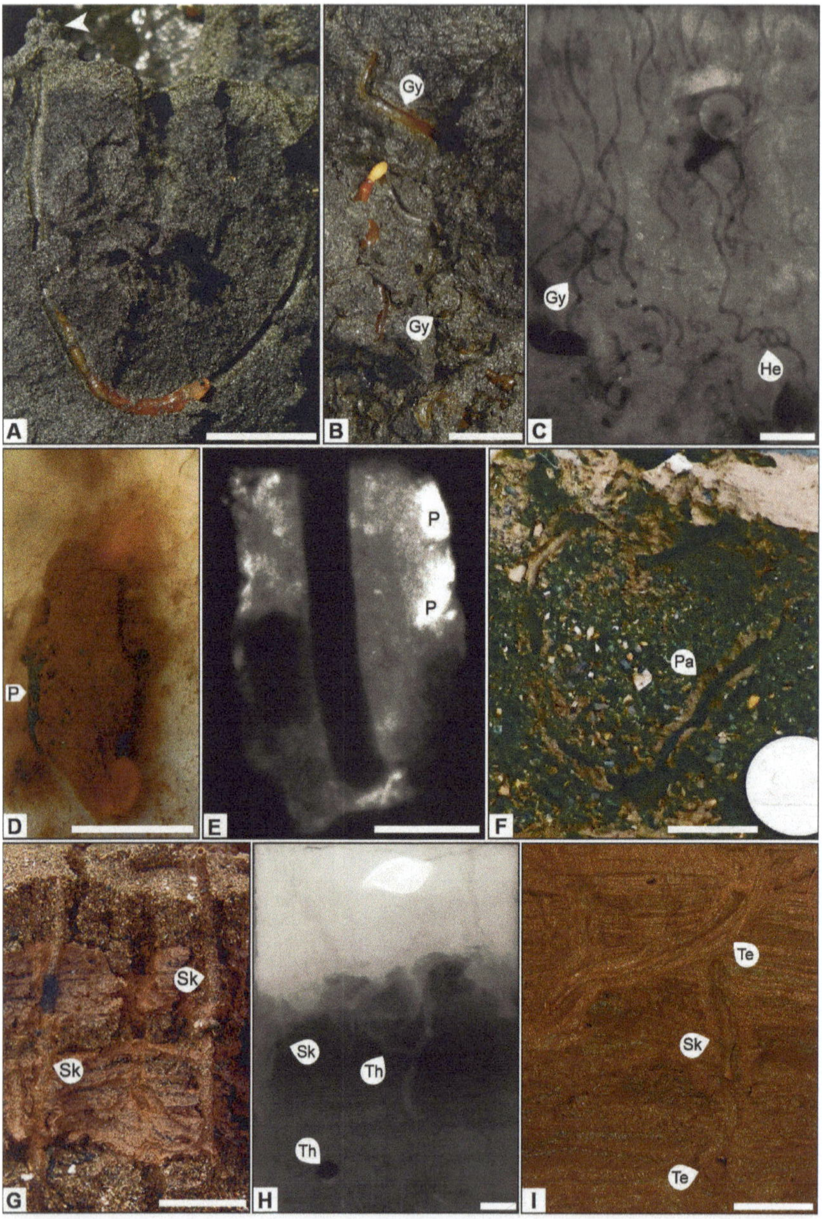

FIGURE 2 Photos (A, B, D, F, G, I) and X-ray radiographs (C, E, H) of vermiform animal burrows. White bar in all photos is 2 cm long. (A) *Abarenicola pacifica* in its U-shaped burrow (*Arenicolites*). (B) and (C) Spiral burrows constructed by the hemichordate, *Dolichoglossus pusillus*. These burrows are most akin to *Gyrolithes* (Gy) and *Helicodromites* (He). (D) and (E) Thick mud linings of terebellid polychaete burrows constructed in subaqueous gravel-dominated sediments. Note the pebbles (P) on the outside of the burrow linings. Terebellid burrows are comparable to

and harvest bacteria (Aller and Yingst, 1978; Bromley, 1990). A last variant of sessile vermiform animals is the development of spiral morphologies (Fig. 2B and C). In neoichnological studies of marine vermiform animals, true spiral morphologies have been attributed to capitellid polychaetes (Howard et al., 1975; Powell, 1977) and paraonid polychaetes (Risk and Tunnicliffe, 1978). A vertical to horizontal helical morphology is also observed at the base of hemichordate worm burrows, mainly those produced by *Saccoglossus* sp. (Figs. 1 and 2B, C; Gingras et al., 1999; Hauck et al., 2009). Variations in the morphology of motile worm burrows are also noted, and include vertical readjustment of burrows (Figs. 1 and 2I), variability in two-dimensional and three-dimensional burrow networks, and backfilling of burrows. Gingras et al. (2008a) discuss at length variations in motile vermiform burrows.

Although deep-water settings are not particularly well studied, oceanographic research has yielded a range of trace-fossil analogs (e.g., Heezen and Hollister, 1971). In these settings, tracks and burrows are reported from photographic images and from box cores, but the tracemakers commonly remain unseen. For example, Hersey (1968) shows a range of modern traces that might be comparable to *Planolites* and *Neonereites*, but the tracemakers are commonly out of view. Several studies document the ichnology of Holocene deep-sea sediments from box cores, and a range of recent burrows are reported, including *Zoophycos* and *Spirophyton* (e.g., Löwemark and Schäfer, 2003; Wetzel, 2008; Wetzel and Werner, 1981) and *Asterosoma* (Ekdale and Lewis, 1991). Although it is generally accepted that a vermiform morphology is suited to the generation of these traces—echiurans and priapulids may be ascribed to such traces—the tracemakers have not been observed in clear association with the biogenic structure. On the other hand, Vannier et al. (2010) used aquarium experiments to show that priapulids were a feasible tracemaker for the commonly deep-water-associated *Treptichnus*.

2.2 Bivalves

Traces produced by deeply burrowing bivalves (burrow depth >5 cm) have a higher preservation potential and, therefore, are ichnologically more important than traces produced by shallowly burrowing bivalves. Burrows and trails made by gastropods and shallow-burrowing bivalves (e.g., *Clinocardium* sp.) are abundant but have a much lower preservation potential due to their near-surface

Cylindrichnus and *Rosselia* (if vertical) and *Asterosoma* (if horizontal). (F) Thick mud lining on the wall of a nereid polychaete burrow constructed in a mixed sand and fine-gravel substrate. This burrow morphology is best described as *Palaeophycus*-like (Pa). (G) Lined *Skolithos* (Sk) traces made by the head-down deposit feeder *Clymenella torquata* (a maldanid polychaete). (H) Trace assemblage including *Skolithos* and *Thalassinoides* (Th) generated by the polychaete *Nereis* sp. Also evident in the radiograph is a crustacean-generated *Thalassinoides* burrow (lower left). (I) *Teichichnus* (Te) and *Skolithos* (Sk) traces generated by the polychaete *Nereis* sp.

position and likely erosion. Deep-burrowing bivalve structures are mainly generated by filter feeders, where the burrows—collectively referred to as *Siphonichnus* (Stanistreet et al., 1980)—comprise two parts: a lower disruption associated with the bivalve and an upper elongate, vertical disruption made by the bivalve's siphon.

The lower part of *Siphonichnus* includes the shell cavity and any retrusive spreite below the shell (e.g., Fig. 3; Reineck and Singh, 1980; Schäfer, 1972). The type of bivalve and its size, the rate and duration of sediment accumulation on a substrate, and the type of substrate burrowed into contribute to the substantial variability observed in the burrows produced. For most bivalves burrowing in soft substrates, the animals vertically readjust their burrow in response to sediment deposition and erosion. The ability to readjust within the sediment tends to decrease for deep-burrowing large bivalves, such as *Tresus nuttalli* (Fig. 4A) and *Mya arenaria* (Figs. 3 and 4B–D) that cannot substantially readjust their burrow once they reach adult size. The resultant structure comprises a mold or empty cavity of the bivalve with minimal sediment disruption above or below the shell (Fig. 4A). If there is no sediment disruption around the bivalve, the basal resting trace is called *Lockeia* (Fig. 4B). Small- and mid-sized bivalves, such as *Macoma* sp. (Fig. 3) and *Nuttallia obscurata* (Fig. 4E), are capable of moving short vertical and horizontal distances within the sediment (up to 40 cm horizontally in the case of *Macoma* spp.; Powilleit et al., 2009). In stiffgrounds, firmgrounds, and hardgrounds, pholadid bivalves produce clavate structures that widen with depth due to the growth of the animal (Fig. 4F). These burrows are permanent and cannot be readjusted upward. Consequently, burrowing in these substrates only occurs in settings characterized by very low sedimentation rates.

The upper part of bivalve-generated burrows includes the siphon trace (one or two siphons) that extends from the body cavity to the sediment/water interface, and any protrusive spreiten above the shell. For deep-burrowing bivalves, the siphon trace (or traces) is (are) relatively large and straight. Adjacent to the siphon trace of large- and mid-sized, deep-burrowing bivalves, it is common to observe downward-deflected sediment lamination (e.g., *M. arenaria*; Figs. 3 and 4D). These laminae record the position of the funnel that developed around the siphon hole at the sediment/water interface (Fig. 4C) or form as sediment caves into the siphon cavity as the animal makes small adjustments to its position in the subsurface. Small bivalves also have siphons, although the siphon traces are temporary structures and are unlikely to produce downward deflecting lamination. For small- and mid-sized bivalves (Fig. 4E), it is also possible to produce convex spreiten that are cross-cut by the siphon trace. These spreiten record the downward movement of the bivalve to maintain its position relative to the sediment/water interface (Stanistreet et al., 1980).

Biogenic structures produced by the lateral movement of shallowly burrowing bivalves and gastropods are also common (e.g., *Mercenaria mercenaria*, Fig. 3). The horizontal movement of bivalves is recorded either as a furrow, as the bivalve uses its foot to pull itself through the sediment, or as a series

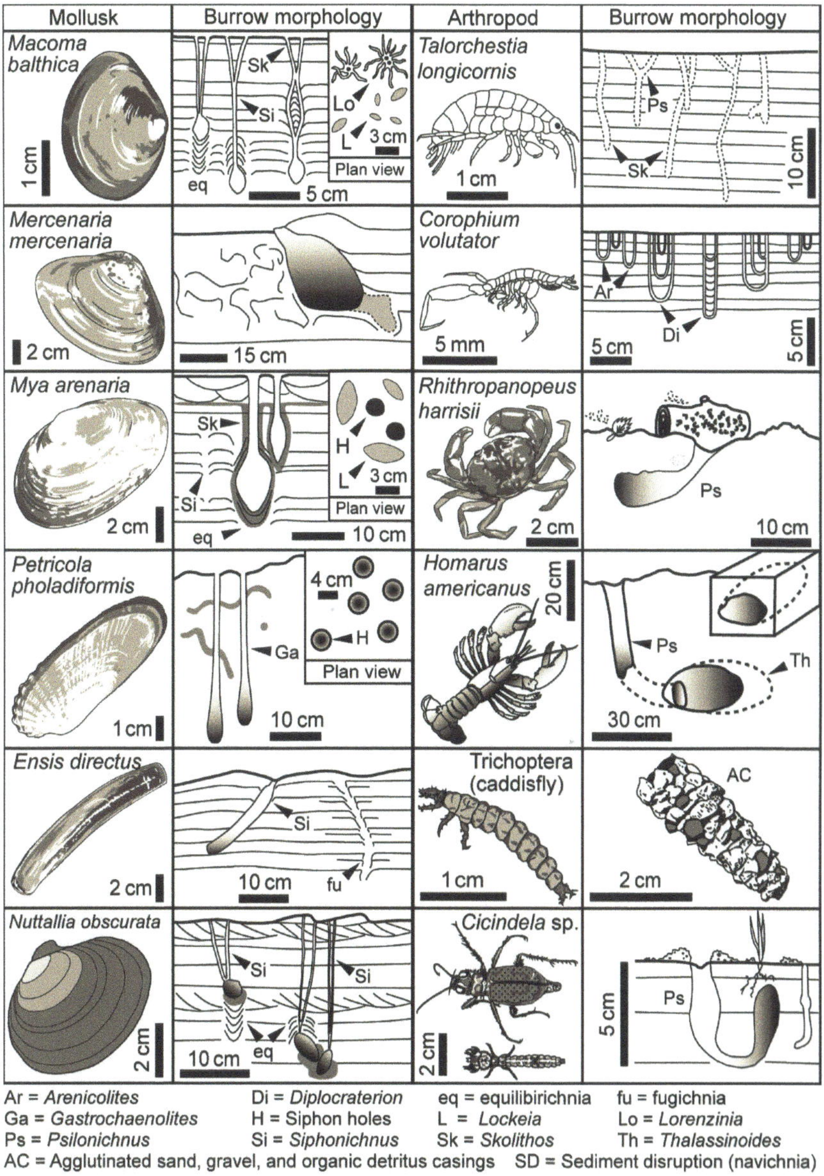

FIGURE 3 Common burrowing bivalves and arthropods and their burrow morphologies. Trace names are assigned based on morphological similarities between the burrows and common trace fossils. All animals burrow in supratidal (e.g., Trichoptera and *Cicindela* spp.), intertidal, and shallow subtidal settings. *Modified after Hauck et al. (2009) and Dashtgard and Gingras (2005).*

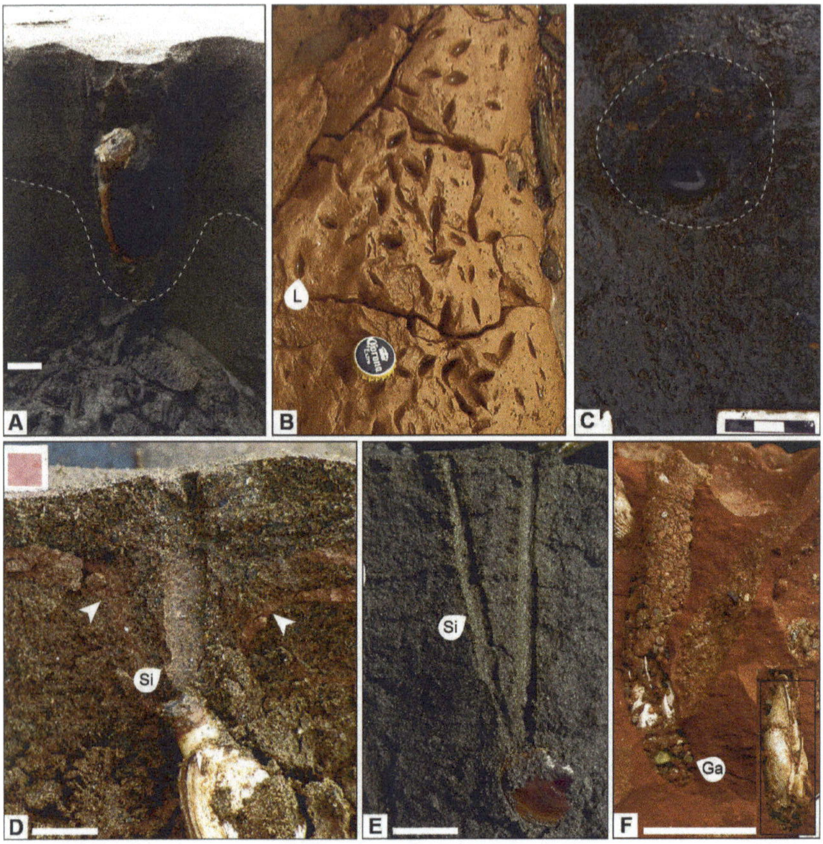

FIGURE 4 Burrowing bivalves and the resultant traces. White scale bar is 2 cm. (A) Downward-deflected sediment lamination (equilibrichnia; outlined by the white dashed line) around the bivalve *Tresus nuttalli*. (B) *Lockeia* (L) on a bedding plane, generated by *Mya arenaria* in mud. Bottle cap is 2.6 cm. (C) Concentric sediment lamination (dashed white line) around the siphon hole of *M. arenaria* in sandy mud. Scale bar is in centimeters. (D) Single-siphon *Siphonichnus* (Si) made by *M. arenaria* in sand. Note the downward-deflected mud laminae around the siphon shaft. (E) Dual-siphon *Siphonichnus* made by *Nuttallia obscurata* in sand. (F) Sand-filled *Gastrochaenolites* (Ga) made by *Petricola pholadiformis* (inset photo) in sandstone.

of *en-echelon* chevrons where the shell was pushed apart to drive the animal forward. The movement of shallowly burrowing bivalves and gastropods can cause significant disruption in the upper few centimeters of the sediment.

2.3 Crustaceans

The range of burrow morphologies produced by crustaceans rivals that of vermiform animals. Crustacean burrows can be unlined or lined with fecal material. Unlike vermiform animals, however, crustacean-burrow diameters are commonly variable along their length. In marine settings, the dominant burrowing crustaceans are isopods, amphipods, and decapods (Figs. 3 and 5). Isopods typically construct trackways (e.g., *Cruziana*, *Isopodichnus*) and very shallow burrows as they scavenge for food (Griffith and Telford, 1985; Hauck et al., 2008). As a result of the shallow burrow depths, the preservation potential of isopod burrows is low.

Amphipods have adapted to living in a range of marginal-marine and marine settings, although only a few extant species are known to burrow. Within the marginal-marine realm, burrowing amphipods are evident from the sandy backshore to the seaward end of the intertidal zone. In sandy backshores, Talitrid amphipods often feed on seaweed and organic detritus washed up by the waves. These animals excavate unlined, vertical burrows, akin to *Skolithos*. Talitrid burrows range up to 25 cm deep (Figs. 3 and 5A; Dashtgard and Gingras, 2005), but are typically less than 5 cm deep.

In the intertidal zone, two burrowing amphipods are commonly found: *Corophium* spp. and *Haustorius* spp. *Corophium volutator* (Fig. 5C and D) is a small mud-loving amphipod which occurs in population densities of up to 63,000/m^2 (Pearson and Gingras, 2006; Thurston, 1990). It constructs a U-shaped tube that promotes water circulation, enabling the amphipod to breathe, surface-deposit feed, and evacuate waste material within the safety of its burrow (Ingle, 1966). *Corophium arenarium* is commonly found in sandy substrates in the intertidal zone, where it excavates a single vertical shaft that provides access to the sediment/water interface for surface-deposit feeding, and allows water to flow freely through the burrow walls (Bromley, 1990; Ingle, 1966). Haustoriid amphipods, commonly referred to as digger amphipods, move through the sediment by excavating and backfilling their burrow as they move. The smaller species of this group are only a few millimeters long and tend to contribute to the formation of cryptobioturbation. Large specimens disrupt the sediment, producing a visible trail of deformed sediment in their wake (Howard and Elders, 1970).

Of the decapods, thalassinid shrimp are the most well-known burrowers (Fig. 5E and F). The common element of most thalassinid shrimp burrows is a vertical shaft that connects to the sediment/water interface and branches at depth (Atkinson and Nash, 1990; Bromley, 1990; Giffis and Chavez, 1988; Gingras et al., 2000a; Shinn, 1968; Swinbanks and Luternauer, 1987). The upper part of the burrow may split, forming a small Y-shape. Deeper in the burrow, it is common to observe tiered and box-work networks (Atkinson and Nash, 1990; Bromley and Frey, 1974; Giffis and Chavez, 1988; Gingras et al., 2008a; Shinn, 1968). In general, thalassinid shrimp use the vertical shaft to maintain a connection to the sediment/water interface and the basal network for deposit feeding. Some genera, including *Upogebia* spp. (Fig. 5E), maintain a broader Y-shaped burrow that the animal employs for filter feeding, and where the descending branch is used as a refuge from predation (Dworschak, 1982; Stevens, 1929). Thalassinid shrimp are also known to make helical burrows (i.e., *Gyrolithes*; Dworschak and Rodrigues, 1997; Wetzel et al., 2010), where the spiral ramp is either used for the easy maintenance of their domiciles or

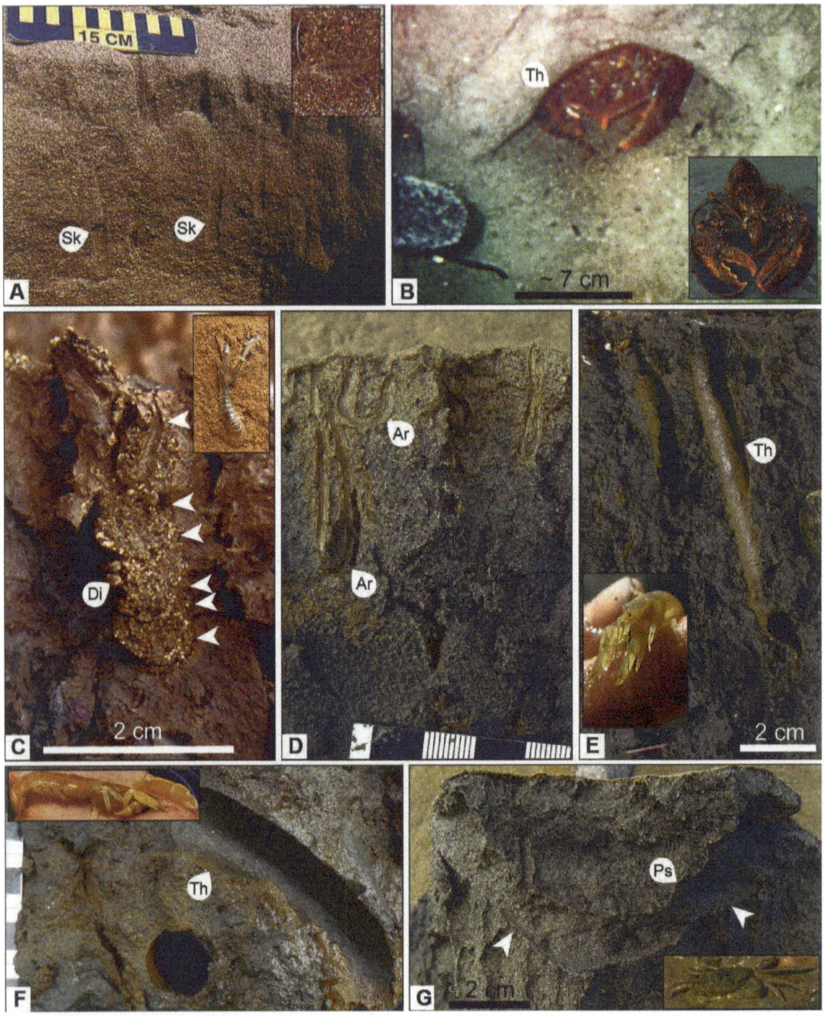

FIGURE 5 Crustacean burrows and their tracemakers (inset photos). (A) Unlined *Skolithos* (Sk) in sand at the intertidal/supratidal transition. Burrows are made by the amphipod, *Talorchestia longicornis*. (B) Large-diameter *Thalassinoides* (Th) produced by the lobster *Homarus americanus* in firm mud in 10 m water depth (photo courtesy of Tyler Hauck). (C) *Diplocraterion* (Di) produced by the amphipod *Corophium volutator* in firm mud. (D) *Corophium* sp. generated *Arenicolites* (Ar) in fine-grained sand. Scale bar is in centimeters. (E) Smooth-walled *Thalassinoides* made by the shrimp *Upogebia pugentensis* in sand. (F) Smooth-walled *Thalassinoides* made by the shrimp *Neotrypaea californiensis* in firm mud. Scale bar is in centimeters. (G) *Psilonichnus* (Ps, white arrows) produced by the crab *Hemigrapsus* sp. in sand.

employed in settings where high population densities restrict the lateral extent of individual burrows.

Other burrowing decapods include crabs and lobsters (Fig. 3). Modern crab burrows have many characteristics that generally are not shared by other crustaceans. These include comparatively simple architectures (U-, Y-, and J-shaped), open apertures and irregular ovoid cross-sections (Figs. 3 and 5). Crab burrows are normally unlined (Atkinson, 1974; Curran and White, 1991; Farrow, 1971) and are best described as *Psilonichnus* (Frey and Pemberton, 1987; Gingras et al., 2000a; Pemberton et al., 2001). Moreover, crab burrows are domiciles, as opposed to combined feeding structures, and lack complex basal geometries. The architectures observed in crab traces are similar to those of lobster traces, although lobster burrows normally display more shallow, subapertural branching (Fig. 5B; Farmer, 1974; Rice and Chapman, 1971) and possess more irregular burrow diameters (Fig. 3). No widely used ichnogenus resembles the irregular burrows constructed by lobsters; however, these traces are most similar to *Psilonichnus* and *Thalassinoides*.

2.4 Echinoderms

Mobile echinoderms include starfish, brittle stars, irregular and regular echinoids, and sea cucumbers. All these animals are known to leave distinctive traces in modern sediments, although only spatangoid (irregular) echinoids and some groups of sea cucumbers burrow deeply into soft substrates (Fig. 6).

Spatangoid echinoids burrow in both sandy and muddy substrates. These animals move through the sediment by passing grains around their bodies using their tube feet (Bromley, 1990; Bromley and Asgaard, 1975; Kanazawa, 1992; Nichols, 1959). In muddy sediments, spatangoid echinoids burrow only a few centimeters below the surface. In sand, they burrow deeper, up to 20 cm in the case of *Echinocardium cordatum* (Fig. 6). The back-filled echinoid tunnels can be broadly classified as *Scolicia* or *Beaconites* (Fu and Werner, 2000), although *Scolicia* should show a preserved axial canal near the center of the trace fossil (Plaziat and Mahmoudi, 1988; Smith and Crimes, 1983). Interfacial preservation of *Scolicia* leads to the production of *Taphrhelminthopsis*, which has recently been observed on submarine videos of the Fraser Delta front, British Columbia, Canada. Sand dollars are also known to tilt into the sediment (in the same manner as brittle stars), such that part of their body is buried and part extends above the sediment/water interface (O'Neill, 1978). This behavior is likely to produce small chevron-shaped disruptions that reflect the orientation of the brittle stars and sand dollars.

Sea cucumbers probably produce a broader range of trace fossils than are presently ascribed to them in the ichnological literature. They are observed to make crude, sinuous trails at the sediment/water interface and are known to make large, if crude, bow-shaped *Arenicolites* and *Diplocraterion* in intertidal settings (Howard, 1968; Pearse, 1908). In deeper water settings, sea cucumbers

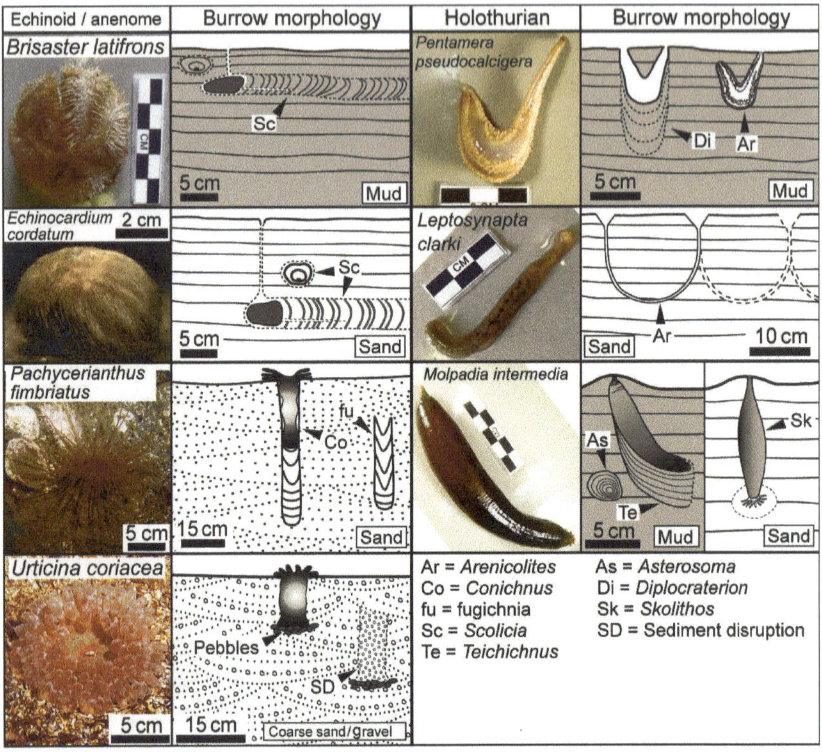

FIGURE 6 Examples of burrowing echinoderms, cnidarians, and holothurians, and their burrow morphologies. Most photos of animals were taken in the Strait of Georgia, Canada. Trace names are assigned based on morphological similarities between the burrows and common trace fossils. The photos of *Brisaster latifrons*, *Pentamera pseudocalcigera*, *Leptosynapta clarki*, and *Molpadia intermedia* were provided by Korhan Ayranci. Burrow morphology is derived from a variety of sources: *B. latifrons* from Goldring et al. (2007); *E. cordatum* from Bromley and Asgaard (1975) and Kanazawa (1992); *Pachycerianthus* is modeled after Schäfer (1972) and Bromley (1996); *Utricina* from Sanamyan and Sanamyan (2006) and McDonald (2010); *P. pseudocalcigera* from Howard (1968) and Gingras et al. (2008a); *L. clarki* from (Myers, 1977b); and, *M. intermedia* from the authors research and from Schäfer (1972), Rhoads (1974), and Bromley (1996). Living position of *M. intermedia* in mud is deduced from ongoing research in the Strait of Georgia, Canada.

produce *Arenicolites*, *Skolithos* (Bromley, 1996; Heezen and Hollister, 1971; Myers, 1977a,b; Rhoads, 1974; Rhoads and Young, 1971; Schäfer, 1972), and concentrically laminated, *Asterosoma*-like and spreitenated, *Teichnichnus*-like traces (Fig. 6; Ginsburg et al., 1966; Zhang et al., 2008). *Arenicolites* and *Diplocraterion*-like traces are produced by suspension-feeding holothurians, whereas traces similar to *Skolithos*, *Asterosoma*, and *Teichnichnus* are made by holothurians employing deposit-feeding strategies. Recognition of holothurian-generated burrows requires additional research; our present understanding of the burrows made by these animals is limited and does not enable recognition of the paleo-infaunal community.

2.5 Sponges and Sea Anemones

In soft sediments, sponge traces are mostly limited to cup-shaped resting marks that are best described as *Bergaueria*-like traces. These structures range in size from centimeter- to decimeter-scale and generally do not exceed a few centimeters in depth. Sponges may also be responsible for generating the deep-water graphoglyptid *Paleodictyon*. *Paleodictyon* was previously attributed to the activity of worms (Ekdale, 1980; Seilacher, 1977) or xenophyophores (Swinbanks, 1982); however, computer modeling and analysis of modern *Paleodictyon* suggest that this trace may, in fact, be a body fossil of a hexactinellid sponge adapted to living in soft sediments (Rona and Merrill, 1978; Rona et al., 2009).

Some sea anemones of the order Ceriantharia can produce large, penetrative biogenic structures (Bromley, 1996; Gingras et al., 2008a; Ricketts et al., 1985; Schäfer, 1972). The burrowing anemone, *Cerianthus*, for example, occurs on both the Pacific and Atlantic coasts of North America. The animal may reach 70 cm in length and burrows deeply below the sediment/water interface (Hargitt, 1907). The animal survives by extending its tentacles and mouth above the sediment/water interface to feed and retracts its body to nearly the full depth of the burrow to take shelter. The iterative probing up through the sediment by burrowing anemones produces a collapsed, chevron-shaped trace fossil that is conical in cross-section (e.g., *Pachycerianthus fimbriatus*, Fig. 6). *Cerianthus* is also very capable of reestablishing its position with respect to the sediment/water interface, such that its dwelling trace can show notable aggradation in response to sedimentation events (Shinn, 1968). *Urticina coriacea* is another burrowing anemone common on the west coast of North America and Europe (Lamb and Hanby, 2005) in permanently subtidal (to 50 m deep) and euhaline waters. It is generally smaller than *Cerianthus*, up to 10 cm in diameter and 15 cm long, and lives in coarse sand and gravel (Fig. 6; McDonald, 2010). Unlike *Cerianthus*, however, *U. coriacea* attaches itself to gravel below the sediment surface, likely resulting in limited vertical shifting of its burrow (Fig. 6).

3. ENVIRONMENTAL STRESSES AND THEIR IMPACT OF NEOICHNOLOGY

Studies in neoichnology aid in establishing the association between environmental stresses and the resultant bioturbate texture. Indeed, neoichnological studies led to the recognition of ichnological models for oxygenation (e.g., Rhoads and Morse, 1971) and for salinity (e.g., Howard and Frey, 1975). The interpretation of archetypal ichnofacies, too, was profoundly influenced by the recognition that ichnofacies are recurrent in modern sedimentary environments (e.g., Ekdale, 1980; Frey and Dashtgard, 2011; Fu and Werner, 2000; Pemberton and Frey, 1985; Rona et al., 2009; Wetzel and Werner, 1981). This recurrence enabled parameterization of associated environmental characteristics.

A wide range of both physical (e.g., grain size, substrate mobility) and chemical (e.g., salinity, water oxygen content) stresses affect burrowing animals, and there is a vast volume of literature discussing these effects (e.g., Gingras et al., 2011; Gray and Elliott, 2009; Pearson, 2001). However, significantly less is known about the impacts of physico-chemical stresses on the burrowing behavior of animals and on the resultant ichnological character. Figure 7 is a synthesis of how physico-chemical stresses impact neoichnological distributions.

Of the physical stresses, grain size, substrate mobility/sedimentation rate (related to hydraulic energy), duration of subaerial exposure (in intertidal environments), substrate consistency, and turbidity represent the main controls on the character of the resultant trace assemblage. Grain size has a major impact in that many animals preferentially burrow in either sand or mud (Bromley, 1996). As grain size increases from sand to gravel, there is a marked decrease in the diversity of animals that burrow in these substrates, and there is an increased occurrence of lined burrows (Dashtgard et al., 2008). Burrow linings are more commonly employed in coarser grained sediments as animals attempt to maintain the environmental conditions within the burrow.

The sedimentation rate exerts a major control on burrowing animals and on the resultant trace assemblage. In particular, the rate of bedform migration, combined with the net accumulation of sediment, can severely restrict the number and type of animals that inhabit a substrate (Dashtgard, 2011a; Wheatcroft, 1990). Low to moderate sedimentation rates serve to attenuate, but not eliminate, the bioturbation intensity. An example includes the study of modern tidally influenced point bars and their adjacent intertidal flats at Willapa Bay, Washington, USA (Gingras et al., 1999). X-radiographs from the point bars at Willapa Bay indicate that the sediments are sparsely to moderately bioturbated. Associated intertidal-zone sediments, which are characterized by very low sedimentation rates, are typically intensely bioturbated.

In the intertidal zone, burrowing infauna are subjected to desiccation, and oxygen and salinity stresses associated with subaerial exposure. With increasing duration of exposure, fewer animals inhabit the substrate, and there is shift in the type of animals toward crustaceans and away from polychaetes (Dashtgard, 2011a; Gingras et al., 1999; Häntzschel, 1939; Swinbanks and Murray, 1981).

Nearly all animals described in the previous section burrow in softgrounds (see Gingras et al., 2000b for quantified definitions of soupgrounds, softgrounds, and hardgrounds), yet with variations in substrate consistency, there are significant shifts in the diversity and density of traces. In soupgrounds, the degree of bioturbation decreases exponentially and there is a shift away from large, shelled animals toward small, shelled animals and polychaetes (Dashtgard et al., 2008; Ekdale, 1985). Moreover, due to their low cohesiveness, soupgrounds heal following the passage of an animal (Uchman and Pervesler, 2006), which does not promote the preservation of a bioturbate texture (Ekdale, 1985; O'Brien, 1987). In firm muds, unlined burrows are typically

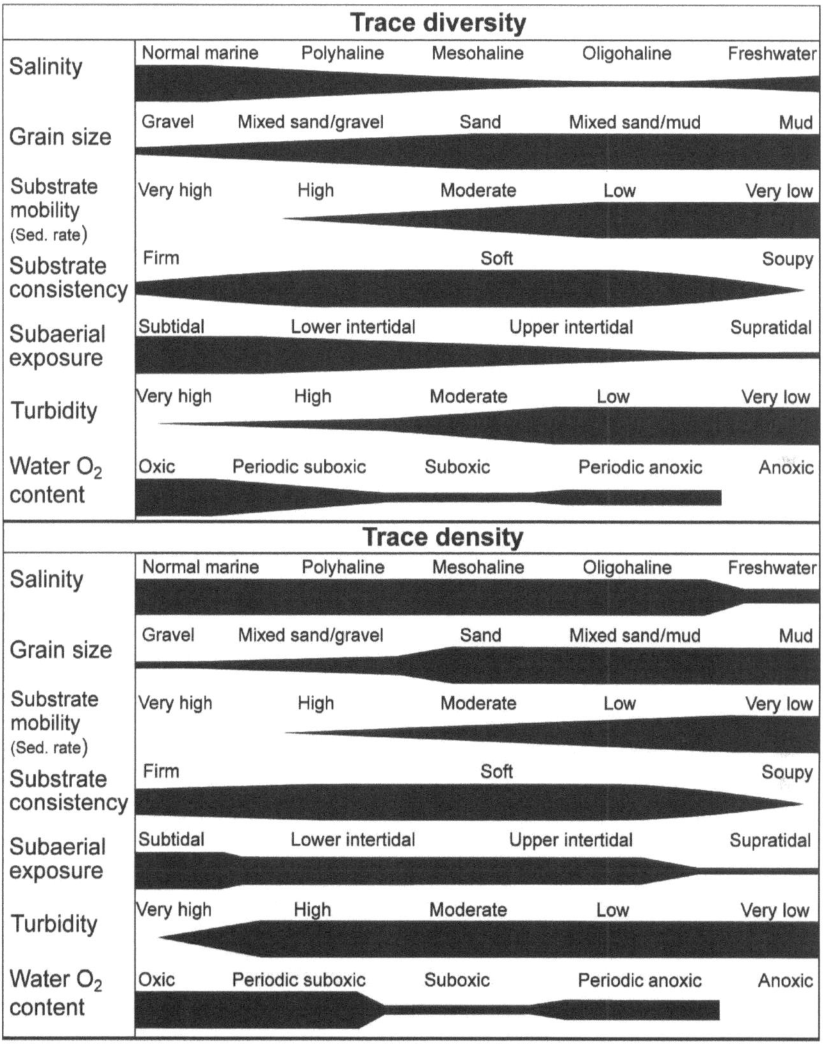

FIGURE 7 Summary chart of the effects of various physico-chemical stresses on trace diversity and trace density. The thickness of the bar reflects the relative density or diversity of burrowing where thicker = more. Note that transitions in bar thickness are more abrupt in the burrow density diagram; this reflects the ability of specific animals to thrive in settings considered stressful by most infauna (e.g., *Corophium volutator* in the upper estuary; Howard et al., 1975). *Data is based on recent neoichnological studies conducted by the authors.*

observed and burrowing is dominated by crustaceans and bivalves and, uncommonly, by polychaetes (e.g., *Polydora* spp.; Dashtgard et al., 2008; Gingras et al., 2001; Yang et al., 2009). The limitations of bioturbation into firmgrounds are that the firm surface must be exposed for a substantial amount of time to permit animal recruitment, and overly compacted, but uncemented, sediment generally substantially reduces burrowing (Gingras et al., 2001). Hardground surfaces are typically bored by bivalves and gastropods capable of rasping on rock, and by animals (e.g., sponges) that chemically dissolve carbonate (Bromley and Asgaard, 1993). Very low sedimentation rates are a prerequisite for boring into cemented substrates.

Turbidity is a significant stress in marginal-marine settings, where there is a large influx and flocculation of mud. Increasing turbidity forces suspension-feeding animals (e.g., most burrowing bivalves) to abandon a substrate (Dashtgard, 2011b; Morton, 1971; Officer et al., 1982), although a minimal amount of suspended organics is necessary for suspension-feeding organisms to receive food. Surface-deposit feeders can continue to thrive in settings with high turbidity, but eventually the amount of particulate matter in the water column overwhelms all burrowing animals (Fig. 7). Moreover, high turbidity promotes the development of soupgrounds, further impeding the ability of infauna to survive in these settings (e.g., Fly River Delta; Dalrymple and Choi, 2007).

In addition to the physical parameters that affect infauna, chemical stresses play a dominant role in limiting the diversity and density of bioturbation. Three main stresses are considered in the chemical stress category: salinity, oxygen content of the water column, and oxygen content of pore water. Pollution, another chemical stress that has received substantial attention (e.g., Gray and Elliott, 2009), is not considered, as the applicability to the rock record of pollution-infauna research is not well established at this time.

With decreasing salinity, from normal marine salinity (35 practical salinity units = psu) to freshwater, there is a corresponding decrease in trace diversity, but not necessarily in bioturbation intensity (Buatois et al., 2005; Gingras et al., 1999; Howard et al., 1975). There is also a significant decrease in the size of burrows with decreasing salinity, partly reflecting the higher density of juvenile *versus* mature animals in low-salinity settings (Chapman and Brinkhurst, 1981; Hauck et al., 2009; Gingras et al., 2012).

The oxygen content of the water column affects infaunal distributions, where periodic influxes of suboxic water into environments dominated by oxygenated water limit animal diversity and periodic anoxic conditions severely limit animal distributions and burrow densities (Fig. 7). However, depending on the persistence of suboxia or anoxia, animals with low O_2 tolerance (e.g., *Travisia pupa*; Manwell, 1960) can completely take over the biome, producing a trace assemblage that is similar to low-salinity (brackish-water) environments. In persistently low-oxygen settings, such as near deep-sea vents (>2,500 m water depth), chemosymbiosis between invertebrates and sulfur-oxidizing chemoautotrophic bacteria has been reported, indicating that low-oxygen levels can sustain some infauna (Bromley, 1996; Cavanaugh, 1994; Frey, 1968). Ichnologically, environments dominated by persistent suboxic waters are characterized by limited bioturbation, whereas persistent anoxia completely halts animal burrowing (Belley et al., 2010; Demaison and Moore, 1980; Hunt, 1996; Wignall and Myers, 1988).

The oxygen content of pore water is linked to the grain size, where the depth of the oxic/anoxic boundary is largely determined by the grain size. In mud, the O_2-content decreases rapidly below the sediment surface, and the zone of oxic sediments is commonly only a few millimeters thick. In sand, the oxic/suboxic boundary may be several meters below the surface and can be shifted downward through the activity of burrowing infauna. However, with increasing organic material incorporated into sandy substrates, the oxic/suboxic boundary rapidly shallows to within a few millimeters to centimeters of the sediment/water interface (Zorn et al., 2006). With a decrease in the oxygen content of pore water in the sediment, there is a corresponding shift from subsurface mining to burrows that maintain an open connection to the sediment surface.

4. SUMMARY AND CONCLUSIONS

Invertebrates in marginal-marine and marine environments produce characteristic traces that can be directly correlated to trace fossils preserved in sedimentary media (Figs. 1 to 6). These traces are a preserved record of the infaunal community that existed at the time of sediment deposition and subsequent colonization. All major invertebrate groups (annelids, bivalves, crustaceans, cnidarians, echinoderms, and sponges) and many less common groups (e.g., hemichordates, nemerteans) have member species that produce biogenic sedimentary structures or traces. As such, animal/trace relationships defined in modern settings (neoichnology) provide a mechanism to partially reconstruct the health of paleomarine ecosystems.

Various physical and chemical properties of the sediment and the environment affect the neoichnological signature of the sediment (Fig. 7). Physical environmental conditions that affect infauna include sediment grain size, the sedimentation rate, subaerial exposure, substrate consistency and turbidity in the water column. Chemical stresses that affect infauna include salinity and oxygen content of the water column and of the pore water. By recognizing and quantifying how neoichnological distributions change with increased or decreased physical and chemical stresses, it may be possible to use these learnings to derive important information about the paleodepositional environment of sedimentary rocks that can not be derived through sedimentological analysis.

Neoichnological studies continue to shed light on the range of infauna inhabiting depositional settings, the physico-chemical stresses that affect them, and the resulting biogenic structures they leave. In the next few decades, renewed research interest in neoichnology promises to advance our understanding of infaunal responses to their environment and to elucidate changes in burrowing behavior and burrow distributions that result. These modern studies have provided and will continue to provide proxies for semiquantification of physico-chemical stresses in sedimentary media and for establishing the diversity of paleoecosystems that otherwise would be based solely on conjecture and interpretation. Neoichnological research holds the promise of significantly improving the overall utility of ichnology in the high-resolution interpretation of the rock record, providing a foundation for interpretation of the paleoichnological results and the advancement and refinement of both ichnofacies and ichnofabric methodologies.

ACKNOWLEDGMENTS

S. E. D.'s research was generously supported by the Natural Sciences and Engineering Council of Canada (NSERC, Grant Number 341789), Nexen Inc., Imperial Oil Ltd., Statoil and Suncor Energy. M. K. G.'s research was supported by grants from NSERC (Grant Number 0001380) and ongoing support from Nexen Inc., ConocoPhillips Canada, Devon Energy Canada, BP Canada, Statoil and Shell. Thanks to Tyler Hauck for permission to reproduce the impressive artwork in Figs. 1 and 3 and to Korhan Aryranci for many of the photos and images included in Fig. 6. We wish to thank A. Martin, R. Bromley, and D. Knaust for their reviews of this chapter.

REFERENCES

Aller, R.C., Yingst, J.Y., 1978. Biogeochemistry of tube-dwellings: a study of the sedentary polychaete *Amphitrite ornata* (Leidy). J. Mar. Res. 36, 201–254.

Atkinson, R.J.A., 1974. Behavioural ecology of the mud-burrowing crab *Goneplax rhomboides*. Mar. Biol. 25, 239–252.

Atkinson, R.J.A., Nash, R.D.M., 1990. Some preliminary observations on the burrows of *Callianassa subterranea* (Montagu) (Decapoda: Thalassinidea) from the west coast of Scotland. J. Nat. Hist. 24, 403–413.

Belley, R., Archambault, P., Sundby, B., Gilbert, F., Gagnon, J.-C., 2010. Effects of hypoxia on benthic macrofauna and bioturbation in the estuary and Gulf of St. Lawrence, Canada. Cont. Shelf Res. 30, 1302–1313.

Bromley, R.G., 1990. Trace Fossils: Biology and Taphonomy. Unwin Hyman, London, 280 pp.

Bromley, R.G., 1996. Trace Fossils: Biology, Taphonomy and Applications. Chapman & Hall, London, 361 pp.

Bromley, R.G., Asgaard, U., 1975. Sediment structures produced by a spatangoid echinoid: a problem of preservation. Bull. Geol. Soc. Denmark 24, 261–281.

Bromley, R.G., Asgaard, U., 1993. Two bioerosion ichnofacies produced by early and late burial associated with sea level change. Geol. Rundsch. 82, 276–280.

Bromley, R.G., Frey, R.W., 1974. Redescription of the trace fossil *Gyrolithes* and taxonomic evaluation of *Thalassinoides*, *Ophiomorpha*, and *Spongeliomorpha*. Bull. Geol. Soc. Denmark 23, 311–335.

Buatois, L.A., Gingras, M.K., MacEachern, J., Mángano, M.G., Zonneveld, J.-P., Pemberton, S.G., Netto, R.G., Martin, A., 2005. Colonization of brackish-water systems through time: evidence from the trace-fossil record. Palaios 20, 321–347.

Cavanaugh, C.M., 1994. Microbial symbiosis: patterns of diversity in the marine environment. Am. Zool. 34, 79–89.

Chapman, P.M., Brinkhurst, R.O., 1981. Seasonal changes in interstitial salinities and seasonal movements of subtidal benthic invertebrates in the Fraser River estuary, B.C. Estuar. Coast. Shelf Sci. 12, 49–66.

Clifton, H.E., Thompson, J.K., 1978. *Macaronichnus segregatis*: a feeding structure of shallow marine polychaetes. J. Sediment. Petrol. 48, 1293–1302.

Curran, H.A., White, B., 1991. Trace fossils of shallow subtidal to dunal ichnofacies in Bahamian Quaternary carbonates. Palaios 6, 498–510.

Dafoe, L.T., Gingras, M.K., Pemberton, S.G., 2008. Determining *Euzonus mucronata* burrowing rates with application to ancient *Macaronichnus segregatis* trace makers. Ichnos 15, 78–90.

Dalrymple, R.W., Choi, K.S., 2007. Morphologic and facies trends through the fluvial-marine transition in tide-dominated depositional systems: a schematic framework for environmental and sequence-stratigraphic interpretation. Earth-Sci. Rev. 81, 135–174.

Dashtgard, S.E., 2011a. Linking invertebrate burrow distributions (neoichnology) to physico-chemical stresses on a sandy tidal flat: implications for the rock record. Sedimentology 58, 1303–1325.

Dashtgard, S.E., 2011b. Neoichnology of the lower delta plain: Fraser River Delta, British Columbia, Canada: implications for the ichnology of deltas. Palaeogeogr. Palaeoclimatol. Palaeoecol. 307, 98–108.

Dashtgard, S.E., Gingras, M.K., 2005. The temporal significance and distribution of bioturbation in coarse-grained upper foreshore and backshore deposits: Waterside Beach, New Brunswick, Canada. Palaios 20, 589–595.

Dashtgard, S.E., Gingras, M.K., Pemberton, S.G., 2008. Grain-size controls on the occurrence of bioturbation. Palaeogeogr. Palaeoclimatol. Palaeoecol. 257, 224–243.

Demaison, G.J., Moore, G.T., 1980. Anoxic environments and oil source bed genesis. Org. Geoch. 2, 9–31.

Dworschak, P.C., 1982. The biology of *Upogebia pusilla* (Petagna) (Decapoda, Thalassinidea) I. The burrows. Mar. Ecol. 4, 19–43.

Dworschak, P.C., Rodrigues, S.de A., 1997. A modern analogue for the trace fossil *Gyrolithes*: burrows of the thalassinidean shrimp *Axianassa australis*. Lethaia 30, 41–52.

Ekdale, A.A., 1980. Graphoglyptid burrows in modern deep-sea sediment. Science 207, 304–306.

Ekdale, A.A., 1985. Paleoecology of the marine endobenthos. Palaeogeogr. Palaeoclimatol. Palaeoecol. 50, 63–81.

Ekdale, A.A., Lewis, D.W., 1991. Trace fossils and paleoenvironmental control of ichnofacies in a Quaternary gravel and loess fan delta complex, New Zealand. Palaeogeogr. Palaeoclimatol. Palaeoecol. 81, 253–279.

Farmer, A.S.D., 1974. Burrowing behavior of the Norway lobster, *Nephrops norvegicus* (L.) (Decapoda: Nephropidae). Estuar. Coast. Mar. Sci. 2, 49–54.

Farrow, G.E., 1971. Back-reef and lagoonal environments of Aldabra Atoll distinguished by their crustacean burrows. Symp. Zool. Soc. London 28, 455–500.

Frey, R.W., 1968. The Lebensspuren of some common marine invertebrates near Beaufort, North Carolina. 1, Pelecypod burrows. J. Paleontol. 42, 570–574.

Frey, R.W., Pemberton, S.G., 1987. The *Psilonichnus* ichnocoenose and its relationship to adjacent marine and nonmarine ichnocoenoses along the Georgia coast. Bull. Can. Petrol. Geol. 35, 333–357.

Frey, S.E., Dashtgard, S.E., 2011. Sedimentology, ichnology, and hydrodynamics of strait-margin, sand-and-gravel beaches and shorefaces: Juan de Fuca Strait, British Columbia, Canada. Sedimentology 58, 1326–1346.

Fu, S., Werner, F., 2000. Distribution, ecology, and taphonomy of the organism trace, *Scolicia*, in Northeast Atlantic deep-sea sediments. Palaeogeogr. Palaeoclimatol. Palaeoecol. 156, 289–300.

Giffis, R.B., Chavez, F.L., 1988. Effects of sediment type on burrows of *Callianassa californiensis* Dana and *C. gigas* Dana. J. Exp. Mar. Biol. Ecol. 117, 239–253.

Gingras, M.K., Pemberton, S.G., Saunders, T., Clifton, H.E., 1999. The ichnology of modern and Pleistocene brackish-water deposits at Willapa Bay, Washington: variability in estuarine settings. Palaios 14, 352–374.

Gingras, M.K., Hubbard, S.M., Pemberton, S.G., Saunders, T., 2000a. The significance of Pleistocene *Psilonichnus* at Willapa Bay, Washington. Palaios 15, 142–151.

Gingras, M.K., Pemberton, S.G., Saunders, T., 2000b. Firmness profiles associated with tidal-creek deposits: the temporal significance of *Glossifungites* assemblages. J. Sediment. Res. 70, 1017–1025.

Gingras, M.K., Pemberton, S.G., Saunders, T., 2001. Bathymetry, sediment texture, and substrate cohesiveness; their impact on modern *Glossifungites* trace assemblages at Willapa Bay, Washington. Palaeogeogr. Palaeoclimatol. Palaeoecol. 169, 1–21.

Gingras, M.K., Dashtgard, S.E., MacEachern, J.A., Pemberton, S.G., 2008a. Biology of shallow-marine ichnology: a modern perspective. Aquat. Biol. 2, 255–268.

Gingras, M.K., Pemberton, S.G., Dashtgard, S.E., Dafoe, L., 2008b. How fast do invertebrates burrow? Palaeogeogr. Palaeoclimatol. Palaeoecol. 270, 280–286.

Gingras, M.K., MacEachern, J.A., Dashtgard, S.E., 2011. Process ichnology and the elucidation of physico-chemical stress. Sediment. Geol. 237, 115–134.

Gingras, M.K., MacEachern, J.A., Dashtgard, S.E., Zonneveld, J.-P., Schoengut, J., Ranger, M.J., Pemberton, S.G., 2012. Estuaries. In: Knaust, D., Bromley, R.G. (Eds.), Trace Fossils as Indicators of Sedimentary Environments. Developments in Sedimentology, vol. 64. Elsevier, Amsterdam, pp. 463–505.

Ginsburg, R.N., Bernard, H.A., Moody, R.A., Daigle, E.G., 1966. The shell method of impregnating cores of unconsolidated sediments. J. Sediment. Petrol. 36, 1118–1125.

Goldring, R., Cadee, G.C., Pollard, J.E., 2007. Climatic control of marine trace fossil distribution. In: Miller III, W. (Ed.), Trace Fossils. Concepts, Problems, Prospects. Elsevier, Amsterdam, pp. 159–171.

Gray, J.S., Elliott, M., 2009. Ecology of Marine Sediments: From Science to Management. Oxford University Press, Oxford, 225 pp.

Griffith, H., Telford, M., 1985. Morphological adaptations to burrowing in *Chiridotea coeca* (Crustacea Isopoda). Biol. Bull. 168, 296–311.

Häntzschel, W., 1939. Tidal flat deposits (Wattenschlick). In: Trask, P.D. (Ed.), Recent Marine Sediments. AAPG, Tulsa, pp. 195–206.

Hargitt, C.W., 1907. Notes on the behavior of sea-anemones. Biol. Bull. 12, 274–284.

Hauck, T.E., Dashtgard, S.E., Gingras, M.K., 2008. Relationships between organic carbon and Fodinichnia morphology. Palaios 23, 336–343.

Hauck, T.E., Dashtgard, S.E., Pemberton, S.G., Gingras, M.K., 2009. Brackish-water ichnological trends in a microtidal barrier island-embayment system, Kouchibouguac National Park, New Brunswick, Canada. Palaios 24, 478–496.

Heezen, B.C., Hollister, C.D., 1971. The Face of the Deep. Oxford University Press, New York, 659 pp.

Hersey, J.B., 1968. Deep Sea Photography (Oceanographic Study). Johns Hopkins University, Baltimore, 310 pp.

Howard, J.D., 1968. X-ray radiography for examination of burrowing in sediments by marine invertebrate organisms. Sedimentology 11, 249–258.

Howard, J.D., Elders, C.A., 1970. Burrowing patterns of haustoriid amphipods from Sapelo Island, Georgia. Geol. J., Spec. Iss. 3, 243–262.

Howard, J.D., Frey, R.W., 1975. Regional animal-sediment characteristics of Georgia estuaries. Senckenberg. Marit. 7, 33–107.

Howard, J.D., Elders, C.A., Heinbokel, J.F., 1975. Animal-sediment relationships in estuarine point bar deposits, Ogeechee River-Ossabaw Sound, Georgia. Senckenberg. Marit. 7, 181–203.

Hunt, J.M., 1996. Petroleum Geochemistry and Geology. W.H. Freeman, New York, 743 pp.

Ingle, R.W., 1966. An account of the burrowing behaviour of the amphipod *Corophium arenarium* Crawford (Amphipoda: Corophiidae). Ann. Mag. Nat. Hist. 9, 309–317.

Kanazawa, K., 1992. Adaptation of test shape for burrowing and locomotion in spatagonoid echinoids. Palaeontology 35, 733–750.

Lamb, A., Hanby, B., 2005. Marine Life of the Pacific Northwest: A Photographic Encyclopedia of Invertebrates, Seaweeds and Selected Fishes. Harbour Publishing, Madeira Park, 398 pp.

Löwemark, L., Schäfer, P., 2003. Ethological implications from a detailed X-ray radiograph and ^{14}C study of the modern deep-sea *Zoophycos*. Palaeogeogr. Palaeoclimatol. Palaeoecol. 192, 101–121.

MacEachern, J.A., Bann, K.L., Pemberton, S.G., Gingras, M.K., 2007. The ichnofacies paradigm: high-resolution paleoenvironmental interpretation of the rock record. In: MacEachern, J.A., Bann, K.L., Gingras, M.K., Pemberton, S.G. (Eds.), Applied Ichnology. SEPM Short Course Notes 52, pp. 27–64.

Manwell, C., 1960. Histological specificity of repiratory pigments-1. Comparisons of the coelom and muscle hemoglobins of the polychaete worm Travisia pupa and the echiuroid worm Arhynchite pugettensis. Comp. Biochem. Physiol. 1, 267–276.

McDonald, G., 2010. Intertidal Invertebrates of the Monterey Bay Area, California. Long Marine Laboratory, University of California, Santa Cruz. http://www2.ucsc.edu/seymourcenter/Inverts/index.html.

Morton, B., 1971. Studies on the biology of *Dreissena polymorpha* Pall.; v. Some aspects of filter-feeding and the effect of micro-organisms upon the rate of filtration. J. Mollusc. Stud. 39, 289–301.

Myers, A.C., 1977a. Sediment processing in a marine subtidal sandy bottom community: I. Physical aspects. J. Mar. Res. 35, 609–632.

Myers, A.C., 1977b. Sediment processing in a marine subtidal sandy bottom community: II. Biological aspects. J. Mar. Res. 35, 633–647.

Nichols, D., 1959. Changes in the Chalk heart-urchin *Micraster* interpreted in relation to living forms. Philos. Trans. R. Soc. Lond., B, Biol. Sci. 242, 347–437.

O'Brien, N.R., 1987. The effects of bioturbation on the fabric of shale. J. Sediment. Petrol. 57, 449–455.

Officer, C.B., Smayda, T.J., Mann, R., 1982. Benthic filter feeding: a natural eutrophication control. Mar. Ecol. Progr. Ser. 9, 203–210.

O'Neill, P.L., 1978. Hydrodynamic analysis of feeding in sand dollars. Oecologia 34, 157–174.

Pearse, A.S., 1908. Observations on the behavior of the holothurian, Thyonebriareus (Leseur). Biol. Bull. Mar. Biol. Lab., Woods Hole 15, 259–288.

Pearson, T.H., 2001. Functional group ecology in soft-sediment marine benthos: the role of bioturbation. In: Gibson, R.N., Barnes, M., Atkinson, R.J.A. (Eds.), Oceanography and Marine Biology Annual Review. Taylor and Francis, New York, pp. 233–267.

Pearson, N.J., Gingras, M.K., 2006. An ichnological and sedimentological facies model for muddy point-bar deposits. J. Sediment. Res. 76, 771–782.

Pemberton, S.G., Frey, R.W., 1985. The *Glossifungites* Ichnofacies: modern examples from the Georgia Coast, U.S.A. In: Curran, H.A. (Ed.), Biogenic Structures: Their Use in Interpreting Depositional Environments. SEPM Spec. Publ., vol. 35, pp. 237–259.

Pemberton, S.G., Spila, M., Pulham, A.J., Saunders, T., MacEachern, J.A., Robbins, D., Sinclair, I.K., 2001. Ichnology & Sedimentology of Shallow to Marginal Marine Systems: Ben Nevis & Avalon Reservoirs, Jeanne d'Arc Basin. Geol. Ass. Can., Short Course Notes 15, 343 pp.

Plaziat, J.-C., Mahmoudi, M., 1988. Trace fossils attributed to burrowing echinoids: a revision including new ichnogenus and ichnospecies. Geobios 21, 209–233.

Powell, E.N., 1977. The relationship of the trace fossil *Gyrolithes* (=*Xenohelix*) to the family Capitellidae (Polychaeta). J. Paleontol. 51, 552–556.

Powilleit, M., Graf, G., Kleine, J., Riethmüller, R., Stockmann, K., Wetzel, M.A., Koop, J.H.E., 2009. Experiments on the survival of six brackish macro-invertebrates from the Baltic Sea after dredge spoil coverage and its implications for the field. J. Mar. Syst. 75, 441–451.

Reineck, H.-E., Singh, I.B., 1980. Depositional Sedimentary Environments. Springer, Berlin, 551 pp.

Rhoads, D.C., 1967. Biogenic reworking of intertidal and subtidal sediments in Barnstable Harbor and Buzzards Bay, Massachusetts. J. Geol. 75, 461–476.

Rhoads, D.C., 1974. Organism-sediment relations on the muddy sea floor. Oceanogr. Mar. Biol. Ann. Rev. 12, 263–300.

Rhoads, D.C., Morse, J.W., 1971. Evolution and ecologic significance of oxygen-deficient marine basins. Lethaia 4, 413–428.

Rhoads, D.C., Young, D.K., 1971. Animal-sediment relations in Cape Cod Bay, Massachusetts. II. Reworking by *Molpadia oolitica* (Holothuroidea). Mar. Biol. 11, 255–261.

Rice, A.L., Chapman, C.J., 1971. Observations on the burrows and burrowing behavior of two mud-dwelling decapod crustaceans, *Nephrops norvegicus* and *Goneplax rhomboides*. Mar. Biol. 10, 330–342.

Ricketts, E.F., Calvin, J., Hedgpeth, J.W., Phillips, D.W., 1985. Between Pacific Tides. Stanford University Press, Stanford, 680 pp.

Rijken, M., 1979. Food and food uptake in *Arenicola marina*. Neth. J. Sea Res. 13, 406–421.

Risk, M.J., Tunnicliffe, V.J., 1978. Intertidal spiral burrows; *Paraonis fulgens* and *Spirophanes wigleyi* in the Minas Basin, Bay of Fundy. J. Sediment. Petrol. 48, 1287–1292.

Rona, P.A., Merrill, G.F., 1978. A benthic invertebrate from the Mid-Atlantic Ridge. Bull. Mar. Sci. 28, 371–375.

Rona, P.A., Seilacher, A., deVargas, C., Gooday, A.J., Bernhard, J.M., Bowser, S., Vetriani, C., Wirsen, C.O., Mullineaux, L., Sherrell, R., Grassle, J.F., Low, S., Lutz, R.A., 2009. *Paleodictyon nodosum*: a living fossil on the deep-seafloor. Deep-Sea Res. 2: Top. Stud. Oceanogr. 56, 1700–1712.

Sanamyan, N.P., Sanamyan, K.E., 2006. The genera *Urticina* and *Cribrinopsis* (Anthozoa: Actiniaria) from the north-western Pacific. J. Nat. Hist. 40, 359–393.

Schäfer, W., 1972. Ecology and Paleoecology of Marine Environments. Oliver and Boyd, Edinburgh, 568 pp.

Seilacher, A., 1953. Über die Methoden der Palichnologie 1: Studien zur Palichologie. N. Jb. Geol. Paläont. 96, 421–452.

Seilacher, A., 1964. Biogenic sedimentary structures. In: Imbrie, J., Newell, N. (Eds.), Approaches to Paleoecology. Wiley, New York, pp. 296–316.

Seilacher, A., 1977. Pattern analysis of *Paleodictyon* and related trace fossils. In: Crimes, T.P., Harper, J.C. (Eds.), Trace Fossils 2. Geol. J., Spec. Iss. 9, pp. 289–334.

Shinn, E.A., 1968. Burrowing in recent lime sediments of Florida and the Bahamas. J. Paleontol. 42, 879–894.

Smith, A.B., Crimes, T.P., 1983. Trace fossils formed by heart urchins—a study of *Scolicia* and related traces. Lethaia 16, 79–92.

Stanistreet, I.G., Le Blanc Smith, G., Cadle, A.B., 1980. Trace fossils as sedimentological and paleoenvironmental indices in the Ecca Group (Lower Permian) of the Transvaal. Trans. Geol. Soc. South Africa 83, 333–344.

Stevens, B.A., 1929. Ecological observations on Callianassidae of Puget Sound. Ecology 10, 399–405.

Swinbanks, D.D., 1982. *Paleodictyon*: the traces of infaunal xenophyophores? Science 218, 47–49.

Swinbanks, D.D., Luternauer, J.L., 1987. Burrow distribution of Thalassinidean shrimp on a Fraser Delta tidal flat, British Columbia. J. Paleontol. 61, 315–332.

Swinbanks, D.D., Murray, J.W., 1981. Biosedimentological zonation of Boundary Bay tidal flats, Fraser River Delta, British Columbia. Sedimentology 28, 201–237.

Thurston, H., 1990. Tidal Life: A Natural History of the Bay of Fundy. Nimbus Publishing, Halifax, 167 pp.

Uchman, A., Pervesler, P., 2006. Surface Lebensspuren produced by amphipods and isopods (crustaceans) from the Isonzo Delta tidal flat, Italy. Palaios 21, 384–390.

Vannier, J., Calandra, I., Gaillard, C., Zylinska, A., 2010. Priapulid worms: pioneer horizontal burrowers at the PreCambrian-Cambrian boundary. Geology 38, 711–714.

Wetzel, A., 2008. Recent bioturbation in the deep South China Sea: a uniformitarian ichnologic approach. Palaios 23, 601–615.

Wetzel, A., Werner, F., 1981. Morphology and ecological significance of *Zoophycos* in deep-sea sediments off NW Africa. Palaeogeogr. Palaeoclimatol. Palaeoecol. 32, 185–212.

Wetzel, A., Tjallingii, R., Stattegger, K., 2010. *Gyrolithes* in Holocene incised-valley fill deposits, offshore southern Vietnam. Palaios 25, 239–246.

Wheatcroft, R.A., 1990. Preservation potential of sedimentary event layers. Geology 18, 843–845.

Wignall, P.B., Myers, K.J., 1988. Interpreting benthic oxygen levels in mudrocks: a new approach. Geology 16, 452–455.

Yang, B.C., Dalrymple, R.W., Gingras, M.K., Pemberton, S.G., 2009. Autogenic occurrence of *Glossifungites* Ichnofacies: examples from wave-dominated, macrotidal flats, southwestern coast of Korea. Mar. Geol. 260, 1–5.

Zhang, G., Uchman, A., Chodyń, R., Bromley, R.G., 2008. Trace fossil *Artichnus pholeoides* igen. isp. nov. in Eocene turbidites, Polish Carpathians: possible ascription to holothurians. Acta Geol. Polon. 58, 75–86.

Zorn, M.E., Lalonde, S.V., Gingras, M.K., Pemberton, S.G., Konhauser, K.O., 2006. Microscale oxygen distribution in various invertebrate burrow walls. Geobiology 4, 137–145.

Part II

Continental and Glacial Systems

Chapter 11

Glacial Environments

Renata G. Netto,*[,1] Jacob S. Benner,[†] Luis A. Buatois,[‡] Alfred Uchman,[§] M. Gabriela Mángano,[‡] John C. Ridge,[†] Vaidotas Kazakauskas[¶] and Algirdas Gaigalas[||,✠]

*Geology Graduate Program, Universidade do Vale do Rio dos Sinos, São Leopoldo, Rio Grande do Sul, Brazil, [†]Department of Earth and Ocean Sciences, Tufts University, Medford, Massachusetts, USA, [‡]Department of Geological Sciences, University of Saskatchewan, Saskatoon, Saskatchewan, Canada, [§]Institute of Geological Sciences, Jagiellonian University, Kraków, Poland, [¶]Institute of Geology and Geography, Vilnius, Lithuania, [||]Department of Geology and Mineralogy, Vilnius University, Čiurlionio, Vilnius, Lithuania

[1]Corresponding author: e-mail: nettorg@unisinos.br

1. INTRODUCTION

Studies using ichnology to improve paleoecological and paleoenvironmental interpretations in glacial environments are still in their infancy, and most reports have focused on the characterization of individual trace fossils and trace-fossil assemblages (e.g., Anderson, 1970, 1975a,b, 1976, 1981; Benner et al., 2008; Dias-Fabrício and Guerra-Sommer, 1989; Fernandes et al., 1987; Gaigalas and Uchman, 2004; Gibbard and Dreimanis, 1978; Gibbard and Stuart, 1974; Knecht et al., 2009; Marques-Toigo et al., 1989; Netto and Goso, 1998; Savage, 1970, 1971; Uchman et al., 2008, 2009; Walter, 1985, 1986). However, the number of studies utilizing trace fossils for paleoenvironmental and paleoecological interpretations in glacial sediments and sedimentary rocks has grown in the past decade (e.g., Balistieri and Netto, 2002; Balistieri et al., 2002, 2003; Benner et al., 2009; Buatois and del Papa, 2003; Buatois and Mángano, 1992, 1993, 1995a,b, 2003; Buatois et al., 2006, 2010; Gandini et al., 2007; Isbell et al., 2001; Lermen, 2006; Netto et al., 2009; Pazos, 2000, 2002a,b; Pazos et al., 2007; Schatz et al., 2011a,b; Trewin, 2000; Trewin et al., 2002; Walter and Suhr, 1998).

Glacial environments are complex, as the rapid and spatially variable processes common in these environments are seen rarely in other sedimentary systems (Menzies, 2002). The dynamic interaction of glaciers and ice sheets with fluvial, eolian, lacustrine, and marine settings generates a great variety of landforms and facies associations (Hambrey, 1994). The movement of ice sheets

✠ Deceased.

Developments in Sedimentology, Vol. 64. http://dx.doi.org/10.1016/B978-0-444-53813-0.00011-3

during glaciations has strongly impacted the original landscape, promoting surficial weathering and deep erosion. During interglacial periods, immense meltwater discharges influenced sedimentation (Buatois et al., 2006, 2010) in addition to exerting erosional forces. Drastic oscillations in climate (Balistieri, 2003; Buatois et al., 2006, 2010; Netto et al., 2009) and the resulting change in the land surface of glaciated regions has impacted the distribution, abundance and, subsequently, evolution of biota through time.

Large-scale glaciations occur about every 200 million years and five inferred global-scale glaciations are known to have impacted the biotic realm (e.g., Clark et al., 2009; Ehlers and Gibbard, 2004; Eyles, 1993; Hoffman et al., 1998; Isbell et al., 2003a,b; Le Heron, 2010; Rieu et al., 2007). Trace fossils from glacial environments are well documented in the Gondwana, and the Quaternary ice ages, and case studies of those glacial periods are presented herein. Thus, the objectives of this chapter are to discuss the main characteristics of trace-fossil assemblages from glacial environments, to evaluate the main abiotic stress factors that constrained tracemakers' impact on trace-fossil abundance and diversity, and to characterize the overall trace-fossil distribution in glacial settings. Inherently, the discussion must be limited to those glacial environments which best preserve trace fossils, and while that includes primarily aquatic depositional environments, some emergent and fully terrestrial environments are considered as well.

2. ECOLOGICAL AND ENVIRONMENTAL CONSTRAINTS IN GLACIAL SETTINGS

The exposure of new environments after the culmination of a glacial period stimulates dispersion of biota from glacial refugia and the establishment of new communities. Ecologic studies on biotic communities from Antarctica suggested that temporary freshwater ecosystems surrounded by ice might have provided oases for a few aquatic species adapted to extreme conditions (Brinck, 1966). In other cases, ice-free high-elevation areas may have provided refuge for certain terrestrial organisms. These resilient species may opportunistically recolonize areas when new habitats become available during glacial retreat. An understanding of this process is critical to the interpretation of the ichnological record of formerly glaciated regions, as environmental constraints placed strict controls on successional patterns in these environments.

2.1 Environmental Stress in Glacial Settings

Physical environmental stress is predominant in glacial environments, and seasonal or episodic variations in physical processes can be extreme and abrupt. The impact of rapid environmental changes upon terrestrial and subaquatic freshwater fauna from glacial foreland environments is amplified relative to subaquatic fauna of brackish and full saline glaciomarine settings. Meltwater input during deglaciation reduces salinity in shallow-marine settings, resulting in conditions that favor opportunistic (*r*-selected) organisms. The endobenthic fauna shares the same characteristics as that commonly observed in other brackish-water settings. Temperature does not seem to play a role in constraining this opportunistic burrowing infauna. In addition, extreme meltwater discharge during deglaciation may result in the establishment of freshwater conditions in the innermost zones of fjords.

The retreat and advance of valley glaciers and ice sheets impact the biota in different ways. During ice retreat, meltwater discharge rapidly inundates exposed, isostatically depressed areas in the glacier foreland, forming either shallow lakes and humid marginal substrates or deep valley lakes, depending on topography, as well as outwash plains. Shallow aquatic habitats quickly develop microbial mats, attracting organisms from surrounding biotic oases. However, the low humidity and high evaporation rates, typical of glacial climates in polar regions, and the relative aridity of glacial settings in temperate zones of lower latitudes, significantly reduce the water table following meltwater discharges, making the shallower lakes ephemeral and exposing littoral zones of perennial lakes (see Netto et al., 2009 for further discussion). Extensive mats formed by shallow microbial communities are exposed during these dry periods, but the leather-like structure of the mat surface retains moisture for a longer period (Noffke, 2010), preserving the colony and providing a source of food to terrestrial mat grazers and decomposers (generally "myriapods", after Kaufmann, 2001). During winter periods, the available area for epigean communities is reduced significantly, as water surfaces freeze, reducing food, light, and oxygen supply to benthic communities. Thus, life conditions are tightly linked with physical glacial processes, and opportunistic taxa tend to be the most successful as they are able to migrate quickly to and from refugia.

In fjords, the freshwater discharge, the sedimentation rate, the water turbidity, the oxygen content, the substrate, and the storm activity each play a role in controlling the distribution of benthic organisms (Buatois and Mángano, 2011). A high rate of sedimentation typifies the fjord environment and is a result of high fluvial input, mass transport, eolian transport, and input from wave and tidal erosion (Syvitski et al., 1987). High sedimentation rates strongly affect the epifauna and limit bioturbation in the innermost zone of the fjord. In addition, fjord waters contain high concentrations of fine-grained particles that preclude establishment of suspension feeders (Feder and Matheke, 1980). As a result, fjord sediments tend to be bioturbated preferentially in the outermost zones, which are dominated by horizontal feeding traces of deposit and detritus feeders; vertical burrows of suspension feeders are typically absent. The concentration of dissolved oxygen in fjord bottom waters is highly variable (Syvitski et al., 1987). The bottom of some fjords, particularly those in enclosed basins, may be characterized by oxygen depletion. Because the redox boundary is very close to the sediment/water interface, shallow-tier structures of small deposit feeders tend to dominate. Under anoxic conditions, bioturbation is suppressed. In other cases, turbidity currents and hyperpycnal flows may supply oxygen to the fjord floor, resulting in intense postevent bioturbation (e.g., Schatz et al., 2011b). The degree of substrate consolidation may be highly variable in fjords (e.g., Schatz et al., 2011a). Muddy soupgrounds are rather common close to the glacier margin, limiting colonization by epifaunal organisms (Eyles et al., 1992). Softground communities with more diverse epifauna and infauna tend to occur distally, while the proximal fluid muds appear to be unbioturbated (Buatois and Mángano, 2011), although this may be due to the low preservation potential of biogenic structures. Firmground suites tend to be associated with current-winnowed substrates (Eyles et al., 1992). Finally, many fjords are subjected to intense wave action, resulting in a deeply emplaced storm-wave base and repeated storms impart a strong influence on benthic communities in relatively deep water (Buatois and Mángano, 2011; Syvitski et al., 1987).

2.2 Postglacial Colonization of Terrestrial Environments

Studies concerning the faunal primary succession in deglaciated terrestrial areas are scarce. Kaufmann (2001) analyzed the terrestrial invertebrate succession in an alpine glacier foreland during the summer–fall period based on the epigean fauna. Despite the fact that the area in front of the glacier had been ice-free for the past 9500 years, the disturbed surface of the outwash plain is much younger. Colonization of the outwash surface began in sunny, stabilized areas and communities from within outwash plains were younger than those from terminal moraine slopes with developed soils. Predators were the first colonizers, while herbivores and decomposers appeared later.

Recolonization of the glacier foreland occurred comparatively faster than in other barren environments (e.g., newly emerged islands), and can be compared with recolonization of volcanic deposits, where a mobile and tolerant invertebrate fauna migrates from adjacent areas to explore and establish breeding populations in newly deposited material (Kaufmann, 2001).

The terrestrial invertebrate fauna was composed chiefly of arthropods, mainly insects (e.g., ants, flies, beetles), arachnids (e.g., spiders), and "myriapods" (centipedes and millipedes; Kaufmann, 2001). Carnivores, like carabeid and staphylinid beetles, and lycopsid and linyphiid spiders were the pioneer species, and despite of the initial lack of vegetation, rhizophagous elaterid beetles were also present. Myriapods colonized later, when communities became established. Centipedes (Chilopoda) appeared first, with the herbivorous beetles and detritivorous millipedes (Diplopoda) appearing later. Lumbricidae were scarce in the glacier foreland but common in adjacent grassland areas. Mobile flying insects, especially dipterans, were widespread in all areas, but dipteran larvae were absent in the youngest sites and in the outwash plain. According to Kaufmann (2001), the biomass produced in the glacier foreland was sufficient to sustain large populations, but there was no clear evidence that the communities were supported by local primary production or that resources were provided by eolian input. The presence of collembolans suggested that algae could be the base of the web food, but no infaunal invertebrates were detected in soils. Considering the pattern of succession, Kaufmann (2001) concluded that decomposers, like dipteran larvae and millipedes, might have facilitated ecological succession where well-established communities existed. The presence of vegetation made no significant difference to the colonization pattern of pioneering invertebrates, which suggests a strong predatory or scavenging component to the pioneer assemblage and possible allogenic inputs.

2.3 Postglacial Colonization of Freshwater Environments

Biodiversity in glacial lakes and streams is low compared with typical temperate and tropical freshwater environments. Biodiversity increases from deep to shallow lakes and from upstream to downstream settings (Likens, 2010). Fishes, when present, are the dominant predator of the glaciolacustrine ecosystem, while the primary benthic community is composed chiefly of worm-like organisms (e.g., oligochaetes, nematodes, rotifers, nematomorphs, and insect larvae), crustaceans, diminutive molluscs, bacteria, algae, and a few aquatic vascular plants (in photic zones).

As only cold-adapted or cold-tolerant species can inhabit these settings, productivity is low and food webs are simple (Likens, 2010). Glacial lakes are typically oligotrophic to ultraoligotrophic, and the development of a pioneer community cannot always depend on planktonic primary producers due to large amounts of suspended glacial rock flour and low nutrient density. In some arctic lacustrine food webs, the benthos is the primary source of organic carbon (Sierszen et al., 2003). Despite low productivity, benthic organisms provide other animals, such as predators, access to environments otherwise too harsh to occupy.

For those glacial freshwater environments linked to the coastline by river systems, offshore freshwater refugia located on exposed continental shelves can play a role in successful recolonization. In addition to terrestrial freshwater habitats, coastal-marine areas may have been available to freshwater fish as glacial meltwaters altered ocean water salinity nearshore (Power, 2002). Power (2002) postulated that during postglacial extensions of fish ranges, proglacial lakes would have been uninhabitable and were too much covered by ice during most of the year. Using Arctic rivers as analogs for what fish may have encountered during range extension (high sediment loads, unstable bed conditions, and frequent meltwater flooding), Powers (2002) argued that proglacial conditions would have been difficult to overcome for migrating fish. These extreme conditions would have required the ability to exploit a wide range of niches, as primary productivity would have been much reduced. In one study of a recently (within the past 40 years) deglaciated area in Alaska, it was found that, despite initial instability of flow and stream channel morphology, the watershed was already heavily colonized by salmonids and even the slimy sculpin, *Cottus cognatus* (Milner and York, 2001). Upstream lake and kettle pond habitats provided the stable environments needed for reproduction and rearing.

Dorava and Milner (2000) evaluated the role of lake regulation in enhancing downstream habitat for migratory species in the Cook Inlet watershed, Alaska. Glacial drainages downstream of lakes had enhanced productivity for a number of reasons: (1) lakes provide supplemental flow in winter and mitigate flood effects in summer, (2) lakes act as settling areas for excess sediment thereby keeping downstream gravel habitat free of fines, (3) lakes act as stable deep-water environments for young salmon, and (4) lakes warm water to suitable temperatures for habitation (5–15 °C). Upstream of lakes, close to glacial margins, habitats were considered too unstable for colonization to take place (Dorava and Milner, 2000).

2.4 Glacial and Postglacial Colonization of Marine Environments

The impact of meltwater discharge in modern fjord benthic invertebrate communities was analyzed by Fetzer et al. (2002), and it was revealed that the total abundance of individuals increased with distance from the fjord head, while diversity was highest close to the glacier. Opportunistic and juvenile deposit-feeding polychaetes were the most abundant taxa, followed by bivalves and crustaceans. However, Fetzer et al. (2002) observed that the abundance of young crustaceans rapidly decreased toward the fjord mouth, possibly in response to stronger bottom currents. Juvenile suspension-feeding bivalves were less disturbed by meltwater discharge but were more vulnerable to currents close to the glacier. Thus, a concentration of burrows made by polychaetes and crustaceans is expected in shallow glaciomarine settings, while bivalve burrows might be common in deeper zones.

This is consistent with the fossil record in glaciomarine settings, which shows dominance of burrows attributed to polychaetes and arthropods and scarce occurrences of bivalve biogenic structures (see Supplementary Table 2 in http://booksite.elsevier.com/9780444538130). Fetzer et al. (2002) also concluded that hydrographic factors mainly are responsible for the benthic community distribution in shallow glaciomarine settings, while grain size and related properties may play a more important role in deeper zones.

Goldring et al. (2004, 2007) summarized work in the climatic control of Cenozoic to Pleistocene marine trace-fossil distribution. While that work did not directly address glacial depositional systems, trends in trace-fossil distribution from cold to warm climates were observed that allowed the recognition of an "Arctic" trace-fossil zone in which annelids and molluscs were the main bioturbators, with an absence of spatangoid burrows and crustacean-related burrows, such as *Thalassinoides* and *Ophiomorpha*.

3. TRACE-FOSSIL ASSEMBLAGES FROM ANCIENT GLACIAL ENVIRONMENTS

3.1 Bioturbated Deposits in Glacial Settings

Diamictites and rhythmites (e.g., varves) are the most characteristic and dominant modes of sedimentation in past glacial environments (e.g., Donovan and Pickerill, 2007; Eyles, 1993; Eyles and Eyles, 1992; Jensen and Wulff-Pedersen, 1996; Thunell et al., 1995). These deposits are directly related to glaciers, which deliver a large volume of sediment to marine and terrestrial/lacustrine basins, forming glacioterrestrial and glaciomarine systems tracts (Eyles and Eyles, 1992). The glacioterrestrial realm is composed of subglacial, supraglacial, glaciolacustrine, and glaciofluvial depositional systems, which are mainly expressed in low-relief glacioterrestrial settings. Glaciomarine settings may be dominated by glacial processes (ice proximal) or marine processes (ice distal). The influence of meltwater input in glaciomarine settings depends on the regional climate regime and controls chemical and biogenic processes. Most glacial deposits in low-relief glaciomarine settings are concentrated on continental shelves, the ice-proximal zone bearing the coarse-grained facies and the fine-grained facies deposited in the ice-distal zone. Fjords characterize the high-relief glaciomarine settings, where the ice-proximal deposits remain on the continental slope and the ice-distal deposits extend to the basin as turbidites (Eyles and Eyles, 1992). Stratigraphic discontinuities are common, as erosion plays an important role in these systems.

Laminated rhythmites are associated with glaciolacustrine deposits and shallow tide-dominated and prodeltaic glaciomarine deposits (e.g., Brodzikowski and van Loon, 1991; Domack, 1984; Mackiewicz et al., 1984; Powell, 1981, 1983; Smith et al., 1990). Millimeter- to centimeter-scale depositional layers, that represent annual cycles known as varves, are the best-known glacigenic rhythmites. However, rhythmites formed in glacially influenced environments do not necessarily represent annual seasonality, but rather result from recurrent episodic meltwater discharge events that form distal turbidite deposits in distinct settings (e.g., Brodzikowski and van Loon, 1991; Eyles and Eyles, 1992; Talbot and Allen, 1996).

Trace fossils in ancient glacial deposits have been recorded chiefly in fine-grained rhythmic deposits and, more rarely, in massive mudstones. They represent colonization after ice retreat, possibly in response to meltwater pulses, being well preserved in regular, varve-like rhythmites from the Upper Carboniferous–Lower Permian of Gondwana and Northern Hemisphere Pleistocene glacial deposits. They represent the bulk of the known glacial trace-fossil record.

The Late Paleozoic Gondwana Ice Age is well represented in the sedimentary record of Bolivia, in the Andean basins of Argentina (Tarija, Calingasta-Uspallata, and Paganzo Basins), in Brazil (Paraná Basin), in South Africa (Karoo Basin), and also in Antarctica and Australia (see Buatois et al., 2006,

2010 and references therein). Most trace-fossil assemblages occur within post-glacial facies and have been recorded in Brazilian, South African, and Argentinian Basins (see references in Supplementary Tables 1 and 2 in http://booksite.elsevier.com/9780444538130), which represent deposition in coastal lakes and fjords that opened toward a shallow sea. The Quaternary Ice Age sedimentary record is well exposed in the Northern Hemisphere, with sedimentary facies similar to those recorded from rocks of the Late Paleozoic Gondwana Ice Age. The preservation of Quaternary rhythmites is extensive and they provide geological evidence that support the interpretation of a seasonal (winter–summer) control of depositional cycles. Trace fossils from Quaternary glacigenic deposits have been recorded chiefly in varved glaciolacustrine sediments from the northern hemisphere, in Austria, Canada, England, Finland, Germany, Lithuania, Poland, Sweden, and the USA (see Uchman et al., 2008, 2009 and references therein; Benner et al., 2008, 2009; Fliri et al., 1970; Knecht et al., 2009).

Despite occurring in disparate time periods, glacial trace-fossil assemblages share some characteristics. The same overall trace-fossil composition has been recorded in glaciolacustrine (or closely related settings strongly influenced by freshwater due to meltwater discharge) deposits of both ages (see Supplementary Table 1 in http://booksite.elsevier.com/9780444538130). Typically these deposits are dominated by horizontal biogenic structures including arthropod trackways, trails, furrows, and resting traces; fish swimming trails; and regularly sinuous to randomly meandering tiny shallow burrows produced by worm-like animals. Glaciolacustrine trace-fossil assemblages occur almost exclusively within and on bedding planes, show a relatively low ichnodiversity, high abundance, and patchy distribution (Uchman et al., 2008, 2009). Aquatic and terrestrial suites may occur on the same bedding plane in palimpsest-type preservation, the latter generally superimposing the former, suggesting a diminished water table and community succession (Netto et al., 2009). Despite the significant mammal body fossils from the Cenozoic to early Holocene, their tracks and trackways have not been recorded from glaciogenic deposits until the present day (Martin, 2009).

3.2 Glaciolacustrine Trace-Fossil Assemblages

The pattern of biogenic structures observed on the bedding planes of ancient ice-proximal glaciolacustrine deposits, the common presence of fish swimming trails, and the patchy distribution of trace fossils, are consistent with the opportunistic colonization pattern observed in modern glaciolacustrine settings. Interestingly, modern pioneer invertebrate communities had the same small-scale spatial heterogeneity and strong correlation with environmental conditions as the older, established sites; a fact that may explain the observed similarity of freshwater invertebrate trace-fossil assemblages across different ages and basins (see Supplementary Table 1 in http://booksite.elsevier.com/9780444538130) and references therein.

The freshwater trace-fossil assemblages recorded in both Late Paleozoic Gondwana and Quaternary varve-like rhythmites are characterized by the presence of two pervasive ichnocoenoses, one dominated by arthropod trackways, and another by non-specialized trails and shallow horizontal burrows, well preserved in broad, shallow glacial basins. The arthropod trackway-dominated ichnocoenosis is the most common in the fossil record, being attributed to the activity of merostomes, isopod crustaceans, and apterygote insects (Fig. 1). A third ichnocoenosis, dominated by fish trace fossils, is well preserved in deep, narrow Pleistocene glacial lake basins (see Supplementary Table 1 in http://booksite.elsevier.com/9780444538130).

In glacial Gondwana deposits, trackways include ichnospecies of *Maculichna*, *Umfolozia*, and *Protichnites* (the last being less common; Fig. 1A–C), but *Kouphichnium* isp. and *Orchesteropus atavus* (Fig. 1D and G) also occur in the Falkland and Paganzo basins, respectively, as well as *Broomichnium permianum* (formerly *Quadrispinichnia parvia*; see Benner et al., 2008 for further discussion) in the Karoo Basin (see Supplementary Table 1 in http://booksite.elsevier.com/9780444538130). Crustacean resting traces are more rare, being represented by *Gluckstadtella cooperi* in the Paraná and Karoo basins (Fig. 1E and F), and by *Kingella natalensis*, exclusive to the Karoo Basin until now. Insect resting traces are represented by *Tonganoxichnus* isp. (Fig. 1I), recorded in the Paraná Basin. A particular suite composed exclusively of myriapod trackways (*Diplichnites gouldi*) and trails (*Diplopodichnus biformis*) also occurs in the Paraná Basin (Fig. 1A, H, and J), being the most conspicuous trace-fossil assemblage in thin-bedded rhythmites of the Paraná Basin (Balistieri et al., 2002, 2003; Buatois et al., 2006, 2010; Nogueira and Netto, 2001; see Netto et al., 2009 for further discussion). Bilobate trace fossils, such as *Cruziana* and *Rusophycus*, may be locally dominant (Schatz et al., 2011a). The Pleistocene glacial trackway assemblage includes *Glaciichnium liebegastensis* and *Warvichnium ulbrichi* as dominant ichnotaxa, but *Steinsfjordichnus* isp., *Lusatichnium slavensis*, *Irichnus* isp., and other arthropod trackways also occur (see Uchman et al., 2008, 2009 and references therein; Supplementary Table 1 in http://booksite.elsevier.com/9780444538130; Fig. 2).

The non-specialized trails and shallow burrows ichnocoenosis is composed chiefly of *Cochlichnus*, *Gordia*, *Helminthoidichnites*, and *Mermia* (=*Gordia carickensis* in Pleistocene records) ichnospecies in both Late Paleozoic Gondwana and Quaternary ice-age deposits (Supplementary Table 1 in http://booksite.elsevier.com/9780444538130). This ichnocoenosis seems have been more diverse in the Gondwana record, in which *Helminthopsis* and slightly deeper-tier burrows of deposit feeders (*Hormosiroidea meandrica*, *Treptichnus pollardi*, *Treptichnus* isp., *Planolites* isp.) and, very rarely, suspension feeders (*Palaeophycus* isp.) also occur (Supplementary Table 1 in http://booksite.elsevier.com/9780444538130; Fig. 3A, D–F, I, and J). Accessory components, including arthropod locomotion (*Cruziana problematica*, *Glaciichnium* isp.), and resting traces (*Rusophycus* isp., *R. carbonarius*), molluscan-type trails and burrows (*Archaeonassa* and *Didymaulichnus*, the latter which can also be produced by arthropods, *Lockeia* and *Protovirgularia* ichnospecies), non-specialized vertical burrows (*Arenicolites* isp., *Diplocraterion* isp., *Skolithos* isp.), and fish trails (*Undichna* isp.), have also been described in Late Paleozoic

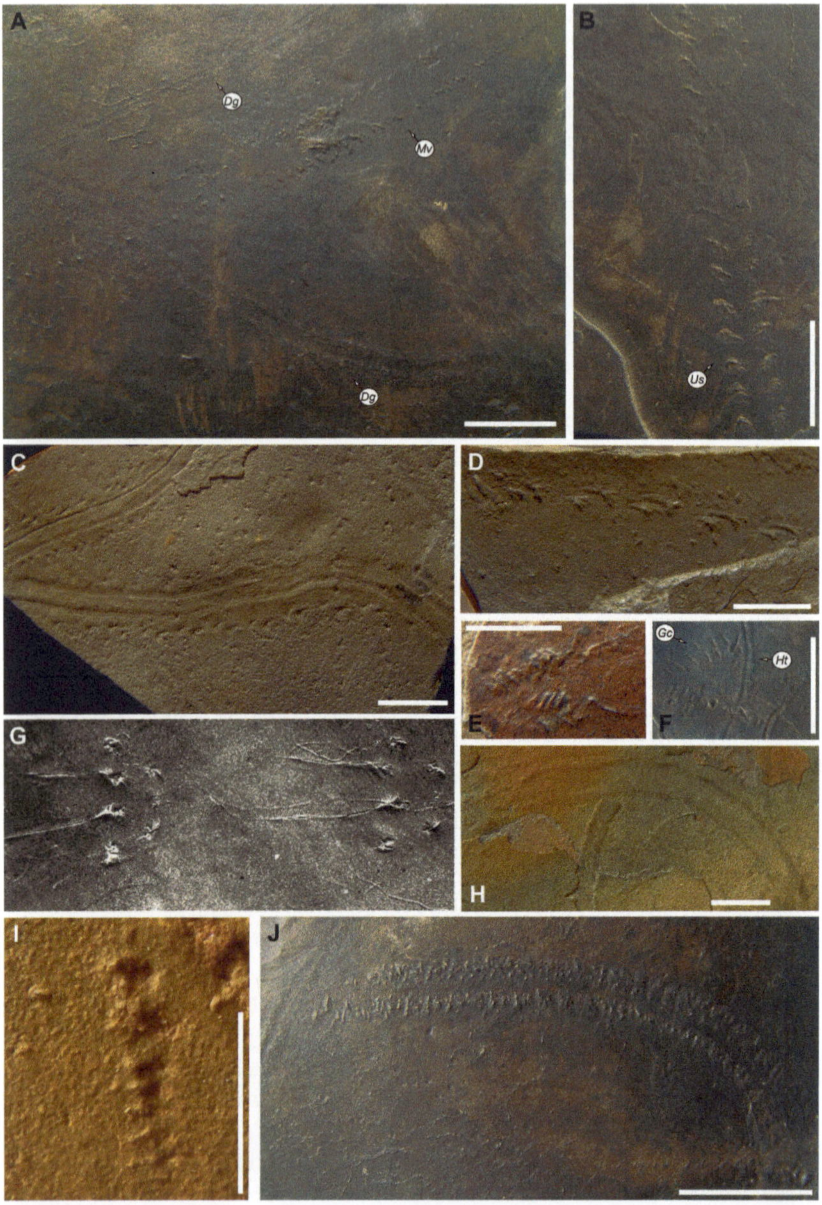

FIGURE 1 Trace fossils from Late Carboniferous–Early Permian terrestrial and lacustrine glacigenic deposits of Gondwanaland basins. (A) *Maculichna varia* (*Mv*), cross-cut by *Diplichnites gouldi* (*Dg*). (B) *Umfolozia sinuosa* (*Us*). (C) *Protichnites* isp. (D) *Kouphichnium* isp. (E) and (F) *Gluckstadtella cooperi*. (G) *Orchesteropus atavus*. (H) *Diplopodichnus biformis*. (I) *Tonganoxichnus* isp. (J) *Diplichnites gouldi*. All specimens occur in the Itararé Group thin-bedded rhythmites (Paraná Basin, southern Brazil) except (G), which comes from the Guadancol Formation (Paganzo Basin, north-western Argentina). Specimens (C) and (F) are preserved as negative epireliefs, whereas all other specimens are preserved as positive hyporeliefs. Scale bars = 10 mm.

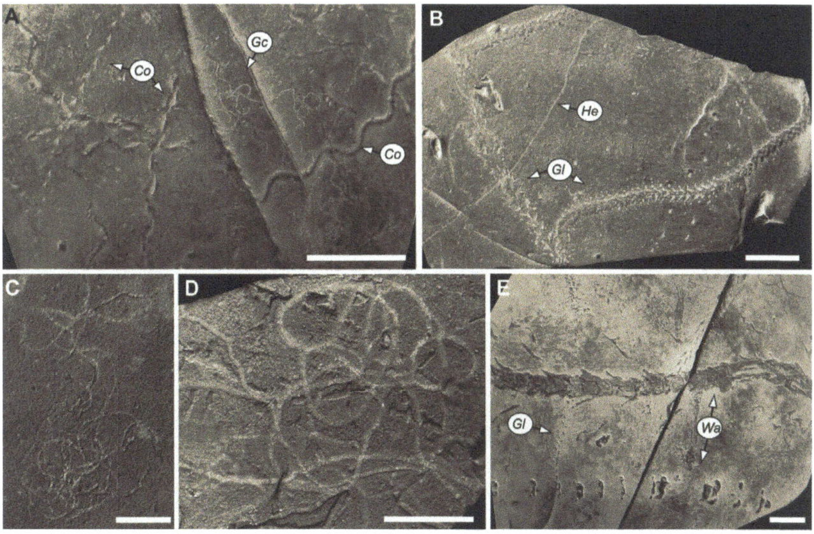

FIGURE 2 Some trace fossils from Pleistocene varved lacustrine sections south of Kaunas, Lithuania. (A) Three different *Cochlichnus anguineus* (*Co*) and *Gordia* (=*Mermia*) *carickensis* (*Gc*); Ppr-2-31 (lower parting surface), Pašaminė section. (B) *Glaciichnium liebegastensis* (*Gl*) and *Helminthoidichnites* isp. (*He*); Ppr-13-14, Pašaminė section. (C) *Gordia carickensis*, Balbieriškis section. (D) *Gordia* isp., Balbieriškis section. (E) *Warvichnium ulbrichi* (*Wa*) and *Glaciichnium liebegastensis* (*Gl*); Ppr-2-41, Pašaminė section. Scale bars = 10 mm, except in (C) (=5 mm).

Gondwana varve-like rhythmites (Supplementary Table 1 in http://booksite.elsevier.com/9780444538130; Fig. 3B, C, G, H, and K–M).

This ichnofauna occurs exclusively on parting surfaces of most rhythmite laminae, including microlaminae of proximal varves. The bioturbated varves, that compose the European Pleistocene glacigenic deposits, reflect events during the summer and on the transitions to clay laminae interpreted as winter laminae. The colonization surfaces are only 0.1–0.3 mm thick and the trace fossils are less common in silt microlaminae within clay (winter) laminae, suggesting that the tracemakers were active mostly during summers and transitional periods. Cryptic bioturbation, marked by reorientation of clay-mineral flakes, is noted in some varved sediment (O'Brien and Pietraszek-Matter, 1998). Preservation potential was highest when the tracemaker moved on or through the thin silt couplet. However, trace fossils are not always present in such situations. Intervals with increased abundance and diversity of trace fossils may reflect warming during interstadials and retreating ice cover (Netto et al., 2009; Uchman et al., 2008).

Trace fossils produced by vertebrates and invertebrates also occur in the Pleistocene varved sediments of the Connecticut and Merrimack valleys (USA; Benner et al., 2008, 2009; Knecht et al., 2009), revealing a general pattern of benthic ecosystem development in deep, narrow glacial lake basins (Table 1). The assemblage is dominated by fish trace fossils represented by *Broomichnium flirii*, *Piscichnus* isp., *Undichna bina*, *U. britannica*, *U. simplicitas*, *U. unisulca*, and other piscine trace fossils. *Cochlichnus* isp. is the only

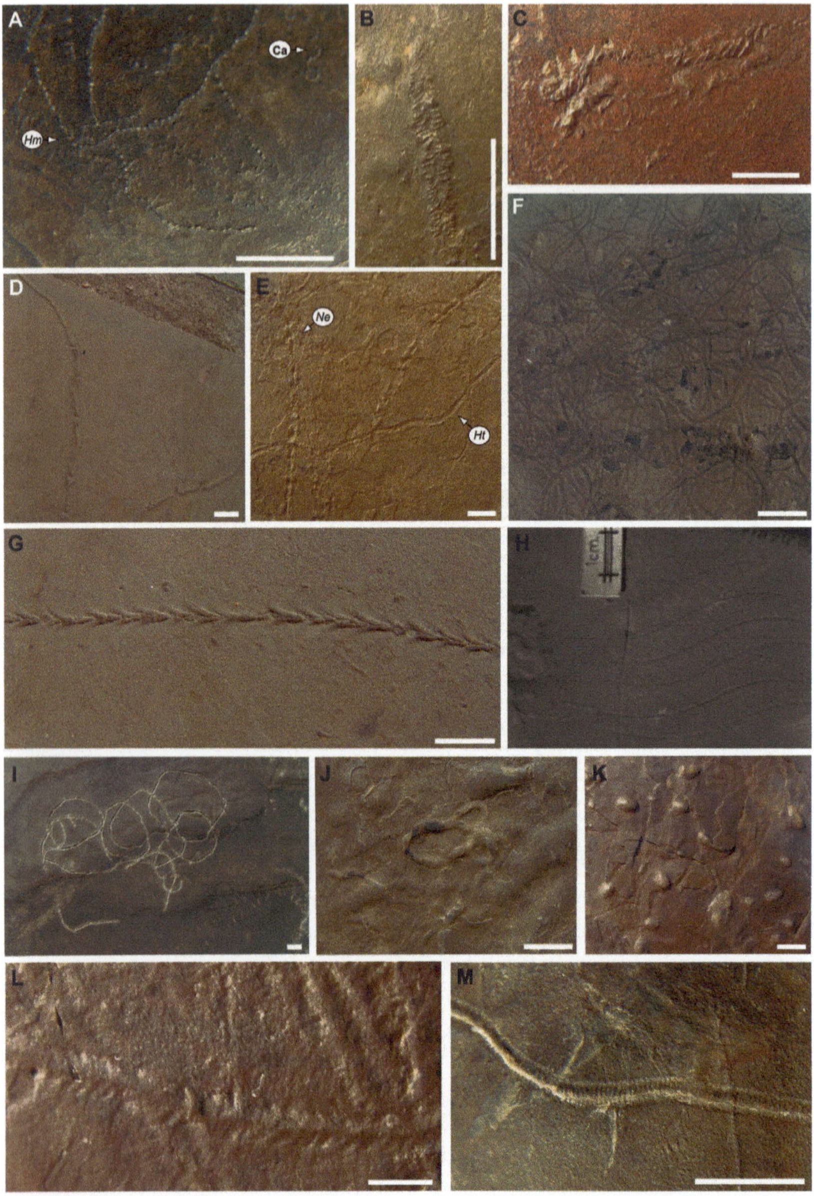

FIGURE 3 Trace fossils from Late Carboniferous–Early Permian lacustrine and marginal-marine glacigenic deposits of Gondwanaland basins. All specimens occur in the Itararé Group thin-bedded rhythmites (Paraná Basin, southern Brazil) except (H), which comes from the Agua Colorada Formation (Paganzo Basin, north-western Argentina). (A) *Cochlichnus anguineus* (*Ca*) and *Hormosiroidea meandrica* (*Hm*). (B) *Cruziana problematica.* (C) *Rusophycus* cf. *carbonarius.* (D) *Treptichnus pollardi.* (E) *Nereites* isp. (*Ne*) and *Helminthopsis* isp. (*Ht*). (F) *Helminthoidichnites tenuis.* (G) *Glaciichnium* isp. (H) *Undichna insolentia.* (I) *Gordia* (=*Mermia*) *carickensis.* (J) *Gordia marina.* (K) *Lockeia siliquaria.* (L) and (M) *Protovirgularia* isp. Specimens (E), (I), and (M) are preserved as negative epireliefs, whereas all other specimens are preserved as positive hyporeliefs. Scale bars = 10 mm.

TABLE 1 Summary of First Occurrences for Each Ichnospecies or Unnamed Type from Connecticut River Valley, USA Varves

			Cochlichnus	*Undichna*				*Broomichnium*	Arthropod	*Piscichnus*
	Site	Varves	isp.	*unisulca*	*bina*	*britannica*	*simplicitas*	*flirii*	trace fossils	isp.
North	PAS-2	7320–7763	7320	7350		7352	7362	7352	7530	
	WR-1	7134–7607	7134	7478		>7350		~7360	>7500	
	NEW	6980–7765	6980	7022		7281	7027	7359	>7500	7610
	MER-2	6150–6210	6124	6209						
	MER-1	6100–6200	6100	6168		6124				
	WAL	5938–5960	5938	5967	5939	5938				
	CB	5930–6100	5930	?						
South	RR	5130–5220	5130	5141	5219					
	KF-C	3820–4150	3820							
	CHIC	3586–3660	3586	d	d	d	d	d	d	d

First occurrence of each reported as the New England Varve Chronology (NEVC; Ridge, 2003, 2004) varve year in which it was found or a reference to the oldest varve year from which it could have been taken (>, ~). Cells have been shaded with color according to the ecological stage; the assemblage represents blue, stage 1; green, stage 2; yellow, stage 3; red, stage 4. See text for explanation.

representative of the non-specialized trails and shallow burrows ichnocoenosis, while *Surculichnus bifurcauda* and other arthropod trace fossils compose the arthropod trackway-dominated ichnocoenosis (see Supplementary Table 1 in http://booksite.elsevier.com/9780444538130).

3.3 Ichnocoenoses and Ichnofacies of Glaciolacustrine Rhythmites

The general structure of the trace-fossil assemblage and the ichnotaxa recorded in the varve-like rhythmite deposits formed by deglaciation during Gondwana and Holarctic-Antarctic ice ages are consistent with those of freshwater ichnofacies. The study developed by Uchman et al. (2008, 2009) in the Lithuanian varves reveals that the lateral distribution of trace fossils is patchy and can change from abundant to absent at distances of 15–20 m without any difference in lithology (Fig. 4). The denser patches may have developed around more nutritionally dense locales (Uchman et al., 2009). The tracemakers moved intensively within a patch or rarely between the patches (cf. Koy and Plotnick, 2007). Densely looping trails can point to activity within highly nutritional patches, whereas less tightly looping to winding forms (transitions to *Helminthoidichnites*) can indicate activity within less nutritional areas or searches for a new patch. A change in trace-fossil composition at adjacent levels may reflect changes in the activity of particular tracemakers. *Cochlichnus* and *Glaciichnium* commonly occur alone, but rarely together. Probably, tracemakers of *Glaciichnium* (*Asellus* or a similar isopod crustacean) and *Cochlichnus* (dipteran larvae or nematodes) were the most successful in the Quaternary, being opportunistic colonizers of new substrates under stressing environmental conditions (Uchman et al., 2008).

The lacustrine varve chronology of Pleistocene glacial Lake Hitchcock (Connecticut Valley, USA) spans >4000 varve years (see Ridge, 2003, 2004 and references therein) and was recently recalibrated (J. Ridge, personal communication, 2010). The vertical distribution of its trace-fossil content allowed Benner et al. (2009) to reconstruct the biotic occupation history of the deep glacial lakes. The initial occupation stage of the lake bottom is represented by a *Cochlichnus* assemblage preserved in oldest varves, closest to receding ice and nearest underlying till. *Cochlichnus* isp. (Fig. 5) in two forms is present in the first varves overlying till at all localities in which till can be located, indicating the pioneering life habit of the causative organism(s). Stage 2 is identified by the presence of the fish-produced trace fossil, *Undichna* (Fig. 5H), in addition to *Cochlichnus*. Stage 2 assemblages persist for circa 1.7 ka, after which a new benthic fish-produced trace fossil, *Broomichnium flirii* (Fig. 5G; see Benner et al., 2008), and a variety of arthropod-produced trace fossils (Fig. 5B–E) including *Surculichnus bifurcauda* (Fig. 5A; see Knecht et al., 2009) are added, producing a Stage 3 assemblage. The shift from a depauperate Stage 2 to a Stage 3 assemblage, rich in arthropod trace fossils and having a higher diversity of fish trace

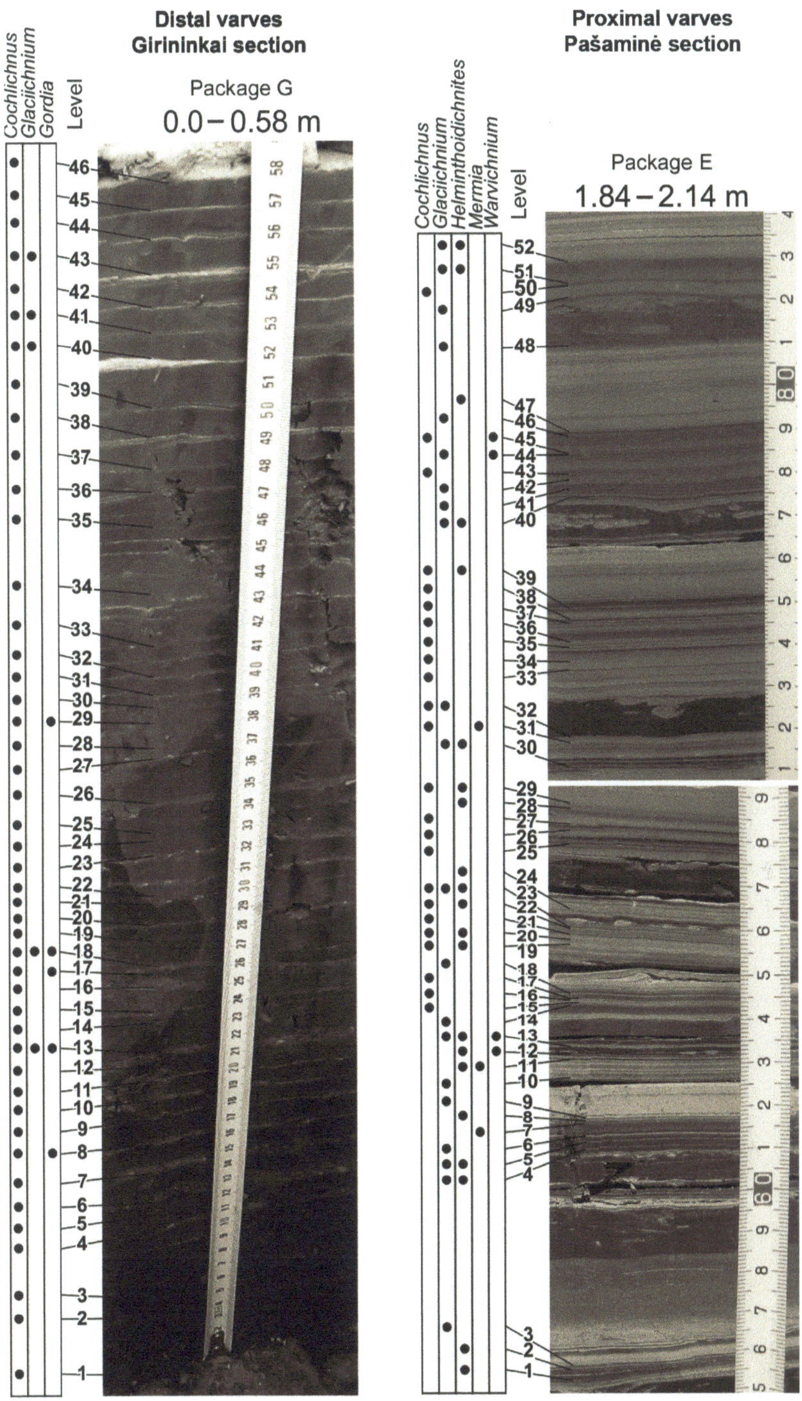

FIGURE 4 Vertical distribution of trace fossils in varved clays of the Late Pleistocene Girininkai and the Pašaminė sections, south of Kaunas, Lithuania.

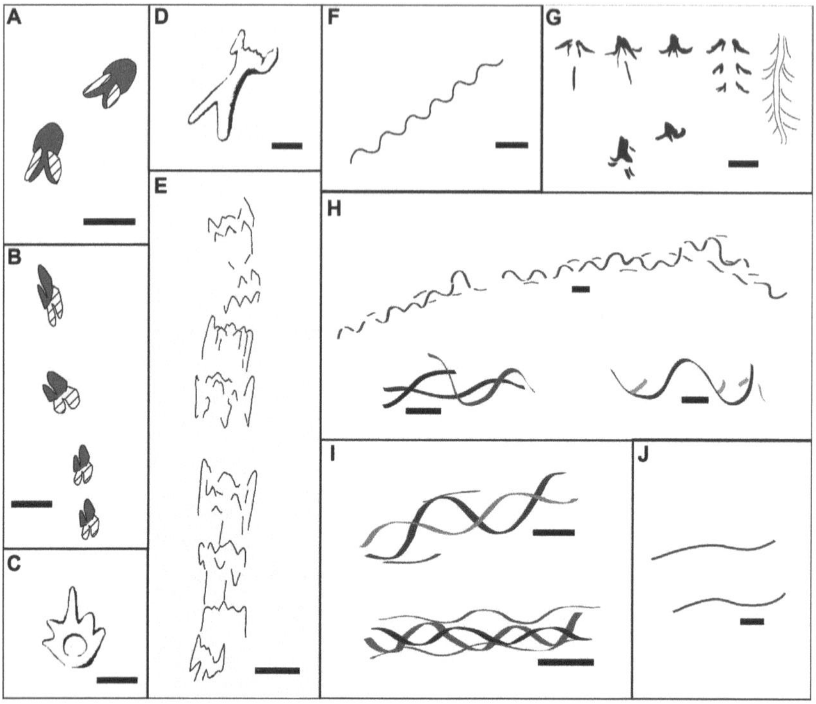

FIGURE 5 Simplified drawings based on actual specimens of trace fossils from Connecticut River valley (USA) varves. (A) *Surculichnus bifurcauda* in epirelief. Gray area is negative relief; hatched area is positive relief. (B) Unknown arthropod trace fossil (UATF) 4 in epirelief. Gray area is negative relief; hatched area is positive relief. (C) UATF 3 in positive hyporelief (cast). (D) UATF 2 in positive hyporelief (cast). (E)'UATF 1 in positive hyporelief (cast). (F) *Undichna unisulca.* (G) Five morphotypes of *Broomichnium flirii.* (H) Three morphotypes of *Undichna britannica.* (I) Two morphotypes of *Undichna simplicitas.* (J) *Undichna bina.* Scale bars in (A), (D), (G–J = 2 cm); (B, C, and E = 1 cm); (F = 10 cm). See Benner et al. (2009) for detailed explanations and interpretations of each trace fossil.

fossils, is coincident with a sedimentologically corroborated shift from mostly glacial input to the lake to mostly non-glacial inputs (Benner et al., 2009). The last identified stage in the succession, Stage 4 (NEVC year 7610; ~13.51 ka BP), coincides with a sedimentological shift to higher percentages of sand deposition, probably indicating shallower lake environments as water levels dropped or as the basin filled. These contain an abundance of fish breeding pits, *Piscichnus*, in addition to arthropod and other fish-produced trace fossils. The apparent lack of small invertebrate traces (i.e., *Cochlichnus*) in Stage 4 varves may be a result of taphonomic factors (poor preservation in sand) rather than actual abundance of the tracemaker. It is possible that the rate of stage transition (succession) increased as time passed in the valley after a significant reservoir of organisms ready to inhabit newly exposed water ways had accumulated in ice-free areas. Close observation of species assemblages in modern late-stage glaciated valley lakes may help to elucidate these processes.

Trace-fossil diversity rarely exceeds three ichnotaxa within individual laminae in ice-distal or ice-proximal varves, especially in the Quaternary deposits. Uchman et al. (2009) found that trace-fossil diversity in a Lithuanian varve sequence was slightly higher in more ice-proximal varves (five ichnogenera) compared with that of ice-distal varves (three ichnogenera) (Fig. 4). Benner et al. (2009) found the opposite—that ichnogeneric diversity increased as varves became more ice distal (Table 1) and contained a higher percentage of non-glacial sediment input. In fact, trace fossils probably made by adult arthropod crustaceans were not present until non-glacial inputs were the majority of sediment delivered to the basin (Knecht et al., 2009). The different conclusions of the two studies may be a result of variable preservation and/or lack of bedding-plane exposure in soft, clay-rich ice-distal varves, or extreme patchiness in the Lithuanian varve record.

Walter (1985) and Walter and Suhr (1998) proposed a model of trace-fossil distribution in a proglacial lake. They distinguished three communities: the repichnia ichnocoenosis with atypically "stretched" *Cochlichnus* (Heftstich type) within sandy varves from the epilimnion of a deltaic environment; the cursichnia ichnocoenosis (renamed as the *Glaciichnium* ichnocoenosis by Uchman et al., 2008) with arthropod trackways *Glaciichnium*, *Warvichnium*, and *Lusatichnium*, typical of proximal varves within the metalimnion; and the *Cochlichnus* ichnocoenosis with common, regular *Cochlichnus* and rare small trace fossils, as well as grazing traces, such as *Gordia* and *Helminthoidichnites*, typical of more distal varves within the hypolimnion. Decreasing oxygen content from the epilimnion to the hypolimnion was considered to be the controlling factor in trace-fossil distribution. The bathymetric control in this model was questioned by Uchman et al. (2008, 2009), because trace fossils of different zones predicted by the model occur within adjacent laminae in intervals that represent short periods of time (Fig. 4). This fact led Uchman et al. (2008, 2009) to suppose that rather than oxygenation, food density or other factors controlled the ichnocoenosis instead of water depth. Further, Benner et al. (2009) noted that a *Cochlichnus* ichnocoenosis occurs in the first varves deposited over till, indicating a most ice-proximal position and potentially shallow water depths.

The arthropod-dominated trace-fossil ichnocoenosis recorded in Gondwanan glacial rhythmites was interpreted by Balistieri (2003) and Netto et al. (2009) as representative of activity in shallow and marginal littoral zones of lakes and proximal areas flanking fjord valleys during climatic amelioration periods that follow deglaciation cycles. These authors assumed that this ichnocoenosis represents an atypical occurrence of the *Scoyenia* Ichnofacies, which always superimposes the non-specialized trail and shallow burrows ichnocoenosis (equivalent to the *Cochlichnus* ichnocoenosis from Pleistocene glacial deposits), representative of the *Mermia* Ichnofacies, in palimpsest preservation (Balistieri, 2003; Buatois et al., 2006, 2010; Netto et al., 2009). The abundance of sedimentary structures induced by the presence of microbial mats formed possibly by cyanobacteria in the Gondwanan bioturbated rhythmites also

FIGURE 6 Trace fossils from Late Carboniferous–Early Permian glaciomarine deposits of Gondwanaland basins. Specimens in (A) to (F) occur in the Itararé Group massive siltstones (Paraná Basin, south Brazil), which characterizes firmgrounds. (A) *Palaeophycus striatus*. (B) *Palaeophycus tubularis*. (C) *Gyrolithes*-like burrows. (D) *Rhizocorallium* isp. (dashed lines representing the inferred U-shaped burrow end). (E) and (J) *Diplocraterion* isp. (F) *Thalassinoides* isp. (G) *Chondrites* isp. (H) *Phycosiphon* isp. (I) *Thalassinoides-Planolites-Palaeophycus-Teichichnus* composite ichnofabric. Specimens in (G) to (J) occur in fine-grained heterolithic deposits of San Gregorio Formation (North Uruguayan Basin, (G) and (H)), and Rio do Sul Formation (Itararé Group, Paraná Basin, southern Brazil). All specimens preserved in full relief. Scale bars = 10 mm.

supports the shallow condition of these ancient glacial lakes (Netto et al., 2009; Noffke, 2010). This model was compared by Netto et al. (2009) with modern postglacial landscapes in Alaska and Antarctica and reinforced by the common occurrence of freshwater ponds and lakes that develop in shallow depressions excavated by ice-mass movements in marginal-marine plains in high-latitude settings (Horne and Goldman, 1994). Several of these lakes overlie permafrost, which melts out during summer, promoting lake drainage. According to Talbot and Allen (1996), cool glacial conditions in polar regions create arid conditions and low lake levels in high-latitude temperate settings. Lake-level changes are controlled dominantly by climate oscillation, which controls the hydrologic budget. Thus, in shallow lakes, reduction of water input and concomitant increases in water loss may reduce the water level, giving rise to wetlands, or cause them to dry up entirely.

Thus, rather than a distinctive “glacial ichnofacies”, ichnofaunas from glacial and periglacial settings characterize particular suites of the *Mermia* Ichnofacies. Suites of *Scoyenia* Ichnofacies, dominated by arthropod trackways (typical of margins of closed lakes, Buatois and Mángano, 2004), also occur, being well documented in the Gondwana record (Buatois et al., 2010; Netto et al., 2009; Fig. 7).

3.4 Glaciomarine Trace-Fossil Assemblages

The ichnofauna of shallow glaciomarine Gondwanaland is preserved in fine- to very fine-grained sandstones and siltstones with flaser and wavy bedding; is mainly composed of ?*Arenicolites*, *Chondrites*, *Diplocraterion*, *Palaeophycus*, *Phycosiphon*, *Planolites*, *Rhizocorallium*, and *Thalassinoides* (see Supplementary Table 2 in http://booksite.elsevier.com/9780444538130; Fig. 6G–J); and is well represented in the Rio do Sul (Paraná Basin) and San Gregorio (North Uruguayan Basin) Formations (Balistieri, 2003; Buatois et al., 2006, 2010). Massive siltstones locally contain sharply bounded burrows (*Thalassinoides* isp., *Diplocraterion* isp., *Palaeophycus* isp., *P. striatus*, ?*Rhizocorallium* isp., and *Gyrolithes*-like burrows; Fig. 6A–F) representative of the *Glossifungites* Ichnofacies (Balistieri and Netto, 2002), suggesting transgressive erosional exhumation and firmground colonization in fjord valley flanks. Glaciomarine trace fossils are small if compared with equivalent ichnofaunas from normal-marine settings, and comprise non-specialized feeding burrows produced by trophic generalists. The overall structure and composition of the assemblage are consistent with impoverished *Cruziana* Ichnofacies suites normally found in brackish-water settings (e.g., Buatois et al., 2005). Dropstones and diamictites with faceted clasts occur throughout most of the succession and thick deposits of fossiliferous marine shales record periods of maximum flooding in the Gondwanaland glacial environment.

The most diverse marine ichnofauna reported in glacially influenced Gondwanaland deposits is preserved in lower shoreface storm sediments from the Talchir Formation in central and eastern India (Bhattacharya and

Bhattacharya, 2007; Sarkar et al., 2009; see Supplementary Table 2 in http://booksite.elsevier.com/9780444538130). Despite high ichnodiversity, the trace fossils are sporadically distributed, and the overall small size of the burrows, the absence of body fossils, and the dominance of worms as tracemakers suggest brackish-water conditions, probably due to influx of glacial meltwater during climatic amelioration (e.g., Fielding et al., 2006; Virtasalo et al., 2006).

Deep glaciomarine bioturbated deposits have been reported only from the Cenozoic record, both in offshore and deep turbidite deposits. *Phymatoderma*, *Thalassinoides*, ?*Nereites*, and *Planolites* were reported by Uchman and Gaździcki (2010) in offshore and deeper-water deposits of the Early Miocene Cape Melville Formation (King George Island, Antarctica), while a more diverse assemblage composed of *Arenicolites*, *Chondrites*, *Diplocraterion*, *Ophiomorpha*, *Rhizocorallium*, *Thalassinoides*, and *Zoophycos* had been reported in the deep-water active-margin turbidites and debris-flow deposits and prograding continental slope mass-flow deposits from the Late Miocene Yakataga Formation (Gulf of Alaska; Armentrout, 1979; Eyles et al., 1992; see Supplementary Table 2 in http://booksite.elsevier.com/9780444538130). Shallow-marine deposits are also present in the Yataga Formation, accumulated in a broad shelf with a high sedimentation rate (including abundant coarse ice-raft debris), but the *Skolithos* Ichnofacies is absent due to the predominance of unstable, muddy substrates (Eyles et al., 1992). Trace fossils from the Yataga Formation were grouped in nine assemblages that were referred to particular depositional settings and types of substrate in submarine channel, slope, shelf boulder pavement, coquina, storm sands, and shelf postglacial muds. Schatz et al. (2011b) analyzed biogenic structures in Holocene cores from three fjords (Maktak, Coronation, and North Pangnirtung) of Baffin Island (Arctic Canada), comparing this information with photos taken from bottom camera stations. The abundance and diversity of biogenic structures was controlled by the proximity to the ice, physical disturbances (e.g., turbidity flows), and ice drift. Sediments have been extremely reworked, especially further away from the fjord head.

Polychaete borings also occur in Quaternary glaciomarine environments. *Caulostrepsis* and *Maeandropolydora* were recorded in early Holocene marine terrace boulders of NE Spitsbergen by Hanken et al. (2012). Earlier, Aitken and Risk (1988) presented borings in Pleistocene-recent shells and limestone clasts from the Arctic Canada. Research on microbioerosion along the Swedish coast (e.g., Wisshak and Rüggeberg, 2006; Wisshak et al., 2005) suggested that some ichnotaxa (e.g., *Flagrichnus baiulus*) occur only in non-tropical settings (Wisshak and Porter, 2006).

3.5 Ichnocoenoses and Ichnofacies of Glaciomarine Rhythmites

Trace fossils reported from tidally influenced and shallow, wave-dominated to deep glaciomarine settings generally compose suites representative of the *Cruziana* and (to a lesser degree) *Glossifungites* Ichnofacies (see Supplementary

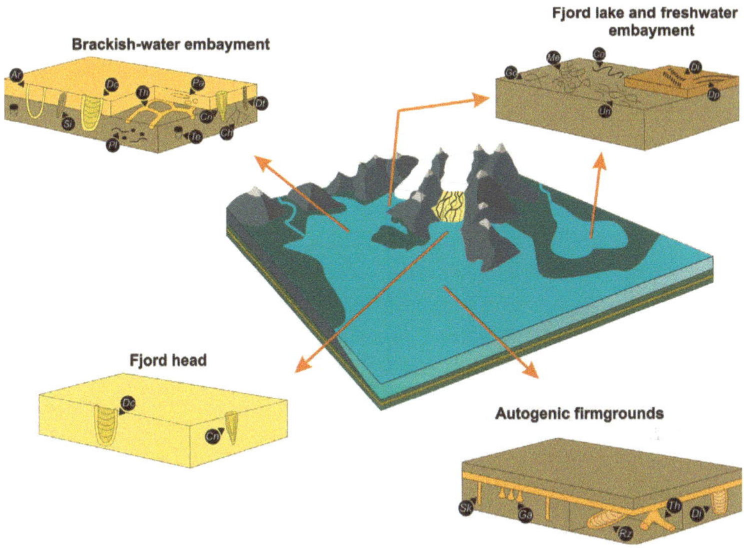

FIGURE 7 General trace-fossil distribution in glacial environments. *Ar*: *Arenicolites*; *Co*: *Cochlichnus*; *Ch*: *Chondrites*; *Cn*: *Conichnus*; *Di*: *Diplichnites*; *Dc*: *Diplocraterion*; *Dp*: *Diplopodichnus*; *Dt*: *Diopatrichnus*; *Ga*: *Gastrochaenolites*; *Go*: *Gordia*; *Me*: *Mermia*; *Pa*: *Palaeophycus*; *Pl*: *Planolites*; *Rh*: *Rhizocorallium*; *Sk*: *Skolithos*; *Si*: *Siphonichnus*; *Te*: *Teichichnus*; *Th*: *Thalassinoides*; *Un*: *Undichna*. *After Buatois and Mángano (2011).*

Table 2 in http://booksite.elsevier.com/9780444538130 for ichnotaxa and references; Fig. 7). They generally show a low degree of bioturbation, impoverished diversity, and diminutive size, when compared with ichnofaunas from normal-saline environments. These features are typical of brackish-water ichnofaunas (e.g., Buatois et al., 2005). However, the presence of truly marine, deep-water ichnogenera (e.g., *Phymatoderma*, *Zoophycos*) suggest the establishment of the archetypical *Cruziana* Ichnofacies in deep-water settings, impoverished due to lowered salinity owing to dilution by glacial meltwaters. The occurrence of the *Cruziana* Ichnofacies in turbidites, instead of the expected *Nereites* Ichnofacies, can be explained by the downslope transport of food by turbiditic currents and mass flows (Eyles et al., 1992).

4. ICHNOLOGY OF RECENT GLACIAL ENVIRONMENTS

Little ichnological research has been performed in modern glacial environments, all of them concentrated in the Arctic regions. The record of bioturbation in glaciolacustrine and terrestrial environments includes chironomid traces from sediments of a Norwegian lake in contact with ice (Duck and McManus, 1984), vertebrate and invertebrate traces of a fluvial pointbar in northern Alaska compared to *Treptichnus*, *Cochlichnus*, *Archaeonassa* (*Aulichnites* in Martin, 2009), and *Helminthoidichnites* (Martin, 2009), and a low-diversity *Palaeophycus*-dominated trace-fossil assemblage with rare *Arenicolites* from postglacial lacustrine settings in the Archipelago Sea (northern Baltic Sea), which is replaced by a higher diversity brackish-water assemblage with *Arenicolites*, *Lockeia*, *Planolites*, and *Teichichnus*, a consequence of marine incursions (Virtasalo et al., 2006).

Burrowing faunas were also reported from glaciomarine settings, in tidal flats of Baffin Island in Arctic Canada (Aitken et al., 1988) and among soupground, softground, and firmground communities of the subtidal and intertidal zones of the Arctic Canada fjords (Dale et al., 1989). These faunas include the anemone *Cerianthus*; the polychaetes *Nereis pelagicus*, *Polydora quadrilobata*, *Pholoe minuta*, and *Arenicola marina*; the bivalves *Portlandia arctica*, *Hiatella arctica*, *Macoma balthica*, and *Mya truncata*; and the priapulid worm *Priapulus caudatus* (Dale et al., 1989). Onuphid polychaete traces associated with gravel-sized dropstones, irregular echinoid furrows, probable burrows of the shrimp *Axius*, and burrowing anemones, compose the biogenic sedimentary structures observed in the substrate of the Baffin Island Fjord (Syvitski et al., 1989).

5. CONCLUDING REMARKS AND PERSPECTIVES

The available data on trace fossils from glacial environments, initially thought to be fossil-poor, have provided more information about the inhabitants of ancient glacial settings than expected. The combination of high sedimentation rates on episodic or annual bases with well-constrained fossil occurrence data has provided a high-quality glaciolacustrine and shallow glaciomarine trace-fossil record. Terrestrial and glaciolacustrine ichnofaunas are dominated by arthropod trackways and shallow horizontal burrows made by worm-like organisms, characterizing *Mermia*, *Scoyenia*, or mixed *Scoyenia-Mermia* ichnofacies assemblages. Glaciomarine ichnofaunas from fjords are highly impacted by meltwater discharge. As a consequence, freshwater conditions prevail in some fjord settings during deglaciation, allowing for the establishment of suites ascribed to these ichnofacies. Under brackish-water conditions, suites illustrate the impoverished *Cruziana* Ichnofacies. The *Skolithos* Ichnofacies is typically suppressed, most likely due to water turbidity.

Extreme environmental conditions dominate in glacial environments, constraining biota diversity and abundance. Amelioration of these conditions during interstadial or interglacial stages may be the most important controlling factor in the establishment of biotic communities, which rapidly occupy newly formed and available habitats, as evidenced by trace fossils within the earliest varves of proglacial lakes. A number of factors control reinhabitation of formerly glaciated valleys by organisms previously confined to refugia during glacial intervals. The dominance of horizontal burrows might be a response to permafrost conditions, constraining the biological activity only in the superficial layers of the substrate. The timing of reinhabitation of aquatic environments, however, depends first upon the presence of inhabitable water ways. For pioneering aquatic organisms that can withstand harsh environmental conditions including perennial ice, low water temperatures, high turbidity, and dissolved solids, only topographic barriers, in the form of spillways, stand in the way.

Paleobiogeographical and paleoecological inferences using bedding-plane trace fossils are tenuous and require identification of tracemakers—a challenge given that the tracemakers are not deep infauna and do not produce trace fossils characteristic of their particular trophic position or of their anatomy with regularity. Because of the abundance of well-preserved bedding-plane trace fossils within the Quaternary glacial rhythmites and the ability to utilize records of modern species occurrences as a starting point, identification in some cases has been more than speculative. Comparisons can be made between the assemblages of the Late Paleozoic glacial strata and those of Cenozoic deposits. The replacement of merostome and myriapod trace fossils of the Paleozoic trace-fossil assemblages by subaquatic crustaceans in the Cenozoic glaciolacustrine deposits is probably a consequence of evolution, as most merostome arthropods were extinct at the end of the Paleozoic. However, ichnofacies relationships and ecological niche occupation remain similar between the old and the young deposits, which are an indication of the constancy of biotic reaction to glacial events through time.

ACKNOWLEDGMENTS

The research summarized in this chapter was supported by the Brazilian National Council for Scientific and Technological Development (CNPq grants 479457/2007-7, 503487/2007-4, 305208/2010-1, and 401826/2010-4 to R. G. N.), Tufts University Faculty (Research Award to J. S. B. and J. C. R.), National Science Foundation (Award 0639830 to J. C. R.), Natural Sciences and Engineering Research Council (NSERC) (Discovery Grants 311727-05/08 to M. G. M. and 311726-05/08 to L. A. B.), Jagillonian University (DS funds to A. U.). The critical reviews made by J. Isbell, H. Walter, and P. Suhr were capital to improve this chapter.

REFERENCES

Aitken, A.E., Risk, M.J., 1988. Biotic interactions revealed by macroborings in Arctic bivalve molluscs. Lethaia 21, 339–350.

Aitken, A.E., Risk, M.J., Howard, J.D., 1988. Animal-sediment relationships on a subartic intertidal flat, Pangnitung Fiord, Baffin Island, Canada. J. Sediment. Petrol. 58, 969–978.

Anderson, A.M., 1970. An analysis of supposed fish trails from interglacial sediments in the Dwyka Series, near Vryheid, Natal. Proceedings of the Second Gondwana Symposium, 637–645.

Anderson, A., 1975a. Turbidites and arthropod trackways in the Dwyka glacial deposits (Early Permian) of southern Africa. Trans. Geol. Soc. S. Afr. 78, 265–273.

Anderson, A., 1975b. Limulid trackways in the Late Paleozoic Ecca sediments and their palaeoenvironmental significance. S. Afr. J. Sci. 71, 249–251.

Anderson, A.M., 1976. Fish trails from the Early Permian of South Africa. Palaeontology 19, 397–409.

Anderson, A., 1981. The *Umfolozia* arthropod trackways in the Permian Dwyka and Ecca Series of South Africa. J. Paleontol. 55, 84–108.

Armentrout, J., 1979. Glacial-marine trace- and body-fossil associations, Upper Yakataga Formation (Pliocene-Pleistocene), Gulf of Alaska. AAPG Bull. 63, 411.

Balistieri, P.R.M.N., 2003. Paleoicnologia da Porção Superior do Grupo Itararé na Região de Mafra (SC): Limitações Paleoecológicas, Paleoambientais e Estratigráficas. Ph.D. Thesis, Universidade do Vale do Rio dos Sinos, São Leopoldo, RS, 128 pp.

Balistieri, P.R.M.N., Netto, R.G., 2002. A *Glossifungites* suite in deposits of the Itararé Group (Upper Carboniferous–Lower Permian of Paraná Basin) at Mafra region, north of Santa Catarina State, Brazil: ichnotaxonomy, and paleoecological and stratigraphical constraints. Acta Geol. Leopold. 55, 91–106.

Balistieri, P.R.M.N., Netto, R.G., Lavina, E.L.C., 2002. Ichnofauna from the Upper Carboniferous–Lower Permian rhythmites from Mafra, Santa Catarina State, Brazil: ichnotaxonomy. Rev. Bras. Paleont. 4, 13–26.

Balistieri, P.R.M.N., Netto, R.G., Lavina, E.L.C., 2003. Icnofauna de ritmitos do topo da Formação Mafra (Permo-Carbonífero da Bacia do Paraná) em Rio Negro, Estado do Paraná (PR), Brasil. In: Buatois, L.A., Mángano, M.G. (Eds.), Icnología: hacia una convergencia entre geología y biología. Publicación Especial de la Asociación Paleontológica Argentina, vol. 9, pp. 131–139.

Benner, J.S., Ridge, J.C., Taft, N.K., 2008. Late Pleistocene freshwater fish (Cottidae) trackways from New England (USA) glacial lakes and a reinterpretation of the ichnogenus *Broomichnium* Kuhn. Palaeogeogr. Palaeoclimatol. Palaeoecol. 260, 375–388.

Benner, J.S., Ridge, J.C., Knecht, R.J., 2009. Timing of post-glacial reinhabitation and ecological development of two New England, USA, drainages based on trace fossil evidence. Palaeogeogr. Palaeoclimatol. Palaeoecol. 272, 212–231.

Bhattacharya, B., Bhattacharya, H.N., 2007. Implications of trace fossil assemblages from Late Palaeozoic glaciomarine Talchir Formation, Raniganj Basin, India. Gondwana Res. 12, 509–524.

Brinck, P., 1966. Animal invasion of Glacial and Late Glacial terrestrial environments in Scandinavia. Oikos 17, 250–266.

Brodzikowski, K., van Loon, A.J., 1991. Glacigenic Sediments. Developments in Sedimentology, vol. 49. Elsevier, Amsterdam, 671 pp.

Buatois, L.A., del Papa, C.E., 2003. Trazas fósiles de la Formación Tarija, Carbonífero Superior del norte argentino: Aspectos icnológicos de la transgresión postglacial en el oeste del Gondwana. In: Buatois, L.A., Mángano, M.G. (Eds.), Icnología: hacia una convergencia entre geología y biología. Publicación Especial de la Asociación Paleontológica Argentina, vol. 9, pp. 119–130.

Buatois, L.A., Mángano, M.G., 1992. Abanicos sublacustres, abanicos submarinos o plataformas glacimarinas? Evidencias icnológicas para una interpretación paleoambiental del Carbonífero de la cuenca Paganzo. Ameghiniana 29, 323–335.

Buatois, L.A., Mángano, M.G., 1993. Trace fossils from a Carboniferous turbiditic lake: implications for the recognition of additional nonmarine ichnofacies. Ichnos 2, 237–258.

Buatois, L.A., Mángano, M.G., 1995a. Postglacial lacustrine event sedimentation in an ancient high mountain setting: the Carboniferous Lake Malanzán from western Argentina. J. Paleolimnol. 12, 1–22.

Buatois, L.A., Mángano, M.G., 1995b. Sedimentary dynamics and evolutionary history of a Late Carboniferous Gondwanic lake in northwestern Argentina. Sedimentology 42, 415–436.

Buatois, L.A., Mángano, M.G., 2003. Caracterización icnológica y paleoambiental de la localidad tipo de *Orchesteropus atavus*, Huerta de Huachi, provincia de San Juan, Argentina. Ameghiniana 40, 53–70.

Buatois, L.A., Mángano, M.G., 2004. Animal-substrate interactions in freshwater ecosystems: applications of ichnology in facies and sequence stratigraphic analysis of fluvio-lacustrine successions. In: McIlroy, D. (Ed.), The Application of Ichnology to Palaeoenvironmental and Stratigraphic Analysis. Geol. Soc. London, Spec. Publ., vol. 228, pp. 311–333.

Buatois, L.A., Mángano, M.G., 2011. Ichnology. Organism-Substrate Interactions in Space and Time. Cambridge University Press, Cambridge, 358 pp.

Buatois, L.A., Gingras, M.K., MacEachern, J., Mángano, M.G., Zonneveld, J.-P., Pemberton, S.G., Netto, R.G., Martin, A.J., 2005. Colonization of brackish-water systems through time: evidence from the trace-fossil record. Palaios 20, 321–347.

Buatois, L.A., Netto, R.G., Mángano, M.G., Balistieri, P.R.M.N., 2006. Extreme freshwater release during the late Paleozoic Gondwana deglaciation and its impact on coastal ecosystems. Geology 34, 1021–1024.

Buatois, L.A., Netto, R.G., Mángano, M.G., 2010. Ichnology of late Paleozoic post-glacial transgressive deposits in Gondwana: reconstructing salinity conditions in coastal ecosystems affected by strong meltwater discharge. In: López-Gamundí, O.R., Buatois, L.A. (Eds.), Late Paleozoic Glacial Events and Postglacial Transgressions in Gondwana. GSA Spec. Paper 468, 149–173.

Clark, P.U., Dyke, A.S., Shakun, J.D., Carlson, A.E., Clark, J., Wohlfarth, B., Mitrovica, J.X., Hostetler, S.W., McCabe, A.M., 2009. The last glacial maximum. Science 325, 710–714.

Dale, J.E., Aitken, A.E., Gilbert, R., Risk, M.J., 1989. Macrofauna of Canadian Arctic fjords. Mar. Geol. 85, 331–358.

Dias-Fabrício, M.E., Guerra-Sommer, M., 1989. Síntese dos estudos icnológicos do Grupo Itararé no Rio Grande do Sul. Pesquisas 22, 71–88.

Domack, E.W., 1984. Rhythmically bedded glaciomarine sediments on Whidbey Island, Washington. J. Sediment. Petrol. 54, 589–602.

Donovan, S.K., Pickerill, R.K., 2007. Dropstones: their origin and significance: a comment. Palaeogeogr. Palaeoclimatol. Palaeoecol. 131, 175–178.

Dorava, J.M., Milner, A.M., 2000. Role of lake regulation on glacier-fed rivers in enhancing salmon productivity: the Cook Inlet watershed, southcentral Alaska, U.S.A. Hydrol. Proc. 14, 3149–3159.

Duck, R.W., McManus, J., 1984. Traces produced by chironomid larvae in sediments of an ice-contact proglacial lake. Boreas 13, 89–93.

Ehlers, J., Gibbard, P.L., 2004. Quaternary Glaciations—Extent and Chronology. Elsevier, Amsterdam, 488 pp.

Eyles, N., 1993. Earth's glacial record and its tectonic setting. Earth-Sci. Rev. 35, 1–248.

Eyles, N., Eyles, C.H., 1992. Glacial depositional systems. In: Walker, R.G., James, N.P. (Eds.), Facies Models: Response to Sea Level Changes. Geol. Ass. Can., Geotext 1, pp. 73–100.

Eyles, N., Vossler, V.M., Lagoe, M.B., 1992. Ichnology of a glacially-influenced continental shelf and slope; the late Cenozoic Gulf of Alaska (Yakatanga Formation). Palaeogeogr. Palaeoclimatol. Palaeoecol. 94, 193–221.

Feder, H.M., Matheke, G.E.M., 1980. Distribution, adundance, community structure and tropic structure of the benthic infauna of the northeast Gulf of Alaska. Institute of Marine Science Report R78–8, University of Alaska, Fairbanks, 209 pp.

Fernandes, A.C.S., Carvalho, I.S., Netto, R.G., 1987. Comentários sobre os traços fósseis do paleolago de Itu, São Paulo. Atas 6° Simpósio Regional de Geologia, Sociedade Brasileira de Geologia, Rio Claro, SP 1, pp. 297–311.

Fetzer, I., Lønne, O.J., Pearson, T., 2002. The distribution of juvenile benthic invertebrates in an arctic glacial fjord. Polar Biol. 25, 303–315.

Fielding, C.R., Bann, K.L., MacEachern, J.A., Tye, S.C., Jones, B.G., 2006. Cyclicity in the nearshore marine to coastal, Lower Permian, Pebbley Beach Formation, southern Sydney Basin,

Australia: a record of relative sea-level fluctuations at the close of the Late Paleozoic Gondwanan ice age. Sedimentology 53, 435–463.

Fliri, F., Bortenschlager, S., Felber, H., Heissel, W., Hilscher, H., Resh, W., 1970. Der Bänderton von Baumkirchen (Inntal, Tirol). Eine neue Schlüsselstelle zur Kenntnis der Würm-Vereisung der Alpen. Z. Gletscherk. Glazialgeol. 6, 5–35.

Gaigalas, A., Uchman, A., 2004. Trace fossils from Upper Pleistocene varved clays S of Kaunas, Lithuania. Geologija 45, 16–26.

Gandini, R., Netto, R.G., Souza, P.A., 2007. Paleoicnologia e a palinologia dos ritmitos do Grupo Itararé na pedreira de Águas Claras (Santa Catarina, Brasil). Gaea 3, 47–59.

Gibbard, P.L., Dreimanis, A., 1978. Trace fossils from the Late Pleistocene glacial lake sediments in southwestern Ontário, Canadá. Can. J. Earth Sci. 15, 1967–1976.

Gibbard, P.L., Stuart, A.J., 1974. Trace fossils from proglacial lake sediments. Boreas 3, 69–74.

Goldring, R., Cadée, G.C., D'Alessandro, A., de Gibert, J.M., Jenkins, R., Pollard, J.E., 2004. Climatic control of trace fossil distribution in the marine realm. In: McIlroy, D. (Ed.), The Application of Ichnology to Palaeoenvironmental and Stratigraphic Analysis. Geol. Soc. London, Spec. Publ., vol. 228, pp. 77–92.

Goldring, R., Cadée, G.C., Pollard, J.E., 2007. Climatic control of marine trace fossil distribution. In: Miller III, W. (Ed.), Trace Fossils. Concepts, Problems, Prospects. Elsevier, Amsterdam, pp. 159–171.

Hambrey, M., 1994. Glacial Environments. CRC Press, London, 302 pp.

Hanken, N.-M., Uchman, A., Jakobsen, S.L., 2012. Late Pleistocene–Early Holocene polychaete borings in NE Spitsbergen and their palaeoecological and climatic implications: an example from the Basissletta area. Boreas 41, 42–55.

Hoffman, P.F., Kaufman, A.J., Halverson, G.P., Schrag, D.P., 1998. A Neoproterozoic Snowball Earth. Science 281, 1342–1346.

Horne, A.J., Goldman, C.R., 1994. Limnology. McGraw-Hill, New York, 576 pp.

Isbell, J.L., Miller, M.F., Babcock, L.E., Hasiotis, S.T., 2001. Ice-marginal environment and ecosystem prior to initial advance of the late Paleozoic ice sheet in the Mount Butters area of the central Transantarctic Mountains, Antarctica. Sedimentology 48, 953–970.

Isbell, J.L., Miller, M.F., Wolfe, K.L., Lenaker, P.A., 2003a. Timing of late Paleozoic glaciation in Gondwana: was glaciation responsible for the development of northern hemisphere cyclothems? In: Chan, M.A., Archer, A.W. (Eds.), Extreme Depositional Environments: Mega End Members in Geologic Time. GSA Spec. Paper 370, pp. 5–24.

Isbell, J.L., Lenaker, P.A., Askin, R.A., Miller, M.F., Babcock, L.E., 2003b. Reevaluation of the timing and extent of late Paleozoic glaciation in Gondwana: Role of the Transantarctic Mountains. Geology 31, 977–980.

Jensen, P.A., Wulff-Pedersen, E., 1996. Glacial or non-glacial origin for the Bigganjargga tillite, Finnmark, Northern Norway. Geol. Mag. 133, 137–145.

Kaufmann, R., 2001. Invertebrate succession on an alpine glacier foreland. Ecology 82, 2261–2278.

Knecht, R.J., Benner, J.S., Rogers, D.C., Ridge, J.C., 2009. *Surculichnus bifurcauda* n. igen., n. isp., a trace fossil from Late Pleistocene glaciolacustrine varves of the Connecticut River Valley, USA, attributed to notostracan crustaceans based on neoichnological experimentation. Palaeogeogr. Palaeoclimatol. Palaeoecol. 272, 232–239.

Koy, K., Plotnick, R.E., 2007. Theoretical and experimental ichnology of mobile foraging. In: Miller III, W. (Ed.), Trace Fossils. Concepts, Problems, Prospects. Elsevier, Amsterdam, pp. 428–441.

Le Heron, D.P., 2010. Trace fossils on a Late Ordovician glacially striated pavement in Algeria. Palaeogeogr. Palaeoclimatol. Palaeoecol. 297, 138–143.

Lermen, R., 2006. Assinaturas Icnológicas em Depósitos Glaciogênicos do Grupo Itararé no RS. M. Sc. Thesis, Universidade do Vale do Rio dos Sinos, São Leopoldo, RS, 84 pp.

Likens, G.E., 2010. Encyclopedia of Island Waters—Lake Ecosystem Ecology. Academic Press/ Elsevier, Amsterdam, 463 pp.

Mackiewicz, N.E., Powell, R.D., Carlson, P.R., Bruce, F.M., 1984. Interlaminated ice-proximal glacimarine sediments in Muir Inlet, Alaska. Mar. Geol. 57, 113–147.

Marques-Toigo, M., Dias-Fabrício, M.E., Guerra-Sommer, M., Cazzulo-Klepzig, M., Piccoli, A.E. M., 1989. Afloramentos da área de Trombudo Central, Permiano Inferior, Santa Catarina: palinologia, icnologia e sedimentologia. In: Anais do 11° Congresso Brasileiro de Paleontologia, Sociedade Brasileira de Paleontologia, Curitiba, PR, 1, pp. 125–150.

Martin, A.J., 2009. Neoichnology of an Arctic fluvial point part, North Slope, Alaska (USA). Geol. Quart. 53, 383–396.

Menzies, J., 2002. Modern and Past Glacial Environments. Butterworth-Heinemann Ltd, Oxford, 543 pp.

Milner, A.M., York, G.S., 2001. Factors influencing fish productivity in a newly formed watershed in Kenai Fjords National Park, Alaska. Arch. Hydrobiol. 151, 627–647.

Netto, R.G., Goso, C.A.A., 1998. Icnología de la Formación San Gregorio (Pérmico Inferior), en el área de los Cerros Guazunambí, Cuenca Norte Uruguaya. Actas del II Congreso Uruguayo de Geología, Punta del Este 1, pp.188–189.

Netto, R.G., Balistieri, P.R.M.N., Lavina, E.L., Silveira, D.M., 2009. Ichnological signatures of shallow freshwater lakes in the glacial Itarare Group (Mafra Formation, Upper Carboniferous-Lower Permian of Parana Basin, S Brazil). Palaeogeogr. Palaeoclimatol. Palaeoecol. 272, 240–255.

Noffke, N., 2010. Geobiology: Microbial Mats in Sandy Deposits from the Archean Era to Today. Springer, New York, 194 pp.

Nogueira, M.S., Netto, R.G., 2001. Icnofauna da Formação Rio do Sul (Grupo Itararé, Permiano da bacia doParaná) na Pedreira Itaú-Itauna, Santa Catarina, Brasil. Acta Geologica Leopoldensia 52/53: 397–406.

O'Brien, N.R., Pietraszek-Matter, S., 1998. Origin of the fabric of laminated fine-grained glaciolacustrine deposits. J. Sediment. Res. 68, 832–840.

Pazos, P.J., 2000. Trace fossils and fácies in glacial to postglacial deposits from the Paganzo Basin (Late Carboniferous), Central Precordillera, Argentina. Ameghiniana 37, 23–38.

Pazos, P.J., 2002a. Paleoenvironmental framework of the glacial-postglacial transition (Late Paleozoic) in the Paganzo-Calingasta Basin (Southern South America) and the Great Karoo-Kalahari Basin (Southern Africa): ichnological implications. Gondwana Res. 5, 619–640.

Pazos, P.J., 2002b. The Late Carboniferous glacial to postglacial transition: facies and sequence stratigraphy, Western Paganzo Basin, Argentina. Gondwana Res. 5, 467–487.

Pazos, P.J., di Pasquo, M., Amenabar, C.R., 2007. Trace fossils of the glacial to post-glacial transition in the El Imperial Formation (Upper Carboniferous), San Rafael Basin, Argentina. In: Bromley, R.G., Buatois, L.A., Mángano, M.G., Genise, J.F., Melchor, R.N. (Eds.), Sediment-Organism Interactions: A Multifaceted Ichnology. SEPM Spec. Publ., vol. 88, pp. 137–147.

Powell, R.D., 1981. A model for sedimentation by tidewater glaciers. Ann. Glac. 2, 129–134.

Powell, R.D., 1983. Glacial marine sedimentation processes and lithofacies of temperate tidewater glaciers, Glacier Bay, Alaska. In: Molnia, B.F. (Ed.), Glacial-Marine Sedimentation. Plenum, New York, pp. 185–232.

Power, G., 2002. Charrs, glaciations and seasonal ice. Env. Biol. Fishes 64, 17–35.

Ridge, J.C., 2003. The last deglaciation of the northeastern United States: a combined varve, paleomagnetic, and calibrated ^{14}C chronology. In: Hart, J.P., Cremeens, D.L. (Eds.), Geoarchaeology of Landscapes in the Glaciated Northeast. New York State Museum, Bull. 497, pp. 15–45.

Ridge, J.C., 2004. The Quaternary glaciation of western New England with correlations to surrounding areas. In: Ehlers, J., Gibbard, P.L. (Eds.), Quaternary Glaciations—Extent and Chronology, Part II: North America. Elsevier, Amsterdam. Dev. Quat. Sci. 2b, pp. 163–193.

Rieu, R., Allen, P.A., Plotze, M., Pettke, T., 2007. Climatic cycles during a Neoproterozoic "snowball" glacial epoch. Geology 35, 299–302.

Sarkar, S., Ghosh, S.K., Chakraborty, C., 2009. Ichnology of a Late Paleozoic ice-marginal shallow marine succession: Talchir Formation, Satpura Gondwana Basin, central India. Palaeogeogr. Palaeoclimatol. Palaeoecol. 283, 28–45.

Savage, N.M., 1970. A preliminary note on arthropod trace fossils from the Dwyka series. Proceedings of the 2nd International Union of Geological Sciences Gondwana Symposium, Pretoria, South Africa, 627–635.

Savage, N.M., 1971. A varvite ichnocoenosis from the Dwyka Series of Natal. Lethaia 4, 217–233.

Schatz, E., Mángano, M.G., Buatois, L.A., Limarino, C.O., 2011a. Life after the Late Paleozoic ice age: trace fossils from glacially influenced deposits in a Late Carboniferous fjord-valley of western Argentina. J. Paleontol. 85, 502–518.

Schatz, E.R., Aitken, A., Mángano, G., 2011b. Ichnology of glaciomarine sediments: Maktak, Coronation and North Pangnirtung Fjords, Baffin Island, Canada. Geohydro 2011. Joint Meeting of the Canadian Quaternary Association and the Canadian Chapter of the International Association of Hydrogeologists. Quebec City, Quebec.

Sierszen, M.E., McDonald, M.E., Jensen, D.A., 2003. Benthos as the basis for arctic food webs. Aquatic Ecol. 37, 437–445.

Smith, N.D., Phillips, A.C., Powell, R.D., 1990. Tidal drawdown: a mechanism for producing cyclic sediment laminations in glaciomarine deltas. Geology 18, 10–13.

Syvitski, J.P., Burrell, D.C., Skei, J.M., 1987. Fjords: Processes and Products. Springer, New York, 379 pp.

Syvitski, J.P.M., Farrow, G.E., Atkinson, R.J.A., Moore, P.G., Andrews, J.T., 1989. Baffin Island Fjord macrobenthos: bottom communities and environmental significance. Arctic 42, 232–247.

Talbot, R.R., Allen, P.A., 1996. Lakes. In: Reading, H.G. (Ed.), Sedimentary Environments. Processes, Facies and Stratigraphy. Blackwell Publishing, Oxford, pp. 83–124.

Thunell, R.C., Tappa, E., Anderson, D.M., 1995. Sediment fluxes and varve formation in Santa Barbara Basin, offshore California. Geology 23, 1083–1086.

Trewin, N.H., 2000. The ichnogenus *Undichna*, with examples from the Permian of the Falkland Islands. Palaeontology 43, 979–997.

Trewin, N.H., MacDonald, D.I.M., Thomas, C.G.C., 2002. Stratigraphy and sedimentology of the Permian of the Falkland Islands: lithostratigraphic and palaeoenvironmental links with South Africa. J. Geol. Soc. London 159, 5–19.

Uchman, A., Gaździcki, A., 2010. *Phymatoderma melvillensis* isp. n. and other trace fossils from the Cape Melville Formation (Lower Miocene) of King George Island, Antarctica. Polar Res. 31, 83–99.

Uchman, A., Gaigalas, A., Kazakauskas, V., 2008. Trace fossils from the Upper Pleistocene glaciolacustrine laminated sediments of Lithuania. Geologija 50, 212–226.

Uchman, A., Kazakauskas, V., Gaigalas, A., 2009. Trace fossils from Late Pleistocene lacustrine varve sediments in eastern Lithuania. Palaeogeogr. Palaeoclimatol. Palaeoecol. 272, 199–211.

Virtasalo, J.J., Kotilainen, A.T., Gingras, M.K., 2006. Trace fossils as indicators of environmental change in Holocene sediments of the Archipelago Sea, northern Baltic Sea. Palaeogeogr. Palaeoclimatol. Palaeoecol. 240, 453–467.

Walter, H., 1985. Zur Ichnologie des Pleistozäns von Liebegast. Freiberger Forschungsh. C400, 101–116.

Walter, H., 1986. Arthropodenfärten in eiszitlichen Bändertonen bei Liebegast, Kreis Hoyerswerda. Natur Landschaft Bez. Cottbus 8, 51–58.

Walter, H., Suhr, P., 1998. Lebensspuren aus kaltzeitlichen Bändersedimenten des Quartärs. Abh. Staatl. Mus. Min. Geol. Dresden 43 (44), 311–328.

Wisshak, M., Porter, D., 2006. The new ichnogenus *Flagrichnus*: a paleoenvironmental indicator for cold-water settings? Ichnos 13, 135–145.

Wisshak, M., Rüggeberg, A., 2006. Colonisation and bioerosion of experimental substrates by benthic foraminifera from euphotic to aphotic depths (Kosterfjord, SW-Sweden). Facies 52, 1–17.

Wisshak, M., Gektidis, M., Freiwald, A., Lundälv, T., 2005. Bioerosion along a bathymetric gradient in cold-temperate setting (Kosterfjord, SW Sweden): an experimental study. Facies 51, 93–117.

Chapter 12

Fluvial Environments

Ricardo N. Melchor,*[,1] Jorge F. Genise,[†] Luis A. Buatois[‡] and Aldo M. Umazano[§]
*INCITAP (CONICET and Universidad Nacional de La Pampa), Santa Rosa, La Pampa, Argentina,
[†]CONICET, División Icnología, Museo Argentino de Ciencias Naturales, Buenos Aires, Argentina,
[‡]Department of Geological Sciences, University of Saskatchewan, Saskatoon, Canada,
[§]Universidad Nacional de La Pampa, Santa Rosa, La Pampa, Argentina
[1]Corresponding author: e-mail: rmelchor@exactas.unlpam.edu.ar

1. INTRODUCTION

The number of studies on the ichnology of fluvial deposits has increased significantly since the early 1980s. Trace fossils in fluvial deposits were considered undiagnostic and their distribution as not related to any particular environment (e.g., Picard and High, 1973). These earlier studies have been influenced by the notion of the *Scoyenia* Ichnofacies as being broadly defined and undistinctive (Seilacher, 1977). Also, it has been argued that the *Scoyenia* Ichnofacies reflects a community of organisms of low diversity, the preservation potential of which is very low (Ekdale et al., 1984). In the past decades, many studies have been directed toward the revision and redefinition of the *Scoyenia* Ichnofacies and to distinguish it from other associated archetypal trace-fossil assemblages that may occur in fluvial and other continental deposits (Buatois and Mángano, 1995, 1998; Frey et al., 1984; Genise et al., 2000, 2010). The possibility of recognizing continental ichnofacies and the ichnofacies concept itself have been questioned (Goldring, 1993; Hasiotis, 2004), although ichnofacies are well established and of common usage (Bromley et al., 2007; MacEachern et al., 2007).

The biological literature on the burrowing and nesting behavior and description of modern structures of arthropods and other organisms associated with fluvial environments is fairly large, but until recently such studies have not been formally incorporated within the background of ichnology. Relatively few contributions on traces from fluvial environments within the conceptual framework of ichnology exist (De, 2002; Laporte and Behrensmeyer, 1980; Lawfield and Pickerill, 2006; Mángano et al., 1996; Martin, 2009; Melchor et al., 2010; Pryor, 1967; Ratcliffe and Fagerstrom, 1980; Smith and Heim, 1971). The diversity of trace fossils that can be found in fluvial environments is reasonably large, and they can potentially contribute to the diagnosis of the paleoenvironment

Developments in Sedimentology, Vol. 64. http://dx.doi.org/10.1016/B978-0-444-53813-0.00012-5

(Buatois and Mángano, 1996, 2004, 2007, 2011; Pollard, 1988). The purpose of this chapter is to evaluate the link between trace-fossil assemblages and ichnofacies with different fluvial subenvironments, with emphasis on the utility of trace fossils in the paleoenvironmental analysis.

2. APPROACH

We have attempted to revise the present knowledge on the sedimentary facies distribution and ichnofacies assignment of published assemblages of trace fossils including invertebrate, vertebrate, and plant trace fossils. The number of ichnogenera recorded in fluvial deposits is over 170 (and more than 20 trace-fossil types are left in open nomenclature); their identification requires special skills and training that are, to some extent, more properly conducted by specialists. In the description of any trace-fossil assemblages, a careful ichnotaxonomical assignment of every trace-fossil type is highly desirable. However, for the practical purposes of this chapter, essentially aimed toward sedimentologists, a more simple classification scheme is adopted, largely based on morphological attributes that can be easily distinguished in the field. This classification scheme (Table 1) is not comprehensive, rather it is intended to cover only the ichnotaxa recorded in fluvial facies, although it can be extended to include other trace-fossil types and environments (see also the proposal by Knaust, 2012). Each category is equivalent to one or more ichnogenera, and no distinction at the ichnospecies level is considered. Some trace-fossil categories adopted herein have been used in an informal way in the ichnological literature, and most of them are proposed herein (Table 1).

The fundamental biogenic structures that are recognized include burrows, trails, tracks, trackways, chambers, nests, rhizoliths and other root structures, and coprolites (Frey and Pemberton, 1985; Klappa, 1980). Bite and gnaw structures on bone and borings in wood do occur in fluvial successions, but they have been used mostly for taphonomical and paleoecological analyses (e.g., Roberts et al., 2007; Uchman et al., 2007), though their applicability in facies analysis is not yet clear. In consequence, these structures are not considered further in this chapter. For tetrapod trackways, the classification is based on a hierarchy of features starting by the quadrupedal or bipedal locomotion of the producer of the trackway (Q, quadrupedal; B, bipedal), then the overall track outline (rounded, triangular), the number of digits on manus and pes tracks (manus-pes for quadrupedal, alternative number of digits impressions for bipedal tracks are separated by a dash), digit morphology (sl, slender; tk, thick; st, stout; d, dot-like), and pace angulation (pa). As shorthand for a category of tetrapod tracks, a formula or abbreviation is used that is composed of the following terms: nature of trackway in capital letters + morphology of footprints in lower case + number of digits manus/number of digits pes + digit morphology + pace angulation (e.g., "Q rounded 5/5"). The basic components of this formula are the nature of the trackway and the number of digits in manus and pes (Table 1). Although

TABLE 1 Morphology of Trace Fossils Found in Fluvial Successions and Ichnogenera that can be Included in Each Category

Category name/ abbreviation	Description	Included ichnogenera
Simple horizontal trail	Simple burrow/trails on bedding plane	*Cochlichnus, Mermia, Gordia, Helminthoidichnites, Helminthopsis, Archaeonassa, Spirodesmos, "Hormosiroidea"*
Sinusoidal trail	One or more sinusoidal trails on bedding plane	*Undichna*
Bilobed burrow/trail	Bilobed burrow/trail on bedding plane	*Cruziana, Didymaulichnus, Diplopodichnus, Didymaulyponomos*
Arthropod trackway	Two or more parallel, linear, or sinuous rows composed of discrete simple traces on bedding plane	*Diplichnites, Umfolozia, Hexapodichnus, Stiallia, Protichnites, Striatichnium, Monomorphichnus, Palmichnium, Merostomichnites*, "scratch traces"
Trough-like relief	Isolated trough-like relief, recording to some extent the latero-ventral morphology of the producer	*Rusophycus, Lockeia, Gluckstadtella, Selenichnites, Svalbardichnus*
Horizontal burrow	Simple horizontal burrow with massive filling	*Palaeophycus, Planolites, Fucusopsis, Tumblagoodichnus, Ctenopholeus*
Burrow system	Horizontal burrow system	*Spongeliomorpha, "Thalassinoides"*
Meniscate burrow	Simple meniscate burrow, surface ornamentation, and a discrete wall may be present	*Scoyenia, Taenidium* (=*Naktodemasis*), *Beaconites, Ancorichnus*
Simple vertical burrow	Simple burrow at high angle to bedding	*Skolithos* (=*Tigillites*), sediment displacement shaft, escape traces
J-, U-, or Y vertical burrow	J-, U-, or Y-shaped burrow at high angle to bedding	*Arenicolites, Psilonichnus, Diplocraterion, Polykladichnus, Camborygma, Loloichnus, Scaphichnium, Macanopsis, Cornulatichnus, Capayanichnus, Spirographites* (=*Lunulichnus*), *Tambia*, "staphyllinid burrow"

Continued

TABLE 1 Morphology of Trace Fossils Found in Fluvial Successions and Ichnogenera that can be Included in Each Category—Cont'd

Category name/ abbreviation	Description	Included ichnogenera
J- or U-shaped burrow, low angle	J- or U-shaped burrow at low angle to bedding	*Katbergia, Fuersichnus, Rhizocorallium*, ephemeroptera burrows
Burrow with pelletal fill	Groups of pellets filling a burrow or chamber	*Edaphichnium, Castrichnus*, "pelleted burrow"
Chambers and chambered burrow systems	Isolated or interconnected chambers with burrows or burrow systems of much smaller size	*Cellicalichnus, Dagnichnus, Pallichnus, Fictovichnus, Termitichnus, Krausichnus, Vondrichnus, Fleaglellius, Parowanichnus, Masrichnus, Monesichnus, Uruguay, Coprinisphaera, Eatonichnus, Palmiraichnus, Teisseirei, Corimbatichnus, Ellipsoideichnus, Rebuffoichnus, Feoichnus*
Borings in chamber walls	Borings in the constructed wall of trace fossils	*Tombownichnus, Lazaichnus*
Bird tracks (B 3–4 sl)	Bipedal trackway, anisodactyl, either lacking a web trace, palmate or semipalmate, tri- or tetradactyl footprints with slender digit imprints	*Gruipeda, Avipeda, Fuscinapeda, Sarjeantopodus, Aquatilavipes, Ignotormis, Barrosopus, Phoenicopterichnum*, bird footprints assigned to Tinamidae, Anatidae, Charadriidae, Laridae, and Rheidae
B 3–4 tk	Bipedal, tri- to tetradactyl, long thick digits, phalangeal pads, claw imprints	*Grallator, Saurexallopus, Neotrisauropodus, Wildeichnus, Sarmientichnus, Dinehichnus*, tyrannosaur footprints
B 3 st	Bipedal, tridactyl, short stout digit impressions, claw imprints	*Hadrosauropodus, Amblydactylus, Limayichnus*, Ornithopod tracks, iguanodontid footprint
Q 2/2	Quadrupedal, bilobed impression, homopodial, digitigrade	*Entelodontipus, Odocoileinichnum, Lamaichnum, Megalamaichnum, Pecoripeda, Gambapes*, bovid footprints

TABLE 1 Morphology of Trace Fossils Found in Fluvial Successions and Ichnogenera that can be Included in Each Category—Cont'd

Category name/ abbreviation	Description	Included ichnogenera
Q 3/3 st	Quadrupedal, homopodial, tridactyl short thick digit impressions	*Macrauchenichnus*, *Eumacrauchenichnus*, *Plagiolophustipus*, Equioidea footprints, rhinocere footprints
Q 3/3 high pa	Quadrupedal, heteropodous, tridactyl, high pace angulation (150–170°), size: 20–40 mm	*Delatorrichnus*
Q 4/5	Quadrupedal, homopodial or heteropodous, tetradactyl manus, pentadactyl pes, low pace angulation, size: less than 135 mm	*Limnopus*, *Palaeosauropus*, *Baropezia*, *Batrachichnus* (=*Cursipes*, *Dromillopus*), "salamander tracks"
Q rounded 5/5	Quadrupedal, rounded impression, pentadactyl, short digit imprints, homopodial	*Dicynodontipus* (=*Gallegosichnus*), "*Chelichnus*"
Q oval	Quadrupedal, oval to kidney-shaped tracks, usually lacking digit impressions	*Venatoripes*, *Neomegatherichnum*, kidney-like footprints, oval impressions
Q rounded	Quadrupedal, rounded, homopodial, usually lacking digit impressions	*Stegomastodonichnum*, *Hippipeda* footprints
Q 5/5 sl	Quadrupedal, heteropodous, pentadactyl, manus and pes long and slender, claw-bearing digits, which are distally curved inward, pace angulation 70–130°, size: 10–80 mm	*Rhynchosauroides*, *Dromopus*, *Tambachichnium*, *Rotodactylus*, *Varanopus*, *Gilmoreichnus*, *Hylopus*, *Hyloidichnus*, *Erpetopus*
Q 5/5 tk	Quadrupedal, heteropodous, pentadactyl, plantigrade, pace angulation (80–130°), size: up to 400 mm	*Dimetropus*, *Pseudobradypus*, *Ichniotherium*, *Amphisauropus*, *Capitosauroides*, *Pachypes*, *Asperipes*
Q 5/5 d	Quadrupedal, manus and pes represented sets of dot-like impressions with a curved arrangement	*Episcopocus*, *Pentasauropus*
Q 5/5 high pa	Quadrupedal, homopodial, pentadactyl, high pace angulation, less than 20 mm in size	*Ameghinichnus*

Continued

TABLE 1 Morphology of Trace Fossils Found in Fluvial Successions and Ichnogenera that can be Included in Each Category—Cont'd

Category name/ abbreviation	Description	Included ichnogenera
Q 4/3	Quadrupedal heteropodous, tetradactyl manus, tridactyl pes	*Stegopodus*, "small heteropod footprint"
Q triangular 4/4	Quadrupedal, tetradactyl, homopodial, triangular sole pad, oval digit imprints	*Bestiopeda, Chelipus, Pumaeichnum, Pehuencoichnum*
Chirotheriid tracks	Quadrupedal, tetradactyl manus, pentadactyl pes, resembling a human hand	*Chirotherium, Brachychirotherium, Isochirotherium, Synaptichnium, Evazoum, Tetrasauropus*
Sauropod track	Quadrupedal, crescentic manus, triangular pes	*Brontopodus, Parabrontopodus*, "sauropod tracks"
Large vertebrate burrow	Large burrow at low to high angle to bedding, showing bilobed floor and convex ceiling, and/or helicoidal path, and/or complex structure (chambers)	*Ichnogyrus, Daimonelix, Alezichnos, Kladosystemites, Polychoredrites, Katarrhedrites*, large burrow, burrow with bilobed bottom, possible lungfish aestivation burrow, burrow and terminal chamber, platter structure
Tetrapod swim traces	Groups of three to five elongate and parallel traces, that may or may not be a trackway	*Characichnos, Hatcherichnus, Lunichnium, Chelonipus*, "swim-trace"
Vertebrate coprolites	Roughly cylindrical mass of sediment, with subcircular cross section, inorganic inclusions, one or more tapering end	*Heteropolacopros, Dicynodontocopros*, "carnivore coprolites"
Indentation structure	Group of laminae with a local concave upward curvature covered by horizontal laminae	Footprints in cross section, concave-up deformation structures
Root structures	Rhizoliths of different morphology (e.g., shallow networks, and taproots of medium-sized plants and trees), isolated or connected to others, or to stump casts	Rhizoliths, stump casts (including *Socialites*)

This scheme is partially based on Genise (2004) for insect trace fossils, other trace-fossil categories adopted herein have been used in an informal way in the ichnological literature. Abbreviations for tetrapod trackways: Q, quadrupedal; B, bipedal; 2/2 to 5/5, the number of digits on manus and pes tracks, respectively; digit morphology, sl, slender; tk, thick; st, stout; d, dot; pa, pace angulation. Triangular, oval, and rounded refers to track outline. The column of "Included Ichnogenera" also contains informal names when no ichnogeneric assignation was made in the original work. See text for further explanation!

each category is defined on the basis of the morphology of the trace fossil, some interpretative names of common use in the literature are adopted (e.g., "arthropod trackway", "bird track", "sauropod track").

This simple classification scheme is not intended to replace the existing taxonomic approach for invertebrate and vertebrate trace fossils. Instead, by reducing the number of categories, it may facilitate recognition of recurrent associations of trace fossils that must be contrasted with definite ichnotaxonomic assignments. In addition, biogenic structures that cannot be or were not identified ichnotaxonomically in the literature can be compared with a trace-fossil category. However, unequivocal comparison of a trace-fossil assemblage with an ichnofacies is possible only through ichnotaxonomic assignment of the component trace fossils.

About 250 case studies describing fluvial trace-fossil assemblages in a definite facies context and with a clear ichnotaxonomical assignment were analyzed for this chapter. A detail of the included case studies can be found as Supplementary Material in http://booksite.elsevier.com/9780444538130. The database is biased toward the description of invertebrate trace-fossil assemblages (52%), a lower number of studies describe vertebrate and invertebrate trace fossils in conjunction (33%), and the less common studies are those that describe vertebrate trace fossils in a definite facies context (15%). There are many papers that are concerned with the description of tetrapod tracks and trackways that are not accompanied by a description of the sedimentary facies, and are thus excluded from this analysis. Only trace-fossil assemblages that were recovered from inland fluvial facies are considered, omitting those that appear in deltaic, estuarine, and other coastal settings. In the analysis of each case study, we distinguished the trace fossils found in channel belts and floodplain facies. Channel-belt deposits belong to the zone of influence of free lateral migration of a river (Bridge and Leeder, 1979) and include channel-bar and channel-fill deposits. The term "channel" *sensu lato* is also used to refer to "channel belts", which is the common usage of the term in the sedimentological literature.

The different trace-fossil assemblages from fluvial sediments were assigned to channel-belt, floodbasin-pond, crevasse-splay, and pedogenized floodplain (paleosol) deposits according to the interpretation of the authors of each case study. Trace-fossil assemblages included under pedogenized floodplain are those that are hosted in rocks that exhibit evidence of pedogenic processes, irrespective of the subaqueous origin of the sediments where the paleosol developed (e.g., crevasse-splay levee, channel).

3. ICHNOFACIES

The recognition of archetypal, recurrent, or Seilacherian ichnofacies in the continental realm has been the subject of substantial debate. Invertebrate archetypal ichnofacies for continental deposits have received general acceptance in the scientific community, in spite of having been proposed only since the 1990s (Buatois and

Mángano, 1995; Genise et al., 2000; Smith et al., 1993). The invertebrate ichnofacies recognized in fluvial deposits include the *Scoyenia*, *Mermia*, *Coprinisphaera*, and *Skolithos* ichnofacies (e.g., Buatois and Mángano, 2007). To this scheme can be added the recently proposed *Celliforma* Ichnofacies and the potentially distinctive and redefined *Termitichnus* Ichnofacies (Genise et al., 2010).

In contrast, the proposal of vertebrate ichnofacies or the incorporation of vertebrate trace fossils to a unified ichnofacies scheme has been controversial (MacEachern et al., 2012). The development of the concept of vertebrate ichnofacies is in an early stage, and there is still much debate and disagreement on the nature of each ichnofacies, the criteria used in its definition, and even the usefulness of the concept in vertebrate ichnology (e.g., Santi and Nicosia, 2008). The principles utilized for recognition of potential vertebrate ichnofacies are not the same as those applied to the establishment of invertebrate ichnofacies. These differences have been recognized, and a suggestion was made to distinguish two kinds of ichnofacies (Hunt and Lucas, 2005): one that reflects behavioral interactions between the organisms and the substrate (ethoichnofacies) and another that attempts to relate tracks and traces to the taxonomy of the producer (biotaxoichnofacies). Lockley and coworkers (Lockley, 2007; Lockley et al., 1994) recognized a number of vertebrate ichnocoenoses as candidates for vertebrate ichnofacies; although most of the ichnofacies were established on the basis of few case studies, their recurrence in time is limited and, in many cases, the ichnofacies also display a restricted geographical distribution (Melchor et al., 2006). Hunt and Lucas (2007) revised the proposal of Lockley et al. (1994) and presented a number of newly defined vertebrate ichnofacies that are composed of individual ichnocoenoses (many of them comparable with the "ichnofacies" of Lockley et al., 1994). The vertebrate ichnofacies of Hunt and Lucas (2007) were conceived as parallel to invertebrate archetypal ichnofacies. Although a unification of the ichnofacies scheme is considered desirable, we follow herein the proposal of vertebrate ichnofacies by Hunt and Lucas (2007). This proposal, even if it is too broad and displays internal inconsistencies, is considered a useful template for further discussion and refinement. This scheme is preferred over the Lockley et al. (1994) proposal because, in the latter, the proposed ichnofacies have a high ichnostratigraphical component (time dependence), except in the case of the "shorebird ichnofacies".

The vertebrate ichnofacies of Hunt and Lucas (2007) that are found in fluvial deposits include the *Grallator* and *Batrachichnus* ichnofacies (Hunt and Lucas, 2007; Lockley, 2007; Lockley et al., 1994). Other proposed ichnofacies that may occur in fluvial deposits lack the recurrence in time and space, to be considered as considered as archetypal, and are probably best viewed as particular ichnocoenoses: these are the *Characichnos* or *Hatcherichnus* (Hunt and Lucas, 2007; Lockley et al., 2010), *Chelonipus* (Lockley and Foster, 2006), and *Brontopodus* (Hunt and Lucas, 2007) "ichnofacies". Although the studies on vertebrate coprolites and the application of ichnotaxonomic names to coprolites are in their infancy, a coprofacies has also been recognized (Hunt et al., 1994).

In particular, the *Heteropolacopros* and *Dicynodontocopros* ichnofacies were proposed for fluvial successions (Hunt et al., 2007).

In the next paragraphs, the expression in fluvial deposits of the previously mentioned ichnofacies or ichnocoenoses is analyzed using selected fluvial trace-fossil assemblages from the literature (Table 2; Fig. 1). Trace-fossil assemblages are included within ichnofacies using the dominant trace fossils and not by means of a single or subordinate component (opposite to the proposal by Keighley and Pickerill, 2003). Some trace-fossil assemblages are too simple to be assigned with confidence to a single ichnofacies or cannot be satisfactorily fitted within the existing ichnofacies scheme (both accounting for about 20% of the analyzed trace-fossil assemblages). Some of the trace-fossil assemblages can be considered the work of a coeval community of organisms or ichnocoenoses in the sense of Bromley (1990, 1996), but in most cases, this is uncertain.

3.1 *Scoyenia* Ichnofacies

Most of the trace-fossil assemblages surveyed (62 cases, 33%) can be assigned with confidence to the *Scoyenia* Ichnofacies (Fig. 1A). These assemblages have been identified as belonging to this ichnofacies because of the dominance of meniscate burrows (Fig. 2A and F) and/or the presence of arthropod trackways (Fig. 2B), burrow systems (Fig. 2C and E), trough-like reliefs (Fig. 2D), and bilobed burrows/trails (e.g., Bromley and Asgaard, 1979; MacNaughton and Pickerill, 1995; Melchor et al., 2006; Morrissey and Braddy, 2004; Stanley and Fagerstrom, 1974). Assemblages containing only vertebrate tracks, but lacking any of the diagnostic components mentioned above, cannot be compared satisfactorily with the *Scoyenia* Ichnofacies and are treated separately in this chapter. Individual *Scoyenia* assemblages usually exhibit a low ichnodiversity, averaging 5 and ranging from 1 to 12 trace-fossil types. They occur in similar proportions in both in fluvial channel belts and overbank settings (mainly crevasse splay and floodbasin ponds). Considering the different river types, these assemblages are more common in braided and ephemeral river systems. The *Scoyenia* Ichnofacies reflects a superposition of suites of trace fossils associated with changes in substrate consistency and water saturation, generally coupled to water-level changes (compare Fig. 2E). Additionally, assemblages dominated or only composed of meniscate burrows from poorly developed and/or Paleozoic paleosols (e.g., ichnocoenosis G of Keighley and Pickerill, 2003) can also be included in this ichnofacies.

3.2 *Coprinisphaera* Ichnofacies

Some of the assemblages that were used to propose the *Coprinisphaera* Ichnofacies are from fluvial deposits (Genise et al., 2000), including the well-known examples from the Eocene Asencio Formation of Uruguay (e.g., Genise et al., 2004), the Paleocene–Eocene Claron Formation of United States (Bown et al.,

TABLE 2 Expression of Ichnofacies Found in Fluvial Deposits

Ichnofacies	Description	Included ichnogenera	Implications
Scoyenia	Diagnostic components: Predominance of simple meniscate burrows, arthropod trackways, trough-like reliefs, and bilobed burrows/trails. Other components that may be present but are not diagnostic: simple vertical burrows, simple horizontal burrows, diverse bipedal and quadrupedal tetrapod trackways, J-, U-, or Y-shaped burrows at high angle to bedding, horizontal burrows systems, and swim traces. Dominance of invertebrate traces. Ichnodiversity is generally low (average 5 morphological types) although some trace fossils may be abundant (particularly arthropod trackways in Paleozoic associations).	Characteristic: *Scoyenia, Taenidium, Beaconites, Diplichnites, Diplopodichnus, Cruziana, Rusophycus, Spongeliomorpha, Skolithos, Fuersichnus* Accessory: *Capayanichnus, Didymaulichnus, Grallator, Batrachichnus, Limnopus, Hyloidichnus, Dimetropus, Ichniotherium, Gilmoreichnus, Amphisauropus, Limayichnus, Anomoepus, Bothriodontipus, Delatorrichnus, Moyenisauropus*, "ornithopod tracks", "sauropod tracks"	Sandy to argillaceous sediments deposited in low-energy subaqueous areas that are periodically emerged. Typical environments include bar tops in fluvial channels, crevasse splays, and floodbasin ponds. Also present in poorly developed or Paleozoic paleosols.
Coprinisphaera	Mainly chambers and chambered trace fossils (mostly assigned to coleopterans and bees), commonly associated with rhizoliths, J-, U-, or Y-shaped vertical burrows, large vertebrate burrows, meniscate burrows, and burrows with pelletal fill. High-diversity trace-fossil assemblages (7–14 trace-fossil morphologies).	Characteristic: *Coprinisphaera, Teisseirei, Eatonichnus, Rebuffoichnus, Celliforma, Palmiraichnus, Uruguay, Ellipsoideichnus, Corimbatichnus* Accessory: *Feoichnus, Cellicalichnus* (except *C. meniscatus* and *C. antiquus*), *Parowanichnus, Beaconites, Tombownichnus, Lazaichnus*	Paleosols developed on former channel-belt deposits and overbank facies. Paleosols of different types and degree of development. Habitats dominated by low vegetation.

Celliforma	Dominance of chambers and chambered trace fossils assigned to bees and wasps. Associated components are rhizoliths, burrows and chambers with pelletal filling, J-, U-, or Y-shaped burrows, vertebrate coprolites, and large vertebrate burrows. Moderate ichnodiversity (average 5, range from 3 to 7 trace-fossil morphologies).	Characteristic: *Celliforma*, *Rebuffoichnus*, *Cellicalichnus* (except *C. meniscatus* and *C. antiquus*), *Fictovichnus* Accessory: *Teisseirei*, *Pallichnus*, *Edaphichnium*, *Macanopsis*, rhizoliths	Carbonate-rich paleosols developed on fine-grained overbank deposits, under semiarid (and seasonal) to arid climate. Indicative of scrubs and woodlands.
Termitichnus	Dominance of chambers and chambered trace fossils assigned to termites. Other components are J-shaped burrows, rhizoliths, large vertebrate burrows, and pelleted burrow fillings. Ichnodiversity ranges from 2 to 14.	Characteristic: *Termitichnus*, *Krausichnus*, *Vondrichnus*, *Fleaglellius* Accessory: *Macanopsis*, *Masrichnus*, *Edaphichnium*, rhizoliths, large vertebrate burrows	Paleosols of closed forests under warm and humid conditions. Moderately developed paleosols on channel-belt deposits.
Mermia	Predominantly horizontal to subhorizontal grazing and feeding traces (simple burrows/trails, simple horizontal burrows), subordinate sinusoidal trails and locally arthropod trackways and horizontal root structures. Low-diversity assemblages (up to five morphotypes).	Characteristic: *Archaeonassa*, *Cochlichnus*, *Undichna*, *Helminthoidichnites*, *Helminthopsis* Accessory: *Didymaulichnus*, *Palaeophycus*, *Planolites*, *Ctenopholeus*, *Diplichnites*, reed-like traces	Sandy to argillaceous sediments deposited in subaqueous settings of floodbasin ponds. Typically current-rippled fine-grained sandstone or siltstone, parallel-laminated mudstone.

Continued

TABLE 2 Expression of Ichnofacies Found in Fluvial Deposits—Cont'd

Ichnofacies	Description	Included ichnogenera	Implications
Skolithos	Vertical cylindrical burrows and more rarely Y, or U-shaped vertical burrows. Other associated trace-fossil types may include bilobed burrows/trails, horizontal burrows, simple horizontal trails, and meniscate burrows. Common monospecific or low-ichnodiversity assemblages (maximum 4 trace-fossil types). Dense ichnofabrics occur, composed of vertical burrows.	Characteristic: *Skolithos, Polykladichnus, Arenicolites,* "escape traces" Accessory: *Cruziana, Archaeonassa* (=*Aulichnites*), *Palaeophycus, Taenidium, Planolites*	Cross-bedded to planar-laminated sandstone and ripple cross laminated sandstone and siltstone. Typically on top of fluvial channel bars or crevasse splays. Possible origins are subaerial exposure by burrowing insects or colonization of high-energy bars. Also developed on sandy substrates of crevasse-splay origin.
Grallator	Dominantly bird tracks and bipedal, tri- to tetradactyl tracks with thick digits (B 3–4 tk). Other accessory trace-fossil morphologies are (in order of abundance) simple horizontal burrows, simple burrows/trails, arthropod trackways, bipedal tridactyl tracks with short stout digits, burrow systems, meniscate burrows, and a high variety of vertebrate tracks. Commonly a mixture of tetrapod tracks and invertebrate traces, where the former are dominant. Medium–high diversity ichnofaunas (5–8 trace-fossil types) with bipedal tridactyl tracks usually dominant.	Characteristic: *Grallator, Anchisauripus, Kayentapus, Eubrontes, Gruipeda, Dinehichnus, Avipeda, Fuscinapeda, Sarjeantopodus, Aquatilavipes, Ignotormis, Barrosopus, Alaripeda* Accessory: *Hadrosauropus, Anomoepus, Parabrontopodus, Dicynodontipus, Macrauchenichnus, Characichnos, Palaeophycus, Planolites, Diplichnites, Spongeliomorpha, Scoyenia, Taenidium, Lockeia, Rusophycus*	Cross-bedded sandstones, ripple cross-laminated sandstones and siltstones, massive to parallel-laminated mudstones. Typically present on top of exhumed fluvial bars, sandy crevasse splays, or floodbasin ponds.

Batrachichnus	Dominantly quadrupedal trackways with tetradactyl manus, pentadactyl pes and low pace angulation (Q 4/5); quadrupedal heteropodous trackways with pentadactyl manus and pes, showing long and slender digit impressions with claw traces (Q 5/5 sl); and quadrupedal heteropodous trackways, with pentadactyl plantigrade imprints (Q 5/5 tk); and/or chirotheriid tracks. Accessory trace-fossil types are arthropod trackways, meniscate burrows, simple trails, horizontal burrows, bilobed burrows/trails, trough-like reliefs, and diverse swim traces.	Characteristic: *Batrachichnus, Amphisauropus, Ichniotherium, Limnopus, Dimetropus, Pseudobradypus, Hylopus, Dromillopus, Cursipes, Chirotherium, Isochirotherium, Brachychirotherium, Synaptichnium* Accessory: *Characichnos, Capitosauroides, Scoyenia, Taenidium, Diplopodichnus, Paleohelcura, Diplichnites, Cruziana, Rusophycus, Helminthoidichnites, Palaeophycus.*	Thin, parallel-laminated or ripple cross-laminated sandstone or siltstone interbedded with mudstone, common mud cracks, and raindrop imprints. Typically occurring on distal crevasse-splay deposits, also present in channel-fill deposits and on top of channel bars.

Modified from Buatois and Mángano (1995, 2007), Genise et al. (2000, 2010), and Hunt and Lucas (2007).

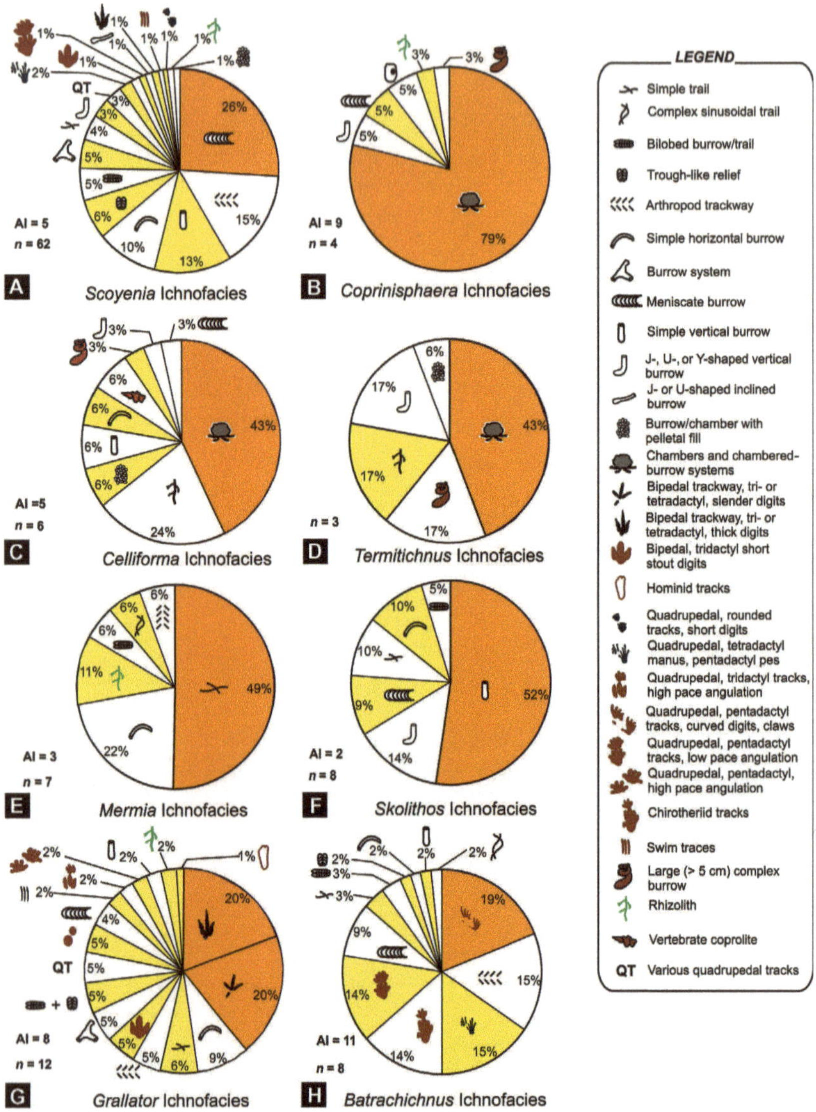

FIGURE 1 Pie diagrams summarizing the relative proportion of trace-fossil types in some trace-fossil assemblages representative of contrasting ichnofacies (see also Table 2). AI, average ichnodiversity of individual assemblages; *n*, number of trace-fossil assemblages. (A) *Scoyenia* Ichnofacies. (B) *Coprinisphaera* Ichnofacies. (C) *Celliforma* Ichnofacies. (D) *Termitichnus* Ichnofacies. (E) *Mermia* Ichnofacies. (F) *Skolithos* Ichnofacies. (G) *Grallator* Ichnofacies. (H) *Batrachichnus* Ichnofacies. In the *Scoyenia* Ichnofacies, QT (quadrupedal trackways) includes the following trace-fossil types: chirotheriid tracks, sauropod tracks, Q 5/5 sl, Q 3/3 high pa, and Q triangular 4/4. In the *Grallator* Ichnofacies, QT: Q 4/3, Q 4/5, Q 2/2, Q 3/3 st (see Table 1).

FIGURE 2 Some representative trace fossils of the *Scoyenia* Ichnofacies. (A) *Scoyenia* (meniscate burrow, black arrows) cut by a mud-crack fill (white arrow). Preserved as positive hyporelief. (B) Two specimens (arrows) of *Siskemia* (arthropod trackway), preserved in negative epirelief. (C) *Spongeliomorpha* (burrow system, arrows) in positive hyporelief. (D) *Rusophycus* (trough-like relief) indicated by arrow, negative epirelief. (E) Bedding plane with mud-draped ripples (white arrows), mud cracks (black arrows), simple horizontal burrows (shb), possible *Scoyenia* (Sc) and *Spongeliomorpha* (Sp), and *Capayanichnus* (Ca). Note cross-cutting relationships between the trace fossils and sedimentary structures. (F) Two specimens (arrows) of the ichnogenus *Taenidium* (meniscate burrow), endorelief. The examples are from the Permian Carapacha Formation, La Pampa, Argentina (A, B, D); the Early Jurassic Santo Domingo Formation, La Rioja Province, Argentina (C and F); and the Oligocene–Miocene Vinchina Formation (E).

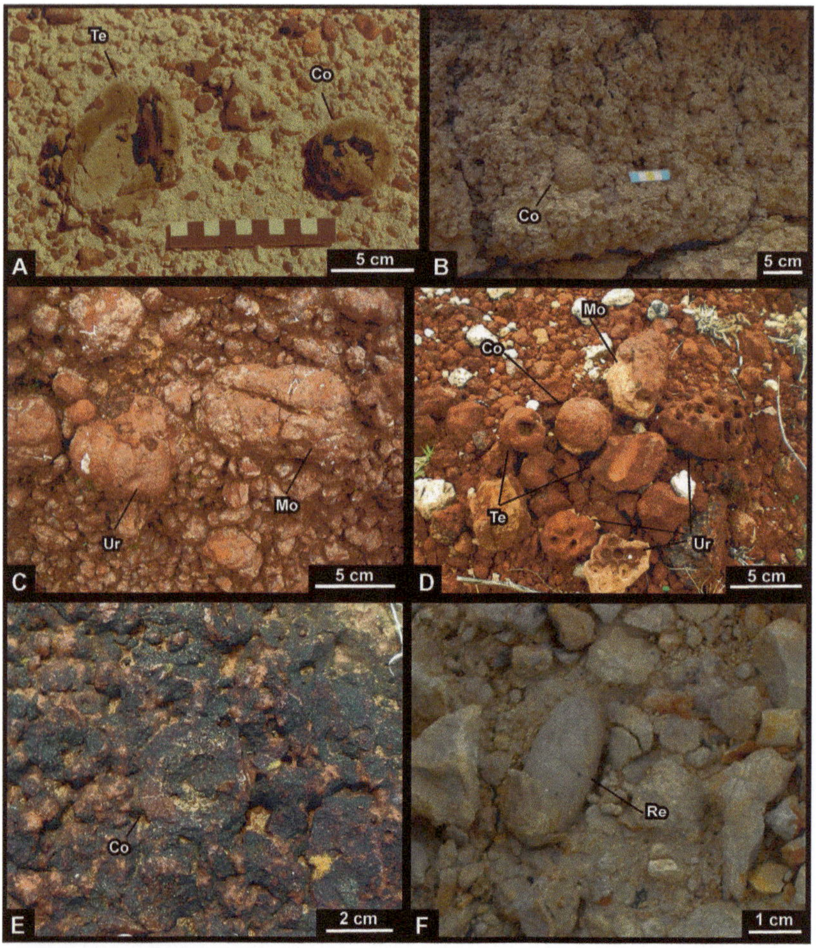

FIGURE 3 Examples of characteristic trace fossils of the *Coprinisphaera* (A–E) and *Celliforma* Ichnofacies (F). (A) *Teisseirei* and *Coprinisphaera*. (B) *Coprinisphaera*. (C) *Uruguay* and *Monesichnus*. (D) *Coprinisphaera*, *Monesichnus*, *Teisseirei*, and many *Uruguay* specimens. (E) Isolated *Coprinisphaera* in cross section. (F) *Rebuffoichnus*. The examples are from paleosol profiles of the Oligocene Upper Puesto Almendra Member of the Sarmiento Formation in Chubut province, Argentina (A and B); from lateritic nodular horizons (C and D) and lateritic hardpan (E) of the Eocene Asencio Formation in Uruguay; and from the Late Cretaceous Bajo Barreal Formation in Chubut province, Argentina (F). Co, *Coprinisphaera*; Mo, *Monesichnus*; Re, *Rebuffoichnus*; Te, *Teisseirei*; Ur, *Uruguay*.

1997), and the Eocene–Miocene Sarmiento Formation of Argentina (e.g., Bellosi et al., 2010). These assemblages display an absolute dominance of chambers and chambered burrow systems (e.g., *Coprinisphaera*, *Teisseirei*, *Eatonichnus*, *Rebuffoichnus*, *Feoichnus*, some ichnospecies of *Cellicalichnus*, *Celliforma*, *Palmiraichnus*, *Uruguay*, and rare *Krausichnus*; see Fig. 3A–E), and a secondary participation of J-, U-, or Y-shaped vertical burrows

(*Loloichnus*), meniscate burrows (*Beaconites*), large vertebrate burrows, perforations in the wall of other trace fossils (*Lazaichnus*, *Tombownichnus*), and rhizoliths (Fig. 1B). These assemblages ($n=4$) display a moderate average ichnodiversity (9 trace-fossil morphologies) and are essentially produced by insects, crayfishes, plants, and tetrapods. The *Coprinisphaera* Ichnofacies characterizes paleosols developed in paleoecosystems of herbaceous communities, particularly grass-dominated habitats (Genise et al., 2000).

3.3 *Celliforma* Ichnofacies

The recently proposed *Celliforma* Ichnofacies is found in calcretes and carbonate-rich paleosols developed in palustrine carbonates. It is typified by a recurrent association of insect trace fossils produced by bees and wasps, particularly *Celliforma*, *Cellicalichnus*, and *Rebuffoichnus* (Genise et al., 2010). At least six fluvial trace-fossil assemblages can be assigned to the *Celliforma* Ichnofacies (e.g., Martin and Varricchio, 2011; Retallack, 1984). As with the *Coprinisphaera* Ichnofacies, the dominant trace-fossil morphology is formed by chambers and chambered burrow systems (e.g., *Cellicallichnus*, *Celliforma*, *Rebuffoichnus*, *Teisseirei*, *Fictovichnus*, *Pallichnus*; see Fig. 3F). Secondary components are rhizoliths (including shallow networks and taproots of medium-sized plants and trees); burrows and chambers with pelletal fill (*Edaphichnium*); J-, U-, or Y-shaped burrows (*Macanopsis*); vertebrate coprolites; and large vertebrate burrows (Fig. 1C). These assemblages can be distinguished from those assigned to the *Coprinisphaera* Ichnofacies by the dominance of *Celliforma*, *Rebuffoichnus*, and *Cellicalichnus*, and by a lower ichnodiversity. The *Celliforma* Ichnofacies would be indicative of a drier climate and lower vegetation coverage than those represented by the *Coprinisphaera* Ichnofacies, particularly for scrubs and woodlands (Genise et al., 2010).

3.4 *Termitichnus* Ichnofacies

The original proposal of the *Termitichnus* Ichnofacies (Smith et al., 1993) was revised by Genise et al. (2000), who suggested it to be retained for assemblages dominated by termite nests in paleosols of closed forest ecosystems under warm and humid conditions. Although this ichnofacies is potentially distinctive, only a few cases are known at present, including three from fluvial settings (Fig. 1D). The best known case study concerns the Oligocene–Miocene Jebel Qatrani Formation of Egypt (Bown, 1982; Genise and Bown, 1994), and other possible examples are the Plio–Pleistocene Upper Siwalik Subgroup of India ("ichnofacies D" of Tandon and Naug, 1984) and the Pleistocene–Holocene Orange River deposits of South Africa (Miller and Mason, 2000). These assemblages also display a dominance of chambers and chambered burrow systems (*Termitichnus*, *Krausichnus*, *Vondrichnus*, *Fleaglellius*, *Masrichnus*; see Fig. 4A and B), J-, U-, or Y-shaped vertical burrows (*Macanopsis*), rhizoliths, large vertebrate burrows, and burrows with pelletal fill (*Edaphichnium*).

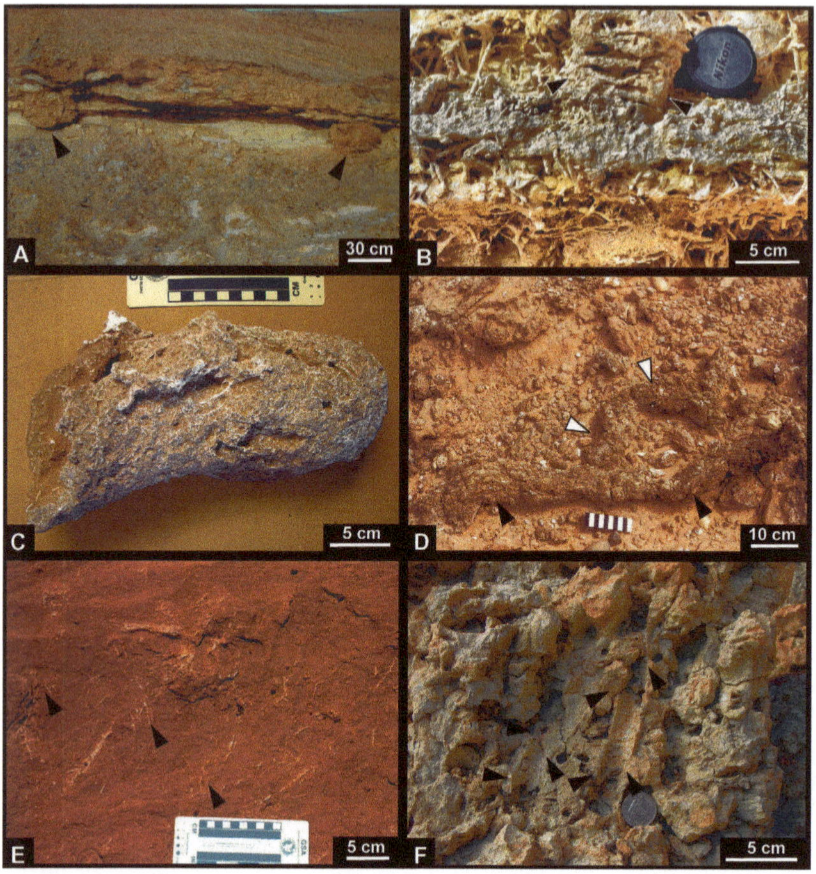

FIGURE 4 Examples of trace fossils from the *Termitichnus* Ichnofacies (A and B) and other trace fossils from paleosol assemblages of uncertain ichnofacies (C–F). (A) Two *Termitichnus* specimens below a cornice (arrows). (B) Dense rhizoliths of different diameters and *Fleaglellius* (arrows). (C) Rounded end of a large vertebrate burrow cast with abundant rhizoliths. (D) Subhorizontal branched siliceous rhizoliths (arrows). (E) Profuse small siliceous rhizoliths (some arrowed) in a vertical outcrop face. (F) Intensely bioturbated ichnofabric with at least two discrete *Loloichnus* specimens (J-, Y-shaped vertical burrow; arrows). The examples are from paleosols of the Oligocene Jebel Qatrani Formation in Egypt (A, B); from a silicified calcrete and silcrete horizons of the Middle Triassic Tarjados Formation in La Rioja province, Argentina (C–E); and from a volcaniclastic paleosol of the Early Cretaceous Punta del Barco Formation in Santa Cruz province, Argentina (F).

3.5 *Mermia* Ichnofacies

A few trace-fossil assemblages in fluvial deposits can be assigned to the *Mermia* Ichnofacies (Buatois and Mángano, 2002, 2004) and are represented by seven cases (Fig. 1E). This ichnofacies is represented by low-diversity associations of simple burrows/trails (Fig. 5A and B), complex sinusoidal trails, arthropod trackways, and/or horizontal burrows, almost exclusively found in

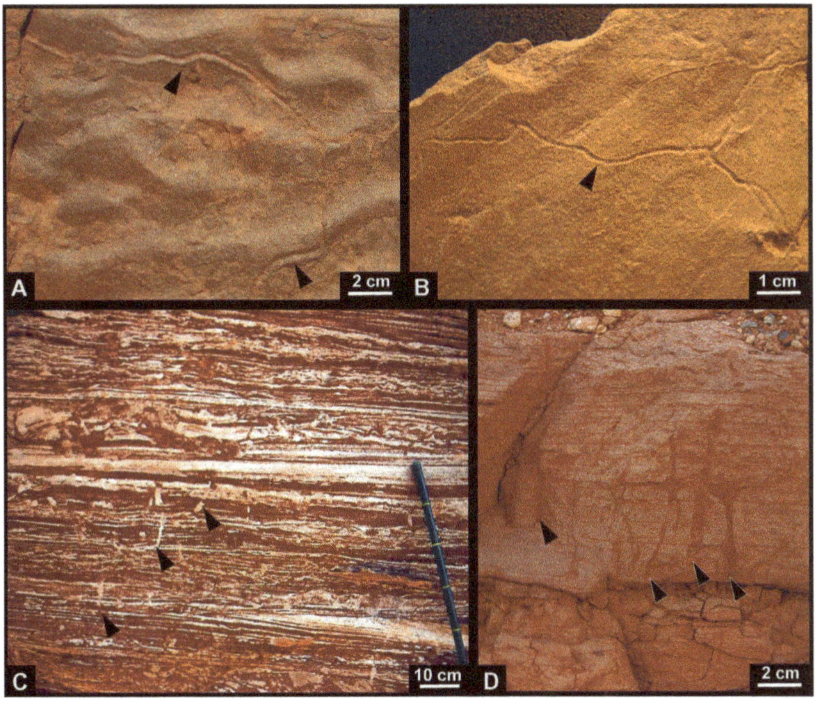

FIGURE 5 Examples of trace fossils from the *Mermia* Ichnofacies (A and B) and *Skolithos* Ichnofacies (C and D). (A) *Archaeonassa* (simple horizontal trail) on top of a rippled-bedding plane. (B) *Helminthoidichnites* (simple horizontal trail) preserved in negative epirelief. (C) and (D) Several *Skolithos* specimens (simple vertical burrow) in cross section (some of the burrows are indicated by arrows). Photographs are from the Oligocene Vinchina Formation in La Rioja province, Argentina (A); from the Permian Carapacha Formation in La Pampa province, Argentina (B); from the Middle Triassic Tarjados Formation in La Rioja province, Argentina (C); and from the Early Jurassic Santo Domingo Formation in La Rioja province, Argentina (D).

floodbasin-pond facies. These assemblages are typically subaqueous and reflect simple feeding or grazing strategies in temporary water bodies.

3.6 *Skolithos* Ichnofacies

The *Skolithos* Ichnofacies (originally defined in the marine realm) has been tentatively recognized in fluvial channel-belt deposits and high-energy lacustrine settings (Buatois and Mángano, 1998, 2004, 2007; Melchor et al., 2003). These ichnofaunas are commonly monospecific, dominated by simple vertical burrows and may result in ichnofabrics with a high degree of bioturbation (piperock). This ichnofacies is recognized by the dominance of simple vertical burrows (Fig. 5C and D) and Y, or U-shaped vertical burrows, although bilobed burrows/trails, horizontal burrows, simple horizontal trails, and meniscate burrows may be present (Fig. 1F). These assemblages were interpreted as reflecting colonization of active channels, and the vertical burrows were tentatively considered as domiciles of suspension feeders (Buatois and Mángano, 2004, 2007). Alternatively, similar assemblages may be interpreted as subaerial in origin, implying channel abandonment or exploitation of sandy crevasse deposits (Netto, 2007; Schlirf et al., 2001). The latter interpretation is supported by the observation of similar dense vertical to oblique burrows produced by staphylinid beetles in modern abandoned anabranches and braided bars during low-water stages of the Platte and Loup rivers, Nebraska (Smith and Heim, 1971).

3.7 *Grallator* Ichnofacies

The recently proposed *Grallator* Ichnofacies (Hunt and Lucas, 2007) can be distinguished in fluvial facies (e.g., Gierliński et al., 2004; Johnson, 1986; Krapovickas et al., 2009), although the typical occurrence is from lakeshore facies (Fig. 1G). This ichnofacies is typified by the abundance of bipedal tridactyl or tetradactyl tracks of avian and non-avian theropods (bird tracks and bipedal tri-tetradactyl trackways with thick digits; see Fig. 6A and B). Other undiagnostic components are bipedal trackways with thick short digit imprints, simple horizontal burrows, simple trails, arthropod trackways, burrow systems, meniscate burrows, and swim traces. The *Grallator* Ichnofacies was proposed only for tetrapod tracks, but here we include the associated invertebrate trace fossils in order to provide a more complete picture of the animal community. In consequence, the examples of the *Grallator* Ichnofacies can be envisaged as sharing components with the *Scoyenia* Ichnofacies (meniscate burrows, arthropod trackways). The ichnofaunas are of moderate to high diversity and almost invariably contain invertebrate trace fossils (which are less diverse than vertebrate trace fossils) in addition to tetrapod tracks. The observed ichnodiversity ranges from 3 to 13 and averages 8 trace-fossil morphologies.

3.8 *Batrachichnus* Ichnofacies

The *Batrachichnus* Ichnofacies was defined by Hunt and Lucas (2007) as ichnofaunas in which the majority of tracks are of quadrupedal carnivores with moderate to high diversity (four to eight ichnogenera). The authors did not propose morphological attributes that can be used for distinguishing carnivore tracks. In consequence, the recognition of this ichnofacies relies essentially on the interpretation of the possible producer of the tracks and/or on the presence of some of the typical ichnogenera (e.g., *Batrachichnus*, *Amphisauropus*, *Ichniotherium*, chirotheriid tracks; see Fig. 6C–E). Considering the examples mentioned by Hunt and Lucas (2006, 2007), at least eight trace-fossil assemblages from fluvial deposits belong to this ichnofacies (Fuglewicz et al., 1990; Lucas et al., 2004; Mack et al., 2003; Melchor and Sarjeant, 2004). Most of these assemblages also contain invertebrate trace fossils that are typical of the *Scoyenia* Ichnofacies (mainly meniscate burrows and various arthropod

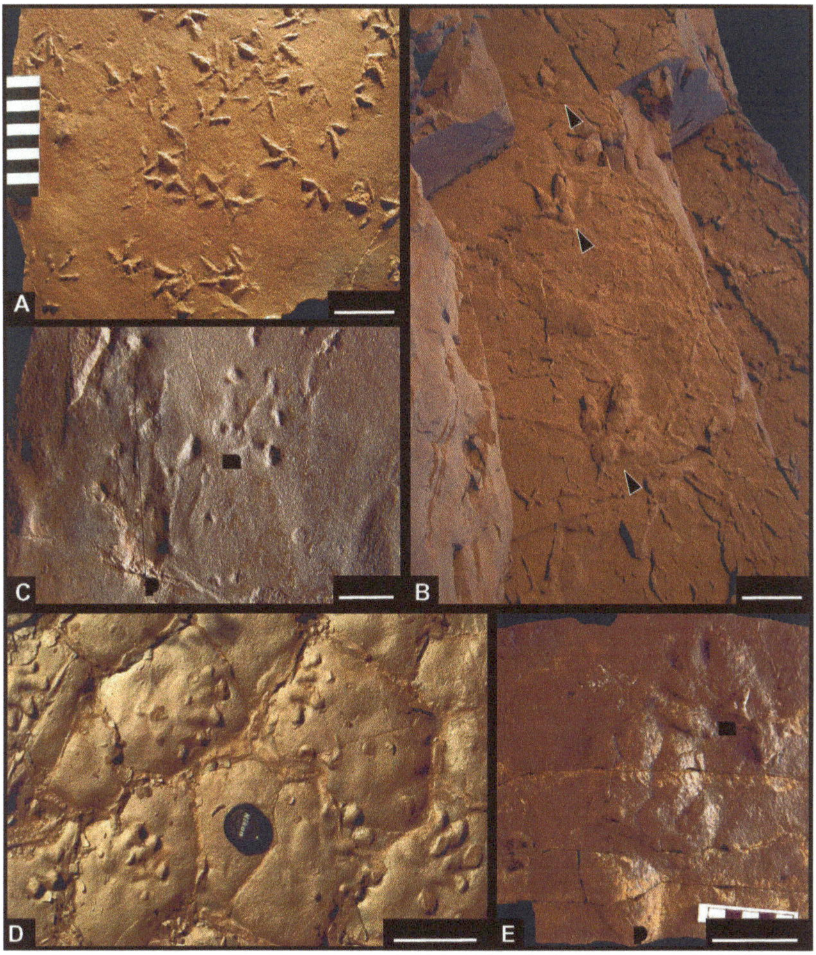

FIGURE 6 Trace fossils that can be assigned to the *Grallator* (A and B), *Batrachichnus* (C and E), and *Brontopodus* Ichnofacies (D). (A) Abundant bird-like tracks (B 3–4 sl) of the ichnogenus *Gruipeda* preserved as positive hyporeliefs. (B) Bipedal trackway composed of large tridactyl footprints (B 3–4 tk, arrows), preserved as positive hyporeliefs. (C) Manus-pes set of the ichnogenus *Batrachichnus* (Q 4/5), negative epirelief. (D) Several *Ichniotherium* footprints (Q 5/5 tk) in a mud-cracked sandstone sole (positive hyporelief). (E) Manus-pes set of the ichnogenus *Brachychirotherium* (chirotheriid tracks), preserved as positive hyporelief. Photographs are from the Early Jurassic Santo Domingo Formation, La Rioja province, Argentina (A); from the Oligocene Vinchina Formation in La Rioja province, Argentina (B); from the Permian Carapacha Formation in La Pampa province, Argentina (C); from the Permian Tambach Sandstone, Germany (D); and from the Middle Triassic Tarjados Formation, La Rioja province, Argentina (E). m, manus; p, pes.

trackways). The *Batrachichnus* Ichnofacies, as represented in fluvial deposits (Fig. 1H), is dominated by different quadrupedal trackways (Q 4/5, Q 5/5 h, Q 5/5 large, or chiroteriid tracks; see Table 1). Associated undiagnostic trace fossils include arthropod trackways, meniscate burrows, simple horizontal trails, horizontal burrows, bilobed burrows/trails, trough-like reliefs, and diverse swim-trace fossils. The overall diversity of trace-fossil types is moderate to high (average 8, range from 4 to 14). Quadrupedal trackways account for most of this diversity (average 5, range 4–7).

3.9 Additional Vertebrate Ichnofacies

Other proposed vertebrate ichnofacies that are represented by fluvial trace-fossil assemblages, but include a few case studies, are the *Characichnos* or *Hatcherichnus*, *Chelonipus*, and *Brontopodus* "ichnofacies". The *Characichnos* "ichnofacies" was proposed to refer to assemblages that contain a majority of vertebrate swim-trace fossils, mostly groups of parallel scratch traces and sinusoidal trails from lacustrine shorelines (Hunt and Lucas, 2007). From a conceptual point of view, a considerable overlap with the *Mermia* Ichnofacies is envisaged, as sinusoidal trace fossils are a diagnostic component of it. At least two of the assemblages of this ichnofacies have been recorded from fluvial facies. One of them is composed of a single manus-pes track impression assigned to *Hatcherichnus sanjuanensis* associated with sauropod tracks on fluvial channel fills of the Jurassic Morrison Formation (Foster and Lockley, 1997; Hunt and Lucas, 2007; Lockley et al., 2010). The other case study is from the Permian Carapacha Formation of Argentina, where swim traces are a minor component of the vertebrate trace-fossil assemblage (Melchor and Sarjeant, 2004). In consequence, its assignment to the *Characichnos* Ichnofacies (Hunt and Lucas, 2006) seems unsupported. At present, the definition and status of the *Hatcherichnus* ichnocoenosis or ichnofacies are unclear (see the discussion in Lockley et al., 2010).

The *Chelonipus* "Ichnofacies" was proposed to distinguish turtle swim traces in fluvial channel deposits on the basis of a few cases of Late Triassic to Cretaceous age (Lockley and Foster, 2006). The only occurrence from inland fluvial deposits described in detail is from the Jurassic Morrison Formation, and the remaining cases in similar facies are from the Late Triassic of Europe (cf. Lockley and Foster, 2006).

The *Brontopodus* Ichnofacies of Hunt and Lucas (2007) was proposed to characterize moderately diverse ichnofaunas with a majority of tracks of terrestrial, quadrupedal herbivores, with a small proportion of terrestrial carnivore tracks. Typical environments are coastal plains, siliciclastic and carbonate-marine shorelines, and lacustrine shorelines. No objective criterion for the distinction of herbivore tracks was proposed. At least five of the case studies of this ichnofacies belong to fluvial deposits and the dominant track types are Q 5/5 sl and Q 5/5 st (Table 1). Three of the case studies are from Permian successions of Italy, Germany, and South Africa (see Hunt and Lucas, 2006). The remaining ichnofaunas are from Cretaceous units of Mongolia and China (Currie et al., 2003; Matsukawa et al., 1995). These ichnofaunas display significant variability in composition and were assigned by Hunt and Lucas (2006, 2007) to three different ichnocoenoses.

The *Heteropolacopros* Ichnofacies is characterized by the presence of microspiral heteropolar coprolites, assigned to xenacanth sharks or lungfish, which occur in Permian–Triassic fluvial redbeds, although a more restricted facies discrimination is not possible at present. Large herbivore (probably produced by dicynodonts) coprolites from swampy environments are the typical components of the *Dicynodontocopros* Ichnofacies (Hunt et al., 2007).

3.10 Potential Ichnofacies from Paleosols

Some trace-fossil assemblages ($n = 40$) recorded essentially in floodplain sediments with low to moderate pedogenesis are dominated by rhizoliths or other root structures (including networks of small roots, medium-sized taproots, and tree-sized taproots; see Fig. 4D and E), vertical J, and Y-shaped burrows (Fig. 4F), large vertebrate burrows (Fig. 4C), meniscate burrows, and burrows with pelletal fill. These components are shared with the *Coprinisphaera* and *Celliforma* ichnofacies but insect trace fossils are absent or rare in these assemblages. The mentioned trace-fossil assemblages cannot be included in any of the recognized ichnofacies, and it is not clear if they meet some criteria for the establishment of a new ichnofacies (temporal and geographic recurrence and definite environmental meaning). Instead, there seems to be at least three distinct groups of assemblages that are distinguished by the dominance of large vertebrate burrows and trace fossils attributed to crayfishes and root structures.

The first group is typified by the occurrence of large vertebrate burrows, mostly exhibiting a helical pattern with a terminal chamber (*Daimonelix*) or a low-angle ramp tunnel with a rounded end (that may or may not be enlarged). Associated trace fossils are rhizoliths, meniscate burrows, and rare insect trace fossils (ant nests and possible dung beetle burrows). These examples commonly appear in well-drained soils, with carbonate nodules or nodular horizons, developed in an arid or semiarid climate (e.g., Gobetz and Martin, 2006; Groenewald et al., 2001; Martin and Bennett, 1977; Sidor et al., 2008; Smith, 1987; Smith et al., 2010).

The second group is characterized by ichnogenera assigned to crayfishes (*Loloichnus*, *Camborygma*, *Dagnichnus*, *Katbergia*, and two ichnospecies of *Cellicallichnus*) associated with rhizoliths and meniscate burrows (*Taenidium*, *Beaconites*). Some of the case studies come from paleosols that exhibit evidence for a fluctuating water table and were developed in a warm climate (Bedatou et al., 2008; Genise et al., 2008; Hasiotis and Honey, 2000).

The third group is typified by trace-fossil assemblages almost exclusively composed of root structures, including rhizoliths and stump casts. A comprehensive classification for rhizolith types is available (Klappa, 1980), but a common and uniform descriptive terminology has not been adopted. Rhizoliths have not received an ichnotaxonomic treatment to date, in part because plant trace fossils are not recognized in the International Code of Botanical Nomenclature (Bertling et al., 2006). Bockelie (1994) proposed a morphological classification of rhizoliths in core, and he was able to recognize "root facies" in Jurassic non-marine and marginal-marine successions. The idea of defining a rhizolith ichnofacies is tempting because there are trace-fossil assemblages dominated by root traces that show more recurrence in time and space than any other terrestrial one. The indication of subaerial exposure of this ichnofacies (except for root structures of swampy vegetation), and other inferences that can be made from the type of roots, will be very valuable.

4. APPLICATION OF ICHNOFACIES AND ICHNOFABRICS TO THE INTERPRETATION OF FLUVIAL SEDIMENTARY FACIES

The identification of key elements of an ichnofacies may help with the sedimentological and paleoenvironmental interpretation of a lithofacies. Some of the ichnofacies recognized in fluvial deposits (particularly archetypal invertebrate ichnofacies) are useful for distinguishing processes or subenvironments. The *Scoyenia* Ichnofacies develops on a substrate that is periodically submerged and exposed, which is also reflected by the associated physical sedimentary structures (e.g., mud cracks, raindrop imprints, water-level marks, ripple marks, rill structures). The presence of the *Scoyenia* Ichnofacies may be used to delineate channel-bar tops or the margins of flood-basin ponds, including temporary pools associated with crevasse-splay processes. Buatois and Mángano (2002, 2004) proposed that the *Scoyenia* Ichnofacies includes a "predesiccation suite" with structures lacking ornamentation (e.g., *Taenidium*, *Beaconites*) and developed in a soft substrate and a "desiccation suite" typified by trace fossils with striated walls (e.g., *Scoyenia*, *Spongeliomorpha*) that cross-cut the former, suggesting firmer substrates. Predesiccation and desiccation suites can be reliably distinguished only using cross-cutting relationships between different ichnotaxa, whereas criteria such as the relative degree of definition of burrow margins are less useful (Melchor et al., 2006).

The identification of the *Mermia* Ichnofacies, and probably also the *Characichnos* ichnocoenosis, suggests low-energy water-lain substrates that never were exposed to air, a situation that is restricted in fluvial settings to floodbasin ponds. Associated sedimentary structures reflect settling from suspension (parallel-laminated mudstone) or weak currents (current-rippled fine-grained sandstone or siltstone).

The identification of the *Coprinisphaera*, *Celliforma*, and *Termitichnus* ichnofacies is indicative of soil development related to herbaceous communities, scrubs and woodlands, and closed forests, respectively. However, the assignment of a trace-fossil assemblage to any of these ichnofacies requires a detailed ichnotaxonomical assessment. The identification of an insect trace fossil (chambers and chambered burrow systems) is an evidence that the hosting sedimentary deposit was exposed subaerially (Genise, 2004) and, consequently, that it can be recognized as a paleosol (Retallack, 2001).

The *Skolithos* Ichnofacies in fluvial settings can be related to the tops of emergent channel bars or sandy crevasse splays that were colonized by insects. In consequence, it may mark variations in fluvial channel discharge. A similar interpretation can be applied to occurrences of the *Grallator* and *Batrachichnus* ichnofacies on top of channel bars, although these ichnofacies are also present in floodplain settings (crevasse splays and floodbasin ponds). An alternative interpretation for the *Skolithos* Ichnofacies is that it represents colonization of fluvial bars in a high-energy setting.

The *Hatcherichnus* and *Chelonipus* ichnocoenoses may be indicative of permanent subaqueous substrates deposited under moving water (lotic settings), particularly in channel facies. No change in fluvial discharge is implied by the presence of these swim-trace fossils in main channel-fill deposits. The meaning of the *Brontopodus* Ichnofacies for sedimentological analysis of fluvial successions is at present unclear.

Retallack (1990, 2001), Bockelie (1994), Kraus and Hasiotis (2006), and Genise et al. (2011) identified different morphologies of root traces that can be linked to well drained (deep water table), more poorly drained (intermediate water table), and poorly drained (shallow water table) paleosol profiles (see also Cohen, 1982). Different types of root traces would increase the value of a potential rhizolith ichnofacies as a paleoenvironmental indicator.

4.1 Composite Ichnofacies

The recognition of significant participation of elements of two ichnofacies in a specific sedimentary facies may be useful for the refinement of the facies analysis in fluvial successions. These examples can be considered as a mixture of two ichnofacies (e.g., the *Mermia* and *Scoyenia* ichnofacies) or a composite ichnofacies (Buatois and Mángano, 2002, 2004; Keighley and Pickerill, 2003; Kim et al., 2005). Another possible example is the change or overprinting of trace-fossil assemblages that characterize the *Scoyenia* and *Celliforma* ichnofacies with increasing degree of subaerial exposure (Genise et al., 2010). In a floodplain pond, short periods of emergence punctuated by flooding are common. In this scenario, a trace-fossil assemblage typical of the *Scoyenia* Ichnofacies may develop. If the time of subaerial exposure was long enough to provide the well aerated and bare soils that hymenopterans need to nest (Genise, 2004), the previous assemblage might be replaced or overprinted by that typical of the *Celliforma* Ichnofacies (Genise et al., 2010).

These composite or mixed assemblages hold the potential of reflecting subtle environmental changes that commonly cannot be inferred from physical sedimentary structures alone. The recognition of ichnocoenoses potentially representative of composite *Mermia–Scoyenia* Ichnofacies has been documented in modern pointbars and bars in braided channels (Lawfield and Pickerill, 2006; Martin, 2009), which are essentially composed of simple horizontal trails (*Mermia* Ichnofacies representatives) and vertebrate tracks (purported *Scoyenia* Ichnofacies component). The potential of preservation of these ichnocoenoses is low, as they can be destroyed due to desiccation after exposure or fluvial reworking during periods of peak-flow discharge. A similar trace-fossil assemblage was described from Oligocene fluvial ponds of Switzerland (Uchman et al., 2004) and type 1 paleosurfaces from Permian crevasse-channel deposits of South Africa (Smith, 1993). In these cases, the vertebrate tracks (purported *Scoyenia* component) are best compared with a vertebrate ichnofacies (like the *Grallator* or *Brontopodus* Ichnofacies), because vertebrate tracks alone are not considered herein as diagnostic of the *Scoyenia* Ichnofacies. The *Mermia* components are represented by simple horizontal trails, arthropod trackways, and horizontal burrows. Additional fossil examples of composite *Mermia–Scoyenia* Ichnofacies from fluvial facies are characterized by a similar participation of simple horizontal trails or burrows and complex sinusoidal trails, the *Mermia* components, associated with meniscate burrows, arthropod trackways, trough-like relief, and simple vertical burrows, the *Scoyenia* components (Głuszek, 1995; Kim et al., 2002; Metz, 1992; ichnocoenosis F of de Gibert and Sáez, 2009; Keighley and Pickerill, 2003). In these cases, the overprinting of significant numbers of components of the *Mermia* and *Scoyenia* ichnofacies suggests transient subaqueous deposits with periodical subaerial exposure.

Another example of mixed or composite assemblages was described by Smith et al. (1993) from the Late Pleistocene of Namibia. These authors identified a soft-sediment assemblage composed of *Planolites*, *Skolithos*, and *Taenidium*, which can be compared to the *Scoyenia* Ichnofacies (*Arenicolites* Ichnofacies of those authors), overprinted by three types of rhizoliths that might indicate the presence of the potential rhizolith ichnofacies (*Termitichnus* Ichnofacies of those authors). A detailed example of the utility of a trace-fossil analysis combined with physical sedimentary structures and observation of modern processes is the description of fluvial paleosurfaces from the Permian Beaufort Group by Smith (1993).

4.2 Stratigraphical Applications

The Neogene, dominantly fluvial Vinchina Basin of north-western Argentina provides a good example of contrasting ichnocoenoses found in fluvial and shallow-lacustrine intervals. The most common trace fossils in the basin are vertical I, or J-shaped large burrows with striated walls (ichnogenus *Capayanichnus*) that are recognized in channel-belt and crevasse-splay deposits of a variety of river systems. *Capayanichnus* is commonly accompanied by burrow systems (*Spongeliomorpha*), meniscate burrows (*Scoyenia*, *Taenidium*), and vertebrate trackways, composing assemblages that can be easily compared with the *Scoyenia* Ichnofacies (Melchor et al., 2010). A 65-m-thick interval of shallow-lacustrine, wave-dominated deposits, that are intercalated in fluvial deposits of the upper member of the Vinchina Formation, exhibits a different trace-fossil assemblage dominantly composed of palmate bird tracks (*Phoenicopterichnum*) and bird tracks without web traces, and subordinate trough-like reliefs (*Lockeia*, *Rusophycus*) (Melchor et al., 2012). The lacustrine assemblage can be compared with the *Grallator* Ichnofacies or the shorebird ichnofacies of Lockley et al. (1994).

Another example comes from the Early Permian of New Mexico (Hunt et al., 2005). In this case, different paleoenvironmental settings are represented by ichnocoenosis assigned to the *Batrachichnus* and *Brontopodus* ichnofacies. Three ichnocoenoses are distinguished on the basis of the dominant ichnotaxon, namely *Ichniotherium* in inland/distal alluvial-fan settings, *Amphisauropus* in alluvial-plain settings, and *Batrachichnus* and *Dimetropus* in coastal/tidal-flat settings (Hunt et al., 2005).

Recognition of trace fossils typical of the *Scoyenia* and other ichnofacies has been used in large-scale stratigraphic studies since its original proposal by Seilacher (1967). In an example that involves fluvial facies, Siggerud and Steel (1999) recognized a number of marine and continental ichnofabrics in an Eocene fan-delta complex from Spain. The identification of an ichnofabric dominated by meniscate (*Scoyenia*, *Taenidium*) and vertical burrows (*Skolithos*) in red siltstone, mudstone, and fine-grained sandstone helped in the identification of sequence boundaries.

The application of ichnology in sequence stratigraphy of fluvial successions is still in its infancy, although a general model has been proposed (Buatois and Mángano, 2004, 2007). According to this model, the distribution of trace fossils in fluvial successions most likely reflects changes at the scale of depositional systems which, in turn, may be related to the stacking of system tracts. During the early lowstand systems tract, widespread erosion, high energy, and high sedimentation rates are conducive to channel amalgamation and extensive reworking of fluvial deposits, conditions that are detrimental for the formation and/or preservation of biogenic structures in fluvial settings. Interfluve areas may be delineated by paleosols, containing root and insect trace fossils, and potentially represented by the *Coprinisphaera*, *Celliforma*, and *Termitichnus* ichnofacies. During the late lowstand, higher accommodation leads to the presence of increasingly isolated fluvial channels encased in fine-grained material. The development of widespread overbank areas tends to increase the preservation potential of biogenic structures, which are typically represented by the *Scoyenia* Ichnofacies and associated vertebrate ichnofacies, namely, the *Grallator* and the *Batrachichnus* ichnofacies. Stagnant water bodies (e.g., marshes, floodbasins) are widespread if the rate of accommodation exceeds the sediment supply, signaling transgressive conditions; the *Mermia* Ichnofacies may develop at this time. Shanley et al. (1992) noted that tidal influence and brackish-water conditions may extend for tens of kilometers inland during a transgression. If this is the case, fluvial deposits may become interbedded with marginal-marine strata displaying ichnological evidence of marine influence (e.g., *Ophiomorpha*, *Teichichnus*, logs containing *Teredolites*). During the subsequent highstand, increased sediment supply and decreased fluvial accommodation are conducive to deltaic progradation and increased establishment of channel bodies, leading to the reestablishment of the *Scoyenia* Ichnofacies.

4.3 Ichnofabric Studies of Fluvial Successions

In recent years, the ichnofabric approach has slowly expanded from marine case studies to continental examples, including fluvial successions. Much progress has been attained in the field of paleosol ichnofabrics, and a conceptual and methodological framework for their analysis is now available (Bedatou et al., 2009; Genise et al., 2004). Bedatou et al. (2009) described four recurrent crayfish and earthworm-dominated complex paleosol ichnofabrics from Late Jurassic to Early Cretaceous volcaniclastic deposits of Patagonia. In comparison with paleosol ichnofabrics, those subaquatic ichnofabrics from channels, crevasse splays, and floodbasins are much simpler. The most typical fluvial ichnofabric consists of meniscate burrows (typically *Beaconites* or *Taenidium*) in abandoned channel and overbank deposits. Morrissey and Braddy (2004) provided a detailed study in the Devonian of Wales and suggested that this ichnofabric records colonization of subaerially exposed sediment as a response to seasonal desiccation, with animals excavating the substrate to aestivate or molt at water-table level. Buatois et al. (2007) documented this ichnofabric in channel and crevasse-splay sandstone and floodplain mudstone, the latter being pervasively bioturbated. In this case, the depth and intensity of bioturbation of the main-channel and crevasse sandstones are a reflection of the time between depositional events (i.e., the colonization window). Main-channel and crevasse sandstones underlying thick packages of bioturbated overbank mudstone are thoroughly bioturbated, as a response to prolonged periods of low-energy sediment fallout between crevassing events. On the other hand, the lowest degree of bioturbation is detected in amalgamated channel sandstones underlying thin intervals of overbank mudstone, revealing short colonization windows due to high-frequency depositional episodes. The ichnofabric of meniscate burrows has been recognized in both outcrop and core (Fig. 7A–C).

5. DEPOSITIONAL ENVIRONMENTS

In this section, we will first examine the broad differences between channel-belt and floodplain trace-fossil assemblages and discuss the possible reasons for this contrast. We will then examine the composition of trace-fossil assemblages of channel-belt and floodplain subenvironments, and different river types.

5.1 Comparison of Channel-Belt and Floodplain Trace-Fossil Assemblages

The existing information on trace-fossil assemblages of fluvial channels and floodplain deposits suggests a dominance of very simple trace-fossil types, mainly of invertebrate origin (Fig. 8). This dominance of invertebrate trace fossils may be a reflection of greater diversity of invertebrate tracemakers. In both settings, the trace-fossil types are essentially similar: more than half of the recorded trace-fossil types are (in approximate order of abundance) meniscate

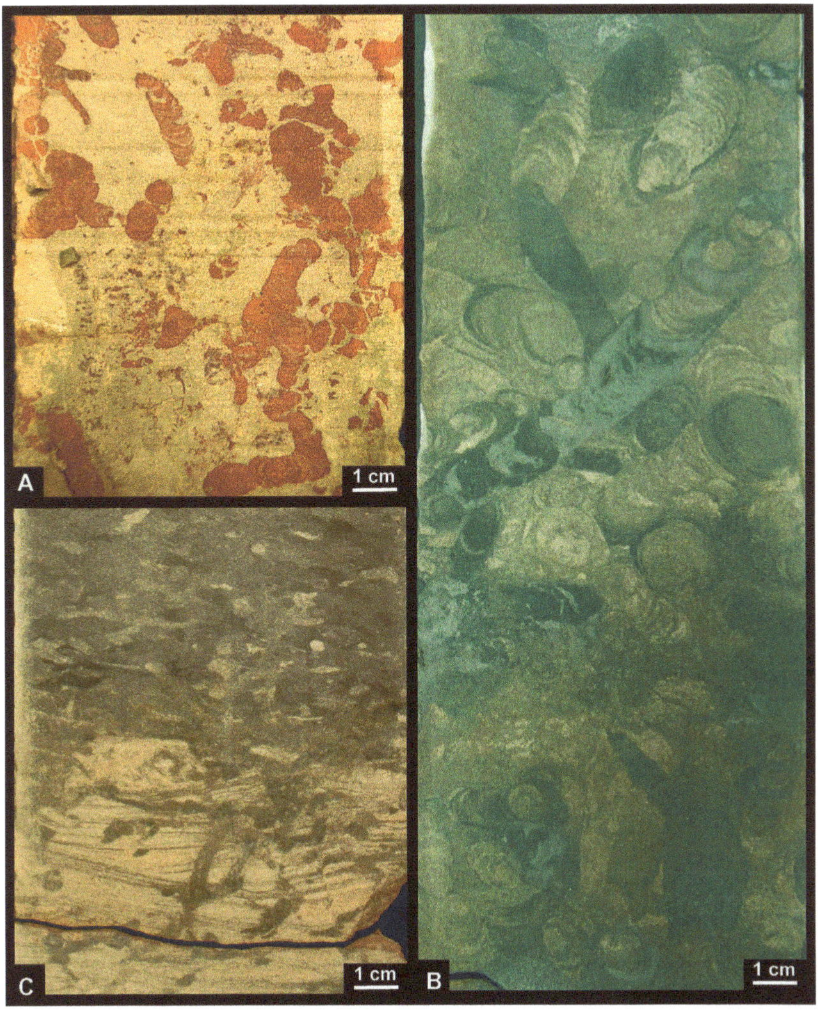

FIGURE 7 Core expression of the *Taenidium* ichnofabric in fluvial channels of subaerial delta-plain deposits, Oligocene–Middle Miocene, Guafita Formation, Apure, Venezuela. (A) and (B) Meniscate trace fossils in floodplain siltstone. (C) Crevasse-splay sandstone (lower interval) and floodplain siltstone (upper interval), displaying colonization from the top of the sandstone layer into the crevasse-splay deposit and totally reworked floodplain deposits.

burrows, arthropod trackways, horizontal burrows, simple vertical burrows, simple horizontal trails, J-, U-, or Y-shaped burrows, rhizoliths, bipedal tri- or tetradactyl footprints with thick digits, bilobed burrows/trails, and trough-like reliefs (Fig. 8A). It is not clear if the dominance of invertebrate trace fossils is a consequence of a bias in observation or if it reflects actual differences in biodiversity or abundance of tracemakers and/or contrasting preservation of invertebrate and vertebrate trace fossils. The major differences between the analyzed assemblages from channel-belt and floodplain settings are in abundance

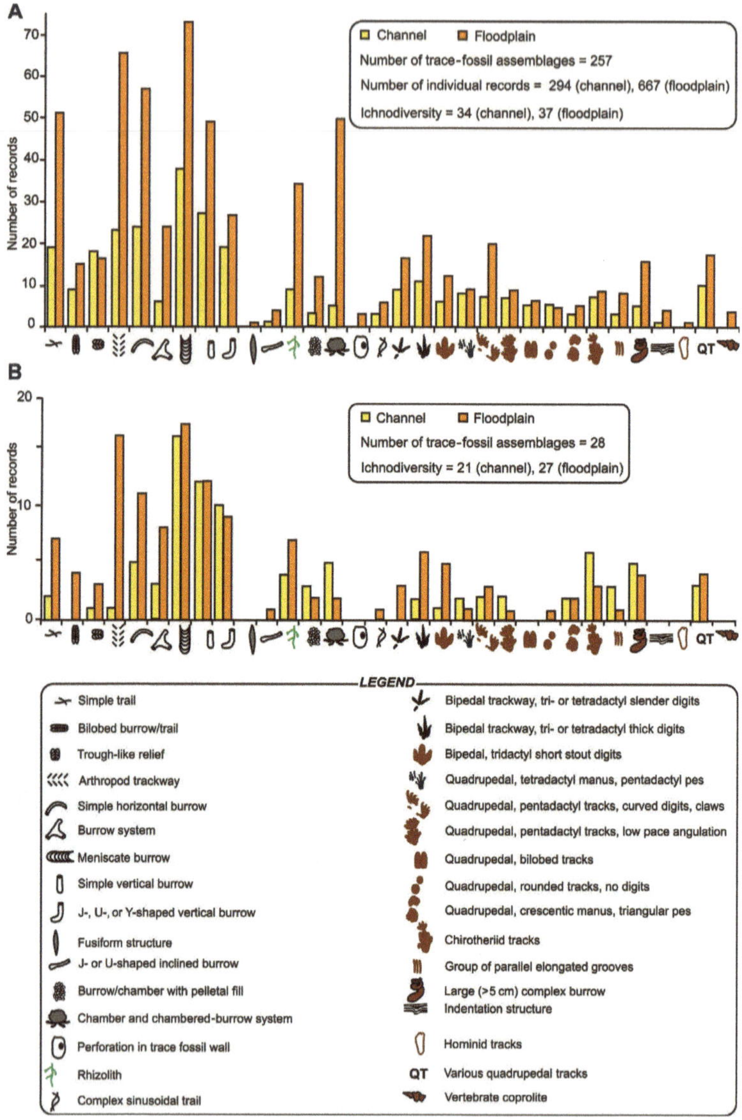

FIGURE 8 Histograms showing the participation of main trace-fossil types in channel and floodplain deposits. (A) All analyzed assemblages. (B) Only including case studies with trace-fossil assemblages from both channel and floodplain facies.

and diversity of trace-fossil types. In floodplain settings, trace fossils are twice as abundant as those found in channel bars and channel fills. Buatois and Mángano (2004, 2007, 2011) noted that the most diverse and abundant trace-fossil assemblages are those of overbank settings, although current analysis suggests that the abundance of trace fossils from channels is larger than previously perceived. Overall, there is a similar number of trace-fossil types that can appear in floodplain facies as in channel facies (Fig. 8A). When considering individual trace-fossil assemblages, it is noteworthy that the majority come from floodplain deposits (175 trace-fossil assemblages against 82 from channels), probably reflecting contrasting preservational conditions. For example, Melchor et al. (2010) found eight trace-fossil assemblages in channel-belt deposits and 16 in floodplain deposits from the more than 5-km-thick dominantly fluvial succession of the Neogene Vinchina Formation of north-western Argentina. The bulk of channel deposits is typically devoid of trace fossils, and most of them are concentrated on particular surfaces and stratigraphic intervals. In fluvial channels, the more common trace-fossil types are meniscate burrows and simple vertical burrows, followed by arthropod trackways, J-, Y-, or U-shaped vertical burrows, simple horizontal burrows, trough-like reliefs, and simple horizontal trails with a subordinate participation of a variety of vertebrate trace fossils. Vertebrate traces are mostly bird-like bipedal trackways and trackways of tri-tetradactyl bipeds with thick digits, and several quadrupedal trackways (the more common are Q 4/5, Q 5/5 sl, Q 5/5 st, chirotheriid tracks; Fig. 8A). In spite of the relatively large variety of trace-fossil types that may appear in a fluvial channel, the average ichnodiversity of individual assemblages from channel deposits is fairly similar to that found in floodplain deposits (four *versus* five trace-fossil morphologies, respectively). About one-third of the trace-fossil assemblages from channel-belt deposits are composed of one or two trace-fossil morphologies. Some of these cases contain either only discrete meniscate burrows (e.g., Graham and Pollard, 1982; Hubert and Dutcher, 2010; Morrissey and Braddy, 2004; Pollard, 1976; Squires and Advocate, 1984) or a highly bioturbated ichnofabric composed of meniscate burrows (Buatois et al., 2007); other cases are represented by simple horizontal and/or vertical burrows (Gibert and Sáez, 2009; Kim et al., 2002; Sarkar and Chaudhuri, 1992) or trackways of bipedal or quadrupedal animals (e.g., Chiappe et al., 2004; Difley and Ekdale, 2002; Platt and Meyer, 1991). A particular case of a low-diversity assemblage from channel facies was described by Cohen (1982) from the Pleistocene of Kenya, where vertical and oblique rhizoliths dominate in multistory ephemeral channels.

Trace-fossil assemblages that display a high ichnodiversity (up to 11 trace-fossil morphologies) in fluvial channel belts are not unusual. Most of them have only few, simple invertebrate traces (simple horizontal trails, meniscate burrows, arthropod trackways) and a more diverse group of tetrapod tracks. Between them, we can distinguish a conspicuous group of Late Paleozoic ichnofaunas that exhibits a dominance of Q 4/5, Q 5/5 sl, and Q 5/5 st (e.g., Mack et al., 2003; Van Allen et al., 2005). Additional case studies are known from the Lower Jurassic of Lesotho (Smith et al., 2009) and from the Miocene of Argentina (Krapovickas et al., 2009). The South African example was recorded from outstanding exposures of low-angle accretion surfaces on pointbar deposits from a meandering river system. Although a revision of the tetrapod ichnotaxa is pending, the authors described simple horizontal trails (*Cochlichnus*, "*Hormosiroidea meandrica*") and meniscate burrows (*Taenidium*), accompanied by five tetrapod-track types (total ichnodiversity = 8) from channel-bar surfaces. The tetrapod tracks are assigned to bipedal tri-tetradactyl tracks with thick digits (*Neotrisauropodus*), bipedal tridactyl tracks with stout digits (*Anomoepus*, *Moyenisauropus*), quadrupedal pentadactyl digitigrade tracks (*Episcopopus*), and chirotheriid tracks. In this case, an analysis of the best preserved trackways helped in the reconstruction of substrate consistency and of the paleolandscape (Smith et al., 2009). The example from the Miocene of Argentina was recorded from the top of sand bars in an anastomosed dryland river. The ichnofauna (ichnodiversity = 10) is dominated by eight types of avian and mammalian tracks (many of them left in open nomenclature) and subordinate simple trails (*Helminthopsis*) and horizontal burrows (*Palaeophycus*). Vertebrate tracks include three bird-like footprint types (including *Fuscinapeda*) and five mammal footprint types (including *Macrauchenichnus* and *Venatoripes*). Krapovickas et al. (2009) emphasized the importance of seasonal flooding events for preservation of tetrapod tracks on bars of anastomosed fluvial systems in a semiarid climate.

A comparison between trace-fossil assemblages from floodplain and associated channel-belt deposits from the same lithostratigraphic units (28 case studies) may be more helpful for highlighting the differences between both settings (Fig. 8B). These cases would reflect the contrasting behavior of biological communities under similar physical constraints in a fluvial system. In most of these case studies, a higher (up to six times) ichnodiversity is recorded in overbank facies than in associated channel facies (e.g., D'Alessandro et al., 1987; Kim et al., 2002; Melchor et al., 2006). In a few cases, the ichnodiversity is greater in channel facies (e.g., Aramayo and Bocanegra, 2003; Fuglewicz et al., 1990). MacNaughton and Pickerill (1995) explained the occurrence of a moderately diverse ichnofauna in Triassic braided-fluvial channels from Canada, as a consequence of higher moisture in coarse-grained channel fluvial deposits in a semiarid climate. In this example, there is no record of overbank trace fossils. The selected trace-fossil assemblages from floodplain and channel-belt deposits from the same lithostratigraphic units (Fig. 8B) underscore a similar participation in both facies of meniscate burrows, simple vertical burrows, and J-, Y-, or U-shaped vertical burrows, which seems to contradict previous inferences about the dominance of vertical burrows in channel facies and horizontal burrows in overbank facies (Maulik and Chaudhuri, 1983). However, for some trace fossils, this relationship can apply. The trace fossil *Capayanichnus vinchinensis* from the Neogene of Argentina preferentially occurs in abandoned channel deposits with a dominantly vertical to oblique orientation and in crevasse-splay facies with a prevailing horizontal orientation (Melchor et al., 2010). This particular distribution and burrow orientation was interpreted as reflecting a marked dependence of the producer of the local water table or moist substrates. Some other notorious differences between channel and floodplain in these selected cases are the larger participation of arthropod trackways, and the apparent restriction of bird tracks and complex sinusoidal trails to overbank facies (Fig. 8B).

5.2 Contrasting Trace-Fossil Preservation in Channel-Belt and Overbank Facies

The differences in ichnodiversity and abundance of trace fossils from channel and overbank facies are related to contrasting taphonomic pathways. Channels are the *loci* of strong currents that typically erode previously deposited sediments producing truncation surfaces of different hierarchy that are typical of channel belts. Even in ephemeral rivers or creeks, the traces formed in channel facies have almost zero preservation potential, as shown by observations in the ephemeral Atuel River near Algarrobo del Águila town, La Pampa province, Argentina (Fig. 9A). The low-sinuosity channel belt shows point and longitudinal bars (Fig. 9A–C). The climate is temperate semiarid, and the river is dry for most of the year. The observations were conducted every 2 weeks from September 2008 to March 2009, at the surface and in shallow pits dug in the channel bottom and a longitudinal bar. The observations include two intervals with water currents (September to October 2008 and December 2008 to January 2009) and two dry intervals (end of November 2008 and February to March 2009). During the dry intervals, the discharge decreased and the remaining water was concentrated in small ponds in the river bed. The generation of shallow horizontal mole-cricket burrows, bird and tetrapod tracks, and plant colonization of channel floors (Fig. 9E) was observed in channel floors and longitudinal bars. Wind ripples (Fig. 9F), laminated mudstones in pools, mud cracks, and raindrop imprints were also recorded. In the section logged in March 2009, there were no deposits of the dry intervals in the longitudinal bar pit and an erosive surface marked the erosion that occurred during the January 2009 current. In the neighboring channel floor pit, a thin mudstone intercalation is the only indication of the dry periods (Fig. 9D). None of the observed biogenic structures were preserved in the analyzed sections. When buried under younger sediments, an adequate interpretation of this channel belt will be aided by an architectural-element analysis. This example also suggests that the absence of trace fossils in fluvial channel floors does not necessarily imply a permanent discharge.

In a compilation of the facies distribution of different track types from 50 localities of the Jurassic Morrison Formation, Foster and Lockley (2006) found different abundances of sauropod and theropod tracks. Sauropod tracks are more abundant in floodplain mudstone, whereas theropod tracks are dominant on the upper surfaces of sandstone formed as bars or crevasse-splay deposits. Sauropod tracks are preserved in floodplain mudstone as natural sandstone casts (some deeply impressed) connected to sandstone overhangs that cover floodplain mudstone. Although a minor paleoenvironmental distinction between both types of tracks is possible, Foster and Lockley (2006) suggested that

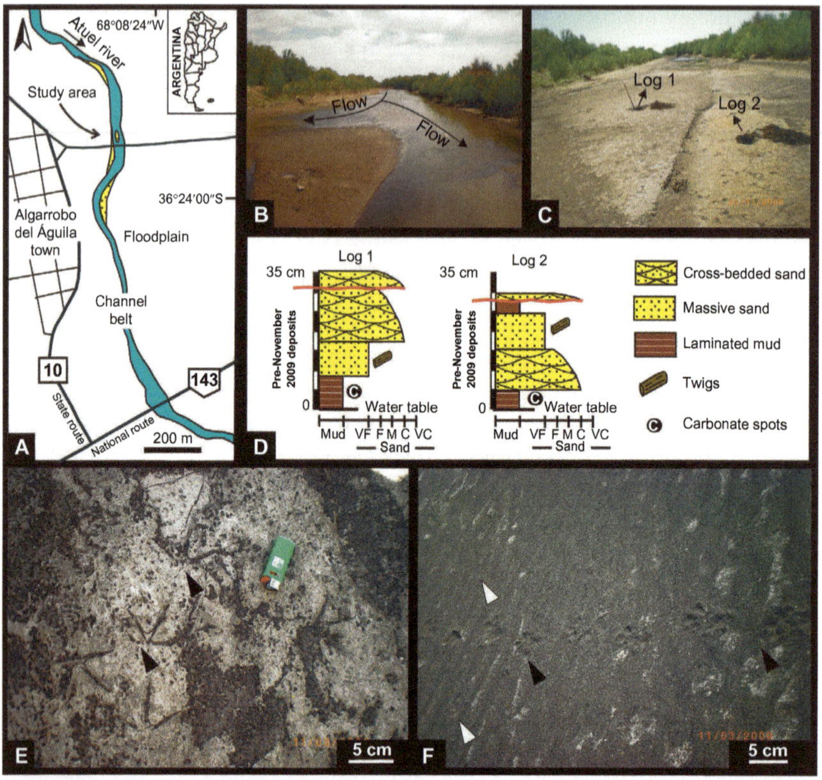

FIGURE 9 Sediments of channel-belt deposits of the Atuel river near Algarrobo del Águila town (36° 23′ 52.3″ S, 67° 08′ 25.9″ W), La Pampa province, Argentina (A–D) and modern traces found in its channel bed (E–H). (A) Location map showing the channel-belt deposits (blue), channel bars (yellow) and study area. (B) and (C) View of the channel belt and emergent longitudinal bar under current conditions (B) and during dry periods (C). Note also the location of shallow pits in figure C. Channel-belt width is 28 m. (D) Measured sections on the bar top (log 1) and the channel floor (log 2) in a dry period. (E) Bird tracks on a mud-draped channel floor (two arrowed). (F) Sand deposits in channel bed showing eolian ripples (white arrows) and mammal trackway (black arrows).

theropod tracks might be less commonly preserved and recorded in this type of outcrop. An additional example of contrasting preservation of groups of vertebrate trace fossils in floodplain sediments is the Late Cretaceous ichnofauna of Mongolia described by Currie et al. (2003). The footprints are exclusively made by large animals (mostly hadrosaurs, but also tyrannosaur theropods and titanosaurs) that sunk more than 1 m into the mud of ephemeral lakes and swales of meandering rivers (compaction was not considered). The absence of small footprints and the preservation of delicate invertebrate trace fossils on bedding planes was interpreted as an indication that the sediments may have been covered by enough water to prevent smaller animals from leaving their footprints (Currie et al., 2003). A similar contrasting preservation of herbivorous (ornithopod) and carnivorous (theropod) dinosaur tracks has been described for seasonal wetland facies of the Maastrichtian of Canada (Nadon, 1993). In this example, herbivorous dinosaur tracks are deeply impressed (to 85 cm) and carnivorous dinosaur tracks are shallowly impressed, suggesting that carnivorous dinosaurs ventured into the wetlands only during the dry season.

Trace fossils display a better preservation potential in overbank areas, particularly in pond and crevasse-splay facies. These different preservational conditions are probably linked to their particular sedimentary dynamics. After peak discharge and consequent breaching of channel margins, large volumes of sandy bedload are dumped on the flanking floodplains. The damp emergent crevasse-splay sediments that cover a large area proximal to the river channel are a very suitable substrate for track production and preservation. Trace preservation is enhanced by a muddy veneer that accumulates on the sand surface, commonly on former swales where water ponded temporarily.

5.3 Trace Fossils in Channel-Belt Deposits

Most of the trace-fossil assemblages in channel facies occur on the top of, or in the upper part of, fluvial bars, although trace fossils can also appear in inactive or abandoned channels, bottom of channels, and between deposits of channel bars. Many assemblages of inactive or abandoned channels have been interpreted as a channel-related pond formed after abandonment of the active channel (e.g., Buatois et al., 2007; Głuszek, 1995; Kim et al., 2002; Tandon and Naug, 1984). The presence of rhizoliths and mud cracks at the bottom of the main channel indicates ephemeral or intermittent river discharge (e.g., Bridge, 2003, 2006). However, other trace-fossil types than rhizoliths can be useful to infer discontinuous channel discharge, taken isolated or in conjunction with physical sedimentary structures, if found at the bottom or between channel bars. For example, diverse mammal (rhinoceros, entelodontid, and camel) footprints were described from the base of Oligocene channel deposits of the United States (Chaffee, 1943), equoid footprints (*Plagiolophustipus*) occur at the bottom of Oligocene meandering-channel deposits in Spain (Astibia et al., 2007), and sauropod dinosaur tracks on the sole of Maastrichtian anastomosed-channel sandstone in Utah, United States (Difley and Ekdale, 2002). In addition, the presence of burrows and rhizoliths at the base of a volcaniclastics-rich channel-fill deposit from the lower member of the Cretaceous Bajo Barreal Formation of Argentina (Fig. 10) was also used to suggest an ephemeral character (Umazano, 2009). The lower part of some channel-belt deposits of this formation exhibits a highly bioturbated interval (Fig. 10A and B) that displays an ichnofabric dominated by simple burrows (Fig. 10C) and small rhizoliths (Fig. 10D). These examples require an initial river incision and subaerial exposure of sediment or ponding of water in the river bed. In the case of the Cretaceous Bajo Barreal Formation, the ponded interval allowed for the development of a paleosol in the river bed.

Trace-fossil assemblages that were preserved between channel bars are also useful for paleoenvironmental interpretation of these facies (Dalla Vecchia and Rustioni, 1996; Melchor, 2009; Melchor et al., 2010; Williamson and Lucas,

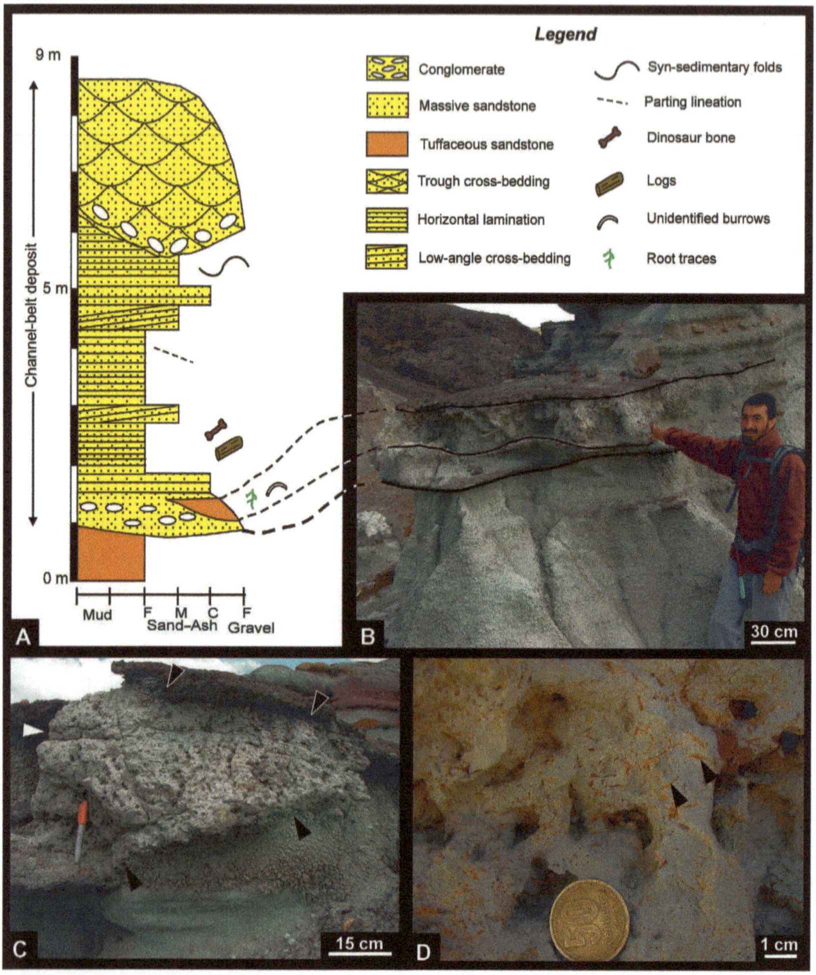

FIGURE 10 Example of pedogenic features within a channel-belt deposit of the Late Cretaceous Bajo Barreal Formation, Chubut province, Argentina. (A) Log of the 7.5-m-thick channel-belt deposit. The lower part of the channel belt contains a lenticular intercalation of fine-grained tuffaceous sandstone with a highly bioturbated ichnofabric. (B) The lower part of the channel-belt deposit and correlation of the lower bounding surface of the channel (thick line) and bounding surfaces of the bioturbated interval (thin lines) with the log of figure A. (C) Detail of the highly bioturbated interval showing the transition to the underlying gravelly sandstone and of the sharp contact with the overlying coarse-grained sandstone (black arrows). Note the relict primary lamination (white arrow). (D) Rhizoliths (some arrowed) in a close-up of the bioturbated interval.

1996). The example described by Melchor (2009) from the Neogene Río Negro Formation of Argentina is composed of a moderate density of bird tracks on the bounding surface of a 3-m-thick coset of trough cross-bedded sandstone composing a channel-belt deposit. These tracks indicate a cessation in water discharge of the river, ponds in the dry channel, and settling of fine-grained sediments, which was the focus of foraging activities by shorebirds.

Not all trace fossils that occur within or at the base of fluvial channel deposits are indicative of ephemeral flow. Two trace-fossil assemblages from the Upper Jurassic of United States (Foster and Lockley, 1997; Lockley and Foster, 2006), involving essentially swim traces in channel facies, are good examples of alternative interpretations. Foster and Lockley (1997) described purported crocodile tracks (*Hatcherichnus sanjuanensis*) associated with possible tail-drag traces at the base of a cross-bedded sandstone set that lies 6 m above the base of a 12-m-thick channel deposit. In this example, inferences about the producer (habitat of crocodiles) and indications of possible swimming or basking behavior, in addition to the absence of emergence and/or desiccation features, suggest that river discharge was essentially uninterrupted. The remaining example from the Upper Jurassic of the United States includes vertebrate tracks from the base and lower part of a channel complex (Lockley and Foster, 2006).

Another example, where trace fossils of fluvial-channel deposits give some indication about changing paleoenvironmental conditions, was described by Hubert and Dutcher (2010). These authors found several tens of oriented subhorizontal meniscate burrows at the top of a trough cross-bedded set of pebbly sandstone from the Triassic of the United States, which were interpreted as reflecting escape traces, probably in response to a rising water table.

5.4 Floodplain Subenvironments

Most of the analyzed overbank ichnofaunas are assigned to three settings: crevasse splay, floodbasin pond, and the pedogenized floodplain. There are a few cases of assemblages from levee deposits (Groenewald et al., 2001; Hasiotis and Honey, 2000; Kim et al., 2002; Tandon and Naug, 1984), which are herein subsumed under pedogenized floodplain. The last setting is represented by a range of floodplain and channel-belt deposits that exhibit different degrees of development of pedogenic features.

The floodplain ichnofaunas share various simple invertebrate burrow types with a roughly similar proportion in each subenvironment (horizontal burrows, burrow systems, meniscate burrows, simple vertical burrows, and J-, U-, or Y-shaped burrows; Fig. 11). However, the floodplain ichnofaunas of these subenvironments can be easily discriminated.

Pedogenized floodplain trace-fossil assemblages are particularly distinctive. They are characterized by chambers and chambered burrow systems (e.g., *Coprinisphaera*, *Rebuffoichnus*, *Cellicalichnus*, *Termitichnus*), rhizoliths and other root structures, burrows with pelletal fill (e.g., *Edaphichnium*, *Castrichnus*), large vertebrate burrows (e.g., *Alezichnos*, *Daimonelix*, L, and G-type burrows), and vertebrate coprolites. These trace fossils are characteristic of ichnofacies that occur in paleosols (*Coprinisphaera*, *Celliforma*, *Termitichnus*, and other potential ichnofacies discussed previously). J-, U-, or Y-shaped vertical burrows (*Loloichnus*, *Camborygma*, *Capayanichnus*, *Macanopsis*) and perforations in the wall of chambered trace fossils (*Lazaichnus* and *Tombownichnus*) can also be present.

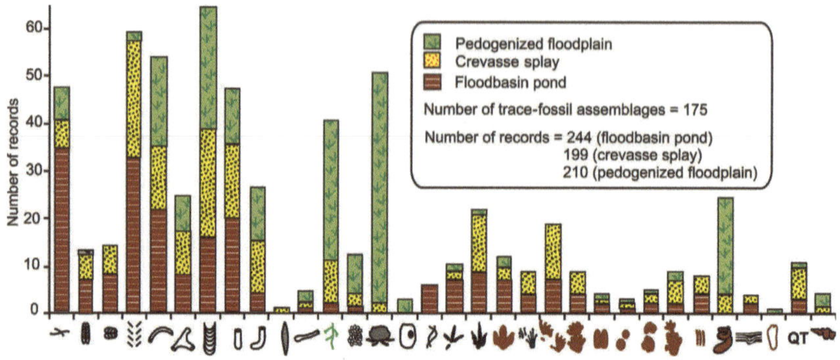

FIGURE 11 Histogram showing the participation of trace-fossil types in several floodplain subenvironments (pond, crevasse splay, and pedogenized floodplain). References as for Fig. 8.

Some examples of trace-fossil assemblage representatives of different ichnofacies that belong to the pedogenized floodplain setting are from the Eocene of Uruguay (Genise, 2004), the Late Oligocene of United States (Retallack et al., 2000), the Oligocene–Miocene of Egypt (Bown, 1982; Genise and Bown, 1994), and the Oligocene of Argentina (Bellosi et al., 2010).

Several trace-fossil morphologies that are typical of the *Mermia* Ichnofacies occur preferentially in floodbasin-pond assemblages, including simple horizontal trails (e.g., *Archaeonassa*, *Helminthoidichnites*, *Helminthopsis*, *Cochlichnus*), complex sinusoidal trails (e.g., *Undichna*), and tetrapod swim traces (e.g., *Characichnos*). Bird tracks (e.g., *Gruipeda*, *Aquatilavipes*, *Ignotormis*, *Barrosopus*) and arthropod-produced trace fossils, mainly trackways (e.g., *Diplichnites*, *Umfolozia*, *Protichnites*), trough-like reliefs (e.g., *Rusophycus*, *Selenichnites*, *Gluckstadtella*), and bilobed burrows/trails (*Cruziana*), also occur preferentially in pond facies. Most tetrapod-track types seem to be equally represented in floodbasin-pond and crevasse-splay facies, but they are very rare in paleosols. The previously mentioned track types include bipedal tri- or tetradactyl trackways with thick digits (*Grallator*, *Anchisauripus*), bipedal tridactyl tracks with stout digits ("iguanodontid tracks"), Q 4/5 (*Batrachichnus*), Q 5/5 sl (*Rhynchosauroides*, *Dromopus*, *Gilmoreichnus*), Q 5/5 st (*Ichniotherium*, *Amphisauropus*), quadrupedal bilobed tracks (*Entelodontipus*, "bovid tracks"), sauropod tracks (*Brontopodus*), and chirotheriid tracks (*Chirotherium*, *Brachychirotherium*, *Tetrasauropus*). Some selected examples of typical floodbasin-pond assemblages are from (1) the Early Devonian Peel Sandstone Group (Isle of Man), dominated by arthropod trackways (Crowley et al., 2009); (2) the Late Carboniferous Tupe Formation (Argentina), essentially composed of horizontal burrows and simple trails (Buatois and Mángano, 2002); and (3) the Late Cretaceous Anacleto Formation (Argentina), characterized by three ichnotaxa of bird footprints associated with meniscate burrows and simple trails and burrows.

Except for the differences mentioned above, trace-fossil assemblages from crevasse-splay deposits appear to lack distinctive features. The crevasse-splay facies seems to have a better preservation potential of vertebrate tracks (e.g., Fuglewicz et al., 1990; Melchor et al., 2010; Smith, 1993). In a detailed study on the ichnotaxonomy and facies distribution of vertebrate tracks from Early Triassic fluvial rocks of Poland, Fuglewicz et al. (1990) recognized that 64% of the footprints appear in a distal crevasse-splay facies, 23% in a proximal crevasse-splay facies, and only 13% in channel bars.

Some of the trace-fossil assemblages from crevasse-splay deposits contain only, or are dominated by, arthropod trackways (mostly *Diplichnites*). These examples range in age from Devonian to Triassic (e.g., Briggs et al., 1984; Lucas et al., 2006; Neef, 2004). Smith (1993) found that the arthropod trackway *Umfolozia* was characteristically linked to proximal crevasse-splay settings in the Permian Beaufort Group, but this possible link between facies and a definite ichnotaxon has not been recorded in other lithostratigraphic units.

5.5 Comparison of Trace-Fossil Assemblages from Different River Systems

In this chapter, we have adopted a division of fluvial systems in braided, meandering, anastomosed, and ephemeral rivers. Braided and meandering channels may be recognized objectively as end members of a fluvial channel spectrum. Since braided and meandering rivers may have a permanent or intermittent discharge, the division of fluvial systems adopted herein, especially the consideration of ephemeral rivers as a distinct category, is not completely objective but reflects the characterization of the paleoenvironment proposed by the various authors. Architectural element analysis of fluvial successions accompanying the description of trace-fossil assemblages is very rare in the literature, Zavala et al. (1993) and Krapovickas et al. (2009) being some exceptions.

The database includes only nine trace-fossil assemblages from anastomosed river-deposits (e.g., Difley and Ekdale, 2002; Johnson, 1986; Marriott et al., 2009; Nadon, 1993), and for this reason, they are analyzed separately. The majority of trace fossils in anastomosed river-deposits occurs in channel and floodbasin-pond deposits, and there is no description of assemblages from pedogenized-overbank deposits. Channel deposits contain simple trails and horizontal and meniscate burrows, but the most remarkable components are tetrapod tracks, particularly bird tracks and large dinosaur (sauropod) tracks. Nadon (1993) suggested that large tetrapod tracks have a higher potential of preservation in anastomosed river-deposits because of the abundance of soft substrates to record the tracks and the occurrence of annual flooding that rapidly buries the footprints.

A comparison of the trace-fossil assemblages from meandering, braided, and ephemeral river-facies (Fig. 12) allows recognizing some general differences between them. The trace-fossil assemblages that were not assigned originally to a specific river system were excluded from this analysis ($n=42$; see Supplementary Material in http://booksite.elsevier.com/9780444538130). There are a number of simple trace-fossil morphologies that exhibit a roughly similar proportion in every river type, namely, simple horizontal trails,

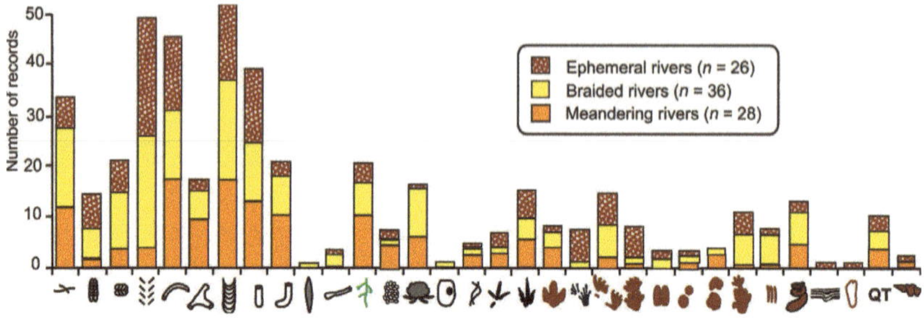

FIGURE 12 Histogram with the absolute abundance of trace-fossil types in meandering, braided, and ephemeral river-systems (including channel-belt and floodplain facies). References as for Fig. 8.

horizontal burrows, meniscate burrows, simple vertical burrows, bipedal tri- or tetradactyl trackways with thick digits, and various quadrupedal trackways. Other trace-fossil morphologies are more common in braided and ephemeral rivers than in meandering rivers, including arthropod trackways and other arthropod-produced traces (bilobed burrows/trails, trough-like reliefs), Q 4/5, Q 5/5 sl, Q 5/5 st, Q 2/2, tetrapod swim traces, and chirotheriid tracks. These trace-fossil morphologies are essentially impressions on bedding planes, and their preservation is probably favored by large areas of channel belts and flooded overbanks in braided and ephemeral rivers containing damp substrates with adequate water saturation for casting of footprints. Similar trace-fossil assemblages in ephemeral and braided rivers may also be linked to the braided pattern commonly found in ephemeral rivers. On the other hand, some trace-fossil types are more common in meandering rivers, namely, rhizoliths, burrows with pelletal fill, large vertebrate burrows, burrow systems, J-, Y-, or U-shaped vertical burrows, bipedal tridactyl trackways with stout digits, and sauropod tracks. The three first trace-fossil types are commonly associated with pedogenesis and might be linked to a greater stability of channel margins and development of paleosols in meandering rivers.

6. CONCLUSIONS

The ichnofacies recognized in fluvial deposits (Table 2) include archetypal or Seilacherian ichnofacies, mostly involving invertebrate trace fossils (*Scoyenia*, *Mermia*, *Coprinisphaera*, *Skolithos*), two incipient trace-fossil assemblages that are probably archetypal (*Celliforma* and *Termitichnus* ichnofacies), and purported archetypal vertebrate ichnofacies (*Batrachichnus* and *Grallator* ichnofacies). Other recently proposed vertebrate ichnofacies, as defined at present, lack the recurrent character to be considered as archetypal (*Hatcherichnus*, *Characichnos*, *Chelonipus*, and *Brontopodus*).

This review emphasizes some key and useful relationships between ichnofacies or trace-fossil assemblages and fluvial subenvironments. The *Scoyenia* Ichnofacies (typified by meniscate burrows, arthropod trackways, trough-like reliefs, and bilobed burrows/trails) is useful for distinguishing emergent channel-bar tops or the margins of floodbasin ponds. It is recommended that tetrapod tracks are not considered as a key component of the *Scoyenia* Ichnofacies; instead they are better compared with a tetrapod ichnofacies. The *Mermia* Ichnofacies (simple horizontal trails, vertebrate swim traces), and probably also the *Characichnos* Ichnofacies (tetrapod swim traces) are typical of floodbasin ponds. Other analogous ichnocoenosis that also contain swim traces but are restricted to subaqueous lotic settings (permanent channels) are the *Chelonipus* and *Hatcherichnus* ichnocoenoses.

The identification of chambers and chambered trace fossils (commonly as dominant components) is indicative of the presence of the *Coprinisphaera*, *Celliforma*, and *Termitichnus* ichnofacies, which were developed in soils supporting different plant communities (herbaceous, scrubs and woodlands, and closed forests, respectively). Several trace-fossil assemblages from paleosols developed in fluvial deposits that lack insect trace fossils cannot be included at present in existing ichnofacies. However, a group of these assemblages is characterized by large burrows of vertebrate origin that occur in well-drained calcareous paleosols, and another group of assemblages is typified by burrows or chambers of possible crayfish origin in paleosols with fluctuating water table. Furthermore, the potential rhizolith ichnofacies, dominated by root traces, is recurrent enough and can yield significant paleoenvironmental inferences to be considered as a tempting idea to explore.

The *Skolithos* Ichnofacies (simple vertical burrows and J, or U-shaped vertical burrows) is related with the top of fluvial bars, although the subaerial or subaqueous nature of the trace fossils must be confirmed in each case. If related with subaerial exposure of fluvial bars, it may mark interruptions in fluvial channel discharge or low flow-regime conditions. The *Grallator* (bird tracks, bipedal tri- or tetradactyl tracks with thick digits) and *Batrachichnus* ichnofacies (Q 4/5, Q 5/5 sl, Q 5/5 st, and chirotheriid tracks) appear on top of channel bars, crevasse splays, and margins of floodbasin ponds.

A major challenge in continental ichnology is the agreement upon a unified ichnofacies scheme that would be able to integrate invertebrate, vertebrate, and plant trace fossils. Alternatively, it is possible that a consensus will never be reached and tetrapod ichnofacies will consolidate as a separate and parallel scheme to archetypal invertebrate ichnofacies.

One of the major contributions of the ichnofacies analysis to the paleoenvironmental interpretation of fluvial successions is the recognition of composite or mixed assemblages that contain significant participation of two ichnofacies. These composite ichnofacies (e.g., *Mermia–Scoyenia*, *Scoyenia–Celliforma*) holds the potential of reflecting subtle environmental changes that commonly cannot be inferred from physical sedimentary structures alone. Vertical replacement of ichnofacies in a fluvial basin can be useful for distinguishing paleoenvironmental changes and can also help in the recognition of larger-scale stacking patterns in sequence-stratigraphic analysis.

The current knowledge on fluvial ichnofabrics is very limited, although ichnofabric studies on continental settings have a large potential for aiding paleoenvironmental, stratigraphical, and paleoecological studies. Application of a definite descriptive methodology may be used to recognize particular ichnofabrics in different fluvial subenvironments and to link the composing trace fossils with their potential producers. Paleosol ichnofabrics exhibit a greater complexity and variety of trace fossils than in other fluvial subaquatic subenvironments. In consequence, they hold a larger potential for paleoenvironmental and paleoecological studies.

Channel-belt and floodplain facies exhibit a dominance of very simple trace-fossil morphologies, essentially of invertebrate origin. The abundance and diversity of trace fossils is much higher in overbank facies than in channel-belt deposits. These differences are linked to contrasting taphonomical conditions in both settings. Coarse-grained channel deposits are mostly devoid of trace fossils, which are concentrated in specific intervals and bedding planes (mostly bar tops and abandoned-channel fills). Rhizoliths and/or tetrapod tracks (that might be associated with mud cracks and/or mud veneers) found at the bottom of channel deposits or between channel bars are good clues to intermittent or ephemeral channel discharge. However, the low preservation potential of these features in a modern ephemeral creek suggests that their absence does not imply permanent discharge, as commonly assumed.

Floodplain ichnofaunas are more abundant and diverse than channel-belt ichnofaunas. Trace-fossil assemblages produced under subaerial conditions, such as those of pedogenized floodplains, are quite different from those showing no or minimal subaerial exposure, such as pond and crevasse splays. Pedogenized floodplain (paleosol) trace-fossil assemblages are typified by chambers and chambered burrow systems, rhizoliths and other root structures, burrows with pelletal fill, large vertebrate burrows, and vertebrate coprolites, whereas vertebrate tracks are rare or absent. Pond deposits commonly contain simple horizontal trails, complex sinusoidal trails, tetrapod swim traces, bird tracks, arthropod trackways, trough-like reliefs, and bilobed burrows/trails. Crevasse-splay trace-fossil assemblages do not show clear distinguishing features, although tetrapod-track preservation is enhanced in this setting.

Trace-fossil assemblages from anastomosed, braided, meandering, and ephemeral rivers display some differences in composition and abundance, especially when considering floodplain ichnofaunas. Anastomosed rivers appear to preserve trace fossils preferentially in channel facies, where bird tracks and large dinosaur (sauropod) tracks are more common. The trace fossils from ephemeral and braided deposits can be distinguished from those of meandering rivers by the larger proportion of structures preserved on bedding planes, including arthropod trackways and other arthropod-produced traces (bilobed burrows/trails, trough-like reliefs), Q 4/5, Q 5/5 sl, Q 5/5 st, Q 2/2, and chirotheriid tracks. Meandering rivers contain a larger proportion of trace fossils related to pedogenized floodplain settings (e.g., rhizoliths, burrows with pelletal fills, large vertebrate burrows, J-, Y-, or U-shaped vertical burrows, burrow systems), perhaps reflecting a higher potential for better developed soils in this type of rivers.

ACKNOWLEDGMENTS

Dirk Knaust and Richard Bromley are thanked for the invitation to contribute to this volume. The volume editors and Nigel Mountney make constructive reviews that improved the chapter. Eduardo Bellosi kindly read critically a former version of this chapter. This research was funded by grants PIP 8010164 from CONICET (Argentina) to R. N. M. and FONCYT-PICT 07/1972 of the ANPCYT (Argentina) to J. F. G.

REFERENCES

Aramayo, S.A., Bocanegra, L., 2003. Icnofacies de *Scoyenia* en la Formación Río Limay (Grupo Neuquén, Cretácico tardío) Provincia de Neuquén, Argentina. In: Buatois, L.A., Mángano, M.G. (Eds.), Icnología: Hacia una Convergencia entre Geología y Biología. Asociación Paleontológica. Argentina. Publicación Especial, vol. 9, pp. 43–48.

Astibia, H., Suberbiola, X.P., Payros, A., Murelaga, X., Berreteaga, A., Baceta, J.I., Badiola, A., 2007. Bird and mammal footprints from the Tertiary of Navarre (western Pyrenees). Ichnos 14, 175–184.

Bedatou, E., Melchor, R.N., Bellosi, E., Genise, J.F., 2008. Crayfish burrows from Late Jurassic-Late Cretaceous continental deposits of Patagonia: Argentina. Their palaeoecological, palaeoclimatic and palaeobiogeographical significance. Palaeogeogr. Palaeoclimatol. Palaeoecol. 257, 169–184.

Bedatou, E., Melchor, R.N., Genise, J.F., 2009. Complex palaeosol ichnofabrics from Late Jurassic-Early Cretaceous volcaniclastic successions of Central Patagonia, Argentina. Sediment. Geol. 218, 74–102.

Bellosi, E.S., Laza, J.H., Sánchez, M.V.,Genise, J.F., 2010. Ichnofacies analysis of the Sarmiento Formation (Middle Eocene–Early Miocene) at Gran Barranca, central Patagonia. In: Madden, R.H., Carlini, A.A., Vucetich, M.G., Kay, R.F. (Eds.), The Paleontology of Gran Barranca. Evolution and Environmental Change through the Middle Cenozoic of Patagonia. Cambridge University Press, Cambridge, pp. 302–312.

Bertling, M., Braddy, S.J., Bromley, R.G., Demathieu, G.D., Genise, J.F., Mikuláš, R., Nielsen, J.K., Nielsen, K.S.S., Rindsberg, A.K., Schlirf, M., Uchman, A., 2006. Names for trace fossils: a uniform approach. Lethaia 39, 265–286.

Bockelie, J.F., 1994. Plant roots in core. In: Donovan, S.K. (Ed.), The Palaeobiology of Trace Fossils. Wiley, Chichester, pp. 177–199.

Bown, T.M., 1982. Ichnofossils and rhizoliths of the nearshore fluvial Jebel Qatrani Formation (Oligocene), Fayum Province, Egypt. Palaeogeogr. Palaeoclimatol. Palaeoecol. 40, 255–309.

Bown, T.M., Hasiotis, S.T., Genise, J.F., Maldonado, F., Brouwers, E.M., 1997. Trace fossils of Hymenoptera and other insects, and paleoenvironments of the Claron Formation (Paleocene and Eocene), southwestern Utah. US Geol. Surv., Bull. 2153, 41–58.

Bridge, J.S., 2003. Rivers and Floodplains: Forms, Processes and Sedimentary Record. Blackwell Scientific Publications, Oxford, 491 pp.

Bridge, J.S., 2006. Fluvial facies models: recent developments. In: Posamentier, H.W., Walker, R.G. (Eds.), Facies Models Revisited. SEPM Spec. Publ., vol. 84, 85–170.

Bridge, J.S., Leeder, M.R., 1979. A simulation model of alluvial stratigraphy. Sedimentology 26, 617–644.

Briggs, D.E.G., Plint, A.G., Pickerill, R.K., 1984. *Arthropleura* trails from the Westphalian of eastern Canada. Palaeontology 27, 843–855.

Bromley, R.G., 1990. Trace Fossils. Biology and Taphonomy. Unwin Hyman, London, 280 pp.

Bromley, R.G., 1996. Trace Fossils: Biology, Taphonomy and Applications. Chapman & Hall, London, 361 pp.

Bromley, R.G., Asgaard, U., 1979. Triassic freshwater ichnocoenoses from Carlsberg Fjord, East Greenland. Palaeogeogr. Palaeoclimatol. Palaeoecol. 28, 39–80.

Bromley, R.G., Buatois, L.A., Genise, J.F., Labandeira, C.C., Mángano, M.G., Melchor, R.N., Schlirf, M., Uchman, A., 2007. Comments on the paper "Reconnaissance of Upper Jurassic Morrison Formation ichnofossils, Rocky Mountain Region, USA: Paleoenvironmental, stratigraphic, and paleoclimatic significance of terrestrial and freshwater ichnocoenoses" by Stephen T. Hasiotis. Sediment. Geol. 200, 141–150.

Buatois, L.A., Mángano, M.G., 1995. The paleoenvironmental and paleoecological significance of the lacustrine *Mermia* Ichnofacies: an archetypical subaqueous nonmarine trace fossil assemblage. Ichnos 4, 151–161.

Buatois, L.A., Mángano, M.G., 1996. Icnología de ambientes continentales: Problemas y perspectivas. In: Melchor, R.N. (Ed.), Primera Reunión Argentina de Icnología. Asociación Paleontológica, Publicación Especial, vol. 4, Argentina, pp. 5–30.

Buatois, L.A., Mángano, M.G., 1998. Trace fossil analysis of lacustrine facies and basins. Palaeogeogr. Palaeoclimatol. Palaeoecol. 140, 367–382.

Buatois, L.A., Mángano, M.G., 2002. Trace fossils from Carboniferous floodplain deposits in western Argentina: implications for ichnofacies models of continental environments. Palaeogeogr. Palaeoclimatol. Palaeoecol. 183, 71–86.

Buatois, L.A., Mángano, M.G., 2004. Animal-substrate interactions in freshwater environments: applications of ichnology in facies and sequence stratigraphic analysis of fluvio-lacustrine successions. In: McIlroy, D. (Ed.), The Application of Ichnology to Palaeoenvironmental and Stratigraphic Analysis. Geol. Soc. London, Spec. Publ., vol. 228, pp. 311–333.

Buatois, L.A., Mángano, M.G., 2007. Invertebrate ichnology of continental freshwater environments. In: Miller III, W. (Ed.), Trace Fossils. Concepts, Problems, Prospects. Elsevier, Amsterdam, pp. 285–323.

Buatois, L.A., Mángano, M.G., 2011. Ichnology: Organism-Substrate Interactions in Space and Time. Cambridge University Press, Cambridge, 358 pp.

Buatois, L.A., Uba, C.E., Mángano, M.G., Hulka, C., Heubeck, C., 2007. Deep bioturbation in continental environments: evidence from Miocene fluvial deposits of Bolivia. In: Bromley, R.G., Buatois, L.A., Mángano, M.G., Genise, J.F., Melchor, R.N. (Eds.), Sediment-Organism Interactions: A Multifaceted Ichnology. SEPM Spec. Publ., vol. 88, pp. 123–136.

Chaffee, R.G., 1943. Mammal footprints from the White River Oligocene. Notulae Naturae, Acad. Nat. Sci. Philadelphia 116, 1–13.

Chiappe, L.M., Schmitt, J.G., Jackson, F.D., Garrido, A., Dingus, L., Grellet-Tinner, G., 2004. Nest structure for sauropods: sedimentary criteria for recognition of dinosaur nesting traces. Palaios 19, 89–95.

Cohen, A.S., 1982. Paleoenvironments of root casts from the Koobi Fora Formation, Kenya. J. Sediment. Petrol. 52, 401–414.

Crowley, S.F., Higgs, K.T., Piper, J.D.A., Morrissey, L.B., 2009. Age of the Peel Sandstone Group, Isle of Man. Geol. J. 44, 57–78.

Currie, P.J., Badamgarav, D., Koppelhus, E.B., 2003. The first Late Cretaceous footprints from the Nemegt locality in the Gobi of Mongolia. Ichnos 10, 1–13.

D'Alessandro, A., Ekdale, A.A., Picard, M.D., 1987. Trace fossils in fluvial deposits of the Duchesne River Formation (Eocene), Uinta Basin, Utah. Palaeogeogr. Palaeoclimatol. Palaeoecol. 61, 285–301.

Dalla Vecchia, F.M., Rustioni, M., 1996. Mammalian trackways in the Conglomerato di Osoppo (Udine, NE Italy) and their contribution to its age determination. Mem. Soc. Geol. 48, 221–232.

De, C., 2002. Continental mayfly burrows within relict-ground in inter-tidal beach profile of Bay of Bengal coast: a new ichnological evidence of Holocene marine transgression. Curr. Sci. 83, 64–67.

de Gibert, J.M., Sáez, A., 2009. Paleohydrological significance of trace fossil distribution in Oligocene fluvial-fan-to-lacustrine systems of the Ebro Basin, Spain. Palaeogeogr. Palaeoclimatol. Palaeoecol. 272, 162–175.

Difley, R.L., Ekdale, A.A., 2002. Footprints of Utah's last dinosaurs: track beds in the Upper Cretaceous (Maastrichtian) North Horn Formation of the Wasatch Plateau, central Utah. Palaios 17, 327–346.

Ekdale, A.A., Bromley, R.G., Pemberton, S.G., 1984. Ichnology: The Use of Trace Fossils in Sedimentology and Stratigraphy. SEPM Short Course Notes 15, 1–316.

Foster, J.R., Lockley, M.G., 1997. Probable crocodilian tracks and traces from the Morrison Formation (Upper Jurassic) of eastern Utah. Ichnos 5, 121–129.

Foster, J., Lockley, M.G., 2006. Preservation and abundance patterns in the ichnological record of the Morrison Formation (Upper Jurassic, North America). New Mexico Mus. Nat. Hist. Sci., Bull. 41, 203–216.

Frey, R.W., Pemberton, S.G., 1985. Biogenic structures in outcrops and cores. I. Approaches to ichnology. Bull. Can. Petrol. Geol. 33, 72–115.

Frey, R.W., Pemberton, S.G., Fagerstrom, J.A., 1984. Morphological, ethological, and environmental significance of the ichnogenera *Scoyenia* and *Ancorichnus*. J. Paleontol. 58, 511–528.

Fuglewicz, R., Ptaszyński, T., Rdzanek, K., 1990. Lower Triassic footprints from the Świętokrzyskie (Holly Cross) Mountains, Poland. Acta Palaeont. Polon. 35, 109–164.

Genise, J.F., 2004. Ichnotaxonomy and ichnostratigraphy of chambered trace fossils in palaeosols attributed to coleopterans, ants and termites. In: McIlroy, D. (Ed.), The Application of Ichnology to Palaeoenvironmental and Stratigraphic Analysis. Geol. Soc. London, Spec. Publ., vol. 228, pp. 419–453.

Genise, J.F., Bown, T.M., 1994. New trace fossils of termites (Insecta: Isoptera) from the Late Eocene-Early Miocene of Egypt, and the reconstruction of ancient isopteran social behavior. Ichnos 3, 155–183.

Genise, J.F., Bellosi, E.S., González, M.G., 2004. An approach to the description and interpretation of ichnofabrics in palaeosols. In: McIlroy, D. (Ed.), The Application of Ichnology to Palaeoenvironmental and Stratigraphic Analysis. Geol. Soc. London, Spec. Publ., vol. 228, pp. 355–382.

Genise, J.F., Bedatou, E., Melchor, R.N., 2008. Terrestrial crustacean breeding trace fossils from the Cretaceous of Patagonia (Argentina): palaeobiological and evolutionary significance. Palaeogeogr. Palaeoclimatol. Palaeoecol. 264, 128–139.

Genise, J.F., Bellosi, E.S., Verde, M., González, M.G., 2011. Large ferruginized palaeorhizospheres from a Paleogene lateritic profile of Uruguay. Sediment. Geol. 240, 85–96.

Genise, J.F., Mángano, M.G., Buatois, L.A., Laza, J.H., Verde, M., 2000. Insect trace fossil associations in paleosols: The *Coprinisphaera* Ichnofacies. Palaios 15, 49–64.

Genise, J.F., Melchor, R.N., Bellosi, E.S., Verde, M., 2010. Invertebrate and vertebrate trace fossils from continental carbonates. In: Alonso-Zarza, A.M., Tanner, L. (Eds.), Carbonates in Continental Settings: Facies, Environments and Processes. Developments in Sedimentology, vol. 61. Elsevier, Amsterdam, pp. 319–369.

Gierliński, G., Pieńkowski, G., Niedźwiedzki, G., 2004. Tetrapod track assemblage in the Hettangian of Soltyków, Poland, and its paleoenvironmental background. Ichnos 11, 195–213.

Głuszek, A., 1995. Invertebrate trace fossils in the continental deposits of an Upper Carboniferous coal-bearing succession, Upper Silesia, Poland. Stud. Geol. Polon. 108, 171–202.

Gobetz, K.E., Martin, L.D., 2006. Burrows of a gopher-like rodent, possibly *Gregorymys* (Geomyoidea: Geomyidae: Entoptychtinae), from the early Miocene Harrison Formation, Nebraska. Palaeogeogr. Palaeoclimatol. Palaeoecol. 237, 305–314.

Goldring, R., 1993. Ichnofacies and facies interpretation. Palaios 8, 403–405.

Graham, J.R., Pollard, J.E., 1982. Occurence of the trace fossil *Beaconites antarcticus* in the Lower Carboniferous fluviatile rocks of county Mayo, Ireland. Palaeogeogr. Palaeoclimatol. Palaeoecol. 38, 257–268.

Groenewald, G.H., Welman, J., MacEachern, J.A., 2001. Vertebrate burrow complexes from the Early Triassic *Cynognathus* Zone (Driekoppen Formation, Beaufort Group) of the Karoo Basin, South Africa. Palaios 16, 148–160.

Hasiotis, S.T., 2004. Reconnaissance of Upper Jurassic Morrison Formation ichnofossils, Rocky Mountain Region, USA: paleoenvironmental, stratigraphic, and paleoclimatic significance of terrestrial and freshwater ichnocoenoses. Sediment. Geol. 167, 177–268.

Hasiotis, S.T., Honey, J.G., 2000. Paleohydrologic and stratigraphic significance of crayfish burrows in continental deposits: examples from several Paleocene Laramide basins in the Rocky Mountains. J. Sediment. Res. 70, 127–139.

Hubert, J.F., Dutcher, J.A., 2010. *Scoyenia* escape burrows in fluvial pebbly sand: upper Triassic Sugarloaf Arkose, Deerfield Rift Basin, Massachusetts, USA. Ichnos 17, 20–24.

Hunt, A.P., Lucas, S.G., 2005. Tetrapod ichnofacies and their utility in the Paleozoic. Pennsylvanian Footprints in the Black Warrior Basin of Alabama. In: Buta, R.J., Rindsberg, A.K., Kopaska-Merkel, D.C. (Eds.), Pennsylvanian Footprints in the Black Warrior Basin of Alabama. Alabama Paleont. Soc., Birmingham, Mon. 1, pp. 113–119.

Hunt, A.P., Lucas, S.G., 2006. Permian tetrapod ichnofacies. In: Lucas, S.G., Cassinis, G., Schneider, J.W. (Eds.), Non-Marine Permian Biostratigraphy and Biochronology. Geol. Soc. London, Spec. Publ., vol. 265, pp. 137–156.

Hunt, A.P., Lucas, S.G., 2007. Tetrapod ichnofacies: a new paradigm. Ichnos 14, 59–68.

Hunt, A.P., Chin, K., Lockley, M.G., 1994. The palaeobiology of vertebrate coprolites. In: Donovan, S. (Ed.), The Palaeobiology of Trace Fossils. Wiley, Chichester, pp. 221–240.

Hunt, A.P., Lucas, S.G., Spielmann, J.A., 2005. Paleoenvironmental transects and tetrapod biotaxonichnofacies in New Mexico. New Mexico Mus. Nat. Hist. Sci., Bull. 31, 49–51.

Hunt, A.P., Lucas, S.G., Spielmann, J.A., Lerner, S.J., 2007. A review of vertebrate coprolites of the Triassic with descriptions of new Mesozoic ichnotaxa. New Mexico Mus. Nat. Hist. Sci., Bull. 41, 88–99.

Johnson, K.R., 1986. Paleocene bird and amphibian tracks from the Fort Union Formation, Bighorn Basin, Wyoming. Contr. Geol. 24, 1–10.

Keighley, D.G., Pickerill, R.K., 2003. Ichnocoenoses from the Carboniferous of eastern Canada and their implications for the recognition of ichnofacies in nonmarine strata. Atlant. Geol. 39, 1–22.

Kim, J.Y., Keighley, D.G., Pickerill, R.K., Hwang, W., Kim, K.S., 2005. Trace fossils from marginal lacustrine deposits of the Cretaceous Jinju Formation, southern coast of Korea. Palaeogeogr. Palaeoclimatol. Palaeoecol. 218, 105–124.

Kim, J.Y., Kim, K.S., Pickerill, R.K., 2002. Cretaceous nonmarine trace fossils from the Hasandong and Jinju formations of the Namhae area, Kyongsangnamdo, southeast Korea. Ichnos 9, 41–60.

Klappa, C.F., 1980. Rhizoliths in terrestrial carbonates: classification, recognition, genesis and significance. Sedimentology 27, 613–629.

Knaust, D., 2012. Trace-fossil systematics. In: Knaust, D., Bromley, R.G. (Eds.), Trace Fossils as Indicators of Sedimentary Environments. Developments in Sedimentology, vol. 64. Elsevier, Amsterdam, pp. 79–101.

Krapovickas, V., Ciccioli, P.L., Mángano, M.G., Marsicano, C.A., Limarino, C.O., 2009. Paleobiology and paleoecology of an arid-semiarid Miocene South American ichnofauna in anastomosed fluvial deposits. Palaeogeogr. Palaeoclimatol. Palaeoecol. 284, 129–152.

Kraus, M.J., Hasiotis, S.T., 2006. Significance of different modes of rhizolith preservation to interpreting paleoenvironmental and paleohydrologic settings: examples from Paleogene paleosols, Bighorn Basin, Wyoming, U.S.A. J. Sediment. Res. 76, 633–646.

Laporte, L.F., Behrensmeyer, A.K., 1980. Tracks and substrate reworking by terrestrial vertebrates in Quaternary sediments of Kenya. J. Sediment. Res. 50, 1337–1346.

Lawfield, A.M.W., Pickerill, R.K., 2006. A novel contemporary fluvial ichnocoenose: unionid bivalves and the *Scoyenia-Mermia* ichnofacies transition. Palaios 21, 391–396.

Lockley, M., 2007. A tale of two ichnologies: the different goals and potentials of invertebrate and vertebrate (tetrapod) ichnotaxonomy and how they relate to ichnofacies analysis. Ichnos 14, 39–57.

Lockley, M.G., Foster, J., 2006. Dinosaur and turtle tracks from the Morrison Formation (Upper Jurassic) of Colorado National Monument, with observations on the taxonomy of vertebrate swim tracks. New Mexico Mus. Nat. Hist. Sci., Bull. 36, 193–198.

Lockley, M.G., Hunt, A.P., Meyer, C.A., 1994. Vertebrate tracks and the ichnofacies concept: implications for palaeocology and palichnostratigraphy. In: Donovan, S.K. (Ed.), Palaeobiology of Trace Fossils. Wiley, Chichester, pp. 241–268.

Lockley, M.G., Lucas, S.G., Milàn, J., Harris, J.D., Avanzini, M., Foster, J.R., Spielmann, J.A., 2010. The fossil record of crocodylian tracks and traces: an overview. In: Milàn, J., Lucas, S.G., Lockley, M.G., Spielmann, J.A. (Eds.), Crocodyle Tracks and Traces. New Mexico Mus. Nat. Hist. Sci., Bull. 51, pp. 1–13.

Lucas, S.G., Lerner, A.J., Hunt, A.P., 2004. Permian tetrapod footprints from the Lucero uplift, central New Mexico, and Permian footprint biostratigraphy. New Mexico Mus. Nat. Hist. Sci., Bull. 25, 291–300.

Lucas, S.G., Lerner, A.J., Spielmann, J.A., Hunt, A.P., 2006. Invertebrate ichnofossils from the Upper Triassic Bull Canyon Formation, east-central New Mexico. New Mexico Mus. Nat. Hist. Sci., Bull. 36, 118–121.

MacEachern, J.A., Bann, K.L., Gingras, M.K., Zonneveld, J.-P., Dashtgard, S.E., Pemberton, S.G., 2012. The ichnofacies paradigm. In: Knaust, D., Bromley, R.G. (Eds.), Trace Fossils as Indicators of Sedimentary Environments. Developments in Sedimentology, vol. 64. Elsevier, Amsterdam, pp. 103–138.

MacEachern, J.A., Pemberton, S.G., Gingras, M.K., Bann, K.L., 2007. The ichnofacies paradigm: a fifty-year retrospective. In: Miller III, W. (Ed.), Trace Fossils. Concepts, Problems, Prospects. Elsevier, Amsterdam, pp. 52–77.

Mack, G.H., Leeder, M., Perez-Arlucea, M., Bailey, B.D.J., 2003. Early Permian silt-bed fluvial sedimentation in the Orogrande basin of the Ancestral Rocky Mountains, New Mexico, USA. Sediment. Geol. 160, 159–178.

MacNaughton, R.B., Pickerill, R.K., 1995. Invertebrate ichnology of the nonmarine Lepreau Formation (Triassic), southern New Brunswick, eastern Canada. J. Paleontol. 69, 160–171.

Mángano, M.G., Buatois, L.A., Claps, G.L., 1996. Grazing trails formed by soldier fly larvae (Diptera: Stratiomyidae) and their paleoenvironmental and paleoecological implications for the fossil record. Ichnos 4, 163–167.

Marriott, S.B., Morrissey, L.B., Hillier, R.D., 2009. Trace fossil assemblages in Upper Silurian tuff beds: evidence of biodiversity in the Old Red Sandstone of southwest Wales, UK. Palaeogeogr. Palaeoclimatol. Palaeoecol. 274, 160–172.

Martin, A.J., 2009. Neoichnology of an Arctic fluvial point bar, North Slope, Alaska (USA). Geol. Quart. 53, 383–396.

Martin, A.J., Varricchio, D.J., 2011. Paleoecological utility of insect trace fossils in dinosaur nesting sites of the Two Medicine Formation (Campanian), Choteau, Montana. Hist. Biol. 23, 15–25.

Martin, L.D., Bennett, D.K., 1977. The burrows of the Miocene beaver *Palaeocastor*, Western Nebraska, U.S.A. Palaeogeogr. Palaeoclimatol. Palaeoecol. 22, 173–193.

Matsukawa, M., Futakami, M., Lockley, M.G., Chen, P., Chen, J., Cao, Z., Bolotsky, U.L., 1995. Dinosaur footprints from the Lower Cretaceous of eastern Manchuria, northeastern China; implications for the recognition of an ornithopod ichnofacies in East Asia. Palaios 10, 3–15.

Maulik, P.K., Chaudhuri, A.K., 1983. Trace fossils from continental Triassic red beds of the Gondwana sequence, Pranhita-Godavari Valley, South India. Palaeogeogr. Palaeoclimatol. Palaeoecol. 41, 17–34.

Melchor, R.N., 2009. Bird tracks preserved in fluvial channel facies of the Río Negro Formation (Neogene), La Pampa province, Argentina. Ameghiniana 46, 209–214.

Melchor, R.N., Bellosi, E., Genise, J.F., 2003. Invertebrate and vertebrate trace fossils from a Triassic lacustrine delta: the Los Rastros Formation, Ischigualasto Provincial Park, San Juan, Argentina. In: Buatois, L.A., Mángano, M.G. (Eds.), Icnología: Hacia una Convergencia entre Geología y Biología. Asociación Paleontológica, Argentina, Publicación Especial, vol. 9, pp. 17–33.

Melchor, R.N., Sarjeant, W.A.S., 2004. Small amphibian and reptile footprints from the Permian Carapacha Basin, Argentina. Ichnos 11, 57–78.

Melchor, R.N., Bedatou, E., de Valais, S., Genise, J.F., 2006. Lithofacies distribution of invertebrate and vertebrate trace-fossil assemblages in an Early Mesozoic ephemeral fluvio-lacustrine system from Argentina: implications for the *Scoyenia* Ichnofacies. Palaeogeogr. Palaeoclimatol. Palaeoecol. 239, 253–285.

Melchor, R.N., Genise, J.F., Farina, J.L., Sánchez, M.V., Sarzetti, L., Visconti, G., 2010. Large striated burrows from fluvial deposits of the Neogene Vinchina Formation, La Rioja, Argentina: a crab origin suggested by neoichnology and sedimentology. Palaeogeogr. Palaeoclimatol. Palaeoecol. 291, 400–418.

Melchor, R.N., Cardonatto, M.C., Visconti, G., 2012. Palaeonvironmental and palaeoecological significance of flamingo-like footprints in shallow-lacustrine rocks: an example from the Oligocene-Miocene Vinchina Formation, Argentina. Palaeogeogr. Palaeoclimatol. Palaeoecol. 315–316, 181–198.

Metz, R., 1992. Trace fossils from the Lower Jurassic nonmarine Towaco Formation, New Jersey. Northeastern Geology 14, 29–34.

Miller, W.R., Mason, T.R., 2000. *Stellavelum arborensis* igen., isp. nov., *Stellavelum uncinum* igen., isp. nov. and *Termitichnus namibiensis* isp. nov.; new ichnofossils from Cenozoic deposits of Namaqualand, South Africa. Ichnos 7, 195–215.

Morrissey, L.B., Braddy, S.J., 2004. Terrestrial trace fossils from the Lower Old Red Sandstone, southwest Wales. Geol. J. 39, 315–336.

Nadon, G.C., 1993. The association of anastomosed fluvial deposits and dinosaur tracks, eggs, and nests: implications for the interpretation of floodplain environments and a possible survival strategy for ornithopods. Palaios 8, 31–44.

Neef, G., 2004. Devonian arthropod trackways from the fluvial Ravendale Formation, western New South Wales. Alcheringa 28, 401–402.

Netto, R.G., 2007. *Skolithos*-dominated piperock in non-marine environments: an example from the Triassic Caturrita Formation, southern Brazil. In: Bromley, R.G., Buatois, L.A., Mángano, M.G., Genise, J.F., Melchor, R.N. (Eds.), Sediment-Organism Interactions: A Multifaceted Ichnology. SEPM Spec. Publ., vol. 88, pp. 109–121.

Picard, M.D., High, L.R.J., 1973. Sedimentary Structures of Ephemeral Streams. Developments in Sedimentology, vol. 17. Elsevier, Amsterdam, 223 pp.

Platt, N.H., Meyer, C.A., 1991. Dinosaur footprints from the Lower Cretaceous of northern Spain: their sedimentological and palaeoecological context. Palaeogeogr. Palaeoclimatol. Palaeoecol. 85, 321–333.

Pollard, J.E., 1976. A problematic trace fossil from the Tor Bay Breccias of south Devon. Proc. Geol. Ass. 87, 105–108.

Pollard, J.E., 1988. Trace fossils in coal-bearing sequences. J. Geol. Soc. London 145, 339–350.

Pryor, W.A., 1967. Biogenic directional features on several recent point-bars. Sediment. Geol. 1, 235–245.

Ratcliffe, B.C., Fagerstrom, J.A., 1980. Invertebrate lebensspuren of Holocene flood plains: their morphology, origin and paleoecological significance. J. Paleontol. 54, 614–630.

Retallack, G.J., 1984. Trace fossils burrowing beetles and bees in an Oligocene paleosol, Badlands National Park, South Dakota. J. Paleontol. 58, 571–592.

Retallack, G.J., 1990. Soils of the Past. An Introduction to Paleopedology. Unwin Hyman, London, 520 pp.

Retallack, G.J., 2001. Soils of the Past. An Introduction to Paleopedology. Blackwell Science, Oxford, 404 pp.

Retallack, G.J., Bestland, E.A., Fremd, T.J., 2000. Eocene and Oligocene paleosols of central Oregon. GSA Spec. Pap. 344, 1–192.

Roberts, E.M., Rogers, R.R., Foreman, B.Z., 2007. Continental insect borings in dinosaur bone: examples from the Late Cretaceous of Madagascar and Utah. J. Paleontol. 81, 201–208.

Santi, G., Nicosia, U., 2008. The ichnofacies concept in vertebrate ichnology. In: Avanzini, M., Petti, F.M. (Eds.), Italian Ichnology. Studi Trentini di scianze Naturali, Acta Geol. 83, pp. 223–229.

Sarkar, S., Chaudhuri, A.K., 1992. Trace fossils in Middle to Late Triassic fluvial redbeds, Pranhita-Godavari valley, south India. Ichnos 2, 7–19.

Schlirf, M., Uchman, A., Kümmel, M., 2001. Upper Triassic (Keuper) non-marine trace fossils from the Haßberge area (Franconia, south-eastern Germany). Paläont. Z. 75, 71–96.

Seilacher, A., 1967. Bathymetry of trace fossils. Mar. Geol. 5, 413–428.

Seilacher, A., 1977. Use of trace fossil assemblages for recognizing depositional environments. In: Basan, P.B. (Ed.), Trace Fossils Concepts. SEPM Short Course Notes 5, pp. 185–200.

Shanley, K.W., McCabe, P.J., Hettinger, R.D., 1992. Tidal influence in Cretaceous fluvial strata from Utah, USA: a key to sequence stratigraphic interpretation. Sedimentology 39, 905–930.

Sidor, C.A., Miller, M.F., Isbell, J.L., 2008. Tetrapod burrows from the Triassic of Antarctica. J. Vert. Paleont. 28, 277–284.

Siggerud, E.I.H., Steel, R.J., 1999. Architecture and trace-fossil characteristics of a 10,000-20,000 year, fluvial-to-marine sequence, SE Ebro Basin, Spain. J. Sediment. Res. 69, 365–383.

Smith, R.M.H., 1987. Helical burrow casts of therapsid origin from the Beaufort Group (Permian) of South Africa. Palaeogeogr. Palaeoclimatol. Palaeoecol. 60, 155–169.

Smith, R.M.H., 1993. Sedimentology and ichnology of floodplain paleosurfaces in the Beaufort Group (Late Permian), Karoo sequence, South Africa. Palaios 8, 339–357.

Smith, N.D., Heim, F.J., 1971. Biogenic reworking of fluvial sediments by staphylinid beetles. J. Sediment. Petrol. 41, 598–602.

Smith, R.M.H., Mason, T.R., Ward, J.D., 1993. Flash-flood sediments and ichnofacies of the Late Pleistocene Homeb Silts, Kuiseb River, Namibia. Sediment. Geol. 85, 579–599.

Smith, R.M.H., Marsicano, C.A., Wilson, J.A., 2009. Sedimentology and paleoecology of a diverse Early Jurassic tetrapod tracksite in Lesotho, Southern Africa. Palaios 24, 672–684.

Smith, J.J., Platt, B.F., Ludvigson, G.A., Thomasson, J.R., 2010. Ant-nest ichnofossils in honeycomb calcretes, Neogene Ogallala Formation, High Plains region of western Kansas, U.S.A. Palaeogeogr. Palaeoclimatol. Palaeoecol. 308, 383–394.

Squires, R.L., Advocate, D.M., 1984. Meniscate burrows from Miocene lacustrine-fluvial deposits, Diligencia Formation, Orocopia Mountains, southern California. J. Paleontol. 58, 593–597.

Stanley, K.O., Fagerstrom, J.A., 1974. Miocene invertebrate trace fossils from a braided river environment, western Nebraska, U.S.A. Palaeogeogr. Palaeoclimatol. Palaeoecol. 15, 63–82.

Tandon, S.K., Naug, B., 1984. Facies-trace fossil relationships in a Plio-Pleistocene fluvial sequence—the upper Siwalik Subgroup, Punjab sub-Himalaya, India. Palaeogeogr. Palaeoclimatol. Palaeoecol. 47, 277–299.

Uchman, A., Pika-Biolzi, M., Hochuli, P.A., 2004. Oligocene trace fossils from temporary fluvial plain ponds: an example from the Freshwater Molasse of Switzerland. Eclogae Geol. Helv. 97, 133–148.

Uchman, A., Gaigalas, A., Melešytė, M., Kazakauskas, V., 2007. The trace fossil *Asthenopodichnium lithuanicum* isp. nov. from Late Neogene brown-coal deposits, Lithuania. Geol. Quart. 51, 329–336.

Umazano, A.M., 2009. Sedimentación fluvial en ambientes volcaniclásticos cretácicos de la Cuenca San Jorge occidental (Grupo Chubut), Argentina. Ph.D. Thesis, Universidad Nacional de San Luis, San Luis, 226 pp.

Van Allen, H.E.K., Calder, J.H., Hunt, A.P., 2005. The trackway record of a tetrapod community in a Walchian conifer forest from the Permo-Carboniferous of Nova Scotia. New Mexico Mus. Nat. Hist. Sci., Bull. 30, 322–332.

Williamson, T.E., Lucas, S.G., 1996. Mammal footprints from the Miocene-Pliocene Ogallala Formation, eastern New Mexico. New Mexico Geol. 18, 1–5.

Zavala, C.A., Santiago, M.F., Amaolo, G.E., 1993. Depositos fluviales en la Formacion Tunas (Pérmico). Cuenca Paleozoica de Ventania, Provincia de Buenos Aires. Rev. Assoc. Geol. Argentina 48, 307–316.

Chapter 13

Lacustrine Environments

Jennifer J. Scott,*,‡,1 Luis A. Buatois† and M. Gabriela Mángano†
**Department of Earth & Atmospheric Sciences, University of Alberta, Edmonton, Alberta, Canada, †Department of Geological Sciences, University of Saskatchewan, Saskatoon, Saskatchewan, Canada, ‡Present address: Department of Earth Sciences, Mount Royal University, Calgary, Alberta, Canada*
[1]Corresponding author: e-mail: jescott@mtroyal.ca

1. INTRODUCTION

Lakes are extremely complex systems in which animal and plant distribution is controlled by a delicate interplay of abiotic (e.g., energy, light, oxygen, temperature, salinity, substrate, and nutrients) and biotic factors (e.g., competition, grazing, predation, and symbiosis) (Buatois and Mángano, 2011; Cohen, 2003; Miller and White, 2007; Renaut and Gierlowski-Kordesch, 2010). Lacustrine systems differ from oceans in several ways, including the smaller volumes of sediment and water in lakes, the direct link between lake level and sediment supply, and the fact that shoreline migration may be due to not only progradation but also due to withdrawal of water (Bohacs et al., 2000). The vertical relationships of lithofacies commonly reflect abrupt changes, particularly when the changes are due to tectonics or climate-controlled fluctuating water levels. The chemistry of groundwater and lake water can be highly variable, even within a single basin, and can change over relatively short time periods. In general, the impacts of environmental changes in continental basins are more pronounced than in oceanic basins. Similarly, the variability of organisms living in adjacent environments can lead to pronounced lateral heterogeneity in trace-fossil assemblages.

Ichnology can be applied to the interpretation of sedimentary successions in lake basins at several scales. Detailed information about sedimentary processes and substrate conditions can be interpreted using the evidence provided by biogenic traces, together with sedimentology. Features showing substrate consistency (e.g., deeply impressed vertebrate footprints), changing sedimentation rates (e.g., escape traces), and depositional hiatuses (e.g., intensely bioturbated horizons), for example, increase the ability to interpret sedimentary processes affecting individual beds and bedsets. By considering the specific types of biogenic structures produced (e.g., vertebrate footprints, insect nests), it may be possible to recognize whether a substrate was subaerially exposed, to interpret the relative duration of

Developments in Sedimentology, Vol. 64. http://dx.doi.org/10.1016/B978-0-444-53813-0.00013-7

exposure and the depth to the water table, which can approximate the extent of lake-level changes (e.g., de Gibert and Sáez, 2009; Genise et al., 2004). By recognizing biogenic structures that are produced subaqueously and cross-cut by those produced subaerially, for example, a change in substrate conditions can be identified (e.g., Buatois et al., 1997; Kim et al., 2002; Metz, 1996).

Recurrent assemblages of plant and animal trace fossils through geological time (i.e., ichnofacies) provide a guide to recognizing the sets of environmental controls to which the organisms responded, and hence to the general depositional setting. A consideration of the "taphonomic pathways" of these assemblages leads to a better understanding of changing conditions (Buatois and Mángano, 2004, 2011). Through the more detailed recognition of recurrent associations of biogenic structures (i.e., at the scale of suites, ichnofabrics, or ichnocoenoses), together with an analysis of their vertical distribution within a sedimentary succession, ichnology can be applied to the reconstruction of the depositional system (e.g., Melchor, 2004; Melchor et al., 2006; Porter and Gallois, 2008; Voigt and Hoppe, 2010). Tiering within a single infaunal community may be uncommon in lake basins, except in well-oxygenated freshwater lakes (e.g., McCall and Tevesz, 1982) and in well-developed paleosols with insects that burrow and nest at different depths (e.g., Genise et al., 2004; Hasiotis, 2007). Palimpsest and/or exhumed substrates may be particularly common in frequently exposed lake-marginal deposits where net sedimentation rates are low and ichnofacies or suites are commonly overprinted (e.g., Buatois and Mángano, 2004, 2009; Scott et al., 2009; Uchman and Álvaro, 2000). These types of substrates represent surfaces that are important for the sequence stratigraphy of lake basins and for deciphering the dynamic history of the lake.

The sedimentary environments in lake basins are dictated by complex sets of controls, many of which also directly impact the biodiversity (numbers of species, abundance) of microorganisms, plants, and animals within the basin (Fig. 1; e.g., Carroll and Bohacs, 2001; Cohen, 2003; Gierlowski-Kordesch and Park, 2004; Renaut and Gierlowski-Kordesch, 2010; Talbot and Allen, 1996). Deposition in lake basins is closely related to both the climate and the tectonic setting, through factors such as the precipitation/evaporation ratio, the size and shape of the basin, and variable subsidence rates (Bohacs et al., 2000). The stratigraphic packaging of the basin fill is a direct response to the relationship between the amount of sediment plus water input to the basin and the potential accommodation, or the spatial volume at elevations lower than basin-bounding structural or geomorphological features (Carroll and Bohacs, 1999). The distribution of terrestrial *versus* aquatic settings within a lake basin is dependent on this relationship, resulting in lake basins that can be either relatively full with water or relatively empty, and which can vary through time.

The tectonic setting can also influence whether the inflow to the lake is provided by rivers and/or springs, with specific structural features in the basin controlling the distribution of the fluvio-lacustrine systems and spring-fed wetlands, for example, as well as the spill points over which lake waters can flow out of the basin (e.g., Pietras and Carroll, 2006). The gradient of the basin floor

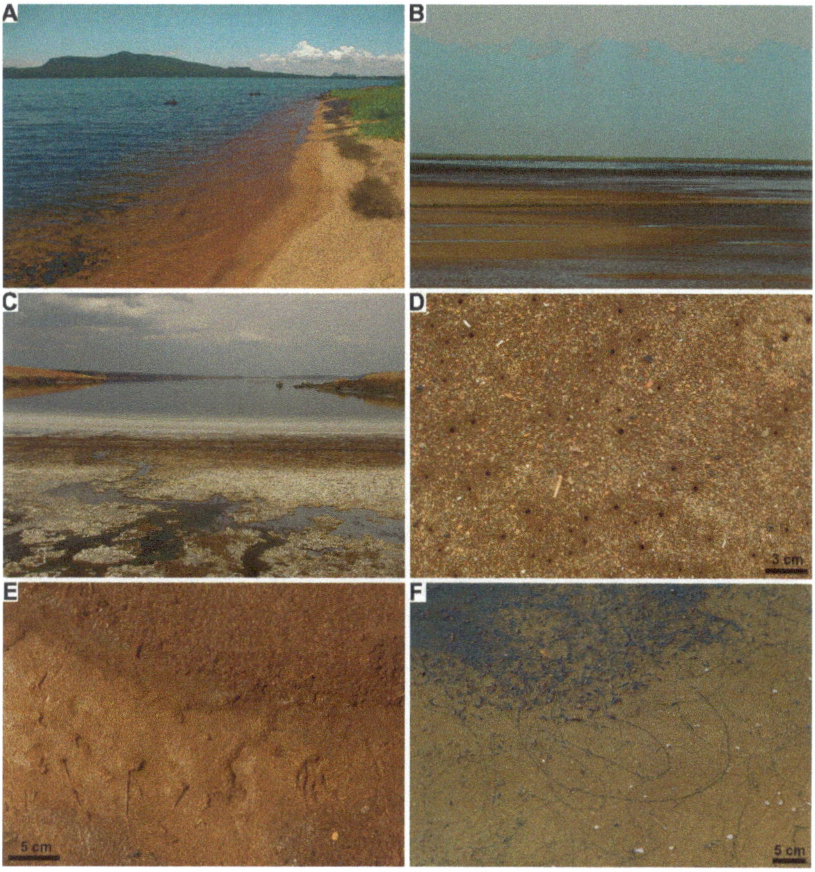

FIGURE 1 Modern lakes and animal traces produced in littoral to eulittoral zones. (A) Freshwater sandy shoreline of north-eastern Lake Victoria, Kenya. (B) Extensive sandy mud flats of the eastern Great Salt Lake, Utah, USA. (C) Organic-rich hypersaline mud flats at Nasikie Engida (Little Magadi), Kenya. (D) Plan view of circular vertical burrows produced by tiger beetle larvae in eulittoral sandy swash zone deposits, saline Lake Bogoria, Kenya. Photograph courtesy of R. Renaut. (E) Branched, surface-tunnel networks of crickets and a mongoose footprint in eulittoral sandy mud flats, saline Lake Bogoria, Kenya. (F) Surface trails of dipteran larvae in very shallow water at saline Lake Magadi, Kenya.

influences the depth of the water table in surrounding terrestrial environments, determines the extent of lake-level fluctuations in lake margins, influences the extent of oxygenated shallow lacustrine waters, and partly determines the depth of the lake. Additionally, the bedrock geology within the catchment may strongly influence clastic and chemical sedimentation as well as the abiotic factors that impact the behavior and distribution of organisms, such as pH, ionic concentrations, salinity, and turbidity of inflow waters (Gierlowski-Kordesch, 2010; Renaut and Gierlowski-Kordesch, 2010).

The zonation scheme of environments within lake basins presented here (Table 1) allows the general characterization of the sets of conditions that determine the nature of sedimentation and the typical biological assemblages present

TABLE 1 A Lake Basin Zonation Scheme for the Interpretation of Lake-Basin Sedimentary Successions with Trace Fossils

<table>
<tr><td rowspan="5">Hierarchical classification of lake-basin zones</td><td colspan="9">Lake-basin zonation</td></tr>
<tr><td colspan="4">Basin margin to lake margin</td><td colspan="5">Lake margin to basin center</td></tr>
<tr><td colspan="2">Terrestrial</td><td colspan="3">Lake margin</td><td colspan="4">Lacustrine</td></tr>
<tr><td>Margin</td><td colspan="6">Zone of lake-level rise-and-fall (including ephemeral lakes)</td><td colspan="2">Permanently lacustrine</td></tr>
<tr><td>Margin</td><td>Terrestrial</td><td colspan="2">Supralittoral</td><td>Eulittoral</td><td colspan="2">Littoral</td><td>Sublittoral</td><td>Profundal</td></tr>
<tr><td></td><td colspan="9">Depositional environments</td></tr>
<tr><td>Wind/wave-influenced lacustrine environments</td><td>Terrestrial</td><td>Eolian</td><td>Backshore (lagoon, flats, and/or marsh)</td><td>Beach</td><td>Swash zone (shoals, spits)</td><td colspan="2">Upper shoreface or high-energy bench platform (above normal wave base)</td><td>Lower shoreface or bench slope</td><td>Offshore (below storm wave base; below chemocline/thermocline)</td></tr>
<tr><td>Alluvial to fluvio-lacustrine environments</td><td>Alluvial fans</td><td>Alluvial floodplain</td><td>Upper delta-plain</td><td>Delta-plain</td><td>Lower delta-plain</td><td>Proximal delta-front</td><td>Distal delta-front</td><td>Prodelta</td><td>Profundal (turbidites, underflow deposits)</td></tr>
<tr><td rowspan="2">Low-energy lacustrine environments</td><td rowspan="2">Terrestrial</td><td colspan="4" rowspan="2">Wetlands and/or mud flats (palustrine)
(not applicable for high-gradient or coincident margins)</td><td colspan="2">Low-energy littoral or low-energy bench platform</td><td rowspan="2">Sublittoral slope or low-energy bench slope</td><td rowspan="2">Profundal (suspension fall-out deposition)</td></tr>
<tr><td>Emergent vegetation</td><td>Submerged vegetation</td></tr>
<tr><td></td><td colspan="5">← Increasing duration of periods of subaerial exposure</td><td colspan="2">Occasionally subaerially exposed</td><td colspan="2">Not normally subaerially exposed</td></tr>
<tr><td></td><td colspan="9">Controls affecting sedimentation, biogenic activity, and production and preservation of biogenic structures by lake-basin zone</td></tr>
<tr><td rowspan="2">Basin-scale controls</td><td colspan="9">Geological controls: Tectonic setting and activity; structural features, bedrock geology and sediment composition, subsidence rate, basin shape and size, gradient, base level, preexisting geomorphologic features, autogenic gradient-related features (e.g., benches, deltas)</td></tr>
<tr><td colspan="9">Climatic controls: Climatic setting, mean annual temperatures, seasonality; precipitation/evapotranspiration ratio, temperature and humidity extremes, climatic cycles and climatic disturbances, paleoclimatic history</td></tr>
</table>

TABLE 1 A Lake Basin Zonation Scheme for the Interpretation of Lake-Basin Sedimentary Successions with Trace Fossils—Cont'd

	Controls affecting sedimentation, biogenic activity, and production and preservation of biogenic structures by lake-basin zone—Cont'd					
	Environmental controls: Abundance and types of food resources, nutrient availability, interconnectedness and heterogeneity of habitats, sedimentary disturbances by erosion or deposition, sediment texture, degree of habitat stability and temporal variability					
	Biological controls: Evolutionary pathways of organisms, dispersal mechanisms of organisms, complexity of food webs, types of animals, body plans, behavior and life history of organisms, interspecific competition and interaction, chance					
	Terrestrial	**Supralittoral**	**Eulittoral**	**Littoral**	**Sublittoral**	**Profundal**
Hydrochemical controls	Ionic composition, ionic concentration, and pH of surface and shallow ground waters		Mixing of phreatic, vadose zone, and lake waters	Ionic composition, ionic concentration, and pH of lake waters, sublacustrine spring waters, and porewaters of lacustrine sediments		
	Temperature of surface, vadose zone, and subaerial spring waters		(Mixing zone)	Temperature of lake waters and sublacustrine spring waters		
	Redoximorphic conditions of sedimentary profile			Dissolved oxygen concentrations in lake waters, oxygenation and redox conditions of lacustrine sediments		
	Concentrations of soil gases (O_2, CO_2, NH_4), alkalinity of pore waters	Concentrations of CO_2 and HCO_3^- (alkalinity) in vadose-zone pore waters and surface waters		Concentrations of CO_2, HCO_3^-, and CO_3^{2-} (alkalinity) in lake and sublacustrine spring waters		
	Climate-related variability of temperature of surface and vadose-zone waters			Temperature and chemical gradients of lake waters (causing density stratification)		
	Lateral variability in hydrochemical characteristics of surface waters and vadose-zone pore waters		(Mixing zone)	Lateral variability of hydrochemical characteristics of lake waters (related to subbasin configuration, sublacustrine springs, and mixing)		
	Solar radiation; shade; air and ground temperatures			Solar radiation, air and lake water temperatures		
	Depth to water table	Capillary evaporation		Direct evaporation		
	Moisture conditions, drainage (permeability and porosity)	Substrate consolidation and degree of induration, substrate cohesiveness and saturation, drainage		Substrate consolidation and degree of induration, substrate cohesiveness		
	Nutrient cycling by infaunal and epifaunal air-breathing organisms			Nutrient cycling by infaunal and epifaunal benthic aquatic organisms		

Continued

TABLE 1 A Lake Basin Zonation Scheme for the Interpretation of Lake-Basin Sedimentary Successions with Trace Fossils—Cont'd

<table>
<tr><th></th><th>Terrestrial</th><th>Supralittoral</th><th>Eulittoral</th><th colspan="2">Littoral</th><th>Sublittoral</th><th>Profundal</th></tr>
<tr><td rowspan="10">Physical controls (including hydrodynamic, substrate-related, and climate-related controls)</td><td>Wind energy and deflation or eolian deposition</td><td>Wind energy, deflation or eolian deposition, fetch and storm wave energy</td><td colspan="3">Fetch and wave energy, daily to seasonal wind-direction variability, variable barometric pressure and wave heights</td><td>Fetch and storm wave energy</td><td>Lake depth</td></tr>
<tr><td></td><td colspan="5">Frequency and intensity of storms</td><td>Event sedimentation (tempestites, varves)</td></tr>
<tr><td colspan="4"></td><td colspan="3">Currents (due to wind, density gradients, atmospheric pressure gradients)</td></tr>
<tr><td colspan="3">Sediment load, discharge, and current energy in fluvial and deltaic channels</td><td colspan="3">Hypopycnal, homopycnal, or hyperpycnal flows in deltaic settings; sediment load, discharge, and current energy in channels</td><td>Hyperpycnal flows and turbidites, sediment load and discharge</td></tr>
<tr><td colspan="3"></td><td colspan="4">Turbidity</td></tr>
<tr><td colspan="5"></td><td colspan="2">Mixing regime</td></tr>
<tr><td>Surface water availability, fluvial flooding cycles</td><td colspan="3">Frequency of lake-level fluctuations, duration of periods of subaerial exposure, effects of short-term and long-term climate cycles on water input and output</td><td colspan="2">Exposure of lake floor due to longer-term climate cycles or tectonic activity</td><td>Exposure due to major basin-scale changes</td></tr>
<tr><td colspan="3">Solar radiation; shade; air and ground temperatures</td><td colspan="4">Solar radiation, air and lake water temperatures</td></tr>
<tr><td>Depth to water table</td><td colspan="2">Capillary evaporation</td><td colspan="4">Direct evaporation</td></tr>
<tr><td>Moisture conditions, drainage (permeability and porosity)</td><td colspan="2">Substrate consolidation and degree of induration, substrate cohesiveness and saturation, drainage</td><td colspan="4">Substrate consolidation and degree of induration, substrate cohesiveness</td></tr>
<tr><td rowspan="5">Biological controls</td><td>Productivity of soil microorganisms</td><td>Productivity of benthic and soil microorganisms</td><td>Mixed</td><td colspan="3">Productivity of benthic and planktonic organisms (depth of light penetration)</td><td>Benthic and planktonic organisms, decay</td></tr>
<tr><td colspan="2">Respiration and gas release by soil organisms</td><td colspan="4">Respiration and gas release by semi-aquatic and aquatic organisms</td><td>Respiration/gas release by aquatic organisms</td></tr>
<tr><td colspan="5">Presence of vegetation, shade from vegetation, types of vegetation</td><td colspan="2">Deposition of plant-derived detritus</td></tr>
<tr><td></td><td colspan="3">Trampling of substrate by vertebrates</td><td colspan="3"></td></tr>
<tr><td colspan="3">Nutrient cycling by infaunal and epifaunal air-breathing organisms</td><td colspan="4">Nutrient cycling by infaunal and epifaunal benthic aquatic organisms</td></tr>
</table>

The presence of the zones and their lateral extent vary with the unique characteristics of each lake basin (e.g., gradient, lake depth) and the features related to lake-basin type (e.g., extent and frequency of lake-level fluctuations). Each lake basin may contain one or more of the depositional systems listed depending on the lake-basin size and morphometry. Carbonate and siliciclastic systems are not subdivided in this classification. Ephemeral lakes are present within the zone of lake-level rise and fall. Springs may be present in any zone. *This scheme was constructed with reference to Ward (1992), Williams (1998), Wetzel (2001), Cohen (2003), Schowalter (2006), Buatois and Mángano (2007), Miller and White (2007), Hasiotis (2007), and Renaut and Gierlowski-Kordesch (2010).*

across a lake basin. Depending on the mixing regime of the lake and on whether it is hydrologically open or closed, among other controls, the environmental characteristics affecting sedimentation and biogenic activity within each zone may differ significantly. The particular environmental conditions between zones in wave/wind-influenced lacustrine depositional systems, alluvial to fluvio-lacustrine depositional systems, or low-energy lacustrine depositional systems are extremely variable. Nevertheless, several parameters affect lakes of all types as well as the distribution and behavior of organisms and the production and preservation of trace fossils (Table 1).

The lake-type model of Carroll and Bohacs (1999) and Bohacs et al. (2000) recognizes three types of lake basins according to the input of sediment and water to the basin and the accommodation potential of the basin: overfilled, balanced-fill, and underfilled. In overfilled lake-type basins, the input of water and sediment exceeds the potential accommodation, and lakes overflow into an adjacent basin. Balanced-fill lake-type basins are characterized by rates of sediment and water supply approximately in balance with potential accommodation, so that balanced-fill lake types periodically shift between being hydrologically open or closed. In underfilled lake-type basins, the potential accommodation exceeds the amount of sediment and water input. This relationship influences the dominant sedimentary facies, the stratal packaging of parasequences, the dominant types of organic matter, and the distribution and flow characteristics of source rocks, reservoirs, and seals (Bohacs et al., 2000; Carroll and Bohacs, 2001). This relationship also determines the distribution of habitats and the main limiting environmental controls on trace-producing organisms.

Buatois and Mángano (2004, 2007, 2009) integrate the lake-type basin model with the ichnofacies model of lacustrine and lake-margin environments. This approach is followed here in order to show how the dynamic interactions between lake-basin sedimentary environments and trace-producing plants and animals can be applied to reconstructing depositional systems and the sequence-stratigraphic interpretation of the basin fill. The extreme complexity and variability of lacustrine systems should always be considered when applying ichnology to the reconstruction of lake-basin successions. However, the lake-type model permits the detection of several general ichnological trends and the placement of documented case studies within an already existing, useful stratigraphic model. Although terrestrial and fluvial systems may be widespread in lake basins, especially in underfilled lake basins, this chapter focuses on the aquatic zones of lakes, as well as the zone of typical lake-level rise and fall (Table 1).

2. ICHNOLOGY

2.1 Biogenic Activity and Structures in Lake Basins

The diversity and distribution of organisms within and surrounding a lake are controlled by numerous abiotic and biotic factors (Table 1; e.g., Cohen, 2003; Hammer, 1986; Miller and White, 2007; Wetzel, 2001). Lakes contain a large variety of benthic organisms capable of producing traces, such as annelids (e.g., oligochaetes), aquatic to semiaquatic insects (e.g., dipterans, coleopterans), crustaceans (e.g., decapods, amphipods), and molluscs (e.g., bivalves, gastropods) (Cohen, 2003; White and Miller, 2008). Vertebrates that inhabit lacustrine settings, including fish, amphibians, and aquatic reptiles, may also interact with the substrate. In subaerially exposed lake-margin settings, trace-producing organisms include semiaquatic to terrestrial insects (e.g., crickets, earwigs, beetles, ants, termites), arachnids (e.g., spiders, mites), crustaceans (e.g., crayfish), and molluscs (e.g., land snails). Vertebrate diversity may be high in lake margins, even around saline lakes, and include trace producers such as tetrapods (e.g., mammals, reptiles, amphibians) and birds (e.g., shorebirds, flamingos). Plants that produce biogenic structures occur in the shallow, oxygenated littoral zone of lakes (e.g., sedges), as well as in lake margins with shallow water tables (e.g., grasses, sedges) and terrestrial and fluvial settings near the lake (e.g., shrubs and trees, ferns, herbaceous plants). The diversity and distribution of plants and their traces are also controlled by similar environmental parameters as invertebrates and vertebrates (e.g., salinity, oxygen, sediment texture and drainage; Cohen, 1982; Kraus and Hasiotis, 2006).

In modern freshwater lakes, the density and diversity of the aquatic benthic fauna usually reach a maximum in the oxygenated sublittoral zone, referred to as the concentration zone by Mackie (2001). In most lakes, this zone occurs between depths of 2 and 4 m, but may be as deep as 18 m in large well-mixed lakes (Mackie, 2001; Martin et al., 2005; White and Miller, 2008). Lakes with high productivity of planktonic microorganisms (e.g., warm tropical lakes, nutrient-rich lakes), and/or high salinities, are generally not well oxygenated, restricting the biodiversity to the littoral to sublittoral zones (Hammer, 1986). In lake-marginal settings with harsh conditions (e.g., high temperature, high salinity), tracemakers may concentrate in localities where conditions (e.g., available freshwater, food resources) are more suitable (e.g., Gierlowski-Kordesch, 1991; Scott et al., 2007, 2009). Hydrochemical conditions, including the availability of free oxygen, the pH, ionic concentrations, temperature, and drainage, may be the primary factors affecting the presence and behavior of infaunal and epifaunal organisms in both aquatic lacustrine and subaerial lake-margin environments with shallow water tables. Energy levels, sedimentation rate, turbidity, and substrate conditions may be secondarily important for the diversity and distribution of tracemaking organisms, but these factors are certainly important for the production and preservation of trace fossils.

The types of traces common in lakes and lake margins that are produced by invertebrates in either subaqueous or subaerial substrates include simple surface trails, open surface tunnels, trackways, burrows backfilled with waste material and/or excavated sediment, branching 3D burrow systems, and vertical burrows (Fig. 1). Vertebrates produce structures such as swim traces, footprints and trackways, trampled surfaces, nest-mounds, and burrows. Vertebrate traces may be present in the profundal and sublittoral zones of a lake, or shallow lacustrine and lake-margin areas, and may be produced either subaqueously or subaerially, depending on the tracemaker, their behavior, and the environmental setting. Insect nests and mammal burrows may also be present in lacustrine or lake-margin deposits if lake levels drop sufficiently following sedimentation (e.g., Genise et al., 2010; Melchor et al., 2002; Scott et al., 2009; Smith and Mason, 1998).

Neoichnological studies of lacustrine and lake-marginal environments have documented many types of biogenic structures produced by a variety of continental organisms (e.g., Chamberlain, 1975; Clark and Ratcliffe, 1989; Garcia and Niell, 1991; Genise et al., 2009; Gingras et al., 2007; Hasiotis and Mitchell, 1993; Knecht et al., 2009; Metz, 1987, 1990; Ratcliffe and Fagerstrom, 1980; Uchman, 2005), although much work remains to characterize the plethora of trace types produced by plants, invertebrates, and vertebrates. Studies that consider the distribution of different infaunal and epifaunal biological communities and the resultant traces produced and potentially preserved in terms of environmental controls are comparatively rare (e.g., Cohen, 1982; Cohen et al., 1991, 1993; Fisher, 1982; Hamer and Sheldon, 2010; McCall and Tevesz, 1982; Scott et al., 2007, 2009). Biological studies of epifaunal and infaunal animals in lacustrine and lake-margin settings (e.g., oligochaetes, insect larvae, bivalves) are numerous (e.g., Fisher and Beeton, 1975; Knisley, 1984; McCall et al., 1986), although in many cases the different research focuses of ichnologists and biologists make the translation of the results necessary.

Although it is sometimes assumed that little has been published on the trace-fossil record of lakes, recent compilations reveal a large volume of literature on lacustrine and lake-margin ichnofaunas (Buatois and Mángano, 2007; Minter et al., 2007). This reflects not only ecological aspects, but also that traces emplaced in lacustrine sediments have the highest preservation potential of all continental trace fossils (Buatois and Mángano, 2011). The highest preservation potential of biogenic structures within lakes is in low-energy areas, such as those that are only weakly affected by wave action. In deep lake environments, alternation of very fine-grained sands and muds deposited from underflow or turbidity currents, or the seasonal to short-term variations in sedimentation rates or sediment composition (e.g., varves), promotes the preservation of delicate and tiny surface trails, as well as very shallow-tier burrows and insect trackways (Benner et al., 2009; Buatois and Mángano, 1998, 2007; Netto et al., 2012; Uchman et al., 2009). In low-energy shoreline deposits, trace-fossil preservation commonly results from the rapid influx of sand via non-erosive sheet floods (e.g., Minter et al., 2007; Zhang et al., 1998). A unique preservational pathway, particularly common in underfilled lakes, involves the rapid cementation of substrates in carbonate and/or saline, alkaline lakes (e.g., by calcite, dolomite, zeolites; Scott et al., 2008; Scrivner and Bottjer, 1986). In subaerially exposed substrates, however, superficial traces are particularly susceptible to destructive processes (e.g., deflation, salt efflorescence) where they are not protected by clay drapes, surface crusts, or microbial mats (e.g., Cohen et al., 1991; Scott et al., 2010).

Many biogenic structures produced in lake basins have not yet been formally defined. Differences between invertebrate ichnotaxa typical of marine environments and similar trace fossils in continental settings (e.g., *Taenidium*, *Spongeliomorpha*), and their implied behavioral and ecological significances, may lead to conceptual ambiguities where ichnological analyses are not accompanied by detailed accounts of associated lithofacies. Also, the supposed presence in continental settings of typical marine ichnogenera (e.g., *Paleodictyon*, *Nereites*, *Scolicia*, *Chondrites*) has been a matter of confusion (e.g., Archer and Maples, 1984; Bohacs et al., 2007b; Hasiotis, 2002, 2004; Hu et al., 1998; Kim et al., 2005; Pickerill, 1990; Smith et al., 1982). Careful reanalysis indicates that the forms reported in continental rocks do not display the diagnostic features of corresponding marine ichnogenera, being relatively simple in morphology (Buatois and Mángano, 2007). A similar problem results from adopting an environmentally based ichnotaxonomy, such as the use of *Isopodichnus* for continental examples of *Cruziana* and *Rusophycus* (Trewin, 1976) or *Naktodemasis* for continental examples of *Taenidium* (Smith et al., 2008). If the depositional environment is used as a basis for ichnotaxonomy, the use of trace fossils as an aid in sedimentary facies analysis would become pervasively affected by circular reasoning.

Vertebrate footprints tend to be named according to the interpreted tracemaker, which can cause difficulties in recognizing recurrent paleoecological patterns through geological time (Hunt and Lucas, 2007). In defense of this practice, it should be noted that most vertebrate tracemakers do produce morphologically distinct footprints from other vertebrates at the level of genus to family. In some cases, the morphological variants present in vertebrate tracks coincide with a higher level of taxonomic resolution than is possible with invertebrate traces (Voigt et al., 2007). Similarly, some of the complex structures produced by Cenozoic social insects may be attributable to tracemakers at the level of subfamily or even genus (e.g., *Macrotermes*).

Some trace-fossil assemblages in lake basins are recurrent throughout geological time. In part due to the effects of hydrochemical conditions and other physical parameters that affect sedimentation as well as the distribution and behavior of organisms, these recurrent assemblages may be closely associated with lithofacies that typify depositional systems in different lake-basin zones (Table 1; e.g., Melchor et al., 2006). Whenever there is a hiatus between initial deposition and the production of traces within a substrate, a change in environmental conditions between the two events becomes more likely as the length of time increases, particularly if the change is related to subaerial exposure of a substrate deposited subaqueously. Thus, the close association of trace-fossil suites with particular lithofacies and sedimentological features (e.g., desiccation cracks) is more probable for non-pedogenically altered deposits, which do not typically undergo large gaps in time between the initial deposition and the recorded biogenic activity.

The recognition of hiatuses may be one of the most applicable aspects of continental ichnology, by facilitating the recognition of stratigraphic surfaces and changing sedimentary processes where sediments themselves are not recorded. With increasing frequency and/or duration of periods of subaerial exposure, the production of overprinted ichnofacies (or suites) within a single substrate becomes more likely. In contrast to the "composite ichnofacies" model of Keighley and Pickerill (2003), the recognition of overprinted ichnofacies in a single substrate aids the recognition of dynamic, changing environmental conditions (e.g., de Gibert and Sáez, 2009; Metz, 1996; Scott et al., 2009; Uchman and Álvaro, 2000). Where sedimentation rates are higher, changing sedimentological and ichnological features through a stratigraphic succession signify changing environmental conditions through time.

2.2 The Ichnofacies Model in Lake Basins

Three main ichnofacies that characterize continental environments through geological time are widely accepted (*Mermia*, *Scoyenia*, *Coprinisphaera*). In addition, the recently proposed *Celliforma* Ichnofacies qualifies as a Seilacherian ichnofacies (Genise et al., 2010). The *Termitichnus* Ichnofacies *sensu stricto* requires additional examples (Genise et al., 2010), while the eolian *Octopodichnus–Entradichnus* Ichnofacies is in need of refinement considering the shared features in Paleozoic and post-Paleozoic examples (Buatois and Mángano, 2011; Ekdale et al., 2007; Hunt and Lucas, 2007). Determining the significance of each of the continental ichnofacies and recognizing new ichnofacies remain major goals for continental ichnology, and the state of the ichnofacies model in continental environments presently is in flux.

Although ichnofacies are associated with sets of environmental controls characteristic of particular depositional settings, trace-fossil types may be present in several environments of a lake basin. The disjunction in time between deposition and biogenic activity by terrestrial organisms in pedogenically modified substrates causes the traces produced to be more closely associated with paleosol features than a particular depositional environment (Genise et al., 2000, 2004; Hasiotis, 2007). In subaerially exposed fluvio-lacustrine and lake-margin settings, which may be frequently flooded, pedogenesis is typically very weak. Thus, the ichnofacies that characterize this zone are typically more closely associated with the depositional conditions and particular lithofacies.

2.2.1 The Mermia *Ichnofacies*

The *Mermia* Ichnofacies (Fig. 2), formally proposed by Buatois and Mángano (1995, 1998), dominantly comprises simple grazing trails and shallow, simple infaunal deposit-feeding burrows associated with permanently subaqueous, oxygenated, low-energy, freshwater lakes, represented by thinly laminated, sometimes heterolithic, fine-grained deposits. Subordinate features include insect trackways and bilobate structures (e.g., *Cruziana*). This ichnofacies has subsequently been recognized in other environments, including short-lived

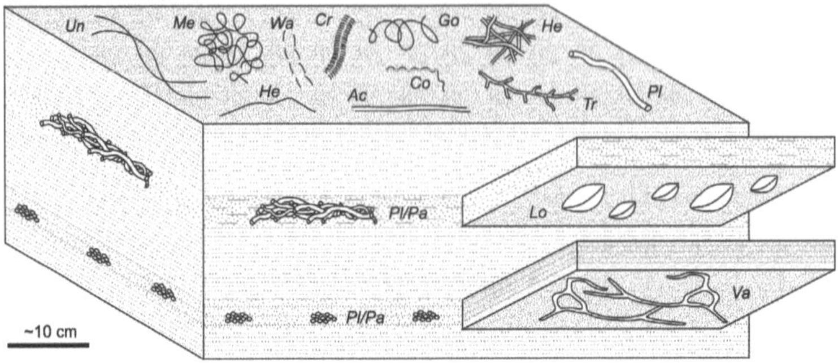

FIGURE 2 Box diagram showing common ichnogenera of the *Mermia* Ichnofacies. The ichnofacies can be informally grouped by similar morphologies, sizes, and orientations that show common associations between ichnotaxa (see Table 2). Surface trails (Group 1): *Me*, *Mermia*; *He*, *Helminthoidichnites*; *Go*, *Gordia*; *Co*, *Cochlichnus*; *Ac*, *Archaeonassa*. Arthropod trails and trackways (Group 2): *Wa*, *Warvichnium*; *Cr*, *Cruziana*. Horizontal infaunal burrows and surface tunnels (Group 3): *Pl*, *Planolites*; *Pa*, *Palaeophycus*; *Tr*, *Treptichnus*; *Va*, *Vagorichnus*. Resting and feeding traces (Group 4): *Lo*, *Lockeia*. Fish swim and resting traces (Group 5): *Un*, *Undichna*.

freshwater ponds on alluvial plains (Buatois and Mángano, 2002, 2004; Melchor et al., 2006), very shallow films of water in low-energy lake margins (e.g., Metz, 1996, 2000), and littoral zones of hypersaline lakes in regions with freshwater input (Scott, 2010). In permanently subaqueous zones of lakes, this ichnofacies is typically exclusive of elements of the *Scoyenia* Ichnofacies and is associated with fine-grained deposits in low-energy sublittoral to profundal zones. In shallow littoral zones or ephemeral water bodies, the *Mermia* Ichnofacies may be closely associated with other trace-fossil types (e.g., meniscate-backfilled burrows, vertebrate footprints), as well as a greater variety of sedimentary structures (e.g., ripple cross-lamination) and indicators of subaerial exposure (e.g., desiccation cracks). In these cases, the *Mermia* Ichnofacies signifies standing or gently flowing water, with traces produced in very shallow subaqueous to saturated substrates. Some examples show the *Mermia* Ichnofacies in settings influenced by glacial runoff, both in continental (e.g., Benner et al., 2009; Uchman et al., 2009) and freshwater-influenced marginal-marine environments (e.g., Buatois et al., 2006, 2010; Netto et al., 2012).

Bivalve locomotion and resting traces (*Lockeia*) are also produced in littoral to sublittoral zones of freshwater lakes (e.g., McCall and Tevesz, 1982; White and Miller, 2008) and are presently included in the *Mermia* Ichnofacies. The backfilled feeding structures of some modern bivalves, such as *Pisidium* (White and Miller, 2008), are as yet undescribed in ichnological terms. Other trace types that have been recorded in low- to moderate-energy littoral to sublittoral and profundal zones include those produced by oligochaetes (incipient cf. *Polykladichnus* isp. and cf. *Labyrintichnus* isp.) and insects such as chironomid larvae or mayfly nymphs (diminutive *Polykladichnus*, *Skolithos*, *Arenicolites*, *Beaconites*) (e.g., Charbonneau and Hare, 1998; Gingras et al., 2007;

McCall and Tevesz, 1982; Uchman and Álvaro, 2000; White and Miller, 2008). These vertical to inclined biogenic structures in lacustrine settings do not correspond to the *Mermia* Ichnofacies (White and Miller, 2008). In contrast, they may be best assigned to the continental *Skolithos* Ichnofacies (Fig. 3), based on their morphology and vertical orientation. These diminutive types of unbranched, branched, and U-shaped vertical burrows were assigned to the *Arenicolites* suite by Bromley and Asgaard (1979), who later erected the *Arenicolites* Ichnofacies (Bromley and Asgaard, 1991). However, the differentiation between the *Arenicolites* Ichnofacies and the *Skolithos* Ichnofacies at present is not possible because the traces are the same morphologically.

With further documentation of case studies from fully lacustrine settings, it may be possible to define new ichnofacies and/or subdivide the *Mermia* Ichnofacies and to clarify the significance of the diminutive trace types of the *Skolithos* Ichnofacies in these settings. For instance, Voigt and Hoppe (2010) recognize the mutual exclusion of assemblages of simple grazing trails from high-density examples of 3D burrow networks in the sublittoral zone of a freshwater lake. In order to recognize some of the variability within the ichnofacies most common in lake basins, it is possible to group ichnogenera based on similarities in their morphology, size, orientation, and preservational style (Table 2). These groups are informal and should not be considered as subdivisions of ichnofacies, but they are useful for comparisons among the dominant assemblages preserved in different lake types. This approach may also provide a way to discuss some of the discrepancies between biological and ichnological data, between modern and fossilized traces in lake basins, and between examples associated with marginal-marine, freshwater-influenced marine, fully lacustrine, lake margin and alluvial environments.

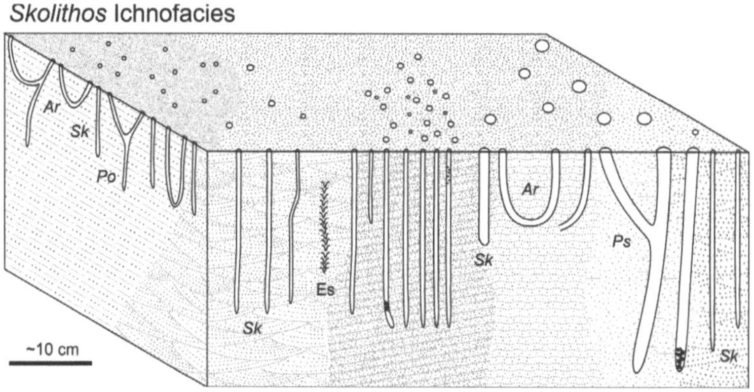

FIGURE 3 Box diagram showing the continental *Skolithos* Ichnofacies. The ichnofacies can be informally grouped by similar morphologies, sizes, and orientations that show common associations between ichnotaxa (see Table 2). Diminutive branched and unbranched vertical burrows (Group 1): *Sk*, *Skolithos*; *Ar*, *Arenicolites*; *Po*, *Polykladichnus*. Medium-sized vertical and U-shaped traces (Group 2): *Sk*, *Skolithos*; *Ar*, *Arenicolites*; Es, escape traces. Large-sized branched and unbranched vertical burrows (Group 3): *Sk*, *Skolithos*; *Ps*, *Psilonichnus*.

TABLE 2 Examples of Ichnogenera Characteristic of the Continental *Mermia, Skolithos,* and *Scoyenia* Ichnofacies

Ichnofacies	Group	General description	Example ichnogenera
Mermia	1	Surface trails	*Helminthoidichnites, Cochlichnus, Gordia, Mermia*
	2	Arthropod trails and trackways	*Glaciichnium, Warvichnium, Diplichnites, Cruziana*
	3	Horizontal infaunal burrows and surface tunnels	*Planolites, Treptichnus, Vagorichnus*
	4	Resting traces	*Lockeia*
	5	Fish swim and resting traces	*Undichnia, Piscichnus*
Skolithos	1	Diminutive branched and unbranched vertical burrows (~1–3 mm in diameter)	*Skolithos, Polykladichnus, Arenicolites*
	2	Medium-sized vertical burrows and U-shaped burrows (~4–9 mm in diameter)	*Skolithos, Arenicolites,* escape traces
	3	Large-sized branched and unbranched vertical burrows (~1–2 cm in diameter)	*Skolithos, Psilonichnus, Polykladichnus*
Scoyenia	1	Bird tracks	*Gruipeda, Avipeda*
	2	Tetrapod and bipedal dinosaur tracks	*Varanopus, Grallator, Amblydactylus, Felipeda*
	3	Insect and other arthropod trackways	*Siskemia, Merostomichnites, Mirandaichnium, Diplichnites*
	4a	Horizontal, shallow-tier burrows (predesiccation suite)	*Taenidium, Beaconites, Planolites, Palaeophycus*
	4b	Horizontal, shallow-tier burrows (postdesiccation suite)	*Spongeliomorpha, Scoyenia*
	5	Vertical, deeper-tier burrows	*Taenidium, Skolithos, Planolites*
	6	3D branching burrow boxworks	*Camborygma, Thalassinoides, Spongeliomorpha, Attaichnus*

The ichnofacies have been informally grouped according to similar morphologies, sizes, behavior, orientation, and preservational style. These groups are not formal subdivisions of the ichnofacies. However, they are useful for summarizing different compositions in assemblages of the case studies used here to compare between different lake-type basins. These groupings may potentially be useful to help recognize different sets of environmental factors (e.g., oxygenation, salinity) that control ichnodiversity and assemblage composition. However, the presence of more than one group within an assemblage does not directly indicate ichnodiversity because there may be only one trace-fossil type present in each group within an assemblage, or there may be several trace types present within a single group of an assemblage.

2.2.2 The continental *Skolithos* Ichnofacies

The *Skolithos* Ichnofacies (Fig. 3), found in sandy, shallow lacustrine and fluvio-lacustrine environments, is similar to the *Skolithos* Ichnofacies typical of marine wave-influenced shorelines (Buatois and Mángano, 2004, 2011). It is dominated by simple vertical dwelling burrows and U-shaped burrows produced in soft substrates and is often associated with cross-bedded and/or ripple cross-laminated sandstones of moderate- to high-energy settings. Subaqueous producers of *Skolithos* and *Arenicolites* in sandy substrates of lacustrine and fluvial environments may include oligochaetes, bivalves, and/or possibly amphipod crustaceans (cf. Dashtgard and Gingras, 2005, 2012; cf. McCall and Tevesz, 1982). Low-diversity assemblages dominated by simple vertical burrows may be produced subaerially in substrates that are sandy and soft or fine grained and cohesive, and even slightly indurated firmgrounds. Tiger beetle (Carabidae: Cicindelinae) larvae, for example, are typical producers of high-density vertical burrows in areas where there is little vegetation (e.g., Knisley, 1984), characterized by frequent flooding by lake or river waters, or frequent erosional and depositional events. Open vertical burrows (e.g., *Skolithos*, *Psilonichnus*) possibly produced by spiders, beetles, and/or decapod crustaceans may be found in supralittoral to terrestrial deposits such as shoal bars exposed during periods of lower lake levels, eolian dunes (Ahlbrant et al., 1978; Ekdale et al., 2007), or abandoned fluvial channels (Fitzgerald and Barrett, 1986; Netto, 2007). Large (~1 cm or more in diameter) and medium-sized (~4–9 mm in diameter) examples are consistently found in substrates that are subject to frequent disturbances by erosion, deposition, or flooding and exposure. These subaerially produced traces share similarities with the *Psilonichnus* Ichnofacies as described for marginal-marine supralittoral zones by Frey and Pemberton (1987), who recognized tiger beetle larvae, spiders, and ghost crabs as the most common producers of vertical burrows in that setting.

2.2.3 The *Scoyenia* Ichnofacies

The *Scoyenia* Ichnofacies (Fig. 4), typical of low-energy environments with frequently flooded and subaerially exposed substrates such as lake margins, dominantly comprises horizontal burrows that are commonly meniscate-backfilled (e.g., *Beaconites*, *Scoyenia*), arthropod trackways and trails (e.g., *Diplichnites*, *Cruziana*), and vertebrate footprints (Buatois and Mángano, 1998; Frey et al., 1984; Minter et al., 2007). Vertical burrows and simple surface trails and tunnels (e.g., *Vagorichnus*, *Treptichnus*) are common subordinate components (e.g., Gierlowski-Kordesch, 1991; Metz, 1996, 2000). Boxwork or 3D burrow systems (e.g., *Thalassinoides*, *Spongeliomorpha*, *Attaichnus*), produced subaerially by earwigs, crickets, and/or ants, as well as large 3D burrow systems produced by decapod crustaceans (e.g., *Camborygma*), are commonly associated with other trace types of the *Scoyenia* Ichnofacies produced in lake-margin areas (e.g., Kim et al., 2002; Metz, 2000; Scott et al., 2009). These traces, however, probably signify slightly deeper water tables than an assemblage

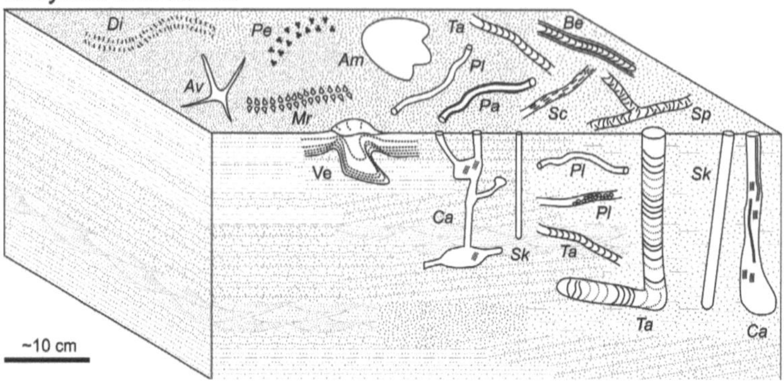

FIGURE 4 Box diagram showing common ichnogenera of the *Scoyenia* Ichnofacies. The ichnofacies can be informally grouped by similar morphologies, sizes, and orientations that show common associations between ichnotaxa (see Table 2). Bird footprints (Group 1): *Av*, *Avipeda*. Tetrapod and bipedal dinosaur tracks (Group 2): *Am*, *Amblydactylus*. Arthropod trackways (Group 3): *Di*, *Diplichnites*; *Pe*, *Paleohelcura*; *Mr*, *Merostomichnites;* Ve, vertebrate tracks in cross section. Horizontal, shallow-tier burrows (predesiccation suite; Group 4a): *Ta*, *Taenidium*; *Be*, *Beaconites*; *Pl*, *Planolites*; *Pa*, *Palaeophycus*. Horizontal, shallow-tier burrows (post-desiccation suite; Group 4b): *Sc*, *Scoyenia*; *Sp*, *Spongeliomorpha*. Vertical, deeper-tier burrows (Group 5): *Ta*, *Taenidium*; *Sk*, *Skolithos*. 3D branching burrow networks (Group 6): *Ca*, *Camborygma*.

dominated by horizontal backfilled burrows (Hamer and Sheldon, 2010; Hasiotis, 2007; Hasiotis and Mitchell, 1993). The transition from proximal littoral to terrestrial areas within the zone of lake-level rise and fall, with progressively deeper water tables and decreasing frequencies of flooding cycles toward the basin margin, may have an important effect on the makeup of the *Scoyenia* Ichnofacies on the scale of suites or ichnocoenoses.

Likewise, vertebrate footprint assemblages, which are typically preserved in low-energy lake margins, also show a characteristic zonation from the edge of the water body to drier substrates, in terms of both their composition and morphology or taphonomy (Cohen et al., 1991, 1993; Hunt and Lucas, 2007; Lockley, 1986; Lockley et al., 1994; Smith, 1993). Five vertebrate ichnofacies were proposed by Hunt and Lucas (2007): (1) *Characichnos*, comprising vertebrate swim traces within standing water bodies; (2) *Grallator*, comparable to the shorebird ichnofacies of Lockley et al. (1994), comprises high-density tridactyl footprints of bipedal animals along shorelines; (3) *Brontopodichnus* and (4) *Batrachichnus*, comprise impressions produced by quadrupedal herbivores and carnivores in lake-margin areas; and (5) *Chelichnus*, being characterized by trackways of amphibians and reptiles produced in eolian environments. Although some workers consider vertebrate-footprint assemblages as separate entities from invertebrate trace fossils present in the same deposits, this simply derives from different research focuses (Lockley, 2007). The treatment of vertebrate trace fossils together with invertebrate trace fossils allows for (1) the refined interpretation of paleoenvironments and ichnological models in the continental realm (Lucas et al., 2011; Melchor et al., 2006, 2012); (2) the recognition of subaerial exposure in low-diversity assemblages containing simple vertical burrows or simple surface trails and burrows, from which the lake-basin zone may otherwise be difficult to discern; and (3) the recognition of stratigraphic surfaces that represent changing conditions in both lake margin (e.g., Prince and Lockley, 1989) and marginal-marine settings (Phillips et al., 2007).

2.2.4 Terrestrial Trace-Fossil Assemblages

Terrestrial trace-fossil assemblages, typical of alluvial and eolian environments, may be present within the zone of lake-level rise and fall if water tables are lowered sufficiently for terrestrial organisms to colonize the substrate. Common traces include vertebrate burrows, ant nests, termite nests and tunnels, beetle burrows and brood balls, and burrows, nests, and cells of wasps and bees. In lake basins, low-diversity assemblages of termite traces, exclusive of cells and brood balls, may even occur in lake-margin to basin-center successions in arid regions (e.g., Smith and Mason, 1998) or where lake levels have dropped for extended periods of time (e.g., Scott et al., 2009). Assemblages dominated by dung beetle nests (e.g., *Coprinisphaera*) or bee cells (e.g., *Celliforma*) appear to be present mainly near the basin margin or in deposits that have undergone moderate to extensive pedogenesis.

Carbonate palustrine environments are low-energy environments with fluctuating water levels and groundwater tables, and are characterized by pedogenic modification of shallow lacustrine to littoral marsh carbonates (e.g., Alonso-Zarza, 2003; Platt, 1989). They contain trace-fossil assemblages typified by bee cells (e.g., *Celliforma*) and other insect burrows and chambers (e.g., *Skolithos*, *Taenidium*, *Teisseirei*; Edwards et al., 1998; Genise et al., 2010; Melchor et al., 2002). Deposits from palustrine environments often contain both freshwater and terrestrial gastropod assemblages, together with weakly developed paleosols and root traces (e.g., Alonso-Zarza, 2003). Melchor et al. (2002) demonstrated the mutual exclusion of very low-diversity assemblages of (1) *Skolithos* in fine-grained carbonates that were relatively waterlogged, when compared to (2) the overlying, pedogenically modified tuffaceous carbonates with abundant *Celliforma*, and (3) a relatively well-developed paleosol with root traces and meniscate-backfilled burrows. This example shows the transition from wetter to drier to vegetated substrates in a vertical succession representing changing conditions in an environment characterized by fluctuating water tables (Melchor et al., 2002).

3. AN INTEGRATED LAKE-TYPE BASIN AND LACUSTRINE ICHNOLOGY MODEL

Depositional environments in lake basins are diverse and are distributed according to the interaction of numerous abiotic factors (e.g., Cohen, 2003; Renaut and Gierlowski-Kordesch, 2010; Wetzel, 2001). The lake-type model of Carroll and Bohacs (1999) and Bohacs et al. (2000) permits the conceptual simplification of lakes based on their prevalent facies associations. The model is useful to summarize the factors that influence the composition and distribution of traces produced by plants and animals. However, few researchers in continental ichnology have considered the basin-scale distribution of deposits that preserve particular trace-fossil assemblages, and much work remains to determine how ichnology can be applied to the interpretation of lake-basin depositional systems at this scale. Nevertheless, numerous documented case studies can be considered within the framework of the lake-type model and improve the present understanding of how organism/sediment interactions in lacustrine basins can be applied to paleoenvironmental reconstructions and stratigraphy (Buatois and Mángano, 2009).

When applying the lake-type basin model, it is important to distinguish between the accommodation *potential* in the lake basin and the accommodation within the water body itself, related to the depth of the lake (Bohacs et al., 2003). It is the accommodation potential of the basin relative to the amount of sediment and water that enters the basin that determines whether a lake is open or closed, whereas the accommodation within the lake itself influences sedimentation. In each of the lake-type basins, the lakes may be deep or shallow, depending on the elevation difference between the basin floor and the sill height, and the depth of the water body relative to this difference. This strongly influences the mixing regime and oxygenation of lacustrine waters and sediments.

In overfilled lake-type basins, terrestrial environments and trace-fossil assemblages produced in drier, terrestrial environments are of limited extent because the basin is relatively filled with water and water tables are high in marginal environments (Bohacs et al., 2007b). The presence of terrestrial and lake-marginal environments in balanced-fill lake-type basins is variable, depending on whether the lake-basin is at or overflowing the basin sill and relatively full or well below the basin sill with widespread subaerially exposed areas. In underfilled lake-type basins, terrestrial and lake-marginal environments are relatively widespread and may form thick packages of sediments toward the basin margins. However, large-scale fluctuations of the lake level also occur in underfilled basins, and lacustrine sediments may form thick vertical successions near the basin margins if they were deposited during high lake levels that remained below the sill height. Lacustrine transgressions may also be due to tectonic activity affecting localized areas with higher subsidence rates and may be best observed in underfilled basins. These factors together determine the distribution and stacking patterns of sedimentary deposits, as well as the composition and distribution of biogenic structures in lake basins.

3.1 Overfilled Lake-Type Basins

Overfilled lake-type basins (Fig. 5) are characterized by fluvio-lacustrine deposits in freshwater lakes with open hydrology, which develop when the rate of sediment and water input exceeds the potential accommodation in the basin (Bohacs et al., 2000, 2003; Carroll and Bohacs, 1999). Parasequences are typically indistinct and are mainly progradational due to relatively constant lake levels that are controlled by the sill over which lake waters flow out of the basin. From a biological perspective, open lakes represent less stressful, more stable ecosystems that tend to support a diverse fauna in relatively stable lakes with low salinity and relatively high nutrient inputs (e.g., Buatois and Mángano, 2004, 2007, 2009; Gierlowski-Kordesch and Park, 2004; Gore, 1989). Sedimentary environments of overfilled basins are characterized by (1) alluvial systems with channel avulsion, high water tables, and abundant vegetation (in older sediments: peat or coal deposits); (2) deltaic deposits; and (3) relatively well-oxygenated, sublittoral to profundal zones (Bohacs et al., 2000).

The most important controls on trace-fossil distribution in permanent subaqueous areas of overfilled lakes are relatively high oxygen contents of lake waters, energy, stable food supply and abundant nutrients, and generally soft to soupy substrates (Buatois and Mángano, 2007, 2009). Cold conditions may also be an important control in glacial lakes. Although overfilled lakes are not salinity stratified, deep overfilled lakes may be stratified due to temperature differences between surface waters and deep waters (Cohen, 2003). Fluvial discharge into overfilled lakes may trigger underflow currents that oxygenate lake bottoms and which may contribute to the turnover of a stratified water body. Resedimentation (e.g., slumping) of material supplied by the fluvial system may also promote bottom oxygenation, which allows the establishment of epifaunal and infaunal communities (Buatois and Mángano, 1998). Abundant organic detritus derived from land plants is brought to overfilled lakes by fluvial systems (Bohacs et al., 2000; Carroll and Bohacs, 2001) and may be introduced into deep lake waters by underflows and turbidity currents, increasing the food supply for infaunal deposit-feeding and grazing communities. High sedimentation rates, high water tables in subaerially exposed lake-marginal areas, and unstable shallow lacustrine substrates are also expected to influence trace-fossil assemblages by precluding the presence of trace types produced in firm substrates (Bohacs et al., 2007b; Buatois and Mángano, 2004).

The *Mermia* Ichnofacies characterizes the low-energy, permanent subaqueous zones (sublittoral, profundal) of overfilled lakes (Figs. 5 and 6; Buatois and Mángano, 2009). Simple grazing (e.g., *Gordia*) and sediment/water interface feeding (e.g., *Vagorichnus*) trace types are commonly dominant (Fig. 6E and F), with fish swim traces (e.g., *Undichna*), arthropod trackways (e.g., *Diplichnites*, *Glaciichnium*), and shallow simple burrows and trails (e.g., *Planolites*, *Palaeophycus*) also being present (Benner et al., 2009; Buatois et al., 1995, 1996; Netto et al., 2012; Uchman et al., 2009). In oxygenated lakes, examples of the *Mermia* Ichnofacies are typically of moderate to high diversity, with a range of behavior and trophic levels represented (e.g., grazers, detritus feeders, deposit feeders, predators). The relative simplicity of ichnofaunas from low-energy, permanently subaqueous lacustrine zones (i.e., *Mermia* Ichnofacies) is in sharp contrast with fully marine ichnofaunas (e.g., *Cruziana* and *Nereites* ichnofacies), even in well-oxygenated waters with abundant and consistent food supplies. No

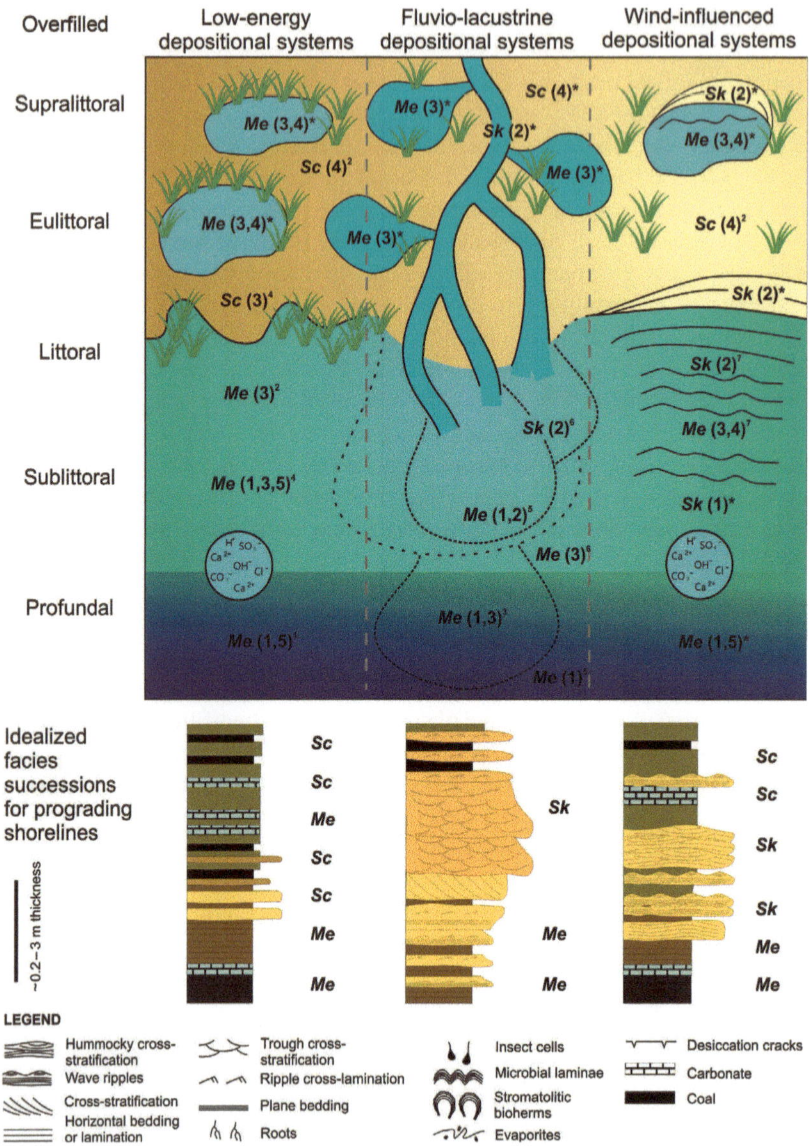

FIGURE 5 Schematic diagram of profundal to supralittoral zones in overfilled lake-type basins showing approximate environmental distributions of ichnofacies and idealized vertical distribution of facies in prograding stratigraphic successions. Dominant controls in overfilled lake-type basins are well-oxygenated lake waters, soft substrates, and high water tables in subaerially exposed settings. Laterally adjacent environments along strike and along depositional dip both form vertical successions under changing conditions. The idealized sections show vertical successions along depositional dip. Examples of ichnofacies shown are based on case studies (referenced here) and predicted assemblages (asterisks). A legend for Figs. 5, 7 and 9 is included here. Abbreviations: *Me*, *Mermia*, *Sk*, *Skolithos*, *Sc*, *Scoyenia*. The numbers following ichnofacies abbreviations refer to the groupings in Table 2. *References:* [1]*Benner et al. (2009)*, [2]*Bohacs et al. (2007b)*, [3]*Buatois et al. (1996)*, [4]*Netto et al. (2009)*, [5]*Uchman et al. (2009)*, [6]*Voigt and Hoppe (2010), and* [7]*this chapter, Fig. 6.*

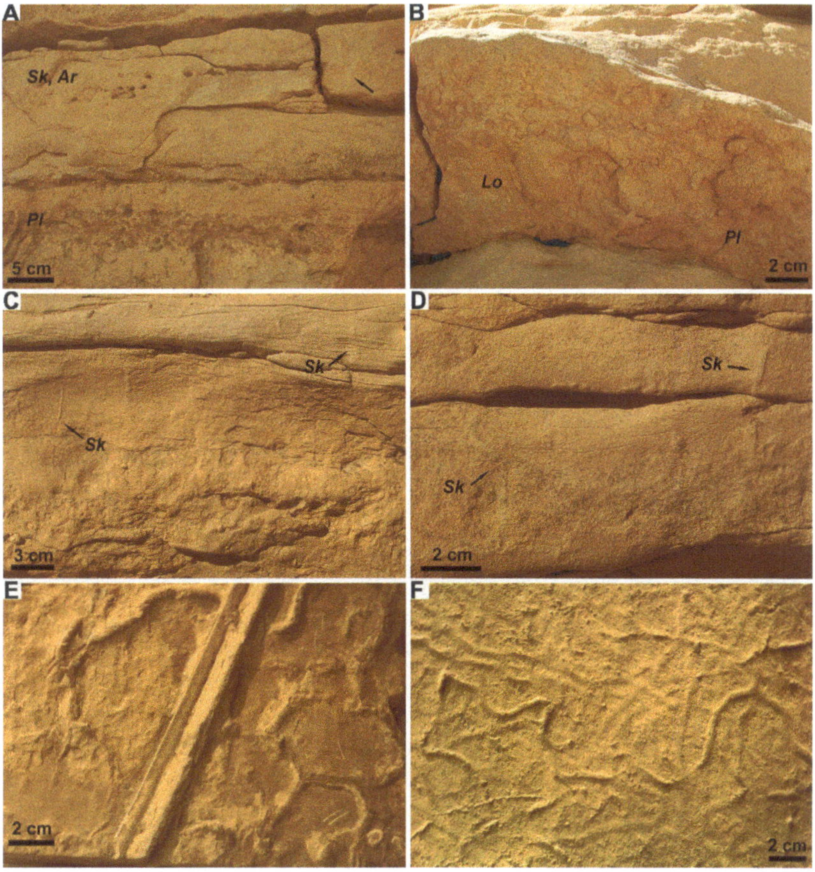

FIGURE 6 Trace fossils in overfilled lake-type basins. (A–D) Examples from the Tipton Shale Member (Scheggs Bed) of the Eocene Green River Formation, Wyoming, USA. (A) and (B) Intensely bioturbated horizons of horizontally bedded sublittoral zone or lower-energy littoral-zone sandstones, showing infaunal burrows (*Pl*, *Planolites*) and resting traces of bivalves (*Lo*, *Lockeia*) of the *Mermia* Ichnofacies. Vertical burrows (*Sk*, *Skolithos*; *Ar*, *Arenicolites*) of the *Skolithos* Ichnofacies are present in the overlying low-angle cross-bedded and laminated sandstones of the littoral zone in (A), (C), and (D). (E) and (F) Examples of the *Mermia* Ichnofacies preserved at the bases of turbidites in the Jurassic Anyao Formation, Jiyuan–Yima Basin, central China. (E) *Vagorichnus anyao* cross-cutting a tool mark at the base of a turbidite. (F) Superimposed specimens of *Paracanthoraphe togwunia* at the base of a turbidite.

archetypal trace-fossil associations can be defined at present to distinguish between the low-energy littoral to profundal zones of lakes. However, Voigt and Hoppe (2010) recognized a distinct assemblage of high-density, shallow-penetrative deposit-feeding burrows in laminated mudstones of the sublittoral to profundal transition of an overfilled lake in Triassic deposits of Kyrgyzstan, Central Asia.

Higher energy shoreface to foreshore and fluvio-deltaic areas of overfilled lakes are characterized by low-density, indistinct vertical burrows (*Skolithos*), escape traces, and U-shaped burrows (*Arenicolites*) produced in soft substrates (Figs. 5 and 6A–D; Bohacs et al., 2007b; Buatois and Mángano, 2004; Voigt and Hoppe, 2010). Areas of rapid sedimentation near the mouth of distributary channels tend not to be burrowed or are sparsely bioturbated. This assemblage is closely comparable to the *Skolithos* Ichnofacies of marine nearshore settings. High-density simple burrows (*Planolites*, *Palaeophycus*) may be common along horizons representing short-term periods of quiescence associated with the deposition of plant material (Fig. 6A–C). Gastropod and bivalve coquinas are also common components (Bohacs et al., 2000). Sediments deposited in quieter waters of the littoral zone may also preserve meniscate-backfilled burrows (e.g., *Taenidium*), arthropod trackways (e.g., *Diplichnites*), bilobate trails (*Cruziana*), and bivalve resting traces (*Lockeia*).

3.2 Balanced-Fill Lake-Type Basins

Balanced-fill lake-type basins (Fig. 7) are characterized by dominantly "fluctuating profundal" facies associations, representing low-energy sublittoral to profundal zones with fine-grained carbonate and siliciclastic deposits and high levels of organic carbon preserved in stratified lake deposits (e.g., oil shale) (Bohacs et al., 2000; Carroll and Bohacs, 2001). In balanced-fill lake-type basins, the input of sediment and water is approximately equal to the accommodation potential. This relationship fluctuates depending on climatic and tectonic controls, and with changing drainage captures from upstream basins (Carroll et al., 2008). These types of lakes periodically overflow into an adjacent basin, with lake levels typically at or near the elevation of the basin-bounding sill (Bohacs et al., 2003). During periods when the lake is hydrologically closed and lake levels are below the sill, salinity may become elevated, the lake waters may become stratified and oxygen depleted, and carbonate and siliciclastic muds accumulate in aggradational packages in the basin center. Lake levels can fluctuate due to minor climatic variations, with widespread subaerially exposed areas (depending on the gradient) and rarely, balanced-fill lakes may dry up almost completely. During periods when the lake is hydrologically open, lake waters become fresher, better oxygenated, and progradational packages of sediments are deposited at the basin margins where lake levels are stable.

Trace fossils produced in balanced-fill basins appear to reflect these variable abiotic conditions and may be closely associated with particular lithofacies associations (Fig. 7). Salinity, substrate, and oxygen are the main limiting factors in balanced-fill lakes. Salinity plays a main role during periods when the lake is hydrologically closed. Saline conditions ($\sim$20–50 g/L) are detrimental to lacustrine faunas (e.g., benthic invertebrates, fish), and density stratification between bottom waters and freshwater inflow is more likely, leading to low-oxygen conditions in sublittoral to profundal zones (e.g., Cohen, 2003; Hammer, 1986).

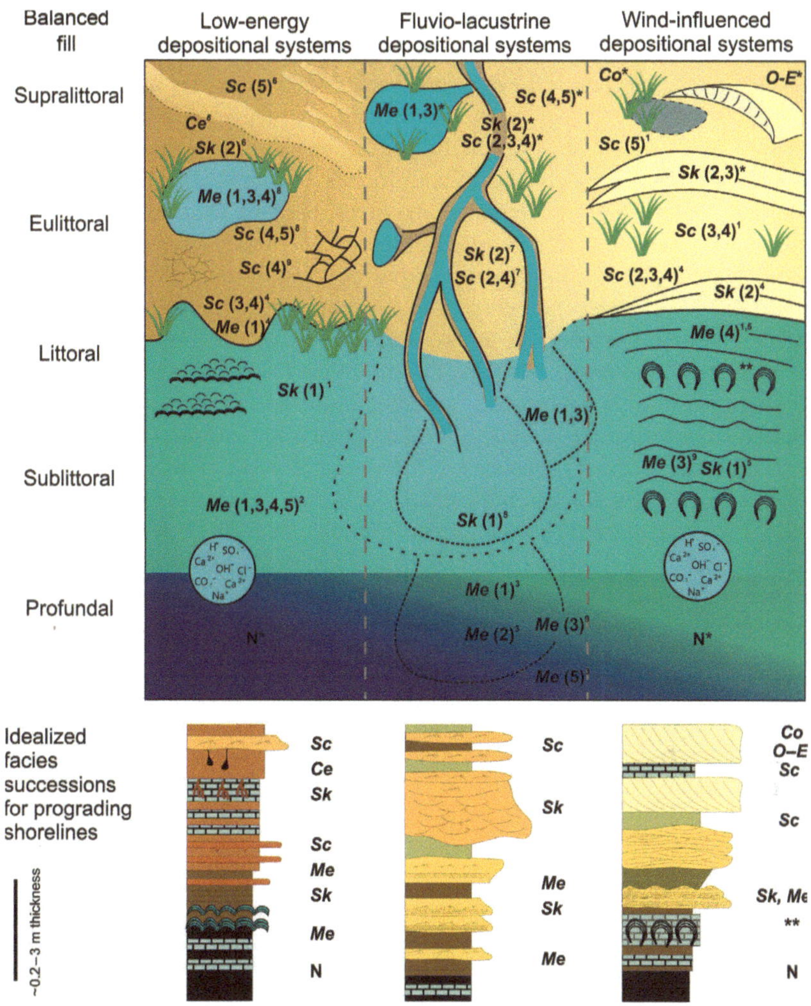

FIGURE 7 Schematic diagram of profundal to supralittoral zones in balanced-fill lake-type basins showing approximate environmental distributions of ichnofacies and idealized vertical distribution of facies in prograding stratigraphic successions. Dominant controls in balanced-fill lake-type basins are variably oxygenated to poorly oxygenated lake waters and substrates, fluctuating salinities and frequently flooded and subaerially exposed substrates, especially in basin-marginal settings. Laterally adjacent environments along strike and along depositional dip both form vertical successions under changing conditions. The idealized sections show vertical successions along depositional dip. Examples are based on case studies (referenced here) and predicted assemblages (asterisks). Abbreviations: *Me*, *Mermia*; *Sk*, *Skolithos*; *Sc*, *Scoyenia*; *Ce*, *Celliforma*; *Co*, *Coprinisphaera*; *O–E*, *Octopodichnus–Entradaichnus*; N, no trace fossils. The numbers following ichnofacies abbreviations refer to the groupings in Table 2. *References:* [1]*Bromley and Asgaard (1979)*, [2]*Buatois et al. (2000)*, [3]*de Gibert et al., 2000*, [4]*Kim et al. (2005)*, [5]*Magyar et al. (2006)*, [6]*Melchor et al. (2002)*, [7]*Melchor (2004)*, [8]*(Metz, 1996)*, [9]*Uchman et al. (2007)*. Note that caddisfly cases (**, *Terrindusia*) may be present in stromatolitic bioherms in high-energy shorelines (e.g., Leggitt et al., 2007), but their potential assignment to an ichnofacies has not yet been determined.

During periods when lake levels are near but below the sill height, shorelines around balanced-fill lakes may fluctuate, leading to the frequent flooding and subaerial exposure of the eulittoral zone and the presence of the *Scoyenia* Ichnofacies in low-energy areas (Fig. 8). In contrast to underfilled basins, these settings are typically near the basin margins. Horizontal simple and branching structures produced in softgrounds (e.g., *Planolites*, *Taenidium*) and firmgrounds (e.g., *Scoyenia*, *Spongeliomorpha*) are common (e.g., Bromley and Asgaard, 1979; Gierlowski-Kordesch, 1991), and vertebrate footprints may be preserved. During periods when lake levels are well below the sill height, desiccation of lacustrine and lake-margin sediments is pervasive and firmgrounds become widespread. These deposits tend to host the *Scoyenia* Ichnofacies. In particular, the firmground suite of this ichnofacies, containing striated trace fossils or vertically oriented structures, may help to delineate parasequences through the recognition of subaerial exposure (Fig. 8F).

High-energy, wave-influenced carbonate shorelines of balanced-fill lake types preserve stromatolitic bioherms, which may be encrusted by caddisfly cases (Leggitt et al., 2007), and/or preserve embedment structures of organisms dwelling within the bioherms (Lamond and Tapanila, 2003). Wave-influenced siliciclastic littoral zones may preserve high-density examples of simple burrows (*Planolites*), which appear to be mainly present during relatively freshwater phases or in well-mixed areas of balanced-fill lakes. Similar lithofacies of the littoral to sublittoral zones that were deposited during saline phases or in poorly mixed lake waters do not preserve evidence of infaunal activity. During highstand progradation of deltaic systems, elements of the *Skolithos* Ichnofacies may occur in delta mouth-bar deposits (Bromley and Asgaard, 1979; Mángano et al., 1994, 2000).

Although freshwater conditions and softgrounds are more common during high lake levels, dysaerobic conditions may prevail in poorly mixed, deeper lake waters, imparting a stress factor on lacustrine biotas. As a result, ichnofaunas from turbidite systems in balanced-fill lakes are less abundant and diverse than those from overfilled lake turbidites, if present at all (e.g., Uchman et al., 2007). Low-diversity, low-abundance examples of simple grazing trails or fish swim traces (*Undichna*; Fig. 8A) of the *Mermia* Ichnofacies are present in laminated, fine-grained lacustrine sediments, which reflect low-oxygen conditions of the sublittoral to profundal zones (Magyar et al., 2006; Uchman et al., 2007). Delta front deposits in these stratified lakes may preserve a low-diversity assemblage of the *Mermia* Ichnofacies (Melchor, 2004). During periods of better oxygenation of lake waters, the *Mermia* Ichnofacies may increase in diversity, with a variety of trace types including grazing trails, deposit-feeding burrows, and arthropod trackways and trails associated with the delta front (Melchor, 2004) to distal prodelta (Fig. 8B; de Gibert et al., 2000). In some cases, vertical burrows may occur in sandy tempestites interbedded with mudstones and appear to represent the temporary oxygenation of the lake bottom due to storms (Magyar et al., 2006).

FIGURE 8 Trace fossils in balanced-fill lake-type basins. (A) and (B) Examples of profundal trace fossils of the *Mermia* Ichnofacies in the Cretaceous of Montsec, Spain. Photographs courtesy of J. de Gibert. (A) *Undichna britannica* in lacustrine shales. (B) The arthropod trackway *Hamipes didactylus* on a bedding plane of a sandstone interpreted as an underflow deposit. (C) *Palaeophycus tubularis* from shallow-lacustrine deposits of the Triassic Tanzhuang Formation, Jiyuan-Yima Basin, central China. (D–F) Examples from the Tipton Shale Member (Rife Bed) of the Eocene Green River Formation, Wyoming, USA. (D) Medium-sized *Arenicolites* of the *Skolithos* Ichnofacies produced in sandy soft sediments of either the eulittoral zone or backshore deposits of the supralittoral zone. Arrows show the U-shaped lower portion of the traces. (E) Example of a horizontal, large-sized meniscate-backfilled trace fossil (*Taenidium*; arrow) of the *Scoyenia* Ichnofacies produced in a soft substrate of the eulittoral zone. (F) Vertical backfilled burrow of the *Scoyenia* Ichnofacies (*Taenidium*; arrow) with sharp burrow boundaries in subaerially exposed high-energy littoral-zone sandstones.

3.3 Underfilled Lake-Type Basins

Underfilled lake-type basins (Fig. 9) are those in which the amount of sediment and water input is less than the potential accommodation, and they are characterized by evaporitic lithofacies associations (e.g., carbonates, evaporite

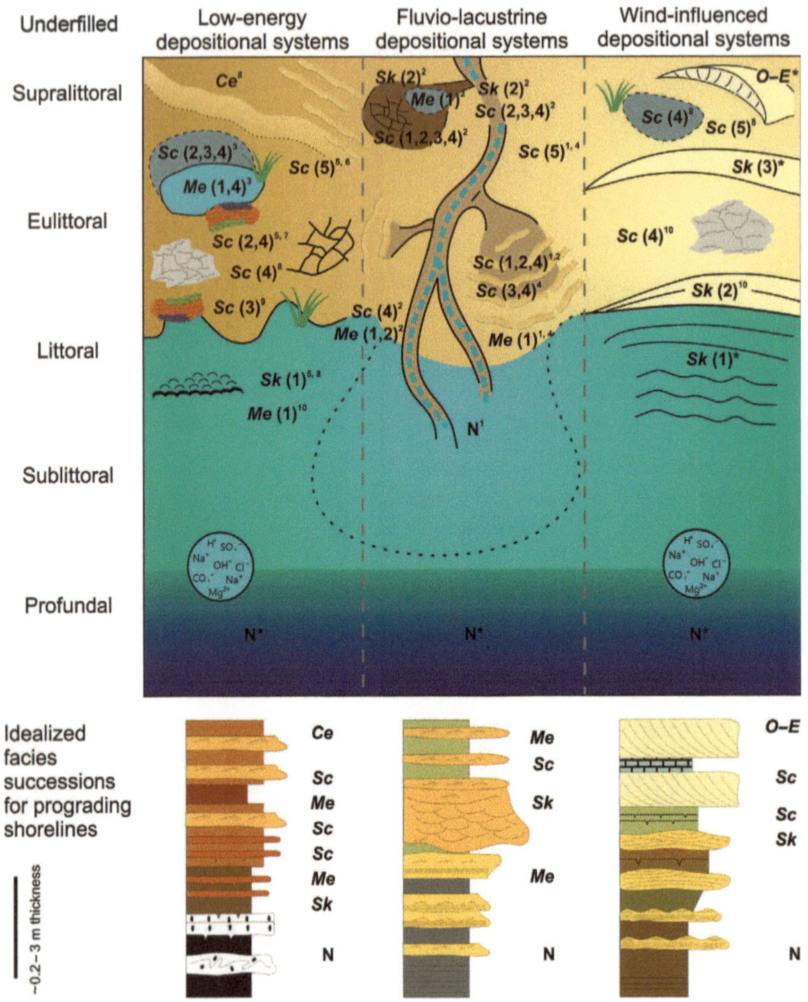

FIGURE 9 Schematic diagram of profundal to supralittoral zones in underfilled lake-type basins showing approximate environmental distributions of ichnofacies and idealized vertical distribution of facies in prograding stratigraphic successions. Dominant controls in underfilled lake-type basins are high salinities, substrate consolidation, patchy input of freshwater, and relatively low water tables in basin-marginal settings. Laterally adjacent environments along strike and along depositional dip both form vertical successions under changing conditions. The idealized sections show vertical successions along depositional dip. Examples are based on case studies (referenced here) and predicted assemblages (asterisks). Abbreviations: *Me*, *Mermia*; *Sk*, *Skolithos*; *Sc*, *Scoyenia*; *Ce*, *Celliforma*; *Co*, *Coprinisphaera*; *O–E*, *Octopodichnus–Entradaichnus*; N, no trace fossils. The numbers following ichnofacies abbreviations refer to the groupings in Table 2. *References:* [1]*de Gibert and Sáez (2009)*, [2]*Melchor et al. (2006)*, [3]*Minter et al. (2007)*, [4]*Pollard et al. (1982)*, [5]*Rodríguez-Aranda and Calvo (1998)*, [6]*Scott et al. (2009)*, [7]*Smith and Mason (1998)*, [8]*Uchman and Álvaro (2000)*, [9]*Zhang et al. (1998)*, [10]*this chapter, Fig. 10.*

minerals) deposited in hydrologically closed, saline to hypersaline water bodies (Bohacs et al., 2000). Siliciclastic lithofacies may be abundant toward the basin margins, but lacking in basin-center successions (e.g., Smoot, 1983). Parasequences form mainly by aggradation rather than by shoreline progradation (Bohacs et al., 2000). Accommodation within an underfilled lake itself is usually limited because of shallow lake waters. Lake levels fluctuate frequently in response to climate variability and tectonic activity, so that shoreline deposits may show extreme displacement across the basin. Lowstand deposition is characterized by evaporite accumulation in the zones of maximum subsidence; while in the remaining zones sediments that accumulated during the previous highstand undergo desiccation, pedogenesis, and reworking (Bohacs et al., 2000, 2007a; Smoot, 1983). Well-mixed, shallow lakes in underfilled basins may not experience density stratification, but deeper saline lakes are usually stratified due to salinity or temperature differences between the epilimnion and the hypolimnion (Renaut and Gierlowski-Kordesch, 2010). They represent very stressful ecosystems, in which faunal diversity can be very low (e.g., Hammer, 1986; Williams, 1998; Williams et al., 1990).

Salinity, which can also contribute to low oxygen levels, is the most important limiting factor on biodiversity within and around lakes of underfilled basins, leading to either the absence of, or impoverished expressions of, the *Mermia* Ichnofacies in eulittoral to sublittoral zones (Buatois and Mángano, 2007). Trace fossils are not expected from profundal deposits. Suites of the *Mermia* Ichnofacies are commonly of very low diversity, are mainly restricted to grazing trails (e.g., *Helminthoidichnites*), and are most common in short-lived, ephemeral water bodies on low-energy lake-margin areas or in shallow littoral to eulittoral lake waters (Fig. 10A; Scott, 2010). The patchy distribution of high-density examples of the *Mermia* Ichnofacies is controlled by the localized availability of relatively freshwater, either by spring-fed inputs or in ephemeral pools fed by rain or overbank flows. Salinity-tolerant animals, that survive within saline to hypersaline lakes (e.g., chironomid fly larvae), may produce high-density, low-diversity assemblages of diminutive vertical burrows (e.g., *Polykladichnus*, *Arenicolites*, *Skolithos*) and possibly meniscate-backfilled burrows (small *Beaconites*), in littoral to sublittoral zones with soft substrates (Ortí et al., 2003; Rodríguez-Aranda and Calvo, 1998; Uchman and Álvaro, 2000). Deposits in littoral to eulittoral zones may also preserve high-density, low-diversity assemblages of bird footprints and trample-grounds produced within water-oversaturated, very shallow lacustrine substrates, attributable to the shorebird (Lockley et al. 1994) or the *Grallator* Ichnofacies (Hunt and Lucas 2007).

Substrate consolidation is a main control on trace-fossil morphology in lake margins of underfilled lake types. The presence of bioglyphs (e.g., striations) and well-defined burrow margins and track morphology are evidence of firmground development due to desiccation, early cementation of the substrate, or exhumation of relatively well-indurated surfaces where net sedimentation rates

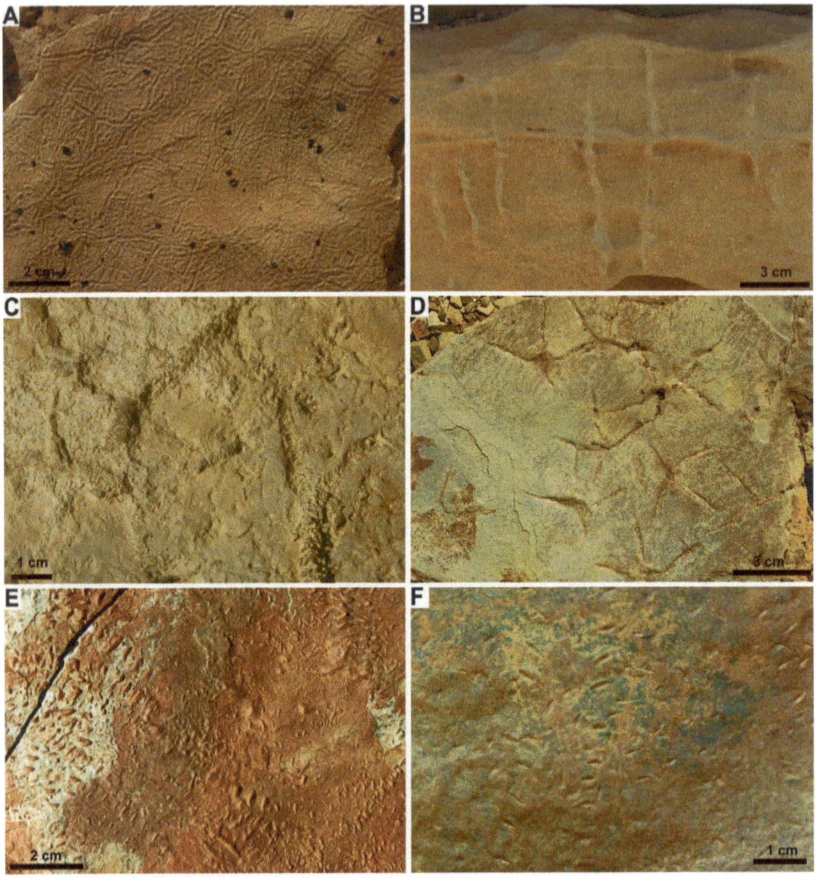

FIGURE 10 Trace fossils in underfilled lake-type basins. (A–D) Examples from the Wilkins Peak Member of the Eocene Green River Formation, Wyoming, USA. (A) High-density, low-diversity expression of the *Mermia* Ichnofacies in low-energy littoral-zone deposits. (B) Medium-sized vertical burrows of the *Skolithos* Ichnofacies in wave-influenced littoral to eulittoral zone deposits. (C) Horizontal, meniscate-backfilled burrows of the *Scoyenia* Ichnofacies in low-energy, subaerially exposed mud flats of the eulittoral zone. (D) Shorebird footprints of the Scoyenia Ichnofacies in lake marginal facies of the Wilkins Peak Member, Eocene Green River Formation. (E) and (F) Examples from the Permian of the Paganzo Basin, Argentina. (E) "Tracked omission surface" showing high-density arthropod trackways of the *Scoyenia* Ichnofacies. (F) Close-up of a "tracked omission surface", illustrating the *Scoyenia* Ichnofacies.

are low. The *Scoyenia* Ichnofacies is present in eulittoral to supralittoral zones, where lake-level fluctuations cause the frequent flooding and subaerial exposure of both, high-energy sandy beach deposits and low-energy fine-grained mud flats (Figs. 9 and 10). Insect trackways, burrows and vertebrate footprints are typical components (Fig. 10; e.g., Knaust and Hauschke, 2005; Lucas et al., 2011; Minter et al., 2007; Zhang et al., 1998). Lake-level rise may lead to increases in species diversity around lake margins in response to a decrease in salinity (due to freshwater input), while low diversities are maintained in the center of the lake due to continued chemical stratification (Cohen, 2003). Areas with localized freshwater input (e.g., spring-fed pools and marshes, delta plains) favor the concentration of many species of insects, mammals, birds, and reptiles. Local development of microbial mats, associated with hypersaline conditions and/or hot springs, may favor burrowing by certain insects (e.g., staphylinid and heterocerid beetles). Matgrounds help to stabilize the substrate or contribute to its early cementation, increasing the preservation potential of both vertebrate and invertebrate biogenic structures (Scott et al., 2007).

The *Scoyenia* Ichnofacies is widespread around lakes in underfilled basins. Lake-margin assemblages mostly comprise root traces of saline-tolerant sedges and grasses, arthropod trackways (e.g., *Diplichnites*, *Umfolozia*), meniscate traces (e.g., *Scoyenia*, *Taenidium*), bilobate traces (e.g., *Cruziana*, *Rusophycus*), chironomid dipteran, coleopteran, and annelid dwelling and feeding traces (e.g., *Labyrintichnus*), and vertebrate trackways produced by sauropsids, amphibians, dinosaurs, birds, and mammals. Due to the influence of salt efflorescence and other destructive taphonomic effects in subaerially exposed, hypersaline settings, poorly preserved vertebrate footprints and infaunal burrows may be the only preserved evidence of the *Scoyenia* Ichnofacies, especially in areas where capillary evaporation of saline pore waters occurs (e.g., Cohen et al., 1991; Rodríguez-Aranda and Calvo, 1998; Scott et al., 2010). High-density, monospecific to low-diversity assemblages dominated by sharp-walled vertical burrows (*Skolithos*) in subaerial substrates of both high-energy and low-energy eulittoral areas without the influence of freshwater are attributable to the *Skolithos* Ichnofacies.

Progressively desiccated, fine-grained substrates typically preserve better defined trackways emplaced in dewatered sediment, which commonly crosscut less-defined imprints that were formed in less-firm substrates (Buatois et al., 1997; Zhang et al., 1998). The density of trackways may be high, with forming arthropod-tracked omission surfaces, or those representing a period of non-deposition (Fig. 10E; e.g., Minter et al., 2007; Zhang et al., 1998) or high-density tetrapod tracksites (e.g., Farlow and Galton, 2003; Szajna and Hartline, 2003). Some omission surfaces may represent sequence boundaries expressed by coplanar surfaces, or those representing both lowstand and subsequent flooding. Lake-level fluctuations are conducive to complex cross-cutting relationships and trace-fossil suite overprinting, particularly where the surfaces involve more than one transgressive/regressive cycle (Scott et al., 2009).

Relatively extreme variations in lake levels and water-table depths in underfilled basins can lead to the presence of traces typically associated with terrestrial environments within lacustrine and lake-margin deposits. Depending on the conditions of the initial depositional environment, terrestrial trace-fossil assemblages (e.g., *Celliforma*, *Termitichnus*) may overprint previously emplaced lacustrine or lake-margin suites, or they may be emplaced within lacustrine deposits that do not preserve other evidence of biogenic activity. In both cases, the drop in lake level, and thus the water table depths within the zone of lake-level rise and fall, is evidenced by the traces (e.g., de Gibert and Sáez, 2009; Uchman and Álvaro, 2000). These substrates are particularly valuable for recognizing stratigraphic surfaces (e.g., sequence boundaries, flooding surfaces) in lacustrine successions (e.g., Buatois and Mángano, 2009; Scott et al., 2009).

4. CONCLUSIONS

Trace-fossil assemblages show overlap between the overfilled, balanced-fill, and underfilled lake types. It is the diversity within ichnofacies, the substrate-controlled characteristics of the traces, and the lateral and vertical distributions of ichnofacies and trace-fossil suites that provide important information for the interpretation of sedimentary processes, depositional settings, environmental conditions, hiatuses, and the architecture of the lake-basin fill. By considering an integrated lake-type model, the understanding of the depositional processes operating within a lake basin, and how they influence the composition, distribution, and behavior of biological communities is improved. Importantly, the lateral distribution and relative areal extent of lake-basin zones (Table 1) varies between the different lake-type basins, as does the degree of lateral heterogeneity of environmental conditions and the frequency of environmental change in an area through time.

The position within a basin, whether it is a basin-margin or basin-center locality for example, is an essential consideration for applying ichnology to lake-basin stratigraphy, and also for a better understanding of the potential limiting controls (e.g., oxygenation, salinity) that influence the precise makeup of an ichnofacies or suite. Although balanced-fill and underfilled lake types preserve both softground and firmground expressions of the *Scoyenia* Ichnofacies in lake margins, for example, these types of depositional environments are commonly present in basin-center areas if the basin is underfilled, but more typically in basin-margin areas if it is a balanced-fill basin. Knowledge of the structural setting of the studied locality can be important for helping to explain the variability in environmental controls represented by different compositions of an ichnofacies along strike, due to the influence of structural features on the distribution of springs and ephemeral streams or perennial rivers, for instance. The frequency and extent of lake-level fluctuations varies between the lake types, which influences not only the distribution of biogenic structures emplaced during or shortly after sedimentation but also the degree of overprinting within palimpsest substrates and the frequency of lake marginal to terrestrial trace types present within lacustrine sediments.

The integration of the lake-type model with the ichnology of lake basins provides a framework from which to better understand the various expressions of ichnofacies, including their composition, diversity, and distribution. Likewise, the applicability of ichnology to lake-basin stratigraphy appears to be extensive, through helping to recognize changing conditions and hiatuses as well as important sequence-stratigraphic surfaces. More case studies that consider lake-basin sedimentary environments in detail, and also trace-fossil distributions at the basin scale, are needed to test and improve the model and help to move lake-basin ichnology forward as an applied discipline useful for stratigraphy, sedimentology, and paleoecology. Neoichnological studies that articulate both the behavioral ecology of trace making organisms (i.e., autecology) and the environmental ecology of trace-producing communities (i.e., synecology) continue to be essential for understanding organism/sediment interactions in continental environments. Taphonomic studies that analyze sedimentary processes and consider the initial and preserved distributions of biogenic structures in different depositional settings will help to better apply neoichnology to understand the rock record.

ACKNOWLEDGMENTS

We are grateful for discussions with Robin Renaut, Bernie Owen, Mike Smith and Alan Carroll, which contributed to this work. Sincere thanks are also extended to Robin Renaut for input on Table 1 and co-supervision of the doctoral thesis of J. J. S. and to Jordi de Gibert and Robin Renaut for sharing some of the photographs used here. Funding was provided by Natural Sciences and Engineering Research Council of Canada (NSERC) grants to J. J. S. (PGS-D scholarship, PDF fellowship), L. A. B. (Discovery Grant # 311726-05 and -08), M. G. M. (Discovery Grant 311727-05 and -08), and Robin Renaut (Research Grant RG629-03). Fieldwork was partially funded by the International Association of Sedimentologists, the Geological Society of America, and the American Association of Petroleum Geologists. The helpful reviews by Elizabeth Gierlowski-Kordesch and Jörg Schneider are very much appreciated. Sincere thanks are also extended to editors Dirk Knaust and Richard Bromley for their careful reviews and the invitation to contribute to this volume.

REFERENCES

Ahlbrant, T.S., Andrews, S., Gwynne, D.T., 1978. Bioturbation in eolian deposits. J. Sediment. Petrol. 48, 839–848.

Alonso-Zarza, A.M., 2003. Palaeoenvironmental significance of palustrine carbonates and calcretes in the geological record. Earth-Sci. Rev. 60, 261–298.

Archer, A.W., Maples, C.G., 1984. Trace-fossil distribution across a marine-to-nonmarine gradient in the Pennsylvanian of southwestern Indiana. J. Paleontol. 58, 448–466.

Benner, J.S., Ridge, J.C., Knecht, R.J., 2009. Timing of post-glacial reinhabitation and ecological development of two New England, USA, drainages based on trace fossil evidence. Palaeogeogr. Palaeoclimatol. Palaeoecol. 272, 212–231.

Bohacs, K.M., Carroll, A.R., Neal, J.E., Mankiewicz, P.J., 2000. Lake-basin type, source potential, and hydrocarbon character: an integrated sequence-stratigraphic–geochemical framework. In: Gierlowski-Kordesch, E.H., Kelts, K.R. (Eds.), Lake Basins through Space and Time. AAPG Studies in Geology, vol. 46, pp. 3–34.

Bohacs, K.M., Carroll, A.R., Neal, J.E., 2003. Lessons from large lake systems: thresholds, nonlinearity, and strange attractors. In: Chan, M.A., Archer, A.W. (Eds.), Extreme Depositional Environments: Mega End Members in Geologic Time. GSA Spec. Pap. 370, pp. 75–90.

Bohacs, K.M., Grabowski Jr., G., Carroll, A.R., 2007a. Lithofacies architecture and variations in expression of sequence stratigraphy within representative intervals of the Green River Formation, Greater Green River Basin, Wyoming and Colorado. Mountain Geol. 44, 39–60.

Bohacs, K.M., Hasiotis, S.T., Demko, T.M., 2007b. Continental ichnofossils of the Green River and Wasatch Formations, Eocene, Wyoming: a preliminary survey, proposed relation to lake-basin type, and application to integrated paleo-environmental interpretation. Mountain Geol. 44, 79–108.

Bromley, R., Asgaard, U., 1979. Triassic freshwater ichnocoenoses from Carlsberg Fjord, East Greenland. Palaeogeogr. Palaeoclimatol. Palaeoecol. 28, 39–80.

Bromley, R., Asgaard, U., 1991. Ichnofacies: a mixture of taphofacies and biofacies. Lethaia 24, 153–163.

Buatois, L.A., Mángano, M.G., 1995. The paleoenvironmental and paleoecological significance of the lacustrine *Mermia* Ichnofacies: an archetypical subaqueous nonmarine trace fossil assemblage. Ichnos 4, 151–161.

Buatois, L.A., Mángano, M.G., 1998. Trace fossil analysis of lacustrine facies and basins. Palaeogeogr. Palaeoclimatol. Palaeoecol. 183, 71–86.

Buatois, L.A., Mángano, M.G., 2002. Trace fossils from Carboniferous floodplain deposits in western Argentina: implications for ichnofacies models of continental environments. Palaeogeogr. Palaeoclimatol. Palaeoecol. 183, 71–86.

Buatois, L.A., Mángano, M.G., 2004. Animal-substrate interactions in freshwater environments: applications of ichnology in facies and sequence stratigraphic analysis of fluvio-lacustrine successions. In: McIlroy, D. (Ed.), The Application of Ichnology to Palaeoenvironmental and Stratigraphic Analysis. Geol. Soc. London, Spec. Publ., vol. 228, pp. 311–333.

Buatois, L.A., Mángano, M.G., 2007. Invertebrate ichnology of continental freshwater environments. In: Miller III, W. (Ed.), Trace Fossils. Concepts, Problems, Prospects. Elsevier, Amsterdam, pp. 285–323.

Buatois, L.A., Mángano, M.G., 2009. Applications of ichnology in lacustrine sequence stratigraphy: potential and limitations. Palaeogeogr. Palaeoclimatol. Palaeoecol. 272, 127–142.

Buatois, L.A., Mángano, M.G., 2011. Ichnology: Organism-Substrate Interactions in Space and Time. Cambridge University Press, Cambridge, 358 pp.

Buatois, L.A., Mángano, M.G., Wu, X., Zhang, G., 1995. *Vagorichnus*, a new ichnogenus for feeding burrow systems and its occurrence as discrete and compound ichnotaxa in Jurassic lacustrine turbidites of Central China. Ichnos 3, 265–272.

Buatois, L.A., Mángano, M.G., Wu, X., Zhang, G., 1996. Trace fossils from Jurassic lacustrine turbidites of the Anyao Formation (central China) and their environmental and evolutionary significance. Ichnos 4, 287–303.

Buatois, L.A., Jalfin, G., Aceñolaza, F.G., 1997. Permian nonmarine invertebrate trace fossils from southern Patagonia, Argentina: ichnologic signatures of substrate consolidation and colonization sequences. J. Paleontol. 71, 324–336.

Buatois, L.A., Mángano, M.G., Wu, X., Zhang, G., 2000. Jurassic lake deposits from the Anyao Formation, Central China. In: Gierlowski-Kordesch, E., Kelts, K. (Eds.), Lake Basins through Space and Time. AAPG Studies in Geology, vol. 46, pp. 189–194.

Buatois, L.A., Netto, R.G., Mángano, M.G., Balistieri, P.R.M.N., 2006. Extreme freshwater release during the late Paleozoic Gondwana deglaciation and its impact on coastal ecosystems. Geology 34, 1021–1024.

Buatois, L.A., Netto, R., Mángano, M.G., 2010. Ichnology of late Paleozoic post-glacial transgressive deposits in Gondwana: reconstructing salinity conditions in coastal ecosystems affected by strong meltwater discharge. In: Lopez Gamundi, O., Buatois, L.A. (Eds.), Late Paleozoic Glacial Events and Postglacial Transgressions in Gondwana. GSA Spec. Pap. 468, pp. 149–173.

Carroll, A.R., Bohacs, K.M., 1999. Stratigraphic classification of ancient lakes: balancing tectonic and climatic controls. Geology 27, 99–102.

Carroll, A.R., Bohacs, K.M., 2001. Lake-type controls on petroleum source rock potential in nonmarine basins. AAPG Bull. 85, 1033–1053.

Carroll, A.R., Doebbert, A.C., Booth, A.L., Chamberlain, C.P., Rhodes-Carson, M.K., Smith, M.E., Johnson, C.M., Beard, B.L., 2008. Capture of high-altitude precipitation by a low-altitude Eocene lake, western U.S. Geology 36, 791–794.

Chamberlain, C.K., 1975. Recent lebensspuren in nonmarine aquatic environments. In: Frey, R.W. (Ed.), The Study of Trace Fossils. Springer, New York, pp. 431–458.

Charbonneau, P., Hare, L., 1998. Burrowing behavior and biogenic structures of mud-dwelling insects. J. N. Am. Benthol. Soc. 17, 239–249.

Clark, G.R., Ratcliffe, B.C., 1989. Observations on the tunnel morphology of *Heterocerus brunneus* Melsheimer (Coleoptera: Heteroceridae) and its paleoecological significance. J. Paleontol. 63, 228–232.

Cohen, A.S., 1982. Paleoenvironments of root casts from the Koobi Fora Formation, Kenya. J. Sediment. Petrol. 52, 401–414.

Cohen, A.S., 2003. Paleolimnology. Oxford University Press, Oxford, 500 pp.

Cohen, A., Lockley, M., Halfpenny, J., Michel, A.E., 1991. Modern vertebrate track taphonomy at Lake Manyara, Tanzania. Palaios 6, 371–389.

Cohen, A., Halfpenny, J., Lockley, M., Michel, A.E., 1993. Modern vertebrate tracks from Lake Manyara, Tanzania and their paleobiological implications. Paleobiology 19, 433–458.

Dashtgard, S.E., Gingras, M.K., 2005. The temporal significance of bioturbation in backshore deposits: Waterside Beach, New Brunswick, Canada. Palaios 20, 589–595.

Dashtgard, S.E., Gingras, M.K., 2012. Marine invertebrate neoichnology. In: Knaust, D., Bromley, R. G. (Eds.), Trace Fossils as Indicators of Sedimentary Environments. Developments in Sedimentology, vol. 64. Elsevier, Amsterdam, pp. 273–295.

de Gibert, J.M., Sáez, A., 2009. Paleohydrological significance of trace fossil distribution in Oligocene fluvial-fan-to-lacustrine systems of the Ebro Basin, Spain. Palaegeogr. Palaeoclimatol. Palaeoecol. 272, 162–175.

de Gibert, J.M., Fregenal-Martínez, M.A., Buatois, L.A., Mángano, M.G., 2000. Trace fossils and their palaeoecological significance in Lower Cretaceous lacustrine conservation deposits, El Montsec, Spain. Palaegeogr. Palaeoclimatol. Palaeoecol. 156, 89–101.

Edwards, N., Jarzembowski, E.A., Pain, T., Daley, B., 1998. Cocoon-like trace fossils from the lacustrine-palustrine Bembridge Limestone Formation (Late Eocene), southern England. Proc. Geol. Ass. 109, 25–32.

Ekdale, A.A., Bromley, R.G., Loope, D.B., 2007. Ichnofacies of an ancient erg: a climatically influenced trace fossil association in the Jurassic Navajo sandstone, southern Utah, USA. In: Miller III, W. (Ed.), Trace Fossils. Concepts, Problems, Prospects. Elsevier, Amsterdam, pp. 562–574.

Farlow, J.O., Galton, P.M., 2003. Dinosaur trackways of Dinosaur State Park, Rocky Hill, Connecticut. In: Letourneau, P.M., Olsen, P.E. (Eds.), The Great Rift Valleys of Pangea in Eastern North America. Sedimentology, Stratigraphy, and Paleontology, vol. 2. Columbia University Press, New York, pp. 248–263.

Fisher, J.B., 1982. Effects of macrobenthos on the chemical diagenesis of freshwater sediments. In: McCall, P.L., Tevesz, M.J.S. (Eds.), Animal-Sediment Relations: The Biogenic Alteration of Sediments. Topics in Geobiology, vol. 2. Academic Press, New York, pp. 177–218.

Fisher, J.A., Beeton, A.M., 1975. The effect of dissolved oxygen on the burrowing behaviour of *Limnodrilus hoffmeisteri* (Oligochaeta). Hydrobiologia 47, 273–290.

Fitzgerald, P.G., Barrett, P.J., 1986. *Skolithos* in a Permian braided river deposit, southern Victoria Land, Antarctica. Palaeogeogr. Palaeoclimatol. Palaeoecol. 52, 237–247.

Frey, R.W., Pemberton, S.G., 1987. The *Psilonichnus* ichnocoenose, and its relationship to adjacent marine and nonmarine ichnocoenoses along the Georgia coast. Bull. Can. Petrol. Geol. 35, 333–357.

Frey, R.W., Pemberton, S.G., Fagerstrom, J.A., 1984. Morphological, ethological and environmental significance of the ichnogenera *Scoyenia* and *Ancorichnus*. J. Paleontol. 58, 511–528.

Garcia, C.M., Niell, F.X., 1991. Burrowing beetles of the genus *Bledius* (Staphylinidae) as agents of bioturbation in the emergent areas and shores of an athalassic inland lake (Fuente de Piedra, southern of Spain). Hydrobiology 215, 163–173.

Genise, J.F., Mángano, G.M., Buatois, L.A., Laza, J.H., Verde, M., 2000. Insect trace fossil associations in paleosols: the *Coprinisphaera* Ichnofacies. Palaios 15, 49–64.

Genise, J.F., Bellosi, E.S., Gonzalez, M.G., 2004. An approach to the description and interpretation of ichnofabrics in palaeosols. In: McIlroy, D. (Ed.), The Application of Ichnology to Palaeoenvironmental and Stratigraphic Analysis. Geol. Soc. London, Spec. Publ., vol. 228, pp. 355–382.

Genise, J.F., Melchor, R.N., Archangelsky, M., Bala, L.O., Straneck, R., De Valais, S., 2009. Application of neoichnological studies to behavioural and taphonomic interpretation of fossil bird-like tracks from lacustrine settings: the Late Triassic–Early Jurassic? Santo Domingo Formation, Argentina. Palaeogeogr. Palaeoclimatol. Palaeoecol. 272, 143–161.

Genise, J.F., Melchor, R.N., Bellosia, E.S., Verde, M., 2010. Invertebrate and vertebrate trace fossils from continental carbonates. In: Alonso-Zarza, A.M., Tanner, L.H. (Eds.), Carbonates in Continental Settings: Facies, Environments and Processes. Developments in Sedimentology, vol. 61. Elsevier, Amsterdam, pp. 319–362.

Gierlowski-Kordesch, E., 1991. Ichnology of an ephemeral lacustrine/alluvial plain system: Jurassic East Berlin Formation, Hartford Basin, USA. Ichnos 1, 221–232.

Gierlowski-Kordesch, E., 2010. Lacustrine carbonates. In: Alonso-Zarza, A.M., Tanner, L.H. (Eds.), Carbonates in Continental Settings: Facies, Environments and Processes. Developments in Sedimentology, vol. 61. Elsevier, Amsterdam, pp. 2–101.

Gierlowski-Kordesch, E., Park, L.E., 2004. Comparing species diversity in the modern and fossil record of lakes. J. Geol. 112, 703–717.

Gingras, M.K., Lalond, S.V., Amskold, L., Konhauser, K., 2007. Wintering chironomids mine oxygen. Palaios 22, 433–438.

Gore, P.J.W., 1989. Toward a model for open- and closed-basin deposition in ancient lacustrine sequences: the Newark Supergroup (Triassic-Jurassic), eastern North America. Palaeogeogr. Palaeoclimatol. Palaeoecol. 70, 29–51.

Hamer, J.M.M., Sheldon, N.D., 2010. Neoichnology at lake margins: implications for paleo-lake systems. Sediment. Geol. 228, 319–327.

Hammer, T.U., 1986. Saline Lake Ecosystems of the World. Dr. W. Junk Publishers, Dordrecht, 616 pp.

Hasiotis, S.T., 2002. Continental trace fossils. SEPM Short Course Notes 51, 1–132.

Hasiotis, S.T., 2004. Reconnaissance of Upper Jurassic Morrison Formation ichnofossils, Rocky Mountain Region, USA: paleoenvironmental, stratigraphic, and paleoclimatic significance of terrestrial and freshwater ichnocoenoses. Sediment. Geol. 167, 177–268.

Hasiotis, S.T., 2007. Continental ichnology: fundamental processes and controls on trace fossil distribution. In: Miller III, W. (Ed.), Trace Fossils. Concepts, Problems, Prospects. Elsevier, Amsterdam, pp. 268–284.

Hasiotis, S.T., Mitchell, C.E., 1993. A comparison of crayfish burrow morphologies: Triassic and Holocene fossil, paleo- and neo-ichnological evidence, and the identification of their burrowing signatures. Ichnos 2, 291–314.

Hu, B., Wang, G., Goldring, R., 1998. *Nereites* (or *Neonereites*) from Lower Jurassic lacustrine turbidites of Henan, Central China. Ichnos 6, 203–209.

Hunt, A.P., Lucas, S.G., 2007. Tetrapod ichnofacies: a new paradigm. Ichnos 14, 59–68.

Keighley, D.G., Pickerill, R.K., 2003. Ichnocoenoses from the Carboniferous of eastern Canada and their implications for the recognition of ichnofacies in nonmarine strata. Atlant. Geol. 39, 1–22.

Kim, J.Y., Kim, K.-S., Pickerill, R.K., 2002. Cretaceous nonmarine trace fossils from the Hasandong and Jinju Formations of the Namhae area, Kyongsangnamdo, southeast Korea. Ichnos 9, 41–60.

Kim, J.Y., Keighley, D.G., Pickerill, R.K., Hwang, W., Kim, K.-S., 2005. Trace fossils from marginal lacustrine deposits of the Cretaceous Jinju Formation, southern coast of Korea. Palaeogeogr. Palaeoclimatol. Palaeoecol. 218, 105–124.

Knaust, D., Hauschke, N., 2005. Living conditions in a Lower Triassic playa system of Central Germany: evidence from ichnofauna and body fossils. Hallesches Jahrb. Geowiss. B19, 95–108.

Knecht, R.J., Benner, J.S., Rogers, D.C., Ridge, J.C., 2009. *Surculichnus bifurcauda* n. igen., n. isp., a trace fossil from Late Pleistocene glaciolacustrine varves of the Connecticut River Valley, USA, attributed to notostracan crustaceans based on neoichnological experimentation. Palaeogeogr. Palaeoclimatol. Palaeoecol. 272, 232–239.

Knisley, C.B., 1984. Ecological distribution of tiger beetles (Coleoptera: Cicindelidae) in Colfax County, New Mexico. Southwest. Nat. 29, 93–104.

Kraus, M.J., Hasiotis, S.T., 2006. Significance of different modes of rhizolith preservation to interpreting paleoenvironmental and paleohydrologic settings: examples from Paleogene paleosols, Bighorn Basin, Wyoming, U.S.A. J. Sediment. Res. 76, 633–646.

Lamond, R.E., Tapanila, L., 2003. Embedment cavities in lacustrine stromatolites: evidence of animal interactions from Cenozoic carbonates in U.S.A. and Kenya. Palaios 18, 445–453.

Leggitt, V.L., Biaggi, R.E., Buchheim, H.P., 2007. Palaeoenvironments associated with caddisfly-dominated microbial-carbonate mounds from the Tipton Shale Member of the Green River Formation: Eocene Lake Gosiute. Sedimentology 54, 661–699.

Lockley, M.G., 1986. The paleobiological and paleoenvironmental importance of dinosaur footprints. Palaios 1, 37–47.

Lockley, M., 2007. A tale of two ichnologies: the different goals and potentials of invertebrate and vertebrate (tetrapod) ichnotaxonomy and how they relate to ichnofacies analysis. Ichnos 14, 39–57.

Lockley, M.G., Lucas, S.G., Hunt, A.P., 1994. Vertebrate tracks and the ichnofacies concept: implications for paleoecology and palichnostratigraphy. In: Donovan, S. (Ed.), The Palaeobiology of Trace Fossils. Wiley and Sons, New York, pp. 241–268.

Lucas, S.G., Voigt, S., Lerner, A.J., Nelson, J.W., 2011. Late Early Permian continental ichnofauna from Lake Kemp, north-central Texas, USA. Palaeogeogr. Palaeoclimatol. Palaeoecol. 308, 395–404.

Mackie, G.L., 2001. Applied Aquatic Ecosystem Concepts. Kendall/Hunt Publishing Company, Dubuque, 744 pp.

Magyar, I., Müller, P.M., Sztanó, O., Babinszki, E., Lantos, M., 2006. Oxygen-related facies in Lake Pannon deposits (Upper Miocene) at Budapest-Köbánya. Facies 52, 209–220.

Mángano, M.G., Buatois, L.A., Wu, X., Sun, J., Zhang, G., 1994. Sedimentary facies, depositional processes and climatic controls in a Triassic lake, Tanzuang Formation, western Henan Province, China. J. Paleolimnol. 11, 41–65.

Mángano, M.G., Buatois, L.A., Wu, X., Sun, J., Zhang, G., 2000. Triassic lacustrine sedimentation from the Tanzhuang Formation, Jiyuan-Yima Basin, southeastern China. In: Gierlowski-Kordesch, E., Kelts, K. (Eds.), Global Geological Record of Lake Basins. AAPG Studies in Geology, vol. 46, pp. 133–140.

Martin, P., Boes, X., Goddeeris, B., Fagel, N., 2005. A qualitative assessment of the influence of bioturbation in Lake Baikal sediments. Glob. Planet. Change 46, 87–99.

McCall, P.L., Tevesz, M.J.S., 1982. The effects of benthos on physical properties of freshwater sediments. In: McCall, P.L., Tevesz, M.J.S. (Eds.), Animal-Sediment Relations: The Biogenic Alteration of Sediments. Topics in Geobiology, vol. 2. Academic Press, New York, pp. 105–176.

McCall, P.L., Matisoff, G., Tevesz, M.J.S., 1986. The effects of a unionid bivalve on the physical, chemical, and microbial properties of cohesive sediments from Lake Erie. Am. J. Sci. 286, 127–159.

Melchor, R.N., 2004. Trace fossil distribution in lacustrine deltas: examples from the Triassic rift lakes of the Ischigualasto–Villa Unión basin, Argentina. In: McIlroy, D. (Ed.), The Application of Ichnology to Palaeoenvironmental and Stratigraphic Analysis. Geological Society London, Spec. Publ., vol. 228. pp. 335–354.

Melchor, R.N., Genise, J.F., Miquel, S., 2002. Ichnology, sedimentology and paleontology of Eocene calcareous paleosols from a palustrine sequence, south west La Pampa, central Argentina. Palaios 17, 16–35.

Melchor, R.N., Bedatou, E., De Valais, S., Genise, J.F., 2006. Lithofacies distribution of invertebrate and vertebrate trace-fossil assemblages in an Early Mesozoic ephemeral fluvio-lacustrine system from Argentina: implications for the *Scoyenia* Ichnofacies. Palaeogeogr. Palaeoclimatol. Palaeoecol. 239, 253–285.

Melchor, R.N., Genise, J.F., Buatois, L.A., Umazano, A.M., 2012. Fluvial environments. In: Knaust, D., Bromley, R.G. (Eds.), Trace Fossils as Indicators of Sedimentary Environments. Developments in Sedimentology, vol. 64. Elsevier, Amsterdam, pp. 329–378.

Metz, R., 1987. Sinusoidal trail formed by a recent biting midge (Family Ceratopogonidae): trace fossil implications. J. Paleontol. 61, 312–314.

Metz, R., 1990. Tunnels formed by mole crickets (Orthoptera: Gryllotalpidae): paleoecological implications. Ichnos 1, 139–141.

Metz, R., 1996. Newark basin ichnology: the Late Triassic Perkasie Member of the Passaic Formation, Sanatoga, Pennsylvania. Northeast. Geol. Environ. Sci. 18, 118–129.

Metz, R., 2000. Triassic trace fossils from lacustrine shoreline deposits of the Passaic Formation, Douglasville, Pennsylvania. Ichnos 7, 253–266.

Miller, M.F., White, D.S., 2007. Ecological and evolutionary controls on the composition of marine and lake ichnofacies. In: Miller III, W. (Ed.), Trace Fossils. Concepts, Problems, Prospects. Elsevier, Amsterdam, pp. 531–544.

Minter, N.J., Krainer, K., Lucas, S.G., Braddy, S.J., Hunt, A.P., 2007. Palaeoecology of an Early Permian playa lake trace fossil assemblage from Castle Peak, Texas, USA. Palaeogeogr. Palaeoclimatol. Palaeoecol. 246, 390–423.

Netto, R.G., 2007. *Skolithos*-dominated piperock in nonmarine environments: an example from the Triassic Caturrita Formation, southern Brazil. In: Bromley, R.G., Buatois, L.A., Mángano, M.G., Genise, J.F., Melchor, R.N. (Eds.), Sediment-Organism Interactions: A Multifaceted Ichnology. SEPM Spec. Publ., vol. 88, pp. 109–121.

Netto, R.G., Balistieri, P.R.M.N., Lavina, E.L.C., Silveira, D.M., 2009. Ichnological signatures of shallow freshwater lakes in the glacial Itararé Group (Mafra Formation, Upper Carboniferous–Lower Permian of Paraná Basin, S Brazil). Palaeogeogr. Palaeoclimatol. Palaeoecol. 272, 240–255.

Netto, R.G., Benner, J.S., Buatois, L.A., Uchman, A., Mángano, M.G., Ridge, J.C., Kazakauskas, V., Gaigalas, A., 2012. Glacial environments. In: Knaust, D., Bromley, R.G. (Eds.), Trace Fossils as Indicators of Sedimentary Environments. Developments in Sedimentology, vol. 64. Elsevier, Amsterdam, pp. 299–327.

Ortí, F., Rosell, L., Anadón, P., 2003. Deep to shallow lacustrine evaporites in the Libros Gypsum (southern Teruel Basin, Miocene, NE Spain): an occurrence of pelletal gypsum rhythmites. Sedimentology 50, 361–386.

Phillips Jr., P.L., Ludvigson, G.A., Joeckel, R.M., González, L.A., Brenner, R.L., Witzke, B.J., 2007. Sequence stratigraphic controls on synsedimentary cementation and preservation of dinosaur tracks: example from the Lower Cretaceous, (Upper Albian) Dakota Formation, Southeastern Nebraska, U.S.A. Palaeogeogr. Palaeoclimatol. Palaeoecol. 246, 367–389.

Pickerill, R.K., 1990. Nonmarine *Paleodictyon* from the Carboniferous Albert Formation of southern New Brunswick. Atlant. Geol. 26, 157–163.

Pietras, J.T., Carroll, A.R., 2006. High-resolution stratigraphy of an underfilled lake basin: Wilkins Peak Member, Eocene Green River Formation, Wyoming, U.S.A. J. Sediment. Res. 76, 1197–1214.

Platt, N.H., 1989. Lacustrine carbonates and pedogenesis: sedimentology and origin of palustrine deposits from the Early Cretaceous Rupelo formation, W Cameros Basin, N Spain. Sedimentology 36, 665–684.

Pollard, J.E., Steel, R.J., Undersrud, E., 1982. Facies sequences and trace fossils in lacustrine/fan delta deposits, Hornelen Basin (M. Devonian), western Norway. Sediment. Geol. 32, 63–87.

Porter, R.J., Gallois, R.W., 2008. Identifying fluvio-lacustrine intervals in thick playa-lake successions: an integrated sedimentology and ichnology of arenaceous members in the mid-late Triassic Mercia Mudstone Group of south-west England, UK. Palaeogeogr. Palaeoclimatol. Palaeoecol. 270, 381–398.

Prince, N.K., Lockley, M.G., 1989. The sedimentology of the Purgatoire tracksite region, Morrison Formation of southeastern Colorado. In: Gillette, D.D., Lockley, M.G. (Eds.), Dinosaur Tracks and Traces. Cambridge University Press, Cambridge, pp. 155–163.

Ratcliffe, B.C., Fagerstrom, J.A., 1980. Invertebrate lebenspuren of Holocene floodplain: their morphology, origin, and paleoecological significance. J. Paleontol. 54, 614–630.

Renaut, R.W., Gierlowski-Kordesch, E., 2010. Lakes. In: James, N., Dalrymple, R. (Eds.), Facies Models 4. Geol. Ass. Can., St. John's, Geotext 6, pp. 541–575.

Rodríguez-Aranda, J.P., Calvo, J.P., 1998. Trace fossils and rhizoliths as a tool for sedimentological and palaeoenvironmental analysis of ancient continental evaporite successions. Palaeogeogr. Palaeoclimatol. Palaeoecol. 140, 383–399.

Schowalter, T.D., 2006. Insect Ecology: An Ecosystem Approach. Elsevier/Academic Press, Amsterdam/Boston, 572 pp.

Scott, J.J., 2010. Saline lake ichnology: composition and distribution of Cenozoic traces in the saline, alkaline lakes of the Kenya Rift Valley and Eocene Green River Formation, U.S.A. Unpublished. Ph.D. Thesis, University of Saskatchewan, 547 pp.

Scott, J.J., Renaut, R.W., Owen, R.B., Sarjeant, W.A.S., 2007. Biogenic activity, trace formation, and trace taphonomy in the marginal sediments of saline, alkaline Lake Bogoria, Kenya Rift Valley. In: Bromley, R.G., Buatois, L.A., Mángano, M.G., Genise, J.F., Melchor, R.N. (Eds.), Sediment-Organism Interactions. A Multifaceted Ichnology. SEPM Spec. Publ., vol. 88, pp. 311–332.

Scott, J.J., Renaut, R.W., Owen, R.B., 2008. Preservation and paleoenvironmental significance of a footprinted surface on the Sandai Plain, Lake Bogoria, Kenya Rift Valley. Ichnos 15, 208–231.

Scott, J.J., Renaut, R.W., Buatois, L.A., Owen, R.B., 2009. Biogenic structures in exhumed surfaces around saline lakes: an example from Lake Bogoria, Kenya Rift Valley. Palaeogeogr. Palaeoclimatol. Palaeoecol. 272, 176–198.

Scott, J.J., Renaut, R.W., Owen, R.B., 2010. Taphonomic controls on animal tracks at saline, alkaline Lake Bogoria, Kenya Rift Valley: impact of salt efflorescence and clay mineralogy. J. Sediment. Res. 80, 639–665.

Scrivner, P.J., Bottjer, D.J., 1986. Neogene avian and mammalian tracks from Death Valley National Monument, California: their context, classification, and preservation. Palaeogeogr. Palaeoclimatol. Palaeoecol. 57, 285–331.

Smith, R.M.H., 1993. Sedimentology and ichnology of floodplain paleosurfaces in the Beaufort Group (Late Permian), Karroo Sequence, South Africa. Palaios 8, 339–357.

Smith, R.M.H., Mason, T.R., 1998. Sedimentary environments and trace fossils of Tertiary oasis deposits in the central Namib Desert, Namibia. Palaios 13, 547–559.

Smith, P.R., Harper, A.S., Wood, M.F., 1982. Nonmarine trace fossils in the Mio-Pliocene Ridge Basin Group, southern California. In: Cromwell, J.C., Link, M.H. (Eds.), Geologic History of Ridge Basin, Southern California. SEPM, Pacific Section, pp. 253–258.

Smith, J.J., Hasiotis, S.T., Woody, D.T., Kraus, M.J., 2008. *Naktodemasis bowni*: new ichnogenus and ichnospecies for adhesive meniscate burrows (AMB), and paleoenvironmental implications, Paleogene Willwood Formation, Bighorn Basin, Wyoming. J. Paleontol. 82, 267–278.

Smoot, J.P., 1983. Depositional subenvironments in an arid closed basin: the Wilkins Peak Member of the Green River Formation (Eocene), Wyoming, U.S.A. Sedimentology 30, 801–827.

Szajna, M.J., Hartline, B.W., 2003. A new vertebrate footprint locality from the Late Triassic Passaic Formation near Birdsboro, Pennsylvania. In: Letourneau, P.M., Olsen, P.E. (Eds.), The Great Rift Valleys of Pangea in Eastern North America. Sedimentology, Stratigraphy, and Paleontology, vol. 2. Columbia University Press, New York, pp. 264–272.

Talbot, M.R., Allen, P.A., 1996. Lakes. In: Reading, H.G. (Ed.), Sedimentary Environments: Processes, Facies and Stratigraphy. Blackwell, Oxford, pp. 83–124.

Trewin, N.H., 1976. *Isopodichnus* in a trace fossil assemblage from the Old Red Sandstone. Lethaia 9, 29–37.

Uchman, A., 2005. Treptichnus-like traces made by insect larvae (Diptera: Chironomidae, Tipulidae). In: Buta, R.J., Rindsberg, A.K., Kopaska-Merkel, D.C. (Eds.), Pennsylvanian Footprints in the Black Warrior Basin of Alabama. Alabama Paleontol. Soc., Monogr. 1, pp. 143–146.

Uchman, A., Álvaro, J.J., 2000. Non-marine invertebrate trace fossils from the Tertiary Calatayud-Teruel Basin, NE Spain. Rev. Esp. Paleontol. 15, 203–218.

Uchman, A., Nemec, W., Ilgar, A., Messina, C., 2007. Lacustrine trace fossils and environmental conditions in the early Miocene Ermenek Basin, southern Turkey. Ann. Soc. Geol. Polon. 77, 123–139.

Uchman, A., Kazakauskas, V., Gaigalas, A., 2009. Trace fossils from Late Pleistocene varved lacustrine sediments in eastern Lithuania. Palaeogeogr. Palaeoclimatol. Palaeoecol. 272, 199–211.

Voigt, S., Hoppe, D., 2010. Mass occurrence of penetrative trace fossils in Triassic lake deposits (Kyrgyzstan, central Asia). Ichnos 17, 1–11.

Voigt, S., Berman, D.S., Henrici, A.C., 2007. First well-established track-trackmaker association of Paleozoic tetrapods based on *Ichniotherium* trackways and diadectid skeletons from the Lower Permian of Germany. J. Vert. Paleontol. 27, 553–570.

Ward, J.V., 1992. Aquatic Insect Ecology: 1. Biology and Habitat. John Wiley & Sons, New York, 438 pp.

Wetzel, R.G., 2001. Limnology: Lake and River Ecosystems. Academic Press, London, 1006 pp.

White, D.S., Miller, M.F., 2008. Benthic invertebrate activity in lakes: linking present and historical bioturbation patterns. Aquat. Biol. 2, 269–277.

Williams, W.D., 1998. Salinity as a determinant of the structure of biological communities in salt lakes. Hydrobiologia 381, 191–201.

Williams, W.D., Boulton, A.J., Taaffe, R.G., 1990. Salinity as a determinant of salt lake fauna: a question of scale. Hydrobiologia 197, 257–266.

Zhang, G., Buatois, L.A., Mángano, M.G., Aceñolaza, F.G., 1998. Sedimentary facies and environmental ichnology of a ?Permian playa-lake complex in western Argentina. Palaeogeogr. Palaeoclimatol. Palaeoecol. 138, 221–243.

Chapter 14

Eolian Environments

Allan A. Ekdale*,[1] and Richard G. Bromley†
*Department of Geology and Geophysics, University of Utah, Salt Lake City, Utah, USA,
†Geological Museum—SNM, Copenhagen, Denmark
[1]Corresponding author: e-mail: a.ekdale@utah.edu

1. INTRODUCTION

On his epic voyage home to Greece following the Trojan War, Odysseus met and became fast friends with Aeolus, keeper of the four winds. Aeolus presented his Greek friend with an ox-hide bag tied with a silver cord, which contained all the necessary breezes to carry Odysseus' ships toward their beloved homeland. But while at sea, the unruly crew ripped open the bag and unleashed all the howling winds at once, thus creating a great tempest that blew them hopelessly off course (from Homer's *Odyssey*, Chapter X).

Winds have been blowing across Earth's surface since our planet's origin, and the geologic record of windy environments is contained within eolianites, aptly named after the mythical custodian of all winds. Deciphering the direction, intensity, and constancy of ancient wind regimes from eolianites is a challenge to continental sedimentologists because sedimentary structures and bed forms in such deposits can be quite confusing. Deciphering the ecology of ancient wind-blown environments from eolianites is a challenge to paleontologists because fossils in such deposits are notoriously sparse. Ichnologic evidence may provide welcome assistance to both.

2. ICHNOFACIES

2.1 Ichnology in Eolian Environments

A sand dune is a subaerial substrate on the move. It is at once an erosional environment and a depositional environment. The gently dipping, windward, and sometimes partially vegetated stoss side of a dune is a site of uphill sediment transport by saltation (or less commonly by traction or suspension). The more steeply dipping, leeward, and generally unvegetated slip face side of a dune is a site of sedimentation, largely by grain flow and/or avalanche depositional mechanisms. The dune moves grain by grain, thus providing a mobile substrate

Developments in Sedimentology, Vol. 64. http://dx.doi.org/10.1016/B978-0-444-53813-0.00014-9

for inhabitants of the dune field. Rarely is the stoss preserved in ancient eolianites; it is the cross-stratified slip face deposits that make up the bulk of ancient eolian sandstone beds.

Tracemaking animals that inhabit the dynamic environment of modern and ancient dune fields consist almost exclusively of arthropods (chiefly insects and arachnids) and vertebrates (chiefly reptiles and mammals). Thus, studies in eolian ichnocoenoses necessarily include attention to both invertebrate and vertebrate traces.

Although the obligatory arid conditions that characterize active dune environments typically preclude high population densities, dune communities may exhibit a surprising diversity of desert organisms and their traces (Fig. 1). It may seem counterintuitive that in modern dunes, the most distinct and most easily identifiable traces occur in sand in the erosional stoss microenvironment, which is preserved only very rarely in the geologic record. Instead, the traces produced by animals in sand in the depositional slip face microenvironment have a much greater preservation potential, but they are typically deformed by collapse and localized slumping.

Plants are generally restricted to the stoss side of a dune, so plant trace fossils (rhizoliths) in eolianites are to be expected only in situations where the stoss facies of stabilized dunes are preserved. Typically, rhizoliths are observed more commonly in coastal dune deposits rather than in inland erg deposits (Curran, 2007; Curran and White, 2001). Animal traces in the stoss sands may include both domichnia and repichnia (and rarely pascichnia), whereas in the slip face sands, the traces are primarily repichnia.

2.2 Ichnofacies in Eolian Paleoenvironments

At present, there are three named ichnofacies that pertain specifically to trace fossil associations in ancient eolianites—the *Entradichnus* Ichnofacies (Ekdale et al., 2007), *Octopodichnus* Ichnofacies (Hunt and Lucas, 2007), both of which are characterized by traces of invertebrate animals, and the *Chelichnus* Ichnofacies (Hunt and Lucas, 2007), which is characterized by vertebrate tracks. To our knowledge, there is no named ichnofacies pertaining specifically to eolianites infested by pervasive rhizomorphs, which often create strikingly prominent ichnofabrics, nor do we suggest that a new ichnofacies should be named for such plant-dominated situations. In fact, for the sake of broader utility and simplicity in ichnology, it seems reasonable that the three aforementioned ichnofacies, as well as rhizomorph-dominated ichnofabrics in eolianites, all should be contained within a single ichnofacies designation, namely the *Entradichnus* Ichnofacies.

The *Chelichnus* Ichnofacies was proposed "for ichnofaunas that have a low diversity (less than five ichnogenera) of tetrapod tracks in which manual and pedal tracks are equant in shape with short digit impressions and in shape often have subequal manual and digit impressions" (Hunt and Lucas, 2007: 66).

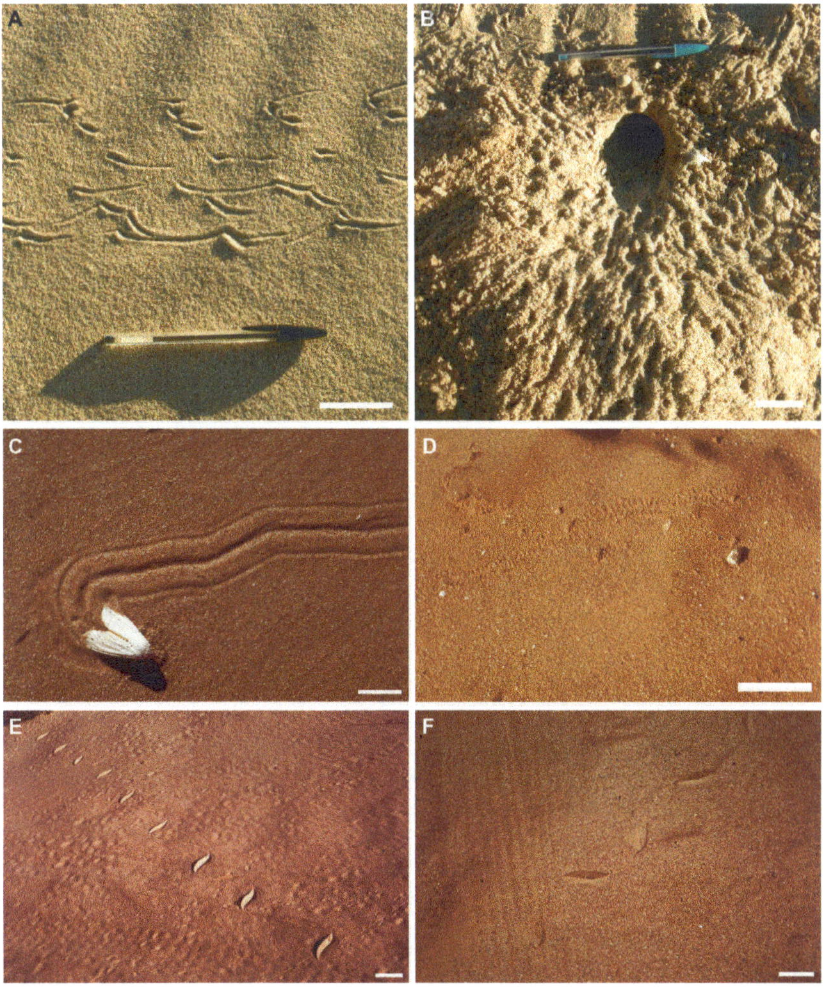

FIGURE 1 Traces in modern coastal dunes. Ephemeral traces of some modern organisms on dune surfaces along the coast of the northern Gulf of California near Puerto Peñasco, Sonora, Mexico. (A) Trackway of a large fiddler crab (*Uca princeps*; scale bar = 5 cm). (B) Burrow opening of a large fiddler crab (*U. princeps*; scale bar = 5 cm). (C) Trail of a small moth (scale bar = 2 cm). (D) Trackway of a small beetle (scale bar = 3 cm). (E) Trail of a sidewinder rattlesnake (*Crotalus cerastes*; scale bar = ca. 15 cm). (F) Trail of a gopher snake (*Pituophis catenifer*; scale bar = ca. 10 cm).

The ichnogenera *Chelichnus* and *Basilichnium* (both tetrapod footprints) are characteristic. The authors regard this trace-fossil association as representative of eolian dune paleoenvironments, but their definition of this ichnofacies is quite restrictive, and they note that it applies only to rocks in three geologic periods (Permian through Jurassic).

The *Octopodichnus* Ichnofacies was proposed for a trace-fossil association that "consists exclusively of a low diversity of walking traces (repichnia) of arthropods, including spiders and scorpions that are preserved in well-sorted, medium-grained sandstones" in eolian dune settings (Hunt and Lucas, 2007: 63). The ichnogenera *Octopodichnus* and *Paleohelcura* (both arachnid trackways) are characteristic, but their delicate nature (shallow surficial impressions in the sand) makes it possible for them to be preserved only under special conditions. The authors report this ichnofacies in Permian and Jurassic eolianites, and they note that these invertebrate traces commonly co-occur in the same strata with the vertebrate traces that typify the *Chelichnus* Ichnofacies.

The *Entradichnus* Ichnofacies was proposed "for the recurrent trace fossil assemblage that typifies sparsely vegetated and unvegetated eolian dune fields in arid climatic settings" (Ekdale et al., 2007: 570). The characteristic ichnogenera (all of them invertebrate burrows) include *Arenicolites*, *Digitichnus*, *Entradichnus*, *Palaeophycus*, *Planolites*, *Skolithos*, and *Taenidium*. The authors cite numerous examples of this ichnofacies in eolianites of Permian, Triassic, Jurassic, Tertiary, and Quaternary age on several continents around the world, most of which include the eponymous ichnogenus *Entradichnus* in the ichnoassemblage (Ekdale and Picard, 1985). Buatois et al. (1998) report that other less common ichnogenera, including the surficial trackways *Octopodichnus* and *Paleohelcura*, also are part of this recurrent trace-fossil association. Vertebrate tracks and trackways in eolian deposits are not excluded.

We regard the *Entradichnus* Ichnofacies of Ekdale et al. (2007) as the primary ichnofacies encompassing the spectrum of trace-fossil assemblages that reflect eolian depositional environments throughout the geological record. The much more narrowly defined trace-fossil assemblages that characterize the *Octopodichnus* and *Chelichnus* ichnofacies of Hunt and Lucas (2007) are subsumed within the *Entradichnus* Ichnofacies.

3. DEPOSITIONAL ENVIRONMENTS

3.1 Trace Fossils in Coastal Dunes

Coastal dune fields generally are localized in bands paralleling marine shorelines, and the component sand grains are reworked from beaches and tidal flats into small, often well-vegetated dunes. They typically intergrade with the adjacent beach and tidal-flat facies, so the occurrence of occasional shell fragments of marine organisms should not be unexpected in coastal-dune deposits; but the trace-fossil associations are distinctly terrestrial.

Quaternary eolianites in the Bahamas are composed entirely of carbonate grains, and they occur in coastal dunes that intergrade with beach and other shallow-marine calcarentic deposits (Curran and White, 2001; Knaust et al., 2012). In this humid, tropical setting, rhizomorphs (i.e., plant-root penetration structures) are ubiquitous and often are so pervasive that they obscure original stratification and all other primary bedding features. The most prominent rhizomorphs in Bahamian eolianites are identical to modern roots of railroad vines (*Ipomoea pes-caprae*) and bay geraniums (*Ambrosia hispida*) growing in the same area today (Curran, 2007: fig. 14.7).

Animal trace fossils in the Bahamian eolianites are mainly attributed to the work of burrowing insects (Curran, 2007) and land-dwelling hermit crabs (Walker et al., 2003). Although vertical tubes (*Skolithos linearis*) and irregular tunnels made by solitary insects are common, the most striking trace fossils are those attributed to the work of social insects. "Cluster burrows" are dense aggregations of large, vertical shafts that are interpreted as the products of brooding/hatching activities of sphecid (digger) wasps (Curran, 2007: fig. 14-8A and B). "Stellate burrows" are closely packed vertical shafts that radiate subhorizontally at the sediment surface, and they are interpreted as the products of the organized nesting behavior of halictid (sweat) bees.

Quaternary eolianites along the eastern coast of the Yucatan Peninsula, Mexico, resemble those in the Bahamas in being composed entirely of carbonate grains. They too are often typified by dense aggregations of rhizomorphs (McKee and Ward, 1983: figs. 10, 12 and 33), but burrows of insects or other animals are uncommon. Similar rhizomorph-dominated ichnofabrics in coastal eolianites have been reported in other parts of the world (Bird, 1972; Glennie and Evamy, 1968; MacKenzie, 1964).

Quaternary eolianites along the northeastern coastline of the Gulf of California in Sonora, Mexico, are composed of mixed carbonate/siliciclastic sand (Phelps, 2002). The climate was arid, and plant cover is sparse; so rhizomorphs are rare in the eolianites. Burrows are common and diverse, but surficial tracks and trackways are almost totally absent from bed surfaces. Most of the small burrows in the Pleistocene eolianites can be attributed with confidence to particular taxa of arthropods, including both insects (mostly cockroaches and beetles) and arachnids (spiders and scorpions), which inhabit the modern dunes in the same area today (Fig. 2A and B). Meniscate burrows (*Taenidium*) are the most abundant trace fossils, and they were made mainly by desert roaches (*Arenivaga investigata*, Family Polyphagidae, Order Blattaria). Non-meniscate burrows (*Planolites* and *Skolithos*) are very common, and they were created mostly by various taxa of beetles (Order Coleoptera), including scarabs (*Cyclocephala*, Family Scarabaeidae), darkling beetles (Family Tenebrionidae), and hister beetles (*Saprinus*, Family Histeridae). Some of the *Skolithos* burrows may have been made by crevice weaver spiders (*Kukulcania utahana*, Family Filistatidae, Order Araneae), which construct short, web-lined, vertical shafts in the modern dunes. Small spiral burrows (*Gyrolithes*) are abundant in the modern dunes and fairly common in the Pleistocene eolianites. They clearly were produced by burrowing scorpions (Family Vaejovidae, Order Scorpiones), several genera of which inhabit the dunes today.

Vertebrate trace fossils also occur in the Quaternary eolianites of Sonora, Mexico. Trackways of lizards, rodents, and lagomorphs and sinuous trails of snakes (rattlers, sidewinders, and gopher snakes) are commonly observed in the sands of the modern dunes, but surficial tracks and trails are rarely preserved as trace fossils. Phelps (2002) reported a rabbit trackway preserved on the surface of one Pleistocene eolianite bed. Vertebrate burrows, however, are locally common in the Pleistocene eolianites, often in direct association with the arthropod burrows (Fig. 2C and D). They are highly branched, horizontal tunnel systems that most likely were excavated by burrowing rodents, but the exact taxa of rodents responsible for the burrows are undetermined at this time. In terms of geometry alone, such burrows would be assigned to *Thalassinoides*, but to avoid obvious confusion with the widespread crustacean burrows that are universally recognized as *Thalassinoides* in marine environments, we choose not to refer the Pleistocene rodent burrows to an ichnogenus at this time.

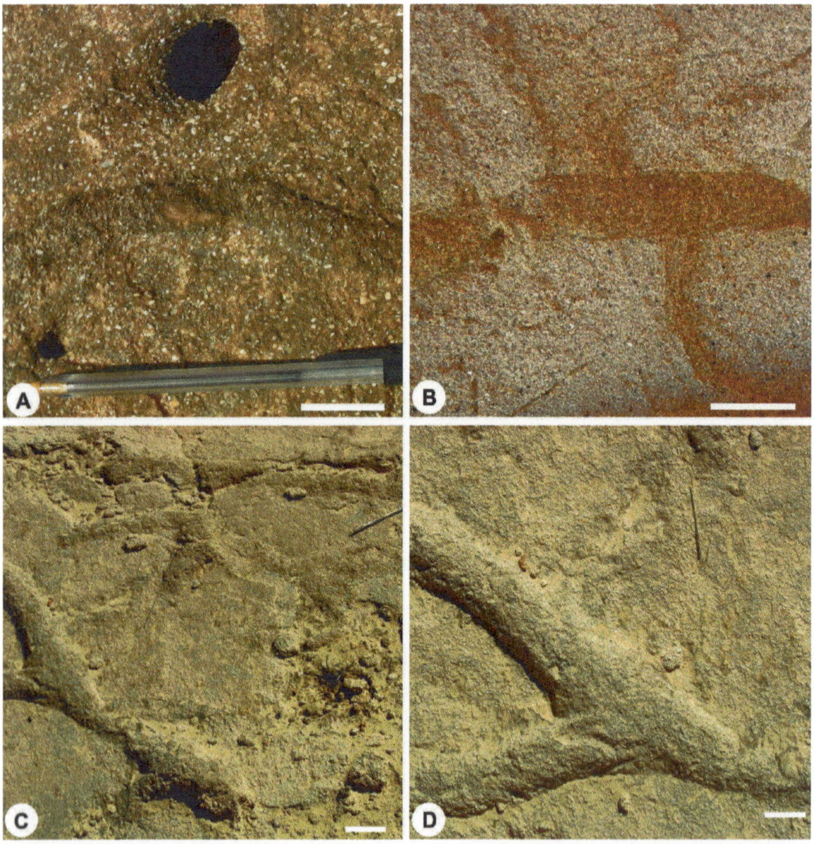

FIGURE 2 Trace fossils in ancient coastal dunes. (A, B) Invertebrate (probably insect) burrows preserved in Pleistocene eolianite near Puerto Peñasco, Sonora, Mexico. Horizontal meniscate burrow (*Taenidium*) and open shaft (*Skolithos*) in plan view (A, scale bar = 15 cm) and horizontal semimeniscate burrow (*Entradichnus*) in plan view (B, scale bar = 3 cm). (C, D) Vertebrate (probably rodent) burrow preserved in Pleistocene eolianite near Puerto Peñasco, Sonora, Mexico. Horizontal branching burrow in plan view (C, scale bar = 5 cm) and close-up view of same (D, scale bar = 3 cm).

3.2 Trace Fossils in Inland Ergs

Inland ergs (vast expanses of high, largely unvegetated, sand dunes) occur in interior locales on several continents today. Some of the most extensive modern ergs include the well-known Sahara of northern Africa, Namib of southwestern Africa, Negev in Israel, and Atacama in Chile and Peru. The world's largest erg today is the Rub'al Khali (Empty Quarter), which covers more than 650,000 km^2 in the southern Arabian Peninsula with vast expanses of sand dunes exceeding 250 m in height. Smaller ergs in western North America, which have been studied extensively by sedimentologists and desert ecologists, include Great Sand Dunes in Colorado, Little Sahara and Coral Pink Sand Dunes in Utah, White Sands in New Mexico, southern Mojave Desert in California, and Gran Desierto de Altar in Sonora, Mexico. Similarly, western North America was a region of widespread ergs off and on throughout the Permian, Triassic, and Jurassic periods (Kocurek and Dott, 1983; Peterson, 1988). These ancient eolianites have been studied extensively by sedimentologists, many of whom have noted common tetrapod tracks and other trace fossils in those deposits (Fig. 3).

Vertebrate footprints have been widely reported in eolianites, and dinosaur tracks and trackways in dune deposits are especially well documented in the scientific literature and widely publicized in the popular media. Images of dinosaurs and evidence of dinosaur lifestyles fascinate paleontologists and nonscientists alike, so it is no wonder that dinosaur footprints are among the most frequently reported trace fossils in eolian deposits.

Jurassic ergs in the southwestern Unites States, particularly in southern Utah and northern Arizona, were favored habitats of dinosaurs and other primitive tetrapods. Their bones are exceedingly rare, but their tracks are amazingly abundant (Hamblin and Foster, 2000; Lockley, 1998; Lockley and Hunt, 1995; Lockley et al., 2007; Loope, 2006b; Milàn and Loope, 2007; Rainforth and Lockley, 1996; Seiler and Chan, 2008; Thulborn, 1990). Well-preserved footprints provide direct evidence of the various kinds of desert-dwelling dinosaurs in the Jurassic, as well as information about their size, foot morphology, and even skin texture. Their trackways offer information about the dinosaurs' gait and perhaps even speed as they traversed the vast Jurassic dune fields. The trace fossil *Grallator*, a three-toed track generally attributed to theropod dinosaurs, is one of the most common ichnogenera in the eolianites (Fig. 4). As theropods were voracious predators, their apparent abundance in the Jurassic ergs demonstrates that there must have been a substantial number of prey animals inhabiting those environments also.

Preservation of tetrapod tracks in eolian deposits includes both surface impressions and undertracks. As the bulk of preserved eolian deposits represent the cross-bedded slip faces of dunes, vertebrate tracks typically exhibit some deformation features. Loope (2006b) described the details of preservation modes of abundant tetrapod (dinosaur and therapsid) tracks in grain-flow layers of slip face deposits

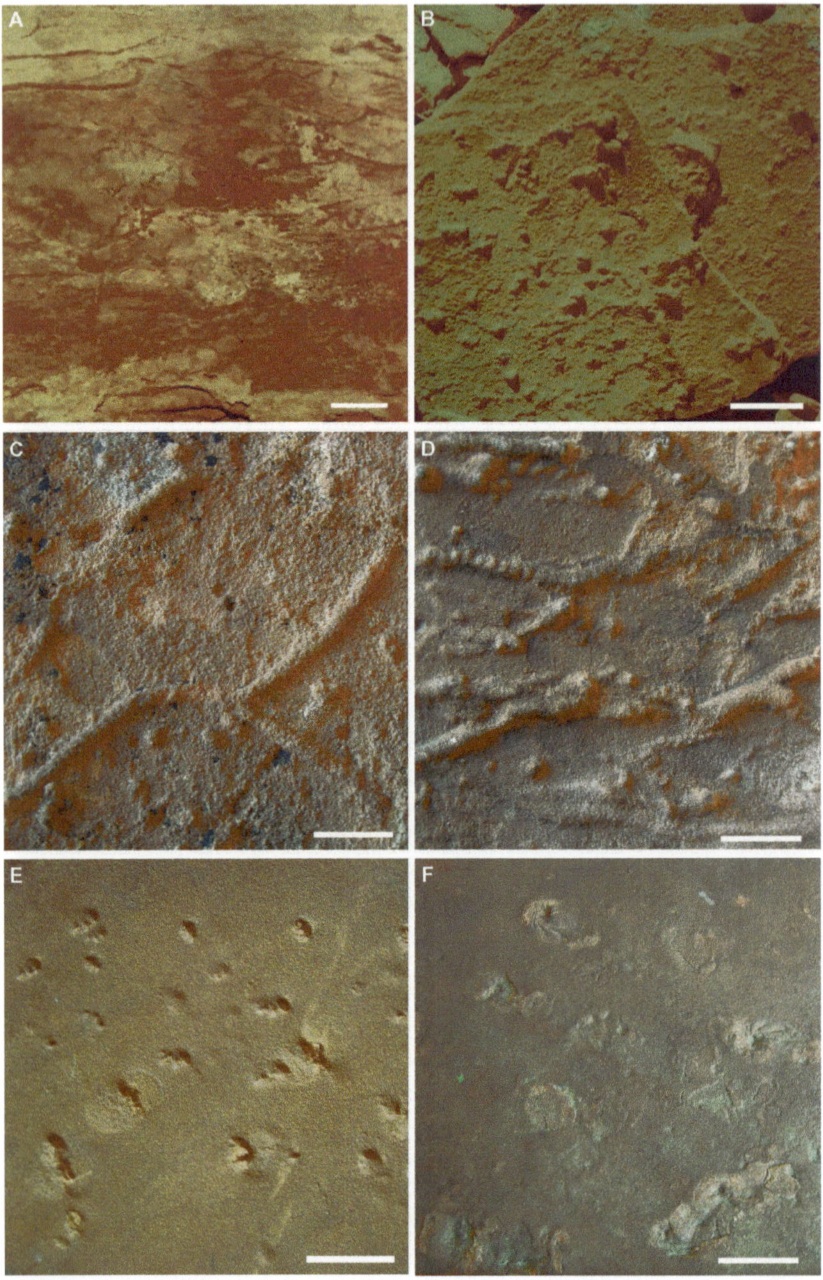

FIGURE 3 Trace fossils in ancient inland erg deposits. (A–D) Invertebrate (probably insect) trace fossils in the Middle Jurassic Entrada Sandstone, San Juan County, Utah. *Entradichnus* meniscus in outcrop (A, scale bar = 5 cm). *Pustulichnus gregarious* in outcrop (B, scale bar = 5 cm). Close-up views of *E. meniscus* and *P. gregarious* (C, D, scale bars = 10 cm). (E, F) Vertebrate trace fossils (*Brasilichnium*) in the Lower Jurassic Nugget Sandstone, Summit County, Utah. Well-defined trackway (showing toe imprints) in plan view of a tetrapod, probably synapsid (E, scale bar = 5 cm). Deformed (slurried) tracks of a tetrapod, probably synapsid (F, scale bar = 5 cm).

FIGURE 4 Trace fossils in ancient inland erg deposits. (A, B) Vertebrate trace fossils in the Lower Jurassic Navajo Sandstone, near Coyote Buttes in Kane County, Utah (photographs courtesy of Marjorie A. Chan). Small, three-toed, dinosaur tracks in concave epirelief (A, scale = 10 cm), and small, three-toed, dinosaur track that has been replaced by dark brown iron oxide (B, pencil = 15 cm long). (C) Large, three-toed, dinosaur track in concave epirelief in the Lower Jurassic Navajo Sandstone, in the San Rafael Swell in Emery County, Utah (shoe = 30 cm long).

in the Lower Jurassic Navajo Sandstone in southern Utah. He demonstrated that the tracks (*Grallator* and *Brasilichnium*) were emplaced deeply in dry sand and subsequently buried by thin grain-flow avalanches on the dune foresets. His analysis suggests not only that Jurassic vertebrates commonly walked on the steeply dipping lee side of dunes but also that their tracks on that side had a high preservation potential. This conclusion effectively contradicts earlier assumptions by previous workers that the steep angles of repose and cohesion-less sand grains of dune slip faces generally precluded good preservation of tracks.

In the Middle Jurassic Entrada Sandstone in southern Utah, Milàn and Loope (2007) examined large numbers of three-toed theropod dinosaur tracks (*Megalosauripus* and *Therangospodus*) in various preservation states, resulting from variations in the gait of the dinosaurs and a range of paleoslopes of the dune foresets on which the dinosaurs were walking. Many of the tracks are very well preserved and show distinct anatomical details of the foot, such as pads on the individual digits and claw impressions, and several examples contain the original sediment infill remaining inside the tracks. In some other examples, the sediment infill was eroded away, partly exposing the base of the track along with a set of concentric circles extending up to 20 cm outward from the track, reflecting the outward deformation of the sand when the heavy animal stepped in it. In still other examples, the track was eroded away completely, leaving just the set of concentric circles in the sand, and the authors point out that such a preservation mode might be easily mistaken for non-biogenic deformation of the sediment rather than the dinosaur trace fossils that it actually represents.

3.3 Uniformitarian Approach to Eeolian Ichnofacies

Because of the general lack of body fossils in eolianites, a uniformitarian approach of seeking modern analogs to link a trace fossil with the tracemaker is obligatory. Virtually all the tracemaking animals in modern dune fields are arthropods (mostly insects and arachnids) and vertebrates (mostly reptiles and mammals), so attribution of traces to the tracemakers is a fairly straightforward accomplishment in geologically young eolianites, but a uniformitarian interpretation becomes more and more tenuous as older and older eolianites are examined.

Although there have been many studies of the ecology of organisms that inhabit sandy deserts today, there have been relatively few studies that focus specifically on the preservable traces that they make in modern sand dunes (Eiseman and Charney, 2010). Ahlbrandt et al. (1978) described the common traces created by present-day dune dwellers, mostly from examples in the Great Sand Dunes of southern Colorado, and their observations have been widely cited in the literature as modern analogs for many trace fossils in ancient eolianites.

Ichnology in most Late Cenozoic eolianites can be linked directly with ichnology in modern dunes. Curran (2007) easily found the link between modern and Pleistocene rhizomorphs (at the genus and species level) in the Bahamas, and Phelps (2002) confidently linked modern and Pleistocene arthropod burrows with their burrowers (at the genus and species level in many cases) in Mexico. Fornós et al. (2002) correctly identified the genus and species of extinct goats that produced fossilized footprints in Pleistocene eolianites on the island of Mallorca, and Milàn et al. (2007) reasonably assigned fossilized footprints in Pleistocene eolianites on the island of Rhodes to the work of elephants and camels. Roberts et al. (2008) were easily able to attribute abundant fossil trackways in Pleistocene eolianites in South Africa to extant African elephants, and Roberts (2008) found footprints that obviously were made by prehistoric humans in Pleistocene eolian deposits at two sites in South Africa that have been dated as 120,000 years old.

Trace fossils in Mesozoic eolianites sometimes can benefit from analogs with modern arthropod tracks and burrows (Albers, 1975; Ekdale and Picard, 1985; Ekdale et al., 2007). However, this is not the case with many vertebrate trace fossils, such as footprints that were made by dinosaurs and other extinct tetrapods that have no close modern relatives (Gillette and Lockley, 1989; Haubold and Katzung, 1978; Lockley, 1991; Lockley and Hunt, 1995; Lockley and Meyer, 2000).

Late Paleozoic (Carboniferous and Permian) eolianites likewise contain fossil invertebrate tracks that may be usefully interpreted by direct analogy with modern insect and arachnid traces (Braddy, 1995; Brady, 1939; McKee, 1944, 1947). However, Late Paleozoic tetrapods included large labyrinthodont amphibians and primitive stem reptiles that are far removed from their modern vertebrate cousins (Haubold, 1971; Haubold et al., 1995a,b; Henderson, 1924; McKee, 1945; McKeever, 1994; McKeever and Haubold, 1996), so attempts to identify the producers of Late Paleozoic vertebrate trackways by analogy with extant tetrapods may not be especially meaningful.

Ichnology of Early Paleozoic (pre-Carboniferous) eolianites is tenuous indeed because land-dwelling faunas in those early times were very primitive relative to modern terrestrial faunas. MacNaughton et al. (2002) reported the remarkable preservation of abundant trackways in a Late Cambrian–Early Ordovician eolianite in the Potsdam Sandstone Group in southern Ontario. The eolian interpretation of the very well-sorted, fine- to coarse-grained, quartz arenite rocks is based on large-scale trough cross-bedding, steeply inclined foreset beds, adhesion marks, and wind-ripple marks. The trackways, which are assigned to the ichnogenera *Protichnites* and *Diplichnites*, are preserved mainly as undertracks on toeset and bottomset surfaces of the eolian sandstone beds, and they are quite large—8–13 cm wide and reaching lengths as much as 2.6 m. The appendage impressions are up to 18 mm in diameter and 5 mm deep in the sandstone surface. These are the oldest known trace fossils of animals (probably myriapod-like arthropods) walking on land, and they indicate that the creatures crawled around quite extensively on the dry sand of the dunes.

Although eolian deposits have been described in Precambrian strata in various parts of the world (Eriksson and Simpson, 1998), ichnology of Precambrian eolianites—for obvious reasons—is non-existent.

3.4 Paleoclimatic Significance of Eolian Ichnofacies

Eolianites, especially those that formed in large inland ergs, are obvious indicators of special paleoclimatic conditions because they depend upon high winds, low precipitation, and a copious source of sand on land. Typically, but not invariably, they form in regions of seasonal high temperatures. Widespread eolianites characterized the arid inland region of the southwestern United States periodically from the Late Paleozoic to the Middle Mesozoic (Blakey et al., 1988; Kocurek and Dott, 1983; Peterson, 1988). In many, if not most, cases, the eolianites display evidence of monsoonal regimes, in which there were pronounced seasonal (annual) changes in wind direction and intensity, as well as precipitation and air temperature (Chan and Archer, 1999, 2000; Loope and Rowe, 2003; Loope et al., 2001, 2004).

Trace fossils play an important role in interpreting the paleoclimatic conditions that are recorded in eolianites because they are direct reflections of the episodic activity of organisms in the dunes throughout the year. Ekdale et al. (2007) described the invertebrate ichnocoenoses in monsoonal deposits of the Jurassic Navajo Sandstone in southern Utah. They found that the invertebrate ichnocoenoses represented a community of shallow-burrowing, solitary (non-social) insects that kept pace with the migrating dunes and flourished during pluvial intervals that brought higher than usual precipitation to the Navajo erg.

4. ICHNOFABRICS

Because most trace fossils in eolianites are horizontal repichnia or very shallow cubichnia, ichnofabrics in vertical cross section typically are poorly developed or entirely absent. Coastal dunes, that are stabilized by dense vegetation, may exhibit an ichnofabric dominated by pervasive rhizomorphs (Fig. 5) or by highly localized social insect burrows, such as the cluster burrows and stellate burrows in Quaternary eolianites in the Bahamas (Curran, 2007). However, in unstabilized, mobile sand dunes, one does not normally encounter complex, composite ichnofabrics (*sensu* Bromley and Ekdale, 1986), or high ichnofabric indices (*sensu* Droser and Bottjer, 1986). Permanent or semi-permanent domichnia or fodinichnia, which typically compose tiered ichnofabrics in other, less-mobile substrates, are rare to absent in active dune environments.

There is one very interesting ichnofabric that is absolutely unique to the loose sand substrate of mobile, subaerial dunes, but it is neither widely recognized nor well understood. A number of vertebrate dune dwellers are highly adapted to moving through loose, dry sand in the very same manner as many aquatic organisms swim through water, namely by muscular undulations of their elongate bodies. This process of infaunal locomotion through unconsolidated sediment is known as sand swimming, and it is a behavior exhibited by

FIGURE 5 Plant trace fossils in ancient eolian deposits. Rhizoliths (plant-root traces with a thick diagenetic halo) in Late Pleistocene eolianite of the South Alghero coast, Sardinia, Italy (photographs courtesy of Dirk Knaust). Rooted horizon in vertical outcrop section (A, image = 1 m wide), and bedding-plane view of three rhizoliths in cross section (B, image = 20 cm wide).

several taxa of desert lizards and snakes (Hembree and Hasiotis, 2007; Maladen et al., 2009).

Dry sand is a granular substrate that displays fluid-like properties because the lack of interstitial moisture eliminates intergranular cohesion and permits grains to be moved independently from one another. Experimental studies of sand-swimming reptiles—and robots—in laboratory situations reveal the locomotory mechanisms employed by the animals (Goldman and Hu, 2010; Hembree and Hasiotis, 2007; Maladen et al., 2009).

Using high-speed X-ray imaging techniques in a laboratory setting, Maladen et al. (2009) studied the infaunal locomotory behavior of the African sandfish (*Scincus scincus*), a lizard that inhabits dry dunes of the Sahara Desert today. The authors observed that the lizard rapidly digs into the sand using its limbs and an undulatory wriggling motion, but once beneath the sediment surface, the lizard's limbs are no longer employed for swimming through the unconsolidated substrate. The limbs are placed tight against the body, effectively achieving a snake-like form, and the animal propels itself through the sand via large-amplitude axial undulations, reaching remarkable speeds of as much as 10 cm/s (which extrapolates to 360 m/h) inside the substrate. An explanation of the physics of such sand swimming is based on the hypothesis that grain–grain and grain–animal friction determine the thrust and drag forces that are produced during propulsion of the animal, and a correlation between the wave frequency of body oscillations and swimming speed can be calculated. Although it might seem counterintuitive, the authors observed that the swimming speed of the lizard increases with the degree of close-packing of the granular sediment—the firmer the sand, the faster the animal swims through the sediment.

In order to assess the ichnologic implications of sand swimming, Hembree and Hasiotis (2007) experimented with the Kenyan sand boa (*Eryx colubrinus*) in laboratory situations. Like the sandfish lizard, the sand boa efficiently propels itself through a dry, granular substrate without creating open burrows, because the sand simply collapses behind the animal as it passes through. The authors observed the snakes moving through both fine- and medium-grained quartz and carbonate sand in dry aquaria containing alternating layers of different colored grains. Because the snakes propelled themselves via lateral undulations, the collapsing sand behind the snakes resulted in minimal compaction of the sediment, and an open burrow would remain only if the sand was moist.

The biogenic structures created by sand boas burrowing in the lab were varied, and in general, they resulted in ichnofabrics without distinct burrows that could be assigned directly to particular ichnogenera (Hembree and Hasiotis, 2007). The upper 4–6 cm of sand was homogenized by burrowing, and the degree of bioturbation, manifested as the deflection of sediment laminae, decreased downward to a maximum depth of 22 cm in the sediment. The sedimentary structures created by the snakes included down-tapering conical structures, vertical shafts, and sinuous tunnels. No burrow linings, meniscate backfill, or permanently open burrows were produced.

Sand swimming is not a commonly observed habit of mammals in desert dunes. However, one species that does employ such behavior is the Grant's golden mole (*Eremitalpa granti*), an insectivore which inhabits the driest parts of the Namib Desert in southern Africa (Fielden et al., 1990; Perrin and Fielden, 1999; Seymour et al., 1998). Not to be confused with a true mole (Family Talpidae), the golden mole (Family Chrysochloridae) is thought to be more closely related to tenrecs. The golden mole is essentially blind, and although it forages for termites on the sand surface at night, it is known to swim quite effectively beneath the surface of the dry sand during the day in search of termites to eat. Unlike sandfish lizards and sand boas, the Grant's golden mole swims through the substrate by using its clawed feet, and undulatory body motions apparently do not play a big role in propelling it forward. Sand collapses behind the animal as it moves, leaving no open or walled burrow in its wake, and so primary stratification of sand is disrupted in a major way. However, little or nothing is known of the ichnofabric details created by the animal's movements inside the sand dunes.

The ichnofabric implications of sand-swimming behavior in dry dunes in the geological record are intriguing, but they have not been investigated thoroughly. Although vertebrate burrows have been described in some Mesozoic eolianites (Loope, 2006a, 2008), they exhibit features such as distinct walls and/or meniscate backfill that are inconsistent with those resulting from sand swimming. Vertebrate burrows are essentially unknown in pre-Mesozoic eolianites.

Because the effect of sand swimmers on the unconsolidated substrate is manifested primarily in the form of disturbed laminae and irregular collapse structures, it is quite possible that a sand-swimming ichnofabric would be misidentified as a non-biogenic, soft-sediment deformation fabric. If sand-swimming ichnofabrics can be recognized with confidence in ancient eolianites, there will be important paleobiologic, sedimentologic, and paleoclimatic implications. Sand swimmers today bury themselves to escape the intense daytime heat (often up to 50 °C) on dry dune surfaces where there is no benefit of shade from leafy plants, but these animals are not inactive inside the dunes. In fact, they are known to move rapidly and efficiently through the sand, presumably in search of infaunal insects for a meal.

Evidence of sand swimming probably occurs unrecognized in many eolianite units, especially geologically young ones, and it simply remains for geologists to try to understand what they see in the rocks. Recognition and description of sand-swimming ichnofabrics in the geological record have the potential of revealing new knowledge and insights regarding life in ancient deserts.

5. CONCLUSIONS

One of the great challenges to ichnologists is that we often are forced to study the product in the absence of the producer. This is especially true with eolianites. Trace fossils occur but are scarce in dune deposits. Body fossils are as exceedingly rare in dune deposits as to be essentially non-existent. Thus without trace fossils, the task of interpreting the paleoecology of eolian paleoenvironments would be as hopeless as Odysseus' fleet being tossed about at the mercy of the four winds.

The main take-home points of this chapter can be summarized briefly by the following points:

1. Ancient dune deposits are generally devoid of body fossils, but trace fossils—although sparse—do occur in sufficient quantity in some eolianites to lend crucial help in interpreting the paleoecology of the paleoenvironment. The *Entradichnus* Ichnofacies, including ichnogenera attributed to both invertebrate and vertebrate animals, represents all the various recurrent assemblages of trace fossils that characterize coastal and inland dune paleoenvironments.
2. Much of what is known about modern traces of dune dwellers comes from observations made on the gently sloping, erosional, stoss side of the dune, whereas the part of the dune that actually becomes preserved in the fossil record is the steeply dipping, depositional, slip face of the dune. Thus, misleading interpretations may result when seeking modern analogs for ancient trace fossils. The stoss tends to exhibit more diverse, more delicate, and less-deformed repichnial, pascichnial, and domichnial traces than does the slip face, where most of the traces are repichnia that exhibit deformation by sediment slippage and collapse.
3. There is trace-fossil evidence of occupation of the eolian environment by some of the first land animals (arthropods) in the Early Paleozoic, perhaps as early as the Late Cambrian. This suggests that at least some early life forms on land were sufficiently hearty to withstand the harsh vicissitudes of blowing winds and shifting sands.
4. Ichnological data in eolianites lend important information to paleoclimatic studies because the trace-fossil occurrences provide evidence of the types of organisms inhabiting the dunes and possibly also the seasonal nature of their activity in monsoonal climates.
5. A unique ichnofabric in modern eolian deposits that has been unrecognized in ancient eolianites heretofore is the distinctive pattern of disruption of stratification resulting from sand-swimming behavior of infaunal vertebrates moving around inside the dry sand of the dunes. Although a sand-swimming ichnofabric has not been reported in the rock record, it must be there, so ichnologists would be well advised to be on the lookout for such an ichnofabric when examining ancient eolian sequences.

ACKNOWLEDGMENTS

This chapter benefited from constructive comments by Ken Glennie and Dirk Knaust.

REFERENCES

Ahlbrandt, T.S., Andrews, S., Gwynne, D.T., 1978. Bioturbation in eolian deposits. J. Sediment. Petrol. 48, 839–848.

Albers, S.H., 1975. Paleoenvironment of the Upper Triassic-Lower Jurassic(?) Nugget(?) Sandstone near Heber, Utah. M.Sc. Thesis, University of Utah, Salt Lake City, 94 pp.

Bird, E.C.F., 1972. Ancient soils at Diamond Bay, Victoria. Victorian Nat. 89, 349–353.

Blakey, R.C., Peterson, F., Kocurek, G.C., 1988. Synthesis of Late Paleozoic and Mesozoic eolian deposits of the western interior of the United States. Sediment. Geol. 56, 3–125.

Braddy, S.J., 1995. The ichnotaxonomy of the invertebrate trackways of the Coconino Sandstone (Lower Permian), northern Arizona. In: Lucas, S.G., Heckert, A.B. (Eds.), Early Permian Footprints and Facies. New Mexico Mus. Nat. Hist. Sci., Bull., vol. 6, pp. 219–224.

Brady, L.F., 1939. Tracks in the Coconino Sandstone compared with those of small living arthropods. Plateau 12, 32–34.

Bromley, R.G., Ekdale, A.A., 1986. Composite ichnofabrics and tiering of burrows. Geol. Mag. 123, 59–65.

Buatois, L.A., Mángano, G.M., Genise, J.F., Taylor, T.N., 1998. The ichnological record of the continental invertebrate invasion: evolutionary trends in environmental expansion, ecospace utilization, and behavioral complexity. Palaios 13, 217–240.

Chan, M.A., Archer, A.W., 1999. Spectral analysis of eolian foreset periodicities: implications for Jurassic decadal-scale paleoclimatic oscillators. Palaeoclimates 3, 239–255.

Chan, M.A., Archer, A.W., 2000. Cyclic eolian stratification on the Jurassic Navajo Sandstone, Zion National Park: periodicities and implications for paleoclimate. In: Sprinkel, D.A., Chidsey Jr., T.C., Anderson, P.B. (Eds.), Geology of Utah's Parks and Monuments. Utah Geol. Ass., Publ. 28, pp. 607–618.

Curran, H.A., 2007. Ichnofacies, ichnocoenoses, and ichnofabrics of Quaternary shallow-marine to dunal tropical carbonates: a model and implications. In: Miller III, W. (Ed.), Trace Fossils. Concepts, Problems, Prospects. Elsevier, Amsterdam, pp. 232–247.

Curran, H.A., White, B., 2001. Ichnology of Holocene carbonate eolianites in the Bahamas. In: Abegg, F.E., Harris, P.M., Loope, D.B. (Eds.), Modern and Ancient Carbonate Eolianites. SEPM Spec. Publ., vol. 71, pp. 47–56.

Droser, M.L., Bottjer, D.J., 1986. A semiquantitative field classification of ichnofabric. J. Sediment. Petrol. 56, 558–559.

Eiseman, C.S., Charney, N.D., 2010. Tracks and Signs of Insects and Other Invertebrates: A Guide to North American Species. Stackpole Books, Harrisburg, 582 pp.

Ekdale, A.A., Picard, M.D., 1985. Trace fossils in a Jurassic eolianite, Entrada Sandstone, Utah, USA. In: Curran, H.A. (Ed.), The Use of Biogenic Structures in Interpreting Depositional Environments. SEPM Spec. Publ., vol. 35, pp. 3–12.

Ekdale, A.A., Bromley, R.G., Loope, D.B., 2007. Ichnofacies of an ancient erg: a climatically influenced trace fossil association in the Jurassic Navajo Sandstone, southern Utah, USA. In: Miller III, W. (Ed.), Trace Fossils. Concepts, Problems, Prospects. Elsevier, Amsterdam, pp. 562–574.

Eriksson, K.A., Simpson, E.L., 1998. Controls on spatial and temporal distribution of Precambrian eolianites. Sediment. Geol. 120, 275–294.

Fielden, L.J., Waggoner, J., Perrin, M.R., Hickman, G.C., 1990. Feeding ecology and foraging behaviour of the Namib Desert golden mole, *Eremitalpa granti namibensis* (Chrysochloridae). J. Zool. Lond. 220, 367–389.

Fornós, J.J., Bromley, R.G., Clemmensen, L.B., Rodriguyez-Perea, A., 2002. Tracks and trackways of *Myotragus balearicus* Bate (Artiodactyla, Caprinae) in Pleistocene eolianites from Mallorca (Balearic Islands, western Mediterranean). Palaeogeogr. Palaeoclimatol. Palaeoecol. 180, 277–313.

Gillette, D.D., Lockley, M.G. (Eds.), 1989. Dinosaur Tracks and Traces. Cambridge University Press, Cambridge, 454 pp.

Glennie, K.W., Evamy, B.D., 1968. Dikaka—plants and plant root structures associated with aeolian sand. Palaeogeogr. Palaeoclimatol. Palaeoecol. 23, 77–87.

Goldman, D.I., Hu, D.L., 2010. Wiggling through the world: the mechanics of slithering locomotion depend on the surroundings. Am. Sci. 98, 314–323.

Hamblin, A.H., Foster, J.R., 2000. Ancient animal footprints and traces in the Grand Staircase-Escalante National Monument, south-central Utah. In: Sprinkel, D.A., Chidsey Jr., T.C., Anderson, P.B. (Eds.), Geology of Utah's Parks and Monuments. Utah Geol. Ass., Publ. 28, pp. 557–568.

Haubold, H., 1971. Ichnia amphiborum et reptilorum fossilium. In: Kuhn, O. (Ed.), Handbuch der Paläoherpetologie 18. Gustav Fischer, Stuttgart, pp. 1–124.

Haubold, H., Katzung, G., 1978. Palaeoecology and palaeoenvironments of tetrapod footprints from the Rotliegend (Lower Permian) of central Europe. Palaeogeogr. Palaeoclimatol. Palaeoecol. 23, 307–323.

Haubold, H., Hunt, A.P., Lucas, S.G., Lockley, M.G., 1995a. Wolfcampian (Early Permian) vertebrate tracks from Arizona and New Mexico. New Mexico Mus. Nat. Hist. Sci., Bull. 6, 135–165.

Haubold, H., Lockley, M.G., Hunt, A.P., Lucas, S.G., 1995b. Lacertoid footprints from Permian dune sandstones. Cornberg and DeChelly Sandstones. New Mexico Mus. Nat. Hist. Sci., Bull. 6, 235–244.

Hembree, D.I., Hasiotis, S.T., 2007. Biogenic structures produced by sand-swimming snakes: a modern analogue for interpreting continental ichnofossils. J. Sediment. Res. 77, 389–397.

Henderson, J., 1924. Footprints in Pennsylvanian sandstones of Colorado. J. Geol. 32, 226–229.

Hunt, A.P., Lucas, S.G., 2007. Tetrapod ichnofacies: a new paradigm. Ichnos 14, 59–68.

Knaust, D., Curran, H.A., Dronov, A., 2012. Shallow-marine carbonates. In: Knaust, D., Bromley, R.G. (Eds.), Trace Fossils as Indicators of Sedimentary Environments. Developments in Sedimentology, vol. 64. Elsevier, Amsterdam, pp. 703–750.

Kocurek, G., Dott Jr., R.H., 1983. Jurassic paleogeography and paleoclimate of the central and southern Rocky Mountains regions. In: Reynolds, M.W., Dolly, E.D. (Eds.), Mesozoic Paleogeography of the West-central United States. Rocky Mountain Section SEPM, Denver, pp. 101–116.

Lockley, M.G., 1991. Tracking Dinosaurs: A New Look at an Ancient World. Cambridge University Press, Cambridge, 238 pp.

Lockley, M.G., 1998. The vertebrate track record. Nature 369, 429–432.

Lockley, M.G., Hunt, A.H., 1995. Dinosaur Tracks and Other Fossil Footprints of the Western United States. Columbia University Press, New York, 338 pp.

Lockley, M.G., Meyer, C., 2000. Dinosaur Tracks and Other Fossil Footprints of Europe. Columbia University Press, New York, 323 pp.

Lockley, M.G., Mitchell, L., Odier, G.P., 2007. Small theropod track assemblages from Middle Jurassic eolianites of eastern Utah: paleoecological insights from dune ichnofacies in a transgressive sequence. Ichnos 14, 131–142.

Loope, D.B., 2006a. Burrows dug by large vertebrates into rain-moistened Middle Jurassic sand dunes. J. Geol. 114, 753–762.

Loope, D.B., 2006b. Dry-season tracks in dinosaur-triggered grainflows. Palaios 21, 132–142.

Loope, D.B., 2008. Life beneath the surfaces of active Jurassic dunes: burrows from the Entrada Sandstone of south-central Utah. Palaios 23, 411–419.

Loope, D.B., Rowe, C.M., 2003. Long-lived pluvial episodes during deposition of the Navajo Sandstone. J. Geol. 111, 223–232.

Loope, D.B., Rowe, C.M., Joekel, R.M., 2001. Annual monsoonal rains recorded by Jurassic dunes. Nature 412, 64–66.

Loope, D.B., Steiner, M.B., Rowe, C.M., Lancaster, N., 2004. Tropical westerlies over Pangaean sand seas. Sedimentology 51, 315–322.

MacKenzie, F.T., 1964. Geometry of Bermuda calcareous dune cross-bedding. Science 144, 1449–1450.

MacNaughton, R.B., Cole, J.M., Dalrymple, R.W., Braddy, S.J., Briggs, D.E.K., Lukie, T.D., 2002. First steps on land: arthropod trackways in Cambrian-Ordovician eolian sandstone, southeastern Ontario, Canada. Geology 30, 391–394.

Maladen, R.D., Ding, Y., Li, C., Goldman, D.I., 2009. Undulatory swimming in sand: subsurface locomotion of the sandfish lizard. Science 325, 314–318.

McKee, E.D., 1944. Tracks that go uphill. Plateau 16, 61–72.

McKee, E.D., 1945. Small scale structures in the Coconino Sandstone of northern Arizona. J. Geol. 53, 313–325.

McKee, E.D., 1947. Experiments on the development of tracks in fine cross-bedded sand. J. Sediment. Petrol. 17, 23–28.

McKee, E.D., Ward, W.C., 1983. Eolian environment. In: Scholle, P.E., Bebout, D.G., Moore, C.H. (Eds.), Carbonate Depositional Environments. AAPG Mem. 33, pp. 131–170.

McKeever, P.J., 1994. A new vertebrate trackway from the Permian of Dumfries and Galloway, Scott. J. Geol. 30, 11–14.

McKeever, P.J., Haubold, H., 1996. Reclassification of vertebrate trackways from the Permian of Scotland and related forms from Arizona and Germany. J. Paleontol. 70, 1011–1022.

Milàn, J., Loope, D.B., 2007. Preservation and erosion of theropod tracks in eolian deposits: examples from the Middle Jurassic Entrada Sandstone, Utah, U.S.A. J. Geol. 115, 375–386.

Milàn, J., Bromley, R.G., Titschack, J., Theodoru, G., 2007. A diverse vertebrate ichnofanua from a Quaternary eolian oolite, Rhodes, Greece. In: Bromley, R.G., Buatois, L.A., Mángano, G., Genise, J.F., Melchoir, R.N. (Eds.), Sediment-Organism Interactions: A Multifaceted Ichnology. SEPM Spec. Publ., vol. 88, pp. 333–343.

Perrin, M.R., Fielden, L.J., 1999. *Eremitalpa granti*. Mamm. Spec. 629, 1–4.

Peterson, F., 1988. Pennsylvanian to Jurassic eolian transportation systems in the western United States. Sediment. Geol. 56, 207–260.

Phelps, W.T., 2002. Comparative ichnology of Pleistocene eolianites and modern coastal dunes, Puerto Peñasco, Sonora, Mexico. M.Sc. Thesis, University of Utah, Salt Lake City, 99 pp.

Rainforth, E.C., Lockley, M.G., 1996. Tracking life in a Lower Jurassic desert: vertebrate tracks and other trace from the Navajo Sandstone. Mus. Northern Arizona Bull. 60, 285–289.

Roberts, D.L., 2008. Last Interglacial hominid and associated vertebrate trackways in coastal eolianites, South Africa. Ichnos 15, 190–207.

Roberts, D.L., Bateman, M.D., Murray-Wallace, C.V., Carr, A.S., Holmes, P.J., 2008. Last Interglacial fossil elephant trackways dated by OSL/AAR in coastal eolianites, Still Bay, South Africa. Palaeogeogr. Palaeoclimatol. Palaeoecol. 257, 261–279.

Seiler, W.M., Chan, M.A., 2008. A wet interdune trampled surface in the Jurassic Navajo Sandstone, Coyote Buttes, Arizona: rare preservation of multiple track types and tail traces. Palaios 23, 700–710.

Seymour, R.S., Withers, P.C., Weathers, W.W., 1998. Energetics of burrowing, running, and free-living in the Namib Desert golden mole (*Eremitalpa namibensis*). J. Zool. Lond. 244, 107–117.

Thulborn, T., 1990. Dinosaur Tracks. Chapman and Hall, London, 410 pp.

Walker, S.E., Holland, S.M., Gardiner, L., 2003. *Coenobichnus currani* (new ichnogenus and ichnospecies): fossil trackway of a land hermit crab, Early Holocene, San Salvador, Bahamas. J. Paleontol. 77, 576–582.

Part III

Shallow-Marine Siliciclastic Systems

Chapter 15

Rocky Shorelines

Jordi M. de Gibert,✠ Rosa Domènech[1] and Jordi Martinell
Departament d'Estratigrafia, Paleontologia I Geociències Marines, Universitat de Barcelona, Barcelona, Spain
[1]*Corresponding author: e-mail: rosa.domenech@ub.edu*

1. INTRODUCTION

Rocky coasts account for more than 33% of the world's coastline today (Reading and Collinson, 1996). Their morphology is controlled by the intervention of three major processes (Anthony, 2008; Trenhaile, 1987): (1) mechanical action, (2) weathering, and (3) biological erosion. Thus, benthic life plays a very active role in shaping these environments through the action of bioeroders (e.g., Spencer and Viles, 2002). The identification of ancient rocky shores is important as they serve as reliable markers of paleoshorelines. Bioerosion assemblages play a key role for rocky-coast recognition (e.g., de Gibert et al., 1998) as borings are cavities in the rockground, which have a higher preservation potential than any other component of the biota of these environments.

The geological significance of rocky-shore bioerosion was evinced as early as in Charles Lyell's "Principles of Geology" (Lyell, 1830), who started his seminal masterpiece with a reproduction of the abundant bivalve borings found in the columns of the temple of Serapis (Pozzuoli, Italy), and used them as a proof of relative sea-level changes (Fig. 1). Despite this pioneering contribution, fossil rocky shores were considered uncommon (e.g., Boucot, 1981). Nevertheless, thanks to the studies carried out during the past two decades, in particular, those of Markes E. Johnson and collaborators (Johnson, 1988a,b, 1992, 2006; Johnson and Baarli, 1999), it is clear today that rocky-shore surfaces and associated deposits can be found throughout the geological column, commonly in relation to transgressive episodes.

2. MODERN ROCKY-SHORE HABITATS AND INHABITANTS

2.1 Rocky-Shore Habitats

As Johnson and Baarli (1999) highlighted, the land/sea interface is the most definitive ecological limit of all existing on Earth. The transition between these two contrasting settings is sharp and clearly juxtaposes different physical,

✠ Deceased.

Developments in Sedimentology, Vol. 64. http://dx.doi.org/10.1016/B978-0-444-53813-0.00015-0

FIGURE 1 Bioerosion on the columns of the temple of Serapis (Pozzuoli, Italy). (A) Original drawing from Charles Lyell's "Principles of Geology" (Lyell, 1830). (B) Close-up of one column with *Gastrochaenolites*. Photograph courtesy of Ester Farinati.

chemical, and biological environments. This transition occurs in diverse coastal environments, sandy and rocky shores being prominent among them. Sandy shores represent areas with net sediment accumulation, whereas rocky shores are essentially erosional. Rocky coasts range from vertical to subvertical cliffs to horizontal abrasion platforms, including extremely varied morphologies as a result of their lithology, particular environmental constraints, and erosional processes. Although rocky shores are fundamentally erosional, they may be associated with a suite of sedimentary facies deposited generally under high-energy conditions (Anthony, 2008).

The rocky-shore habitat is characterized by the hard character of the substrate available to organism colonization and by its littoral position. The latter implies zones of variable subaerial exposure, and also common high-energy, unpredictable conditions due to waves, storms, and/or tides. Nevertheless, although it can be said that rocky shores are rather hostile environments subjected to harsh and varying environmental conditions, numerous organisms are capable of colonizing and thriving on them.

2.2 Rocky-Shore Inhabitants

Modern rocky-shore biota includes a wide range of organisms, many of them specifically adapted to live in this habitat (e.g., Lewis, 1964; Little and Kitching, 1996; Southward, 1965). Even the supralittoral area hosts biota adapted to the sporadic arrival of water. Sessile epilitho-episkeletobionts and endolitho-endoskeletobionts (terminology of Taylor and Wilson, 2002) are the most characteristic rocky-shore inhabitants, but vagile fauna is also present (Figs. 2 and 3).

The main groups of organisms that dwell in recent rocky coasts are the following (see Little and Kitching, 1996 for an exhaustive revision): (1) microorganisms and algae, including cyanobacteria, fungi, foraminifers, diatoms, lichens, as well as green (laminarians), brown, and coralline red algae; (2) plants, most important in the highest, never submerged parts, where vegetation is usually dominated by halophytes, but also present in the submerged habitats; (3) sponges, fouling and boring forms dominant; (4) worms, mostly sabellid and spirorbid polychaetes; (5) cnidarians, in particular, non-skeletal anthozoans such as anemones, but also corals mostly in tropical seas; (6) arthropods, such as vagile decapod crabs, isopods, and amphipods, as well as fouling acrothoracican, lepadomorph, and balanomorph barnacles; (7) mollusks, including limpets, snails, chitons, bivalves (mytilids, oysters, pectinids), and octopuses; (8) bryozoans, with forms encrusting organic (shells, algae) and inorganic (rock) substrates; (9) echinoderms, represented by vagile regular sea urchins, starfish, and sea cucumbers; and (10) chordates, encompassing ascidians, fish, birds, and mammals (sirenids and pinnipeda).

The distribution of benthic organisms is not random, and ecological zonation is evident in virtually all modern rocky shores (Little and Kitching, 1996). Limits between zones are usually sharp and continuous, although it is difficult to establish an archetypal zonation due to the high variability of physical and biological constraints. There are two fundamental parameters affecting the distribution of the biota: (1) the ability to resist desiccation, particularly on coasts with tides and (2) the competition among species. Overall, a lower (sublittoral) fringe with a preponderance of green algae, accompanied by numerous groups of invertebrates and fish, is generally present. Above it, the foreshore (midlittoral) zone is characterized by the growth of crustose coralline algae and the frequent presence of byssate (mussels) and boring (lithophagous) bivalves, barnacles (Balanidae), snails, chitons, crabs, sponges, sea urchins, starfish, etc. The littoral zone is inhabited by species able to survive prolonged periods outside the water (algae, limpets, barnacles, chitons), and some vagile arthropods. Finally, the supralittoral fringe (or splash zone) is the highest one, wetted by only a few tides in the year, or by waves or splash, and is dominated by lichens and algae, periwinkle snails (littorinids), and barnacles (Chthamalidae).

Many species are obligate members of rocky-coast communities, unable to live elsewhere in offshore environments, whereas others enjoy a wider range of habitats, intertidal or subtidal. Many rocky-shore organisms show defensive mechanisms that help them to survive in such a hostile physical environment. Desiccation in supra- and intertidal areas is a major problem to solve. Epilithobionts with protective skeletal structures (barnacles, gastropods, chitons) thrive on the surfaces periodically exposed to the atmosphere. High-energy conditions require fixation to the rock, either by cementing, as in coralline algae, vermetid gastropods, oysters, or barnacles, or by byssal filaments, as in mussels. An endolithic or semi-endolithic mode of life also provides protection against both, the physical environment and the predators (Fig. 3). This is the strategy used by

FIGURE 2 Modern rocky shores of the Catalan Mediterranean coast (NE Spain). (A) Supralittoral zone of a rocky shore in Tarragona. Approximate width of photograph: 4 m. (B) Sublittoral zone of a

some sponges (clionaids), polychaete worms or lithophagous bivalves, that bore permanent dwellings into the substrate, and some regular echinoids, which excavate depressions in it. Searching for protection in existing cavities is also a valid mechanism employed by cryptic taxa that seek sporadic or permanent shelter in caves and fissures (crabs, brachiopods, and bryozoans).

Animals living on rocky coasts display three main feeding strategies: grazing, filter-feeding, and predation. Vagile epilithozoans (whelk gastropods, crabs, and starfish) are mostly carnivorous, preying mainly on mollusks. Sessile epilithozoans (barnacles, mussels, oysters, sabellid, and spirorbid polychaetes) are filter-feeders, except for the anemones, which actively capture live small organisms.

2.3 Preservation Potential of Rocky-Shore Biota in the Fossil Record

As stated above, rocky-shore environments are basically erosional and associated with high-energy, unpredictable conditions. This scenario is not favorable to the preservation of inhabiting biotas in the geological record. Among the diversity of forms described in the previous section, it is easy to understand that only those capable of modifying the substrate (bioeroders) or of strongly attaching to it (cementers) would have a good chance to be preserved as *in-situ* fossils. Other organisms with hard skeleton can be preserved only as transported elements in high-energy sedimentary deposits located in the vicinity of the rocky shore.

Ancient rocky shores therefore typically preserve two types of fossils: bioerosional trace fossils and shells of encrusting epilithobionts. Among the latter, red algae, cirripedes, and oysters are common, particularly in the Cenozoic (e.g., de Gibert et al., 1998). In contrast to borings, they have a higher potential of being destroyed during the biostratonomic phase of fossilization or even after burial. Thus, bioerosion structures are the most common fossils in ancient rocky shores and, in consequence, they are paramount for their identification and interpretation. Even among bioerosion structures, the preservation of superficial structures (e.g., grazing traces such as *Gnathichnus* or *Radulichnus*) is

rocky shore in Palamós at a water depth of −30 m with the branching gorgonian coral (Cnidarian) *Paramuricea clavata*. Approximate width of photograph: 80 cm. (C) Midlittoral (=littoral) zone (−2 m) with the mussel *Mytilus galloprovincialis* and the fish *Sarpa salpa* at Port de la Selva. Approximate width of photograph: 30 cm. (D) A variety of encrusting organisms in the sublittoral zone (−25 m) at Illes Medes, showing two demosponges, *Phorbas tenacior* (blue) and *Antho inconstans* (orange), the dendrophyllid coral *Leptosammia pruvoti* (yellow), and the branching bryozoan *Myriapora truncata* (orange). Approximate width of photograph: 20 cm. (E) Sublittoral zone with the bryozoan *Sertella septentrionales* and the demosponge *Spirastrella cuncatrix* (red). Approximate width of photograph: 20 cm. (F) Sublittoral zone (−30 to 40 m) at Cap de Creus with the demosponges *Phorbas tenacior* (bluish) and *Antho inconstans* (orange), the corals *Corallium rubrum* (red), and *Leptosammia pruvoti* (yellow). Approximate width of photograph: 30 cm. Photographs (B–E), courtesy of Roger de Marfà; (F), courtesy of Creu Palacín.

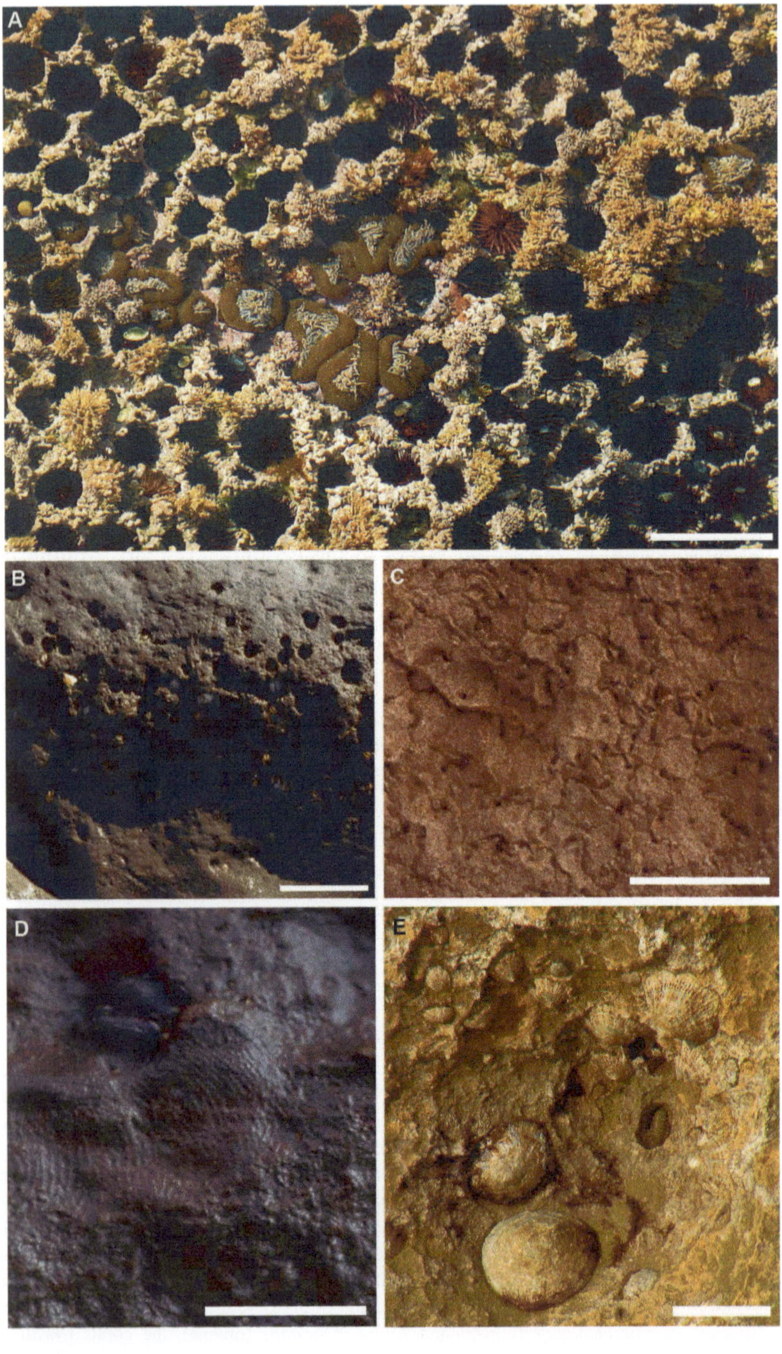
A
B
C
D
E

uncommon, and rocky shore trace-fossil assemblages are generally dominated by borings produced by endolithobionts.

3. MAIN BIOEROSION ICHNOTAXA

Both micro- and macrobioerosion occur in rocky shores. Nevertheless, little work has been done on the first, compared to the attention paid to macroborings, which are obviously more conspicuous and easily identifiable. Thus, this review focuses on macroscopic bioerosion structures, which are usually without great difficulty assignable to a variety of groups of invertebrates. They are described here according to their abundance in ancient rocky shores.

3.1 Bivalve Borings

Several families of bivalves, such as Pholadidae, Gastrochaenidae, and Mytilidae, along with some genera in other families, are active borers in modern marine environments (Kleemann, 1973; Warme, 1975). They produce flask-shaped borings with a narrow aperture and a larger ovoid chamber known from the fossil record as *Gastrochaenolites* (Kelly and Bromley, 1984) (Figs. 3B and 4B). This trace fossil became common in the Jurassic (Bromley, 2004) and thus, its fossil record parallels that of boring bivalves. Paleozoic *Gastrochaenolites* are known from the Ordovician (Wilson and Palmer, 1988; Benner et al., 2004; Ekdale and Bromley, 2001) and Carboniferous (Wilson and Palmer, 1998), but the identity of their tracemakers is still unknown. *Gastrochaenolites* is a ubiquitous element in Mesozoic and Cenozoic rocky-shore trace-fossil assemblages (Fig. 5), generally being the most conspicuous and abundant trace fossil. As bivalves include both chemical and mechanical boring representatives, their bioerosion structures may be present in substrates of a variety of lithologies.

3.2 Sponge Borings

Entobia is the most widespread and characteristic ichnogenus attributed to sponges (Fig. 4A). This complex boring consists of an anastomosing network of canals and chambers with numerous apertures connecting the system to the substrate surface (Bromley and D'Alessandro, 1984). Species belonging to the family Clionaidae are the dominant endolithic sponges today, although

FIGURE 3 Modern bioerosion in littoral settings. (A) Midlittoral platform with abundant *Paracentrotus lividus* within their borings, and some abandoned cavities occupied by sea anemones, Bakio, Basque Country, N Spain. Scale bar = 10 cm. Photograph courtesy of Oriol Ferrer. (B) Cryptoendolithic bivalves occupying previously bored *Gastrochaenolites*, Monte León, Puerto Santa Cruz, Patagonia, Argentina. Scale bar = 2 cm. (C) Bioerosion produced by the periwinkle gastropod *Littorina neritoides* in the supralittoral zone, Salou, NE Spain. Scale bar = 2 cm. (D) *Radulichnus*-like traces produced by the gastropod *Patella* sp., Monte León, Puerto Santa Cruz, Patagonia, Argentina. Scale bar = 2 cm. (E) Bioerosion produced by *Patella caerulea* and the polyplacophoran *Chiton olivaceus*, Tarragona, NE Spain. Scale bar = 2 cm.

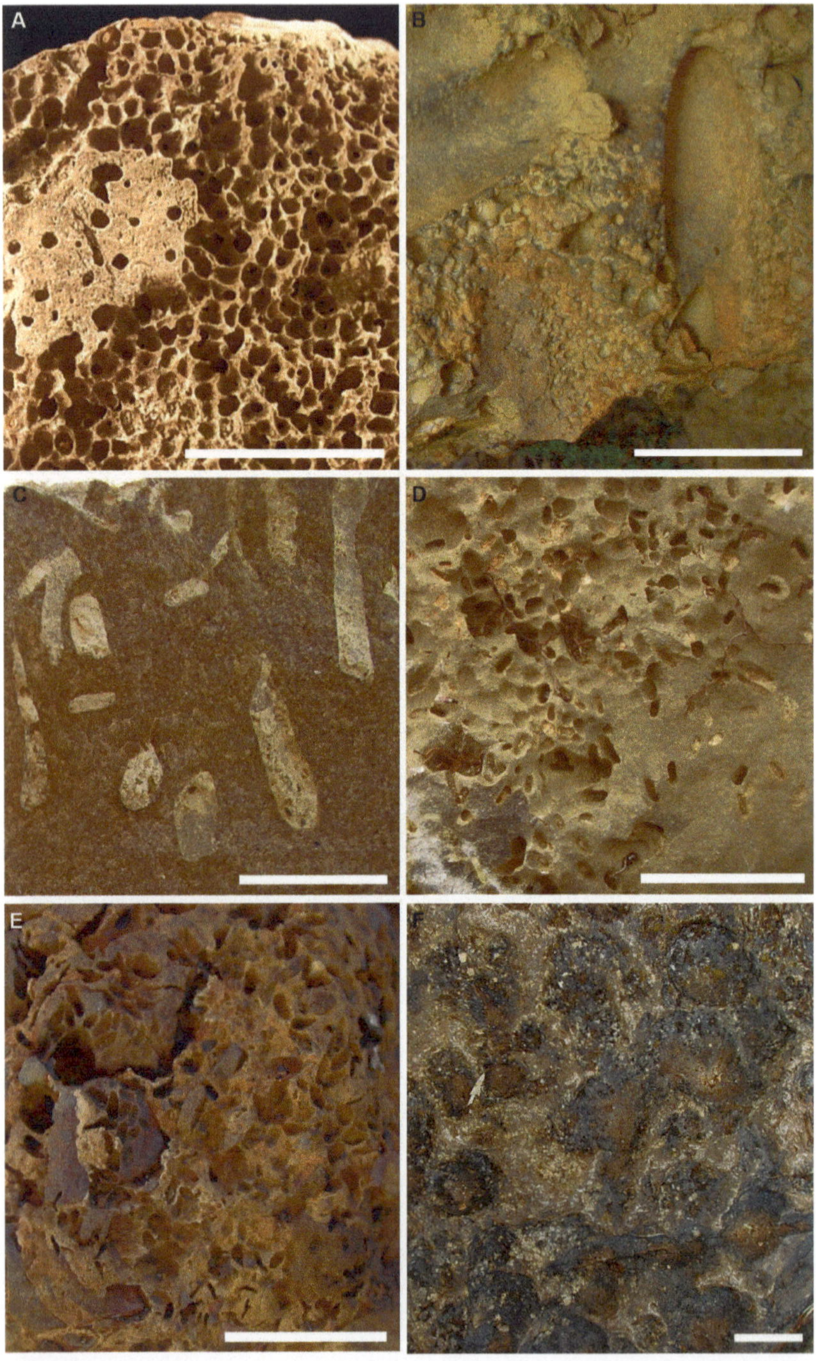

FIGURE 4 Rocky-shore trace fossils. (A) *Entobia* isp. in a clast from the Pliocene Baix Llobregat Basin, NE Spain. (B) *Gastrochaenolites torpedo* and *Entobia* isp. from the Pliocene Baix Ebre Basin, NE Spain. (C) *Trypanites weisei*, Pliocene, Arroyo Vaquero, Málaga, S Spain. (D) *Caulostrepsis* isp., Pliocene, Arroyo Vaquero, Málaga, S Spain. (E) *Rogerella* isp., Pliocene, Almeria, SE Spain. (F) *Circolites*. El Canuto, Málaga, S Spain. Scale bars = 2 cm.

some other taxa are known to produce *Entobia*-like borings (Bromley, 2004). Sponges are chemical borers, and their borings are restricted to carbonate substrates. *Entobia* is present in almost all described Cenozoic carbonate rocky shores, but their presence in Mesozoic rocky shores is limited to the Cretaceous (Fig. 5).

3.3 Worm Borings

A variety of worm borings are known in fossil rocky shores, the most common being *Trypanites*, *Maeandropolydora*, and *Caulostrepsis*. *Trypanites* (Fig. 4C) is a simple, cylindrical boring similar to those produced today by some sipunculans and polychaetes (Bromley, 1978). It has a long fossil record from the Cambrian (James et al., 1977) to the Holocene. It is often the only bioerosion structure in Paleozoic rocky shores, but their presence in Mesozoic and Cenozoic assemblages is, in most cases, a secondary component (Fig. 5). *Caulostrepsis* (Fig. 4D) and *Maeandropolydora* (Bromley and D'Alessandro, 1983) are two ichnogenera generally attributed to the work of spionid polychaetes (Domènech et al., 2008; Voigt, 1965). *Caulostrepsis* is a U-shaped boring, while *Maeandropolydora* consists of winding cylindrical tubes sometimes connected to pouch-like structures. Both ichnogenera are well known from the Mesozoic and Cenozoic, but some occurrences have been reported in the Paleozoic (Bromley, 2004). They are present in many Neogene rocky shores (Fig. 5), where they occur as secondary components together with *Gastrochaenolites* and *Entobia*. They may be found in the rockground or, more often, affecting the shells of epilithic mollusks.

3.4 Echinoid Borings

Some regular echinoids are important rock borers in very shallow waters today (Fig. 3A). They typically inhabit shallow, bowl-shaped depressions, which they bore mechanically by using the jaws and spines (Martinell, 1981; Warme, 1975). This allows for bioerosion of lithologically diverse substrates. Echinoid borings are known since the Jurassic (Bromley, 2004), and Mikuláš (1992) erected the ichnogenus *Circolites* (Fig. 4F) to name this type of trace fossil. *Circolites* has been identified in Cretaceous and Neogene rocky shores (Fig. 5).

3.5 Cirripede Borings

Acrothoracican cirripedes bore small, narrow, pouch-shaped cavities with narrow apertures (Seilacher, 1969). They are assigned to the ichnogenus *Rogerella* (Bromley and D'Alessandro, 1987) (Fig. 4E), which ranges from the Devonian to the Holocene (Bromley, 2004). Although not very abundant, *Rogerella* has been described from Paleozoic (Carboniferous), Mesozoic (Cretaceous), and Neogene rocky shores (Fig. 5).

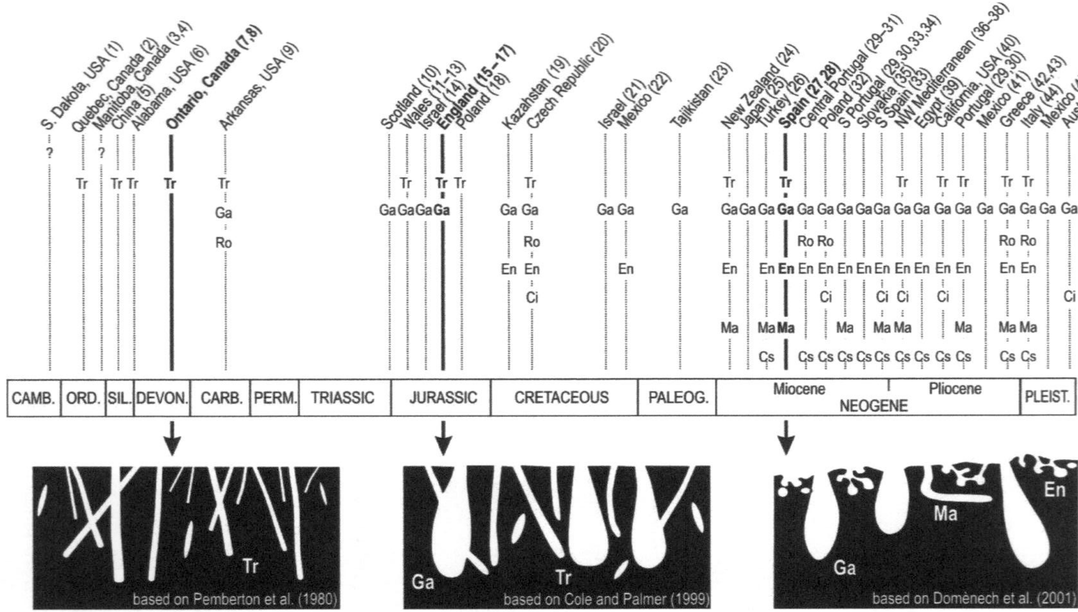

FIGURE 5 Composition of rocky-shore trace-fossil assemblages through the Phanerozoic based on bibliographic data. Only the most common ichnogenera are included; some assignments have been made based on pictures and descriptions. The drawings in the lower part illustrate three selected assemblages, which are indicated in bold. References used: (1) *Johnson et al. (2010)*, *(2) Desrochers and James (1988)*, *(3) Johnson and Baarli (1987)*, *(4) Johnson et al. (1988)*, *(5) Rong and Johnson (1996)*, *(6) Baarli et al. (1992)*, *(7) Kobluk et al. (1977)*, *(8) Pemberton et al. (1980)*, *(9) Webb (1994)*, *(10) Farris et al. (1999)*, *(11) Ager (1986)*, *(12) Fletcher (1988)*, *(13) Johnson and McKerrow (1995)*, *(14) Wilson et al. (2005)*, *(15) De La Beche (1846)*, *(16) Bromley (1975)*, *(17) Cole and Palmer (1999)*, *(18) Radwanski (1959)*, *(19) Pianovskaya and Hecker (1966)*, *(20) Mikuláš (1992)*, *(21) Lewy (1985)*, *(22) Lescinski et al. (1991)*, *(23) Hecker (1965)*, *(24) Campbell et al. (2004)*, *(25) Masuda (1968)*, *(26) Uchman et al. (2002)*, *(27) de Gibert et al. (1996)*, *(28) Domènech et al. (2001)*, *(29) da Silva et al. (1995)*, *(30) da Silva et al. (1999)*, *(31) Santos et al. (2010)*, *(32) Radwanski (1970)*, *(33) Santos et al. (2008)*, *(34) Cachão et al. (2009)*, *(35) Radwanski (1968)*, *(36) Martinell and Domènech (1986)*, *(37) Martinell and Domènech (1995)*, *(38) de Gibert et al. (1998)*, *(39) Aigner (1983)*, *(40) Watkins (1990)*, *(41) Ledesma-Vázquez and Johnson (1994)*, *(42) Bromley and Asgaard (1993a)*, *(43) Bromley and Asgaard (1993b)*, *(44) Bromley and D'Alessandro (1987)*, *(45) Zwiebel and Johnson (1995)*, *(46) Libbey and Johnson (1997)*, *(47) Semeniuk and Johnson (1985)*. Abbreviations used for ichnogenera: ?, unknown; Tr, *Trypanites*; Ga, *Gastrochaenolites*; Ro, *Rogerella*; En, *Entobia*; Ci, *Circolites*; Ma, *Maeandropolydora*; Cs, *Caulostrepsis*.

4. TRACE-FOSSIL ASSEMBLAGES AND ICHNOFACIES

Figure 5 shows a collection of published fossil rocky shores with information on bioerosion structures including examples from the Ordovician to the Pleistocene. An analysis of these data reveals two major controls on the composition of trace-fossil assemblages: substrate and time (evolution). In this section, we examine these two aspects to later discuss the assemblages in the context of the archetypal ichnofacies model.

4.1 Substrate Influence on Rocky-Shore Assemblages

Organisms may bore into hard lithic substrates by chemical and/or mechanical means (e.g., Ekdale et al., 1984: table X-2; Bromley, 1992: table 2). Those that use chemical techniques are capable of dissolving $CaCO_3$, and as a consequence, they are largely restricted to carbonate substrates. Thus, while sponges are obligatory carbonate borers, echinoids or decapods, which bore by scrapping off the substrate, are not limited by the chemical composition of the substrate. Among bivalves and worms, different taxa bore chemically, mechanically, or by combining both techniques. Most fossil rocky-shore bioerosion assemblages described in the literature are developed on carbonate rocky coasts where both chemical and mechanical borers were capable of bioeroding. Nevertheless, a few examples are known in other lithologies. Pitted surfaces have been described as bioerosion on rocky-shore quartzite boulders from the Late Cambrian (Johnson et al., 2010) and Late Ordovician–Early Silurian (Johnson and Baarli, 1987; Johnson et al., 1988). In contrast, clear *Trypanites* are known from Paleozoic carbonate rockgrounds. Most striking differences are found between carbonate and non-carbonate rocky shores in the Cenozoic. Studies carried out on examples from the Miocene (igneous rocks, Masuda, 1968), Pliocene (siltstones, Ledesma-Vázquez and Johnson, 1994), and Pleistocene (andesite, Libbey and Johnson, 1997; Zwiebel and Johnson, 1995) reported only the presence of bivalve borings, whereas the sponge boring *Entobia* is typically absent in non-carbonate rocky shores. Nevertheless, *Entobia* may be present in shells of encrusting organisms, as in the case of some Pliocene, shale-constituted rocky shores in the NW Mediterranean (de Gibert et al., 1998). Some reports on the bioerosion of modern rocky shores made of igneous rock record the presence of echinoid, crustacean, worm, and bivalve borings (Allouc et al., 1996; Fischer, 1981).

4.2 Evolution of Rocky-Shore Bioerosion

The oldest reported bioerosion traces on rocky shores are structures described as such by Johnson et al. (2010) on Late Cambrian quartzite boulders. Nevertheless, the oldest unambiguous evidence of macrobioerosion in rocky shores was reported from the Middle Ordovician (Desrochers and James, 1988). These authors documented a monotypic *Trypanites* assemblage, which is very comparable to other Paleozoic examples. A higher diversity is only reported by Webb

(1994) and Wilson and Palmer (1998) from boulders associated to a Carboniferous rocky shore. Thus, Paleozoic rockground bioerosion is essentially typified by low-diversity (monotypic) assemblages of worm borings, maybe produced by sipunculids (e.g., Pemberton et al., 1980). The Carboniferous example may suggest the incorporation of new boring taxa to Paleozoic rocky-shore communities, although more examples would be needed to confirm this tendency.

No Permian or Triassic bioerosion in rocky shores is known to the authors to have been reported, which does not allow the assessment of the possible impact of the end-Permian extinction on coastal borers. On the contrary, several Jurassic examples are known (Fig. 5), which are characterized by bivalve borings (*Gastrochaenolites*), which may be associated with *Trypanites*. The assemblages known from the Cretaceous take on a more modern aspect. The sponge boring *Entobia* is present in most cases, while at least one example (Mikuláš, 1992) has diversity comparable to Neogene and modern assemblages, including not only bivalve and sponge borings but also echinoid and cirripede trace fossils.

There is a considerable gap of information concerning Paleogene rocky shores (Johnson, 1988a). Nevertheless, Neogene and Pleistocene records of bioerosion are very abundant (Fig. 5 includes only some examples). Trace-fossil assemblages are the same as those seen in modern coasts and can be typified by the presence of bivalve and sponge borings (*Gastrochaenolites* and *Entobia*, respectively), often accompanied by worm borings (*Maeandropolydora* and *Caulostrepsis*), echinoid bioerosional bowls (*Circolites*), and cavities produced by acrothoracican cirripedes (*Rogerella*). This general increase in diversity is even more remarkable in some examples, such as in the Pliocene of Greece (Bromley and Asgaard, 1993a,b) or the Plio-Pleistocene of Italy (Bromley and D'Alessandro, 1987), where other trace fossils such as *Gnathichnus*, *Radulichnus*, and '*Conchotrema*' (=*Talpina*) also occur.

This trend in the diversification of rocky-shore bioerosion during the Phanerozoic was already acknowledged by de Gibert et al. (1998), and it shows a clear parallelism with the pattern exhibited by boring assemblages in marine hardgrounds in general (Palmer, 1982; Taylor and Wilson, 2003; Wilson and Palmer, 1990, 1992) and also to the general trend to increasing diversity of rocky-shore biotas shown by Johnson and Baarli (1999).

4.3 Rocky-Shore Ichnofacies

Bioerosional trace-fossil assemblages were not incorporated into the Seilacherian or archetypal ichnofacies model until Frey and Seilacher (1981) defined the *Trypanites* Ichnofacies. These authors characterized it as constituted by borings, raspings, and gnawings formed in a variety of hard substrates, including not only rocky coasts, beachrocks, and sublittoral hardgrounds but also organic substrates such as shell, bone, and wood. Later, Bromley et al. (1984) erected the *Teredolites* Ichnofacies for marine woodgrounds. Thus, the *Trypanites*

Ichnofacies was left to typify marine-bioerosion assemblages in rocky, sedimentary, and skeletal substrates.

In their study of bioerosion in the Late Pliocene of Rhodes, Bromley and Asgaard (1993a) proposed the existence of two different ichnofacies: (1) *Entobia* Ichnofacies, dominated by deep-tier dwelling borings found in rocky-shore settings; (2) *Gnathichnus* Ichnofacies, constituted mainly by epigenic and shallow-tier trace fossils seen in shells and lithoclasts found in deeper water sediments. They suggested that these two ichnofacies could be either subdivisions or even substitutes of the *Trypanites* Ichnofacies. Later, de Gibert et al. (1998, 2007) demonstrated the recurrence of these two assemblages since the Jurassic and validated them as archetypal ichnofacies with paleoenvironmental significance.

Thus, there are two archetypal ichnofacies available for rocky-shore settings: *Trypanites* and *Entobia* (de Gibert et al., 1998, 2007). Both are characterized by deeply penetrating dwelling borings (domichnia). Based on that, MacEachern et al. (2007) considered both ichnofacies to be equivalent. Nevertheless, as pointed out by de Gibert et al. (2007), there are important particularities in the *Entobia* Ichnofacies that reflect compositional and structural evolutionary changes of endolithic rocky-shore communities. Our proposal here is to keep the *Trypanites* Ichnofacies to encompass all bioerosional trace-fossil assemblages constituted by dwelling borings resulting from long-term colonization in permanent or semipermanent hard substrates, such as those of rocky shores, some hardgrounds, gravel pavements, and reefs, in contrast to the *Gnathichnus* Ichnofacies (see de Gibert et al., 2007), which consists mostly of epigenic and shallow-tier traces in ephemeral substrates such as shellgrounds, and some hardgrounds and gravels. Nevertheless, this *Trypanites* Ichnofacies in Phanerozoic rocky shores includes two clear and distinct recurrent assemblages (Fig. 5). The first corresponds to low-diversity bioerosion associations constituted by vertical to subvertical borings and typifies most Paleozoic and Jurassic rocky-shore trace-fossil assemblages and also some younger occurrences in non-carbonate substrates (Fig. 5). Meanwhile, the other one (the *Entobia* Ichnofacies of Bromley and Asgaard, 1993a) is characteristic of Cretaceous to modern carbonate rocky shores, and it is characterized by a higher diversity with both shallow- and deep-tier borings. The differences between both are linked to the nature of the substrate and, most importantly, to the increasing complexity of marine benthic hardground communities after the Mesozoic Marine Revolution.

5. CASE STUDIES

5.1 Early Devonian, Ontario

Among published descriptions of Paleozoic rocky shores, Kobluk et al. (1977) and Pemberton et al. (1980) provide interesting details on bioerosional trace fossils. This case refers to the Silurian–Devonian disconformity in southern Ontario (Canada), where the Early Devonian (Emsian) Oriskany and Springvale sands disconformably overlie the Upper Silurian dolomitic Bertie Formation. A second disconformity exists between the two Devonian sandy units. The top surface of the Bertie Formation has a local undulatory topography of up to 1 m with the presence of some dolomitic blocks. The dolomitic substrate bears a system of joints and fractures, which were widened by dissolution where Springvale erosion removed the Oriskany sands. Other evidences of exposure and karstification include vug and cavern porosity. *Trypanites* borings penetrate the Silurian dolomites in the top horizontal discontinuity and in the walls of open fractures of both, the pre- and the post-Oriskany disconformity. They are tubular, rectilinear to gently curved, 2–3 mm in diameter and 10–30 mm long. This morphology is comparable to the borings of some modern sipunculid worms (Pemberton et al., 1980). Algal microborings originate from the walls of *Trypanites*. These bioerosional features were formed during transgression following substrate emersion and karstification. No other evidence of marine submersion, such as epilithozoan skeletons, was found.

5.2 Middle Jurassic, England

In the Mendip Hills, west of Frome (Somerset), Carboniferous (Dinantian) limestones are overlain by Jurassic units (Wall and Jenkyns, 2004). One of the first researchers studying this unconformity was De La Beche (1846), who recognized borings and provided some of the first illustrations of fossil bioerosion on rocky shores (Fig. 6A and B). Bromley (1975: fig. 18.6) pointed out the

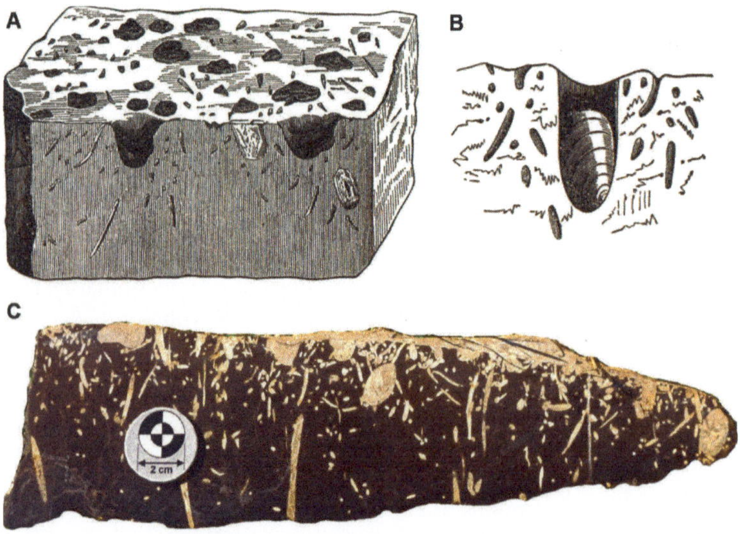

FIGURE 6 Jurassic rocky-shore trace fossils from England. (A, B) Original figures by De La Beche (1846) showing *Trypanites* and *Gastrochaenolites*. (C) Section of a Carboniferous limestone with Jurassic *Trypanites* and *Gastrochaenolites*.

presence of *Gastrochaenolites* and *Trypanites* not only in the rockground itself but also in hardgrounds found within the Middle Jurassic (Bajocian–Bathonian) Upper Inferior Oolite overlying the Carboniferous limestones. A more detailed study of the rocky-shore bioerosion in the area was carried out by Cole and Palmer (1999). These authors characterized the assemblage as dominated by *Trypanites weisei* (diameter 0.7–3.2 mm, maximum length 130 mm), which are generally very abundant, reaching densities of more than 100,000 borings per m^2. Larger worm borings (*Trypanites fosteryeomani*) and bivalve-produced cavities (*Gastrochaenolites*) are also common (Fig. 6C). Oysters (*Liostrea*) are found encrusting the substrate, while the bivalve *Hiatella* and some gastropods are found within the borings, probably indicating a nestling habit.

5.3 Miocene, Spain

Domènech et al. (2001) published a detailed account on the bioerosion found in rocky shores at several localities in the Camp de Tarragona Basin in NE Spain. These rocky shores were formed during the Langhian–Serravalian (Middle Miocene) transgression by the flooding of a previously existing topography constituted by Mesozoic carbonates. In several localities, the transgressive surface displays a planar morphology parallel to the overlying Miocene beds, suggesting an interpretation as a wave-cut abrasion platform (Fig. 7A). Bioerosion is intense and dominated by *Gastrochaenolites* and *Entobia* (Fig. 7B–D). Two ichnospecies of *Gastrochaenolites* were recognized, *G. lapidicus* and *G. torpedo*, together with *Phrixichnus* isp. Sponge borings (*Entobia*) occupy the area between *Gastrochaenolites*. Some *Maeandropolydora decipiens* (Fig. 7D) are occasionally found. Bivalve borings are the most conspicuous feature on the surface, occurring often in densities exceeding 1,500 individuals per m^2 (Fig. 7C). They are often truncated, most likely due to the weakening of the substrate due to sponge activity, more intense as transgression progressed. The rock substrate is generally covered by heterometric breccias of a variable thickness, followed by coquines or calcarenites (Ardenya Unit). The clasts in the breccias are dolomitic and are also intensely bored with the same ichnotaxa (plus *Trypanites rectus* in one locality), but *Entobia* dominates here, while *Gastrochaenolites* is much more rare except in big boulders. This is due to higher instability of such mobile substrates that favor rapid growing borers such as sponges. No epilithobionts have been found either on the rockground or on the clasts.

6. CONCLUSIONS

Rocky shores in the fossil record are found associated with regional unconformities. The right paleoenvironmental scenario for their formation and preservation occurs during major transgressive events capable of flooding rocky

FIGURE 7 Miocene bioerosion in rocky shores of the Camp de Tarragona Basin, NE Spain. (A) Very flat boundary between the Jurassic dolomites and the Miocene breccias and calcarenites, interpreted as a wave-cut abrasion platform at Punta del Miracle. (B) Close-up of the abrasion platform with *Gastrochaenolites*, *Entobia* and *Maeandropolydora*. (C) Intense bioerosion of *Gastrochaenolites* and *Entobia* at the same locality. (D) *Gastrochaenolites* and *Entobia* in vertical section penetrating the Jurassic dolomites and infilled by Miocene sediments, Punta Savinosa. Scale bar = 2 cm. Photograph courtesy of Zain Belaústegui.

paleoreliefs devoid of sedimentary cover. Marine evidence of submersion (bioerosion, body fossils, tidal, or wave erosion) commonly overprints other features formed during the period of subaerial exposure (e.g., karstification in carbonate rocks) preceding the transgression. Although it is possible to identify sedimentary deposits linked to rocky shores, these are essentially erosional environments, in which bioerosion is often the best aid for their identification in the fossil record. Bioeroded rocky-shore surfaces can be extensive and, given the adequate outcrop conditions, can be recognized and correlated through different localities in a basin, thus contributing to its paleogeographic reconstruction. Such large surfaces are often time-averaged as a consequence of the overprinting of succeeding communities as the transgression proceeded. Bioerosion can be also a very precise indicator of maximum sea level in those areas where borings disappear above a certain altitude, a feature that can be related with other morphological characteristics indicative of sea level, such as erosional notches. Identifying these paleopositions of sea level may be important to interpret eustatic sea-level changes or relative changes due to tectonic activity.

Rocky-shore bioerosion assemblages are typified by the *Trypanites* Ichnofacies characterized by the presence of deep-tier dwelling borings produced by a variety of invertebrates (sponges, worms, bivalves, cirripedes, and echinoids) seeking refuge from hostile conditions and predators. Two clearly distinct recurrent assemblages can be recognized when analyzing trace-fossil associations in rocky shores throughout the Phanerozoic. They do not correlate with significant paleoenvironmental differences but rather reflect variations in the complexity of the ecological structure of the benthic community related to evolutionary aspects or substrate limitations. Thus, trace-fossil assemblages in Paleozoic rocky shores previous to the Jurassic diversification of borers, but also some younger occurrences on non-carbonate substrates, exhibit a low diversity of vertical, deep-penetrating borings (*Trypanites* and *Gastrochaenolites*). Meanwhile, more diverse assemblages consisting of shallow- and deep-tier structures, most typically *Entobia* and *Gastrochaenolites*, are found in Cretaceous and Cenozoic rocky shores on carbonate substrates.

ACKNOWLEDGMENTS

This research was funded by the Spanish Ministry of Research and Innovation as a contribution to the projects CGL 2007-60507/BTE and CGL 2010-15047/BTE.

REFERENCES

Ager, D., 1986. A reinterpretation of the basal "Littoral Lias" of the Vale of Glamorgan. Proc. Geol. Assoc. 97, 29–35.

Aigner, T., 1983. A Pliocene cliff-line around the Giza Pyramids Plateau, Egypt. Palaeogeogr. Palaeoclimatol. Palaeoecol. 42, 313–322.

Allouc, J., Le Campion, T., Leung Task, D., 1996. La bioérosion des substrats magmatiques en millieu littoral: l'exemple de la presqu'ile du Cap Vert (Sénégal Occidental). Geobios 29, 485–502.

Anthony, E.J., 2008. Rocky shores. In: Anthony, E.J. (Ed.), Shore Processes and Their Palaeo-environmental Applications. Developments in Marine Geology, vol. 4. Elsevier, Oxford, pp. 369–399.

Baarli, B.G., Brande, S., Johnson, M.E., 1992. Proximality trends in the Red Mountain Formation (Lower Silurian) of Birmingham, Alabama. Okla. Geol. Surv. Bull. 145, 1–13.

Benner, J.S., Ekdale, A.A., de Gibert, J.M., 2004. Macroborings (*Gastrochaenolites*) in Lower Ordovician hardgrounds of Utah: sedimentologic, paleoecologic, and evolutionary implications. Palaios 19, 543–550.

Boucot, A.J., 1981. Principles of Benthic Marine Palaeoecology. Academic Press, New York, 463 pp.

Bromley, R.G., 1975. Comparative analysis of fossil and recent echinoid bioerosion. Palaeontology 18, 725–739.

Bromley, R.G., 1978. Bioerosion of Bermuda reefs. Palaeogeogr. Palaeoclimatol. Palaeoecol. 23, 169–197.

Bromley, R.G., 1992. Bioerosion: eating rocks for fun and profit. In: Maples, C.G., West, R.R. (Eds.), Trace Fossils. Short Courses in Paleontology, vol. 5, pp. 121–129.

Bromley, R.G., 2004. A stratigraphy of marine bioerosion. In: McIlroy, D. (Ed.), The Application of Ichnology to Palaeoenvironmental and Stratigraphic Analysis. Geol. Soc. London, Spec. Publ., vol. 228. pp. 455–479.

Bromley, R.G., Asgaard, U., 1993a. Endolithic community replacement on a Pliocene rocky coast. Ichnos 2, 93–116.

Bromley, R.G., Asgaard, U., 1993b. Two bioerosion ichnofacies produced by early and late burial associated with sea-level change. Geol. Rundsch. 82, 276–280.

Bromley, R.G., D'Alessandro, A., 1983. Bioerosion in the Pleistocene of southern Italy: ichnogenera *Caulostrepsis* and *Maeandropolydora*. Riv. Ital. Paleontol. Strat. 89, 283–309.

Bromley, R.G., D'Alessandro, A., 1984. The ichnogenus *Entobia* from the Miocene, Pliocene and Pleistocene of southern Italy. Riv. Ital. Paleontol. Strat. 90, 227–296.

Bromley, R.G., D'Alessandro, A., 1987. Bioerosion of the Plio-Pleistocene transgression of southern Italy. Riv. Ital. Paleontol. Strat. 93, 379–442.

Bromley, R.G., Pemberton, S.G., Rahmani, R.A., 1984. A Cretaceous woodground: the *Teredolites* ichnofacies. J. Paleontol. 58, 488–498.

Cachão, M., da Silva, C.M., Santos, A., Domèmech, R., Martinell, J., Mayoral, E., 2009. The bioeroded megasurface of Oura (Algarve, south Portugal): implications for the Neogene stratigraphy and tectonic evolution of southwest Iberia. Facies 55, 213–225.

Campbell, K.A., Grant-Mackie, J.A., Buckeridge, J.S., Hudson, N., Alfaro, A.C., Hoverd, J., Morgan, S., Horne, N., Banfield, A., 2004. Paleoecology of an early Miocene, rapidly submerging rocky shore, Motuketekete Island, Hauraki Gulf, New Zealand. N. Z. J. Geol. Geophys. 47, 731–748.

Cole, A.R., Palmer, T.J., 1999. Middle Jurassic worm borings, and a new giant ichnospecies of *Trypanites* from the Bajocian/Dinantian unconformity, southern England. Proc. Geol. Assoc. 110, 203–209.

da Silva, C.M., Cachão, M., Martinell, J., Domènech, R., 1995. Estruturas bioerosivas como indicadores de paleolitorais rochosos. O exemplo do Miocénico da Foz da Fonte (Sesimbra, Portugal). Dados preliminares. Universidade do Porto-Facultade de Ciéncias. Mus. Lab. Min. Geol. Mem. 4, 133–137.

da Silva, C.M., Cachão, M., Martinell, J., Domènech, R., 1999. Bioerosional evidence of rocky palaeoshores in the Neogene of Portugal: environmental and stratigraphical significance. Bull. Geol. Soc. Denmark 45, 156–160.

de Gibert, J.M., Martinell, J., Domènech, R., 1996. El Mioceno marino entre las playas de L'Arrabassada y El Miracle (Tarragona): aspectos paleontológicos e implicaciones sedimentológicas. Acta Geol. Hisp. 29, 133–148.

de Gibert, J.M., Martinell, J., Domènech, R., 1998. *Entobia* ichnofacies in fossil rocky shores, Lower Pliocene, Northwestern Mediterranean. Palaios 13, 476–487.

de Gibert, J.M., Domènech, R., Martinell, J., 2007. Bioerosion in shell beds from the Pliocene Roussillon Basin, France: implications for the (macro)bioerosion ichnofacies model. Acta Palaeontol. Polon. 52, 783–798.

De La Beche, H.T., 1846. On the formation of the rocks of south Wales and south western England. Mem. Geol. Surv. Great Britain 1, 1–296.

Desrochers, A., James, N.P., 1988. Early Paleozoic surface and subsurface paleokarst: Middle Ordovician carbonates, Mingan Islands, Quebec. In: James, N.P., Choquette, P.W. (Eds.), Paleokarst. Springer, New York, pp. 183–210.

Domènech, R., de Gibert, J.M., Martinell, J., 2001. Ichnological features of a marine transgression: Middle Miocene rocky shore of Tarragona, Spain. Geobios 34, 99–107.

Domènech, R., Martinell, J., de Porta, J., 2008. Bioerosión por poliquetos espiónidos (Polychaeta, Spionidae) en moluscos marinos del Cuaternario caribeño de Colombia. Rev. Acad. Colomb. Cienc. 32, 411–419.

Ekdale, A.A., Bromley, R.G., 2001. Bioerosional innovation for living in carbonate hardgrounds in the Early Ordovician of Sweden. Lethaia 34, 1–12.

Ekdale, A.A., Bromley, R.G., Pemberton, S.G., 1984. Ichnology: The Use of Trace Fossils in Sedimentology and Stratigraphy. SEPM Short Course Notes 15, 317 pp.

Farris, M.A., Oates, M.J., Torrens, H.S., 1999. New evidence on the origin and Jurassic age of palaeokarst and limestone breccias, Loch Slapin, Isle of Skye. Scott. J. Geol. 35, 25–29.

Fischer, R., 1981. Bioerosion of basalt of the Pacific Coast of Costa Rica. Senck. Marit. 13, 1–41.

Fletcher, C.J.N., 1988. Tidal erosion, solution cavities and exhalative mineralisation associated with the Jurassic unconformity at Ogmore, South Glamorgan. Proc. Geol. Assoc. 99, 1–14.

Frey, R.W., Seilacher, A., 1981. Uniformity in marine invertebrate ichnology. Lethaia 13, 183–207.

Hecker, R.F., 1965. Introduction to Palaeoecology. Elsevier, New York, 166 pp.

James, N.P., Kobluk, D.R., Pemberton, S.G., 1977. The oldest macroborers: lower Cambrian of Labrador. Science 197, 980–983.

Johnson, M.E., 1988a. Why are ancient rocky shores so uncommon? J. Geol. 96, 469–480.

Johnson, M.E., 1988b. Hunting for ancient rocky shores. J. Geol. Educ. 36, 147–154.

Johnson, M.E., 1992. Studies of ancient rocky shores: a brief history and annotated bibliography. J. Coast. Res. 8, 797–812.

Johnson, M.E., 2006. Uniformitarianism as a guide to rocky-shore ecosystems in the geological record. Can. J. Earth Sci. 43, 1119–1147.

Johnson, M.E., Baarli, B.G., 1987. Encrusting corals on a latest Ordovician to earliest Silurian rocky shore, south-west Hudson Bay, Manitoba, Canada. Geology 15, 15–17.

Johnson, M.E., Baarli, B.G., 1999. Diversification of rocky-shore biotas through geologic time. Geobios 32, 257–273.

Johnson, M.E., McKerrow, W.S., 1995. The Sutton stone: an Early Jurassic rocky shore deposit in South Wales. Palaeontology 38, 529–541.

Johnson, M.E., Skinner, D.F., MacLeod, K.G., 1988. Ecological zonation during the carbonate transgression of a Late Ordovician rocky shore (Northeastern Manitoba, Hudson Bay, Canada). Palaeogeogr. Palaeoclimatol. Palaeoecol. 65, 93–114.

Johnson, M.E., Wilson, M.A., Redden, J.A., 2010. Borings in quartzite surf boulders from the Upper Cambrian basal Deadwood Formation, Black Hills of South Dakota. Ichnos 17, 48–55.

Kelly, S.R.A., Bromley, R.G., 1984. Ichnological nomenclature of clavate borings. Palaeontology 27, 793–807.

Kleemann, K.H., 1973. Der Gesteinsabbau durch Ätzmuscheln an Kalkküsten. Oecologia 13, 377–395.

Kobluk, D.R., Pemberton, S.G., Karolyi, M., Risk, M.J., 1977. The Silurian–Devonian disconformity in southern Ontario. Bull. Can. Petrol. Geol. 25, 1157–1186.

Ledesma-Vázquez, J., Johnson, M.E., 1994. Late Pliocene abrasion platform from the Cantil Costero Formation of Baja California. Cienc. Mar. 20, 139–157.

Lescinski, H.L., Ledesma-Vázquez, J., Johnson, M.E., 1991. Dynamics of Late Cretaceous rocky shores (Rosario Formation) from Baja California, Mexico. Palaios 6, 126–141.

Lewis, J.R., 1964. The Ecology of Rocky Shores. English Universities Press, London, 323 pp.

Lewy, Z., 1985. Paleoecological significance of Cretaceous bivalve borings from Israel. J. Paleontol. 59, 643–648.

Libbey, L.K., Johnson, M.E., 1997. Upper Pleistocene rocky shores and intertidal biotas at Playa La Palmita (Baja California Sur, Mexico). J. Coast. Res. 13, 216–225.

Little, C., Kitching, J.A., 1996. The Biology of Rocky Shores. Oxford University Press, Oxford, 240 pp.

Lyell, C., 1830. Principles of Geology, Being an Attempt to Explain the Former Changes of the Earth's Surface, by Reference to Causes Now in Operation, vol. 1. John Murray, London, 511 pp.

MacEachern, J.A., Pemberton, S.G., Gingras, M.K., Bann, K.L., 2007. The ichnofacies paradigm: a fifty-year retrospective. In: Miller III, W. (Ed.), Trace Fossils. Concepts, Problems, Prospects. Elsevier, Amsterdam, pp. 52–77.

Martinell, J., 1981. Actividad erosiva de *Paracentrotus lividus* (Lmk.) (Echinodermata, Echinoidea) en el litoral gerundense. Oecol. Aquat. 5, 219–225.

Martinell, J., Domènech, R., 1986. Actividad bioerosiva en el plioceno marino del Empordà (Catalunya). Paleontol. Evol. 20, 247–251.

Martinell, J., Domènech, R., 1995. Bioerosive structures on the Pliocene rocky shores of Catalonia (Spain). Rev. Esp. Paleontol. 10, 37–44.

Masuda, K., 1968. Sandpipes penetrating igneous rocks in the environs of Sendai, Japan. Trans. Proc. Palaeontol. Soc. 72, 351–361.

Mikuláš, R., 1992. Early Cretaceous borings from Stramberk (Czechoclovakia). Čas. Min. Geol. 37, 297–323.

Palmer, T., 1982. Cambrian to Cretaceous changes in hardground communities. Lethaia 13, 309–323.

Pemberton, S.G., Kobluk, D.R., Yeo, R.K., RIisk, M.J., 1980. The boring *Trypanites* at the Silurian-Devonian disconformity in southern Ontario. J. Paleontol. 54, 1258–1266.

Pianovskaya, L.A., Hecker, R.T., 1966. Rocky shores and hardground of the Cretaceous and Palaeogene seas in the central Kizil-Kum and their inhabitants. In: Organisms and Environment in the Geological Past. A Symposium. Nauka, Moscow, pp. 222–245 (in Russian).

Radwanski, A., 1959. Littoral structures (cliff, clastic dikes and veins, and borings of *Potamilla*) in the high-tatric Lias. Acta Geol. Polon. 9, 270–280.

Radwanski, A., 1968. Tortonian cliff deposits at Zahorska Bystrica near Bratislava (Southern Slovakia). Bull. Acad. Polon. Sci., Ser. Sci. Geol. Geogr. 16, 97–102.

Radwanski, A., 1970. Dependence of rock-borers and burrowers on the environmental conditions within the Tortonian littoral zone of Southern Poland. In: Crimes, T.P., Harper, J.C. (Eds.), Trace Fossils. Geol. J., Spec. Iss, vol. 9, pp. 371–390.

Reading, H.G., Collinson, J.D., 1996. Clastic coasts. In: Reading, H.G. (Ed.), Sedimentary Environments. Processes, Facies and Stratigraphy. Blackwell, Oxford, pp. 154–231.

Rong, J., Johnson, M.E., 1996. A stepped karst unconformity as an Early Silurian rocky shoreline in Guizhou Province (South China). Palaeogeogr. Palaeoclimatol. Palaeoecol. 121, 115–129.

Santos, A., Mayoral, E., da Silva, C.M., Cachão, M., Domènech, R., Martinell, J., 2008. Trace fossil assemblages on Miocene rocky shores of southern Iberia. In: Wisshak, M., Tapanila, L. (Eds.), Current Developments in Bioerosion. Erlangen Earth Conference Series. Springer, Berlin, pp. 431–450.

Santos, A., Mayoral, E., da Silva, C.M., Cachão, M., Kullberg, J.C., 2010. *Trypanites* ichnofacies: palaeoenvironmental and tectonic implications. A case study from the Miocene disconformity at Foz da Fonte (Lower Tagus Basin, Portugal). Palaeogeogr. Palaeoclimatol. Palaeoecol. 292, 35–43.

Seilacher, A., 1969. Paleoecology of boring barnacles. Am. Zool. 9, 705–719.

Semeniuk, V., Johnson, D.P., 1985. Modern and Pleistocene rocky shore sequences along the carbonate coastlines, south-western Australia. Sediment. Geol. 44, 225–261.

Southward, A.J., 1965. Life on the Sea-Shore. Harvard University Press, Cambridge, 153 pp.

Spencer, T., Viles, H., 2002. Bioconstruction, bioerosion and disturbance on tropical coasts: coral reefs and rocky limestone shores. Geomorphology 48, 23–50.

Taylor, P.D., Wilson, M.A., 2002. A new terminology for marine organisms inhabiting hard substrates. Palaios 17, 522–525.

Taylor, P.D., Wilson, M.A., 2003. Palaeoecology and evolution of marine hard substrate communities. Earth-Sci. Rev. 62, 1–103.

Trenhaile, A.S., 1987. The Geomorphology of Rock Coasts. Clarendon Press, Oxford, 370 pp.

Uchman, A., Demírcan, H., Toker, V., Derman, S., Sevim, S., Szulc, J., 2002. Relative sea-level changes recorded in borings from a Miocene rocky shore of the Mut Basin, southern Turkey. Ann. Soc. Geol. Polon. 72, 263–270.

Voigt, E., 1965. Über parasitische Polychaeten in Kreide-Austern sowie einige andere in Muschelschalen bohrende Würmer. Paläont. Z. 39, 193–211.

Wall, G.R.T., Jenkyns, H.C., 2004. The age, origin and tectonic significance of Mesozoic sediment-filled fissures in the Mendip Hills (SW England): implications for extensional models and Jurassic sea-level curves. Geol. Mag. 141, 471–504.

Warme, J.E., 1975. Borings as trace fossils, and the processes of marine bioerosion. In: Frey, R.W. (Ed.), The Study of Trace Fossils. Springer, New York, pp. 181–227.

Watkins, S.R., 1990. Paleoecology of a Pliocene rocky shore-line, Salton Trough Region, California. Palaios 5, 167–175.

Webb, G.E., 1994. Paleokarst, paleosol, and rocky shore deposits at the Mississippian-Pennsylvanian unconformity, northwestern Arkansas. GSA Bull. 106, 634–648.

Wilson, M.A., Palmer, T.J., 1988. Nomenclature of a bivalve boring from the Upper Ordovician of the Midwestern United States. J. Paleontol. 62, 306–308.

Wilson, M.A., Palmer, T.J., 1990. A review of evolutionary trends in carbonate hardground communities. In: Miller III, W. (Ed.), Paleocommunity Temporal Dynamics: The Long-term Development of Multispecies Assemblies. Paleontol. Soc., Spec. Publ., vol. 5, pp. 137–152.

Wilson, M.A., Palmer, T.J., 1992. Hardgrounds and Hardground Faunas. Institute of Earth Studies Publications, University of Wales, Aberystwyth, vol. 9, pp. 1–131.

Wilson, M.A., Palmer, T.J., 1998. The earliest *Gastrochaenolites* (Early Pennsylvanian, Arkansas, USA): an upper Paleozoic bivalve boring? J. Paleontol. 72, 769–772.

Wilson, M.A., Wolfe, K.R., Avni, Y., 2005. Development of a Jurassic rocky shore complex (Zohar Formation, Makhtesh Qatan, southern Israel). Isr. J. Earth Sci. 54, 171–178.

Zwiebel, J.A., Johnson, M.E., 1995. Late Pleistocene Mytilid and Petricolid bivalves from the open rocky shores of Pacific Baja California (Mexico): unusual preservation of macrofossils. J. Coast. Res. 11, 704–716.

Chapter 16

Estuaries

Murray K. Gingras,*,[1] James A. MacEachern,† Shahin E. Dashtgard,† John-Paul Zonneveld,* Jesse Schoengut,‡ Michael J. Ranger§ and S. George Pemberton*

*Department of Earth and Atmospheric Sciences, University of Alberta, Edmonton, Alberta, Canada, †Department of Earth Sciences, Simon Fraser University, Burnaby, British Columbia, Canada, ‡Canadian Natural Resources Limited Calgary, Alberta, Canada, §808 West Chestermere Drive, Chestermere, Alberta, Canada

[1]Corresponding author: e-mail: mgingras@ualberta.ca

1. INTRODUCTION

Our understanding of the ichnology of modern and ancient estuaries has considerably advanced since the 1970s. The foundation of estuary ichnology is built on the extensive neoichnological studies conducted in estuaries of Georgia, USA (Dorjes and Howard, 1975; Howard and Frey, 1975; Howard et al., 1975; Mayou and Howard, 1975). This is notwithstanding an immensely important body of work produced previously from North Sea sites (e.g., Häntzschel, 1939; Reineck, 1956), focused on marginal-marine neoichnology. In 1982, the neoichnological observations from the North Sea and the estuaries of Georgia were used to formulate the brackish-water ichnological model, providing the initial key for the identification of estuaries in the rock record (cf. Pemberton et al., 1982).

The ichnological model for estuaries developed alongside a rapidly expanding ability to identify tidally influenced sedimentation, in concert with a growing understanding of seismic data and sequence stratigraphy (Bubb and Hatlelid, 1977; Dobrin, 1976; Mitchum, 1977; Mitchum and Vail, 1977; Vail and Mitchum, 1977; Van Wagoner et al., 1987). From this stratigraphic research came a growing awareness that incised valleys were cut during relative lowstands of sea level but were dominantly filled with strata deposited during the subsequent marine transgression (e.g., Allen and Posamentier, 1993; Dalrymple and Zaitlin, 1994; MacEachern et al., 2012). Basal transgressive strata represent the leading edge of marine incursion, and as such, ichnologically discernible, brackish-water deposits are strongly associated with many incised valley-fill deposits (e.g., Beynon et al., 1988; Karvonen, 1989; MacEachern and Gingras, 2007; MacEachern and Pemberton, 1994; Pemberton et al., 1982; Rahmani,

Developments in Sedimentology, Vol. 64. http://dx.doi.org/10.1016/B978-0-444-53813-0.00016-2

1984). Broadly speaking, the geological criteria for recognition of fossil estuaries established in these early works are still used today.

Since the late 1980s, several studies have demonstrated the usefulness of ichnology in the identification of ancient estuary deposits. The most successful efforts have combined detailed sedimentological and stratigraphic observations with ichnological data, in order to provide evidence for (1) brackish-water sedimentation, (2) tidally influenced sedimentation, (3) the presence of incised valleys, and (4) landward (i.e., transgressive) shifts of sedimentary environments (e.g., Bann et al., 2004; Buatois et al., 1998; Gingras and MacEachern, 2012; Hubbard et al., 2004; Karvonen, 1989; MacEachern and Gingras, 2007; MacEachern and Pemberton, 1994; MacEachern et al., 1992; Pattison, 1992; Pattison and Walker, 1994, 1998; Pemberton et al., 1992; Ranger and Pemberton, 1992; Rossetti and Santos, 2004; Savrda and Nanson, 2003; Wetzel et al., 2010; Yang et al., 2007). The role of ichnology in establishing these important characteristics of estuaries is discussed below. Additionally, various case studies are used to emphasize the importance of trace-fossil occurrences and distributions in identifying and classifying the deposits of estuaries.

The term "estuary" has been applied in a number of ways. The Latin term for estuary is *aestuarium*, which refers to tidal marshes and inlets and *aestus estus* is Latin for tide, so the root is strongly linked to tidal activity. In the biological sciences, "estuary" simply refers to any physiographic zone characterized by the presence of brackish water. In the earth sciences, it is more common to refer to brackish-water settings as "estuarine", while reserving "estuary" for coastal geomorphological entities. Pritchard (1967) defined "estuary" as a physiographically restricted coastal body of water with rivers or streams flowing into it and a connection to the ocean. Wolanski (2007) refined this view, suggesting that an estuary is "*a semi-enclosed body of water connected to the sea as far as the tidal limit or the salt-intrusion limit and receiving freshwater runoff; however, the freshwater inflow may not be perennial, the connection to the sea may be closed for part of the year and tidal influence may be negligible*". Although these definitions can be useful, they are not practical definitions for the identification of estuaries in the rock record. After all, fluvial or tidal-inlet deposits may not be demonstrable in the rock record as that is dependent on their preservation potential.

The above definitions are notwithstanding the sequence-stratigraphic context of estuaries, wherein estuaries are the depositional system represented by incised valley fill (IVF). In these cases, the concept of estuary is tied to stratigraphic characteristics that include a valley floor represented by either a composite subaerial unconformity (SU) and transgressive surface or a composite SU and tidal-ravinement surface (TRS; MacEachern et al., 2012). Brackish-water-associated strata dominantly represent the valley fill. Another important criterion is that estuaries demonstrably receive sediment from both fluvial and tidal sources (i.e., Dalrymple et al., 1992). Boyd et al. (2006) nicely summarize these aspects of estuaries with their definition: "*... estuaries, as defined geologically* *here, are transgressive in nature. They receive sediment from both fluvial and marine sources, commonly occupy the seaward portion of a drowned valley, contain facies influenced by tide, wave, and fluvial processes, and are considered to extend from the landward limit of tidal facies at their heads to the seaward limit of coastal facies at their mouths*". We are in agreement with this definition but modify the definition to say "from the landward limit of *persistently tidally modulated facies* to the seaward limit of coastal facies". This allows the practitioner to use readily identifiable evidence for the presence of tidally influenced sedimentation that may include one or more of the following observations: (1) the presence of brackish-water bioturbation in the landward part of the estuary, as the presence of a typical brackish-water assemblage requires larval recruitment from the marine realm and thereby depends upon landward transport of larvae by tidal currents; (2) the occurrence of intertidal-flat deposits, which have a characteristic ichnological signature; (3) the presence of bioturbated inclined heterolithic stratification (IHS), which is consistent with landward larval recruitment and may provide evidence for salinity-associated flocculation processes; and/or (4) the preservation of tidal sedimentary structures. The latter observation is the most commonly used criterion for recognizing paleo-estuary deposits but is likely to be the most difficult

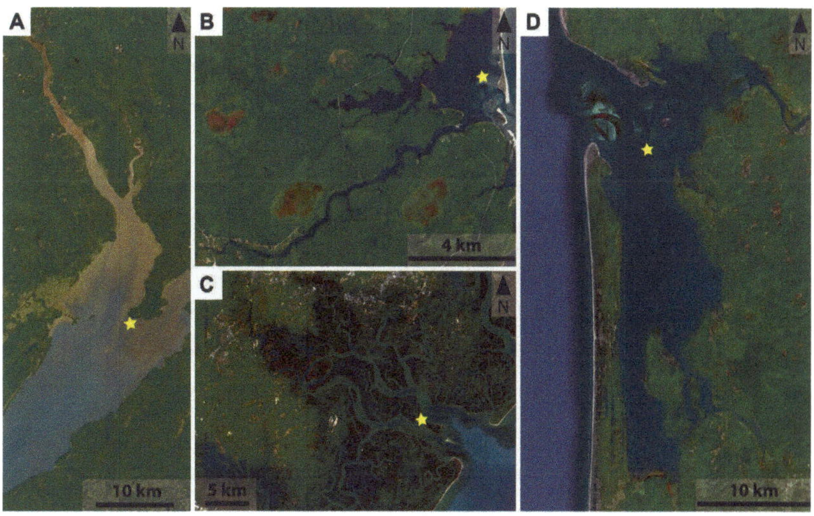

FIGURE 1 Location of modern estuaries referred to in this chapter. Each satellite image shows the estuary to the approximate tidal limit. The yellow stars show the locations of the reported coordinates. Images courtesy of Google Earth, 2011. (A) Chignecto Bay, Atlantic Canada (45°43′5.66″N, 64°32′22.31″W), a large, tide-dominated estuary. (B) Kouchibouguac-River estuary, Atlantic Canada (46°50′42.44″N, 64°55′30.75″W), a wave-dominated estuary. (C) Ogeechee River estuary (31°51′41.88″N, 81°5′0.62″W), a mixed-energy estuary. (D) Willapa Bay, Washington, USA (46°38′57.00″N, 124°0′10.70″W), a mixed-energy estuary.

to observe in limited datasets (e.g., core). These features are consistently observed in modern estuaries from the zone of abrupt sinuosity decrease and channel-width expansion (see Fig. 1) to the estuary mouth. Within estuaries, the following trends are noted. Tidal sedimentary structures decrease in abundance landward and may be essentially absent in the inner estuary. Brackish-water trace-fossil assemblages are increasingly marine in character seaward, and bioturbated IHS are nominally limited to the inner estuary and, more rarely, the fluvio-tidal transition (e.g., Clifton et al., 1976; Dalrymple et al., 1992; Gingras et al., 1999; Hovikoski et al., 2008; Lanier and Tessier, 1998; Martinius and Van den Berg, 2011). Intertidal-flat deposits may be expressed throughout the estuary, but they are most ichnologically distinctive in the inner and middle estuary.

A strong influence on the distribution of estuary tracemakers is the range of estuary morphologies, which are bracketed by the sedimentological end members referred to as wave- and tide-dominated estuaries (Fig. 1A and B); between the two end members reside "mixed" estuaries (Fig. 1C and D). Finally, the state of estuarine fill (filled *versus* unfilled) has been used to classify estuaries more specifically (e.g., Roy, 1994), although we consider these criteria to reflect the later stage or stages of estuary filling, and assert that our ichnological criteria for assisting in the identification of estuaries apply to the inferred sediment and the terminal estuary fill.

2. THE ICHNOLOGICAL IDENTIFICATION OF ESTUARIES

The identification of estuaries in the geological record is consistent with at least four ichnologically or sedimentologically discernible conditions. The first is the presence of brackish-water sedimentation, which is strongly associated with diminutive trace fossils in low-diversity associations. Secondly, tidally influenced sedimentation, although not limited to estuaries, provides evidence for sedimentation in marginal-marine settings. Thirdly, the presence of a transgressed valley floor may be ichnologically discernible by occurrences of omission suites such as those of the *Glossifungites* or *Trypanites* ichnofacies. Finally, transgressive backstepping of sedimentary environments can be easily documented using ichnological datasets. It is important to recognize that not all of the above characteristics may be recognized within a dataset. For example, the transgressed base of the estuary is not always demarcated by omission suites, discerning a tidal signal from core datasets may be challenging, and the outer part of the estuary—and thereby stratigraphically higher units—may not display brackish-water trace-fossil assemblages. However, the more of the criteria that can be established for the system as a whole, the more probable is the interpretation of estuary.

2.1 Brackish-Water Trace-Fossil Assemblages

Brackish-water trace-fossil assemblages are recognized on the basis of characteristic combinations of trace fossils, trace-fossil sizes, assemblage diversities, and distribution trends (Fig. 2). The fundamental characteristics of the brackish-water model were developed by Pemberton et al. (1982) and refined by Beynon et al. (1988). Working in the Mannville Group of the oil sands deposits of NE Alberta, they demonstrated that large parts of the succession were dominated by estuarine facies. Several important characteristics reported by Pemberton et al. (1982) are now routinely used as evidence for brackish-water sedimentation. These include the following:

1. *The identification of ethological generalist behaviors derived from the marine realm.* In brackish-water deposits, trace fossils record behaviors of animals that exploit food resources under a diverse range of environmental conditions (Fig. 2). These types of trace fossils are referred to as "facies-crossing" elements. Owing to the rich food resource associated with sedimentation in tidally influenced settings, traces that are the product of rapid once-over harvesting of food within the sediment are common (e.g., *Planolites*, *Thalassinoides*, and *Protovirgularia*). Likewise, burrow morphologies that suit head-down deposit-feeding, interface deposit-feeding, and/or subordinate suspension-feeding behaviors (e.g., *Skolithos*, *Cylindrichnus*, *Arenicolites*, *Gyrolithes*, and *Siphonichnus*) are exceedingly common in brackish-water strata. The behaviors employed in estuaries are primarily imported from adjacent marine environs; thus trace fossils not normally associated with freshwater (e.g., *Thalassinoides*-, *Cylindrichnus*-, *Arenicolites*-, and *Gyrolithes*-like traces) are common in low-salinity (i.e., < 10 psu [practical salinity units]) waters (e.g., Gingras et al., 1999).
2. *Trace fossils are comparably diminutive in brackish water.* In brackish-water deposits, diminution of trace fossils is interpreted to result from size selection and physiological duress, associated with low and/or fluctuating salinity (Fig. 2). A decrease in size may occur for two reasons: (1) a high body surface/volume ratio of the organism can be desirable for osmotic and ionic regulation (e.g., Moore and Francis, 1985); (2) owing to physiological stress, energy expenditure toward regulatory processes slows animal growth rates and reduces survival rates (e.g., Holste et al., 2009; Spaargaren, 1979, 1995). A high surface-area-to-volume ratio is required by animals that move solutes (such as salt or oxygen) or heat energy across their body membranes. We refer to this as "facilitative diminution", which is prevalent in brackish-water settings (Gingras et al., 1999; MacEachern and Gingras, 2007). Diminution in response to physiological stress can be referred to as "enforced diminution" and can be ascribed to animals living in low- and fluctuating-salinity settings, as well as very warm waters, both of which lead to sustained high metabolism in animals (Remane and Schlieper, 1972;

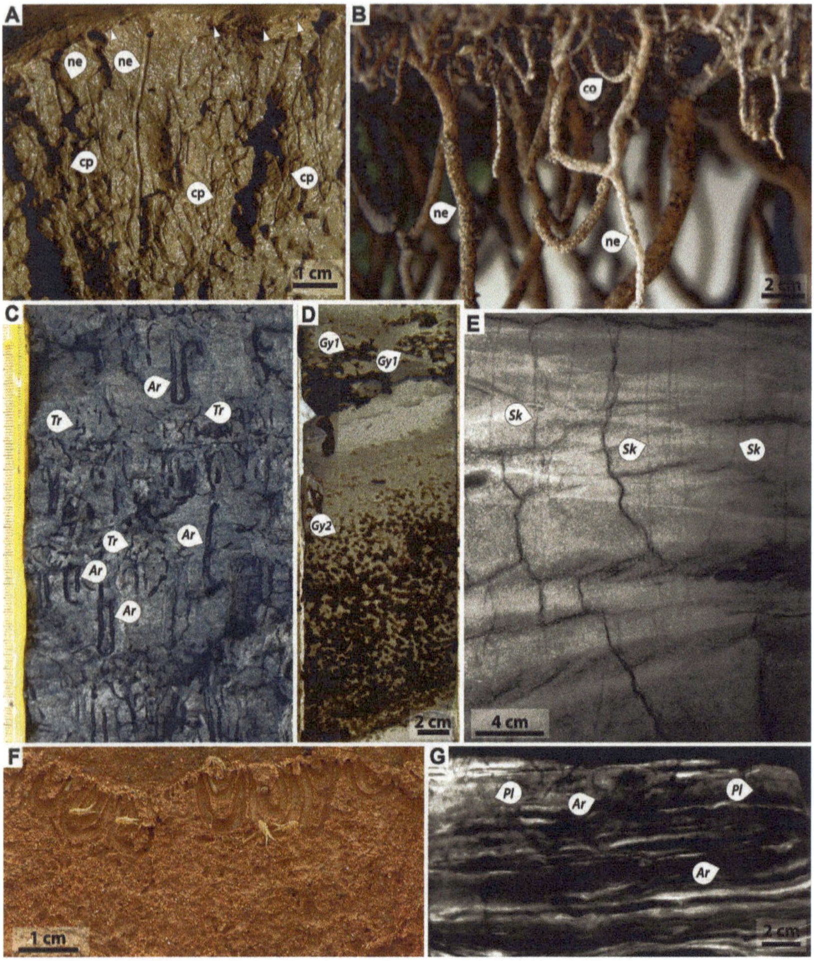

FIGURE 2 Examples of brackish-water assemblages of trace fossils. (A) Nereid (ne) and capitellid (cp) polychaete burrows that resemble the ichnogenera *Skolithos* (*Sk*) and *Trichichnus*. Gironde Estuary, France. (B) Resin cast of a low-diversity trace assemblage comprising *Polykladichnus* made by nereid (ne) polychaetes. Kouchibouguac, New Brunswick, Canada. (C) Low-diversity suite of *Arenicolites* (*Ar*) and *Trichichnus* (*Tr*) from a tidally influenced bar deposit of the Fraser River delta. (D) Comparably robust *Gyrolithes* (*Gy1*) and more gracile examples (*Gy2*) in inclined heterolithic stratification (IHS) of the Cretaceous McMurray Formation. (E) *Skolithos* in cross-bedded sands of the inner Ogeechee estuary, Georgia, USA. (F) Monospecific assemblage of *Arenicolites*, Bay of Fundy, Canada. (G) Low-diversity trace assemblage comprising diminutive traces (*Pl*, *Planolites*; *Ar*, *Arenicolites*) from an upper point bar, inner estuary, Willapa Bay, Washington, USA.

Roy and Martien, 2001; Spaargaren, 1979). Enforced diminution also may induce shortened lifecycles, thereby promoting elevated mortality rates. Such conditions lead to a predominance of juvenile tracemakers (Chapman and Brinkhurst, 1981) that produce comparably small traces.

3. *Assemblage diversities in brackish-water strata are substantially lower than in equivalent marine strata* (Fig. 2). Trace-fossil diversities are considerably lower in brackish-water strata than in marine units. Lower behavioral diversity may be ascribed to the importance of ethologically flexible feeding strategies needed to rapidly harvest the rich food resources in dynamic sedimentary conditions (see 1). Another consideration is that, for modern settings, these robust ethologies are mostly used by opportunistic, commonly brackish-water tolerant animals (e.g., nereid polychaetes, the bivalve *Macoma balthica*, arthropods such as *Corophium volutator*, and the lugworm *Arenicola marina*). As such, the traces produced by animals employing these generalist behaviors are rapidly and widely distributed in marginal-marine settings.
4. *Common presence of high population densities.* Although the diversity of trace fossils is lower in brackish-water settings, trace fossils are commonly present in high population densities, that is, with high bioturbation indices (BI) (Fig. 2A–D; MacEachern and Gingras, 2007; Pemberton et al., 1982). This characteristic is exemplified by the occurrence of mono-ichnospecific trace-fossil assemblages in highly burrowed strata (i.e., BI = 4 or 5). There are at least two likely reasons for this pattern. First, due to the presence of tidal transport from the marine realm and low overall energies in many brackish-water settings, food resources are comparably abundant on and in the sediment. As well, favorable physiological or behavioral adaptations by some animals enable them to outcompete others under stressful conditions (see 3). The successful animals are thereby able to flourish in the environment, with minimal interference from competitors.
5. *Brackish-water environments tend to promote infaunal over epifaunal lifestyles.* One reason for this may be that living within the sediment provides protection against high-frequency (daily or semi-daily) salinity fluctuations (Chapman, 1981; Knox, 1986). Interstitial waters show much more uniform salinities throughout the tidal cycle because they are buffered from the more variable surface waters by the sediment body. Infaunal lifestyles are also a consequence of the abundance of food resources in such regimes, encouraging deposit-feeding within the sediment (Gingras et al., 1999).
6. *Longitudinal trends in brackish-water settings are indicative of landward freshening of the depositional waters* (Fig. 3). Although not specifically outlined by Pemberton et al. (1982), it is derivative of their work (2 and 3). A similar trend was identified by Howard et al. (1975) and Howard and Frey (1975) from their work on the Ogeechee River estuary. Hauck et al. (2009) showed correspondence between the function ([maximum burrow size observed] × [diversity of macroscopic infauna]) and mean salinity within the modern estuary, Kouchibouguac Bay, New Brunswick, Canada. Similarly, Gingras et al. (1999) document a progressive diminution and diversity reduction at Willapa Bay, Washington. This work suggests that both animal sizes and burrow diversities show a crude relationship to

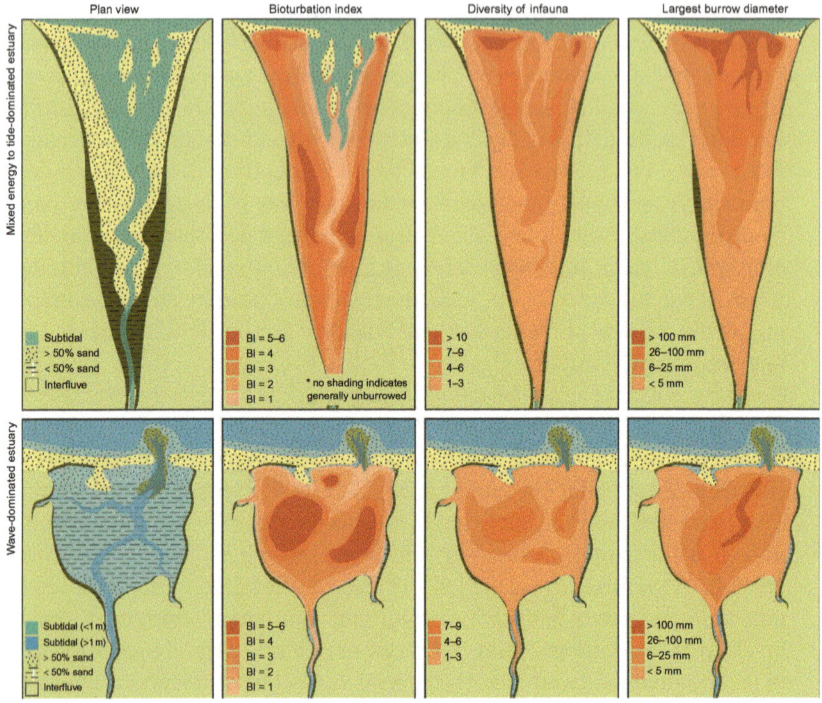

FIGURE 3 Conceptual framework for the distribution of trace-fossil characteristics in mixed-energy to tide-dominated (top row) and wave-dominated (bottom row) estuary settings. Broadly speaking, the mixed to tidal framework is based on personal observations and data from the Willapa and Tillamook bays, the Ogeechee River estuary and the Bay of Fundy. The wave-dominated framework is based on the Kouchibouguac-River estuary. These distributions are most sensitive to the amount of freshwater flux into the system: a higher fluvial discharge can significantly freshen the estuary.

salinity and that their product provides a logarithmic relationship. Although similar efforts have not been attempted in the stratigraphic record, this type of data-intensive analysis has excellent potential for the identification of various marginal-marine settings.

Regarding criteria 2 and 3, it can be challenging to establish what is meant by the terms "diminutive trace fossil" and "low-diversity assemblage". In practice, it is difficult to recognize trends in diminution and diversity, unless a fully marine baseline can be established from contemporaneous rocks in the associated sedimentary basin (cf. MacEachern and Gingras, 2007). In other words, workers must establish a case for relatively lower salinity as opposed to determining absolute salinity. This practice is important. The basin from which the "marine" ichnological suite is defined, may itself have contained brackish water (e.g., the Baltic Sea; Virtasalo et al., 2011), and the salinity of the basin may decrease rapidly into the adjoining bays and estuaries.

2.2 Tides and Trace Fossils

A growing dataset of tidally generated sedimentary structures has also been used to establish a link between trace-fossil assemblages and estuary deposits. Relying on the work of sedimentologists studying tidal deposits, ichnological workers were able to make fundamental observations that linked trace fossil distributions to the presence of tidal bundles and a broad range of rhythmic sedimentary patterns generally referred to as tidal rhythmites (*sensu* Dalrymple et al., 1991; Kvale et al., 1989; Rahmani, 1988; Williams, 1989). Examples of this integration include Bann et al. (2004), Buatois et al. (1998), MacEachern and Gingras (2007), MacEachern and Pemberton (1994), and MacEachern et al. (1992).

Recently, there has been increased interest in the use of trace fossils to identify tidally influenced sedimentation (e.g., Desjardins et al., 2012; Gingras and MacEachern, 2012). Tides impose several important characteristics within estuaries, including fluctuating depositional energies, landward transport of marine-derived food resources, and the mixing of marine and continental waters (discussed previously). These characteristics influence the distribution of infauna and, in some cases, promote key morphological aspects of the preserved traces.

The record of regularly fluctuating depositional energies is preserved in tidal settings as rhythmic bedding. Similarly, the infilling of trace fossils can also result from tidal sedimentation referred to as tubular tidalites by Gingras and MacEachern (2012) (Fig. 4B). Tubular tidalites contain several sedimentary couplets that show regular waxing and waning of depositional currents, as well as gradual thickening and thinning of the infill laminations. It is important to note that complete neap-spring cycles are generally not visible in core or outcrop examples; thus the identification of tubular tidalites is challenging. Generally, the fewer the number of preserved laminae, the greater the likelihood of misinterpretation.

It is unclear whether specific animal behaviors, as reflected by ichnogenera, are directly influenced by tidal currents. For example, it is tempting to infer that the U-shape of *Arenicolites*—a trace fossil that is common in tidally influenced environments—facilitates feeding and defecation in switching tidal currents. Evidence for such rheotactic behavior, however, requires a substantial dataset, and in the case of tidal currents and burrow alignment, this has yet to be demonstrated from modern *Arenicolites* tracemakers.

The comparably rich food resource present in tidal settings has a great influence on trace distributions in estuaries. Food is concentrated in estuaries in two ways: (1) suspension settling during slackwater periods, when plankton (i.e., algae, diatoms, phytoplankton) are able to settle on the estuary floor and on intertidal flats; and (2) straining of food particles (e.g., algae, diatoms, larvae) in intertidal positions, where water drains through the sediment during the final

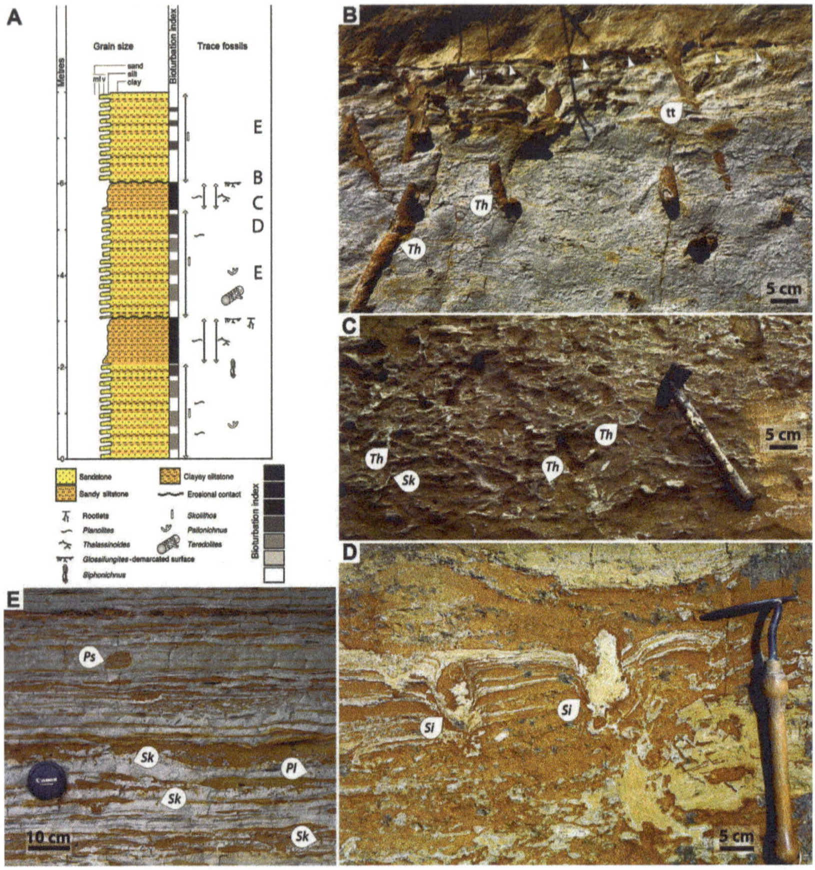

FIGURE 4 Pleistocene examples from Willapa Bay, Washington, USA (see Gingras et al., 2000). (A) Typical vertical succession observed in a tidally influenced bay and in estuary settings (e.g., Anima et al., 1989; Clifton, 1983; Gingras et al., 1999; Pearson and Gingras, 2006). (B) An erosional channel base (white arrows) demarcated by a suite attributable to the *Glossifungites* Ichnofacies, comprising firmground *Thalassinoides* and a tubular tidalite infill (tt). (C and D) Moderately to intensely bioturbated intertidal strata: in these examples, the outer tidal flat is characterized by *Siphonichnus* (*Si*) and the inner tidal flat by *Thalassinoides* (*Th*). (E) Subtidal laterally accreted units. These beds are characteristically rarely burrowed; however, moderate bioturbation may be present with either the sand- or the mud-dominated members of the IHS. In this example, *Psilonichnus* (*Ps*) and *Skolithos* (*Sk*) are indicated.

stages of the falling tide (summarized in Gingras and MacEachern, 2012). This and other factors lead to a preponderance of surface and subsurface deposit-feeding and infaunal lifestyles (discussed above) and also result in a characteristic distribution of trace fossils (Fig. 4A). Coupled to this, due to the presence of stronger currents in the deeper parts of channels, tidally influenced bars (i.e., point bars and longitudinal bars) characteristically display an increasing-upward bioturbate character (Gingras and MacEachern, 2012; MacEachern and Gingras, 2007).

Thomas et al. (1987) presented conceptual interpretations for a range of inclined sand/mud alternations, referred to as IHS. Subsequent studies have shown a strong association between IHS and tidally influenced bar deposits (Fenies and Faugeres, 1998; Geier, 1995; Gingras et al., 1999; Hovikoski et al., 2008; Leckie and Singh, 1991; Sisulak and Dashtgard, 2012; Van den Berg, 1981). The presence of brackish-water trace-fossil assemblages coupled with IHS (Figs. 3D and 4E) has, since, become virtually synonymous with sedimentation within estuaries (see MacEachern and Gingras, 2007 for a review). This is, of course, an oversimplification, as similarly bioturbated IHS is present in the distal parts of delta distributary channels (e.g., Sisulak and Dashtgard, 2012) and in other tidal channels as well (e.g., Choi et al., 2004; Dalrymple and Choi, 2007; Pearson and Gingras, 2006).

2.3 Ichnological Evidence for Transgressive Incised Valley Fills

Trace fossils have proven to be useful in sequence-stratigraphic studies of incised valleys. Discussion of ichnological applications in sequence-stratigraphic analysis has centered around the use of substrate-controlled ichnofacies as a means to identify and interpret the origin of strata-bounding discontinuities. This includes occurrences of *Glossifungites* Ichnofacies-demarcated discontinuities and, to a lesser degree, omission suites attributable to the *Teredolites* and *Trypanites* ichnofacies (Figs. 4B and 5A–C). The conditions for the generation of omission suites marking stratigraphic discontinuities and their association with transgressed sequence boundaries (transgressive surfaces/composite subaerial unconformity, TS/SU) are outlined in MacEachern et al. (2012) and are not recounted here. The established association of both the *Glossifungites* and *Teredolites* ichnofacies with TS/SU is of importance because these stratigraphic surfaces—if they are mappable and suggest a valley form—may represent the floor and margins (i.e., the container) of the estuary. However, where there are true freshwater fluvial deposits within the valley fill, the container will not be marked by omission suites of the *Glossifungites*, *Teredolites*, or *Trypanites* ichnofacies.

Another application of estuary-associated trace-fossil assemblages is to demonstrate the persistent transgressive nature of the estuary fill (summarized in Fig. 5). Taken as a whole, trace fossils provide the required information to determine whether stratigraphic successions have a transgressive or a regressive character (e.g., Bann et al., 2004; MacEachern and Pemberton, 1994; MacEachern et al., 2010; McIlroy, 2007; McIlroy et al., 2005; Pemberton et al., 1992). Recognizing the stacking pattern of estuary deposits hinges on the workers' ability to establish proximal and distal trends based on trace-fossil size and diversity and to differentiate ichnogenera that are consistent with marine, brackish-water, and freshwater settings (see discussion above). Although persistently

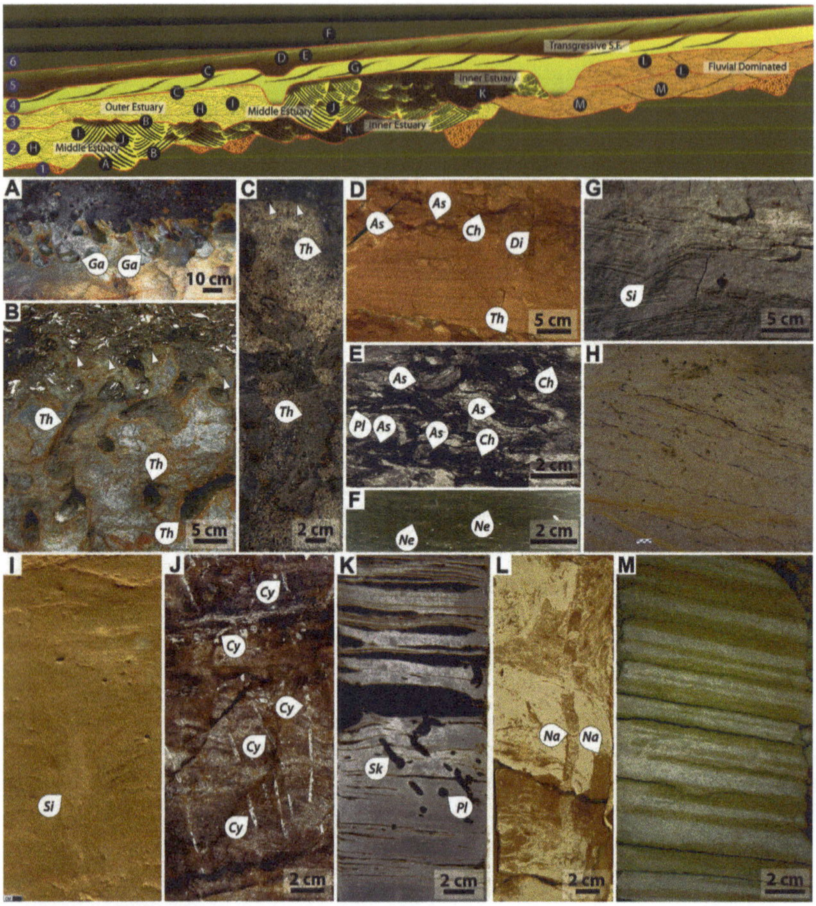

FIGURE 5 Stratigraphic application of trace fossils in estuaries (left is oceanward and right is landward). The uppermost panel shows a schematic interpretation of a transgressively filled estuary. This differs from the models presented by Dalrymple et al. (1992) and Martinius and Van den Berg (2011) in that it specifically recognizes bioturbated IHS belts and assigns them to the inner estuary. The schematic model also attempts to encapsulate a backstepping stratigraphic framework. Genetic stratigraphic units are bound by red lines (TS/SU) and six phases of transgression are shown (numbered 1 through 6 with blue markers on left of schematic). Unit 1 is fluvial. Units 2 and 3 are estuarine. Units 4–6 are shoreface through to offshore. The facies backstep and so the middle estuary zone is labeled in phases 2 and 3 for reference. Images (A) through (M) are indicated in their schematic position by their letter on the black circles. (A) *Gastrochaenolites* (*Ga*) at the base of a channelized Pleistocene TS/SU (Coos Bay, Oregon, USA). (B) *Thalassinoides* (*Th*) at the base of a tidal channel (Pleistocene, Willapa Bay, Washington, USA). (C) *Thalassinoides* (*Th*) at the base of a shoreface-associated marine flooding surface (WRS/SU) (Centenario Formation, Cretaceous, Argentina). (D) Proximal offshore characterized by hummocky cross-stratification (HCS), *Asterosoma* (*As*), *Diplocraterion* (*Di*), *Thalassinoides* (*Th*) and *Chondrites* (*Ch*) (Clearwater Formation, Cretaceous, Alberta, Canada). (E) Offshore characterized by HCS, *Asterosoma* (*As*), *Planolites* (*Pl*), and *Chondrites* (*Ch*) (Jurassic, Germany). (F) Distal offshore characterized by *Nereites* (*Ne*) (Jurassic, Germany). (G) *Siphonichnus* (*Si*) with small-scale HCS, interpreted as shoreface

subaerially exposed coastal deposits (i.e., salt-marsh, soil-forming, or paludal environments) may be highly burrowed (e.g., Dashtgard and Gingras, 2005a), their associated fluvial deposits are normally unburrowed.

Many marine trace fossils do not typically occur in brackish-water environments. The ability to identify characteristic "marine" trace fossils, that are not associated with brackish water, is critical. Such trace fossils include all graphoglyptids, and the common marine ichnogenera *Nereites*, *Zoophycos*, *Spirophyton*, *Chondrites*, *Rhizocorallium*, *Asterosoma*, *Phoebichnus*, robust *Rosselia*, and large-diameter *Ophiomorpha*. At least since the Mesozoic, brackish-water strata commonly contain *Planolites*/*Teichichnus*-dominated assemblages, *Cylindrichnus*/*Skolithos*-dominated assemblages, horizons dominated by *Gyrolithes*, or zones containing abundant *Arenicolites*. *Siphonichnus* can be locally common. In more distal locales (i.e., closer to the "marine" basin), brackish-water deposits may contain comparably small examples of *Rosselia*, *Thalassinoides*, or *Ophiomorpha*. In proximal positions (i.e., within the inner estuary), the trace-fossil assemblage tends to be progressively dominated by *Planolites* and/or small *Skolithos*. Although subaqueous fluvial deposits are normally unburrowed, trace fossils such as rare *Planolites*, *Skolithos*, *Camborygma*, and *Protovirgularia* may be observed. Additionally, Buatois et al. (1997) documented the presence of insect tracks on the intertidal portion of a tidally influenced fluvial bar in Carboniferous strata of the Tonganoxie Sandstone, Kansas, USA.

By recognizing the proximal and distal shifts between stratigraphic levels, a framework for overall transgression or regression can be established. In transgressive regimes, comparably distal trace-fossil suites are positioned immediately above proximal assemblages. In marginal-marine settings, trace-fossil suites reflect highly variable depositional conditions and heterogeneous distributions of infauna (e.g., Dashtgard, 2011a,b). Correspondingly, the sequence-stratigraphic resolution resulting from this ichnological approach can be of high fidelity.

3. TRACE DISTRIBUTIONS WITHIN WAVE- AND TIDE-DOMINATED ESTUARIES

The distribution of food resources and salinity within estuaries is passively related with one another through their codependence on the dynamics of the tidal or tidal–fluvial system. Research in modern estuaries suggests that, in these settings, the distribution of food resources and brackish water, coupled

---

(McMurray Formation, Cretaceous, Alberta, Canada). (H) Current reversals on sand waves from a tidal-inlet deposit (Pleistocene, Willapa Bay, Washington, USA). (I) *Siphonichnus* (*Si*) in a sand-dominated tidal-bar deposit (Pleistocene, Willapa Bay, Washington, USA). (J) *Cylindrichnus* (*Cy*) in a sand-dominated tidal-bar deposit (McMurray Formation, Cretaceous, Alberta, Canada). (K) *Skolithos* (*Sk*) and *Planolites* in a sand-mud IHS from an inner estuary tidal-bar deposit (McMurray Formation, Cretaceous, Alberta, Canada). (L) Illuviated, pedogenically altered medium with insect-generated trace fossils *Naktodemasis* (McMurray Formation, Cretaceous, Alberta, Canada). (M) Cross-bedded fluvial sandstone (Centenario Formation, Cretaceous, Argentina).

with grain-size and sedimentation rates, overwhelmingly dictates the resulting longitudinal distributions of tracemaking organisms and their biogenic structures (e.g., Gingras et al., 1999; Hauck et al., 2009; Hertweck, 1992; Hertweck et al., 2007; Howard and Frey, 1975; Hubbard et al., 2004; Noffke et al., 2009; Fig. 3). This is in marked contrast with the dominant controls on ichnology that prevail in more wave-exposed settings (i.e., the shoreface), wherein infaunal distributions are mainly influenced by the magnitude of wave energy and frequency and magnitude of storm events, as well as sedimentation rate, grain-size availability, and grain-size distributions (Pemberton et al., 2012).

Below, we draw from modern and ancient examples of tidally influenced, marginal-marine depositional environments, with increasing tidal energy relative to wave and fluvial energy. From the modern, we consider a wave-dominated estuary (Kouchibouguac, Canada), mixed-energy estuaries (Willapa Bay and the Ogeechee River Estuary, USA), and a tide-dominated estuary (Chignecto Bay, Canada), where discrete ichnological distribution patterns can be discerned. From the rock record, we compare the wave-dominated Viking Formation (Albian) of Alberta, Canada; a mixed(?) estuary from the Montney Formation (Triassic) of Alberta, Canada; and a tide-dominated(?) example from the McMurray Formation (Aptian to Albian) of Alberta, Canada.

Included in the depositional environments are a range of tidally influenced subenvironments, including tidal flats, tidal channels, shoals, ebb- and flood-tidal deltas, and salt marshes. It is emphasized here that these examples show only a small part of the broad range of estuary occurrences. The character of the estuary is heavily influenced by wave energy, volume of the tidal prism, fluvial flux, and the available sediment sources. As such, although the examples below likely provide useful comparisons, they fall short of defining an all-encompassing "estuary ichnology".

3.1 Wave-Dominated Estuaries

Wave-dominated estuaries constitute one end member in the continuum of estuary types proposed by Dalrymple et al. (1992). Wave-dominated estuaries typically display a well-demarcated tripartite facies zonation (inner, middle, and outer estuary, or alternatively, bay-head delta, central basin, and estuary mouth) recognized by earlier workers (e.g., Dalrymple et al., 1992; Reinson, 1980; Reinson et al., 1988).

Bay-head delta deposits encompass sediments supplied to the estuary via fluvial influx. Generally, the low wave energies and very weak tidal flux in the central basin result in the bay-head delta developing a river-dominated, digitate morphology. Distributary channels are locally developed in association with the delta front. Under conditions of low fluvial flux, the tidal flow may impinge landward of the bay-head delta, but for most wave-dominated estuaries, this effect is minimal.

Central-basin deposits are laid down in relatively shallow, standing bodies of water, which receive sediment both from the river at the landward end and from tidal exchange at the mouth of the system. Such bodies of water tend to be fairly small so that fair-weather waves and storm waves are less effective at reworking the substrate. As such, central basins tend to be characterized by muddier deposits, and sand interbeds—which are associated with the flood-tidal delta, deposited during river-flood events, or generated during storms—are predominantly oscillation ripple laminated, with rare current ripples and thin beds of low-angle undulatory parallel lamination (micro-HCS). Tidal ranges are typically low in such settings so that, although tidal flats may form along bay margins, they tend to be thin and their deposits difficult to recognize.

Some wave-dominated estuaries are essentially filled, such that their central basin consists of salt marshes that are dissected by tidal channels. The salt-marsh deposits are preserved in the geological record as illuviated, root-bearing strata that cap burrowed subtidal facies. The salt marsh can be thick, because of vertical aggradation due to accumulation during relative sea-level rise.

Estuary-mouth deposits are sand dominated. Wave-dominated estuaries develop on strongly wave-swept coasts, where barrier complexes can be readily established and maintained. The tidal energy is generally sufficient to breach the barrier locally, forming tidal inlets. The tidal energy is high where it is constricted through the inlet but dissipates rapidly as it enters the central basin. This decelerating flow commonly builds a flood-tidal delta complex. Intermittent storms may result in washover events over the barrier into the central basin.

The above tripartite zonation is the result of pronounced energy partitions that yield discrete substrate characteristics and are considered diagnostic of wave-dominated estuaries. The clear relationship between resultant sediment calibers and sedimentary process is extensively discussed in the geological literature (e.g., Dalrymple et al., 1990, 1992; Roy et al., 1980; Visher and Howard, 1974), as is the distribution of trace fossils (MacEachern and Gingras, 2007; MacEachern and Pemberton, 1994; Pemberton et al., 1992). The relatively low volumes of water exchanged in wave-dominated estuaries—due to a small tidal prism—greatly influence the biological dynamics of microtidal bays within the estuary. Microtidal-bay water may be prone to partial thermal and salinity stratification; thus, bottom waters can become eutrophic during neap tides. Freshets (seasonal high-discharge, fluvial events) can markedly freshen the bay, as freshwaters are sequestered within the middle, lower-energy parts of the bay. These factors combine to make conditions unattractive for long-term colonization, and intervals of unburrowed sediments in these settings are therefore common (Gingras et al., 1999; Hauck et al., 2009).

3.2 Wave-Dominated Estuary Case Study: Kouchibouguac Bay, New Brunswick, Canada

Kouchibouguac Bay is situated on the Northumberland Strait in the southern Gulf of St. Lawrence. Kouchibouguac Bay comprises 29 km of arcuate barrier islands fronting several estuaries (Fig. 1B). The climate of New Brunswick is cool temperate. The lagoons and estuaries represent preglacial valleys drowned during the Holocene transgression. Kouchibouguac Bay is microtidal, with mean and maximum tidal ranges of 0.67 and 1.25 m, respectively (Davidson-Arnott and Greenwood, 1974; McCann and Bryant, 1972).

3.2.1 *Valley Margins and Substrate-Controlled Suites*

In Kouchibouguac Bay, animals inhabit exposed cohesive lagoonal and salt-marsh deposits. A depauperate assemblage of burrowers is present therein, producing burrow suites that are analogous to the *Glossifungites* Ichnofacies. The two dominant burrowers in muddy firmgrounds are the bivalve *Petricola pholadiformis* (Fig. 6A) and the very small polychaete *Polydora ligni*. The lobster *Homarus americanus* has also been observed, producing large-diameter firmground tunnels in channel flanks and bases. A bay-margin assemblage attributable to the *Glossifungites* Ichnofacies is not observed.

3.2.2 *Trace-Fossil Distributions by Subenvironment*

Hauck et al. (2009) show that the microtidal wave-dominated estuary at Kouchibouguac Bay displays a predictable ichnological distribution along its length. Within the inner estuary, salinities range between 1 and 10 psu, and the ichnological character comprises a low-diversity assemblage of biogenic structures, mainly *Skolithos*, *Palaeophycus*, and *Arenicolites* (Fig. 6E and F). Moreover, endobenthos in the inner estuary dominantly reside in channel-bar tops and in small intertidal flats developed along the banks of the brackish-water reach of the river (Fig. 6E and F). Subtidal settings of the inner estuary are characteristically unburrowed into the fluvial reaches.

Toward the middle part of the Kouchibouguac estuary, the ambient salinity is approximately 25 psu and a moderately diverse suite of traces is present, including *Psilonichnus*, *Gyrolithes*, *Planolites*, *Palaeophycus*, *Thalassinoides*, and *Siphonichnus*. Bioturbation in the middle estuary is sporadically distributed, ranging from BI = 1–3 in the tidal channels to BI = 3–5 in the subtidal flats (Fig. 6C and D). Notably, subtidal flats represent the spatially dominant part of the middle estuary, suggesting that bioturbated media of low to moderate trace diversity constitute the main sedimentary facies therein.

In the outer estuary (the most marine-influenced part of the system), animal distributions are very patchy (BI = 1–3) (Fig. 6B). Ten "ichnogenera" are observed, with *Skolithos*, *Siphonichnus*, *Arenicolites*, and *Polykladichnus* being generally dominant. Cryptobioturbation and equilibrichnia are also common, with *Psilonichnus*, *Planolites*, *Thalassinoides*, and *Palaeophycus* present more rarely.

3.3 Mixed-Energy Estuaries

Mixed-energy estuaries can be both strongly wave- and tide-influenced with variable fluvial influence. A reduced tide *versus* wave energy promotes effective estuary mouth-bar development and limits tidal exchange between the

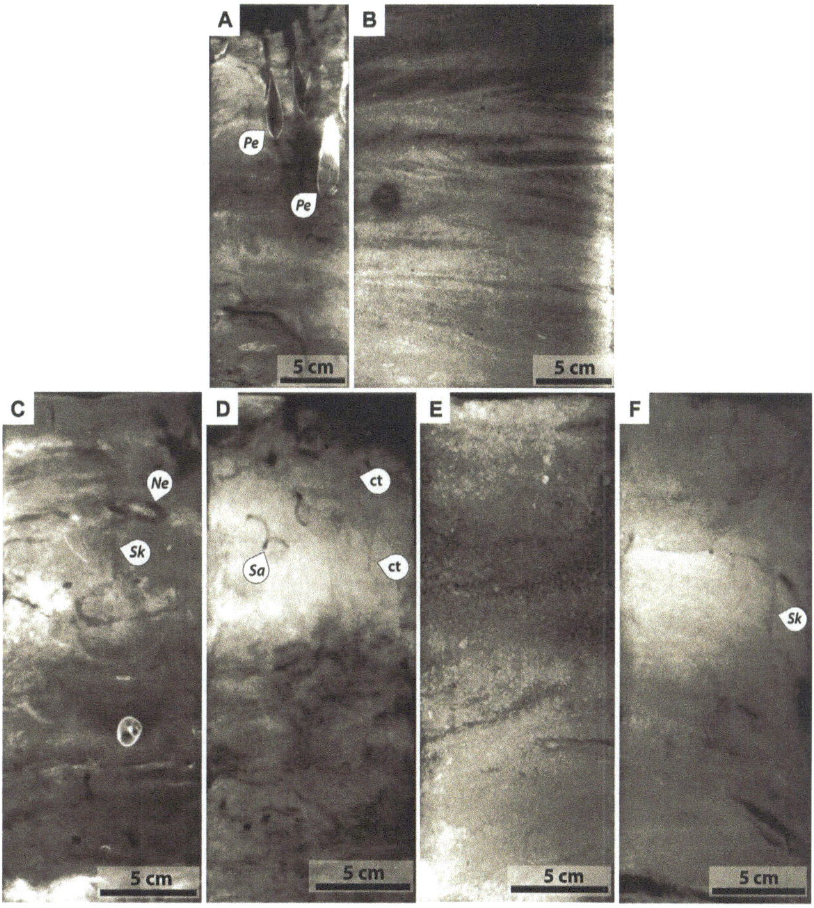

FIGURE 6 X-radiographs of modern estuary deposits in Kouchibouguac Bay, New Brunswick, Canada. (A) The *Gastrochaeonlites*-shaped traces are associated with salt-marsh (bay-margin) firmgrounds. The tracemaker is *Petricola pholadiformis* (*Pe*). (B) Herringbone cross-lamination in a tidal-inlet deposit. (C) Nereid polychaetes (ne) contribute to the emplacement of *Skolithos*, *Polykladichnus*, and *Palaeophycus* in otherwise bioturbated middle estuary deposits. (D) Highly burrowed middle estuary with *Saccoglossus* (*Sa*) and capitellid burrows indicated. (E) Cross-bedded sand from a fluvial influenced inner estuary channel. (F) Burrowed bar top (*Arenicolites* and *Skolithos* visible) from the inner estuary.

central basin and the ocean or seaway. Higher tidal energies induce an increased hydraulic exchange between the bay and the open ocean, such that barriers across the mouth of the estuary are dissected by tidal inlets.

Within mixed-energy estuaries, the increased importance of tidal processes can have the effect of substantially attenuating the salinity within the inner and middle estuary. Sedimentological and ichnological characteristics are, therefore, spread over larger geographical zones, particularly parallel to the axis of the estuary, and with increasing tidal influence, discrete sedimentary and biological boundaries become uncommon. As such, clear tripartite divisions are not developed, and the transitions through the inner, middle, and outer estuary are gradational. Nevertheless, these estuaries also display discernible physiographic zones.

The inner estuary is fluvially influenced and receives most of its sediment from the fluvial reaches. The tidally influenced channels may possess a meander form, and they expand in width and depth basinward. Salinities within the inner estuary are typically very low.

The middle estuary receives sediment from both the estuary mouth and the fluvial end of the system. As a result of tidal/fluvial interactions, facies successions of the transition zone from the inner estuary to the middle estuary are markedly heterolithic. The channels' meander form is attenuated in the middle estuary, and their channel widths and depths continue to increase basinward. Salinities tend to be variable, depending on the fluvial flux into the estuary. Within the middle estuary, the tidal-current energies tend to be low to moderate compared to the outer estuary. This permits the emplacement and preservation of a range of biogenic sedimentary structures.

The outer estuary is characterized by wide and deep, sand-dominated tidal channels. The channels are gently sinuous to straight, and the dominant mode of sediment storage is as tidal dunes and longitudinal tidal bars. Most of the sediment is derived from the tidal inlet. Although salinities can be high, energetic tidal currents inhibit sediment colonization and trace preservation.

3.4 Mixed-Energy Example: Willapa Bay, Washington, USA

Willapa Bay is located in the southwestern corner of Washington, USA. The bay is separated from the Pacific Ocean by a 27-km long spit (North Beach Peninsula, Fig. 1D). Owing to sediment supplied from the Columbia River, the spit is progradational and is constructed by high-energy waves of the Pacific Ocean. Willapa Bay is a mesotidal estuary, with a tidal range of 2–3.4 m. The local climate is temperate.

The estuary sits within a Pleistocene-aged incised valley, entrenched into Eocene basalts and sandstones (summarized in Anima et al., 1989; Clifton, 1983; Clifton et al., 1976). A stratigraphic record of three or more stacked incised valleys is preserved within the main valley. Substantive stratigraphic, sedimentological, and ichnological studies have been conducted at Willapa Bay, and the area is particularly well known.

3.4.1 Valley Margins and Substrate-Controlled Suites

A striking heterogeneity in omission suites attributable to the *Glossifungites* Ichnofacies is documented from these modern firmgrounds (Fig. 7A and B). This variability is related to intertidal zonation, sediment texture, the absolute firmness of the firmground (substrate consistency), and the presence or absence of a soft-sediment veneer (Gingras et al., 2001).

In the middle estuary, valley-margin expansion is ongoing; wave and tidal erosion are continually exposing compacted Pleistocene strata. These locales typically reside within the intertidal zone, and a zonation of firmground burrowers is readily observed. Shallow subtidal and lower to middle intertidal firmgrounds are preferentially colonized by crustaceans or pholadiid bivalves. These surfaces characteristically possess firmground *Thalassinoides*-, *Gastrochaenolites*-, and rare *Psilonichnus*-like traces. The upper intertidal zone is dominated by small polychaetes (*Polydora*), which produce diminutive *Diplocraterion*- and *Arenicolites*-like traces (Fig. 7A). Firm, sand-dominated substrates contain *Gastrochaenolites*-like burrows (Fig. 7B). Finally, a *Thalassinoides*-dominated suite is locally observed at the base of tidal channels (e.g., Fig. 3B). In Pleistocene units associated with the bay, nearly every observed TS/SU is demarcated by *Thalassinoides*, *Psilonichnus*, or, more rarely, *Gastrochaenolites*, attributable to the *Glossifungites* Ichnofacies (Gingras et al., 1999).

3.4.2 Trace-Fossil Distributions by Subenvironment

Willapa Bay displays notably different trace-fossil distributions than are observed in purely wave-dominated estuaries (Fig. 3). The fluvial reaches and bay-head delta are gravel- and sand-dominated and are typically unburrowed. Owing to a low degree of fluvial influence, the inner estuary exhibits salinities between 0 and 17 psu and is characterized by 1–3 m thick, mud-dominated, subtidal point-bar deposits. The point bars contain thinly bedded, mud-dominated (70–95% mud) IHS, which contain biogenic sedimentary structures comparable to *Palaeophycus*, *Polykladichnus*, *Arenicolites*, *Diplocraterion*, and *Skolithos*. Burrows are generally small (<2 mm in diameter), and animal distributions are patchy, leading to spatially variable bioturbation (BI = 1–3, Fig. 7C). Bioturbation is present down to the channel thalweg. Moderately burrowed (BI = 3–4) intertidal-flat deposits overlie the point-bar deposits of the inner estuary, with suites dominated by small (<2 mm in diameter) *Skolithos*, *Arenicolites*, and *Diplocraterion*. Bioturbation intensities increase upward from BI = 1–2 in the lower point bar to BI = 3–4 at the bar top.

In the middle estuary, salinities are generally 20–25 psu and traces such as *Skolithos*, *Planolites*, *Palaeophycus*, *Thalassinoides*, *Siphonichnus*, *Arenicolites*, *Cylindrichnus*, and *Piscichnus* are present (Fig. 7D). Traces are larger (2–20 mm in diameter) than those of the inner estuary. The intertidal flats of the middle estuary are typically intensely bioturbated (BI = 5–6). Due to physical reworking in the channels, the abundance of traces and animal diversities is markedly lower than in the intertidal flats (BI = 1–2). The subtidal deposits are dominated by *Thalassinoides*, *Skolithos*, *Planolites*, *Teichichnus*, and shallow motility traces of bivalves and, potentially, cryptobioturbation. Bioturbation

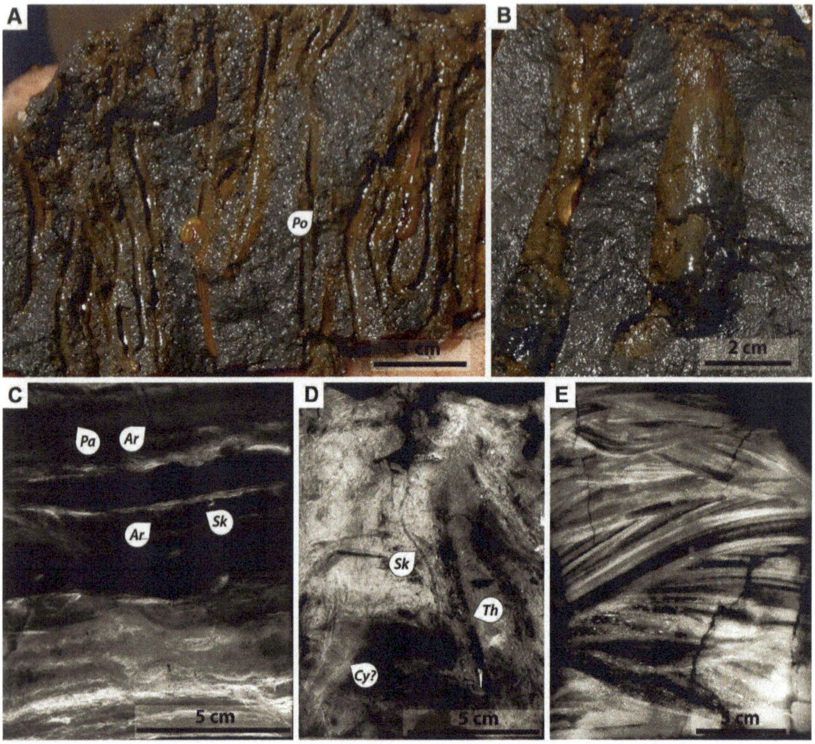

FIGURE 7 Photos (in color) and X-radiographs (gray scale) of modern estuary deposits of Willapa Bay, Washington, USA. (A) *Arenicolites* and *Diplocraterion* in a Pleistocene firmground. The tracemaker *Polydora* is indicated as *Po*. (B) *Gastrochaenolites*-shaped trace in a Pleistocene firmground. The tracemaker was *Petricola pholadiformis*. (C) Inner-estuary bioturbate texture showing a low-diversity and diminutive trace assemblage dominated by *Palaeophycus* (*Pa*), *Arenicolites* (*Ar*), and *Skolithos* (*Sk*). (D) The middle estuary shows larger trace fossils such as *Skolithos* (*Sk*), *Thalassinoides* (*Th*), and *Cylindrichnus* (*Cy*). (E) In the channels of the outer estuary, cross-bedded sands are the dominant lithofacies.

intensities increase upward from BI = 1–2 in the lower point bar to BI = 5–6 at the bar top.

At Willapa Bay, the outer estuary salinities are essentially fully marine (27–30 psu). Nevertheless, rapid sediment transport and bedform migration resulting from high-wave and tidal-current energies tend to inhibit infaunal colonization, such that the density of biogenic structures in the outer estuary is surprisingly low (generally BI = 0–1, rarely BI = 2; Fig. 7E). Sheltered tidal-flat deposits situated near the bay mouth are similar to those of the middle estuary (i.e., intensely bioturbated; BI = 5–6) but are invariably dominated by the *Thalassinoides*-making sand shrimp *Neotrypaea californiensis*. In contrast, the surface sediments on tidal flats formed along the exposed spit platforms are too unstable (rapidly shifting) for most infauna. As such, most units are essentially unburrowed or support, at most, a low-density population of sand shrimp or razor clams. In contrast, opheliid polychaetes, such as *Ophelia* and *Euzonus*, can thrive in this zone, and thus the trace fossil *Macaronichnus segregatis* may be present.

3.5 Mixed-Energy Example: Ogeechee River Estuary, Georgia, USA

The Ogeechee River–Ossabaw Sound estuary is partially barred, with a well-developed tidal inlet (Fig. 1C). Owing to a large tidal prism, wave-generated shoals at the mouth of the bay are dissected by deep and broad tidal channels. Mean high tide is approximately 2.1 m. The climate is warm temperate.

Dorjes and Howard (1975) reported a mappable, tripartite zonation within the Ogeechee River–Ossabaw Sound system. Follow-up work by the authors of this chapter expands on Dorjes and Howard's observations. As with the Willapa Bay example, the transitions through the inner, middle, and outer estuary are gradational.

3.5.1 Valley Margins and Substrate-Controlled Suites

In the Ogeechee River estuary, animals inhabit exhumed, cohesive salt-marsh deposits. A depauperate suite of biogenic structures is present, including *Psilonichnus* (constructed by the shore crab, *Uca* sp.) and small *Arenicolites* (generated by *Corophium volutator*). Throughout the estuary, tidal channels cut deeply into very firm Pleistocene muds or muddy sands. Pleistocene firmgrounds are consistently burrowed by *Corophium* and *Polydora*, which construct *Arenicolites*- and *Diplocraterion*-like burrows into the cut banks and bases of channels of the inner and middle estuary. Similar settings and firmground exposures in the outer estuary are dominated by *Thalassinoides* (Fig. 8).

3.5.2 Trace-Fossil Distributions by Subenvironment

As with the other modern examples, the inner estuary of the Ogeechee is flanked by expansive salt-marsh deposits. In the Georgia coastal area, the salt marshes commonly contain *Psilonichnus* produced by shore crabs. Here, salinities vary between 1 and 15 psu. The intertidal sediments are highly burrowed (BI = 3–5). Much of the sediment reworking in this zone is due to amphipods, which produce cryptobioturbation (Howard and Frey, 1975). *Skolithos* is also common and occurs in high densities in lower intertidal and subtidal zones. Low sedimentation rates permit moderate degrees of bioturbation on the subtidal point bars (approximately BI = 2–3), but near the channel thalweg bioturbation is sparse (BI = 0–1).

There is a marked difference in the degree of bioturbation between point-bar (BI = 1–3) and intertidal-flat deposits (BI = 3–5) within the middle estuary, where salinities generally range from 15 to 25 psu. An increase in the diversity of infauna

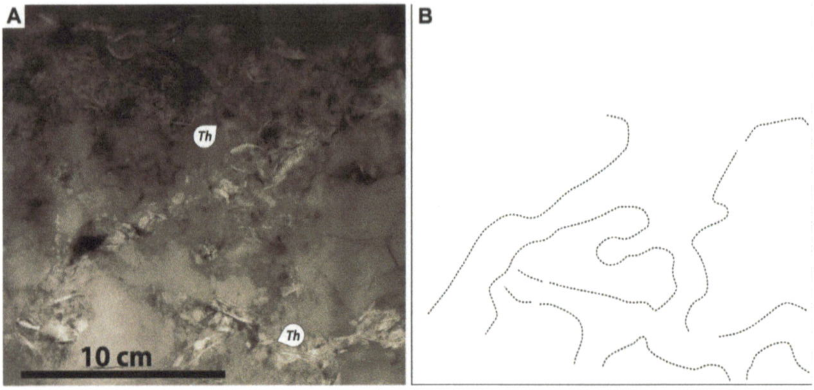

FIGURE 8 (A) X-radiograph of a modern estuary-mouth deposit of Ogeechee River, Georgia, USA. In this area, the tidal-inlet channel has cut down into palimpsest muddy sands, and a firmground *Thalassinoides*-dominated suite attributable to the *Glossifungites* Ichnofacies is present. Shell detritus infills the shrimp burrows. (B) Line tracing of the burrows.

is denoted by the presence of *Thalassinoides*-, *Skolithos*-, *Palaeophycus*-, and *Planolites*-like structures (e.g., Fig. 2E), as well as cryptobioturbation.

The outer estuary of the Ogeechee system is dominated by polyhaline waters and contains the highest diversity of infaunal species. The major bioturbating elements of the community are capable of producing cryptic mottling, as well as biogenic sedimentary structures including *Skolithos*, *Thalassinoides* (Fig. 7), *Palaeophycus*, *Planolites*, *Arenicolites*, *Diplocraterion*, *Gyrolithes*, *Polykladichnus*, and *Chondrites*. Despite the heightened diversity of the community, the outer estuary shows reduced bioturbation intensities overall, which is attributed to increased rates of sediment deposition and bedform migration. The degree of bioturbation is lower in both the intertidal flats (BI = 2–3) and subtidal environments (BI = 1–2).

3.6 Tide-Dominated Estuaries

A result of the presence of a very large tidal prism is that tide-dominated estuaries tend to be large and that their physiographic boundaries are greatly attenuated (Dalrymple et al., 1992). Facies changes are gradational over a scale of several kilometers to tens of kilometers. Saltwater incursion can extend over long distances inland—although this is greatly influenced by fluvial discharge—and vigorous tidal currents can effectively disrupt the vertical stratification of the water column within the estuary.

Most commonly, the inner estuary of tide-dominated estuaries is dominated by salt marshes, muddy intertidal substrates, and variably muddy or sandy subtidal sediments. Oligohaline or mesohaline waters are typically present. Muddy intertidal flats are commonly bioturbated; however, high current energies within the channels tend to limit the degree and diversity of bioturbation in the subtidal inner estuary.

The middle parts of tide-dominated estuaries characteristically display a variably muddy intertidal zone transitional with a sand-dominated outer intertidal flat. The subtidal channels are likely to be sand dominated—large longitudinal tidal sand-bars are common within and between channels—but mud can still be transported seaward from the inner estuary. The water is mesohaline. As with the inner estuary, the subtidal deposits are typically unburrowed. Owing to the presence of waves and variable tidal currents, the sandy intertidal flats and estuary-margin shorefaces are sporadically burrowed, although bioturbation is pervasive in the often muddy bay-margin zones.

The outer estuary is broadly similar to the middle estuary, but is characterized by broad and deep tidal channels, wherein bioturbation has a low preservation potential. Under conditions of persistent exposure to waves, the intertidal flat is replaced by estuary-margin shoreface deposits (Dalrymple et al., 2011). Estuary-margin shorefaces and intertidal flats are, nevertheless, sporadically bioturbated.

3.7 Tide-Dominated Example: Chignecto Bay, Bay of Fundy, Atlantic Canada

Chignecto Bay is one of two subbasins at the head of the Bay of Fundy. It is a tide-dominated estuary with a tidal range of 10–13 m (Dashtgard et al., 2007; Desplanque and Mossman, 2001) and occurs in a cool temperate climatic zone (Fig. 1A). Due to the tidal mixing in the channels, the salinity is generally between 15 and 20 psu. The depositional history of Chignecto Bay records the postglacial history of the bay over the past 13.5 ka. The maximum lowstand was reached approximately 6.5 ka ago, at which point Chignecto Bay was a wave-dominated estuary subjected to microtidal conditions (Amos et al., 1991; Dashtgard et al., 2007). Tidal amplification occurred over the past 7 ka (Grant, 1970; Scott and Greenberg, 1983; Shaw et al., 2002), and strongly macrotidal to megatidal conditions have only persisted for the past 4 ka (Scott and Greenberg, 1983).

3.7.1 Valley Margins and Substrate-Controlled Suites

Chignecto Bay is currently undergoing a transgression, mainly resulting from tidal amplification in the inner Bay of Fundy. During maximum lowstand, Chignecto Bay was barred and the system corresponded to a wave-dominated estuary with a quiet central basin, wherein mud was deposited in predominantly subtidal positions (Dashtgard et al., 2007). When the relative sea level rose and tidal amplification led to transgression of the gravel barrier at the mouth of the bay, the system evolved from a wave-dominated estuary to a tide-dominated estuary. Mud deposits were subjected to erosion and served to provide the main source of fine-grained material to the tidal flats in the middle and inner estuary (Amos, 1987; Amos and Asprey, 1979; Amos et al., 1991).

Progradation of the tidal flats and salt marshes is common in the inner and middle estuary, as an abundant sediment supply coupled with flow deceleration

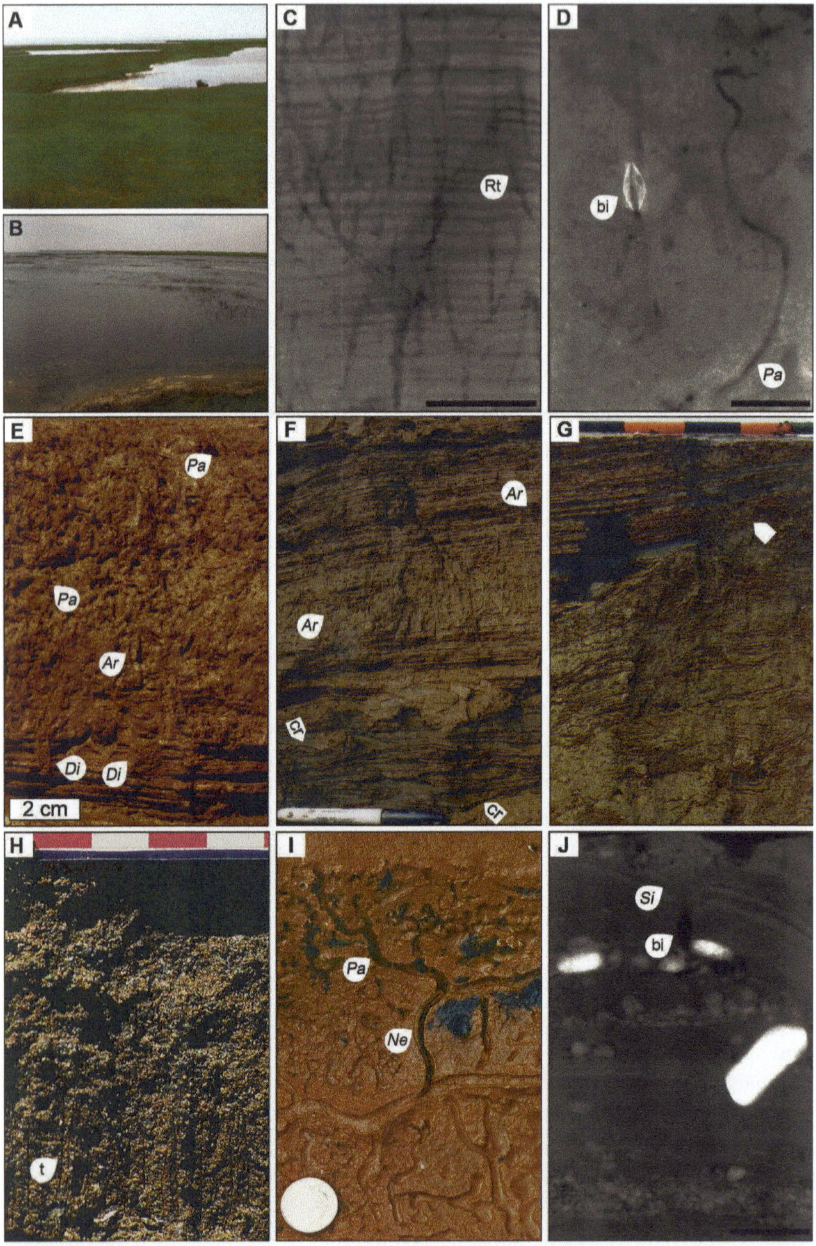

FIGURE 9 Photos (in color) and X-radiographs (gray scale) of inner, middle, and outer estuary deposits from Chignecto Bay, inner Bay of Fundy, Canada. (A) An example of the vegetated platform of salt marshes that ring the estuary. Note the shallow pools that commonly develop on the salt-marsh surface. (B) The vegetated platform submerged during spring high tide. The platform is typically fully submerged for 5–6 days per month. (C) X-radiograph of parallel-laminated,

onto the tidal flats promotes mud deposition. By contrast, in the outer estuary, much of the coastline is transgressive, and salt-marsh and glaciomarine deposits are commonly eroded and exposed on the sea floor, making them available for subsequent colonization. The range of burrowing fauna is remarkably consistent throughout the embayment, with a dominance of *Corophium volutator* (an amphipod) in intertidal mud beds and a range of sessile and motile subsurface deposit-feeding and surface deposit-feeding polychaetes (e.g., *Nereis* sp., *Clymenella torquata*) in all deposit types (Dashtgard and Gingras, 2005a; Dashtgard et al., 2008; Hauck et al., 2008; Pearson and Gingras, 2006; Pearson et al., 2007). In the outer estuary, where suspended sediment concentrations are lower and substrates are typically sand- or gravel-dominated, bivalves (e.g., *Mya arenaria, Ensis directus*), sediment-swimming polychaetes (e.g., *Nephtys* sp.), and sessile deposit-feeding polychaetes (terebellids) are increasingly common (Dashtgard et al., 2008).

3.7.2 Trace-Fossil Distributions by Subenvironment

There is no strong longitudinal (ichnological) gradation through the inner and middle estuary of Chignecto Bay. This is most likely because the rivers are small and deliver too little freshwater even during peak discharge. The inner and middle estuary of Chignecto Bay is fringed by a heavily rooted salt marsh that abruptly grades into well-developed, typically broad, muddy tidal flats. The salt marsh comprises vegetated platforms, shallow pools, and tidal channels that drain the marsh (Fig. 9A and B), of which the vegetated platforms are nearly devoid of bioturbation (Fig. 9C). Salt-marsh pools are either unbioturbated in landward positions or intensely bioturbated in locations where they commonly receive influxes of oxygenated water (Fig. 9D). The tidal channels are bioturbated where sedimentation rates are moderate to low (Dashtgard and Gingras, 2005a). Tidal flats are generally unburrowed in the wave-winnowed and sand-

silt-dominated salt-marsh deposits with abundant root traces (Rt). Black scale bar is 5 cm. (D) Intensely bioturbated mud deposited in a salt-marsh pool near the marsh-intertidal zone contact. This pool is regularly inundated by marine water. Note the presence of the bivalve (bi) *Mya arenaria* and polychaete-generated *Palaeophycus* (*Pa*). Scale bar is 5 cm. (E)–(G) Interbedded laminated and current-rippled beds and burrowed beds of the middle estuary mud flats. Laminated deposits represent winter deposition when bioturbation is severely restricted, while intensely bioturbated units represent summer deposits (Pearson and Gingras, 2006). The range of traces in these deposits is produced by a low-diversity suite of infauna, mainly consisting of *Corophium* and *Nereis*. Traces observed in these photos include *Arenicolites* (*Ar*), *Diplocraterion* (*Di*), *Palaeophycus*, and cryptobioturbation (cr). The pen in (F) is 13.5 cm long. The orange and black bands on the meter stick in (G) are 10 cm each. The white arrow in (G) marks the basal scour surface of a tidal channel cut into horizontally bedded and intensely bioturbated tidal-flat deposits (images provided by I. Armitage). (H) Small-diameter threadworm burrows (t) in gravelly sand of the outer estuary. Scale in centimeters. (I) Permanent dwelling of the trophic-generalist *Nereis* (*Ne*) in muddy gravel. This trace is best described as *Palaeophycus*. Coin is 2.1 cm in diameter. (J) X-radiograph of a bivalve and its trace (*Siphonichnus*, *Si*) in gravelly sand. Scale bar is 5 cm.

dominated innermost 50 m from the mean high-tide level. Exceptions to this generalization include relatively large (2–3 cm long, 1 cm wide), burrowing amphipods with a propensity for making short, wide-diameter *Skolithos*-like traces (up to 20 cm long). In high burrow densities, these can lead to the development of cryptobioturbation (Dashtgard and Gingras, 2005b). In the upper intertidal flats, *Siphonichnus* and *Arenicolites* are common (BI = 2–3). In the middle and lower intertidal zones of muddy tidal flats, the diversity of the infaunal community is generally low, although the intensity of burrowing is high (BI = 3–4, Fig. 9E–G; Pearson et al., 2007; Thurston, 1990). Typical burrows include *Arenicolites*, *Palaeophycus*, *Teichichnus*, and *Polykladichnus*. The outer part of muddy tidal flats and subtidal channels is reworked daily, and infauna is likely to be rarely present.

In the outer estuary of Chignecto Bay, the muddy tidal flats are replaced by sand-dominated deposits, and the tidal flats are replaced by tidally modulated shorefaces (TMS; Dashtgard et al., 2006, 2009). Bioturbation is patchy in the middle and lower intertidal zones of TMS (BI = 0–4; Fig. 9H–J). Where mud beds are present, *Siphonichnus*, *Arenicolites*, and *Diplocraterion* typically dominate, and bioturbation values range from BI = 1 to 4 (Dashtgard et al., 2008). In sand-dominated substrates, *Skolithos*-, *Polykladichnus*-, *Palaeophycus*-, and rare *Siphonichnus*-like structures occur with bioturbation values of BI = 0–2. The animals residing in both the sandy and muddy substrates tend to live in low-diversity communities and represent a mixture of small crustaceans and bivalves, as well as large polychaetes. In the lower intertidal and subtidal zones, *Skolithos*-, *Cylindrichnus*-, and *Palaeophycus*-like traces dominate.

In addition to TMS, the channels and subtidal portions of the outer estuary of Chignecto Bay are dominantly erosional (Amos and Asprey, 1979; Amos et al., 1991). The deposits under the subtidal outer estuary are largely eroded by strong tidal currents and have a low preservation potential. Large subtidal sand bars are mapped in this part of the bay, yet little is known of the ichnological character of these deposits. Rare *Skolithos*-generating *Cerebratulus*, fugichnia- and *Siphonichnus*-producing razor clams have been observed. The resulting bioturbation intensities are inferred to be low (BI = 0–1).

4. ANCIENT EXAMPLES—TRACE-FOSSIL DISTRIBUTION

4.1 Wave-Dominated Estuaries in the Viking Formation (Albian), Alberta, Canada

Wave-dominated estuarine incised valleys have been described from a number of Cretaceous stratigraphic units of Alberta, Canada, but the best-studied examples are from the Albian Viking Formation (Fig. 10). The Viking Formation contains at least five petroleum-producing deposits that are interpreted to reflect estuarine IVF successions. The most extensively studied Viking IVF resides in the Crystal Field (Pattison, 1991; Pattison and Walker, 1994; Reinson et al.,

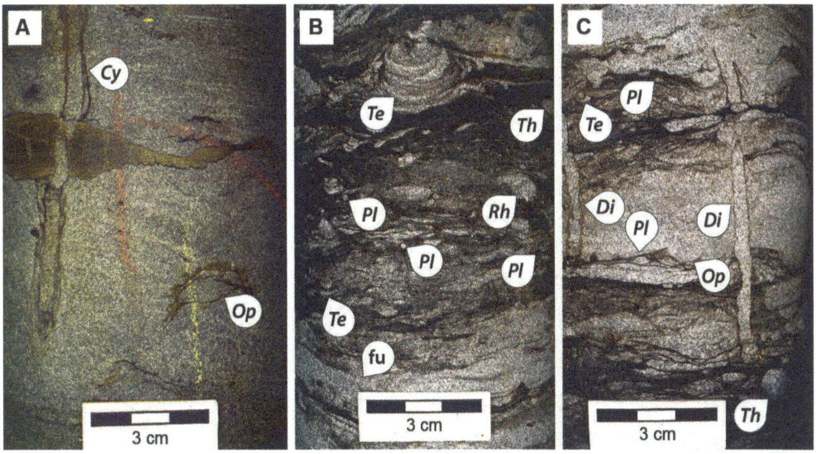

FIGURE 10 Proximal-to-distal distribution of trace-fossil suites from the wave-dominated estuarine complex of the Early Cretaceous (Albian) Viking Formation in the Crystal Field. (A) Sandstone facies from the bay-head delta complex toward the landward end of the estuary. The medium-grained sandstone contains disseminated carbonaceous detritus, sideritized mudstone layers, and generally shows BI = 0–2. The photo shows well-developed *Cylindrichnus* (*Cy*) and *Ophiomorpha* (*Op*). (B) Wavy-bedded heterolithic interval consisting of thoroughly bioturbated (BI = 4) sandy mudstones and fine-grained oscillation-and current-ripple-laminated sandstone with rare, thin dark, largely unburrowed carbonaceous fissile mudstone interbeds. Trace fossils are abundant, but suites generally show an overall low diversity. Dominant elements comprise *Teichichnus* (*Te*), *Planolites* (*Pl*), *Rhizocorallium* (*Rh*), *Thalassinoides* (*Th*), and fugichnia (fu). (C) Sporadically bioturbated (BI = 1–3/4) muddy sandstone from the estuary mouth complex, showing more robust traces. The suite comprises *Diplocraterion* (*Di*), *Ophiomorpha* (*Op*), *Thalassinoides* (*Th*), *Planolites* (*Pl*), and *Teichichnus* (*Te*).

1988), although studies have also focused on the Willesden Green Field (Boreen and Walker, 1991), the Sundance and Edson fields (Pattison and Walker, 1998), and Cyn-Pem (Pattison, 1991). These studies concentrated on the facies architecture, sequence-stratigraphic framework, and mapped distributions of facies. Early studies addressed the ichnological expression of the IVFs (Pattison, 1992; Pemberton et al., 1992), but these studies were based on incomplete assemblages from a small number of cores. An ichnological summary of all facies in Viking IVFs was provided in MacEachern and Pemberton (1994) and expanded upon the light of other brackish-water assemblages as well as marine mudstones from the Viking Formation (MacEachern et al., 1999). MacEachern and Gingras (2007) provided an updated evaluation of ichnological suites of wave-dominated estuaries of the Viking Formation. The ichnological demarcation of Viking valley-margins and internal discontinuities were also addressed in MacEachern and Pemberton (1994), MacEachern et al. (1992), and Pemberton and MacEachern (1995). A summary of the discontinuities within the sequence-stratigraphic framework is presented in MacEachern et al. (2012).

4.1.1 Valley Margins and Substrate-Controlled Suites

During relative sea-level fall, valleys are incised into underlying successions forming an SU and served as zones of sediment bypass. During late lowstand and early transgression, however, aggradation and eventually retrogradation dominate the valley-fill architectures. In the Crystal Field, the valley floors and margins are demarcated by firmground omission suites, indicating that the initial preserved facies of the valley were marine or marginal-marine in origin, indicating a transgressive surface developed on the SU (TS/SU). The TS is non-appreciably erosional within the bay and along the bay margins, but ranges from a TRS near the mouth of the system, where it is associated with the tidal inlets, to a wave-ravinement surface (WRS) once the valley is filled and transgression has breached the barrier fronting the system.

Infaunal colonization of the TS/SU may yield suites attributable to all three substrate-controlled ichnofacies, depending upon the character of the substrate that was exhumed. In the Crystal Viking IVF, however, only suites of the *Glossifungites* Ichnofacies are preserved along the valley margins and consist of firmground *Diplocraterion*, *Thalassinoides*, *Rhizocorallium*, *Gastrochaenolites*, and *Skolithos*. The valleys are excavated into coarsening-upward, regionally extensive, fully marine parasequences of the underlying highstand systems tract (Pattison and Walker, 1994). These stacked parasequences consist of thoroughly bioturbated silty mudstones, sandy mudstones, and muddy sandstones, containing fully marine, high-diversity ichnological suites attributable to distal-to-proximal expressions of the *Cruziana* Ichnofacies.

4.1.2 Trace-Fossil Distributions by Subenvironment

The bay-head delta complex (i.e., inner estuary) is characterized by sediments supplied by the river and deposited into the central basin as lobate, typically river-dominated deltaic wedges (Fig. 10). Facies are largely characterized by interstratified planar-laminated and current-rippled sandstones, as well as moderately burrowed sandstones and muddy sandstones (Pattison and Walker, 1994). The sandstones are associated with the delta front and the distributary channels. Mudstone interlaminae and interbeds are common. Bioturbation intensities are generally low (BI = 0–2), and burrowed horizons are sporadically distributed. The ichnological suites are dominated by *Planolites*, *Palaeophycus*, *Skolithos*, *Ophiomorpha*, and fugichnia, with subordinate *Teichichnus*, *Arenicolites*, *Cylindrichnus*, *Rosselia*, and *Diplocraterion* (Fig. 10). Rare suites may contain isolated occurrences of *Conichnus*, *Bergaueria*, *Asterosoma*, *Schaubcylindrichnus freyi*, *Lockeia*, *Macaronichnus*, *Thalassinoides*, and *Siphonichnus* (MacEachern and Gingras, 2007).

The central-basin complex (Fig. 10B) represents the standing body of water lying behind the barrier (middle estuary) and ranges from successions that are sand dominated (bay margin) to those that are mud dominated (deeper bay). Whether sand or mud dominated, the central-basin deposits consist of two regularly interbedded subfacies. The most distinctive subfacies comprises moderately to intensely bioturbated (BI = 2–5), interstratified sandy mudstones with largely unburrowed, dark, fissile mudstone drapes and thin sandstone stringers (Fig. 10). This heterolithic subfacies is characterized by wavy bedding with oscillation ripples and rare current and combined flow ripples (MacEachern and Pemberton, 1994; Pattison, 1992). The trace fossils are more uniformly distributed than elsewhere in the estuary, are typically moderately diverse, and are characterized by a predominance of structures inferred to reflect deposit-feeding and dwelling behaviors. The trace-fossil suites are dominated by *Planolites*, *Teichichnus*, and *Schaubcylindrichnus freyi*, with subordinate numbers of *Palaeophycus*, *Siphonichnus*, *Lockeia*, *Chondrites*, *Thalassinoides*, and diminutive *Rosselia*. Uncommon components include *Ophiomorpha*, *Diplocraterion*, *Cylindrichnus*, *Rhizocorallium*, *Phycosiphon*, and diminutive *Asterosoma*. The reduced size of many biogenic structures, coupled with the presence of synaeresis cracks, supports the interpretation of persistently fluctuating salinities and brackish-water conditions.

The estuary-mouth complex (outer estuary) comprises sand derived from alongshore transport of sediment along the barrier margin and from tidal exchange through the tidal inlet. On the estuary side of the barrier, most sand deposition reflects washover events, flood-tidal delta accumulation, and tidal-inlet deposits (Pattison and Walker, 1994). These sandstones are typically the most marine influenced in the estuary, and unless sedimentation rates are high, they tend to be the most intensely bioturbated facies within the IVF. Successions generally display interstratified muddy sandstones and horizontal parallel-laminated, planar cross-stratified, current-ripple laminated, and trough-cross-stratified sandstones (Fig. 10C). Mudstone interlaminae are common, with moderate numbers of mudstone interbeds that are locally siderite cemented. The bioturbation intensities are generally higher than in the other associated estuarine deposits (BI = 1–4), although burrowed zones are sporadically distributed, due to episodic deposition. The suites tend to be diverse but are dominated by *Planolites*, *Ophiomorpha*, *Teichichnus*, and *Palaeophycus*. Subordinate components comprise *Arenicolites*, *Schaubcylindrichnus freyi*, *Rosselia*, *Diplocraterion*, *Thalassinoides*, *Siphonichnus*, and fugichnia. Very minor elements include *Asterosoma*, *Skolithos*, *Cylindrichnus*, *Chondrites*, *Bergaueria*, *Macaronichnus*, and *Lockeia*. The trace fossils tend to be comparatively robust, which is attributed to higher and more uniform salinities.

4.2 Mixed(?) Estuary, Montney Formation (Triassic), Alberta, Canada

The Montney Formation was deposited on the west coast of the Pangaean supercontinent during the Early Triassic (basal Induan to end Olenekian). Within the Western Canada Sedimentary Basin, the Paleozoic/Mesozoic transition records a regional lowstand with concomitant subaerial exposure and development of an erosional unconformity in all but the westernmost locales. The basal beds of the Montney Formation reflect a regional marine transgression and a shift of the shoreline hundreds of kilometers to the east. The Montney Formation records several regional fluctuations in relative sea level, permitting subdivision of this unit into several unconformity-bound sequences.

4.2.1 Valley Margins and Substrate-Controlled Suites

The sequence boundary in the Montney Formation occurs approximately between the Induan and Olenekian stages (Davies et al., 1997; Kendall, 1999; Markhasin, 1998; Moslow and Davies, 1997; Panek, 2000). This surface, referred to as the mid-Montney sequence boundary, resulted in significant erosional incision into an older clastic ramp succession. In the Simonette–Kaybob area of west-central Alberta, this erosional incision includes several incised paleovalley complexes (Buatois et al., 2005; Markhasin, 1998). A low-density, low-diversity firmground omission suite consisting of *Skolithos*, *Thalassinoides*, and rare *Rhizocorallium* of the *Glossifungites* Ichnofacies characterizes the basal contact of the estuarine IVF succession.

4.2.2 Trace-Fossil Distributions by Subenvironment

Planar- to wavy-parallel-laminated siltstones with thin, sharp-based sandstone beds are interpreted as subtidal bayfill and storm-washover complexes (Fig. 11B–E). These beds contain a low-diversity, locally high-density assemblage consisting of *Phycosiphon*, *Gyrochorte*, *Lingulichnus*, *Planolite*s, and *Trichichnus* (Fig. 11). Heterolithic interlaminated siltstone and very fine-grained sandstone beds are characterized by low-angle inclined surfaces and display low-diversity but locally high-density assemblages consisting of *Phycosiphon*, *Conichnus*, *Lingulichnus*, *Palaeophycus*, *Planolites*, *Psilonichnus*, *Skolithos*, and *Thalassinoides*. These heterolithic facies are interpreted as intertidal-flat/estuarine-bar deposits (Fig. 11F and G).

Lingulide brachiopod dwelling traces (*Lingulichnus*) are notably abundant in the Montney estuarine deposits (Fig. 11B and C) and occur in adjacent shoreface successions as well. Lingulide brachiopods, both modern and ancient, exhibit wide environmental tolerances and, thus, occur in settings that range from distal offshore through shoreface and intertidal flat (Emig, 1997; Zonneveld and Pemberton, 2003; Zonneveld et al., 2007). Consequently, the presence of *Lingulichnus* in Montney estuarine successions is not diagnostic. Notably however, the Montney *Lingulichnus* are smaller within estuary-fill successions and diminish in size to the east/southeast (paleo-upstream).

4.3 Tide-Dominated(?) Estuary, McMurray Formation (Aptian to Albian), Alberta, Canada

The McMurray Formation was deposited in a series of North–South trending valleys during the Aptian and Albian transgression of the Boreal Sea, which inundated the setting from North to South. The drainage was complex, and it is likely

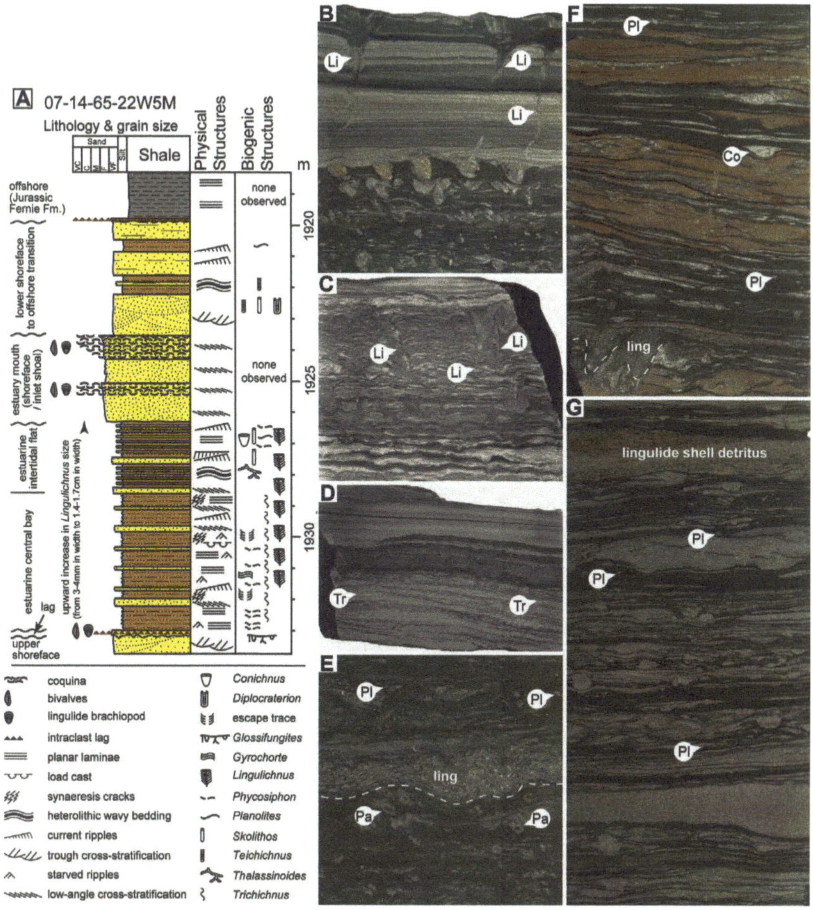

FIGURE 11 Trace fossils from estuarine successions in the Early Triassic Montney Formation. (A) Description of the cored Montney interval in well 07-14-65-22W6 of the Simonette-Kaybob area, showing the distribution of physical and biogenic sedimentary structures. (B) Sharp-based sandstone bed from the bay-center succession. Note the *Lingulichnus* (*Li*) and lingulide escape traces that move through the event bed. Note also the pyritized burrow fill in the abandoned horizon beneath the event bed. Well Asplund Creek 04-22-66-23W6, 1980.0 m. (C) Interbedded siltstone and very fine-grained sandstone with abundant *Lingulichnus* (*Li*), well 7-14-65-22W6, 1928.3 m. (D) Laminated siltstone with abundant tiny vertical traces (cf. *Trichichnus*, *Tr*), well 7-14-65-22W6, 1928.3 m. (E) Bioturbated siltstone with sharp-based (load-casted) bioclastic sandstone bed (above dashed line). The siltstone contains a low-diversity assemblage of *Palaeophycus* (*Pa*) and *Planolites* (*Pl*). Asplund Creek 04-22-66-23W6, 1975.3 m. (F) Low-angle, heterolithic siltstone and very fine-grained sandstone containing *Conichnus* (*Co*), *Planolites* (*Pl*), rare *Skolithos*, and escape traces. This succession was deposited in an intertidal-bar setting. Asplund Creek 04-22-66-23W6, 1980.5 m. (G) Low-angle, heterolithic siltstone and very fine-grained sandstone containing two different sizes of *Planolites* (*Pl*). The low-angle cross-stratified sandstone near the top of the succession contains abundant, admixed, sand-sized bioclastic detritus. This succession was deposited in an intertidal-bar setting. Asplund Creek 04-22-66-23W6, 1982.8 m.

that some aspects of the McMurray fill are deltaic and others are estuarine. We focus here on the bioturbated units within the McMurray Formation that have been interpreted to have an estuarine affinity. The flooding of the McMurray subbasin led to the establishment of a large complex of estuarine and deltaic deposits that are observable and mapped over an area exceeding 16,000 km^2. Owing to the presence of notable bitumen resources, the deposit is well known from outcrop as well as from a very large core dataset. Although the stratigraphy is exceedingly complex, it is clear that estuary-associated valley fills represent a large proportion of the strata assigned to the McMurray Formation.

Trace fossils are common in many facies of the McMurray Formation. Pemberton et al. (1982) showed that impoverished trace-fossil suites associated with the ubiquitous IHS, defining tidally influenced point-bar deposits, indicated a brackish-water, estuary environment. Later work (e.g., Ranger and Gingras, 2008) differentiated outer and inner estuary deposits, each characterized by discrete ichnological assemblages.

4.3.1 Valley Margins and Substrate-Controlled Suites

McMurray Formation strata sit variably on argillaceous limestones and calcareous shales of the Devonian. Regionally, this erosional discordance represents an angular unconformity, and the valleys were carved into more recessive levels. Evidence for trace-fossil omission suites are rare in the lower levels of the McMurray Formation. However, parasequences at the top of the McMurray Formation can be identified by the presence of rare, low-diversity firmground suites of *Thalassinoides* and *Gastrochaenolites*, attributable to the *Glossifungites* Ichnofacies. The Clearwater Formation lies on top of the McMurray Formation, and the contact between the two formations is easily identified by the presence of firmground suites of *Diplocraterion*, *Thalassinoides*, *Rhizocorallium*, and *Skolithos*. Within the Clearwater Formation, there is a switch to marine sedimentation and bioturbation associated with shoreface and deltaic depositional environments, indicating that the ongoing relative sea-level rise led to the filling of the subbasin.

4.3.2 Trace-Fossil Distributions by Subenvironment

No bay-head delta deposits are observed in the estuary portions of the McMurray Formation. Rather, there appears to be a gradational change from fluvially influenced sedimentation to tidally influenced sedimentation in a northward direction.

In landward deposits, the strata are current-ripple laminated to cross-stratified. Mudstone beds and mudstone drapes are absent to moderately abundant, and intraformational mudclasts and carbonaceous detritus are locally common. Importantly, bioturbation is rare (BI = 1; rarely BI = 2), and trace fossils (*Planolites* and *Skolithos*) are sporadically distributed. Local vertical and lateral trends in trace-fossil distributions have not been discerned. These deposits are likely associated with the tidally influenced part of the fluvial system.

The inner estuary deposits in the McMurray Formation are dominated by brackish water (i.e., oligohaline to perhaps mesohaline). Strata of the inner estuaries characteristically comprise IHS. The impoverished trace-fossil assemblages that are characteristic of the IHS have been well documented (Pemberton et al., 1982; Ranger and Pemberton, 1992) and consist of abundant but generally diminutive and monospecific assemblages of *Planolites*, *Skolithos*, and *Cylindrichnus*, or spiral forms such as *Gyrolithes* and undiagnosed microhelical forms (Figs. 3D and 4K). Trace fossils are typically restricted to either the sandstone or the mudstone member of the IHS, which is potentially indicative of seasonal larval recruitment and colonization (Gingras et al., 2011). Within the IHS units, the bioturbation intensities characteristically increase upward: BI = 0 at the base of IHS successions, BI = 2–3 throughout much of the medial part of the IHS (Fig. 3D and K), and locally grading to BI = 5 at the top of the succession. This vertical distribution is interpreted to represent lowered sedimentation rates upward and the transition to bioturbated intertidal-flat-associated strata.

The middle- to outer-estuary deposits of the McMurray Formation are characterized by amalgamated decimeter- to meter-bedded, high-angle cross-stratified sandstone, which is interpreted to represent laterally accreted mid-channel tidal bars. The cross-bedded sandstones contain a sparse, low-diversity marine-influenced trace-fossil assemblage. Considering the high-energy stress indicated by the physical structures, trace fossils within the cross-stratified sands are, not surprisingly, uncommon. Nonetheless, conical plugs assigned to *Conichnus* and *Siphonichnus* are locally present, but rare. Thin, mud-draped hiatal surfaces, that locally cap cross-stratified bedsets, contain a very low-diversity trace-fossil assemblage consisting of *Cylindrichnus* and/or *Skolithos*. The trace-fossil evidence (the largest trace fossils observed and the local presence of large *Siphonichnus* and *Conichnus*) as well as the presence of higher-energy physical sedimentary structures are most consistent with meso- to macrotidal regimes. The differentiation between middle and outer estuary remains unclear in the McMurray subbasin.

5. DISCUSSION

Although the ichnological content varies from estuary to estuary, trends in trace-fossil distributions are discernible. Important commonalities that relate to our ability to identify estuary deposits in the rock record can, therefore, be established. First, in all of our above-reported examples, the trace-fossil assemblage is influenced in landward locales by the presence of brackish water. As such, a salinity gradient is always evident at the estuary scale, and this can be established through simple observations of trace-fossil sizes, diversities, and types, particularly when compared to their fully marine counterparts (cf. MacEachern and Gingras, 2007). The second major concordance is that firm substrates are well colonized by low-diversity assemblages of infauna in brackish-water zones, and so channel and valley margins should normally contain omission suites of trace fossils that demarcate stratigraphic levels of transgression across the SU. Another commonality exhibited by many of our examples is the upward increase of bioturbation intensities and diversities, which is a pattern associated with tidally influenced point bars and, to a lesser degree, longitudinal bars.

In short, our modern and ancient analogs demonstrate that estuaries can be ichnologically identified in the rock record. Importantly, it is apparent that the type of estuary (wave dominated through to tide dominated) can be understood, especially if taken in the context of the more general sedimentary facies associations. There is some question as to whether or not tide-dominated estuaries can be discerned from tide-dominated deltas. To answer this question, a set of neoichnological studies in tide-dominated deltas need to be conducted; for the time being, no detailed characterization exists. We predict that limitations on larval recruitment in fluvially dominated settings will lead to ichnological impoverishment basinward of the inner estuary and that, at least on the map scale, the ichnology of the two systems will be discrete.

For finer-scale subdivisions, it is useful to contrast the ichnology of the various estuaries. In particular, the sharp ichnological gradients and the presence of more marine conditions in the outer estuary of wave-dominated estuaries set them apart from the mixed-energy and tide-dominated estuaries. The most important difference between wave- and tide-dominated estuaries is that tide-dominated settings experience hydraulic energy conditions that far exceed those imposed by tidal currents in either wave-dominated or mixed-energy settings. A notable characteristic of tide-dominated estuaries is the enormous volume of their tidal prisms. Large tidal prisms have the effect of effectively mixing bay waters, such that water stratification is absent and lateral salinity changes are gradual. Salinity stratification, eutrification, and rapid salinity changes are, by contrast, more readily associated with the wave-dominated end members. Larval recruitment is also profoundly influenced by volumetrically large tide-water exchange, in that animal spat are more evenly distributed over larger areas in macrotidal estuaries than in their microtidal counterparts (e.g., Ayata et al., 2009; Bentley and Pacey, 1992). The gradient of processes and salinities along the length of macrotidal estuaries leads to a broad distribution of ichnological suites that may differ little from the middle to the outer parts of the system. Finally, the resulting biogenic structures in tide-dominated estuaries reflect organism responses that are more profoundly influenced by the grain size of the substrate, sedimentation rates, turbidity of the water column, and the overall hydraulic energy than are those of organisms that inhabit wave-dominated estuaries.

6. CONCLUSIONS

Although there are a number of definitions for an estuary that are appropriate depending on their context, we have chosen to emphasize a definition that suits the identification of estuaries in the stratigraphic record as IVFs. The main criteria for the identification of such IVF estuaries include the following:

1. The estuary fill is commonly encapsulated within a valley, wherein the shared contact represents a composite subaerial unconformity and a transgressive surface (TS/SU), or a subaerial unconformity and a tidal-ravinement surface (TRS/SU). This criterion may be established ichnologically through the repeated identification of omission suites of trace fossils, commonly representative of the *Glossifungites* or *Trypanites* ichnofacies. Note that the SU does not host suites attributable to the *Glossifungites* Ichnofacies.
2. Brackish-water-associated strata dominantly represent the estuarine IVF. By recognizing trends in trace-fossil sizes and diversities, and by comparing the observed assemblages to their fully marine counterparts as a base line, this criterion can be established ichnologically.
3. Estuaries are filled during transgression, which requires the recognition of proximal and distal trends of trace-fossil size and diversity, as well as the differentiation of trace-fossil suites that are consistent with fully marine, brackish-water, and freshwater fluvial settings.
4. Estuaries receive sediment from both fluvial and tidal sources. Tidal currents influence the distribution of infauna largely through the transport of larvae. The presence of a brackish-water fauna indicates larval transport into the estuary and can be used as evidence for landward-directed suspended-load tidal transport, which may or may not be accompanied by tidal bedload. Additionally, certain ethologies can be used to infer the presence of tides.

Ichnological data are clearly important in the identification of estuaries. In spite of the surprising range of sedimentological variability observed in estuaries, the ichnological range is comparatively more constrained and therefore distinctive. Trace-fossil assemblages are ideally suited to the recognition of brackish-water conditions. The tidal bars within the estuaries tend to show trends of upward-increasing bioturbation. Finally, trace-fossil omission suites are common along the erosionally exhumed valley margins of estuaries.

ACKNOWLEDGMENTS

M. K. G.'s research is supported by an NSERC Discovery Grant (No. 238530) and ongoing support from Nexen Inc., ConocoPhillips Canada, Devon Energy Canada, BP Canada, Statoil and Shell. S. E. D.'s research is supported through an NSERC Discovery Grant (No. 341789) and funding from Nexen Inc., Imperial Oil Ltd., Statoil and Suncor. J. A. M. is funded through NSERC Discovery Grant No. 184293.

REFERENCES

Allen, G.P., Posamentier, H.W., 1993. Sequence stratigraphy and facies model of an incised valley fill: the Gironde Estuary, France. J. Sediment. Petrol. 63, 378–391.

Amos, C.F., 1987. Fine-grained sediment transport in Chignecto Bay, Bay of Fundy, Canada; dynamics of turbid coastal environments. Cont. Shelf Res. 7, 1295–1300.

Amos, C.L., Asprey, K.W., 1979. Geophysical and sedimentary studies in the Chignecto Bay System, Bay of Fundy-A progress report. Curr. Res. B, Geol. Surv. Can., Paper 79-lB, 245–252.

Amos, C.L., Tee, K.T., Zaitlin, B.A., 1991. The post-glacial evolution of Chignecto Bay, Bay of Fundy, and its modern environment of deposition. In: Smith, D.G., Reinson, G.E., Zaitlin, B.A., Rahmani, R.A. (Eds.), Clastic Tidal Sedimentology. Can. Soc. Petrol. Geol., Mem. 16, pp. 59–89.

Anima, R.J., Clifton, H.E., Phillips, R.L., 1989. Comparison of modern and Pleistocene estuarine facies in Willapa Bay, Washington. In: Reinson, G.E. (Ed.), Modern and Ancient Examples of Clastic Tidal Deposit: A Core and Peel Workshop. Can. Soc. Petrol. Geol., Second International Research Symposium on Clastic Tidal Deposits, pp. 1–19.

Ayata, S., Ellien, C., Dumas, F., Dubois, S., Thiebaut, E., 2009. Modelling larval dispersal and settlement of the reef-building polychaete *Sabellaria alveolata*: role of hydroclimatic processes on the sustainability of biogenic reefs. Cont. Shelf Res. 29, 1605–1623.

Bann, K.L., Fielding, C.R., MacEachern, J.A., Tye, S.C., 2004. Differentiation of estuarine and offshore marine deposits using integrated ichnology and sedimentology: Permian Pebbley Beach Formation, Sydney Basin, Australia. In: McIlroy, D. (Ed.), The Application of Ichnology to Palaeoenvironmental and Stratigraphic Analysis. Geol. Soc. London, Spec. Publ., vol. 228, pp. 179–211.

Bentley, M., Pacey, A., 1992. Physiological and environmental control of reproduction in polychaetes. Oceanogr. Mar. Biol. Annu. Rev. 30, 443–481.

Beynon, B.M., Pemberton, S.G., Bell, D.D., Logan, C.A., 1988. Environmental implications of ichnofossils from the Lower Cretaceous Grand Rapids Formation, Cold Lake Oil Sands Deposit: sequences, stratigraphy, sedimentology: surface and subsurface. Can. Soc. Petrol. Geol., Mem. 15, 275–289.

Boreen, T., Walker, R.G., 1991. Definition of allomembers and their facies assemblages in the Viking Formation, Willesden Green area, Alberta. Bull. Can. Petrol. Geol. 39, 123–144.

Boyd, R., Dalrymple, R.W., Zaitlin, B.A., 2006. Estuarine and incised-valley facies models. In: Posamentier, H.W., Walker, R.G. (Eds.), Facies Models Revisited. SEPM Spec. Publ., vol. 84, pp. 171–235.

Buatois, L.A., Mangano, M.G., Maples, C.G., Lanier, W.P., 1997. The paradox of nonmarine ichnofaunas in tidal rhythmites: integrating sedimentologic and ichnologic data from the Late Cretaceous of eastern Kansas, USA. Palaios 12, 467–481.

Buatois, L.A., Mangano, M.G., Maples, C.G., Lanier, W.P., 1998. Ichnology of an Upper Carboniferous fluvio-estuarine paleovalley: the Tonganoxie Sandstone, Buildex Quarry, eastern Kansas, USA. J. Paleontol. 72, 152–180.

Buatois, L.A., Gingras, M.K., MacEachern, J., Mángano, M.G., Zonneveld, J.-P., Pemberton, S.G., Netto, R.G., Martin, A., 2005. Colonization of brackish-water systems through time: evidence from the trace fossil record. Palaios 20, 321–347.

Bubb, J.N., Hatlelid, W.G., 1977. Seismic stratigraphy and global changes of sea level. Part 10. Seismic recognition of carbonate buildups. In: Payton, C.E. (Ed.), Seismic Stratigraphy: Applications to Hydrocarbon Exploration. AAPG Mem. 26, pp. 185–204.

Chapman, P.M., 1981. Measurements of the short-term stability of interstitial salinities in subtidal estuarine sediments. Estuar. Coast. Shelf Sci. 12, 67–81.

Chapman, P.M., Brinkhurst, R.O., 1981. Seasonal changes in interstitial salinities and seasonal movements of subtidal benthic invertebrates in the Fraser River estuary, B.C. Estuar. Coast. Shelf Sci. 12, 49–66.

Choi, K.S., Dalrymple, R.W., Chun, S.S., Kim, S.P., 2004. Sedimentology of modern, inclined heterolithic stratification (IHS) in the macrotidal Han River Delta, Korea. J. Sediment. Res. 74, 677–689.

Clifton, H.E., 1983. Discrimination between subtidal and intertidal facies in Pleistocene deposits, Willapa Bay, Washington. J. Sediment. Petrol. 53, 353–369.

Clifton, H.E., Phillips, R.L., Scheihing, J.E., 1976. Modern and ancient estuarine-fill facies, Willapa Bay, Washington. AAPG Bull. 60, 657–658.

Dalrymple, R.W., Choi, K.S., 2007. Morphologic and facies trends through the fluvial-marine transition in tide-dominated depositional systems: a schematic framework for environmental and sequence-stratigraphic interpretation. Earth-Sci. Rev. 81, 135–174.

Dalrymple, R.W., Zaitlin, B.A., 1994. High-resolution sequence stratigraphy of a complex, incised valley succession, Cobequid Bay-Salmon River Estuary, Bay of Fundy, Canada. Sedimentology 41, 1069–1091.

Dalrymple, R.W., Knight, R.J., Zaitlin, B.A., Middleton, G.V., 1990. Dynamics and facies model of a macrotidal sand-bar complex, Cobequid Bay-Salmon River Estuary (Bay of Fundy). Sedimentology 37, 577–612.

Dalrymple, R.W., Makino, Y., Zaitlin, B.A., 1991. Temporal and spatial patterns of rhythmite deposition on mud flats in the macrotidal Cobequid Bay-Salmon River Estuary, Bay of Fundy, Canada. In: Smith, D.G., Reinson, G.E., Zaitlin, B.A., Rahmani, R.A. (Eds.), Clastic Tidal Sedimentology. Can. Soc. Petrol. Geol., Mem. 16, pp. 137–160.

Dalrymple, R.W., Zaitlin, B.A., Boyd, R., 1992. Estuarine facies models: conceptual basis and stratigraphic implications. J. Sediment. Petrol. 62, 1130–1146.

Dalrymple, R.W., MacKay, D.A., Ichaso, A.A., Choi, K.S., 2011. Processes, morphodynamics, and facies of tide-dominated estuaries. In: Davis, R., Dalrymple, R.W. (Eds.), Principles of Tidal Sedimentology. Springer, Dordrecht, pp. 79–107.

Dashtgard, S.E., 2011a. Neoichnology of the lower delta plain: Fraser River Delta, British Columbia, Canada: implications for the ichnology of deltas. Palaeogeogr. Palaeoclimatol. Palaeoecol. 307, 98–108.

Dashtgard, S.E., 2011b. Linking invertebrate burrow distributions (neoichnology) to physicochemical stresses on a sandy tidal flat: implications for the rock record. Sedimentology 58, 1303–1325.

Dashtgard, S.E., Gingras, M.K., 2005a. Facies architecture and ichnology of Recent salt-marsh deposits: Waterside Marsh, New Brunswick, Canada. J. Sediment. Res. 75, 596–607.

Dashtgard, S.E., Gingras, M.K., 2005b. The temporal significance of bioturbation in backshore deposits: Waterside Beach, New Brunswick, Canada. Palaios 20, 589–595.

Dashtgard, S.E., Gingras, M.K., Butler, K.E., 2006. Sedimentology and stratigraphy of transgressive, muddy gravel beach: Waterside Beach, Bay of Fundy, Canada. Sedimentology 53, 279–296.

Dashtgard, S.E., White, R.O., Butler, K.E., Gingras, M.K., 2007. Effects of relative sea level change on the depositional character of an embayed beach, Bay of Fundy, Canada. Mar. Geol. 239, 143–161.

Dashtgard, S.E., Gingras, M.K., Pemberton, S.G., 2008. Grain-size controls on the occurrence of bioturbation. Palaeogeogr. Palaeoclimatol. Palaeoecol. 257, 224–243.

Dashtgard, S.E., Gingras, M.K., MacEachern, J.A., 2009. Tidally modulated shorefaces. J. Sediment. Res. 79, 793–807.

Davidson-Arnott, R.G.D., Greenwood, B., 1974. Bedforms and structures associated with bar topography in the shallow-water wave environment, Kouchibouguac Bay, New Brunswick, Canada. J. Sediment. Petrol. 44, 698–704.

Davies, G.R., Moslow, T.F., Sherwin, M.D., 1997. The Lower Triassic Montney Formation, west-central Alberta: an issue focused on the study of Triassic of the Western Canada Sedimentary Basin. Bull. Can. Petrol. Geol. 45, 474–505.

Desjardins, P.R., Buatois, L.A., Mángano, M.G., 2012. Tidal flats and subtidal sand bodies. In: Knaust, D., Bromley, R.G. (Eds.), Trace Fossils as Indicators of Sedimentary Environments. Developments in Sedimentology, vol. 64. Elsevier, Amsterdam, pp. 529–562.

Desplanque, C., Mossman, D.J., 2001. Bay of Fundy tides. Geosci. Can. 28, 1–11.

Dobrin, M.B., 1976. Seismic stratigraphy, new kind of geophysical interpretation. AAPG Bull. 60, 2179.

Dorjes, J., Howard, J.D., 1975. Fluvial-marine transition indicators in an estuarine environment, Ogeechee River-Ossabaw Sound. Senckenberg. Marit. 7, 137–179.

Emig, C.C., 1997. Ecology of inarticulated brachiopods. In: Kaesler, R.L. (Ed.), Part H, Brachiopoda. Treatise on Invertebrate Paleontology 1, Geological Society of America and Paleontological Institute, Boulder and Lawrence, pp. 473–495.

Fenies, H., Faugeres, J., 1998. Facies and geometry of tidal channel-fill deposits (Arcachon Lagoon, SW France). Mar. Geol. 150, 131–148.

Geier, M.H., 1995. Ichnology, sedimentology and stratigraphy of the ostracode zone, Bellshill Lake-Killam area, east-central Alberta. M.Sc. Thesis, University of Alberta, 204 pp.

Gingras, M.K., MacEachern, J.A., 2012. The tidal ichnology of shallow-water clastic settings. In: Davis, R., Dalrymple, R.W. (Eds.), Principles of Tidal Sedimentology. Springer, Dordrecht, pp. 57–78.

Gingras, M.K., Pemberton, S.G., Saunders, T., Clifton, H.E., 1999. The ichnology of modern and Pleistocene brackish-water deposits at Willapa Bay, Washington: variability in estuarine settings. Palaios 14, 352–374.

Gingras, M.K., Hubbard, S.M., Pemberton, S.G., Saunders, T., 2000. The significance of Pleistocene *Psilonichnus* at Willapa Bay, Washington. Palaios 15, 142–151.

Gingras, M.K., Pemberton, S.G., Saunders, T., 2001. Bathymetry, sediment texture, and substrate cohesiveness: their impact on modern *Glossifungites* trace assemblages at Willapa Bay, Washington. Palaeogeogr. Palaeoclimatol. Palaeoecol. 169, 1–21.

Gingras, M.K., MacEachern, J.A., Dashtgard, S.E., 2011. Process ichnology and the elucidation of physico-chemical stress. Sediment. Geol. 237, 115–134.

Grant, D.R., 1970. Recent coastal submergence of the Maritime provinces, Canada. Can. J. Earth Sci. 7, 676–689.

Häntzschel, W., 1939. Die Lebens-Spuren von *Corophium volutator* (Pallas) und ihre paläontologische Bedeutung. Senckenbergiana 21, 215–227.

Hauck, T.E., Dashtgard, S.E., Gingras, M.K., 2008. Relationships between organic carbon and pascichnia morphology in intertidal deposits: Bay of Fundy, New Brunswick, Canada. Palaios 23, 336–343.

Hauck, T.E., Dashtgard, S.E., Pemberton, S.G., Gingras, M.K., 2009. Brackish-water ichnological trends in a microtidal barrier island-embayment system, Kouchibouguac National Park, New Brunswick, Canada. Palaios 24, 478–496.

Hertweck, G., 1992. Distribution patterns of benthos and lebensspuren in the Spiekeroog backbarrier tidal flat area (southern North Sea): modern and ancient clastic tidal deposits. Courier Forsch. Inst. Senck. 151, 40–42.

Hertweck, G., Wehrmann, A., Liebezeit, G., 2007. Bioturbation structures of polychaetes in modern shallow marine environments and their analogues to *Chondrites* group traces. Palaeogeogr. Palaeoclimatol. Palaeoecol. 245, 382–389.

Holste, L., John, M.A.S., Peck, M.A., 2009. The effects of temperature and salinity on reproductive success of *Temora longicornis* in the Baltic Sea: a copepod coping with a tough situation. Mar. Biol. 156, 527–540.

Hovikoski, J., Rasanen, M., Gingras, M., Ranzi, A., Melo, J., 2008. Tidal and seasonal controls in the formation of late Miocene inclined heterolithic stratification deposits, western Amazonian foreland basin. Sedimentology 55, 499–530.

Howard, J.D., Frey, R.W., 1975. Introduction. In: Howard, J.D., Frey, R.W. (Eds.), Estuaries of the Georgia Coast, U.S.A. Sedimentology and Biology. Senckenberg. Marit., vol. 7, pp. 1–31.

Howard, J.D., Elders, C.A., Heinbokel, J.F., 1975. Animal-sediment relationships in estuarine point bar deposits, Ogeechee River-Ossabaw Sound. In: Howard, J.D., Frey, R.W. (Eds.), Estuaries of the Georgia Coast, U.S.A. Sedimentology and Biology. Senckenberg. Marit. vol. 7, pp. 181–203.

Hubbard, S.M., Gingras, M.K., Pemberton, S.G., 2004. Palaeoenvironmental implications of trace fossil in estuary deposits of the Cretaceous Bluesky Formation, Cadotte region, Alberta, Canada. Fossils Strata 51, 68–87.

Karvonen, R.L., 1989. The Ostracode Member east-central Alberta: an example of an estuarine valley fill deposit. In: Reinson, G.E. (Ed.), Modern and Ancient Examples of Clastic Tidal Deposits: A Core and Peel Workshop. Can. Soc. Petrol. Geol., Second International Research Symposium on Clastic Tidal Deposits, pp. 105–116.

Kendall, D.R., 1999. Sedimentology and Stratigraphy of the Lower Triassic Montney Formation, Peace River Basin, Subsurface of Northwestern Alberta. M.Sc. Thesis, University of Calgary, 368 pp.

Knox, G., 1986. Estuarine Ecosystems, vol. 1. CRC Press, Boca Raton, 304 pp.

Kvale, E.P., Archer, A.W., Johnson, H.R., 1989. Daily, monthly, and yearly tidal cycles within laminated siltstones of the Mansfield Formation (Pennsylvanian) of Indiana. Geology 17, 365–368.

Lanier, W.P., Tessier, B., 1998. Climbing-ripple bedding in the fluvio-estuarine transition: a common feature associated with tidal dynamics (modern and ancient analogues). SEPM Spec. Publ., vol. 61, 109–117.

Leckie, D.A., Singh, C., 1991. Estuarine deposits of the Albian Paddy Member (Peace River Formation) and lowermost Shaftesbury Formation, Alberta, Canada. J. Sediment. Petrol. 61, 825–849.

MacEachern, J.A., Gingras, M.K., 2007. Recognition of brackish-water trace fossil suites in the Cretaceous Western Interior Seaway of Alberta, Canada. In: Bromley, R.G., Buatois, L.A., Mángano, G., Genise, J.F., Melchor, R.N. (Eds.), Sediment-Organism Interactions: A Multifaceted Ichnology. SEPM Spec. Publ., vol. 88, pp. 50–59.

MacEachern, J.A., Pemberton, S.G., 1994. Ichnological aspects of incised-valley fill systems from the Viking Formation of the Western Canada sedimentary basin, Alberta, Canada. In: Dalrymple, W., Boyd, R., Zaitlin, B.A. (Eds.), Incised-Valley Systems: Origin and Sedimentary Sequences. SEPM Spec. Publ., vol. 51, pp. 129–157.

MacEachern, J.A., Bechtel, D.J., Pemberton, S.G., 1992. Ichnology and sedimentology of transgressive deposits, transgressively related deposits and transgressive systems tracts in the Viking Formation of Alberta. In: Pemberton, S.G. (Ed.), Applications of Ichnology to Petroleum Exploration—A Core Workshop. SEPM Core Workshop Notes 17, pp. 251–290.

MacEachern, J.A., Stelck, C.R., Pemberton, S.G., 1999. Marine and marginal marine mudstone deposition: paleoenvironmental interpretations based on the integration of ichnology, palynology and foraminiferal paleoecology. In: Bergman, K.M., Snedden, J.W. (Eds.), Isolated Shallow Marine Sand Bodies: Sequence Stratigraphic Analysis and Sedimentologic Interpretation. SEPM Spec. Publ., vol. 64, pp. 205–225.

MacEachern, J.A., Pemberton, S.G., Gingras, M.K., Bann, K.L., 2010. Ichnology and facies models. In: James, N., Dalrymple, R. (Eds.), Facies Models 4. Geol. Ass. Can., St. John's, Geotext 6, pp. 19–58.

MacEachern, J.A., Dashtgard, S.E., Knaust, D., Bann, K.L., Pemberton, S.G., 2012. Sequence stratigraphy. In: Knaust, D., Bromley, R.G. (Eds.), Trace Fossils as Indicators of Sedimentary Environments. Developments in Sedimentology, vol. 64. Elsevier, Amsterdam, pp. 157–194.

Markhasin, B., 1998. Sedimentology and Stratigraphy of the Lower Triassic Montney Formation, Subsurface of Northwestern Alberta. M.Sc. Thesis, University of Calgary, 153 pp.

Martinius, A.W., Van den Berg, J.H., 2011. Atlas of Sedimentary Structures in Estuarine and Tidally Influenced River Deposits of the Rhine-Meuse-Scheldt System. Their Application to the Interpretation of Analogous Outcrop and Subsurface. EAGE Publications, Houten, 298 pp.

Mayou, T.V., Howard, J.D., 1975. Animal-sediment relationships of a salt marsh estuary, Doboy Sound. In: Howard, J.D., Frey, R.W. (Eds.), Estuaries of the Georgia Coast, U.S.A. Sedimentology and Biology. Senckenberg. Marit, vol, 7, pp. 205–236.

McCann, S.B., Bryant, E.A., 1972. Beach changes and wave conditions, New Brunswick. Proc. Coast. Eng. Conf. 13, 1293–1304.

McIlroy, D., 2007. Ichnology of a macrotidal tide-dominated deltaic depositional system: Lajas Formation, Neuquen Province, Argentina. In: Bromley, R.G., Buatois, L.A., Mángano, G., Genise, J.F., Melchor, R.N. (Eds.), Sediment-Organism Interactions: A Multifaceted Ichnology. SEPM Spec. Publ., vol. 88, pp. 195–211.

McIlroy, D., Flint, S., Howell, J.A., Timms, N., 2005. Sedimentology of the tide-dominated Jurassic Lajas Formation, Neuquen Basin, Argentina. In: Veiga, G.D., Spalletti, L.A., Howell, J.A., Schwarz, E. (Eds.), The Neuquen Basin, Argentina: A Case Study in Sequence Stratigraphy and Basin Dynamics. Geol. Soc. London, Spec. Publ., vol. 252, pp. 83–107.

Mitchum Jr., R.M., 1977. Seismic stratigraphy and global changes of sea level, Part 11. Glossary of terms used in seismic stratigraphy. In: Payton, C.E. (Ed.), Seismic Stratigraphy: Applications to Hydrocarbon Exploration. AAPG Mem. 26, pp. 205–212.

Mitchum Jr., R.M., Vail, P.R., 1977. Seismic stratigraphy and global changes of sea level, Part 7. Seismic stratigraphic interpretation procedure. In: Payton, C.E. (Ed.), Seismic Stratigraphy: Applications to Hydrocarbon Exploration. AAPG Mem. 26, pp. 135–143.

Moore, P., Francis, C., 1985. On the water relations and osmoregulation of the beach-hopper *Orchestia gammarellus* (Pallas) (Crustacea: Amphipoda). J. Exp. Mar. Biol. Ecol. 94, 131–150.

Moslow, T.F., Davies, G.R., 1997. Turbidite reservoir facies in the Lower Triassic Montney Formation, west-central Alberta: an issue focused on the study of Triassic of the Western Canada Sedimentary Basin. Bull. Can. Petrol. Geol. 45, 507–536.

Noffke, A., Hertweck, G., Kroncke, I., Wehrmann, A., 2009. Particle size selection and tube structure of the polychaete *Owenia fusiformis*. Estuar. Coast. Shelf Sci. 81, 160–168.

Panek, R., 2000. The Sedimentology and Stratigraphy of the Lower Triassic Montney Formation in the Subsurface of the Peace River Area, Northwestern Alberta. M.Sc. Thesis, University of Calgary, 275 pp.

Pattison, S., 1991. Sedimentology and Allostratigraphy of Regional, Valleyfill, Shoreface and Transgressive Deposits of the Viking Formation (Lower Cretaceous), Central Alberta. Ph.D. Thesis, University of Calgary, 380 pp.

Pattison, S.A.J., 1992. Recognition and interpretation of estuarine mudstones (central basin mudstones) in the tripartite valley-fill deposits of the Viking Formation, central Alberta. In:

Pemberton, S.G. (Ed.), Applications of Ichnology to Petroleum Exploration—A Core Workshop. SEPM Core Workshop Notes 17, pp. 223–249.

Pattison, S.A.J., Walker, R.G., 1994. Incision and filling of a lowstand valley: late Albian Viking Formation at Crystal, Alberta, Canada. J. Sediment. Res. 64, 365–379.

Pattison, S.A.J., Walker, R.G., 1998. Multiphase transgressive filling of an incised valley and shoreface complex, Viking Formation, Sundance-Edson area, Alberta. Bull. Can. Petrol. Geol. 46, 89–105.

Pearson, N.J., Gingras, M.K., 2006. An ichnological and sedimentological facies model for muddy point-bar deposits. J. Sediment. Res. 76, 771–782.

Pearson, N.J., Gingras, M.K., Armitage, I.A., Pemberton, S.G., 2007. Significance of Atlantic sturgeon feeding excavations, Mary's Point, Bay of Fundy, New Brunswick, Canada. Palaios 22, 457–464.

Pemberton, S.G., MacEachern, J.A., 1995. The sequence stratigraphic significance of trace fossils: examples from the Cretaceous foreland basin of Alberta, Canada. In: Van Wagoner, J.C., Bertram, G. (Eds.), Sequence Stratigraphy of Foreland Basin Deposits: Outcrop and Subsurface Examples from the Cretaceous of North America. AAPG Mem. 64, pp. 429–475.

Pemberton, S.G., Flach, P.D., Mossop, G.D., 1982. Trace fossils from the Athabasca Oil Sands, Alberta, Canada. Science 217, 825–827.

Pemberton, S.G., Reinson, G.E., MacEachern, J.A., 1992. Comparative ichnological analysis of late Albian estuarine valley-fill and shelf-shoreface deposits, Crystal Viking Field, Alberta. In: Pemberton, S.G. (Ed.), Applications of Ichnology to Petroleum Exploration—A Core Workshop. SEPM Core Workshop Notes 17, pp. 291–317.

Pemberton, S.G., MacEachern, J.A., Dashtgard, S.E., Bann, K.L., Gingras, M.K., Zonneveld, J.-P., 2012. Shorefaces. In: Knaust, D., Bromley, R.G. (Eds.), Trace Fossils as Indicators of Sedimentary Environments. Developments in Sedimentology, vol. 64. Elsevier, Amsterdam, pp. 563–604.

Pritchard, D.W., 1967. What is an estuary? Physical viewpoint. In: Lauff, G.H. (Ed.), Estuaries. American Association for the Advancement of Science 3, Washington, pp. 3–5.

Rahmani, R.A., 1984. Estuaries on mesotidal shorelines of a Late Cretaceous epicontinental seaway, Canadian Western Interior, Drumheller, Alberta. In: Stott, D.F., Glass, D.J. (Eds.), The Mesozoic of Middle North America. Can. Soc. Petrol. Geol., Mem. 9, pp. 561–572.

Rahmani, R.A., 1988. Estuarine tidal channel and nearshore sedimentation of a Late Cretaceous epicontinental sea, Drumheller, Alberta, Canada. In: De Boer, P.L., Van Gelder, A., Nio, S.D. (Eds.), Tide-influenced sedimentary environments and facies. Symposium on Clastic Tidal Deposits, Utrecht, pp. 433–471.

Ranger, M., Gingras, M., 2008. Depositional history of the McMurray Formation, a model driven by outcrop observation. Can. Soc. Petrol. Geol. Res. 35, 17–18.

Ranger, M.J., Pemberton, S.G., 1992. The sedimentology and ichnology of estuarine point bars in the McMurray Formation of the Athabasca oil sands deposit, northeastern Alberta, Canada. In: Pemberton, S.G. (Ed.), Applications of Ichnology to Petroleum Exploration—A Core Workshop. SEPM Core Workshop Notes 17, pp. 401–421.

Reineck, H.-E., 1956. Der Wattenboden und das Leben im Wattenboden, ein geologischer Streifzug. Natur Volk 86, 268–284.

Reinson, G.E., 1980. Variations in tidal-inlet morphology and stability, northeast New Brunswick. In: McCann, S.B. (Ed.), The Coastline of Canada. Geological Survey of Canada, Paper 80-10, pp. 23–39.

Reinson, G.E., Clark, J.E., Foscolos, A.E., 1988. Reservoir geology of Crystal Viking Field, Lower Cretaceous estuarine tidal channel-bay complex, south-central Alberta. AAPG Bull. 72, 1270–1294.

Remane, A., Schlieper, C., 1972. The Biology of Brackish Water. Wiley Interscience, New York, 372 pp.

Rossetti, D.F., Santos Jr., A.E., 2004. Facies architecture in a tectonically influenced estuarine incised valley fill of Miocene age, northern Brazil. J. South Am. Earth Sci. 17, 267–284.

Roy, P.S., 1994. Holocene estuary evolution. Stratigraphic studies from southeastern Australia. In: Dalrymple, R.W., Boyd, R., Zaitlin, B.A. (Eds.), Incised-Valley Systems. Origin and Sedimentary Sequences. SEPM Spec. Publ., vol. 51, pp. 241–263.

Roy, K., Martien, K., 2001. Latitudinal distribution of body size in north-eastern Pacific marine bivalves. J. Biogeogr. 28, 485–493.

Roy, P.S., Thom, B.G., Wright, L.D., 1980. Holocene sequences on an embayed high-energy coast: an evolutionary model: shallow marine processes and products. Sediment. Geol. 26, 1–19.

Savrda, C.E., Nanson, L.L., 2003. Ichnology of fair-weather and storm deposits in an Upper Cretaceous estuary (Eutaw Formation, western Georgia, USA). Palaeogeogr. Palaeoclimatol. Palaeoecol. 202, 67–83.

Scott, D.B., Greenberg, D.A., 1983. Relative sea-level rise and tidal development in the Fundy tidal system. Can. J. Earth Sci. 20, 1554–1564.

Shaw, J., Gareau, P., Courtney, R.C., 2002. Palaeogeography of Atlantic Canada 13-0 kyr. Q. Sci. Rev. 21, 1861–1878.

Sisulak, C.F., Dashtgard, S.E., 2012. Seasonal controls on the development and character of inclined heterolithic stratification in a tide-influenced, fluvially dominated channel: Fraser River, Canada. J. Sediment. Res. 82, 244–257.

Spaargaren, D.H., 1979. A comparison of the composition of physiological saline solutions with that of calculated Pre-Cambrian seawater. Comp. Biochem. Physiol. 63A, 319–323.

Spaargaren, D.H., 1995. A functional model for describing the responses to salinity stress in aquatic animals. Comp. Biochem. Physiol. 111A, 501–506.

Thomas, R.G., Smith, D.G., Wood, J.M., Visser, J., Calverley-Range, E.A., Koster, E.H., 1987. Inclined heterolithic stratification: terminology, description, interpretation and significance. Sediment. Geol. 53, 123–179.

Thurston, H., 1990. Tidal Life: A Natural History of the Bay of Fundy. Camden House Publishing, Camden East, 167 pp.

Vail, P.R., Mitchum Jr., R.M., 1977. Seismic stratigraphy and global changes of sea level. Part 1. Overview. In: Payton, C.E. (Ed.), Seismic Stratigraphy: Applications to Hydrocarbon Exploration. AAPG Mem. 2, pp. 51–52.

Van den Berg, J.H., 1981. Rhythmic seasonal layering in a mesotidal channel fill sequence, Oosterschelde Mouth, The Netherlands. In: Nio, S.-D., Shüttenhelm, R.T.E., Van Weering, T.C.E. (Eds.), Holocene Marine Sedimentation in the North Sea Basin. IAS Spec. Publ. 5. Blackwell Publishing Ltd, Oxford, pp. 147–159.

Van Wagoner, J.C., Mitchum Jr., R.M., Posamentier, H.W., Vail, P.R., 1987. Seismic stratigraphy interpretation using sequence stratigraphy. Part 2. Key definitions of sequence stratigraphy. In: Bally, A.W. (Ed.), Atlas of Seismic Stratigraphy. AAPG Studies in Geology, vol. 27, pp. 11–14.

Virtasalo, J.J., Bonsdorff, E., Moros, M., Kabel, K., Kotilainen, A.T., Ryabchuk, D., Kallonen, A., Hämäläinen, K., 2011. Ichnological trends along an open-water transect across a large marginal-marine epicontinental basin, the modern Baltic Sea. Sediment. Geol. 241, 40–51.

Visher, G.S., Howard, J.D., 1974. Dynamic relationship between hydraulics and sedimentation in the Altamaha Estuary. J. Sediment. Petrol. 44, 502–521.

Wetzel, A., Tjallingii, R., Stattegger, K., 2010. *Gyrolithes* in Holocene estuarine incised-valley fill deposits, offshore southern Vietnam. Palaios 25, 239–246.

Williams, G.E., 1989. Tidal rhythmites: geochronometers for the ancient Earth-Moon system. Episodes 12, 162–171.

Wolanski, E., 2007. Estuarine Ecohydrology. Elsevier, Amsterdam, 157 pp.

Yang, B., Dalrymple, R.W., Gingras, M.K., Chun, S., Lee, H., 2007. Up-estuary variation of sedimentary facies and ichnocoenoses in an open-mouthed, macrotidal, mixed-energy estuary, Gomso Bay, Korea. J. Sediment. Res. 77, 757–771.

Zonneveld, J.-P., Pemberton, S.G., 2003. Ichnotaxonomy and behavioral implications of lingulide-derived trace fossils from the Lower and Middle Triassic of Western Canada. Ichnos 10, 25–39.

Zonneveld, J.-P., Beatty, T.W., Pemberton, S.G., 2007. Lingulide brachiopods and the trace fossil *Lingulichnus* from the Triassic of Western Canada: implications for faunal recovery after the end-Permian mass extinction. Palaios 22, 74–97.

Chapter 17

Deltas

Nicola S. Tonkin[1,2]
Department of Earth Sciences, Memorial University, St. John's, Newfoundland and Labrador, Canada
[1]Corresponding author: e-mail: nicola.tonkin-normore@woodside.com.au
[2]Current address: Woodside Energy Ltd., Perth, Australia

1. INTRODUCTION

Subaqueous deltaic environments in marine basins are shaped by the interactions of fluvial and marine hydrodynamic processes as well as the geometry of the recipient basin. Fluvial, tidal, and wave-dominated or -influenced deltas produce distinctive coastal morphologies, as well as having diagnostic lateral and stratigraphic facies trends. The matrix of possible combinations of these parameters is evident in the resultant sedimentology and ichnology. Sedimentological analysis is used as the primary tool to define deltaic deposits and their subenvironments. Ichnological characteristics may provide an excellent additional tool in recognition of deltaic subenvironments. Integration of sedimentology and ichnology enables the construction of realistic paleoenvironmental models, incorporating physico-chemical parameters such as salinity, hydrodynamic energy, sedimentation rate, oxygenation, substrate consistency, turbidity, light, and temperature. Such insights can be important in appreciating the facies shifts associated with deltaic processes that are not evident from the study of physical sedimentary structures alone. The ichnological characteristics of a deltaic succession can be used to infer the relative importance of fluvial, tidal, or wave processes and can aid in recognition of the key stages in the delta development (Galloway, 1975). The regressive (progradational) and transgressive phases of the deltaic cycle are associated with changes in relative sea level and sediment supply (Boyd et al., 1992; Galloway, 1975; Roberts, 1997).

A delta is defined as the—partially subaerial—site of sediment deposition where a river enters a standing body of water (Galloway, 1975). Deltas classically form discrete shoreline protuberances, where the alluvial system entering the sedimentary basin supplies sediment more rapidly than it can be redistributed away from the distributary mouth, by the predominant basinal processes of waves and/or tides (Boyd et al., 1992; Galloway, 1975; Orton and Reading, 1993). Ancient deltaic successions, as seen in outcrop and core, are recognized as coarsening- and

Developments in Sedimentology, Vol. 64. http://dx.doi.org/10.1016/B978-0-444-53813-0.00017-4

shallowing-upward progradational packages of siliciclastic rocks. Deltaic successions are abundant and widely distributed throughout geological time (see Supplementary Table 1: http://booksite.elsevier.com/9780444538130).

Deltaic research, from a geological perspective, began in the late 1880s and focused on the facies architecture, sedimentary processes, and stratigraphic evolution of deltaic sedimentary successions (Barrell, 1912; Gilbert, 1885; see historical review in Le Blanc, 1975). Microfossils and macrofossils are routinely used as proxies for deltaic paleoenvironmental reconstruction, including their use in defining the extent of marine influence and brackish to freshwater fluvial conditions (e.g., Beerbower, 1961; Coleman and Wright, 1975; Galloway, 1975; Horowitz, 1966; Murphy and Schlanger, 1962). Ichnology was not an integral part of deltaic studies until the 1960s, this coinciding with the promotion of ichnological research (see Baucon et al., 2012). The integration of ichnology into facies analysis has been used to help understand both modern and ancient deltas since the late 1960s (Allen, 1970; Fisher and McGowen, 1969; Oomkens, 1970). Several recent studies of ancient deltas have used ichnology as a tool to classify and define subenvironments and facies within deltaic systems (see Supplementary Table 1: http://booksite.elsevier.com/9780444538130), although further research is required to fully understand these dynamic environments.

The sedimentology of deltaic environments is comparatively well understood, and a significant amount of published literature is available (reviews in Bhattacharya, 2006, 2010; Reading and Collinson, 1996). The wealth of studies of modern deltaic systems such as the Mississippi River Delta (review in Roberts, 1997) stems from their importance as sites of agriculture and trade. In addition, ancient deltas host significant hydrocarbon reserves (Whateley and Pickering, 1989). Ichnology can, with care, be used to help determine delta type, define subenvironments, and characterize depositional facies, when combined with sedimentological datasets (e.g., MacEachern et al., 2005; McIlroy, 2004b; Pemberton et al., 2001). This chapter focuses on the subaqueous marine component of the deltaic system and aims to clarify the application of ichnology in ancient deltas and discuss its use as a tool for study of deltaic paleoenvironments and facies analysis.

2. CLASSIFICATION OF DELTAIC SYSTEMS

Modern deltas can be classified in terms of (1) the dominant process (fluvial, tidal, or wave energy; see Galloway, 1975); (2) grain size (see Orton and Reading, 1993); (3) delta morphology (Weise, 1980); (4) type of feeder; (5) water depth; and (6) mouth-bar type (see Postma, 1990). Characterization of ancient deltas is somewhat more challenging, as the deltaic coastline morphology (e.g., elongate, lobate) can usually only be studied in 3D seismic. Outcrop and 2D seismic datasets cannot fully characterize the geomorphology of a delta unless the limit of progradation of a deltaic package can be mapped. Deltaic sediments have significant potential to be reworked by subsequent deltaic processes (e.g., ravinement during abandonment or through erosion by subsequent distributary channels; see McIlroy et al., 2005). Recent classifications subdivide wave-dominated deltas into symmetric, asymmetric, and deflected morphologies (Bhattacharya and Giosan, 2003). Deltas have also been described in terms of river mouth process and dispersal into (1) hypopycnal (buoyancy- and friction-dominated deltas), (2) hyperpycnal (inertia-dominated deltas), and (3) homopycnal (friction-dominated deltas) (see Bhattacharya, 2006; Wright, 1977), which affects the mouth-bar development (Postma, 1990).

The ichnology of the three end members of the deltaic classification of Galloway (1975) is discussed herein (Supplementary Table 1: http://booksite.elsevier.com/9780444538130 and Fig. 1). Most deltas encompass a combination of processes (Galloway, 1975; Supplementary Table 1: http://booksite.elsevier.com/9780444538130). Detailed sedimentological descriptions and facies analysis of fluvial-dominated, wave-dominated, and tide-dominated ancient deltaic successions are not covered in this chapter, as several excellent reviews exist in the current literature (see Bhattacharya, 2010; Dalrymple, 2010; Reading and Collinson, 1996).

2.1 Bioturbation Intensity and Ichnodiversity

The intensity of bioturbation (BI) and ichnological diversity in ancient deltaic settings are useful first-order tools in recognition of deltaic facies, facies association, and the dominant sedimentary processes. BI is a measure of the destruction

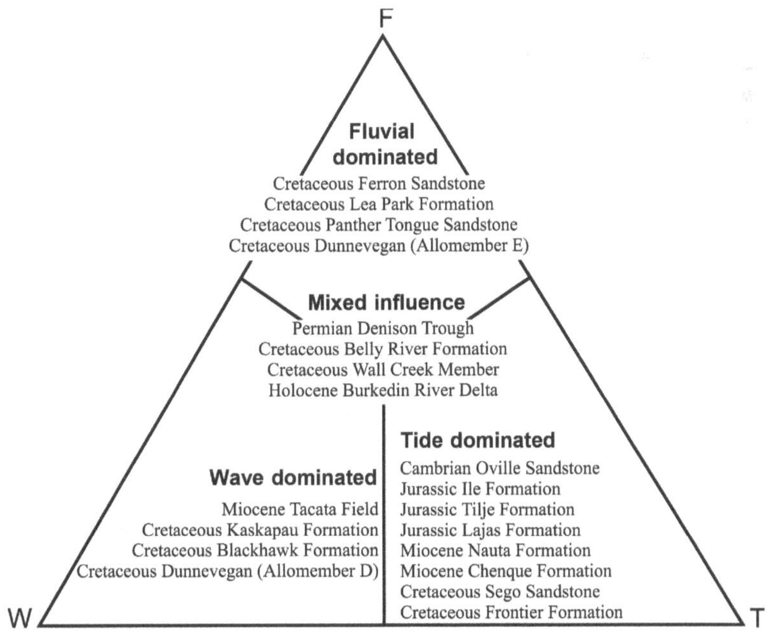

FIGURE 1 Tripartite classification of ancient deltas based on the dominant processes of fluvial, tidal, and wave energy (based on Galloway, 1975). Mixed-influence deltas lie in the central area of the triangle. Twenty-two examples from refereed publications on fine-grained ancient deltas are used (see Supplementary Table 1: http://booksite.elsevier.com/9780444538130).

of the primary sedimentological fabric (Knaust, 2012; Taylor and Goldring, 1993). The bioturbation index reaches from BI = 0, which equates to zero bioturbation and extends up to BI = 6, which indicates 100% complete biological reworking (e.g., homogenization) of the primary sedimentary fabric. The measurement of ichnodiversity (number of ichnotaxa per unit) is influenced by (1) taphonomy (preservation), (2) sediment textural and mineralogical maturity, (3) colonization history, (4) BI, and (5) the taxonomic philosophy of the investigator (sedimentologist or specialist ichnologist; e.g., "lumper" or "splitter").

2.2 Ichnofabric and Ichnofacies Analysis

The ichnofabric is a characteristic approach used to categorize the ichnology, diversity, bioturbation level, and colonization order of bioturbated beds (see Ekdale et al., 2012; McIlroy, 2008; Taylor et al., 2003 for detailed methodology). Ancient deltas can also be categorized using Seilacherian ichnofacies (see MacEachern et al., 2007a, 2012 for detailed methodology). Ancient fluvial-, tide-, and wave-dominated deltas may contain archetypal, stressed, proximal, distal, and mixed expressions of the *Skolithos* and *Cruziana* ichnofacies (Supplementary Table 1: http://booksite.elsevier.com/9780444538130). In addition, "*Glossifungites* surfaces" are recognized at the bases of distributary channels (McIlroy, 2004b) and at sequence-stratigraphic surfaces with submarine erosion such as the transgressive surface of erosion (Gingras et al., 2002; MacEachern and Gingras, 2008; Pemberton et al., 1992). The *Curvolithus* Ichnofacies characterizes shallow deltaic or marginal-marine environments with rapid deposition in quiet-water settings (Lockley et al., 1987), in contrast to rapid deposition in high-energy settings that characterize the *Skolithos* Ichnofacies. Alternatively, the study of ichnofabrics (e.g., McIlroy, 2004b, McIlroy et al., 2005) can be incorporated with sedimentology, and the results presented as an ichnological analysis (see McIlroy, 2008).

3. COMMON DELTAIC TRACE FOSSILS

The assemblage or suite of trace fossils in an ichnofabric or ichnofacies, when considered in their sedimentological context, may be used to refine subenvironments of marine deltas. The most abundant trace fossils in ancient marine deltas are also the most widely distributed in siliciclastic environments. *Planolites*, *Thalassinoides*, *Palaeophycus*, and *Teichichnus* are documented in all deltaic types and occur in all subaqueous deltaic facies associations (lower delta plain, delta front, and prodelta; see Supplementary Table 1: http://booksite.elsevier.com/9780444538130 and Fig. 2).

The diversity of a trace-fossil assemblage (ichnodiversity) should not be directly linked to species diversity or abundance on the sea floor. Factors that may obscure or enhance apparent ichnodiversity are (1) preservational biases (taphonomy), which favor mid- to deep-tier traces (mixed layer may be removed by erosion); (2) burrow overprinting (e.g., palimpsest surfaces);

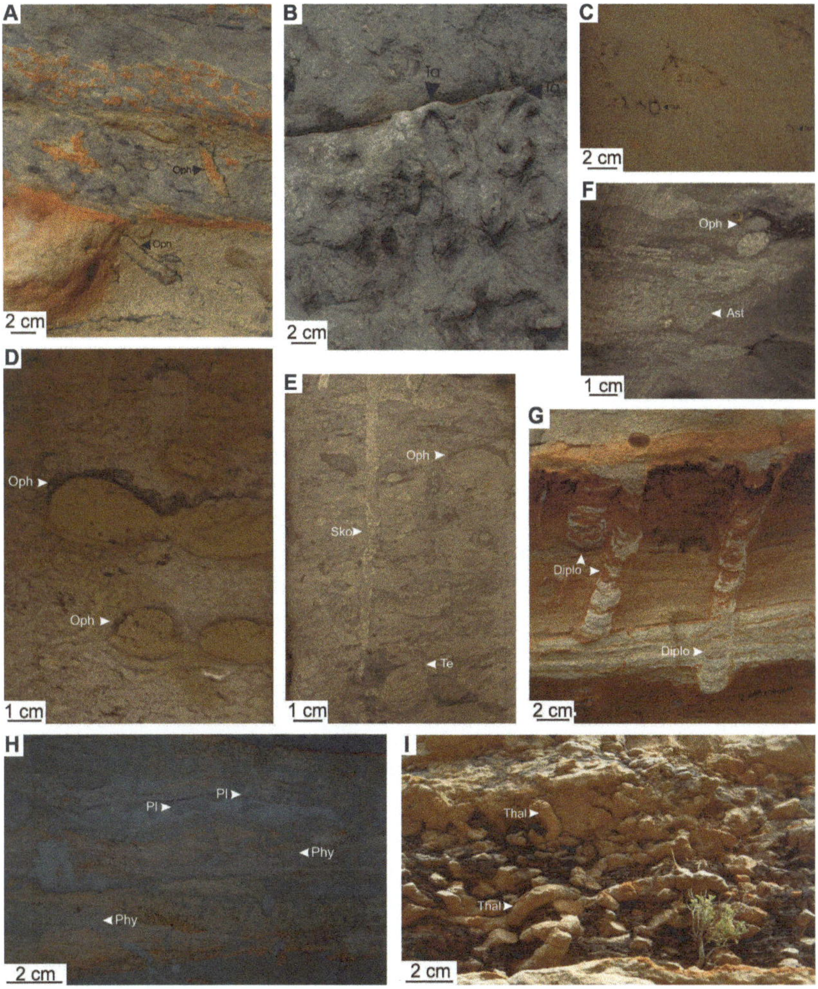

FIGURE 2 Common deltaic trace fossils. (A–C) Tidal-deltaic facies, Middle Jurassic Lajas Formation, Neuquén Basin, Argentina. (A) *Ophiomorpha* (Oph) in delta-front facies. (B) *Taenidium* (Ta). (C) *Thalassinoides* (Th) at a parasequence boundary. (D–F) Wave-influenced facies, Cretaceous Ben Nevis Formation, offshore Newfoundland, Canada. (D) *Ophiomorpha* (Oph). (E) *Ophiomorpha* (Oph), *Skolithos* (Sko), and *Teichichnus* (Te). (F) *Asterosoma* (Ast) and *Ophiomorpha* (Oph). (G–H) Gilbert-type delta, Pleistocene, Conway Flats, North Canterbury, New Zealand. (G) *Diplocraterion* (Diplo). (H) Burrow mottling, *Planolites* (Pl) and *Phycosiphon* (Phy). (I) *Thalassinoides* (Th) in channel facies, Cretaceous Blackhawk Formation, Book Cliffs, USA.

(3) the ability for one organism to create multiple burrow morphologies (behavioral plasticity) and to create extensive burrow networks (e.g., *Thalassinoides*); and (4) changes in ichnodiversity through geologic time (e.g., colonization of brackish-water environments, Buatois et al., 2005). While ichnodiversity cannot be used as a direct indicator of species richness, it is a valid proxy for

diversity of behavior and it is extremely useful to document changes throughout a deltaic succession from, for example, prodelta to delta front. The presence or absence of monotaxic assemblages is also highly significant. Ichnofacies composed of a single ichnotaxon is usually indicative of a "stressed" paleoenvironment, associated with high energy, high rates of sedimentation, and lowered salinity (e.g., distributary channels and mouth bars at the proximal delta front).

4. CONTROLS ON DELTAIC ICHNOLOGY

The controls on benthic ecology in marine subaqueous environments are physicochemical parameters including (1) hydrodynamic energy (fluvial, tidal, and wave); (2) salinity; (3) oxygen; (4) sedimentation rate; (5) substrate consistency; and (6) turbidity, light, and temperature. Biological parameters include the salinity tolerance of the tracemaker, burrow morphology, duration of colonization window, food supply, and trophic/feeding mode (see reviews in Gingras et al., 2011; MacEachern et al., 2007b; McIlroy, 2004a; Taylor et al., 2003). Reworking of fluvial-derived sediments at the coastline by oceanographic processes (e.g., wave and tidal) controls the intensity, diversity, and distribution of trace fossils in deltaic environments. The main paleoenvironmental controls on bioturbation in deltaic settings are sedimentation rate, salinity, and hydrodynamic energy.

4.1 Hydrodynamic Energy and Response of Bioturbators

The distribution of trace fossils is intimately linked to hydrodynamic energy in all depositional settings, including deltaic systems. High sedimentation rates and erosion are commonly associated with intense hydrodynamic energy in distributary channel and delta-front facies. These facies are characterized by vertical burrows conventionally interpreted as being made by suspension-feeding organisms (e.g., *Diplocraterion* and *Thalassinoides*; Fig. 2G and I). Trace-fossil diversity and intensity typically increase away from the source of intense hydrodynamic energy (e.g., distributary mouth and shoreface). High-energy subenvironments are associated with the largest grain size, so their geometry and architecture are of particular interest in hydrocarbon reservoir characterization (Moiola et al., 2004). Rapid sedimentation associated with event beds can take place in otherwise quiescent conditions, for example, hyperpycnal-derived beds of sand or mud in fluvial-dominated deltas (Olariu et al., 2010) and storm-induced turbidite-like sand-rich beds in the prodelta of (wave-dominated) deltas (Pattison, 2005).

4.1.1 Lower Delta Plain

In the lower delta plain, distributary channels are the conduits through which fluvial sediment is delivered to the receiving sedimentary basin. The branching patterns and abundance of channels at the lower delta plain to proximal delta front vary between delta types (Olariu and Bhattacharya, 2006), although it is unclear as to how these data can be used geologically, as most deltaic deposits are cut by non-contemporaneous distributary channels during progradation (Zaitlin et al., 1994). All distributary channels are, at least periodically, zones of relatively high, albeit variable, energy. The funnel-shaped morphology of distributary channels, and their direct connection to the ocean, means that the currents in the channels are most likely to be tidally modulated (Brettle et al., 2002; Dalrymple and Choi, 2007). Along-strike variability in deltaic successions can be highly significant, especially in the more proximal environments of the lower delta plain, where quiescent interdistributary areas (e.g., swamp, salt marshes, lakes, bays, lagoons, or tidal flats) are cut by high-energy distributary channels, although interdistributary areas of higher energy are common in wave-dominated systems (e.g., beaches; Boyd, 2010).

Typical trace fossils in the high-energy environments of distributary channels and lower delta-plain deposits are *Diplocraterion*, *Macaronichnus*, *Skolithos*, *Arenicolites*, and *Ophiomorpha* (Fig. 2A, C, and F). As such, the deposits have an ichnological assemblage comparable to the *Skolithos* Ichnofacies of Seilacher (1967).

4.1.2 Delta Front

At the delta front, tidal and wave energy in the receiving basin significantly control the sea floor environment, as fluvial energy dissipates. Seaward of the distributary mouth bar, hydrodynamic energy dissipates rapidly. This dissipation of hydrodynamic energy results in trace-fossil diversity and intensity Increasing seaward.

At the distal delta front of fluvial-dominated deltas, reworking of sediment by tides and waves is low. In wave-dominated delta fronts, physical and biogenic sedimentary structures have similar associations as those seen on non-deltaic coastlines such as beaches and strandplains. In large deltas, such coastal facies are contemporaneous with, and lateral to, deposition on the active delta lobe (Bhattacharya and Giosan, 2003).

In the moderately high-energy environments of the delta front, seaward of the distributary mouth and above storm-wave base, distal mouth-bar deposits are dominated by the trace fossils of sediment-feeding tracemakers such as *Teichichnus*, *Asterosoma*, *Ophiomorpha*, *Thalassinoides*, *Rosselia*, and *Chondrites* (equivalent to the proximal *Cruziana* Ichnofacies; Fig. 2F and I).

4.1.3 Prodelta

Hypopycnal and hyperpycnal flows from the distributary are the primary dispersal systems of mud and sand to the prodelta (see Fig. 3A–C). Delivery of mud to the distal delta can also be attributed to rapid post-event-bed deposition in quiescent hydrodynamic conditions, or flocculation and suspended sediment settling in quiescent conditions (e.g., slack-water tidal systems; Ichaso and Dalrymple, 2009; Lamb and Parsons, 2005; McIlroy et al., 2005; Sheremet et al.,

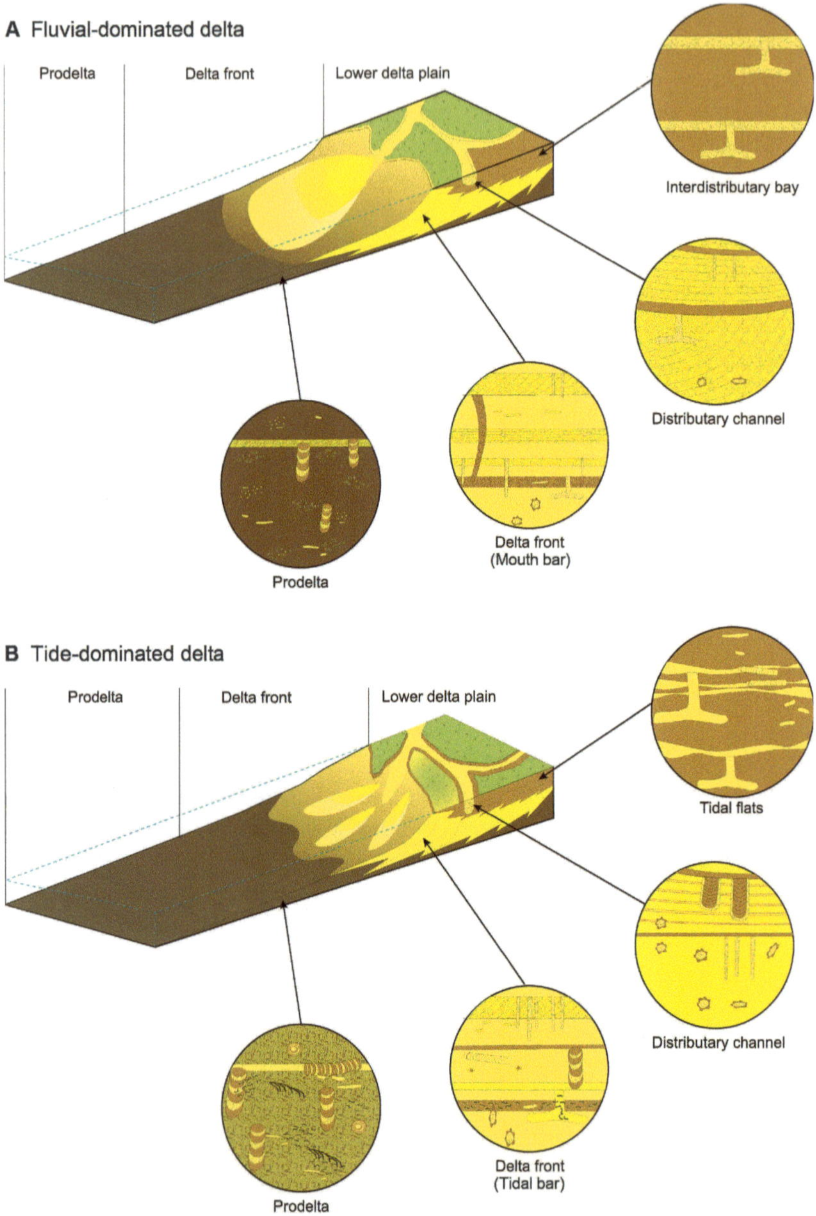
A Fluvial-dominated delta
Prodelta
Delta front
Lower delta plain
Interdistributary bay
Distributary channel
Delta front
(Mouth bar)
Prodelta
B Tide-dominated delta
Prodelta
Delta front
Lower delta plain
Tidal flats
Distributary channel
Delta front
(Tidal bar)
Prodelta

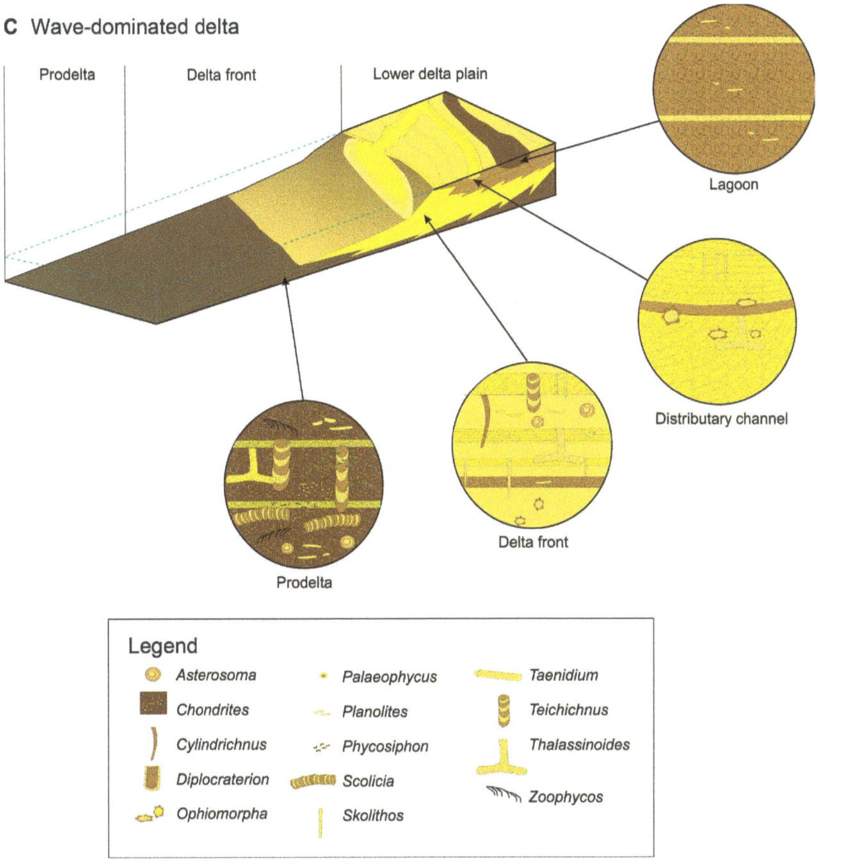

FIGURE 3 Deltaic ichnology facies models, including block diagram of an idealized environment and corresponding idealized log through an interdistributary area (swamp, tidal flat, and lagoon) distributary channel, delta front, and prodelta facies. (A) Fluvial-dominated ancient river delta, (B) tide-dominated ancient river delta, and (C) wave-dominated ancient river delta. *Based on Allen (1970).*

2005; Uncles et al., 2006). Fluid-mud deposits are often characterized as unbioturbated structureless mudstone beds. Sandstone interbeds in the prodelta can be wave induced, as sand stored in the proximal delta is transported seaward with enhanced river discharge or wave and tidal currents (Dalrymple and Choi, 2007; Willis et al., 1999).

In the distal delta front below the fair-weather wave base and in the prodelta, deposits typically contain marine trace fossils, for example, *Thalassinoides*, *Phycosiphon*, *Chondrites*, *Zoophycos*, *Scolicia*, and *Palaeophycus* (equivalent to the distal *Cruziana* and *Zoophycos* ichnofacies; Fig. 2H and I).

4.2 Salinity and Salinity Tolerance of Tracemakers

Deltas form at the junction between a river and a receiving basin. The receiving basin, into which deltaic systems develop, can be grouped into marine, brackish/marginal-marine, and freshwater or hypersaline environments (see Melchor et al., 2012), but in this chapter, only subaqueous deltaic-marine environments will be discussed. Biological diversity and organism size are thought to covary with salinity. The salinity-specific effects on infaunal organisms are less well constrained. Rapid salinity fluctuations are somewhat buffered in pore-water systems due to slow rates of diffusion (see discussion in McIlroy, 2004a), and most infaunal, especially shelled, organisms have adaptations to protect themselves against rapid salinity changes. Ichnological assemblages of exclusively freshwater environments are usually devoid of complex feeding burrows and contain no diagnostically marine ichnotaxa (Melchor et al., 2012). Brackish-water assemblages are in general environmentally unstable, and salinity is seldom demonstrably the first-order control on ecology (temperature, sedimentation rate, substrate consistency, and turbidity are generally all involved).

4.2.1 Salinity Wedge

Buoyant freshwater sits above saline water (the saltwater wedge), and the wedge laden with fine suspended sediment extends landward, trapping silts and clays in the lower delta plain, until fluvial discharge is high enough to push the wedge seaward into the receiving marine basin (Dalrymple and Choi, 2007). The limit of tidal influence (tidal reach) is the boundary between the lower and upper delta plain (Coleman and Prior, 1982) and is usually associated with tidal to fluvial facies. The salinity wedge is critically important in facies analysis of tide-dominated ancient deltaic successions (Ichaso and Dalrymple, 2009; McIlroy, 2004b; Willis et al., 1999). Fluvial influence can also extend seaward of the distributary mouth as hypopycnal, homopycnal, or hyperpycnal flows (Bhattacharya and MacEachern, 2009; Wright, 1977). As a consequence of the salinity wedge, marine influence from the receiving basin can extend into the delta plain, and consequently the range of marine trace fossils follows a similar trend.

4.2.2 Lower Delta Plain and Delta Front

Fluvial, brackish, and marine ichnological assemblages are distinctly different, with some wide-ranging marine ichnotaxa being extremely tolerant of low-salinity conditions or salinity variations. It has been suggested that the trace fossils tolerant of salinity changes, such as *Ophiomorpha* or *Thalassinoides* in monotaxic assemblages, could be used as indicators of penetration of the salinity wedge (Swinbanks and Luternauer, 1987). In tidal-deltaic settings, marine and brackish salinity can extend into the lower delta plain, extending the landward range of some trace fossils. Typical brackish-water ichnological assemblages contain *Skolithos*, *Thalassinoides*, *Planolites*, *Lockeia*, *Palaeophycus*, and low-diversity assemblages (see Supplementary Table 1: http://booksite.elsevier.com/

9780444538130; reviews in Buatois et al., 2005; Pemberton et al., 1992). In fluvial-dominated deltas, salinity decreases markedly due to the dominance of fluvial outflow, resulting in freshwater, or brackish-water, lower delta plain ichnological assemblages with low BI (e.g., BI = 0–2; see Supplementary Table 1: http://booksite.elsevier.com/9780444538130).

Depauperate assemblages are commonly considered to be indicative of brackish-water settings, but an integrated approach incorporating careful sedimentary facies analysis along with consideration of ichnofaunal distributions is probably needed (cf. McIlroy, 2004a). Persistently brackish paleoenvironments, in association with deltaic deposition, are only really expected in distributary channels, near-channel paleoenvironments, and potentially in restricted lagoons (Bhattacharya, 2010).

4.2.3 Prodelta

The salinity in the prodeltaic settings is predominantly marine, with episodic brackish-water conditions during sediment-laden discharge from riverine outflow, particularly in fluvial-dominated deltaic systems. This gives prodeltaic facies a distinctly marine ichnological assemblage with high diversity and variable BI.

4.3 Sedimentation Rate and the Colonization Window

The deceleration of fluvial currents upon meeting standing marine waters causes rapid deposition at the basin margin to produce deltas. The fate of that sediment is variable, from instantaneous and permanent deposition to erosion and redistribution by waves and tides, sometimes in a multicyclic fashion. Deltaic coastlines are all characterized by comparatively high rates of sediment supply. The sediment source in deltas is characteristically fluvial, and rates of shoreline progradation are largely controlled by a combination of sediment supply (usually high), accommodation space (usually low), and the relative importance of river, tide, and wave processes acting at the coastline (Orton and Reading, 1993; Roberts, 1997).

Shallow-marine successions with intense bioturbation indicate slow rates of sedimentation relative to the rate of bioturbation. In order to assess the importance of any given highly bioturbated bed or bedset, it is important to be able to calibrate the intensely bioturbated unit with its unbioturbated counterpart. In deltaic successions, some facies are persistently highly bioturbated, indicating either low rates of sedimentation or optimal conditions for bioturbation, or a combination of both. In deltaic depositional systems, the distal delta front to prodelta and interdistributary bay fills commonly meet these criteria.

4.3.1 Lower Delta Plain

In the lower delta-plain facies, sedimentation rates are highly variable. Interdistributary areas have low net accumulation rates and may have high to complete BI, but many contain only low-diversity assemblages (e.g., abandoned channel fills with *Ophiomorpha*; McIlroy et al., 2005). Conversely, distributary channel facies are characterized by episodic high rates of sedimentation and overall low to moderate net accumulation, comprising a highly variable trace-fossil distribution with dominantly non-bioturbated beds.

4.3.2 Delta Front and Prodelta

Sedimentation rates are typically highest in distal mouth-bar deposits and within terminal distributary channels (*sensu* Olariu and Bhattacharya, 2006) of the proximal delta-front facies. In areas with rapid deposition, the colonization window is open only for short periods of time, inhibiting colonization of the ancient sea floor and establishment of a benthic epifaunal and infaunal assemblage. Net sediment accumulation on the delta front is typically rapid, successions are often devoid of trace fossils or sparsely bioturbated with low-diversity assemblages, and primary sedimentary fabrics remain intact and preserved. Limited bioturbation may in part be a result of high sedimentation rates but also of paleoenvironmental stress (e.g., salinity fluctuations) at the distributary mouth. Conversely, in areas of infrequent event-bed deposition on the distal delta front and the prodelta, BI and ichnodiversity are higher. The colonization window in the distal prodelta is usually long enough open for biogenic reworking of sediments. In the predominantly quiescent setting of the distal delta front, bioturbation commonly disturbs and largely obscures primary sedimentary structures, reflected in intense bioturbation. The low paleoenvironmental stress in the distal delta front and the high organic content commonly result in diverse ichnological assemblages of trace fossils.

4.3.3 Event-Bed Deposition and the Colonization Window

Fluvial input into the marine basins can be variable and linked to seasonal storms. In such cases, deposition is generally episodic and event-bed deposition is likely to dominate the delta front and prodelta settings. This inter-bedding of fair-weather and event-bed deposits may produce "lam-scram" fabrics of alternating intense to low bioturbation (Howard, 1972). The thickness of event beds is an important limiting factor on benthic ecology, as it may smother existing infaunal communities, effectively causing defaunation if the pre-event community is unable to escape to the new sediment/water interface (Pollard et al., 1993; Wheatcroft and Drake, 2003). If colonization from below is not possible, post-depositional recolonization by juveniles or adult organisms is possible (Taylor and Goldring, 1993), although the new sea floor substrate may not be entirely hospitable to deposit-feeding organisms due to a lack of deposited organic matter (see discussion in Herringshaw et al., 2010). *Macaronichnus*, *Skolithos*, *Diplocraterion*, *Arenicolites*, *Rosselia*, and *Cylindrichnus* are often found in association with shifting substrates and event-bed deposition (Fig. 2G; Ekdale and Lewis, 1991; Gingras et al., 1998) and have the ability to adjust to the new sediment/water interface. Sandstone-infilled *Thalassinoides* burrows in a mudstone matrix are often interpreted to be associated with hiatal "*Glossifungites* surfaces" and are characteristic of the *Glossifungites* Ichnofacies (Pemberton and Frey, 1985), which are commonly interpreted as markers for key stratal surfaces of allocyclic significance (Fig. 2I; MacEachern et al., 2007c, 2012).

4.4 Oxygenation

The oxygenation of both interstitial pore waters and sea-water is relatively common first-order controls on trace-fossil distribution in subaqueous environments. Well-oxygenated sea-water may encourage intense bioturbation and deep burrowing. Deep or shallow infaunal habitation and colonization of the shallow, mid, and deep tiers (vertical partitioning of the substrate) can often be used as an indicator of well-oxygenated bottom waters and pore waters in the substrate (Bromley, 1996).

4.4.1 Lower Delta Plain

Some deltaic environments may have high concentrations of sedimentary organic matter (e.g., swamps, tidal flats, marshes, lagoons, and bay fills), which create oxygen-poor pore-water environments. Endobenthic activity in such organic-rich sediments is seldom limited by the absence of pore-water oxygen, which can be introduced by bioirrigation from the overlying water column. Bioirrigation stimulates microbial growth in the burrow wall (see Herringshaw et al., 2010). It seems likely that organisms have endosymbionts that thrive in low-oxygen sediment (e.g., some echinoids; Bromley, 1996) and produce ichnofabrics with suppressed BI and diversity, independent of the bathymetry (Ekdale and Mason, 1988). Tidal-flat facies containing solely *Ophiomorpha* or *Thalassinoides* may be indicative of a stressed paleoenvironment where pore-water oxygenation fluctuates with tidal exposure (Fig. 2B, C, and I; Swinbanks and Luternauer, 1987).

4.4.2 Delta Front and Prodelta

Deltaic systems, being areas of dynamic coastal processes, are generally characterized by strong bottom currents. These currents are commonly a combination of fluvial, wave, and tidal processes, which keep mixed the water column and bottom water. Consequently, sea-water oxygenation is not generally a controlling factor in the occurrence and distribution of trace fossils in deltaic environments, especially in the delta front and prodeltaic environments.

Localized areas of bottom-water dysoxia (e.g., interdistributary areas) are associated with a low ichnodiversity (e.g., monotaxic trace-fossil assemblages), a low BI, and poorly developed tiering profiles. Oxygenated sea-water, however, is the "norm" in all deltaic types, although pore waters in sediments rich in organic matter may be sulfidic and hostile to endobenthos.

4.5 Substrate Consistency and Benthic Colonization Style

Substrate consistency on the sea floor in ancient deltaic setting is commonly soft to loose, with occasional soup- and firmgrounds (Goldring, 1995). The colonization of soft- and looseground substrates in the lower delta plain to prodelta is interpreted as reflecting relatively continuous deposition or aggradation, whereas development of a firmground can be indicative of a hiatal surface produced by erosion or non-deposition in the deltaic complex (e.g., Fig. 2I; *Glossifungites* surface). Firmground surfaces in the delta-front setting are commonly colonized by *Thalassinoides* and may result from a hiatus or colonization of buried dewatered cohesive mudstones (e.g., concealed firmgrounds; Bromley, 1996). Soupgrounds are generally found in association with fluvial- and tide-dominated deltas, especially where hyperpycnal flows cause deposition of fluid-mud.

4.6 Turbidity, Light, and Temperature

Turbidity of the water column is a result of suspended sediment load. A high turbidity is common in many deltaic environments, including buoyant plumes emanating from the distributary mouth into the basin. Turbidity may reduce the BI and diversity in deltaic environments (MacEachern et al., 2005) and can also reduce light penetration to the sediment/water interface, which may have a feedback into microbial food productivity for tracemaking organisms. Seasonal temperature fluctuations of the water column triggered by flood or changes in ocean circulation may place stress on tracemakers, reducing the BI and ichnodiversity (e.g., fluvial-dominated and intertidal deltaic facies). Turbidity, light, and temperature are not considered major controlling factors for the ichnological distribution in the deltaic realm.

5. ICHNOLOGY OF DELTAIC END MEMBERS

The complex interplay between fluvial, wave, and tidal processes, and their fluctuating environmental parameters of hydrodynamic energy and sedimentation rates are common to all deltas. Subtle variations in the relative parameters enable characteristic facies successions and ichnological assemblages to be defined. Salinity profiles down-dip (e.g., a salt-wedge in a tide-dominated delta or flood discharge in a fluvial-dominated delta) are distinctive for each deltaic type and their corresponding subenvironment (see Supplementary Table 1: http://booksite.elsevier.com/9780444538130 and Fig. 3). Distributary channels are commonly fresh to brackish and are characterized by non-bioturbated or sparsely bioturbated facies. Delta-front facies in all three end-member deltaic types demonstrate the largest ichnological variability. This reflects fundamental geomorphological and sedimentological paleoenvironmental differences. Bioturbation intensity and ichnodiversity typically increase seaward of the distributary mouth (see Fig. 3). In prodelta facies, the contrasts between intensities and trace-fossil diversity in the four deltaic types appear to even out (Fig. 3).

5.1 Fluvial-Dominated Deltaic Ichnology

Bioturbation intensity and diversity of trace-fossil assemblages are reduced in the lower delta plain and delta front of fluvial-dominated delta successions (BI = 0–4; see Supplementary Table 1: http://booksite.elsevier.com/9780444538130 and Fig. 3A). The highly variable ichnodiversity and BI of these facies are a direct result of the variable fluvial discharge and associated high sedimentation rates at, and near, the distributary mouth. Seaward of the distributary mouth, the fluvial influence dissipates and prodeltaic deposits contain highly diverse ichnofaunal assemblages with moderate to intense bioturbation (BI = 3–6; see Supplementary Table 1: http://booksite.elsevier.com/9780444538130 and Fig. 3A). Typically, localized monotaxic assemblages of *Ophiomorpha* or *Palaeophycus* or *Teredolites* are prevalent in the distributary channels, interdistributary areas, and proximal delta front. Low-diversity assemblages containing *Planolites*, *Skolithos*, *Lockeia*, *Teichichnus*, *Diplocraterion*, and *Cylindrichnus* characterize the delta front (Fielding, 2010; Gingras et al., 1998; Moiola et al., 2004; Olariu et al., 2005).

Depauperate assemblages in fluvial-dominated deltas are a consequence of seasonally high rates of sedimentation and sustained brackish-water conditions during high river discharge. Other factors that may suppress the ichnological diversity and BI in fluvial-dominated deltas are the prevalence of seasonal hypopycnal and subordinate hyperpycnal flows (associated with a high fluvial discharge) which create a higher water turbidity within the water column and at the sea-floor (MacEachern et al., 2005). It is important to note that all five fluvial-dominated deltaic examples are from Cretaceous rocks deposited in interior seaways, so that further research is required to give a balanced argument (see Supplementary Table 1: http://booksite.elsevier.com/9780444538130 and Figs. 1 and 3A).

5.2 Tide-Dominated Deltaic Ichnology

The marine signature in tide-dominated deltaic successions is able to extend well into the deltaic plain; therefore, moderate to complete bioturbation intensities (up to BI = 6) exist in all subenvironments as a result of the tidal range (see Fig. 3B). A high trace-fossil diversity persists in this facies association, notably higher than in wave-dominated, fluvial-dominated, and mixed-influence ancient deltas (see Supplementary Table 1: http://booksite.elsevier.com/9780444538130). The salt-water wedge increases the marine ichnological signature landward, up the distributary channels into the lower delta plain. Therefore, distributary channels, tidal channels, and tidal flats in tide-dominated deltas have distinct facies, mud-rich heterolithic sandstones with moderate intensity and diversity. The delta front is characterized by moderate- to high-diversity trace-fossil assemblages with a strongly heterolithic nature. The prodeltaic facies is similar to the facies of other end-member deltaic types of fluvial- and wave-dominated deltas but is characterized by variable intensities and high-diversity trace-fossil assemblages. Euryhaline tracemakers of *Planolites*, *Teichichnus*, *Phycosiphon*, *Rosellia*, *Asterosoma*, *Chondrites*, *Taenidium*, *Ophiomorpha*, *Palaeophycus*, *Nereites*, *Scolicia*, and *Diplocraterion* characterize the delta and prodeltaic facies of tide-dominated ancient deltas (Carmona et al., 2009; Ichaso and Dalrymple, 2009; McIlroy, 2004b; McIlroy et al., 2005; Rebata-H et al., 2006; Willis and Gabel, 2001).

5.3 Wave-Dominated Deltaic Ichnology

High diversity of trace-fossil assemblages and moderate to intense bioturbation intensities (BI = 3–5; see Supplementary Table 1: http://booksite.elsevier.com/9780444538130 and Fig. 3C) characterize the delta front and prodelta of wave-dominated ancient deltaic successions. Ancient wave-dominated river deltas developed shoreface facies and may be difficult to distinguish from non-deltaic shorelines unless the large-scale spatial context is well delineated by seismic data, good outcrop exposures or closely spaced, reliably correlated core, or well-log data (Hampson and Howell, 2005). In the wave-dominated prodeltaic facies, the BI and ichnodiversity are the highest of all deltaic types as a function of extremely well mixed and highly oxygenated waters that have a full marine signature, creating optimal conditions for colonization of the ancient sea floor. Trace-fossil assemblages in the lower delta plain to delta front result from euryhaline tracemakers of *Planolites*, *Ophiomorpha*, *Teichichnus*, *Skolithos*, *Phycosiphon*, *Helminthopsis*, *Thalassinoides*, *Cylindrichnus*, *Arenicolites*, *Rosellia*, *Asterosoma*, *Chondrites*, *Rhizocorallium*, *Palaeophycus*, *Diplocraterion*, and *Macaronichnus* (Buatois et al., 2008; Charvin et al., 2010; Gingras et al., 1998; Reid and Pemberton, 2007). In the prodelta, marine tracemakers such as those of *Zoophycos* and *Scolicia* are found. In wave-dominated deltas, the lower delta plain and proximal delta front to prodelta may be indistinguishable from non-deltaic successions (e.g., shorefaces) unless a fluvial point source can be recognized and the spatial context is constrained by seismic or continuous outcrop exposures.

6. CONCLUSIONS

The dominant physico-chemical controls in ancient river deltas are the sedimentation rate and the salinity; these parameters control the spatial distribution of trace fossils. Bioturbated fluvial-dominated and tidal-dominated ancient deltaic successions contain mostly of trace fossils made by euryhaline organisms, while wave-dominated deltaic successions contain a mix of both euryhaline and stenohaline organisms. Deltaic successions are typically stressed or mixed expressions of *Skolithos* and *Cruziana* ichnofacies, with *Cruziana* elements becoming more distinct in the distal delta front to prodelta. The dynamic combination of fluvial, tidal, and wave processes creates distinct environments and subenvironments with corresponding ichnological signatures (see Fig. 1 and Supplementary Table 1: http://booksite.elsevier.com/9780444538130).

Fluvial-dominated deltas have a weaker ichnological marine signature in the lower delta plain and proximal delta-front environments than their marine and tidal counterparts. Fresh- and brackish-water trace-fossil assemblages with absent to high BI and reduced diversity (BI = 0–4; Supplementary Table 1: http://booksite.elsevier.com/9780444538130) dominate in proximal settings. The ichnodiversity and BI of assemblages increase markedly (BI = 3–6; Supplementary Table 1: http://booksite.elsevier.com/9780444538130) seaward of fluvial influence in the marine distal portions of the fluvial delta complex (Fig. 3A).

In tide-dominated deltaic successions, the marine ichnological signature extends into the lower delta plain as a result of the salinity wedge (Fig. 3B). High-diversity trace-fossil assemblages of brackish to marine trace fossils persist throughout the successions, while the BI is high but variable (BI = 0–6; Supplementary Table 1: http://booksite.elsevier.com/9780444538130).

In wave-dominated deltas, the marine signature is also suppressed in the lower delta plain, with absent to moderate BI (BI = 0–3; Supplementary Table 1: http://booksite.elsevier.com/9780444538130). In the strongly marine, high-diversity and intensity assemblages (BI = 0–5; Supplementary Table 1: http://booksite.elsevier.com/9780444538130) of the delta front and prodelta, it is often difficult to distinguish non-deltaic shoreface from offshore successions (Fig. 3C), unless three-dimensional data are available.

REFERENCES

Allen, J.R.L., 1970. Sediments of the modern Niger delta, a summary and review. In: Morgan, J.P. (Ed.), Deltaic Sedimentation, Modern and Ancient, SEPM Spec. Publ., vol. 15, pp. 138–151.

Barrell, J., 1912. Criteria for the recognition of ancient delta deposits. GSA Bull. 23, 377–446.

Baucon, A., Bordy, E., Brustur, T., Buatois, L.A., Cunningham, T., De, C., Duffin, C., Felletti, F., Gaillard, C., Hu, B., Hu, L., Jensen, S., Knaust, D., Lockley, M., Lowe, P., Mayor, A., Mayoral, E., Mikuláš, R., Muttoni, G., Neto de Carvalho, C., Pemberton, S.G., Pollard, J., Rindsberg, A.K., Santos, A., Seike, K., Song, H., Turner, S., Uchman, A., Wang, Y., Yi-ming, G., Zhang, L., Zhang, W., 2012. A history of ideas in ichnology. In: Knaust, D., Bromley, R.G. (Eds.), Trace Fossils as Indicators of Sedimentary Environments. Developments in Sedimentology, vol. 64. Elsevier, Amsterdam, pp. 3–43.

Beerbower, J.R., 1961. Origin of cyclothems of the Dunkard Group (Upper Pennsylvanian-Lower Permian) in Pennsylvania, West Virginia, and Ohio. GSA Bull. 72, 1029–1050.

Bhattacharya, J.P., 2006. Deltas. In: Posamentier, H.W., Walker, R.G. (Eds.), Facies Models Revisited. SEPM Spec. Publ., vol. 84, pp. 237–292.

Bhattacharya, J.P., 2010. Deltas. In: James, N.P., Dalrymple, R.W. (Eds.), Facies Models 4. Geol. Ass. Can., St. John's. Geotext 6, pp. 233–264.

Bhattacharya, J.P., Giosan, L., 2003. Wave-influenced deltas: geomorphological implications for facies reconstruction. Sedimentology 50, 187–210.

Bhattacharya, J.P., MacEachern, J.A., 2009. Hyperpycnal rivers and prodeltaic shelves in the Cretaceous Seaway of North America. J. Sediment. Res. 79, 184–209.

Boyd, R., 2010. Transgressive wave-dominated coasts. In: James, N.P., Dalrymple, R.W. (Eds.), Facies Models 4. Geol. Ass. Can., St. John's, Geotext 6, pp. 265–294.

Boyd, R., Dalrymple, R.W., Zaitlin, B.A., 1992. Classification of coastal sedimentary environments. Sediment. Geol. 80, 139–150.

Brettle, M.J., McIlroy, D., Davies, S.J., Elliott, T., Waters, C.N., 2002. Identifying cryptic tidal influences within deltaic successions: an example from the Marsdenian (Namurian) interval of the Pennine Basin, UK. J. Geol. Soc. London 159, 379–391.

Bromley, R.G., 1996. Trace Fossils: Biology, Taphonomy and Applications. Chapman & Hall, London, 361 pp.

Buatois, L.A., Gingras, M.K., MacEachern, J., Mángano, M.G., Zonneveld, J.-P., Pemberton, S.G., Netto, R.G., Martin, A., 2005. Colonization of brackish-water systems through time: evidence from the trace-fossil record. Palaios 20, 321–347.

Buatois, L.A., Santiago, N., Parra, K., Steel, R., 2008. Animal–substrate interactions in an Early Miocene wave-dominated tropical delta: delineating environmental stresses and depositional dynamics (Tácata Field, Eastern Venezuela). J. Sediment. Res. 78, 458–479.

Carmona, N.B., Buatois, L.A., Ponce, J.J., Mángano, M.G., 2009. Ichnology and sedimentology of a tide-influenced delta, Lower Miocene Chenque Formation, Patagonia, Argentina: trace-fossil distribution and response to environmental stresses. Palaeogeogr. Palaeoclimatol. Palaeoecol. 273, 75–86.

Charvin, K., Hampson, G.J., Gallagher, K.L., Labourdette, R., 2010. Intra-parasequence architecture of an interpreted asymmetrical wave-dominated delta. Sedimentology 57, 760–785.

Coleman, J.M., Prior, D.B., 1982. Deltaic environments. In: Scholle, P.A., Spearing, D.R. (Eds.), Sandstone Depositional Environments. AAPG Mem. 31, pp. 139–178.

Coleman, J.M., Wright, L.D., 1975. Major river deltas: variability of processes and sandbodies. In: Broussard, M.L. (Ed.), Deltas, Models for Exploration. Houston Geol. Soc., Mem. 13, pp. 99–149.

Dalrymple, R.W., 2010. Tidal depositional systems. In: James, N.P., Dalrymple, R.W. (Eds.), Facies Models 4. Geol. Ass. Can., St. John's, Geotext 6, pp. 201–231.

Dalrymple, R.W., Choi, K., 2007. Morphologic and facies trends through the fluvial-marine transition in tide-dominated depositional systems: a schematic framework for environmental and sequence stratigraphic interpretation. Earth-Sci. Rev. 81, 135–174.

Ekdale, A.A., Lewis, D.W., 1991. Trace fossils and paleoenvironmental control of ichnofacies in a late Quaternary gravel and loess fan delta complex, New Zealand. Palaeogeogr. Palaeoclimatol. Palaeoecol. 81, 253–279.

Ekdale, A.A., Mason, T.R., 1988. Characteristic trace-fossil associations in oxygen-poor sedimentary environments. Geology 16, 720–723.

Ekdale, A.A., Bromley, R.G., Knaust, D., 2012. The ichnofabric concept. In: Knaust, D., Bromley, R.G. (Eds.), Trace Fossils as Indicators of Sedimentary Environments. Developments in Sedimentology, vol. 64. Elsevier, Amsterdam, pp. 139–155.

Fielding, C.R., 2010. Planform and facies variability in asymetric deltas: facies analysis and depositional architecture of the Turonian Ferron Sandstone in the Western Henry Mountains, south-central Utah, USA. J. Sediment. Res. 80, 455–479.

Fisher, W.L., McGowen, J.H., 1969. Depositional systems in the Wilcox Group of Texas and their relation to occurrence of oil and gas. AAPG Bull. 53, 30–54.

Galloway, W.E., 1975. Process framework for describing the morphologic and stratigraphic evolution of deltaic depositional systems. In: Broussard, M.L. (Ed.), Deltas, Models for Exploration. Houston Geol. Soc., Mem. 13, pp. 87–98.

Gilbert, G.K., 1885. The topographic features of lake shores. U.S. Geol. Surv., 5th Ann. Rep. (1883–1884), 69–123.

Gingras, M.K., MacEachern, J.A., Pemberton, S.G., 1998. A comparative analysis of the ichnology of wave- and river-dominated allomembers of the Upper Cretaceous Dunvegan Formation. Bull. Can. Petrol. Geol. 46, 51–73.

Gingras, M.K., MacEachern, J.A., Dashtgard, S.E., 2011. Process ichnology and the elucidation of physico-chemical stress. Sediment. Geol. 237, 115–134.

Gingras, M.K., Räsänen, M., Pemberton, S.G., Romero, L.P., 2002. Ichnology and sedimentology reveal depositional characteristics of bay-margin parasequences in the Miocene Amazonian Foreland Basin. J. Sediment. Res. 72, 871–883.

Goldring, R., 1995. Organisms and substrate: response and effect. In: Bosence, D.W.J., Allison, P.A. (Eds.), Marine Palaeoenvironmental Analysis from Fossils. Geol. Soc. London, Spec. Publ., vol. 83, pp. 151–180.

Hampson, G.J., Howell, J.A., 2005. Sedimentologic and geomorphic characterization of ancient wave-dominated shorelines: examples from the Late Cretaceous Blackhawk Formation, Book Cliffs, Utah. In: Giosan, L., Bhattacharya, J.P. (Eds.), River Deltas: Concepts, Models and Examples. SEPM Spec. Publ., vol. 83, pp. 133–154.

Herringshaw, L.G., Sherwood, O., McIlroy, D., 2010. Ecosystem engineering by bioturbating polychaetes in event bed microcosms. Palaios 25, 46–58.

Horowitz, D.H., 1966. Evidence for deltaic origin of an Upper Ordovician sequence in the central Appalachians. In: Shirley, M.L., Ragsdale, J.A. (Eds.), Deltas in Their Geologic Framework. Houston Geol. Soc., pp. 159–169.

Howard, J.D., 1972. Trace fossils as criteria for recognizing shorelines in stratigraphic record. In: Rigby, J.K., Hamblin, W.K. (Eds.), Recognition of Ancient Sedimentary Environments. SEPM Spec. Publ., vol. 16, pp. 215–225.

Ichaso, A.A., Dalrymple, R.W., 2009. Tide- and wave-generated fluid mud deposits in the Jurassic Tilje Formation, offshore Norway. Geology 37, 539–542.

Knaust, D., 2012. Methodology and techniques. In: Knaust, D., Bromley, R.G. (Eds.), Trace Fossils as Indicators of Sedimentary Environments. Developments in Sedimentology, vol. 64. Elsevier, Amsterdam, pp. 245–271.

Lamb, M.P., Parsons, J.D., 2005. High-density suspensions formed under waves. J. Sediment. Res. 75, 386–397.

Le Blanc, R.J., 1975. Significant studies of modern and ancient deltaic sediments. In: Broussard, M.L. (Ed.), Deltas, Models for Exploration. Houston Geol. Soc., pp. 13–85.

Lockley, M.G., Rindsberg, A.K., Zeiler, R.M., 1987. The paleoenvironmental significance of the nearshore *Curvolithus* ichnofacies. Palaios 2, 255–262.

MacEachern, J.A., Gingras, M.K., 2008. Recognition of brackish-water trace fossil suites in the Cretaceous Western Interior Seaway of Alberta, Canada. In: Bromley, R.G., Buatois, L.A., Mángano, G., Genise, J.F., Melchor, R.N. (Eds.), Sediment-Organism Interactions: A Multifaceted Ichnology. SEPM Spec. Publ., vol. 88, pp. 149–194.

MacEachern, J.A., Bann, K.L., Bhattacharya, J.P., Howell, C.D., 2005. Ichnology of deltas; organism responses to the dynamic interplay of rivers, waves, storms, and tides. In: Giosan, L., Bhattacharya, J.P. (Eds.), River Deltas: Concepts, Models and Examples. SEPM Spec. Publ., vol. 83, pp. 49–85.

MacEachern, J.A., Dashtgard, S.E., Knaust, D., Catuneanu, O., Bann, K.L., Pemberton, S.G., 2012. Sequence stratigraphy. In: Knaust, D., Bromley, R.G. (Eds), Trace Fossils as Indicators of Sedimentary Environments. Developments in Sedimentology, vol. 64. Elsevier, Amsterdam, pp. 157–194.

MacEachern, J.A., Gingras, M.K., Bann, K., Pemberton, S.G., 2007a. The ichnofacies paradigm: high-resolution paleoenvironmental interpretation of the rock record. In: MacEachern, J.A., Pemberton, S.G., Gingras, M.K., Bann, K. (Eds.), Applied Ichnology. SEPM Short Course Notes 52, pp. 27–65.

MacEachern, J.A., Gingras, M.K., Bann, K., Pemberton, S.G., Reich, L.T., 2007b. Applications of ichnology to high-resolution genetic stratigraphic paradigms. In: MacEachern, J.A., Pemberton, S.G., Gingras, M.K., Bann, K. (Eds.), Applied Ichnology. SEPM Short Course Notes 52, pp. 95–131.

MacEachern, J.A., Pemberton, S.G., Bann, K., Gingras, M.K., 2007c. Departures from the archetypal ichnofacies: effective recognition of environmental stress in the rock record. In: MacEachern, J.A., Pemberton, S.G., Gingras, M.K., Bann, K. (Eds.), Applied Ichnology. SEPM Short Course Notes 52, pp. 65–95.

McIlroy, D., 2004a. A review of ichnological concepts, methodologies, applications and frontiers. In: McIlroy, D. (Ed.), The Application of Ichnology to Palaeoenvironmental and Stratigraphic Analysis. Geol. Soc. London, Spec. Publ., vol. 228, pp. 3–27.

McIlroy, D., 2004b. Ichnofabrics and sedimentary facies of a tide-dominated delta; Jurassic Ile Formation of Kristin Field, Haltenbanken, offshore mid-Norway. In: McIlroy, D. (Ed.), The Application of Ichnology to Palaeoenvironmental and Stratigraphic Analysis. Geol. Soc. London, Spec. Publ., vol. 228, pp. 237–272.

McIlroy, D., 2008. Ichnological analysis. The common ground between ichnofacies workers and ichnofabric analysts. Palaeogeogr. Palaeoclimatol. Palaeoecol. 270, 332–338.

McIlroy, D., Flint, S., Howell, J.A., Timms, N., 2005. Sedimentology of the tide-dominated Jurassic Lajas Formation, Neuquén Basin, Argentina. In: Veiga, G.D., Spalletti, L.A., Howell, J.A., Schwarz, E. (Eds.), The Neuquén Basin, Argentina: A Case Study in Sequence Stratigraphy and Basin Dynamics. Geol. Soc. London, Spec. Publ., vol. 252, pp. 83–107.

Melchor, R.N., Genise, J.F., Buatois, L.A., Umazano, A.M., 2012. Fluvial environments. In: Knaust, D., Bromley, R.G. (Eds.), Trace Fossils as Indicators of Sedimentary Environments. Developments in Sedimentology, vol. 64. Elsevier, Amsterdam, pp. 329–378.

Moiola, R.J., Welton, J.E., Wagner, J.B., Fearn, L.B., Farrell, M.E., Enrico, R.J., Echols, R.J., 2004. Integrated analysis of the Upper Ferron deltaic complex, Southern Castle Valley, Utah. In: Chidsey Jr., T.C., Adams, R.D., Morris, T.H. (Eds.), Analog for Fluvial-Deltaic Reservoir Modeling: Ferron Sandstone of Utah. AAPG Studies in Geology 50, pp. 79–92.

Murphy, M.A., Schlanger, S.O., 1962. Sedimentary structures in Cretaceous Ilhas and Sho Sebastiho Formations, Reconcavo Basin, Brazil. AAPG Bull. 46, 457–477.

Olariu, C., Bhattacharya, J.P., 2006. Terminal distributary channels and delta front architecture of river-dominated delta systems. J. Sediment. Res. 76, 212–233.

Olariu, C., Bhattacharya, J.P., Xu, X., Aiken, C.L.V., Zeng, X., McMechan, G.A., 2005. Integrated study of ancient deltafront deposits, using outcrop, ground-penetrating radar, and three-dimensional photorealistic data; Cretaceous Panther Tongue sandstone, Utah, USA. In: Giosan, L., Bhattacharya, J.P. (Eds.), River Deltas: Concepts, Models and Examples. SEPM Spec. Publ., vol. 83, pp. 155–177.

Olariu, C., Steel, R.J., Petter, A.L., 2010. Delta-front hyperpycnal bed geometry and implications for reservoir modeling: Cretaceous Panther Tongue delta, Book Cliffs, Utah. AAPG Bull. 94, 819–845.

Oomkens, E., 1970. Depositional sequences and sand distribution in the postglacial Rhone delta complex. In: Morgan, J.P. (Ed.), Deltaic Sedimentation Modern and Ancient. SEPM Spec. Publ., vol. 15, pp. 198–212.

Orton, G.J., Reading, H.G., 1993. Variability of deltaic processes in terms of sediment supply, with particular emphasis on grain size. Sedimentology 40, 475–512.

Pattison, S.A.J., 2005. Isolated highstand shelf sandstone body of turbiditic origin, lower Kenilworth Member, Cretaceous Western Interior, Book Cliffs, Utah. USA. Sediment. Geol. 177, pp. 131–144.

Pemberton, S.G., Frey, R.W., 1985. The *Glossifungites* ichnofacies: modern examples from the Georgia Coast, USA. In: Curran, H.A. (Ed.), Biogenic Structures: Their Use in Interpreting Depositional Environments. SEPM Spec. Publ., vol. 35, pp. 237–259.

Pemberton, S.G., Frey, R.W., Ranger, M.J., MacEachern, J., 1992. The conceptual framework of ichnology. In: Pemberton, S.G. (Ed.), Applications of Ichnology to Petroleum Exploration—A Core Workshop. SEPM Core Workshop Notes 17, pp. 1–32.

Pemberton, S.G., Spila, M., Pulham, A.J., Saunders, T., MacEachern, J.A., Robbins, D., Sinclair, I.K., 2001. Ichnology & Sedimentology of Shallow to Marginal Marine Systems. Geol. Ass. Canada, Short Course Notes 15, 343 pp.

Pollard, L.E., Goldring, R., Buck, S.G., 1993. Ichnofabrics containing *Ophiomorpha*: significance in shallow-water facies interpretation. J. Geol. Soc. Lond. 150, 149–164.

Postma, G., 1990. Depositional architecture and facies of river and fan deltas: a synthesis. In: Colella, A., Prior, D.B. (Eds.), Coarse-Grained Deltas. IAS Spec. Publ., vol. 10, pp. 13–27.

Reading, H.G., Collinson, J.D., 1996. Clastic coasts. In: Reading, H.G. (Ed.), Sedimentary Environments and Facies. Blackwell Science, Oxford, pp. 154–231.

Rebata-H, L.A., Gingras, M.K., Räsänen, M.E., Barberi, M., 2006. Tidal channel deposits on a delta plain from the Upper Miocene Nauta Formations, Maranon Foreland sub-basin, Peru. Sedimentology 53, 971–1013.

Reid, S.A., Pemberton, S.G., 2007. Facies architecture of the Doe Creek Member of the Kaskapau Formation, NW Alberta: utilizing ichnology to enhance deltaic depositional models. In: MacEachern, J.A., Pemberton, S.G., Gingras, M.K., Bann, K. (Eds.), Applied Ichnology. SEPM Short Course Notes 52, pp. 280–297.

Roberts, H.H., 1997. Dynamic changes of the Holocene Mississippi River delta plain; the delta cycle. J. Coast. Res. 13, 605–627.

Seilacher, A., 1967. Bathymetry of trace fossils. Mar. Geol. 5, 413–428.

Sheremet, A., Mehta, A.J., Liu, B., Stone, G.W., 2005. Wave-sediment interaction on a muddy inner shelf during Hurricane Claudette. Estuar. Coast. Shelf Sci. 63, 225–233.

Swinbanks, D.D., Luternauer, J.L., 1987. Burrow distribution of thalassinidean shrimp on a Fraser Delta tidal flat, British Columbia. J. Paleontol. 61, 315–332.

Taylor, A.M., Goldring, R., 1993. Description and analysis of bioturbation and ichnofabric. J. Geol. Soc. Lond. 150, 141–148.

Taylor, A.M., Goldring, R., Gowland, S., 2003. Analysis and application of ichnofabrics. Earth-Sci. Rev. 60, 227–259.

Uncles, R.J., Stephens, J.A., Law, D.J., 2006. Turbidity maximum in the macrotidal, highly turbid Humber Estuary, UK: flocs, fluid mud, stationary suspensions and tidal bores. Estuar. Coast. Shelf Sci. 67, 30–52.

Weise, B.R., 1980. Wave-dominated deltaic systems of the Upper Cretaceous San Miguel Formation, Maverick Basin, South Texas. Report of Investigations107, Texas Bureau of Economic Geology, Austin, Texas, 39 pp.

Whateley, M.K.G., Pickering, K.T., 1989. Deltas: sites and traps for fossil fuels. Geol. Soc. London, Spec. Publ., vol. 41, 360 pp.

Wheatcroft, R.A., Drake, D.E., 2003. Post-depositional alteration and preservation of sedimentary event layers on continental margins. I. The role of episodic sedimentation. Mar. Geol. 199, 123–137.

Willis, B.J., Gabel, S.L., 2001. Sharp-based, tide-dominated deltas of the Sego Sandstone, Book Cliffs, Utah, USA. Sedimentology 48, 479–506.

Willis, B.J., Bhattacharya, J.P., Gabel, S.L., White, C.D., 1999. Architecture of a tide-influenced river delta in the Frontier Formation of central Wyoming, USA. Sedimentology 46, 667–688.

Wright, L.D., 1977. Sediment transport and deposition at river mouths: a synthesis. GSA Bull. 88, 857–868.

Zaitlin, B.A., Dalrymple, R.W., Boyd, R., 1994. The stratigraphic organization of incised-valley systems associated with relative sea-level change. In: Dalrymple, R.W., Boyd, R., Zaitlin, B.A. (Eds.), Incised-Valley Systems: Origin and Sedimentary Sequences. SEPM Spec. Publ., vol. 51, pp. 45–60.

Zonneveld, J.-P., Gingras, M.K., Beatty, T.W., Bottjer, D.J., Chaplin, J.R., Greene, S.E., Martindale, R.C., Mata, S.A., McHugh, L.P., Pemberton, S.G., Schoengut, J.A., 2012. Mixed siliciclastic/carbonate systems. In: Knaust, D., Bromley, R.G. (Eds.), Trace Fossils as Indicators of Sedimentary Environments. Developments in Sedimentology, vol. 64. Elsevier, Amsterdam, pp. 807–834.

Chapter 18

Tidal Flats and Subtidal Sand Bodies

Patricio R. Desjardins,*[,1] Luis A. Buatois[†] and M. Gabriela Mángano[†]
*Shell International Exploration and Production Inc., Houston, Texas, USA, [†]Department of Geological Sciences, University of Saskatchewan, Saskatoon, Saskatchewan, Canada
[1]Corresponding author: e-mail: p.desjardins@shell.com

1. INTRODUCTION

During the past two decades, the number of publications dealing with the ichnology of wave-dominated shallow-marine systems has outpaced those on tide-dominated settings. In addition, most of the research dealing with tide-influenced or tide-dominated environments is actually focused on marginal-marine brackish-water environments, such as estuaries, bays, and deltas (e.g., Carmona et al., 2009; Gingras and MacEachern, 2012; Gingras et al., in press; MacEachern and Gingras, 2007; McIlroy, 2007) and, even more recently, tidally influenced and modulated shorefaces (Dashtgard et al., in press), rather than open-marine tide-dominated settings. Despite these shortcomings, a number of key contributions have outlined some of the most relevant characteristics of ichnofaunas in fully marine environments where tides play a major role in mobilizing and depositing sediments, namely, tidal flats, subtidal sandbar complexes, and tidal dunes and sand ridges in open shelf settings (e.g., Desjardins et al., 2010b, 2012a; Mángano and Buatois, 1999, 2004a,b; Mángano et al., 2002a; Schäfer, 1972).

A major difference between wave-dominated and tide-dominated systems is that the former display a clear onshore–offshore trend in energy conditions (Buatois and Mángano, 2011; Pemberton et al., 2012), whereas the energy gradient in tide-dominated shallow-marine environments is not straightforward (Yoshida et al., 2007). Despite all these complexities, tide-dominated shorelines, in particular those developed in macrotidal settings, can be subdivided into supratidal, upper, middle, and lower intertidal, and subtidal environments (Klein, 1971, 1977), the associated depositional profile displaying clear ichnological trends (Buatois and Mángano, 2011; Mángano and Buatois, 1999,

Developments in Sedimentology, Vol. 64. http://dx.doi.org/10.1016/B978-0-444-53813-0.00018-6

2004b). In the subtidal region, energy-related and other ecological parameters can also vary across a single sand body, which leads to localized colonization trends (Desjardins et al., 2012a).

The focus of this study is placed on macrotidal (tidal range >4 m) open-marine settings, in which tidal currents are the dominant sedimentary processes. Excluded are the ichnofaunas of tidally influenced or tidally dominated marginal-marine settings, such as estuaries, deltas, or interdistributary bays, which are typically of lower diversity in brackish-water environments (see Gingras et al., 2012), or contain freshwater trace fossils as in the case of fluvio-estuarine transitions (see Buatois et al., 1997; Mángano and Buatois, 2004b). In addition to the classic zonation for intertidal settings, the environmental framework for tidal shelves adopted in this study is that of Desjardins et al. (2010a), where the position of the effective wave bases (*sensu* Flemming, 2005) serves to separate three main environmental zones: (1) the outer shelf below the effective storm-wave base, (2) the inner shelf above the effective storm-wave base and below the effective fair-weather wave base, and (3) the shallow subtidal environment above the effective fair-weather wave base and below the low-tide line, bathymetrically equivalent of the shoreface as commonly defined for wave-dominated settings (Plint, 2010). The intertidal and supratidal environments lie landward of the shallow subtidal environment (Fig. 1).

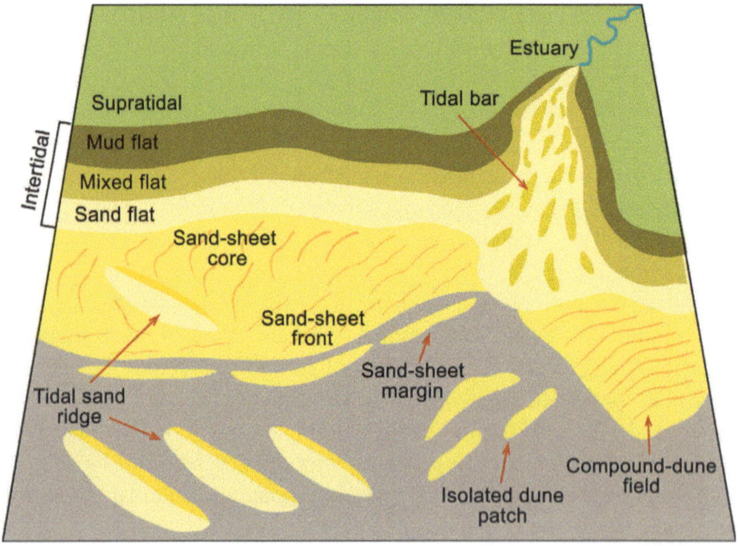

FIGURE 1 Sedimentary environments and sand-body distribution in tide-dominated shallow-marine settings.

2. SUPRATIDAL AND INTERTIDAL ENVIRONMENTS

In contrast to wave-dominated shorelines, which coarsen landward, tidal flats in general coarsen seaward parallel to an increase in tidal energy. In tide-dominated shorelines, a typical tidal-flat profile in a landward direction consists of a lower intertidal sand flat, a middle intertidal mixed (sand and mud) flat, and an upper intertidal mud flat. Landward of the mud flat, supratidal salt marshes may be present (Figs. 1 and 2; Dalrymple, 2010; Klein, 1977). The intertidal zone represents an extremely harsh ecosystem in which few species are able to inhabit the entire tidal range and, accordingly, different animal communities live in different areas within the tidal flat (Reise, 1985). Benthic faunas in supratidal to intertidal settings are mostly controlled by the interplay of (1) salinity, (2) sediment mobility and hydrodynamic energy, (3) time of exposure to subaerial conditions, (4) temperature, (5) substrate type, and (6) food supply.

Salinity strongly controls the type of organisms and associated biogenic structures of tidal flats, the emplacement of which may range from the open shoreline to the fluvio-estuarine transition, encompassing highly variable salinity conditions, from fully marine to brackish water and freshwater (see Mángano and Buatois, 2004b). Even in the case of tidal flats directly facing the open sea, salinity fluctuations may be significant. Periodical fluctuations in salinity are typically related to periodical emersions and submersions of the intertidal zone, in combination with wind and solar radiation. In addition, seasonal

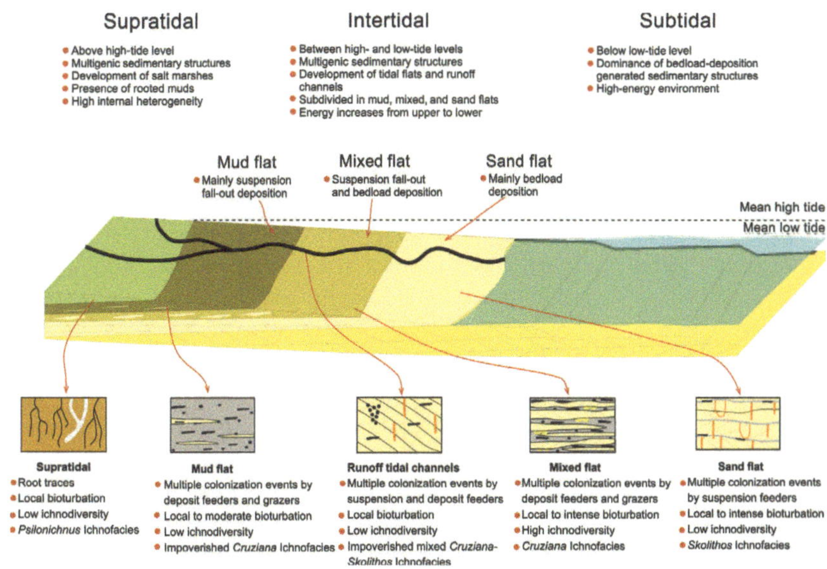

FIGURE 2 Integrated sedimentological and ichnological model for intertidal to supratidal environments.

rainfall and drainage from the continent affect both the salinity and the position of the water table. As a result of the complex hydrological conditions of tidal flats, a number of behavioral strategies are adopted for protection. For example, the amphipod *Corophium* is a good osmoregulator that can tolerate salinities between 2‰ and 47‰. On the other hand, inhabiting a burrow or taking temporary refuge in the sediment provides effective protection against strong salinity variations (Sanders et al., 1965). Also, horizontal migration represents a strategy to minimize the dramatic salinity shifts in the upper intertidal zone. In general, salinity tolerance controls the zonal distribution of intertidal animals, with euryhaline species being particularly abundant in the upper intertidal zone (Newell, 1979).

The hydrodynamic energy in intertidal zones, which results from the local interplay of tides, waves, and currents, controls sediment mobility and bedform development. High-energy zones of tidal flats are commonly characterized by low-diversity assemblages of suspension feeders or passive predators represented, such as *Diplocraterion* or *Skolithos* (e.g., Cornish, 1986; Mángano and Buatois, 2004a; Simpson, 1991). In moderate- to low-energy coastal settings, ichnodiversity increases and assemblages tend to be dominated by horizontal traces of deposit feeders, detritus feeders, and grazers (e.g., Mángano et al., 2002a). Scouring on the tidal flat, resulting from either tides or storms, is conducive to erosional truncation of burrows, as illustrated by specimens of *Lockeia* forming composite surfaces and revealing successive colonization events by a bivalve fauna (Mángano et al., 1998).

The time of exposure is particularly significant on tidal flats, becoming a limiting factor particularly in the upper intertidal zone (Mángano et al., 2002a). Many organisms have developed biological rhythms (e.g., quasi-tidal, quasi-semilunar rhythms) of vertical or horizontal migration, controlled by tidal cyclicity in order to cope with the stress associated with subaerial exposure (Palmer, 1995). Some species such as the crab *Sesarma reticulatum* hide in their burrows during low tide and are active during high tide (Palmer, 1967; Seiple, 1981). The isopod *Eurydice pulchra* lives buried in the sand flat during emersion but rises into the water column with the flood tide to swim at the water's edge in order to feed on epifauna, infauna, and detritus. During the ebb tide, it retreats seaward and reburies itself for protection (Warman et al., 1991). In general, marine invertebrate surface activity on the tidal flat is typically more intense during high tide, whereas a relatively large number of semiterrestrial and terrestrial animals (e.g., many terrestrial crabs and the intertidal beetle *Thalassotrechus barbarae*) may display an activity peak during low-tide emersion (Palmer, 1995; Pieńkowski, 1983; Vader, 1964).

Tidal flats commonly experience rapid changes in temperature as a result of periodical subaerial exposure, which represents a limiting factor particularly, but not exclusively, in the upper intertidal zone (Mángano et al., 2002a). In tropical environments, the upper intertidal zone is extremely inhospitable for marine organisms due to very high temperatures, long periods of exposure and abnormal salinities. As a result, the highest density of biogenic structures in tropical tidal flats is found in the lower intertidal zone (Terwindt, 1988). In colder areas, tidal flats may exhibit a high density of biogenic structures in the upper intertidal zone (e.g., Yeo and Risk, 1981). Bathymetric displacement of certain species along latitudinal gradients is common, as illustrated by the bivalve *Gemma gemma* that lives in intertidal areas of northern North America and in subtidal areas further south in order to avoid the hazards of high temperatures (Green and Hobson, 1970; Reise, 1985).

The substrate is a controlling factor operating at different scales, affecting ichnology from overall trace-fossil distribution to trace-fossil morphology (Mángano et al., 2002a). The substrate can be characterized in terms of composition and degree of consolidation, both features influencing the distribution and preservation of biogenic structures. Vertical and horizontal differences in the degree of substrate consolidation influence the diversity, abundance, and distribution of intertidal organisms (Newell, 1979; Reise, 1985). The pore-fluid content within the sediment can vary during a tidal cycle, the tidal-flat surface at low tide being characterized by the presence of tidal pools within a commonly emerged area, resulting in a range of substrate conditions along an isochronous surface. As a result, organisms moving across the tidal flat may produce biogenic structures that display a considerable morphological variability, as illustrated by the bivalve trail *Protovirgularia* (Mángano et al., 1998). Erosional exhumation of firm sediment layers or even hardgrounds due to tidal scouring (Willis, 2005) is quite common in tide-dominated settings, resulting in the presence of certain elements of the *Glossifungites* Ichnofacies (Buatois and Mángano, 2011).

Food supply displays a large-scale gradient along a transect from the supratidal to lower intertidal region, with abundant organic particles concentrated on the sediment surface in more proximal positions and abundant organic particles in the water column in more agitated seaward settings. In particular, suspension-feeder burrows are common in high-energy sand flats (e.g., Mángano and Buatois, 2004a). Also, it has been suggested that some ichnotaxa (e.g., *Syringomorpha*) may be associated with the exploitation of meiofauna and epigranular bacteria that coat sand grains (Mángano and Buatois, 2004a). In addition to general trends, food supply is strongly variable at a smaller scale, resulting in high spatial heterogeneity (Buatois and Mángano, 2011; Mángano et al., 2002a). Water entering through ripple troughs and exiting through the ripple slopes and crests, driven by wave pressure gradients (e.g., Webb, 1969), results in a resorting of sediment with small grains and organic detritus being drawn into the troughs. The distribution of organics at the scale of microtopography accounts for the aggregation of meiofauna and invertebrate grazers in ripple troughs (Jansson, 1967; Newell, 1979). The grazing trail *Psammichnites grumula* has been observed forming guided meanders along ripple troughs in sand-flat deposits (Mángano et al., 2002b).

2.1 Supratidal Marshes and Mangroves

Along modern temperate shorelines, the supratidal area may be vegetated by salt marshes, the cordgrass *Spartina* being one of the most widespread plants, their root traces being typically pervasive in such deposits (Fig. 2; e.g., Basan and Frey, 1977; Edwards and Frey, 1977; Montague et al., 1981; Pomeroy et al., 1981). Along tropical shorelines, mangroves tend to be common, allowing the formation of extensive root networks of, for example, *Avicennia*, *Rhizopora*, and *Sonneratia* (Fig. 3A; Cadée, 1998). Animal traces include elements of the *Psilonichnus* Ichnofacies, with gastropods and crustaceans (mainly crabs) forming some of the most important marine elements (Frey and Pemberton, 1987; Fig. 2). The beetle *Bledius arenarius* typically constructs dwelling burrows (Schäfer, 1972). Insects are by far the most important terrestrial

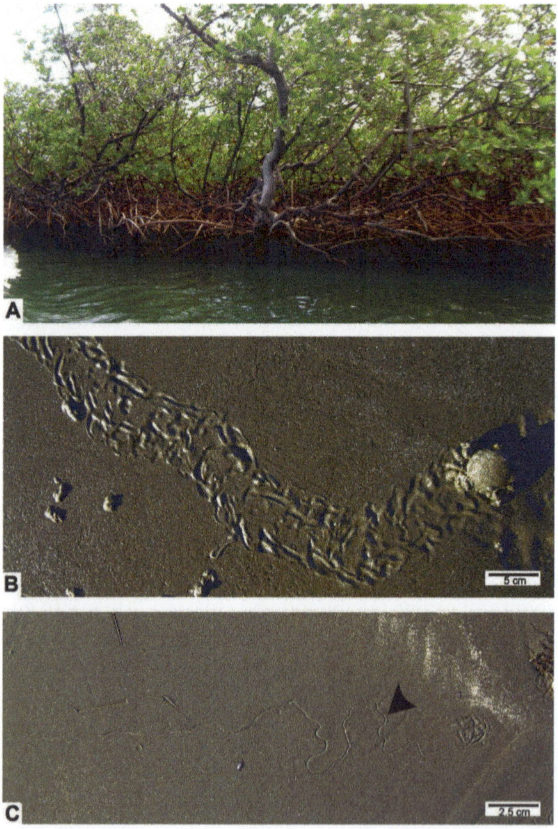

FIGURE 3 Modern mangrove and mud flats. (A) Modern mangrove. Gran Mangle, Dominican Republic. (B) Crab trackway on a modern mud flat near Estancia Maria Luisa, Tierra del Fuego, southern Argentina. (C) Grazing trail on a modern mud flat between Cabo Irigoyen and Cabo San Pablo, Tierra del Fuego, southern Argentina.

invertebrate representatives, with trackways, burrows, trails and particularly nests becoming dominant landward, delineating the transition into terrestrial ichnofacies, such as *Scoyenia* and *Coprinisphaera* (Buatois and Mángano, 1995; Genise et al., 2000). Among the vertebrates, mammals, reptiles, and birds produce a wide variety of structures, including burrows and trackways (Baucon, 2008; Frey and Pemberton, 1986). Ichnological observations in supratidal deposits are essentially restricted to modern environments, there being a remarkable dearth of studies in ancient deposits. In addition, given the low preservation potential of many of the supratidal biogenic structures (root traces and crustacean burrows being the exception), it is still premature to provide an ichnological classification of associated ancient analogs.

2.2 Mud Flats

Mud flats constitute the upper zone of tidal flats, depositional processes being dominated by the fallout of suspended sediment comprising sortable silts, flocs, and aggregates (Chang et al., 2007; Flemming, 2012), as well as biodeposition due to the production of fecal pellets and pseudofeces (Fig. 2; Augustinus, 2002; De Boer, 1998; Potter et al., 2005). Mud-flat deposits are dominated by parallel-laminated or massive mudstone, lenticular-bedded siltstone and very fine-grained sandstone being less common (Fig. 4A). Salinity shifts, together with changes in exposure and temperature, are typically dramatic in the upper intertidal zone (Newell, 1979; Reise, 1985).

Modern mud flats are intensely bioturbated, polychaetes, gastropods, and bivalves being common producers (Reineck and Singh, 1980; Schäfer, 1972). However, limited lithological contrast commonly precludes visualization and preservation of biogenic structures (Mángano et al., 2002a). Accordingly, discrete trace fossils are relatively rare and an indistinct mottled texture is commonly observed instead, it being locally accompanied by more distinct polychaete and crustacean burrows (Baucon, 2008). In the fossil record, elements of the *Cruziana* Ichnofacies may be preserved where lithological interfaces are available (e.g., Mángano et al., 2002a). Typical components are *Cruziana*, *Rusophycus*, *Palaeophycus*, *Psammichnites*, *Lockeia*, and *Protovirgularia*. The presence of these ichnogenera may either reflect a wide environmental adaptation of these bioturbators, or the short-term incursion into this zone by faunal elements of the middle and lower intertidal zones (Mángano et al., 2002a). Migrations may be regulated by tidal cyclicity in connection with the search for food. Also, invertebrate trackways and simple grazing trails, such as *Helminthopsis* and *Helminthoidichnites*, are common (Figs. 3B, C and 4B); their preservation potential is low, but it may increase in the presence of associated microbial mats (e.g., Desjardins et al., 2010a). The mud flat also hosts vertebrate trackways that illustrate some of the ichnocoenoses of the *Brontopodus* and *Batrachichnus* ichnofacies of Hunt and Lucas (2007).

FIGURE 4 Early Cambrian St. Piran Formation, Gog Group, southern Rocky Mountains, Canada. (A) Vertical succession of two intertidal parasequences composed of mixed-flat and mud-flat deposits. (B) Mud flat bedding-plane view (top) containing wrinkle marks (W), associated with the grazing trace fossils *Helminthoidichnites* isp. (*He*) and *Helminthopsis tenuis* (*Hp*).

2.3 Mixed Flats

Mixed flats occur across the transition between sand flats (<5% mud content) and mud flats proper (>95% mud content), and are characterized by alternating bedload sedimentation and fallout from suspension (Figs. 2 and 4A; Flemming, 2012). Deposits typically consist of thinly interbedded wave- and current-ripple cross-laminated, very fine-grained sandstone and massive or parallel-laminated mudstone. Heterolithic tidal bedding is typical, being mostly represented by sand/mud rhythmites and flaser or wavy bedding (Fig. 5A). Flat-topped ripples and washout structures may also be present.

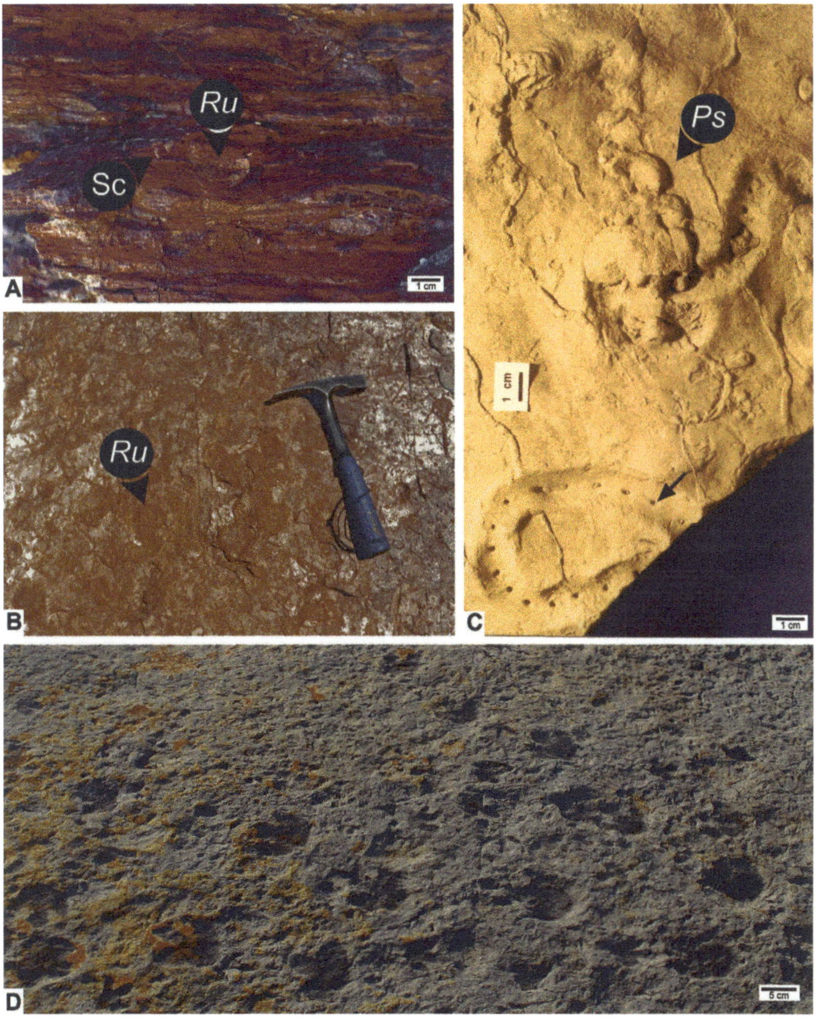

FIGURE 5 Mixed tidal flat. (A) Cross-section view of *Rusophycus leifeirikssoni* (*Ru*) in mixed tidal-flat deposits. Note associated syneresis cracks (Sc). Angosto del Morro de Chucalezna, Quebrada de Huamahuaca, northwest Argentina. (B) Bedding-plane view (top) of a cluster of *R. leifeirikssoni* in mixed tidal-flat deposits. Early to Middle Cambrian, Campanario Formation, Mesón Group, Angosto del Morro de Chucalezna, Quebrada de Huamahuaca, northwest Argentina. Hammer is 33.5 cm long. (C) *Psamichnites grumula* preserved on a sandstone base. Note siphon traces (arrow). Late Carboniferous, Stull Shale, Kanwaka Formation, Waverly fossil site, Kansas, USA. (D) Dinosaur tracks in tidal-flat deposits. View of a sandstone surface (top) with large ornithopod trackways (*Caririchnium leonardii*) and small theropod trackways (*Magnoavipes loewi*). Early Cretaceous, Dakota Group, Alameda Avenue, west of Denver, Colorado, USA.

The intensity of bioturbation is typically moderate (Reineck and Singh, 1980). Alternation of sandstone and mudstone layers enhances preservation of horizontal interface traces, such as those that typify the *Cruziana* Ichnofacies. Common components are *Cruziana*, *Rusophycus*, *Psammichnites*, *Protovirgularia*, *Lockeia*, *Palaeophycus*, *Planolites*, *Helminthopsis*, *Helminthoidichnites*, and *Bergaueria* (Fig. 5B and C). Clusters of *Rusophycus* are common in Early Paleozoic tidal-flat deposits (Fig. 5B; Mángano and Buatois, 2004a). Vertebrate trackways are typically preserved in sandy layers of the mixed intertidal flats. Mesozoic examples include spectacular dinosaur tracks, commonly forming megatrack sites (Fig. 5D; e.g., Avanzini et al., 2006; Lockley et al., 1992). Microbial mats may be extensive, allowing a number of interactions, including protection and undermat mining (Baucon, 2008). Insect and polychaete burrows are common, decapod burrows being particularly concentrated along mat margins (Baucon, 2008). Substrate stabilization by microbial activity has also been recognized in the stratigraphic record (e.g., Gerdes et al., 1985; Noffke, 2010; Schieber, 2004), as also illustrated by the presence of wrinkle marks associated with relict ripple troughs (e.g., Mángano et al., 2002a).

2.4 Sand Flats

Sand flats occupy the lower zones of tidal flats, and are dominated by bedload transport of sand-sized sediment (Fig. 2). They are the most variable intertidal areas in terms of both sedimentary facies differentiation and trace-fossil content, this variability being essentially controlled by the intensity of tidal currents in combination with wave action (Mángano and Buatois, 2004b; Reineck and Singh, 1980). In macrotidal settings characterized by high-current velocities, migration of large-scale bedforms (i.e., two-dimensional and three-dimensional dunes) is the dominant process (Boyd et al., 2006; Dalrymple, 2010; Dalrymple and Rhodes, 1995). Deposits typically consist of medium- to thick-bedded, trough and planar cross-bedded coarse- to fine-grained sandstone, and medium- to very fine-grained sandstone with parallel lamination formed in the upper flow regime (Fig. 6A; Dalrymple, 2010; Dalrymple and Choi, 2007; Dalrymple et al., 1990). Current-rippled fine-grained sandstone is less common. Under these conditions, the lower intertidal zone is very difficult to distinguish from subtidal areas. In lower-energy micro- to mesotidal settings, ripples are the dominant bedforms, and deposits typically consist of current- and wave-rippled, cross-laminated, very fine- and fine-grained sandstone, locally containing mud drapes and flaser bedding. Flat-topped ripples, washout structures, erosional remnants and pockets, and wrinkle marks are common (Fig. 6B).

On high-energy sand flats, rapidly migrating bedforms generally preclude intense bioturbation (e.g., Baucon, 2008; Mángano and Buatois, 2004a; Reineck and Singh, 1980). Where present, ichnofaunas are dominated by vertical burrows of suspension feeders or passive predators (e.g., *Skolithos*, *Ophiomorpha*, *Arenicolites*, and *Diplocraterion*, illustrating the *Skolithos* Ichnofacies; Fig. 6C–E).

FIGURE 6 Sand flat. (A) Vertical section of a sand-flat deposit with erosional base overlying subtidal sandstone. Early Cambrian St. Piran Formation, Gog Group, southern Rocky Mountains, Canada. (B) Flat-topped ripples and relict troughs. Late Carboniferous Stull Shale, Kanwaka Formation, Waverly fossil site, Kansas, USA. (C) Cross-section view of deep *Arenicolites* isp. Early to Middle Cambrian Campanario Formation, Mesón Group, Angosto de Perchel, Quebrada de Huamahuaca, northwest Argentina. Lens cap = 5.5 cm in diameter. (D) Bedding-plane view (top) of *Diplocraterion* isp. Early to Middle Cambrian Campanario Formation, Mesón Group, Quebrada de Moya, Quebrada de Huamahuaca, northwest Argentina. Coin = 1.5 cm in diameter. (E) Cross-section view of *Syringomorpha* pipe rock. Early to Middle Cambrian Campanario Formation, Mesón Group, Angosto del Morro de Chucalezna, Quebrada de Huamahuaca, northwest Argentina. Lens cap = 5.5 cm in diameter.

These assemblages commonly reflect short-term colonization windows along dune reactivation surfaces (Mángano et al., 1996; Pollard et al., 1993). In Cambrian examples, the ichnogenus *Syringomorpha* may occur in high densities in rippled sandstone (Fig. 6E; Mángano and Buatois, 2004a). Mud drapes formed during slack water may contain *Planolites*. In modern examples of sand flats, the polychaete *Arenicola marina* and the bivalve *Mya arenaria* typically produce significant bioturbation (Reineck and Singh, 1980; Schäfer, 1972).

On low-energy sand flats, relatively quiet conditions, coupled with very short periods of subaerial exposure, allow the development of a diverse resident fauna. These tidal flats are typically dominated by high-diversity assemblages of horizontal burrows and trails of deposit and detritus feeders, as well as predators, namely, *Cruziana*, *Rusophycus*, *Asteriacites*, *Pentichnus*, *Psammichnites*, *Curvolithus*, *Nereites*, *Lockeia*, *Protovirgularia*, *Palaeophycus*, and *Planolites* (Fig. 7A–D). This assemblage illustrates the *Cruziana* Ichnofacies (Mángano and Buatois, 2004b; Mángano et al., 2002a).

2.5 Tidal Channels and Creeks

Tidal-flat deposits are commonly dissected by a network of meandering tidal channels and creeks (also known as gullies) that migrate across the intertidal zone, producing lateral accretion in point bars and resulting in inclined heterolithic stratification (Fig. 2; Thomas et al., 1987). In the muddy upper intertidal zone, channels are small to medium size, but in the lower sandy areas, they tend to coalesce to form wider and deeper channels (Dalrymple, 2010). The

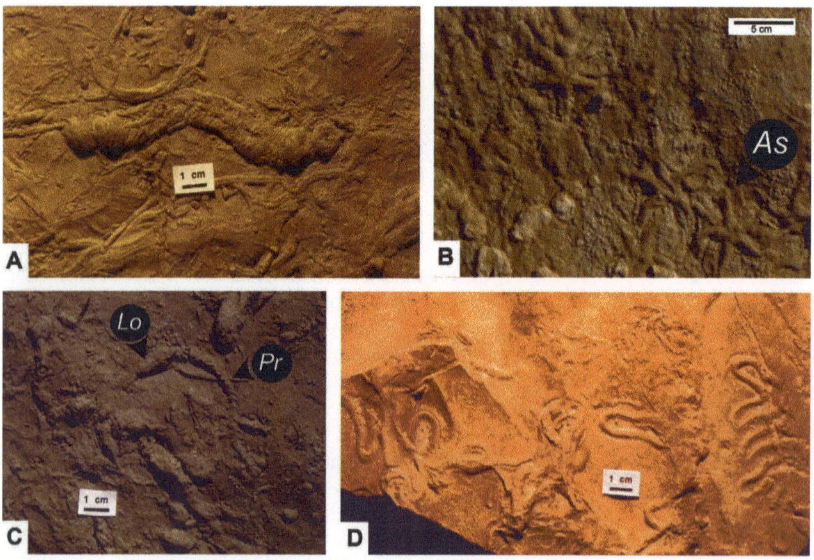

FIGURE 7 Sand flat containing the *Cruziana* Ichnofacies. Traces are preserved at the soles of sandstone beds. Late Carboniferous Stull Shale, Kanwaka Formation, Waverly fossil site, Kansas, USA. (A) *Nereites imbricata*. (B) *Asteriacites lumbricalis* (*As*). (C) Intergradation between *Lockeia ornata* (*Lo*) and *Protovirgularia rugosa* (*Pr*). (D) *Psammichnites implexus*.

degree of bioturbation in the point bars is lower than in tidal-flat deposits, most likely reflecting higher rates of sedimentation along unstable channel margins (e.g., Gingras et al., 1999; Reineck and Singh, 1980). Bivalve escape traces may be present as a result of rapid sedimentation in channels and large gullies (Reineck and Singh, 1980). Overall, tidal-creek deposits tend to be slightly more bioturbated than larger-scale tidal channels, being benthic organisms remarkably different from those in the adjacent tidal flats (Reineck and Singh, 1980). Associated exhumed compacted sediments may contain dense populations of the crustacean *Corophium volutator* (Dörjes, 1970; Reineck and Singh, 1980). Intense bioturbation by decapods and polychaetes has been observed along the margins of tidal creeks dissecting matgrounds (Baucon, 2008).

3. SUBTIDAL ENVIRONMENTS

The subtidal zone either occurs immediately seaward of the intertidal area or is separated from it by sandy barriers or barrier islands. It is a high-energy environment characterized by high-current velocities (Figs. 1 and 2; Dalrymple, 2010). In macrotidal settings, strong tidal currents tend to be the dominant sedimentary process, thus large-scale bedforms develop in the subtidal region. However, it should be noted that under microtidal or mesotidal conditions this may not be the case, as for example in regions of the southern North Sea, which are dominated by wave action weakening the tidal-current influence (Antia et al., 1994). In zones of actively migrating large-scale bedforms, conditions are not favorable for benthos colonization and, as a result, few species are able to survive in this setting (Wilson, 1982, 1986). Modern studies and information from the stratigraphic record indicate that faunal diversity increases toward areas with smaller bedforms, and on the outer shelf where dunes tend to be replaced by small ripples and interbeds with mud laminae. Parallel to this trend in faunal diversity, variations in trophic types are apparent, with suspension feeders being dominant in high-energy, shallow subtidal environments, and deposit and detritus feeders becoming more abundant toward the outer shelf (Buatois and Mángano, 2011; Desjardins et al., 2012a; Wilson, 1982). Benthic faunas in tide-dominated subtidal sand bodies are mostly controlled by the interplay of (1) salinity, (2) sediment mobility and hydrodynamic energy, (3) water turbidity, (4) substrate type, and (5) food supply.

Shelf sand bodies are typically emplaced under normal-marine conditions and, therefore, salinity fluctuation is not an important control factor. However, in brackish-water subtidal settings, such as deltaic or estuarine tidal bars (Fig. 1), salinity dilution may significantly affect benthic faunas (e.g., Carmona et al., 2009).

Sediment mobility in the form of bedform migration essentially controls the duration of the colonization window (*sensu* Pollard et al., 1993). Under conditions of continuous migration, the colonization window is closed or of short duration, preventing bioturbation or allowing burrowing by opportunistic organisms adapted to high-energy conditions, respectively. More extensive bioturbation typically results from longer-term colonization windows associated with discontinuous migration or dormant forms (Desjardins et al., 2010b).

Water turbidity significantly influences the trophic type (Buatois and Mángano, 2011). Clogging by abundant clay and silt particles in suspension in the water column commonly precludes the establishment of suspension feeders (Perkins, 1974). Suspension-feeding is particularly inhibited in deltaic settings (Gingras et al., 1998; MacEachern et al., 2005) due to the presence of buoyant mud plumes extending in a seaward direction, particularly under hypopycnal conditions (Bates, 1953; Kineke et al., 1996). In cases of extreme water turbidity, primary production can be severely inhibited, resulting in an impoverishment not only of suspension feeders, but also of deposit and detritus feeders (Leithold and Dean, 1998).

As in the case of intertidal settings, both the grain size and the degree of consolidation of the substrate play an important role in controlling the subtidal ichnofaunas. Some ichnotaxa tend to be restricted to certain types of substrate, such as *Skolithos* and *Ophiomorpha* to sand-rich and *Planolites* to mud-rich heterolithic intervals. Accumulation of fluid mud is quite common in tide-dominated settings, imparting a substrate stress by reducing boundary shear stress, which prevents benthic organisms from constructing permanent structures or from actively backfilling tunnels (Buatois and Mángano, 2011; MacEachern et al., 2005; Schieber, 2003). Fluid-mud deposits are typically non-bioturbated or, more rarely, contain "mantle and swirl" biogenic structures (Bhattacharya and MacEachern, 2009; Schieber, 2003).

The type of organic matter essentially controls feeding strategies. Suspension-feeder burrows, such as *Skolithos*, tend to dominate under conditions of relatively high energy conducive to a high content of suspended organic particles in the water column (e.g., Mángano and Buatois, 2004a; Mángano et al., 1996). Mud and silt layers rich in organic matter are typically bioturbated by deposit and detritus feeders (e.g., *Planolites*, *Cruziana*).

Five broad categories of subtidal sand bodies have recently been defined on the basis of sedimentological and ichnological criteria: (1) compound dune fields, (2) sand sheets, (3) sand ridges, (4) isolated dune patches, and (5) tidal bars (Fig. 1; Desjardins et al., 2012a).

3.1 Compound Dune Fields

Compound dune fields comprise trains of large dunes with superimposed smaller dunes, the crests of which are mainly oriented perpendicular to the predominant tidal-current direction, and therefore accrete in a forward direction (Fig. 8; e.g., Ashley, 1990). This type of sand body was originally described from the Eocene Baronia Formation in Spain by Mutti et al. (1985) as "tidal bars" but

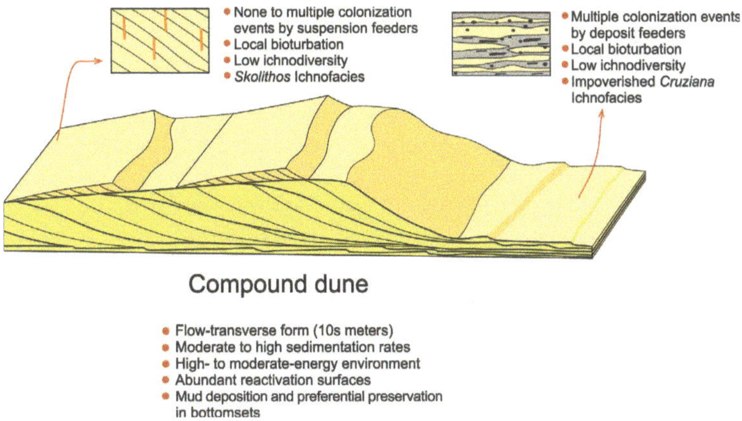

FIGURE 8 Integrated sedimentological and ichnological model for subtidal compound dunes.

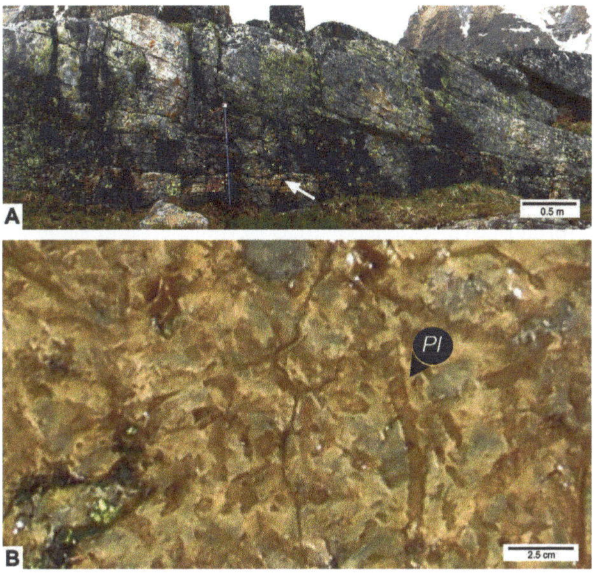

FIGURE 9 Compound dune-field deposits. Early Cambrian St. Piran Formation, Gog Group, southern Rocky Mountains, Canada. (A) Compound cross-stratified sandstone. (B) *Planolites beverleyensis* (*Pl*) on a bedding surface at bottomset of A (white arrow). Note that the high-energy foresets are non-bioturbated.

have subsequently been reinterpreted as representing compound dunes within a strait or seaway (Dalrymple and Choi, 2007; Dalrymple et al., 2003). In compound dune-field deposits, the dip direction of internal discontinuities is the same as that of the cross-stratified sets, and the resulting sandstone packages display thickening- and coarsening-upward trends. Mud drapes are common locally.

These deposits are typically non-bioturbated (Fig. 9A) or contain a few vertical burrows (e.g., *Skolithos*, *Ophiomorpha*), representing the *Skolithos* Ichnofacies. Thin-bedded horizontal bottomsets may contain only scattered horizontal burrows (e.g., *Planolites*; Fig. 9B). Compound dune deposits may gradationally overlie heterolithic intervals containing a diverse ichnofauna, dominated by horizontal trails and burrows mostly produced by deposit-feeding organisms such as *Rusophycus*, *Cruziana*, *Planolites*, *Phycodes*, *Asterosoma*, and *Helminthoidichnites*, illustrating the archetypal *Cruziana* Ichnofacies (Desjardins et al., 2012a).

The lack of bioturbation in the sand body indicates moderate to high sedimentation rates and erosion resulting from continuous and rapid bedform migration. Flocculated clay and mud aggregates recorded in the mud drapes may show restricted colonization by suspension feeders. Bioturbation in the bottomset deposits suggests that the colonization window was relatively long. The associated intensely bioturbated deposits containing the *Cruziana* Ichnofacies reveal preferential preservation, low-energy conditions, and a more or less continuous colonization window in a fully marine environment.

3.2 Sand Sheets

Sand sheets consist of compound dunes but differ from compound dune fields because the former cover extensive areas of the shelf and show a decrease of bedform sizes as a product of a streamwise change in the current velocity across the sand sheet (Desjardins et al., 2012a). Local tidal flow-field velocities mainly control the dune patterns in sand sheets, although other factors such as grain size and water depth also play important additional roles in determining their final configuration (Flemming, 2000). The typical modern sand-sheet examples are those of the North Sea and Irish Sea, which migrate under the influence of strong tidal currents forming sediment transport paths that extend over hundreds of square kilometers, developing suites of bedforms in the course that decrease in size in conjunction with reductions in the current energy (Belderson et al., 1982; Van Landeghem et al., 2009). Along such transport paths, bedforms are arranged in three zones: (1) a zone of large compound dunes forming the sand-sheet core, (2) a zone of small compound dunes forming the sand-sheet front, and (3) a zone of small dunes and sand ripples forming the sand-sheet margin (Fig. 1; Desjardins et al., 2010a). Where sand supply is abundant within areas of large compound dunes, tidal sand ridges may develop, whereas areas of highest energy are commonly characterized by sand ribbons, furrows, and gravel lags, but sand ridges can also develop in these zones (Stride et al., 1982).

Desjardins et al. (2010a) proposed a facies model for sand sheets based on observations in the Early Cambrian Gog Group, western Canada, characterizing each subzone in terms of its ichnology and sedimentology (Figs. 10 and 11A).

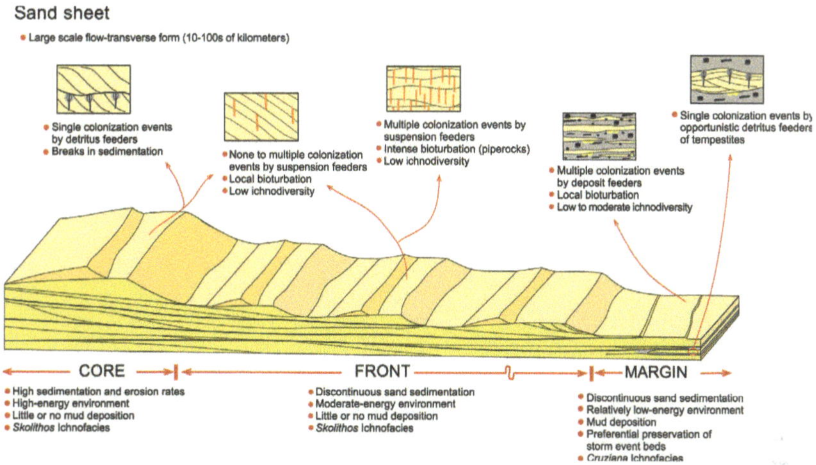

FIGURE 10 Integrated sedimentological and ichnological model for sand sheets.

The sand-sheet core comprises medium- to thick-bedded, cross-stratified sets of planar, trough and compound cross-stratified sandstone. The core of the sand sheet is mostly unburrowed or locally contains *Skolithos* and large *Rosselia*. The sand-sheet front is characterized by thin- to medium-bedded, planar and compound cross-stratified sandstone (Fig. 11C). Some cross-stratified beds display scattered *Skolithos* (Fig. 11B). However, pipe-rock ichnofabrics riddled with *Skolithos linearis* are commonly developed within these deposits (Fig. 11F). Locally, concave-up or sigmoidal surfaces within compound cross-stratified packages merge downwards into intercalated ripple cross-laminated sandstone and mudstone. These represent the toes of compound dunes, which contain simple feeding trace-fossils such as *Planolites* and *Teichichnus* (Fig. 11C–E). The sand-sheet margins, off the flanks of the complex, comprise interbedded, very thin- to thin-bedded, ripple cross-laminated sandstone and mudstone, hummocky cross-stratified sandstone, and local thin- to medium-bedded, planar cross-stratified sandstone. These heterolithic deposits contain a trace-fossil fauna dominated by *Planolites*, *Teichichnus*, and locally also various ichnospecies of *Rusophycus*. Hummocky cross-stratified sandstones tend to contain scattered small *Rosselia* (Fig. 11G).

Sand-sheet deposits exhibit predictable trends in trace-fossil distributions along the sediment transport path characterized by decreasing energy conditions (Fig. 10; Desjardins et al., 2010b). High-energy conditions associated with continuous bedform migration, coupled with intense erosion and high sedimentation rates, commonly preclude bioturbation in the core of the sand sheet. Local occurrences of *Skolithos* record the sporadic development of colonization windows represented by suspension feeders of the *Skolithos* Ichnofacies. However, the presence of *Rosselia* in the high-energy facies indicates prolonged periods of dune inactivity and the presence of suspended mud in the core of the complex

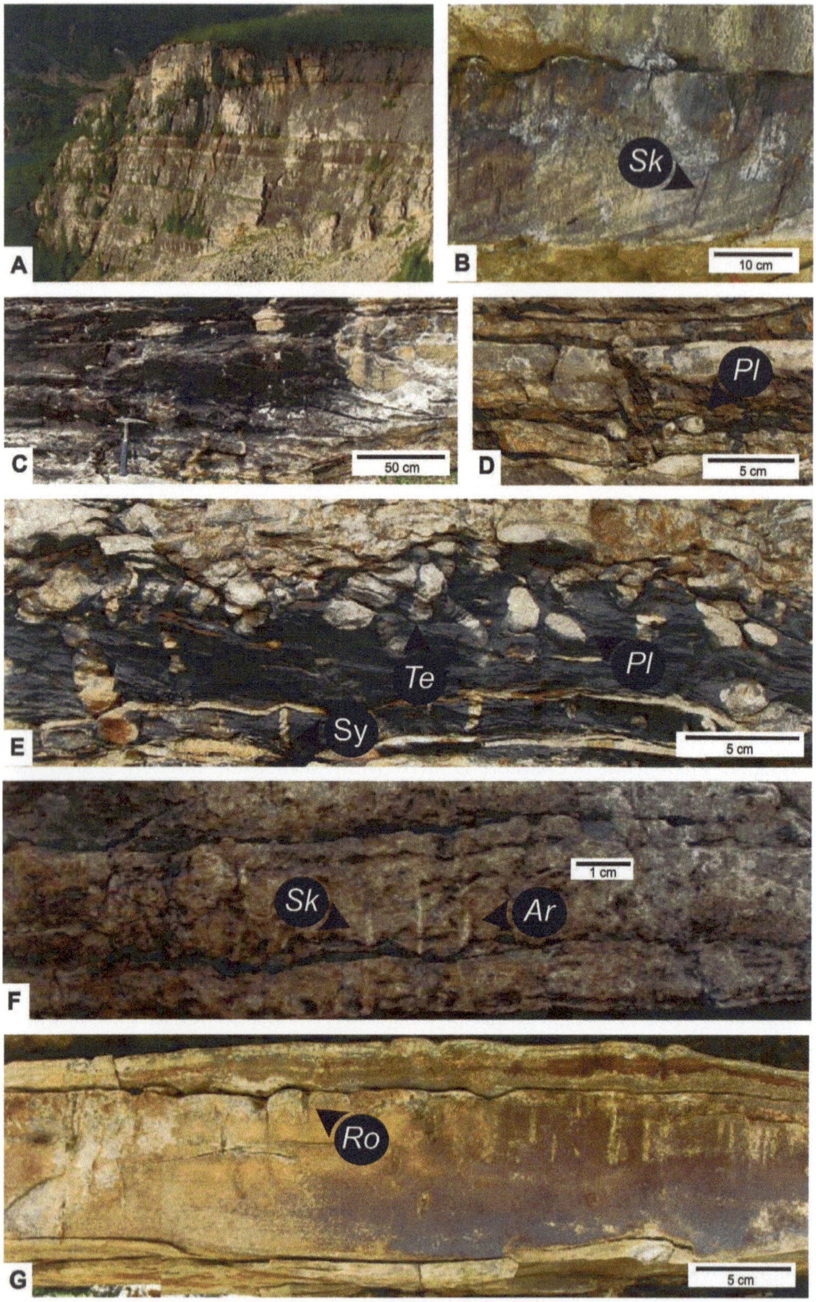

FIGURE 11 Sand-sheet deposits. Early Cambrian Lake O'Hara Member, St. Piran Formation, Gog Group, southern Rocky Mountains, Canada. (A) General view. (B) Cross-section view of *Skolithos linearis* (*Sk*) extending down from the upper surfaces of the foresets of a planar cross-stratified sandstone. (C) Compound cross-stratified, fine- to medium-grained sandstone showing sigmoidal-shaped surfaces separating different cross-stratified beds. (D) and (E) Cross-section view of *Planolites montanus* (*Pl*) in intercalated rippled cross-laminated sandstone and mudstone at bottomsets of a compound cross-stratified sandstone. (F) Cross-section view of a pipe rock comprising *Skolithos linearis* (*Sk*) and *Arenicolites* isp. (*Ar*). (G) Hummocky cross-stratified sandstone containing several specimens of *Rosselia* isp., which extend down from the upper surface.

associated with flooding events. The sand-sheet front deposits are characterized by smaller bedforms containing composite types of burrows (i.e. resulting from multiple bioturbation events), with vertical burrows penetrating from various colonization surfaces to develop pipe rocks associated with multiple pauses in sedimentation. Multiple colonization events of sandy substrates by suspension-feeding organisms are typical of the *Skolithos* Ichnofacies. Ichnofabrics of the mud-rich bottomsets contain a deposit-feeding fauna represented by *Planolites* and *Teichichnus*, recording an impoverished *Cruziana* Ichnofacies. Low ichnodiversity, even in the low-energy facies, is attributed to soupy substrate conditions. Sand-sheet margin deposits are characterized by alternating episodic sandstone sedimentation and mud deposition under relatively low-energy conditions. The presence of trace fossils of deposit feeders, which fed on organic particles that fell out of suspension and accumulated on the sea floor, is characteristic of lower-energy conditions and the development of the *Cruziana* Ichnofacies. On the other hand, *Rosselia* within hummocky cross-stratified sandstone reflects colonization after storms.

3.3 Tidal Sand Ridges

Sand ridges are elongated sand bodies that are larger and geographically more stable in comparison to dunes, being oriented at an oblique angle to the strongest current (Figs. 1 and 12). A sand ridge is composed of dunes and compound

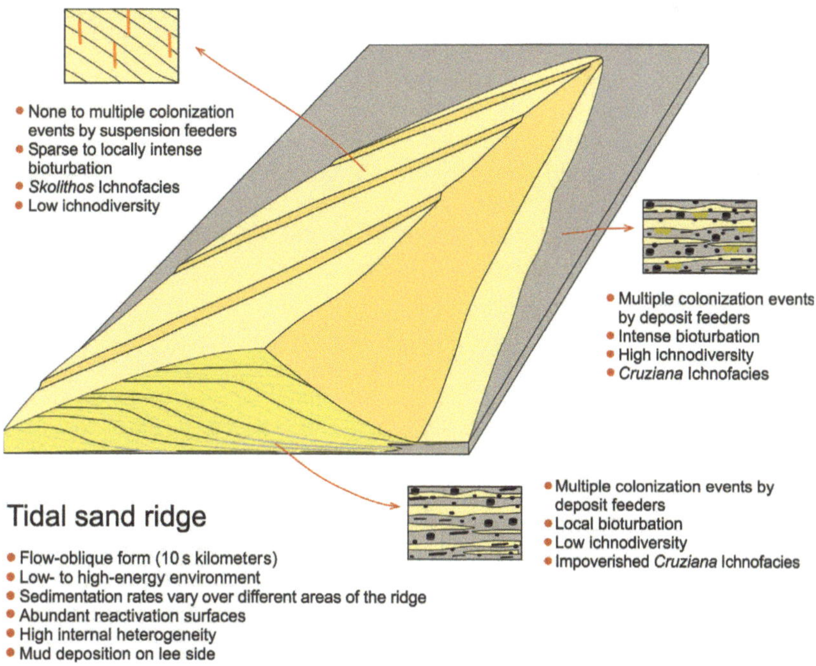

FIGURE 12 Integrated sedimentological and ichnological model for tidal sand ridges.

dunes of varied forms and sizes which migrate both on the stoss and lee sides of the ridge. On modern shelves they typically occur in localized groups (Snedden and Dalrymple, 1999). Four conditions are required for sand-ridge formation: (1) an initial irregularity, commonly a transgressed coastal sand body (e.g., Penland et al., 1988; Posamentier, 2002; Snedden and Dalrymple, 1999; Snedden et al., 1999; Swift and Field, 1981), (2) sufficient sand supply, (3) a current capable of moving sand, and (4) sufficient time for the sand to be molded into a ridge. Tidal sand ridges in the North Sea were formed when glacioeustatic lowstand deposits were reworked by tidal currents (Belderson et al., 1982). Sand ridges in the East China Sea (southern Yellow Sea) also developed during the Pleistocene–Holocene transgression. These overlie deltaic deposits (Berné et al., 2002; Wang et al., 2012) and become younger toward the present location of the Yangtze River due to continued transgression (Liu et al., 1998).

The internal structures of tidal sand ridges comprise smaller-scale cross-stratified beds deposited mainly on the lee sides (Fig. 13A). Paleocurrent directions determined from individual cross-stratified beds are bipolar but oriented at an angle to the direction of sand-ridge accretion. Thickening- and coarsening-up intervals are common and record progradation of dunes on the flanks of the lower-energy lee-sides and in the swales between the ridges. The trace-fossil content is mainly restricted to horizontal and subhorizontal structures in interbedded sandstone and mudstone belonging to *Cruziana*, *Dimorphichnus*, *Planolites* (typically *P. montanus*), *Palaeophycus*, *Phycodes*, *Rusophycus*, and *Teichichnus* (typically *T. rectus*). *Skolithos linearis* is present in cross-stratified beds.

Sand ridges display clear ichnological trends perpendicular to the ridge axes. Bioturbation is virtually absent in the cross-stratified beds that make up the cores of the tidal sand ridges, and only scarce *Skolithos linearis* are present in medium- to thick-bedded cross-stratified sandstones, clearly indicating closed to short-term colonization windows linked to rapidly migrating bedforms (Fig. 13B). The lee-side deposits of sand ridges are characterized by increased heterogeneity as recorded by pervasive interfingering of muddy layers (Fig. 13C). These heterolithic deposits mark the transition from the *Skolithos* Ichnofacies to a depauperate *Cruziana* Ichnofacies, being characterized by low-diversity suites dominated by *Planolites* and *Teichichnus*. Low ichnodiversity and sparse bioturbation most probably reflect soupy substrates in connection with fluid mud. The shelf deposits adjacent to the sand ridges consist of mud-dominated heterolithic successions, containing abundant arthropod (e.g., *Cruziana*, *Dimorphichnus*, and *Rusophycus*) and worm (*Planolites*, *Palaeophycus*, and *Phycodes*) trace fossils (Fig. 13D and E). This association is characteristic of low-energy conditions and abundant organic particles settling from suspension onto the sea floor, thereby allowing the development of the archetypal *Cruziana* Ichnofacies (Desjardins et al., 2012a).

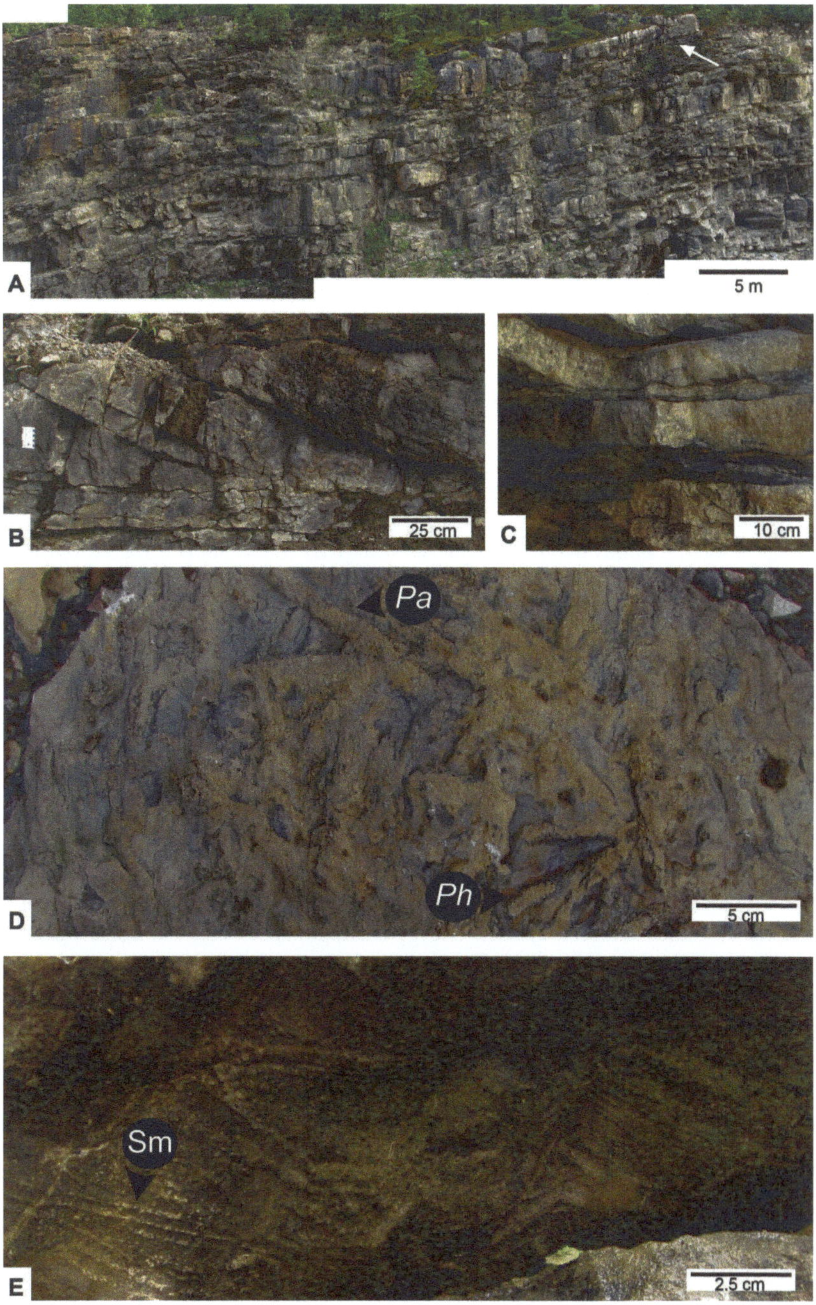

FIGURE 13 Sand-ridge deposits. Early Cambrian Wiwaxy Peaks Member, St. Piran Formation, Gog Group, southern Rocky Mountains, Canada. (A) Sand-ridge lateral-accretion deposits exhibiting very thick-bedded, compound cross-stratified sandstone. (B) Non-bioturbated, compound cross-stratified sandstone at the core of the sand ridge. (C) Intercalated sandstone and mudstone along sand- ridge flanks with local *Planolites* isp. (D) Bedding-plane view (base) of *Palaeophycus* isp. (*Pa*) and *Phycodes* isp. (*Ph*). (E) Bedding-plane view (base) of *Dimorphichnus*-like scratch traces (Sm).

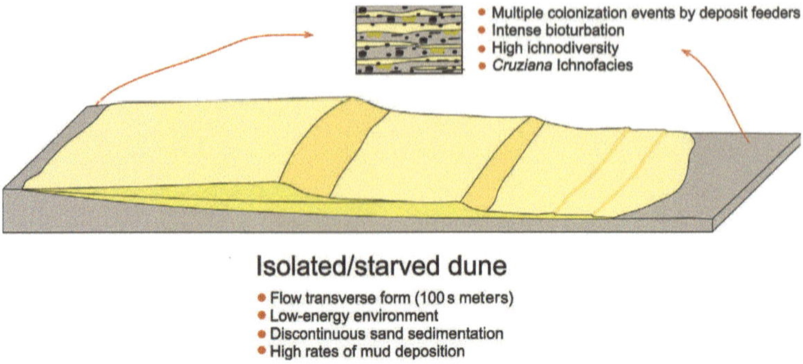

FIGURE 14 Integrated sedimentological and ichnological model for isolated dune patches.

3.4 Isolated Dune Patches

Isolated dune patches occur in sand-starved areas of the shelf affected by tidal currents under conditions of very low sediment supply (Figs. 1 and 14). They typically develop during transgressions when shallow-water sands are redistributed toward the open shelf. This type of deposit is characterized by lenticular, medium- to thick-bedded, low-angle compound cross-stratified packages (Fig. 15A). Sigmoidal-shaped reactivation surfaces truncating the foresets are common and grade laterally into intensively bioturbated, very thin- to thin-bedded, ripple cross-laminated sandstone and mudstone displaying lenticular bedding.

The sandy cross-stratified portions of the isolated dune patches are generally not bioturbated, probably because of the high mud contents in the water column, which may have prevented the establishment of the typical suspension-feeding faunas adapted to relatively high-energy settings. However, the interfingering muddy deposits typically contain a high diversity of trace fossils characteristic of the archetypal *Cruziana* Ichnofacies. The trace-fossil fauna consists of deposit-feeder trace fossils produced by both arthropods (e.g., *Cruziana*, *Diplichnites*, and *Rusophycus*) and worms (e.g., *Helminthopsis*, *Palaeophycus*, *Phycodes*, *Planolites*, *Teichichnus*, and *Trichophycus*; Fig. 15B–E). The high ichnodiversity and degree of bioturbation reflect the existence of continuous colonization windows in mud-dominated low-energy zones of the inner shelf (Desjardins et al., 2012a).

3.5 Tidal Bars

Bars are longer-lived sand bodies than subaqueous dunes (Dalrymple and Rhodes, 1995). They include flow-transverse, flow-oblique, and longitudinally oriented sand bodies, commonly having dunes superimposed on them. Tidal bars are typically developed within distributary channels, along delta fronts and in the outer parts of estuaries (Dalrymple and Choi, 2007), but do not occur on the shelf proper because, in the offshore, tidal currents are not constrained by geomorphological features. As in the case of tidal sand ridges, the migration of

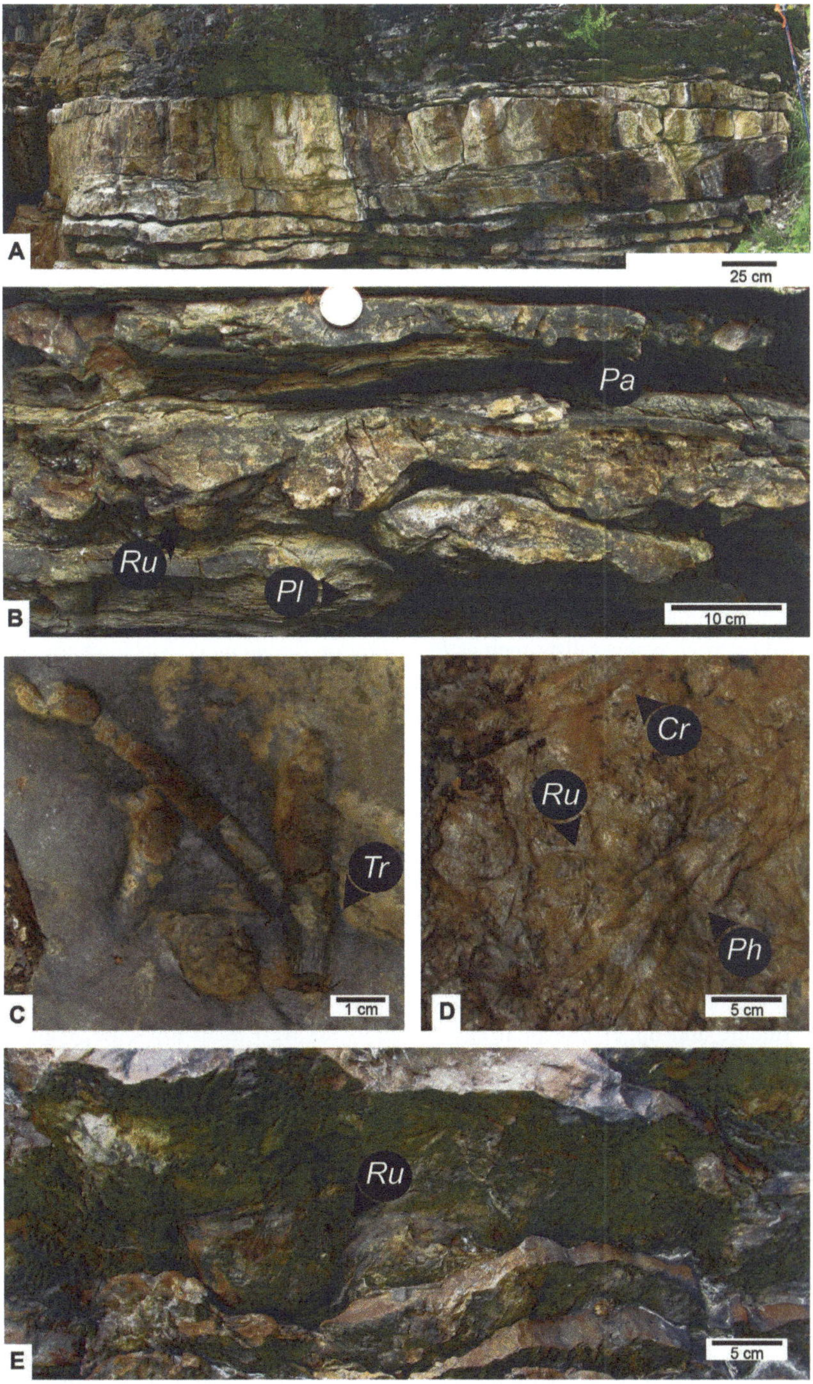

FIGURE 15 Isolated dune-patch deposits. Early Cambrian Lake Louise Formation, Gog Group, southern Rocky Mountains, Canada. (A) Compound cross-stratified sandstone encased within intercalated mudstone and sandstone. (B) Cross-section view of intercalated thin-bedded sandstone and mudstone containing relatively deep *Rusophycus* isp. (*Ru*), local *Planolites montanus* (*Pl*) and *Palaeophycus* isp. (*Pa*). (C) Bedding-plane view (base) of *Trichophycus venosus* (*Tr*). (D) Bedding-plane view (base) of *Cruziana* isp. (*Cr*), *Rusophycus* isp. (*Ru*) and *Phycodes* isp. (*Ph*). (E) Deep *Rusophycus* isp. (*Ru*) at the soles of sandstone beds.

tidal bars is slightly oblique, but in the direction of the dominant current. Migration proceeds by erosion of their stoss side by the dominant current and deposition on the opposite lee side. This leads to the preferential preservation of the down-current flank of the bar, where sedimentation is dominated by the subordinate current. As a result, the small cross-beds within the large-scale lateral accretion deposits are preferentially oriented in the direction of the regional weaker current (Dalrymple and Choi, 2007: fig. 29). The lateral-accretion deposits of tidal bars typically overlie the erosion surface produced by the migration of the thalweg of the adjacent channel (e.g., Yu et al., 2012). Because the water depth and the current speed both decrease upward from the thalweg toward the bar crest, the bedset thicknesses decrease and grain sizes decrease upward, except where fluid mud occurs at the channel bottom (Dalrymple et al., 2003). This contrasts with the tidal sand ridges occurring in unconfined areas of open shelves.

Tidal bars, which typically develop in restricted brackish-water settings (e.g., estuaries, funnel-shaped bays), tend to contain low-diversity ichnofaunas dominated by trophic generalists such as *Planolites* or *Teichichnus* (MacEachern and Gingras, 2007; MacEachern and Pemberton, 1994; Mángano and Buatois, 2004a) and are addressed by Gingras et al. (2012).

4. ICHNOFACIES TRENDS

The ichnofacies gradient across tide-dominated shorelines is to some extent opposite to that observed in wave-dominated shoreface environments. Near the shoreline, the overall tidal energy increases from the supratidal to the shallow subtidal. Whereas the *Skolithos* Ichnofacies tends to dominate lower intertidal and shallow subtidal deposits, the *Cruziana* Ichnofacies is best developed in middle and upper intertidal environments, as well as in low-energy areas of the inner shelf (Desjardins et al., 2012a; Mángano and Buatois, 2004b; Mángano et al., 2002a). The decrease in energy toward the upper intertidal is accompanied by a decrease in sand content, sediment oxygen content, and mobility of the substrate. Thus, muds and organic particles are prone to settle from suspension as the energy decreases toward the upper intertidal zone. However, the highest concentration (mass per unit volume) of mud and organic particles occurs in the zone of mixed sediments (muddy sand–sandy mud), although the highest contents (mass per unit mass) of mud and organics occurs in pure mud in the uppermost intertidal area (Flemming, 2012). Accordingly, the highest trace-fossil diversity tends to occur in the mixed-flat area or, in the case of lower-energy systems, in the sand flat, with ichnofaunas belonging to the *Cruziana* Ichnofacies. The decrease in suspended organic particles toward the upper intertidal is a consequence of the reduction in the thickness of water column toward the high-tide line (i.e., the total amount of suspended material available for sedimentation gets progressively smaller; Flemming, 2012).

However, in the shallow subtidal and shelf environments, different tidal sand-body types have different ichnological signatures owing to variations in substrate type, sediment mobility, water turbidity, type of organic matter (suspended *versus* deposited), and salinity. In the shallow subtidal and on the inner shelf, the *Skolithos* Ichnofacies tends to dominate the high-energy deposits formed by subaqueous dunes and bars, the degree of bioturbation by these suspension-feeders reflecting well-developed colonization windows. Commonly, muddy substrates at the toes of dunes are characterized by an impoverished detritus-feeding ichnofauna (impoverished *Cruziana* Ichnofacies). If the sand bodies are emplaced on the inner shelf, a diverse deposit-feeder ichnofauna develops along the margins of sand sheets, compound dune fields, and tidal sand-ridges (*Cruziana* Ichnofacies). However, if the sand bodies are developed in restricted areas such as bays, delta fronts, or estuaries, they tend to contain low-diversity ichnofaunas dominated by trophic generalists (Gingras et al., 2012; Tonkin, 2012).

5. SEQUENCE STRATIGRAPHY

Parasequences from wave- and tide-dominated settings contrast in their internal bedset stacking pattern as a result of markedly different hydrodynamic energy gradients. Parasequences in wave-dominated environments are characterized by a coarsening-upward trend, whereas those representing shorelines adjacent to tidal flats have a fining-upward character, recording the transitions from sand flats to mixed flats and mud flats (Fig. 16B; Klein, 1977; Van Wagoner et al.,

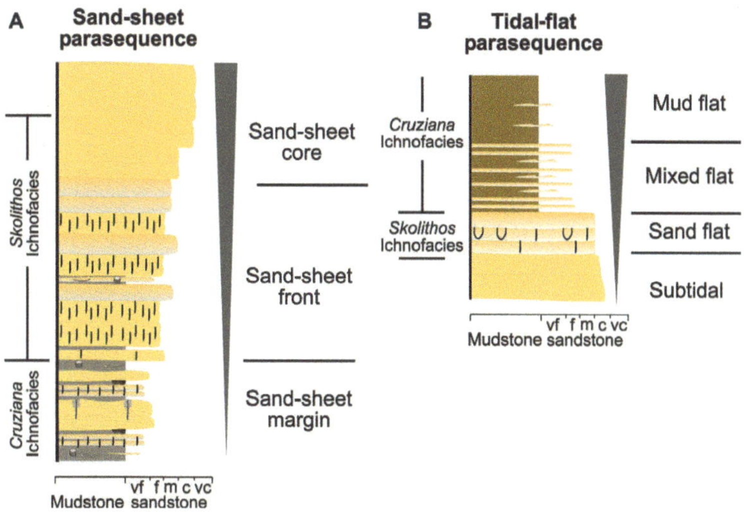

FIGURE 16 Variability of tide-dominated parasequences. (A) Coarsening-upward sand-sheet parasequence. (B) Fining-upward tidal-flat parasequence. See Mángano and Buatois (2004b) for details on the zonation of the tidal-flat parasequence.

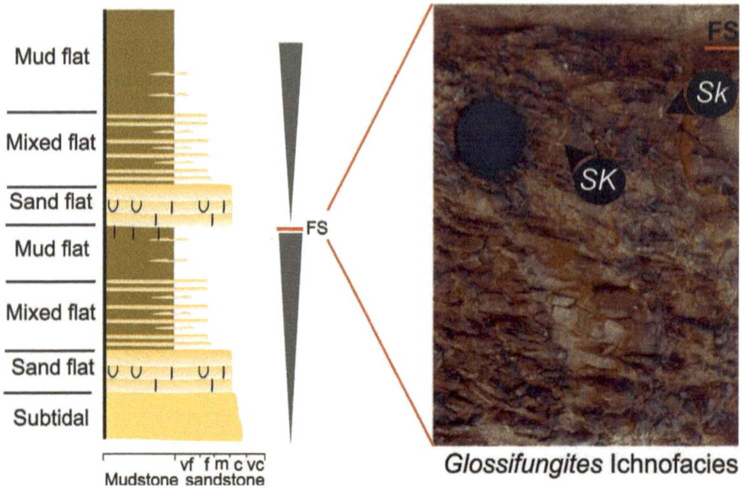

FIGURE 17 Cross-section view of *Glossifungites* Ichnofacies with *Skolithos* (*Sk*) below tidal-flat flooding-surfaces. Photo is from the Early to Middle Cambrian Campanario Formation, Mesón Group, Purmamarca town, Quebrada de Huamahuaca, northwest Argentina.

1990). The two types of parasequences also differ in their ichnological signatures. Wave-dominated parasequences display an overall vertical transition from the *Cruziana* to the *Skolithos* Ichnofacies, whereas tide-dominated parasequences exhibit a vertical replacement of the *Skolithos* by the *Cruziana* Ichnofacies (Buatois and Mángano, 2011; Mángano and Buatois, 1999, 2004b).

On the other hand, parasequences developed in the subtidal areas of a tide-dominated shelf are mainly controlled by the migration and accretion of spatially extensive sand bodies. For example, in sand sheets, parasequences are characterized by the vertical transition from intercalated very thin- to thin-bedded, very fine- to fine-grained sandstone and mudstone, hummocky cross-stratified, fine-grained sandstone, and highly bioturbated *Skolithos* pipe rock to non-bioturbated planar cross-stratified, fine- to medium-grained sandstone of the sand-sheet core (Fig. 16A; Desjardins et al., 2010b).

The distinctive lithological and ichnological characteristics of parasequences developed in intertidal and subtidal areas of tide-dominated environments may be used for the recognition of sequence-stratigraphically significant surfaces. The lowermost interval of a sandy tidal-flat parasequence is characterized by sharp-based sandstone, which overlies the truncated mud-flat deposit of the previous parasequence. This high-energy flooding surface is commonly delineated by firmground *Skolithos* of the *Glossifungites* Ichnofacies (Fig. 17). Desjardins et al. (2012b) recognized that the parasequences at the top of the Lake O'Hara Member of the Gog Group were truncated at a pipe-rock interval by an erosion surface and that the expected capping cross-stratified sandstone was missing. On top of this surface, fining-upward tidal-flat parasequences occur (Fig. 6A). Because the missing facies would have recorded progradation due to gradual shallowing, this surface implies an erosional event. The facies jump was interpreted to represent a forced regression and the surface was interpreted as a response to sea-level fall.

6. CONCLUSIONS

Although still lagging behind our knowledge of wave-dominated marine settings, the ichnology of tide-dominated open-marine environments has received increased attention during the past decade. As a result, an integrated ichnological and sedimentological model that allows the delineation of supratidal, intertidal (mud, mixed, and sand flats), and shallow subtidal subenvironments has been proposed. Integration of sedimentological and ichnological datasets also helps to characterize compound dune fields, sand sheets, sand ridges, isolated dune patches, and tidal bars. The integration of this information into a sequence-stratigraphic framework, in turn, allows the distinction of two types of parasequences. The first type is related to tidal-flat progradation and is reflected by the replacement of the *Skolithos* Ichnofacies by the *Cruziana* Ichnofacies within fining-upward successions. The second type relates to the progradation of a sand sheet and is characterized by the transition from the *Skolithos* Ichnofacies to non-bioturbated high-energy sandstones of the sand-sheet core as represented in coarsening-upward successions. Tidal-flat parasequences are typically characterized by high-energy flooding surfaces delineated by the *Glossifungites* Ichnofacies.

ACKNOWLEDGMENTS

We thank Dirk Knaust and Richard Bromley for their willingness and patience to include this chapter as part of their book, Brian Pratt for his valuable feedback on the Gog Group, and Burghard Flemming for his insightful comments which have considerably improved this chapter. Financial support for this study was provided by Natural Sciences and Engineering Research Council (NSERC) Discovery Grants 311727-05/08 and 311726-05/08 awarded to Mángano and Buatois, respectively.

REFERENCES

Antia, E., Flemming, B., Wefer, G., 1994. Transgressive facies sequence of a high energy, wave-tide-storm-influenced shoreface: a case study of the East Frisian barrier islands (southern North Sea). Facies 30, 15–24.

Ashley, G.M., 1990. Classification of large-scale subaqueous bedforms: a new look at an old problem. J. Sediment. Petrol. 60, 160–172.

Augustinus, P., 2002. Biochemical factors influencing deposition and erosion of fine-grained sediment. In: Healy, T., Wang, Y., Healy, J.A. (Eds.), Muddy Coasts of the World: Processes, Deposits, and Function. Proc. Mar. Sci., vol. 4, pp. 203–228.

Avanzini, M., Piubelli, D., Mietto, P., Roghi, G., Romano, R., Masetti, D., 2006. Lower Jurassic (Hettangian-Sinemurian) dinosaur track megasites, southern Alps, northern Italy. New Mexico Mus. Nat. Hist. Sci., Bull. 37, 207–216.

Basan, P.B., Frey, R.W., 1977. Actual-palaeontology and neoichnology of salt marshes near Sapelo Island, Georgia. In: Crimes, P., Harper, J.C. (Eds.), Trace Fossils 2. Geol. J., Spec. Iss. 9, pp. 41–69.

Bates, C.C., 1953. A rational theory of delta formation as exemplified by the present-day Mississippi Delta. J. Sediment. Petrol. 23, 132–133.

Baucon, A., 2008. Neoichnology of a microbial mat in a temperate, siliciclastic environment: Spiaggia al Bosco (Grado, Northern Adriatic, Italy). In: Avanzini, M., Petti, F. (Eds.), Italian Ichnology. Studi Trent. Sci. Nat., Acta Geol., vol. 83, pp. 183–203.

Belderson, R.H., Johnson, M.A., Kenyon, N.H., 1982. Bedforms. In: Stride, A.H. (Ed.), Offshore Tidal Sands: Processes and Deposits. Chapman & Hall, New York, pp. 27–57.

Berné, S., Vanner, P., Guichard, F., Lericolais, G., Liu, Z., Trentesaux, A., Yin, P., Yi, H., 2002. Pleistocene forced regressions and tidal sand ridges in the East China Sea. Mar. Geol. 188, 293–315.

Bhattacharya, J.P., MacEachern, J.A., 2009. Hyperpycnal rivers and prodeltaic shelves in the Cretaceous seaway of North America. J. Sediment. Res. 79, 184–209.

Boyd, R., Dalrymple, R.W., Zaitlin, B.A., 2006. Estuarine and incised-valley facies models. In: Posamentier, H.W., Walker, R. (Eds.), Facies Models Revisited. SEPM Spec. Publ., vol. 84, pp. 175–240.

Buatois, L.A., Mángano, M.G., 1995. The paleoenvironmental and paleoecological significance of the lacustrine *Mermia* ichnofacies: an archetypical subaqueous nonmarine trace fossil assemblage. Ichnos 4, 151–161.

Buatois, L.A., Mángano, M.G., 2011. Ichnology: Organism-Substrate Interactions in Space and Time. Cambridge University Press, Cambridge, 358 pp.

Buatois, L.A., Mángano, M.G., Maples, C.G., Lanier, W.P., 1997. The paradox of nonmarine ichnofaunas in tidal rhythmites: integrating sedimentologic and ichnologic data from the Late Carboniferous of eastern Kansas, USA. Palaios 12, 467–481.

Cadée, G.C., 1998. Influence of benthic fauna and microflora. In: Eisma, D. (Ed.), Intertidal Deposits: River Mouths, Tidal Flats, and Lagoons. CRC Press, Boca Raton, pp. 383–402.

Carmona, N.B., Buatois, L.A., Ponce, J.J., Mángano, M.G., 2009. Ichnology and sedimentology of a tide-influenced delta, Lower Miocene Chenque Formation, Patagonia, Argentina: trace-fossil distribution and response to environmental stresses. Palaeogeogr. Palaeoclimatol. Palaeoecol. 273, 75–86.

Chang, T.S., Flemming, B.W., Bartholomä, A., 2007. Distinction between sortable silts and aggregates in muddy intertidal sediments of the East Frisian Wadden Sea, southern North Sea. Sediment. Geol. 202, 453–463.

Cornish, F.G., 1986. The trace-fossil *Diplocraterion*; evidence of animal-sediment interactions in Cambrian tidal deposits. Palaios 5, 478–491.

Dalrymple, R.W., 2010. Tidal depositional systems. In: James, N., Dalrymple, R. (Eds.), Facies Models 4. Geol. Ass. Can., St. John's, Geotext 6, pp. 201–231.

Dalrymple, R.W., Choi, K., 2007. Morphologic facies trends through the fluvial-marine transition in tide-dominated depositional systems: a schematic framework for environmental and sequence-stratigraphic interpretation. Earth-Sci. Rev. 81, 135–174.

Dalrymple, R.W., Rhodes, R.N., 1995. Estuarine dunes and bars. In: Perillo, G.M.E. (Ed.), Geomorphology and Sedimentology of Estuaries. Developments in Sedimentology, vol. 53. Elsevier, Amsterdam, pp. 359–406.

Dalrymple, R.W., Knight, R.J., Zaitlin, B.A., Middleton, G.V., 1990. Dynamics and facies model of a macrotidal sand-bar complex, Cobequid Bay-Salmon River estuary (Bay of Fundy). Sedimentology 37, 577–612.

Dalrymple, R.W., Baker, E.K., Harris, P.T., Hughes, M., 2003. Sedimentology and stratigraphy of a tide-dominated foreland-basin delta (Fly River, Papua New Guinea). In: Sidi, F.H., Nummedal, D., Imbert, P., Darman, H., Posamentier, H. (Eds.), Tropical Deltas of Southeast Asia—Sedimentology, Stratigraphy and Petroleum Geology. SEPM Spec. Publ., vol. 76, pp. 147–173.

Dashtgard, S.E., MacEachern, J.A., Frey, S.E., Gingras, M.K., in press. Tidal effects on the shoreface: towards a conceptual framework. Sediment. Geol. doi:10.1016/j.sedgeo.2011.05.007.

De Boer, P.L., 1998. Intertidal sediments: composition and structure. In: Eisma, D. (Ed.), Intertidal Deposits: River Mouths, Tidal Flats and Coastal Lagoons. CRC Press, Boca Raton, pp. 912–921.

Desjardins, P.R., Pratt, B.R., Buatois, L.A., Mángano, M.G., 2010a. Stratigraphy and sedimentary environments of the Lower Cambrian Gog Group in the southern Rocky Mountains of western Canada: evolution of transgressive sandstones on a broad continental margin. Bull. Can. Petrol. Geol. 58, 1–37.

Desjardins, P.R., Mángano, M.G., Buatois, L.A., Pratt, B.R., 2010b. *Skolithos* pipe rock and associated ichnofabrics from the southern Rocky Mountains, Canada: colonisation trends and environmental controls in an Early Cambrian sand-sheet complex. Lethaia 43, 507–528.

Desjardins, P.R., Buatois, L.A., Pratt, B.R., Mángano, M.G., 2012a. Sedimentological–ichnological model for tide-dominated shelf sandbodies: lower Cambrian Gog Group of western Canada. Sedimentology 59, 1452–1477.

Desjardins, P.R., Buatois, L.A., Pratt, B.R., Mángano, M.G., 2012b. Forced-regressive tidal flats: response to falling sea level in tide-dominated settings. J. Sediment. Res. 82, 149–162.

Dörjes, J., 1970. Das Watt als Lebensraum. In: Reineck, H.-E. (Ed.), Das Watt, Ablagerungs- und Lebensraum. Kramer, Frankfurt, pp. 71–105.

Edwards, J.M., Frey, R.W., 1977. Substrate characteristics within a Holocene salt marsh, Sapelo Island, Georgia. Senckenberg. Marit. 9, 215–259.

Flemming, B.W., 2000. The role of grain size, water depth and flow velocity as scaling factors controlling the size of subaqueous dunes. In: Trentesaux, A., Garlan, T. (Eds.), Marine Sandwave Dynamics. 23–24 March 2000, University of Lille 1 (France), Proceedings, pp. 55–60.

Flemming, B.W., 2005. The concept of wave base: fact and fiction. In: Haas, H., Ramseyer, K., Schlunegger, F. (Eds.), Abstracts. Sediment 2005, 18–19 July, Gwatt, Lake Thun, Switzerland. Schriftenr. Deutsche Ges. Geowiss. 38, 57 pp.

Flemming, B.W., 2012. Siliciclastic back-barrier tidal flats. In: Davis Jr., R.A., Dalrymple, R.W. (Eds.), Principles of Tidal Sedimentology. Springer, Dordrecht, pp. 231–267.

Frey, R.W., Pemberton, S.G., 1986. Vertebrate lebensspuren in intertidal and supratidal environments, Holocene barrier islands. Georgia. Senckenberg. Marit. 18, 45–95.

Frey, R.W., Pemberton, S.G., 1987. The *Psilonichnus* ichnocoenose, and its relationship to adjacent marine and nonmarine ichnocoenoses along the Georgia coast. Bull. Can. Petrol. Geol. 35, 333–357.

Genise, J.F., Mángano, M.G., Buatois, L.A., Laza, J., Verde, M., 2000. Insect trace fossil associations in paleosols: the *Coprinisphaera* ichnofacies. Palaios 15, 33–48.

Gerdes, G., Krumbein, W.E., Reineck, H.-E., 1985. The depositional record of sandy, versicolored tidal flats (Mellum Island, southern North Sea). J. Sediment. Petrol. 55, 265–278.

Gingras, M.K., MacEachern, J.A., 2012. Tidal ichnology of shallow-water clastic settings. In: Davis, Jr., R.A., Dalrymple, R.W. (Eds.), Principles of Tidal Sedimentology. Springer, Dordrecht, pp. 57–77.

Gingras, M.K., MacEachern, J.A., Pemberton, S.G., 1998. A comparative analysis of the ichnology of wave- and river-dominated allomembers of the Upper Cretaceous Dunvegan Formation. Bull. Can. Petrol. Geol. 46, 51–73.

Gingras, M.K., Pemberton, S.G., Saunders, T., Clifton, H.E., 1999. The ichnology of modern and Pleistocene brackish-water deposits at Willapa Bay, Washington: variability in estuarine settings. Palaios 14, 352–374.

Gingras, M.K., MacEachern, J.A., Dashtgard, S.E., Zonneveld, J.-P., Schoengut, J., Ranger, M.J., Pemberton, S.G., 2012. Estuaries. In: Knaust, D., Bromley, R.G. (Eds.), Trace Fossils as Indicators of Sedimentary Environments. Developments in Sedimentology, vol. 64. Elsevier, Amsterdam, pp. 463–505.

Gingras, M.K., MacEachern, J.D., Dashtgard, S.E., in press. The potential of trace fossils as tidal indicators in bays and estuaries. Sediment. Geol. doi:10.1016/j.sedgeo.2011.05.007.

Green, R.H., Hobson, K.D., 1970. Spatial and temporal structure in a temperate intertidal community with special emphasis on *Gemma gemma* (Pelecypoda: Mollusca). Ecology 51, 999–1011.

Hunt, A., Lucas, S., 2007. Tetrapod ichnofacies: a new paradigm. Ichnos 14, 59–68.

Jansson, J.D., 1967. The significance of grain size and pore water content for the interstitial fauna of sandy beaches. Oikos 18, 311–322.

Kineke, G.C., Sternberg, R.W., Trowbridge, J.H., Rockwell, G.W., 1996. Fluid-mud processes on the Amazon continental shelf. Cont. Shelf Res. 16, 667–696.

Klein, G., 1971. A sedimentary model for determining paleotidal range. GSA Bull. 82, 2585–2592.

Klein, G., 1977. Clastic Tidal Facies. CEPCO, Champaign, 149 pp.

Leithold, E.L., Dean, W.E., 1998. Depositional processes and carbon burial on a Turonian prodelta at the margin of the Western Interior Seaway. In: Dean, W.E., Arthur, M.A. (Eds.), Stratigraphy and Palaeoenvironments of the Cretaceous Western Interior seaway, USA. SEPM Concepts in Sedimentology and Palaeontology, vol. 6, pp. 189–200.

Liu, Z.X., Dias, D.X., Berné, S., Yang, W.K., Marsset, T., Tang, Y.X., Bourillet, J.F., 1998. Tidal depositional systems of China's continental shelf, with special reference to the eastern Bohai Sea. Mar. Geol. 145, 225–268.

Lockley, M.G., Holbrook, J., Hunt, A.P., Matsukawa, M., Meyer, C., 1992. The dinosaur freeway: a preliminary report on the Cretaceous megatracksite, Dakota Group, Rocky Mountain Front Range and High Plains, Colorado, Oklahoma and New Mexico. In: Flores, R. (Ed.), Mesozoic of the Western Interior. SEPM Mid-year Meeting, Fieldtrip Guidebook, pp. 39–54.

MacEachern, J.A., Gingras, M., 2007. Recognition of brackish-water trace fossil assemblages in the Cretaceous western interior seaway of Alberta. In: Bromley, R.G., Buatois, L.A., Mángano, G., Genise, J.F., Melchor, R.N. (Eds.), Sediment-Organism Interactions: A Multifaceted Ichnology. SEPM Spec. Publ., vol. 88, pp. 149–194.

MacEachern, J.A., Pemberton, S.G., 1994. Ichnological aspects of incised valley fill systems from the Viking Formation of the Western Canada Sedimentary Basin, Alberta, Canada. In: Boyd, R., Zaitlin, B.A., Dalrymple, R. (Eds.), Incised Valley Systems—Origin and Sedimentary Sequences. SEPM Spec. Publ., vol. 51, pp. 129–157.

MacEachern, J.A., Bann, K.L., Bhattacharya, J.P., Howell Jr., C.D., 2005. Ichnology of Deltas; Organism Responses to the Dynamic Interplay of Rivers, Waves, Storms, and Tides. SEPM Spec. Publ., vol. 83, 49–85.

Mángano, M.G., Buatois, L.A., 1999. Ichnofacies models in Early Paleozoic tide-dominated quartzites: onshore-offshore gradients and the classic Seilacherian paradigm. Acta Univ. Carol. 43, 151–154.

Mángano, M.G., Buatois, L.A., 2004a. Reconstructing Early Phanerozoic intertidal ecosystems: ichnology of the Cambrian Campanario Formation in northwest Argentina. In: Webby, B.D., Mángano, M.G., Buatois, L.A. (Eds.), Trace Fossils in Evolutionary Palaeoecology. Fossils Strata 51, pp. 17–38.

Mángano, M.G., Buatois, L.A., 2004b. Ichnology of Carboniferous tide-influenced environments and tidal flat variability in the North American Midcontinent. In: McIlroy, D. (Ed.), The Application of Ichnology to Palaeoenvironmental and Stratigraphic Analysis. Geol. Soc. London, Spec. Publ., vol. 228, pp. 157–178.

Mángano, M.G., Buatois, L.A., Aceñolaza, G.F., 1996. Trace fossils and sedimentary facies from an Early Ordovician tide-dominated shelf (Santa Rosita Formation, northwest Argentina): implications for ichnofacies models of shallow marine successions. Ichnos 5, 53–88.

Mángano, M.G., Buatois, L.A., West, R.R., Maples, C.G., 1998. Contrasting behavioral and feeding strategies recorded by tidal-flat bivalve trace fossils from the Upper Carboniferous of eastern Kansas. Palaios 13, 335–351.

Mángano, M.G., Buatois, L.A., West, R.R., Maples, C.G., 2002a. Ichnology of a Pennsylvanian Equatorial Tidal Flat: The Stull Shale Member at Waverly, Eastern Kansas. Kansas Geol. Surv. 245, pp. 1–133.

Mángano, M.G., Buatois, L.A., Rindsberg, A.K., 2002b. Carboniferous *Psammichnites*: systematic re-evaluation, taphonomy and autecology. Ichnos 9, 1–22.

McIlroy, D., 2007. Ichnology of a macrotidal tide-dominated deltaic depositional system: Lajas Formation, Neuquén Province, Argentina. In: Bromley, R.G., Buatois, L.A., Mángano, G., Genise, J.F., Melchor, R.N. (Eds.), Sediment-Organism Interactions: A Multifaceted Ichnology. SEPM Spec. Publ., vol. 88, pp. 195–211.

Montague, C.L., Bunker, S.M., Haines, E.B., Pace, M.L., Wetzel, R.L., 1981. Aquatic macroconsumers. In: Pomeroy, L.R., Wiegert, R.G. (Eds.), The Ecology of Salt Marsh. Springer, New York, pp. 69–85.

Mutti, E., Rosell, J., Allen, G.P., Fonnesu, F., Sgavetti, M., 1985. The Eocene Baronia tide-dominated delta-shelf system in the Ager Basin. In: Miall, A.D., Rosell, J. (Eds.), 6th European Regional Meeting. IAS, Lleida, Spain, Excursion Guide-Book, pp. 579–600.

Newell, R.C., 1979. Biology of Intertidal Animals. Marine Ecological Surveys Ltd., Favershem, Kent, 781 pp.

Noffke, N., 2010. Geobiology. Microbial Mats in Sandy Deposits from the Archaean Era to Today. Springer, Heidelberg, 194 pp.

Palmer, J.D., 1967. Daily and tidal components in the persistent rhythmic activity of the crab *Sesarma*. Nature 215, 64–66.

Palmer, J.D., 1995. The Biological Rhythms and Clocks of Intertidal Animals. Oxford University Press, New York, 217 pp.

Pemberton, S.G., MacEachern, J.A., Dashtgard, S.E., Bann, K.L., Gingras, M.K., Zonneveld, J.-P., 2012. Shorefaces. In: Knaust, D., Bromley, R.G. (Eds.), Trace Fossils as Indicators of Sedimentary Environments. Developments in Sedimentology, vol. 64. Elsevier, Amsterdam, pp. 563–604.

Penland, S., Boyd, R.L., Suter, J.R., 1988. Transgressive depositional systems of the Mississippi delta plain: a model for barrier shoreline and shelf sand development. J. Sediment. Petrol. 58, 932–949.

Perkins, E.J., 1974. The Biology of Estuaries and Coastal Waters. Academic Press, London, 678 pp.

Pieńkowski, M.H., 1983. Surface activity of some intertidal invertebrates in relation to temperature and the foraging behaviour of their shorebird predators. Mar. Ecol. Prog. Ser. 11, 141–150.

Plint, A.G., 2010. Wave- and storm-dominated shoreline and shallow-marine system. In: James, N.P., Dalrymple, R.W. (Eds.), Facies Models 4. Geol. Ass. Can., St. John's, Geotext 6, pp. 167–200.

Pollard, J.E., Goldring, R., Buck, S.G., 1993. Ichnofabrics containing *Ophiomorpha*: significance in shallow-water facies interpretation. J. Geol. Soc. London 150, 149–164.

Pomeroy, L.R., Darley, W.M., Dunn, E.L., Gallagher, J.L., Haines, E.B., Whitney, D.M., 1981. Primary production. In: Pomeroy, L.R., Wiegert, W. (Eds.), Ecology of a Salt Marsh. Springer, New York, pp. 39–67.

Posamentier, H.W., 2002. Ancient shelf ridges—a potentially significant component of the transgressive systems tract: case study from offshore northwest Java. AAPG Bull. 86, 75–106.

Potter, P.E., Maynard, J.B., Depetris, P.J., 2005. Mud and Mudstone: Introduction and Overview. Springer, Berlin, 297 pp.

Reineck, H.-E., Singh, I.B., 1980. Depositional Sedimentary Environments. Springer, Berlin, 551 pp.

Reise, K., 1985. Tidal Flat Ecology: An Experimental Approach to Species Interactions. Ecological Studies 54. Springer, Berlin, 191 pp.

Sanders, H.L., Mangelsdorf, P.C., Hampson, G.R., 1965. Salinity and faunal distribution in the Pocasset River, Massachusetts. Limnol. Oceanogr. 10, R216–R229.

Schäfer, W., 1972. Ecology and Palaeoecology of Marine Environments. Chicago University Press, Chicago, 568 pp.

Schieber, J., 2003. Simple gifts and buried treasures; implications of finding bioturbation and erosion surfaces in black shales. Sediment. Rec. 1, 4–8.

Schieber, J., 2004. Microbial mats in the siliciclastic rock record. In: Eriksson, P.K., Altermannn, D.R., Nelson, D.R., Mueller, W.E., Catuneanu, O. (Eds.), The Precambrian Earth: Tempos and Events. Elsevier, Amsterdam, pp. 663–673.

Seiple, W., 1981. The ecological significance of the locomotor activity rhythms of *Sesarma cinereum* and *Sesarma reticulatum*. Crustaceana 40, 5–15.

Simpson, E.L., 1991. An exhumed, Lower Cambrian tidal flat: the Antietam Formation, central Virginia, U.S.A. In: Smith, D.G., Reinson, G.E., Zaitlin, B.A., Rahmani, R.A. (Eds.), Clastic Tidal Sedimentology. Can. Soc. Petrol. Geol., Mem. 16, pp. 123–134.

Snedden, J.W., Dalrymple, R.W., 1999. Modern shelf sand ridges: from historical perspective to a unified hydrodynamic and evolutionary model. In: Bergman, K.M., Snedden, J.W. (Eds.), Isolated Shallow Marine Sand Bodies: Sequence Stratigraphic Analysis and Sedimentologic Interpretations. SEPM Spec. Publ., vol. 64, pp. 13–28.

Snedden, J.W., Kreisa, R.D., Tillman, R.W., Culver, S.J., Schweller, W.J., 1999. An expanded model for modern shelf sand ridge genesis and evolution on the New Jersey Atlantic shelf. In: Bergman, K.M., Snedden, J.W. (Eds.), Isolated Shallow Marine Sand Bodies: Sequence Stratigraphic Analysis and Sedimentologic Interpretations. SEPM Spec. Publ., vol. 64, pp. 147–163.

Stride, A.H., Belderson, R.H., Kenyon, N.H., Johnson, M.A., 1982. Offshore tidal deposits: sand sheet and sand bank facies. In: Stride, A.H. (Ed.), Offshore Tidal Sands: Processes and Deposits. Chapman & Hall, New York, pp. 95–125.

Swift, D.J.P., Field, M.F., 1981. Evolution of a classic ridge field, Maryland sector, North American inner shelf. Sedimentology 28, 461–482.

Terwindt, J.H.J., 1988. Palaeo-tidal reconstructions of inshore tidal depositional environments. In: De Boer, P.L., Van Gelder, A., Nio, S.D., Dordrecht, D. (Eds.), Tide-Influenced Sedimentary Environments. Reidel Publishing, Dordrecht, pp. 233–263.

Thomas, R.G., Smith, D.G., Wood, J.M., Visser, J., Caverley-Range, E.A., Koster, E.H., 1987. Inclined heterolithic stratification: terminology, description, interpretation and significance. Sediment. Geol. 53, 123–179.

Tonkin, N., 2012. Deltas. In: Knaust, D., Bromley, R.G. (Eds.), Trace Fossils as Indicators of Sedimentary Environments. Developments in Sedimentology, vol. 64. Elsevier, Amsterdam, pp. 507–528.

Vader, W.J.M., 1964. A preliminary investigation into the reactions of the infauna of the tidal flats to tidal fluctuations in water level. Neth. J. Sea Res. 2, 189–222.

Van Landeghem, K.J.J., Wheeler, A.J., Mitchell, N.C., Sutton, G., 2009. Variations in sediment wave dimensions across the tidally dominated Irish Sea, NW Europe. Mar. Geol. 263, 108–119.

Van Wagoner, J.C., Mitchum, R.M., Campion, K.M., Rahmanian, V.D., 1990. Siliciclastic Sequences, Stratigraphy in Well Logs, Cores, and Outcrops. AAPG, Methods in Exploration Series 7, 55 pp.

Wang, Y., Zhang, Y., Zou, X., Zhu, D., Piper, D., 2012. The sand ridge field of the South Yellow Sea: origin by river-sea interaction. Mar. Geol. 291–294, 132–146.

Warman, C.G., O'Hare, T.J., Naylor, E., 1991. Vertical swimming in wave-induced currents as a control mechanism of intertidal migration by a sand-beach isopod. Mar. Biol. 111, 49–50.

Webb, J.E., 1969. Biologically significant properties of submerged marine sands. Proc. Roy. Soc. London 174, 355–402.

Willis, B.J., 2005. Deposits of tide-influenced river deltas. In: Giosan, L., Bhattacharya, J.P. (Eds.), River Deltas—Concepts, Models and Examples. SEPM Spec. Publ., vol. 83, pp. 87–129.

Wilson, J.B., 1982. Shelly faunas associated with temperate offshore tidal deposits. In: Stride, A.H. (Ed.), Offshore Tidal Sands. Chapman and Hall, London, pp. 126–171.

Wilson, J.B., 1986. Faunas of tidal current and wave-dominated continental shelves and their use in the recognition of storm deposits. In: Knight, R.J., McLean, J.R. (Eds.), Shelf Sands and Sandstones. Can. Soc. Petrol. Geol., Mem. 11, pp. 313–326.

Yeo, R.K., Risk, M.J., 1981. The sedimentology, stratigraphy, and preservation of intertidal deposits in the Minas Basin system, Bay of Fundy. J. Sediment. Petrol. 51, 245–260.

Yoshida, S., Steel, R.J., Dalrymple, R.W., 2007. Changes in depositional processes; an ingredient in a new generation of sequence stratigraphic models. J. Sediment. Res. 77, 447–460.

Yu, Q., Wang, Y., Gao, S., Flemming, B., 2012. Modeling the formation of a sand bar within a large funnel-shaped, tide-dominated estuary: Qiantangjiang Estuary, China. Mar. Geol. 299–302, 63–76.

Chapter 19

Shorefaces

S. George Pemberton,*[,1] James A. MacEachern,[†] Shahin E. Dashtgard,[†] Kerrie L. Bann,[‡] Murray K. Gingras* and John-Paul Zonneveld*

*Ichnology Research Group, Department of Earth & Atmospheric Sciences, University of Alberta, Edmonton, Alberta, Canada, [†]Applied Research in Ichnology and Sedimentology (ARISE), Department of Earth Sciences, Simon Fraser University, Burnaby, British Columbia, Canada, [‡]Ichnofacies Analysis Inc., Calgary, Alberta, Canada

[1]Corresponding author: e-mail: george.pemberton@ualberta.ca

1. INTRODUCTION

The marine shoreface complex consists of a steeply seaward-sloping sediment ramp (1:200) that extends from then low-tide mark to a nearly flat (1:2000) inner shelf or basin floor, typically corresponding to approximately 10–20 m of water depth (Clifton, 2003; Immenhauser, 2009; Johnson, 1919). Strandplains are more typical of regressive coasts; however, shorefaces may occur on both regressive and transgressive coasts (Fig. 1). The shoreface setting is dominated by wave energy and, as a result of decreasing wave interaction with the substrate in a seaward direction, shows a pronounced basinward fining. The shoreface is classically divided into three subenvironments (from seaward to landward): the lower, the middle, and the upper shoreface. The boundaries between these zones, and between the lower shoreface and the offshore, are not always clearly defined (Reinson, 1984). For example, with greater storm domination, the effective wave base is lowered to depths approaching the storm-wave base. As a result, erosively amalgamated tempestites erroneously attributed to "lower shoreface" conditions may actually have been deposited well below fair-weather wave base. Although the large-scale regional context may vary, the specific subenvironments of the shoreface are not significantly different whether they occur as part of a strandplain, barrier island, or wave/storm-dominated delta free from interference from active distributary channels (cf. MacEachern and Bann, 2008; MacEachern et al., 2005; Tonkin, 2012). The shoreface grades distally into offshore units and landward into foreshore deposits. A complete shoreface progradational succession reflects offshore through to foreshore environments; consequently, these adjacent environments bear inclusion in any discussion of shoreface deposits *per se*.

Developments in Sedimentology, Vol. 64. http://dx.doi.org/10.1016/B978-0-444-53813-0.00019-8

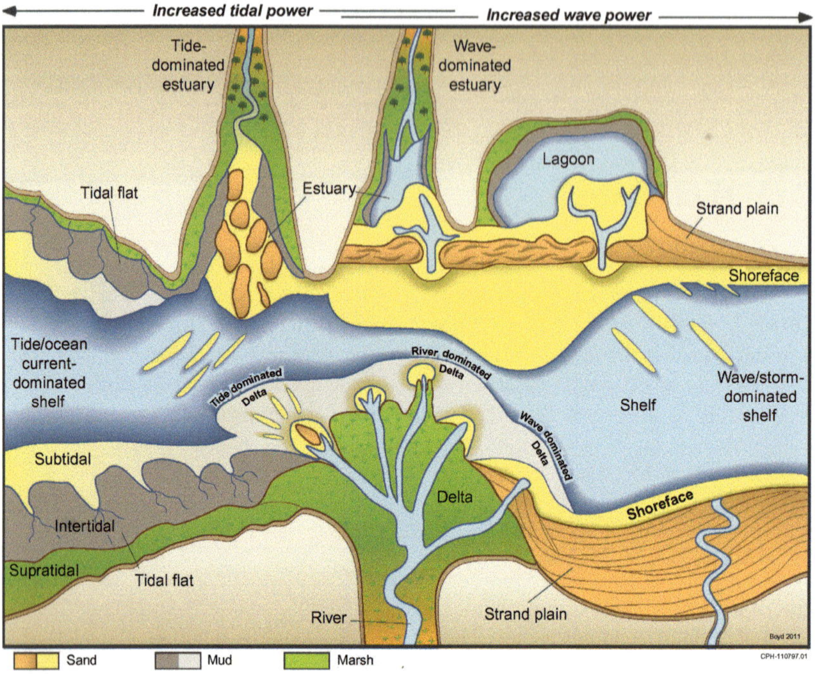

FIGURE 1 Siliciclastic marginal-marine and shallow-marine subenvironments of deposition under varying conditions of tidal power and wave power (*Boyd, 2010; courtesy of Ron Boyd*). *Figure originally based on Heward (1981) and Boyd et al. (1992).*

A long history of trace-fossil analysis of shoreface systems has generated a robust ichnological model of their various subenvironments (Fig. 2). This model builds upon a variety of general models and approaches for using trace fossils in the identification and interpretation of specific sedimentary environments (e.g., Bann et al., 2004; Gingras et al., 1998; Howard and Frey, 1984; MacEachern and Pemberton, 1992; MacEachern et al., 1999b; Pemberton and Wightman, 1992; Pemberton et al., 1992a,b,d; Pemberton et al. (1992c) Savrda and Bottjer, 1989; Seilacher, 1967, Taylor et al., 2003; Vossler and Pemberton, 1988, 1989; see Fig. 3). Animal/sediment interactions are well documented from the more uniform conditions associated with strandplain shorefaces, and the shoreface ichnofacies model is robust (Figs. 2 and 3; e.g., Bann and Fielding, 2004; Bann et al., 2008; Frey, 1990; Howard and Frey, 1984; Howard and Reineck, 1981; MacEachern and Bann, 2008; MacEachern and Pemberton, 1992; MacEachern et al., 1999a,b; Pemberton and Frey, 1984, Pemberton and MacEachern, 1995; Vossler and Pemberton, 1988). In deltaic settings, the dynamic interplay of fluvial influx, fluvial discharge types, tidal energy, wave action, and storms have a strong effect on animal/sediment characteristics (Tonkin, 2012). Such interactions, at least as they affect infaunal diversity, abundance, feeding strategy, and overall behaviors, are generally poorly

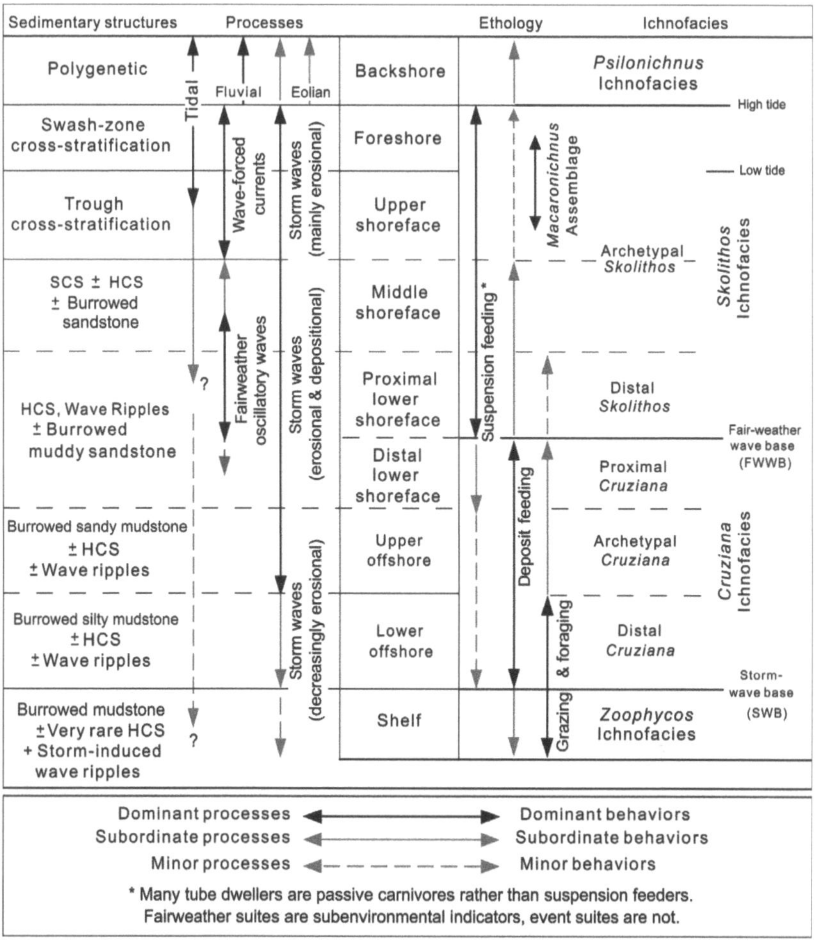

FIGURE 2 Integrated ichnological/sedimentological model for shoreface settings, with positions of the subenvironments. *Modified after MacEachern and Bann (2008), and MacEachern et al. (1999b).*

understood. As such, earlier models appear to have incorporated ichnological–sedimentological characteristics from deltaic successions (particularly wave-dominated deltaic complexes) into the shoreface model (e.g., markedly heterolithic successions in basinal "offshore" positions better characterized as prodelta, abundant mudstone interbeds in proximal "middle shoreface" positions more accurately identified as delta front; cf. MacEachern and Bann, 2008). However, studies on the ichnology of ancient deltaic successions have dramatically increased in number recently and the resulting deltaic ichnological models highlight their deviation from shoreface facies characteristics (e.g., Bann and Fielding, 2004; Bann et al., 2008; Bhattacharya and MacEachern, 2009; Buatois et al., 2008, 2011, 2012; Coates and MacEachern, 2007; Gingras et al., 1998; Hansen and

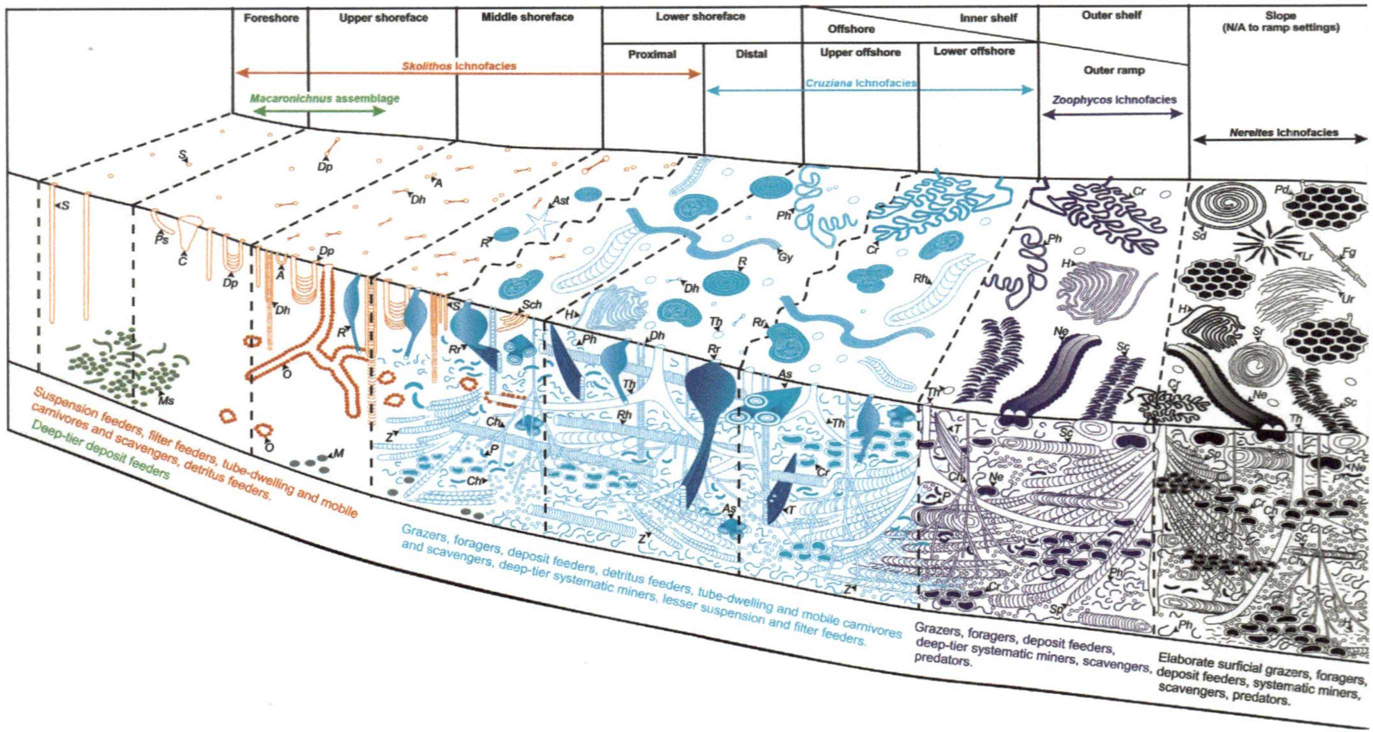

FIGURE 3 Schematic distribution of common ichnogenera, marine ichnofacies, and prevailing animal behaviors along a proximal-to-distal trend. Abbreviations: *A*, *Arenicolites*; *As*, *Asterosoma*; *Ast*, *Asteriacites*; *C*, *Conichnus*; *Ch*, *Chondrites*; *Cr*, *Cosmorhaphe*; *Dh*, *Diplocraterion habichi*; *Dp*, *Diplocraterion parallelum*; *Fg*, *Fustiglyphus*; *Gy*, *Gyrochorte*; *H*, *Helminthopsis*; *Lr*, *Lorenzinia*; *M*, *Macaronichnus* isp.; *Ms*, *Macaronichnus segregatis*; *Ne*, *Nereites*; *O*, *Ophiomorpha*; *P*, *Planolites*; *Pa*, *Palaeophycus*; *Pd*, *Paleodictyon*; *Ph*, *Phycosiphon*; *Ps*, *Psilonichnus*; *R*, *Rosselia socialis*; *Rh*, *Rhizocorallium*; *Rr*, *Rosselia rotatus*; *S*, *Skolithos*; *Sc*, *Scolicia*; *Sch*, *Schaubcylindrichnus coronus*; *Sd*, *Spirodesmos*; *Sp*, *Spirophyton*; *Sr*, *Spirorhaphe*; *T*, *Teichichnus*; *Th*, *Thalassinoides*; *Ur*, *Urohelminthoida*; *Z*, *Zoophycos*. *Diagram adapted from Seilacher (1967), modified after MacEachern and Bann (2008).*

MacEachern, 2007; MacEachern and Bann, 2008; MacEachern et al., 2005, 2007; McIlroy, 2004, 2007; Moslow and Pemberton, 1988; Raychaudhuri and Pemberton, 1992; Siggerud and Steel, 1999).

More recently, interest has focused on tidal effects on otherwise wave-dominated shorefaces (Ainsworth et al., 2008; Dashtgard et al., 2006, 2009, in press; Desjardins et al., 2012; Frey and Dashtgard, 2011; Vakarelov et al., in press). Dashtgard et al. (in press) present models for two tidal shorefaces: tide-influenced shorefaces (TIS) and tidally modulated shorefaces (TMS). TIS are developed in settings with strong tidal currents, such as straits and bays. TMS are developed in settings with large tidal ranges (>8 m macrotidal and megatidal settings), wherein wave regimes are displaced along the depositional profile as a consequence of the rising and falling of tides. Although these models are new and relatively untested, work by Ainsworth et al. (2008) and Vakarelov et al. (in press) suggests that the criteria for recognizing tidal shorefaces are valid. Ongoing work on these models promises to develop ichnological and sedimentological criteria for distinguishing shoreface through to tidal-flat successions that occur along the continuum of wave-tidal coastal environments.

2. SHOREFACE SUBENVIRONMENTS

Coastal morphodynamics are subject to a wide range of variables depending upon whether the shoreline is prograding due to net-sediment accumulation or due to transgression. Estuaries and barrier-island systems, for example, are generally a function of transgression, whereas shore-attached strandplains and deltas are more apt to develop along prograding coastlines (Fig. 1). Nevertheless, the shoreface and adjacent environments, whether associated with progradation or erosional shoreface retreat, can be grouped into three depositional complexes: the offshore complex, the lower–middle shoreface complex, and the upper shoreface–foreshore–backshore complex. Each complex corresponds to discrete combinations of physical processes and concomitant faunal responses, leading to distinct combinations of primary physical and biogenic sedimentary structures.

2.1 Offshore Complex

Many sedimentologists adhere to the paradigm set by Reading and Collinson (1996) who placed the offshore regime below effective (fair-weather) wave base, but above maximum (storm-weather) wave base. The distribution of biogenic structures in the offshore is largely dictated by availability of food resources and substrate consistency. In these regimes, food particles are deposited onto the bed and substrates are generally cohesive, leading to increased interspecific competition and a corresponding increase in specialized feeding strategies. The resulting ichnological suites are highly diverse and dominated by deposit-feeding and grazing structures, attributable to the *Cruziana* Ichnofacies (Fig. 3).

The offshore is commonly dominated by fair-weather deposition, although in strongly storm-dominated regimes, the offshore complex may consist of shaly zones interbedded with thin, relatively unburrowed tempestites (e.g., Frey, 1990; Howard and Frey, 1984; Pemberton and Frey, 1984; Pemberton and MacEachern, 1997; Vossler and Pemberton, 1989). Fair-weather deposits are characterized by pervasive burrowing (BI = 4–5), manifest by high diversities and uniform distributions of ichnogenera. The suites lack domination by individual forms and possess ichnogenera characteristic of equilibrium behaviors, consistent with fully marine, unstressed settings (cf. Jumars, 1993; Pianka, 1970). The offshore is commonly subdivided into lower and upper subzones (e.g., MacEachern and Bann, 2008; MacEachern and Pemberton, 1992).

2.1.1 Lower Offshore

The lower offshore, as defined here, comprises dark, typically organic-rich, silty shale, commonly thoroughly homogenized via biogenic reworking (Fig. 4A–C). Fine- to very fine-grained sandstone beds are present, though in limited amounts. Thin sandstone beds are predominantly intensely bioturbated. In thicker beds, remnant undulatory parallel laminae are preserved (Fig. 4D). Thicker sandstone beds tend to show sharp, erosional bases, with moderately to well-developed parallel lamination. Most burrowing is restricted to the tops of the laminated sandstone beds, interpreted as distal tempestites. These distal storm beds are deposited well below fair-weather wave base, such that fair-weather physical processes are unable to modify them; such beds, therefore, have a high preservation potential (Dott, 1983, 1988; Wheatcroft, 1990).

Fair-weather deposits of the lower offshore commonly display abundant and uniformly distributed bioturbation, characterized by a high-diversity ichnological suite that corresponds to a distal expression of the *Cruziana* Ichnofacies. Trace-fossil suites comprise a mixture of deposit-feeding, and to a lesser degree, grazing/foraging traces (Fig. 4). Most of these traces show minimal vertical penetration of the bed. Under such circumstances, distal storm beds may show little biogenic reworking, except near their tops, even if the tempestite is relatively thin (e.g., Pemberton and MacEachern, 1997). In low-energy settings, ichnological suites in this zone may be characterized by a proximal (shallow-water) expression of the *Zoophycos* Ichnofacies dominated by grazing, foraging, and mining traces (Fig. 3).

Common ichnotaxa in lower offshore deposits include *Helminthopsis*, *Phycosiphon*, *Chondrites*, *Zoophycos*, *Schaubcylindrichnus freyi*, *Planolites*, *Rosselia socialis*, *Ancorichnus capronus*, *Ophiomorpha annulata*, *Cylindrichnus concentricus*, and diminutive *Asterosoma* and *Thalassinoides* (e.g., Fig. 4A–C). Less common forms include small *Cosmorhaphe* (e.g., Fig. 4C and D), *Scolicia*, *Teichichnus*, and *Palaeophycus*.

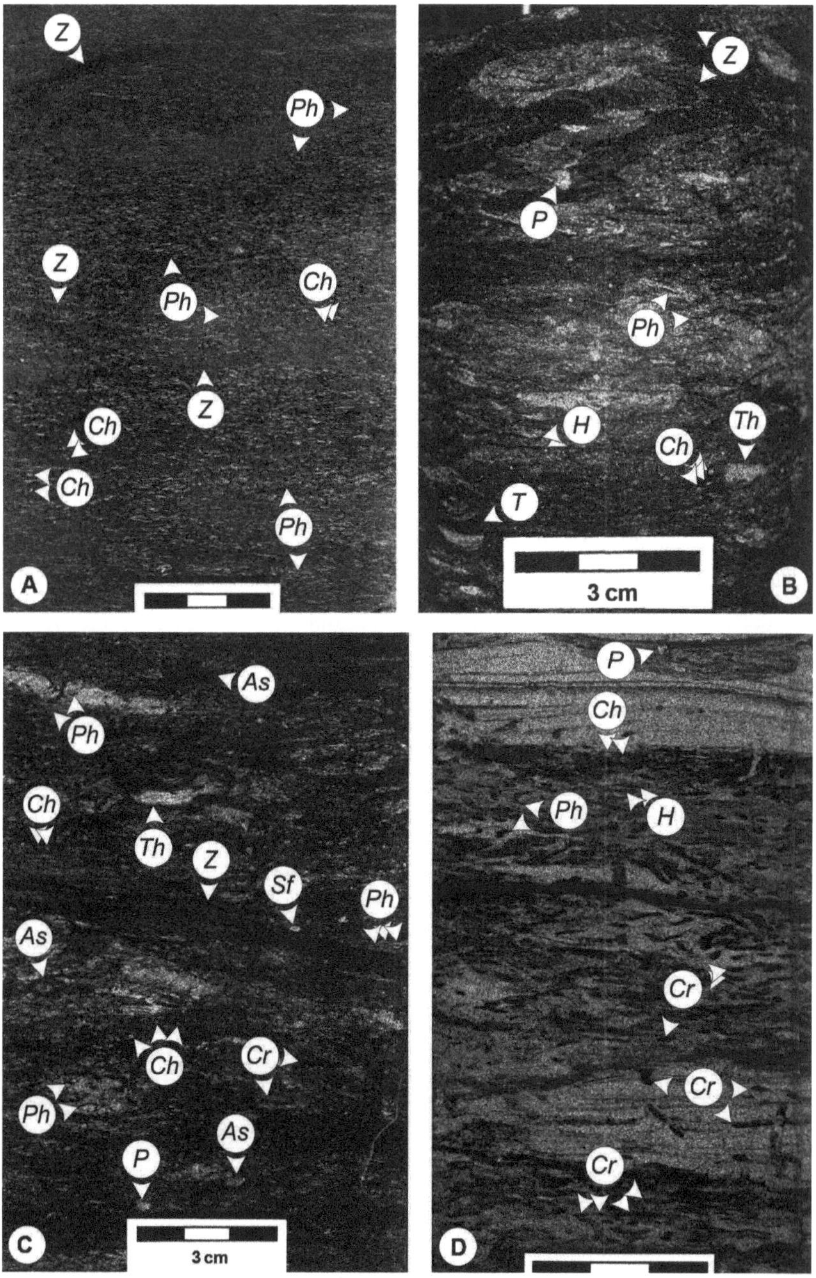

FIGURE 4 Lower offshore silty mudstones with ichnological suites consistent with a distal expression of the *Cruziana* Ichnofacies. (A) Unit shows BI = 5, with abundant *Phycosiphon* (*Ph*), with *Chondrites* (*Ch*), and rare, muddy *Zoophycos* (*Z*). (B) Unit shows BI = 5, with robust *Zoophycos* (*Z*), *Helminthopsis* (*H*), *Planolites* (*P*), *Phycosiphon* (*Ph*), *Chondrites* (*Ch*), *Teichichnus* (*T*), and *Thalassinoides* (*Th*). (C) Unit shows BI = 5, with *Asterosoma* (*As*), *Thalassinoides* (*Th*), *Zoophycos* (*Z*) (example cut oblique to the chevrons), *Phycosiphon* (*Ph*), *Chondrites* (*Ch*), *Schaubcylindrichnus freyi* (*Sf*), and *Planolites* (*P*). (D) Bioturbated silty mudstone showing BI = 5, interbedded with thin, wavy parallel-laminated sandstone beds with BI = 0–2. Laminated sandstone beds correspond to distal tempestites. Trace fossils include *Phycosiphon* (*Ph*), *Cosmorhaphe* (*Cr*), *Helminthopsis* (*H*), *Planolites* (P), and *Chondrites* (*Ch*). Scale bars in centimeters.

2.1.2 Upper Offshore

The upper offshore is considerably more sedimentologically and ichnologically variable than the lower offshore, attributable to the greater degree of storm and wave interaction with the substrate. The interval commonly consists of intensely bioturbated silty and very fine- to fine-grained sandy mudstones, locally interbedded with thin, very fine- to fine-grained sandstones (Fig. 5). The sandstone beds range in internal character from complete homogenization to erosionally based beds possessing remnant low-angle ($< 15°$) parallel laminae, oscillation, or combined-flow ripples. These beds are interpreted as tempestites, with local preservation of waning-flow deposits. Less intensely burrowed upper offshore intervals may show more regular interbedding of shale- and storm-generated sandstone beds, probably reflecting stronger storm domination, higher storm frequency, and/or high sedimentation rates. This has been referred to as the "offshore transition", although the zone does not properly refer to a depositional subenvironment (cf. MacEachern and Bann, 2008). The primary physical structures of the sandstone and siltstone beds are consistent with tempestites deposited in a setting lying seaward of fair-weather wave base (cf. Aigner, 1985; Pemberton and Frey, 1984; Pemberton and MacEachern, 1997; Saunders, 1989; Seilacher and Aigner, 1991; Snedden and Nummendal, 1991). Many of the tempestites record the waning stages of storm action on the substrate, characterized by normal grading, oscillation rippled tops, and suspension fall-out of silts and muds following storm abatement. Tempestites generally tend to show moderate to intense degrees of burrowing near their tops. Distal tempestites deposited in the upper offshore typically display a higher degree of biogenic reworking than the thinner storm beds of the lower offshore (cf. Pemberton and MacEachern, 1997; Fig. 5B and D). This reflects the deeper vertical penetration of the substrate by the more robust benthic organisms that inhabit the upper offshore (Dörjes and Hertweck, 1975).

As in the lower offshore silty shale beds, burrowing in the upper offshore is typically intense, with a high diversity and uniform distribution of ichnogenera (Fig. 5A and C). Ichnological suites correspond to a diverse archetypal expression of the *Cruziana* Ichnofacies. The bulk of the suite consists of deposit-feeding traces such as *Planolites*, *Teichichnus*, *Chondrites*, *Thalassinoides*, small *Asterosoma*, and *Rhizocorallium*. Subordinate trace fossils in the suite include passive carnivore traces such as *Schaubcylindrichnus freyi*, *S. coronus*, and *Palaeophycus*; detritus-feeding traces such as *Cylindrichnus* and *Rosselia*; and bivalve feeding- and resting-traces such as *Siphonichnus* and *Lockeia*. Grazing/foraging traces such as *Helminthopsis*, *Phycosiphon*, *Cosmorhaphe*, and *Zoophycos* are also locally abundant. Vertical dwelling traces such as *Ophiomorpha*, *Arenicolites*, *Skolithos*, and *Diplocraterion habichi* are associated with tempestite beds and represent opportunistic colonization of the sandy substrates following storms (e.g., Pemberton and MacEachern, 1997; Pemberton et al., 1992b; Vossler and Pemberton, 1988, 1989). These organisms are

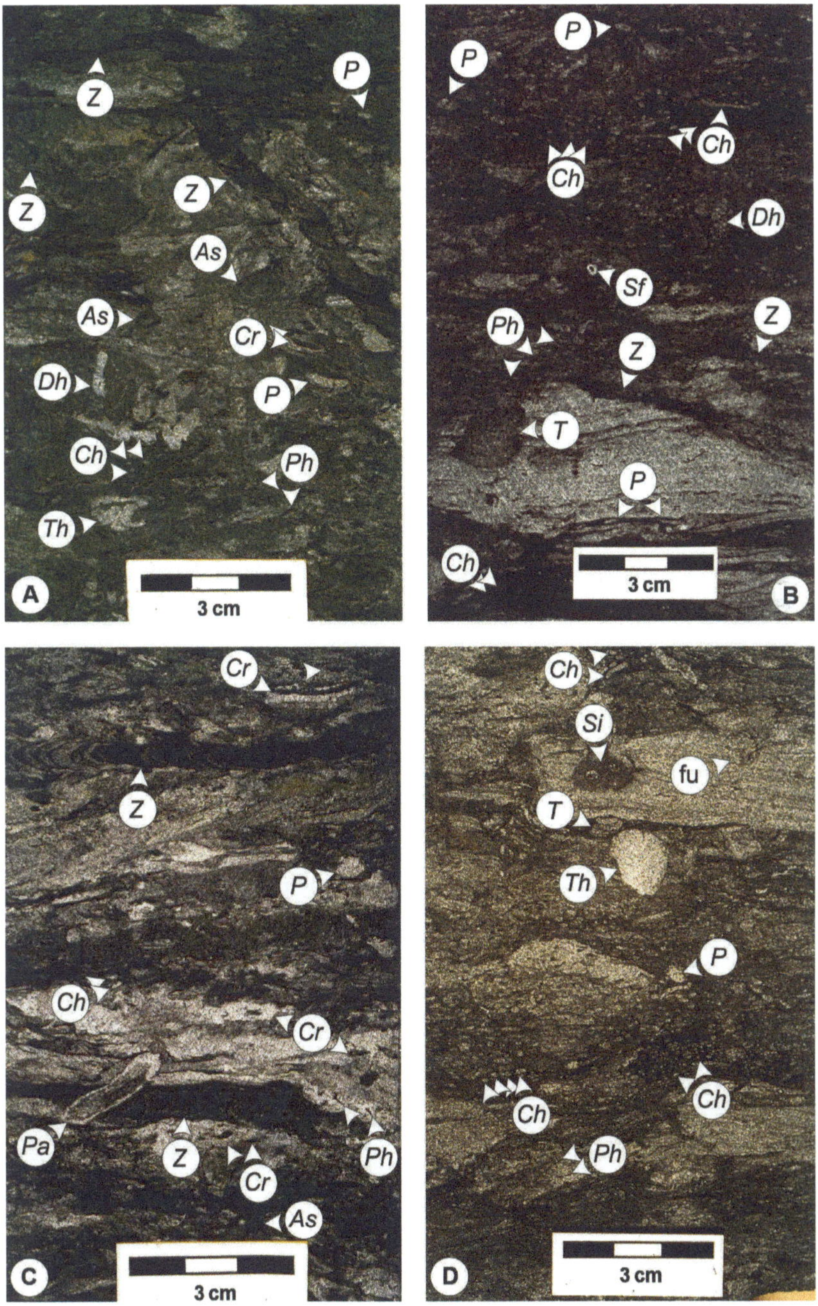

FIGURE 5 Upper offshore sandy mudstones containing ichnological suites consistent with the archetypal expression of the *Cruziana* Ichnofacies. (A) BI = 5, with *Zoophycos* (*Z*), *Planolites* (*P*), *Asterosoma* (*As*), *Diplocraterion habichi* (*Dh*), *Chondrites* (*Ch*), *Cosmorhaphe* (*Cr*), *Phycosiphon* (*Ph*), and *Thalassinoides* (*Th*). (B) BI = 4, with a remnant, wavy parallel-laminated sandstone bed interpreted as a tempestite. Trace-fossil suite consists of *Planolites* (*P*), *Chondrites* (*Ch*), *Diplocraterion habichi* (*Dh*), *Phycosiphon* (*Ph*), *Schaubcylindrichnus freyi* (*Sf*), *Zoophycos* (*Z*), and *Teichichnus* (*T*). (C) BI = 5 with *Cosmorhaphe* (*Cr*), *Planolites* (*P*), *Chondrites* (*Ch*), *Palaeophycus heberti* (*Pa*), *Phycosiphon* (*Ph*), *Zoophycos* (*Z*), and *Asterosoma* (*As*). (D) BI = 4, with a remnant, parallel-laminated sand bed interpreted as a tempestite. Trace-fossil suite includes *Siphonichnus* (*Si*), *Planolites* (*P*), *Chondrites* (*Ch*), *Teichichnus* (*T*), *Phycosiphon* (*Ph*), *Thalassinoides* (*Th*), and fugichnia (fu).

replaced by the resident community during extended periods of fair-weather and the resultant burrows are commonly overprinted. Tempestites may also contain fugichnia (escape traces), recording the disturbance of stratification by animals buried by the rapidly deposited sediment, as they attempted to reach the new sediment/water interface (Fig. 5D). The distribution of trace fossils records the typical character of ecological stresses imparted by episodic depositional events (cf. Bann et al., 2008; MacEachern and Bann, 2008; MacEachern, et al., 2007; Pemberton and MacEachern, 1997; Pemberton et al., 1992b). Trace fossils such as *Helminthopsis*, *Phycosiphon*, *Planolites*, and *Teichichnus* record the reestablishment of grazing and deposit-feeding infauna following the event and typically occur near the tops of sandstone beds or within the mudstones capping them. The low abundance of suspension-feeding traces suggests that most sandstone beds were quickly buried by silt and mud following the depositional event, precluding colonization by organisms that favor sandy substrates.

2.2 Lower–Middle Shoreface Complex

2.2.1 Lower Shoreface

The lower shoreface begins at the lower limit of the fair-weather (effective) wave base, where offshore processes continue to dominate deposition (Reinson, 1984). Waves constitute the dominant physical process controlling deposition in the interval, and most physical sedimentary structures reflect storm deposition, such as hummocky cross-stratification (HCS; cf. Dott and Bourgeois, 1982; Duke, 1985), rare swaley cross-stratification (SCS; cf. Leckie and Walker, 1982), and quasi-planar lamination (QPL; cf. Arnott, 1993). Intervals without significant erosional amalgamation may show waning-stage oscillation or combined-flow ripple-lamination capping HCS and QPL. SCS characterizes strongly erosionally amalgamated bed sets. Fair-weather-generated oscillation ripples may be present but are relatively uncommon because of bioturbation. The intensity of burrowing within lower shoreface deposits is highly variable, depending upon the degree of storm influence (Fig. 6). High-intensity and high-frequency storm events favor minimal preservation of burrowing, whereas low-intensity and infrequent storms permit thorough homogenization of the tempestites and accumulation of thick fair-weather intervals (e.g., Fig. 6D).

The trace-fossil suites represent a proximal expression of the *Cruziana* Ichnofacies (cf. MacEachern and Bann, 2008; Fig. 3). Deposit-feeding and mobile carnivore traces of the *Cruziana* Ichnofacies include such traces as robust *Rosselia*, *Asterosoma*, *Teichichnus*, *Cylindrichnus*, *Schaubcylindrichnus*, *Planolites*, *Thalassinoides*, *Ancorichnus*, *Phoebichnus*, *Siphonichnus*, *Macaronichnus*, *Chondrites*, and *Rhizocorallium* (Fig. 6). Rarer grazing/foraging traces, largely confined to muddier zones, are also locally present and include *Phycosiphon* and very rare *Zoophycos*, *Scolicia*, and *Aulichnites.*

FIGURE 6 Lower shoreface muddy sandstones containing ichnological suites consistent with proximal expressions of the *Cruziana* Ichnofacies and distal expressions of the *Skolithos* Ichnofacies. (A) A general photo of lower to middle shoreface deposits, with the proximal part of the lower shoreface near the base of the photo. The red grains are siderite-cemented burrows overwhelmingly dominated by *Ophiomorpha borneensis*. (B) Closeup of part of an *O. borneensis* (*Ob*) burrow network. (C) Pervasively bioturbated (BI = 5) muddy sandstone of the distal part of the lower shoreface, with *Diplocraterion* (*D*), *Asterosoma* (*As*), *Planolites* (*P*), *Phoebichnus* (*Pb*), *Chondrites* (Ch), *Phycosiphon* (*Ph*), and *Thalassinoides* (*Th*), recording a proximal expression of the *Cruziana* Ichnofacies. (D) Laminated-to-burrowed bedding in the lower shoreface. Remnant laminated sandstone at the base corresponds to a tempestite with *Macaronichnus* isp. (*M*). The burrowed fair-weather bed overlies the tempestite and shows BI = 5, with *Ophiomorpha irregulaire* (*Oi*), *Skolithos* (*S*), *Chondrites* (in muddy interlaminae), *Planolites* (*P*), and *Thalassinoides* (*Th*), attributable to a distal expression of the *Skolithos* Ichnofacies. Scale bar = 3 cm.

Lower shoreface suites also contain subordinate numbers of permanent dwellings such as horizontal *Ophiomorpha borneensis* (Fig. 6A and B), *O. nodosa*, *O. irregulaire* (Fig. 6C), as well as rare *O. annulata*, *Skolithos*, *Conichnus*, *Diplocraterion habichi*, *D. parallelum* (Fig. 6D), *Bergaueria*, and the passive carnivore trace *Palaeophycus tubularis*. In addition, escape traces (fugichnia) and those of various opportunistic organisms may also be common, generally representing colonization of storm bed tops.

The dominance of deposit-feeding behaviors, with significant contributions of permanent dwellings associated with suspension-feeding organisms, suggests that persistent wave shoaling (i.e., above fair-weather wave base) is important in winnowing mud and moving it offshore, in suspending food particles above the bed and in shifting sand at the sediment/water interface. The resulting fair-weather trace-fossil suite corresponds to a proximal expression of the *Cruziana* Ichnofacies (cf. MacEachern and Bann, 2008; Fig. 3).

Storm events impart considerable control on the character of the lower shoreface. Storms tend to uproot, destroy, and/or bury resident (fair-weather) benthic communities (Frey, 1990; Pemberton and Frey, 1984; Pemberton et al., 1992b). During poststorm recovery, initial colonization of the tempestite generally records opportunistic dwelling/suspension-feeding structures, which do not typically reflect the original resident community. Continued fair-weather conditions may see a return to the original trace-fossil suite (see Section 3 in the section on storms). Depending upon the degree of storm dominance, the preserved record of the lower shoreface can be quite variable (Dashtgard et al., in press; MacEachern and Pemberton, 1992; Plint, 2010; Walker and Plint, 1992).

Some thin tempestites may escape thorough biogenic reworking under conditions of high-frequency storms, moderate- to low-energy storm conditions, and high sedimentation rates (cf. Wheatcroft, 1990). Saunders and Pemberton (1986) and Saunders (1989) noted that thin storm beds common in the lower shoreface deposits of the Appaloosa Sandstone of the Bearpaw–Horseshoe Canyon Formation transition, Alberta, permitted *Rhizocorallium jenense*, *Teichichnus rectus*, and *Rosselia socialis* to display reequilibration to each successive sediment/water interface. *R. socialis* reflects this reequilibration by vertical stacking of individual burrows. Similar observations were made in the Pleistocene of Japan by Nara (1997) and the Permian of the Sydney Basin, Australia by Bann et al. (2008).

2.2.2 Middle Shoreface

The middle shoreface extends over the zone of shoaling and initial breaking of waves (Reinson, 1984) and is characterized by high oscillatory wave energy. Longshore bars are commonly present near the upper portion of the middle shoreface within intermediate (barred) states of shoreface morphodynamics

(Thom et al., 1986; Wright et al., 1979). Preserved deposits of longshore bars are rarely intercalated within middle shoreface intervals, as under fair-weather conditions bars migrate landward and weld to the foreshore (Davidson-Arnott and Greenwood, 1976; Davis, 1978). Middle shoreface sandstone beds tend to be well sorted, winnowed of fines and fine- to medium-grained, with only minor shale, siltstone, and shell layers (Fig. 7). Preserved structures are predominantly low-angle, wedge-shaped sets of parallel laminae, constituting SCS and lesser HCS or QPL. Oscillation-ripple laminae, combined flow-ripple laminae, and rarer trough cross-stratification are locally present. Storms have a strong influence on the middle shoreface, stronger than in the lower shoreface, and may constitute much of its depositional record. The SCS or QPL tempestite beds are typically erosionally amalgamated, becoming progressively more cannibalistic (Aigner and Reineck, 1982) with increased shallowing and associated enhancement of storm-induced scouring. As a result, the degree of bioturbation is highly variable but generally quite low.

Ichnologically, the middle shoreface corresponds to a dominance of suspension-feeding behaviors (constituting the *Skolithos* Ichnofacies; Fig. 3) over deposit-feeding ethologies (the *Cruziana* Ichnofacies), although some traces of the latter continue to persist as subordinate elements (e.g., *Macaronichnus*, *Rosselia*, *Cylindrichnus*) (Howard, 1972, 1975). The bulk of the suite consists of ichnogenera such as *Skolithos*, *Conichnus*, *Diplocraterion*, *Ophiomorpha* (commonly vertical components), *Arenicolites*, *Bergaueria*, *Schaubcylindrichnus coronus*, and *Palaeophycus* (Fig. 7). Howard (1971a,b, 1975) and others have noted that *Ophiomorpha* systems tend to shift from predominantly horizontal to vertical orientations as energy levels increase. This shift in orientation may accompany the transition from the lower to middle shoreface. Truncation of burrows and fugichnia may be commonly associated with amalgamation of tempestites (Fig. 7A).

Cruziana Ichnofacies elements include robust *Rosselia* (*R. socialis* dominates, Fig. 7D), *Cylindrichnus*, and *Siphonichnus*. Frey (1990) noted that *Cylindrichnus* and *Rosselia* become more steeply inclined as energy levels increase. This increase in inclination may also accompany the transition from the lower to middle shoreface, mimicking the trend observed for *Ophiomorpha*.

As in the lower shoreface, variability in storm intensity and frequency strongly affects the sedimentological and ichnological character of the middle shoreface. As storm effects become more pronounced, the diversity and abundance of deposit-feeding traces declines markedly. This appears to be attributable to the abundance of well-winnowed sand and the general paucity of deposited food for tracemakers of ichnogenera common to the *Cruziana* Ichnofacies. The preservation potential of biogenic structures in the middle shoreface is intrinsically associated with the character of tempestite accumulation. In storm-dominated conditions, as tempestites become progressively more cannibalistic, only deeply penetrating traces (e.g., *Ophiomorpha* and fugichnia) may be preserved (see, for instance, Fig. 7A–C). It is not uncommon for storm-dominated middle shoreface

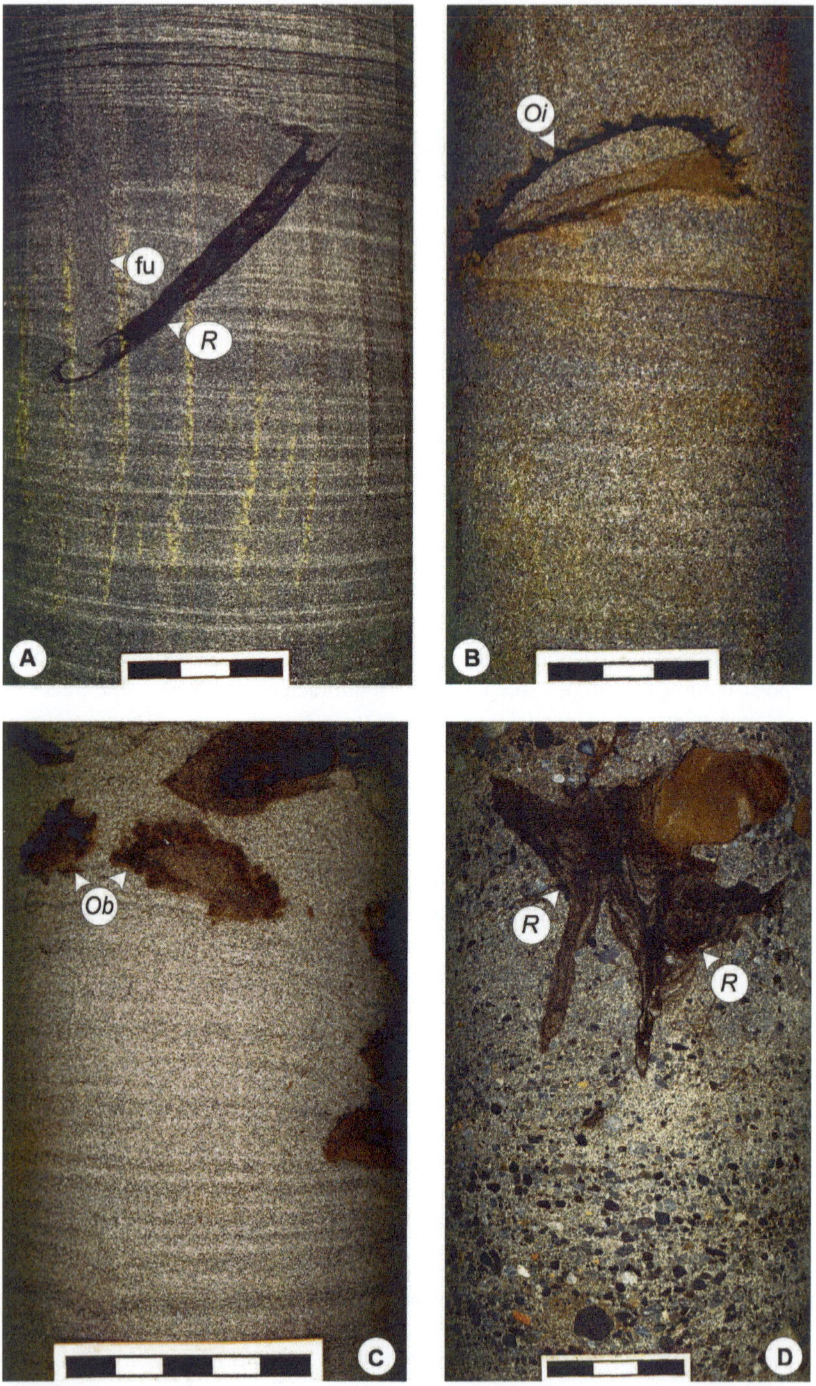

FIGURE 7 Middle shoreface sandstones displaying low-intensity and sporadically distributed bioturbation. (A) HCS showing erosional amalgamation and truncation of the dwelling tube of *Rosselia* (*R*). Note the fugichnia associated with the truncated *Rosselia*. The upper part of the lower sandstone bed shows crypto bioturbation. (B) Thick-bedded tempestite with opportunistic colonization of the upper surface with robust *Ophiomorpha irregulaire* (*Oi*). (C) Thick-bedded HCS with

sandstones to show little or no burrowing (cf. MacEachern and Pemberton, 1992). Unfortunately, discerning whether their absence corresponds to high intensity storms (greater erosional amalgamation) or to high-frequency storms (minimal time for reestablishment of benthic communities) is virtually impossible.

Like the offshore complex, the lower to middle shoreface complex consists of sediments accumulated during both fair-weather and storm conditions. In contrast to the offshore, however, the preserved depositional record of the lower–middle shoreface complex of many successions is characterized by a predominance of tempestite beds, with fair-weather sediments constituting a common, though subordinate, component of the interval (Kumar and Sanders, 1976; Niedoroda et al., 1984).

2.3 Upper Shoreface–Foreshore–Backshore Complex

2.3.1 Upper Shoreface

The upper shoreface is situated in the high-energy build-up and surf zones and lies between the breaker zone and the low-tide mark (Clifton et al., 1971; Davidson-Arnott and Greenwood, 1976; Greenwood and Mittler, 1985; Hunter et al., 1979). The zone corresponds to the "Outer Rough", "Outer Planar", and possibly part of the "Inner Rough" zones of Clifton et al. (1971), where the bulk of sediment transport is related to multidirectional current flow (Davis, 1978). Within the upper shoreface, wave- and storm-driven currents paralleling the shoreline (e.g., longshore drift) interact with shore-normal currents generated by translatory flow (combined-flow oscillation and current interference) associated with plunging waves, producing multidirectional, sinuous-crested subaqueous dunes (Figs. 8A, B and 9A).

Under intermediate barred states (cf. Davidson-Arnott and Greenwood, 1976; Greenwood and Davidson-Arnott, 1979; Hunter et al., 1979; Thom et al., 1986; Wright et al., 1979), wave- and storm-induced currents run obliquely shoreward over the bar, turn shore-parallel along the longshore troughs, and ultimately return seaward through rip channels. Greenwood and Mittler (1985) pointed out that there is generally little depositional record of a bar complex, based on their study of the Kouchibouguac Bay barred system of New Brunswick. Preserved sediments correspond to the fill of the longshore troughs and those associated with the seaward slope of the bar itself.

Hunter et al. (1979) studied a barred coastline in Oregon and pointed out that, under a barred configuration, an erosional discontinuity, reflecting the base of the deepest rip channels or longshore troughs, tends to separate the

---

opportunistic colonization of the upper surface with robust *O. borneensis* (*Ob*). Rare *O. irregulaire* (*Oi*) are present as well. Note the subtle fugichnion (fu) in the lower left side of the core. (D) Granule-bearing sandstone with low-angle planar-parallel laminae, interpreted as large-scale SCS. Isolated zones of *Rosselia socialis* (*R*) represent tempestite colonization. Scale bars in centimeters.

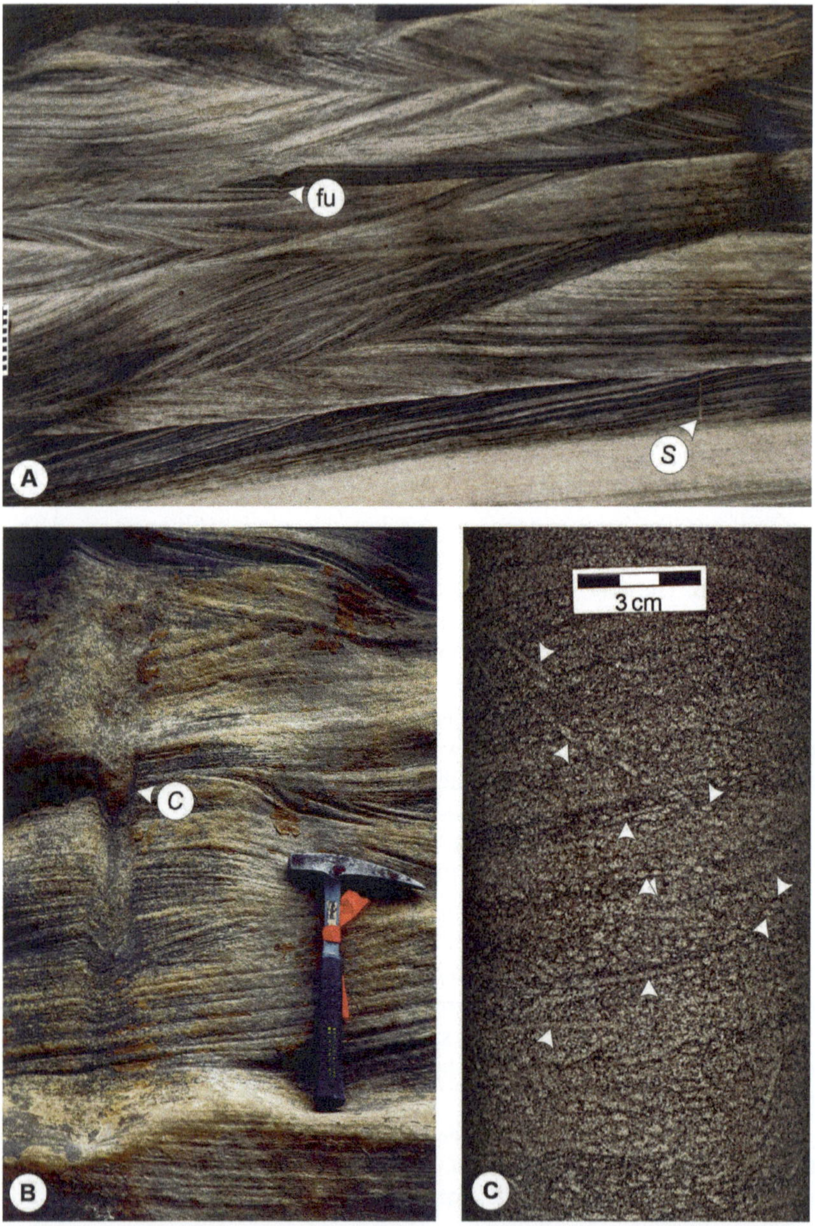

FIGURE 8 Upper shoreface sandstones. (A) Generally unburrowed multidirectional trough cross-stratification. Note the single *Skolithos* (*S*) in the lower right of the photo and an isolated escape structure (fu) in the middle left of the image. Scale bar = 15 cm. (B) Close-up of trough cross-stratified sandstones with a single, robust *Conichnus conicus* (*C*). (C) Cross-stratified sandstones with BI = 5, consisting of *Macaronichnus segregatis* (arrows) in a "toe-of-the-beach" association. The zone occurs near the top of the upper shoreface near the foreshore boundary.

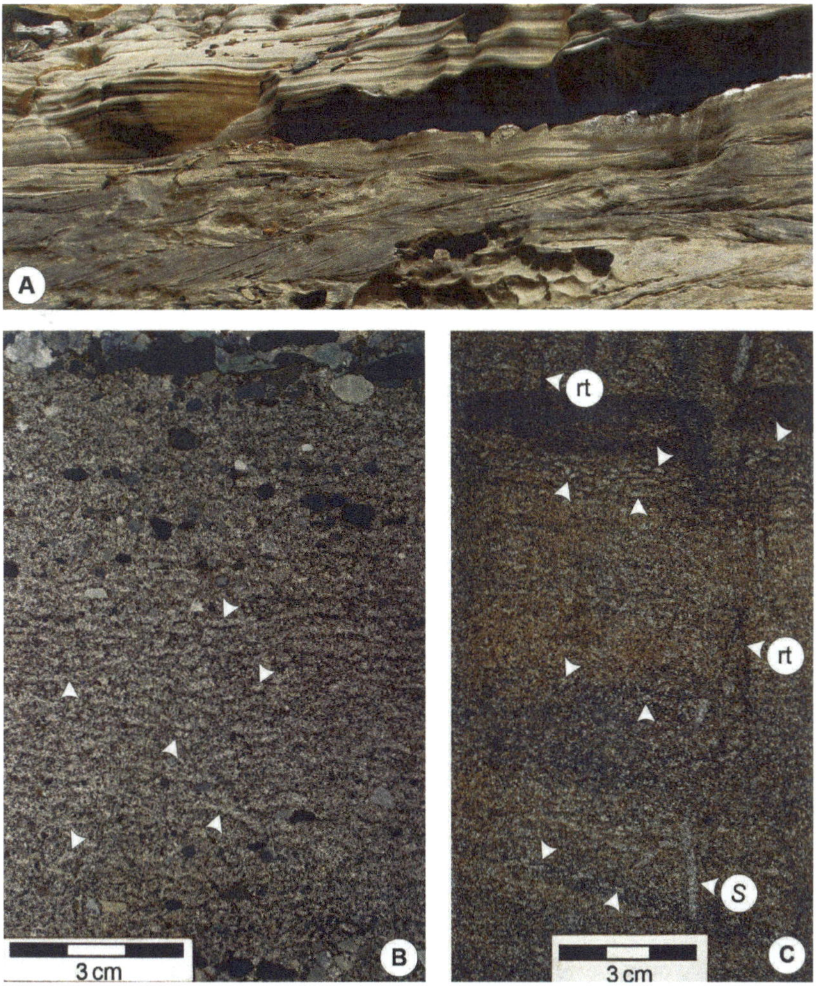

FIGURE 9 Foreshore deposits. (A) Upper shoreface cross-stratified sandstones abruptly overlain by low-angle planar-parallel-stratified sandstones reflecting swash-zone cross-stratification. (B) Pervasively burrowed (BI = 5) sandstone with a monospecific suite of *Macaronichnus segregatis* (arrows) with beds of well-segregated granule and fine-pebble conglomerate. (C) Cross-stratified to planar-bedded sandstone of the foreshore, with *M. segregatis* (arrows) and *Skolithos* (*S*), locally cross-cut by roots (rt) that subtend from overlying coastal-plain deposits.

lower–middle shoreface complex from the upper shoreface–foreshore complex. Lateral shift of these channels, coupled with seaward progradation of the entire shoreface system, facilitates the widespread development of this erosional surface. A similar relationship was noted from barred systems in Eastern Long Island, New York (Shipp, 1984) and Kouchibouguac Bay, New Brunswick (Greenwood and Mittler, 1985), and was applied to the Cadotte Member of the Peace River Formation of Alberta (Rahmani and Smith, 1988).

Upper shoreface deposits are distinctive in the dominance of multidirectional trough cross-stratification, typically in 15–45 cm thick sets, interbedded with low-angle bidirectional planar cross-bedded sets (Davidson-Arnott and Greenwood, 1976; Reinson, 1984; Roy et al., 1980; Thom et al., 1986). Sandstone units are typically well sorted, well winnowed, medium grained, or coarser. Conglomerate and pebbly sandstone intervals tend to have a more massive appearance and contain low-angle planar-stratified beds. Heavy-mineral concentrations may exceed those into the foreshore zone (Howard and Frey, 1984). Wave-induced liquefaction may locally produce convolute bedding and dewatering structures.

Trace fossils are locally common but rarely abundant, and diversities are characteristically low. Continuously migrating bedforms present a major ecological problem for endobenthic organisms; consequently, there is a general scarcity of animals that are capable of constructing permanent domiciles under such conditions (Howard, 1972, 1975; Saunders, 1989). Those biogenic structures that are formed have a low preservation potential and, in general, only deeply penetrating structures are preserved (Howard and Frey, 1984). The main ichnological elements correspond to the *Skolithos* Ichnofacies (Fig. 8) and are characterized by the ichnogenera *Skolithos*; heavily lined, vertical, or deeply penetrating *Ophiomorpha nodosa* and *Conichnus conicus* (Fig. 8B). Under very high-energy conditions, *Macaronichnus* isp. and *M. segregatis* are also locally abundant (Fig. 8C), such as in the Bluesky Formation (Moslow and Pemberton, 1988), Cadotte Member (MacEachern and Pemberton, 1992), the Appaloosa Sandstone of the Bearpaw-Horseshoe Canyon transition (Saunders, 1989; Saunders and Pemberton, 1986), and the Blood Reserve Sandstone (Nadon, 1988) of the Western Canada Sedimentary Basin. *Macaronichnus* is typically more common near the upper shoreface/foreshore contact (Pemberton et al., 2001; Saunders, 1989; Saunders and Pemberton, 1986; Saunders et al., 1994). Under more reflective shoreface states, characterized by steeper depositional profiles and a larger grain size, the zone may occur lower down in the upper shoreface. The Cadotte Member contains upper shoreface intervals consisting of conglomerates interbedded with medium- to coarse-grained sandstone beds, many of which possess intensely burrowed zones of *M. segregatis*, demonstrating its high-energy affinity (Saunders et al., 1994).

Unlike the lower and middle shoreface, storm effects in the upper shoreface are generally reflected by ridge and runnel systems rather than major depositional events. Ridge and runnel systems indicate erosion of the beachface and transport of sediments to the shoreface, related to short-period waves and wave setup. Poststorm recovery is manifest by landward migration of ridges and longshore bars, which weld to the beachface in response to long-period fair-weather waves (Davies et al., 1971).

2.3.2 Foreshore

The foreshore is confined to the intertidal zone occupying the area of wave swash. It also corresponds to the littoral zone in areas where tidal activity is minimal (e.g., Baltic Sea and Mediterranean Sea). The swash and backwash mechanism produces distinctive low-angle seaward-dipping planar-parallel laminations, typically in well-sorted, very well-winnowed, medium- to coarse-grained sandstone (Fig. 9A). Locally, however, textures may incorporate up to cobble-sized particles. Stratification occurs as wedge-shaped sets with internal stratification parallel to subparallel to the lower set contacts, commonly referred to as swash-zone stratification. The set boundaries are not truncated, except by storm erosion. Angular discordance of bed sets reflects the changing slope of the prograding beachface during accretionary phases (Reinson, 1984). Some convolute laminae and flame structures may also be present (Howard and Frey, 1984), possibly reflecting wave-induced liquefaction during intense storm activity.

As sediment calibers increase to pebble and cobble sizes, internal stratification becomes indistinct and is visible mainly as clast segregation (Fig. 9B). Stratification, where visible, remains low-angle planar parallel. Conglomerates are mostly clast-supported with a sand or granule matrix, although open framework (no matrix) conglomerates are preserved locally. Unlike alluvial conglomerates, foreshore conglomerates tend to display a good lateral continuity of bedding (Clifton, 1969) and clasts may vary considerably in size over the entire foreshore interval, but are very well sorted within each bed (MacEachern and Hobbs, 2004). Elongate clasts may be imbricated, with long axes parallel to depositional strike. Non-resistant clasts tend to be absent due to the intensity of wave-induced erosion.

The diversity and abundance of trace fossils is low, due to the harsh environmental conditions, continuously shifting substrates and overall low preservation potential (Howard and Frey, 1984). Most elements reflect the *Skolithos* Ichnofacies, particularly the ichnotaxa *Conichnus conicus* (dwelling structure of sea anemones), *Ophiomorpha nodosa* (vertical components), *Skolithos linearis*, and lesser *Diplocraterion* and *Arenicolites*, indicating the predominance of suspension-feeding behaviors within deeply penetrating domiciles (Fig. 3). Some passively predaceous organisms inhabit the foreshore, which may produce *Palaeophycus* and possibly *Schaubcylindrichnus*, but these traces are shallowly penetrating and, as such, are rarely preserved (Howard and Frey, 1984).

In contrast to the lower–middle shoreface complex, the upper shoreface–foreshore complex shows virtually no storm-induced deposition. Storm events are characterized by pronounced erosion in this zone, where short-period waves cut ridge and runnel systems and plane off the accretionary profile of the beachface (Fox and Davis, 1978; Kumar and Sanders, 1976; Niedoroda et al., 1984; Swift et al., 1985). The bulk of the eroded sediment is transported basinward to the lower–middle shoreface complex. Much of it may be returned to the beach in the form of longshore bars, which migrate landward and weld to the beachface in response to the long-period waves generated during fair-weather conditions (Cook and Gorsline, 1972; Davidson-Arnott and Greenwood, 1976). Some sediment, however, is permanently lost to the lower–middle shoreface, where it is mobilized during storms into tempestites. It is not uncommon for upper shoreface–foreshore complexes to acquire an intermediate, barred state during extended storm seasons and return to largely unbarred reflective states during fair-weather conditions (Thom et al., 1986; Wright et al., 1979).

The presence of discrete zones of *Macaronichnus segregatis* is highly significant to the interpretation of the nearshore-to-foreshore complex succession as upper shoreface and foreshore deposits (Saunders, 1989; Saunders and Pemberton, 1986). Clifton and Thompson (1978) originally described and defined *M. segregatis* and attributed it to the deposit-feeding behavior of the opheliid polychaete *Ophelia limicina*. In addition, the modern polychaete *Euzonus mucronata*, also of the family *Ophelidae*, makes similar traces (Pemberton et al., 2001). Both of these modern tracemakers ingest sand as a means of exploiting an epigranular food source of bacteria and other grain-attached organic materials and subsequently excrete the "cleansed" sand to form the fill of the structure. The polychaetes preferentially ingest quartz and feldspar sand and avoid mica flakes and mafic grains; consequently, the fill of the structure tends to be lighter in color than the surrounding host sediment.

In the intertidal and innermost surf zones of wave-exposed beaches, however, high oxygen windows occur as permanent, hydrodynamically sustained features. There, oxygenated surface waters can circulate several meters into the sand, well beneath the reach of surface wave disturbance. This is accompanied by the filtering of tremendous volumes of dissolved and finely particulate organic matter, virtually all of which is mineralized at depth, on passage through the porous framework of the beach (McLachlan et al., 1985). Under these conditions, the more familiar, physically controlled habitat at the surface is adjoined at depth by a second habitat. This second habitat is very stable, predictable, and possesses a potentially abundant and constantly replenishing food supply. In modern oxygen-windowed beaches, only meiofauna and opheliid polychaetes, such as *Euzonus mucronata*, have thus far been found to exploit this unique, deep-sand habitat. In the Cadotte Member and other boreal beaches of the Early Cretaceous, this is confirmed by the presence of *Macaronichnus* (Fig. 9B and C). In these ancient settings, the deep-window habitat was shared by this specially adapted species of macrofaunal deposit feeder, most probably opheliid or opheliid-like polychaetes similar to those in modern equivalents (Clifton and Thompson, 1978; Saunders, 1989). The ability to exploit deep habitats accounts for the preservation of their burrows in high-energy, nearshore sand and sandy conglomerates (Clifton, 1988; Clifton and Thompson, 1978; Saunders, 1989).

In the normal Seilacherian ichnofacies scheme, the foreshore zone should be dominated by the dwelling burrows of suspension feeders that characterize the

Skolithos Ichnofacies (see Figs. 2 and 3). However, the *M. segregatis* suite, reflecting horizontal, intrastratal sediment-ingesting behaviors, seems to characterize higher-energy upper shoreface and foreshore environments where domiciles cannot be established, owing to shifting substrates and difficulty of filter-feeding because of intense wave action (e.g., Aguirre et al., 2010; Saunders et al., 1994; Seike, 2007, 2009).

2.3.3 Backshore

The backshore represents a mixture of marine, marginal-marine, and non-marine conditions. Typical environments include the beach backshore, coastal dunes, washover fans, and supratidal flats. Frey and Pemberton (1987) indicated that such environments are subject to extreme variations in energy levels, sediment types, and physical and biogenic sedimentary structures. These settings are also subject to temporal fluctuations between fresh and saline water. Marine processes generally dominate during spring tides and storm surges, whereas eolian processes and subaerial exposure of supratidal flats predominate during neap tides and non-storm periods.

The topographically elevated position of this setting and its extended periods of subaerial exposure preclude colonization by most benthic marine animals. The only persistent, notable exceptions are amphibious crabs of the family Ocypodidae and amphipods of the family Talitridae. Ocypodid crabs, which include both scavengers and surficial deposit feeders, typically excavate J-, Y-, or U-shaped dwelling burrows referable to the tracefossils *Psilonichnus* (Fig. 10A) and *Polykladichnus* (Fig. 10B). Talitrid amphipods construct vertical burrows up to 25 cm deep in sand at the landward limit of the foreshore and feed on seaweed washed up on the beach (Dashtgard and Gingras, 2005). Other biogenic structures are generated by essentially terrestrial organisms and include the vertical shafts of insects and spiders; the horizontal tunnels of other insects and tetrapods; and the ephemeral tracks, trails, and fecal pellets of insects, reptiles, birds, and mammals. Plant roots constitute another biogenic structure common to the *Psilonichnus* Ichnofacies (Fig. 10C) and commonly penetrate into the underlying foreshore deposits (Fig. 9C). The types of plants able to exploit these substrates range from intertidal halophytes on the distal margins of some washover fans to maritime and terrestrial grasses, weeds, vines, shrubs, bushes, and trees on dunes.

3. SHOREFACE VARIABILITY

Except for textural variations, the upper shoreface–foreshore complex remains relatively consistent in character over a wide range of shoreface morphodynamic states. There are few facies clearly indicative of barred conditions, as most depositional bodies have a low preservation potential (Greenwood and Mittler, 1985), and those that are preserved closely resemble the facies

FIGURE 10 Backshore succession. (A) *Ocypoda quadrata* (red arrow), an inferred tracemaker of *Psilonichnus* (*Ps*), in a backshore coastal dune complex. Facies also show *Planolites* (*P*), fugichnia (fu), and fecal pellets (pe) in the burrow fills. Roots (rt) are from present-day vegetation and cross-cut the Pleistocene interval. (B) Y-branched dwelling trace *Polykladichnus* (*Pk*), attributed to crustacean-generated backshore burrowing. (C) Backshore coastal eolianite and washover sandstones, cross-cut by roots (rt).

of non-barred systems. As well, the preserved record of the upper shoreface–foreshore complex reflects fair-weather conditions and is characterized by ichnogenera of the *Skolithos* Ichnofacies. In contrast, the lower– middle shoreface complex contains the highest degree of facies variability, both sedimentologically and ichnologically (MacEachern and Pemberton, 1992). This part of the shoreface may range from virtually unburrowed HCS, SCS, or QPL sandstone to thoroughly bioturbated muddy and silty sandstone, because successions consist of both storm deposits and fair-weather accumulation under shoaling waves. The lower–middle shoreface complex straddles the transition from proximal expressions of the *Cruziana* Ichnofacies to the *Skolithos* Ichnofacies, further enhancing its variability. Although the offshore also receives sediment during both storm and fair-weather conditions, the setting is generally fair-weather dominated and resulting trace-fossil suites are representative of the *Cruziana* Ichnofacies (Figs. 2 and 3).

Storms and, to a lower degree, fair-weather waves are the dominant physical processes operating on the lower and middle shoreface. As these deposits may constitute the bulk of the ancient record of this portion of the shoreface (Elliott, 1986; Kumar and Sanders, 1976; Reinson, 1984; Swift et al., 1985), it follows that much of the observed trace-fossil variability may be attributable to variability in the intensity and frequency of storm activity. Fluctuations in the sedimentation rate are intimately associated with the episodic character of storm deposition and this appears to control the ultimate preserved expression of lower and middle shoreface.

Storm-induced variability in the preserved record of the lower and middle shoreface is associated with the degree of storm influence (i.e., the interplay of storm erosion *versus* tempestite deposition). By contrast, foreshore and upper shoreface settings are characterized by preferential erosion rather than deposition as a result of storms; most facies successions show erosionally bounded fair-weather deposits. In general, it is possible to identify three principal "types" of lower and middle shoreface successions: those that are strongly storm-dominated (high energy), those that are moderately affected by storms (intermediate energy), and those that are only weakly affected by storms or dominated by fair-weather deposition (low energy) (Dashtgard et al., in press; MacEachern and Pemberton, 1992; MacEachern et al., 1999b). A complete intergradation exists, both along depositional strike (depending on the regional paleogeography) and vertically through the facies succession (related to shallowing and increasing effectiveness of storm action on the bed). Much of the confusion that exists in defining the character of the lower and middle shoreface, as well as in determining the boundaries between the two subenvironments, appears to revolve around the variability imposed by the degree of storm dominance and preservability of the fair-weather deposits (MacEachern and Pemberton, 1992).

The preserved record of strongly storm-dominated shorefaces is characterized by stacked, erosionally amalgamated tempestites (HCS, SCS, and QPL) with few preserved biogenic structures (Fig. 11). The preserved record of moderately storm-dominated shorefaces is characterized by stacked tempestites with preserved burrowed tops and/or fair-weather beds (Fig. 12). Such facies show a laminated-to-burrowed appearance (so-called lam-scram; MacEachern and Pemberton, 1992). The preserved record of weakly storm-affected shorefaces is characterized by little or no preserved tempestites and a succession dominated by fair-weather deposits that impart a thoroughly bioturbated appearance (Fig. 13).

FIGURE 11 Storm-dominated shoreface successions. (A) The Eagle Sandstone shoreface succession of Montana, USA, consisting of amalgamated tempestites. (B) The Virgelle Member of the Milk River Formation at Writing-On-Stone Provincial Park, Alberta, showing amalgamated HCS and SCS. (C) Core-box shots of amalgamated HCS and SCS in the lower- to middle-shoreface succession of the Cadotte Member of the Peace River Formation, Alberta. Base (B) is to the lower left and top (T) is to the upper right. Note the general absence of macroscopic burrowing. Scale bar = 15 cm. (D) Cryptically bioturbated sandstone near the base, erosionally overlain by HCS tempestites (arrow) in the Early Cretaceous Cadotte Member, Alberta, Canada. Scale bar = 5 cm.

4. TIDAL EFFECTS ON THE SHOREFACE

Tidal effects on the shoreface, and how these are manifesting sedimentologically and ichnologically, have only recently received attention. What is known today is almost assuredly going to evolve with time. To date, only a few studies have been conducted on the sedimentology and ichnology of these systems (e.g., Ainsworth et al., 2008; Dashtgard et al., 2006, 2009, in press; Frey and Dashtgard, 2011; Vakarelov et al., in press); yet, these studies have identified viable criteria for recognizing tidal shorefaces (Dashtgard et al., in press). Tides can be manifest in one of two ways across the shoreface: either as direct controls on sediment deposition or, with increasing tidal range, as tidal modulation.

Direct tidal controls (i.e., tide-influenced shorefaces or TIS) are observed in settings with strong tidal currents and intermittent slack-water periods, wherein strong tidal currents, running parallel to the coastline, dominate sediment transport and deposition in the lower shoreface and, to a lesser extent, the middle shoreface (Dashtgard et al., in press; Frey and Dashtgard, 2011). In the upper shoreface and foreshore of TIS, the wave-orbital velocities and wave-generated current velocities exceed the tidal current velocities, such that the upper shoreface and foreshore are sedimentologically and ichnologically similar to their fair-weather wave-dominated and storm-dominated (non-tidal) counterparts.

For TIS, tidal controls on sediment deposition are best expressed in the lower shoreface and offshore. Across offshore through middle shoreface zones, grain sizes either remain constant or increase slightly in the offshore direction, and the mud content of the sediment is very low (Frey and Dashtgard, 2011). In the lower shoreface and offshore, sand accumulates as current-generated structures and plane beds, although these structures are commonly obliterated by pervasive bioturbation. Burrows are predominantly vertical (Fig. 14) and form suites attributable to the archetypal *Skolithos* Ichnofacies. Toward the middle shoreface, there is increased preservation of wave-generated sedimentary structures (e.g., SCS and HCS), with relatively limited preservation of bioturbated fair-weather beds (Fig. 14). Burrows observed in modern middle shoreface deposits of TIS are also typical elements of the *Skolithos* Ichnofacies and include *Arenicolites*, *Planolites*, *Skolithos*, and *Conichnus* (Fig. 14). The upper shoreface and foreshore of TIS are sedimentologically and ichnologically similar to their wave-dominated and storm-dominated counterparts and commonly possess the *Macaronichnus segregatis* "toe-of-the-beach" assemblage (cf. Saunders et al., 1994), developed between the upper shoreface and foreshore (Vakarelov et al., in press).

Tidal modulation, unlike direct tidal controls on sediment deposition, refers to the lateral translation of wave zones across the shoreface profile in response to the rising and falling of tides. In microtidal and lower mesotidal settings, the translation of wave zones along the depositional dip is minimal and probably falls within the range of variability imposed by changing wave conditions. With increasing tidal range (macrotidal and megatidal settings), however, the proximal–distal displacement of wave zones is pronounced, and it is not uncommon to subaerially expose upper, middle and possibly lower shoreface sediments at low tide (Dashtgard and Gingras, 2007; Masselink and Hegge, 1995; Masselink and Short, 1993; Masselink and Turner, 1999; Short, 1999). The sedimentological and ichnological expression of such significant tidal ranges are distinct and the resulting shoreface character is referred to as a "tidally modulated shoreface" (Dashtgard et al., 2009, in press).

TMS constitute the tidal end member of the wave-tidal spectrum for shoreface settings. TMS are identified by (1) common interbedding of tidal sedimentary structures down the shoreface profile, (2) the notable absence of an obvious middle shoreface, (3) an anomalously thick foreshore, and (4) the presence of a low-diversity and low-density trace-fossil suite of simple vertical and

FIGURE 12 Intermediate storm-influenced lower and middle shoreface successions, characterized by laminated-to-burrowed bedding. (A) Outcrop expression of a sandy tempestite containing HCS and fugichnia (fu), which is burrowed at its top with *Skolithos* (*S*) and *Ophiomorpha borneensis*

FIGURE 13 Weakly storm-affected shoreface successions, typified by intense bioturbation. (A) Crudely bedded, bioturbated muddy sandstone passing into burrowed silty sandstone of the lower shoreface and middle shoreface, respectively, Fall River Formation, Montana. (B) Core boxes with a thoroughly bioturbated succession, showing pervasively bioturbated (BI = 5) sandy mudstones of the upper offshore, grading into bioturbated (BI = 5) muddy sandstones of the lower shoreface and into intensely bioturbated (BI = 4–5) silty sandstones of the middle shoreface. A marine flooding surface (black arrow) is overlain by burrowed sandy mudstones and thin tempestites of the upper offshore. The interval preserves only a single tempestite near the base of the fourth column from the bottom of the interval. Scale bar = 15 cm.

horizontal burrows (Dashtgard et al., 2009). First, sedimentary structures produced by shoaling waves, breakers and surf, and swash-backwash are interbedded across much of the TMS depositional profile, as a result of the lateral shift of wave zones during the tidal cycle (Fig. 15). With increasing storm-wave influence on TMS, sedimentary structures produced by storm waves (i.e., HCS and SCS) are increasingly dominant throughout the lower and upper shoreface (Ainsworth et al., 2008). Second, as a result of the regular lateral shifting of wave zones, the sedimentological and ichnological criteria

(*Ob*), and passes into pervasively bioturbated (BI = 5) sandstone of the proximal lower shoreface. (B) Box cores with a laminated-to-burrowed sandstone succession, consisting of bioturbated (BI = 5) muddy sandstones interstratified with decimeter-scale HCS sandstones, passing into less thoroughly burrowed (BI = 3–4) silty sandstones interbedded with thicker HCS sandstones. Interval is interpreted to reflect lower shoreface deposits passing into middle shoreface units. Scale bar = 15 cm. (C) Close-up photo of burrowed (BI = 5) lower shoreface silty sandstone with *Rosselia* (*R*), *Chondrites* (*Ch*), and *Skolithos* (*S*), erosionally overlain by wavy parallel-laminated sandstone, interpreted as a tempestite.

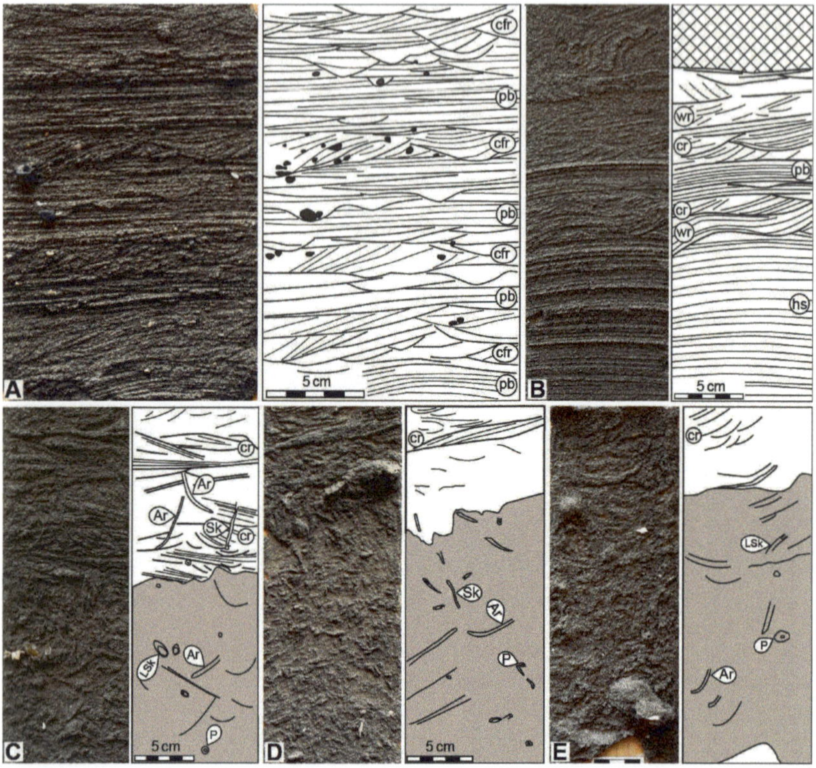

FIGURE 14 Resin peels and line interpretations from three modern tide-influenced shorefaces (TIS) in the Juan de Fuca Strait, Canada. *Images from Frey and Dashtgard (2011), and Dashtgard et al. (in press).* For all images, the landward direction is to the left of the page and the oceanward is to the right. (A) Upper shoreface to foreshore transition: rhythmically interbedded plane-bedded and combined flow/current-rippled sand with isolated pebbles (dark gray) and shell fragments (light gray). Note the absence of bioturbation. (B) Middle to upper shoreface transition: wave-rippled and current-ripped sand interbedded with plane-bedded (possibly hummocky cross-stratified) sand. These sediments overlie hummocky cross-stratified sand. Note the lack of bioturbation in this core. Cross-hatch pattern defines area of distorted core. (C) Lower shoreface: current-rippled and plane-bedded medium-grained sand (BI = 2) overlying intensely bioturbated (BI = 5–6) sand (gray pattern). (D) Lower shoreface: intensely bioturbated (BI = 5–6), medium-grained sand below a current-rippled surface. Burrows are mainly simple vertical and horizontal tubes. (E) Upper offshore: partial preservation of current ripples near top of core, with increasing sediment reworking with depth. Bioturbation index values increase from BI = 2–3 near the top of the core to BI = 6 near the base. Acronyms used for sedimentary structures: current ripples (cr), combined-flow ripples (cfs), hummocky cross-stratification (hs), plane bedding (pb), and wave ripples (wr). Acronyms used for identifiable traces: *Arenicolites* (Ar), *Planolites* (P), *Skolithos* (Sk), and lined-*Skolithos* (lSk). Scale bar = 3 cm.

normally used to identify the middle shoreface (e.g., SCS, trace-fossil suites of the archetypal *Skolithos* Ichnofacies) become interbedded with structures typically attributed to the upper shoreface and/or lower shoreface. Consequently, a "middle shoreface" zone cannot be differentiated in the TMS, and the lower shoreface appears to grade landward into the upper shoreface.

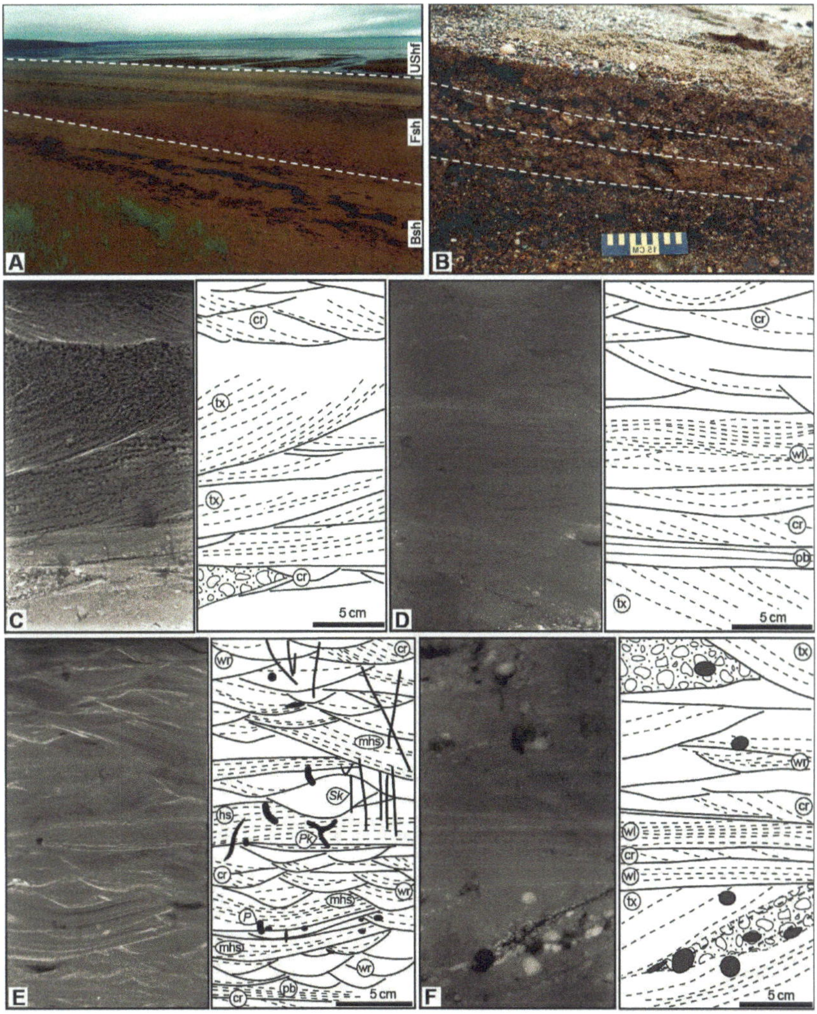

FIGURE 15 Photos, X-ray radiographs, and line interpretations of a modern tidally modulated shoreface (TMS) from the Bay of Fundy, Canada. *Images from Dashtgard et al. (2009), and Dashtgard et al. (in press).* For all X-ray radiographs and line drawings, the landward direction is to the left of the page and the oceanward is to the right. (A) Photo of a TMS taken during the falling tide, showing the position of the backshore (Bsh), foreshore (Fsh), and upper shoreface (UShf). The foreshore dips seaward at 2–4° and is 100–200 m wide. (B) Trench through beds of pebbly sand and gravel that dip seaward (to the right). (C) Proximal upper shoreface: onshore-directed trough cross-beds in sand, interbedded with ebb-oriented current ripples. (D) Distal upper shoreface to proximal lower shoreface: trough cross-bedded and current-rippled sand with interbedded wavy-parallel-laminated sand (possibly swaley cross-stratification). (E) Proximal lower shoreface: complex vertical succession of wave ripples, current ripples, microhummocky cross-stratification, plane beds, and hummocky cross-stratification. Sediment is muddy sand (lighter colors indicative of higher mud content). Bioturbation is low (BI = 1–2) and is dominated by vertical structures (e.g., *Polykladichnus*, *Skolithos*). (F) Proximal lower shoreface: wavy-parallel laminae and current ripples in sand, interbedded with trough cross-bedded pebbly sand. Acronyms used for sedimentary structures: current ripples (cr), microhummocky cross-stratification (mhs), plane beds (pb), trough cross-beds (tx), wave ripples (wr), and wavy-parallel laminae (wl). Acronyms used for identifiable traces: *Planolites* (*P*), *Polykladichnus* (*Pk*), and *Skolithos* (*Sk*).

The third criterion for recognizing TMS is the anomalously large thickness of the foreshore and, to a lesser extent, the change in dip angle between the foreshore and the upper shoreface, particularly in gravel-prone settings. TMS foreshores are typically 4–5 m thick (Dashtgard et al., 2009, in press; Vakarelov et al., in press), which is at least twice that of a microtidal, fair-weather wave-dominated shoreface. Although foreshore thicknesses are significantly less than the vertical tidal range under which TMS develop, the overthickened foreshore does suggest an increased tidal range. For sandy foreshores, the contact between the foreshore and the upper shoreface is gradational and the depositional angle of beds is reduced (Ainsworth et al., 2008; Vakarelov et al., in press). A sharp change in dip angles between the foreshore and upper shoreface is common in composite sand-and-gravel beaches, wherein the foreshore is gravelly and the upper shoreface is sandy (Clifton, 1981; Frey and Dashtgard, 2011; Jennings and Shulmeister, 2002; Kirk, 1980). It is worth mentioning that distinctive sedimentary structures recording wave–tide interaction are documented in tidal flats (e.g., wave-generated tidal bundles, tidally modulated HCS; e.g., Yang et al., 2008a,b). Could it perhaps be possible that the same or similar structures are generated on TMS? Even if not, TMS might form a continuum with storm/wave-modulated tidal flats, each containing subtle sedimentological and ichnological evidence for mixed process regime (Fig. 16).

The remaining criteria for recognizing TMS are ichnological. For low and moderately storm-influenced TMS in the rock record, the lower shoreface is characterized by low BI values in sandstone beds and a dominance of deep-penetrating traces such as *Ophiomorpha* and *Thalassinoides* (Ainsworth et al., 2008). In the lower shoreface of modern TMS, the trace suite is dominated by vertical suspension- and surface-deposit-feeding traces, as well as horizontal subsurface-deposit-feeding traces. Traces observed include *Skolithos*, *Arenicolites*, *Planolites*, *Siphonichnus*, *Palaeophycus*, and *Polykladichnus* (Fig. 15). BI values range between 0 and 3. The upper shoreface also exhibits burrowing with the same range of traces as seen in the lower shoreface, yet the density of burrowing is significantly reduced (BI=0–2). The foreshore is unburrowed. Although our knowledge of the ichnological character of TMS is limited to only a few examples, it appears that TMS can be characterized, ichnologically, by a low-diversity and low-density suite consisting of simple vertical and horizontal traces. Trace diversities remain consistent across the upper and lower shoreface, whereas bioturbation intensities decrease landward.

5. SUMMARY

The integration of ichnology and sedimentology provides a high-resolution depositional model for identifying wave-dominated foreshore through to offshore subenvironments, whether they occur on strandplain, barrier islands, or even locations in wave/storm-dominated deltas removed from the interference of distributary channels (Fig. 17). Wave-dominated (non-tidal) shorefaces can be grouped into three main depositional complexes, each of which responds to a different set of

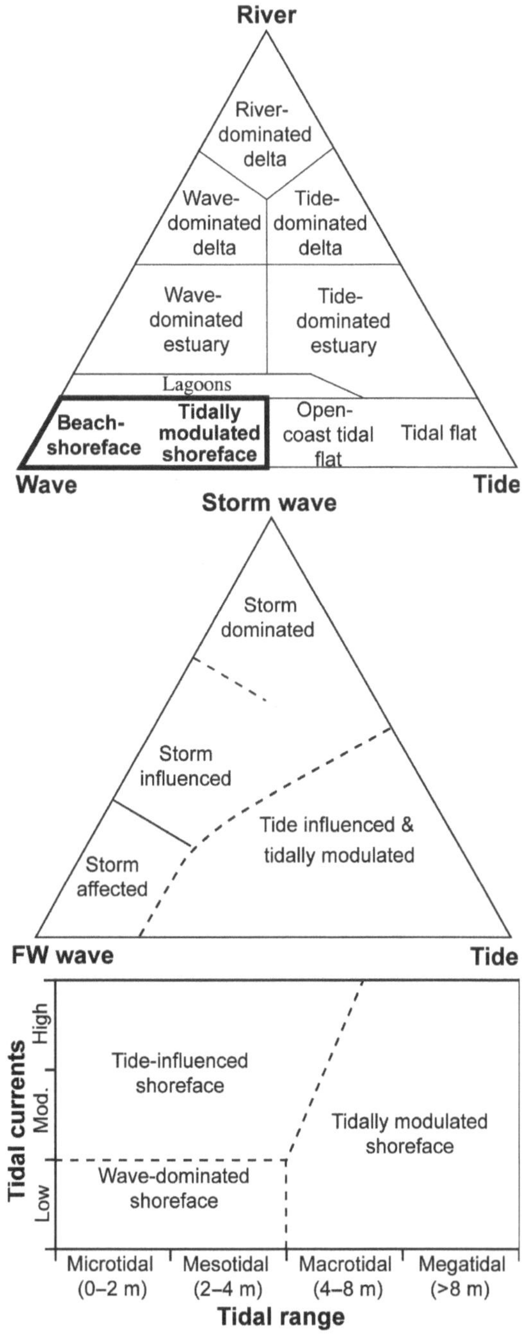

FIGURE 16 Conceptual model proposed for shoreface deposits. The top ternary diagram is modified after Dalrymple et al. (1992) and Dashtgard et al. (2009), showing the relative position of tidally modulated shorefaces (the tidal end member of shorefaces) with respect to wave, tide, and river influence on sediment deposition. The middle ternary diagram is developed strictly for shorefaces and considers the three main influences on sediment deposition in these settings: fair-weather (FW) waves, storm waves, and tides. The lower diagram considers only tidal processes (currents and tidal range) and their influences on shoreface character. *Modified after Dashtgard et al. (in press).*

FIGURE 17 Split core schematic diagram, showing the idealized transition from shelf through to foreshore deposits of a prograding shoreface succession (after MacEachern et al., 2010).

physical parameters and displays discrete ichnological characteristics. Within each depositional complex, the subenvironments can be reasonably differentiated through a combination of physical and biogenic features.

The offshore complex is typically dominated by fair-weather conditions and is characterized by ichnological suites attributable to the distal to archetypal *Cruziana* Ichnofacies. Such suites display a predominance of deposit-feeding traces and grazing/foraging traces, with rare dwelling traces of suspension feeders. Storm effects are variable but ineffective in removing fair-weather suites except under exceptionally high degrees of storm domination. The offshore typically displays an equilibrium suite of trace fossils, with variable overprinting by opportunistic suites related to initial tempestite colonization.

The lower–middle shoreface complex is affected by both storm and fair-weather conditions, and hence, the preserved record shows a high degree of variability. Fair-weather (equilibrium) trace-fossil suites range from those characteristic of proximal expressions of the *Cruziana* Ichnofacies in the lower shoreface to the archetypal *Skolithos* Ichnofacies in the middle shoreface (Figs. 2 and 3).

The upper shoreface–foreshore complex is also quite variable, but this is mainly a consequence of sediment texture and morphodynamic configuration of the shoreline. Preserved facies are remarkably similar over a wide range of shoreface states and accumulate in response to fair-weather conditions. Storm events are non-depositional and strongly erosive. In dissipative, lower energy systems, the trace-fossil suites are generally sparse and correspond to the archetypal *Skolithos* Ichnofacies. In high-energy reflective systems, the upper shoreface–foreshore transition may contain zones of *Macaronichnus segregatis*.

The main variations in the preserved record of a shoreface succession are reflected in the lower–middle shoreface complex and can be attributed to variations in overall storm intensity, storm frequency, and relative water depth. Three main "types" of shoreface successions can be recognized, although a complete continuum exists. Erosionally amalgamated tempestites with minimal preserved biogenic structures constitute the lower–middle shoreface complex of strongly storm-dominated systems. Moderately storm-dominated systems lead to lower–middle shoreface complexes that consist of stacked, erosionally based storm beds with bioturbated tops consisting of opportunistic trace-fossil suites, cross-cut by resident (fair-weather) suites. Weakly storm-affected shorefaces constitute the lowest energy and most dissipative conditions, with the lower–middle shoreface complex characterized by thoroughly bioturbated muddy and silty sandstones recording nearly continuous fair-weather accumulation.

Discrimination between lower and middle shoreface deposits may be possible on an ichnological basis, provided well-developed fair-weather suites are preserved. The boundary between the lower and middle shoreface corresponds to the transition from suites attributable to the *Cruziana* Ichnofacies to those that correspond to the *Skolithos* Ichnofacies (MacEachern and Pemberton, 1992; MacEachern et al., 1999b). Under progressively higher degrees of storm domination, fair-weather suites are insufficiently preserved to accurately resolve the two subenvironments ichnologically. In such scenarios, physical characteristics such as coarsening of sediment texture, decrease in thickness of fair-weather deposits, and replacement of lower energy HCS beds by SCS beds may have to be employed. In the case of QPL-bearing intervals, it is unclear whether such a distinction is possible. In any case, vertical and lateral transitions between shoreface "types" may occur, depending upon relative water depth, storm frequency, storm intensity, and shoreline orientation in relation to dominant wave attack. These conditions serve to vary the character of contemporaneous lower and middle shoreface deposits along depositional strike, which may make differentiation of these two subenvironments problematic.

Two types of tidal shorefaces also have been defined, depending upon the tidal hydrodynamic conditions. TIS develop on coastlines affected by strong tidal currents and are considered typical of strait margins and embayed coastlines. Sedimentological and ichnological criteria considered indicative of TIS are confined to the lower shoreface and offshore as well as, to a lesser extent, the middle shoreface. Across all three of these zones, grain sizes either remain constant or increase slightly in the offshore direction and mud content of the sediment is very low (Dashtgard et al., in press; Frey and Dashtgard, 2011). Sand accumulates as current-generated structures and plane beds in the lower shoreface and offshore, although these structures are commonly obliterated by pervasive bioturbation. Burrowing across the shoreface and into the offshore is dominated by vertical structures (Fig. 14C–E) that are typical of the archetypal *Skolithos* Ichnofacies, marking a basinward shift in its paleodepositional position. At the transition from the upper shoreface to the foreshore of the TIS, it is common to find the *Macaronichnus segregatis* "toe-of-the-beach" assemblage (cf. Saunders et al., 1994; Vakarelov et al., in press).

TMS constitute the tidal end member of the wave-tidal spectrum for shoreface settings (Figs. 15 and 16). TMS are identified by (1) common interbedding of sedimentary structures down the shoreface profile; (2) the notable absence of an obvious middle shoreface zone; (3) an anomalously thick foreshore interval; and (4) the presence of a low-diversity and low-density trace-fossil suite, consisting of simple vertical and horizontal burrows (Dashtgard et al., 2009). Trace diversities remain consistent across the upper and lower shoreface, whereas bioturbation intensities decrease landward.

ACKNOWLEDGMENTS

The authors would like to thank the Natural Science and Engineering Research Council of Canada (NSERC) for research funding; Discovery Grant 184293 to J. A. M., Discovery Grant 341789 to S. E. D., Discovery Grant 0001380 to M. K. G., Discovery Grant 371662 to J.-P. Z., and Discovery Grant A0816 to S. G. P.. S. G. P. would like to thank the Canada Research Chairs program for their support of his research. We are grateful to Ron Boyd for supplying the copy of Figure 1. The chapter benefited from a thoughtful review by Gary Hampson and comments from Richard Bromley and Dirk Knaust. We thank them for their insights that helped to improve the quality of this chapter.

REFERENCES

Aguirre, J., de Gibert, J.M., Puga-Bernabéu, A., 2010. Proximal–distal ichnofabric changes in a siliciclastic shelf, Early Pliocene, Guadalquivir Basin, southwest Spain. Palaeogeogr. Palaeoclimatol. Palaeoecol. 291, 328–337.

Aigner, T., 1985. Storm depositional systems: dynamic stratigraphy in modern and ancient shallow-marine sequences. Lect. Notes Earth Sci. 3, 174 pp.

Aigner, T., Reineck, H.E., 1982. Proximality trends in modern storm sands from the Helgoland Bight (North Sea) and their implications for basin analysis. Senckenberg. Marit. 14, 183–215.

Ainsworth, R.B., Flint, S.S., Howell, J., 2008. Predicting coastal depositional style: influence of basin morphology and accommodation/sediment supply regime within a sequence stratigraphic framework. In: Hampson, G.J., Steel, R.J., Burgess, P.B., Dalrymple, R.W. (Eds.), Recent Advances in Models of Siliciclastic Shallow-Marine Stratigraphy. SEPM Spec. Publ., vol. 90, pp. 237–263.

Arnott, R.W.C., 1993. Quasi-planar-laminated sandstone beds of the lower Cretaceous Bootlegger Member: North-Central Montana: evidence of combined flow sedimentation. J. Sediment. Petrol. 63, 488–494.

Bann, K.L., Fielding, C.R., 2004. An integrated ichnological and sedimentological comparison of non-deltaic shoreface and subaqueous delta deposits in Permian reservoir units of Australia. In: McIlroy, D. (Ed.), The Application of Ichnology to Palaeoenvironmental and Stratigraphic Analysis. Geol. Soc. London, Spec. Publ., vol. 228, pp. 273–310.

Bann, K.L., Fielding, C.R., MacEachern, J.A., Tye, S.C., 2004. Differentiation of estuarine and offshore marine deposits using integrated ichnology and sedimentology: Permian Pebbley Beach Formation, Sydney Basin, Australia. In: McIlroy, D. (Ed.), The Application of Ichnology to Palaeoenvironmental and Stratigraphic Analysis. Geol. Soc. London, Spec. Publ., vol. 228, pp. 179–211.

Bann, K.L., Tye, S.C., MacEachern, J.A., Fielding, C.R., Jones, B.G., 2008. Ichnological and sedimentologic signatures of mixed wave- and storm-dominated deltaic deposits: examples from the early Permian Sydney Basin, Australia. In: Hampson, G.J., Steel, R.J., Burgess, P.M., Dalrymple, R.W. (Eds.), Recent Advances in Models of Siliciclastic Shallow Marine Stratigraphy. SEPM Spec. Publ., vol. 90, pp. 293–332.

Bhattacharya, J.P., MacEachern, J.A., 2009. Hyperpycnal rivers and prodeltaic shelves in the Cretaceous seaway of North America. J. Sediment. Res. 79, 184–209.

Boyd, R.L., 2010. Transgressive wave dominated coasts. In: James, N.P., Dalrymple, R.W. (Eds.), Facies Models 4. Geol. Ass. Can., St. John's, Geotext 6, pp. 265–294.

Boyd, R.L., Dalrymple, R.W., Zaitlin, B.A., 1992. Classification of clastic coastal depositional environments. Sediment. Geol. 80, 139–150.

Buatois, L.A., Santiago, N., Parra, K., Steel, R., 2008. Animal-substrate interactions in an Early Miocene wave-dominated tropical delta: delineating environmental stresses and depositional dynamics (Tácata field, eastern Venezuela). J. Sediment. Res. 78, 458–479.

Buatois, L.A., Saccavino, L.L., Zavala, C., 2011. Ichnologic signatures of hyperpycnal flow deposits in Cretaceous river-dominated deltas, Austral Basin, southern Argentina. In: Slatt, R.M., Zavala, C. (Eds.), Sediment Transfer from Shelf to Deep Water—Revisiting the Delivery System. AAPG Studies in Geology, vol. 61, pp. 153–170.

Buatois, L.A., Santiago, N., Herrera, M., Plink-Björklund, P., Steel, R., Espin, M., Parra, K., 2012. Sedimentological and ichnological signatures of changes in wave, river and tidal influence along a Neogene tropical deltaic shoreline. Sedimentology 59, 1568–1612.

Clifton, H.E., 1969. Beach lamination: nature and origin. Mar. Geol. 7, 553–559.

Clifton, H.E., 1981. Progradational sequences in Miocene shoreline deposits, southeastern Caliente Range, California. J. Sediment. Petrol. 51, 165–184.

Clifton, H.E., 1988. Sedimentologic approaches to paleobathymetry with applications to the Merced Formation of central California. Palaios 3, 507–522.

Clifton, H.E., 2003. Coastal sedimentary facies. In: Middleton, G.V. (Ed.), Encyclopedia of Sediments and Sedimentary Rocks. Kluwer Academic Publishers, Dordrecht, pp. 149–157.

Clifton, H.E., Thompson, J.K., 1978. *Macaronichnus segregatis*—a feeding structure of shallow marine polychaetes. J. Sediment. Petrol. 48, 1293–1302.

Clifton, H.E., Hunter, R.E., Phillips, R.L., 1971. Depositional structures and processes in the non-bar high energy nearshore. J. Sediment. Petrol. 41, 651–670.

Coates, L., MacEachern, J.A., 2007. The ichnological signatures of river- and wave-dominated delta complexes: differentiating deltaic from non-deltaic shallow marine successions, Lower Cretaceous Viking Formation and Upper Cretaceous Dunvegan Formation, west-central Alberta. In: MacEachern, J.A., Bann, K.L., Gingras, M.K., Pemberton, S.G. (Eds.), Applied Ichnology. SEPM Short Course Notes 52, pp. 227–254.

Cook, D.O., Gorsline, D.S., 1972. Field observations of sand transport by shoaling waves. Mar. Geol. 13, 31–55.

Dalrymple, R.W., Zaitlin, B.A., Boyd, R., 1992. Estuarine facies models: conceptual basis and stratigraphic implications. J. Sediment. Petrol. 62, 1130–1146.

Dashtgard, S.E., Gingras, M.K., 2005. The temporal significance and distribution of bioturbation in coarse-grained upper foreshore and backshore deposits: Waterside Beach, New Brunswick, Canada. Palaios 20, 589–595.

Dashtgard, S.E., Gingras, M.K., 2007. Tidal controls on the morphology and sedimentology of gravel-dominated deltas and beaches: examples from the megatidal Bay of Fundy, Canada. J. Sediment. Res. 77, 1063–1077.

Dashtgard, S.E., Gingras, M.K., Butler, K.E., 2006. Sedimentology and stratigraphy of a transgressive, muddy gravel beach: Waterside Beach, Bay of Fundy, Canada. Sedimentology 53, 279–296.

Dashtgard, S.E., Gingras, M.K., MacEachern, J.A., 2009. Tidally modulated shorefaces. J. Sediment. Res. 79, 793–807.

Dashtgard, S.E., MacEachern, J.A., Frey, S.E., Gingras, M.K. in press. Tidal effects on the shoreface: towards a conceptual framework. Sediment. Geol. http://dx.doi.org/10.1016/j.sedgeo.2010.09.006.

Davidson-Arnott, R.G.D., Greenwood, B., 1976. Facies relationships in a barred coast, Kouchibouguac Bay, New Brunswick, Canada. In: Davis Jr., R.A., Ethridge, R.L. (Eds.), Beach and Nearshore Sedimentation. SEPM Spec. Publ., vol. 24, pp. 149–168.

Davies, D.K., Ethridge, F.G., Berg, R.R., 1971. Recognition of barrier environments. AAPG Bull. 55, 550–565.

Davis Jr., R.A., 1978. Beach and nearshore zone. In: Davis Jr., R.A. (Ed.), Coastal Sedimentary Environments. Springer, New York, pp. 237–285.

Desjardins, P.R., Buatois, L.A., Mángano, M.G., 2012. Tidal flats and subtidal sand bodies. In: Knaust, D., Bromley, R.G. (Eds.), Trace Fossils as Indicators of Sedimentary Environments. Developments in Sedimentology, vol. 64. Elsevier, Amsterdam, pp. 529–562.

Dörjes, J., Hertweck, G., 1975. Recent biocoenoses and ichnocoenoses in shallow-water marine environments. In: Frey, R.W. (Ed.), The Study of Trace Fossils. Springer, New York, pp. 459–491.

Dott Jr., H.R., 1983. Episodic sedimentation—how normal is average? How rare is rare? Does it matter? J. Sediment. Petrol. 53, 5–23.

Dott Jr., H.R., 1988. An episodic view of shallow marine classic sedimentation. In: Debor, P.L., Van Gelder, A., Nio, S.D. (Eds.), Tide-Influenced Sedimentary Environments and Facies. D. Reidel Publishing, Dordrecht, pp. 3–12.

Dott Jr., R.H., Bourgeois, J., 1982. Hummocky stratification: significance of its variable bedding sequences. GSA Bull. 93, 663–680.

Duke, W.L., 1985. Hummocky cross-stratification, tropical hurricanes, and intense winter storms. Sedimentology 32, 167–194.

Elliott, T., 1986. Siliciclastic shorelines. In: Reading, H.G. (Ed.), Sedimentary Environments and Facies. Blackwell, Oxford, pp. 155–188.

Fox, W.T., Davis Jr., R.A., 1978. Seasonal variation in beach erosion and sedimentation on the Oregon coast. GSA Bull. 89, 1541–1549.

Frey, R.W., 1990. Trace fossils and hummocky cross-stratification, Upper Cretaceous of Utah. Palaios 5, 203–218.

Frey, S.E., Dashtgard, S.E., 2011. Sedimentology, ichnology and hydrodynamics of strait-margin, sand and gravel beaches and shorefaces: Juan de Fuca Strait, British Columbia, Canada. Sedimentology 58, 1326–1346.

Frey, R.W., Pemberton, S.G., 1987. The *Psilonichnus* ichnocoenose and its relationship to adjacent marine and nonmarine ichnocoenoses along the Georgia coast. Bull. Can. Petrol. Geol. 35, 333–357.

Gingras, M.K., MacEachern, J.A., Pemberton, S.G., 1998. A comparative analysis of the ichnology of wave- and river-dominated allomembers of the Upper Cretaceous Dunvegan Formation. Bull. Can. Petrol. Geol. 46, 51–73.

Greenwood, B., Davidson-Arnott, R.G.D., 1979. Sedimentation and equilibrium in wave formed bars: a review and case study. Can. J. Earth Sci. 16, 312–332.

Greenwood, B., Mittler, P.R., 1985. Vertical sequence and lateral transitions in the facies of a barred nearshore environment. J. Sediment. Petrol. 55, 366–375.

Hansen, C.D., MacEachern, J.A., 2007. Application of the asymmetric delta model to along-strike facies variations in a mixed wave- and river-influenced delta lobe, Upper Cretaceous Basal Belly River Formation, central Alberta. In: MacEachern, J.A., Bann, K.L., Gingras, M.K., Pemberton, S.G. (Eds.), Applied Ichnology. SEPM Short Course Notes 52, pp. 255–271.

Heward, A.P., 1981. A review of wave-dominated clastic shoreline deposits. Earth-Sci. Rev. 17, 223–276.

Howard, J.D., 1971a. Comparison of the beach-to-offshore sequence in modern and ancient sediments. In: Howard, J.D., Valentine, J.W., Warme, J.E. (Eds.), Recent Advances in Paleoecology and Ichnology. Short Course Lecture Notes, American Geological Institute, pp. 148–183.

Howard, J.D., 1971b. Trace fossils as paleoecological tools. In: Howard, J.D., Valentine, J.W., Warme, J.E. (Eds.), Recent Advances in Paleoecology and Ichnology. Short Course Lecture Notes, American Geological Institute, pp. 184–212.

Howard, J.D., 1972. Trace fossils as criteria for recognizing shorelines in stratigraphic record. In: Rigby, J.K., Hamblin, W.K. (Eds.), Recognition of Ancient Sedimentary Environments. SEPM Spec. Publ., vol. 16, pp. 215–225.

Howard, J.D., 1975. The sedimentological significance of trace fossils. In: Frey, R.W. (Ed.), The Study of Trace Fossils. Springer, New York, pp. 131–146.

Howard, J.D., Frey, R.W., 1984. Characteristic trace fossils in nearshore to offshore sequences. Upper Cretaceous of east central Utah. Can. J. Earth Sci. 21, 200–219.

Howard, J.D., Reineck, H.-E., 1981. Depositional facies of a high energy beach-to-offshore sequence: comparison with low energy sequence. AAPG Bull. 65, 807–830.

Hunter, R.E., Clifton, H.E., Phillips, R.L., 1979. Depositional processes, sedimentary structures, and predicted vertical sequences in barred nearshore systems, southern Oregon coast. J. Sediment. Petrol. 49, 711–728.

Immenhauser, A., 2009. Estimating palaeo-water depth from the physical rock record. Earth-Sci. Rev. 96, 107–139.

Jennings, R., Shulmeister, J., 2002. A field based classification scheme for gravel beaches. Mar. Geol. 186, 211–228.

Johnson, D.W., 1919. Shore Processes and Shoreline Development. John Wiley and Sons, New York, 587 pp.

Jumars, P.A., 1993. Concepts in Biological Oceanography. Oxford University Press, New York, 348 pp.

Kirk, R.M., 1980. Mixed sand and gravel beaches: morphology, processes and sediments. Progr. Phys. Geogr. 4, 189–210.

Kumar, N., Sanders, J.E., 1976. Characteristics of shoreface storm deposits: modern and ancient examples. J. Sediment. Petrol. 46, 145–162.

Leckie, D.A., Walker, R.G., 1982. Storm- and tide-dominated shorelines in Cretaceous Moosebar-Lower Gates interval—outcrop equivalents of Deep Basin gas trap in Western Canada. AAPG Bull. 66, 138–157.

MacEachern, J.A., Bann, K.L., 2008. The role of ichnology in refining shallow marine facies models. In: Hampson, G., Steel, R., Burgess, P., Dalrymple, R. (Eds.), Recent Advances in Models of Siliciclastic Shallow-Marine Stratigraphy. SEPM Spec. Publ., vol. 90, pp. 73–116.

MacEachern, J.A., Hobbs, T.W., 2004. The ichnological expression of marine and marginal marine conglomerates and conglomeratic intervals, Cretaceous Western Interior Seaway, Alberta and northeastern British Columbia. In: Moslow, T., Zonneveld, J.-P. (Eds.), Marine Conglomerates. Bull. Can. Soc. Petrol. Geol., vol. 52, pp. 77–104.

MacEachern, J.A., Pemberton, S.G., 1992. Ichnological aspects of Cretaceous shoreface successions and shoreface variability in the Western Interior Seaway of North America. In: Pemberton, S.G. (Ed.), Applications of Ichnology to Petroleum Exploration. SEPM Core Workshop Notes 17, pp. 57–84.

MacEachern, J.A., Stelck, C.R., Pemberton, S.G., 1999a. Marine and marginal marine mudstone deposition: paleoenvironmental interpretations based on the integration of ichnology, palynology and foraminiferal paleoecology. In: Bergman, K.M., Snedden, J.W. (Eds.), Isolated Shallow Marine Sand Bodies: Sequence Stratigraphic and Sedimentologic Interpretation. SEPM Spec. Publ., vol. 64, pp. 205–225.

MacEachern, J.A., Zaitlin, B.A., Pemberton, S.G., 1999b. A sharp-based sandstone succession of the Viking Formation, Joffre Field, Alberta, Canada: criteria for recognition of transgressively incised shoreface complexes. J. Sediment. Res. 69, 876–892.

MacEachern, J.A., Bann, K.L., Bhattacharya, J.P., Howell, C.D., 2005. Ichnology of deltas: organism responses to the dynamic interplay of rivers, waves, storms and tides. In: Giosan, L., Bhattacharya, J.P. (Eds.), River Deltas: Concepts, Models and Examples. SEPM Spec. Publ., vol. 83, pp. 49–85.

MacEachern, J.A., Pemberton, S.G., Bann, K.L., Gingras, M.K., 2007. Departures from the archetypal Ichnofacies: effective recognition of environmental stress in the rock record. In: MacEachern, J.A., Bann, K.L., Gingras, M.K., Pemberton, S.G. (Eds.), Applied Ichnology. SEPM Short Course Notes 52, pp. 65–93.

MacEachern, J.A., Pemberton, S.G., Gingras, M.K., Bann, K.L., 2010. Ichnology and facies models. In: James, N.P., Dalrymple, R.W. (Eds.), Facies Models 4. Geol. Ass. Can., St. John's, Geotext 6, pp. 19–58.

Masselink, G., Hegge, B.J., 1995. Morphodynamics of meso- and macrotidal beaches. Examples from central Queensland. Mar. Geol. 129, 1–23.

Masselink, G., Short, A.D., 1993. The effect of tidal range on beach morphodynamics and morphology: a conceptual beach model. J. Coast. Res. 9, 785–800.

Masselink, G., Turner, I., 1999. The effects of tides on beach morphodynamics. In: Short, A.D. (Ed.), Handbook of Beach and Shoreface Morphodynamics. John Wiley & Sons, Toronto, pp. 204–229.

McIlroy, D., 2004. Ichnofabrics and sedimentary facies of a tide-dominated delta: Jurassic Ile Formation of Kristin Field, Haltenbanken, Offshore Mid-Norway. In: McIlroy, D. (Ed.), The Application of Ichnology to Palaeoenvironmental and Stratigraphic Analysis. Geol. Soc. London, Spec. Publ., vol. 228, pp. 237–272.

McIlroy, D., 2007. Ichnology of a macrotidal tide-dominated deltaic depositional system: Lajas Formation, Neuquén Province, Argentina. In: Bromley, R.G., Buatois, L.A., Genise, J., Mángano, M.G., Melchor, R. (Eds.), Sediment-Organism Interactions: A Multifaceted Ichnology. SEPM Spec. Publ., vol. 88, pp. 195–211.

McLachlan, A., Eliot, I.G., Clarke, D.J., 1985. Water filtration through reflective microtidal beaches and shallow sublittoral sands and its implications for an inshore ecosystem in Western Australia. Estuar. Coast. Shelf Sci. 21, 91–104.

Moslow, T.F., Pemberton, S.G., 1988. An integrated approach to the sedimentological analysis of some Lower Cretaceous shoreface and delta front sandstone sequences. In: James, D.P., Leckie, D.A. (Eds.), Sequences, Stratigraphy, Sedimentology: Surface and Subsurface. Can. Soc. Petrol. Geol., Mem. 15, pp. 337–386.

Nadon, G., 1988. Tectonic Controls on Sedimentation within a Foreland Basin: The Bearpaw, Blood Reserve, and St. Mary River Formations, Southwestern Alberta. Can. Soc. Petrol. Geol., Calgary, Alberta Field Guide, 85 pp.

Nara, M., 1997. High-resolution analytical method for event sedimentation using *Rosselia socialis*. Palaios 12, 489–494.

Niedoroda, A.W., Swift, D.J.P., Hopkins, T.S., Ma, C.-M., 1984. Shoreface morphodynamics on wave-dominated coasts. In: Greenwood, B., Davis Jr., R.A. (Eds.), Hydrodynamics and Sedimentation in Wave-Dominated Coastal Environments. Marine Geology, vol. 60, pp. 331–334.

Pemberton, S.G., Frey, R.W., 1984. Ichnology of storm-influenced shallow marine sequence: Cardium Formation (Upper Cretaceous) at Seebe, Alberta. In: Stott, D.F., Glass, D.L. (Eds.), The Mesozoic of Middle North America. Can. Soc. Petrol. Geol., Mem. 9, pp. 281–304.

Pemberton, S.G., MacEachern, J.A., 1995. The sequence stratigraphic significance of trace fossils: examples from the Cretaceous foreland basin of Alberta, Canada. In: Van Wagoner, J.C., Bertram, G. (Eds.), Sequence Stratigraphy of Foreland Basin Deposits—Outcrop and Subsurface Examples from the Cretaceous of North America. AAPG Mem. 64, pp. 429–475.

Pemberton, S.G., MacEachern, J.A., 1997. The ichnological signature of storm deposits: the use of trace fossils in event stratigraphy. In: Brett, C.E. (Ed.), Paleontological Event Horizons—Ecological and Evolutionary Implications. Columbia University Press, New York, pp. 73–109.

Pemberton, S.G., Wightman, D.M., 1992. Ichnological characteristics of brackish water deposits. In: Pemberton, S.G. (Ed.), Application of Ichnology to Petroleum Exploration. A Core Workshop. SEPM Core Workshop Notes 17, pp. 141–167.

Pemberton, S.G., MacEachern, J.A., Frey, R.W., 1992a. Trace fossil facies models: environmental and allostratigraphic significance. In: Walker, R.G., James, N.P. (Eds.), Facies Models: Response to Sea Level Change. Geol. Ass. Can., St. John's, Geotext 1, pp. 47–72.

Pemberton, S.G., MacEachern, J.A., Ranger, M.J., 1992b. Ichnology and event stratigraphy: the use of trace fossils in recognizing tempestites. In: Pemberton, S.G. (Ed.), Application of Ichnology to Petroleum Exploration. A Core Workshop. SEPM Core Workshop Notes 17, pp. 85–117.

Pemberton, S.G., Ranger, M.J., MacEachern, J.A., 1992c. The conceptual framework of ichnology. In: Pemberton, S.G. (Ed.), Application of Ichnology to Petroleum Exploration. A Core Workshop. SEPM Core Workshop Notes 17, pp. 1–32.

Pemberton, S.G., Van Wagoner, J.C., Wach, G.D., 1992d. Ichnofacies of a wave-dominated shoreline. In: Pemberton, S.G. (Ed.), Application of Ichnology to Petroleum Exploration. A Core Workshop. SEPM Core Workshop Notes 17, pp. 339–382.

Pemberton, S.G., Spila, M., Pulham, A.J., Saunders, T., MacEachern, J.A., Robbins, D., Sinclair, I.K., 2001. Ichnology & Sedimentology of Shallow to Marginal Marine Systems: Ben Nevis & Avalon Reservoirs, Jeanne d'Arc Basin. Geol. Ass. Can., Short Course Notes 15, 343 pp.

Pianka, E.R., 1970. On r and K selection. Am. Nat. 104, 592–597.

Plint, A.G., 2010. Wave- and storm-dominated shoreline and shallow-marine systems. In: James, N.P., Dalrymple, R.W. (Eds.), Facies Models 4. Geol. Ass. Can., St. John's, Geotext 6, pp. 167–199.

Rahmani, R.A., Smith, D.G., 1988. The Cadotte Member of northwestern Alberta: a high energy barred shoreline. In: James, D.P., Leckie, D.A. (Eds.), Sequences, Stratigraphy, Sedimentology: Surface and Subsurface. Can. Soc. Petrol. Geol., Mem. 15, pp. 431–438.

Raychaudhuri, I., Pemberton, S.G., 1992. Ichnologic and sedimentologic characteristics of open marine to storm dominated restricted marine settings within the Viking/Bow Island formations, south-central Alberta. In: Pemberton, S.G. (Ed.), Application of Ichnology to Petroleum Exploration. A Core Workshop. SEPM Core Workshop Notes 17, pp. 119–139.

Reading, H.G., Collinson, J.D., 1996. Clastic coasts. In: Reading, H.G. (Ed.), Sedimentary Environments: Processes, Facies and Stratigraphy. Blackwell Science, Oxford, pp. 154–231.

Reinson, G.E., 1984. Barrier-island and associated strand-plain systems. In: Walker, R.G. (Ed.), Facies Models. Geoscience Canada, Reprint Series 1. Geol. Ass. Can., St. John's, pp. 119–140.

Roy, P.S., Thom, B.G., Wright, L.D., 1980. Holocene sequences on an embayed high-energy coast: an evolutionary model. Sediment. Geol. 26, 1–19.

Saunders, T., 1989. Trace fossils and sedimentology of a Late Cretaceous progradational barrier island sequence: Bearpaw-Horseshoe Canyon Formation transition, Dorothy, Alberta. M.Sc. Thesis, University of Alberta, 187 pp.

Saunders, T., Pemberton, S.G., 1986. Trace fossils and sedimentology of the Appaloosa Sandstone: Bearpaw-Horseshoe Canyon Formation transition, Dorothy, Alberta. Can. Soc. Petrol. Geol., Canada Field Trip Guide Book, 117 pp.

Saunders, T., MacEachern, J.A., Pemberton, S.G., 1994. Cadotte Member Sandstone: progradation in a boreal basin prone to winter storms. In: Pemberton, S.G., James, D.P., Wightman, D.M. (Eds.), Mannville Core Conference. Can. Soc. Petrol. Geol., Calgary, Alta, pp. 331–350.

Savrda, C.E., Bottjer, D.J., 1989. Trace fossil model for reconstructing oxygenation histories of ancient marine bottom waters: application to Upper Cretaceous Niobrara Formation, Colorado. Palaeogeogr. Palaeoclimatol. Palaeoecol. 74, 49–74.

Seike, K., 2007. Palaeoenvironmental and palaeogeographical implications of modern *Macaronichnus segregatis*-like traces in foreshore sediments on the Pacific coast of central Japan. Palaeogeogr. Palaeoclimatol. Palaeoecol. 252, 497–502.

Seike, K., 2009. Influence of beach morphodynamics on the distributions of the opheliid polychaete *Euzonus* sp. and its feeding burrows on a sandy beach: paleoecological and paleoenvironmental implications for the trace fossil *Macaronichnus segregatis*. Palaios 24, 799–808.

Seilacher, A., 1967. Bathymetry of trace fossils. Mar. Geol. 5, 413–428.

Seilacher, A., Aigner, T., 1991. Storm deposition at the bed, facies and basin scale: the geologic perspective. In: Einsele, G., Ricken, W., Seilacher, A. (Eds.), Cycles and Events in Stratigraphy. Springer, Berlin, pp. 249–267.

Shipp, R.C., 1984. Bedforms and depositional sedimentary structures of a barred nearshore system, Eastern Long Island, New York. Mar. Geol. 60, 235–259.

Short, A.D. (Ed.), 1999. Handbook of Beach and Shoreface Morphodynamics. John Wiley & Sons, Toronto, 379 pp.

Siggerud, E.I., Steel, R., 1999. Architecture and trace-fossil characteristics of a 10,000-20,000 year, fluvial to marine sequence, SE Ebro basin, Spain. J. Sediment. Res. 69, 36–383.

Snedden, J.W., Nummendal, D., 1991. Origin and geometry of storm-deposited sand beds in modern sediments of the Texas continental shelf. In: Swift, D.J.P., Oertel, G.F., Tillman, R.W., Thorne, J.A. (Eds.), Shelf Sand and Sandstone Bodies: Geometry, Facies and Sequence Stratigraphy. IAS Spec. Publ., vol. 14, pp. 283–308.

Swift, D.J.P., Niedoroda, A.W., Vincent, C.E., Hopkins, T.S., 1985. Barrier island evolution, middle Atlantic shelf, U.S.A. Part I: shoreface dynamics. In: Oertel, G.F., Leatherman, S.P. (Eds.), Barrier Islands. Mar. Geol., vol. 63, pp. 331–361.

Taylor, A., Goldring, R., Gowland, S., 2003. Analysis and application of ichnofabrics. Earth-Sci. Rev. 60, 227–259.

Thom, B.G., Roy, P.S., Short, A.D., Hudson, J., Davis Jr., R.A., 1986. Modern coastal and estuarine environments of deposition in southeastern Australia. In: Guide for Excursion 4A, 12th International Sedimentological Congress, Canberra, 279 pp.

Tonkin, N.S., 2012. Deltas. In: Knaust, D., Bromley, R.G. (Eds.), Trace Fossils as Indicators of Sedimentary Environments. Developments in Sedimentology, vol. 64. Elsevier, Amsterdam, pp. 507–528.

Vakarelov, B.K., Ainsworth, R.B., MacEachern, J.A., in press. Recognition of wave-dominated, tide-influenced shoreline systems in the rock record: variations from a microtidal shoreline model. Sediment. Geol. http://dx.doi.org/10.1016/j.sedgeo.2011.03.004.

Vossler, S.M., Pemberton, S.G., 1988. *Skolithos* in the Upper Cretaceous Cardium Formation: an ichnofossil example of opportunistic ecology. Lethaia 21, 351–362.

Vossler, S.M., Pemberton, S.G., 1989. Ichnology and palaeoecology of offshore siliciclastic deposits in the Cardium Formation (Turonian, Alberta, Canada). Palaeogeogr. Palaeoclimatol. Palaeoecol. 74, 217–229.

Walker, R.G., Plint, A.G., 1992. Wave- and storm-dominated shallow marine systems. In: Walker, R.G., James, N.P. (Eds.), Facies Models: Response to Sea Level Change. Geol. Ass. Can., St. John's, Geotext 1, pp. 219–238.

Wheatcroft, R.A., 1990. Preservation potential of sedimentary event layers. Geology 18, 843–845.

Wright, L.D., Chappell, J., Thom, B.G., Bradshaw, M.P., Cowell, P., 1979. Morphodynamics of reflective and dissipative beach and inshore systems: southeastern Australia. Mar. Geol. 32, 105–140.

Yang, B.C., Dalrymple, R.W., Chun, S.S., Johnson, M.F., Lee, H.J., 2008a. Tidally modulated storm sedimentation on open-coast tidal flats, southwestern Korea. In: Hampson, G.J., Steel, R.J., Burgess, P.M., Dalrymple, R.W. (Eds.), Recent Advances in Models of Siliciclastic Shallow-Marine Stratigraphy. SEPM Spec. Publ., vol. 90, pp. 161–176.

Yang, B.C., Gingras, M.K., Pemberton, S.G., Dalrymple, R.W., 2008b. Wave-generated tidal bundles as an indicator of wave-dominated tidal flats. Geology 36, 39–42.

Part IV

Deep-Marine Siliciclastic Systems

Chapter 20

Slopes

Stephen M. Hubbard,*,[1] James A. MacEachern† and Kerrie L. Bann‡
*Department of Geoscience, University of Calgary, Calgary, Canada, †Department of Earth Sciences, Simon Fraser University, Burnaby, Canada, ‡Ichnofacies Analysis Inc., Calgary, Canada
[1]Corresponding author: e-mail: shubbard@ucalgary.ca

1. INTRODUCTION

Deposits of the continental slope, and of other slopes associated with high-relief basin margins, contain the record of sedimentological processes critical to the understanding of sediment transport from continents into deep-sea basins. In many instances, slopes can be considered primarily as zones of sedimentary bypass, wherein coarse-grained sediment derived from the continent is funneled through conduits such as submarine canyons, ultimately accumulating as submarine fans on the basin-floor (Fig. 1). Extensive and regular mass-wasting events also play key roles in sediment remobilization and bypass on slopes. On many margins, contour currents are also significant, and help to shape the seascape. It is important to recognize that slopes can be as diverse as many better-documented depositional realms (e.g., marginal-marine settings, or basin-floors) and that, in many instances, accretion of slopes by widespread suspension settling of fine-grained particulate debris constitutes the dominant process.

Due to their prevalence as zones of sedimentary bypass or of suspended sediment settling, deposits of slopes are commonly dominated by fine-grained sediment. Major erosion surfaces that are excavated into fine-grained substrata and subsequently overlain by fine-grained material, are subtle, but may comprise the only record of high-energy processes responsible for the transport of sands and gravels across this depositional setting. In many instances then, slope regimes do not host extensive coarse-grained sedimentary accumulations, and this, in turn, has dictated the quality and abundance of data available for their study in two ways: slope deposits (1) have been traditionally under-represented in drill cores acquired for hydrocarbon exploration because they were typically considered to lack significant reservoir potential until the last two decades; and (2) are commonly poorly exposed in outcrop, owing to their recessive weathering character. By contrast, the deposits of more broadly sand-prone sedimentary environments of continental shelf or basin-floor affinity were often more extensively drilled or are better exposed in outcrop. For these

Developments in Sedimentology, Vol. 64. http://dx.doi.org/10.1016/B978-0-444-53813-0.00020-4

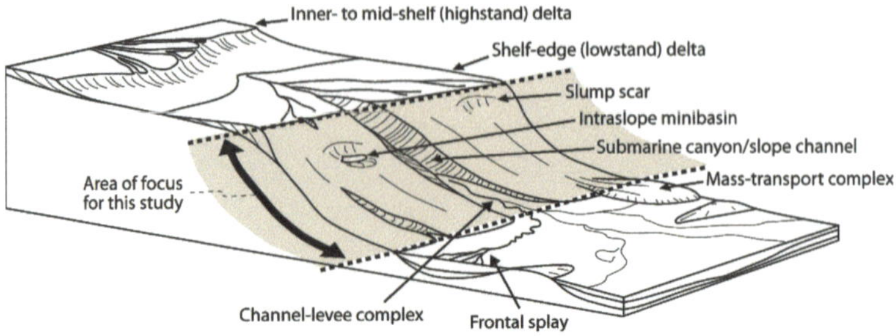

FIGURE 1 Schematic overview of a continental margin highlighting the various subenvironments. *Modified from Posamentier and Walker (2006).*

reasons, there has been a relative paucity of ichnological studies focused on the deposits of slope environments.

The integration of ichnological and sedimentological data is necessary in order to construct robust depositional models for ancient sedimentary successions. Over the past two decades, substantial sedimentological and geomorphological analyses of slope units have led to a superior understanding of these settings, including the dominant physical processes (e.g., Hodgson et al., 2011; Normark et al., 2009; Paull et al., 2005; Posamentier and Kolla, 2003; Prather et al., 1998). These studies provide the framework for integrating trace-fossil data, which will lead, ultimately, to the development of more robust facies models suited for interpretations of slope stratigraphy. In his classical work, Seilacher (1967) considered the *Zoophycos* Ichnofacies to be characteristic of the zone between the area of wave-influence (i.e., the shelf) and turbidite sedimentation (i.e., the basin-floor). Of course, we now better appreciate the breadth of sedimentary processes that impact the slope, including both depositional and bypassing turbidity currents. In popular bathymetric schemes, the *Zoophycos* Ichnofacies is typically portrayed as an intermediary between the *Cruziana* Ichnofacies and the *Nereites* Ichnofacies (cf. MacEachern et al., 2007a), a perception that does the ichnofacies a disservice. Reevaluation of the *Zoophycos* Ichnofacies by Seilacher (1978) and Frey and Seilacher (1980) showed that one of the major environmental controls typifying the *Zoophycos* Ichnofacies is lowered oxygen levels, associated with organic debris accumulating in largely persistent low-energy regimes. Such conditions are commonly characteristic of distal (outer) shelf settings as well as slope settings that lack persistent oceanic currents. In more dynamic slope environments, however, the resulting ichnological suites are more diverse, but commonly remain within the ethological constraints of the *Zoophycos* Ichnofacies.

It is now widely recognized that gravity flows, including turbidity currents, shape the bathyal realm (cf. Posamentier and Walker, 2006). Recent studies have revealed a marked variability in the character of bathyal depositional subenvironments (Fig. 1). The focus of this review includes submarine canyons, intraslope minibasins, slope channels, and levee complexes.

The recognition of substantial intraslope sand accumulations has led to a revolution in slope sedimentology and stratigraphy over the past two decades, spawned in large part by the development of predictive models by researchers in the hydrocarbon industry (e.g., Prather et al., 1998). Intraslope minibasins have been shown to be characterized by sand-rich fills in many instances and formed from a combination of processes such as salt withdrawal, mass-wasting, extensional rifting, and/or growth-fault development. Conduits, such as submarine canyons and gullies, also have been shown to contain ponded sandstone in many cases (e.g., Anderson et al., 2006; Paull et al., 2005). Additionally, channel-levee complexes and other channel subenvironments are common on the slope, representing the proximal *loci* of sand deposition associated with expansive basin-floor fan complexes (Fig. 1).

Acknowledging the variability in continental slopes and their deposits, this chapter focuses on areas of the slope associated with high current energies, many of which are sand prone (Fig. 1). This approach fosters the incorporation of high-quality subsurface and outcrop data. Furthermore, the integration of ichnological observations and interpretations complements developing sedimentological models and aims to be particularly useful to hydrocarbon explorationists and sedimentologists attempting to discern slope deposits and their stratigraphic architecture in the rock record. Of note, Wetzel et al. (2008) focused on bioturbation associated with contour-current-dominated slope deposits and correspondingly, this topic is not addressed in detail herein.

2. THE DISTRIBUTION OF INFAUNA ON SLOPES: INSIGHTS FROM MODERN BASIN MARGINS

The distribution of sea-bottom-dependant (epibenthic and endobenthic) organisms, both on the seabed and within the substrate in the Earth's oceans and peripheral seas, has been extensively documented (e.g., Heezen and Hollister, 1971; Sanders, 1968). Although not all of these organisms leave distinctive traces, their distribution provides insights into physical and chemical conditions across the bathyal realm. Tracemakers are variable, including holothurians, echinoids, polychaetes, echiurians, enteropneusts, gastropods, molluscs, seapens, and decapods (Heezen and Hollister, 1971). Overall faunal densities are typically observed to decrease with increasing water depth across basin margin slopes (e.g., Flach and Thomsen, 1998; Rowe et al., 1982; Stora et al., 1999), although in some instances, perturbations in this trend are notable (e.g., Gage et al., 2000). Species diversities can mimic this downward-decreasing trend of biomass abundance (e.g., Stora et al., 1999; Vetter and Dayton, 1998), but there are numerous exceptions. In certain instances, species diversity *increases* with water depth (e.g., Rowe et al., 1982; Vetter and Dayton, 1998), and other slopes are characterized by a gradual increase in overall species diversity to the mid-slope position, followed by a persistent decrease in this trend toward the abyssal depths (e.g., Gage et al., 2000; Stuart et al., 2003).

Faunal densities and diversities on slopes are significantly influenced by their proximity to sediment point sources on the adjacent shelf and shoreline, as well as by the presence of sediment/current conduits, such as submarine canyons (Cartes et al., 2010; Griggs et al., 1969; Morse and Beazley, 2008; Rowe, 1971; Vetter and Dayton, 1998). Factors that influence faunal distribution trends include strong currents, varied oxygenation of bottom and interstitial waters, focused sedimentation and erosion, and elevated organic enrichment (e.g., via primary production at the water surface and transported terrestrial detritus; Cartes et al., 2010; Gerino et al., 1999; Josselyn et al., 1983; Morse and Beazley, 2008; Rowe, 1971; Vetter and Dayton, 1998; Vetter et al., 2010; Wetzel, 2008).

Parameters controlling the distribution, diversity, and abundance of benthic organisms on the slope are numerous, as is typical of any depositional setting. Consequently, no single physico-chemical factor can be used to explain the diversity and abundance trends observed in the bathyal realm (Levin and Gage, 1998). Physical and chemical factors such as current energy, sedimentation rate, mass-wasting, pressure, light, substrate particle size, substrate consistency, oxygenation, food availability, temperature, and turbidity, as well as biological controls such as productivity, predation, and competitive displacement, can vary with depth and influence feeding behaviors (Carney, 2005; Flach and Thomsen, 1998; Gage and Tyler, 1991; Levin and Gage, 1998; Morse and Beazley, 2008; Puig et al., 2001; Rogers, 2000; Savrda et al., 1984; Stuart et al., 2003; Young and Richardson, 1998). In the vicinity of submarine canyons or gullies, increased food resources, oxygenation, and sedimentation rates commonly lead to conditions on the slope akin to those of shelf or coastal settings (Blake and Hilbig, 1994; Rhoads and Hecker, 1994). Thus, many slopes on the margins of continents or peripheral seas like the Mediterranean are characterized by species normally associated with shelf environments (e.g., Rhoads and Hecker, 1994; Stora et al., 1999; Wetzel, 2008). The resulting variability in organism distribution trends observed on basin-margin slopes indicates that species diversity and abundance develop in response to local conditions and that a fundamental response to water depth does not exist (Gage et al., 2000; Heezen and Hollister, 1971).

3. ICHNOFACIES IN BATHYAL SETTINGS

The ichnological characterization of ancient slope deposits constitutes a highly underdeveloped research topic. Likewise, even the basic sedimentological criteria defining slope deposits are vague compared to the adjacent shelf and basin-floor counterparts, often requiring expansive outcrops in order to recognize the large-scale features that might typically be regarded as diagnostic, such as mass-transport complexes (e.g., Armitage et al., 2009). Over the past two decades, high-resolution bathymetric images and 3D seismic surveys have enlightened sedimentologists and stratigraphers as to the characteristics and complexity of slope deposits, inspiring their recognition in both the subsurface and outcrops. Many of these intervals may otherwise have been misclassified as basin-floor fan or shelf deposits (cf. Gjelberg et al., 2001; Shultz and Hubbard, 2005; Sømme et al., 2009).

Despite the challenges associated with slope-deposit recognition, a number of studies have provided tremendous insight into ichnological trends on high-relief slope-to-basin-floor transects in ancient basins (e.g., Crimes, 1977; Heard and Pickering, 2008; Savrda et al., 2001). Many studies of shelf strata, as well as others focused on abyssal fan deposits, have incorporated analyses of slope units (e.g., Eyles et al., 1992; Uchman, 2009). Furthermore, some studies have focused on ichnological components from bathyal deposits (e.g., Bromley et al., 2003; Miller and Vokes, 1998; Savrda, 2012). In general, suites from these higher-relief slope regimes vary markedly in diversity at any one locality. Ambient suites yield reasonably diverse associations that fall into either distal expression of the *Cruziana* Ichnofacies or to high-diversity expressions of the *Zoophycos* Ichnofacies. These are commonly interstratified with low-diversity opportunistic suites associated either with turbiditic sands, turbiditic muds, or contourite-(oceanic-current)-generated muddy heterolithic facies. Although pelagic to hemipelagic deposition is not uncommon in slope regimes, ichnological associations lack the graphoglyptid suites diagnostic of the *Nereites* Ichnofacies. Graphoglyptid-bearing suites are generally regarded to represent the stable environmental conditions associated with the abyssal plain and probably represent compelling evidence of basin-floor positions (cf. Uchman and Wetzel, 2011, 2012). Trace fossils typical of facies deposited on the slope are numerous and include such elements as *Arenicolites*, *Asterosoma*, *Chondrites*, *Cosmorhaphe*, *Cylindrichnus*, *Gyrochorte*, *Diplocraterion*, *Helminthopsis*, *Neonereites*, *Nereites*, *Ophiomorpha*, *Palaeophycus*, *Phycodes*, *Phycosiphon*, *Planolites*, *Rhizocorallium*, *Rosselia*, *Scolicia*, *Skolithos*, *Spirophyton*, *Teichichnus*, *Thalassinoides*, and *Zoophycos* (Fig. 2).

3.1 *Zoophycos* Ichnofacies

The *Zoophycos* Ichnofacies is typically considered intermediate between ichnofacies typical of inner shelf or offshore deposits (i.e., the *Cruziana* Ichnofacies) and abyssal deposits (i.e., the *Nereites* Ichnofacies). Correspondingly, deposits of the bathyal zone are most commonly ascribed to the *Zoophycos* Ichnofacies (cf. MacEachern et al., 2007a, 2010; Seilacher, 1967). Low-energy shelves, particularly those that are mud-prone, however, are also typified by the *Zoophycos* Ichnofacies (cf. Pemberton et al., 2012; MacEachern et al., 2012). Shelfal expressions of the *Zoophycos* Ichnofacies, however, tend to comprise low-diversity suites of grazing and deposit-feeding structures, dominated by low-energy and/or facies-crossing components common to both the *Cruziana* Ichnofacies and the *Zoophycos* Ichnofacies (e.g., *Zoophycos*, *Helminthorhaphe*, *Phycosiphon*, *Cosmorhaphe*, *Chondrites*, *Planolites*, and *Thalassinoides*).

Higher diversity slope trace-fossil suites attributable to the *Zoophycos* Ichnofacies commonly comprise grazing structures and deposit-feeding structures produced by vagile and sessile deposit-feeding organisms. Overall intensities of bioturbation (as expressed with the Bioturbation Index, BI) are highly variable (BI = 0–5), depending upon the overall sedimentation rate and intercalation of gravity-flow mudstones. Ambient suites are typically characterized by high

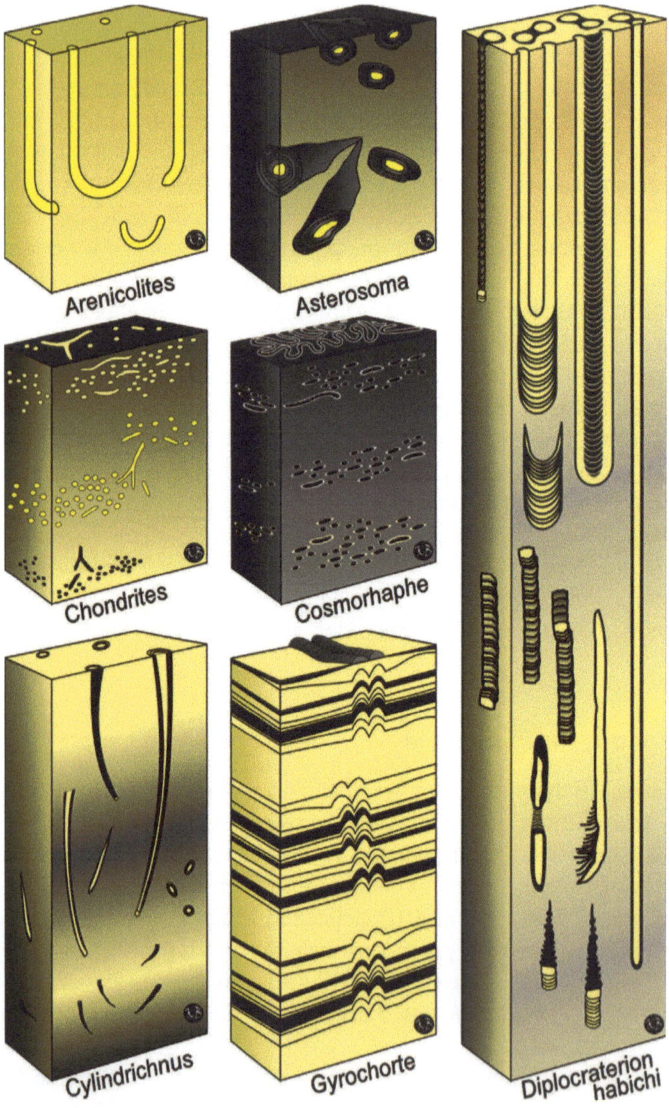

FIGURE 2 Trace fossils common to deep-water slope strata.

abundances and moderate to high diversities. Abundances and diversities can decline markedly in the presence of active turbiditic or contouritic sedimentation. Correspondingly, slope deposits tend to show marked changes in ichnological fabrics at the bed and bedset scale, reflecting *in loco* changes in depositional dynamics. Both deep-tier traces (e.g., *Thalassinoides*, *Zoophycos*, and *Chondrites*) and shallow-tier traces (e.g., *Cosmorhaphe*, *Helminthorhaphe*,

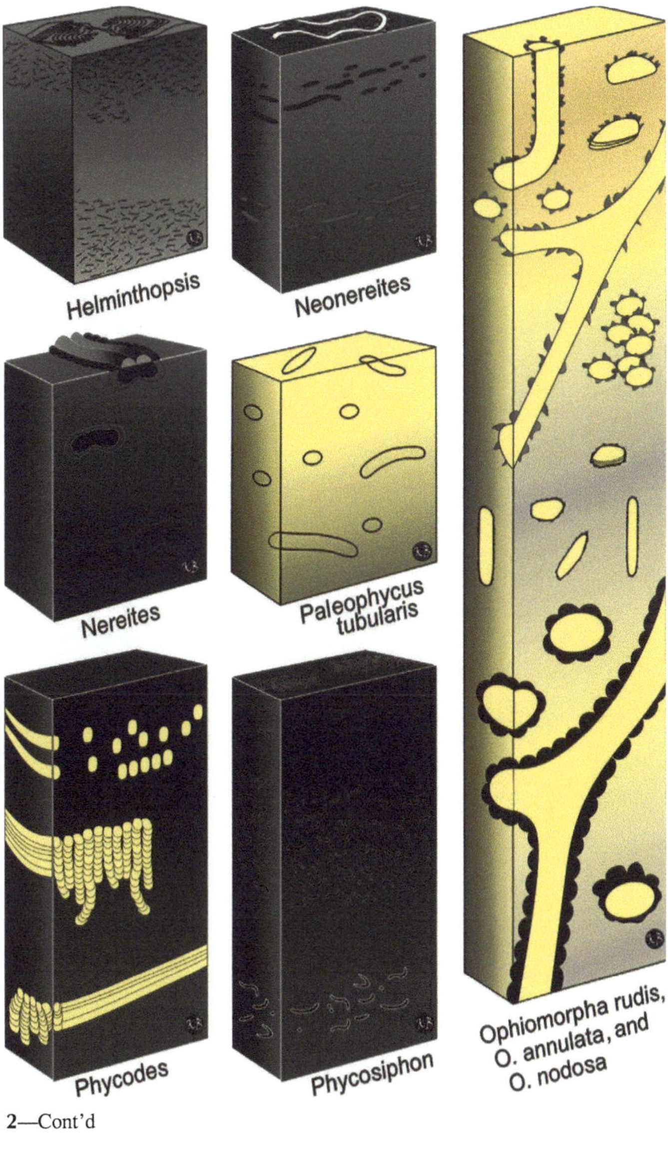

FIGURE 2—Cont'd

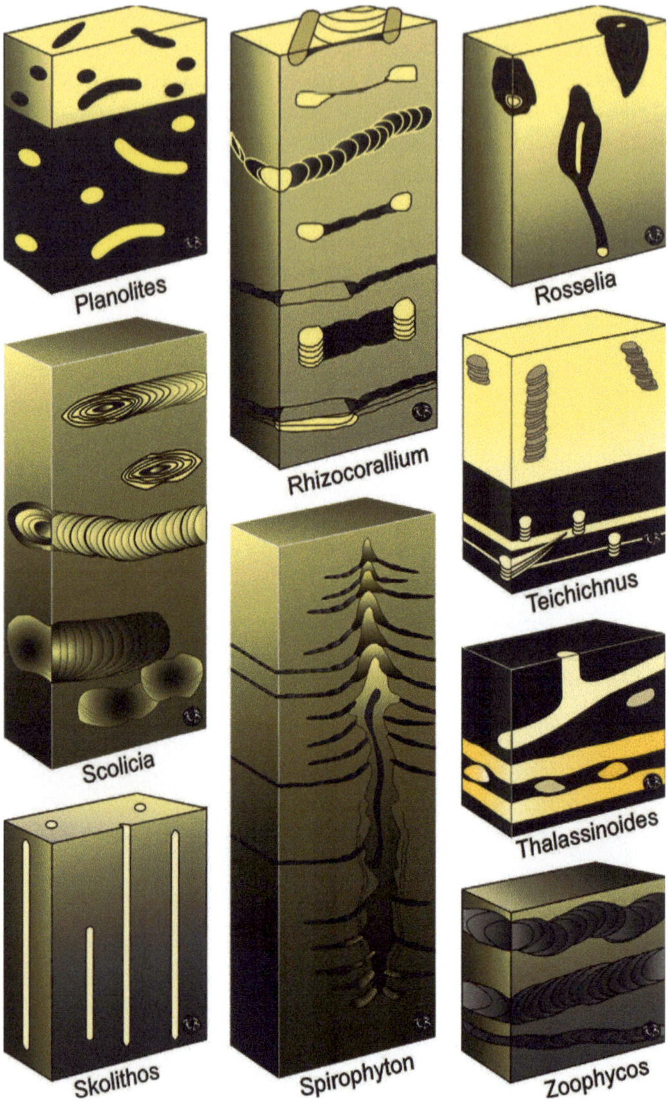

FIGURE 2—Cont'd

and *Phycosiphon*) represent important constituents. *Zoophycos*, *Spirophyton*, horizontal to subhorizontal meniscate mobile deposit-feeding traces such as *Scolicia*, *Taenidium*, and *Lophoctenium*, and traces that result from surface foraging such as *Neonereites*, are also characteristic.

A primary physico-chemical control on the archetypal *Zoophycos* Ichnofacies has typically been considered lowered oxygen levels, associated with muddy cohesive substrates (Frey and Seilacher, 1980; Seilacher, 1978). Therefore, it is not surprising that ancient deposits associated with the ichnofacies include those attributed to shelf and muddy slope settings, where the accumulation of abundant organic matter from suspension are characteristic (e.g., Olivero and Gaillard, 1996; Wetzel, 1984). However, as the supply of oxygen and food to the substrate increase, departures from this archetypal expression are to be expected. The resulting high diversity and high abundance suites may indicate parts of the slope swept by various oceanic currents (e.g., contour currents and density currents).

3.2 *Skolithos* Ichnofacies

Conduits for currents and sediment transport across continental slopes are associated with deposits typified by trace-fossil suites notably distinct from those of the background slope. Trace fossils more commonly attributed to high-energy, shallow-marine settings comprise important constituents of suites, manifested by localized to semi-regional occurrences of the *Skolithos* and *Glossifungites* ichnofacies (e.g., Hubbard and Shultz, 2008; Kane et al., 2007; Shultz and Hubbard, 2005).

Sandy, shifting (non-cohesive) substrates associated with persistently high energy, impacted by frequent and abrupt shifts in sedimentation and erosion, are more typical of nearshore sedimentary environments. However, a comparable set of physical conditions can also be achieved in some slope regimes, particularly submarine canyons, gullies, and other deep-marine channels (cf. Crimes, 1970; Frey, 1971; Frey and Howard, 1970; Frey et al., 1990; Hayward, 1976; Ineson, 1987). Food and oxygenated waters are transported throughout bathyal depths, particularly in areas where canyon heads are located adjacent to fluvio-deltaic point sources (e.g., Cartes et al., 2010; Morse and Beazley, 2008). Submarine canyon, intraslope minibasin, and lower-slope channel deposits have been shown to locally exhibit typically impoverished ichnological suites commonly assigned to the *Skolithos* Ichnofacies, which include *Skolithos*, *Arenicolites*, *Ophiomorpha nodosa*, *O. rudis*, *Diplocraterion*, *Cylindrichnus*, and *Rosselia* (e.g., Crimes, 1970, 1977; Crimes et al., 1981; Cummings and Hodgson, 2011; Heard and Pickering, 2008; Kane et al., 2007; Kern and Warme, 1974; Shultz and Hubbard, 2005).

3.3 *Glossifungites* Ichnofacies

Passively filled burrows excavated into firm, erosionally exhumed substrates are common on slopes, particularly in areas dominated by recurring, high-energy currents (e.g., Anderson et al., 2006; Callow et al., 2012a; Di Celma et al., 2010; Eyles et al., 1992; Hayward, 1976; Hubbard and Shultz, 2008; MacEachern et al., 2007b; Pemberton and MacEachern, 2005; Savrda et al., 2001). Such firmground suites are attributable to the *Glossifungites* Ichnofacies.

The pioneering work of Hayward (1976) featured the recognition of bioglyph-bearing (scratch-marked) firmground burrows (e.g., *Rhizocorallium*, *Skolithos* [*Tigillites*], and *Thalassinoides*) excavated into the walls of an Early Miocene submarine canyon succession. Similar firmground suites have been identified from submarine canyon margins in the West Ahken Field of the Nile Delta, and in the Mississippian Canyon, Gulf of Mexico (e.g., MacEachern et al., 2007b). Additionally, contour currents were deemed important in the generation of a firm substrate on the Pliocene Atlantic-margin continental-slope off the east coast of North America, showing that like marginal-marine manifestations of the ichnofacies, bathyal examples could be associated with widespread, sequence-stratigraphically significant discontinuities (Savrda et al., 2001). Hubbard and Shultz (2008) proposed a list of potential mechanisms for widespread firmground generation in the deep sea, with substrate exhumation most likely linked to (1) widespread turbidity current erosion, (2) large-scale slumps, and (3) contour currents (cf. Savrda et al., 2001).

Channelized turbidity currents and slumps are commonly focused, leading to localized effects on basin-margin slopes. Larger spatial extents may be anticipated with interconnected slope-channel systems. Nevertheless, erosion and colonization of firm substrates may impact the sea floor over limited areas, such that the stratigraphic significance of the *Glossifungites* Ichnofacies is likely to be minimal. Indeed, submarine canyon examples of this ichnofacies are, themselves, spatially limited (e.g., Anderson et al., 2006; Di Celma et al., 2010; Hayward, 1976).

An alternative utility of passively filled firmground burrows in the bathyal realm could be as indicators of sediment bypass. Specifically, sand-infilled burrows (in some instances) record the passage of coarse-grained material through an erosional conduit such as a submarine canyon or channel, without deposition of a coarse-grained lag. The slope is commonly not a locus of deposition for coarse-grained detritus; background strata and conduit fill typically consist of conformable siltstone and mudstone. Alternatively, channels may be filled with siltstone- and mudstone-dominated mass-transport deposits (Fig. 3). In either case, the only evidence of coarse-grained sediment in the system may be recorded within the burrow fills (Fig. 3). As such, such trace fossils could be considered "tubular turbidites", in a manner similar to the record of storm events preserved in "tubular tempestites" (Tedesco and Wanless, 1991).

3.4 *Cruziana* Ichnofacies

The archetypal *Cruziana* Ichnofacies is typical of fully marine settings characterized by unconsolidated cohesive to semi-cohesive muddy substrates generated in areas of low to moderate energy. Common depositional settings include the distal lower shoreface, offshore, and inner shelf (MacEachern and Bann, 2008; MacEachern et al., 2007a; Pemberton et al., 1992). The archetypal expression of the

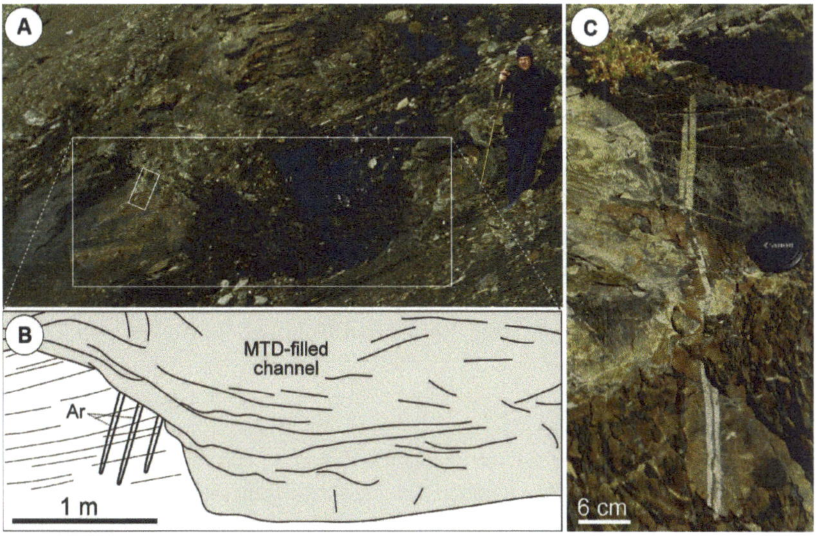

FIGURE 3 Trace fossils as indicators of sediment bypass (tubular turbidites) in slope strata. (A) Photograph and (B) line-drawing of a slope channel filled with fine-grained mass-transport deposits (MTD) in the Late Cretaceous Cerro Toro Formation, Chile (Hubbard et al., 2008). *Arenicolites* (*Ar*) subtend from the margin of the channel into background slope strata. (C) Close-up of *Arenicolites* from area outlined by small box in (A). Despite the fact that the fill of the channel is devoid of sandstone, the trace fossils were passively filled by sand from bypassing turbidity currents.

ichnofacies is represented by high-diversity suites corresponding to a predominance of deposit-feeding strategies and subordinate grazing behaviors.

Trace-fossil suites from units deposited in moderate to low-energy slope environments are commonly dominated by similar deposit-feeding ethologies, leading to their assignment to the archetypal *Cruziana* Ichnofacies or to distal expressions of the *Cruziana* Ichnofacies (cf. MacEachern and Bann, 2008; MacEachern et al., 2007a). Slope environments include areas of relative quiescence, at least periodically, particularly on the upper to middle slope (e.g., Ballance et al., 1984; Savrda et al., 2001), within channels (e.g., Eyles et al., 1992; Phillips et al., 2011), interchannel areas (e.g., Buck and Bottjer, 1985), areas influenced by low-energy bottom currents (e.g., Eyles et al., 1992; Wetzel et al., 2008), confining margins of intraslope topographic depressions (e.g., Phillips et al., 2011), and intraslope minibasins (e.g., Shultz and Hubbard, 2005).

Occurrences of suites attributable to the *Cruziana* Ichnofacies are somewhat impoverished compared to those typifying the offshore and distal lower shoreface. Trace-fossil diversities are typically less than 6 recognizable ichnogenera in most deposits. The most commonly documented trace fossils include *Chondrites*, *Palaeophycus*, *Planolites*, *Taenidium*, *Teichichnus*, *Thalassinoides*, and *Zoophycos*. *Asterosoma*, *Helminthoida*, *Ophiomorpha*, *Phycodes*, *Phycosiphon*, *Rhizocorallium*, and *Schaubcylindrichnus* are locally important, for instance, in the Late Cenozoic Yakataga Formation, Alaska (Eyles et al., 1992), the Pleistocene of the New Jersey margin (Savrda et al., 2001), and the Late Cretaceous Tres Pasos Formation of Chile (Shultz and Hubbard, 2005). Slope examples of the *Cruziana* Ichnofacies are most common in areas where oxygen and food are readily distributed to deeper-water settings, such as tectonically active slopes associated with narrow shelves, areas where seasonal coastal down-welling of cold waters is significant, glacially influenced slopes, slopes dominated by strong contour currents, regions where surface-water productivity is high, and areas proximal to conduits that focus currents and transfer sediment from the shelf to the basin-floor (e.g., Buck and Bottjer, 1985; Eyles et al., 1992; Ineson, 1987; Savrda et al., 2001; Shultz and Hubbard, 2005; Wetzel, 2008).

3.5 *Nereites* Ichnofacies

As described above, slope environments are dynamic, commonly characterized by quiescent conditions interrupted by mass-wasting, strong currents, and recurring high-energy, erosive sediment-gravity flows in some instances. Owing to these factors, coupled with a relative abundance of food resources, deposit- and filter-feeding organisms are expected locally on the slope. Correspondingly, suites attributable to the *Nereites* Ichnofacies, and specifically graphoglyptids, are not widespread in deposits of slope affinity. The *Nereites* Ichnofacies is common to settings characterized by continuous yet exceedingly slow rates of sediment deposition from suspension. These conditions may be interrupted by periodic out-sized gravity flows, but their frequency must be less than that required for substrate recolonization by the ambient benthic community. Conditions favorable for trace-fossil suites attributable to the *Nereites* Ichnofacies are common to basin-floor depositional environments (e.g., Crimes et al., 1981; Leszczyński and Seilacher, 1991; Uchman, 2001; Uchman and Wetzel, 2011, 2012). However, deep-sea drill cores have shown that the ichnofacies is largely confined to fine-grained substrates that include distal turbidites; it is largely absent from great expanses of the abyssal plain where sedimentation is more or less constant and low (e.g., Scholle et al., 1983). It is probable that, in many instances, the lack of lithological variability within slope strata strongly impacts graphoglyptid preservation (Seilacher, 1977; Uchman, 2001), and where they do occur, such homogeneity has a negative impact on our ability to recognize them.

Slope settings, where suites attributable to the *Nereites* Ichnofacies have been documented, include distal levees associated with large channel systems, channel-margin deposits, minibasins characterized by regular turbidity current input, and the channel-lobe transition zone (e.g., Callow et al., 2012a; Cummings and Hodgson, 2011; Heard and Pickering, 2008; Hubbard and Shultz, 2008; Kane et al., 2007). These environments of deposition are all present in close proximity to basin-floor depozones, where the *Nereites* Ichnofacies is considered diagnostic (Seilacher, 1967). Unfortunately, previous work has not focused on the range of BI values, uniformity of burrow distributions, and proportions of ichnological elements, and this remains an important area of future research. The *Nereites* Ichnofacies is not observed in upper slope or shelf deposits (cf. MacEachern et al., 2007a).

4. SLOPE DEPOSITIONAL ENVIRONMENTS—CASE STUDIES

4.1 Submarine Canyon Fill, Early Miocene Tirikohua Formation, Northland, New Zealand

Early Miocene strata of the Waitakere Group crop out on the west coast of Northland, New Zealand (Fig. 4), and contain two submarine canyons that were cut into bathyal volcanogenic turbidites of the Nihotupu Formation (Hayward, 1976). Canyon margins are overlain by outer neritic to upper bathyal canyon-floor deposits of the Tirikohua Formation. Paleobathymetric assessments have been made on the basis of molluscan and foraminiferal assemblages (Murray, 1973). The canyon margins are interpreted to have been incised during a relative base-level fall associated with tectonic uplift and block faulting of the margins of an inter-arc basin. Some 30 km from these localities, toward Auckland, nearly time-equivalent deposits of the Waitemata flysch possess suites of the archetypal *Nereites* Ichnofacies (including graphoglyptids), consistent with basin-floor depositional positions (cf. Uchman and Wetzel, 2012; Wetzel and Uchman, 2012).

The underlying Nihotupu Formation consists of weakly and sporadically burrowed (BI = 0–3, mainly BI = 0–2) turbiditic mudstones and sandstones, as well as conglomeratic debris-flow deposits, all of volcanogenic origin. Andesitic pillow piles also are locally present. Biogenically reworked units are generally isolated and local. Ichnogenera identified by Hayward (1976) are limited to *Skolithos* (called '*Tigillites*' but synonymized with *Skolithos* by Alpert (1974) and presently broadly accepted), *Thalassinoides*, *Planolites*, and *Nereites* (=*Scalarituba*). In addition, *Chondrites* occurs at the Tirikohua Point locality.

4.1.1 Submarine Canyon Margin

The submarine canyons in the area range from 600 to 2000 m wide and 80 to 150 m deep. The canyon margins correspond to a sequence boundary (Fig. 4A) and are commonly demarcated by widespread palimpsest ichnological suites, representing firmground substrate conditions (Fig. 4B–D). No true borings were identified from the surface (Hayward, 1976). Firmground burrows are present along overhangs, locally possess bioglyphs (scratch marks), and are passively infilled with sand from the overlying Tirikohua Formation. The sequence boundary shows BI = 1–3, with firmground *Rhizocorallium*, *Thalassinoides*, *Spongeliomorpha*, and *Skolithos* of the *Glossifungites* Ichnofacies. A few simple inclined to horizontal tubes may be attributable to firmground *Palaeophycus*. Studies of the modern Scripps Submarine Canyon (Warme et al., 1971) and some New England canyons (Dillon and Zimmerman, 1970) display similar ichnological expressions (cf. Hayward, 1976).

FIGURE 4 Ichnological characterization of a submarine canyon margin, Early Miocence strata of Northland, New Zealand. (A) General exposure of the canyon margin sequence boundary (yellow arrows) at Tirikohua Point. (B) Firmground *Rhizocorallium* (*Rh*) attributable to the *Glossifungites* Ichnofacies. The omission suite is associated with the sequence boundary shown in (A) excavated

4.1.2 Canyon-Floor Fill

The overlying Tirikohua Formation is interpreted to record canyon-floor deposits that accumulated above the sequence boundary in outer neritic to upper bathyal environments (Hayward, 1976). The succession consists of proximal, thick-bedded turbiditic sandstones and siltstones, intercalated with grain-flow derived conglomerates. Trace fossils are absent from the well-rounded cobble to pebble conglomerates but are sporadically distributed (BI = 0–3) in sandstone and siltstone units. The proximal sandy turbidites of the canyon floor contain allochthonous wood fragments with abundant *Teredolites*, presumably reflecting basinward transport of bored logs from shallow-water settings (Fig. 4E). *In-situ* trace fossils include fugichnia, *Ophiomorpha* (*O. rudis*), *Skolithos*, *Thalassinoides*, *Planolites*, *Scolicia*, and *Nereites* (=*Scalarituba*). Much of the suite records post-turbidite recolonization rather than bioturbation of the substrate during persistent periods of quiescence. The predominance of dwellings and deposit-feeding structures indicates that the slope setting was exposed to a persistent supply of oxygen and food, which was interspersed by periodic sediment gravity flows as the submarine canyons filled (Hayward, 1976).

4.2 Slope Deposits, Middle Eocene Ainsa-Jaca Basin, Spain

A spectrum of deposits that record deep-water environments is preserved in Middle Eocene strata in the Ainsa-Jaca Basin of the Spanish Pyrenees (Heard and Pickering, 2008; Uchman, 2001). The sedimentological and architectural characterization of these deep-water units is well established and provides a framework for ichnological analyses (e.g., Clark and Pickering, 1996; Mutti and Normark, 1987). Upper basin slope and slope gully through to fan-fringe and basin-floor deposits have been reported from the succession (Heard and Pickering, 2008), although only the most proximal of these are discussed herein: basin slope, slope gully, and canyon fill (Fig. 5).

In general, proximal slope deposits are characterized by low-diversity trace-fossil suites, dominated by post-depositional forms (Heard and Pickering, 2008). Mudstone- and siltstone-dominated ambient basin-slope deposits are highly bioturbated (up to BI = 3), with an overall trace-fossil diversity of 14 ichnogenera (Fig. 5). Of these, 10 forms constitute the post-depositional elements associated with the sandstone beds, and two forms are regarded as pre-depositional graphoglyptids (i.e., *Belorhaphe zickzack* and *Paleodictyon strozzii*). Basin-slope deposits are cross-cut by erosional gullies, filled with sand-dominated facies characterized by trace-fossil diversities that range from 5 to

---

into muddy beds of the Nihotupu Formation. (C) Firmground suite consisting of *Rhizocorallium* (*Rh*) and *Thalassinoides* (*Th*) marking the same discontinuity as shown in (B). (D) Firmground *Thalassinoides* (*Th*) marking the same discontinuity as shown in (B) and (C). (E) Burrowed sandstones of the Tirikohua Formation showing allochthonous wood containing *Teredolites* (*Ter*) and *in-situ Ophiomorpha rudis* (*Or*) and *Scolicia* (*Sc*).

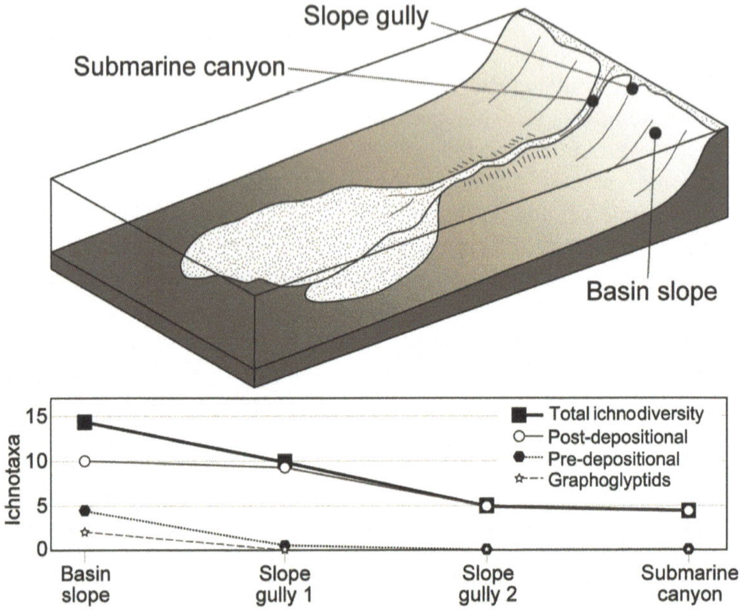

FIGURE 5 Summary of trace-fossil distribution within background slope, slope gully, and submarine canyon deposits of the Eocene Ainsa Basin, Spain *(upper schematic redrawn, and data compiled, from Heard and Pickering, 2008).*

10 ichnogenera (Fig. 5). Post-depositional trace fossils are associated with BI=2–3. A single submarine canyon-fill succession was assessed by Heard and Pickering (2008) and consists of mass-transport deposits associated with a post-depositional suite of *Ophiomorpha annulata, O. rudis*, *Thalassinoides suevicus*, and *Thalassinoides* isp.

Heard and Pickering (2008) speculate that the trace-fossil suites associated with high-energy canyon- and gully-fills are largely a consequence of (1) the low preservation potential of pre-depositional forms in these conduits due to erosion and (2) the lack of stable conditions in these settings, which favors opportunistic burrowing strategies. Their comprehensive analysis shows that proximal fine-grained slope units, characterized by active turbiditic accumulation, possess relatively impoverished trace-fossil assemblages that lack the diverse graphoglyptid-rich suites typifying lithologically similar facies in basin-floor fan subenvironments (cf. Uchman and Wetzel, 2012).

4.3 Ambient Slope Deposits, Cretaceous Nise Formation, Norwegian Shelf

The Late Cretaceous Nise Formation of offshore Norway forms part of the K85–K90 Sequence (Vergara et al., 2001), and is well developed in the Vøring and Møre basins. In these Norwegian Sea basins, the Late Jurassic basement structural template controlled Early Cretaceous deep-water sedimentary systems in a manner similar to the North Sea Basin. In the Vøring Basin, the Nise Formation may exceed 1200 m in thickness and is mainly interpreted as deep-sea submarine fan deposits (e.g., Fjellanger et al., 2005; Knaust, 2009; Martinsen et al., 2005). Successions in those localities are dominated by sediment gravity-flow deposits, with thoroughly bioturbated zones interstratified with bedded turbidites and other mass-flow units.

By contrast, the Nise Formation in the Ellida area (6405/7-1), lying further to the south within the Vøring Basin, possesses few turbidites or turbiditic/contouritic mudstones. Facies associations there are dominated by pervasively bioturbated silty mudstones, sandy mudstones, and muddy sandstones (Fig. 6). These successions are interpreted to record upper to lower slope settings, dominated by background or ambient (non-event) conditions. These ambient facies contrast markedly with the few intercalated turbidite and contourite heterolithic mudstone units.

4.3.1 Event-Generated Heterolithic Facies

The event-style deposits in the Ellida well are characterized by thin (cm-scale) heterolithic sandstone/siltstone and claystone composite bedsets (e.g., Fig. 6B), as well as graded to stratified lamina-sets displaying a characteristic pinstriped appearance. The sandstone layers are of mm-scale, and are delicately parallel laminated with minor current ripples. Traction structures are generally attributed to dilute turbidity currents, although some may record contour-current reworking. Sandstones are characterized by BI = 0–1, with isolated fugichnia and *Phycosiphon*, and normally grade to largely unburrowed silty claystones to silt-poor claystone layers, interpreted as rapidly emplaced T_e turbidite divisions. Trace fossils are sporadically distributed (BI = 0–2) and occur as thin bands of diminutive *Chondrites*, small- and large-diameter *Planolites*, and minor *Phycosiphon*. Locally, mud-filled *Zoophycos*, *Schaubcylindrichnus* (*S. freyi*), and diminutive *Asterosoma* are also present. Overall, trace-fossil suites are characterized by a low diversity and record rapid reoccupation of newly available muddy substrates by opportunists following event sedimentation. Prevailing elements are predominantly shallow-tier forms.

4.3.2 Ambient, Bioturbated Mudstone Facies

Ambient (background) facies of the Nise Formation, by contrast, consist of thoroughly biogenically admixed clay, silt and sand, predominantly expressed as silty and sandy mudstones (Fig. 6A and B) that locally pass into muddy sandstones (Fig. 6C and D). Sandy and silty mudstones are pervasively bioturbated (BI = 4–5), and no primary sedimentary structures are preserved. The ichnological suite comprises *Phycosiphon*, *Chondrites*, *Planolites*, *Zoophycos*, robust *Scolicia*, *Helminthopsis*, *Cosmorhaphe*, *Taenidium*, *Thalassinoides*, *Schaubcylindrichnus* (*S. freyi*), and *Rhizocorallium*. Locally, these suites also include isolated *Asterosoma*, *Spirophyton*, *Phoebichnus*, and *Nereites*. These ambient suites constitute highly diverse expressions of the *Zoophycos* Ichnofacies. The bioturbated mudstone facies are interpreted to reflect the slope margin

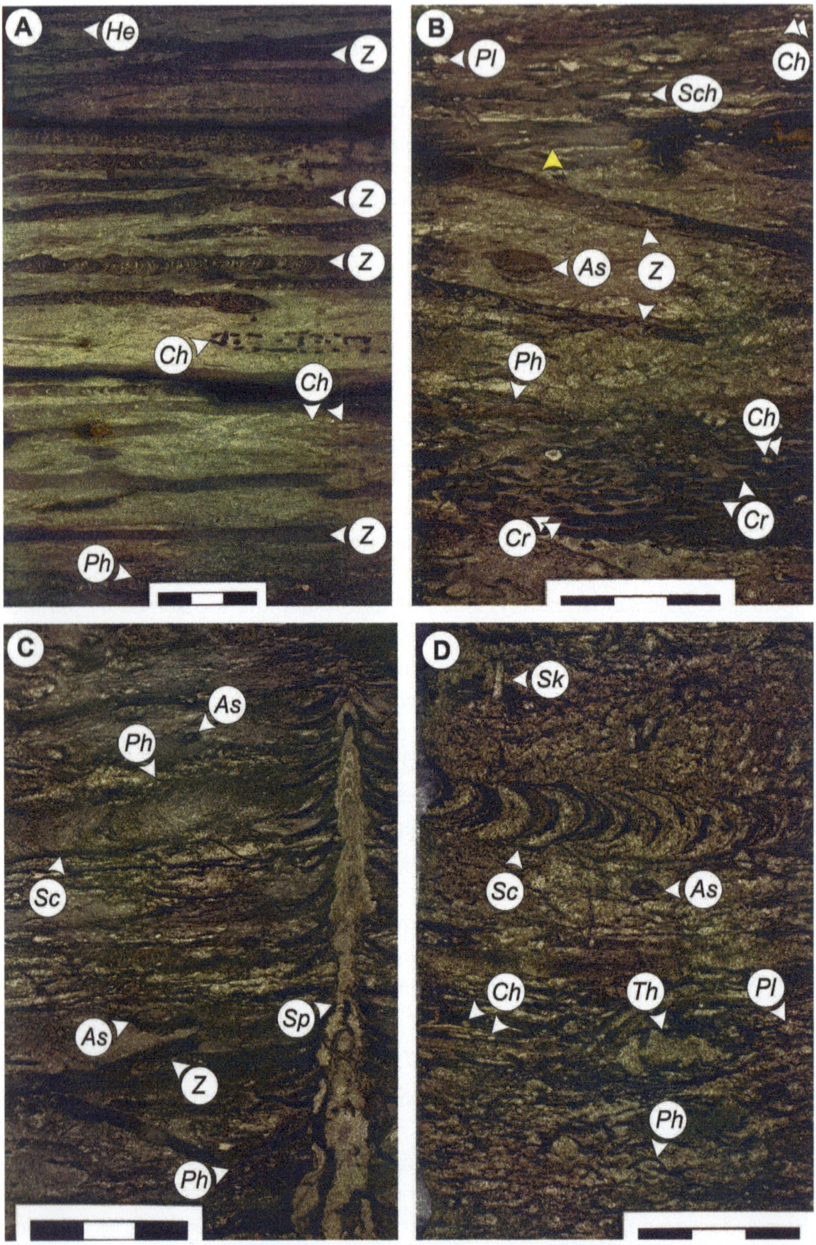

FIGURE 6 Overview of ambient slope facies from the Cretaceous Nise Formation of the Norwegian Sea. Each scale-bar shows 3 cm. (A) Bioturbated (BI = 5) silty mudstone showing pervasively burrowed background facies containing *Helminthopsis* (*He*) and *Phycosiphon* (*Ph*), overprinted by deeper-tier *Chondrites* (*Ch*) and *Zoophycos* (*Z*). (B) Bioturbated sandy mudstones showing general BI = 5, with a zone of BI = 4 toward the top. Yellow arrow points to remnant Bouma T_e mud drape

to upper slope but could extend to lower slope positions, particularly where oceanic currents were sufficiently energetic to supply food and oxygen to the bed. Such pervasive and uniformly bioturbated mudstones are consistent with settings prone to low sedimentation rates largely unaffected by turbiditic events.

4.3.3 Ambient, Bioturbated Muddy Sandstone

Pervasively bioturbated (BI = 5–6) muddy and locally glauconitic sandstones are also common in the Nise Formation. Sand ranges from lower fine to upper medium in grain size. Trace fossils are uniformly distributed and overprint one another. The suite is dominated by *Scolicia*, *Zoophycos*, *Asterosoma*, *Thalassinoides*, *Chondrites*, *Planolites*, *Phycosiphon*, *Palaeophycus*, *Helminthopsis*, *Cosmorhaphe*, and *Schaubcylindrichnus* (*S. freyi*), with rare *Ophiomorpha* (*O. rudis*) as well as very rare *Nereites* and *Spirophyton*. The presence of robust *Scolicia*, muddy *Zoophycos*, small *Asterosoma*, and possible *Nereites* marks this as a sandy, high-diversity expression of the archetypal *Zoophycos* Ichnofacies. The substrate consistency and grain size of the facies probably limited the abundance of surface grazing structures, which lends the unit an appearance more proximal than is likely warranted by its actual depositional position. Nevertheless, the abundance of grazing structures in these sandy intervals is atypical of shallow-marine sandy settings. Sandstone-dominated environments in more proximal positions (e.g., inner shelf and distal shoreface) typically display suites attributable to the archetypal *Cruziana* Ichnofacies, proximal expressions of the *Cruziana* Ichnofacies and distal expressions of *Skolithos* Ichnofacies (cf. MacEachern and Bann, 2008; MacEachern et al., 2007a, 2012).

Distal depositional positions are supported by the high abundance of grazing and low-energy deposit-feeding traces, despite the coarser grain sizes within the facies. Additionally, few of the biogenic structures present in the suite facilitate thorough churning of the sediment; bioturbation intensities of BI = 5–6, therefore, indicate that the deposits accumulated through slow, gradual sedimentation (cf. Wetzel et al., 2008). This may be distinctive of ambient slope settings, wherein oceanic currents (geostrophic, contour, etc.) are responsible for slow transport of material as bedload and are also a source of food and oxygen. Much of the mud admixed within the sand may have been introduced biogenically (e.g., fecal strings, fecal pellets, and mucous linings), rather than through suspended sediment settling.

---

containing isolated burrows that cross-cut it from higher levels. The ambient suite is diverse, and includes *Cosmorhaphe* (*Cr*), *Phycosiphon* (*Ph*), *Zoophycos* (*Z*), *Asterosoma* (*As*), *Chondrites* (*Ch*), and *Planolites* (*P*). (C) Bioturbated muddy sandstone showing BI = 5. The ichnological suite includes *Phycosiphon* (*Ph*), *Asterosoma* (*As*), *Spirophyton* (*Sp*), *Scolicia* (*Sc*), and *Zoophycos* (*Z*) and is attributable to the *Zoophycos* Ichnofacies. (D) Bioturbated (BI = 5) muddy sandstone with *Skolithos* (*Sk*), *Scolicia* (*Sc*), *Asterosoma* (*As*), *Thalassinoides* (*Th*), *Chondrites* (*Ch*), and abundant *Phycosiphon* (*Ph*) attributed to the *Zoophycos* Ichnofacies.

4.4 Slope Minibasin Deposits, Cretaceous Tres Pasos Formation, Chile

Deposits of a slope system are preserved in the Late Cretaceous Tres Pasos Formation of southern Chile, recording infill of the Magallanes foreland basin (Hubbard et al., 2010). A series of slope clinoforms, characterized by hundreds of meters of bathymetric relief and lengths >25 km, are well exposed in outcrop (Fig. 7A). The upper parts of the clinoforms (i.e., the shelfal top-set deposits) are characterized by shelf-edge deltaic complexes, with trace-fossil suites primarily assigned to the *Skolithos* and *Cruziana* ichnofacies (Covault et al., 2009; Hubbard et al., 2010). Slope strata are dominated by mudstone, punctuated locally by sandstone deposits. A sand-prone deposit at El Chingue Bluff, ~60 m thick and characterized by upward coarsening and bed thickening, is present along a clinoform surface (Shultz and Hubbard, 2005). This anomalous unit is interpreted to record progradation of turbiditic deposits into a growth-fault-controlled intraslope minibasin. The minibasin formed in a middle- to lower-slope position.

Thinly interbedded turbidites (5–10 cm thick; cf. Bouma, 1962) yield two distinctive trace-fossil suites (Fig. 7B–E). Mudstone layers characterized by Bouma T_d and T_e divisions contain *Chondrites*, *Planolites*, and *Thalassinoides* (Fig. 7B). Sandstone beds possess *Arenicolites*, *Glockerichnus*, *Ophiomorpha annulata*, *Palaeophycus*, *Phycosiphon*, *Planolites*, *Spirophyton*, *Taenidium*, *Teichichnus*, *Thalassinoides*, and *Zoophycos* (Fig. 7C–E). Shultz and Hubbard (2005) assigned these trace-fossil suites to the *Zoophycos* Ichnofacies. The variation in trace-fossil assemblages was attributed to the changes in food and oxygen availability in the slope environment, with more abundant resources introduced to the setting by higher energy turbidity currents.

Thick-bedded, high-density turbidites (cf. Lowe, 1982) are increasingly significant toward the top of the succession. *Chondrites*, *Glockerichnus*, *Planolites*, *Spongeliomorpha*, and *Thalassinoides* are moderately abundant within thin siltstone interbeds (Fig. 7B and C). Biogenic reworking of turbiditic sandstone is reflected by the presence of *Arenicolites*, *Chondrites*, *Gordia*, *Ophiomorpha annulata*, *O. nodosa*, *Palaeophycus*, *Phycodes*, *Phycosiphon*, *Planolites*, *Skolithos*, *Taenidium*, *Teichichnus*, *Thalassinoides*, and *Zoophycos* (Fig. 7D). This suite is assigned to the archetypal *Cruziana* Ichnofacies, although a series of trace-fossil forms typical of the *Skolithos* Ichnofacies (e.g., *Arenicolites*, *O. nodosa*, and *Skolithos*) are notable in discrete beds (Shultz and Hubbard, 2005). In general, the trace-fossil suite is more comparable to ichnofacies typical of shelf and distal shoreface environments. Organisms survived in slope settings that contained sufficient oxygen levels and food resources yet were also adaptable to recurring high-energy sediment gravity flows and associated sediment influx (cf. Crimes, 1977). Shultz and Hubbard (2005) compared this adaptation to fluctuating physical conditions to that experienced by organisms in shelf settings that are subjected to periodic storms (e.g., Frey and Pemberton, 1987; Howard and Frey, 1984; Pemberton and MacEachern, 1997).

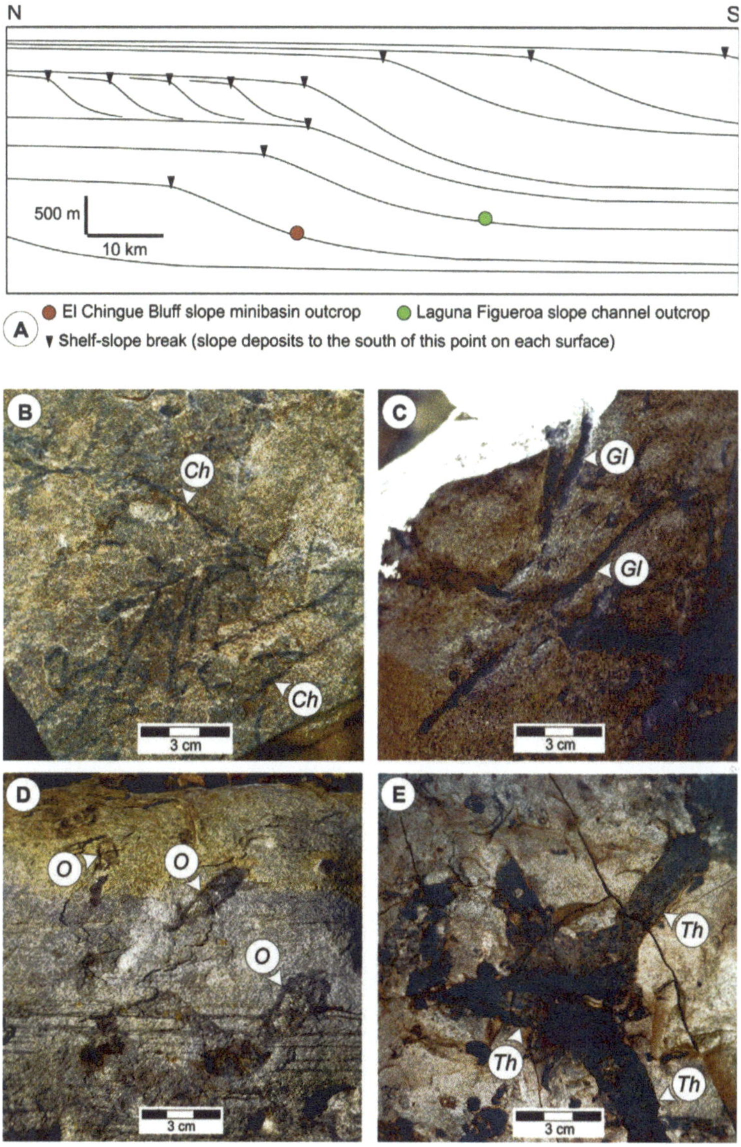

FIGURE 7 Summary of slope-minibasin deposits in the Late Cretaceous Tres Pasos Formation at El Chingue Bluff, Magallanes Basin, Chile. (A) Depositional dip profile context for the outcrops of the formation featured in this study *(modified from Hubbard et al., 2010)*. The deposits comprise low- to moderate-diversity trace-fossil suites within turbiditic sandstone and mudstone, including (B) *Chondrites* (*Ch*) on the top of a turbidite bed, (C) *Glockerichnus* (*Gl*) on a turbidite sole, (D) *Ophiomorpha* (*O*) in a turbidite bed characterized by Bouma $T_{b\text{–}c}$ divisions, and (E) *Thalassinoides* (*Th*) exposed on the base of a thick-bedded turbidite.

4.5 Lower-Slope Channel Deposits, Cretaceous Tres Pasos Formation, Chile

Channel deposits in the Tres Pasos Formation are present toward the toe of a slope clinoform (Fig. 7A), first described by Fildani et al. (2009) and Hubbard et al. (2010). Individual channels are 300–400 m wide and 15–25 m thick (Fig. 8). The basal scour surfaces are commonly draped with siltstone, which record deposition from the tails of high-energy turbulent currents that largely bypassed the channel setting (cf. Mutti and Normark, 1987). The main

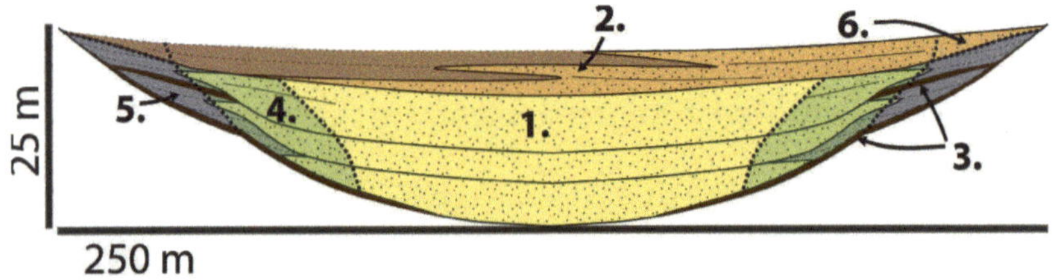

	1. Axis	2. Axial abandonment	3. Bypass drape	4. Proximal margin	5. Distal margin	6. Distal abandonment
Physical sedimentary structures	Thick-bedded turbidites (T_{a-b})	Thick- to thin-bedded turbidites (T_{a-d})	Thin-bedded turbidites (T_{b-e})	Thin-bedded turbidites (T_{a-e})	Thin-bedded turbidites (T_{a-e})	Thin-bedded turbidites (T_{a-e})
BI	sst = 0 ms = 0	sst = 0–5 ms = 0–4	sst = 0–1 ms = 1–3	sst = 0–2 ms = 0–2	sst = 0–5 ms = 0–3	sst = 0–2 ms = 0–1
Ichnogenera	NA	sst: *Op* ms: *Th*	sst: *Pl m., Ph* ms: *He, Neo, Pl, Sk*	palimpsest: Di h., Th	sst: *Ast, Gyr, Op, Pa t., Pa h., Ph, Pl m., Sk, Te* ms: *Co, He, Neo, Pl, Sc, Sk, Th*	sst: *Ar, Ast, Di h., Gyr, fu, Neo, Op, Pa t., Pl, Pl m., Rh, Ro, Sc, Sk, Te* ms: *Ast, Neo, Pl, Sk, Te, Th*
Ethology	NA	Common to abundant opportunists	Uncommon grazers and deposit feeders	Common opportunists (palimpsest only)	Rare to common grazers and opportunists	Rare grazers, common and diverse deposit feeders and opportunists

FIGURE 8 Slope-channel model (top) and physical and biogenic structure overview for the Late Cretaceous Tres Pasos Formation at Laguna Figueroa, Magallanes Basin, Chile. The low-sinuosity channels are characterized by approximately symmetric margins, as shown in the strike-oriented cross-section. Intrachannel subenvironments can be discerned by trace-fossil suites; trace fossils are more abundant upward and at the edges of channel fills (i.e., most abundant and diverse in distal abandonment deposits). Trace-fossil abbreviations: *Ar*, *Arenicolites*; *Ast*, *Asterosoma*; *Co*, *Cosmorhaphe*; *Di h*, *Diplocraterion habichi*; fu, fugichnia; *Gyr*, *Gyrochorte; He*, *Helminthopsis*; Ne, *Neonereites*; *Op*, *Ophiomorpha*; *Pa t*, *Palaeophycus tubularis*; *Pa h*, *P. heberti*; *Ph*, *Phycosiphon*; *Pl m*, *Planolites montanus*; *Pl*, *Planolites*; *Rh*, *Rhizocorallium*; *Ro*, *Rosselia*; *Sk*, *Skolithos*; *Sc*, *Scolicia*; *Te*, *Teichichnus*; *Th*, *Thalassinoides*.

constituents of the channel fill can be broken up into axial deposits, which are dominated by amalgamated sandstone, and margin facies, largely made up of thinly interbedded sandstone and siltstone (Fig. 8). Marginal facies can be subdivided into "proximal margin" (relative to the channel axis) and "distal margin" associations. Both margin facies consist of non-amalgamated sandstone beds; however, sandstone beds are significantly thinner in the distal margin deposits. The transition from channel axis to channel margin records the shift from high- to low-energy turbulent sediment transport. The upper part of the channel fill is characterized by abandonment deposits, which are manifest by a thinning- and fining-upward stratal package formed through the waning of turbidity current input. Axial abandonment facies consist of non-amalgamated, thick-bedded turbidites. Laterally, toward the channel margins, these beds pass gradationally into thinly interbedded turbidity-current deposits (distal abandonment), which are distinguished from the distal-margin deposits by an increase in organic detritus and trace-fossil diversity (Fig. 8).

Trace-fossil suites in the channel fill record rare to common grazing traces (e.g., *Neonereites*), uncommon to common deposit-feeding traces (e.g., *Asterosoma*, *Gyrochorte*, and *Teichichnus*), and common to abundant opportunistic strategies (e.g., *Diplocraterion*, *Ophiomorpha* and *Thalassinoides*) (Figs. 8 and 9). Amalgamated channel-axis sandstones are generally devoid of bioturbation, with palimpsest firmground suites consisting of *Diplocraterion habichi* and *Thalassinoides* confined to autogenic scour surfaces within the channel fill in the proximal margin setting (Fig. 9B). Bioturbation indices are highly variable, ranging from BI = 0–5 across the various intrachannel facies (Fig. 8). Trace-fossil diversity, however, varies among the intrachannel facies, depending upon current energies as well as temporal and spatial variations in erosion. In particular, a systematic increase in trace-fossil diversity can be observed from the channel axis to the channel margins. Similarly, trace-fossil diversities increase upward through the channel-fill succession. As such, the highest diversity trace-fossil suites are notable in distal abandonment deposits (Fig. 9C–F). Bypass drape-deposits, which are associated with the highest energy flows that passed through the channels, are characterized by the lowest diversity trace-fossil suites and tend to lack elements reflecting opportunistic ethologies (Fig. 9G). Numerous intrachannel subenvironments are characterized by similar appearing thin-bedded turbidites (cf. Bouma, 1962; Fig. 9). Ichnological analysis is more effective than the physical sedimentology in differentiating these facies types in the Tres Pasos Formation.

4.6 Levee Deposits, Nanaimo Group, Gulf Islands, British Columbia

The Late Cretaceous Nanaimo Group of the Gulf Islands consists of a thick (>4000 m) succession of siliciclastic slope-fan deposits that accumulated in the foreland basin setting of the Strait of Georgia during encroachment of Vancouver Island to the mainland of British Columbia (Mustard, 1994).

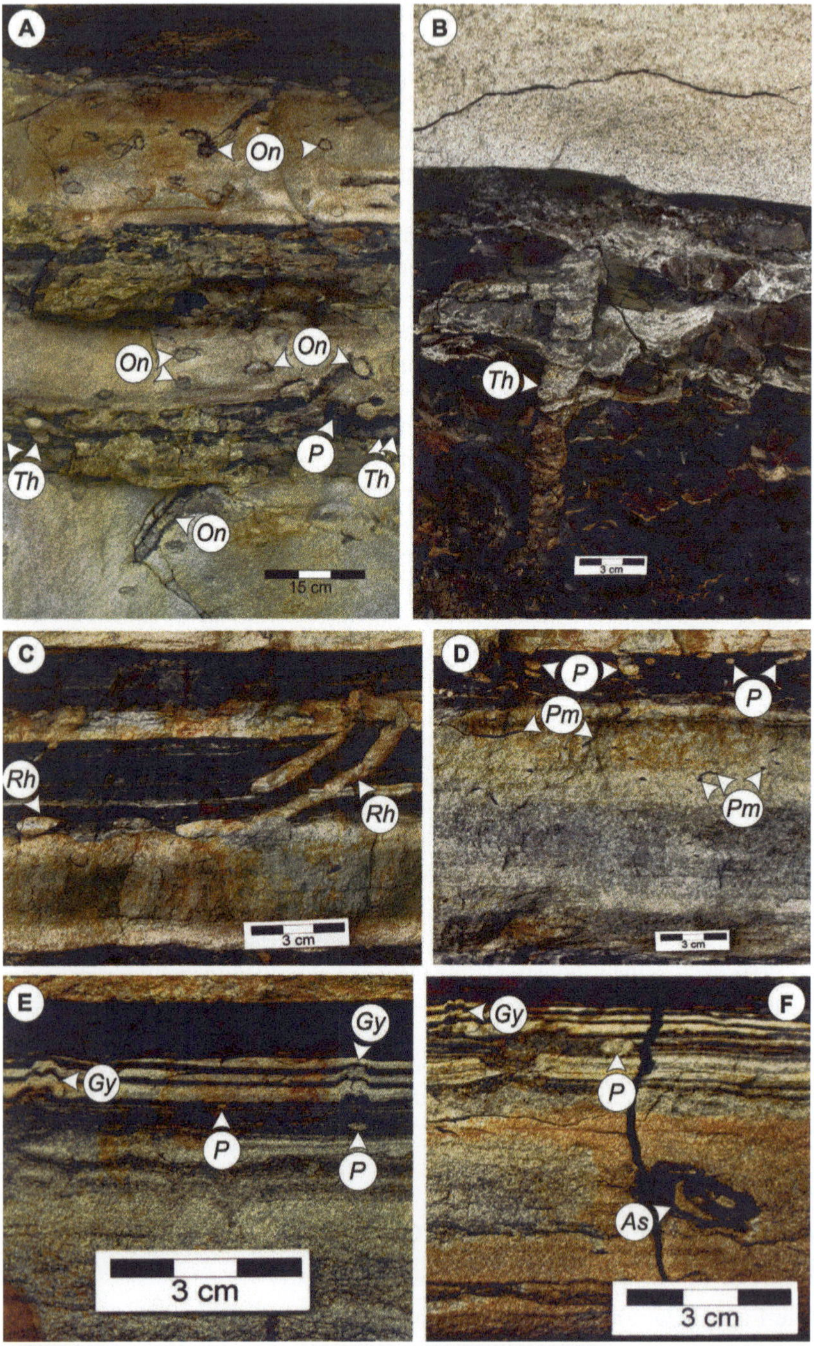
A
On
On
On
P
Th
Th
On
15 cm
B
Th
3 cm
C
Rh
Rh
3 cm
D
P
P
Pm
Pm
3 cm
E
Gy
Gy
P
P
3 cm
F
Gy
P
As
3 cm

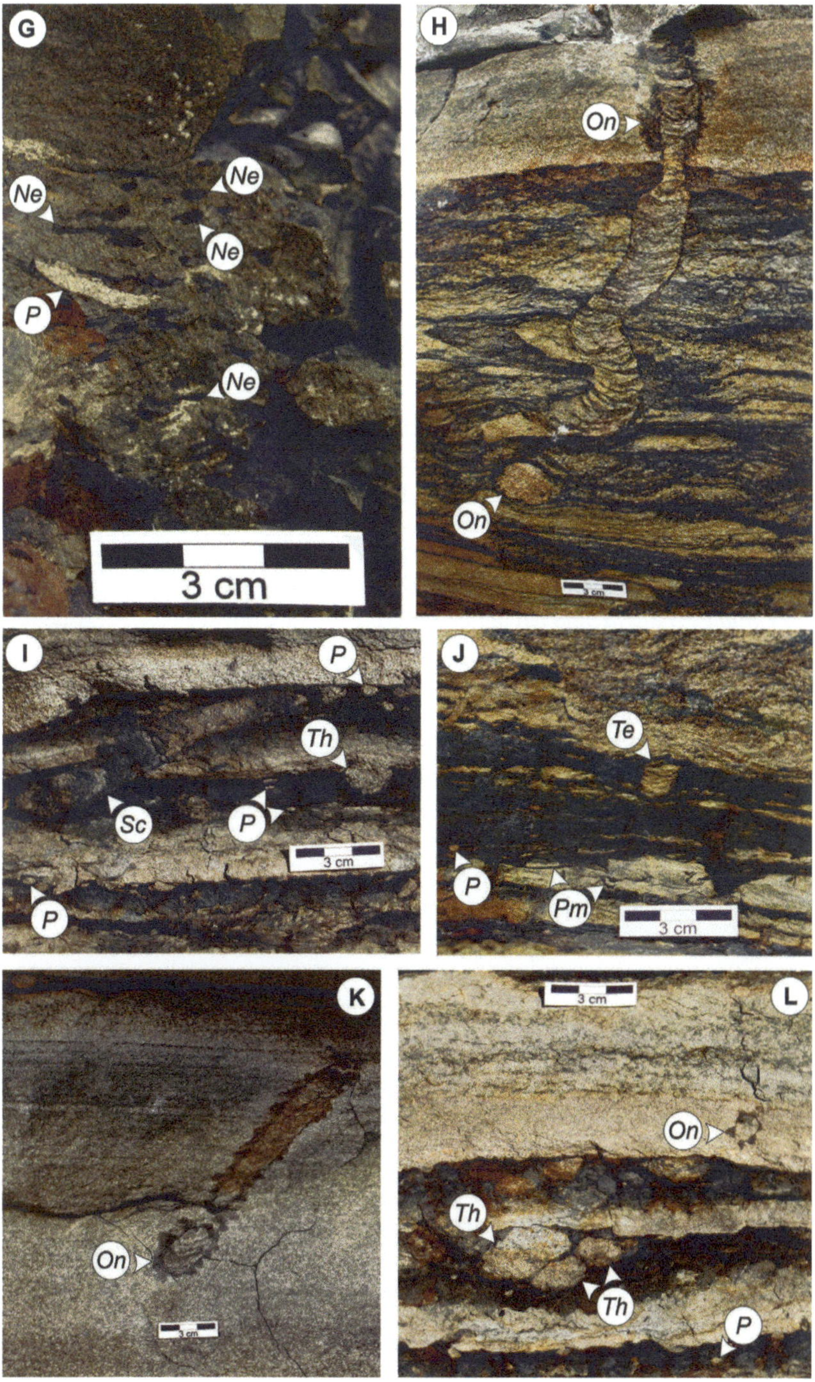
G
Ne
Ne
Ne
P
Ne
3 cm
H
On
On
3 cm
I
P
Th
Sc
P
3 cm
P
J
Te
P
Pm
3 cm
K
On
3 cm
L
3 cm
On
Th
Th
P

Well-exposed successions of channel-levee deposits locally crop out along the intertidal zone of several of the southern Gulf Islands.

Channel deposits consist of amalgamated, thick-bedded (meters to tens of meters) granule, pebble, and cobble conglomerates that are interstratified with normally graded, massive to planar-laminated medium- to coarse-grained sandstones. A few beds toward the top of the channel fill succession show dm-scale trough cross-stratification. Most of the channel-fill deposits are unburrowed (BI=0). Sandstone beds locally display BI=1, manifest by isolated, thin-walled *Ophiomorpha* and small-diameter *Skolithos*.

Levee deposits (Fig. 10) display sharp-based, thin- to moderately thin-bedded (cm- to dm-scale), low-density turbidites ($T_{b–e}$ divisions of Bouma, 1962). Most levee turbidites display convolute bedding (Fig. 10B and F), with rarer aggradational current ripples (Fig. 10E), and uncommon mudstone rip-up clasts. Load structures are locally associated with some basal contacts (Fig. 10C). Slumped intervals also are prevalent, associated with downslope displacement and rotational folding of beds. Boudinage structures are rare.

Levee units typically display BI=0–1, with isolated zones of BI=2–3. Biogenic structures include common fugichnia, lesser *Planolites*, *Thalassinoides*, *Macaronichnus*, *Chondrites*, and *Phycosiphon*, as well as very rare *Asterosoma*, *Taenidium*, *Scolicia* and an unidentified vertical burrow with fecal-pellets packed into meniscae (possibly attributable to *Compaginatichnus*; cf. Pickerill, 1989; Fig. 10F). Suites show a predominance of infaunal responses to event deposition (e.g., escape or rare equilibrichnia), and opportunistic colonization of the bed by largely mobile deposit-feeders (e.g., *Planolites*, *Macaronichnus*, and *Scolicia*). Subordinate behaviors are reflected by sessile deposit-feeding fauna (e.g., *Thalassinoides*, *Chondrites*, and *Phycosiphon*). Such associations are consistent with high sedimentation rates and variable depositional energies. Longer pauses between turbidite emplacements on the levees led to horizons conducive to sessile deposit-feeding, although dominated by the more opportunistic elements (e.g., *Phycosiphon*) of that behavioral group. *Chondrites* constitutes a subordinate element.

FIGURE 9 Trace fossils in slope-channel deposits of the Late Cretaceous Tres Pasos Formation at Laguna Figueroa, Magallanes Basin, Chile. (A) Axial abandonment deposits, BI=3–4, with an opportunistic trace-fossil suite consisting of *Ophiomorpha nodosa* (*On*), *Thalassinoides* (*Th*), and *Planolites* (*P*). (B) Proximal-margin deposits with uncommon *Thalassinoides* (*Th*). (C–F) Distal-abandonment deposits with a relatively diverse assemblage of trace fossils representing the *Cruziana* Ichnofacies. BI ranges from 0 to 4. Ichnogenera include *Rhizocorallium* (*Rh*), sand-filled *Planolites* (*P*), small, silt-filled *Planolites montanus* (*Pm*), *Gyrochorte* (*Gy*), and *Asterosoma* (*As*). (G) Bypass-drape deposits, BI=1–3, with a low-diversity suite of trace fossils consisting of *Planolites* (*P*) and *Neonereites* (*Ne*). (H–L) Distal-margin deposits, BI=0–4, containing a moderately diverse trace-fossil assemblage dominated by opportunistic traces including *Ophiomorpha nodosa* (*On*) with distinctive laminated fill, sand-filled *Planolites* (*P*), silt-filled *Planolites montanus* (*Pm*), *Teichichnus* (*Te*), and *Thalassinoides* (*Th*). *Scolicia* (*Sc*) occurs locally.

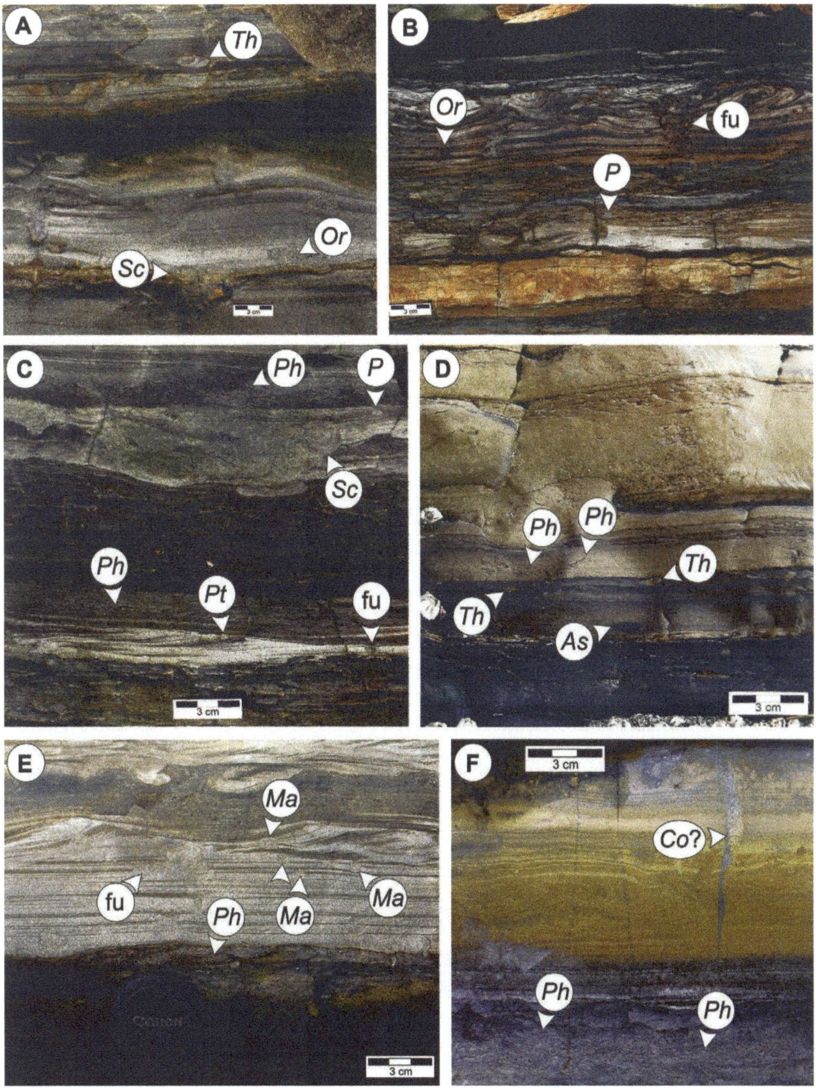

FIGURE 10 Trace fossils in levee deposits of the Late Cretaceous Nanaimo Group, Gulf Islands, British Columbia. (A) Thin-bedded turbidites with wavy parallel laminae. Beds show BI = 0–2, with isolated trace fossils such as *Thalassinoides* (*Th*), *Scolicia* (*Sc*), and *Ophiomorpha rudis* (*Or*). (B) Turbidite bed with well-developed convolute bedding at its top. Unit shows BI = 0–1, with isolated fugichnia (fu), *Ophiomorpha rudis* (*Or*), and *Planolites* (*P*). (C) Thin-bedded levee turbidites, with the lower unit showing well-developed current ripples and the upper sandstone characterized by loading into underlying mudstone. The mudstone is unburrowed. The sandstones contain isolated fugichnia (fu) and *Scolicia* (*Sc*). The silty turbidite caps contain *Planolites* (*P*), *Palaeophycus tubularis* (*Pt*), and *Phycosiphon* (*Ph*). (D) Thin- to thick-bedded turbidites with *Asterosoma* (*As*) (probably allochthonous), *Thalassinoides* (*Th*), and *Phycosiphon* (*Ph*). (E) Well-developed Bouma $T_{b\text{–}e}$ turbidite containing aggradational current ripples. The unit shows BI = 2, with isolated fugichnia

5. DISCUSSION AND CONCLUSIONS

Beyond the shelf-break, the deep-water realm is characterized by a variable seascape that consists of subaqueous canyons, sinuous channels, slump scars, and minibasins, which contribute to substantial topographic rugosity (Fig. 1). The seascape morphology is influenced by numerous factors, including tectonic setting, oceanic currents, fluctuations in sea-level, and sediment input. Sedimentological and stratigraphical analyses of slope deposits over the past decade have led to the development of increasingly realistic and complex facies models (e.g., Hodgson and Flint, 2005; Hubbard et al., 2011; Wynn et al., 2007). Recent ichnological work has tended toward helping to better establish the utility of trace-fossil analysis for recognizing the deposits of these highly variable slope subenvironments in the rock record (e.g., Heard and Pickering, 2008). Although there have been substantial advances to date, we have only just started and undoubtedly there are numerous discoveries yet to be made that will help future sedimentary geologists to better resolve ancient slope successions. Slope ichnology arguably can be considered a field of research with tremendous potential at this time.

The trace-fossil assemblages of slope deposits show a marked variability, controlled by physico-chemical parameters such as current energy, sedimentation rate, slope instability, the character of the substrate, oxygen and food availability, and water turbidity. Background (ambient) slope deposits are dominated by deposit-feeding and grazing structures, although diversities appear to vary from setting to setting (Figs. 5 and 6). Ambient slope environments are subject to slow, continuous suspension settling of fine-grained material, with associated oceanic currents, turbidity currents (low- or high-density), and/or mass-wasting events. Bioturbation can be thorough and pervasive in successions typified by gradual sedimentation (Fig. 6), but considerably lower and sporadically distributed in event bed-dominated intervals (Fig. 5). An additional challenge, associated with characterizing slope trace-fossil assemblages, derives from the fact that many slope settings are typified by bedding-parallel trace fossils. Hence, cross-sectional perspectives of slope deposits, such as those observed in cores, will likely result in underrepresented trace-fossil diversity measurements, affecting forms attributed to grazing behaviors, as well as graphoglyptids (cf. Knaust, 2009; MacEachern et al., 2010).

Conduits that cross-cut the slope, including canyons, gullies, and channels, are commonly characterized by increased and replenished oxygenation and more abundant food resources (e.g., Blake and Hilbig, 1994; Rhoads and Hecker, 1994). Although large gravity flows are expected to significantly

(fu) and *Macaronichnus* (*Ma*). The underlying siltstone contains abundant *Phycosiphon* (*Ph*), and locally, *Chondrites* and *Planolites* as well. (F) Levee sandstone erosionally overlying pervasively burrowed siltstone. The siltstone shows BI = 3–4 with abundant *Phycosiphon* (*Ph*). The overlying turbiditic sandstone shows convolute bedding and fecal-pellet filled meniscate burrow tentatively identified as *Compaginatichnus* (*Co*?).

influence canyon environments, it is important to consider that major events often recur at a frequency of decades or centuries (e.g., Romans et al., 2009), rates at least an order of magnitude lower than bed recolonization by benthic organisms (e.g., Grassle, 1977). For long periods, submarine conduits are commonly associated with focused and sustained sediment-poor and oxygen- and nutrient- rich currents (e.g., upwelling currents). Therefore, it is likely that these conduits are ideal areas for infaunal activity for extended periods. The paucity of trace fossils in many ancient channel-fill successions probably reflects their preservation potential, in part owing to pronounced erosion from the rare, outsized events (Fig. 8). Furthermore, although numerous interpretations of ancient deep-water strata consider punctuated sedimentation as a major deterrent to biogenic reworking, many of the deposit-feeding ethologies recorded in ancient slope deposits suggest that organisms were likely able to withstand, or recover (as a community) relatively quickly from turbidity current sedimentation. As a result of all these factors, trace-fossil assemblages (mostly impoverished) that are more typical of shallow-marine environments are commonly observed in bathyal deposits (e.g., Figs. 7–9). Although it is possible that some organisms residing on the slope have been transported there via turbidity currents (cf. Crimes, 1977; Föllmi and Grimm, 1990; Wetzel, 1984), it is more likely that the environmental conditions themselves can sustain a so-called shallow-water community in many instances. Deep erosion, associated with submarine conduits, can expose semi-consolidated substrates, and therefore, substrate-controlled trace-fossil associations attributable to the *Glossifungites* Ichnofacies are also common and should be expected in channel fill and margin deposits (e.g., Figs. 3 and 4; Anderson et al., 2006; Callow et al., 2012a; Hayward, 1976; MacEachern et al., 2007b).

Channel-levee complexes are an important component of submarine fan and slope systems, and the ichnology of levees is a burgeoning research topic (e.g., Fig. 10; Callow et al., 2012a,b; Cummings and Hodgson, 2011; Heard and Pickering, 2008; Hubbard and Shultz, 2008; Kane et al., 2007). In general, levee deposits are characterized by higher trace-fossil diversities relative to those of many other slope environments, including abundant graphoglyptids in some instances. For example, in Cretaceous-Eocene levee units of the Basque Basin in northern Spain, Cummings and Hodgson (2011) recognized 30 distinct ichnotaxa, including 17 pre-depositional forms and 12 post-depositional forms. Of these, 12 forms were considered graphoglyptids. Although trace-fossil suites with common graphoglyptids are not featured in this review, it is apparent that the lower bathyal realm is host to deposits where they are at least locally abundant (e.g., Heard and Pickering, 2008). In comparison to basin-floor fan deposits, however, graphoglyptids are commonly lacking in slope deposits due to unstable environmental conditions (e.g., currents, sedimentation) and the presence of relatively abundant food resources on the slope. Organisms that adapt specialized feeding strategies, like many that are attributed to the construction of graphoglyptids, are out-competed on the slope by more mobile organisms, many of which employ opportunistic feeding strategies. Ichnogenera typical of suites attributed to the *Nereites* Ichnofacies generally reflect organism behaviors that attempt to optimize sea-floor coverage in response to resource deficiency.

Since the pioneering work of Seilacher (1967), sedimentologists have attempted to use trace fossils and trace-fossil suites as a proxy for interpreting paleo-water depths. Physical and chemical parameters do not change consistently with water depth from basin to basin, nor even along an individual basin's margin. Factors, such as basin circulation, water stratification, shelf width, and the details of sediment supply to the basin (e.g., point source *versus* line source), impart significant influences on infaunal distributions of modern slopes, and without doubt, in ancient slope deposits as well. The consideration of trace-fossil suites that can be assigned to a series of ichnofacies, including the *Zoophycos*, *Cruziana*, *Skolithos*, *Glossifungites*, and *Nereites* ichnofacies, is a reflection of the varied conditions that influence organisms across the slope.

Abandoning attempts to tie trace-fossil suites to paleobathymetry (cf. Frey et al., 1990), high-resolution ichnological analyses have more potential for developing means to differentiate subtle facies changes in ancient slope deposits. Such an approach is common in the analysis of marginal- and shallow-marine deposits (e.g., Bann et al., 2004; Frey and Pemberton, 1987; MacEachern et al., 2005). In one such example of this type of investigation in bathyal strata, Cummings and Hodgson (2011) document systematic changes in trace-fossil assemblages across levees relative to their proximity to a channel margin (e.g., increases in graphoglyptids distally). Overall, linking subtle facies and trace-fossil shifts across the deposits of slope subenvironments has been rather poorly assessed until recently (e.g., Callow et al., 2012a,b; Greene et al., 2012; Phillips et al., 2011). Future ichnological work, linked closely with the stratigraphic architecture of slope deposits, will undoubtedly uncover insights into a range of paleoenvironmental conditions and establish the utility of deep-water ichnology in high-resolution slope-facies modeling.

ACKNOWLEDGMENTS

S. M. H. graciously acknowledges the support of Chevron Energy Technology Company, ConocoPhillips Company, Marathon Oil Corporation, Talisman Energy Inc., Rohöl-Aufsuchungs AG, and the Natural Sciences and Engineering Research Council of Canada (NSERC) for research on slope deposits. J. A. MacEachern acknowledges funding via NSERC Discovery Grant 184293. Reviews by Drs. Ian Kane, Dirk Knaust, and Richard Bromley improved the clarity of the chapter and are greatly appreciated.

REFERENCES

Alpert, S.P., 1974. Systematic review of the genus *Skolithos*. J. Paleontol. 48, 661–669.

Anderson, K.S., Graham, S.A., Hubbard, S.M., 2006. Facies, architecture and origin of a reservoir-scale sand-rich succession within submarine canyon fill: insights from Wagon Caves Rock (Paleocene), Santa Lucia Range, California. J. Sediment. Res. 76, 819–838.

Armitage, D.A., Romans, B.W., Covault, J.A., Graham, S.A., 2009. The influence of mass transport deposit surface topography on the evolution of turbidite architecture: the Sierra Contreras, Tres Pasos Formation (Cretaceous), southern Chile. J. Sediment. Res. 79, 287–301.

Ballance, P.F., Gregory, M.R., Gibson, G.W., Chaproniere, G.C.H., Kadar, A.P., Sameshima, T., 1984. A late Miocene and early Pliocene upper shelf-to-slope sequence of calcareous fine sediment from the Pacific margin of New Zealand. In: Stow, D.A.V., Piper, D.J.W. (Eds.), Fine-Grained Sediments: Deep Water Processes and Facies. Geol. Soc. London, Spec. Publ., vol. 15, pp. 331–342.

Bann, K.L., Fielding, C.R., MacEachern, J.A., Tye, S.C., 2004. Differentiation of estuarine and offshore marine deposits using integrated ichnology and sedimentology: Permian Pebbley Beach Formation, Sydney Basin, Australia. In: McIlroy, D. (Ed.), The Application of Ichnology to Palaeoenvironmental and Stratigraphic Analysis. Geol. Soc. London, Spec. Publ., vol. 228, pp. 179–211.

Blake, J.A., Hilbig, B., 1994. Dense infaunal assemblages on the continental slope off Cape Hatteras, North Carolina. Deep-Sea Res. II 41, 875–899.

Bouma, A.H., 1962. Sedimentology of Some Flysch Deposits. Elsevier, Amsterdam, 168 pp.

Bromley, R.G., Uchman, A., Gregory, M., Martin, A.J., 2003. *Hillichnus lobosensis* igen. et isp. nov., a complex trace fossil produced by tellinacean bivalves, Paleocene, Monterey, California, U.S.A. Palaeogeogr. Palaeoclimatol. Palaeoecol. 192, 157–186.

Buck, S.P., Bottjer, D.J., 1985. Continental slope deposits from a Late Cretaceous, tectonically active margin, Southern California. J. Sediment. Petrol. 55, 843–855.

Callow, R.H.T., McIlroy, D., Kneller, B., Dykstra, M., 2012a. Integrated ichnological and sedimentological analysis of a Late Cretaceous submarine channel-levee system: The Rosario Formation, Baja California, Mexico. Mar. Petrol. Geol. (in press). DOI: 10.1016/j.marpetgeo.2012.02.001.

Callow, R.H.T., McIlroy, D., Kneller, B., Dykstra, M., 2012b. Ichnology and palaeoenvironments of a slope-channel-levee system: the Late Cretaceous Rosario Formation, Baja California, Mexico. Ichnos (in press).

Carney, R.S., 2005. Zonation of deep biota on continental margins. Oceanogr. Mar. Biol. Ann. Rev. 43, 211–278.

Cartes, J.E., Fanelli, E., Papiol, V., Maynou, F., 2010. Trophic relationships at intrannual spatial and temporal scales of macro and megafauna around a submarine canyon off the Catalonian coast (western Mediterranean). J. Sea Res. 63, 180–190.

Clark, J.D., Pickering, K.T., 1996. Architectural elements and growth patterns of submarine channels: application to hydrocarbon exploration. AAPG Bull. 80, 194–221.

Covault, J.A., Romans, B.W., Graham, S.A., 2009. Outcrop expression of a continental margin-scale shelf-edge delta from the Cretaceous Magallanes Basin, Chile. J. Sediment. Res. 79, 523–539.

Crimes, T.P., 1970. The significance of trace fossils in sedimentology, stratigraphy and palaeoecology with examples from Lower Palaeozoic strata. In: Crimes, T.P., Harper, J.C. (Eds.), Trace Fossils. Geol. J., Spec. Iss. 3, pp. 101–125.

Crimes, T.P., 1977. Trace fossils of an Eocene deep-sea fan, northern Spain. In: Crimes, T.P., Harper, J.C. (Eds.), Trace Fossils 2. Geol. J., Spec. Iss. 9, pp. 71–90.

Crimes, T.P., Goldring, R., Homewood, P., van Stuijvenberg, J., Winkler, W., 1981. Trace fossil assemblages of deep-sea fan deposits, Gurnigel and Schlieren flysch (Cretaceous-Eocene), Switzerland. Eclogae Geol. Helv. 74, 953–995.

Cummings, J.P., Hodgson, D.M., 2011. Assessing controls on the distribution of ichnotaxa in submarine fan environments, the Basque Basin, northern Spain. Sediment. Geol. 239, 162–187.

Di Celma, C., Cantalamessa, G., Didaskalou, P., Lori, P., 2010. Sedimentology, architecture, and sequence stratigraphy of coarse-grained, submarine canyon fills from the Pleistocene (Gelasian-Calabrian) of the Peri-Adriatic basin, central Italy. Mar. Petrol. Geol. 27, 1340–1365.

Dillon, W.P., Zimmerman, H.B., 1970. Erosion by biological activity in two New England Submarine Canyons. J. Sediment. Petrol. 40, 542–547.

Eyles, N., Vossler, S.M., Lagoe, M.B., 1992. Ichnology of a glacially influenced continental shelf and slope: the Late Cenozoic Gulf of Alaska (Yakataga Formation). Palaeogeogr. Palaeoclimatol. Palaeoecol. 94, 193–221.

Fildani, A., Hubbard, S.M., Romans, B.W., 2009. Stratigraphic evolution of deep-water architecture: examples of controls and depositional styles from the Magallanes Basin, Chile. SEPM Fieldtrip Guidebook 10, 73 pp.

Fjellanger, E., Surlyk, F., Wamsteeker, L.C., Midtun, T., 2005. Upper Cretaceous basinfloor fans in the Vøring Basin, Mid Norway shelf. In: Wandas, B. (Ed.), Onshore-Offshore Relationships on the North Atlantic Margin. NPF Spec. Publ., vol. 12, Elsevier, Amsterdam, pp. 135–164.

Flach, E., Thomsen, L., 1998. Do physical and chemical factors structure the macrobenthic community at a continental slope in the NE Atlantic? Hydrobiologia 375 (376), 265–285.

Föllmi, K.B., Grimm, K.A., 1990. Doomed pioneers: gravity-flow deposition and bioturbation in marine oxygen-deficient environments. Geology 18, 1069–1072.

Frey, R.W., 1971. Ichnology—the study of fossil and recent lebensspuren. In: Perkins, B.F. (Ed.), Trace Fossils: A Field Guide to Selected Localities in Pennsylvanian, Permian, Cretaeous, and Tertiary rocks of Texas and Related Papers. Louisiana State University, Miscellaneous Publication 71-1, pp. 91–125.

Frey, R.W., Howard, J.D., 1970. Comparison of Upper Cretaceous ichnofaunas from siliceous sandstones and chalk, Western Interior Region, U.S.A. In: Crimes, T.P., Harper, J.C. (Eds.), Trace Fossils. Geol. J., Spec. Iss. 3, pp. 141–166.

Frey, R.W., Pemberton, S.G., 1987. The *Psilonichnus* ichnocoenose, and its relationship to adjacent marine and nonmarine ichnocoenoses along the Georgia coast. Bull. Can. Petrol. Geol. 35, 333–357.

Frey, R.W., Seilacher, A., 1980. Uniformity in marine invertebrate ichnology. Lethaia 13, 183–207.

Frey, R.W., Pemberton, S.G., Saunders, T.D.A., 1990. Ichnofacies and bathymetry: a passive relationship. J. Paleontol. 64, 155–158.

Gage, J.D., Tyler, P.A., 1991. Deep-Sea Biology. Cambridge University Press, Cambridge, 520 pp.

Gage, J.D., Lamont, P.A., Kroeger, K., Paterson, G.L.J., Gonzalez Vecino, J.L., 2000. Patterns in deep-sea macrobenthos at the continental margin: standing crop, diversity and faunal change on the continental slope off Scotland. Hydrobiologia 440, 261–271.

Gerino, M., Stora, G., Weber, O., 1999. Evidence of bioturbation in the Cap-Ferret Canyon in the deep northeastern Atlantic. Deep-Sea Res. II 46, 2289–2307.

Gjelberg, J.G., Enoksen, T., Kjærnes, P., Mangerud, G., Martinsen, O.J., Roe, E., Vågnes, E., 2001. The Maastrichtian and Danian depositional setting, along the eastern margin of the Møre Basin (mid-Norwegian Shelf): implications for reservoir development of the Ormen Lange field. In: Martinsen, O.J., Dreyer, T. (Eds.), Sedimentary Environments Offshore Norway–Paleozoic to Recent. Norwegian Petrol. Soc., Spec. Publ., vol. 10, pp. 421–440.

Grassle, J.F., 1977. Slow recolonization of deep-sea sediment. Nature 265, 618–619.

Greene, T.J., Gingras, M.K., Gordon, G.S., McKeel, D.R., 2012. The significance of deep-water cryptic bioturbation in slope-channel massive sand deposits of the lower Rio Dell Formation, Eel River Basin, California. Mar. Petrol. Geol. 29, 152–174.

Griggs, G.B., Carey, A.G., Kulm, L.D., 1969. Deep-sea sedimentation and sediment-fauna interaction in Cascadia Channel and on Cascadia Abyssal Plain. Deep-Sea Res. 16, 157–170.

Hayward, B.W., 1976. Lower Miocene bathyal and submarine canyon ichnocenoses from Northland, New Zealand. Lethaia 9, 149–162.

Heard, T.G., Pickering, K.T., 2008. Trace fossils as diagnostic indicators of deep-marine environments, Middle Eocene Ainsa-Jaca basin, Spanish Pyrenees. Sedimentology 55, 809–844.

Heezen, B.C., Hollister, C.D., 1971. The Face of the Deep. Oxford University Press, Toronto, 659 pp.

Hodgson, D.M., Flint, S.S. (Eds.), 2005. Submarine Slope Systems: Processes and Products. Geol. Soc. London, Spec. Publ., vol. 244, 220 pp.

Hodgson, D.M., Di Celma, C.N., Brunt, R.L., Flint, S.S., 2011. Submarine slope degradation and aggradation and the stratigraphic evolution of channel-levee systems. J. Geol. Soc. 168, 625–628.

Howard, J.D., Frey, R.W., 1984. Characteristic trace fossils in nearshore to offshore sequences, Upper Cretaceous of east-central Utah. Can. J. Earth Sci. 21, 200–219.

Hubbard, S.M., Shultz, M.R., 2008. Deep burrows in submarine fan-channel deposits of the Cerro Toro Formation (Cretaceous), Chilean Patagonia: implications for firmground development in the deep-sea. Palaios 23, 223–232.

Hubbard, S.M., Romans, B.W., Graham, S.A., 2008. Deep-water foreland basin deposits of the Cerro Toro Formation, Magallanes Basin, Chile: architectural elements of a sinuous axial channel belt. Sedimentology 55, 1333–1359.

Hubbard, S.M., Fildani, A., Romans, B.W., Covault, J.A., McHargue, T., 2010. High-relief slope clinoform development: insights from outcrop, Magallanes Basin, Chile. J. Sediment. Res. 80, 357–375.

Hubbard, S.M., Romans, B.W., Fildani, A., 2011. The stratigraphic evolution of deep-water architecture. Mar. Petrol. Geol. 28, 609–611.

Ineson, J.R., 1987. Trace fossils from a submarine fan-slope apron complex in the Cretaceous of James Ross Island, Antarctica. Br. Antarct. Surv., Bull. 74, 1–16.

Josselyn, M.N., Cailliet, G.M., Niesen, T.M., Cowen, R., Hurley, A.C., Connor, J., Hawes, S., 1983. Composition, export and faunal utilization of drift vegetation in the Salt River Submarine Canyon. Estuar. Coastal Shelf Sci. 17, 447–465.

Kane, I.A., Kneller, B.C., Dykstra, M., Kassem, A., McCaffrey, W.D., 2007. Anatomy of a submarine channel-levee: an example from Upper Cretaceous slope sediments, Rosario Formation, Baja California, Mexico. Mar. Petrol. Geol. 24, 540–563.

Kern, J.P., Warme, J.E., 1974. Trace fossils and bathymetry of the Upper Cretaceous Point Loma Formation, San Diego, California. GSA Bull. 85, 893–900.

Knaust, D., 2009. Characterisation of a Campanian deep-sea fan system in the Norwegian Sea by means of ichnofabrics. Mar. Petrol. Geol. 26, 1199–1211.

Leszczyński, S., Seilacher, A., 1991. Ichnocoenoses of a turbidite sole. Ichnos 1, 293–303.

Levin, L.A., Gage, J.D., 1998. Relationships between oxygen, organic matter and the diversity of bathyal macrofauna. Deep-Sea Res. II 45, 129–163.

Lowe, D.R., 1982. Sediment gravity flows: II. Depositional models with special reference to the deposits of high-density turbidity currents. J. Sediment. Petrol. 52, 279–297.

MacEachern, J.A., Bann, K.L., 2008. The role of ichnology in refining shallow marine facies models. In: Hampson, G., Steel, R., Burgess, P., Dalrymple, R. (Eds.), Recent Advances in Models of Siliciclastic Shallow-Marine Stratigraphy. SEPM Spec. Publ., vol. 90, pp. 73–116.

MacEachern, J.A., Bann, K.L., Bhattacharya, J.P., Howell, C.D., 2005. Ichnology of deltas: organism responses to the dynamic interplay of rivers, waves, storms and tides. In: Bhattacharya, J.P., Giosan, L. (Eds.), River Deltas: Concepts, Models and Examples. SEPM Spec. Publ., vol. 83, pp. 49–85.

MacEachern, J.A., Pemberton, S.G., Gingras, M.K., Bann, K.L., 2007a. The ichnofacies concept: a fifty-year retrospective. In: Miller III, W. (Ed.), Trace Fossils. Concepts, Problems, Prospects. Elsevier, Amsterdam, pp. 50–75.

MacEachern, J.A., Pemberton, S.G., Gingras, M.K., Bann, K.L., Dafoe, L.T., 2007b. Use of trace fossils in genetic stratigraphy. In: Miller III, W. (Ed.), Trace Fossils. Concepts, Problems, Prospects. Elsevier, Amsterdam, pp. 105–128.

MacEachern, J.A., Pemberton, S.G., Gingras, M.K., Bann, K.L., 2010. Ichnology and facies models. In: Dalrymple, R.W., James, N.P. (Eds.), Facies Models 4. Geol. Ass. Can., St. John's, Geotext 6, pp. 19–58.

MacEachern, J.A., Bann, K.L., Gingras, M.K., Zonneveld, J.-P., Dashtgard, S.E., Pemberton, S.G., 2012. The ichnofacies paradigm. In: Knaust, D., Bromley, R.G. (Eds.), Trace Fossils as Indicators of Sedimentary Environments. Developments in Sedimentology, vol. 64. Elsevier, Amsterdam, pp. 103–138.

Martinsen, O.J., Lien, T., Jackson, C., 2005. Cretaceous and Palaeogene turbidite systems in the North Sea and Norwegian Sea Basins: source, staging area and basin physiography controls on reservoir development. In: Doré, A.G., Vining, B.A. (Eds.), Petroleum Geology: North-West Europe and Global Perspectives, Proceedings of the Sixth Petroleum Geology Conference. Geol. Soc. London, pp. 1147–1164.

Miller III, W., Vokes, E.H., 1998. Large *Phymatoderma* in Pliocene slope deposits, northwestern Ecuador: associated ichnofauna, fabrication, and behavioral ecology. Ichnos 6, 23–45.

Morse, J.W., Beazley, M.J., 2008. Organic matter in deepwater sediments of the Northern Gulf of Mexico and its relationship to the distribution of benthic organisms. Deep-Sea Res. II 55, 2563–2571.

Murray, J.W., 1973. Distribution and Ecology of Living Benthic Foraminiferids. Heinemann, London, 274 pp.

Mustard, P.S., Mustard, P.S., 1994. The Upper Cretaceous Nanaimo Group, Georgia Basin. In: Monger, J.W.H. (Ed.), Geology and Geological Hazards of the Vancouver Region, Southwestern British Columbia. Geol. Surv. Can., Bull. 481, pp. 29–95.

Mutti, E., Normark, W.R., 1987. Comparing examples of modern and ancient turbidite systems: problems and concepts. In: Legget, J.K., Zuffa, G.G. (Eds.), Deep Water Clastic Deposits: Models and Case Histories. Graham and Trotman, London, pp. 1–38.

Normark, W.R., Paull, C.K., Caress, D.W., Ussler, W., Sliter, R., 2009. Fine-scale relief related to Late Holocene channel shifting within the floor of the upper Redondo Fan, offshore southern California. Sedimentology 56, 1690–1704.

Olivero, D., Gaillard, C., 1996. Paleoecology of Jurassic *Zoophycos* from south-eastern France. Ichnos 4, 249–260.

Paull, C.K., Mitts, P., Ussler, W., Keaten, R., Greene, H.G., 2005. Trail of sand in upper Monterey Canyon: Offshore California. GSA Bull. 117, 1134–1145.

Pemberton, S.G., MacEachern, J.A., 1997. The ichnological signature of storm deposits: the use of trace fossils in event stratigraphy. In: Brett, C.E., Baird, G.C. (Eds.), Paleontological Events. Stratigraphic, Ecological and Evolutionary Implications. Columbia University Press, New York, pp. 73–109.

Pemberton, S.G., MacEachern, J.A., 2005. Significance of ichnofossils in applied stratigraphy. In: Koutsoukos, E.A. (Ed.), Applied Stratigraphy. Springer, Dordrecht, pp. 281–302.

Pemberton, S.G., MacEachern, J.A., Frey, R.W., 1992. Trace fossil facies models: environmental and allostratigraphic significance. In: Walker, R.G., James, N.P. (Eds.), Facies Models: Response to Sea-Level Change. Geol. Ass. Can., St. John's, Geotext 1, pp. 47–72.

Pemberton, S.G., MacEachern, J.A., Dashtgard, S.E., Bann, K.L., Gingras, M.K., Zonneveld, J.-P., 2012. Shorefaces. In: Knaust, D., Bromley, R.G. (Eds.), Trace Fossils as Indicators of Sedimentary Environments. Developments in Sedimentology, vol. 64. Elsevier, Amsterdam, pp. 563–604.

Phillips, C., McIlroy, D., Elliott, T., 2011. Ichnological characterization of Eocene/Oligocene turbidites from the Gres d'Annot Basin, French Alps, SE France. Palaeogeogr. Palaeoclimatol. Palaeoecol. 300, 67–83.
Pickerill, R.K., 1989. *Compaginatichnus*: a new ichnogenus from Ordovician flysch of Eastern Canada. J. Paleontol. 63, 913–919.
Posamentier, H.W., Kolla, V., 2003. Seismic geomorphology and stratigraphy of depositional elements in deep-water settings. J. Sediment. Res. 73, 367–388.
Posamentier, H.W., Walker, R.G., 2006. Deep-water turbidites and submarine fans. In: Posamentier, H.W., Walker, R.G. (Eds.), Facies Models Revisited. SEPM Spec. Publ., vol. 84, pp. 397–520.
Prather, B.E., Booth, J.R., Steffens, G.S., Craig, P.A., 1998. Classification, lithologic calibration, and stratigraphic succession of seismic facies of intraslope basins, deep-water Gulf of Mexico. AAPG Bull. 82, 701–728.
Puig, P., Company, J.B., Sardà, F., Palanques, A., 2001. Responses of deep-water shrimp to intermediate nepheloid layer detachments on the northwestern Mediterranean continental margin. Deep-Sea Res. I 48, 2195–2207.
Rhoads, D.C., Hecker, B., 1994. Processes on the continental slope off North Carolina with special reference to the Cape Hatteras region. Deep-Sea Res. II 41, 965–980.
Rogers, A.D., 2000. The role of the oceanic oxygen minima in generating biodiversity in the deep sea. Deep Sea Res. II 47, 119–148.
Romans, B.W., Normark, W.R., McGann, M.M., Covault, J.A., Graham, S.A., 2009. Coarse-grained sediment delivery and distribution in the Holocene Santa Monica Basin, California: implications for evaluating source-to-sink flux at millennial time scales. GSA Bull. 121, 1394–1408.
Rowe, G.T., 1971. Observations on bottom currents and epibenthic populations in Hatteras Submarine Canyon. Deep-Sea Res. 18, 569–581.
Rowe, G.T., Polloni, P.T., Haedrich, R.L., 1982. The deep-sea macrobenthos on the continental margin of the northwest Atlantic Ocean. Deep-Sea Res. 29, 257–278.
Sanders, H.L., 1968. Marine benthic diversity: a comparative study. Am. Naturalist 102, 243–282.
Savrda, C.E., 2012. Chalk and related deep-marine carbonates. In: Knaust, D., Bromley, R.G. (Eds.), Trace Fossils as Indicators of Sedimentary Environments. Developments in Sedimentology, vol. 64. Elsevier, Amsterdam, pp. 777–806.
Savrda, C.E., Bottjer, D.J., Gorsline, D.S., 1984. Development of a comprehensive oxygen-deficient marine biofacies model: evidence from Santa Monica, San Pedro, and Santa Barbara Basins, California Continental Borderland. AAPG Bull. 68, 1179–1192.
Savrda, C.E., Krawinkel, H., McCarthy, F.M.G., McHugh, C.M.G., Olson, H.C., Mountain, G., 2001. Ichnofabrics of a Pleistocene slope succession, New Jersey margin: relations to climate and sea-level dynamics. Palaeogeogr. Palaeoclimatol. Palaeoecol. 171, 41–61.
Scholle, P.A., Arthur, M.A., Ekdale, A.A., 1983. Pelagic environment. In: Scholle, P.A., Bebout, D.G., Moore, C.H. (Eds.), Carbonate Depositional Environments. AAPG Mem. 33, pp. 619–691.
Seilacher, A., 1967. Bathymetry of trace fossils. Mar. Geol. 5, 413–428.
Seilacher, A., 1977. Pattern analysis of *Paleodictyon* and related trace fossils. In: Crimes, T.P., Harper, J.C. (Eds.), Trace Fossils 2. Geol. J., Spec. Iss. 9, pp. 289–334.
Seilacher, A., 1978. Use of trace fossils for recognizing depositional environments. In: Basan, P.B. (Ed.), Trace Fossil Concepts. SEPM Short Course 5, pp. 175–201.
Shultz, M.R., Hubbard, S.M., 2005. Sedimentology, stratigraphic architecture, and ichnology of gravity-flow deposits partially ponded in a growth-fault bounded slope mini-basin, Tres Pasos Formation, southern Chile. J. Sediment. Res. 75, 440–453.

Sømme, T.O., Helland-Hansen, W., Martinsen, O.J., Thurmond, J.B., 2009. Predicting morphological relationships and sediment partitioning in source-to-sink systems. Basin Res. 21, 361–387.

Stora, G., Bourcier, M., Arnoux, A., Gerino, M., Le Campion, J., Gilbert, F., Durbec, J.P., 1999. The deep-sea macrobenthos on the continental slope of the northwestern Mediterranean Sea: a quantitative approach. Deep-Sea Res. I 46, 1339–1368.

Stuart, C.T., Rex, M.A., Etter, R.J., 2003. Large-scale spatial and temporal patterns of deep-sea benthic species diversity. In: Tyler, P.A. (Ed.), Ecosystems of the Deep Oceans. Elsevier, Amsterdam, pp. 295–311.

Tedesco, L.P., Wanless, H.R., 1991. Generation of sedimentary fabrics and facies by repetitive excavation and storm infilling of burrow networks, Holocene of south Florida and Caicos Platform, British West Indies. Palaios 6, 326–343.

Uchman, A., 2001. Eocene flysch trace fossils from the Hecho Group of the Pyrenees, northern Spain. Beringeria 28, 3–41.

Uchman, A., 2009. The *Ophiomorpha rudis* ichnosubfacies of the *Nereites* ichnofacies: characteristics and constraints. Palaeogeogr. Palaeoclimatol. Palaeoecol. 276, 107–119.

Uchman, A., Wetzel, A., 2011. Deep-sea ichnology: the relationships between depositional environment and endobenthic organisms. In: Hüneke, H., Mulder, T. (Eds.), Deep-Sea Sediments. Developments in Sedimentology, vol. 63. Elsevier, Amsterdam, pp. 517–556.

Uchman, A., Wetzel, A., 2012. Deep-sea fans. In: Knaust, D., Bromley, R.G. (Eds.), Trace Fossils as Indicators of Sedimentary Environments. Developments in Sedimentology, vol. 64. Elsevier, Amsterdam, pp. 643–672.

Vergara, L., Wreglesworth, I., Trayfoot, M., Richardsen, G., 2001. The distribution of Cretaceous and Paleocene deep-water reservoirs in the Norwegian Sea basins. Petrol. Geosci. 7, 395–408.

Vetter, E.W., Dayton, P.K., 1998. Macrofaunal communities within and adjacent to a detritus-rich submarine canyon system. Deep-Sea Res. II 45, 25–54.

Vetter, E.W., Smith, C.R., De Leo, F.C., 2010. Hawaiian hotspots: enhanced megafaunal abundance and diversity in submarine canyons on the oceanic islands of Hawaii. Mar. Ecol. 31, 183–199.

Warme, J.E., Scanland, T.B., Marshall, N.F., 1971. Submarine canyon erosion: contribution of marine rock burrowers. Science 173, 1127–1129.

Wetzel, A., 1984. Bioturbation in deep-sea fine-grained sediments: influence of sediment texture, turbidite frequency and rates of environmental change. In: Stow, D.A.V., Piper, D.J.W. (Eds.), Fine-Grained Sediments: Deep Water Processes and Facies. Geol. Soc. London, Spec. Publ., vol. 15, pp. 595–608.

Wetzel, A., 2008. Recent bioturbation in the deep South China Sea: a uniformitarian ichnologic approach. Palaios 23, 601–614.

Wetzel, A., Uchman, A., 2012. Hemipelagic and pelagic basin plains. In: Knaust, D., Bromley, R.G. (Eds.), Trace Fossils as Indicators of Sedimentary Environments. Developments in Sedimentology, vol. 64. Elsevier, Amsterdam, pp. 673–702.

Wetzel, A., Werner, F., Stow, D.A.V., 2008. Bioturbation and biogenic sedimentary structures in contourites. In: Rebesco, M., Camerlenghi, A. (Eds.), Contourites. Developments in Sedimentology, vol. 60. Elsevier, Amsterdam, pp. 183–202.

Wynn, R.B., Cronin, B.T., Peakall, J., 2007. Sinuous deep-water channels: genesis, geometry and architecture. Mar. Petrol. Geol. 24, 341–387.

Young, D.K., Richardson, M.D., 1998. Effects of waste disposal on benthic faunal succession on the abyssal seafloor. J. Mar. Systems 14, 319–336.

Chapter 21

Deep-Sea Fans

Alfred Uchman*[,1] and Andreas Wetzel[†]
*Institute of Geological Sciences, Jagiellonian University, Kraków, Poland, [†]Geologisch-Paläontologisches Institut der Universität, Basel, Switzerland
[1]Corresponding author: e-mail: alfred.uchman@uj.edu.pl

1. INTRODUCTION

Bioturbational structures and trace fossils are important components of turbidites since late Cambrian to early Ordovician times, when endobenthic animals began to colonize the deep-sea floor more intensively (e.g., Crimes and Fedonkin, 1994). The variable but still poorly understood burrowing activity in the abyssal habitat was and is influenced by many environmental factors, such as pore water oxygenation, trophic level, and sediment properties.

Research into deep-sea ichnology developed more slowly than work in shallow-water settings, because of a general lack of knowledge about the deep-sea environments until the 1950s. In the nineteenth century, trace fossils occurring in deep-marine deposits were often described as algae (fucoids) without any recognition of their true origin (e.g., Heer, 1876/1877; von Fischer-Ooster, 1858). Once the true nature of trace fossils had been recognized (Nathorst, 1881), they were increasingly studied (Abel, 1935; Richter, 1928; Seilacher, 1953, 1955; see Uchman, 2007a, for partial review). The potential use of bioturbational structures as ecological indicators dramatically increased with the study of modern environments about 50 years ago.

In the following, an overview is given of the ichnology of deep-marine settings that are considerably affected by turbidite deposition, and a state-of-the-art compilation is provided regarding the interpretation of the main trace-fossil groups. Additionally, some ichnological processes and phenomena in the deep sea are illustrated by some examples.

2. CLASSIFICATION OF DEEP-SEA TRACE FOSSILS

In the fossil record, two main behavioral and, hence, often variably preserved groups of trace fossils can be distinguished with respect to the time of sedimentation of the bed in which they occur. Trace fossils that are produced on or very close to the sediment surface become preserved only when they are partly

Developments in Sedimentology, Vol. 64. http://dx.doi.org/10.1016/B978-0-444-53813-0.00021-6

scoured and cast with sediment. For such burrows, Ksiazkiewicz (1954) introduced the term pre-depositional , because they were produced prior to the deposition of a turbidite. The trace fossils within previously deposited beds are named post-depositional (see also Leszczynski, 1993a; Seilacher, 1962). As a rule, the post-depositional suite consists of deeply penetrating trace fossils, whereas the pre-depositional suite includes highly organized, often delicate burrows for which Fuchs (1895) coined the term graphoglyptids . The pre-depositional trace fossils are normally produced by shallow burrowing organisms which are mostly mud dwellers (Kern, 1980).

In most cases, the producers of deep-sea trace fossils are unknown, and commonly they are referred to as worms of unknown taxonomic affinity. However, some burrows are formed by crustaceans (e.g., *Ophiomorpha*, *Thalassinoides*), bivalves (e.g., *Protovirgularia*), or echinoids (*Scolicia*), based on the recognition of characteristic features.

The ethological interpretation of trace fossils is based on the analysis of their functional morphology. The basic ethological categories were introduced by Seilacher (1953) and new ones were subsequently added (see Rindsberg, 2012, for a review). In the deep sea, the most common categories are the following:

- **Pascichnia:** mostly horizontal burrows produced by vagile deposit-feeding organisms (e.g., *Scolicia*, *Protovirgularia*).
- **Fodinichnia:** traces produced during localized deposit-feeding (e.g., *Zoophycos*).
- **Agrichnia:** delicate, mostly shallow, regularly patterned burrow systems showing a range of morphologically varying meanders (e.g., *Helminthorhaphe*), spirals (*Spirorhaphe*), rosettes (*Lorenzinia*), or nets (*Paleodictyon*), produced for the trapping or farming of microbes or other very small organisms; most of the trace fossils within this category are termed graphoglyptids (Fuchs, 1895; Miller, 1991b; Seilacher, 1977a; Uchman, 2003).
- **Chemichnia:** structures produced by organisms feeding on chemosymbiotic microbes; the organisms maintain a connection to oxygenated water but penetrate into anoxic sediments rich in sulfides or ammonium that are required for microbe-feeding (e.g., *Chondrites*, *Trichichnus*).
- **Domichnia:** open burrows used as dwelling structures (e.g., *Ophiomorpha rudis*); transitional forms to fodinichnia or even to agrichnia are common (the case of *O. rudis*; see Cummings and Hodgson, 2011a; Uchman, 2009).
- **Repichnia:** structures recording the movement of organisms (e.g., *Helminthopsis*).

Some of the deep-marine trace fossils display a very complex morphology, suggesting that they may have been produced by more than one ethological activity, for instance, *Hillichnus*, which is interpreted as a bivalve locomotion and feeding trace (Bromley et al., 2003). A complex morphology is ascribed to either the living strategy or the various behavioral activities of the tracemaker that respond to short-term environmental fluctuations. There are, however, more guesses than proofs in this matter (for a discussion, see Miller, 2007).

Numerous studies on deep-sea trace fossils with significant systematic descriptions are available, not only for the Paleozoic, for instance by Delgado (1910), Pfeiffer (1968), Osgood (1970), Chamberlain (1971), Pickerill (1980), Benton (1982a), Stepanek and Geyer (1989), Fillion and Pickerill (1990), Crimes and Crossley (1991), McCann (1993), Orr (1995, 1996), and references therein, Orr et al. (1996); but also for the Mesozoic and Cenozoic, for instance by Sacco (1888), Squinabol (1890), Azpeitia Moros (1933), Crimes (1973, 1977), Ksiazkiewicz (1977), Crimes et al. (1981), McCann and Pickerill (1988), Leszczynski and Seilacher (1991), Miller (1991a, 1993), Uchman (1995a, 1998, 1999, 2001), Buatois et al. (2001), and references therein.

3. PRE-DEPOSITIONAL TRACE FOSSILS

Most of the pre-depositional forms are represented by graphoglyptids, which are preserved often as semi-reliefs resulting from the scouring and casting of shallow burrow systems; most of them are interpreted as agrichnia. They are common, but never abundant on the soles of sandy turbidites. The tubes are commonly less than 2 mm wide. For a discussion of graphoglyptids, the reader is referred to Seilacher (1977a, 2007), Miller (1991b), and Uchman (1995a, 1998, 2003).

Cosmorhaphe (Fig. 1A) is a hypichnial, semicircular ridge forming first- and second-order, regular meanders.

Helminthorhaphe (Fig. 1B) is a hypichnial, semicircular ridge forming only first-order, regular, high-amplitude meanders.

Lorenzinia (Fig. 1C) is a hypichnial structure composed of semicircular bars and knobs radially arranged around a central empty area.

Megagrapton (Fig. 1D) is composed of hypichnial semicircular ridges, forming an irregular net.

Paleodictyon (Fig. 1H) is composed of hypichnial semicircular ridges forming a regular hexagonal net. The meshes of the net are usually 1 to 35, exceptionally up to 140 mm wide.

Spirorhaphe (Fig. 1F) is a hypichnial semicircular ridge forming a spiral with a loop in the centre.

Spirophycus (Fig. 1G) is a horizontal ridge bent at one end in a spiral. This non-graphoglyptid trace fossil is probably a preservational variant of *Nereites*, while scoured and cast. For further details, see Ksiazkiewicz (1977) and Uchman (1998).

Scolicia strozzii (=*Taphrhelminthopsis*; Fig. 1E) is a bilobate hypichnial smooth ridge with a central furrow. At least in Late Jurassic and younger sediments, it is interpreted as a washed-out and cast shallow irregular echinoid burrow (Uchman, 1995a).

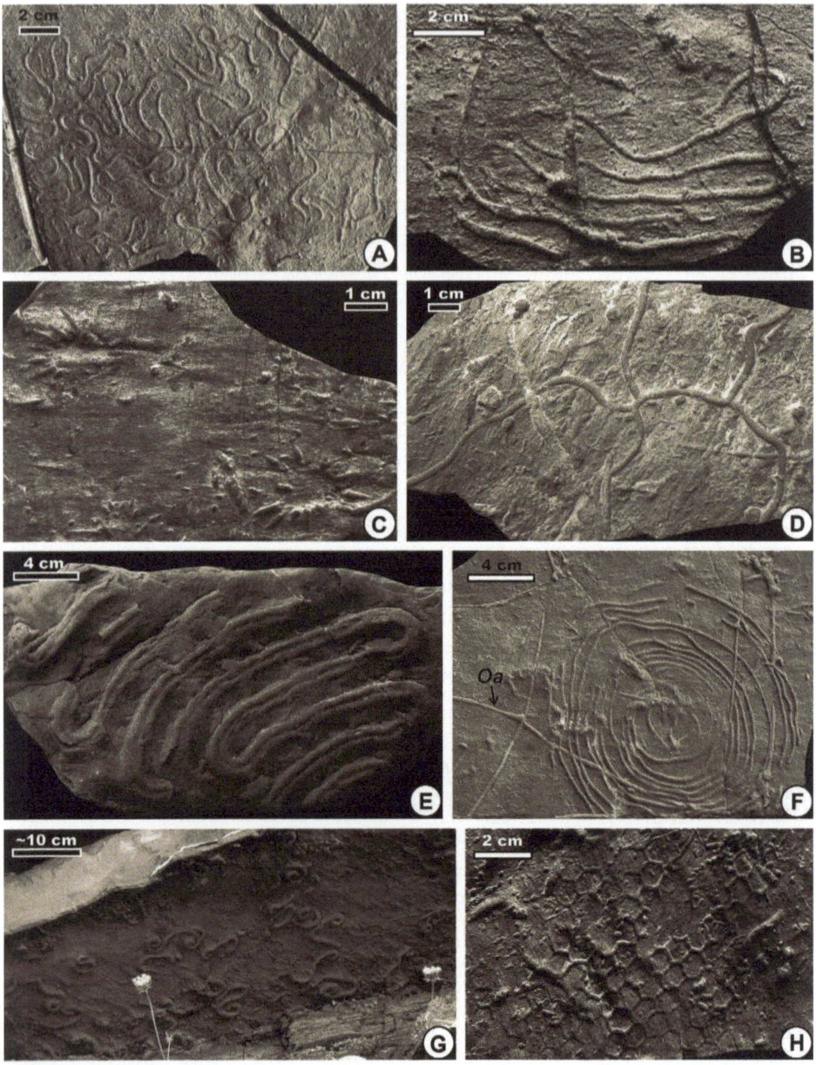

FIGURE 1 Some pre-depositional trace fossils, preserved in hypichnial semi-relief. (A) *Cosmorhaphe sinuosa* (Azpeitia Moros), Słopnice, Ropianka Formation (Senonian Paleocene), Magura Nappe. (B) *Helminthorhaphe japonica* Tanaka. Słopnice, Ropianka Formation (Senonian Paleocene), Magura Nappe, Polish Carpathians. (C) *Lorenzinia nowaki* (Ksiazkiewicz). Sub-Magura Beds (Eocene), Kłodne, Magura Nappe, Polish Carpathians. (D) *Megagrapton irregulare* Ksiazkiewicz, Sromowce Beds (Coniacian Lower Campanian), Jaworki, Potok Skalski section, Polish Carpathians. (E) *Scolicia strozzii* (Savi and Meneghini), Senonian Paleocene flysch of the Ropianka Formation, Polish Carpathians, locality unknown. (F) *Spirorhaphe involuta* (De Stefani) and *Ophiomorpha annulata* (Ksiazkiewicz) (*Oa*), Middle Eocene flysch, about 3 km east of Zumaya, northern Spain. (G) *Spirophycus bicornis* (Heer), Middle Eocene flysch, about 3 km east of Zumaia, northern Spain. (H) *Paleodictyon majus* Meneghini in Peruzzi, Zbludza, Beloveza Formation (Eocene), Magura Nappe, Polish Carpathians.

4. POST-DEPOSITIONAL TRACE FOSSILS

Chondrites (Figs. 2A, H and 5A C) is a three-dimensional, regularly branching tunnel system consisting of an open connection to the surface and numerous tunnels that ramify at angles of 30 60° to form a dendritic pattern. *Chondrites* is interpreted as a deep-tier chemichnion (for discussion, see Fu, 1991; Uchman, 1999; Wetzel, 2008).

Dictyodora (Fig. 2B) is a meandering to looping deep-tier pascichnion composed of a basal subcylindrical structure and a dorsal crest. *Dictyodora* is reported only from Paleozoic deep-sea sediments. For discussion, see Benton and Trewin (1980).

Nereites (Fig. 2C) is a winding to regularly meandering, more or less horizontal pascichnion, consisting of a median back-filled tunnel enveloped by an even to lobate zone of reworked sediment. Synonyms of *Nereites* are post-depositional *Helminthoida* and probably pre-depositional *Spirophycus* (Uchman, 1995a). Modern *Nereites missouriensis* is produced in spatial relation to the redox boundary, $\sim$1 2 cm above the Fe Mn-stained horizon (Wetzel, 2002).

Ophiomorpha (Figs. 1F and 2E, G) comprises variably oriented, branched cylinders, locally with a granulated wall interpreted to be produced by crustaceans. In post-Tithonian deep-sea sediments, *O. annulata* and *O. rudis* occur. *O. nodosa* is typical of shallow-marine sediments. For discussion, see Tchoumatchenco and Uchman (2001), Uchman (2001, 2009), and Cummings and Hodgson (2011a).

Palaeophycus is a branched or unbranched, smooth or ornamented, lined, essentially cylindrical, predominantly horizontal burrow of varying diameter; the fill is passive, typically structureless and of the same lithology as the host rock. These burrows are interpreted as pascichnia or domichnia. For discussion, see Pemberton and Frey (1982) and Keighley and Pickerill (1995).

Phycosiphon (Figs. 2D and 5C) consists of compound antler-like arranged series of spreiten. A spreite is a laminated biogenic structure composed of closely spaced successive tunnel margins produced as the burrow shifts laterally through the sediment. The width of a single spreite is 1 5 mm, and the diameter of the marginal tube is 0.5 1 mm. The marginal tubes show a thin lining. *Phycosiphon* is interpreted as a fodinichnion. For discussion, see Wetzel and Bromley (1994), Naruse and Nifuku (2008), and Bednarz and McIlroy (2009).

Planolites (Figs. 2H and 5C) consists of rarely branched, straight or gently curved, inclined or horizontally oriented, cylindrical tunnels, commonly filled with sediment of different lithology. Differences in diameter, penetration depth, and sediment infill may allow different types to be distinguished (Wetzel, 1981). Active filling of the burrow is indicated by grain-size differences in relation to the surrounding sediment and relatively close packing of the infill. *Planolites* is interpreted as a pascichnion. For further details, see Pemberton and Frey (1982) and Keighley and Pickerill (1995).

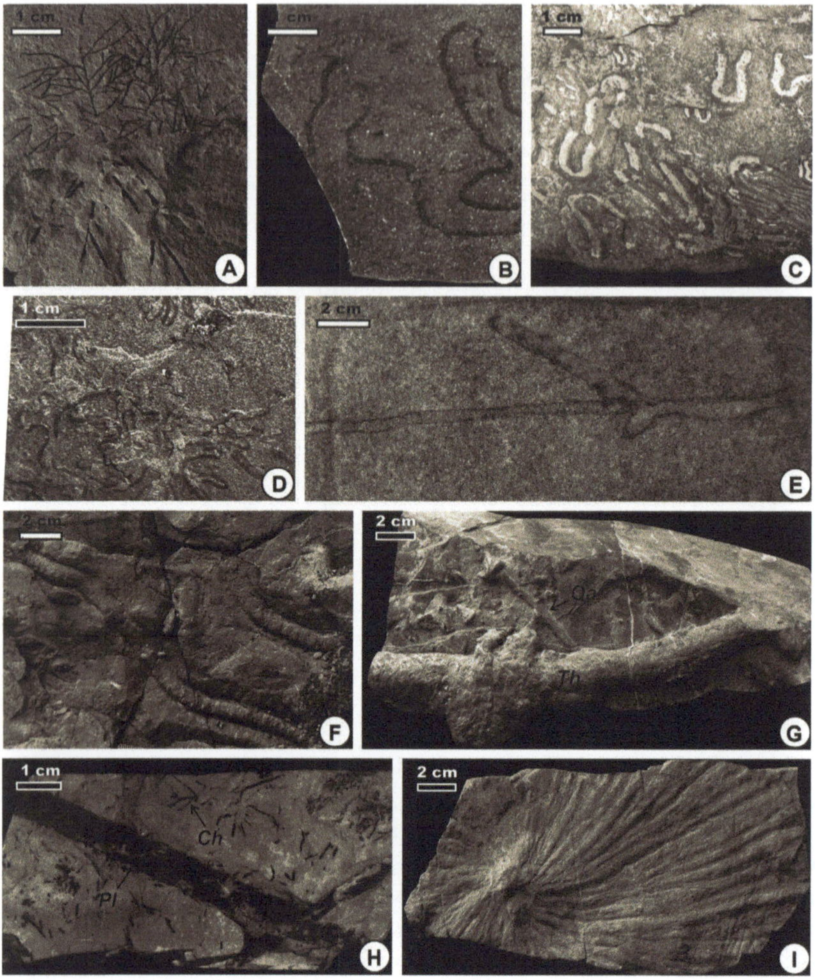

FIGURE 2 Some post-depositional trace fossils. (A) *Chondrites intricatus* (Brongniart), endichnial full reliefs in turbiditic marlstone, parting surface, Krzeczkowa, Holownia Siliceous Marls (Turonian Lower Santonian), Skole Nappe, Polish Carpathians. (B) *Dictyodora liebeana* Geinitz in horizontal section. Turbiditic mudstone, Kulm facies, Jivova Quarry, Moravia, Czech Republic. (C) *Nereites irregularis* Schafhautl, parting horizontal surface of a muddy/silty turbidite, top view, Szczawa-Głebieniec, Ropianka Formation (Paleocene), Magura Nappe, Polish Carpathians. (D) *Phycosiphon incertum* von Fischer-Ooster, endichnial full relief on a parting surface at the top of a sandy turbidite, Zbludza, Beloveza Beds (Eocene), Magura Unit, Polish Carpathians. (E) *Ophiomorpha annulata* (Ksiazkiewicz), full reliefs in a turbiditic sandstone, Beloveza Beds (Eocene), Naszacowice, Magura Unit, Polish Carpathians. (F) *Scolicia prisca* Quatrefages, weathered endichnial full relief at the top of a turbiditic sandstone, field photograph, Łabowa, Beloveza Beds (Eocene), Magura Nappe, Polish Carpathians. (G) *Thalassinoides* isp. (*Th*) and *Ophiomorpha annulata* (Ksiazkiewicz) (*Oph*), full reliefs on the lower surface of a sandy turbidite, Lubomierz, Beloveza Beds (Eocene), Magura Nappe, Polish Carpathians. (H) *Chondrites intricatus* (Brongniart) (*Ch*) and *Planolites* isp. (*Pl*) reworked by *Chondrites* isp., parting surface of a marly mudstone, Muntigl, Atlenbacher Schichten (Maastrichtian Paleocene), Rhenodanubian Flysch, Austria. (I) *Zoophycos* isp., parting surface of a marly turbidite, Kanina, Ropianka Formation (Senonian Paleocene), Magura Unit, Polish Carpathians.

Scolicia (Figs. 2F and 5B) is a large, bilaterally symmetrical, subcylindrical pascichnion filled with meniscate lamellae, broadly divided into two concave sets. The basal face shows a faint bilobate morphology with two strings and the upper face a slight axial groove. In transverse cross sections, the concentric structure of the bilobate lamellae is the most conspicuous feature. *Scolicia* is produced by *Spatangus*-like irregular echinoids. For discussion, see Uchman (1995a).

Thalassinoides (Figs. 2G and 5A C) is a burrow system consisting of horizontally branched tunnels connected to the surface by more or less vertical shafts; swellings often occur at points of branching. A typical asymmetrical (eccentric) fill structure often results from active filling by the burrowing organism or collapse of the burrow walls. The tubes are 5 20 mm in diameter and smooth-walled. *Thalassinoides* is interpreted to be produced by crustaceans. For further details, see, for example, Frey et al. (1984), Ekdale (1992), and Schlirf (2000).

Trichichnus is a thread-like, rarely branching, cylindrical burrow commonly exceeding 20 cm in length. Its fill is usually pyritized. For further details, see McBride and Picard (1991) and Uchman (1995a, 1999).

Zoophycos (Fig. 2I) is a spreiten structure composed of distinct levels or coilings and/or lobes. The regular fill structure appears as crescent-shaped en-echelon features in vertical section. The presence of open (collapsed) or actively filled marginal tubes indicates lower or higher oxygen contents, respectively, in the surrounding waters (e.g., Wetzel and Werner, 1981). For further details, see Bromley and Hanken (2003).

5. ICHNOFACIES

The ichnofacies model was introduced by Seilacher (1964, 1967; Fig. 3) to distinguish between the different trace-fossil associations that characterize continental, shallow- and deep-marine deposits, so-called molasse-type deposits in

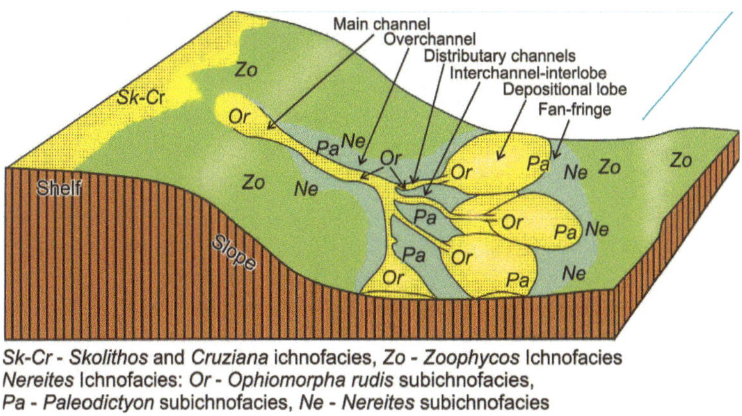

FIGURE 3 Ichnofacies from shelf to deep sea *(modified after Uchman, 2007a).*

shallow foreland basins and turbidite-dominated, so-called flysch deposits in deep-marine basins (i.e., trenches) in the front of orogens. Initially, the resolution of this model was low within deep-marine turbidites, being grouped into the *Nereites* Ichnofacies, which is characterized by graphoglyptids and meandering trace fossils (Seilacher, 1967). Muddy deposits below the upper offshore zone are ascribed to the *Zoophycos* Ichnofacies, which is characterized by the ichnogenera *Zoophycos*, *Phycosiphon*, and *Chondrites*. Traditionally, this ichnofacies is related to slope deposits (Seilacher, 1967). Principal ecological factors, such as organic-matter input, sedimentation rate, grain size, oxygenation, etc., vary significantly in the deep sea, and consequently deep-sea ichnofacies cannot be ascribe to a specific water depth with any precision (Wetzel, 1983).

The ichnofacies are considered to some extent as taphofacies, and thus as dependent on the level of preservation of the trace fossils (e.g., Bromley and Asgaard, 1991). For instance, graphoglyptids characterizing the *Nereites* Ichnofacies require subtle scouring and casting in order to be preserved. Photographs of the modern deep-sea floor, however, document graphoglyptid lebensspuren (e.g., Gaillard, 1991), also in places where their preservation seems almost impossible due to the absence of such processes (Ekdale, 1985; Miller, 1991b). This is probably true for many fossil pelagites not preserving graphoglyptids but trace fossils typical of the *Zoophycos* Ichnofacies (e.g., Wetzel, 2010).

With time, the *Nereites* Ichnofacies has been subdivided into the *Ophiomorpha rudis* subichnofacies for thick-bedded sandstones in channels and proximal lobes in turbidite successions (Heard and Pickering, 2008; Knaust, 2009; Phillips et al., 2011; Uchman, 2001, 2009); the *Paleodictyon* subichnofacies for more sandy, medium-to-thin bedded normal flysch; and the *Nereites* subichnofacies for mud-rich distal flysch (Seilacher, 1974). Roughly, the *Ophiomorpha rudis Paleodictyon Nereites* subichnofacies may express a bathymetric trend from inner to outer fan (Uchman, 2007a) and a non-bathymetric trend from channel axis, levee to overbank or inter-channel areas (Monaco et al., 2010, Olivero et al., 2010; Uchman, 2009). In addition, within sandy flysch deposits, graphoglyptids are frequent (e.g., *Paleodictyon*, *Desmograpton*) and muddy deposits contain relatively more meandering forms. Monaco et al. (2010) noted a higher abundance and diversity of graphoglyptids in detached lobes than in attached lobes, and a reduced ichnodiversity in channelized areas. However, the diversity and composition of the trace-fossil associations are affected by the lithological variability and other ecological (Uchman, 2001) and evolutionary factors (Uchman, 2004a).

The *Nereites* Ichnofacies occurs over a wide bathymetric range. To give an example, the *Paleodictyon* subichnofacies occurs in an Eocene turbidite succession a few tens of meters below tempestites in the Sinop Basin, Turkey (Uchman et al., 2004). This subichnofacies occurs even in tempestite-bearing, deep intra-shelf troughs reported from the Late Cretaceous shelf of Tanzania

(Ernst and Zander, 1993). However, it is questionable whether these cases should be regarded as some odd exceptions or as an indication for a habitat with such conditions. The trace-fossil assemblage of the Early Cretaceous Kamchia Formation, Bulgaria, contains a mixture of forms typical of both the *Nereites* Ichnofacies (*Squamodictyon*) and the *Cruziana* Ichnofacies (*Curvolithus*, *Gyrochorte*). Probably, the sediments of the Kamchia Formation were deposited in an offshore to deeper setting with storm sand layers and marly background sedimentation. It is possible that storm-induced currents transported producers of such traces from the shelf into the deeper sea (Uchman and Tchoumatchenco, 2003; cf. Savrda et al., 2010).

In general, the co-occurrence of deep-sea and shelf trace fossils can be caused by the transport of tracemakers by storm-induced or other downslope currents from shallower to deeper areas. This reasoning has been used to explain the occurrence of typical shallow-marine trace fossils, mostly *Ophiomorpha* and *Thalassinoides*, in deep-sea sediments (Crimes, 1977; Follmi and Grimm, 1990; Wetzel, 1984, 2008). However, most of the producers of so-called shallow-water trace fossils, such as *Ophiomorpha*, appear to be well adapted to the deep-sea environment, living there for a long time in deep tiers. Size analysis of *Ophiomorpha* suggests that the tracemakers cover the whole age spectrum in the deep sea (Uchman, 1995a). Ichnotaxonomic analysis shows distinct differences of the ichnospecies level, for example, between the deep shallow-water forms and true shelf trace-fossils (Uchman, 1995a, 2009). Thus, most of the deep shallow-water forms can be ascribed probably to permanent deep-sea residents.

In addition to using ichnofacies, distinguishing trace-fossil assemblages or ichnocoenoses based on certain facies settings or even sedimentary features of individual beds is also useful. For instance, Crimes and Crossley (1991) noted for the Aberystwyth Grits Formation (Silurian) in Wales that *Paleodictyon* occurs in intervals in which shale is thicker than sandstone, whereas meandering traces (their *Helminthoida crassa*) occur in opposite situations. Crimes (1973) showed that the trace fossils content varies from lithofacies to lithofacies and he found that rosette traces are more abundant in proximal turbidites in the Paleogene in Zumaia (N Spain), while spiral and patterned forms as well as winding and meandering forms are more abundant in the distal facies. Such a distribution was challenged by Uchman (2001) for the Hecho Group (Eocene) in N Spain and by Cummings and Hodgson (2011b), partly for the same area. According to Crimes (1973), the distribution of trace fossils is only partly controlled by preservational factors. While studying the Gurnigel and Schlieren flysch units (Paleocene Eocene) in Switzerland, Crimes et al. (1981) demonstrated that the channelized inner fan facies contain the so-called shallow-marine ichnotaxa (*Arenicolites*, *Skolithos*), but meandering (except for *Helminthopsis*) and network forms are absent. The inter-channel facies contain the radiating, meandering, spiral, and network forms. In the depositional lobes and lobe fringes, trace fossils are more abundant, with a maximum in the channelized parts of depositional lobes, where the shallow-water forms can be still present. Monaco et al. (2010) distinguished five ichnocoenoses in the Oligocene Miocene foredeep and piggy-back basins of the Apennines. They noted that some ichnotaxa are more abundant in certain settings, for example, *Scolicia* in channelized fan lobes and *Arthrophycus strictus* in the middle part of the lobes. It is questionable whether similar relationships can be found in other areas and deposits of different age. Cummings and Hodgson (2011b) considered the ratios of pre- to post-depositional forms as the most powerful tool in the characterization of the trace-fossil distribution in different parts of the depositional system in the Late Cretaceous Eocene deep-sea fan sediments of the Basque Basin, northern Spain. They concluded that the higher diversity and intensity of bioturbation and domination of the pre-depositional forms is typical of marginal parts of channel environments and lobes, while the axial part of channels displays a lower diversity of trace-fossil assemblages dominated by post-depositional forms. In the axial part of lobes, post-depositional forms exclusively occur with a dominance of *Ophiomorpha*. Moreover, fan fringe deposits are characterized by highly diverse trace-fossil assemblages with both pre- and post-depositional forms, while the basin-floor fan is dominated by *Zoophycos*. Such a picture of trace-fossil distribution is in general consistence with the so-far published data, showing the differences between the distal and proximal facies (e.g., Heard and Pickering, 2008; Ksiazkiewicz, 1977, Uchman, 2001) and the subichnofacies trends presented above. Phillips et al. (2011) discovered the highest diversity (but generally low less than 10 ichnogenera) in the Eocene/Oligocene heterolithic facies of the Grès d Annot Basin, SE France, which are interpreted as an accumulation of distal turbidites on a protected slope or channel filling.

Giannetti and McCann (2010) noted a strong dependence of the trace-fossil distribution on lithology, substrate consistency, and food supply in the carbonate-rich turbidite succession of the Upper Paleocene in the Zumaia section, northern Spain. Giannetti (2010) stressed also fluctuations in the rate of sedimentation as a controlling factor for the same deposits.

6. TRACE-FOSSIL ASSEMBLAGES

The composition and diversity of deep-sea trace-fossil assemblages can record paleoenvironmental and ecological conditions. Ekdale (1985) distinguished between K- and r-selected ichnotaxa. K-selected ichnotaxa are produced by animals adapted to a stable environment with low or moderate ecological stress; they usually reproduce slowly. Their diversity is high, but their abundance is relatively low. Graphoglyptids are the best example. Producers of r-selected trace fossils, mostly opportunists, are adapted to instability and high environmental stress. The morphology of these trace fossils is often simple and their diversity is low, but their abundance can be high.

In turbidites, K-selected forms are mostly pre-depositional traces, whereas r-selected forms are normally among the post-depositional burrows (Tunis and

Uchman, 1996b). The ratio between r-selected and K-selected ichnotaxa can vary from formation to formation in one basin in response to environmental changes such as oxygenation, sedimentation rate, and grain size (Tunis and Uchman, 1996a; Uchman, 1991, 1992, 2004b). Commonly, trace-fossil assemblages rich in graphoglyptids occur in thin- to medium-bedded turbidites characterized by greenish to grayish shales, formed under moderately oligotrophic conditions (Miller, 1991b; Tunis and Uchman, 1996b). However, Olivero et al. (2010) noted abundant graphoglyptids in levees of a channel formed in Eocene deeper shelf settings of Tierra del Fuego in Argentina, while Monaco et al. (2010) recorded occurrences in the levees of deep-sea channels and in turbidites along the margin of a submarine high in the Oligocene Miocene turbidites of the Apennines in Italy. All these settings, especially channel areas, are expected to be rich in organic matter transported from the land and shelf, but most of it being refractory while derived from a terrestrial source (e.g., Baudin et al., 2010).

Not all turbidity currents bring enough oxygen allowing colonization of the sea floor by infauna in an area experiencing a general oxygen deficiency. Probably, many turbidites were not colonized because of this reason (Leszczynski, 1991; Uchman, 1991, 1992; Wetzel and Uchman, 2012). The ichnodiversity is low if oxygenation drops below a threshold value. However, when slowly accumulating sediment is very well oxygenated and if only little organic matter is available for the burrowing organisms, diversity can decrease (Leszczynski and Uchman, 1993; Wetzel and Uchman, 2012). This is well demonstrated by the relationship between ichnodiversity and sediment color: the highest diversity occurs in gray sediments, and the lowest in red or black deposits (Leszczynski, 1993b). The diversity can also be lowered by too frequent disturbances by turbidity currents (Leszczynski, 2003). Taking into account observations of the impact of modern physical disturbances on soft-sediment benthos (e.g., Hall, 1994), an influence of this factor is possible.

7. ICHNOFABRICS

Turbidites, like other event beds, are normally colonized from the top. Turbidity currents and other gravity-induced flows can transport tracemakers from shallower to deeper water. These tracemakers can survive transport and produce burrows (Crimes, 1977; Follmi and Grimm, 1990), but this mechanism appears to be overestimated in the interpretation of the ichnological record (see Section 5).

In detail, turbiditic/pelagic or turbiditic/hemipelagic couplets consist of rapidly deposited turbidites and slowly accumulated background sediment. Basically, two intervals can be distinguished in the couplets, a spotty and an elite layer (Uchman, 1999). These correspond roughly to the mixed layer and the transitional layer, respectively, in modern deep-sea sediments (Berger et al., 1979; Bromley, 1990, 1996; Ekdale and Berger, 1978; Ekdale et al., 1984, and references therein). The characteristic spotty layer is entirely bioturbated (Fig. 5A) and it occupies the uppermost part of a turbidite, commonly corresponding to Bouma s $T_d - T_e$ intervals (Fig. 4; Uchman, 1999). Oval spots of different color and sharpness of contours are visible against the indistinctly mottled background. The spots are cross sections of trace fossils, commonly *Planolites* and *Thalassinoides*. In some layers, the color contrast is so low that the layer seems to be structureless. This picture can be altered in thin-bedded turbidites, if some deep-tier trace fossils of the so-called multilayer colonizers *sensu* Uchman (1995b) or crossichnia *sensu* Monaco and Caracuel (2007), such as *Chondrites* and *Ophiomorpha*, penetrate from the overlying bed to the spotty layer of the underlying bed (Rajchel and Uchman, 1998). Their sections are visible as last-generation spots (Fig. 5A), having very distinct margins, strong color contrast, and commonly different lithology from the host rock. The lithology of the spotty layer differs from that of the underlying sediment; it is more fine-grained and shows a different color or tint. In sediments deposited below the Calcite Compensation Depth (CCD), it is free of carbonate, at least in the upper part. Above the CCD, the $CaCO_3$ content in the spotty layer can be higher than in the underlying

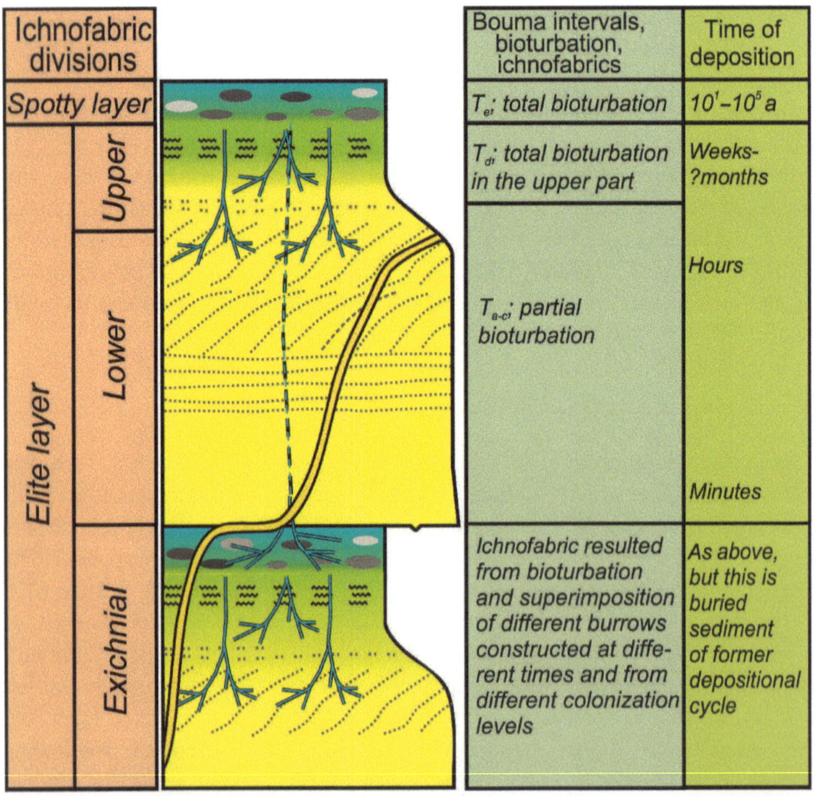

FIGURE 4 Model of ichnofabrics in turbidite/hemipelagite couplets *(modified after Uchman, 1999)*.

siliciclastic deposits if the supply of pelagic carbonate particles is significant. In oligotrophic, well-oxygenated environments, the spotty layer is lighter than the underlying sediments. In eutrophic environments, the spotty layer is commonly darker than the sediment underneath (Wetzel and Uchman, 1998). Some trace fossils of the spotty layer are preserved as pre-depositional forms on the soles of overlying turbidites. They belong to the so-called background ichnofauna (Leszczynski, 1993a), which includes mainly graphoglyptids, which are hardly recognizable in cross section.

The spotty layer is sometimes mistaken for purely pelagic or hemipelagic sediment but in fact represents the mixed layer in modern deep-sea sediments, which in the rock record is not only composed of pelagic material but, due to bioturbation, also of some material derived from turbidity currents (Uchman 1995a, 1999). Pelagic sediment is not only transported downward into the turbidite layer, but turbidite particles are also conveyed upward to the sea floor that is mainly covered with pelagic material. This can be deduced from the fact that, in modern environments, differently colored sediment of the transitional layer is conveyed to the sea floor by burrowers (Ekdale et al., 1984).

The spotty layer cannot be considered as being entirely equivalent to the mixed layer when fossilized, especially with respect to thickness. The formation of the spotty layer in turbidites is probably a dynamic process. Initially, a thin mixed layer is formed at the top of the sediment following turbidite deposition. Trace fossils produced during this phase are hard to recognize due to lithological contrast and do not have sharp outlines. At the same time, pelagic particles accumulate. If the rate of biological reworking exceeds the rate of the accumulation, the pelagic material becomes totally mixed by local, homogenizing processes and non-local, far downward-reaching transport within burrows into turbiditic sediment. The lower boundary of the spotty layer can be either distinct, because of the lithological contrast, or gradational due to intense bioturbation by producers of *Scolicia* or *Ophiomorpha*. The contrast between the spotty layer and the turbiditic mud can be low if they are lithologically similar.

The most notable trace fossils occur below the spotty layer in the elite layer (elite trace fossils *sensu* Ekdale and Bromley, 1991). In most marly turbidites, deep-tier elite *Chondrites*, *Planolites*, *Nereites*, *Phycosiphon*, and *Scolicia* occur (Fig. 5A and B). To estimate the depth of burrowing, the distance to the top of the elite layer should be known. Some burrows penetrate below turbidites and they form the so-called exichnial elite layer (Uchman, 1999) (Fig. 5C).

Ichnofabric analysis is useful in reconstructing the complex colonization history of a single layer in turbidites/hemipelagic sediments. It appears that the cross-cutting relationships between trace fossils result not only from the accretion of sediment and upward movement of tiers such as in (hemi)pelagic sediments, but also from the sequential colonization of event layers over time (Wetzel and Uchman, 2001). Newly deposited turbidites contain relatively well-oxygenated pore water and a relatively high amount of food. Initially,

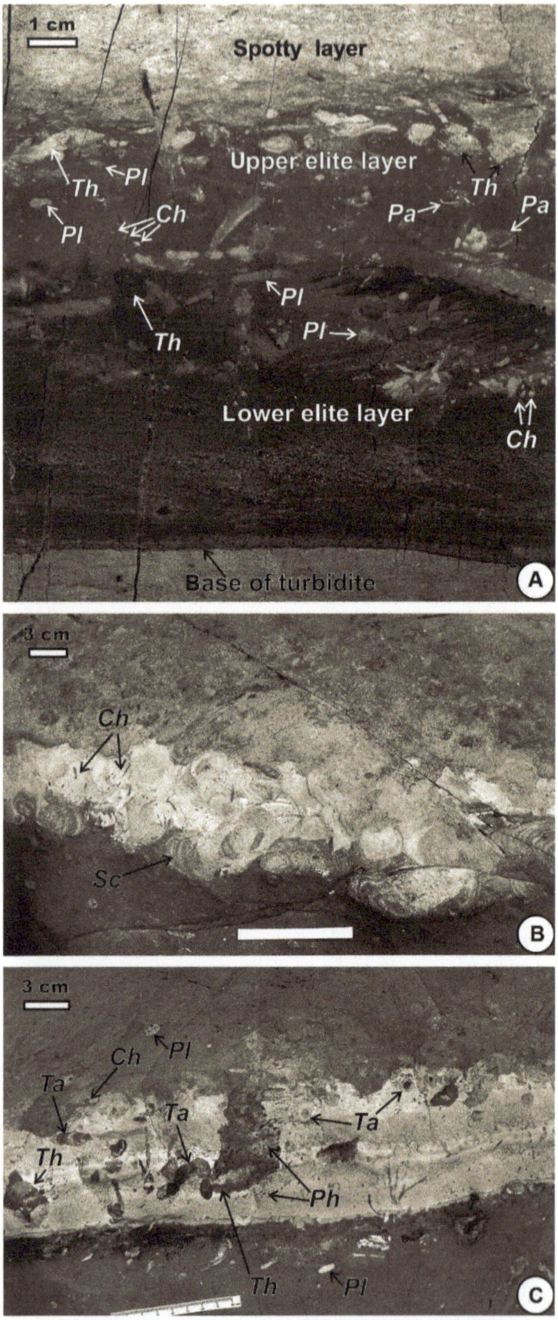

FIGURE 5 Ichnological aspects of turbidites in vertical section. (A) Thick turbidite, Late Cretaceous flysch of the Gropello area, Southern Alps. *Chondrites* (*Ch*), *Planolites* (*Pl*), *Palaeophycus*

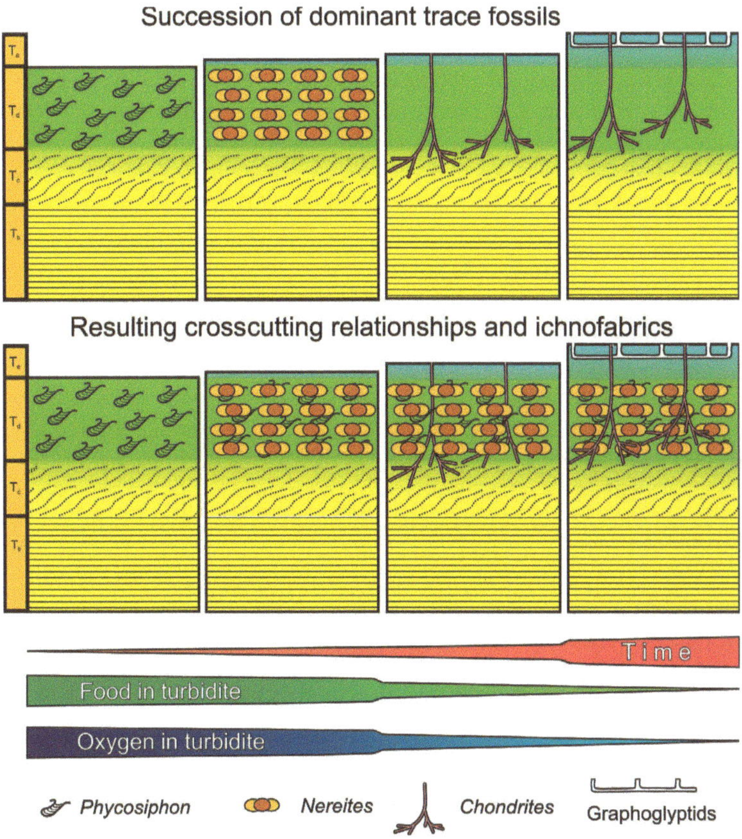

FIGURE 6 Model of sequential colonization in turbidite/hemipelagite couplets according to Wetzel and Uchman (2001). T_b, T_c, T_d, T_e indicate the Bouma intervals (see text). *Modified after Uchman (2007a).*

the sediment is colonized by small opportunistic deposit feeders producing *Phycosiphon*, rather than by the larger sediment-feeding producers of *Nereites* (Fig. 6). Both burrows penetrate the sediment horizontally and have no connection to the sea floor, benefiting from the still oxygenated pore water. When oxygen and food decreases to a level that horizontal mining becomes inefficient, the sediment is colonized by stationary burrowers producing *Chondrites*;

(*Pa*), and *Thalassinoides* (*Th*). (B) Totally bioturbated turbidites in a Neogene slope basin, Eastern Cape; Northern Island, New Zealand. *Scolicia* (*Sc*) cross-cut by *Chondrites* (*Ch*). (C) Totally bioturbated turbidites in a Neogene slope basin, Eastern Cape; Northern Island, New Zealand. *Chondrites* (*Ch*), *Planolites* (*Pl*), *Phycosiphon* (*Ph*), *Taenidium* (*Ta*), and *Thalassinoides* (*Th*). Note light turbidite-derived material as infilling of burrows below the base of the turbidite (exichnial elite layer).

the producing organisms take advantage of chemosymbionts utilizing the gradient between reducing pore water and oxygenated bottom water pumped into the burrow via the connection to the sea floor. During the long periods between the deposition of successive turbidites, commonly hundreds or thousands of years, the slowly deposited background sediment is colonized by graphoglyptids. There are also tracemakers independent of this succession, such as echinoids producing *Scolicia*, and the multilayer colonizers *sensu* Uchman (1995b) which overprint the pre-existing ichnofabric to varying degrees. Such a scenario of sequential colonization, elaborated on the basis of Paleogene deep-marine turbidite successions, is also applicable to older units, including at least part of the Upper Paleozoic (Mikulas et al., 2002).

Certain types of ichnofabrics can characterize different parts of a depositional system. For instance, Knaust (2009) recognized seven types of ichnofabrics in a Campanian deep-sea fan in the Norwegian Sea and related them to amalgamated channels, lobate sand sheets, proximal and distal overbank, fan fringe, and hemipelagic deposits.

8. ICHNOLOGY OF CARBONATE TURBIDITES

Trace-fossil preservation and diversity differs in siliciclastic and carbonate (marly) Cretaceous Paleogene flysch deposits (Giannetti and McCann, 2010; Savrda, 2012; Uchman, 1999). Generally, trace fossils in marly turbidites are more often of post-depositional than of pre-depositional origin. Graphoglyptids are relatively rare, although there are exceptions (e.g., Monaco and Checconi, 2010), and this is probably the result of preservation conditions and benthic food level. For the preservation of pre-depositional trace fossils, particularly graphoglyptids, subtle erosion is necessary, which scours such burrow systems before casting. Late Jurassic and younger marls consist mostly of microfossil remnants (e.g., Tucker, 1974), which behave hydraulically like silt or even clay particles. Therefore, turbidity currents carrying calcareous mud cannot cause erosion similar to that of muddy siliciclastic suspension currents. Thus, marly turbidites mask a shallow-tier ichnofauna. On the other hand, the marly turbidites often display features of eutrophic environments, in which graphoglyptids are less abundant than in oligotrophic environments (Uchman, 2003, 2007b). Only few trace fossils apparently are characteristic of either siliciclastic or marly sediment, respectively. Most probably, *Cladichnus fischeri* is restricted to marly sediments. *Nereites irregularis* is very common in marly turbidites; however, it also occurs in carbonate-free sediments, such as in the variegated shales of the Upper Paleocene Lower Eocene in the Carpathians (Leszczynski and Uchman, 1993). Filling of some late colonizer burrows in calcarenitic tempestites consist of calcilutite. This is interpreted as an effect of the long settling time of carbonate from suspension, which can take months (Miller et al., 2004).

9. TRACE FOSSILS AS INDICATORS OF ECOLOGICAL AND PHYSICAL SEDIMENTARY PARAMETERS

Tracemakers in the deep sea are subjected to various ecological and physical factors, but their influence is recognized not to the same degree. For instance, oxygenation of pore and bottom water is well known, but the influence of temperature is not. The trace-fossil record of environmental factors is also influenced by preservation, which depends on, among other factors, erosion of the sea floor and intensity of bioturbation in deep tiers. Action of currents can be reflected in the morphology and orientation of trace fossils.

The influence of temperature on deep-sea organisms is not really known. In modern environments, the temperature in the deep sea is quite uniform and low, but in the geological past, like the Early Eocene, temperature was higher than 10C (e.g., Brass et al., 1982; Shackleton, 1986). Such changes influence oxygenation, the development of microbes and the turnover of organic matter (e.g., Thomas et al., 2000). The ichnodiversity in the Phanerozoic deep-sea deposits appears to partly reflect temperature changes (Uchman, 2004a), especially the decline of diversity in the Late Carboniferous and Permian related to the Gondwana glaciations, and the increase in diversity during the Early Eocene. This is also valid for shallow-marine trace fossils and burrowing organisms (Goldring et al., 2004).

Oxygenation of sediment is an important parameter controlling burrowing organisms. Normally, there is enough time for the colonization of turbidites during the usually long interturbidite period. Nevertheless, some turbidites or packages of beds are not colonized. This may be related to anoxic events, especially if the interturbiditic background sediments are dark (Uchman, 2004b). At a larger scale, fluctuations in the oxygenation of the sediments influence the trace-fossil diversity in turbidite successions. This is well recorded in the Early Cretaceous flysch in the Polish Carpathians (Uchman, 2004b). In some beds or packages of beds, the ichnofabrics are dominated by one or two trace fossils, such as *Scolicia*, which record the invasion of their producer. This phenomenon is related to the bulldozing effect *sensu* Thayer (1979), who concluded that relatively large burrowers prevent or reduce colonization of the substrate by immobile suspension-feeders. Thus, the term bulldozing effect may be expanded to include the above-described phenomenon in deep-sea environments (Uchman, 1995a).

Trace fossils can also record the erosion caused by turbidity currents, if the tiering pattern can be recognized and calibrated. Then, the minimum depth of erosion below a turbidite can be estimated based on the presence of trace fossils belonging to a specific tier that is preserved as semi-relief on the sole of a turbidite (Wetzel and Aigner, 1986).

Some pre-depositional trace fossils display characteristic asymmetrical scouring, which can be used for the reconstruction of the current direction of a mass flow if better evidence is lacking. The steeper slopes of ridges are always up-current, similarly to flute casts. The orientation of some trace fossils, like *Paleodictyon* meshes, can also indicate the paleocurrent direction (Crimes and Crossley, 1980). In addition, several physical taphonomic features of graphoglyptids (stretching, smoothing, bending, tapering, thinning, and straightening) indicate that the burrow systems were exhumed at least partly by bottom currents prior to turbidite deposition (Monaco, 2008). The direction of the currents can be deduced from such features.

10. CYCLICITY AND SEQUENCE STRATIGRAPHY ASPECTS

Milankovitch cyclicity controls many aspects of sedimentation, and this is especially evident in the pelagic environment. For turbidites, Heard et al. (2008) noted cycles in the intensity of bioturbation of 41 ka and about 112 ka in very-thin- and thin-bedded turbidites of the Eocene Ainsa Basin in the southern Pyrenees, related to glacio-eustasy and related changes of bottom-water oxygenation.

The 600 m thick section of Eocene, mostly thin- to medium-bedded turbidites in the Basque Basin, N Spain, displays changes in ichnodiversity, without any clear relationship with turbidite facies trends. The changes appear to reflect sea-level fluctuations affecting the basin-margin areas as evidenced by sequence-stratigraphic correlations (Rodr guez-Tovar et al., 2010). In this way, sea-level fluctuations influence the trophic conditions, temperature, and oxygenation and thus the ichnofauna. Corresponding trends have also been noted for the benthic foraminifers.

In the Late Paleocene turbidites of the Zumaia section, northern Spain, the lowstand systems tract, and the early transgressive systems tract are characterized by a reduced rate of sedimentation. These intervals display the *Zoophycos* trace-fossil assemblages (*Zoophycos*, *Chondrites*, *Trichichnus*, *Palaeophycus*), while the highstand systems tract, characterized by a high rate of carbonate accumulation, displays the *Palaeophycus tubularis* assemblage (*Palaeophycus tubularis*, *Chondrites*, *Ophiomorpha recta*, *Fucrculosus*, *Nereites*, *Thalassinoides*, *Phycodes*, *Trichichnus*; Giannetti and McCann, 2010; see also Giannetti, 2010).

The influence of cyclicity and eustasy on the ichnology of turbidites is a new aspect that now becomes increasingly studied. Nevertheless, the results suggest that not only sedimentological processes affect trace fossils, but that complex, often multi-feedback mechanisms influence the atmosphere, hydrosphere, and biosphere, and thus also sedimentology and ichnology.

11. EVOLUTIONARY ASPECTS

Tracemakers are, just like other organisms, subject to evolutionary changes. The ichnological record of their changes and the impact on sediments is intriguing. This problem was outlined by Seilacher (1962), but the first attempts to elucidate the evolution of deep-sea trace-fossil communities were made later.

For this purpose, the diversity of trace fossils per formation was analyzed; Seilacher (1974, 1977b) compared 16 formations; Crimes (1974) did so for a similar number; McCann (1990) studied 34 formations, Orr (2001) 50 Ordovician Carboniferous formations, and Uchman (2004a) 151 Phanerozoic flysch formations (Fig. 7).

A sharp increase in the number of ichnogenera in Cretaceous Neogene flysch was described by Crimes (1974), but he emphasized that more detailed studies were needed. An increasing diversity of deep-sea trace fossils during the Phanerozoic, with a rapid acceleration in the Cretaceous, was found by Seilacher (1974, 1976), who subsequently mentioned a gradual increase in the diversity of flysch ichnocoenoses through the Phanerozoic (Seilacher, 1977b), but then returned to his previous idea and proposed a mid-Cretaceous diversity burst (Seilacher, 1978). These findings and hypotheses were tested by McCann (1990), who concluded that the increase of diversity was neither gradual during most of the Phanerozoic nor rapid in the Cretaceous; in addition, he drew attention to a methodological problem. For instance, the Jurassic-Tertiary flysch of the Polish Carpathians had been taken as one formation, although it includes different units deposited in a number of basins (Ksiazkiewicz, 1977). According to Orr (2001: 270), the diversity of deep-sea trace-fossil

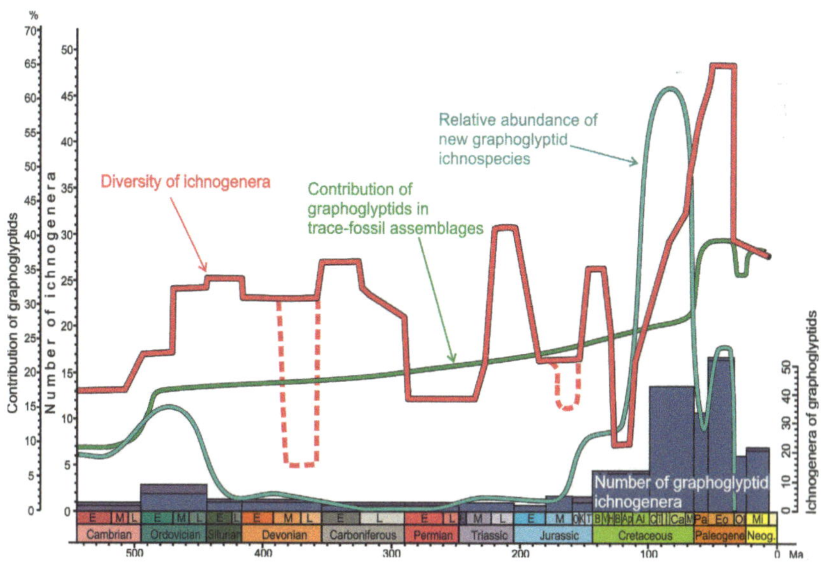

FIGURE 7 Diversity through time of deep-sea ichnogenera (the dashed line shows some less well-documented portions of the curve). The graphs show the contribution of graphoglyptids in trace-fossil assemblages and the relative abundance of new graphoglyptid ichnospecies, whereas the chart shows the number of graphoglyptid ichnogenera in the Phanerozoic (upper portions of the columns express ichnogenera tentatively included in the graphoglyptids). *Based on Uchman (2003, 2004a), with complementary data by Uchman (2007c) and Wetzel et al. (2007), updated.*

assemblages shows a distinct increase from the Cambrian to the Ordovician, but later differences are considered as not obvious .

The diversity-*versus*-time graph (Fig. 7; Uchman, 2004a) indicates that the diversity of deep-sea trace fossils markedly changed through the Phanerozoic, with peaks in the Ordovician Early Silurian and Early Carboniferous, a low in the Permian to early Late Jurassic, a peak in the Tithonian to Aptian, a low in the Albian, and a maximum peak in the Eocene. The contribution of graphoglyptids gradually rose up to the end of the Mesozoic, with a peak in the Paleocene Eocene and a low in the Oligocene. The diversity changes were influenced mostly by the availability of, and competition for, benthic food, the temperature, and oxygenation of bottom water, and also, indirectly, by the varying areas of turbidite sedimentation (i.e., suitable deep-sea habitats). The diversity of deep-water trace fossils was not severely affected by the major biotic crises, such as those at the Ordovician/Silurian, Frasnian/Famennian, Triassic/Jurassic, and Cretaceous/Tertiary boundaries (for details, see Uchman et al., 2005), except for the minor Eocene/Oligocene crisis. However, the Ordovician/Silurian, Cretaceous/Tertiary, and Paleocene/Eocene crises influenced the graphoglyptid ichnodiversity and their relative abundance (Uchman, 2003). The decrease of diversity after the Late Carboniferous was probably caused by the Gondwana glaciations and then reinforced by the Permian/Triassic mass extinction. Recovery took a long time, and diversity remained low until the end of the Jurassic, although the data from Oman show that the diversity in Tethyan refuge areas was already high in the Late Triassic (Wetzel et al., 2007).

The diversity of graphoglyptids (Uchman, 2003) was low from the Cambrian through the Middle Jurassic, with a minor peak in the Ordovician, followed by a drop in the Silurian. In the Late Jurassic, there was a gradual increase. The ichnologic radiation was probably delayed in the Early Cretaceous by the widespread anoxia at that time. Then, a marked rise occurred in the Late Cretaceous, probably during the Turonian. Graphoglyptids display an increase in complexity with time (Seilacher, 1977a), showing a distinct acceleration in the infaunal colonization during the Late Cretaceous, when the farming activity of their tracemakers became a common feeding strategy. The Late Cretaceous radiation of graphoglyptids may be correlated with global changes in the circulation and oceanic settings, resulting in an increasing amount of plankton and organic matter. In the Paleocene, the ichnodiversity decreased, but subsequently increased (to a maximum) in the Eocene. The Eocene or Paleocene peak in graphoglyptid diversity (total number of ichnogenera) and the Eocene peak in the frequency of graphoglyptids may be related to the common occurrence of oceanic oligotrophy in the Late Paleocene and Early Eocene (Uchman, 2003). The Oligocene was marked by a sharp decrease in both graphoglyptid diversity and frequency, followed by an increase in the Miocene (Uchman, 2003). However, the diversity trends are not applicable to every region, and some departures from the model can be noted locally, *viz.*, where different phenomena can control trace-fossil assemblages (e.g., Lopez Cabrera et al., 2008).

Individual ichnotaxa changed their environmental range over time. For example, *Zoophycos* migrated to the deep sea during the Jurassic (Bottjer et al., 1988; Olivero, 2003) and occurs rarely in Miocene shelf sediments (Hantzschel, 1960; Pervesler and Uchman, 2004). *Ophiomorpha* expanded its range, since both *O. rudis* (Ksiazkiewicz 1977) and *O. annulata* (Ksiazkiewicz 1977) occur in deep-marine sediments from the Tithonian onward, thus documenting the invasion of large burrowing crustaceans to the deep sea, together with echinoids producing *Scolicia* (Tchoumatchenco and Uchman, 2001). Additionally, cross-cutting relationships between burrows may change through geological time.

The evolutionary changes of some ichnospecies were so rapid, for example, regarding those belonging to the Paleozoic form *Dictyodora* (Benton, 1982b), that the corresponding traces occur within narrow stratigraphic ranges and thus can be used for biostratigraphic purposes (Uchman, 2004a). For the Paleogene Podhale flysch, for instance, the occurrence of trace fossils in relation to volcanic ash layers clearly documents that the characteristic trace fossils occupy the same stratigraphic position (Pienkowski and Westwalewicz-Mogilska, 1986).

The *Nereites* Ichnofacies has some time constraints. It can be distinguished since the Ordovician, when graphoglyptids colonized the deep sea (Orr, 2001). Prior to that, in the Cambrian, graphoglyptids occurred only in shelf sediments. A specific trace-fossil association, related to shallow burrowers and surface structures associated with microbial mats, occurred in the Vendian Cambrian (Buatois and Mangano, 2003). The *Ophiomorpha rudis* subichnofacies can be distinguished since the Tithonian, when its index ichnotaxon first appeared in flysch deposits (Tchoumatchenco and Uchman, 2001).

12. PERSPECTIVES AND CONCLUSIONS

Trace fossils and ichnofabrics are still underestimated in sedimentological analyses. For instance, detailed divisions of turbiditic facies, such as those by Pickering et al. (1989), do not sufficiently take ichnofabrics into account. Furthermore, there is a considerable gap between deep-sea paleoichnology and neoichnology. The world of modern endobenthic burrowing animals is poorly known and their interrelationships with the fluctuating environmental factors are rarely investigated. Complex neoichnological research in the various deep-sea environments is needed in order to recognize and identify the tracemakers, their general burrowing behavior and distribution. Probably, laboratory experiments using different deep-sea invertebrates and simulation of sedimentary processes will result in significant advances in this matter. Expeditions with ichnological programs using submersible vehicles would be of great help. Proper employment of robots for resin casting, for example, would be a significant step forward in ichnological research in the deep sea.

Other achievements are expected in the further recognition of trace fossils and ichnofabrics, especially in paleoenvironmentally well-recognized facies tracts, and by their comparisons based on standardized data. Ichnological data of deep-sea sediments from some periods are underrepresented. This is especially true for the Cambrian, Devonian, and the Late Carboniferous to Jurassic, which have been studied only in a few places (Uchman, 2004a).

ACKNOWLEDGMENTS

A. U. was supported by the Jagiellonian University (DS funds). A. W. was supported by the Swiss National Science Foundation (several grants). Tommy McCann (Bonn), Richard G. Bromley (Copenhagen), and Dirk Knaust (Stavanger) provided helpful comments and proposed many corrections.

REFERENCES

Abel, O., 1935. Vorzeitliche Lebensspuren. Gustav Fischer, Jena, 644 pp.

Azpeitia Moros, F., 1933. Datos para el estudio paleontologico del flysch de la Costa Cantabrica y de algunos otros puntos de Espana. Bol. Inst. Geol. Min. Esp. 53, 1 65.

Baudin, F., Disnar, J.-R., Martinez, P., Dennielou, B., 2010. Distribution of the organic matter in the channel-levees systems of the Congo mud-rich deep-sea fan (West Africa). Implication for deep offshore petroleum source rocks and global carbon cycle. Mar. Petrol. Geol. 27, 995 1010.

Bednarz, M., McIlroy, D., 2009. Three-dimensional reconstruction of phycosiphoniform burrows: implications for identification of trace fossils in core. Palaeontologica Electronica 12, 15. http://palaeo-electronica.org/2009_3/195/index.html.

Benton, M.J., 1982a. Trace fossils from Lower Palaeozoic ocean-floor sediments of the Southern Uplands of Scotland. Trans. R. Soc. Edinburgh, Earth Sci. 73, 67 87.

Benton, M.J., 1982b. *Dictyodora* and associated trace fossils from the Palaeozoic of Thuringia. Lethaia 15, 115 132.

Benton, M.J., Trewin, N.J., 1980. *Dictyodora* from Silurian of Peeblesshire, Scotland. Palaeontology 23, 501 513.

Berger, W.H., Ekdale, A.A., Bryant, P.F., 1979. Selective preservation of burrows in deep-sea carbonates. Mar. Geol. 32, 205 230.

Bottjer, D.J., Droser, M.L., Jablonski, D., 1988. Palaeoenvironmental trends in the history of trace fossils. Nature 333, 252 255.

Brass, G.W., Southam, J.R., Peterson, W.H., 1982. Warm saline bottom water in the ancient ocean. Nature 296, 620 623.

Bromley, R.G., 1990. Trace Fossils: Biology and Taphonomy. Unwin Hyman, London, 280 pp.

Bromley, R.G., 1996. Trace Fossils: Biology, Taphonomy and Applications. Chapman and Hall, London, 361 pp.

Bromley, R.G., Asgaard, U., 1991. Ichnofacies: a mixture of taphofacies and biofacies. Lethaia 24, 153 163.

Bromley, R.G., Hanken, N.-M., 2003. Structure and function of large, lobed *Zoophycos*, Pliocene of Rhodes, Greece. Palaeogeogr. Palaeoclimatol. Palaeoecol. 192, 79 100.

Bromley, R.G., Uchman, A., Gregory, M., Martin, A.J., 2003. *Hillichnus lobosensis* igen. et isp. nov., a complex trace fossil produced by tellinacean bivalves, Paleocene, Monterey, California, U.S.A. Palaeogeogr. Palaeoclimatol. Palaeoecol. 192, 157 186.

Buatois, L.A., Mangano, M.G., 2003. Early colonization of the deep sea: ichnologic evidence of deep-marine benthic ecology from the Early Cambrian of northwest Argentina. Palaios 18, 572 581.

Buatois, L.A., Mangano, M.G., Sylvester, Z., 2001. A diverse deep-marine ichnofauna from the Eocene Tarcau Sandstone of the eastern Carpathians, Romania. Ichnos 8, 23 62.

Chamberlain, C.K., 1971. Morphology and ethology of trace fossils from the Ouachita Mountains, southeast Oklahoma. J. Paleontol. 45, 212 246.

Crimes, T.P., 1973. From limestones to distal turbidites: a facies and trace fossil analysis in the Zumaya flysch (Paleocene-Eocene), North Spain. Sedimentology 20, 105 131.

Crimes, T.P., 1974. Colonisation of the early ocean floor. Nature 248, 328 330.

Crimes, T.P., 1977. Trace fossils of an Eocene deep-sea fan, northern Spain. In: Crimes, T.P., Harper, J.C. (Eds.), Trace Fossils 2. Geol. J., Spec. Iss. 9, pp. 71 90.

Crimes, P.T., Crossley, J., 1980. Inter-turbidite current orientation from trace fossils with an example from the Silurian flysch of Wales. J. Sediment. Petrol. 50, 821 830.

Crimes, T.P., Crossley, J.D., 1991. A diverse ichnofauna from Silurian flysch of the Aberystwyth Grits Formation, Wales. Geol. J. 26, 27 64.

Crimes, P.T., Fedonkin, M.A., 1994. Evolution and dispersal of deepsea traces. Palaios 9, 74 83.

Crimes, T.P., Goldring, R., Homewood, P., Van Stuijvenberg, J., Winkler, W., 1981. Trace fossil assemblages of deep-sea fan deposits, Gurnigel and Schlieren flysch (Cretaceous-Eocene). Eclogae Geol. Helv. 74, 953 995.

Cummings, J.P., Hodgson, D.M., 2011a. An agrichnial feeding strategy for deep-marine Paleogene *Ophiomorpha* group trace fossils. Palaios 26, 212 224.

Cummings, J.P., Hodgson, D.M., 2011b. Assessing controls on the distribution of ichnotaxa in submarine fan environments, the Basque Basin, Northern Spain. Sediment. Geol. 239, 162 187.

Delgado, J.F.N., 1910. Terrains paleozo ques du Portugal. Etude sur les fossiles des schichtes à nereites de San Domingo et des schichtes à nereite et à graptolites de Barrancos. Com. Serv. Geol. Port. 56, 1 68.

Ekdale, A.A., 1985. Paleoecology of the marine endobenthos. Palaeogeogr. Palaeoclimatol. Palaeoecol. 50, 63 81.

Ekdale, A.A., 1992. Muckraking and mudslinging: the joys of deposit-feeding. In: Maples, C.G., West, R.R. (Eds.), Trace Fossils. Short Courses in Paleontology 5. Paleontological Society, Knoxville, pp. 145 171.

Ekdale, A.A., Berger, W.H., 1978. Deep-sea ichnofacies: modern organism traces on and in pelagic carbonates of the western equatorial Pacific. Palaeogeogr. Palaeoclimatol. Palaeoecol. 23, 268 278.

Ekdale, A.A., Bromley, R.G., 1991. Analysis of composite ichnofabrics: an example in uppermost Cretaceous chalk of Denmark. Palaios 6, 232 249.

Ekdale, A.A., Muller, L.N., Novak, M.T., 1984. Quantitative ichnology of modern pelagic deposits in the abyssal Atlantic. Palaeogeogr. Palaeoclimatol. Palaeoecol. 45, 189 223.

Ernst, G., Zander, J., 1993. Stratigraphy, facies development, and trace fossils of the Upper Cretaceous of southern Tanzania (Kilwa District). In: Abbate, E., Sagri, M., Sassi, F.P. (Eds.), Geology and Mineral Resources of Somalia and Surrounding Regions. Relazioni e Monografie Agrarie Suptropicali e Tropicali, Nuova Serie, vol. 113, pp. 259 278.

Fillion, D., Pickerill, R.K., 1990. Ichnology of the Upper Cambrian? to Lower Ordovician Bell Island and Wabana groups of eastern Newfoundland, Canada. Palaeontogr. Can. 7, 1 119.

Follmi, K.B., Grimm, K.A., 1990. Doomed pioneers: gravity-flow deposition and bioturbation in marine oxygen-deficient environments. Geology 18, 1069 1072.

Frey, R.W., Curran, A.H., Pemberton, G.S., 1984. Trace making activities of crabs and their environmental significance: the ichnogenus *Psilonichnus*. J. Paleontol. 58, 333 350.

Fu, S., 1991. Funktion, Verhalten und Einteilung fucoider und lophocteniider Lebensspuren. Courier Forsch. Inst. Senck. 135, 1 79.

Fuchs, T., 1895. Studien uber Fucoiden und Hieroglyphen. Denkschr. Kais. Akad. Wiss. Wien, Math.-Nat. Kl. 62, 369 448.

Gaillard, C., 1991. Recent organism traces and ichnofacies on the deep-sea floor off New Caledonia, southwestern Pacific. Palaios 6, 302 315.

Giannetti, A., 2010. Influence of climate, sea-level changes and tectonics and ichnoassemblages distribution in carbonate-dominated, deep-marine environment (Upper Paleocene, Zumaya section). Palaeogeogr. Palaeoclimatol. Palaeoecol. 285, 104 118.

Giannetti, A., McCann, T., 2010. The Upper Paleocene of the Zumaya section (northern Spain): review of the ichnological content and preliminary palaeoecological interpretation. Ichnos 17, 137 161.

Goldring, R., Cadee, G.C., D Alessandro, A., de Gibert, J.M., Jenkins, R., Pollard, J.E., 2004. Climatic control on trace fossil distribution in the marine realm. In: McIlroy, D. (Ed.), The Application of Ichnology to Palaeoenvironmental and Stratigraphic Analysis. Geol. Soc. London, Spec. Publ., vol. 228, pp. 77 93.

Hall, S.J., 1994. Physical disturbance and marine benthic communities: life in unconsolidated sediments. Oceanogr. Mar. Biol. Ann. Rev. 32, 179 239.

Hantzschel, W., 1960. Spreitenbauten (*Zoophycos* Massalongo) im Septarienton Nordwest-Deutschlands. Mitt. Geol. Staatsinst. Hamburg 29, 95 100.

Heard, T., Pickering, K.T., 2008. Trace fossils as diagnostic indicators of deep-marine environments, middle Eocene Ainsa-Jaca Basin, Spanish Pyrenees. Sedimentology 55, 809 844.

Heard, T.G., Pickering, K.T., Robinson, S.A., 2008. Milankovitch forcing of bioturbation intensity in deep-marine thin-bedded siliciclastic turbidites. Earth Planet. Sci. Lett. 272, 130 138.

Heer, O., 1876/1877. Flora Fossilis Helvetiae. Die vorweltliche Flora der Schweiz. J. Wurster, Zurich, 182 pp.

Keighley, D.G., Pickerill, R.K., 1995. The ichnotaxa *Palaeophycus* and *Planolites:* historical perspectives and recommendations. Ichnos 3, 301 309.

Kern, J.P., 1980. Origin of trace fossils in Polish Carpathians flysch. Lethaia 13, 347 723.

Knaust, D., 2009. Characterisation of a Campanian deep-sea fan system in the Norwegian Sea by means of ichnofabrics. Mar. Petrol. Geol. 26, 1199 1211.

Ksiazkiewicz, M., 1954. Uziarnienie frakcjonalne i laminowane we fliszu karpackim [Graded and laminated bedding in the Carpathian Flysch]. Rocznik Pol. Tow. Geol. 22, 399 471 (in Polish).

Ksiazkiewicz, M., 1977. Trace fossils in the flysch of the Polish Carpathians. Palaeont. Polon. 36, 1 208.

Leszczynski, S., 1991. Oxygen-related controls on predepositional ichnofacies in turbidites, Guipuzcoan flysch (Albian-lower Eocene), northern Spain. Palaios 6, 271 280.

Leszczynski, S., 1993a. A generalized model for the development of ichnocoenoses in flysch deposits. Ichnos 2, 137 146.

Leszczynski, S., 1993b. Ichnocoenosis *versus* sediment colour in upper Albian to lower Eocene turbidites, Guipuzcoan province, northern Spain. Palaeogeogr. Palaeoclimatol. Palaeoecol. 100, 251 265.

Leszczynski, S., 2003. Bioturbation structures in the Holovnia Siliceous Marls (Turonian Lower Santonian) in Rybotycze (Polish Carpathians). Ann. Soc. Geol. Polon. 73, 103 122.

Leszczynski, S., Seilacher, A., 1991. Ichnocoenoses of a turbidite sole. Ichnos 1, 293 303.

Leszczynski, S., Uchman, A., 1993. Biogenic structures of organic-poor sediments: examples from the Paleogene variegated shales, Polish Outer Carpathians. Ichnos 2, 267 275.

Lopez Cabrera, M.I., Olivero, E.B., Carmona, N.B., Ponce, J.J., 2008. Cenozoic trace fossils of the *Cruziana*, *Zoophycos* and *Nereites* ichnofacies from the Fuegian Andes, Argentina. Ameghiniana 45, 377 392.

McBride, E.F., Picard, M.D., 1991. Facies implications of *Trichichnus* and *Chondrites* in turbidites and hemipelagites, Marnoso-arenacea Formation (Miocene), northern Apennines, Italy. Palaios 6, 281 290.

McCann, T., 1990. Distribution of Ordovician-Silurian ichnofossil assemblages in Wales implications for Phanerozoic ichnofaunas. Lethaia 23, 243 255.

McCann, T., 1993. A *Nereites* ichnofacies from the Ordovician-Silurian Welsh Basin. Ichnos 3, 39 56.

McCann, T., Pickerill, R.K., 1988. Flysch trace fossils from the Cretaceous Kodiak Formation of Alaska. J. Paleontol. 62, 330 347.

Mikulas, R., Lechotsky, T., Babek, O., 2002. Lower Carboniferous ichnofabrics of the Culm facies: a case study of the Moravice Formation (Moravia and Silesia, Czech Republic). Geol. Carpathica 53, 141 148.

Miller III, W., 1991a. Intrastratal trace fossil zonation, Cretaceous flysch in northern California. Ichnos 1, 161 171.

Miller III, W., 1991b. Paleoecology of graphoglyptids. Ichnos 1, 305 312.

Miller III, W., 1993. Trace fossil zonation in Cretaceous turbidite facies, northern California. Ichnos 3, 11 28.

Miller III, W., 2007. Complex trace fossils. In: Miller III, W. (Ed.), Trace Fossils. Concepts, Problems, Prospects. Elsevier, Amsterdam, pp. 458 465.

Miller III, W., Stefani, C., Grandesso, P., 2004. Alternation of ecologic regimes in a deep-marine carbonate basin: calciturbidite trace fossils from the Cretaceous Scaglia Rossa, northeastern Italy. Palaeogeogr. Palaeoclimatol. Palaeoecol. 204, 317 330.

Monaco, P., 2008. Taphonomic features of *Paleodictyon* and other graphoglyptid trace fossils in Oligo-Miocene thin-bedded turbidites, Northern Apennines, Italy. Palaios 23, 667 682.

Monaco, P., Caracuel, J.E., 2007. Il valore stratinomico delle tracce fossili negli strato evento (event bed) del registro geologico: esempi significativi di ichnologia comportamentale dall Italia e dalla Spagna. Studi e Ricerche - Associazione Amici del Museo - Museo Civico G. Zannato 14, 43 60.

Monaco, P., Checconi, A., 2010. Taphonomic aspects of the Miocene ichnofossil-lagerstatte from calcarenite turbiditic beds in the Verghereto Marls Formation (northern Apennines, Italy). Rev. Ital. Paleont. Strat. 116, 237 252.

Monaco, P., Milighetti, M., Checconi, A., 2010. Ichnocoenosis in the Oligocene to Miocene foredeep basins (Northern Apennines, central Italy) and their relation to turbidite deposition. Acta Geol. Polon. 60, 53 70.

Naruse, H., Nifuku, K., 2008. Three-dimensional morphology of the ichnofossils *Phycosiphon incertum* and its implication for paleoslope inclination. Palaios 23, 270 279.

Nathorst, A.G., 1881. Om spar av nagra evertebrede djur och deras paleontologiska betydelse. Kongl. Svenska Vetenskaps-Akademien Handlingar 18, 1 104.

Olivero, D., 2003. Early Jurassic to Late Cretaceous evolution of *Zoophycos* in French Subalpine Basin (southeast France). Palaeogeogr. Palaeoclimatol. Palaeoecol. 192, 59 78.

Olivero, E.B., Lopez Cabrera, M.I., Malumian, N., Torres Carbonell, P.J., 2010. Eocene graphoglyptids from shallow-marine, high-energy, organic-rich, and bioturbated turbidites, Fuegian Andes, Argentina. Acta Geol. Polon. 60, 77 91.

Orr, P.J., 1995. A deep-marine ichnofaunal assemblage from Llandovery strata of the Welsh Basin, west Wales, UK. Geol. Mag. 132, 267 285.

Orr, P.J., 1996. The ichnofauna of the Skiddaw Group (early Ordovician) of the Lake District, England. Geol. Mag. 133, 193 216.

Orr, P.J., 2001. Colonization of the deep-marine environment during the early Phanerozoic: the ichnofaunal record. Geol. J. 36, 265 278.

Orr, P.J., Benton, M., Trewin, N.H., 1996. Deep marine trace fossil assemblages from the Lower Carboniferous of Menorca, Balearic Islands, western Mediterranean. Geol. J. 31, 235 258.

Osgood, R.G., 1970. Trace fossils of the Cincinnati area. Palaeont. Am. 6, 193 235.

Pemberton, G.S., Frey, R.W., 1982. Trace fossil nomenclature and the *Planolites-Palaeophycus* dilemma. J. Paleontol. 56, 843 881.

Pervesler, P., Uchman, A., 2004. Ichnofossils from the type area of the Grund Formation (Miocene, Lower Badenian) in northern Lower Austria (Molasse Basin). Geol. Carpathica 55, 103 110.

Pfeiffer, H., 1968. Die Spurenfossilien des Kulms (Dinants) und Devons der Frankenwalder Querzone (Thuringen). Jb. Geol. 2, 651 717.

Phillips, C., McIlroy, D., Elliott, T., 2011. Ichnological characterization of Eocene/Oligocene turbidites from the Grès d Annot Basin, French Alps, SE France. Palaeogeogr. Palaeoclimatol. Palaeoecol. 300, 67 83.

Pickerill, R.K., 1980. Phanerozoic flysch trace fossil diversity observations based on an Ordovician flysch ichnofauna from the Aroostook-Metapedia Carbonate Belt of northern New Brunswick. Can. J. Earth Sci. 17, 1259 1270.

Pickering, K.T., Hiscott, R.N., Hein, F.J., 1989. Deep Marine Environments: Clastic Sedimentation and Tectonics. Unwin Hyman, London, 416 pp.

Pienkowski, G., Westwalewicz-Mogilska, E., 1986. Trace fossils from the Podhale Flysch Basin, Poland an example of ecologically based lithocorrelation. Lethaia 19, 53 65.

Rajchel, J., Uchman, A., 1998. Ichnological analysis of an Eocene mixed marly-siliciclastic flysch deposits in the Nienadowa Marls Member, Skole Unit, Polish Flysch Carpathians. Ann. Soc. Geol. Pol. 68, 61 74.

Richter, R., 1928. Die fossilen Fahrten und Bauten der Wurmer, ein Uberblick uber ihre biologischen Grundformen und deren geologische Bedeutung. Palaont. Z. 9, 193 241.

Rindsberg, A.K., 2012. Ichnotaxonomy: finding patterns in a welter of information. In: Knaust, D., Bromley, R.G. (Eds.), Trace Fossils as Indicators of Sedimentary Environments. Developments in Sedimentology 64, Elsevier, Amsterdam, pp. 45 78.

Rodr guez-Tovar, F.J., Uchman, A., Payros, A., Orue-Etxebarria, X., Apellaniz, E., Molina, E., 2010. Sea-level dynamics and palaeoecological factors affecting trace fossil distribution in Eocene turbiditic deposits (Gorrondatxe section, N Spain). Palaeogeogr. Palaeoclimatol. Palaeoecol. 285, 50 65.

Sacco, F., 1888. Note di paleoicnologia italiana. Atti Soc. Ital. Sci. Nat. 31, 151 192.

Savrda, C.E., 2012. Chalk and related deep-marine carbonates. In: Knaust, D., Bromley, R.G. (Eds.), Trace Fossils as Indicators of Sedimentary Environments. Developments in Sedimentology 64, Elsevier, Amsterdam, pp. 777 806.

Savrda, C.E., Counts, J.W., Bigham, E., Martin, S., 2010. Ichnology of siliceous facies in the Eocene Tallahatta Formation (eastern United States Gulf coastal plain): implications for depositional conditions, storm processes, and diagenesis. Palaios 25, 642 655.

Schlirf, M., 2000. Upper Jurassic trace fossils from the Boulonnais (northern France). Geol. Palaeont. 34, 145 213.

Seilacher, A., 1953. Studien zur Palichnologie. I. Uber die Methoden der Palichnologie. N. Jb. Geol. Palaont. Abh. 96, 421 452.

Seilacher, A., 1955. Spuren und Fazies im Unterkambrium. In: Schindewolf, O.H., Seilacher, A. (Eds.), Beitrage zur Kenntnis des Kambriums in der Salt Range (Pakistan). Akad. Wiss. Lit. Mainz., Abh. Math.-Nat. Kl. 10, pp. 373 399.
Seilacher, A., 1962. Paleontological studies of turbidite sedimentation and erosion. J. Geol. 70, 227 234.
Seilacher, A., 1964. Biogenic sedimentary structures. In: Imbrie, J., Newell, N. (Eds.), Approaches to Paleoecology. Wiley and Sons, New York, pp. 296 316.
Seilacher, A., 1967. Bathymetry of trace fossils. Mar. Geol. 5, 413 428.
Seilacher, A., 1974. Flysch trace fossils: evolution of behavioural diversity in the deep-sea. N. Jb. Geol. Palaont. Mh. 1974, 223 245.
Seilacher, A., 1976. Evolution von Spuren-Vergesellschaftungen. Zbl. Geol. Palaont. T. 2, 396 402.
Seilacher, A., 1977a. Pattern analysis of *Paleodictyon* and related trace fossils. In: Crimes, T.P., Harper, J.C. (Eds.), Trace Fossils 2. Geol. J., Spec. Iss. 9, pp. 289 334.
Seilacher, A., 1977b. Evolution of trace fossil communities in the deep sea. In: Hallam, A. (Ed.), Patterns of Evolution. Elsevier, Amsterdam, pp. 359 376.
Seilacher, A., 1978. Evolution of trace fossil communities in the deep sea. N. Jb. Geol. Palaont. Abh. 157, 251 255.
Seilacher, A., 2007. Trace Fossil Analysis. Springer, Berlin, 226 pp.
Shackleton, N.J., 1986. Paleogene stable isotope events. Palaeogeogr. Palaeoclimatol. Palaeoecol. 57, 91 102.
Squinabol, S., 1890. Alghe e Pseudoalghe fossili italiane. Atti Soc. Ling. Sci. Nat. Geogr. 1 (29 49), 166 199.
Stepanek, J., Geyer, G., 1989. Spurenfossilien aus dem Kulm (Unterkarbon) des Frankenwaldes. Beringeria 1, 1 55.
Tchoumatchenco, P., Uchman, A., 2001. The oldest deep-sea *Ophiomorpha* and *Scolicia* and associated trace fossils from the Upper Jurassic-Lower Cretaceous deep-water turbidite deposits of SW Bulgaria. Palaeogeogr. Palaeoclimatol. Palaeoecol. 169, 85 99.
Thayer, C.W., 1979. Biological bulldozers and the evolution of marine benthic communities. Nature 203, 458 461.
Thomas, E., Zachos, J.C., Bralower, T.J., 2000. Deep-sea environments on a warm earth: latest Paleocene-early Eocene. In: Huber, B.T., MacLeod, K.G., Wing, S.L. (Eds.), Warm Climates in Earth History. Cambridge University Press, Cambridge, pp. 132 160.
Tucker, M.E., 1974. Sedimentology of pelagic Paleozoic limestones in the Devonian Griotte (South France) and Cephalopodenkalk (Germany). In: Hsu, K.J., Jenkyns, H.C. (Eds.), Pelagic Sediments: On Land and Under the Sea. IAS Spec. Publ., vol. 1, pp. 71 92.
Tunis, G., Uchman, A., 1996a. Trace fossil and facies changes in the Upper Cretaceous-Middle Eocene flysch deposits of the Julian Prealps (Italy and Slovenia): consequences of regional and world-wide changes. Ichnos 4, 169 190.
Tunis, G., Uchman, A., 1996b. Ichnology of the Eocene flysch deposits in the Istria Peninsula, Croatia and Slovenia. Ichnos 5, 1 22.
Uchman, A., 1991. Trace fossils from stress environments in Cretaceous-Paleogene flysch of the Polish Outer Carpathians. Ann. Soc. Geol. Pol. 61, 207 220.
Uchman, A., 1992. An opportunistic trace fossil assemblage from the flysch of the Inoceramian beds (Campanian-Palaeocene), Bystrica Zone of the Magura Nappe, Carpathians, Poland. Cret. Res. 13, 539 547.
Uchman, A., 1995a. Taxonomy and palaeoecology of flysch trace fossils: the Marnoso-arenacea Formation and associated facies (Miocene, Northern Apennines, Italy). Beringeria 15, 3 115.

Uchman, A., 1995b. Tiering patterns of trace fossils in Paleogene flysch deposits of the Carpathians, Poland. Geobios Mem. Spec. 18, 389 394.

Uchman, A., 1998. Taxonomy and ethology of flysch trace fossils: a revision of the Marian Ksiazkiewicz collection and studies of complementary material. Ann. Soc. Geol. Polon. 68, 105 218.

Uchman, A., 1999. Ichnology of the Rhenodanubian Flysch (Lower Cretaceous-Eocene) in Austria and Germany. Beringeria 25, 65 171.

Uchman, A., 2001. Eocene flysch trace fossils from the Hecho Group of the Pyrenees, northern Spain. Beringeria 28, 3 41.

Uchman, A., 2003. Trends in diversity, frequency and complexity of graphoglyptid trace fossils: evolutionary and palaeoenvironmental aspects. Palaeogeogr. Palaeoclimatol. Palaeoecol. 192, 123 142.

Uchman, A., 2004a. Phanerozoic history of deep-sea trace fossils. In: McIlroy, D. (Ed.), The Application of Ichnology to Palaeoenvironmental and Stratigraphic Analysis. Geol. Soc. London, Spec. Publ., vol. 228, pp. 25 139.

Uchman, A., 2004b. Deep-sea trace fossils controlled by palaeo-oxygenation and deposition: an example from the Lower Cretaceous dark flysch deposits of the Silesian Unit, Carpathians, Poland. Fossils Strata 51, 39 57.

Uchman, A., 2007a. Deep-sea ichnology: development of major concepts. In: Miller III, W. (Ed.), Trace Fossils. Concepts, Problems, Prospects. Elsevier, Amsterdam, pp. 248 267.

Uchman, A., 2007b. Deep-sea trace fossils from the mixed carbonate-siliciclastic flysch of the Monte Antola Formation (Late Campanian-Maastrichtian), North Apennines, Italy. Cret. Res. 28, 980 1004.

Uchman, A., 2007c. Trace fossils of the Pagliaro Formation (Paleocene) in the North Apennines, Italy. Beringeria 37, 217 237.

Uchman, A., 2009. The *Ophiomorpha rudis* ichnosubfacies of the *Nereites* ichnofacies: characteristics and constraints. Palaeogeogr. Palaeoclimatol. Palaeoecol. 276, 107 119.

Uchman, A., Tchoumatchenco, P., 2003. A mixed assemblage of deep-sea and shelf trace fossils from the Lower Cretaceous (Valanginian) Kamchia Formation in the Troyan region, central Fore-Balkan, Bulgaria. Ann. Soc. Geol. Polon. 73, 27 34.

Uchman, A., Janbu, N.E., Nemec, W., 2004. Trace fossils in the Cretaceous Eocene flysch of the Sinop-Boyabat Basin, Central Pontides, Turkey. Ann. Soc. Geol. Polon. 74, 197 235.

Uchman, A., Bub k, M., Mikulas, R., 2005. The ichnological record across the Cretaceous/Tertiary boundary in turbiditic sediments at Uzgrun (Moravia, Czech Republic). Geol. Carpathica 56, 57 65.

von Fischer-Ooster, C., 1858. Die fossilen Fucoiden der Schweizer Alpen, nebst Erorterung uber deren geologisches Alter. Huber u. Comp., Bern, 74 pp.

Wetzel, A., 1981. Okologische und stratigraphische Bedeutung biogener Gefuge in quartaren Sedimenten am NW-afrikanischen Kontinentalrand. "Meteor" Forsch.-Ergebn. C34, 1 47.

Wetzel, A., 1983. Biogenic sedimentary structures in a modern upwelling region: NW African continental margin. In: Thiede, J., Suess, E. (Eds.), Coastal Upwelling and Its Sediment Record. Part B: Sedimentary Records of Ancient Coastal Upwelling. Plenum, New York, pp. 123 144.

Wetzel, A., 1984. Bioturbation in deep-sea fine-grained sediments: influence of sediment texture, turbidite frequency and rates of environmental changes. In: Stow, D.A.V., Piper, D.J.W. (Eds.), Fine Grained Sediments: Deep-Water Processes and Facies. Geol. Soc. Spec. Publ., vol. 15, pp. 597 608.

Wetzel, A., 2002. Modern *Nereites* in the South China Sea ecological associations with redox conditions in the sediment. Palaios 17, 507 515.

Wetzel, A., 2008. Recent bioturbation in the deep South China Sea: a uniformitarian ichnologic approach. Palaios 23, 601 615.
Wetzel, A., 2010. Deep-sea ichnology: observations in modern sediments to interpret fossil counterparts. Acta Geol. Polon. 60, 125 138.
Wetzel, A., Aigner, T., 1986. Stratigraphic completeness: tiered trace fossils provide a measuring stick. Geology 14, 234 237.
Wetzel, A., Bromley, R.G., 1994. *Phycosiphon incertum* revisited: *Anconichnus horizontalis* is its junior subjective synonym. J. Paleontol. 68, 1396 1402.
Wetzel, A., Uchman, A., 1998. Trophic level in the deep-sea recorded by ichnofabrics: an example from Palaeogene flysch in the Carpathians. Palaios 13, 533 546.
Wetzel, A., Uchman, A., 2001. Sequential colonization of muddy turbidites: examples from Eocene Beloveza Formation, Carpathians, Poland. Palaeogeogr. Palaeoclimatol. Palaeoecol. 168, 171 186.
Wetzel, A., Uchman, A., 2012. Hemipelagic and pelagic basin plains. In: Knaust, D., Bromley, R.G. (Eds.), Trace Fossils as Indicators of Sedimentary Environments. Developments in Sedimentology 64, Elsevier, Amsterdam, pp. 673 702.
Wetzel, A., Werner, F., 1981. Morphology and ecological significance of *Zoophycos* in deep-sea sediments off NW Africa. Palaeogeogr. Palaeoclimatol. Palaeoecol. 32, 185 212.
Wetzel, A., Blechschmidt, I., Uchman, A., Matter, A., 2007. A highly diverse ichnofauna in late Triassic deep-sea fan deposits of Oman. Palaios 22, 567 576.

Chapter 22

Hemipelagic and Pelagic Basin Plains

Andreas Wetzel*,[1] and Alfred Uchman†
*Geologisch-Paläontologisches Institut der Universität, Basel, Switzerland, †Institute of Geological Sciences, Jagiellonian University, Kraków, Poland
[1]Corresponding author: e-mail: andreas.wetzel@unibas.ch

1. INTRODUCTION

In the first half of the nineteenth century, trace fossils from bathyal and abyssal deposits were described as algae (fucoids), without recognition of their deep-sea origin (e.g., von Fischer-Ooster, 1858). For the following decades, the deep sea was seen often as a mysterious environment because there was no equipment to observe or to study the deep-sea floor directly. Therefore, deep-sea research, including ichnology, lagged behind of investigations done in shallow-water settings.

The potential of trace fossils as ecological indicators markedly increased when modern environments were studied in addition, because of accessibility, first in shallow-marine settings (Richter, 1928; Schäfer, 1956), and later in the deep sea (e.g., Eving and Davis, 1967; Heezen and Hollister, 1971; Hollister et al., 1975; Laughton, 1961). Nonetheless, the beginnings of ichnological research in the modern deep sea dates back to the end of nineteenth century (Fuchs, 1894).

At first, deep-sea trace fossils were related to parameters such as the energy of the environment (turbidites *versus* non-turbidites) and indirectly to bathymetry (Seilacher, 1967). Then the composition of trace-fossil assemblages was considered, starting with turbidite successions (see Uchman and Wetzel, 2012), later in rather continuously accumulating hemipelagic and pelagic deposits (e.g., Chamberlain and Clark, 1973; Reineck, 1973), especially when deep-marine deposits became available through the Deep Sea Drilling Project (e.g., Ekdale, 1977; Warme et al., 1973). Subsequently, environmental factors such as organic-matter content, oxygenation, and sedimentation rate were taken into account (e.g., Byers, 1977; Edwards, 1985; Savrda et al., 2001; Wetzel, 1981, 1984). At the same time, interrelationships between benthic communities and their dependence on environmental parameters were increasingly addressed

Developments in Sedimentology, Vol. 64. http://dx.doi.org/10.1016/B978-0-444-53813-0.00022-8

(e.g., Gage and Tyler, 1991; Lopez et al., 1989; Rex, 1983). Concomitantly, (hemi)pelagic deep-sea deposits were interpreted ecologically, based on their trace-fossil content (Ekdale, 1985; Ekdale and Mason, 1988; Leszczyński and Uchman, 1991, 1993; Bąk, 1995; Wetzel and Uchman, 1998a, 2001; Rodríguez-Tovar and Uchman, 2004; Rodríguez-Tovar et al., 2010).

Bioturbational structures store information about the ecological conditions during their production, but in turn, the endobenthic animals disturb the originally layered sedimentary record. In particular in Pleistocene and Holocene sediments of the ocean basins, bioturbation has been studied to better understand the effects of vertical mixing, so as to improve the accuracy of paleoceanographic and paleoclimatic investigations (e.g., Berger and Heath, 1968; DeMaster et al., 1991; Smith and Rabouille, 2002; Trauth et al., 1997). The analysis of vertical displacement of tracer particles by burrowers led to a better understanding of the bioturbation process and the behavior of the producers (e.g., Kotake, 1989; Löwemark and Werner, 2001; Wetzel, 2002). Similarly, the preservation potential of event layers has been addressed (e.g., Bentley et al., 2006; Ruddiman and Glover, 1982; Wetzel, 2009; Wheatcroft, 1990). In these terms, the recolonization of the sea floor after event bed deposition or an ecological disaster has been evaluated (e.g., Ingole et al., 2005; Thiel, 1992).

Mixing also enhances the flux of oxygen and water into the sediment and extends the zone of organic-matter remineralization and hence plays an important role with respect to global biogeochemical cycles (Chester, 2000). Thus, the bioturbational processes affect diagenesis (e.g., Bromley et al., 1975; Savrda et al., 2010). The purpose of this chapter is to give an overview of the ichnology and the ecological interpretation of hemipelagic and pelagic deposits as well as to provide examples of the ichnological record of some deep-sea processes and phenomena.

2. ICHNOLOGY

2.1 Aspects Regarding the Bioturbated Zone

To quantify burrowing activity, bioturbation has been described as a diffusion-like process comprising an infinite number of small random steps (e.g., Boudreau, 1986a), but also non-local, long-range vertical transport was considered (Boudreau, 1986b; Boudreau and Imboden, 1987). Within the so-called surface mixed layer, the sediment is churned by burrowing organisms so effectively that even short-lived radioisotopes exhibit a roughly constant concentration; below the mixed layer, the radiotracer concentration decreases exponentially (e.g., DeMaster et al., 1991; Fig. 1). For modern, stable oceanic environments, the mean thickness of the mixed layer has been estimated to be in the range of 9.8 ± 4.5 cm (Boudreau, 1998). At a specific site, however, the thickness of the mixed layer was found to be related to the flux of organic matter to the sea floor (e.g., Smith and Rabouille, 2002; Trauth et al., 1997). Therefore,

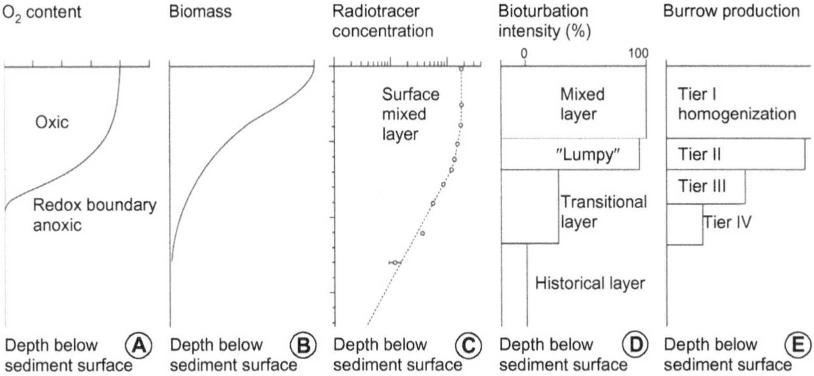

FIGURE 1 Subdivision of the sea floor deposits that are so strongly enriched in organic matter that the pore water becomes anoxic within the bioturbated zone. (A) Oxygen content within the sea floor; the redox boundary marks the transition from oxic to anoxic deposits. (B) Schematic representation of the biomass within the sea floor, normally showing an exponential decrease. (C) Concentration profiles of short-lived radioisotopes, for instance, ^{210}Pb, suggesting a homogeneously mixed layer and no mixing below. (D) Bioturbation intensity within a multilayer model using sediment particles as tracers. (E) Arrangement of tiers based on cross-cutting relationships; traces cross-cutting each other belong to one tier, and traces of the deeper tier cross-cut those of shallower ones; although production of burrows of the deep tier can be too low to affect radioisotope distribution, in fact it dominates the ichnofabric.

it is not surprising that high mixing rates have been determined for areas receiving high amounts of benthic food (e.g., Legeleux et al., 1994).

The mixed-layer concept addresses the turnover of sedimentary particles, but not the aspects of the resulting biogenic fabric. The abundance of macro-organisms decreases with depth in the sediment (e.g., Gage and Tyler, 1991), and therefore, the time required to bioturbate a layer of given thickness increases with depth below the surface, in particular as the endofauna is often patchily distributed (e.g., Jumars and Ekman, 1983). The chance of a trace to become preserved as a trace fossil increases with the burrowing depth of its producer (e.g., Werner and Wetzel, 1982). Consequently, trace fossils have a high preservation potential in the interval that is affected by low rates of bioturbation by deeply penetrating organisms; it is called the "transitional layer" (Berger, 1982).

This purely descriptive two-layer concept has been refined by the tiering concept that takes into account the change of ecological conditions with depth within the sediment. Consequently, the various infaunal organisms utilize different depth intervals (e.g., Gage and Tyler, 1991; Jumars and Ekman, 1983; Jumars et al., 1990). The vertical zonation of animals results in tiered burrows ("*Gefügestockwerke*", Reineck et al., 1967), later called storeys (Wetzel, 1981) or tiers (Ausich and Bottjer, 1982). Each tier comprises a distinct depth interval containing co-occurring traces cross-cutting each other (Fig. 1; Wetzel, 1981, 1984; Gaillard, 1984; Bromley and Ekdale, 1986). In the fossil record, the youngest burrows within a specific horizon can cut through previously produced ones, whereas all those that have been produced later can intersect the oldest bioturbation structures. Thus, tiering can be deduced by the analysis of the cross-cutting relationships and the number of tiers can be established. A cross-cutting relationship can result from the vertical migration of one endofaunal community keeping pace with the sedimentary surface during accumulation (Bromley and Ekdale, 1986; Miller, 1991; Werner and Wetzel, 1982) or from the juxtaposition of burrows of replacing communities while environmental conditions change (e.g., Miller et al., 2004; Savary et al., 2004; Wetzel and Uchman, 2001).

Using the tiering concept for environmental analysis, steady-state conditions are often invoked. However, this is not always likely, as illustrated by the following example. Holocene deep-sea sediments off NW Africa accumulated at a rate of 10 cm/ka and the tiers may be as deep as 50 cm (Werner and Wetzel, 1982: 280). In this case, steady-state ichnofabric formation would need stable conditions for more than 5,000 years, a requirement seldom met if the stable-isotope record in Quaternary deep-sea sediments is analyzed (e.g., Seibold and Berger, 1996). Consequently, it is necessary to evaluate whether the burrowing organisms and the related trace fossils of the deeper tiers are associated with actual conditions or whether they represent responses to past conditions provided by the host sediment. Therefore, the conditions during the formation of deep-tier traces have to be considered by compensating for their penetration depth and estimating the lifetime intervals of their producers (Löwemark, 2007; Wetzel, 1981).

2.2 Bioturbational Structures

Two general types of bioturbational sedimentary structures can be distinguished. (1) Trace fossils exhibit sharp outlines and possess a characteristic recurrent geometry that allows their classification in terms of ichnotaxonomy, whereas (2) biodeformational structures have less distinct outlines and do not display a recurrent geometry that would allow their classification; biodeformational structures are formed within soft to soupy or incoherent sediment (e.g., Wetzel and Uchman, 1998b; Lobza and Schieber, 1999; Leszczyński, 2004 calling them "turbichnia"). They destroy pre-existing sedimentary structures (Fig. 2). An intermediate class represents cryptobioturbation, constituted by small-sized trace fossils in such an overwhelming abundance that it results in a quasi-homogenization of the sediment (e.g., Pemberton et al., 2008).

In the fossil record, two main behavioral and hence often differently preserved groups of trace fossils are distinguished. Trace fossils that are produced on or very close to the sediment surface become preserved only when they are scoured and cast by depositional events; with respect to the event they are predepositional (Książkiewicz, 1954). Especially, the highly organized, often delicate graphoglyptids (Fuchs, 1895) belong to this category. Post-depositional

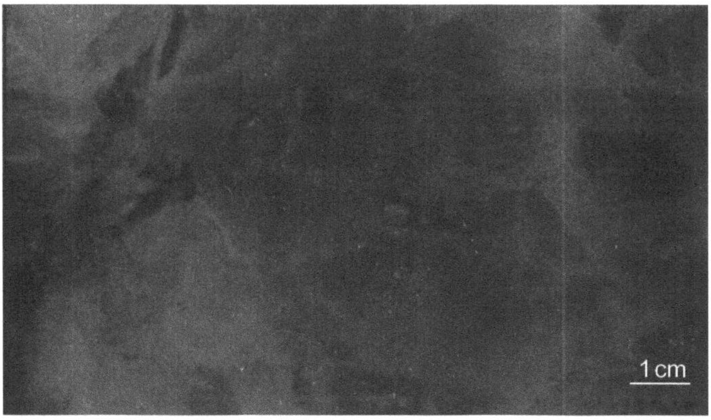

FIGURE 2 Biodeformational structures in modern deep-sea sediments off NW Africa, 23°08.4′ N, 17°44.7′ W, 2096 m water depth (X-ray radiograph negative; core 12379-2, Institute of Geosciences Kiel; see Wetzel, 1981).

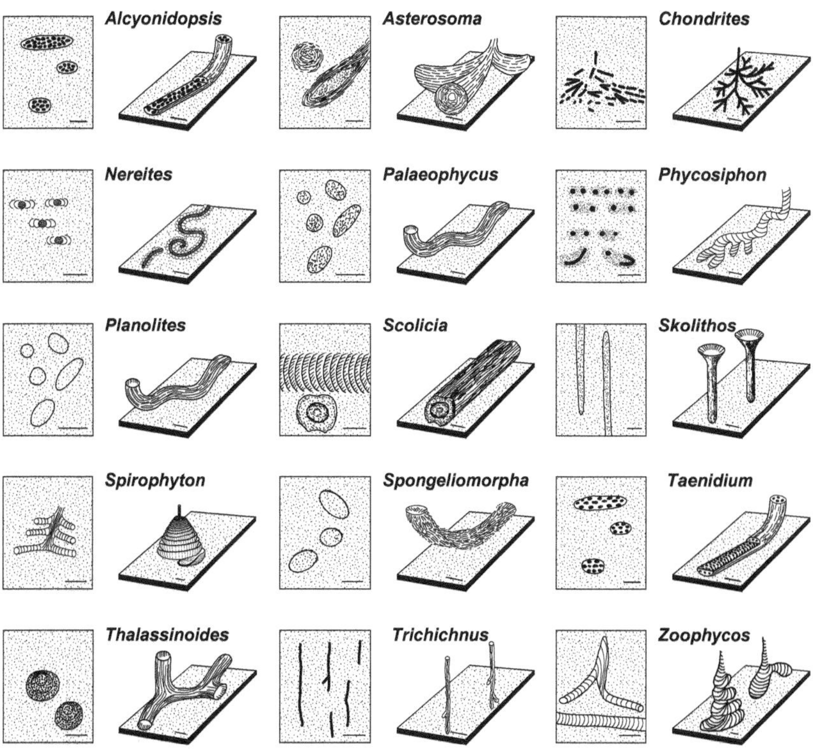

FIGURE 3 Schematic drawings of trace fossils commonly encountered in (hemi)pelagic settings.

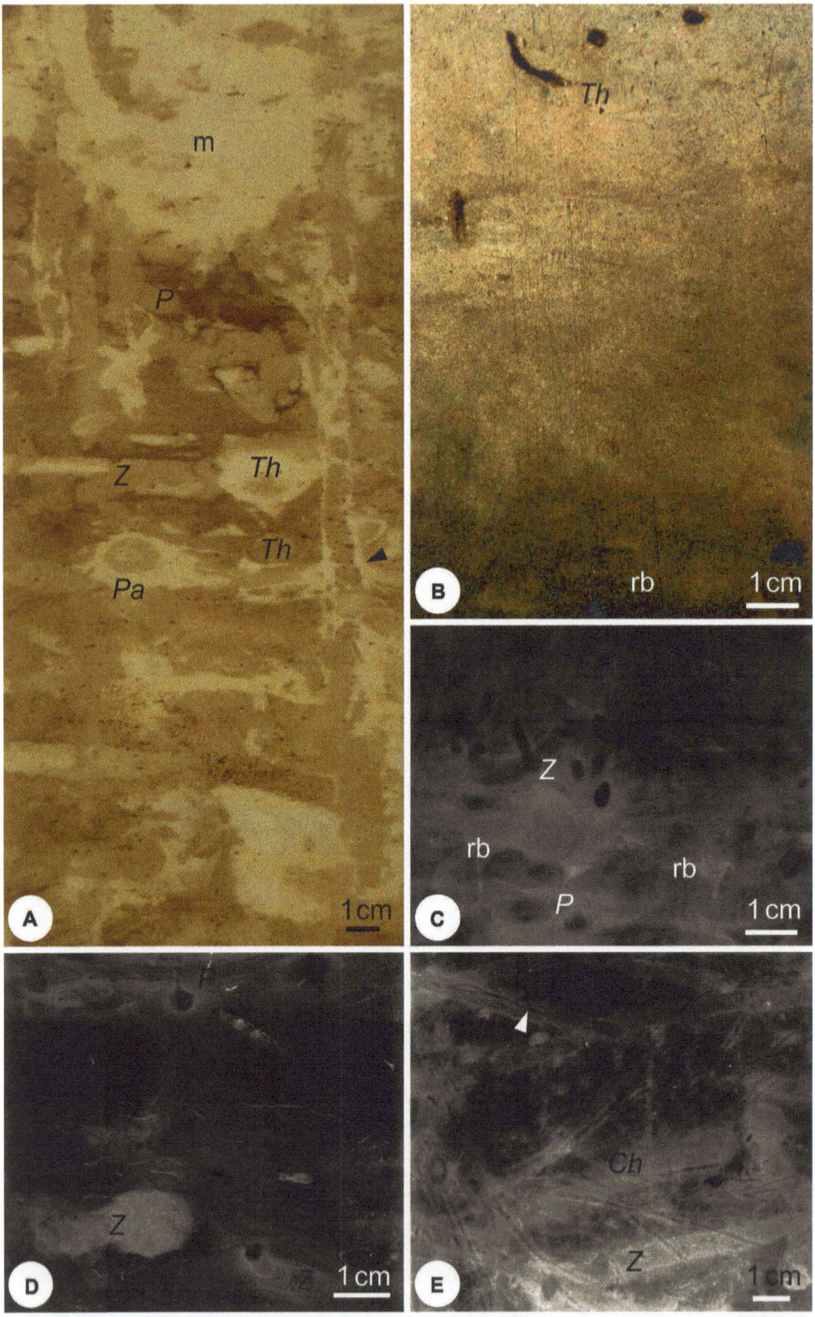

FIGURE 4 Modern examples of burrows in oxic deposits without (or with a very low content of) organic matter. (A) Institute of Geosciences Kiel, core 10148-1 (~8°40′N, ~149°30′W, ~5,100 m water depth, 202–230 cm, fresh core); note the deeply penetrating vertical burrows (arrow);

trace fossils mainly penetrate deeply and cross previously deposited beds (see also Seilacher, 1962). Continuously deposited sediments contain only post-depositional burrows, while graphoglypids are overprinted by organisms burrowing deeper. However, in (hemi)pelagic settings, pre-depositional burrows may be preserved below ash layers (e.g., Manville and Wilson, 2004). Graphoglyptids are addressed in detail by Uchman and Wetzel (2012).

2.3 Systematic Ichnology

Only post-depositional burrows are described briefly here because they commonly occur and dominate in (hemi)pelagic deep-sea deposits (Fig. 3). In most cases, the producers of deep-sea trace fossils are unknown and of unknown taxonomic affinity. In practice, most of the trace fossils mentioned here are observed on cut or parting surfaces as 2D structures, particularly in cores. Recognition of their 3D aspects, being a prerequisite for their proper identification, can be difficult (see Ekdale et al., 1984, for linking 2D views to 3D models).

Alcyonidiopsis is an endichnial, simple, almost straight, strongly flattened cylindrical burrow, mostly a few millimeter wide, filled with very small elongate fecal pellets.

Asterosoma (Fig. 3) consists of elongate, several centimeter long, a few centimeter wide bulbous segments having a terminal or eccentric, laminated fill; the segments are oriented (sub)horizontally and are arranged around a central vertical, somewhat twisted axis. *Asterosoma* occurs in impure mud containing some silt and fine sand.

Chondrites (Figs. 3 and 4–7) represents a regularly bifurcating tunnel system consisting of a (small number of) master shafts open to the surface, which ramify at depth to form a dendritic network.

Nereites (Figs. 3, 5D and 7D) is a usually selectively preserved, winding to regularly meandering, often horizontal trail, consisting of a central back-filled tunnel enveloped by an even to lobate zone of reworked sediment. Commonly, only the external part of the enveloping zone is preserved as a densely packed uni- or multiserial chain of small depressions or pustules.

Palaeophycus (Figs. 3 and 4A) is a typically unbranched, smooth or ornamented, lined, essentially cylindrical, predominantly horizontal burrow of

---

staining resulting from local dissolution and precipitation of iron (see Hartmann, 1979 for details). (B) RV *Sonne* Cruise 114, core 20 (11°54′ N, 118°04′ E, 2,919 m water depth; fresh core). (C) RV *Sonne* Cruise 140, core 03 (15°25′ N, 115°5′ E, 4248 m water depth; X-ray radiograph negative; 1–8 cm). (D) Institute of Geosciences Kiel, core 13209-2 (12°29.2′ N, 20°02.7′ W, 4,713 m water depth; 726–740 cm; X-ray radiograph negative); carbonate shells (white dots) within *Zoophycos* document a temporary surface-feeding of the producer while the sediment was deposited just below the carbonate compensation depth where $CaCO_3$ dissolution occurs on the sediment surface. (E) Institute of Geosciences Kiel, core 10148-1 (~8°40′ N, ~149°30′ W; ~5,100 m water depth; X-ray radiograph negative; 619–630 cm); note that *Chondrites* cross-cut *Zoophycos*. m, mottled homogeneous surface layer; rb, redox boundary (stained by Mn–Fe-oxides; appearing light in X-ray radiograph negatives); *Ch*, *Chondrites*; *Pa*, *Palaeophycus*; *P*, *Planolites*; *Th*, *Thalassinoides*; *Z*, *Zoophycos*.

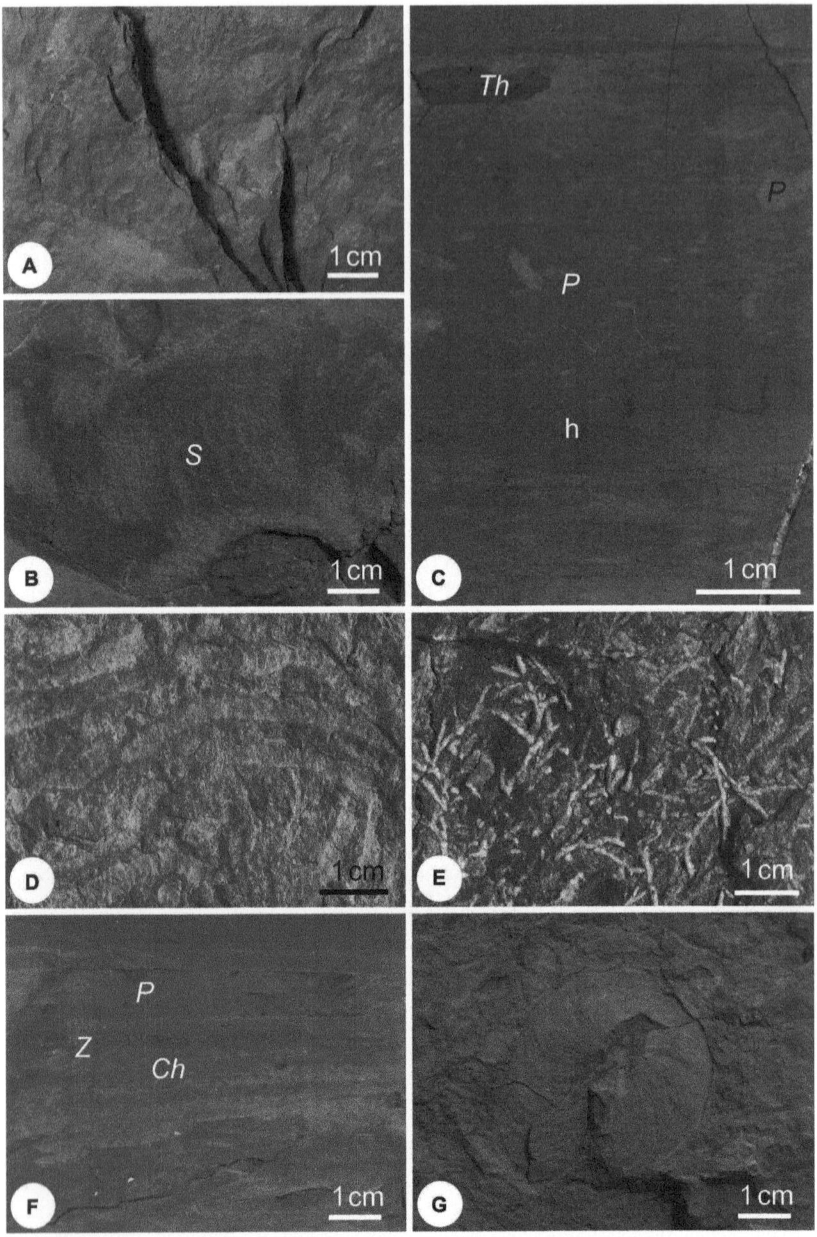

FIGURE 5 Bioturbation in highly oxic deposits not preserving organic matter; the small size of the burrows and the vertically only little extended bioturbated zone imply restricted food availability (for details, see text). (A) Slates without any evident trace fossils (Variegated Shale). (B) *Scolicia* (*S*); the coloration of the sediment suggests burn-down of organic matter while animals irrigated the sediment (Variegated Shale, Late Paleocene–Early Eocene, Słopnice, Carpathians, Poland). (C) Red limestone showing only a few trace fossils, although completely bioturbated (Upper Cretaceous, near Montbovon, Switzerland). (D) *Nereites* (Variegated Shale, Late Paleocene–Early Eocene, Carpathians, Lubomierz, Poland). (E) *Chondrites* (Variegated Shale, Late Paleocene–Early Eocene, Carpathians, Lubomierz, Poland). (F) ODP Site 718C (30°50.4′ N, 67°38.9′ W, 5,124 m water depth, core 64X2, 145–150 cm); note the small size of the traces. (G) *Zoophycos* (Couches rouges, Late Cretaceous, near Montbovon, Switzerland). h, homogeneous; *Ch*, *Chondrites*; *P*, *Planolites*; *S*, *Scolicia*; *Th*, *Thalassinoides*; *Z*, *Zoophycos*.

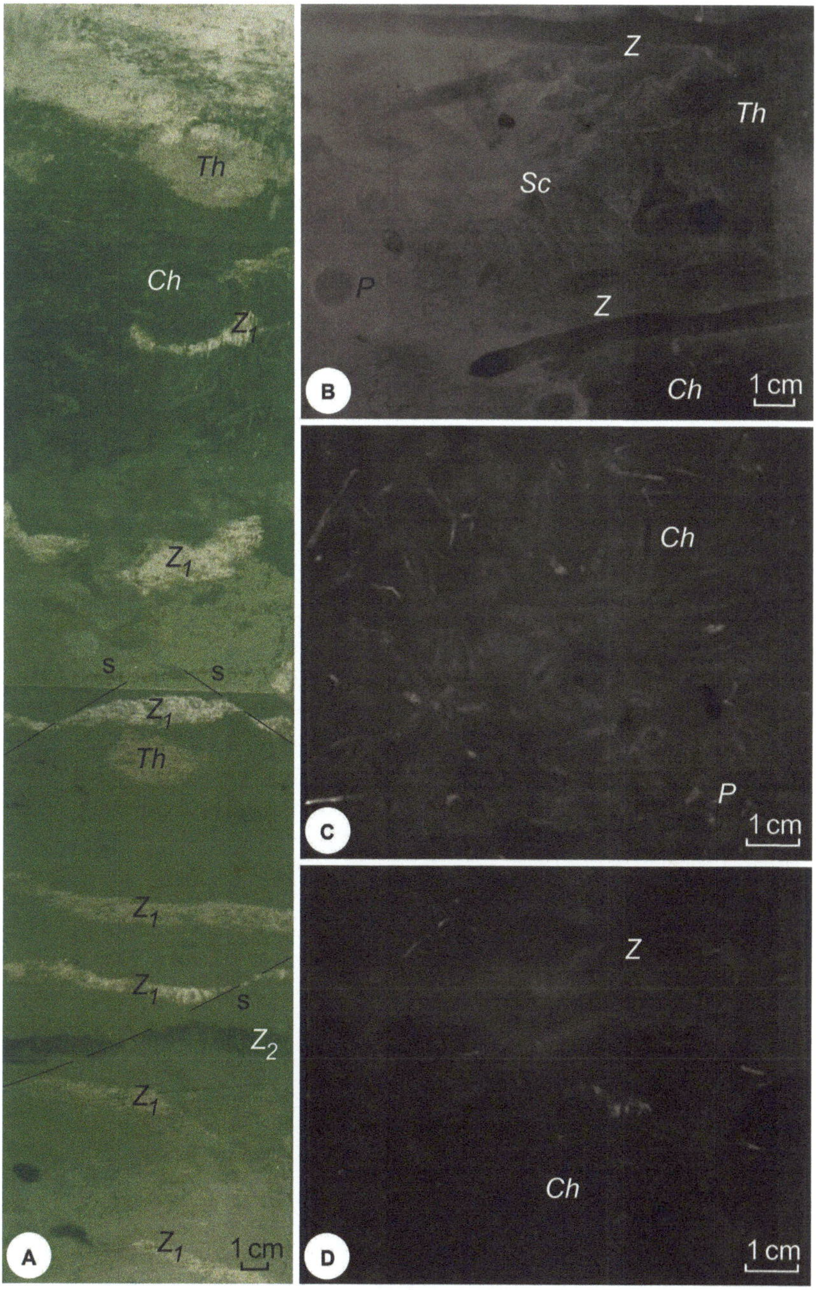

FIGURE 6 Modern examples of burrows in deposits containing organic matter. (A) Core 16867-3 (2.2° S, 5.1° E; 3,894 m water depth; 1,655–1,695 cm), producers of *Zoophycos* (*Z*) displaced material downward as indicated by sediment color; Z_1 from the top carbonate ooze layer, Z_2 from the black mud layer. (B) Core 12337-5 (15°58.4′ N, 18°07.1′ W, 3,082 m water depth; 34–44 cm; X-ray radiograph negative); cross-cutting relationships suggest a four-tier bioturbated zone. (C and D) *Sonne* Cruise 187, core 61-2 (11°25.5′ N, 111°17′ E, 2,226 m water depth); X-ray-radiograph negative; (C) 487–493 cm core depth; (D) 505–512 cm core depth; bioturbation within the oxygen minimum zone; low oxygenation indicated by the high proportion of biodeformational structures, and by the small size of traces. h, homogeneous (biodeformational structures); s, shear plane formed during retrieval of the core; *Ch*, *Chondrites*; *P*, *Planolites*; *Sc*, *Scolicia*; *Th*, *Thalassinoides*; Z, *Zoophycos*.

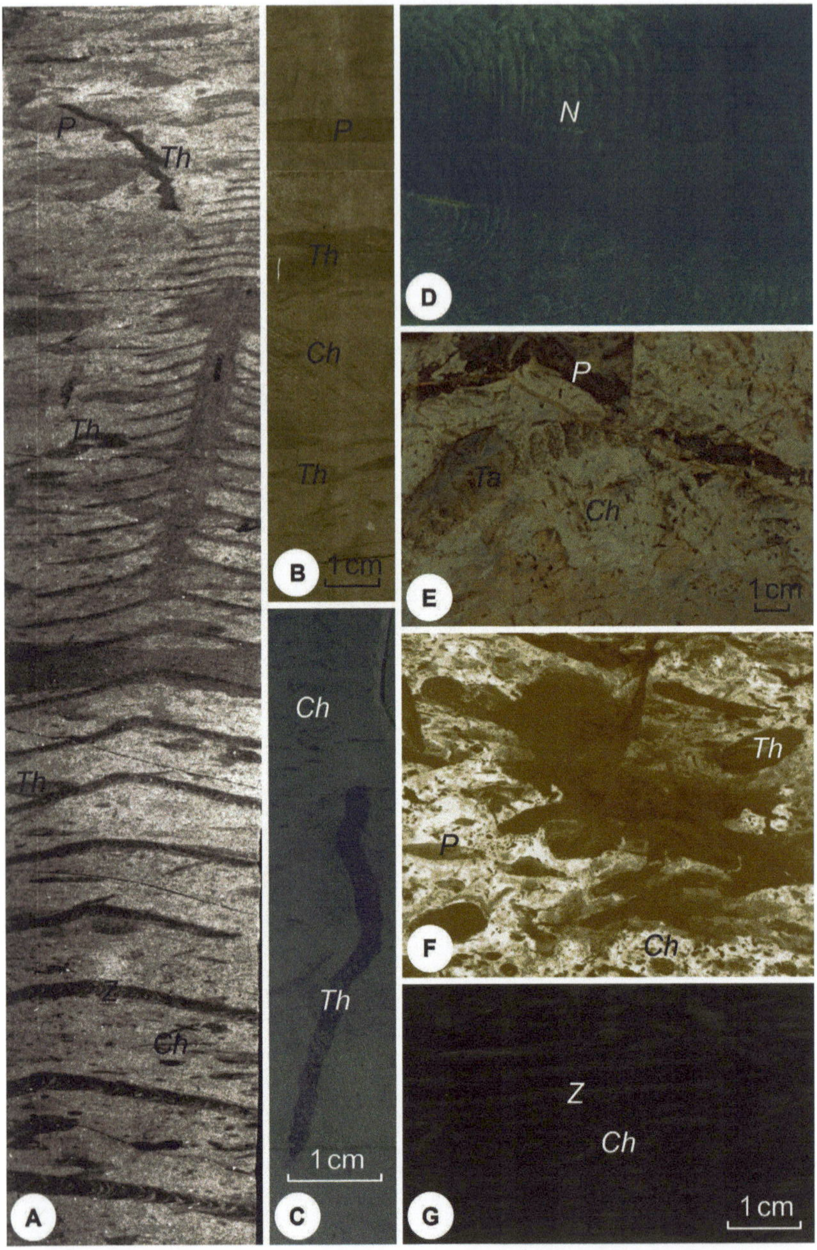

FIGURE 7 Bioturbational structures in deposits containing organic matter. (A–C) DSDP Site 605 (38°45.5′ N, 72°36.6′ W, 2,194 m water depth); (A) core 57-1, 99–124 cm; (B) core 51-2, 88–100 cm; (C) core 63-4, 110–118 cm. (D) Helminthoid Flysch (Weissenburg, Switzerland). (E) Late Cretaceous marls (Fähneren area, Switzerland). (F) So-called *Zoophycos* Beds (Middle Jurassic, Blumenstein, Switzerland). (G) ODP Site 718C (30°50.4′ N, 67°38.9′ W, 5,124 m water depth, core 94X1, 23–26 cm). Multiple cross-cutting relationships suggest a multitiered, vertically extended bioturbated zone. *Ch*, *Chondrites*, *N*, *Nereites irregularis* (formerly *Helminthoida labyrinthica*), *P*, *Planolites*, *Ta*, *Taenidium dieslingi*, *Th*, *Thalassinoides*, *Z*, *Zoophycos*.

variable diameter; the fill is typically structureless and of the same lithology as the host rock.

Phycosiphon (Fig. 3) is an extensive, small-scale spreite trace fossil consisting of repeated narrow, U-shaped lobes enclosing a spreite at millimeter- to centimeter-scale, branching regularly from an axial spreite of similar width. Lobes are protrusive, mainly parallel to bedding.

Planolites (Figs. 3 and 4–7) represents an unlined, rarely branched, straight to tortuous, smooth to annulated, tubular burrow of variable dimensions. The fill is essentially structureless but differs from the host rock.

Scolicia (Figs. 3, 5B and 6B) is a large, bilaterally symmetrical, subcylindrical burrow with meniscate lamellae often divided into two concave sets. In transverse cross section, a concentric structure of bilobate lamellae surrounds an eccentric axis. Today, *Scolicia* is produced by irregular echinoids that prefer to live in sandy to silty sediment.

Skolithos (Figs. 3 and 8B) is a straight, simple, unbranched, normally vertical tube with a uniform diameter (4–10 mm).

Spirophyton (Fig. 3) consists of a spirally coiled outward, slightly bent-upward spreite around a central axis. The diameter of the coils increases downward regularly.

Spongeliomorpha (Fig. 3) is a 3D burrow system composed of cylindrical tubes with Y- and T-shaped branching with enlarged junctions, having mm-long, irregularly oriented casts of scratches (bioglyphs) along the burrow margin.

Taenidium (Figs. 3 and 7E) represents an unlined or very thinly lined, simple, straight to sinuous, cylindrical burrow with a fill of meniscus-shaped segments.

Teichichnus (Fig. 3) is a vertical, blade-like elongate spreite burrow resulting from the upward or downward displacement of the causative tube.

Thalassinoides (Figs. 3, 4–7 and 8B) is an endichnial or hypichnial, slightly winding, semi-circular, straight or curved, branching burrow system having swellings at Y- or T-shaped junctions.

Trichichnus (Fig. 3) represents a vertical to oblique, straight or slightly winding, rarely branched cylindrical burrow, up to 1 mm in diameter.

Zoophycos (Figs. 3, 4–7 and 8) is an endichnial, horizontal to subhorizontal planar spreiten burrow visible as lobes. The lobes are surrounded by a marginal tunnel. Fine spreiten laminae run obliquely according to the concavity of the lobe or as a subcircular, helical coiled structure and may contain poorly outlined pellets. Spreiten laminae are slightly curved and run from the elevated center toward a thin, indistinct marginal tunnel, where they tend to curve.

3. DEPOSITIONAL ENVIRONMENTS AND TRACE FOSSILS

The biogenic sedimentary structures can be used to decipher the environmental conditions by looking at a specific trace fossil characterizing a typical condition (index trace fossil) or at the ichnofabric as such. The combination of both attempts will provide maximum information.

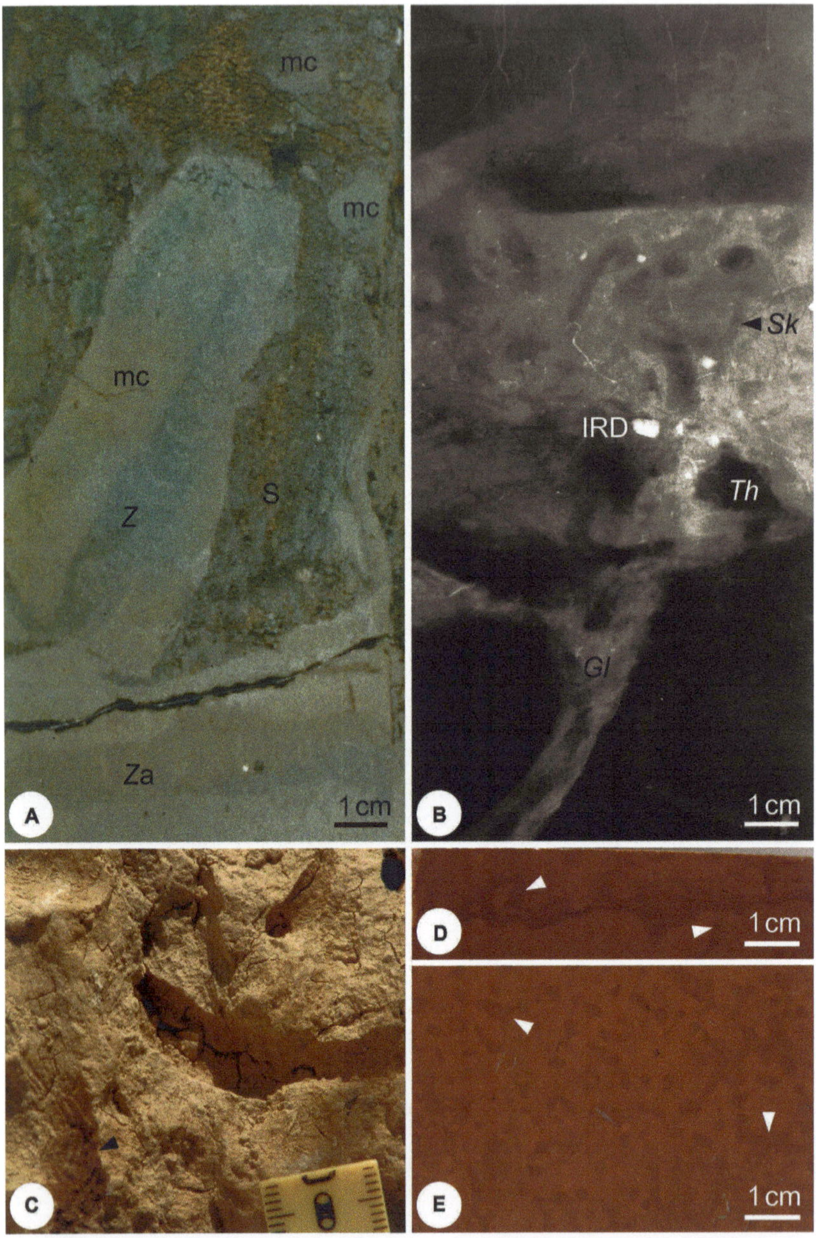

FIGURE 8 Bioturbation structures in bottom-current-affected settings. (A) Hiatus horizon covered with mud pebbles, representing the base of the Late Pleistocene within the Iceland Shetland Channel in 1,100 m water depth. British Geological Survey core 60-05/50 (60°49.7′N, 4°12.5′W), 346–358 cm (for details, see Akhurst et al., 2002). The mud clasts (mc), one with *Zoophycos* (Z), are embedded in coarse sand (S) that in turn rests on mud with autochthonous *Zoophycos* (Z_a). The minimum amount of erosion is 40 cm, based on the burrowing depth of the

3.1 Sediment Types

(Hemi)pelagic sediments may differ regarding their lithology. It is not clear if the lithology itself is a significant factor influencing burrowing organisms. Burrows are better preserved in calcareous than in muddy sediments (Berger et al., 1979). Radiolarites, made of spiny sediment particles, would theoretically influence burrowing organisms and display the same trace-fossil content as the other (hemi)pelagic sediments (e.g., Kakuwa, 2004). The ichnological record is positively or negatively affected by diagenesis because this process can enhance or reduce color differences between burrows and host sediment.

Trace fossils can help to distinguish between deep-water and shelf equivalents of the same (hemi)pelagic facies (e.g., Ekdale and Bromley, 1984; Monaco, 1995). The former authors proposed to distinguish between deep-sea Cretaceous chalk containing *Zoophycos* and shelf chalk with *Thalassinoides*.

3.2 Sediment Consistency

The margin of burrows stores information about the sediment consistency. In addition, the porosity of the substrate during trace-fossil production can be estimated by measuring the actual porosity and the compaction-induced deformation of a trace fossil (Wetzel and Aigner, 1986). Direct observations allow distinction of six categories (Wetzel and Uchman, 1998b).

- **Looseground**: Tubes within loose grains of sandy substrates are stabilized by a mucus lining (e.g., *Skolithos*) or by a pellet-armed wall (e.g., *Ophiomorpha*).
- **Soupground**: Animals "swim" through the sediment producing biodeformational structures and produce burrows having characteristic "mantle and swirl" appearance in cross section (Lobza and Schieber, 1999), or assemble a "floating pile foundation" in suspension-rich settings (Wetzel and Bromley, 1996).
- **Softground**: Thin-walled (e.g., *Palaeophycus*) or pelleted tubes are stable.
- **Stiffground**: Unlined tunnels are stable (e.g., *Thalassinoides*, *Zoophycos*).
- **Firmground**: Animals use body appendages to burrow, and burrow margins often exhibit a sculptured ornamentation (e.g., *Spongeliomorpha*).

Zoophycos producer (Wetzel and Aigner, 1986). (B) Sand (light) resting on mud (dark); sand has filled up the open burrows (*Gl*) typical for *Glossifungites* surfaces; the sand has been intensely bioturbated. X-ray radiograph (negative). Institute of Geosciences Kiel, Germany, core 16384-2 (61°57.55′N, 11°53.15′W), 310–320 cm (Late Pleistocene), southern flank of the Iceland-Færø Ridge at 1255 m water depth. IRD, ice rafted debris; *Th*, *Thalassinoides*; *Sk*, *Skolithos*. (C) Modern hiatus horizon produced by bottom currents that traverse a seamount chain within the South China Sea (14°52.4′N, 118°35.4′E; 3,940 m water depth; core SO 114-16; for details, see Wetzel, 2008). The sediment has been partly eroded and the surface has become stained by Fe- and Mn-oxides. A stiff sediment consistency is indicated by the numerous open burrows and the ornamentation of the burrow margins by claw-sculptures (arrow). (D, E) Vertical (D) and horizontal (E) section through a bottom-current-affected omission surface (Late Cretaceous, Prealpes medianes, Switzerland); interconnected U-shaped tubes (arrows mark deviations) were produced in a stiffened substrate; later overprint by pressure solution.

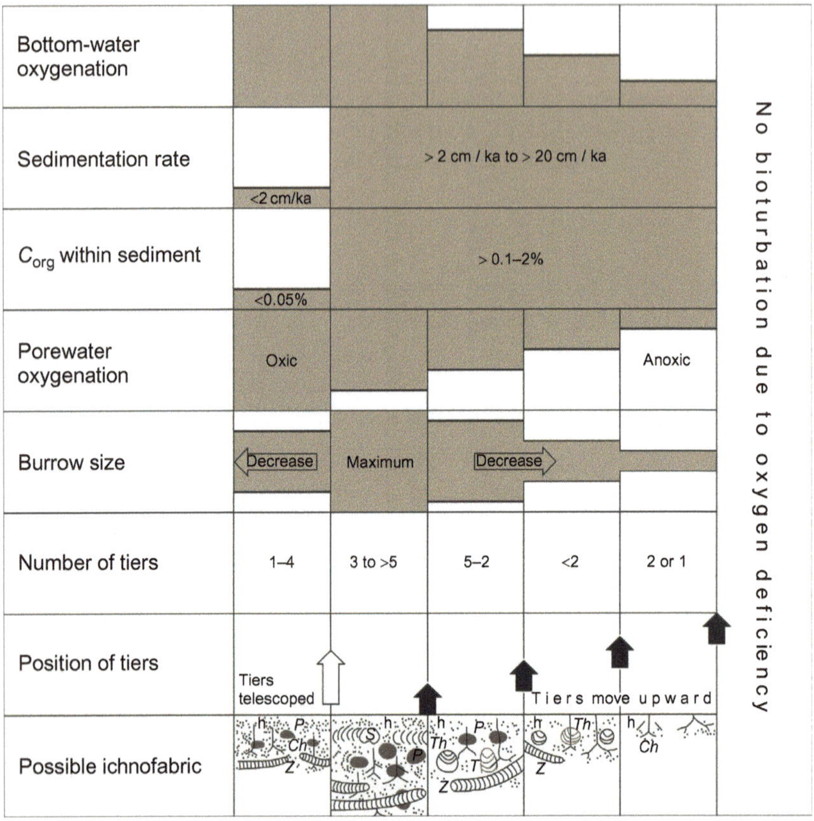

FIGURE 9 Burrowing organisms in deep-marine, (hemi)pelagic settings sensitively respond to environmental changes. The diagram is based on observations in Quaternary sediments in the upwelling off NW Africa. *Modified after Wetzel (1981, 1991, 2010).*

- **Hardground**: Traces cut through particles and cement of indurated sediment (e.g., *Entobia*).

3.3 Hemipelagic and Pelagic Sediments

Trace fossils are suitable to evaluate various ecological parameters of pelagic and hemipelagic sediments that accumulate continuously. If the steady accumulation, however, is interrupted by the deposition of an event layer, the previous tiering structure can become completely or partly fossilized, depending on the erosional power during the deposition of the event bed (see Wetzel and Aigner, 1986). Such "frozen" tiers are useful to calibrate ichnofabrics. The development of an ichnofabric is affected by many interrelated environmental factors, especially organic-matter supply, sedimentation rate, and oxygenation (Fig. 9). As these can be only partially unraveled by ichnofabric analysis, the potential effects are discussed first, and then two standard sediment types with respect to pore-water oxygenation are outlined.

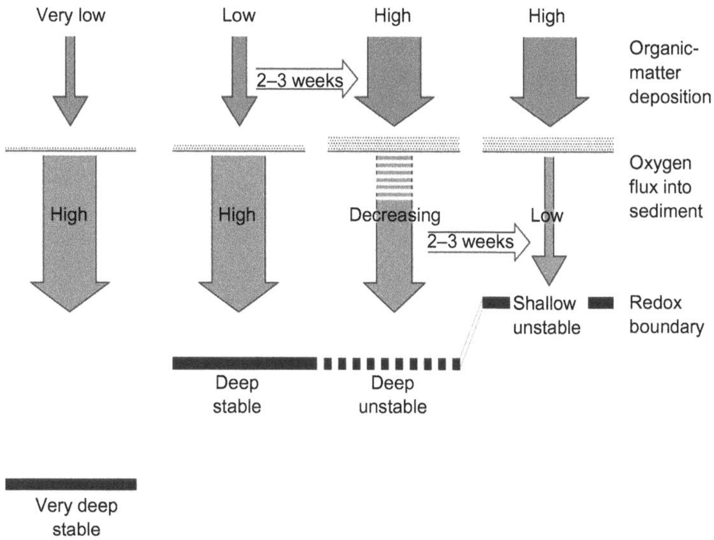

FIGURE 10 Schematic representation of how episodic deposition of organic matter could affect the oxygen flux into the sediment and the temporary position of the redox boundary (*after Wetzel, 2010, slightly changed*).

The sedimentation rate controls the burial of organic matter (see this section below). Therefore, the trophic level of sediments cannot be evaluated from trace fossils for a wide range of sedimentation rates (<3 to >30 cm/ka). Such deposits are completely bioturbated if accumulated under fully oxygenated bottom water. Nonetheless, trends can be found (Figs. 9 and 10). With increasing sedimentation rate, the vertical extension of tiers may increase, and deeply penetrating burrows such as *Thalassinoides* and *Zoophycos* may become dominant in deep tiers. In intermediate tiers, patchy bioturbation by the producers of *Phycosiphon* can occur; furthermore, *Chondrites* and *Teichichnus* may be present. In the case of slowing down sediment input, the penetration depth tends to decrease. At a drastically reduced sediment input, little organic matter is buried and the substrate tends to stiffen. In this case, burrows are less compacted, often sharply walled (stiffground), and passively filled, and they show claw-sculptured ornamentation along their margins (Savrda et al., 2001; Wetzel, 2008).

The organic-matter content of deep-sea deposits accumulating at rates <5 cm/ka depends on the sedimentation rate (Müller and Suess, 1979). Above that value, a dilution effect occurs (Tyson, 2001). In fact, the burial velocity of organic matter influences the exposure time of organic matter to oxygen (e.g., Hartnett et al., 1998). A feedback mechanism complicates this relationship because with increasing benthic food content, both the long-term benthic standing stock and the burrowing activity expressed as the benthic mixing rate increase (e.g., Legeleux et al., 1994). In turn, the oxygenation of the pore water is affected; the higher the burrowing activity, the more extended is the oxygenated zone (e.g., Reimers et al., 1986). Thus, effects of organic-matter burial and sedimentation rate cannot be clearly defined and directly evaluated. However, if the availability of benthic food is very high, biodeformational structures normally >2 cm in diameter dominate the ichnofabric under well-oxygenated conditions. Endobenthic animals appear to burrow without distinct behavioral specialization and hence no trace fossils are produced (Wetzel, 1981, 1991). In modern sediments off NW Africa, biodeformational structures dominate if the C_{org} content is >2%; below this value, trace fossils are present. However, if oxygenation of the bottom water decreases, biodeformational structures may form at lower C_{org} values (see this section below).

For vast areas of the oceans, the organic-matter production fluctuates significantly during the year and during longer time spans in response to the seasonality of insolation, wind stress, and circulation (e.g., Antoine et al., 1996; Lutz et al., 2007). Depending on water depth and of the size of the (mineral) organic aggregates, such organic-rich particles reach the sea floor after 2–4 weeks (e.g., Smith et al., 1996; Wiesner et al., 1996). While settling, a considerable portion of the organic particles is oxidized (e.g., Suess, 1980; Tyson, 2001). Thus, the organic-matter deposition on the sea floor can fluctuate considerably with time (e.g., Lampitt and Antia, 1997; Lutz et al., 2002). The deposition of phytodetritus is a major energy source to the benthos and has been linked to both seasonal patterns of growth and to reproduction (Tyler, 1988) as well as to regional variability in benthic biomass (Thurston et al., 1994). A surplus of benthic food results in enhanced macrobenthic activity after a short-time lag; microbes and other organisms respond immediately (e.g., Gooday and Turley, 1990).

In the short term, following blooms, organic flocks can cover the sea floor (e.g., Smith et al., 2002; Wetzel, 2008). Concomitantly, while oxygen is consumed within this organic-rich layer, the oxygen flux into the sediment decreases rapidly (e.g., Gehlen et al., 1997). Measurements in oceanic deposits imply that the decrease in oxygen flux into the sediment may lag behind organic-matter deposition by 2–3 weeks (e.g., Smith et al., 2002). When the oxygen flux decreases, the redox boundary separating oxic and anoxic deposits moves upward. In this way, the effect of organic-matter deposition propagates within some tens of days into the sediments (Soetaert et al., 1996).

Some trace fossils can indicate such food fluctuation. For instance, in Holocene and Pleistocene times the producer of *Zoophycos* collected food mostly from the sedimentary surface and kept it for periods of low food availability (cache behavior; Bromley, 1991; Miller and D'Alberto, 2001; Löwemark et al., 2006; Wetzel et al., 2011). Similarly, the producers of *Nereites missouriensis* appear to respond to the arrival of organic matter. After upwelling periods, in the central South China Sea the producers of this trace move upward and feed on the surface, and displace some surface sediment downward as evidenced by 1991 ash present in these *Nereites* (Wetzel, 2002). The geometry of different ichnospecies of *Nereites* (following the taxonomy of Uchman, 1995) can be explained appropriately by these observations. If organic-matter deposition fluctuates seasonally, *Nereites missouriensis* is produced in response to fluctuating degrees of oxygenation. However, if organic-matter supply to the sea floor is constant, horizontal *Nereites irregularis* (formerly *Helminthoida labyrinthica*) tends to be formed (for details, see Wetzel, 2010; Fig. 10).

Regarding oxygenation, three different situations are commonly distinguished with respect to the oxygen content of the bottom/pore water (Rhoads and Boyer, 1982): aerobic/oxic (> 1.0 ml O_2/l), dysaerobic/dysoxic (1.0 to 0.3–0.1 ml O_2/l), and anaerobic/anoxic (< 0.3–0.1 ml O_2/l). With respect to bioturbation, it is useful to distinguish within the aerobic setting between (1) oxygenated deposits being low in organic matter, having also oxygenated pore water characterized by a brownish color and (2) partially oxic sediments enriched in organic matter, exhibiting a redox boundary within the bioturbated zone, and hence, color changes from brownish at the top to grayish green further down, and a dark brownish black Fe–Mn-enriched zone occurs at the redox boundary (e.g., Froelich et al., 1979). Under anaerobic conditions, laminated, unbioturbated black mud forms as a rule. The change between dysaerobic and anaerobic conditions can be unraveled by the study of trace fossils (e.g., Rodríguez-Tovar and Uchman, 2010; Uchman et al., 2008).

3.3.1 Oxygenated Deposits without Organic Matter

Because of the low deposition of sediment and/or organic matter, organic material is oxidized within the upper interval of the sea floor to such a high degree that iron remains in the oxidized state, and hence, the deposits exhibit a brownish color (e.g., König et al., 1997). Such deposits contain dissolved oxygen throughout the bioturbated zone. They accumulate very slowly, and because of the extent of oxygen exposure time, they are very low in organic matter (e.g., Mangini et al., 2001). Food is restricted, at least deeper within the sediment, but the near-surface benthic-food level may increase after plankton blooms. The sediment is completely bioturbated and the number of tiers is low, often ≤3. The surface mixed layer is thin and burrows tend to be small (Figs. 9 and 4). The penetration depth of burrows is generally low (up to 10–15 cm). The substrate is soft to stiff. The small size of burrows, the fine grain size and the often small variations in grain size lead to a homogeneous slaty appearance of such deposits when compacted; furthermore, oxic pore water has the tendency to extinguish any initial color differences.

In the rock record in such deep-marine "oceanic red beds" (e.g., Hu et al., 2005), there are several factors making trace fossils often hardly recognizable: (1) burrows are so tiny that they cannot be recognized clearly; (2) the sorting of the sediment is so good that sorting by burrowing does not result in a distinct structure; (3) the sediment is very fine-grained and rather soft and hence burrows were later extinguished by compaction (this is especially true when microfossil shells are dissolved within the sediment); (4) highly oxic conditions prevent the production of a chemical gradient between burrow and surrounding sediment that is sufficiently distinct to survive diagenesis.

In highly oxic mudstones, a low degree of bioturbation and only a few, mainly small trace fossils are commonly observed (Bąk, 1995; Leszczyński and Uchman, 1993; Mikuláš et al., 2009). Commonly, *Chondrites*, *Palaeophycus*, *Phycosiphon*,

Planolites, *Teichichnus*, *Thalssinoides*, and *Zoophycos* occur (e.g., Leszczyński and Uchman, 1991, 1993; Mikuláš et al., 2009; Wetzel and Uchman, 1998b). Many of these trace fossils have an open connection to the sea floor. It is not yet clear whether this is to enhance respiration water supply in the low-permeable mud or to catch food from the surface during bloom times (Wetzel and Uchman, 1998b).

The conditions in such settings are not as stable as usually supposed; subtle color changes occur; these are related to fluctuations in sedimentation rate and/or organic-matter deposition. Environmental changes have been documented by intercalated calcareous layers or distal turbidites (e.g., Leszczyński and Uchman, 1991, 1993; Mikuláš et al., 2009), and hence, it is possible to decipher the regulating factors to some degree. In these intervals, the intensity of bioturbation, size and diversity of trace fossils increases. Evidently, the benthic food-level rose for a short time (e.g., Leszczyński, 1993; Mikuláš et al., 2009). However, the coloration of the sediments suggests that the depositional area remained unstable as episodes of increased food availability alternated with periods of slow sediment accumulation that favored downward oxidation of previously deposited sediment (cf. Jung et al., 1997).

Notably, in some sections of such reddish, deep-marine, fine-grained deposits *Chondrites* was found as one of the most deeply produced traces. This observation implies that *Chondrites* is not strictly indicative of lowered oxygenation. Furthermore, not all producers of *Chondrites* very likely follow a nutritional strategy relying on chemosymbiosis in a reducing host sediment (e.g., Fu, 1991; Seilacher, 1990).

Large *Zoophycos* that occur in reddish hemipelagic deposits (e.g., Miller et al., 2004; Fig. 4) probably record cache behavior as found in modern deep-marine settings (Löwemark et al., 2006; Wetzel et al., 2011). These *Zoophycos* appear to be indicative of very low sedimentation rates, but influenced by seasonal or episodic organic-matter deposition.

3.3.2 Oxygenated Deposits with Preserved Organic Matter

Oxygenated deposits enriched in organic matter can accumulate under well-oxygenated bottom water, but burial of organic matter leads to the complete consumption of dissolved oxygen within the upper part of the bioturbated zone, so that the pore water becomes anoxic. The boundary between oxic and anoxic pore water is defined as the redox boundary, where microbes flourish (e.g., Konhauser, 2007). The deposits below the redox boundary are grayish to green colored, as iron occurs in a reduced state. Under such conditions, even slight changes in the oxygen content of the bottom water can affect the burrowing fauna and the resultant ichnofabric considerably (e.g., Bromley and Ekdale, 1984; Savrda and Bottjer, 1989; Wetzel, 1983, 1991).

Because of high benthic food levels and good oxygen availability, a vertically extended bioturbation zone is found in soft sediments that can occupy an about 1 m thick interval or more (Wetzel, 1981; Figs. 6, 7 and 9). A large number of tiers may develop. The large number of tiers is an expression of efficient fractionation of ecospace where conditions change with depth in sediment and because various food sources are potentially available. Food sources include surface organic matter, detrital organic matter within the sediment, and microbes living along the redox boundary or along burrow walls, where a steep geochemical gradient is maintained. However, the change of one of the above-mentioned factors affects the whole bioturbated zone (Fig. 9). Oxygen-deficiency is considered here in some more detail.

When the oxygen content of the bottom water decreases, the penetration depth, burrow diameter, trace-fossil diversity and number of tiers also decrease and tiers show an order of disappearance, starting with the shallowest (Bromley and Ekdale, 1984; Savrda and Bottjer, 1989): *Planolites—Zoophycos—Thalassinoides—Chondrites* (large)—*Chondrites* (small). Burrows with an open connection to the surface become relatively more frequent while those having no connection to the sea floor disappear (Wetzel, 1983, 1991).

The benthic-community succession paradigm, originally set-up for shallow-water deposits (Rhoads and Boyer, 1982), differs from that trace-fossil approach (Wheatcroft, 1989). Under well-oxygenated equilibrium conditions, intensive bioturbation and particle mixing occurs, and the surface is covered with numerous feeding pits, fecal pellets and tube openings. Such a community changes during deoxygenation to (1) surface-deposit-feeding and suspension-feeding (including production of pellets); over (2) fluid bioturbation pumping water into and out of the bottom through vertical tubes while particle bioturbation is subordinate; and (3) small opportunistic tube dwelling polychaetes at high densities, whereby tubes protect against soupy substrate and poisonous pore water (H_2S); to finally (4) no bioturbation.

The growing number of observations in modern settings exhibits that there is no unequivocal scenario for bioturbation in settings experiencing low oxygenation. For deep-marine assemblages developing under severe hypoxia, food availability remains a significant determinant of animal abundance, as well as depth and type of bioturbational structure produced, whereas oxygen influences patterns of organisms' diversity and dominance (e.g., Levin et al., 2009).

At low-oxygen levels in deep-marine settings, biodeformational structures tend to be produced more commonly by a shallow burrowing fauna as it is the case, for instance, in sediments of the California borderland basins (e.g., Behl, 1995; Edwards, 1985). Therefore, during deoxygenation, the known trend to decreasing size and penetration depth of burrows is supplemented by the trend to decreasing production of burrows with distinct outlines and concomitantly increasing production of biodeformational structures. Even though tube-building organisms are quite abundant at low-oxygen levels, their tubes are normally small in diameter (<1 mm) and short (<1 cm) (Levin et al., 2009). In fact, a stage of cryptobioturbation is approached. Low oxygenation provokes shallow bioturbation, and excessive benthic food content makes behavioral specialization superfluous and maintenance of open tubes in soft sediment inefficient (Figs. 6 and 7). It appears that in oxygen-deficient settings with soft sediments, the production of trace fossils ceases at lower C_{org} values than in oxygenated environments; mainly biodeformational structures are produced instead. Off NW Africa, for instance, under fully oxygenated conditions, biodeformational structures dominate the ichnofabric above ~2% C_{org} (Wetzel, 1981, 1991). In contrast, off central Vietnam, this threshold value is in the range of 0.5–0.7% C_{org} during periods of lowered oxygenation (Wetzel et al., 2011).

In addition to the benthic-food content and oxygenation, other aspects need to be taken into account when investigating ichnofabrics; especially for mud rocks, the original sediment consistency, the role of biodeformational structures, and the degree of compaction need to be considered. The more enriched a muddy sediment is in organic matter, the more water-rich it tends to be when deposited (e.g., Keller, 1982). A rather soft consistency, however, favors the production of biodeformational structures (e.g., Lobza and Schieber, 1999; Wetzel et al., 2011). The higher the porosity of a mud, the higher is the chance that clay minerals become reoriented perpendicular to the pressure during compaction (e.g., Mitchell, 1993). In this way, the originally formed biodeformational structures may become macroscopically pseudolaminated when compacted. Another aspect is that the biodeformational structures are rarely mentioned in ichnofabric analysis (cf. Bromley and Ekdale, 1984; Savrda and Bottjer, 1986).

3.4 Settings Affected by Bottom Currents

The effects of deep-sea bottom currents on the habitat have been studied mainly in modern settings that are influenced by contour currents (for summary, see Fig. 11). Such areas have been investigated in great detail (e.g., Nowell and Hollister, 1985; Stow et al., 2002, and references therein). The mean flow velocity of contour currents is in the range of 10–30 cm/s (McCave et al., 1980). In fact, the current speed can fluctuate within a few days to several weeks from quiescence to benthic storms (>70 cm/s; Richardson et al., 1981). Contour currents may therefore rework the sea floor and take-up particles that are settling down (Fig. 8). Thus, such currents can carry a considerable amount of fine grains including particulate organic matter (e.g., McCave, 1985; Thomsen et al., 2002) and supply food to deep-marine biota (e.g., Thistle et al., 1985). Therefore, the local benthos exhibits a higher population density and biomass than in adjacent, more tranquil settings (e.g., Thistle et al., 1985: 103).

At the contour-current affected HEBBLE (high energy benthic boundary layer experiment) site off Nova Scotia, the macro- and meiofauna are on average extremely abundant for that depth and comparable to sites 2,000 m shallower (Thistle et al., 1985; Thomson et al., 2000). Because of the repeated physical disturbance, the fauna resembles that found in recolonization experiments (Thistle et al., 1991). With respect to behavioral groups, ordinary suspension feeders are not abundant, probably as their filter apparatus can easily be plugged when the suspension concentration is very high for some time (Thistle et al., 1991). Taxa, however, that passively extract drifting particles by maintaining

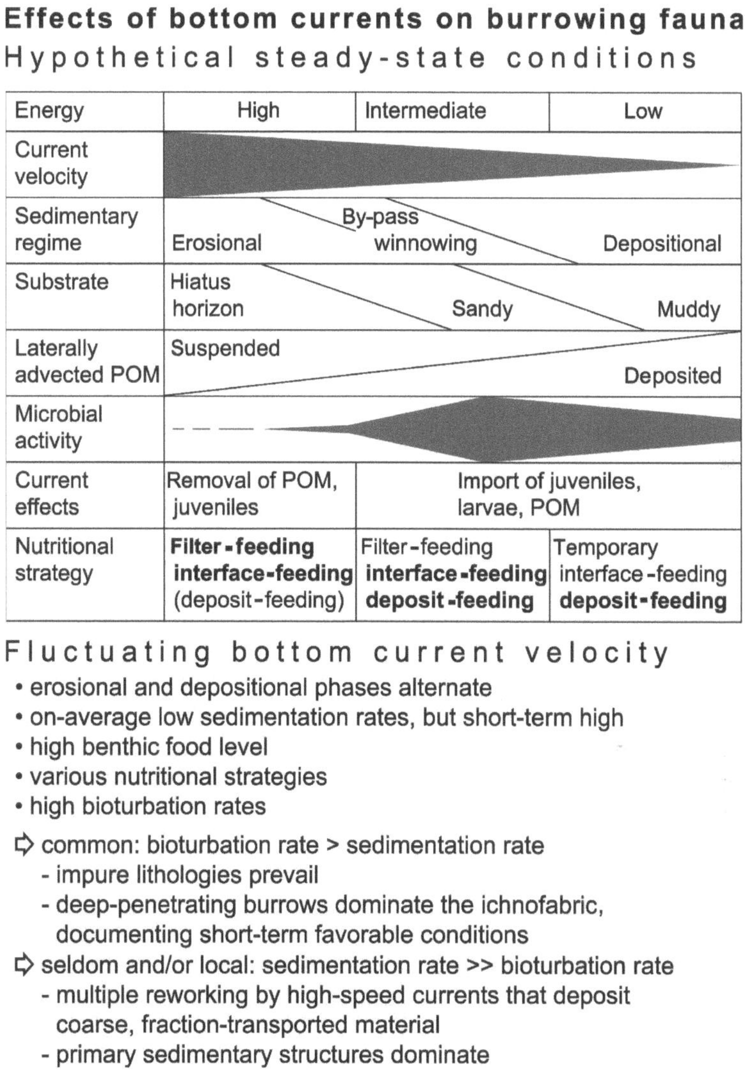

FIGURE 11 Bioturbation and trace fossils in bottom-current-affected settings; POM, particulate organic matter (*after Wetzel et al., 2008, slightly altered*).

a relief on the sea floor, exhibit a pronounced abundance (Aller and Aller, 1986). Similarly, in many organism groups, taxa that can enter the sediment occur more abundantly than at tranquil sites (Thistle et al., 1985: 103).

The high-trophic mode of bioturbation in contourite settings results in an enhanced rate of burrow production, a distinct depth of bioturbation and a specialized behavior. In contrast to numerous deep-sea biological studies, there are only few detailed investigations of bioturbation and ichnofabrics, such as those by Fu and Werner (1994, 2000), Baldwin and McCave (1999), Löwemark et al. (2004), and Wetzel et al. (2008). They mainly exhibit a clear relationship between ichnofabrics and grain size; the effects of the other environmental factors are not unequivocal. Therefore, a clear ichnological criterion recording the contourite mode of bioturbation is not yet available.

4. CONCLUSIONS AND PERSPECTIVES

Trace fossils are still underestimated in sedimentological analyses of (hemi) pelagic sediments. There is still a great gap between deep-sea paleoichnology and neoichnology. The world of modern endobenthic burrowing animals is only little known and their interrelationships to the fluctuating environmental factors are rarely investigated. Multidisciplinary neoichnological research in different deep-sea environments is needed in order to recognize and identify the trace-makers, their burrowing and general behavior, and their distribution in relation to the environmental parameters. The faunal structure in the deep sea and the factors regulating it are also unknown, including the aspects of patchiness. In the rock record, it would be worthwhile to study the spatial distribution and lateral continuity of trace fossils and trace-fossil assemblages in detail to gain additional information.

ACKNOWLEDGMENTS

A. U.'s research benefited financially from the Alexander von Humboldt Foundation, the Jagiellonian University (DS funds), and the Komitet Badań Naukowych (now Ministry of Education and High Schools, Poland) as a 2004–2005 research project. A. W.'s research was supported by the Swiss National Science Foundation and the University of Basel. W. Miller (Arcata, CA, USA) and the editors carefully reviewed the chapter and made helpful suggestions. All these contributions are gratefully acknowledged.

REFERENCES

Akhurst, M.C., Stow, D.A.V., Stoker, M.S., 2002. Late Quaternary glacigenic contourite, debris flow and turbidite process interaction in the Faroe-Shetland Channel, NW European Continental Margin. In: Stow, D.A.V., Pudsey, C.J., Howe, J.A., Faugères, J.-C., Viana, A.R. (Eds.), Deep-Water Contourite Systems: Modern Drifts and Ancient Series, Seismic and Sedimentary Characteristics. Geol. Soc. London, Mem. 22, pp. 73–84.

Aller, J.Y., Aller, R.C., 1986. Evidence for localized enhancement of biological activity with tube and burrow structures in deep-sea sediments at the HEBBLE site, western North Atlantic. Deep-Sea Res. 33, 755–790.

Antoine, D., André, J.-M., Morel, A., 1996. Oceanic primary production 2. Estimation at global scale from satellite (coastal zone color scanner) chlorophyll. Glob. Biogeochem. Cy. 10, 57–69.

Ausich, W.I., Bottjer, D., 1982. Tiering in suspension-feeding communities on soft substrata through the Phanerozoic. Science 216, 173–174.

Bąk, K., 1995. Trace fossils and ichnofabrics in the Upper Cretaceous red deep-sea marly deposits of the Pieniny Klippen Belt, Polish Carpathians. Ann. Soc. Geol. Polon. 64, 81–97.

Baldwin, C.T., McCave, I.N., 1999. Bioturbation in an active deep-sea area: implications for models of trace fossil tiering. Palaios 14, 375–388.

Behl, R.J., 1995. Data report: detailed lithostratigraphy of Hole 893A, Santa Barbara Basin, southern California continental borderland. In: Kennett, J.P., Baldauf, J.G., Lyle, M. (Eds.), Proc. ODP, Scient. Res. 146, pp. 347–351.

Bentley, S.J., Sheremet, A., Jaeger, J.M., 2006. Event sedimentation, bioturbation, and preserved sedimentary fabric: field and model comparisons in three contrasting marine settings. Cont. Shelf Res. 26, 2108–2124.

Berger, W.H., 1982. The benthic interface of deep-sea carbonates: a three-tiered sequence controlled by depth of deposition. In: Fanning, K.A., Manheim, F.T. (Eds.), The Dynamic Environment of the Ocean Floor. Heath and Company, Lexington, pp. 95–114.

Berger, W.H., Heath, G.R., 1968. Vertical mixing in pelagic sediments. J. Mar. Sci. 26, 134–147.

Berger, W.H., Ekdale, A.A., Bryant, P.F., 1979. Selective preservation of burrows in deep-sea carbonates. Mar. Geol. 32, 205–230.

Boudreau, B.P., 1986a. Mathematics of tracer mixing in sediments. I. Spacially dependent diffusive mixing. Am. J. Sci. 286, 161–198.

Boudreau, B.P., 1986b. Mathematics of tracer mixing in sediments. II. Nonlocal mixing and biological conveyor belt. Am. J. Sci. 286, 199–238.

Boudreau, B.P., 1998. Mean mixed depth of sediments: the where-fore and the why. Limnol. Oceanogr. 43, 524–526.

Boudreau, B., Imboden, D.M., 1987. Mathematics of tracer mixing in sediments: III. The theory of nonlocal mixing within sediments. Am. J. Sci. 287, 693–719.

Bromley, R.G., 1991. *Zoophycos*: strip mine, refuse dump, cache or sewege farm? Lethaia 24, 460–462.

Bromley, R.G., Ekdale, A.A., 1984. *Chondrites*: a trace fossil indicator of anoxia in sediments. Science 224, 872–874.

Bromley, R.G., Ekdale, A.A., 1986. Composite ichnofabric and tiering of burrows. Geol. Mag. 123, 49–65.

Bromley, R.G., Schulz, M.-G., Peake, N.B., 1975. Paramoudras: giant flints, long burrows and the early diagenesis of chalk. Det Kong. Danske Vidensk. Selsk. Biol. Skr. 20, 1–31.

Byers, C.W., 1977. Biofacies Patterns in Euxinic Basins: A General Model. SEPM, Spec. Publ., vol. 25, 5–17.

Chamberlain, C.K., Clark, D.L., 1973. Trace fossils and conodonts as evidence for deep-water deposits in the Oquirrah Basin of Central Utah. J. Paleontol. 47, 663–682.

Chester, R., 2000. Marine Geochemistry. Blackwell, Oxford, 506 pp.

DeMaster, D.J., Brewster, D.C., McKee, B.A., Nittrouer, C.A., 1991. Rates of particle scavenging, sediment reworking, and longitudinal ripple formation at the HEBBLE site based on measurements of ^{234}Th and ^{210}Pb. Mar. Geol. 99, 423–444.

Edwards, B.D., 1985. Bioturbation in a dysaerobic, bathyal environment: California borderland. In: Curran, H.A. (Ed.), Biogenic Structures: Their Use in Interpreting Depositional Environments. SEPM Spec. Publ., vol. 35, pp. 309–331.

Ekdale, A.A., 1977. Abyssal trace fossils in worldwide deep sea drilling project cores. In: Crimes, T.P., Harper, J.C. (Eds.), Trace Fossils 2. Geol. J., Spec. Iss. 9, pp. 163–182.

Ekdale, A.A., Bromley, R.G., 1984. Comparative ichnology of shelf-sea and deep-sea chalk. J. Paleontol. 58, 322–332.

Ekdale, A.A., 1985. Trace fossils and mid-Cretaceous anoxic events in the Atlantic Oceans. In: Curran, H.A. (Ed.), Biogenic Structers: Their Use in Interpreting Depositional Environments. SEPM Spec. Publ., vol. 35, 333–342.

Ekdale, A.A., Mason, T.R., 1988. Characteristic trace-fossil association in oxygen-poor sedimentary environments. Geology 16, 720–723.

Ekdale, A.A., Bromley, R.G., Pemberton, S.G., 1984. Ichnology: The Use of Trace Fossils in Sedimentology and Stratigraphy. SEPM Short Course Notes 15, Tulsa, 317 pp.

Eving, M., Davis, R.A., 1967. Lebensspuren photographed on the ocean floor. In: Hersey, J.B. (Ed.), Deep-Sea Photography. John Hopkins University Press, Baltimore, pp. 259–294.

Froelich, P.N., Klinkhammer, G.P., Luedtke, N., Heath, G.R., Cullen, D., Dauphin, P., Hammond, D., Hartman, B., Maynard, V., 1979. Early oxidation of organic matter in pelagic sediments of the eastern equatorial Atlantic: suboxic diagenesis. Geochim. Cosmochim. Acta 43, 1075–1090.

Fu, S., 1991. Funktion, Verhalten und Einteilung fucoider und lophocteniider Lebensspuren. Cour. Forsch. Inst. Senck. 135, 1–79.

Fu, S., Werner, F., 1994. Distribution and composition of biogenic structures on the Iceland-Faeroe Ridge: relation to different environments. Palaios 9, 92–101.

Fu, S., Werner, F., 2000. Distribution, ecology and taphonomy of the organism trace *Scolicia* in northeast Atlantic deep-sea sediments. Palaeogeogr. Palaeoclimatol. Palaeoecol. 156, 289–300.

Fuchs, T., 1894. Über einige von der Österreichischen Tiefsee-Expedition S.M. Schiffes "Pola" in bedeutenten Tiefen gedredschte *Cylindrites*-ähnliche Körper und deren Verwandschaft mit *Gyrolithes*. Denkschr. Kais. Akad. Wiss. Wien, Math.-Nat. Kl. 61, 11–22.

Fuchs, T., 1895. Studien über Fucoiden und Hieroglyphen. Denkschr. Kais. Akad. Wiss. Wien, Math.-Nat. Kl. 62, 369–448.

Gage, J.D., Tyler, P.A., 1991. Deep-Sea Biology: A Natural History of Organisms at the Deep-Sea Floor. Cambridge University Press, Cambridge, 504 pp.

Gaillard, C., 1984. Bioturbation des sédiments pélagiques du Crétacé inférieur dans le Bassin Vocontien (Chaînes Subalpines méridionales, France). Geobios, Mém. Spéc. 8, 205–214.

Gehlen, M., Rabouville, C., Ezat, U., Guidi-Guilvard, L.D., 1997. Drastic changes in deep-sea sediment porewater composition induced by episodic input of organic matter. Limnol. Oceanogr. 42, 980–986.

Gooday, A.J., Turley, C.M., 1990. Responses by benthic organisms to inputs of organic material to the ocean floor: a review. Phil. Trans. R. Soc. Lond. A 331, 119–138.

Hartmann, M., 1979. Evidence for early diagenetic mobilization of trace metals from discolorations of pelagic sediments. Chem. Geol. 26, 277–293.

Hartnett, H.E., Keil, R.G., Hedges, J.I., Devol, A.H., 1998. Influence of oxygen exposure time on organic carbon preservation in continental margin sediments. Nature 391, 572–574.

Heezen, B.C., Hollister, C.D., 1971. The Face of the Deep. Oxford University Press, New York, 659 pp.

Hollister, C.D., Heezen, B.C., Nafe, K.E., 1975. Animal traces on the deep-sea floor. In: Frey, R.W. (Ed.), The Study of Trace Fossils. Springer, New York, pp. 493–510.

Hu, X., Jansa, L., Wang, C., Sarti, M., Bak, K., Wagreich, M., Soták, J., 2005. Upper Cretaceous oceanic red beds (CORB) in the Tethys: occurrences, lithofacies, age and environments. Cret. Res. 26, 3–26.

Ingole, B.S., Pavithran, S., Ansari, Z.A., 2005. Restoration of deep-sea macrofauna after simulated benthic disturbance in the Central Indian Basin. Mar. Geores. Geotechnol. 23, 267–288.

Jumars, P., Ekman, J.E., 1983. Spatial structure within deep-sea benthic communities. In: Rowe, G.T. (Ed.), Deep-Sea Biology. The Sea, vol. 8. Wiley, New York, pp. 399–451.

Jumars, P.A., Mayer, L.M., Deming, J.W., Barros, J.A., Wheatcroft, R.A., 1990. Deep-sea deposit-feeding strategies suggested by environmental and feeding constraints. Phil. Trans. R. Soc. Lond. A 331, 85–101.

Jung, M., Ilmberger, J., Mangini, A., Emeis, K.-C., 1997. Why some Mediterranean sapropels survived burn-down (and others did not). Mar. Geol. 141, 51–60.

Kakuwa, Y., 2004. Trace fossils of the Triassic to Jurassic radiolarian chert successions deposited on the pelagic oceanic bottom. Fossils Strata 51, 58–67.

Keller, G., 1982. Organic matter and the geotechnical properties of submarine sediments. Geo-Mar. Lett. 2, 191–198.

Konhauser, K., 2007. Introduction to Geomicrobiology. Blackwell, Oxford, 425 pp.

König, I., Drodt, M., Suess, E., Trautwein, A.X., 1997. Iron reduction through the tan-green color transition in deep-sea sediments. Geochim. Cosmochim. Acta 61, 1679–1683.

Kotake, N., 1989. Paleoecology of the *Zoophycos* producers. Lethaia 22, 327–341.

Książkiewicz, M., 1954. Uziarnienie frakcjonalne i laminowane we fliszu karpackim [Graded and laminated bedding in the Carpathian Flysch]. Rocznik Pol. Tow. Geol. 22, 399–471 (in Polish).

Lampitt, R.S., Antia, A.N., 1997. Particle flux in deep seas: regional characteristics and temporal variability. Deep-Sea Res. I 44, 1377–1403.

Laughton, A.S., 1961. Photography of the ocean floor. Ann. Rep. Smithsonian Inst. 1960, 319–326.

Legeleux, F., Reyss, J.-L., Schmidt, S., 1994. Particle mixing rates in sediments of the northeast tropical Atlantic from $^{210}Pb_{xs}$, ^{137}Cs, $^{238}Th_{xs}$ and $^{234}Th_{xs}$ downcore distribution. Earth Plan. Sci. Lett. 128, 545–562.

Leszczyński, S., 1993. Ichnocoenosis *versus* sediment colour in upper Albian to lower Eocene turbidites, Guipúzcoan province, northern Spain. Palaeogeogr. Palaeoclimatol. Palaeoecol. 100, 251–265.

Leszczyński, S., 2004. Bioturbation structures of the Kropivnik Fucoid Marls (Campanian-lower Maastrichtian) of the Huwniki–Rybotycze area (Polish Carpathians). Geol. Quart. 48, 35–60.

Leszczyński, S., Uchman, A., 1991. To the origin of variegated shales. Geol. Carp. 42, 279–289.

Leszczyński, S., Uchman, A., 1993. Biogenic structures of organic-poor sediments: examples from the Paleogene variegated shales, Polish Outer Carpathians. Ichnos 2, 267–275.

Levin, L.A., Whitcraft, C.R., Mendoza, G.F., Gonzalez, J.P., Cowie, G., 2009. Oxygen and organic matter thresholds for benthic faunal activity on the Pakistan margin oxygen minimum zone (700-1100 m). Deep-Sea Res. II 56, 449–471.

Lobza, V., Schieber, J., 1999. Biogenic sedimentary structures produced by worms in soupy, soft muds: observations from the Chattanooga Shale (Upper Devonian) and experiments. J. Sediment. Res. 69, 1041–1049.

Lopez, G., Taghon, G., Levinton, J.E., 1989. Ecology of Marine Deposit Feeders. Springer, Berlin, 322 pp.

Löwemark, L., 2007. Importance and usefulness of trace fossils and bioturbation in paleoceanography. In: Miller III, W. (Ed.), Trace Fossils. Concepts, Problems, Prospects. Elsevier, Amsterdam, pp. 413–427.

Löwemark, L., Werner, F., 2001. Dating errors in high-resolution stratigraphy: a detailed X-ray radiograph and AMS-^{14}C study of *Zoophycos* burrows. Mar. Geol. 177, 191–198.

Löwemark, L., Schönfeld, J., Werner, F., Schäfer, P., 2004. Trace fossils as a paleoceanographic tool: evidence from Late Quaternary sediments of the southwestern Iberian margin. Mar. Geol. 204, 27–41.

Löwemark, L., Lin, H.-L., Sarnthein, M., 2006. Temporal variations of the trace fossil *Zoophycos* in a 425 ka long sediment record from the South China Sea: implications for the ethology of the *Zoophycos* producer. Geol. Mag. 143, 105–114.

Lutz, M., Dunbar, R., Caldeira, K., 2002. Regional variability in the vertical flux of particulate organic carbon in the ocean interior. Glob. Biogeochem. Cy. 16, 1037–1055.

Lutz, M.J., Caldeira, K., Dunbar, R.B., Behrenfeld, M.J., 2007. Seasonal rhythms of net primary production and particulate organic carbon flux describe biological pump efficiency in the global ocean. J. Geophys. Res. 112, C10011, 26 pp.

Mangini, A., Jung, M., Laukenmann, S., 2001. What do we learn from peaks of uranium and of manganese in deep sea sediments? Mar. Geol. 177, 63–78.

Manville, V., Wilson, C.J.N., 2004. Vertical density currents: a review of their potential role in the deposition and interpretation of deep-sea ash layers. J. Geol. Soc. London 161, 947–958.

McCave, I.N., 1985. Sedimentology and stratigraphy of box cores from the HEBBLE site on the Nova Scotian continental rise. Mar. Geol. 66, 59–89.

McCave, I.N., Lonsdale, P.F., Hollister, C.D., Gardner, W.D., 1980. Sediment transport over the Hatton and Gardar contourite drifts. J. Sediment. Petrol. 50, 1049–1062.

Mikuláš, R., Skupien, P., Bubík, M., Vašíček, Z., 2009. Ichnology of the Cretaceous Oceanic Red Beds (Outer Western Carpathians, Czech Republic). Geol. Carp. 60, pp. 233–250.

Miller III, W., 1991. Paleoecology of graphoglyptids. Ichnos 1, 305–312.

Miller III, W., D'Alberto, L., 2001. Paleoethologic implications of *Zoophycos* from Late Cretaceous and Paleocene limestones of the Venetian Prealps, northeastern Italy. Palaeogeogr. Palaeoclimatol. Palaeoecol. 166, 237–247.

Miller III, W., Stefani, C., Grandesso, P., 2004. Alternation of ecologic regimes in a deep-marine carbonate basin: calciturbidite trace fossils from the Cretaceous Scaglia Rossa, northeastern Italy. Palaeogeogr. Palaeoclimatol. Palaeoecol. 204, 317–330.

Mitchell, J.K., 1993. Fundamentals of Soil Behavior. Wiley and Sons, New York, 437 pp.

Monaco, P., 1995. Relationships between trace-fossil communities and substrate characteristics in some Jurassic pelagic deposits in the Umbria-Marche basin, central Italy. Geobios 18, 299–311.

Müller, P.J., Suess, E., 1979. Productivity, sedimentation rate and sedimentary organic matter in the oceans. I. Organic carbon preservation. Deep-Sea Res. 26, 1347–1362.

Nowell, A.R.M., Hollister, C.D. (Eds.), 1985. Deep Ocean Sediment Transport. Preliminary Results of the High Energy Benthic Boundary Layer Experiment. Elsevier, Amsterdam, 418 pp.

Pemberton, S.G., MacEachern, J.A., Gingras, M.K., Saunders, T.D.A., 2008. Biogenic chaos: cryptobioturbation and the work of sedimentologically friendly organisms. Palaeogeogr. Palaeoclimatol. Palaeoecol. 270, 273–279.

Reimers, C.E., Fischer, K.M., Merewether, R., Smith Jr., K.L., Jahnke, R.A., 1986. Oxygen microprofiles measured *in situ* in deep ocean sediments. Nature 320, 741–744.

Reineck, H.-E., 1973. Schichtung und Wühlgefüge in Grundproben vor der ostafrikanischen Küste. "Meteor" Forsch.-Ergeb. C16, 67–81.

Reineck, H.-E., Gutmann, W.F., Hertweck, G., 1967. Das Schlickgebiet südlich Helgoland als Beispiel rezenter Schelfablagerungen. Senck. Leth. 48, 219–275.

Rex, M.A., 1983. Geographic patterns of species diversity in the deep-sea benthos. In: Rowe, G.T. (Ed.), Deep-Sea Biology. Wiley, New York, pp. 453–472.

Rhoads, D.C., Boyer, L.F., 1982. The effects of marine benthos on physical properties of sediments: a successional perspective. In: McCall, P.L., Tevesz, M.J.S. (Eds.), Animal-Sediment Relations. Plenum, New York, pp. 3–52.

Richardson, M.J., Wimbush, M., Mayer, L., 1981. Exceptionally strong near-bottom flows on the continental rise of Nova Scotia. Science 213, 887–888.

Richter, R., 1928. Die fossilen Fährten und Bauten der Würmer, ein Überblick über ihre biologischen Grundformen und deren geologische Bedeutung. Paläontol. Z. 9, 193–241.

Rodríguez-Tovar, F.J., Uchman, A., 2004. Trace fossils after the K-T boundary event from the Agost section, SE Spain. Geol. Mag. 141, 429–440.

Rodríguez-Tovar, F.J., Uchman, A., 2010. Ichnofabric evidence for the lack of bottom anoxia during the Lower Toarcian Oceanic Anoxic Event (T-OAE) in the Fuente de la Vidriera section, Betic Cordillera, Spain. Palaios 25, 576–587.

Rodríguez-Tovar, F.J., Uchman, A., Payros, A., Orue-Etxebarria, X., Apellaniz, E., Molina, E., 2010. Sea-level dynamics and palaeoecological factors affecting trace fossil distribution in Eocene turbiditic deposits (Gorrondatxe section, N Spain). Palaeogeogr. Palaeoclimatol. Palaeoecol. 285, 50–65.

Ruddiman, W.F., Glover, L.K., 1982. Mixing of volcanic ash zones in subpolar North Atlantic sediments. In: Scrutton, R., Talwani, M. (Eds.), The Ocean Floor. Wiley, New York, pp. 37–60.

Savary, B., Olivero, D., Gaillard, C., 2004. Calciturbidite dynamics and endobenthic colonization. Example from a Late Barremian (Early Cretaceous) succession of southeastern France. Palaeogeogr. Palaeoclimatol. Palaeoecol. 211, 221–239.

Savrda, C.E., Bottjer, D.J., 1986. Trace-fossil model for reconstruction of paleo-oxygenation in bottom waters. Geology 14, 3–6.

Savrda, C.E., Bottjer, D.J., 1989. Trace-fossil model for reconstructing oxygenation histories of ancient marine bottom waters: application to Upper Cretaceous Niobrara Formation, Colorado. Palaeogeogr. Palaeoclimatol. Palaeoecol. 74, 49–74.

Savrda, C.E., Krawinkel, H., McCarthy, F.M.G., McHugh, C.M.G., Olson, H.C., Mountain, G., 2001. Ichnofabrics of a Pleistocene slope succession, New Jersey margin: relations to climate and sea level dynamics. Palaeogeogr. Palaeoclimatol. Palaeoecol. 171, 41–61.

Savrda, C.E., Counts, J.W., Bigham, E., Martin, S., 2010. Ichnology of siliceous facies in the Eocene Tallahatta Formation (eastern United States Gulf coastal plain): implications for depositional conditions, storm processes, and diagenesis. Palaios 25, 642–655.

Schäfer, W., 1956. Wirkungen der Benthos-Organismen auf den jungen Schichtverband. Senck. Leth. 37, 183–263.

Seibold, E., Berger, W.H., 1996. The Sea Floor. Springer, Berlin, 356 pp.

Seilacher, A., 1962. Paleontological studies of turbidite sedimentation and erosion. J. Geol. 70, 227–234.

Seilacher, A., 1967. Bathymetry of trace fossils. Mar. Geol. 5, 413–428.

Seilacher, A., 1990. Aberration in bivalve evolution related to photo- and chemosymbiosis. Hist. Biol. 3, 289–311.

Smith, C.R., Rabouille, C., 2002. What controls the mixed-layer depth in deep-sea sediments? The importance of POC flux. Limnol. Oceanogr. 47, 418–426.

Smith, C.R., Hoover, D.J., Doan, S.E., Pope, R.H., deMaster, D.J., Dobbs, F.C., Altabet, M.A., 1996. Phytodetritus at the abyssal seafloor across 10° of latitude in the central equatorial Pacific. Deep-Sea Res. II 43, 1309–1338.

Smith Jr., K.L., Baldwin, R.J., Karl, D.M., Boetius, A., 2002. Benthic community responses to pulses in pelagic food supply: North Pacific Subtropical Gyre. Deep-Sea Res. I 49, 971–990.

Soetaert, K., Herman, P.M.J., Middelburg, J.J., 1996. Dynamic response of deep-sea sediments to seasonal variations: a model. Limnol. Oceanogr. 41, 1651–1668.

Stow, D.A.V., Pudsey, C.J., Howe, J.A., Faugères, J.-C., Viana, A.R. (Eds.), 2002. Deep-Water Contourite Systems: Modern Drifts and Ancient Series, Seismic and Sedimentary Characteristics. Geol. Soc. London, Mem. 22, pp. 1–464.

Suess, E., 1980. Particulate organic carbon flux in the oceans—surface productivity and oxygen utilization. Nature 288, 260–263.

Thiel, H., 1992. Deep-sea environmental disturbance and recovery potential. Int. Rev. Ges. Hydrobiol. 77, 331–339.

Thistle, D., Yingst, J.Y., Fauchald, K., 1985. A deep-sea benthic community exposed to strong near-bottom currents on the Scotian Rise (western Atlantic). Mar. Geol. 66, 91–112.
Thistle, D., Ertman, S.C., Fauchald, K., 1991. The fauna of the HEBBLE site: patterns in standing stock and sediment-dynamic effects. Mar. Geol. 99, 413–422.
Thomson, J., Brown, L., Nixona, S., Cook, G.T., MacKenzie, A.B., 2000. Bioturbation and Holocene sediment accumulation fluxes in the north-east Atlantic Ocean (Benthic Boundary Layer experiment sites). Mar. Geol. 169, 21–39.
Thurston, M.H., Bett, B.J., Rice, A.L., Jackson, P.A.B., 1994. Variations in the invertebrate abyssal megafauna in the North Atlantic Ocean. Deep-Sea Res. I 41, 1321–1348.
Trauth, M., Sarnthein, M., Arnold, M., 1997. Bioturbational mixing depth and carbon flux at the seafloor. Paleoceanography 12, 517–526.
Tyler, P.A., 1988. Seasonality in the deep sea. Oceanogr. Mar. Biol. Ann. Rev. 26, 227–258.
Tyson, R.V., 2001. Sedimentation rate, dilution, preservation and total organic carbon: some results of a modelling study. Org. Geochem. 32, 333–339.
Uchman, A., 1995. Taxonomy and palaeoecology of flysch trace fossils: the Marnoso-arenacea Formation and associated facies (Miocene, Northern Apennines, Italy). Beringeria 15, 3–115.
Uchman, A., Wetzel, A., 2012. Deep-sea fans. In: Knaust, D., Bromley, R.G. (Eds.), Trace Fossils as Indicators of Sedimentary Environments. Developments in Sedimentology, vol. 64. Elsevier, Amsterdam, pp. 643–672.
Uchman, A., Bąk, K., Rodríguez-Tovar, F.J., 2008. Ichnological record of deep-sea palaeoenvironmental changes around the Oceanic Anoxic Event 2 (Cenomanian–Turonian boundary): an example from the Barnasiówka section, Polish Outer Carpathians. Palaeogeogr. Palaeoclimatol. Palaeoecol. 262, 61–71.
von Fischer-Ooster, C., 1858. Die fossilen Fucoiden der Schweizer Alpen, nebst Erörterung über deren geologisches Alter. Huber u. Comp., Bern, 74 pp.
Warme, J.E., Kennedy, W.J., Schneidermann, N., 1973. Biogenic sedimentary structures (trace fossils) in LEG 15 Cores. In: Edgar, N.T., Kaneps, A.G., Herring, J.R. (Eds.), Initial Reports of the Deep Sea Drilling Project 15, pp. 813–831.
Werner, F., Wetzel, A., 1982. Interpretation of biogenic structures in oceanic sediments. Bull. l'Inst. Géol. Bassin d'Aquit. 31, 275–288.
Wetzel, A., 1981. Ökologische und stratigraphische Bedeutung biogener Gefüge in quartären Sedimenten am NW-afrikanischen Kontinentalrand. "Meteor" Forsch.-Ergeb. C34, 1–47.
Wetzel, A., 1983. Biogenic sedimentary structures in a modern upwelling region: NW African continental margin. In: Thiede, J., Suess, E. (Eds.), Coastal Upwelling and Its Sediment Record, Part B, Sedimentary Records of Ancient Coastal Upwelling. Plenum, New York, pp. 123–144.
Wetzel, A., 1984. Bioturbation in deep-sea fine-grained sediments: influence of sediment texture, turbidite frequency and rates of environmental changes. In: Stow, D.A.V., Piper, D.J.W. (Eds.), Fine Grained Sediments: Deep-Water Processes and Facies. Geol. Soc. London, Spec. Publ., vol. 15, pp. 597–608.
Wetzel, A., 1991. Ecologic interpretation of deep-sea trace fossil communities. Palaeogeogr. Palaeoclimatol. Palaeoecol. 85, 47–69.
Wetzel, A., 2002. Modern *Nereites* in the South China Sea—ecological association with redox conditions in the sediment. Palaios 17, 507–515.
Wetzel, A., 2008. Recent bioturbation in the deep South China Sea: a uniformitarian ichnologic approach. Palaios 23, 601–615.
Wetzel, A., 2009. The preservation potential of ash layers in the deep-sea: the example of the 1991-Pinatubo ash in the South China Sea. Sedimentology 56, 1992–2009.

Wetzel, A., 2010. Deep-sea ichnology: observations in modern sediments to interpret fossil counterparts. Acta Geol. Pol. 60, 125–138.

Wetzel, A., Aigner, T., 1986. Stratigraphic completeness: tiered trace fossils provide a measuring stick. Geology 14, 234–237.

Wetzel, A., Bromley, R.G., 1996. The ichnotaxon *Tasselia ordamensis* and its junior synonym *Caudichnus annulatus*. J. Paleontol. 70, 523–526.

Wetzel, A., Uchman, A., 1998a. Deep-sea benthic food content recorded by ichnofabrics: a conceptual model based on observations from Paleogene flysch, Carpathians, Poland. Palaios 13, 533–546.

Wetzel, A., Uchman, A., 1998b. Biogenic sedimentary structures in mudrocks – an overview. In: Schieber, J., Zimmerle, W., Sethi, P. (Eds.), Shales and Mudrocks I. Schweizerbart, Stuttgart, pp. 351–369.

Wetzel, A., Uchman, A., 2001. Sequential colonization of muddy turbidites: examples from Eocene Beloveža Formation, Carpathians, Poland. Palaeogeogr. Palaeoclimatol. Palaeoecol. 168, 171–186.

Wetzel, A., Werner, F., Stow, D.A.V., 2008. Bioturbation and biogenic sedimentary structures in contourites. In: Rebesco, M., Camerlenghi, A. (Eds.), Contourites. Developments in Sedimentology, vol. 60. Elsevier, Amsterdam, pp. 183–202.

Wetzel, A., Tjallingii, R., Wiesner, M.G., 2011. Bioturbational structures record environmental changes in the upwelling area off Vietnam (South China Sea) for the last 150,000 years. Palaeogeogr. Palaeoclimatol. Palaeoecol. 311, 256–267.

Wheatcroft, R.A., 1989. Comment on "Characteristic trace-fossil associations in oxygen-poor sedimentary environments". Geology 17, 674.

Wheatcroft, R.A., 1990. Preservation potential of sedimentary event layers. Geology 18, 843–845.

Wiesner, M.G., Zheng, L., Wong, H.K., Wang, Y., Chen, W., 1996. Fluxes of particulate matter in the South China Sea. In: Ittekot, V., Schäfer, P., Honjo, S., Depetris, P.J. (Eds.), Particle Flux in the Ocean. SCOPE Series No. 57, Wiley, New York, pp. 293–312.

Part V

Marine Carbonate Systems

Chapter 23

Shallow-Marine Carbonates

Dirk Knaust,*,[1] H. Allen Curran† and Andrei V. Dronov‡
*Statoil ASA, Stavanger, Norway, †Department of Geosciences, Smith College, Northampton, Massachusetts, USA, ‡Russian Academy of Sciences, Geological Institute, Moscow, Russia
[1]Corresponding author: e-mail: dkna@statoil.com

1. INTRODUCTION

Carbonates comprise only 20–25% of the total volume of all sedimentary rocks (Boggs, 2009). This fact is reflected in the relatively small number of ichnological studies of carbonates compared with siliciclastics, although other aspects (such as sedimentology or diagenesis) are roughly equally covered. Given the importance of carbonates as oil and gas reservoirs and aquifers (about 50% of the world's total proven hydrocarbon reserves are held in carbonate reservoirs), the ichnology of carbonate sedimentary systems has much potential for future investigations.

This chapter deals with the ichnology of shallow-marine carbonates, ranging from marginal-marine (peritidal) environments to the shelf. Trace fossils in carbonate rocks were previously reviewed by Kennedy (1975) and Ekdale et al. (1984), and ichnocoenoses specific to Quaternary, Bahamian-style carbonate environments by Curran (1994, 2007).

A review of literature with significant studies on the ichnology of shallow-marine carbonates shows an unequal representation of different time periods (studies focused solely on individual ichnotaxa are disregarded). Precambrian to Cambrian systems with their microbial mat-related, early bilaterian lifestyles are generally poorly covered (e.g., Weber et al., 2007). In the Phanerozoic, Paleozoic carbonate ichnology is strongly underrepresented, whereas the Mesozoic is relatively well documented. This is especially true for the Silurian to Permian (e.g., Archer, 1984; Knaust, 2009, 2010a; Narbonne, 1984), compared to the relatively intensively studied Ordovician (e.g., Dronov et al., 2002; Fei and Zhang, 2002; Osgood, 1970; Pickerill and Forbes, 1979; Pickerill et al., 1984). In the Mesozoic, several studies deal with the ichnology of Triassic carbonate platforms and ramps (e.g., Diedrich, 2008 and references therein; Knaust, 1998, 2007a,b, 2009, 2010a,b; Knaust and Costamagna, 2012; Rodríguez-Tovar et al., 2007; Twitchett and Wignall, 1996), but many are dedicated to the Jurassic (e.g., Avanzini et al., 1997; de Gibert, 1996; de Gibert

Developments in Sedimentology, Vol. 64. http://dx.doi.org/10.1016/B978-0-444-53813-0.00023-X

and Ekdale, 1999; El-Asa'ad, 1987; Gaillard et al., 1994; Giannetti and Monaco, 2004; Goldring et al., 2005; Monaco, 1995; Monaco and Garassino, 2001; Monaco and Giannetti, 2002; Olóriz and Rodríguez-Tovar, 2002; Schweigert, 1998; Wilson and Palmer, 1994). In contrast, Cretaceous carbonates are scarcely dealt with in ichnological–paleoenvironmental studies (e.g., Husinec and Read, 2011; Mángano and Buatois, 1991; Mezga et al., 2006; Moro et al., 2007; Petti et al., 2008). The Cenozoic again is covered by numerous studies dealing with various environments and topics (e.g., Braithwaite and Talbot, 1972; Cunningham et al., 2009; Curran, 1984, 1994, 2007; Curran and Martin, 2003; Curran and White, 1991; de Gibert and Goldring, 2007; Farrow, 1971; Garrett, 1977; Lukasik and James, 2003; Massari and D'Alessandro, 2010; Pedley, 1992; Shinn, 1968; Tedesco and Wanless, 1991).

Following their main morphological characteristics, carbonate systems can broadly be classified into rimmed carbonate shelf, carbonate ramp (homoclinal and distally steepened), and isolated platform (Read, 1985). Each system has a different position for the carbonate factory and sediment transport pathways, giving reason to the classification into T factory (tropical), C factory (cool-water), and M factory (mud-mound) (Schlager, 2005).

2. ICHNOLOGICAL CHARACTERISTICS OF SHALLOW-MARINE CARBONATE SYSTEMS

The presence, distribution, and quality of trace fossils in shallow-marine carbonates are controlled by a number of criteria which can be grouped into three major categories (Table 1): (1) the faunal composition and distribution of benthic organisms as potential tracemakers, (2) the sediment type and its alteration, and (3) the early-diagenetic history. The interaction of these three criteria can create and convey a specific trace-fossil record which represents only a fraction of the full range of possible outcomes. According to which category played the most important role in the fossilization process, distinct ichnofaunas and trace-fossil associations may represent the same paleoenvironment in different settings. To facilitate evaluation of the trace-fossil record and to employ it for interpretation of paleoenvironments, each category is discussed in more detail before turning to characteristic ichnofacies of shallow-marine carbonates.

2.1 The Composition and Distribution of Benthic Organisms

The location of a given carbonate depositional system in relation to the equator is an important factor controlling its development and the fauna available for modification of the sediment. Warm waters of tropical systems generally contain a higher diversity of organisms than non-tropical, cool-water systems, which may have an effect on the kind and the amount of substrate modification that can occur. Most trace-fossil forms occur in both carbonate and siliciclastic environments, with only a few exceptions of carbonate-specific ichnotaxa that result from particular burrowing or boring techniques adapted to early consolidated substrates such as firmgrounds and hardgrounds. For instance, some borings are driven by chemical means, such as *Trypanites* and *Entobia*, and are typical in carbonate substrates. In general, climatic control has an influence on the distribution of some trace-fossil groups. Decapod crustaceans are important burrowers and are most abundant in carbonate substrates from tropical and subtropical zones (Dworschak, 2000). As such, burrows like *Ophiomorpha*, *Thalassinoides*, *Spongeliomorpha*, *Psilonichnus*, *Pholeus*, and *Macanopsis* may be more common in carbonates than in siliciclastics.

Evolution is an important aspect in the generation of trace fossils, as different tracemakers change throughout the Phanerozoic. A relevant example is the occurrence of crustacean burrows in Mesozoic and Cenozoic deposits, whereas similar trace fossils such as *Thalassinoides* in Paleozoic strata were constructed by trilobites (e.g., Cherns et al., 2006) or unidentified animals. Other trace fossils (e.g., *Zoophycos*) are known from shallow-marine carbonates until the Jurassic but migrate to deeper environments during the Cretaceous and Cenozoic (Seilacher, 2007). On an ichnospecific level, *Rhizocorallium commune* (herein regarded as a senior synonym of *R. irregulare*) occurs from the Cambrian through Holocene and apparently is the product of polychaetes, whereas *R. jenense* (*sensu stricto*) characterizes Triassic and younger deposits and may be the product of arthropods. Radular bite traces known as *"Radulichnus"* are common in Proterozoic and Cambrian matgrounds (Seilacher, 2007) but also appear in the Triassic, where similar environments prevailed after the end-Permian mass extinction (Knaust, 2010b). In contrast, Cretaceous to modern *Radulichnus* occurs in hard substrates (Bromley, 2004).

Given the simple case of a homoclinal ramp, general trends can be observed from proximal (backshore, supra- to intertidal) to distal (shelf, subtidal) settings, although exceptions may occur. These trends include an overall decrease of burrow size and depth, an increase in ichnodiversity and bioturbation intensity, and a change from predominantly horizontal to vertical burrows (Laporte, 1969). The distribution of potential tracemaker benthic organisms is mainly controlled by the availability of organic matter within the substrate as food resource. Suspension feeders (e.g., many bivalves, sponges, brachiopods) prefer a stabilized (firm and hard) substrate in a proximal position (supratidal to shallow subtidal), whereas deposit feeders (e.g., some worms, some echinoids, some bivalves, many gastropods) thrive on and in more distal silty and muddy bottoms where they find a higher organic content.

2.2 Sediment Types

Carbonates differ from siliciclastics in some important ways. Carbonates are predominantly biogenic sediments and as such are mainly generated by organisms and their reworked skeletal components. Such deposits are also more prone to bioerosion by means of chemical dissolution than are siliciclastics.

TABLE 1 **Relevant Differences Between Carbonates and Siliciclastics with Respect to Formation and Preservation of Trace Fossils.** *Modified from Curran (1994)*

	Carbonates	Siliciclastics
Infaunal distribution	1. Favored trace fossils (e.g., crustacean burrows, *Thalassinoides, Ophiomorpha, Psilonichnus, Macanopsis,* and *Pholeus*) due to preferential occurrence of various benthic organisms in tropical and subtropical zones where shallow-marine carbonate environments dominate.	A wide range of ichnotaxa commonly occurs.
	2. Carbonate platforms and ramps comprise broad and very shallow facies belts, often with restricted environments that are favorable for intense bioturbation.	Differentiated depositional systems with relatively small facies belts, heterogeneous bioturbation patterns, and intense bioturbation in offshore sediments.
	3. Given their very low-gradient, carbonate platforms are commonly subject to wide-ranging subaerial exposure, resulting in the development of peritidal ichnofauna and trace fossils formed by plant roots (rhizomorphs or rhizoliths) and prostrate stems.	Rhizomorphs (rhizoliths) can occur, but are not nearly so common.
Nature of sediment	4. Bioerosion is common and related to reefs, bioherms, biostromes, and reworked skeletal components.	Bioerosion is subordinate because organic buildups are much less dominant, and relatively few organisms bioerode non-carbonates.
	5. Borings driven by chemical means dominate (*Trypanites, Entobia*).	Borings with a mechanical or combined mechanical/chemical method occur.
	6. Hardgrounds and firmgrounds common and form rapidly, both have significant degrees and diversity of borer and burrower forms of the *Glossifungites* and *Trypanites* ichnofacies.	Hardgrounds rare, firmgrounds less common.
	7. Preferential preservation of trace fossils (including fecal pellets) due to fine-grained lime mud in combination with microbial mats,	Delicate traces or parts of them are rarely well preserved due to the clastic nature of sediment, reworking, compaction, and late-diagenetic cementation.

TABLE 1 Relevant Differences Between Carbonates and Siliciclastics with Respect to Formation and Preservation of Trace Fossils. ***Modified from Curran (1994)*****—Cont'd**

	Carbonates	Siliciclastics
	biofilms, and early-diagenetic cementation.	
Early diagenesis	8. Rapid lithification; cementation process accentuates trace-fossil definition, that is, differential cementation between burrow matrix and fill materials.	Lithification may never occur or be delayed for a significant amount of geological time.
	9. Normally little to no sediment-color contrast between burrow materials and matrix (except in cases of diagenetic staining), limiting the outlining of trace-fossil forms.	Sediment-color contrast is the norm, commonly outlining trace-fossil forms.
	10. High potential for diagenetic alteration of rocks bearing trace fossils, including the development of ichnogenic macroporosity.	Lower potential for diagenetic alteration and development of ichnogenic macroporosity, although it can be locally important.
General	11. Some depositional environments minor or essentially unrepresented, such as glacial, fluvial, fluvial-deltaic, and deepest sea.	A wider range of depositional environments represented in the rock record.

Marine siliciclastic deposition commonly takes place on a well-defined gradient from the shoreline to the shoreface, across the shelf, and down the slope to a basin floor. In contrast, warm-water carbonate systems are chiefly controlled by the growth of organisms with prime dependence on available light. Consequently, their platforms have low-gradient geometries resulting in a shelf and ramp sea-floor with very low relief and facies development as broad belts.

Given the overall geometry of carbonate platforms with broad facies zonation, favorable conditions for the preservation of traces occur when different factors come together. Many shallow-marine carbonate rocks consist of fine-grained carbonate (micrite) generated from a degradation of skeletal material combined with *in-situ* precipitation which covers extensive areas in tropical and subtropical zones with abundant benthic life. The fine-grained and homogeneous sediment within peritidal facies is typically subject to periodic subaerial exposure, which leads to gradually changing substrate consistencies from soft to stiff and firm over short time intervals (e.g., diurnally). This prograding change in substrate consistency may be documented by continuously changing trace-fossil morphologies and their excellent and detailed preservation, even of meiobenthic trace fossils less than 1 mm in shortest dimension (Knaust, 2007a, 2010b).

Another contribution comes from the microbial mats and biofilms, which are typical in such environments. They cover the sediment surface (including fresh animal traces) and prevent subsequent destruction, creating a slightly reducing microenvironment and in some cases even preventing the oxidization of the tracemakers, in which case they are preserved together with the traces. Rapid coverage by a new layer of fine-grained sediment (obrution) together with early-diagenetic cementation further increase the chance to preserve the traces in an optimal way.

Specific carbonate particles in combination with early-diagenetic cementation and alteration are another aspect of carbonates. A diverse worm and crustacean fauna contributes with major sediment-feeding organisms to bioturbate and rework the sediment, in many places completely, and results in a considerable production of fecal pellets. Such accumulations of microcoprolites are subject to early lithification and may turn into carbonate rocks. Fecal pellets are regarded as trace fossils and provide valuable information about the paleoenvironment (e.g., Knaust, 2008). Moreover, broken shells, ooids, and lithoclasts are commonly affected by microboring organisms that partly or completely destroy the structure of the grains while generating peloids (Wisshak, 2012).

2.3 Early Diagenesis

Taphonomic processes are important for trace-fossil preservation in siliciclastics, especially the casting of traces due to contrasting lithotypes such as sand and mud, whereas grain-size contrasts (e.g., arenite or rudite on carbonate mud) typically play only a minor role in the preservation in carbonates. However, the taphonomic history of trace fossils in carbonates differs from that in siliciclastics by the importance of diagenetic processes that lead to preferential preservation and trace-fossil selection (Narbonne, 1984). The reason for this is sea-floor lithification, and therefore less compaction than in siliciclastics, and a better preservation of sedimentary and biogenic structures. In other circumstances, carbonate cementation and recrystallization are able to alter trace fossils in carbonates, which can lead to a poor level of preservation.

Diagenetic processes on carbonate platforms and ramps play an important role in the transition of sediments to rocks. The different stages of substrate consistency and the timing of the inherent processes are often reflected by significant trace-fossil assemblages (including borings in hardgrounds) and thus can be utilized in the recognition of the kind of substrate, duration of sedimentation, omission, discontinuity surfaces, etc. (Bromley, 1975; Ekdale and Bromley, 2001; Goldring and Kaźmierczak, 1974; Hölder and Hollmann, 1969; Kobluk et al., 1977; Lukasik and James, 2003; Wilson and Palmer, 1994). The taphonomy of trace fossils is controlled by the chemical microenvironment around the burrows and borings, which act as open conduits to the sediment/water interface and facilitate diagenetic processes such as the precipitation of calcite, dolomite, celestite, or pyrite (Bertling, 1999; Gingras et al., 2004).

Furthermore, diagenesis is capable of total alteration of the deposits as well as the development of secondary porosity. Because of common early-diagenetic consolidation of carbonate sediment, burrow-related porosity is an even more important aspect for the quality of hydrocarbon reservoirs and aquifers than in siliciclastic deposits (Cunningham et al., 2009, 2012; Knaust, 2009; Pemberton and Gingras, 2005).

3. ICHNOFACIES

Following the gradient on a carbonate platform (or ramp) from a relative proximal (landward) to distal (seaward) position, there are three ichnofacies of importance for the reconstruction of paleoenvironments (Fig. 1): (1) *Psilonichnus* Ichnofacies, (2) *Skolithos* Ichnofacies, and (3) *Cruziana* Ichnofacies. During periods with strongly reduced or no deposition (condensation, omission) and progressing lithification, each of the before-mentioned ichnofacies can be overprinted by the substrate-controlled (4) *Glossifungites* Ichnofacies, or the (5) *Trypanites* Ichnofacies.

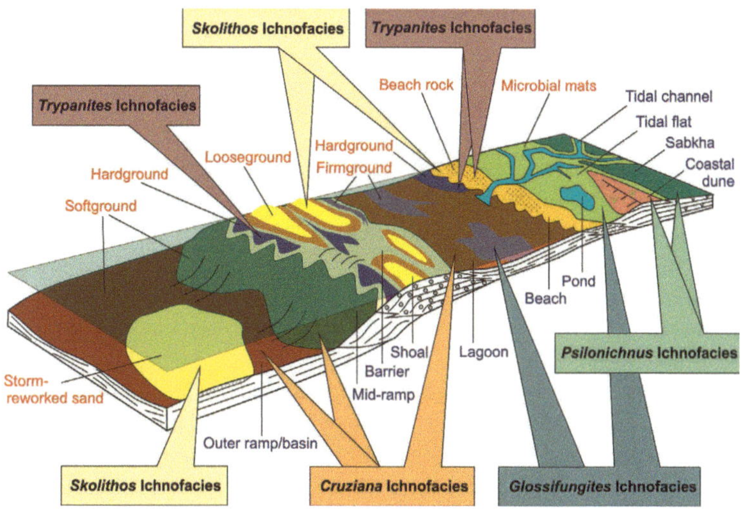

FIGURE 1 Conceptual model of a carbonate ramp with characteristical depositional environments and related ichnofacies (see text for explanations). Note the complexity and resulting variation in ichnofacies occurrence.

3.1 *Psilonichnus* Ichnofacies (Frey and Pemberton, 1987)

The most landward, coastal dunes and supratidal to upper intertidal facies zone is dominated by the *Psilonichnus* Ichnofacies, which characterizes transitional marine/terrestrial environments, particularly the backshore, coastal dunes, washover fans, and supratidal flats (Frey and Pemberton, 1987; Frey et al., 1984; MacEachern et al., 2007, 2012). The ichnogenus *Psilonichnus* comprises predominantly vertical, cylindrical burrows with short horizontal or oblique side branches, with an upper part that is commonly Y-shaped (Fürsich, 1981).

Burrows formed by amphibious ghost crabs (Ocypodidae) typify the *Psilonichnus* Ichnofacies. They must not be confused with similar burrows of the mud shrimp *Upogebia* (Upogebiidae) containing turning chambers in the upper part of the burrow, attributed to the ichnogenus *Parmaichnus* Pervesler and Uchman, 2009, and occurring in intertidal and subtidal areas (Nesbitt and Campbell, 2006; Radwański et al., 2012; see Knaust, 2012). Ichnospecies relevant to the *Psilonichnus* Ichnofacies include *P. upsilon* Frey, Curran and Pemberton, 1984 and *P. tubiformis* Fürsich, 1981 (*P. quietis* Myint, 2001 is probably a junior synonym to *P. tubiformis*). The morphology of *Psilonichnus* is variable, including J-, I-, and U-shapes, depending on ontogenetic stages (Myint, 2001) and environmental conditions (De, 2005).

Other constituents of this ichnofacies include root structures (e.g., rhizomorphs or rhizoliths, overprinting existing invertebrate burrows), various arthropod trace fossils (mainly occurring as repichnia), and vertebrate tracks. The deeply penetrating dwelling burrows of *Psilonichnus* are mainly known from Jurassic and younger deposits. Although the *Psilonichnus* Ichnofacies can be regarded as a softground ichnofacies, carbonate systems are prone to substrate modifications with superimposition of firmground ichnocoenoses of the *Glossifungites* Ichnofacies (*Rhizocorallium jenense* and *Balanoglossites* ichnospecies).

3.2 *Skolithos* Ichnofacies (Seilacher, 1967)

Adjacent to the *Psilonichnus* Ichnofacies (typical of low- to moderate-energy settings), the *Skolithos* Ichnofacies characterizes higher-energy sandy deposits from the beach (intertidal) to environments situated more distally on the platform or ramp (subtidal). Beach foreshore and sandy shoals on the platform margin are predestined as the *Skolithos* Ichnofacies. Opportunistic colonization of event beds (e.g., storm deposits) was referred to the *Arenicolites* Ichnofacies by Bromley and Asgaard (1991).

A low ichnodiversity but high burrow density indicates rapid colonization of sediment surfaces. Vertically oriented or steeply inclined shafts dominate, mainly produced by suspension-feeding organisms, for example, shrimp, polychaetes, and bivalves. *Skolithos* ichnospecies vary in size and comprise simple tubular shafts with or without lining. *Skolithos linearis* is the most common ichnospecies occurring in the *Skolithos* Ichnofacies. The obvious low number of attributes and therefore the broad diagnosis of *Skolithos* lead to the uncomfortable situation that it can be produced by a range of higher organism taxa and even plant roots, a fact that clearly weakens the environmental value of *Skolithos* (Gregory et al., 2006).

Depending on the climate zone, kind of sediment and age, the following ichnotaxa may dominate within the *Skolithos* Ichnofacies or are part of the ichnocoenoses: *Altichnus*, *Arenicolites*, *Diplocraterion*, *Lingulichnus*, *Macaronichnus*, *Ophiomorpha*, *Polykladichnus*, *Piscichnus*, *Psilonichnus tubiformis*, rhizomorphs, escape and equilibrium traces.

3.3 *Cruziana* Ichnofacies (Seilacher, 1967)

The *Cruziana* Ichnofacies is probably the most common and extensive among carbonate systems. It occurs in shallow-marine environments between the fair-weather wave base and the storm-wave base, commonly in a moderate- to low-energy system. The resulting deposition rate is low with the dominant deposition of fine sand to soft lime mud, occasionally punctuated by moderate to high-energy event deposition (e.g., storms).

The trace-fossil association mainly encompasses horizontally oriented burrows and traces produced by deposit-feeding and crawling organisms, accompanied by vertical structures. Given favorable conditions in terms of salinity, oxygen, currents, etc., the ichnodiversity and bioturbation is moderate to high and includes traces of different trophic groups of organisms (mainly arthropods and worms).

Common trace fossils of the *Cruziana* Ichnofacies are *Archaeonassa*, *Arthrophycus*, *Asterosoma*, *Cruziana*, *Curvolithus*, *Diplocraterion*, *Gyrochorte*, *Ophiomorpha*, *Palaeophycus*, *Phycodes*, *Planolites*, *Protovirgularia*, *Rhizocorallium commune*, *Teichichnus*, *Thalassinoides*, and *Zoophycos* (in part).

3.4 *Glossifungites* Ichnofacies (Seilacher, 1967)

The substrate-controlled *Glossifungites* Ichnofacies occurs in semiconsolidated carbonate sediment with stable substrate properties forming stiff- to firmgrounds, which only allow specialized organisms to burrow into it. It is related to omission or erosion surfaces and generally occurs in different environments (Frey and Seilacher, 1980). However, preferred *loci* for substrate dewatering and consolidation on carbonate ramps and platforms are given in shallow subtidal and intertidal to supratidal environments, such as lagoons, beaches, and shoals (Pemberton and Frey, 1985).

Vertical burrows (dwelling structures) of suspension feeders are predominant and occur in moderate to high density, although the ichnodiversity is limited. Numerous burrowing organisms are adapted to colonize omission surfaces, for instance different species of polychaetes, nemerteans, and crustaceans. Given the cohesive stage of the substrate, burrow outlines are sharp and clearly defined if not blurred subsequently by diagenetic processes. The burrows are commonly passively infilled from above with sediment unrelated to the host rock. Early cementation is important and is aided by the chemical microenvironment around the burrows, resulting in the crystallization of calcite, celestite, dolomite, goethite, or pyrite.

The name-bearer trace fossil of the *Glossifungites* Ichnofacies is now regarded as a junior synonym of *Rhizocorallium jenense* (Uchman et al., 2000). Other characteristic trace fossils are *Balanoglossites*, *Gastrochaenolites*, *Pholeus*, *Spongeliomorpha*, and *Thalassinoides*. The former two ichnogenera include both firmground burrows and borings, whereas other tracemakers are capable of burrowing into soft and firm substrates to produce, for instance, *Chondrites*, *Palaeophycus*, *Siphonichnus*, *Lingulichnus*, and *Zoophycos*.

A special expression of the *Glossifungites* Ichnofacies is a firm matground substrate, resulting from the development of algal or microbial mats on the sediment surface. Such conditions are well known from Precambrian and Early Cambrian units, but similar conditions repeatedly occur through the Phanerozoic. The related ichnofauna comprises surface traces of crawling and rasping organisms, which benefit from the organic matter as mat scratchers and digesters, mat stickers, or undermat miners (Seilacher, 2007).

3.5 *Trypanites* Ichnofacies (Frey and Seilacher, 1980)

The substrate-controlled *Glossifungites* and *Trypanites* ichnofacies are intimately associated, as the former has the potential to develop gradually into a hardground of the latter (Bromley, 1975). In contrast to the *Glossifungites* Ichnofacies, the *Trypanites* Ichnofacies is more restricted in its distribution to areas with rapid cementation. Such conditions are common in subtidal and intertidal environments, where this process may lead to an extensive pavement by hardgrounds or beachrock, but the *Trypanites* Ichnofacies also occurs on rocky shores or in reefs. Rapid lithification preferentially takes place in front of sand-shoal complexes (including the shelf break), where the sediment is exposed to wave sweeping and reworking.

Bromley and Asgaard (1993) proposed to replace or subdivide the *Trypanites* Ichnofacies with the *Entobia* Ichnofacies for deep-tier borings with high preservation potential and the *Gnathichnus* Ichnofacies including superficial sculptures resulting from short-term bioerosion. However, *Entobia* and *Gnathichnus* are probably better treated as ichnocoenoses within the *Trypanites* Ichnofacies (MacEachern et al., 2007).

Hardgrounds are often well defined by a clear top surface colonized by epibenthic organisms (e.g., oysters, serpulid polychaetes) and mineral staining. Bioerosion takes place on the hardground surface or from its top, marginal, or undercut surface into the substrate. It comprises a variety of borings mainly produced by suspension-feeding organisms. *Trypanites weisei* from its type area in the Middle Triassic of Germany occurs in micritic limestone assignable to the *Glossifungites* Ichnofacies, where it superimposes a *Balanoglossites* ichnofabric (Knaust, 2008). Its stratigraphical position, sedimentological and ichnological context, and early-diagenetic dolomitization indicate an origin in the lower intertidal zone.

Important trace fossils of the *Trypanites* Ichnofacies are *Caulostrepsis*, *Entobia*, *Gastrochaenolites*, *Gnathichnus*, *Maeandropolydora*, *Rogerella*, and *Trypanites*, as well as numerous microborings.

4. CASE STUDIES

4.1 Ordovician Cool-Water Carbonates: St. Petersburg Region, Russia

During the Ordovician, the Baltic paleocontinent migrated from a subpolar to a subequatorial position within the southern hemisphere (Cocks and Torsvik, 2005). This is reflected in the succession of facies from predominantly siliciclastics (sand and black shale) in the Tremadocian, through cool-water bioclastic wackestones in the Floian to Sandbian, to tropical sabkha dolomites and pelmicrites in the Katian to Hirnantian (Dronov and Rozhnov, 2007). The Ordovician carbonates of the St. Petersburg region (Floian to Katian) reach a thickness of 20–160 m and belong to the relatively shallow-water part of the epicontinental basin of Baltoscandia. The rocks are typically rich in bryozoans, brachiopods, trilobites, ostracods, echinoderms, and conodonts. Shells of gastropods, bivalves, cephalopods, and sponge spicules are locally abundant. In this chapter, cool-water carbonates of Volkhovian (Dapingian) age are described (Fig. 2), which are best exposed and most extensively studied in natural outcrops and quarries in the St. Petersburg region.

Basic ichnological features of these carbonates were described by Dronov et al. (2002) and Dronov and Mikuláš (2010). In general, the ichnodiversity of the Dapingian ramp is relatively low but due to the condensed character of the section, the trace-fossil density can be very high at some levels. The main ichnogenera are *Balanoglossites*, *Bergaueria*, *Chondrites*, *Gastrochaenolites*, *Palaeophycus*, *Thalassinoides*, and *Trypanites*. Major portions of the succession belong to the *Cruziana* Ichnofacies with numerous omission surfaces demarcated by the *Glossifungites* and *Trypanites* ichnofacies.

The well-bedded Dapingian limestone has been extensively quarried for building purposes since the foundation of the town of St. Petersburg in 1703 (Fig. 3). The basements and staircases of most buildings in the historical part of the city are made from certain beds of the "Glauconite Limestone". There is an old tradition among the local quarrymen to give names to distinctive beds and bedsets as well as to some bedding surfaces (Fig. 2). Regional conformity of this nomenclature reflects the lateral persistence of these lithological units. Objectively, this informal terminology is related to some distinctive lithological and ichnological features, such as hardness and homogeneity of the rock, mode of intercalation, distribution of colors as well as intensity of bioturbation, and specifically the ichnotaxa content of each bed. This informal terminology has

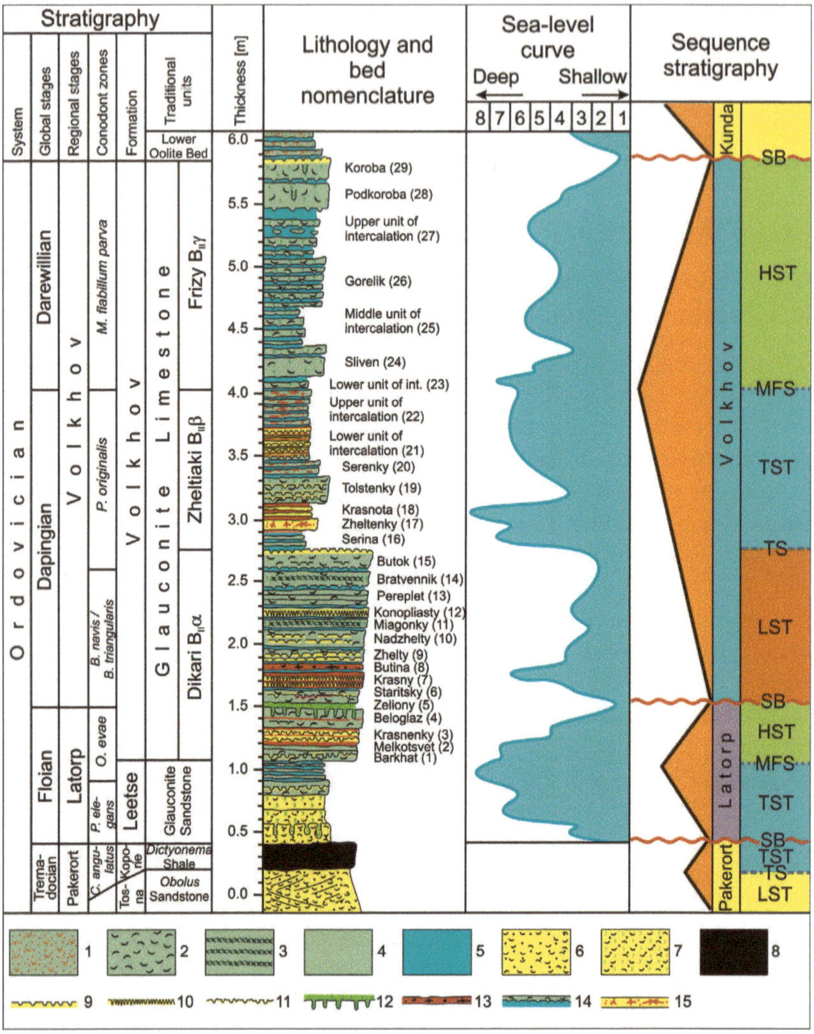

FIGURE 2 Detailed stratigraphy and interpreted sea-level changes in the Volkhovian (Dapingian) of the St. Petersburg region, Russia. (1) Bioclastic wackestone with iron ooids; (2) bioclastic packstone; (3) marlstone with *Thalassinoides* ichnofabric; (4) carbonate mudstone to wackestone; (5) terrigenous siltstone; (6) quartz sandstone; (7) quartz sandstone enriched with glauconite; (8) black shale; (9) hardground surface pitted with *Trypanites*; (10) hardground surfaces with *Balanoglossites* ("Karandashi"); (11) hardground surface with superimposed *Gastrochaenolites*, *Balanoglossites*, and *Trypanites*; (12) hardground surface with well-developed *Gastrochaenolites oelandicus* ("Steklo" surface); (13) marine red bed with *Thalassinoides* and *Chondrites*; (14) *Bergaueria* at the base of the beds; (15) marine red bed with *Chondrites*.

FIGURE 3 Sawn section through the Middle Ordovician Dikari unit in the Babino quarry, approximately 150 km east of St. Petersburg, Russia. The bed numbers correspond to the bed nomenclature of Fig. 2.

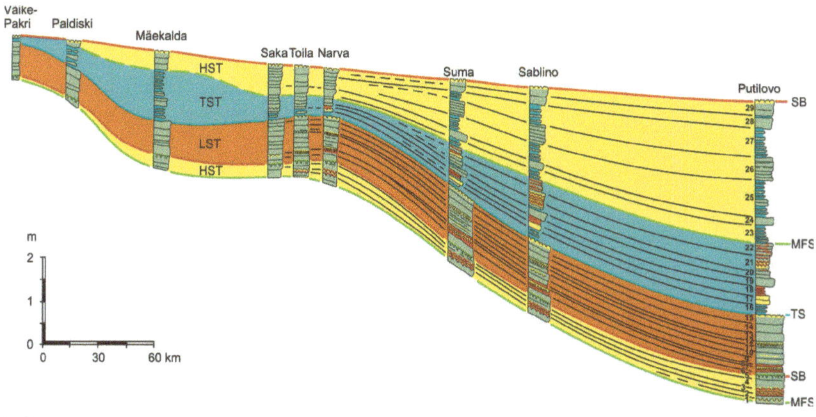

FIGURE 4 Bed-by-bed correlation of sections of the Middle Ordovician Volkhov Formation (Dapingian) along the Baltic-Ladoga Glint Line (Estonia-Russia).

been adopted for subdivision of the succession into numerous elementary units that, to a significant degree, are traceable over the eastern part of the Baltic-Ladoga Glint (a Holocene coast cliff) over a distance of more than 300 km (Jaanusson, 1976; Männil, 1966; Dronov et al., 1996, 2000; Fig. 4). In total, 29 lithostratigraphic units (parasequences) can be recognized in the Volkhov Formation (Figs. 2 and 3).

The Volkhov Regional Stage is interpreted as a single depositional sequence bounded by distinctive erosional surfaces with associated stratigraphic gaps. Sequence boundaries coincide with the hardground surfaces at the base and the top of that sequence (Fig. 2). The lowermost part of the Dikari Limestone is interpreted as a highstand systems tract of the underlying Latorp sequence. The upper part of the Dikari Limestone represents a lowstand systems tract, whereas the Zheltiaki Limestone and the Frizy Limestone represent transgressive and highstand systems tracts, respectively (Dronov and Holmer, 1999; Dronov et al., 2002).

4.1.1 Highstand Systems Tract of the Underlying Latorp Sequence

The lowermost part of the Dikari Limestone is represented by four distinctive beds (Fig. 2). The succession of the beds demonstrates a shallowing-upward trend and is interpreted as a highstand systems tract. On top of the Beloglaz Bed (bed 4 in Fig. 2) rests the Zeliony Bed (bed 5) with a prominent flat hardground surface covered by a thin glauconite veneer. It marks the base of the Middle Ordovician and is interpreted as a transgressive surface and sequence boundary (Fig. 5). The surface is pitted by so-called Amphora-like burrows and/or borings (Orviku, 1960; Vishnjakov and Hecker, 1937), corresponding to *Gastrochaenolites oelandicus* Ekdale and Bromley, 2001 (Fig. 6A and B). The hardground surface at the base of the Volkhov depositional sequence occurs not only in the Syas River area ("Steklo" surface) but also in Estonia ("Pustakkiht"), on the island of Öland in Sweden "Blommiga Bladet", and in

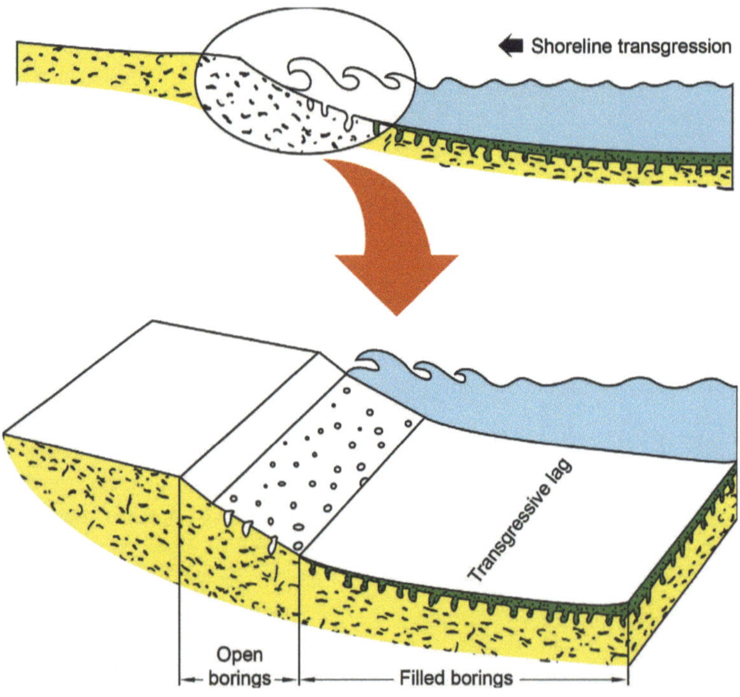

FIGURE 5 Development of the "Steklo" transgressive surface mainly bioeroded with *Gastrochaenolites oelandicus. Modified after Posamentier and Allen (1999).*

the Oslo region in Norway. Everywhere it is recognizable by *G. oelandicus* pits usually filled with glauconite grains.

4.1.2 Lowstand Systems Tract, Upper Part of the Dikari Limestone (BIIα)

The upper part of the Dikari Limestone consists of 10 distinctive beds (Figs. 2 and 3). The Butina Bed is the most remarkable one, comprising 1–5 cm of relatively soft red marlstone with thin *Thalassinoides* forming a network typical for the Central Baltoscandian Confacies belt. This bed is the best marker and may be interpreted as a short-term invasion of relatively deep-water conditions. The Dikari are predominantly of gray bioclastic packstone or grainstone with numerous scattered glauconite grains. Distinctive hardground surfaces, emphasized by yellow goethite impregnation, are abundant at some levels (e.g., the Krasny, Zhelty, Nadzhelty, and Konoplasty beds). Some of these surfaces are pitted with *Balanoglossites* and, more rarely, *Trypanites*. These beds usually consist of four to eight elementary layers, each 3–4.5 cm thick. Most of the layers are distinctly graded. Brachiopods, echinoderms, bryozoans, ostracods, and trilobites are the main fossils. The uppermost bed of the Dikari Limestone (Butok) has a distinctive hardground non-depositional surface at the top, marked by an extensive yellow impregnation down to 2 cm. It is penetrated by shallow vertical borings attributed to *Trypanites* (Fig. 6C) and a complex *Balanoglossites* ichnofabric augmented by the weathering/dissolution of the surface prior to its hardening (Fig. 6D–F). The hardground is interpreted as a transgressive surface. Other ichnogenera of this unit include *Thalassinoides*, *Bergaueria*, and *Palaeophycus*.

4.1.3 Transgressive Systems Tract, Zheltiaki Limestone (BIIβ)

The Zheltiaki Limestone differs from the underlying rocks of the Dikari Limestone in having more argillaceous material within the carbonate rock, the appearance of numerous clay layers, and the variegated, mostly red and yellow, color of the rocks. Glauconite is rare or absent. The faunal assemblages recovered from interbeds of clay are dominated by brachiopods, ostracods, and echinoderms, whereas those from the limestone beds are rather different and contain many trilobites. These differences can be explained by the tempestite origin of the limestone beds. The yellow and red colors of the rocks and the smaller grain size in comparison with the underlying and overlying strata point to a relatively deep-water origin. The Zheltiaki Limestone is subdivided into seven informal lithostratigraphic units (Fig. 2). The units bear rich ichnocomplexes dominated by *Thalassinoides* and *Chondrites* (Fig. 6G). Washed-out surface trace fossils are rare (e.g., *Rusophycus*; Dronov et al., 2002).

4.1.4 Highstand Systems Tract, Frizy Limestone (BIIγ)

The Frizy Limestone consists of intercalations of greenish gray bioclastic limestone and bluish gray clay, both containing scattered glauconite grains. It can be subdivided into seven informal lithostratigraphical units (Fig. 2).

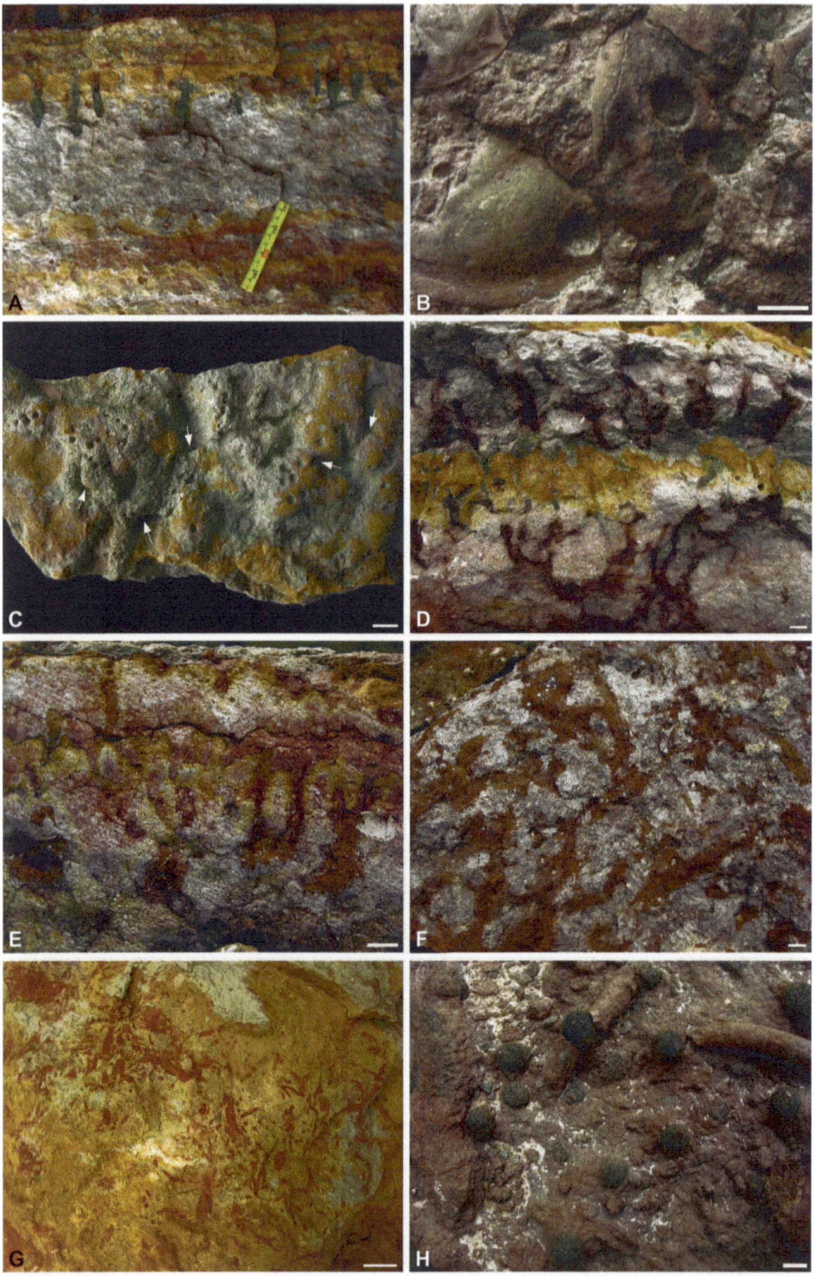

FIGURE 6 Common trace fossils from the Volkhovian (Dapingian) of the St. Petersburg region, Russia. Scale bars = 1 cm. (A) Vertical section of the "Steklo" hardground (Zeliony Bed) with *Gastrochaenolites oelandicus* filled with glauconite grains, Putilovo quarry. (B) *G. oelandicus* penetrating carapaces of the trilobite *Megistaspis estonica*, Zeliony Bed, Kingisepp quarry. (C) Hardground surface in the Zhelty Bed (Dikari unit) with grooves and openings of *Balanoglossites triadicus* (arrows) and *Trypanites* isp. on the ridges, Babino quarry. (D) Vertical section with the same hardground as in C, revealing a complex *B. triadicus* ichnofabric, Babino quarry. (E) Vertical section between the Krasnenky and Melkotsvet beds displaying a firm- to hardground with U- and Y-shaped

A proximal-to-distal trend is clearly recognizable in these storm deposits. The most distal facies are closer to the shore than the red-colored tempestites of the Zheltiaki Limestone. The individual units of the Frizy Limestone contain, in specific colonization horizons, *Thalassinoides* (various patterns and sizes of networks and boxworks), *Palaeophycus* (simple, rarely branched, lined tunnels), *Bergaueria* (Fig. 6H), and rarely *Gastrochaenolites*. Inside the upper unit of intercalation, a "patchy hardground", composed of hardened *Thalassinoides* tunnels, is bioeroded with *Trypanites*. The top of the Koroba Bed is marked by a prominent hardground surface pitted with *Trypanites* and is interpreted as the upper sequence boundary.

4.1.5 Sea-Level Fluctuations

The sediments of the Dapingian succession in the St. Petersburg region were deposited on a cool-water carbonate ramp within a shallow-marine, storm-dominated environment (Dronov, 1998). These conditions reflect short-term sea-level fluctuations where even minor changes in depth can cause an abrupt shift of facies. About eight lithofacies can be discriminated based on their lithology and ichnological content and can be arranged according to the relative depth of their deposition on the carbonate ramp. A sea-level curve has been reconstructed on the basis of the shift of these facies along the storm-dominated ramp profile (Dronov, 1999).

Major rises of relative sea level occurred at the following levels (with reference to the traditional bed nomenclature): (1) Krasnenky, (2) Butina, and (3) Krasnota. All of these events are marked by the appearance of red-colored deposits accumulated in the central part of the basin under relatively deep-water conditions. Important relative sea-level drops occurred at the following levels: (1) "Steklo" surface (base of Volkhov), (2) Butok, (3) Tolstenky, and (4) Koroba. Overall, the sea-level curve is comparable to the one constructed by Nielsen (2004) for the Komstad Limestone in Scania, except for the Middle Volkhov interval, where the data from Russia support the interpretation of a greater water depth than in the Lower Volkhov. As a consequence, the $B_{II}\alpha$/$B_{II}\beta$ boundary in the shallow-water model (Dronov, 1999) is interpreted as a deepening (transgressive) event, whereas the same boundary in the deep-water model (Nielsen, 2004) is interpreted as a shallowing (regressive) event. This discrepancy is related to a different interpretation of facies and stratigraphy; further ichnological studies have the potential to assess the interpretation of these sea-level changes.

B. triadicus burrows/borings, Putilovo quarry. (F) Bedding plane within the Pereplet Bed with a dense network of *B. triadicus*, Putilovo quarry. (G) Zheltenky Bed of the Zheltiaky unit with *Chondrites* isp., Putilovo quarry. (H) Lower bedding surface from the upper member of intercalations of the Frizy unit with glauconitic *Bergaueria* isp. penetrating arcuate burrows cf. *Phycodes*, Putilovo quarry.

4.2 Triassic Epicontinental Carbonates: Germany

The Middle Triassic (Anisian-Ladinian), epicontinental carbonates and evaporites of the Muschelkalk Group in the Germanic Basin were deposited between continental siliciclastics of the Buntsandstein and Keuper groups. The carbonates of the Lower and Upper Muschelkalk contain a moderate to rich ichnofauna, whereas the evaporitic Middle Muschelkalk is principally devoid of trace fossils, except in its lower and upper transition zones (Knaust, 2007b). In contrast to the storm-dominated homoclinal carbonate ramp of the Upper Muschelkalk, the Lower Muschelkalk resulted from deposition on a partly detached and distally steepened carbonate ramp with influence of synsedimentary tectonics along the basin axis (Knaust, 2000). Sedimentation of the Lower Muschelkalk Group (Anisian) took place in a restricted, dysoxic, and evaporitic environment with starved basin conditions. Its ichnology is briefly outlined below.

The vast portion of the low-energy, lime-mud-dominated Lower Muschelkalk Group belongs to the *Cruziana* Ichnofacies, with intercalated high-energy skeletal and Oolite Beds of the *Skolithos* Ichnofacies, and omission surfaces demarcated by the *Glossifungites* Ichnofacies and more rarely *Trypanites* Ichnofacies. Ichnodiversity is moderate, but only a few ichnotaxa repeatedly occur throughout the succession.

The Lower Muschelkalk is built of numerous stacked depositional sequences, most of them showing a shallowing-upward trend (e.g., Knaust, 1998). Such an idealized sequence is typically between 0.5 and 10 m thick and shows a generally upward-decreasing amount of bioturbation from high/moderate to low or zero. It consists of characteristic lithofacies and trace-fossil assemblages reflecting different depositional environments on a prograding carbonate ramp. Depending on the stratigraphic and regional position within the Germanic Basin, different variations of carbonate ramps can be recognized, such as a homoclinal ramp with a very low gradient and siliciclastic influx in parts, a distally steepened ramp with a considerable amount of deformed and reworked sediment, and a detached carbonate ramp with a low gradient and two or more belts of shoals. Taking the detached ramp as an example (as it occurs in the region of central Germany), five lithofacies and related trace-fossil associations occur from the top of the sequence downward (Figs. 7 and 8).

4.2.1 Grainstone/Packstone (Rudite, Arenite): Rare Bioturbation

Beds of bioclastic (bivalves, brachiopods, echinoderms), intraclastic (flat-pebble conglomerates), or oolitic limestone occur on top of shallowing-upward cycles and may contain elements of the *Skolithos* Ichnofacies. These regional marker beds occur above lagoonal or intertidal deposits. Large-scale cross-bedding and vadose diagenetic fabric indicate an origin as high-energy

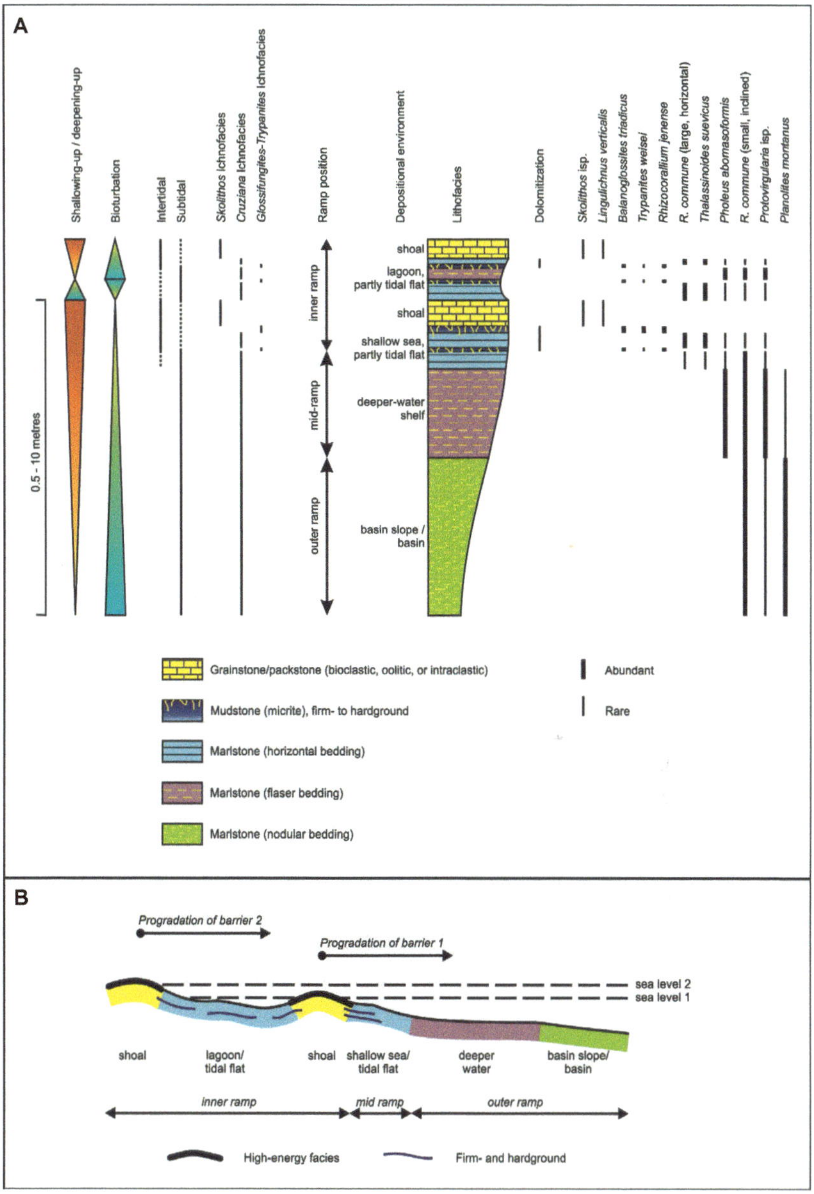

FIGURE 7 The Lower Muschelkalk Group (Middle Triassic, Anisian) in central Germany. (A) Idealized shallowing-upward sequence with the occurrence of characteristic trace-fossil associations. (B) Cartoon with an interpretation of the depositional system following the concept of a detached ramp (Schlager, 2005) with a basinward progradation barrier 1, subsequent minor flooding, and progradation of barrier 2 (*inspired by Burchette et al., 1990*).

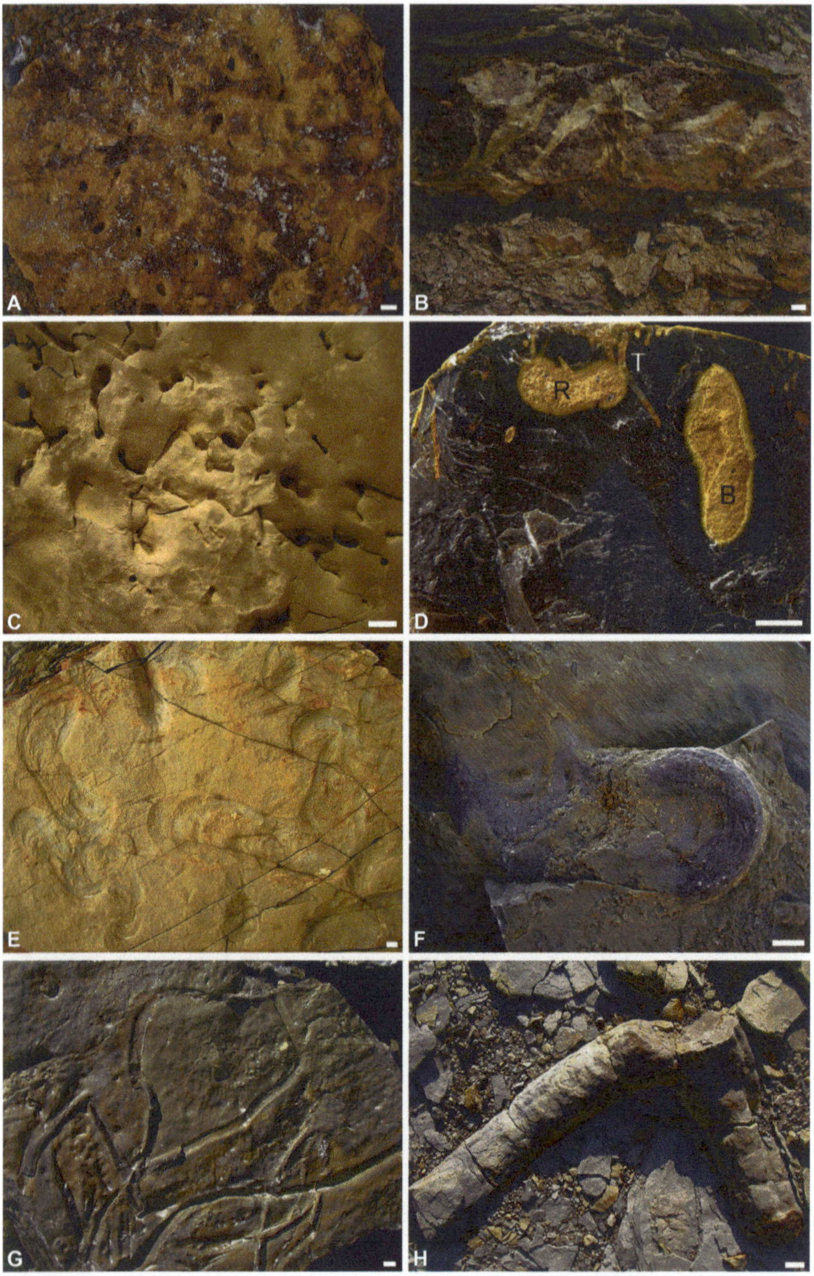

FIGURE 8 Common trace fossils from the Lower Muschelkalk Group (Middle Triassic, Anisian) of Thuringia, central Germany, in bedding plane expression (except B and D, which show vertical sections). For localities and stratigraphical positions see Knaust (2007b). Scale bar = 1 cm. (A) Openings of *Lingulichnus verticalis* in a grainy packstone (arenite) with incipient firmground conditions. *Skolithos* Ichnofacies. Upper *Terebratula* Bed, Plaue quarry. (B) Steeply inclined and dolomite-filled burrows (*Altichnus* isp.) in a bioclastic and intraclastic (rudstone) limestone bed (event bed). *Skolithos* Ichnofacies. f-3 Bed, Jena-Steudnitz quarry. (C) Omission surface on a mudstone (micrite) with abundant openings of steeply inclined and open *R. jenense* burrows. *Glossifungites* Ichnofacies. Upper Wellenkalk, Rittersdorf quarry. (D) Mudstone (micrite) with a firmground

shoals in a shallow subtidal to intertidal environment. Because of the continued redeposition of sediment, bioturbation is rare and encompasses vertically oriented trace fossils such as *Lingulichnus verticalis* (Fig. 8A), *Skolithos* isp., and *Altichnus* isp. (Fig. 8B). Marginal regions of the shoal bodies merge into firm- or hardgrounds and may contain *Balanoglossites triadicus* and *Trypanites weisei*. Minor cycles typically preserve an event bed on the top.

4.2.2 Lime Mudstone (Micrite): B. triadicus-T. weisei *Assemblage*

Discontinuity surfaces with firm- and hardground features are common throughout the Lower Muschelkalk but are best developed as ravinement surfaces just below the grainstone/packstone beds (e.g., Upper Oolite Bed, Lower *Terebratula* Bed). Thus, they occur either in front of the shoals in a very shallow-subtidal to intertidal environment or in the back-shoal lagoon (subtidal). In front of the shoals, intense lithification with the development of thick and multiple hardground layers took place, whereas slightly deeper areas farther offshore and back-barrier lagoonal environments typically contain numerous thin firmground layers. The bioturbation and bioerosion of such lime mudstone layers is moderate to high with a low ichnodiversity, typically with few main components. The continuous spectrum of burrows and borings with sharp margins demarcates firmgrounds with *B. triadicus* and *R. jenense* (*Glossifungites* Ichnofacies, Fig. 8C), which progressively grade into hardgrounds with *T. weisei* (*Trypanites* ichnofacies, Fig. 8D; Knaust, 2008). They are passively filled by marly dolomite and probably originated on a lower tidal flat.

4.2.3 Marly Limestone with Horizontal Bedding: R. commune-Thalassinoides suevicus *Assemblage*

The lower and upper portion of the Lower Muschelkalk succession, where the facies shift from supratidal to inter- and subtidal conditions and back, is characterized by platy limestone. This facies also occurs in the upper part of major

to hardground suite, exhibiting passively dolomite-filled *R. jenense* in cross section (R), a fragment of a *Balanoglossites triadicus* system (B), and superimposed needle-like *Trypanites weisei* borings (T) from the top margin. *Glossifungites–Trypanites* Ichnofacies. Upper Oolite Bed, Jena-Steudnitz quarry (type area and type bed of *T. weisei* and *B. triadicus*). (E) Dolomitic and marly limestone with extensive *R. commune* (large, horizontal) on the bedding surface. *Cruziana* Ichnofacies (intertidal). Basal Muschelkalk, Tannroda-Böttelborn quarry. (F) Marly limestone with a stunt and slightly inclined *R. commune* (small, inclined). *Cruziana* Ichnofacies (subtidal). Upper Wellenkalk, Haufeld quarry. (G) *Protovirgularia* isp. within marly limestone with parallel bedding. *Cruziana* Ichnofacies. Between Lower and Upper *Terebratula* Bed, Plaue quarry. (H) *Pholeus bifurcatus* (incomplete) within flaser-bedded marly limestone. *Cruziana* Ichnofacies. Middle Wellenkalk, Tünschütz quarry.

shallowing-upward cycles, below and between the grainstone/packstone beds (e.g., the Oolite and Schaumkalk Beds). The marly limestone can be dolomitic and contain features of subaerial exposure (e.g., desiccation cracks), thus documenting deposition on low-energy tidal flats. Where dolomitization and exposure features are lacking, rocks of the same texture can be interpreted as shallow subtidal deposits. Inter- and subtidal environments can be differentiated on the basis of the ichnofauna. Bedding planes of the dolomitic marly limestone (intertidal) are covered by large and winding *R. commune* together with the fecal pellets *Coprulus oblongus* (*Cruziana* Ichnofacies, Fig. 8E), resulting from feeding activity, probably by polychaetes. Desiccated parts of the tidal flat were apparently visited by crustaceans, preferably following the weakened cracks between the desiccation polygons and producing *T. suevicus* (*Glossifungites* Ichnofacies). In contrast, the subtidal marly limestone contains more ichnotaxa and is dominated by short and oblique *R. commune* (Fig. 8F) accompanied by *Protovirgularia* isp.

4.2.4 *Marlstone with Flaser Bedding:* Protovirgularia *isp.*-Pholeus abomasofomis *Assemblage*

This lithofacies mainly occurs in the middle part of the cycles (outer ramp) and within the lagoonal facies (inner ramp). Its texture results from undulating and flaser-like micrite layers separated by thin seams of marl. Although belonging to the *Cruziana* Ichnofacies, the local and patchy occurrence of a firm substrate indicates overprinting by the *Glossifungites* Ichnofacies. Bioturbation is moderate to high, dominated by elongated bivalve burrows *Protovirgularia* isp. (with only weakly pronounced chevron ornament due to the stiff to firm substrate, Fig. 8G), the supposed polychaete feeding trace *R. commune* (small, inclined), the crustacean dwelling burrow *P. abomasofomis*, and more rarely, other ichnospecies of *Pholeus* (Fig. 8H; Knaust, 2002).

4.2.5 *Marlstone with Nodular Bedding:* Planolites montanus-R. commune *Assemblage*

This lithofacies occurs in the lower part of shallowing-upward cycles and consists of thinly bedded and alternating mudstone and marlstone. Various disruptions of the bedding result from synsedimentary deformation, reworking, or diagenetic overprint. Many units contain reworked sediment in the form of debris- and mud-flow deposits, where bioturbated sediment became unstable and subject to reworking. Consequently, many reworked clasts are actually disrupted burrows (ichnoclasts). All in all, a distal, outer ramp facies on a distally steepened carbonate ramp can be inferred. Many intervals are highly bioturbated and mainly contain *Planolites montanus*, *R. commune* (small, inclined), and *Protovirgularia* isp. (*Cruziana* Ichnofacies). Intergradations between *P. montanus* and *Protovirgularia* isp. suggest that they were produced by bivalves burrowing in soft *versus* firm substrate.

Each of the three Lower Muschelkalk units contains shallowing-upward cycles and terminates with two or three major carbonate sand bodies at the top. Facies patterns from outer to inner ramp suggest a prograding shoreline, of which the bioclastic *Terebratula* Beds in the middle unit represent shoal deposits in a relatively distal position, whereas the Oolite Beds of the lower unit and the Schaumkalk Beds of the upper unit are proximal. Because the inner ramp facies between the coupled shoal deposits is more restricted (and partly lagoonal) than the mid-ramp facies, it is concluded that drowning of the first barrier resulted in subsequent deposition of lagoonal and tidal deposits, before the now proximally positioned second barrier prograded over the system again (cf. Burchette et al., 1990; see Knaust, 1997).

5. NEOICHNOLOGICAL APPROACHES

5.1 Rimmed Carbonate Platforms: Quaternary of the Bahamas

The islands and shallow subtidal platforms of the Bahama Archipelago are geologically world famous as a natural laboratory for the study of the sedimentology and paleontology of Quaternary carbonate rocks and sediments (see Curran and White, 1995 on Bahamian terrestrial and shallow-marine geology, and Ginsburg, 2001 on Bahamian carbonate platform evolution). Pleistocene and Holocene rocks cap these islands, and their diverse modern carbonate sedimentary environments contain a commonly diagnostic ichnological signature.

The purpose of this section is to provide a brief, informative overview of the ichnology of the Bahamas; in-depth coverage can be found in Curran (1994, 2007) and references therein. A principal benefit gained from study of Bahamian ichnology is insight regarding the modern-to-fossil transition. Given that rocks of the Bahama Islands are geologically young, in many cases one can match modern tracemaker activity to analogs found in the fossil record. Trace fossils discussed here all fall within either the *Skolithos* or the *Psilonichnus* Ichnofacies that characterize the predominately grainstone lithofacies. Bioerosion is an important aspect in reef destruction and rocky-shore erosion and was recently reviewed by Bergman et al. (2010).

Modern carbonate environments and their counterpart units in the Bahamian rock record range from shallow-subtidal, commonly reefal settings, to beach and dune environments, as found along grainy, windward, or leeward island coasts. More protected environments that might exist marginal to coastal embayments, such as tidal flat–lagoon complexes, also are included. Figure 9 illustrates two hypothetical Bahamian islands and the spatial distribution of these coastal environments, with the side panel illustrating five ichnocoenoses recognized by Curran (2007), and each keyed to its respective depositional environment. In addition, Garrett (1977) described several invertebrate communities (nearshore, pond, and levee) from the extensive, mud-dominated tidal flats on the west side of Andros Island that could be regarded as a separate ichnocoenosis.

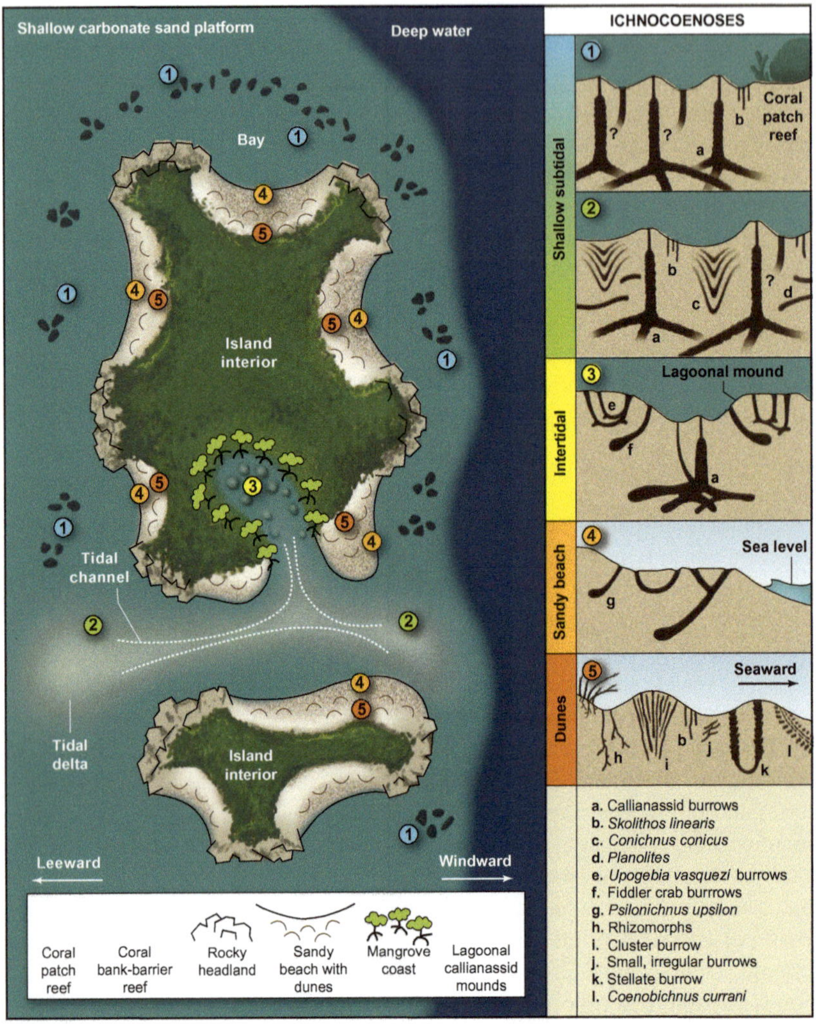

FIGURE 9 Hypothetical aerial views of two Bahamian islands and associated coastal environments. Panel at right illustrates the ichnocoenoses described in this section. Numbers key each ichnoecoenosis to its environment of occurrence and letters indicate traces present. *Modified from Curran (2007).*

An ichnocoenosis represents an ecologically pure assemblage of traces or trace fossils derived from collective activities of a single endobenthic community (Bromley, 1996). In Bahamian-style settings, most trace-fossil assemblages are true ichnocoenoses; the exception is the occurrence of rhizomorphs (or rhizoliths) formed by plant roots. Rhizomorphs are typically an integral, and sometimes dominant part of eolian complexes, but they also can result from plant roots penetrating any preexisting lithofacies during an extended period of subaerial exposure with lowered sea level. Therefore, to at least some degree, rhizomorphs are ubiquitous in Quaternary Bahamian lithofacies.

5.1.1 *The* Skolithos *Ichnofacies*

In Bahamian and similar tropical carbonate settings, the *Skolithos* Ichnofacies, as defined and summarized by Bromley (1996), Pemberton et al. (2001), and McIlroy (2004), includes subtidal to intertidal ichnocoenoses dominated by bioturbation and fossil burrows of callianassid shrimp. Here callianassids commonly have created complex and robust burrow systems that result in distinctive, often maximum intensity ichnofabrics (Fig. 10).

As one continues seaward on the platform shelf, an increase in trails and tracks typically occurs with increasing water depth, suggesting transition to a *Cruziana* Ichnofacies assemblage. This transition has not yet been documented in the Bahamian rock record because of the abrupt shelf-to-slope break at the edges of Bahamian platforms. Accordingly, little is currently known about the deeper-water ichnology of the offshore areas.

Domination by the burrowing activities of endobenthic thalassinidean shrimp, particularly callianassids, of carbonate-sand substrates in tropical,

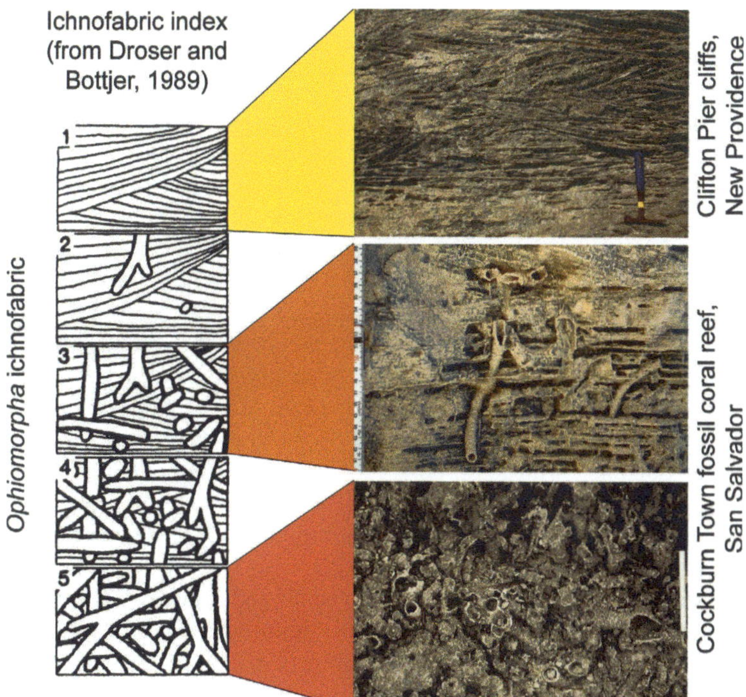

FIGURE 10 *Ophiomorpha* ichnofabrics in Bahamian Late Pleistocene grainstones keyed to the scale of Droser and Bottjer (1989).

shallow subtidal environments is important to both sedimentologists and ichnologists because "they often occur in high densities and influence the whole sedimentology and geochemistry of the seabed" (Dworschak, 2004). With respect to geochemistry, some studies, mostly by marine biologists (e.g., Webb and Eyre, 2004), support this statement, but data are limited. What has been clearly documented is the many effects callianassids have on surrounding carbonate sediments (Tudhope and Scoffin, 1984), including the total transformation of original lithofacies by burrowing activity (Curran, 1994; Tedesco and Wanless, 1991; Fig. 10). Nonetheless, ichnologists and sedimentologists should be mindful that overall knowledge of tropical callianassids remains poorly known with respect to species identification, burrow morphology, and general ecology and sedimentary environment effects.

There are many species of thalassinideans, and the families Callianassidae and Upogebiidae are cosmopolitan, with decreasing species numbers toward the poles. In a survey of extant thalassinidean diversity, Dworschak (2000) recorded 155 species of callianassids and 139 species of upogebiids (species numbers have increased to more than 200; Dworschak, 2010: personal communication) with subsequent new discoveries. Furthermore, most thalassinidean species are found in shallow-marine waters, living at depths of less than 20 m, with by far the highest numbers of species found at tropical to subtropical latitudes.

In the light of these numbers, it is no surprise that the bioturbation and burrows of callianassids (*Ophiomorpha)* dominate the ichnocoenoses of the *Skolithos* Ichnofacies as it occurs in the shallow-subtidal Quaternary grainstones of the Bahamas, south Florida, and other similar tropical carbonate settings. This is important because *Ophiomorpha* ichnofabrics commonly create high levels of permeable ichnogenic macroporosity, permitting host strata to have significant potential as carbonate aquifers and reservoirs (Cunningham et al., 2009, 2012).

The dominant ichnotaxon occurring in Late Pleistocene shallow-subtidal grainstones of the Bahamas and south Florida is *Ophiomorpha* (Ichnocoenosis 1). Bahamian *Ophiomorpha* burrow systems are commonly robust and well preserved (Fig. 10) and have been described in detail (Curran, 1994, 2007; Curran and White, 1991 and references therein). *Ophiomorpha*-bearing grainstones can occur in association with fossil coral reefs and intervene with coral rudstone (Curran, 1994). *S. linearis* is always present but secondary to *Ophiomorpha*, occurring in greatest densities in grainstones formed in the shallowest parts of shallowing-upward sequences.

Ichnocoenosis 2 includes *Conichnus conicus* and *Planolites* in addition to *Ophiomorpha* and *Skolithos linearis*. In Late Pleistocene rocks, *C. conicus* occurs most commonly in grainstones characterized by tabular and trough cross-bedding and interpreted as representing relatively rapid sediment accumulation under shallow-subtidal, shoaling conditions influenced by nearshore and/or tidal currents (Curran and White, 1997; Fig. 11A). *Planolites* typically is less abundant than

C. conicus, but specimens can be well formed and are common. *C. conicus* is generally attributed to the escape-burrowing activities of sea anemones (Shinn, 1968). However, a mechanical origin for *C. conicus* cannot be ruled out, based on the study of Buck and Goldring (2003). Further discussion of the possible modes of origin and significance of *C. conicus* was given by Curran (2007). *Planolites* in Bahamian grainstones and the Miami Limestone can be attributed to the deposit-feeding activity of balanoglossid worms (Curran, 2007).

Ichnocoenosis 3 is characteristic of the intertidal zone of semi-enclosed, tidally influenced, slightly hypersaline lagoonal areas, commonly referred to as "creeks". An example is Pigeon Creek, on San Salvador Island, where the margins of a lagoon are bordered by mangroves that commonly give way to extensive carbonate flats of muddy sand. The ichnology of these intertidal flats and possible trace-fossil analogs and rock-record implications were investigated by Curran and Martin (2003). These flats and other similar areas throughout the Bahamas display a distinctive topography of mounds and craters formed by the callianassid shrimp *Glypturus acanthochirus* (Fig. 11B).

This deep-tier burrower profoundly modifies the intertidal and shallow-subtidal areas that it inhabits and is a true ecosystem engineer (Berkenbusch and Rowden, 2003). *G. acanthochirus* burrows are large, well lined and have a distinctive, downward-spiraling morphology (Curran and Martin, 2003; Dworschak and Ott, 1993). Deep, spiraling burrows from Florida, the Bahamas, and the Caicos Platform and generically attributed to "*Callianassa*" almost certainly also were formed by *G. acanthochirus* and indicate a distribution for this species throughout the wider Caribbean (Shinn, 1968; Tedesco and Wanless, 1991). Farther afield, Farrow (1971) and Braithwaite and Talbot (1972) described casts of large, spiraling callianassid-burrow systems from modern carbonate substrates in the Seychelles Archipelago, Indian Ocean, with the indication being that downward-spiraling morphology is widespread for callianassids in tropical, intertidal, and shallow-subtidal carbonate settings.

With the development of thin "*Schizothrix*" microbial mats, the surfaces of large, coalesced mounds formed by *G. acanthochirus* become stabilized and colonized by other burrowers (Curran and Martin, 2003). The most distinctive and complex burrows are those of the shrimp *Upogebia vasquezi*, which penetrate 10–15 cm into the mound surfaces; they are thickly lined, and have smooth interior and coarsely knobby exterior surfaces (Fig. 11F). Also present on mound surfaces are numerous fiddler crabs burrows. The larger burrows are formed by adult *Uca major*, with burrow openings commonly surrounded by scratch traces and numerous excavation and feeding pellets. Smaller openings indicate juvenile *U. major* burrows and/or other unidentified *Uca* species.

In an island interior facies of the Late Pleistocene Grotto Beach Formation on San Salvador, Curran and Martin (2003) described trace fossils identified as *U. vasquezi* burrows in association with the tops of presumed *G. acanthochirus* burrow shafts. Although not yet formally named, fossil *U. vasquezi* burrows, as characteristic of intertidal, mounded surfaces, should be geologically useful

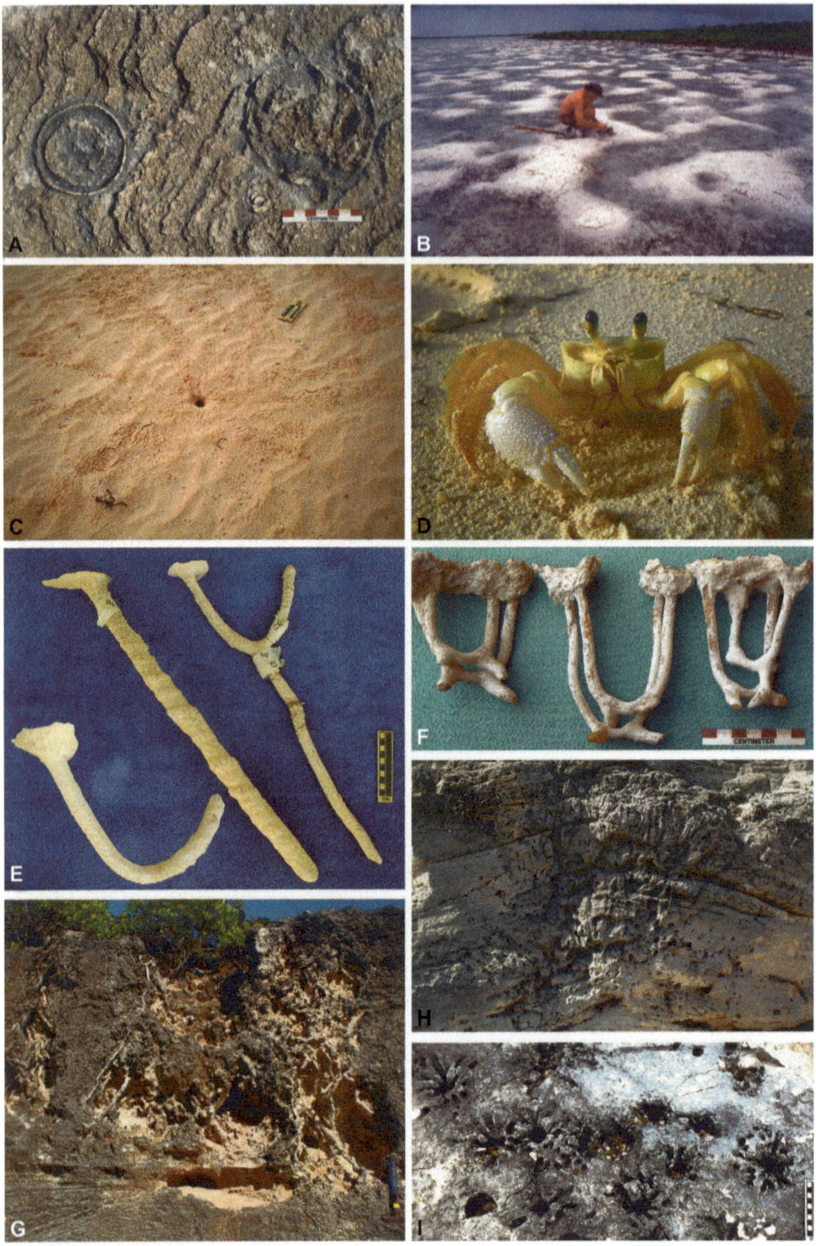

FIGURE 11 Bahamian ichnology. (A) Cross sections of *Conichnus conicus* in cross-bedded Late Pleistocene grainstone (horizontal surface), Harry Cay, Little Exuma. Scale bar = 7 cm. (B) Mounded topography created by the deep-burrowing callianassid *Glypturus acanthochirus* on the wide intertidal zone of Pigeon Creek lagoon, San Salvador. (C) *Ocypode quadrata* burrow opening with radiating trackways, East Beach, San Salvador. Scale bar = 12 cm. (D) Mature *O. quadrata* ghost crab, same locality as C. About 10 cm from side to side. (E) Casts of typical *O. quadrata* burrows made by K. Seike, same location as C. Scale bar = 10 cm. (F) Casts of the complex burrow of the shrimp *Upogebia vasquezi*. One shrimp occupies each tube, a male in one and a female in the other, and the tubes are surrounded by a thick, unified, lining with large pellets on the exterior. (G) Dense occurrence of rhizomorphs in a Late Pleistocene regressive

as indicators of past sea-level position. Fossil fiddler crab burrows have not been reported from Pleistocene carbonate rocks, but, if they should be discovered in the future, they also would be good sea-level position markers.

5.1.2 *The* Psilonichnus *Ichnofacies*

The beach and coastal dune ichnocoenoses of the Bahamas fall within and conform fully to the definition of the *Psilonichnus* Ichnofacies, as conceived by Frey and Pemberton (1987, originally termed the *Psilonichnus* Ichnocoenose). The holotype specimen of *P. upsilon* (Frey et al., 1984) is a burrow likely formed by the ghost crab, *Ocypode quadrata* (Fig. 11C and D), and preserved within Holocene carbonate grainstones of backshore beds on San Salvador.

Ichnocoenosis 4 (Fig. 9) is characteristic of carbonate sand beaches of the Bahamas and elsewhere in the wider Caribbean and beyond, as well as equivalent Holocene strata in the Bahamas and the Pleistocene of Bermuda (Curran, 1994, 2007). This ichnocoenosis is dominated by the distinctive modern burrows of ocypodid crabs (e.g., ghost crabs) and *P. upsilon* (Fig. 11E). In the Bahamas, the analog relationship between *O. quadrata*, its modern burrows, and *P. upsilon* is well established (Curran, 1994; Curran and White, 1991; Frey et al., 1984). The typical range of vertical oblique, J-, U-, and Y-shaped morphologies for *P. upsilon* occurs in the Holocene beach-backshore beds in the Bahamas. Vannini (1980) and De (2005) illustrated an even wider range of morphologies for burrows constructed by species of *Ocypode* in modern siliciclastic beaches.

Beaches in the Bahamas tend to be relatively steep and narrow and usually are backed with heavily vegetated dunes. Burrowing activity of *O. quadrata* is effectively confined to the beach backshore zone. Unlike its distribution in coastal siliciclastic settings, where ghost crabs commonly range well back into the dunes, Bahamian *O. quadrata* normally does not inhabit dunes, owing to the thick vegetation cover which impedes its mobility, and possibly also because of abundant presence of the common land crab, *Gecarcinus lateralis*, within dunal areas (Curran 1994; Curran and White, 1991).

P. upsilon is widespread in Holocene upper foreshore and backshore successions throughout the Bahamas. In contrast, the backshore zone is not well represented in Pleistocene beach to dune sequences, and *P. upsilon* has not been reported from these rocks (Curran, 2007). However, *P. upsilon* is common in a carbonate beach to primary dune coastal exposure of the middle Pleistocene Belmont Formation on Bermuda (Curran, 1994: Fig. 3.11), where its occurrence has real value as a sea-level position indicator. This utilitarian aspect of *P. upsilon* likely can be applied to other similar tropical carbonate settings elsewhere.

---

eolianite, near Whale Point, North Eleuthera. Hammer length = 28 cm. (H) Large cluster burrow in Holocene eolianite, Hanna Bay, San Salvador. Height of burrow complex = 1.4 m. (I) Cross sections of stellate burrows (shafts with radiating short tunnels) exposed on horizontal surface of Holocene eolianite, same location as H. Scale bar = 10 cm.

The original definition of the *Psilonichnus* Ichnofacies included trace-fossil assemblages of coastal eolian dunes (Pemberton and Frey, 1985). A distinctive suite of trace fossils occurs in Bahamian coastal-carbonate eolianites (Ichnocoenosis 5; Curran and White, 2001). In the past, eolianites were thought to contain only sparse ichnofauna at best, usually rare vertebrate trackways. With establishment of the *Entradichnus* Ichnofacies to characterize sparsely vegetated to unvegetated eolian dune fields in arid settings (Ekdale et al., 2007; and/or the *Octopodichnus* Ichnofacies of Hunt and Lucas, 2007, featuring arthropod trackways) and recognition of the Bahamian eolian ichnocoenosis, it is clear that eolianites can have a very real and diverse ichnologic signature (see Ekdale and Bromley, 2012).

Rhizomorphs (or rhizoliths) are a ubiquitous aspect of Bahamian eolianites, with prolific occurrences reaching maximum ichnofabric level and completely obscuring original bedding and textures, particularly in Late Pleistocene eolian sediments deposited during sea regression (Fig. 11G). In Holocene eolianite sequences throughout the Bahamas, thin caliche (micritic) crusts with well-developed rhizomorphs parallel to bedding are common and represent short breaks in the buildup of dunal sands. These horizontal rhizomorphs can be distinctive enough to be attributed to the activity of some common coastal plant species (Curran and White, 2001; White and Curran, 1988).

Various types of arthropods are the likely tracemakers for all five of the animal trace fossils previously described from Bahamian eolianites (Curran, 2007 and earlier references). The largest and most distinctive trace fossils in these eolianites are cluster burrows and stellate burrows (Fig. 11H and I), attributed respectively to brooding and hatching activities of sphecid (digger) wasps and nesting activities of halictid (sweat) bees. Stellate burrows likely can be assigned to the ichnogenus *Cellicalichnus* Genise, 2000. Both forms are common and can be well preserved in Holocene beds throughout the Bahamas and, although less common, also in Late Pleistocene eolianites.

5.2 Homoclinal Carbonate Ramp: Quaternary of the Southern Arabian Gulf

Modern carbonates of the southern region of the Arabian (Persian) Gulf comprise one of the best studied and understood carbonate systems of the world. Although they have been widely used as a key analog for homoclinal carbonate ramps encountered elsewhere (see Alsharhan and Kendall, 2003), care must be taken before rigorous applications are made (e.g., Knaust, 1997; Walkden and Williams, 1998). Despite its well known sedimentology and diagenesis, the ichnological aspect of the Arabian Gulf has been only marginally studied and a comprehensive treatment is still missing. In this section, an overview of the ichnology in relationship to established facies zones is given. A more detailed characterization of the area can be found in Purser (1973) and Riegl et al. (2010).

The well-oxygenated conditions of the Gulf permit abundant benthic life at all depths that are responsible for a relatively high degree of bioturbation and destruction of lamination. Only hypersaline lagoons have local anaerobic conditions within the sediment, as indicated by sulfate reduction and organic-rich layers with pyrite. Al-Ghadban et al. (1998) discussed the amount of bioturbation in the Gulf in relation to water depth (water circulation), light, sedimentation rate, and grain size. They concluded that bioturbation dominates in coarse-grained sediments deposited mainly in the shallow south-western part of the Gulf, where high water agitation, good light penetration, favorable hard substrate, and coarse-grained sediment provide adequate supplies of oxygen and food which allow for growth of various epifaunal and infaunal (burrowing) communities. Benthic organisms with a burrowing behavior include foraminifers, cnidarians, crustaceans, bivalves, echinoderms, sponges, and a diverse worm fauna (Basson et al., 1977; George, 2005; John and George, 2005).

The modern facies pattern along the southern Arabian Gulf consists of a barrier-island system (mainly microtidal) and a coastal plain with sabkha deposits (Evans, 1995). Several localities along the coastline of the United Arab Emirates offer insight into different aspects of modern carbonate deposition (Alsharhan and Kendall, 2002, 2003), of which the facies belts in the area south of Al Aryam Island (Kirkham, 1997, 1998) illustrate well the related traces. On a S-N transect, the five facies belts detailed below can be recognized from the land to the sea.

5.2.1 Supratidal Flat (Sabkha) with Beach Ridges

This broad zone consists of a polygonally cracked halite surface (upper supratidal), a convoluted halite crust (middle supratidal), and a moist carbonate mud and gypsum mush (lower supratidal, cf. Alsharhan and Kendall, 2002). It is relatively barren of traces except for some traces of terrestrial animals, mainly arthropods (ants, termites, spiders, scorpions, and other insects) and vertebrates, concentrated on local, slightly elevated areas with sand accumulation. Stranded Pleistocene beach ridges as a result of the Flandrian transgression (approximately 5000 years BP) build such topographic highs (Kirkham, 1998) and run parallel to the coast in an upper supratidal position (some 2–4 km inland). Weak and patchy lithification of the bioclastic carbonate sand allows some organisms to excavate open burrows. Most obvious are the shallow subterranean nests of termites, about 5 cm in diameter and packed with plant material from the sparse shrub cover (Fig. 12A). The nests are associated with elongated pathways, which emerge from the central chamber and run along the surface. This habitat is also occupied by desert geckos of the genus *Stenodactylus*, which comprise ground-dwelling species. A low level of bioturbation becomes more common in the lower supratidal zone, where sandy and muddy layers originate due to storm washover events.

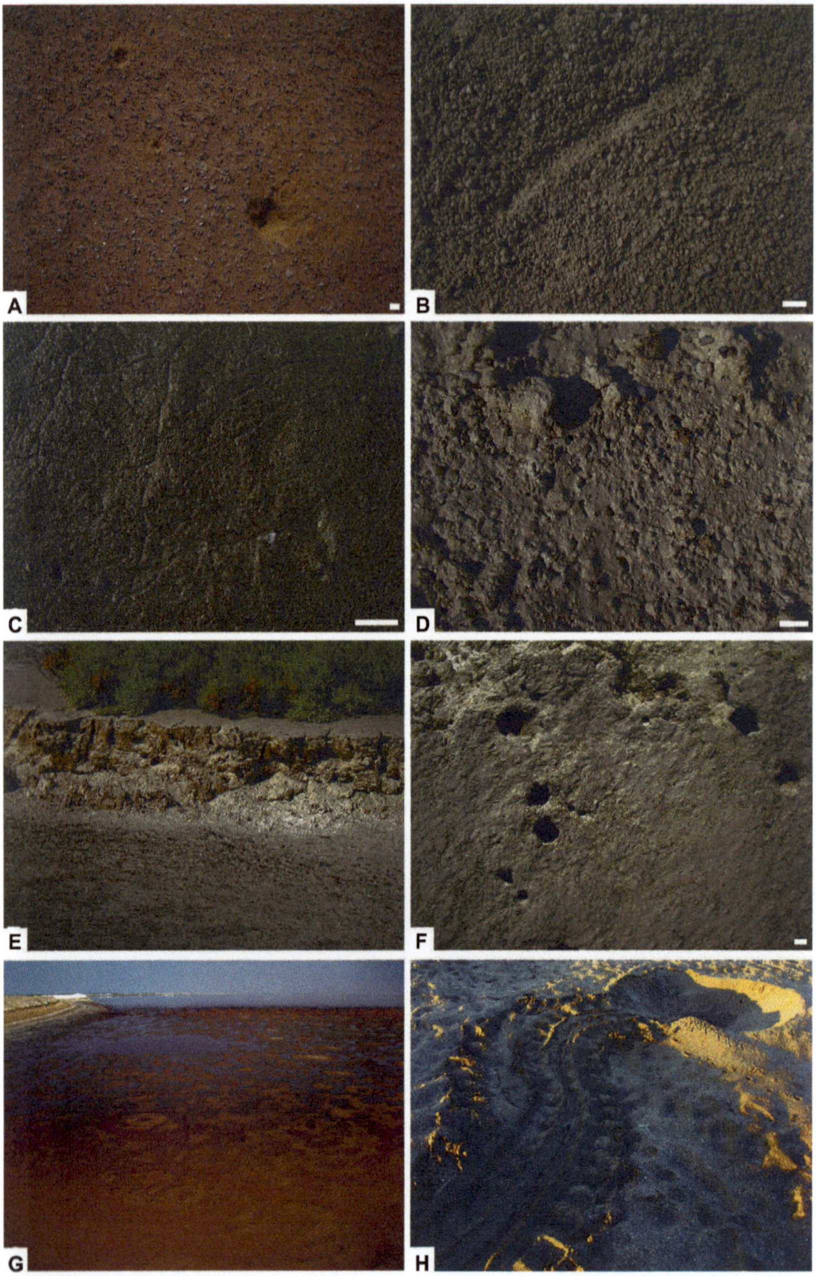

FIGURE 12 Modern ichnology of the tidal flat complex south of Al Aryam Island (southern Arabian Gulf, UAE). Scale bars = 1 cm. (A) Stranded Pleistocene beach ridges with abundant shells of cerithid gastropods on a supratidal flat (sabkha). The two larger burrows were made by termites, whereas the smaller one in between results from ant activities. (B) Upper intertidal sand flat with a groove-like surface trace of the sand-dwelling crab *Scopimera crabricauda* and its pseudo-pellets. (C) Soft mud surface at the margin of a lower intertidal pool with minute trails produced by foraminifers. (D) Firmground surface of a mud-bank in the lower intertidal with intense bioturbation by worms. (E) Margin of a lower intertidal channel (cliff height = approximately 0.3 m). The channel floor shows abundant trackways of birds, whereas the bank acts as a stiff- to firmground and is

5.2.2 Upper Intertidal Microbial/Sand Flats

A broad coastal zone is occupied by microbial (cyanobacterial) mats with different morphologies (Alsharhan and Kendall, 2003; Kendall et al., 2002). Such microbial flats with a leather-like surface develop in a harsh and hypersaline environment, which is uninhabitable for most burrowing organisms and therefore very few traces can be found here. Terrestrial arthropods are temporary visitors of this transition zone and burrow underneath the mats while feeding on decomposing mat material. Sand-filled shrinkage cracks along the margins of mat polygons are preferred sites of moisture and weakness for entering the substrate. Marine colonization by crustaceans and worms occurs in parallel with diurnal inundation of seawater, resulting in a patchy distribution of sand or sandy beaches. Volcano-like mounds, approximately 10 cm high and 4 cm wide and with a 1–3 mm diameter axial canal, occur in the vicinity of tidal channels. The columnar structures consist of alternating biolamination and pellet layers and result from the work of polychaetes within the polygonal algal zone (Shinn, 1972). Amphipods and ghost crabs (*Ocypode saratan*) are marine animals with a semiterrestrial lifestyle and move up from the water and cross the high-tide line to burrow into lower supratidal deposits. Amphipods are known to produce *Skolithos*-, *Arenicolites*-, or *Diplocraterion*-like traces (Gingras et al., 2008), whereas ghost crabs excavate large J-, U-, Y-shaped or complexly branched burrows like *Macanopsis* and *Psilonichnus* (e.g., De, 2005). Besides scavenging, the ghost crabs may also rework organic-rich sand (deposit-feeding) and produce pellets (Fig. 12B). Trackways and resting traces may have a slight chance to become preserved, but this is unlikely for the up to 30 cm high sand towers build by some male ghost crabs to mark their territory (Basson et al., 1977).

5.2.3 Lower Intertidal Sand/Mud Flats with Tidal Channels and Beachrock

The substrate of this zone is extensively reworked by a diverse fauna, dominated by elements of the meiofauna (e.g., foraminifers, nematodes, copepods, polychaetes, and turbellarians; Fig. 12C), several kinds of worms (e.g., polychaetes, nemerteans, and sipunculids; Fig. 12D), molluscs (gastropods, bivalves), crustaceans (e.g., amphipods, isopods, and decapods), insects (e.g., beetles), and birds. Aragonite-cemented coastal sand, composed of cerithid gastropod shells, may turn into well-developed beachrock (Alsharhan and Kendall, 2003;

---

intensively burrowed mainly by crabs (probably *Metopograpsus messor*) and worms. (F) Muddy firmground surface on a lower intertidal mud flat with the paired opening of U- or Y-shaped crab burrows. (G) Lower intertidal mud (salt) flat with circular and bow-shaped feeding traces (approximately 1.2–1.6 m in diameter) of the Greater Flamingo (*Phoenicopterus roseus*). (H) Nesting site and trackway (approximately 0.6 m wide) of the Green Turtle (*Chelonia mydas*). Photo taken at Ras Al Hadd, Oman.

Kendall et al., 1994). These crusts have a patchy cement distribution with numerous crevices, which provide shelter and food supply for a diverse fauna. Semiconsolidated substrate is bioeroded by cyanobacteria, bivalves (*Botula*, *Lithophaga*, and *Gastrochaena*) and the shrimp *Alpheus* (cf. Asgaard et al., 1997; Fischer and Meyer, 1985). Muddy and firm substrate along eroding channel margins is intensively burrowed by crustaceans, polychaetes, echiurans, and other worms (Fig. 12D and E). Lower intertidal mud flats host burrows of deposit-feeding ocypodid crabs (mainly *Macrophthalmus depressus*, see Basson et al., 1977) producing *Macanopsis*-, *Psilonichnus*-, *Skolithos*-, and *Gyrolithes*-like burrow morphologies (Fig. 12F), with the echiuran *Ikeda taenioides* leaving deep subvertical burrows (Hornby, 2005), and with bivalves. A rich association of avifauna traces is present, including trackways of herons, flamingo, plovers, sandpipers, gulls, and terns. On protected mud flats, the Greater Flamingo (*Phoenicopterus roseus*) leaves extensive feeding traces when sifting through the silt while turning its body slowly around (Allen, 1956). The circular and semicircular structures are approximately 1.2–1.6 m in diameter and comprise a central pedestal surrounded by a feeding trough (Fig. 12G). The surface of tidal pools and soupy channel mud banks is covered with minute trails produced by foraminifers and abundant trails of cerithid snails with orientation toward the water.

5.2.4 Subtidal Lagoon with Sand Bars

The lagoonal areas between barrier islands and the tidal flat are mainly sandy and subordinately muddy. Subaqueous cementation is common and leads to the synsedimentary development of extensive hardgrounds, which become bioeroded by algae, grazing snails, endolithic sponges (*Entobia*-like borings), bivalves (*Gastrochaenolites*-like borings), crabs, and stomatopods. Patch reefs consisting of corals and algae grow where a suitable hard substrate is available and are subject to bioerosion not only by the same kinds of organisms but also by sea urchins, polychaetes, sipunculids, and fish. A complex system of tidal currents within the lagoon gives rise to the accumulation of various sand bodies such as bars, spits, and ebb-tidal deltas consisting of ooids. This sandy bottom is inhabited by a diverse meiobenthic fauna in the interstices as well as by macrobenthic animals with a burrowing behavior. Different snails and sea urchins (sand dollars and heart urchins) leave their trails on the sediment surface or dig slightly deeper. Bivalves, crustaceans, and worms all bioturbate the sand, but their burrows are poorly known. Common features are sand cones up to 1 m in height, produced by callianassid shrimp, while excavating an extensive burrow system—known from the fossil record as *Ophiomorpha*, *Spongeliomorpha*, and *Thalassinoides* (see Section 5.1). Burrows made by alpheid shrimp are frequently inhabited by gobies, a small commensal fish. Larger fish (e.g., soles) hide in the sand while waiting for prey, whereas rays form deep plug-shaped feeding excavations due to hydraulic jetting. If fossilized, such traces are known as *Piscichnus*, and numerous of such depressions are nicely exposed on a Pleistocene bedding plane along the northern coast of Futaysi Island (Kirkham and Evans, 2008). The muddy substrate is burrowed by different species of bivalves, gastropods, worms, echinoids, crabs, and shrimp.

5.2.5 Barrier Islands with Sabkha and Leeward Spit–Bar Complexes

The exposed seaward sides of the barrier islands are often fringed with coral reefs, which build similar habitats as the patch reefs in the lagoon. In addition to the above-mentioned fauna, the beach sands are periodically visited by breeding sea turtles, especially the Green Turtle (*Chelonia mydas*, Fig. 12H). The turtles come onshore to lay their eggs in pits and leave impressive trackways in the sand which have potential for preservation.

5.3 Holocene Along the Mussafah Channel near Abu Dhabi

In addition to patchy Pleistocene outcrops (e.g., Evans et al., 2002), the Pleistocene–Holocene sections along the man-made Mussafah Channel SW of the city of Abu Dhabi offer good outcrops for studying the facies zones described in Section 5.2. (Kirkham, 1997, 1998) and their trace fossils. The about 2-m-thick section shows lateral and vertical facies variations but in general consists of the three units described below (Strohmenger et al., 2010).

5.3.1 Pleistocene Reworked Dune Deposits (Eolianites)

The massive to cross-bedded carbonate-rich sandstone crops out at the base of the succession. This rock is affected by high-tide currents and contains numerous modern borings made by crustaceans (crabs and/or shrimp, Fig. 13A). This evidence implies that ancient crustacean structures similar to *Psilonichnus* and *Macanopsis* not only derive from the burrowing activity in sediment but sometimes also could be bioerosion features (cf. Asgaard et al., 1997; Fischer and Meyer, 1985).

5.3.2 Holocene Transgressive Unit (Intertidal to Subtidal)

Above an unconformity, the carbonate succession starts with a thin layer of microbial mat deposits (intertidal) with weak bioturbation, locally overlain by a consolidated and cemented, cerithid-rich grainstone. The latter is a firm- to hardground that builds the lag of a compound channel cutting down into the eolianite. The eroded top surface of the firm- to hardground reveals various cross sections of burrows/borings, many of them with a dumb-bell shape suggesting pouch-shaped excavations similar to the ones of modern *Polydora* (Fig. 13B). Above it follows bioclastic pack-, grain-, and rudstone of tidal channel origin, which interfingers laterally with extensive lagoonal deposits (subtidal) consisting of mud-rich packstone with oxidized rootlets.

FIGURE 13 Ichnology of Pleistocene (A) and Holocene (B–H) deposits from the Mussafah Channel SW of the city Abu Dhabi. (A) Reworked dune deposits (eolianites) with modern crustacean borings in the upper intertidal zone. Cliff height = approximately 0.6 m. (B) Cerithid-rich firm- to hardground (channel lag deposit) with an eroded top surface and numerous cross sections of burrows/borings, probably produced by polychaetes. Scale bar = 1 cm. (C) Lower bedding plane of a beach sand layer (intertidal) with an extensive burrow system of *Thalassinoides paradoxicus*, probably produced by crustaceans. Scale bar = 1 cm. (D) Lagoonal packstone with large lenticular gypsum crystals (L, subtidal), bioturbated beachrock (B, lower intertidal) and microbial mats with gypsum crystals (M, upper intertidal). Scale bar = 20 cm. (E) Top of the lagoonal packstone with

Sections in a most landward position commonly contain beach deposits, some of which are intensively bioturbated. The burrows are about 5 mm in diameter and build a comprehensive system of *Thalassinoides paradoxicus*, maybe the product of crustaceans (Fig. 13C). The burrows promote a selective cementation which creates an ichnofabric in hypichnial preservation.

5.3.3 Holocene Regressive Unit (Subtidal to Supratidal)

The lagoonal, peloidal–skeletal packstone with large lenticular gypsum crystals of the transgressive unit continues into the upper part of the succession. Its sharp top surface exhibits many irregular shafts to crudely Y-shaped burrows (*Parmaichnus* isp.), penetrating deep into the substrate (Fig. 13D and E). They were likely produced by thalassinidean decapods (like *Upogebia*) and indicate lower intertidal conditions (Pervesler and Uchman, 2009). The colonization surface is covered by a 10- to 20-cm-thick rudstone/grainstone with cerithid gastropods and cross-bedding (Fig. 13D). This loosely consolidated sediment in places contains an irregularly branching burrow system (*Balanoglossites ramosus*, Knaust, 2008; Fig. 13F and G); the sediment is interpreted as high-energy tidal channel and beach deposit. Its top surface is locally covered by a thin firm- to hardground representing beachrock (Fig. 13H). The bioclastic packstone/grainstone locally continues toward the top of the section, where it results from extensive beach spit and bar deposition (intertidal), but is less bioturbated and subsequently overprinted by large gypsum crystals (supratidal).

6. CONCLUSIONS

The ichnology of carbonate sedimentary systems is less well studied compared to that of siliciclastic systems. Published ichnological/sedimentological studies of shallow-marine carbonates are biased toward the Mesozoic (especially Jurassic), whereas Precambrian, Paleozoic, and Cenozoic systems are underrepresented.

Shallow-marine carbonates differ from their siliciclastic counterparts by a number of criteria which can be grouped into three categories: (1) The composition and distribution of trace-fossil associations is controlled by the preferential abundance of various benthic organisms in tropical and subtropical zones where shallow-marine carbonate environments are common. This results in favored trace fossils such as crustacean burrows. (2) Given their specific

a dense accumulation of *Parmaichnus* isp., interpreted as the burrows of thalassinidean decapods (e.g., *Upogebia*) in a lower intertidal environment. Scale bar = 1 cm. (F) Cerithid-rich rudstone/grainstone (lower intertidal) with an irregularly branching burrow system of *Balanoglossites ramosus* (vertical section), interpreted as the burrows of polychaetes. Scale bar = 1 cm. (G) As figure F, bedding surface. Scale bar = 1 cm. (H) Top surface of a firm- to hardground (intertidal beachrock) with a *Balanoglossites* ichnofabric below it. Scale bar = 20 cm.

morphology, carbonate ramps and rimmed shelves comprise broad and very shallow facies belts favorable for intense bioturbation, subaerial exposure, and development of peritidal ichnofauna and trace fossils formed by plant roots. (3) Carbonate sediments and their high diagenetic potential have a strong impact on the ichnological signature, resulting in the common occurrence of bioerosion, chemically etched borings, abundance of firm- and hardgrounds, and preferential trace-fossil preservation. Early-diagenetic processes can lead to rapid lithification with trace-fossil accentuation, but can also result in little to no sediment color contrast between burrows and matrix and bears a high potential for diagenetic alteration of trace fossils and hosting rocks.

Three ichnofacies are important for the reconstruction of paleoenvironments on carbonate shelves (or ramps): the *Psilonichnus* Ichnofacies, the *Skolithos* Ichnofacies, and the *Cruziana* Ichnofacies. In addition, the substrate-controlled *Glossifungites* Ichnofacies and the *Trypanites* Ichnofacies can overprint the former ones during periods with stagnant or low net-sediment accumulation (condensation, omission).

The value of ichnological analysis in shallow-marine carbonates is demonstrated in two case studies, the Middle Ordovician cool-water carbonates of the St. Petersburg region in Russia and the Middle Triassic epicontinental carbonates in Germany. Furthermore, neoichnological approaches can set a focus on rimmed carbonate platforms examplified on the Quaternary of the Bahamas, and on homoclinal carbonate ramps such as the Quaternary of the southern Arabian Gulf. These examples, with their distinctive ichnocoenoses, are similar to other carbonate systems elsewhere and may serve as a comparative model for future investigations around the world.

Ichnological studies of modern carbonate environments and their rock-record equivalents are far from complete and have much potential for fruitful future research. Carbonates should not be viewed as the object of a separate subdiscipline of ichnology. Rather, ichnological information from carbonates should be combined with that of siliciclastics to achieve a better, more comprehensive understanding of depositional environments and the sedimentary rock record.

ACKNOWLEDGMENTS

D. K. thanks Statoil for enabling me to perform work in the Middle East and to publish it. H. A. C. thanks Brian White (Smith College), Jim Carew (College of Charleston), John Mylroie (Mississippi State University), Tony Martin (Emory University), and numerous other colleagues who have contributed to my understanding of Bahamian geology and ichnology over several decades, as well as the many Smith College geology students who assisted with field studies. I am grateful to the directors and staff of the Gerace Research Centre for support of my fieldwork on San Salvador and for facilitating much of my work on other islands of the Bahamas. A. V. D. acknowledges financial support from the Russian Foundation for Basic Research (Grant 10-05-00848). Jeff Lukasik (Statoil) kindly reviewed the manuscript and helped to improve this chapter.

REFERENCES

Al-Ghadban, A.N., Abdali, F., Massoud, M.S., 1998. Sedimentation rate and bioturbation in the Arabian Gulf. Environ. Int. 24, 23–31.

Allen, R.P., 1956. The Flamingos: Their Life History and Survival. National Audubon Society, New York, 285 pp.

Alsharhan, A.S., Kendall, C.G.St.C., 2002. Holocene carbonate/evaporates of Abu Dhabi, and their Jurassic ancient analogs. In: Barth, H.-J., Boer, B. (Eds.), Sabkha Ecosystems. Kluwer Academic Publishers, Dordrecht, pp. 187–202.

Alsharhan, A.S., Kendall, C.G.St.C., 2003. Holocene coastal carbonates and evaporites of the southern Arabian Gulf and their ancient analogues. Earth-Sci. Rev. 61, 191–243.

Archer, A.W., 1984. Preservational control of trace-fossil assemblages: Middle Mississippian carbonates of South-Central Indiana. J. Paleontol. 58, 285–297.

Asgaard, U., Bromley, R.G., Hanken, N.-M., 1997. Recent firmground burrows produced by a upogebiid crustacean: palaeontological implications. Cour. Forsch. Inst. Senck. 201, 23–28.

Avanzini, M., Frisia, S., Van den Driessche, K., Keppens, E., 1997. A dinosaur tracksite in an early Liassic tidal flat in northern Italy; paleoenvironmental reconstruction from sedimentology and geochemistry. Palaios 12, 538–551.

Basson, P.W., Burchard Jr., J.E., Hardy, J.T., Price, A.R.G., 1977. Biotopes of the Western Arabian Gulf. Marine Life and environments of Saudi Arabia. Aramco Department of Loss Prevention and Environmental Affairs, Dhahran, 284 pp.

Bergman, K.L., Westphal, H., Janson, X., Poiriez, A., Eberli, G.P., 2010. Controlling parameters on facies geometries of the Bahamas, an isolated carbonate platform environment. In: Westphal, H., Riegl, B., Eberli, G.P. (Eds.), Carbonate Depositional Systems: Assessing Dimensions and Controlling Parameters. Springer, Berlin, pp. 5–80.

Berkenbusch, K., Rowden, A.A., 2003. Ecosystem engineering—moving away from 'just-so' stories. N. Zeal. J. Ecol. 27, 67–73.

Bertling, M., 1999. Taphonomy of trace fossils at omission surfaces (Middle Triassic, East Germany). Palaeogeogr. Palaeoclimatol. Palaeoecol. 149, 27–40.

Boggs, S., 2009. Petrology of Sedimentary Rocks. Cambridge University Press, Cambridge, 600 pp.

Braithwaite, C.J.R., Talbot, M.R., 1972. Crustacean burrows in the Seychelles, Indian Ocean. Palaeogeogr. Palaeoclimatol. Palaeoecol. 11, 265–285.

Bromley, R.G., 1975. Trace fossils at omission surfaces. In: Frey, R.W. (Ed.), The Study of Trace Fossils. Springer, New York, pp. 399–428.

Bromley, R.G., 1996. Trace Fossils: Biology, Taphonomy and Applications. Chapman & Hall, London, 361 pp.

Bromley, R.G., 2004. A stratigraphy of marine bioerosion. In: McIlroy, D. (Ed.), The Application of Ichnology to Palaeoenvironmental and Stratigraphic Analysis. Geol. Soc. London, Spec. Publ., vol. 228, pp. 455–479.

Bromley, R.G., Asgaard, U., 1991. Ichnofacies: a mixture of taphofacies and biofacies. Lethaia 24, 153–163.

Bromley, R.G., Asgaard, U., 1993. Two bioerosion ichnofacies produced by early and late burial associated with sea-level changes. Geol. Rundsch. 82, 276–280.

Buck, S.G., Goldring, R., 2003. Conical sedimentary structures, trace fossils or not? Observations, experiments, and review. J. Sediment. Petrol. 73, 338–353.

Burchette, T.P., Wright, V.P., Faulkner, T.J., 1990. Oolitic sandbody depositional models and geometries, Mississippian of southwest Britain: implications for petroleum exploration in carbonate ramp settings. Sediment. Geol. 68, 87–115.

Cherns, L., Wheeley, J.R., Karis, L., 2006. Tunneling trilobites; habitual infaunalism in an Ordovician carbonate seafloor. Geology 34, 657–660.

Cocks, L.R.M., Torsvik, T.H., 2005. Baltica from the late Precambrian to mid-Palaeozoic times: the gain and loss of terrain's identity. Earth-Sci. Rev. 72, 39–66.

Cunningham, K.J., Sukop, M.C., Huang, H., Alvarez, P.F., Curran, H.A., Renken, R.A., Dixon, J.F., 2009. Prominence of ichnologically influenced macroporosity in the karst Biscayne aquifer: stratiform "super-K" zones. GSA Bull. 121, 164–186.

Cunningham, K.J., Sukop, M.C., Curran, H.A., 2012. Carbonate aquifers. In: Knaust, D., Bromley, R.G. (Eds.), Trace Fossils as Indicators of Sedimentary Environments. Developments in Sedimentology, vol. 64. Elsevier, Amsterdam, pp. 871–898.

Curran, H.A., 1984. Ichnology of Pleistocene carbonates on San Salvador, Bahamas. J. Paleontol. 58, 312–321.

Curran, H.A., 1994. The palaeobiology of ichnocoenoses in Quaternary, Bahamian-style carbonate environments: the modern to fossil transition. In: Donovan, S.K. (Ed.), Palaeobiology of Trace Fossils. John Wiley & Sons, Chichester, pp. 83–104.

Curran, H.A., 2007. Ichnofacies, ichnocoenoses, and ichnofabrics of Quaternary shallow-marine to dunal tropical carbonates: a model and implications. In: Miller III, W. (Ed.), Trace Fossils. Concepts, Problems, Prospects. Elsevier, Amsterdam, pp. 232–247.

Curran, H.A., Martin, A.J., 2003. Complex decapod burrows and ecological relationships in modern and Pleistocene intertidal carbonate environments, San Salvador Island, Bahamas. Palaeogeogr. Palaeoclimatol. Palaeoecol. 192, 229–245.

Curran, H.A., White, B., 1991. Trace fossils of shallow subtidal to dunal ichnofacies in Bahamian Quaternary carbonates. Palaios 6, 498–510.

Curran, H.A., White, B. (Eds.), 1995. Terrestrial and Shallow Marine Geology of the Bahamas and Bermuda. GSA Spec. Pap. 300, 344 pp.

Curran, H.A., White, B., 1997. A *Conichnus conicus*-generated ichnofabric in the late Pleistocene limestones at Clifton Pier, New Providence Island, Bahamas. In: Curran, H.A. (Ed.), Guide to Bahamian Ichnology: Pleistocene, Holocene, and Modern Environments. Bahamian Field Station, San Salvador, pp. 55–61.

Curran, H.A., White, B., 2001. Ichnology of Holocene carbonate eolianites of the Bahamas. In: Abegg, F.E., Harris, P.M., Loope, D.B. (Eds.), Modern and Ancient Carbonate Eolianites: Sedimentology, Sequence Stratigraphy, and Diagenesis. SEPM Spec. Publ., vol. 71, pp. 47–56.

De, C., 2005. Biophysical model of intertidal beach crab burrowing: application and significance. Ichnos 12, 11–29.

de Gibert, J.M., 1996. *Diopatrichnus odlingi* n. isp. (annelid tube) and associated ichnofabrics in the White Limestone (M. Jurassic) of Oxfordshire: sedimentological and palaeoecological significance. Proc. Geol. Ass. 107, 189–198.

de Gibert, J.M., Ekdale, A.A., 1999. Trace fossil assemblages reflecting stressed environments in the Middle Jurassic Carmel Seaway of Central Utah. J. Paleontol. 73, 711–720.

de Gibert, J.M., Goldring, R., 2007. An ichnofabric approach to the depositional interpretation of the intensely burrowed Bateig Limestone, Miocene, SE Spain. Sediment. Geol. 194, 1–16.

Diedrich, C., 2008. Millions of reptile tracks—Early to Middle Triassic carbonate tidal flat migration bridges of Central Europe—reptile immigration into the Germanic Basin. Palaeogeogr. Palaeoclimatol. Palaeoecol. 259, 410–423.

Dronov, A.V., 1998. Stormovaja sedimentatsija v nizhneordovikskikh karbonatno-terrigennykh otlozhenijakh okrestnostej Sankt-Peterburga [Storm sedimentation in the Lower Ordovician carbonate-terrigenous deposits of the St. Petersburg region]. Bulleten' Moskovskogo Obschestva Ispytateley Prirody, Otdelenie Geologicheskoe 73, 43–51 (in Russian).

Dronov, A.V., 1999. Kolebanija urovnja morja v rannem ordovike I ikh otrazhenie v razrezakh vostochnoi chasti glinta [Sea-level changes in the Lower Ordovician and their reflection in the rock record of the eastern part of the Glint line]. Bulleten' Moskovskogo Obschestva Ispytateley Prirody, Otdelenie Geologicheskoe 74, 39–47 (in Russian).

Dronov, A.V., Holmer, L.E., 1999. Depositional sequences in the Ordovician of Baltoscandia. In: Kraft, P., Fatka, O. (Eds.), Quo vadis Ordovician? Act. Univ. Carol., Geol. 43, pp. 133–136.

Dronov, A., Mikuláš, R., 2010. Paleozoic ichnology of St. Petersburg region. In: Excursion Guidebook for the IV Workshop on Ichnotaxonomy, June 21-25, Moscow-St. PetersburgTransactions of the Geological Institute of RAS, vol. 596, pp. 1–70.

Dronov, A., Rozhnov, S., 2007. Climatic changes in the Baltoscandian basin during the Ordovician: sedimentological and palaeontological aspects. Acta Palaeont. Sinica 46 (Suppl.), 108–113.

Dronov, A.V., Savitsky, J.V., Fedorov, P.V., Tsyganova, E.A., 1996. Detailed lithostratigraphy of the Ordovician lower Volkhovian limestone along the eastern part of the Baltic-Ladoga Glint, northwestern Russia. GFF 118, 19–24.

Dronov, A.V., Meidla, T., Ainsaar, L., Tinn, O., 2000. The Billingen and Volkhov Stages in the northern East Baltic: detailed stratigraphy and lithofacies zonation. Proc. Est. Acad. Sci. Geol. 49, 3–16.

Dronov, A.V., Mikuláš, R., Logvinova, M., 2002. Trace fossils and ichnofabrics across the Volkhov depositional sequence (Ordovician, Arenigian of St. Petersburg Region, Russia). J. Czech Geol. Soc. 47, 133–146.

Droser, M.L., Bottjer, D.J., 1989. Ichnofabric of sandstones deposited in high energy nearshore environments: measurement and utilization. Palaios 4, 598–604.

Dworschak, P.C., 2000. Global diversity in the Thalassinidea (Decapoda). J. Crust. Biol. 20, 238–245.

Dworschak, P.C., 2004. Biology of Mediterranean and Caribbean Thalassinidea (Decapoda). In: Tamaki, A. (Ed.), Proceedings of the Symposium on "Ecology of Large Bioturbators in Tidal Flats and Shallow Sublittoral Sediments—From Individual Behavior to Their Role as Ecosystem Engineers". Nagasaki University, Nagasaki, Japan, pp. 15–22.

Dworschak, P.C., Ott, J.A., 1993. Decapod burrows in mangrove-channel and back-reef environments at the Atlantic Barrier Reef, Belize. Ichnos 2, 277–290.

Ekdale, A.A., Bromley, R.G., 2001. Bioerosional innovation for living in carbonate hardgrounds in the Early Ordovician of Sweden. Lethaia 34, 1–12.

Ekdale, A.A., Bromley, R.G., Pemberton, S.G., 1984. Ichnology: the use of trace fossils in sedimentology and stratigraphy. SEPM Short Course Notes 15, 1–316.

Ekdale, A.A., Bromley, R.G., Loope, D.B., 2007. Ichnofacies of an ancient erg: a climatically influenced trace fossil association in the Jurassic Navajo Sandstone, southern Utah, USA. In: Miller III, W. (Ed.), Trace Fossils. Concepts, Problems, Prospects. Elsevier, Amsterdam, pp. 562–574.

Ekdale, A.A., Bromley, R.G., 2012. Eolian environments. In: Knaust, D., Bromley, R.G. (Eds.), Trace Fossils as Indicators of Sedimentary Environments. Developments in Sedimentology, vol. 64. Elsevier, Amsterdam, pp. 419–437.

El-Asa'ad, G.M.A., 1987. Mesozoic trace fossils from Central Saudi Arabia. Arab Gulf J. Sc. Res. Math. Phys. Sci. A5, 205–224.

Evans, G., 1995. The Arabian Gulf: a modern carbonate-evaporite factory; a review. Cuad. Geol. Iberica 19, 61–96.

Evans, G., Kirkham, A., Carter, R.A., 2002. Quaternary development of the United Arab Emirates coast: new evidence from Marawah Island, Abu Dhabi. GeoArabia 7, 441–458.

Farrow, G.F., 1971. Back-reef and lagoonal environments of Aldabra Atoll distinguished by their crustacean burrows. In: Stoddart, D.R., Yonge, M. (Eds.), Regional Variation in Indian Ocean Coral Reefs. Symp. Zool. Soc. London 28, pp. 455–500.

Fei, A., Zhang, Z., 2002. Middle Ordovician bioturbation structures from southeastern margin of Ordos Basin and their environmentary interpretation. J. China Univ. Geosci. 13, 321–329.

Fischer, R., Meyer, W., 1985. Observations on rock boring by *Alpheus saxidomus* (Crustacea: Alpheidae). Mar. Biol. 89, 213–219.

Frey, R.W., Pemberton, S.G., 1987. The *Psilonichnus* ichnocoenose, and its relationship to adjacent marine and nonmarine ichnocoenoses along the Georgia coast. Bull. Can. Petrol. Geol. 35, 333–357.

Frey, R.W., Seilacher, A., 1980. Uniformity in marine invertebrate ichnology. Lethaia 13, 183–207.

Frey, R.W., Curran, H.A., Pemberton, S.G., 1984. Tracemaking activities of crabs and their environmental significance: the ichnogenus *Psilonichnus*. J. Paleontol. 58, 333–350.

Fürsich, F.T., 1981. Invertebrate trace fossils from the Upper Jurassic of Portugal. Comm. Serv. Geol. Port. 67, 53–168.

Gaillard, C., Bernier, P., Gall, J.C., Gruet, Y., Barale, G., Bourseau, J.P., Buffetaut, E., Wenz, S., 1994. Ichnofabric from the Upper Jurassic lithographic limestone of Cerin, southeast France. Palaeontology 37, 285–304.

Garrett, P., 1977. Biological communities and their sedimentary record. In: Hardie, L.A. (Ed.), Sedimentation on the Modern Carbonate Tidal Flats of Northwest Andros Island, Bahamas. Studies in Geology, vol. 22. Johns Hopkins University Press, Baltimore, pp. 124–158.

Genise, J.F., 2000. The Ichnofamily Celliformidae for *Celliforma* and allied ichnogenera. Ichnos 7, 267–282.

George, D., 2005. Marine invertebrates. In: Hellyer, P., Aspinall, S. (Eds.), The Emirates. A Natural History. Trident Press, London, pp. 197–221.

Giannetti, A., Monaco, P., 2004. Burrow decreasing-upward parasequence (BDUP): a case study from the Lower Jurassic of the Trento Carbonate Platform (Southern Alps), Italy. Riv. Ital. Paleontol. Strat. 110, 77–86.

Gingras, M.K., Pemberton, S.G., Muelenbachs, K., Machel, H., 2004. Conceptual models for burrow-related, selective dolomitization with textural and isotopic evidence from the Tyndall Stone, Canada. Geobios 2, 21–30.

Gingras, M.K., Dashtgard, S.E., MacEachern, J.A., Pemberton, S.G., 2008. Biology of shallow marine ichnology: a modern perspective. Aquat. Biol. 2, 255–268.

Ginsburg, R.N. (Ed.), 2001. Subsurface Geology of a Prograding Carbonate Platform Margin, Great Bahama Bank: Results of the Bahamas Drilling Project. SEPM Spec. Publ., vol. 70, 271 pp.

Goldring, R., Kaźmierczak, J., 1974. Ecological succession in intraformational hardground formation. Palaeontology 17, 949–962.

Goldring, R., Taylor, A.M., Hughes, G.W., 2005. The application of ichnofabrics towards bridging the dichotomy between siliciclastic and carbonate shelf facies: examples from the Upper Jurassic Fulmar Formation (UK) and the Jubaila Formation (Saudi Arabia). Proc. Geol. Ass. 116, 235–249.

Gregory, M.R., Campbell, K.A., Zuraida, R., Martin, A.J., 2006. Plant traces resembling *Skolithos*. Ichnos 13, 205–216.

Hölder, H., Hollmann, R., 1969. Bohrgänge mariner Organismen in jurassischen Hart- und Felsböden. N. Jb. Geol. Paläont. Abh. 133, 79–88.

Hornby, R.J., 2005. An intertidal spoon worm (Phylum Echiura) in the United Arab Emirates: occurrence, distribution, taxonomy and ecology. Tribulus 15, 3–8.

Hunt, A.P., Lucas, S.G., 2007. Tetrapod ichnofacies: a new paradigm. Ichnos 14, 59–68.

Husinec, A., Read, J.F., 2011. Microbial laminite *versus* rooted and burrowed caps on peritidal cycles: Salinity control on parasequence development, Early Cretaceous isolated carbonate platform, Croatia. GSA Bull. 123, 1896–1907.

Jaanusson, V., 1976. Faunal dynamics in the Middle Ordovician (Viruan) of Baltoscandia. In: Bassett, M.G. (Ed.), The Ordovician System. Proceedings of the Palaeontological Association Symposium, Birmingham, Cardiff, 1974, pp. 301–326.

John, D., George, D., 2005. The shore & shallow seas. In: Hellyer, P., Aspinall, S. (Eds.), The Emirates. A Natural History. Trident Press, London, pp. 123–131.

Kendall, C.G., St, C., Sadd, J.L., Alsharhan, A., 1994. Holocene marine cement coatings on beach-rocks of the Abu Dhabi coastline (UAE); analogs for cement fabrics in ancient limestones. Carb. Evapor. 9, 119–131.

Kendall, C.G., St, C., Alsharhan, A.S., Cohen, A., 2002. The Holocene tidal flat complex of the Arabian Gulf coast of Abu Dhabi. In: Barth, H.-J., Boer, B. (Eds.), Sabkha Ecosystems. Kluwer Academic Publishers, Dordrecht, pp. 21–35.

Kennedy, W.J., 1975. Trace fossils in carbonate rocks. In: Frey, R.W. (Ed.), The Study of Trace Fossils. Springer, New York, pp. 377–398.

Kirkham, A., 1997. Shoreline evolution, aeolian deflation and anhydrite distribution of the Holocene, Abu Dhabi. GeoArabia 2, 403–416.

Kirkham, A., 1998. A Quaternary proximal foreland ramp and its continental fringe, Arabian Gulf, UAE. In: Wright, V.P., Burchette, T.P. (Eds.), Carbonate Ramps. Geol. Soc. London, Spec. Publ., vol. 149, pp. 15–41.

Kirkham, A., Evans, G., 2008. Giant burrows in the Quaternary Limestones of Futaysi Island and Al Dabb'iya, Abu Dhabi Emirate. Palaeogeogr. Palaeoclimatol. Palaeoecol. 270, 324–331.

Knaust, D., 1997. Die Karbonatrampe am SW-Rand des Persischen Golfes (V.A.E.)—rezentes Analogon für den Unteren Muschelkalk der Germanischen Trias? Greifswalder Geowiss. Beitr. 5, 101–123.

Knaust, D., 1998. Trace fossils and ichnofabrics on the Lower Muschelkalk carbonate ramp (Triassic) of Germany: tool for high-resolution sequence stratigraphy. Geol. Rundsch. 87, 21–31.

Knaust, D., 2000. Signatures of tectonically controlled sedimentation in Lower Muschelkalk carbonates (Middle Triassic) of the Germanic Basin. Zbl. Geol. Paläont., Teil I 1998, 893–924.

Knaust, D., 2002. Ichnogenus *Pholeus* Fiege, 1944, revisited. J. Paleontol. 76, 882–891.

Knaust, D., 2007a. Meiobenthic trace fossils as keys to the taphonomic history of shallow-marine epicontinental carbonates. In: Miller III, W. (Ed.), Trace Fossils. Concepts, Problems, Prospects. Elsevier, Amsterdam, pp. 502–517.

Knaust, D., 2007b. Invertebrate trace fossils and ichnodiversity in shallow-marine carbonates of the German Middle Triassic (Muschelkalk). In: Bromley, R.G., Buatois, L.A., Mángano, M.G., Genise, J.F., Melchor, R.N. (Eds.), Sediment-Organism Interactions: A Multifaceted Ichnology. SEPM Spec. Publ., vol. 88, pp. 221–238.

Knaust, D., 2008. *Balanoglossites* Mägdefrau, 1932 from the Middle Triassic of Germany: part of a complex trace fossil probably produced by burrowing and boring polychaetes. Paläontol. Z. 82, 347–372.

Knaust, D., 2009. Ichnology as a tool in carbonate reservoir characterization: a case study from the Permian–Triassic Khuff Formation in the Middle East. GeoArabia 14, 17–38.

Knaust, D., 2010a. The end-Permian mass extinction and its aftermath on an equatorial carbonate platform: insights from ichnology. Terra Nova 22, 195–202.

Knaust, D., 2010b. Remarkably preserved benthic organisms and their traces from a Middle Triassic (Muschelkalk) mud flat. Lethaia 43, 344–356.

Knaust, D., Costamagna, L.G., 2012. Ichnology and sedimentology of the Triassic carbonates of northwest Sardinia, Italy. Sedimentology 59, 1190–1207.

Knaust, D., 2012. Trace-fossil systematics. In: Knaust, D., Bromley, R.G. (Eds.), Trace Fossils as Indicators of Sedimentary Environments. Developments in Sedimentology, vol. 64. Elsevier, Amsterdam, pp. 79–101.

Kobluk, D.R., Pemberton, S.G., Karolyi, M., Risk, M.J., 1977. The Silurian-Devonian disconformity in southern Ontario. Bull. Can. Petrol. Geol. 25, 1157–1186.

Laporte, L.F., 1969. Recognition of a transgressive carbonate sequence within an epeiric sea: Helderberg Group (Lower Devonian) of New York State. SEPM Spec. Publ., vol. 14, 98–119.

Lukasik, J.J., James, N.P., 2003. Deepening-upward subtidal cycles, Murray Basin, South Australia. J. Sediment. Res. 73, 653–671.

MacEachern, J.A., Pemberton, S.G., Gingras, M.K., Bann, K.L., 2007. The ichnofacies paradigm: a fifty-year retrospective. In: Miller III, W. (Ed.), Trace Fossils. Concepts, Problems, Prospects. Elsevier, Amsterdam, pp. 52–77.

MacEachern, J.A., Dashtgard, S.E., Knaust, D., Catuneanu, O., Bann, K.L., Pemberton, S.G., 2012. Sequence stratigraphy. In: Knaust, D., Bromley, R.G. (Eds.), Trace Fossils as Indicators of Sedimentary Environments. Developments in Sedimentology, vol. 64. Elsevier, Amsterdam, pp. 157–194.

Mángano, M.G., Buatois, L.A., 1991. Discontinuity surfaces in the Lower Cretaceous of the High Andes (Mendoza, Argentina): Trace fossils and environmental implications. J. South Am. Earth Sci. 4, 215–229.

Männil, R., 1966. Istorija razvitija Balijskogo bassejna v ordovike [History of the evolution of the Baltic basin in the Ordovician]. Valgus Publishers, Tallinn, 210 pp (in Russian with English abstract).

Massari, F., D'Alessandro, A., 2010. Icehouse, cool-water carbonate ramps: the case of the Upper Pliocene Capodarso Fm (Sicily): role of trace fossils in the reconstruction of growth stages of prograding wedges. Facies 56, 47–58.

McIlroy, D., 2004. Some ichnological concepts, methodologies, applications and frontiers. In: McIlroy, D. (Ed.), The Application of Ichnology to Palaeoenvironmental and Stratigraphic Analysis. Geol. Soc. London, Spec. Publ., vol. 228, pp. 3–27.

Mezga, A., Meyer, C.A., Tešović, B.C., Bajraktarević, Z., Gušić, I., 2006. The first record of dinosaurs in the Dalmatian part (Croatia) of the Adriatic-Dinaric carbonate platform (ADCP). Cret. Res. 27, 735–742.

Monaco, P., 1995. Relationships between trace-fossil communities and substrate characteristics in some Jurassic pelagic deposits in the Umbria-Marche Basin, Central Italy. Geobios 18, 299–311.

Monaco, P., Garassino, A., 2001. Burrows and body fossil of decapod crustaceans in the Calcari Grigi, Lower Jurassic, Trento platform (Italy). Geobios 34, 291–301.

Monaco, P., Giannetti, A., 2002. Three-dimensional burrow systems and taphofacies in shallowing-upward parasequences, Lower Jurassic carbonate platform (Calcari Grigi, Southern Alps, Italy). Facies 47, 57–82.

Moro, A., Tunis, G., Mezga, A., Tarlao, A., Cosovic, V., 2007. Depositional environments of upper Cenomanian limestones with rudists and dinosaur footprints, Istria, Croatia. In: Scott, R.W. (Ed.), Cretaceous Rudists and Carbonate Platforms; Environmental Feedback. SEPM Spec. Publ., vol. 87, pp. 37–44.

Myint, M., 2001. *Psilonichnus quietis* isp. nov. from the Eocene Iwaki Formation, Shiramizu group, Joban coal field, Japan. Ichnos 8, 1–14.

Narbonne, G.M., 1984. Trace fossils in Upper Silurian tidal flat to basin slope carbonates of Arctic Canada. J. Paleontol. 58, 398–415.

Nesbitt, E.A., Campbell, K.A., 2006. The paleoenvironmental significance of *Psilonichnus*. Palaios 21, 187–196.

Nielsen, A.T., 2004. Ordovician sea level changes: a Baltoscandian perspective. In: Webby, B.D., Paris, F., Droser, M.L., Percival, G. (Eds.), The Great Ordovician Biodiversification Event. Columbia University Press, New York, pp. 84–93.

Olóriz, F., Rodríguez-Tovar, F.J., 2002. Trace-fossils and minor discontinuities in a marl limestone rhythmite, Lower–Middle Kimmeridgian, southern Spain. Geobios 35, 581–593.

Orviku, K.K., 1960. O litostratigraphii volkhovskogo (BII) i kundaskogo (BIII) gorizontov nizhnego ordovika Estonii [On the lithostratigraphy of the Volkhovian (BII) and Kundan (BIII) stages of the lower Ordovician of Estonia]. Eesti NSV Teaduste Akad. Geol. Inst., Uurimused 5, 45–88 (in Russian).

Osgood Jr., R.G., 1970. Trace fossils of the Cincinnati area. Plaeont. Am. 6, 281–444.

Pedley, M., 1992. Bio-retexturing: early diagenetic fabric modifications in outer-ramp settings—a case study from the Oligo-Miocene of the Central Mediterranean. Sediment. Geol. 79, 173–188.

Pemberton, S.G., Frey, R.W., 1985. The *Glossifungites* Ichnofacies: modern examples from the Georgia Coast, U.S.A. In: Curran, H.A. (Ed.), Biogenic Structures: Their Use in Interpreting Depositional Environments. SEPM Spec. Publ., vol. 35, pp. 237–259.

Pemberton, S.G., Gingras, M.K., 2005. Classification and characterizations of biogenically enhanced permeability. AAPG Bull. 89, 1493–1517.

Pemberton, S.G., Spila, M., Pulham, A.J., Saunders, T., MacEachern, J.A., Robbins, D., Sinclair, I.K., 2001. Ichnology & sedimentology of shallow to marginal marine systems: Ben Nevis & Avalon Reservoirs, Jeanne d'Arc Basin. Geol. Ass. Can., Short Course Notes 15, 343 pp.

Pervesler, P., Uchman, A., 2009. A new Y-shaped trace fossil attributed to upogebiid crustaceans from Early Pleistocene of Italy. Acta Palaeont. Polon. 54, 135–142.

Petti, F.M., D'Orazi Porchetti, S., Conti, M.A., Nicosia, U., Perugini, G., Sacchi, E., 2008. Theropod and sauropod footprints in the Early Cretaceous (Aptian) Apenninic carbonate platform (Esperia, Lazio, central Italy): a further constraint on the palaeogeography of the central Mediterranean area. In: Avanzini, M., Petti, F.M. (Eds.), Italian Ichnology. Studi Trentini di Scienze Naturali, Acta Geologica, vol. 83, pp. 323–334.

Pickerill, R.K., Forbes, W.H., 1979. Ichnology of the Trenton Group in the Quebec City area. Can. J. Earth Sci. 16, 2022–2039.

Pickerill, R.K., Fillion, D., Harland, T.L., 1984. Middle Ordovician trace fossils in carbonates of the Trenton Group between Montreal and Quebec City, St. Lawrence Lowland, Eastern Canada. J. Paleontol. 58, 416–439.

Posamentier, H.W., Allen, G.P., 1999. Siliciclastic Sequence Stratigraphy: Concepts and Applications. SEPM Concepts in Sedimentology and Paleontology, vol. 7, 210 pp.

Purser, B.H. (Ed.), 1973. The Persian Gulf, Holocene Carbonate Sedimentation and Diagenesis in a Shallow Epicontinental Sea. Springer, New York, 471 pp.

Radwański, A., Wysocka, A., Górka, M., 2012. Miocene burrows of the Ghost Crab *Ocypode* and their environmental significance (Mykolaiv Sands, Fore-Carpathian Basin, Ukraine). Acta Geol. Polon. 62, 217-229.

Read, J.F., 1985. Carbonate platform facies models. AAPG Bull. 69, 1–21.

Riegl, B., Poiriez, A., Janson, X., Bergman, K.L., 2010. The Gulf: facies belts, physical, chemical, and biological parameters of sedimentation on a carbonate ramp. In: Westphal, H., Riegl, B., Eberli, G.P. (Eds.), Carbonate Depositional Systems: Assessing Dimensions and Controlling Parameters. Springer, Berlin, pp. 145–213.

Rodríguez-Tovar, F.J., Pérez-Valera, F., Pérez-López, A., 2007. Ichnological analysis in high-resolution sequence stratigraphy: The *Glossifungites* ichnofacies in Triassic successions from the Betic Cordillera (southern Spain). Sediment. Geol. 198, 293–307.

Schlager, W., 2005. Carbonate sedimentology and sequence stratigraphy. SEPM Concepts in Sedimentology and Paleontology, vol. 8, 200 pp.

Schweigert, G., 1998. Die Spurenfauna des Nusplinger Plattenkalks (Oberjura, Schwäbische Alb). Stuttgarter Beitr. Naturk. B 262, 1–47.

Seilacher, A., 1967. Bathymetry of trace fossils. Mar. Geol. 5, 413–428.

Seilacher, A., 2007. Trace Fossil Analysis. Springer, Berlin, 226 pp.

Shinn, E.A., 1968. Burrowing in Recent lime sediments of Florida and the Bahamas. J. Paleontol. 42, 879–894.

Shinn, E.A., 1972. Worm and algal-built columnar stromatolites in the Persian Gulf. J. Sediment. Petrol. 42, 837–840.

Strohmenger, C.J., Al-Mansoori, A., Al-Jeelani, O., Al-Shamry, A., Al-Hosani, I., Al-Mehsin, K., Shebl, H., 2010. The sabkha sequence at Mussafah Channel (Abu Dhabi, United Arab Emirates): Facies stacking patterns, microbial-mediated dolomite and evaporite overprint. GeoArabia 15, 49–90.

Tedesco, L.P., Wanless, H.R., 1991. Generation of sedimentary fabrics and facies by repetitive excavation and storm infilling of burrow networks, Holocene of South Florida and Caicos Platform, B.W.I. Palaios 6, 326–343.

Tudhope, A.W., Scoffin, T.P., 1984. The effects of *Callianassa* bioturbation on the preservation of carbonate grains in Davies Reef Lagoon, Great Barrier Reef, Australia. J. Sediment. Petrol. 54, 1091–1096.

Twitchett, R.J., Wignall, P., 1996. Trace fossils and the aftermaths of the Permo-Triassic mass extinction: evidence from northern Italy. Palaeogeogr. Palaeoclimatol. Palaeoecol. 124, 137–151.

Uchman, A., Bubniak, I., Bubniak, A., 2000. The *Glossifungites* ichnofacies in the area of its nomenclatural archetype, Lviv, Ukraine. Ichnos 7, 183–193.

Vannini, M., 1980. Researches on the coast of Somalia, the shore and the dune of Sar Uanle. It. J. Zool. N. Ser. 13, 11–44.

Vishnjakov, S.G., Hecker, R.F., 1937. Sledy razmyva I vnutriplastovye narushenija v glaukonitovykh izvestnjkakh nizhnego silura Leningradskoi oblasti [Traces of erosion and synsedimentary disturbances in the Lower Silurian "Glauconite Limestone" of the Leningrad Region]. Jubilejny sbornik v chest' N.F. Pogrebova, 30–45 (in Russian).

Walkden, G., Williams, A., 1998. Carbonate ramps and the Pleistocene-Recent depositional systems of the Arabian Gulf. In: Wright, V.P., Burchette, T.P. (Eds.), Carbonate Ramps. Geol. Soc. London, Spec. Publ., vol. 149, pp. 43–53.

Webb, A.P., Eyre, B.D., 2004. Effect of natural populations of burrowing thalassinidean shrimp on sediment irrigation, benthic metabolism, nutrient fluxes and denitrification. Mar. Ecol. Progr. Ser. 268, 205–220.

Weber, B., Steiner, M., Zhu, M.-Y., 2007. Precambrian-Cambrian trace fossils from the Yangtze Platform (South China) and the early evolution of bilaterian lifestyles. Palaeogeogr. Palaeoclimatol. Palaeoecol. 254, 328–349.

White, B., Curran, H.A., 1988. Mesoscale physical sedimentary structures and trace fossils in Holocene carbonate eolianites from San Salvador Island, Bahamas. Sediment. Geol. 55, 163–184.

Wilson, M.A., Palmer, T.J., 1994. A carbonate hardground in the Carmel Formation (Middle Jurassic, SW Utah, USA) and its associated encrusters, borers and nestlers. Ichnos 3, 79–87.

Wisshak, M., 2012. Microbioerosion. In: Knaust, D., Bromley, R.G. (Eds.), Trace Fossils as Indicators of Sedimentary Environments. Developments in Sedimentology, vol. 64. Elsevier, Amsterdam, pp. 213–243.

Chapter 24

Reefs and Mounds

Leif Tapanila*,†,1 and Pat Hutchings‡
*Department of Geosciences, Idaho State University, Pocatello, Idaho, USA,
†Earth Sciences Division, Idaho Museum of Natural History, Pocatello, Idaho, USA,
‡Australian Museum, Sydney, New South Wales, Australia
[1]Corresponding author: e-mail: tapaleif@isu.edu

1. INTRODUCTION

Traces in reef environments are produced in hard substrates, including the skeletons of reef constructors, associated fauna, microbial carbonate crusts, and early carbonate cements. A multitude of organisms use mechanical and chemical means to grind, scrape, bore, gouge, and etch hard materials, thus leaving cavities that record their behavior. A separate process of embedment allows symbiotic organisms to be "grown around" by a living host skeletonized animal, thereby making a bioclaustration cavity. These hard-substrate trace fossils in reefs record two fundamental ethologies: dwelling and grazing structures, the former include macroborings (>1 mm), microborings (<1 mm), and bioclaustrations, whereas the latter are simply called "grazings" (de Gibert et al., 2004; Tapanila, 2005).

The diverse fossil record of reef ecosystems is well studied in terms of evolutionary innovation and biodiversity (Kiessling et al., 2010); yet, the literature on fossil reef ichnology is sparse when compared to the vast literature on trace fossils in non-reef depositional settings. Two factors may account for the differences: the hard substrates of reefs restrict the types of traces to borings, etchings, and bioclaustrations, and relegate most soft-sediment burrows to off-reef facies; and observation of traces in reef constructors often requires significant preparation (cutting, thin sections, epoxy casting, etc.), thus making their study more deliberate and laboratory intensive. Hard-substrate traces have received increasing interest from ichnologists (Schönberg and Tapanila, 2006a), but much remains to be understand about the role and distribution of tracemakers in ancient reefs.

Developments in Sedimentology, Vol. 64. http://dx.doi.org/10.1016/B978-0-444-53813-0.00024-1

2. TRACES AND TRACEMAKERS OF THE REEF

2.1 Macroborings and Bioclaustrations

The ichnoguild concept of Bromley (1990, 1996) (see also the guilds for biotaxa: Bambach, 1983) provides an ecological basis for categorizing hard-substrate dwelling structures by combining the traits of ethology, vagility, and trophic level to unite the activities of disparate phyla (Tapanila, 2008b). In this framework, all hard-substrate dwelling cavities made in reef settings are constructed by two superguilds: the heterotrophic endolithic guild, which includes all macroborings, bioclaustrations, and a few heterotrophic microborings; and the autotrophic endolithic guild, which includes all algal and cyanobacterial microborings. The latter comprise a significant aspect of reef ichnology, and they are treated thoroughly by Wisshak (2012).

Heterotrophic endoliths either bore or embed their way into a substrate, although a combination of processes may be used in live skeletal substrates. A wide variety of organisms have been documented as macroborers, including sponges, bivalve molluscs, sipunculans, polychaetes, and phoronids (Bromley, 1978; Hutchings, 1986).

The manifold ecological similarities among animals of the heterotrophic endolithic guild permit some generalizations. The great majority of macroborings and all bioclaustrations are produced for the primary function of dwelling, and are constructed by sessile, soft bodied, suspension feeders (predatory borings, such as *Oichnus*, are important exceptions). Adoption of a lifestyle confined within a hard substrate cavity precludes locomotion, and has minimized the need for a protective exoskeleton, thus, with the exception of the rasping shell of pholadid bivalves, most heterotrophic endoliths have reduced, or lack, external hard parts. As suspension feeders, environmental conditions often dictate where heterotrophic endoliths will flourish. It is generally accepted that increased nutrients (e.g., oligo- and eutrophic waters) benefit endolithic activity, whereas increased siliciclastic input may inhibit some endoliths by reducing the efficiency of feeding currents or by enhancing the burial of their substrate (Edinger and Risk, 1994).

By contrast to their shallow-marine coralline substrates, heterotrophic endoliths are eurytopic with respect to water temperature, turbidity, and light intensity (Cortes and Risk, 1985). Macroborers of dead substrates are not usually limited by substrate type, although factors that increase exposure time of the substrate to the water column, such as substrate height above the sea floor, often encourage greater intensities of bioerosion (Tapanila et al., 2004). Endoliths of live coralline substrates often show some degree of host specificity, especially among embedding symbionts that make bioclaustrations (Tapanila, 2005).

The most recent literature reviews on the history of hard substrate tracemakers and their trace fossils are the following: Taylor and Wilson (2003) and Bromley (2004) for macroborings, Glaub and Vogel (2004) and Wisshak (2012) for microborings, and Tapanila (2005, 2008a) for bioclaustrations.

2.1.1 Worms

The polyphyletic group of morphologically similar "worm" borings refers especially to cavities produced by polychaete annelids and sipunculans, which according to the literature are mainly cosmopolitan in distribution, but are more likely to represent suites of cryptic species. Diverse boring polychaetes are found on modern reefs belonging to several families that are not closely related, suggesting that boring evolved several times during the evolution of this old group. The dominant families with boring representatives are the Eunicidae, Dorvilleidae, Sabellidae, Cirratulidae, and the Spionidae. While the precise methods of boring are still being investigated, two main methods appear to be involved, chemical dissolution and/or mechanical removal of the substrate (Hutchings, 2008). A well-defined succession of borers occurs, with initial colonizers being short-lived species followed by long-lived species (Hutchings, 2008).

As polychaetes are soft-bodied animals, their fossil record is sparse. However, the Dorvilleidae have well-developed jaws that are well represented as scolecodonts, starting in the Early Ordovician (Rouse and Pleijel, 2001) and closely resemble modern-day boring species. The diversity of scolecodonts and hence the clade Eunicida increased dramatically through the Ordovician, followed by a major extinction at the end of the Permian (Szaniawski, 1966). This raises the possibility that boring polychaetes were present in fossil reefs.

The most common fossil worm macroborings have a simple cylindrical form (*Trypanites*; Fig. 1A–C) and may have a distal enlargement (*Palaeosabella*). Although *Trypanites* is the oldest macroboring in the fossil record, appearing in the Early Cambrian (James et al., 1977), Bromley (2004) points out that distinguishing it from *Palaeosabella* requires good exposure of the distal end, which may result in underreporting of *Palaeosabella*. Taylor and Wilson (2003) recognize the earliest *Palaeosabella* in the Ordovician. Sipunculans and polychaetes are most frequently cited as potential tracemakers of *Trypanites* and *Palaeosabella*.

Spionid polychaetes are responsible for modern equivalent structures of *Caulostrepsis* and *Maeandropolydora* (Bromley and D'Alessandro, 1983). Both fossils include a U-shaped pouch, but whereas *Caulostrepsis* is stouter with a single aperture, *Maeandropolydora* has elongated curved tunnels that lead to more than one aperture. The type material of *Caulostrepsis* is Devonian in age, but the origin of *Maeandropolydora* is contentious, with some evidence pointing to a Silurian–Devonian beginning (Bromley, 2004), or as late as the Cretaceous (Taylor and Wilson, 2003).

Talpina is a network of curved, branching tubular cavities that most easily could be mistaken for sponge borings. Phoronid lophophorates generate such cavities today, and their fossil record extends to the Devonian, and they are most common starting in the Jurassic (Taylor and Wilson, 2003).

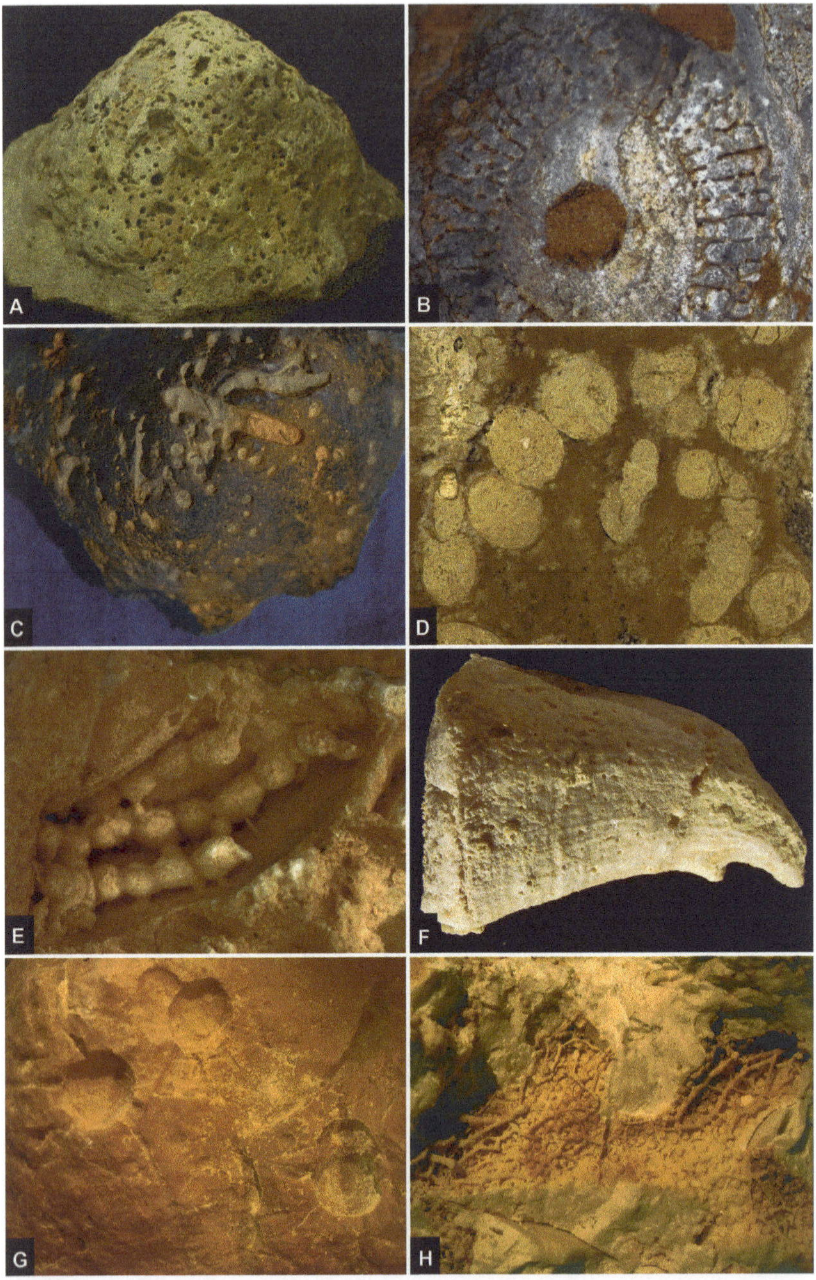

FIGURE 1 Bioerosion in fossil reefs. (A) Monotaxic bioerosion of *Trypanites* in a stromatoporoid skeleton, off-reef facies, Late Silurian, Gotland, Sweden. Width of image is 20 cm. (B) *Trypanites* in early cemented fill within archaeocyath sponge, Early Cambrian, Labrador. Width of image is 1 cm. (C) Three elongate *Petroxestes pera* and several circular *Trypanites* apertures in a stromatoporoid,

Most ancient bioclaustrations are thought to be made by polychaete worms, as are many modern equivalents, for example, Serpulidae and Sabellariidae (Nishi and Nishihira, 1999). Much of the fossil literature on symbiotic bioclaustrations is confined to coral- and sponge-hosted associations. The earliest known coral-hosted bioclaustration, *Chaetosalpinx*, is also the most pervasive ichnogenus of the Paleozoic. *Chaetosalpinx* is a straight cavity formed by the interaction of a symbiont residing for up to 5 years in the skeleton of *Columnopora* tabulate coral hosts from Late Ordovician patch reefs in eastern Canada (Tapanila, 2002). *Helicosalpinx* is a spiraled tube in tabulate corals and stromatoporoids from biostromes of the Late Ordovician to Late Devonian (Tapanila, 2004). *Hicetes* is a complex spiraled U-shaped bioclaustration associated with Devonian *Pleurodictyum* tabulate corals from muddy-bottom facies (Brett and Cottrell, 1982). Bioclaustrations show strong host specificity—from the family to genus level—and their stratigraphic record mirrors that of their preferred host clade (Tapanila, 2004, 2008a).

2.1.2 Bivalves

Within the bivalve molluscs, boring species are found in several families on modern reefs. The families Petricolidae, Pholadidae, and Clavagellidae are represented by only a few species and genera, but by far most species boring into dead coral belong to the Lithophaginae and the Gastrochaenidae. Species restricted to boring into live coral belong to the Mytilidae and some genera of the Lithophaginae. Some species of *Lithophaga* can live in several coral species, whereas others are restricted to a single species. Host specificity among bivalve borers may indicate how they evolved with respect to their host-coral substrates. For example, the composition of corals found in the Atlantic and Indo-Pacific differ with only 7% generic overlap and none at the specific level. The less specialized, dead coral-boring bivalves that occur in several species are restricted to the coral genera that have wide geographical distributions, whereas the boring species that are restricted to live coral are found only in coral species that have limited distributions. This suggests that boring bivalves evolved as the corals themselves diversified (Morton, 1990; Rosen, 1984). In addition to bivalve molluscs, there are some species of modern gastropods (all in the family Coralliophilidae) that can bore into living corals (Soliman, 1969), although fossil gastropod borings have yet to be found.

Petroxestes pera is a slot-shaped boring with a wide aperture found in Ordovician and Silurian stromatoporoids and hardgrounds (Tapanila and Copper,

off-reef facies, Early Silurian, Goéland Member, Jupiter Formation, Anticosti Island, Québec, Canada. Width of image is 8 cm. (D) *Gastrochaenolites* oriented parallel to scleractinian corallites, Oligocene, Lares Quarry, Puerto Rico. Width of image is 9 cm. (E) Multichambered *Entobia retiformis* in reef rubble; Miocene, Rio Guajataca locality, Puerto Rico. Width of image is 1.7 cm. (F) *Entobia* in rudist bivalve, Cretaceous, Huqf area, Oman. Width of image is 7 cm. (G) Large chambers of *Entobia devonica* in a stromatoporoid, Devonian, Iowa. Width of image is 8 cm. (H) Bryozoan borings in scleractinian coral, Miocene, Rio Guajataca locality, Puerto Rico. Width of image 5 cm. (Photos: L. Tapanila, except E, which was provided by D. Knaust.)

2002; Wilson and Palmer, 1988; Fig. 1C). Whitfield's (1893) discovery of *Corallidomus scobina* mytilid bivalves preserved inside *Petroxestes* borings provides unusual, but compelling, evidence for the tracemaker identity (Pojeta and Palmer, 1976). By contrast, the affinity of the earliest *Gastrochaenolites* is not as straightforward.

Gastrochaenolites is an elongate boring that enlarges toward the base, and has been described variously as pouch shaped, flask shaped, or clavate (Kelly and Bromley, 1984; Fig. 1D). The distinct morphology of the cavity corresponds to a tracemaker that grows in width as it deepens its boring into the substrate. The ichnogenus is most frequently cited as a product of bivalve bioerosion, a claim validated by modern activities of gastrochaenid and pholadid bivalves. Fossil borings often preserve the tracemaking bivalve inside the cavity (although "nestling" fauna could be mistaken for a tracemaker: e.g., Kelly, 1980) and in the instance of pholadid mechanical borings, bioglyphic ornament from the rasp-like shell adorn the internal surface of the boring (Evans, 1970). The earliest record of the *Gastrochaenolites* morphology comes from Early Ordovician hardgrounds, although these are unlikely to be the product of bivalves (Benner et al., 2004; Ekdale and Bromley, 2001). *Gastrochaenolites* preserved in reef substrates are absent in the fossil record until the Jurassic.

2.1.3 Sponges

Members of several orders of sponges bore into modern-day coral substrates, including Hadromerida (Clionaidae, perhaps Spirastrellidae, and Alectonidae), Poecilosclerida (Acarnidae, *viz.*, the genus *Zyzzya*), Halichondrida (Halichondriidae, *viz.*, the genus *Amorphinopsis*), and Haplosclerida (Phloeodictyidae, *viz.*, the genus *Aka*) and there are a few "maybe eroders" in other Orders (Hooper and Van Soest, 2002; Christine Schönberg, Australian Institute of Marine Science, personal communication, 2011). Sponges bore by the chemical etching of the substrate and the mechanical removal of the resultant sponge chips (for a review on sponge boring, see Schönberg, 2008). Rates of boring vary according to species of sponge, substrate density (Edinger and Risk, 1997; Highsmith et al., 1983), location, and depth (López-Victoria and Zea, 2005), with most material being removed from massive, less porous coral species (Buddemeier et al., 1974). Rates vary over the lifetime of the sponge colony, with highest rates of growth and erosion exhibited by newly settling larvae or sponge fragments and the lowest rates by established colonies (Neumann, 1966; Rützler, 1975).

Entobia consists of one or more chambers interconnected by narrow tunnels and leading to multiple apertures at the surface of the substrate. Much of the *Entobia* excavation occurs below the substrate surface, which often obscures the magnitude of substrate removal. Several ichnospecies are named under *Entobia* (see reviews by Bromley and D'Alessandro, 1984, 1989), and can be categorized—for the sake of simplicity—within two end-member morphologies: those colonies comprised of many small chambers and those with few large chambers.

Small-chambered *Entobia* fossils (Fig. 1E and F) are frequently considered the work of clionaid sponges (e.g., *E. cretacea*, *E. ovula*, *E. retiformis*, and *E. laquea*). Borings resembling *Entobia* have been ascribed to sponge activity several times from Paleozoic rocks, from Cambrian archaeocyathid reefs (Kobluk, 1981) to Ordovician trilobite carapaces (Lindström, 1979). Despite their presence, small-chambered *Entobia* are either underreported or rare until the Jurassic, when they became a standard component of the bioeroding community (Bromley, 2004).

Large-chambered *Entobia devonica* (Fig. 1G), with a structure analogous to cavities formed by *Aka*, first appear in Devonian reef and peri-reef facies, but disappear following the Frasnian (Schönberg and Tapanila, 2006b; Tapanila, 2006). The large-chambered entobians return by the Early Cretaceous and are part of modern bioerosion assemblages (e.g., *E. solaris* and *E. gigantea*: Bromley and D'Alessandro, 1984; Mikuláš, 1992).

2.1.4 Barnacles and Bryozoans

Twenty-four genera and 102 nominal species of barnacles have been recorded as boring and embedding into live coral, of which 67 species are extant (Ross and Newman, 2002), and traditionally they have been placed within the Pyrgomatidae. A recent molecular study by Simon-Blecher et al. (2007) has suggested that they are not a monophyletic group and at least one genus is far closer to free-living barnacles, suggesting that the ability to bore and embed into coral has evolved more than once in the group.

The ancient record of acrothoracican barnacles starts at least by Devonian time, and may extend back to the Silurian (Bromley, 2004). *Rogerella*, a small pouch-shaped boring with a slit-like aperture, is the most commonly referenced ichnogenus. In fossil reef substrates, barnacle borings are usually less abundant and cause less damage than sponge, worm, and bivalve excavators, but important exceptions may occur in reefs from deeper or cooler water settings (Perry and Bertling, 2000).

Borings by ctenostome bryozoans (Fig. 1H) are known from Late Ordovician to modern carbonate substrates and consist of elongate cavities connected by narrow canals, which correspond to zooids and stolons (e.g., *Ropalonaria*: Pohowsky, 1978). Cheilostome bryozoans produce shallow etchings and pits on carbonate substrates, and are recognized from as early as the Early Cretaceous (e.g., *Leptichnus*: Taylor et al., 1999).

2.2 The Grazers

External erosion or grazing also occurs on both live and dead coral but by different organisms. Scarids or parrotfish, now regarded as belonging to the Labridae family (Cowman et al., 2009), are important grazers on modern-day reefs (Fig. 2C). Recent studies by Bellwood (2003) have shown that they can be divided based on jaw morphology into excavators that remove pieces

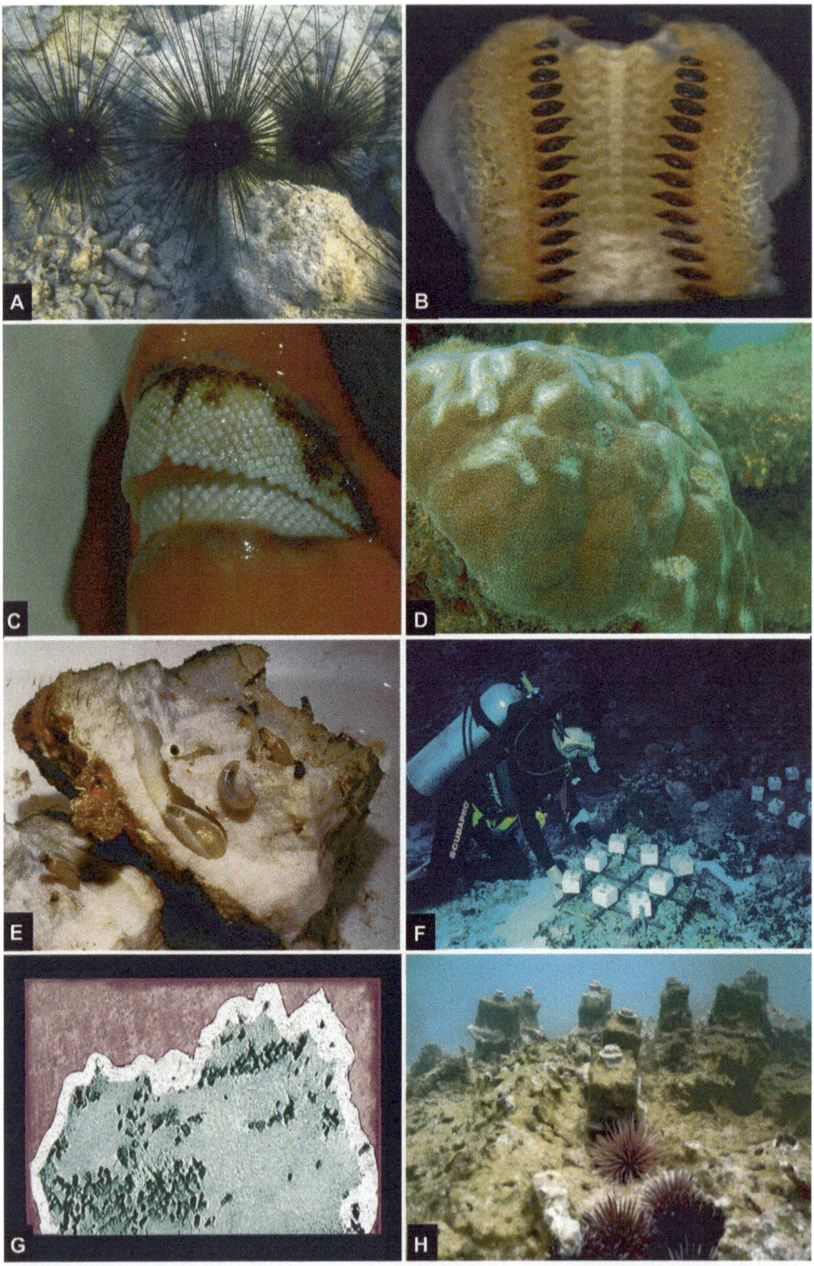

FIGURE 2 Tracemakers and bioerosion in modern reefs. (A) *Diadema setosum*, a grazing echinoid which causes major erosion of reefs in the western Indian Ocean (photo: O. Hoegh-Guldberg). (B) Radula of chiton, *Acanthopleura hirtosa* (photo: Shaw et al., 2010, reprinted with permission). (C) Jaws of *Bolbometopon muricatum* on the outer barrier near Lizard Island (photo: D. Bellwood). (D) Scrape marks of scarids and a boring barnacle embedded in live *Porites* (photo: P. Hutchings).

of the substrate and scrapers that have a non-excavating bite that removes only the surface of the substrate (Fig. 2D). Although most scarids feed on dead coral substrates, the large *Bolbometopon muricatum* is an excavator that feeds heavily on live coral, especially on species of *Acropora*, and this can constitute over 50% of its diet on the Great Barrier Reef. Recent studies in the Caribbean also have suggested that such feeding on live coral is far more widespread than previously thought (Rotjan and Lewis, 2005).

Echinoids are important grazers of dead coral substrate (Asgaard and Bromley, 2008; Peyrot-Clausade et al., 2000) and on reefs that have been impacted by overfishing or eutrophication. Dense populations of echinoids can develop, which leads to high rates of grazing and substantial loss of reef framework (Pari et al., 2002) (Fig. 2A and H). Grazing chitons (Barbosa et al., 2008) and gastropods (Lam et al., 2007; Shafir et al., 2008) may also be locally abundant and graze on dead coral substrate (Fig. 2B).

All these grazers of both live and dead coral are feeding on the endolithic algae that live just underneath the surface of the coral substrate, as well as epilithic algae on the surface of dead coral substrate. Specialized mouth parts include jaws (scarids, Fig. 2C), the grinding plates of Aristotle's lantern (echinoids), and the radulae (chitons and gastropods; Fig. 2B) to scrape off the epilithic algae and the top layers of the substrate, which contains a network of endolithic algae. Deep excavating species of scarid fish have well-developed plates within their pharynx that grind the coral into a fine powder, which helps in breaking down the walls of the algae releasing their cell contents. Fine sediment is defecated and contributes to lagoonal sediments. The specialized mouth parts of echinoids, gastropods, and chitons grind the coral into a fine powder, which releases the algal cells, and it is presumed that they have the necessary enzymes to break down the plant cell walls, as experiments show for fish (Choat et al., 2004).

Damage from grazing invertebrate animals is typically shallow and unlikely to be preserved. Grazing gastropods and chitons leave radular scratches on limestone and skeletal substrates (*Radulichnus*), and regular echinoids can scrape a set of five radial grooves (*Gnathichnus*) using their Aristotle's lantern.

Trace fossils ascribed to grazing by scarid fish have yet to be named formally, but fossils of parrotfish are well preserved on Cenozoic reefs, the oldest ones including a species of *Bolbometopon*, an eroder of Miocene age (Bellwood and Schultz, 1991).

(E) *In-situ* dead coral habitat split open to reveal boring bivalves and sipunculans (photo: P. Hutchings). (F) Experimental study of bioerosion at Osprey Reef, Coral Sea, two replicate grids with newly laid coral blocks to be exposed for varying periods of time (photo: J. Johnson). (G) Diagrammatic representation of a coral block illustrating how the various components of bioerosion (grazing, boring, accretion) are measured by slicing experimental blocks into a series of sections. Knowing the density of the block, measurements can be scaled up to rates per square meter and net rates of bioerosion can be calculated (a, original block; b, accretion; c, block remaining after grazing and boring) (image: K. Attwood). (H) Experimental blocks after 6 months' exposure at Faaa, Tahiti, illustrating extensive grazing by *Echinometra mathaei* (photo: M. Peyrot-Clausade).

3. REEF ICHNOLOGY THROUGH TIME

3.1 Fossil Reefs

Most fossil reef studies are qualitative, but some have attempted quantitative methods to describe and contrast bioerosion with the modern record. This section attempts to summarize key trends in fossil reef ichnology through the Phanerozoic eras.

3.1.1 Paleozoic

Archaeocyathid calcareous sponges are the first metazoan animals to produce reefs, and they preserve the earliest evidence of macrobioerosion in the fossil record (James et al., 1977). Early Cambrian archaeocyathan reefs from Labrador (eastern Canada) contain the simple cylindrical boring *Trypanites*, which penetrates the skeletal fabric of the framebuilders as well as the early cemented carbonate muds that surround and infill the central osculum of the sponges (Fig. 1B).

With few exceptions, *Trypanites* borings are the most common Paleozoic reef (and non-reef) hard-substrate trace fossil. After their first appearance in the Early Cambrian, and an absence during the Middle Cambrian, *Trypanites* are next recorded in Late Cambrian and Early Ordovician hardgrounds (e.g., Chow and James, 1992; Rozhnov, 1994). By the Middle to Late Ordovician, the endolithic habit expands in parallel with the Great Ordovician Biodiversity Event (Webby et al., 2004). Termed the Ordovician Bioerosion Revolution (Wilson and Palmer, 2006), dwelling cavities made by macroborers in reef substrates increase in abundance and intensity during the Middle Ordovician (e.g., Kapp, 1975), and by the end of the Ordovician, the diversity of ichnotaxa includes seven distinct forms. Despite the increased diversity in ichnogenera among various carbonate substrates, reef macrobioerosion remains monotonously dominated by *Trypanites*.

Studies on the Late Ordovician to Early Silurian carbonate ramp preserved on Anticosti Island (eastern Canada) exemplify the early Paleozoic pattern of *Trypanites*-dominated bioerosion. Tapanila et al. (2004) examined more than 2500 coral and stromatoporoid sponge substrates from patch reef and off-reef facies, examining boring frequency and intensity. *Trypanites* accounted for $>99.9\%$ of all macroborings, with rare off-reef occurrences of *Petroxestes pera*. Although an average of 30% of reef constructors had been bored, corals and stromatoporoids living in off-reef settings were more frequently bored (40–50%) and with greater intensity (up to 35 *Trypanites* per 4 cm^2 of substrate surface) (Tapanila et al., 2004). Increased postmortem exposure time above the sediment/water interface was determined to be the most important environmental control to promote bioerosion in a substrate. The relatively lower frequency of borings in reef constructors, as compared to off-reef corals and stromatoporoids, may be due to increased competition for space in the reef from encrusters, coupled with the flatter growth forms that dominate Anticosti's reef substrates. Tapanila et al. (2004) found that, regardless of facies, tabular (flatter) growth forms of corals and stromatoporoids are bored nearly 30% less than high domical morphologies. Ebbestad and Tapanila (2005) further quantified the relationship between *Trypanites* borings and absolute height of a substrate, finding that the boring frequency increases as a logarithmic function of substrate height, with an asymptote (maximum boring frequency per substrate) reached at 12 cm above the sea floor (also see Nield, 1984, for observations on *Trypanites* frequency as a function of substrate topography). The boring frequency of *Trypanites* also shows a positive correlation with the substrate density (the inverse of substrate porosity). Tapanila et al. (2004) speculated that a denser substrate affords the borer greater control on the final morphology of the borehole, as well as increased durability of the substrate. Similar concepts have been suggested for bioerosion of modern reefs (Hutchings, 1986).

Macroborings are notably rare in the Silurian stromatoporoid reefs on Gotland, Sweden (e.g., Hoburgen-type reefs studied by Segars and Liddell, 1988), but in the muddy-bottom Visby beds, *Trypanites*, *Palaeosabella*, and the large *Osprioneides* are abundant (Beuck et al., 2008; Kershaw, 1993). Rugose coral and stromatoporoid reefs of the Middle to Late Devonian are abundant globally (Copper, 2002) and bioerosion is prevalent. In addition to the usual *Trypanites*, a newcomer joins the reef and lagoonal macroboring assemblage: *Entobia devonica* (formerly *Topsentopsis devonica*, see Tapanila, 2006, for a taxonomic redescription). This large, usually single-chambered sponge boring resides 1 cm or more below the outer substrate surface, and therefore most of the erosion by this boring is concealed deep within the substrate. The Frasnian reef complexes of western North America (Alberta to Nevada) tend to preserve *Entobia devonica* more frequently than *Trypanites* borings, whereas Wood's (2000) investigation of the Australian Canning Basin reefs (late Frasnian) only reports a low abundance of *Trypanites* and no *Entobia* borings.

Following the end-Frasnian decline of reef ecosystems, the Paleozoic bioerosion record in reefs similarly decline (Webb, 2002). Scant records of *Trypanites* in microbial buildups and some Carboniferous reefs demonstrate that the endolithic guild remained active, but poorly studied (Wahlman, 2002; Webb, 1987, 1998). Permian reef systems are similarly poorly studied with the exception of Weidlich (1996), who identifies several morphologies of macroboring in rugose corals from Oman. *Trypanites* and some of the earliest small-chambered *Entobia* sponge borings occur frequently in fasciculate rugosans. The observation of *Entobia* near the end of the Permian is significant in that it marks a transition toward the Meso-Cenozoic domination of clionaid borers in reef ecosystems.

3.1.2 Mesozoic–Cenozoic

Perry and Bertling (2000) most recently reviewed the literature on Mesozoic and Cenozoic macrobioerosion in reefs. Following the Permian/Triassic mass extinction, non-scleractinian reefs of the Early to Middle Triassic show little influence from macroborers, and are dominated by borings from fringe elements, the barnacles and phoronids—the latter are unknown in modern reefs, but common off-reef. The rapid diversification of scleractinians in the Middle Triassic establishes their future dominance as reef constructors (Flügel, 2002), while the first fully endolithic bivalves (i.e., lithophagids and gastrochaenids) adapt to a reef destroyer lifestyle (Carter and Stanley, 2004; Flügel and Senowbari-Daryan, 2001). From the Late Triassic onward, bivalves become a significant component of the macroboring community, producing trace fossils belonging to the ichnogenus *Gastrochaenolites*. The few studies on Late Triassic reef bioerosion show boring frequencies on a par with Ordovician–Silurian reef values. For example, 30% of the corals and spongeliomorphids in latest Triassic (Rhaetian) Austrian reefs show evidence of worm and bivalve macroborings (Perry and Bertling, 2000; Stanton and Flügel, 1989). A general reef trend through the Mesozoic is a transfer of dominance from "worm" borers to bivalve borers, although local environmental conditions may upset this trend (Perry and Bertling, 2000). Debate remains over the earliest clionaid sponge, *Entobia*, recognized by small, multichambered borings (Late Permian: Weidlich, 1996; Late Triassic: Perry and Bertling, 2000; Early Jurassic: Fraser et al., 2004), but it is evident that they do not become significant bioeroders until the Late Jurassic.

Jurassic reef studies more frequently comment on bioerosion as an important and common process, with bivalve and worm borings being most abundant, followed by sponge excavations (Perry and Bertling, 2000). Warme (1977) illustrates the great intensity of Middle Jurassic excavation of scleractinian skeletons by deeply penetrating (lower tier) *Gastrochaenolites* and shallow (upper tier) *Entobia*. Grazers, including echnioids, first appear in reefs and associated microbial crusts during the Late Jurassic (Leinfelder, 1986).

Bivalves and worms continue to be the most abundant contributors to coral reef erosion during the Early Cretaceous, although few studies are available to draw upon (Perry and Bertling, 2000). The Late Cretaceous rudist buildups, by contrast, show increased sponge bioerosion, but the relative dominance among bivalves, "worms", and sponges appears to correlate with environmental factors (shelf position and siliciclastic input) rather than an intrinsic quality of the bioeroding community (Perry and Bertling, 2000).

Following the end-Cretaceous mass extinction, trace fossils from reef environments show the same morphologies, demonstrating the resilience and conservatism of the marine bioeroders. Bivalve borings continue to dominate over those made by worms and sponges during the Paleocene–Eocene interval. But, by the Late Oligocene and Miocene, reef bioerosion assemblages become dominated by sponges over bivalves and worms. The zonation of bioerosion across reef facies shows a greater intensity and diversity of bioeroders in back-reef and lagoon facies and lower intensity and diversity at the reef crest and fore-reef sites (Edinger and Risk, 1994; Frost et al., 1983; Perry, 1996; Pleydell and Jones, 1988). The composition and zonation of the reef bioeroding community that is most prevalent in Quaternary oceans was thereby established already roughly 24 million years ago (Perry, 2000; Perry and Bertling, 2000; Perry and Hepburn, 2008).

3.2 Modern Reef Synthesis

Modern reefs, like their fossil predecessors, are dynamic systems that balance reef growth and reef destruction (for a review, see Hutchings, 2011).

3.2.1 Qualitative Studies on Bioerosion

The earliest studies on bioerosion mainly consisted of examining large heads of living coral and documenting the amount of the substrate occupied by boring sponges and noting the presence of bivalve molluscs and sipunculans, and calculating rates based on estimates of the age of the coral (Hudson, 1977; MacGeachy, 1977; Risk and MacGeachy, 1978). While such studies highlighted the widespread occurrence of bioerosion especially at the base of living coral colonies, they failed to recognize the widespread occurrence of bioerosion in dead coral substrates, which makes up the bulk of the coral reefs, and which is subjected to much higher levels of bioerosion. This is because all borers recruit as pelagic larvae (Hutchings, 1986), although sponges can also recruit via fragments of tissue (López-Victoria and Zea, 2004), and as living corals are active carnivores, recruitment is restricted to dead coral or damaged areas of living coral colonies. This means that the highest occurrence of bioerosion occurs in the dead reefal framework (Fig. 2E). Qualitative studies of such habitats reveal the great diversity of borers in such habitats (Hutchings, 1986, 2011) and evidence suggests that the longer the time since death of the coral colony, the greater diversity and biomass of borers. Reef cavities produced from bioeroders promote diversity (Stella et al., 2011), especially among non-boring cryptofaunal organisms (Hutchings, 1986). Other studies have documented the importance of grazers in removing the coral substrate such as echinoids (Asgaard and Bromley, 2008; Peyrot-Clausade et al., 2000), scarid fish (Bellwood, 2003; Rotjan and Lewis, 2005), gastropods (Lam et al., 2007; Shafir et al., 2008), and chitons (Barbosa et al., 2008).

3.2.2 Quantitative Studies on Bioerosion

Until experimental studies were undertaken, rates and agents of macroboring and grazing could not be accurately determined and apportioned to the various groups of agents. Complementary studies were carried out on the Great Barrier Reef, Australia (Kiene and Hutchings, 1994a,b; Osorno et al., 2005) and in French Polynesia (Pari et al., 2002), where experimental blocks of recently killed colonies of *Porites* were used (Fig. 2F). For details of the methodology, see Osorno et al. (2005). These studies allowed developments of balance sheets of net losses and net gains, which could then be compared between habitats, between reefs and geographic areas (Osorno et al., 2005; Pari et al., 2002;

Fig. 2G). During these experiments, which were conducted over several years, microborers were also investigated as their importance was increasingly recognized by how they alter substrate and facilitate recruitment of macroborers and encourage grazing (Chazottes et al., 2002; Tribollet et al., 2002).

All these experiments, by default, were also measuring rates of loss by chemical and physical erosion. Most of the experimental blocks were laid in relatively quiet waters to mitigate chemical and physical erosion, although some at Lizard Island, Great Barrier Reef, were laid on the reef front subjected to South East trade winds (Kiene and Hutchings, 1994a,b).

Grazing rates have been determined for echinoids in the Galapagos (Glynn, 1988; Reaka-Kudla et al., 1996), in French Polynesia (Peyrot-Clausade et al., 2000) and in the Indian Ocean (Peyrot-Clausade et al., 2000); chitons on the Great Barrier Reef (Barbosa et al., 2008), and scarid fish on the Great Barrier Reef (Bellwood and Choat, 1990), in the Indian Ocean (Peyrot-Clausade et al., 2000), and in the Caribbean (Rotjan and Lewis, 2006). Attempting to compare these figures is difficult as different methods have been used, but a substantial loss of reef framework can occur especially from echinoids (see Table 1, especially in the Galapagos and Faaa in Tahiti; Fig. 2H).

Most of the quantitative studies discussed here are limited by their short duration (<5 years) and use of a single type of coral, *Porites*, as the substrate. Rates are known to vary according to the type and density of coral skeleton (Hutchings, 1986). These studies reveal considerable variations within and between reefs, and over time. Boring rates do not continue to increase over time but rather slow down over time as boring communities mature. This is particularly true for sponges (Rützler, 1975; Schönberg, 2008). The agents of boring change over time with a distinct succession of borers, from short-lived polychaete species to long-lived polychaete species, followed by sponges, sipunculans, and bivalves after a delay of 2–3 years. Therefore, caution is warranted before extrapolating these quantitative results through time and by facies.

In spite of these limitations, all quantitative studies show that bioerosion is a major process occurring on modern reefs and that the structure and form of reefs is the result of interactions between reef growth and reef destruction. Just as past climatic and evolutionary episodes have reshaped reef ecosystems over the eras, modern climate change and anthropogenic factors are contributing to the balance between productivity and destruction in reef carbonate budgets.

4. ENVIRONMENTAL CONTROLS ON BIOEROSION

What it means to be a "healthy" reef is challenging to quantify and probably will remain a subjective concept; however, the carbonate budget of a reef system can capture the dynamics of reef expansion, stability, or contraction. Studies on bioerosion from anthropogenically affected reefs *versus* more pristine reefs allow identification of environmental factors that control the bioerosion process.

Studies at Lizard Island, Great Barrier Reef, showed that rates varied across the reef, with subtidal reef slopes and lagoonal sites experiencing higher rates of grazing than deeper sites and reef flats, and these differences were maintained in experimental substrates exposed for 9 years (Kiene and Hutchings, 1994a,b). The density of grazing by scarids in these habitats is suggested to control the bioerosion rates. Recruitment of boring organisms was highly seasonal and variations occurred between years (Hutchings and Murray, 1982), with prevailing wind patterns important in successful larval recruitment. Sediment loads in the water column were also very important for inshore sites, especially those near river mouths experiencing higher levels of boring than grazing compared to sites in clear water (Hutchings et al., 2005; Osorno et al., 2005; Fig. 3). Again, the distribution of grazing fish was important and this was related to the amount of endolithic algae in the surface layers of the substrate. Inshore sites had lower levels of these algae than offshore sites due to levels of sedimentation on the surface of the blocks, which reduced light penetration necessary for algal growth.

Experimental sites subjected to increased nutrients without sediment load, but with healthy grazing fish populations, also exhibited variations in the levels of microborers (Osorno, 2005; Table 1) and no changes in the levels of grazing. In contrast, eutrophic sites also subjected to overfishing experienced high levels of grazing (Pari et al., 2002; Table 1, Fig. 2H). This was because the high levels of nutrients in the water column encouraged dense epilithic algal populations which facilitated the development of large populations of grazing echinoids. Such populations could develop as severe overfishing had occurred and therefore little or no predation of the juvenile echinoids occurred. Grazing levels were so high that little boring occurred in the experimental substrates at this polluted site.

Increasing incidences of bleaching, which may lead to death of coral colonies, are substantially increasing the amount of dead coral substrate available and can therefore lead to localized increases in bioerosion (Eakin, 1992; see Table 1). Substrates heavily bored may be more susceptible to physical removal during storm events, the intensity of which is predicted to increase with climatic warming (Przeslawski et al., 2008). In some cases, though, sponge boring may allow dead coral colonies to bend with the water movement with the presence of the sponge enmeshed in the coral skeleton strengthening the colony, making them less susceptible to being dislodged during storm events (Chaves-Fonnegra and Zea, 2007; López-Victoria and Zea, 2004). Increased ocean acidification is also likely to influence rates and agents of bioerosion with reduced densities of coral substrate (Przeslawski et al., 2008).

A variety of factors such as depth, levels of sedimentation, water quality, bleaching, overfishing, and plagues of Crown of Thorns affect the rates of bioerosion as well as the ratio of grazing to boring, and often these act synergistically and are difficult to separate. In Table 1, the rates of grazing and boring at selected sites subjected to varying environmental conditions are shown,

TABLE 1 Modern Rates of Bioerosion (Grazing and Boring) from Experimental Substrates under a Variety of Environmental Conditions

Environmental conditions	Location	Net rates of bioerosion (grazing + boring) − accretion kg $CaCO_3$ m^{-2} a^{-1}	Rates of grazing kg $CaCO_3$ m^{-2} a^{-1}	Rates of macroboring kg $CaCO_3$ m^{-2} a^{-1}	Rates of microboring kg $CaCO_3$ m^{-2} a^{-1}	Rates of calcification on reef kg $CaCO_3$ m^{-2} a^{-1}	Comments	Exposure duration	Substrate	References
Pristine	Lizard Islands, Great Barrier Reef	1.459 ± 0.679	1.145 ± 0.518	0.391 ± 0.277	0.75 ± 0.03	3.7–4.6 Estimated for windward margin	Marine park, good populations of grazing fish, and limited echinoids	48 months	Blocks of *Porites*	Osorno et al. (2005) and Kinsey (1979)
Pristine	Osprey, Coral Sea	4.918 ± 0.078	4.886 ± 1.995	$0.154 \pm$	1.34 ± 0.74		Open ocean, good populations of grazing fish, limited echinoids	48 months	Blocks of *Porites*	Osorno et al. (2005)
Lagoonal, limited flushing	Tikehau, French Polynesia	2.74 ± 0.79	1.95 ± 0.58	0.78 ± 0.27			No echinoids, medium scarids	5 years	Blocks of *Porites*	Pari et al. (2002)
Closed lagoon	Takapoto, French Polynesia	1.70 ± 0.54	3.16 ± 0.38	0.7 ± 0.11			No echinoids, medium scarids	5 years	Blocks of *Porites*	Pari et al. (2002)
High sediment loads	Snapper-entrance to Daintree River, North Queensland	1.590 ± 0.773	0.590 ± 0.188	1.075 ± 0.659				48 months	Blocks of *Porites*	
Major bleaching event, mortality of 50% corals	Uva Island, Panama	>15.4	N/A	N/A			Includes echinoid and fish, grazers + internal borers, after severe bleaching events	85, 282, 419, 655, 702, and 938 days	Cores of *Porites*	Eakin (1992)

TABLE 1 Modern Rates of Bioerosion (Grazing and Boring) from Experimental Substrates under a Variety of Environmental Conditions—Cont'd

Environmental conditions	Location	Net rates of bioerosion (grazing + boring) − accretion kg $CaCO_3$ m^{-2} a^{-1}	Rates of grazing kg $CaCO_3$ m^{-2} a^{-1}	Rates of macroboring kg $CaCO_3$ m^{-2} a^{-1}	Rates of microboring kg $CaCO_3$ m^{-2} a^{-1}	Rates of calcification on reef kg $CaCO_3$ m^{-2} a^{-1}	Comments	Exposure duration	Substrate	References
Severely impacted by El Niño	Galapagos, Eastern Pacific	25.4	22.8	2.4			Dense populations of echinoids	14.8 months	Blocks of *Porites*	Reaka-Kudla et al. (1996)
Eutrophic waters	Reunion Island, Indian Ocean	1.79–5.37	1.51–4.313	0.28–1.063			Includes micro- and macroborings	12 months	Slices of *Porites*	Chazottes (1996)
Eutrophic waters	Faaa-Tahiti, French Polynesia	10.45	10.38 ± 8.73, increased to 25.14 ± 2.40	0.07 ± 0.02	0.6	Very little coral recruitment	High densities of echinoids	6 months	Blocks of *Porites*	Pari et al. (1998)
Eutrophic waters	Moorea, French Polynesia	1.39 ± 0.3	1.27 ± 0.27	0.12 ± 0.005	0.9–2.6	2.160	High densities of echinoids and scarids	5 years	Blocks of *Porites*	Guillaume (1990) and Pari et al. (2002)
Mildly eutrophic waters	Atimaonao-Tahiti, French Polynesia	1.91 ± 0.37	1.58 ± 0.27	0.33 ± 0.27		Very limited coral recruitment	High densities of echinoids, medium scarids	5 years	Blocks of *Porites*	Pari et al. (2002)

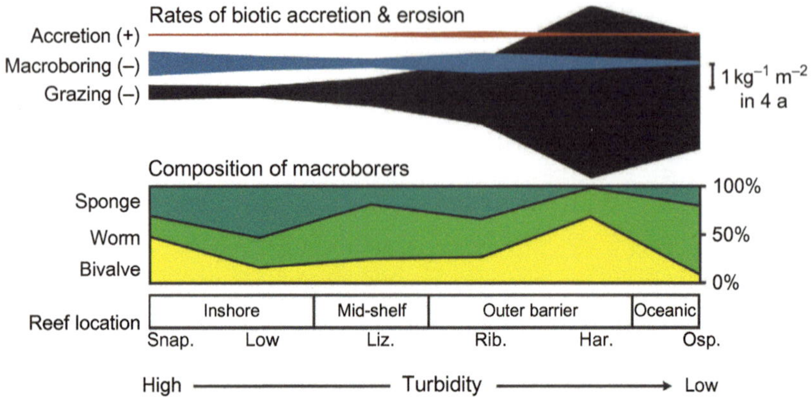

FIGURE 3 Biotic accretion and degradation along an offshore transect of the Great Barrier Reef. Experiments using dead coral substrate exposed over a 4-year period demonstrate the rates of carbonate accretion by algae, and carbonate erosion by grazing and macroboring fauna. Offshore sediment gradient and increasing water clarity extend from the coast at Snapper Island (Snap.) to Osprey Reef (Osp.), Coral Sea. *See Osorno et al. (2005) for full results and map of reef localities.* Abbreviations: Low, Low Isle; Liz., Lizard Island; Rib., Ribbon Reef No. 3; Har., Harrier Reef.

which illustrate that these rates vary considerably, and while rates increase on anthropogenically impacted sites, these are not uniform.

An example of reef recovery from grazing, for example, Crown of Thorns plague, can be seen in areas of the Great Barrier Reef. After the plague has moved on, new coral recruits coming from adjacent reefs can settle on the newly created space and—provided that the water quality is good with low sediment loads and healthy fish populations present—within 5–10 years these reefs can make a full recovery (Emslie et al., 2008). However, if other factors are working against this, such as poor water quality and high sediment loads, recovery will be slow and/or incomplete.

5. CONCLUSIONS

The process of bioerosion is inextricably linked with Phanerozoic reef ecosystems. The intensity and diversity of bioeroding agents has generally increased over time. Dominance among macroboring assemblages has shifted from worm borings (Cambrian–Triassic) to bivalve borings (Jurassic–Eocene) to sponge borings (Oligocene–Recent). Boring has evolved several times within each of the major groups of borers and may have produced similar-looking trace fossils.

The lack of substrate specificity among most bioeroders, and their eurytopic synecology, may be significant contributors to their relative prosperity and behavioral conservatism through times of biotic crisis, at the same time as their reef coralline substrates undergo significant taxonomic turnover. By contrast, host-specific endolithic symbionts show turnover with the disappearance of their host taxa.

Factors that increase bioerosion among heterotrophic endoliths include increased inorganic nutrient supply, decreased siliciclastic input, and increased exposure time of the carbonate substrate to the water column. Bioerosion intensity increases in response to modern climate change and anthropogenic impacts, for example, increasing temperature, alkalinity, eutrophication, rising sea level, overfishing, intensity of storms, and changing ocean currents. As a destructive process of the carbonate budget, rates of bioerosion exceed those of reef carbonate production leading to loss of reefal framework.

Whereas fossil reefs have been subjected to changes in seawater temperature and sea level, the elevated modern rates coupled with overfishing and eutrophication facilitate bioerosion and diminish carbonate productivity, resulting in net contraction of reefs. This new paradigm of human influence on the carbonate budget of reefs is not comparable to historical observations based on the fossil record and we are entering a new phase in reefal budgets and unchartered waters.

ACKNOWLEDGMENTS

L. T. thanks P. Copper, E. Edinger, A. A. Ekdale, and M. Wilson for their encouragement through the years in pursuing the fossil record of endoliths. P. H. would like to thank Dirk Knaust for an invitation to prepare this review. Michelle Yerman helped in the preparation of Fig. 2E. Edinger provided bioeroded specimens from Puerto Rico.

REFERENCES

Asgaard, U., Bromley, R.G., 2008. Echinometrid sea urchins, their trophic styles and corresponding bioerosion. In: Wisshak, M., Tapanila, L. (Eds.), Current Developments in Bioerosion. Erlangen Earth Conference Series. Springer, Berlin, pp. 280–303.

Bambach, R.K., 1983. Ecospace utilization and guilds in marine communities through the Phanerozoic. In: Tevesz, M.J.S., McCall, P.L. (Eds.), Biotic Interactions in Recent and Fossil Benthic Communities. Plenum Press, New York, pp. 719–746.

Barbosa, S.S., Byrne, M., Kelahar, B.P., 2008. Bioerosion caused by foraging of the tropical chiton *Acanthopleura gemmata* at One Tree Reef, southern Great Barrier Reef. Coral Reefs 27, 635–639.

Bellwood, D.R., 2003. Origins and escalation of herbivory in fishes: a functional perspective. Palaeobiology 29, 71–83.

Bellwood, D.R., Choat, J.H., 1990. A functional analysis of grazing in parrot fishes (family Scaridae): the ecological implications. Env. Biol. Fish. 28, 189–214.

Bellwood, D.R., Schultz, O., 1991. A review of the fossil record of the parrotfishes (Labroidei: Scaridae) with a description of a new *Calotomus* species from the Middle Miocene (Badenian) of Austria. Ann. Naturhist. Mus. Wien 92, 55–71.

Benner, J.S., Ekdale, A.A., de Gibert, J.M., 2004. Macroborings (*Gastrochaenolites*) in Lower Ordovician hardgrounds of Utah: sedimentologic, paleoecologic, and evolutionary implications. Palaios 19, 543–550.

Beuck, L., Wisshak, M., Munnecke, A., Freiwald, A., 2008. A giant boring in a Silurian stromatoporoid analysed by computer tomography. Acta Palaeont. Polon. 53, 149–160.

Brett, C.E., Cottrell, J.F., 1982. Substrate specificity in the Devonian tabulate coral *Pleurodictyum*. Lethaia 15, 247–262.

Bromley, R.G., 1978. Bioerosion of Bermuda reefs. Palaeogeogr. Palaeoclimatol. Palaeoecol. 23, 169–197.

Bromley, R.G., 1990. Trace Fossils, Biology and Taphonomy. Unwin Hyman, London, 280 pp.

Bromley, R.G., 1996. Trace Fossils: Biology. Taphonomy and Applications. Chapman and Hall, London, 361 pp.

Bromley, R.G., 2004. A stratigraphy of marine bioerosion. In: McIlroy, D. (Ed.), The Application of Ichnology to Palaeoenvironmental and Stratigraphic Analysis. Geol. Soc. London, Spec. Publ., vol. 228, pp. 455–479.

Bromley, R.G., D'Alessandro, A., 1983. Bioerosion in the Pleistocene of southern Italy: ichnogenera *Caulostrepsis* and *Maeandropolydora*. Riv. It. Paleontol. Strat. 89, 283–309.

Bromley, R.G., D'Alessandro, A., 1984. The ichnogenus *Entobia* from the Miocene, Pliocene and Pleistocene of southern Italy. Riv. It. Paleontol. Strat. 90, 227–296.

Bromley, R.G., D'Alessandro, A., 1989. Ichnological study of shallow marine endolithic sponges from the Italian coast. Riv. It. Paleontol. Strat. 95, 279–340.

Buddemeier, R.W., Maragos, J.E., Knutson, D.W., 1974. Radiographic studies of reef coral exoskeletons: rates and patterns of coral growth. J. Exp. Biol. Ecol. 141, 179–200.

Carter, J.G., Stanley Jr., G.D., 2004. Late Triassic gastrochaenid and lithophaginid borings (Mollusca: Bivalvia) from Nevada (USA) and Austria. J. Paleontol. 78, 230–234.

Chaves-Fonnegra, A., Zea, S., 2007. Observations on reef coral undermining by the Caribbean excavating sponge *Cliona delitrix* (Demospongiae, Hadromerida). In: Custódio, M.R., Hajdu, E., Lôbo-Hajdu, G., Muricy, M. (Eds.), Porifera Research: Biodiversity, Innovation and Sustainability. Proc. 7th Int. Sponge Symp., Rio de Janeiro, pp. 247–254.

Chazottes, V., 1996. Etude expérimentale de la bioérosion et de la sédimentogénèse en milieu récifal: effets de l'eutrophisation (île de la Réunion, Océan Indien). Compt. Rend. Acad. Sci. Paris 323, 133–142.

Chazottes, V., Le Campion-Alsumard, T., Peyrot-Clausade, M., Cuet, P., 2002. The effects of eutrophication-related alterations to coral reef communities on agents and rates of bioerosion (Reunion Island, Indian Ocean). Coral Reefs 21, 375–390.

Choat, J.H., Robbins, W.D., Clements, K.D., 2004. The trophic status of herbivorous fishes on coral reefs II: food processing modes and trophodynamics. Mar. Biol. 145, 445–454.

Chow, N., James, N.P., 1992. Synsedimentary diagenesis of Cambrian peritidal carbonates: evidence from hardgrounds and surface paleokarst in the Port au Port Group, western Newfoundland. Bull. Can. Petrol. Geol. 40, 115–127.

Copper, P., 2002. Silurian and Devonian reefs: 80 million years of global greenhouse between ice ages. In: Kiessling, W., Flügel, E., Golanka, J. (Eds.), Phanerozoic Reef Patterns. SEPM Spec. Publ., vol. 72, pp. 181–238.

Cortes, J., Risk, M., 1985. A reef under siltation stress: Cahuita, Costa Rica. Bull. Mar. Sci. 36, 339–356.

Cowman, P.F., Bellwood, D.R., van Herwerden, L., 2009. Dating the evolutionary origins of wrasse lineages (Labridae) and the rise of trophic novelty on coral reefs. Molec. Phylogen. Evol. 52, 621–631.

de Gibert, J.M., Domènech, R., Martinell, J., 2004. An ethological framework for animal bioerosion trace fossils upon mineral substrates with proposal of a new class, fixichnia. Lethaia 37, 429–437.

Eakin, C.M., 1992. Post-El Niño Panamanian Reefs: less accretion, more erosion and damselfish protection. Proc. 7th Int. Coral Reef Symp., vol. 2, 387–396.

Ebbestad, J.O.R., Tapanila, L., 2005. Non-predatory borings in Phanerotrema (Gastropoda, Early Silurian), Anticosti Island, Québec, Canada. Palaeogeogr. Palaeoclimatol. Palaeoecol. 221, 325–341.

Edinger, E.N., Risk, M.J., 1994. Oligocene-Miocene extinction and geographic restriction of Caribbean corals: roles of turbidity, temperature, and nutrients. Palaios 9, 576–598.

Edinger, E.N., Risk, M.J., 1997. Sponge borehole size as a relative measure of bioerosion and paleoproductivity. Lethaia 29, 275–286.

Ekdale, A.A., Bromley, R.G., 2001. Bioerosional innovation for living in carbonate hardgrounds in the Early Ordovician of Sweden. Lethaia 34, 1–12.

Emslie, M.J., Cheal, A.J., Sweatman, H., Delean, S., 2008. Recovery from disturbance of coral and reef fish communities on the Great Barrier Reef. Australia. Mar. Ecol. Progr. Ser. 371, 177–190.

Evans, J.W., 1970. Palaeontological implications of a biological study of rock boring clams (Family Pholadidae). In: Crimes, T.P., Harper, J.C. (Eds.), Trace Fossils. Geol. J., Spec. Iss. 3, pp. 127–140.

Flügel, E., 2002. Triassic reef patterns. In: Kiessling, W., Flügel, E., Golanka, J. (Eds.), Phanerozoic Reef Patterns. SEPM Spec. Publ., vol. 72, pp. 391–463.

Flügel, E., Senowbari-Daryan, B., 2001. Triassic reefs of the Tethys. In: Stanley Jr., D. (Ed.), The History and Sedimentology of Ancient Reef Systems. Kluwer Academic/Plenum Publishers, New York, pp. 217–249.

Fraser, N.M., Bottjer, D.J., Fischer, A.G., 2004. Dissecting "Lithiotis" bivalves: implications for the Early Jurassic reef eclipse. Palaios 19, 51–67.

Frost, S.H., Harris, P.M., Harbour, J.L., 1983. Late Oligocene reef-tract carbonates, southwestern Puerto Rico. In: Harris, P.M. (Ed.), Carbonate Buildups. SEPM Core Workshop 4, pp. 491–518.

Glaub, I., Vogel, K., 2004. The stratigraphic record of microborings. Fossils Strata 51, 126–135.

Glynn, P.W., 1988. El Nino warming, coral mortality and reef framework destruction by echinoid bioerosion in the eastern Pacific. Galaxea 7, 20–30.

Guillaume, M., 1990. Growth and calcium carbonate production of massive *Porites*. Proc. Int. Soc. Reef Stud., Noumea, 91–98.

Highsmith, R.C., Lueptow, R.L., Schonberg, S.C., 1983. Growth and bioerosion of three massive corals on the Belize barrier reef. Mar. Ecol. Prog. Ser. 13, 261–271.

Hooper, J.N.A., Van Soest, R.W.M., 2002. Systema Porifera: A Guide to the Classification of Sponges. 2 Volumes. Kluwer Academic/Plenum Publishers, New York, 1810 pp.

Hudson, J.H., 1977. Long term bioerosion rates on a Florida reef: new method. Proc. 3rd Int. Coral Reef Symp. vol. 2, 491–498.

Hutchings, P.A., 1986. Biological destruction of coral reefs—a review. Coral Reefs 4, 239–252.

Hutchings, P.A., 2008. Role of polychaetes in bioerosion of coral substrate. In: Wisshak, M., Tapanila, L. (Eds.), Current Developments in Bioerosion. Erlangen Earth Conference Series. Springer, Berlin, pp. 249–264.

Hutchings, P.A., 2011. Bioerosion. In: Hopley, D. (Ed.), Encyclopedia of Modern Coral Reefs: Structure, Form and Process. Springer, Berlin, pp. 139–156.

Hutchings, P.A., Murray, A., 1982. The establishment of polychaete populations to coral substrates at Lizard Island, Great Barrier Reef—an experimental approach. Austr. J. Mar. Freshwater Res. 33, 1029–1037.

Hutchings, P.A., Peyrot-Clausade, M., Osnorno, A., 2005. Influence of land runoff on rates and agents of bioerosion of coral substrates. Mar. Poll. Bull. 51, 438–447.

James, N.P., Kobluk, D.R., Pemberton, S.G., 1977. The oldest macroborers: Lower Cambrian of Labrador. Science 197, 980–983.

Kapp, U.S., 1975. Paleoecology of Middle Ordovician stromatoporoid mounds in Vermont. Lethaia 8, 195–207.

Kelly, S.R.A., 1980. *Hiatella*—a Jurassic squatter? Palaeontology 23, 769–781.

Kelly, S.R.A., Bromley, R.G., 1984. Ichnological nomenclature of clavate borings. Palaeontology 27, 793–807.

Kershaw, S., 1993. Sedimentation control on growth of stromatoporoid reefs in the Silurian of Gotland. Sweden. J. Geol. Soc. London 150, 197–205.

Kiene, W.E., Hutchings, P.A., 1994a. Long-term bioerosion of experimental coral substrates from Lizard Island, Great Barrier Reef. Proc. 7th Int. Coral Reef Congr., Guam, vol. 1, pp. 397–403.

Kiene, W.E., Hutchings, P.A., 1994b. Experimental investigations on patterns in the rates of bioerosion at Lizard Island, Great Barrier Reef. Coral Reefs 13, 91–98.

Kiessling, W., Simpson, C., Foote, M., 2010. Reefs as cradles of evolution and sources of biodiversity in the Phanerozoic. Science 327, 196–198.

Kinsey, D.W., 1979. Carbon turnover and accumulation by coral reefs. Ph.D. Thesis, University of Hawaii, 248 pp.

Kobluk, D.R., 1981. Lower Cambrian cavity-dwelling endolithic (boring) sponges. Can. J. Earth Sci. 18, 972–980.

Lam, K., Shin, P.K.S., Hodgson, P., 2007. Severe bioerosion caused by an outbreak of corallivorous *Drupella* and *Diadema* at Hoi Ha Wan Marine Park, Hong Kong. Coral Reefs 26, 893.

Leinfelder, R.R., 1986. Facies, stratigraphy and paleogeographic analysis of Upper? Kimmeridgian to Upper Portlandian sediments in the environs of Arruda dos Vinhos, Estremadura, Portugal. Münchner Geowiss. Abh. A 7, 1–216.

Lindström, M., 1979. Probable sponge borings in Lower Ordovician limestone of Sweden. Geology 7, 152–155.

López-Victoria, M., Zea, S., 2004. Storm-mediated coral colonization by an excavating Caribbean sponge. Clim. Res. 26, 251–256.

López-Victoria, M., Zea, S., 2005. Current trends of space occupation by encrusting excavating sponges on the Columbian coral reefs. Mar. Ecol. 26, 33–41.

MacGeachy, J.K., 1977. Factors controlling sponge boring in Barbados reef corals. Proc. 3rd Int. Coral Reef Symp., Florida, vol. 2, 478–483.

Mikuláš, R., 1992. Early Cretaceous borings from Stramberk (Czechoslovakia). Casopis Min. Geol. 37, 297–312.

Morton, B., 1990. Corals and their bivalve borers—the evolution of a symbiosis. In: Morton, B. (Ed.), The Bivalvia—Proceedings of a Memorial Symposium in Honour of Sir Charles Maurice Yonge, Edinburgh, 1986. Hong Kong University Press, Honkong, pp. 11–46.

Neumann, A.C., 1966. Observations on coastal erosion in Bermuda and measurements of the boring rate of the sponge, *Cliona lampa*. Limnol. Oceanogr. 11, 92–108.

Nield, E.W., 1984. The boring of Silurian stromatoporoids—towards an understanding of larval behaviour in the *Trypanites* organism. Palaeogeogr. Palaeoclimatol. Palaeoecol. 48, 229–243.

Nishi, E., Nishihira, M., 1999. Use of annual density banding to estimate longevity of infauna of massive corals. Fisher. Sci. 65, 48–56.

Osorno, A., 2005. Impact d'une eutrophisation experimental sur les processus de la bioerosion recifale, One Tree Island, Grande Barriere d'Australie. Ph.D. Thesis, Université de la Méditerranée, Aix–Marseille, 11, 134 pp.

Osorno, A., Peyrot-Clausade, M., Hutchings, P.A., 2005. Patterns and rates of erosion in dead *Porites* across the Great Barrier Reef (Australia) after 2 years and 4 years of exposure. Coral Reefs 24, 292–303.

Pari, N., Peyrot–Clausade, M., Le Campion–Alsumard, T., Hutchings, P.A., Chazottes, V., Golubic, S., Le Campion, T., Fontaine, M.F., 1998. Bioerosion of experimental substrates on high islands and atoll lagoons (French Polynesia) after 2 years of exposure. J. Exp. Mar. Ecol., Progr. Ser. 166, 119–130.

Pari, N., Peyrot–Clausade, M., Hutchings, P.A., 2002. Bioerosion of experimental substrates on high islands and atoll lagoons (French Polynesia) during 5 years of exposure. J. Exp. Mar. Biol. Ecol. 276, 109–127.

Perry, C.T., 1996. Distribution and abundance of macroborers in an upper Miocene reef system, Mallorca, Spain: implications for reef development and framework destruction. Palaios 11, 40–56.

Perry, C.T., 2000. Macroboring of Pleistocene coral communities, Falmouth Formation, Jamaica. Palaios 15, 483–491.

Perry, C.T., Bertling, M., 2000. Spatial and temporal patterns of macroboring within Mesozoic and Cenozoic coral reef systems. In: Insalaco, E., Skelton, P.W., Palmer, T.J. (Eds.), Carbonate Platform Systems: Components and Interactions. Geol. Soc. London, Spec. Publ., vol. 178, pp. 33–50.

Perry, C.T., Hepburn, L.J., 2008. Syn-depositional alteration of coral reef framework through bioerosion, encrustation, and cementation: taphonomic signatures of reef accretion and reef depositional events. Earth-Sci. Rev. 86, 106–144.

Peyrot-Clausade, M., Chabanet, P., Conand, C., Fontaine, M., Letourneur, Y., Harmelin-Vivien, M., 2000. Sea urchin and fish bioerosion on La Réunion and Moorea reefs. Bull. Mar. Sci. 66, 477–485.

Pleydell, S.M., Jones, B., 1988. Borings of various faunal elements in the Oligocene-Miocene Bluff Formation of Grand Cayman, British West Indies. J. Paleontol. 62, 348–367.

Pohowsky, R.A., 1978. The boring ctenostomate Bryozoa: taxonomy and paleobiology based on cavities in calcareous substrata. Bull. Am. Paleontol. 73, 1–192.

Pojeta Jr., J., Palmer, T.J., 1976. The origin of rock boring in mytilacean pelecypods. Alcheringa 1, 167–179.

Przeslawski, R., Ahyong, S., Byrne, M., Wörheide, G., Hutchings, P., 2008. Beyond corals and fish: the effects of climate change on non-coral benthic invertebrates of tropical reefs. Global Change Biol. 14, 1–23.

Reaka-Kudla, M., Feingold, J.S., Glynn, P., 1996. Experimental studies of rapid bioerosion of coral reefs in the Galapagos. Coral Reefs 15, 101–107.

Risk, M.J., MacGeachy, J.K., 1978. Aspects of bioerosion of modern Caribbean reefs. Rev. Biol. Trop. 26, 85–105.

Rosen, B.R., 1984. Reef coral biogeography and climate through the late Cainozoic: just islands in the sun or a critical pattern of islands? In: Brenchley, P. (Ed.), Fossils and Climate. John Wiley and Sons, New York, pp. 201–260.

Ross, A., Newman, W.A., 2002. Coral barnacles: Cenozoic decline and extinction in the Atlantic/East Pacific *versus* diversification in the Indo-West Pacific. Proc. 9th Int. Coral Reef Symp., Bali, Indonesia, pp. 179–184.

Rotjan, R.D., Lewis, S.M., 2005. Selective predation by parrotfishes on the reef coral *Porites astroides*. Mar. Ecol. Progr. Ser. 305, 193–201.

Rotjan, R.D., Lewis, S.M., 2006. Parrotfish abundance and selective corallivory on a Belizean coral reef. J. Exp. Mar. Biol. Ecol. 335, 292–301.

Rouse, G.W., Pleijel, F., 2001. Polychaetes. Oxford University Press, Oxford, 354 pp.

Rozhnov, S.V., 1994. Changes in the hardground at the Cambrian-Ordovician boundary. Paleontol. J. 28, 84–91.

Rützler, K., 1975. The role of burrowing sponges in bioerosion. Oecologia 19, 203–216.

Schönberg, C.H.L., 2008. A history of sponge erosion: from past myths and hypotheses to recent approaches. In: Wisshak, M., Tapanila, L. (Eds.), Current Developments in Bioerosion. Erlangen Earth Conference Series. Springer, Berlin, pp. 165–202.

Schönberg, C.H.L., Tapanila, L., 2006a. Bioerosion research before and after 1996—a discussion on what has changed since the First International Bioerosion Workshop. Ichnos 13, 99–102.

Schönberg, C.H.L., Tapanila, L., 2006b. The bioeroding sponge *Aka paratypica*, a modern trace-making analogue for the Paleozoic ichnogenus *Entobia devonica*. Ichnos 13, 147–157.

Segars, M.T., Liddell, W.D., 1988. Microhabitat analyses of Silurian stromatoporoids as substrata for epibionts. Palaios 3, 391–403.

Shafir, S., Gur, O., Rinkevich, B., 2008. A *Drupella cornus* outbreak in the northern Gulf of Eilat and changes in coral prey. Coral Reefs 27, 379.

Shaw, J.A., Macey, D.J., Brooker, L.R., Clode, P.L., 2010. Tooth use and wear in three iron-biomineralizing mollusc species. Biol. Bull. 218, 132–144.

Simon-Blecher, N., Huchon, D., Achituv, Y., 2007. Phylogeny of coral-inhabiting barnacles (Cirripedia; Thoracica; Pyrgomatidae) based on 12S, 16S and 18S rDNA analysis. Molec. Phylogen. Evol. 44, 1333–1341.

Soliman, G.N., 1969. Ecological aspects of some coral-boring gastropods and bivalves of the north-western Red Sea. Am. Zool. 9, 887–894.

Stanton, R.J., Flügel, E., 1989. Problems with reef models: the Late Triassic Steinplatte "reef" (Northern Alps, Salzburg/Tyrol, Austria). Facies 20, 1–138.

Stella, J., Pratchett, M.S., Hutchings, P., Jones, G.P., 2011. Diversity, importance and vulnerability of coral associated invertebrates. Oceanogr. Mar. Biol. Annu. Rev. 49, 43–116.

Szaniawski, H., 1966. Scolecodonts. In: Jansonius, J., McGregor, D.C. (Eds.), Palynology: Principles and Applications. Am. Ass. Strat. Palynol. Found., vol. 1, pp. 337–354.

Tapanila, L., 2002. A new endosymbiont in Late Ordovician tabulate corals from Anticosti Island, eastern Canada. Ichnos 9, 109–116.

Tapanila, L., 2004. The earliest *Helicosalpinx* from Canada and the global expansion of commensalism in Late Ordovician sarcinulid corals (Tabulata). Palaeogeogr. Palaeoclimatol. Palaeoecol. 215, 99–110.

Tapanila, L., 2005. Palaeoecology and diversity of endosymbionts in Palaeozoic marine invertebrates: trace fossil evidence. Lethaia 38, 89–99.

Tapanila, L., 2006. Devonian *Entobia* borings from Nevada, with a revision of *Topsentopsis*. J. Paleontol. 80, 760–767.

Tapanila, L., 2008a. Direct evidence of ancient symbiosis using trace fossils. In: Kelley, P.H., Bambach, R.K. (Eds.), From Evolution to Geobiology: Research Questions Driving Paleontology at the Start of a New Century. Paleontol. Soc., Pap. 14, pp. 271–287.

Tapanila, L., 2008b. The endolithic guild: an ecological framework for residential cavities in hard substrates. In: Wisshak, M., Tapanila, L. (Eds.), Current Developments in Bioerosion. Erlangen Earth Conference Series. Springer, Berlin, pp. 3–20.

Tapanila, L., Copper, P., 2002. Endolithic trace fossils in Ordovician-Silurian corals and stromatoporoids, Anticosti Island, eastern Canada. Acta Geol. Hisp. 37, 15–20.

Tapanila, L., Copper, P., Edinger, E., 2004. Environmental and substrate control on Paleozoic bioerosion in corals and stromatoporoids, Anticosti Island, eastern Canada. Palaios 19, 292–306.

Taylor, P.D., Wilson, M.A., 2003. Palaeoecology and evolution of marine hard substrate communities. Earth-Sci. Rev. 62, 1–103.

Taylor, P.D., Wilson, M.A., Bromley, R.G., 1999. A new ichnogenus for etchings made by cheilostome bryozoans into calcareous substrates. Palaeontology 42, 595–604.

Tribollet, A., Decherf, G., Hutchings, P.A., Peyrot–Clausade, M., 2002. Spatial large scale variability in bioerosion of experimental coral substrates on the GBR (Australia); importance of microborers. Coral Reefs 21, 424–432.

Wahlman, G.P., 2002. Upper Carboniferous-Lower Permian (Bashkirian-Kungurian) mounds and reefs. In: Kiessling, W., Flügel, E., Golanka, J. (Eds.), Phanerozoic Reef Patterns. SEPM Spec. Publ., vol. 72, pp. 271–338.

Warme, J.E., 1977. Carbonate borers—their role in reef ecology and preservation. In: Frost, S., Weiss, M., Saunders, J. (Eds.), Reefs and Related Carbonates—Ecology and Sedimentology. AAPG, Studies in Geology, vol. 4, pp. 261–279.

Webb, G.E., 1987. Late Mississippian thrombolite bioherms from the Pitkin Formation of northern Arkansas. GSA Bull. 99, 686–698.

Webb, G.E., 1998. Earliest known Carboniferous shallow-water reefs, Gudman Formation (Tn1b), Queensland, Australia: implications for Late Devonian reef collapse and recovery. Geology 26, 951–954.

Webb, G.E., 2002. Latest Devonian and Early Carboniferous reefs: depressed reef building after the Middle Paleozoic collapse. In: Kiessling, W., Flügel, E., Golanka, J. (Eds.), Phanerozoic Reef Patterns. SEPM Spec. Publ., vol. 72, pp. 239–269.

Webby, B.D., Paris, F., Droser, M., Percival, I.G., 2004. The Great Ordovician Biodiversification Event. Columbia University Press, New York, 484 pp.

Weidlich, O., 1996. Bioerosion in Late Permian Rugosa from reefal blocks (Hawasina Complex, Oman Mountains): Implications for reef degradation. Facies 35, 133–142.

Whitfield, R.P., 1893. New genera and species of burrowing fossil bivalve shell. Rep. Geol. Surv. Ohio 7, 492–493.

Wilson, M.A., Palmer, T.J., 1988. Nomenclature of a bivalve boring from the Upper Ordovician of the midwestern United States. J. Paleontol. 62, 306–308.

Wilson, M.A., Palmer, T.J., 2006. Patterns and processes in the Ordovician Bioerosion Revolution. Ichnos 13, 109–112.

Wisshak, M., 2012. Microbioerosion. In: Knaust, D., Bromley, R.G. (Eds.), Trace Fossils as Indicators of Sedimentary Environments. Developments in Sedimentology, vol. 64. Elsevier, Amsterdam, 213–243.

Wood, R., 2000. Palaeoecology of a Late Devonian back reef, Canning Basin, Western Australia. Palaeontology 43, 671–703.

Chapter 25

Chalk and Related Deep-Marine Carbonates

Charles E. Savrda[1]
Department of Geology and Geography, Auburn University, Auburn, Alabama, USA
[1]*Corresponding author: e-mail: savrdce@auburn.edu*

1. INTRODUCTION

This chapter focuses on the ichnology of those marine carbonates that are composed mainly of the remains of calcareous nannoplankton, principally coccolithophores, that is, carbonate ooze, chalk, and nannofossil limestone. Because coccolithophores did not emerge until the Early Jurassic, this class of carbonates has a relatively restricted stratigraphic range. Nonetheless, these deposits are locally important elements of Early Jurassic to modern marine successions; they bear records of paleoenvironmental, paleoceanographic, and paleoclimatic conditions, and they commonly are of economic significance, particularly to the petroleum industry (Scholle, 1977).

Nannofossil ooze and its variably lithified equivalents (chalk and limestone) are commonly referred to as "deep-water" and "pelagic" carbonates, to distinguish them from facies formed in shallow carbonate-platform settings. However, this terminology requires qualification. Through most of post-Triassic time, deposition of carbonate ooze occurred in relatively deep, low-energy marine settings distant from significant sources of siliciclastic sediments but above the carbonate compensation depth (CCD). These settings, typically beneath 1–5 km of water in modern oceans, include continental slopes and rises, abyssal plains, and tops and flanks of ridges and plateaus. However, at times in the late Mesozoic and earliest Tertiary, but particularly during the Cretaceous, comparable coccolith-rich muds also accumulated in relatively shallow settings, including continental shelves and epeiric basins, some of which were influenced substantially by storm-induced or other bottom currents. This incursion of typically deep-water facies into shallower realms is attributed to (1) high sea-level stands, which resulted in siliciclastic sediment starvation (Hancock, 1975), and (2) lower Mg/Ca ratios in Cretaceous sea water, which significantly enhanced coccolith production (Stanley et al., 2005). In most settings, carbonate

Developments in Sedimentology, Vol. 64. http://dx.doi.org/10.1016/B978-0-444-53813-0.00025-3

ooze accumulated slowly as a pelagic rain, typically at rates of a few centimeters per thousand years. Yet, some oozes were redistributed by persistent bottom currents (e.g., Surlyk and Lykke-Andersen, 2007) or rapidly redeposited by mass-movement processes (Bromley and Ekdale, 1987), including turbidity currents (Uchman, 2007), to ocean depths that in some cases were below the CCD. Thus, despite "pelagic" and "deep water" monikers, the carbonate deposits addressed herein accumulated via multiple processes, under a significant range of water depths and in a wide array of marine environments (Fig. 1; Table 1).

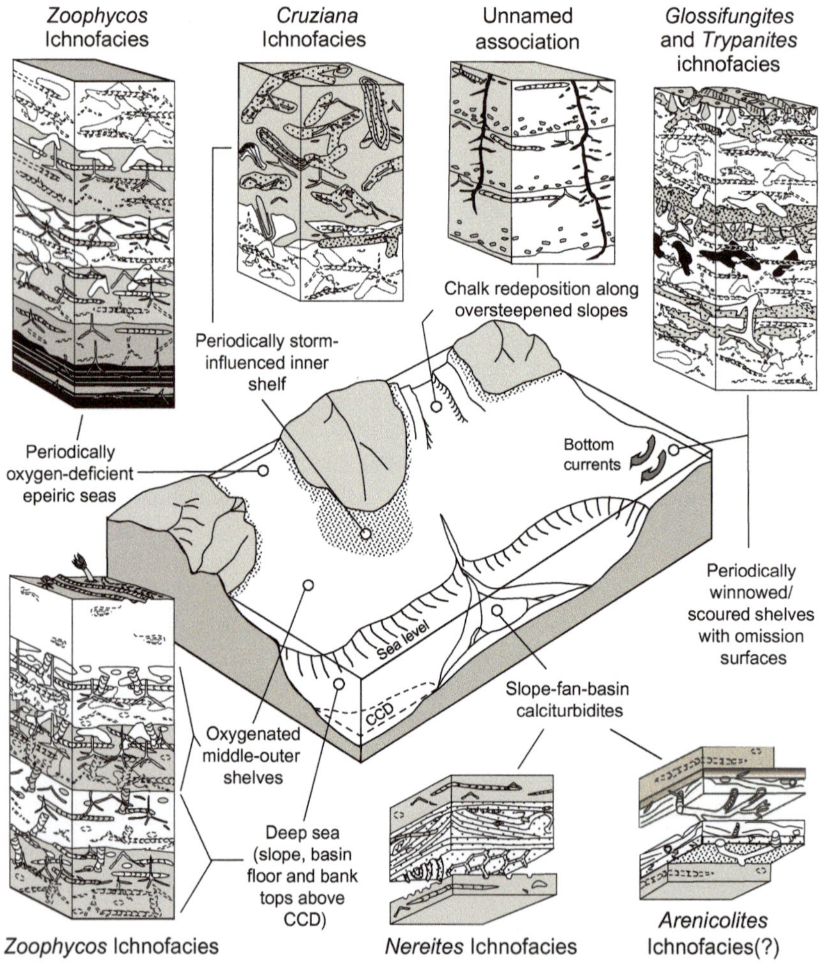

FIGURE 1 Various chalk and deep-marine carbonate depositional regimes and associated ichnofacies.

TABLE 1 Examples of Ichnofacies and Trace-Fossil Assemblages Recognized in Various Chalk or Deep-Water Carbonate Facies

Environmental setting	Substrate	Ichnofacies	Background fabrics	Recurring ichnotaxa	Age, material, and location	Source
Modern deep sea	Softground	*Zoophycos*	Homogeneous, burrow-mottled	Pl, Sk, Ch, Te, Zo?	Recent, box cores, Abyssal Pacific Ocean	Ekdale et al. (1984a)
				Pl, Sk, rare Ch, Zo	Recent, box cores, Abyssal Atlantic Ocean	Ekdale and Berger (1978)
Ancient deep sea	Softground	*Zoophycos*	Homogeneous, burrow-mottled	Pl, Ch, Zo, Te	Cretaceous–Tertiary, DSDP cores, Caribbean Sea	Warme et al. (1973)
				Pl, Ch, Zo, Te, He(?)	Cretaceous–Tertiary, DSDP cores Pacific Ocean	Chamberlain (1975)
				Pl, Ch, Zo, Te	Cretaceous–Tertiary, DSDP cores worldwide	Ekdale (1977)
				Pl, Ch, Sk	Tertiary, DSDP cores, Philippine Sea	Ekdale (1980)
				Pl, Ch, Zo, Te, rare Th	Philippine Sea, DSDP cores, South Atlantic Ocean	Fütterer (1984)
				Pl, Ch, Zo, Te, Sk, Cy, Tr, rare Th	Eocene–Cretaceous, DSDP cores, North Atlantic Ocean	Wetzel (1987)
				Pl, Ch, Zo, rare Sk	Cretaceous–Tertiary, ODP cores, Indian Ocean	Droser and Bottjer (1991)
				Ch, Zo, Te	Cretaceous, Scisti a Fucoidi, Italy	Erba and Premoli-Silva (1994)

Continued

TABLE 1 Examples of Ichnofacies and Trace-Fossil Assemblages Recognized in Various Chalk or Deep-Water Carbonate Facies—Cont'd

Environmental setting	Substrate	Ichnofacies	Background fabrics	Recurring ichnotaxa	Age, material, and location	Source
Outer shelf	Softground	*Zoophycos*	Homogeneous, burrow-mottled	Ch, Zo, Th, Ba	Cretaceous, chalks, England, and France	Kennedy (1970) and Mortimore and Pomerol (1991)
				Pl, Ch, Zo, Ta, Ph, Th	Cretaceous, chalks and marls, Denmark	Ekdale and Bromley (1982, 1991) and Lauridsen et al. (2011)
				Pl, Ch, Zo, Te, Ta, Ph, Th	Cretaceous, Alabama, USA, Demopolis Chalk	Frey and Bromley (1985) and Locklair and Savrda (1998a,b)
				Pl, Zo, Th	Cretaceous, Annona Chalk, Arkansas, USA	Bottjer (1986)
Epeiric basin	Softground	*Zoophycos*	Homogeneous, burrow-mottled	Pl, Ch, Zo, Te, Ta, Th, Tr, As, La	Cretaceous, Niobrara Formation, U.S. Western Interior	Frey (1970) and Savrda (1998a)
				Pl, Ch, Zo, Te, Ta, Th, Tr	Cretaceous, Bridge Creek Limestone, U.S. Western Interior	Hattin (1971) and Savrda (1998b)
Shallower shelf	Softground	*Cruziana*	Homogeneous, burrow-mottled	Th, Ch, Pl	Cretaceous, chalks, England	Kennedy (1975)
				Th, Pl	Cretaceous, Saratoga and Annona Chalks, Arkansas, USA	Bottjer (1985, 1986)
				Th, Ch, Cy, Tr, Te, Sk	Cretaceous, Prairie Bluff Chalk, Alabama, USA	Frey and Bromley (1985)
				Th, Cy, Ch, Pl	Cretaceous, middle and upper Austin Chalk, Texas, USA	Dawson and Reaser (1985, 1990)

TABLE 1 Examples of Ichnofacies and Trace-Fossil Assemblages Recognized in Various Chalk or Deep-Water Carbonate Facies—Cont'd

Environmental setting	Substrate	Ichnofacies	Background fabrics	Recurring ichnotaxa	Age, material, and location	Source
Shallower shelf—cont'd	Firmground	*Glossifungites*	Homogeneous and softground fabrics	Th, Ps	Cretaceous, chalks, Europe	Bromley (1967, 1975) and Kennedy and Garrison (1975)
				Th, Sp	Cretaceous, Arcola Limestone, Prairie Bluff Chalk, Alabama, USA	Frey and Bromley (1985)
				Th	Cretaceous, Annona and Saratoga formations, Arkansas, USA	Bottjer (1985)
				Th, Ps, Rh	Cretaceous, Austin Chalk, Texas, USA	Dawson and Reaser (1990)
	Hardground	*Trypanites*	Soft- and firmground fabrics	Th, Try, Ga, En	Cretaceous, chalks, Europe	Bromley (1967, 1975) and Kennedy and Garrison (1975)
Unstable margins, deep-sea allochthonous	Softground	Unnamed	Homogeneous	Ba, Ch, Zo	Re-deposited Cretaceous, chalks of Europe	Bromley et al. (1975), Nygaard (1983) and Bromley and Ekdale (1987)
		Nereites/ *Arenicolites*(?)	Laminated/ cross-laminated	Cl, Ne, Op, Tr, Ch, Ta, Pl, rare Ph, Zo, and graphoglyptids	Cretaceous, turbidite marls, Monte Antola Formation, Italy	Uchman (2007)
				An(?), Th, Te, Phy, Pl, rare Ch	Cretaceous, calciturbidites, Scaglia Rossa, Italy	Miller et al. (2004)

An, *Ancorichnus*; As, *Asterosoma*; Ba, *Bathichnus*; Ch, *Chondrites*; Cl, *Cladichnus*; Cy, *Cylindrichnus*; En, *Entobia*; Ga, *Gastrochaenolites*; He, *Helminthopsis*; La, *Laevicyclus*; Pl, *Planolites*; Ph, *Phycosiphon*; Phy, *Phycodes*; Ps, *Pseudobilobites*; Ne, *Nereites*; Op, *Ophiomorpha*; Rh, *Rhizocorallium*; Sk, *Skolithos*; Sp, *Spongeliomorpha*; Ta, *Taenidium*; Te, *Teichichnus*; Th, *Thalassinoides*; Tr, *Trichichnus*; Try, *Trypanites*; Zo, *Zoophycos*.
DSDP, Deep Sea Drilling Project; ODP, Ocean Drilling Program.

The composition and texture of modern and ancient nannofossil-rich carbonates vary considerably, reflecting the diversity of depositional regimes in which they formed. While some deposits are nearly wholly composed of coccoliths, nannofossil ooze may be variably admixed with coarser skeletal material, siliciclastic sand or mud, biogenic silica, organic matter, or some combination of these to form various hybrid deposits; for example, calcarenitic chalk, sandy chalk, marly chalk and marl, diatomaceous chalk, etc. Diagenetic mineral assemblages also vary; chalks locally may contain pyrite, glauconite, phosphatic crusts and nodules, and chert or flint concretions, in addition to carbonate cements. Whether primary or diagenetic, compositional differences reflect temporal and spatial changes in physical and chemical conditions. Notably, in many chalk successions, particularly, but not exclusively, those of Cretaceous age, temporal changes in the paleoenvironmental regime are manifested by decimeter-scale variations in composition, texture, and/or degree of lithification. This rhythmic bedding has been attributed to cyclic changes in carbonate production, siliciclastic influx, benthic oxygenation, bottom-current energy, and/or carbonate dissolution (Arthur et al., 1986; Bottjer et al., 1986). Such paleoceanographic cycles, in turn, are linked to climate rhythms mediated by Earth's orbital perturbations—that is, Milankovitch cycles—that in various ways impacted ocean circulation, sediment delivery, and diagenetic regimes in settings characterized by carbonate ooze sedimentation (e.g., Fischer and Schwarzacher, 1984; Gale, 1995; Weedon, 1993).

Temporal and spatial variations in chalk depositional regimes are likewise reflected by ichnological attributes. The ichnology of these deposits has been well documented in studies of modern carbonate ooze as well as in studies of Cretaceous and Tertiary shallow- to deep-water chalk and limestone successions revealed in outcrops, especially those exposed in northern Europe and North America, and in cores recovered during the Deep Sea Drilling Project (DSDP), the Ocean Drilling Program (ODP), and other ocean or continental drilling ventures (Table 1). These investigations indicate that oozes and equivalent sedimentary rocks are characterized by a wide array of ichnofabrics and ichnocoenoses—representing as many as six different ichnofacies—that are indicative of environmental parameters such as general bathymetry, benthic oxygenation, substrate consistency, bottom-current regime, and depositional process.

The carbonate deposits addressed herein may contain trace-fossil suites that are representative of the *Zoophycos*, *Cruziana*, *Glossifungites*, *Trypanites*, *Nereites*, and, possibly, *Arenicolites* ichnofacies (Table 1; Fig. 1). *Zoophycos* Ichnofacies suites are most common; they predominate in modern carbonate oozes and ancient deep-sea, outer-shelf, and epeiric basin chalks. *Cruziana*, *Glossifungites*, and *Trypanites* ichnofacies assemblages, representing soft, firm, and hard substrates, respectively, are largely restricted to chalks and limestones that formed in shallower shelf-sea settings in which the environmental energy was at least episodically higher and, in the case of the latter two ichnofacies, periodical non-deposition and/or winnowing resulted in partial or complete syndepositional cementation at the sea floor. Different trace-fossil suites, including those representing the *Nereites* and *Arenicolites* ichnofacies, developed locally in oozes that were rapidly deposited by turbidity currents or other gravity-flow processes in deeper basinal settings. The goal of this chapter is to review and exemplify ichnofabrics, ichnocoenoses, and ichnofacies expressed in chalk and related carbonates from the complete spectrum of deep-sea and shallower settings.

2. *ZOOPHYCOS* ICHNOFACIES

2.1 Environments and Substrates

Ichnocoenoses representing the *Zoophycos* Ichnofacies predominate in chalks, marly chalks, and marls that accumulated in relatively deep, quiet settings; that is, environments that were below storm-wave base and unaffected by turbidity currents. Such ichnocoenoses are dominant in modern deep-sea oozes and in Cretaceous and Tertiary pelagic carbonate successions recovered in DSDP and ODP cores, but they also are well represented in Cretaceous outer-shelf and epeiric basin chalk successions exposed in North America and Europe (Table 1). Despite their collective grouping into the *Zoophycos* Ichnofacies, ichnocoenoses and ichnofabrics do vary, reflecting differences in benthic conditions as well as in taphonomic influences. Before reviewing this variability, it is helpful to address general traits that are commonly shared among deposits characterized by the *Zoophycos* Ichnofacies.

2.2 Common Ichnofabric Traits

Pelagic deposits assigned to the *Zoophycos* Ichnofacies exhibit three common characteristics. First, with some notable exceptions related to episodes of benthic oxygen-deficiency, these deposits are thoroughly bioturbated. Second, ichnofabrics in these deposits typically comprise two general components: (1) a homogeneous or burrow-mottled background fabric that is overprinted by (2) a low-diversity assemblage of relatively distinct endogenic structures dominated by simple to moderately complex feeding traces, including planar to gently inclined spreiten. Third, cross-cutting relationships among distinct trace fossils are commonly systematic; that is, certain recurring ichnotaxa regularly cross-cut or overprint others.

Thorough bioturbation requires little explanation. Associated deposits accumulated very slowly in quiet settings in which there was ample time for benthic organisms to thoroughly rework sediments many times over.

The two-component aspect of *Zoophycos* Ichnofacies fabrics can be explained by the general burrow stratigraphy that has been recognized in modern deep-sea pelagic oozes (Berger et al., 1979; Ekdale and Berger, 1978;

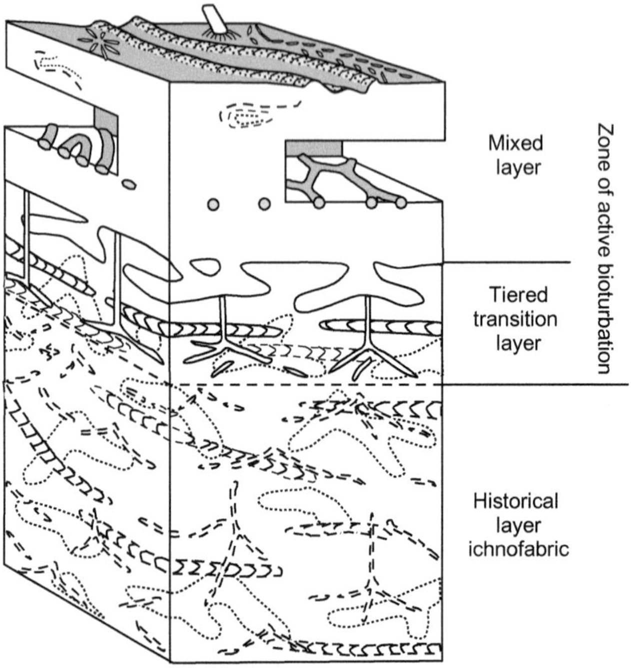

FIGURE 2 Burrow stratigraphy in carbonate ooze and resulting ichnofabric preserved in the stratigraphic record.

Ekdale et al., 1984a). In these substrates, active bioturbation occurs in two layers; a surface mixed layer and a transition layer (Fig. 2). The surface mixed layer is a zone of rapid and continuous homogenization by epibenthic and shallow endobenthic animals. As revealed by sea-floor photographs and in the tops of box cores, these organisms may produce surface tracks and trails, as well as endogenic locomotion, dwelling, feeding, and gardening structures, including graphoglyptids—for example, *Cosmorhaphe*, *Spirorhaphe*, and *Paleodictyon* (e.g., Ekdale et al., 1984a: fig. 7; Gaillard, 1991). The transition layer, the thickness of which varies as a function of environmental factors, is a zone of heterogeneous mixing by organisms that live or feed at greater depths in the substrate. Transition-layer organisms produce suites of endogenic traces that may include open and actively filled burrows or burrow systems, as well as complex spreiten. Notably, the preservation potential of distinct mixed-layer traces is very low. Because mixed-layer sediments are more fluid, open burrows produced therein may collapse shortly after formation, and those that are actively or passively filled may be compacted beyond recognition. Moreover, surface traces and shallow endogenic structures are subsequently overprinted by transition-layer structures as the zone of active bioturbation migrates slowly upward in pace with sedimentation. Hence, ichnofabrics preserved in the stratigraphic record comprise a homogeneous or burrow-mottled background produced during passage through the mixed layer, and an assemblage of preferentially preserved endogenic structures subsequently emplaced during passage through the transition layer (Berger et al., 1979; Fig. 2).

Systematic cross-cutting relationships among recurring ichnotaxa also can be accounted for by sequential biogenic activity in substrates. Transition layers may be tiered (Fig. 2); that is, tracemakers may preferentially occupy distinct depth zones below the mixed layer (Bromley and Ekdale, 1986; Wetzel, 1984). As tiers migrate upward in pace with sedimentation, deeper tier structures sequentially overprint those produced in shallower tiers. Thus, in many cases, cross-cutting relationships observed in *Zoophycos* Ichnofacies reflect the tiering structure of a single ichnocoenosis. In other words, they record the sequential work of depth-stratified tracemakers belonging to a temporally stable endobenthic community. However, in some cases, as will be described below, cross-cutting relationships may reflect the overprinting of one ichnocoenosis by another, in response to changes in environmental conditions and associated benthic communities.

2.3 Ichnocoenoses and Ichnofabrics

Ichnocoenoses vary among modern and ancient pelagic carbonates, mainly reflecting differences in water depth, substrate character, and benthic oxygenation. Most relatively pure pelagic carbonates that accumulated in oxygenated deep-sea settings are characterized by low-diversity ichnocoenoses. Modern carbonate oozes are dominated by *Planolites*, *Chondrites*, *Zoophycos*, and *Teichichnus* (Ekdale and Berger, 1978; Ekdale et al., 1984a), and these same ichnotaxa are nearly ubiquitous in Cretaceous and Tertiary deep-sea chalks and pelagic limestones (Chamberlain, 1975, Ekdale, 1977, Fütterer, 1984, Warme et al., 1973; see Table 1, Figs. 3 and 4). While additional ichnotaxa (e.g., *Trichichnus*, *Cylindrichnus*) also may occur locally, *Thalassinoides* is notably rare or absent in deep-sea ichnocoenoses.

Ichnocoenoses in chalks formed in somewhat shallower, well-oxygenated outer-shelf or distal epeiric basin settings—for example, the Maastrichtian marly chalk beds of Denmark (Ekdale and Bromley, 1991; Lauridsen et al., 2011) and the Campanian Demopolis Chalk of the U.S. Gulf coastal plain (Fig. 5)—are similar to those in deep-sea deposits in that they typically include *Planolites*, *Chondrites*, *Teichichnus*, and *Zoophycos*. However, additional ichnotaxa (e.g., *Taenidium* and *Phycosiphon*) are commonly present and, more important, *Thalassinoides* is nearly ubiquitous (Table 1, Figs. 3A and 5). Indeed, the presence or absence of *Thalassinoides* is a key criterion for distinguishing deep-sea and shelf-sea chalks (Ekdale and Bromley, 1984a). Ichnocoenoses that include *Thalassinoides* but lack *Zoophycos* are generally restricted to even shallower-water chalks and are more allied with the *Cruziana* and *Glossifungites* ichnofacies described in Section 3 and 4.

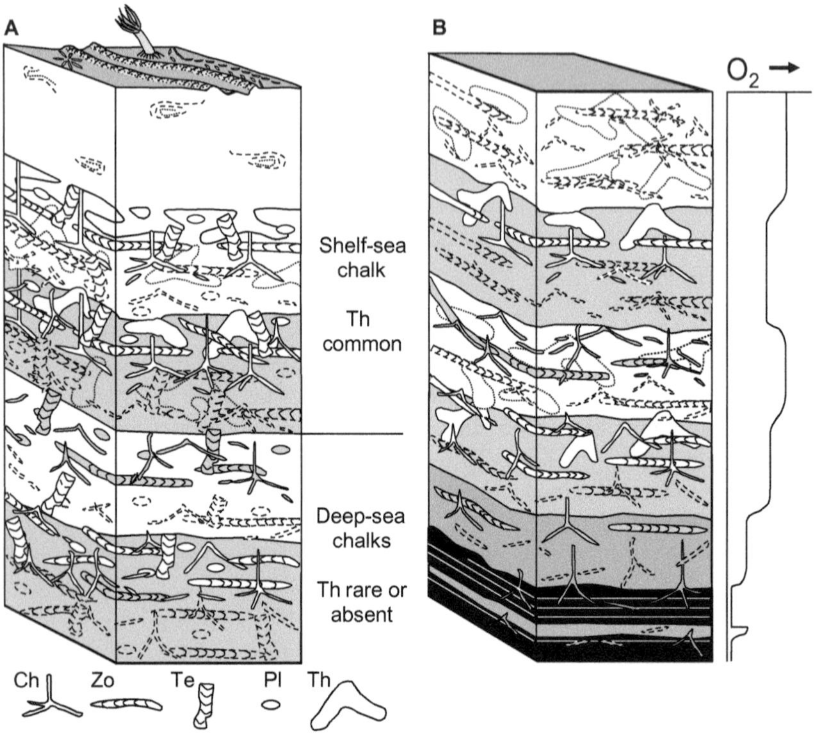

FIGURE 3 Schematic representation of typical ichnofabrics and recurring trace-fossil suites representing the *Zoophycos* Ichnofacies in chalks deposited in (A) well-oxygenated deep-sea *versus* shallow-shelf settings, and (B) variably oxygenated basins (Ch; *Chondrites*, Zo; *Zoophycos*, Te; *Teichichnus*, Pl; *Planolites*, Th; *Thalassinoides*). As shown in both A and B, the expression of the trace fossils is improved in heterogeneous successions at transitions between lighter and darker beds (bed-junction preservation). As depicted in B, systematic changes in trace-fossil diversity, burrow diameters, and depths of burrow penetration with changes in benthic oxygenation result in the stacking of distinct oxygen-related ichnocoenoses. Note in B that burrows representing one ichnocoenosis may overprint laminae or ichnofabrics formed under previous conditions. Examples of deep-sea, shelf, and oxygen-deficient epeiric basin chalk ichnofabrics are provided in Figs. 4, 5 and 6, respectively.

Tiering is particularly well manifested by cross-cutting relationships among recurring ichnotaxa in outer-shelf and epeiric basin chalks (e.g., Ekdale and Bromley, 1982, 1991; Frey and Bromley, 1985; Locklair and Savrda, 1998a,b; Fig. 3). *Thalassinoides* typically is overprinted by all other ichnotaxa and, thus, characterizes a shallow tier. In contrast, *Chondrites* is superimposed on all other structures (e.g., Fig. 5C), reflecting its prevalence in the deepest tier. Other recurring ichnotaxa (*Phycosiphon*, *Planolites*, *Taenidium*, *Teichichnus*, and *Zoophycos*) represent intermediate tiers.

The substrate composition and consistency may have an impact on ichnocoenoses and ichnofabrics. As carbonate contents decrease substantially, either due to increased influx of siliciclastic muds or enhanced carbonate dissolution

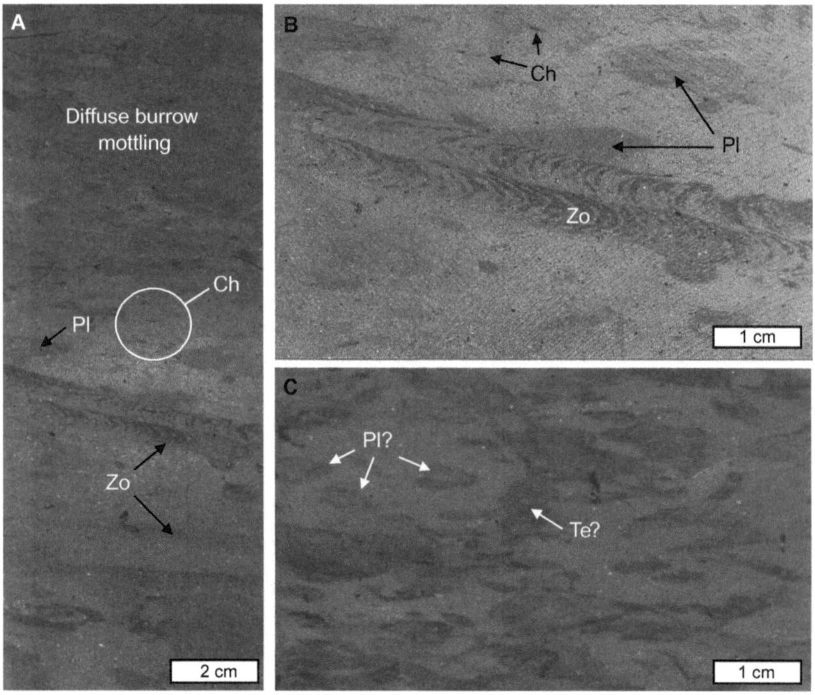

FIGURE 4 Examples of ichnofabrics in deep-sea chalks and marls, Eocene, ODP core, site 1073, Leg 174A, New Jersey margin. (A) Core section showing both diffuse burrow mottling in relatively carbonate-poor mud (top) and distinct trace fossils in chalk (Pl; *Planolites*, Zo; *Zoophycos*, Ch; *Chondrites*). (B) Close-up of the chalk ichnofabric in A. (C) Ill-defined structures in clay-rich ooze. Burrow mottling may include *Planolites* (Pl?) and *Teichichnus* (Te?).

(e.g., in abyssal settings below the lysocline), substrates become less cohesive. As a result of lower sediment shear strengths, even transition-layer structures may be significantly influenced by compaction. Consequently, in modern deep-sea marly oozes and in some Cretaceous and Tertiary marls, ichnofabrics are characterized by lower diversity assemblages of ill defined, typically horizontal structures (e.g., Fig. 4C) that cautiously may be ascribed to *Planolites*, *Chondrites*, or *Thalassinoides* (Ekdale, 1980; Ekdale and Berger, 1978; Ekdale and Bromley, 1982, 1984b; Frey and Bromley, 1985; Wetzel, 1991).

Ichnocoenoses representing the *Zoophycos* Ichnofacies also vary as a function of benthic redox conditions (Figs. 3B and 6). While modern carbonate ooze and most Cretaceous and Tertiary chalks accumulated in oxygenated settings, some pelagic carbonates accumulated beneath dysoxic or anoxic waters. These include some deep-sea carbonates—for example, those deposited during Cretaceous oceanic anoxic events (Ekdale, 1985) or the K/T boundary interval (Droser and Bottjer, 1991)—and chalks and marls deposited in stratified epeiric basins, such as the U.S. Western Interior (Savrda, 1998a,b; see Fig. 6) and the

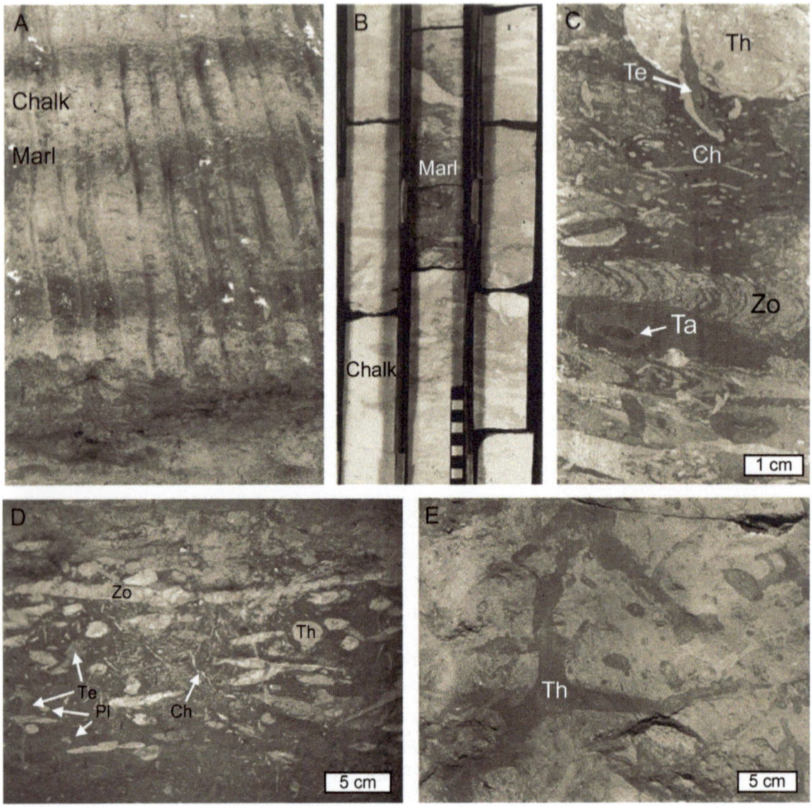

FIGURE 5 *Zoophycos* Ichnofacies manifested in rhythmically bedded, outer-shelf chalk and marl of the Campanian Demopolis Chalk, Alabama, USA (see Locklair and Savrda, 1998a,b). (A) and (B) Decimeter-scale rhythmic alternation of lighter chalks and darker marls, manifested in outcrop (A) and core (B), reflecting siliciclastic dilution cycles mediated by orbital precession. Note the blurred bed boundaries associated with thorough bioturbation and improved visibility of trace fossils at chalk–marl and marl–chalk transitions. Despite variations in the sediment character, chalk and marl beds in the Demopolis Chalk contain the same ichnocoenosis (exposure in A is ~2.5 m high; scale in B is graduated in cm). (C) and (D) Trace fossils expressed in cross-sectional views of marl–chalk transitions in core (C) and outcrop (D). Homogeneous background fabrics are cut by *Thalassinoides* (Th), *Zoophycos* (Zo), *Teichichnus* (Te), *Chondrites* (Ch), *Taenidium* (Ta), *Planolites* (Pl), and *Phycosiphon* (not visible in photograph). (E) Bedding-parallel view of *Thalassinoides* (Th) and other trace fossils in a chalk–marl transition.

Central Graben of the North Sea (Danholt and Surlyk, 2004). Ichnocoenoses developed under progressively lower levels of benthic oxygenation are characterized by progressively lower trace-fossil diversities. Reduction in diversity is generally accompanied by decreases in burrow diameters and bioturbation depths (Figs. 3B and 6). Larger, shallow-tier structures such as *Thalassinoides* are normally the first structures to be excluded from ichnocoenoses as benthic oxygenation decreases (e.g., compare Fig. 6B and C), while smaller diameter, deeper-tier structures normally are the last to disappear before bioturbation ceases altogether (e.g., Fig. 6D). Indeed, ichnofabrics characterized only by *Chondrites* commonly appear immediately below or above laminated, carbonaceous chalks or marls (Bromley and Ekdale, 1984a; Ekdale, 1985). Notably, in chalk–marl successions that accumulated under fluctuating levels of benthic oxygenation, various oxygen-related ichnocoenoses typically overprint one another or laminated intervals with which they may be intercalated (Figs. 3B and 6E). Recognition of stacked oxygen-related ichnocoenoses in these successions provides a means of deciphering basin oxygenation histories (Erba and Premoli-Silva, 1994; Savrda, 1998a,b; Savrda and Bottjer, 1989; Fig. 6E).

2.4 Factors Affecting the Expression of Trace Fossils

The quality of the expression of ichnofabrics and trace fossils in pelagic carbonates, or in any of the carbonate facies described herein, mainly is controlled by (1) the level of textural and compositional contrast between biogenic structures and background sediment, and (2) the extent to which structures have been deformed in depositional and diagenetic regimes. The trace-fossil contrast is governed by factors such as stratigraphic heterogeneity, tracemaker behavior, preferential diagenesis, and weathering. In some relatively pure chalks, the natural trace-fossil contrast is extremely low, and detailed ichnofabric observations may not be possible without employing contrast-enhancing techniques (oil application, high-contrast photography, image analysis, etc.; Bromley, 1981; Bromley and Richter, 1999). The natural contrast commonly is higher in more heterolithic successions containing alternating light and dark beds (e.g., chalks and marly chalks; Figs. 3 and 5); at bed transitions, the structures typically are filled with or comprise sediments that are darker or lighter than the sediments in which they were emplaced (Lauridsen et al., 2011; Locklair and Savrda, 1998a). This type of bed-junction preservation (Simpson, 1957) is biased toward (1) open burrows that are maintained as dwelling or feeding structures but are susceptible to passive filling from above (e.g., *Thalassinoides*) and (2) structures produced by deep-dwelling reverse conveyors that actively transport surface or near-surface sediments into the transition layer (e.g., *Zoophycos*; Kotake, 1989).

Natural enhancement of a subtle primary textural or compositional contrast also may result from early diagenetic processes (Berger et al., 1979). The trace-fossil visibility in chalk may be accentuated by preferential precipitation of early diagenetic minerals in or around biogenic structures (i.e., the diagenetic preservation of Simpson, 1957). For example, in some European shelf-sea chalk successions containing omission surfaces (*Glossifungites* Ichnofacies), elements of *Zoophycos* Ichnofacies ichnofabrics are more readily apparent in diagenetic flint nodules than in unmineralized host chalks (Bromley and Ekdale, 1984b). Weathering commonly has the opposite effect; leaching, oxidation, and formation of superficial crusts diminish the natural trace-fossil contrast in surface exposures (Dawson and Reaser, 1990; Frey and Bromley, 1985).

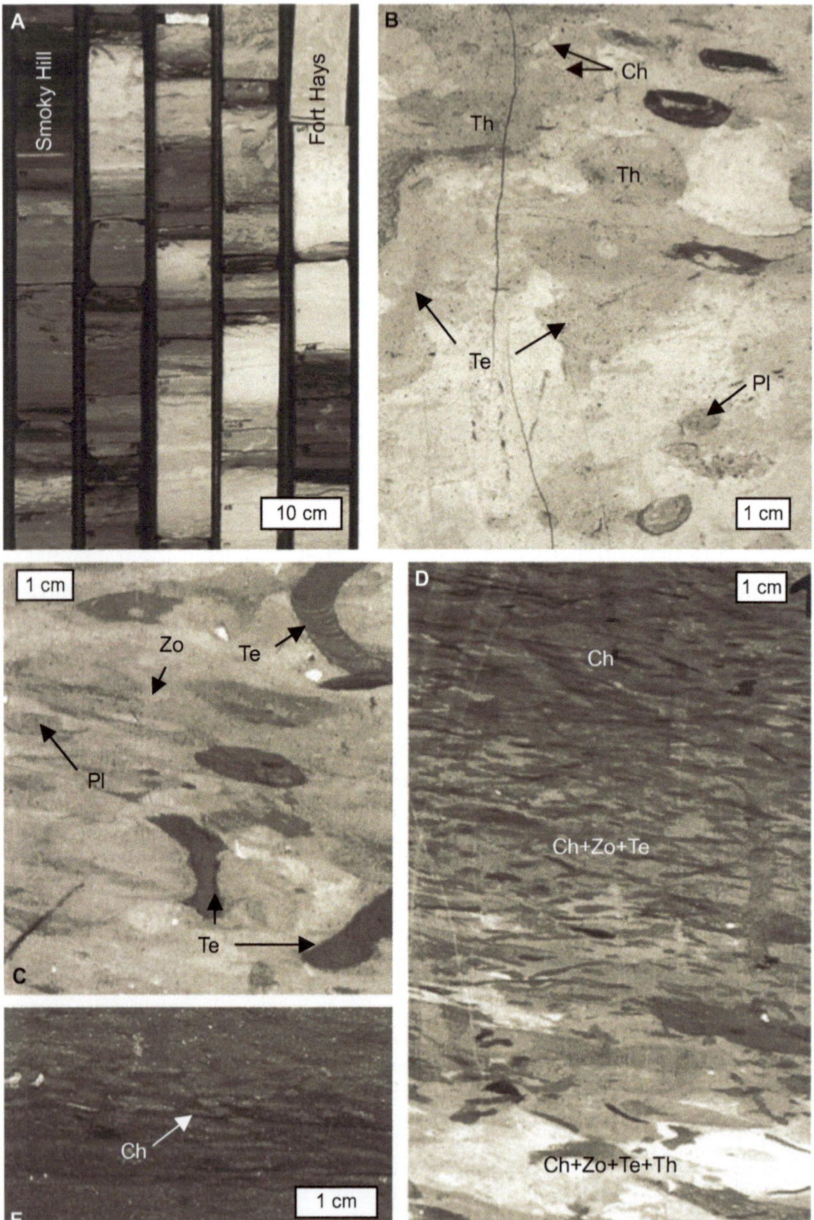

FIGURE 6 Rocks and ichnofabrics representing the *Zoophycos* Ichnofacies in rhythmically bedded, epeiric basin chalks, marls, and calcareous shales of the Turonian-Campanian Niobrara Formation, U.S. Western Interior (Savrda, 1998a). (A) Decimeter-scale rhythmic alternation of lighter chalk (limestone), moderately carbonaceous gray marl (marlstone), and dark carbonaceous calcareous shale manifested in core. Variations in carbonate and organic contents, as well as ichnofabrics,

In addition to the mechanical compaction discussed above, other deformational processes may act to obscure ichnofabric elements in deposits characterized by the *Zoophycos* Ichnofacies. What were once distinct trace fossils may be smeared or truncated as a result of (1) downslope remobilization of carbonate ooze (via slow creep or rapid slumping or flowage; Bromley and Ekdale, 1987) or (2) chemical compaction and formation of solution seams, stylolites (Garrison and Kennedy, 1977; Kennedy and Garrison, 1975; Scholle, 1977), or microstylolites (e.g., Ekdale and Bromley, 1988; Steinich, 1967).

3. *CRUZIANA* ICHNOFACIES

3.1 Environments and Substrates

Ichnofabrics best ascribed to the *Cruziana* Ichnofacies are characteristic of sandy, fossiliferous, commonly glauconitic, and locally phosphatic chalk and marl that accumulated in relatively shallow-shelf settings. Such facies are well represented in Cretaceous shelf-sea chalk successions, including those of Europe (Kennedy, 1975) and North America, such as the Annona and Saratoga chalks of Arkansas (Bottjer, 1985, 1986), the Prairie Bluff Chalk of Alabama (Frey and Bromley, 1985), and the Austin Chalk of Texas (Dawson and Reaser, 1985, 1990; see Table 1). Abundance of siliciclastic sand, concentrations of molluscan skeletal debris, and the localized occurrence of shallow channels and scour surfaces indicate deposition in periodically energetic settings between fair-weather and storm-wave bases. The prevalence of glauconite and the common occurrence of encrusted and bored shells and phosphatic molds indicate relative sediment starvation and stratigraphic condensation.

3.2 Ichnocoenoses and Ichnofabrics

Chalks characterized by the *Cruziana* Ichnofacies are thoroughly bioturbated, except rare vestiges of laminated or cross-laminated sand or calcarenite. As with the *Zoophycos* Ichnofacies, any surface structures and shallow endogenic

reflect combined redox-dilution cycles linked to orbitally driven wet/dry climate cycles. The photograph includes the transition from the limestone-dominated Fort Hays Member to the more heterolithic Smoky Hill Member of the Niobrara Formation. (B) Ichnofabric in relatively pure chalk deposited under well-oxygenated bottom waters. *Thalassinoides* are cut by *Teichichnus* (Te), *Planolites* (Pl), *Zoophycos* (not visible in the photograph), and *Chondrites* (Ch). (C) Marly chalk-reflecting moderate paleo-oxygenation. The trace-fossil assemblage includes *Teichichnus* (Te), *Planolites* (Pl), *Zoophycos* (Zo), and *Chondrites* (not visible in the photograph), but lacks *Thalassinoides*. (D) Ichnofabrics reflecting the progressive decrease in trace-fossil diversity and burrow size associated with a phase of bottom-water deoxygenation. Ch, Te, Th, and Zo are defined as above. (E) Ichnofabric formed in response to increased benthic oxygenation. Note the penetration of lighter *Chondrites* (Ch) into previously unburrowed laminated calcareous shale.

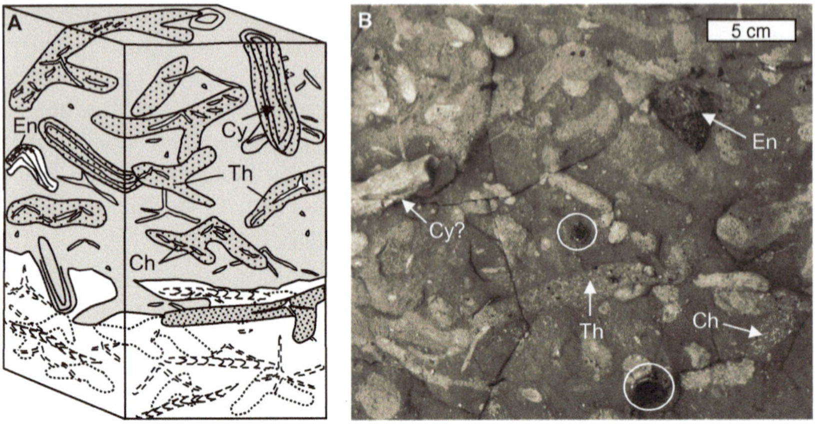

FIGURE 7 *Cruziana* Ichnofacies in typically sandy and shelly shallow-shelf chalks. (A) Schematic representation of common trace fossils, including *Cylindrichnus* (Cy), *Thalassinoides* (Th), and *Chondrites* (Ch). Associated robust shells or phosphatic molds thereof commonly contain *Entobia* (En) and other borings. Intervals characterized by the *Cruziana* Ichnofacies may be interbedded with finer-grained chalk characterized by the *Zoophycos* Ichnofacies (base). (B) *Cruziana* Ichnofacies in sandy, glauconitic chalk of the Maastrichtian Prairie Bluff Chalk, Alabama. The chalk formed in a shallow-shelf setting during a period of sediment starvation. The ichnofabric is dominated by robust *Thalassinoides* (Th), *Cylindrichnus* (Cy?), and *Chondrites* (Ch). The *Chondrites* tracemaker preferentially exploited coarser, glauconitic fills of the larger burrows. Note the *Entobia*-bored (En) oyster and phosphatic molds (circled).

structures that may have been produced in these substrates normally are not preserved owing to heavy overprinting by deeper-tier trace fossils.

Trace-fossil assemblages within sandy chalk successions, including the Maastrichtian Prairie Bluff Chalk exposed in the U.S. Gulf coastal plain, mainly comprise relatively simple deposit-feeding structures but also may include structures produced by sessile suspension feeders. Virtually all assemblages include *Thalassinoides*, most contain *Planolites*, *Cylindrichnus*, and *Chondrites* (Fig. 7), and some locally include *Skolithos*, *Teichichnus*, and *Trichichnus* (Table 1). Notably, *Zoophycos* is conspicuously absent. Shells and phosphate nodules within these deposits commonly contain borings produced by sponges (*Entobia*; Fig. 7) and bivalves (*Gastrochaenolites*).

Comparatively little is known about trace-fossil tiering in these sandier substrates. Where present, *Chondrites* typically cross-cuts all other ichnotaxa and commonly is concentrated in the fills of larger *Thalassinoides* (Kennedy, 1975; Fig. 7B). This indicates that *Chondrites* was deeply emplaced. However, given the environmental dynamics associated with these shallower-water deposits, it is not clear whether the occupation of substrates by the *Chondrites* tracemaker and other organisms were contemporaneous. The observed cross-cutting relationships may reflect the superimposition of two or more temporally disjunct ichnocoenoses rather than true tiering.

3.3 Factors Affecting the Expression of Trace Fossils

The expression of trace fossils in deposits characterized by the *Cruziana* Ichnofacies may be influenced by some of the same factors described in Section 2.3 for the *Zoophycos* Ichnofacies. Sandy chalk successions typically are heterogeneous in terms of texture and composition, for example, in abundance of sand or glauconite. Hence, the visibility of trace fossils, particularly that of relatively robust *Thalassinoides*, is commonly enhanced by bed-junction preservation. The expression of trace fossils also may be enhanced in surface exposures by differential cementation and weathering of burrow fills. Notably, sandier chalk substrates generally were more cohesive than finer-grained pelagic oozes and, hence, trace fossils produced therein were less susceptible to mechanical compaction and other soft-sediment deformational mechanisms.

4. *GLOSSIFUNGITES* ICHNOFACIES

4.1 Environments and Substrates

Trace-fossil assemblages representing the firmground *Glossifungites* Ichnofacies occur locally in deep-sea chalks (e.g., Bromley and Allouc, 1992; see Savrda et al., 2001) but are most common in shelf-sea chalk successions (Bromley, 1967, 1975; Kennedy and Garrison, 1975). Chalk firmgrounds developed in two ways: (1) erosional exhumation of semiconsolidated carbonate ooze in response to significant sea-level changes, and (2) incipient synsedimentary cementation of carbonate ooze associated with periods of increased bottom-current intensity and sediment bypass.

Firmgrounds formed by the first mechanism, referred to here as type 1 firmgrounds, have sequence-stratigraphic significance. They locally occur at the tops of shelf-sea chalk or marl successions that were truncated by coplanar sequence boundaries/transgressive surfaces. Subaerial erosion during sea-level lowstand and subsequent transgressive erosion resulted in the exhumation and marine inundation of dewatered calcareous ooze. This mechanism is likely responsible for the isolated firmground ichnofabrics developed at formational boundaries in Cretaceous Gulf coastal plain successions, for example, at the top of the Austin Chalk in Texas (Dawson and Reaser, 1980; Fürsich et al., 1981), bases of the Saratoga and Annona chalks in Arkansas (Bottjer, 1985), and the top of the Prairie Bluff Chalk in Alabama (Savrda, 1991; Fig. 8A). Comparable firmground ichnofabrics also have been recognized beneath scour surfaces that were cut into Tertiary deep-sea chalk in slope and rise settings (Fig. 8B). Firmground development at these depths is linked to extreme sediment starvation and increased impact of contour currents, both of which were deep-water responses to phases of rapid sea-level rise (Savrda et al., 2001).

Firmgrounds formed by the second mechanism, or type 2 firmgrounds, mark omission surfaces *within* relatively shallow-shelf sea chalk successions.

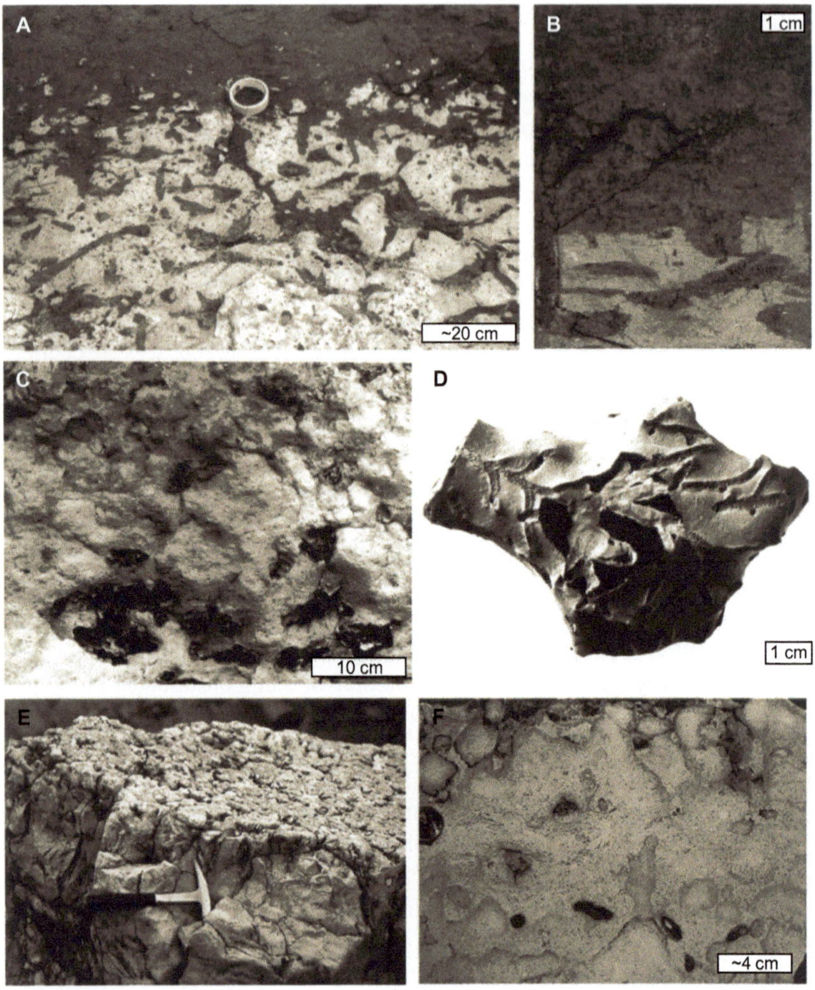

FIGURE 8 Ichnofabrics and trace fossils associated with the firmground *Glossifungites* and hardground *Trypanites* ichnofacies in chalks. (A) Well-defined robust firmground *Thalassinoides* across a coplanar sequence boundary-transgressive surface (type 1 firmground) marking the contact between the Maastrichtian Prairie Bluff Chalk below and sandy glauconitic marls of the Danian Clayton Formation above (western Alabama). (B) Type 1 firmground *Thalassinoides* marking an Eocene transgressive surface bounding diatomaceous chalk (below) and glauconitic sandy mud (above), ODP core, site 1073, Leg 174A, New Jersey margin. (C) Turonian chalk succession exposed near Normandy, France, showing the transition from soft chalk characterized by the *Zoophycos* Ichnofacies to nodular chalk marking an omission surface with the *Glossifungites* (type 2 firmground) and *Trypanites* ichnofacies (topmost part of the photograph). Flint nodules in the lower part preferentially nucleated beneath the omission surface in and around large *Thalassinoides* (*from Ekdale and Bromley, 1984a*). (D) Flint nodule from Danish chalk nucleated in or around a *Thalassinoides* branch juncture and preserving pelletal fill fabric of *Chondrites* on its exterior (*from Bromley and Ekdale, 1984b*). (E) Bored hardground surface, Cretaceous chalk of Normandy, France (*from Scholle et al., 1983*). (F) Polished slab of Cretaceous chalk from southern England with multiple phosphatized and glauconitized, bored hardground surfaces (*from Scholle et al., 1983*).

Omission surfaces characterized by the *Glossifungites* Ichnofacies occur only locally in North American chalk successions, for example, in the Austin Chalk of Texas (Dawson and Reaser, 1990). However, they are particularly common in relatively pure Cretaceous chalk successions in north-western Europe (Bromley, 1967, 1975; Kennedy and Garrison, 1975; Table 1) wherein they define decimeter- to meter-scale scour cycles and are commonly associated with nodular chalk or flints (Figs. 8C and 9).

4.2 Ichnocoenoses and Ichnofabrics

In both type 1 and type 2 chalk firmgrounds, crustaceans were the preeminent tracemakers. Type 1 firmgrounds are dominated by vertically extensive, sharp-walled *Thalassinoides* (Bottjer, 1985; Savrda, 1991; Savrda et al., 2001; Fig. 8A), some of which exhibit well-defined wall ornamentation (e.g., scratch traces, the *Spongeliomorpha* of Frey and Bromley, 1985), but they locally also include *Rhizocorallium* (Dawson and Reaser, 1980).

Ichnocoenoses characterizing type 2 firmgrounds also are dominated by sharp-walled *Thalassinoides* burrow networks but may include other related arthropod traces such as *Pseudobilobites*, *Spongeliomorpha*, and "*Glossifungites*" (=*Rhizocorallium*) (Bromley, 1967, 1975; Bromley and Allouc, 1992; Kennedy and Garrison, 1975; Table 1). *Thalassinoides* burrow morphologies vary depending on the extent of substrate cementation at the time of emplacement. *Thalassinoides* formed during earlier stages of omission—that is, before the onset of syndepositional cementation—are relatively undeformed, or idiomorphic. In contrast, burrow systems formed contemporaneously with incipient cementation and localized growth of hard nodules exhibit more distorted or stenomorphic configurations (Fig. 9), reflecting evolving heterogeneity of the available burrowing space (Bromley, 1967, 1975).

Regardless of the mechanism responsible for formation of firm substrates, the carbonate sediments that comprise them were originally soft oozes. Hence, firmground *Thalassinoides* invariably overprint pre-erosion or preomission softground ichnocoenoses representing the *Zoophycos* or possibly the *Cruziana* ichnofacies. Ichnofabrics of omission surfaces may be particularly complex. In some cases, ichnofabrics record periodical filling and reexcavation of *Thalassinoides* burrows (Bromley, 1967), reflecting the amalgamation of omission surfaces. Moreover, ichnofabrics formed in type 2 firmgrounds may be overprinted by a postlithification omission suite of borings of the *Trypanites* Ichnofacies—if extended omission subsequently led to complete sea-floor cementation—and/or by a postomission softground trace-fossil suite (Fig. 9). The latter suite reflects the resumption of normal sedimentation, filling of irregular firmground burrows with soft ooze, and exploitation of a soft burrow fill by reestablished softground bioturbators. Burrows emplaced within these burrow fills are distorted or xenomorphic (Bromley, 1967) as a result of the irregular geometry of the soft-sediment burrowing space.

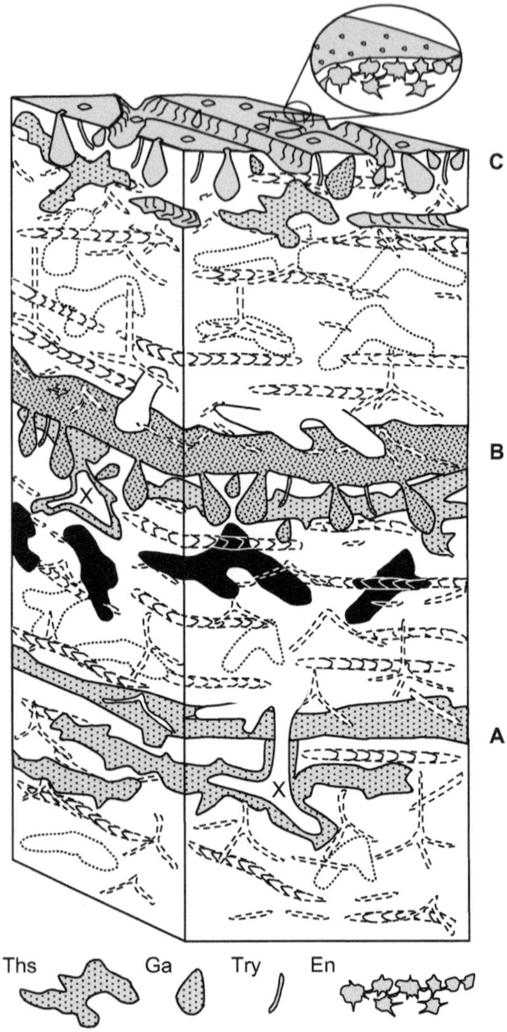

FIGURE 9 Schematic depiction of three chalk omission surfaces (A–C) illustrating the overprinting of softground trace-fossil suites (*Zoophycos* Ichnofacies) by firmground and hardground trace fossils representing the *Glossifungites* and *Trypanites* ichnofacies, respectively. Such a decimeter-scale alternation of softground ichnofabrics and omission assemblages typify intervals of European chalks (Bromley, 1975; Kennedy and Garrison, 1975) and likely reflect climate-driven scour cycles (Bottjer et al., 1986). Omission surface (A) includes only a prelithification firmground suite of robust stenomorphic *Thalassinoides* (Ths), reflecting resumption of sediment before significant sea-floor cementation. Exploitation of softer fills of firmground *Thalassinoides* by subsequently established softground tracemakers may result in distorted, xenomorphic burrow morphologies (X). At omission surfaces (B) and (C), both softground and firmground ichnofabrics are cut by postlithification suites of borings, including *Gastrochaenolites* (Ga), *Trypanites* (Try), and *Entobia* (En). As depicted beneath surface (B), depositional hiatuses may result in dissolution and reprecipitation of silica to form flint nodules, which commonly nucleate in or around crustacean burrows (e.g., *Thalassinoides*) below omission horizons.

4.3 Factors Affecting the Expression of Trace Fossils

Firmground ichnofabrics typically are rather well expressed. *Thalassinoides* are robust, their burrow boundaries are sharp and their burrow fills commonly contrast markedly with the host sediments. Enhanced visibility of burrows related to bed-junction preservation occurs in both types of firmgrounds but is particularly obvious in type 1 firmgrounds, which invariably are overlain by sediments that differ appreciably in texture and/or composition from the host chalk (Fig. 8A). In type 2 firmgrounds, *Thalassinoides* may be further accentuated by the formation of authigenic glauconitic, phosphatic, or ferruginous crusts along the burrow walls, particularly if burrowing was followed by substrate lithification (Bromley, 1967, 1975; Kennedy and Garrison, 1975).

Diagenetic preservation of burrow fills by silica replacement at or beneath omission surfaces, manifested in flint nodules, is a common feature in the Middle and Upper Chalk of the United Kingdom (Fig. 8C and D). This special mode of preservation reflects the precipitation of silica, derived from intrastratal dissolution of biogenic components such as diatoms and sponge spicules, in subsurface redox zones that were stabilized at shallow substrate depths in response to extended periods of non-deposition (Madsen and Stemmerik, 2010). Flint nodules are preferentially nucleated in or around burrow fills, particularly that of *Thalassinoides*, and commonly avail three-dimensional views of the trace fossils they incorporate (Bromley and Ekdale, 1984b; Fig. 8D). Although silicification is associated with omission and in some cases may have selectively preserved firmground *Thalassinoides* (Bromley, 1967; Kennedy and Garrison, 1975), conditions necessary for silica precipitation commonly developed at substrate depths *below* active firmground bioturbation (Fig. 9). In the latter case, flint-nodule growth preferentially preserved components of softground ichnofabrics in chalk subjacent to omission surfaces (Bromley and Ekdale, 1984b; Kennedy and Garrison, 1975).

Other diagenetic processes may have had adverse impacts on the expression of firmground ichnofabrics. As exemplified by some European chalk omission surfaces, particularly those associated with nodular chalks, chemical compaction, and the formation of solution seams around harder nodules locally have produced flaser-bedded chalk fabrics within which burrows may be distorted or obscured (Kennedy and Garrison, 1975).

5. *TRYPANITES* ICHNOFACIES

5.1 Environments and Substrates

The *Trypanites* Ichnofacies develops in lithified carbonate substrates—in rockgrounds, formed by the exhumation of once deeply buried limestones (e.g., Bromley and Asgaard, 1993; Malfait and Van Andel, 1980), and in true hardgrounds, formed by synsedimentary lithification of carbonate ooze. Hardgrounds, like the type 2 firmgrounds from which they developed, result from bottom-current winnowing and sediment bypass and thus are characteristic of omission surfaces (Figs. 8E, F and 9). While such hardgrounds can form at considerable water depths of up to ~4 km for some Quaternary substrates (Bromley and Allouc, 1992), most well-documented examples are those contained in relatively shallow-shelf sea chalk successions of Late Cretaceous age in north-western Europe (Bromley, 1975; Kennedy and Garrison, 1975).

5.2 Ichnocoenoses and Ichnofabrics

Hardground *Trypanites* Ichnofacies assemblages, or postlithification omission suites, comprise bioerosion structures (Bromley, 1967, 1975). Owing to taphonomic biases imposed by extended phases of omission and bioerosion (Bromley, 1994), hardground ichnocoenoses developed in shelf-sea chalks typically are dominated by relatively deeply emplaced borings such as *Entobia*, *Trypanites*, and *Gastrochaenolites* produced by sponges, polychaetes, and bivalves, respectively (Bromley, 1975; Fig. 9). Notably, *Gastrochaenolites* is generally rare or absent in deep-sea hardgrounds (Ekdale et al., 1984b).

Because the sediments that embody chalk hardgrounds originated as soft ooze and subsequently passed through a firmground stage prior to lithification, assemblages of borings invariably overprint earlier formed softground and firmground ichnocoenoses (Fig. 9). Indeed, borings not only penetrate downward across omission surfaces but also commonly were emplaced at various orientations into the hardened, mineralized walls of open firmground burrows of the prelithification suite (Bromley, 1975).

5.3 Factors Affecting the Expression of Trace Fossils

The visibility of bioerosion structures in hardgrounds is generally good. Like firmground burrows, borings have extremely sharp walls, which may be accentuated by diagenetic mineralization (Bromley, 1975; Fig. 8F), and are commonly filled with sediment that contrasts with the host substrate.

6. TRACE FOSSILS IN REMOBILIZED SUBSTRATES

Most of the deposits described above are true pelagites or hemipelagites. However, chalk or limestone that formed via redeposition of pelagic ooze and other carbonates by gravity flows, including turbidity currents, may be important locally.

6.1 Ichnofabrics of Slump, Ooze-Flow, and Debris-Flow Deposits

Structures and fabrics indicative of mass transport of oozes by slumping, gliding, and ooze or debris flow are represented at least locally in Cretaceous and Danian chalk successions throughout northern Europe, reflecting oversteepened slopes along unstable margins (Bromley and Ekdale, 1987; Bromley et al., 1975; Ekdale et al., 2012; Nygaard, 1983). In many cases, ichnofabrics in allochthonous chalk beds simply reflect shearing and truncation of trace-fossil assemblages emplaced prior to sediment remobilization, and these assemblages may be representative of any of the ichnofacies described above (Bromley and Ekdale, 1987). However, some of these beds also may contain, in their upper parts, trace-fossil assemblages emplaced after ooze remobilization. Typified by *Zoophycos*, *Chondrites*, and/or *Trichichnus*, these assemblages resemble *Zoophycos* Ichnofacies suites in deep-sea or oxygen-deficient basin chalks (Ekdale et al., 1984b; Fig. 10A). Multiple, stacked allochthonous chalk beds may be penetrated by narrow, vertically extensive (up to 9 m long), sparsely branched structures that commonly are encased in cylindrical flint nodules. Referred to as *Bathichnus*, these unusually large enigmatic structures have been interpreted as combined feeding-dwelling-escape structures of unknown but long-lived organisms (Bromley et al., 1975). Notably, *Bathichnus* also is recognized in some pelagic successions (Table 1) and, thus, is not necessarily diagnostic of allochthonous chalk deposits.

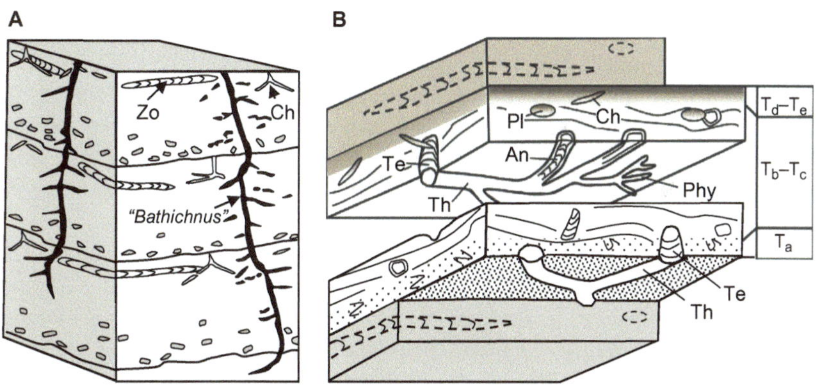

FIGURE 10 Trace-fossil associations in redeposited chalks. (A) Ichnofabrics observed in European chalks deposited by ooze flows or debris flows. The tops of the depositional units may be marked by normal softground trace fossils, most commonly *Chondrites* (Ch) and *Zoophycos* (Zo), while multiple units may be penetrated by vertically extensive, enigmatic structures referred to as *Bathichnus* (see text). The latter structures may be encased in cylindrical flint nodules (not shown). (B) Ichnofabric in a calciturbidite based on observations by Miller et al. (2004) from the Cretaceous Scaglia Rossa of Italy. Unlike siliciclastic turbidites, which are typified by the *Nereites* Ichnofacies, calciturbidites commonly lack pre-depositional traces on their soles, and post-depositional traces are commonly enigmatic. In this example, distinct trace fossils in the middle part of the calciturbidite (divisions T_b and T_c) include compound *Thalassinoides* (Th)—*Phycodes* (Phy)—*Teichichnus* (Te)—*Ancorichnus*? (An) that may have been produced by crustaceans transported by associated turbidity currents. Trace fossils in turbidite divisions T_d and T_e include *Planolites* (Pl) and *Chondrites* (Ch).

6.2 Calciturbidites and the *Nereites* and *Arenicolites* Ichnofacies

Some deeper water carbonate or mixed carbonate/siliciclastic successions, including Cretaceous and Tertiary packages exposed in the Alps, Apennines, and Carpathians, contain limestones, chalks, or marls that were emplaced as calciturbidites (Miller et al., 2004; Uchman, 2007). Notably, the carbonates that comprise turbidites in these successions are of varied origin; some represent redeposited nannofossil ooze (e.g., Uchman, 2007), while others represent allochthonous shallow-platform carbonate (Giannetti and McCann, 2010; Savary et al., 2004). Nonetheless, these calciturbidites share several features, including ichnological attributes that differ markedly from those in siliciclastic turbidites (Uchman and Wetzel, 2012).

With noted exceptions (e.g., Uchman, 2007; Table 1), calciturbidites commonly lack the highly patterned graphoglyptids and other pre-depositional trace fossils that are so diagnostic of the *Nereites* Ichnofacies in their siliciclastic counterparts (D'Alessandro et al., 1986; Giannetti and McCann, 2010; Miller et al., 2004; Savary et al., 2004; Seilacher, 1964; Uchman, 2007). In rare cases, this can be attributed to inhospitable background conditions (e.g., anoxia) that excluded the production of biogenic structures altogether during deposition of interturbidite sediments (D'Alessandro et al., 1986). However, in most cases, background sediments are bioturbated, requiring alternative explanations. According to Uchman (2007), the general paucity of graphoglyptids and other pre-depositional trace fossils in calciturbidites may be related to differences in energy regimes; that is, turbidity currents carrying carbonate mud may be less energetic than those transporting siliciclastic sediments and, hence, they lack the erosive power to exhume, cast, and thereby preserve shallow mixed-layer traces in background sediments.

Calciturbidites typically contain post-depositional trace-fossil assemblages (Giannetti and McCann, 2010; Miller et al., 2004; Savary et al., 2004), the origins of which are enigmatic. In some cases, post-depositional assemblages include *Planolites*, *Chondrites*, *Zoophycos*, *Taenidium*, and *Thalassinoides*, and thus they resemble *Zoophycos* Ichnofacies suites in pelagic deposits. This would suggest that post-depositional suites originated, at least in part, as transition-layer structures of the background ichnocoenoses. However, such assemblages commonly differ from those seen in associated background sediments, and in some examples, there is evidence that post-depositional traces were emplaced *before* background conditions were reestablished (D'Alessandro et al., 1986; Miller et al., 2004). These two observations suggest that the post-depositional traces in some calciturbidites reflect the activities of ephemeral benthic communities of opportunistic or event-transported organisms.

The unique nature of calciturbidite trace-fossil assemblages is exemplified well by the work of Miller et al. (2004). In the deep-water Scaglia Rossa exposed in north-eastern Italy, calciturbites typically include (1) a thin basal, cryptically burrow-mottled grainstone (turbidite division T_a); (2) a thicker

(2.5–4 cm) laminated to cross-laminated packstone to grainstone (divisions T_b and T_c) with an assemblage of discrete trace fossils (*Thalassinoides*, *Teichichnus*, *Phycodes*, and *Ancorichnus*?); and (3) an upper heavily bioturbated wackestone (divisions T_d and T_e) with *Planolites*, *Chondrites*, or *Taenidium* (Fig. 10B). Interbedded background limestones contain poorly preserved burrows, including *Zoophycos*, *Planolites*, and/or *Thalassinoides*, none of which penetrate into subjacent turbidites. Based on these observations, Miller et al. (2004) concluded that the ichnofabrics of these calciturbidites reflect the work of distinct short-lived associations of endobenthic organisms contemporaneous with and immediately following event deposition. Notably, the four ichnotaxa identified in these turbidites (*Ancorichnus*?, *Thalassinoides*, *Teichichnus*, and *Phycodes*) are apparently intimately associated with one another as compound structures (Fig. 10B) and may represent the work of crustacean trace-makers, possibly ones that were transported by associated turbidity currents. Very similar compound structures attributed to allochthonous crustaceans recently have been documented in siliciclastic storm beds in siliceous shelf deposits (Savrda et al., 2010).

So, to what ichnofacies can calciturbidite trace-fossil suites be assigned? In the general absence of pre-depositional traces, assignment to the *Nereites* Ichnofacies is difficult to justify. Given the evidence that post-depositional traces in many calciturbidites reflect unique event-related benthic communities, assignment to a distinct ichnofacies may be favorable. Bromley and Asgaard (1991) and Bromley (1996) argued that suites of post-depositional structures that result from the work of transient opportunistic or pioneer communities in turbidites and storm beds warrant consideration as the *Arenicolites* Ichnofacies. Assignment of calciturbidites to this ichnofacies, after some modification of its diagnostic description to account for a greater variety of traces and associated behaviors, may be most appropriate.

7. SUMMARY

Throughout the post-Triassic, but particularly during the Cretaceous, chalk and related carbonates accumulated in a significantly broad array of deep- to shallow-marine depositional regimes. When integrated with sedimentological observations, trace-fossil assemblages can be employed to discriminate these regimes and the depositional processes that operated therein, for instance, slow pelagic deposition *versus* mass-transport mechanisms. Broad ichnological differences between ichnofacies (e.g., *Zoophycos versus Cruziana* ichnofacies), as well as within ichnofacies (e.g., *Zoophycos* Ichnofacies assemblages with or without *Thalassinoides*), can help recognize general paleobathymetric gradients. Moreover, ichnological variations at the bedding scale (e.g., alternation of low *versus* high diversity *Zoophycos* Ichnofacies assemblages or overprinting of *Zoophycos* Ichnofacies suites by *Glossifungites* and *Trypanites* ichnofacies associations) can be used to decipher short-term changes in benthic oxygenation, bottom-current strength, sedimentation rate, substrate consistency, and early diagenetic regimes that signal paleoceanographic and paleoclimatic dynamics.

The utility of ichnology in chalk facies analysis and sequence stratigraphy has obvious implications for the petroleum industry. However, ichnological investigations may have more direct impacts on petroleum exploration and production. As gauges of benthic paleo-oxygenation, trace fossils can help assess hydrocarbon-source potential. As indicators of diagenetic regimes or depositional processes, such as remobilization of chalks, that govern porosity and permeability, they can help assess the reservoir qualities of chalks and associated carbonates.

The ichnological attributes of chalks, particularly those characterized by the *Zoophycos*, *Cruziana*, *Glossifungites*, and *Trypanites* ichnofacies, seemingly are well studied. Arguably, there is a greater need for additional systematic studies to further document and interpret trace-fossil assemblages preserved in allochthonous chalk facies. These include deep-sea calciturbidites as well as other gravity-flow deposits formed in shallow epeiric basins.

ACKNOWLEDGMENTS

The author's research on chalk and related deposits was supported by the U.S. National Science Foundation, the American Chemical Society—Petroleum Research Fund, and Auburn University.

REFERENCES

Arthur, M.A., Bottjer, D.J., Dean, W.E., Fischer, A.G., Hattin, D.E., Kauffman, E.G., 1986. Rhythmic bedding in Upper Cretaceous pelagic carbonate sequences—varying sedimentary response to climatic forcing. Geology 14, 153–156.

Berger, W.H., Ekdale, A.A., Bryant, P.F., 1979. Selective preservation of burrows in deep-sea carbonates. Mar. Geol. 32, 203–230.

Bottjer, D.J., 1985. Trace fossils and paleoenvironments of two Arkansas Upper Cretaceous discontinuity surfaces. J. Paleontol. 59, 282–298.

Bottjer, D.J., 1986. Campanian-Maastrichtian chalks of southwestern Arkansas: Petrology, paleoenvironments and comparison with other North American and European chalks. Cret. Res. 7, 161–196.

Bottjer, D.J., Arthur, M.A., Dean, W.E., Hattin, D.E., Savrda, C.E., 1986. Rhythmic bedding produced in Cretaceous pelagic carbonate environments: sensitive recorders of climatic cycles. Paleoceanography 1, 467–481.

Bromley, R.G., 1967. Some observations on burrows of thalassinidean Crustacea in chalk hardgrounds. Q. J. Geol. Soc. Lond. 123, 157–182.

Bromley, R.G., 1975. Trace fossils at omission surfaces. In: Frey, R.W. (Ed.), The Study of Trace Fossils. Springer, New York, pp. 399–428.

Bromley, R.G., 1981. Enhancement of visibility of structures in marly chalk: modification of the Bushinsky oil technique. Bull. Geol. Soc. Denmark 29, 111–118.

Bromley, R.G., 1994. The palaeoecology of bioerosion. In: Donovan, S.K. (Ed.), The Palaeobiology of Trace Fossils. Wiley, Chichester, pp. 134–154.

Bromley, R.G., 1996. Trace Fossils: Biology, Taphonomy and Applications. Chapman & Hall, London, 361 pp.

Bromley, R.G., Allouc, J., 1992. Trace fossils in bathyal hardgrounds, Mediterranean Sea. Ichnos 2, 43–54.

Bromley, R.G., Asgaard, U., 1991. Ichnofacies: a mixture of taphofacies and biofacies. Lethaia 24, 153–163.

Bromley, R.G., Asgaard, U., 1993. Endolithic community replacement on a Pliocene rocky coast. Ichnos 2, 93–116.

Bromley, R.G., Ekdale, A.A., 1984a. *Chondrites*: a trace fossil indicator of anoxia in sediments. Science 224, 872–874.

Bromley, R.G., Ekdale, A.A., 1984b. Trace fossil preservation in flint in the European chalk. J. Paleontol. 58, 298–311.

Bromley, R.G., Ekdale, A.A., 1986. Composite ichnofabrics and tiering in burrows. Geol. Mag. 123, 59–65.

Bromley, R.G., Ekdale, A.A., 1987. Mass transport in European Cretaceous chalk: fabric criteria for recognition. Sedimentology 34, 1079–1092.

Bromley, R.G., Richter, B., 1999. Direkte scanning: Ny teknik til illustrering af geologisk material. (Direct scanning: a new technique to illustrate geological material). Geol. Tidsskrift 4, 1–10 [In Danish].

Bromley, R.G., Schultz, M.-G., Peake, N.P., 1975. Paramoudras: giant flints, long burrows and the early diagenesis of chalks. Det Kongelige Danske Videnskabernes Selskab, Biologiske Skrifter 20, 1–31.

Chamberlain, C.K., 1975. Trace fossils in DSDP cores of the Pacific. J. Paleontol. 49, 1074–1096.

D'Alessandro, A., Ekdale, A.A., Sonnino, M., 1986. Sedimentologic significance of turbidite ichnofacies in the Saraceno Formation (Eocene), southern Italy. J. Sediment. Petrol. 56, 294–306.

Danholt, T., Surlyk, F., 2004. Laminated-bioturbated cycles in Maastrichtian chalk of the North Sea: oxygenation fluctuations within the Milankovitch frequency band. Sedimentology 51, 1323–1342.

Dawson, W.C., Reaser, D.F., 1980. *Rhizocorallium* in the upper Austin Chalk: Ellis County, Texas. Texas J. Sci. 32, 207–214.

Dawson, W.C., Reaser, D.F., 1985. Ichnology and Paleoenvironments of the Middle and Upper Austin Chalk (Upper Cretaceous), Northeastern Texas. Transactions, AAPG Southwest Section 1985, pp. 47–67.

Dawson, W.C., Reaser, D.F., 1990. Trace Fossils and Paleoenvironments of Lower and Middle Austin Chalk (Upper Cretaceous), North-Central Texas. Transactions, Gulf Coast Association of Geological Societies 40, pp. 161–173.

Droser, M.L., Bottjer, D.J., 1991. Trace fossils and ichnofabrics in Leg 119 cores. In: Barron, J., Larsen, B. et al., (Eds.), Proceedings of the Ocean Drilling Program, Scientific Results 119 pp. 635–641.

Ekdale, A.A., 1977. Abyssal trace fossils in worldwide Deep Sea Drilling Project cores. In: Crimes, T.P., Harper, J.C. (Eds.), Trace Fossils 2. Geol. J., Spec. Iss. 9, pp. 163–182.

Ekdale, A.A., 1980. Trace fossils in Deep Sea Drilling Project Leg 58 cores. In: de Vries Klein, G., Kobyashi, K. et al., (Eds.), Initial Reports of the Deep Sea Drilling Project 58, pp. 601–605.

Ekdale, A.A., 1985. Trace fossils and Mid-Cretaceous anoxic events in the Atlantic Ocean. In: Curran, H.A. (Ed.), Biogenic Structures: Their Use in Interpreting Depositional Environments. SEPM Spec. Publ., vol. 35, pp. 333–342.

Ekdale, A.A., Berger, W.H., 1978. Deep-sea ichnofacies: modern organism traces on and in pelagic carbonates of the western equatorial Pacific. Palaeogeogr. Palaeoclimatol. Palaeoecol. 23, 263–278.

Ekdale, A.A., Bromley, R.G., 1982. Trace fossils and ichnofabrics in the Kjølby Gaard Marl, uppermost Cretaceous, Denmark. Bull. Geol. Soc. Denmark 31, 107–119.

Ekdale, A.A., Bromley, R.G., 1984a. Comparative ichnology of shelf-sea and deep-sea chalk. J. Paleontol. 58, 322–332.

Ekdale, A.A., Bromley, R.G., 1984b. Cretaceous chalk ichnofacies in northern Europe. Geobios 17 (Suppl. 1), 201–204.

Ekdale, A.A., Bromley, A.A., 1988. Diagenetic microlamination in chalk. J. Sediment. Petrol. 58, 857–861.

Ekdale, A.A., Bromley, R.G., 1991. Analysis of composite ichnofabrics: an example in the uppermost Cretaceous chalk of Denmark. Palaios 6, 232–249.

Ekdale, A.A., Bromley, R.G., Knaust, D., 2012. The ichnofabric concept. In: Knaust, D., Bromley, R.G. (Eds.), Trace Fossils as Indicators of Sedimentary Environments. Developments in Sedimentology, vol. 64. Elsevier, Amsterdam, pp. 139–155.

Ekdale, A.A., Bromley, R.G., Pemberton, S.G., 1984b. Ichnology: The Use of Trace Fossils in Sedimentology and Stratigraphy. SEPM Short Course Notes 15, 317 pp.

Ekdale, A.A., Muller, L.N., Novak, M.T., 1984a. Quantitative ichnology of modern pelagic deposits in the abyssal Atlantic. Palaeogeogr. Palaeoclimatol. Palaeoecol. 45, 189–223.

Erba, E., Premoli-Silva, I., 1994. Orbitally driven cycles in trace-fossil distribution from the Piobbico core (late Albian, central Italy). In: De Boer, P.L., Smith, D.G. (Eds.), Orbital Forcing and Cyclic Sequences. IAS Spec. Publ., vol. 19, Blackwell Scientific, Oxford, pp. 211–225.

Fischer, A.G., Schwarzacher, W., 1984. Cretaceous bedding rhythms under orbital control? In: Berger, A., Imbrie, J., Hays, J., Kukla, G., Saltzman, B. (Eds.), Milankovitch and Climate. NATO ASAI Series C 126. Reidel, Dordrecht, pp. 163–176.

Frey, R.W., 1970. Trace fossils of the Fort Hays Limestone Member of the Niobrara Chalk (Upper Cretaceous), west-central Kansas. Univ. Kansas Paleont. Contr. 53, 1–41.

Frey, R.W., Bromley, R.G., 1985. Ichnology of American chalks: the Selma Group (Upper Cretaceous), western Alabama. Can. J. Earth Sci. 22, 801–828.

Fürsich, F.T., Kennedy, W.J., Palmer, T.J., 1981. Trace fossils at a regional discontinuity surface: the Austin-Taylor (Upper Cretaceous) contact in central Texas. J. Paleontol. 55, 537–551.

Fütterer, D.K., 1984. Bioturbation and trace fossils in deep sea sediments of the Walvis Ridge, southeastern Atlantic, Leg 74. In: Moore, T.C., Rabinowitz, P.D. et al., (Eds.), Initial Reports of the Deep Sea Drilling Project 74, pp. 543–555.

Gaillard, C., 1991. Recent organism traces and ichnofacies on the deep-sea floor off New Caledonia, southwestern Pacific. Palaios 6, 302–315.

Gale, A.S., 1995. Cyclostratigraphy and correlation of the Cenomanian of Europe. In: House, M.R., Gale, A.S. (Eds.), Orbital Forcing Timescale for the Late Cretaceous, pp. 177–197.

Garrison, R.E., Kennedy, W.J., 1977. Origin of solution seams and flaser structures in the Upper Cretaceous chalks of southern England. Sediment. Geol. 19, 107–137.

Giannetti, A., McCann, T., 2010. The Upper Paleocene of the Zumaya section (northern Spain): review of the ichnological content and preliminary palaeoecological interpretation. Ichnos 17, 137–161.

Hancock, J.M., 1975. The petrology of chalk. Proc. Geol. Ass. 86, 499–535.

Hattin, D.E., 1971. Widespread, synchronously deposited, burrow-mottled limestone beds in Greenhorn Limestone (Upper Cretaceous) of Kansas and southeastern Colorado. AAPG Bull. 55, 412–431.

Kennedy, W.J., 1970. Trace fossils in the chalk environment. In: Crimes, T.P., Harper, J.C. (Eds.), Trace Fossils. Geol. J., Spec. Iss. 3, pp. 263–281.

Kennedy, W.J., 1975. Trace fossils in carbonate rocks. In: Frey, R.W. (Ed.), The Study of Trace Fossils. Springer, New York, pp. 377–396.

Kennedy, W.J., Garrison, R.E., 1975. Morphology and genesis of nodular chalks and hardgrounds in the Upper Cretaceous of southern England. Sedimentology 22, 311–386.

Kotake, N., 1989. Paleoecology of the *Zoophycos* producers. Lethaia 22, 379–385.

Lauridsen, B.W., Surlyk, F., Bromley, R.G., 2011. Trace fossils of cyclic chalks, Rørdal Member, Denmark. Cret. Res. 32, 194–202.

Locklair, R.E., Savrda, C.E., 1998a. Ichnology of rhythmically bedded Demopolis Chalk (Upper Cretaceous, Alabama): implications for paleoenvironment, depositional cycle origins, and trace-maker behavior. Palaios 13, 423–438.

Locklair, R.E., Savrda, C.E., 1998b. Ichnofossil tiering analysis of a rhythmically bedded chalk-marl sequence in the Upper Cretaceous of Alabama. Lethaia 31, 311–322.

Madsen, H.B., Stemmerik, L., 2010. Diagenesis of flint and porcellanite in the Maastrichtian chalk at Stevns Klint, Denmark. J. Sediment. Res. 80, 578–588.

Malfait, B.T., Van Andel, T.H., 1980. A modern oceanic hardground on the Carnegie Ridge in the eastern equatorial Pacific. Sedimentology 27, 467–496.

Miller III, W., Stefani, C., Grandesso, P., 2004. Alternation of ecologic regimes in a deep-marine carbonate basin: calciturbidite trace fossils from the Cretaceous Scaglia Rossa, northeastern Italy. Palaeogeogr. Palaeoclimatol. Palaeoecol. 204, 317–330.

Mortimore, R.N., Pomerol, B., 1991. Stratigraphy and eustatic implications of trace fossil events in the Upper Cretaceous Chalk of northern Europe. Palaios 6, 216–231.

Nygaard, E., 1983. *Bathichnus* and its significance in the trace fossil association of Upper Cretaceous chalk, Mors, Denmark. Danm. Geol. Unders., Årbog 1982, 107–113.

Savary, B., Olivero, D., Gaillard, C., 2004. Calciturbidite dynamics and endobenthic colonization: example from a late Barremian (Early Cretaceous) succession in southeastern France. Palaeogeogr. Palaeoclimatol. Palaeoecol. 211, 221–239.

Savrda, C.E., 1991. Ichnology in sequence stratigraphic studies: an example from the Lower Paleocene of Alabama. Palaios 6, 39–53.

Savrda, C.E., 1998a. Ichnocoenoses in the Niobrara Formation: implication for benthic oxygenation histories. In: Dean, W.E., Arthur, M.A. (Eds.), Stratigraphy and Paleoenvironments of the Cretaceous Western Interior Seaway, USA. SEPM Concepts in Sedimentology and Paleontology, vol. 6, pp. 137–151.

Savrda, C.E., 1998b. Ichnology of the Bridge Creek Limestone: evidence for temporal and spatial variations in paleo-oxygenation in the Western Interior seaway. In: Dean, W.E., Arthur, M.A. (Eds.), Stratigraphy and Paleoenvironments of the Cretaceous Western Interior Seaway, USA. SEPM Concepts in Sedimentology and Paleontology, vol. 6, pp. 127–136.

Savrda, C.E., Bottjer, D.J., 1989. Trace fossil model for reconstructing oxygenation histories of ancient marine bottom waters: application to Upper Cretaceous Niobrara Formation, Colorado. Palaeogeogr. Palaeoclimatol. Palaeoecol. 74, 49–74.

Savrda, C.E., Browning, J.V., Krawinkel, H., Hesselbo, S.P., 2001. Firmground ichnofabrics in deep-water sequence stratigraphy, Tertiary clinoform-toe deposits, New Jersey slope. Palaios 16, 294–305.

Savrda, C.E., Counts, J.W., Bigham, E., Martin, S., 2010. Ichnology of siliceous facies in the Eocene Tallahatta Formation (eastern United States Gulf coastal plain): implications for depositional conditions, storm processes, and diagenesis. Palaios 25, 642–655.

Scholle, P.A., 1977. Chalk diagenesis and its relation to petroleum exploration: oil from chalks, a modern miracle? AAPG Bull. 61, 982–1009.

Scholle, P.A., Arthur, M.A., Ekdale, A.A., 1983. Pelagic. In: Scholle, P.A., Bebout, D.G., Moore, C.H. (Eds.), Carbonate Depositional Environments. AAPG Mem. 33, pp. 619–691.

Seilacher, A., 1964. Biogenic sedimentary structures. In: Imbrie, J., Newell, N.D. (Eds.), Approaches to Paleoecology. Wiley, New York, pp. 296–316.

Simpson, S., 1957. On the trace fossil *Chondrites*. Q. J. Geol. Soc. Lond. 112, 475–495.

Stanley, S.M., Ries, J.B., Hardie, L.A., 2005. Seawater chemistry, coccolithophore population growth, and the origin of Cretaceous chalk. Geology 33, 593–596.

Steinich, G., 1967. Sedimentstrukturen der Rügener Schreibkreide. Geologie 16, 570–583.

Surlyk, F., Lykke-Andersen, H., 2007. Contourite drifts, moats and channels in the Upper Cretaceous chalk of the Danish Basin. Sedimentology 54, 405–422.

Uchman, A., 2007. Deep-sea trace fossils from the mixed carbonate-siliciclastic flysch of the Monte Antola Formation (Late Campanian-Masstrichtian), North Apennines, Italy. Cret. Res. 28, 980–1004.

Uchman, A., Wetzel, A., 2012. Deep-sea fans. In: Knaust, D., Bromley, R.G. (Eds.), Trace Fossils as Indicators of Sedimentary Environments. Developments in Sedimentology, vol. 64. Elsevier, Amsterdam, pp. 643–672.

Warme, J.E., Kennedy, W.J., Scheidermann, N., 1973. Biogenic sedimentary structures (trace fossils) in Leg 15 cores. In: Edgar, N.T., Saunders, J.B. et al., (Eds.), Initial Reports of the Deep Sea Drilling Project 15, pp. 813–831.

Weedon, G., 1993. The recognition and stratigraphic implications of orbital-forcing of climate and sedimentary cycles. Sediment. Rev. 1, 31–50.

Wetzel, A., 1984. Bioturbation in deep-sea fine-grained sediments: influence of sediment texture, turbidite frequency and rates of environmental change. In: Stow, D.A.V., Piper, D.J.W. (Eds.), Fine-grained Sediments: Deep Water Processes and Facies. Blackwell, Oxford, pp. 595–608.

Wetzel, A., 1987. Ichnofabrics in Eocene to Maestrichtian sediments from Deep Sea Drilling Project Site 605, off the New Jersey coast. In: Hinte, J.E., Wise Jr., S.W. et al., (Eds.), Initial Reports of the Deep Sea Drilling Project 93, pp. 825–835.

Wetzel, A., 1991. Ecologic interpretation of deep-sea trace fossil communities. Palaeogeogr. Palaeoclimatol. Palaeoecol. 85, 47–69.

Chapter 26

Mixed Siliciclastic/Carbonate Systems

John-Paul Zonneveld,*[,1] Murray K. Gingras,* Tyler W. Beatty,† David J. Bottjer,‡ James R. Chaplin,§ Sarah E. Greene,‡ Rowan C. Martindale,‡ Scott A. Mata,‡ Luke P. McHugh,* S. George Pemberton* and Jesse A. Schoengut*

**Department of Earth and Atmospheric Sciences, University of Alberta, Edmonton, Alberta, Canada, †Department of Geology and Geophysics, University of Calgary, Calgary, Alberta, Canada, ‡Department of Earth Sciences, University of Southern California, Los Angeles, California, USA, §Oklahoma Geological Survey, University of Oklahoma, Norman, Oklahoma, USA*

[1]Corresponding author: e-mail: zonneveld@ualberta.ca

1. INTRODUCTION

Mixed siliciclastic/carbonate depositional systems are distinguished by the pervasive and discrete co-occurrence of both siliciclastic and carbonate lithotypes. Ideally, the lithotypes should be interlaminated, interbedded, or interstratified. Admixture of siliciclastic and carbonate sediment can occur due to a variety of factors, most of which are centered on either spatial variability in the facies distribution (i.e., lateral facies mixing) or temporal variability in the sediment type (i.e., variation in sediment supply/stratigraphic occurrence) (Brett et al., 2010; Budd and Harris, 1990; Lomondo and Harris, 1991; McNeill et al., 2010; Mount, 1984). In most examples, trends of mixing vary strongly along both depositional strike and dip as a function of proximity to sources of carbonate or clastic sediment. In some cases, such as tropical coastal areas with ephemeral fluvial input, the mixing varies strongly in both the geographical sense (proximity to the clastic point source) and the temporal sense (periodicity/seasonality of fluvial flow). In other examples, sediment mixing may occur during rare, high-intensity events (such as storms) and may be limited to discrete horizons within otherwise unmixed successions (Mount, 1984). Finally, sediment mixing may be a function of cycles in the rates of sedimentation, with intercalation of sediment-starved concentrations and obrution deposits (Brett et al., 2010).

Sea-level fluctuations have commonly been implicated in the development of mixed systems. In some cases, such as the South Florida Neogene, changes in

Developments in Sedimentology, Vol. 64. http://dx.doi.org/10.1016/B978-0-444-53813-0.00026-5

basin configuration related to a changing sea level resulted in basin deflection of shallow oceanic currents that dramatically affected the amount of siliciclastic material delivered to depocenters (Guertin et al., 1999; McNeill et al., 2004). In other cases, the influx of siliciclastic material transported basinward during intervals of rapidly dropping sea level inhibited carbonate accumulation (Campbell, 2005; Schlager, 1991). In these systems, lowstand systems tracts are dominated by siliciclastic sedimentation, whereas transgressive and highstand systems tracts are characterized by dominantly carbonate deposition (García-Hidalgo et al., 2007; Southgate et al., 1993).

Mixed siliciclastic/carbonate systems also occur, albeit more rarely, in continental successions. In our experience, these most commonly occur in mixed fluvial/lacustrine and marginal-lacustrine successions. The continental examples discussed herein are limited to examples of these environments.

Even though mixed siliciclastic/carbonate systems are common, arguably as common as the pure end-members, ichnological analyses that have focused specifically on mixed successions are sparse (i.e., Brett et al., 2010; Chaplin, 1996, 2010; Zonneveld et al., 2001). Similar to purely carbonate or siliciclastic systems, the trace-fossil distribution in mixed systems reflects physical and chemical parameters that control the distribution of organisms. Ichnological studies of mixed systems reveal that these systems commonly preserve the comingling of typical clastic and carbonate facies and fabrics. Trace-fossil associations (such as *Glossifungites* assemblages), commonly considered characteristic of clastic successions, may occur in carbonate-dominated mixed systems. Conversely, bored shells, cobbles, or boulders and bored hardground surfaces with variably diverse *Trypanites* assemblages may be locally common in clastics-dominated mixed systems.

This chapter summarizes the current body of literature that deals with the ichnology of mixed siliciclastic/carbonate systems. Although many works that focus on depositional or taphonomic attributes of mixed siliciclastic/carbonate successions make mention of trace fossils, in few of these, it is clear how the mixed lithology nature of the succession influences the trace-fossil distribution. As well, examples of mixed argillaceous calcareous mudstone successions are common in the rock record. Although these are clearly excellent examples of mixed siliciclastic/carbonate successions, there has been no documentation that their ichnotaxonomic composition and diversity differ in any appreciable way from either purely argillaceous or purely calcareous shale successions deposited in identical paleoenvironments. Although we view this as an area in need of further investigation, it is not further explored herein.

This study focuses solely on examples where admixture of clastic and carbonate sediment produces a unique ichnological signature. These are discussed using case studies that provide examples wherein sediment mixing produces unique relationships within and between trace-fossil assemblages. The examples discussed herein, all of which were chosen from Phanerozoic successions in North America (Fig. 1), represent one of two end-member circumstances. In the first, spatial variability/lateral facies mixing resulted in unique relationships between trace-fossil assemblages. This includes examples wherein variability in clastic and carbonate

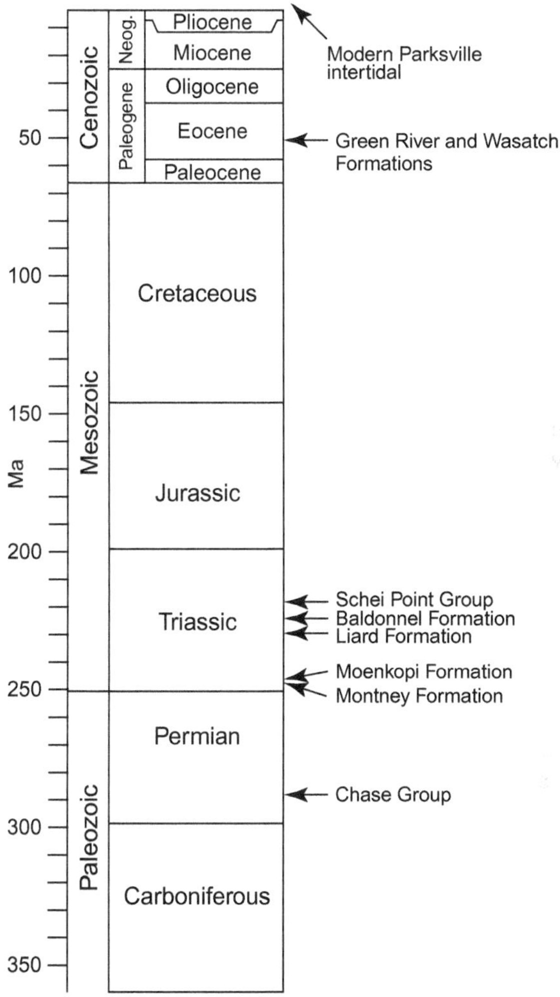

FIGURE 1 Stratigraphic occurrence of mixed siliciclastic/carbonate examples used in this chapter.

sediment supply resulted in co-occurrence of unique trace-fossil assemblages. In the second end-member, temporal variability in the form of stratigraphic occurrence resulted in the vertical juxtaposition of trace-fossil assemblages and ichnofacies.

2. SPATIAL VARIABILITY (LATERAL FACIES MIXING)

Mixed systems that occur through lateral variability in lithofacies distribution are perhaps the most common type of mixed system. In many cases, differences in trace-fossil distributions between carbonate and clastic components are a direct function of grain size and shape and thus commonly reflect the influences of depositional energy and environmental segregation of grain type. Resource distribution may also be unevenly allocated in mixed systems resulting in preferential colonization of one lithotype over another. Finally, consistency or compactional differences in different lithology types may result in ichnotaxonomic segregation as some taxa prefer loose media within which to burrow and others prefer firm or cemented substrates.

Mixed systems occur throughout the geological record; however, they are more common in specific intervals. Arguably, the most notable among these include the lowermost Mesozoic. Examples are derived from western Canada and the western United States. In both of these regions, the Permian was dominated by carbonate deposition. The demise of many of the carbonate-producing taxa coincides in many areas with a shift from carbonate dominance to clastic dominance at the Permian/Triassic boundary. Although syndepositional diagenetic controls have also likely played an important role, the prevalence of bioclastic packstone and grainstone beds in preextinction successions and their absence in similar deposits in postextinction settings were in part a function of the demise of many shell-secreting organisms (i.e., many brachiopod and echinoderm taxa) rather than a shift in environmental setting. Survival and recolonization by survivor taxa as well as the origination and environmental expansion of new forms produced mixed clastic/carbonate successions in many Early and Middle Triassic successions worldwide, of which some examples are included in this section (e.g., Montney Formation, Alberta; Liard Formation, British Columbia; Moenkopi Formation, Nevada and Utah).

2.1 Montney Formation (Early Triassic), Alberta

Shallow- and marginal-marine systems dominated by siliciclastic siltstone and sandstone and bioclastic packstone and grainstone characterize Early Triassic successions in west-central Alberta, Canada (Moslow and Zonneveld, 2005; Orchard and Zonneveld, 2009). These units comprise the Vega and Phroso members of the Sulphur Mountain Formation in the Foothills and Front Ranges outcrop belt and the Montney Formation in the subsurface. Although siltstone and sandstone throughout the entire Montney Formation are dominated by quartz, dolomite (both syndepositional and late diagenetic) is also abundant and, indeed, commonly occurs in nearly equal abundance with quartz. In the middle part of this succession, a bioclastic grainstone/packstone succession known as the Coquinal Dolomite Middle Member (Mackenzie Dolomite Lentil in outcrop) is present. Stratigraphic and sedimentological evidence indicates that the middle part of the Montney Formation records deposition in a proximal ramp setting and includes units interpreted as shoreface, foreshore, and intertidal flat (Moslow and Zonneveld, 2005; Orchard and Zonneveld, 2009).

Shoreface units exhibit gradation from very fine-grained and fine-grained sandstone at their bases to bioclastic sandstone and sandy bioclastic packstone/grainstone at their tops (Moslow and Zonneveld, 2005). Bioclastic units consist of whole and fragmentary bivalves, gastropods, and lingulide brachiopods. Bioturbation within the shoreface sandstone units consists of low-diversity proximal *Cruziana* and *Skolithos* assemblages. Bioturbation is less commonly observed within the bioclastic grainstone and packstone units, although *Skolithos* and *Diplocraterion* have been recognized. In all cases, the bioclastic units represent deposition in wave-dominated settings (upper shoreface and foreshore), and thus, it is likely that the drop in bioturbation intensity is due to deposition within a more proximal setting than coeval siliciclastic facies.

2.2 Moenkopi Formation (Early Triassic), Nevada and Utah

Similar to the Montney Formation, shallow-marine successions in the Moenkopi Formation are characterized by mixed siliciclastic/carbonate deposition that is the result of lateral facies mixing in both nearshore and offshore environments. Bioclastic limestone, characterized by variably abundant crinoid ossicles, echinoid skeletal debris, and bivalves, occurs interstratified with mudstone in proximal offshore environments and gradationally overlays sandstone beds in upper shoreface to tidal-inlet environments. The trace-fossil distribution is strongly controlled by grain size and inferred depositional environment, with proximal offshore/offshore transition sandstone beds containing low-diversity assemblages of dominantly *Arenicolites* and *Planolites*, with rare *Diplocraterion*, *Helminthopsis*, and *Rhizocorallium* (Fig. 2). Shoreface successions of amalgamated and non-amalgamated sandstone beds contain more diverse assemblages that include *Arenicolites*, *Asteriacites*, *Asterosoma*, *Conichnus*, *Diplocraterion*, *Gyrochorte*, *Lockeia*, *Palaeophycus*, *Planolites*, and *Rhizocorallium* (Fig. 2). Bioclastic units contain lower-diversity assemblages, limited primarily to *Planolites* and *Thalassinoides* (Fig. 2).

Offshore bioclastic packstone units consist primarily of disarticulated crinoid ossicles and were deposited as storm-induced concentrations of skeletal debris in proximal offshore through offshore-transition successions and as such were the result of rapid deposition and high-energy conditions. Nearshore bioclastic packstone and grainstone successions, restricted to upper shoreface and tidal-inlet paleoenvironments, lack evidence of bioturbation, and the absence of trace fossils may be due to the coarse bioclastic nature of the deposits, as well as the persistent high-energy conditions present during deposition. Differences in trace-fossil assemblages between carbonate and siliciclastic depositional units are inferred to be the product of both, differences in depositional energy between the two end-members as well as grain-size and grain-shape constraints on bioturbator activity and the preservation of individual trace fossils. Many taxa, such as those that constructed *Diplocraterion* and *Rhizocorallium*, likely avoided substrates characterized by coarse skeletal debris, preferring instead sand-dominated successions, in which it would have been easier to construct burrows. Bedding-plane traces, such as *Asteriacites* and *Gyrochorte*, have a low preservation potential in coarse substrates. Whether the constructors of these traces avoided coarse bioclastic substrates is uncertain, but their preservation potential, and if preserved, recognition potential, is low.

FIGURE 2 Trace fossils in the Virgin Limestone Member of the Moenkopi Formation (Early Triassic), Nevada and Utah, USA. (A) *Asteriacites lumbricalis* on the sole of a sandy storm bed, lower shoreface, Ute, Nevada. Coin = 1.8 cm in diameter. (B) Bedding-plane view of branching *Thalassinoides* within a bioclastic packstone bed, proximal offshore, Beaver Dam Mountains, Utah. The burrow fill comprises disarticulated and loosely packed ossicles of the crinoid *Holocrinus smithi*. Pen = 1.3 cm in diameter. (C) *Rhizocorallium* within amalgamated hummocky cross-stratified sandstone, lower shoreface, Ute, Nevada. (D) Bedding-plane view of branching *Thalassinoides* within a bivalve wackestone, proximal offshore, Lost Cabin Spring, Nevada. The burrow fill consists of densely packed bioclastic debris. (E) Firmground horizon with sharp-walled *Thalassinoides* penetrating the underlying bivalve-wackestone storm bed, offshore transition, Ute, Nevada. The burrow was passively filled with bioclastic debris from overlying bed. Pen = 15 cm long.

2.3 Liard Formation (Middle Triassic), British Columbia

Brachiopod- and echinoderm-dominated biostromes are common within late Ladinian strata of the Liard Formation in northeastern British Columbia, Canada (Zonneveld, 2001; Zonneveld et al., 1997). These biostromes are composed primarily of bioclastic material but occur within dominantly siliciclastic shoreface successions. This material provided the stable substrate that the biostrome taxa colonized. They formed through both allogenic and autogenic taphonomic feedback mechanisms (*sensu* Kidwell and Jablonski, 1983; Zonneveld, 2001).

The base of all biostromes observed consists of a layer of abraded, disarticulated bioclastic detritus interpreted to reflect storm-induced concentrations of shell debris. The Liard biostromes are dominated by organisms that preferred hard or stable substrates such as the terebratulide brachiopod *Aulacothyroides* sp., the articulate crinoid *Isocrinus* sp., the cidaroid echinoid *Miocidaris* sp., and an ostreid/pseudo-ostreid of unknown affinity (Zonneveld, 2001). The architecture of these biostromes is complex and differs between occurrences. In some examples, brachiopod-dominated rudstone occurs interbedded with either calcareous siltstone or medium- to coarse-grained crinoidal grainstone that in turn occurs interbedded with quartzose sandstone. In other occurrences, the crinoidal beds are absent and the brachiopod-dominated rudstone beds occur interbedded with quartzose sandstones.

Trace fossils in biostrome-center deposits are limited to borings such as rare examples of the pedicle-attachment trace *Podichnus* (Zonneveld, 2001) and diminutive, solitary pits (cf. *Rogerella*, Fig. 3). Rare patches of admixed calcareous/quartzose sand occur in biostrome-center deposits, wherein low-diversity trace-fossil assemblages of *Asterosoma*, and *Palaeophycus*, and *Planolites* have been observed (Fig. 3). Where present, the crinoidal grainstone deposits are primarily massive in appearance, but large, dominantly horizontal, meniscate

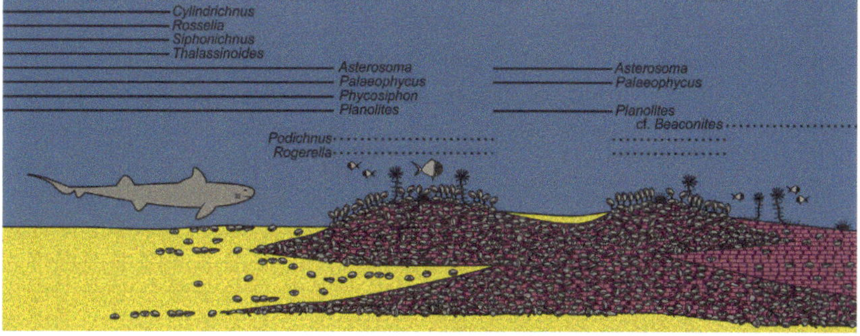

FIGURE 3 Distribution of trace fossils adjacent to brachiopod–echinoderm biostromes, Liard Formation (Middle Triassic), northeastern British Columbia, Canada. Trace-fossil distributions are intended to reflect their distribution at a specific point in time. Within interfingering strata, trace-fossil distributions may be quite complex, but at any point in time, trace distributions follow the distribution illustrated herein. Individual biostromes are 10–100 s of meters across and decimeters to meters thick. *Adapted from Zonneveld (2001).*

burrows (cf. *Beaconites*) do occur (Fig. 3). Calcareous siltstone beds, which commonly occur interbedded with both the crinoidal grainstone and brachiopod rudstone beds, contain a low-diversity trace-fossil assemblage of dominantly horizontal deposit feeders (*Asterosoma* and *Planolites*) as well as scattered crinoid and echinoid skeletal elements, brachiopods (*Aulacothyroides* sp. and *Discinisca* sp.), and decapod crustaceans (*Chimaerastacus* sp.).

Interfingering layers of clastic sandstone and carbonate beds characterize the fringes of the biostromes (Zonneveld, 2001). Clastic sandstone layers contain an abundance of fossils that preferred loose or shifting substrates, such as lingulide brachiopods (*Lingularia* sp.) and a variety of bivalves and gastropods. Where abundant shell debris is present in the sand, the traces are limited to horizontal deposit-feeding and dwelling traces (*Asterosoma*, *Palaeophycus*, *Phycosiphon,* and *Planolites*; Fig. 3). Lateral to the biostromes, in areas devoid of significant shell debris, vertically oriented dwelling traces of infaunal filter feeders (i.e., *Cylindrichnus*, *Lingulichnus*, *Rosselia,* and *Siphonichnus*) occur in roughly equal numbers with those of infaunal deposit feeders (i.e., *Asterosoma*, *Phycosiphon*, and *Planolites*) and carnivores/scavengers (i.e., *Palaeophycus* and *Thalassinoides*; Fig. 3; Zonneveld, 2001).

The Liard biostromes represent "islands" or patches of stable substrate that formed within shifting quartz-dominated sandy shoreface successions and record interaction between characteristic carbonate and siliciclastic trace-fossil assemblages. The compositions of trace-fossil assemblages in these successions are clearly influenced by the nature and distribution of the different sediment types as well as food availability at the interface of siliciclastic and carbonate-dominated systems.

2.4 Baldonnel Formation (Late Triassic), British Columbia

The Carnian Baldonnel Formation in northeastern British Columbia, Canada, is a carbonate-dominated succession with abundant admixed quartz-dominated sand (Martindale et al., 2010; Zonneveld and Orchard, 2002; Zonneveld et al., 2004, 2007). Although this unit contains diverse and locally abundant trace fossils, these traces are commonly subtly displayed and are difficult to differentiate from host lithologies (not an uncommon problem in carbonate-dominated successions). Similar to the pattern observed in the Liard Formation, trace fossils exhibit a unique relationship between *in-situ* carbonate production and trace-fossil assemblage composition.

Small coralliferous patch reefs occur in the lower Baldonnel Formation (Martindale et al., 2010; Zonneveld et al., 2007). These patch reefs occur within a succession of calcareous-quartzose sandstone. Trace fossils within the sandstone successions include *Cylindrichnus*, *Diplocraterion*, *Palaeophycus*, *Planolites*, *Psilonichnus*, *Rhizocorallium*, *Rosselia*, *Siphonichnus*, *Skolithos*, *Scolicia*, *Teichichnus*, and *Thalassinoides.* On the fringes of individual patch reefs, bioclastic grainstone/rudstone interfingers with the sandstone. In these areas trace fossils are limited to *Palaeophycus*, *Planolites*, *Psilonichnus*, and *Teichichnus*. Within the patch reefs, proper trace fossils are rare, possibly partly because extensive recrystallization of reef facies may have obliterated trace fossils that were present. However, in one of the patch reefs, bioclast-filled *Planolites*, *Thalassinoides*, *Skolithos*, and *Scolicia* occur within fine-grained sandstone beds within intrareef cavities. Moreover, within the patch reefs occur numerous hardgrounds, many of which are penetrated by small (0.5–4.0 mm in diameter), simple tubular borings referred to as *Trypanites*.

3. TEMPORAL VARIABILITY (STRATIGRAPHIC MIXING)

Many mixed systems are produced by stratigraphic controls; changing relative sea-level results in changes to sediment sources and concomitant changes in preservation potential of sedimentary successions within a given area. Diagenetic processes commonly operate at different scales and temporal intervals in siliciclastics and carbonates, leading to fundamental differences in preservation and in substrate consistency. Carbonate systems, particularly those deposited in tropical to subtropical settings, are prone to early cementation and thus *Trypanites* assemblages commonly occur (i.e., Blanchon and Perry, 2004; Brett, 1995; Brett et al., 2010; Knaust et al., 2012; McLaughlin et al., 2008; Tapanila et al., 2004; Webb, 1994). *Glossifungites*-demarcated discontinuity surfaces are common in both carbonate and siliciclastics-dominated successions (i.e., Brett et al., 2010; Gingras et al., 2001; Knaust, 2007; McLaughlin et al., 2008; Rodríguez-Tovar and Perez-Valera, 2008; Yang et al., 2009). Within mixed systems, these substrate-controlled ichnofacies may become juxtaposed and commonly occur at significant lithostratigraphic or sequence-stratigraphic boundaries.

3.1 Chase Group (Early Permian), Oklahoma

The Early Permian Chase Group in northcentral Oklahoma, USA, consists of a series of interstratified mixed siliciclastic/carbonate units interpreted to represent major transgressive and regressive events (Chaplin, 1996, 2010; Fig. 4). The regressive components consist of clastics-dominated marginal-marine successions, mainly of red and green, pedogenically altered mudstone (Chaplin, 1996, 2010). Transgressive components consist of marine and marginal-marine bioclastic wackestone, packstone, and grainstone (Chaplin, 1996, 2010). The contact between regressive units and overlying transgressive units comprises *Glossifungites*-demarcated discontinuity surfaces characterized by dense, horizontal networks of *Thalassinoides* (Fig. 5B and C) as well as burrowing bivalves (*Pinna*; Fig. 5A; Chaplin, 1996, 2010). Abundant, large *Rhizocorallium* dominate the top surface of each carbonate succession (Fig. 5D and E). Also present are rarer *Arenicolites*, *Chondrites*, *Diplocraterion*, *Planolites*, and *Teichichnus* (Chaplin, 1996). Although these surfaces represent hiatuses, they are interpreted

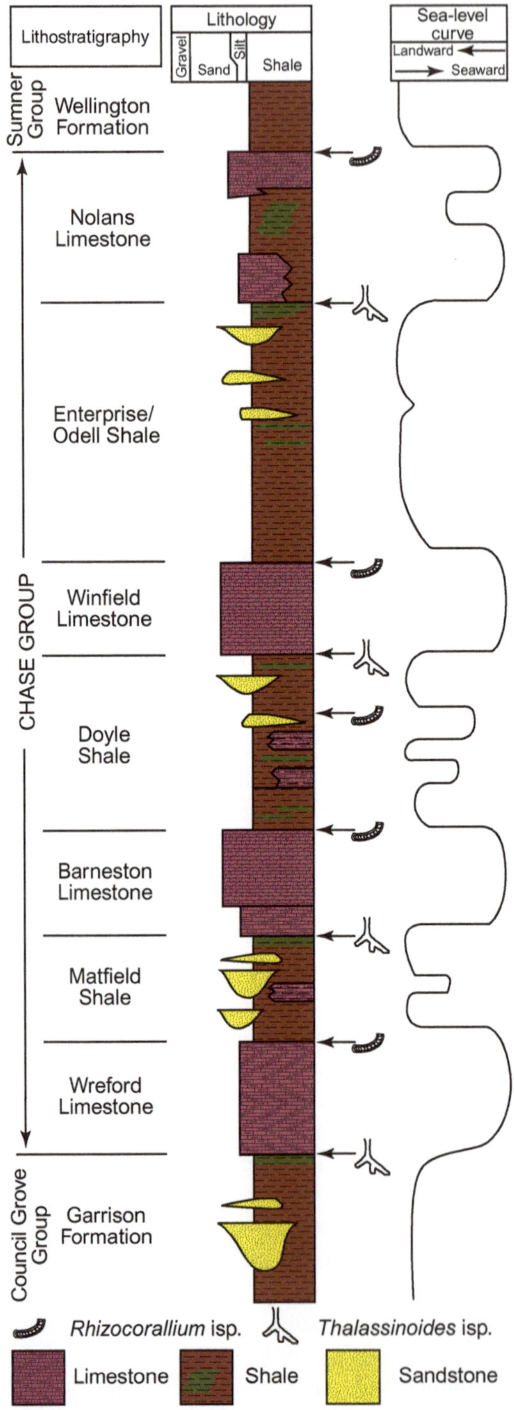

FIGURE 4 Stratigraphic succession in the Chase Group (Early Permian), Oklahoma, USA. *Adapted from Chaplin (1996).*

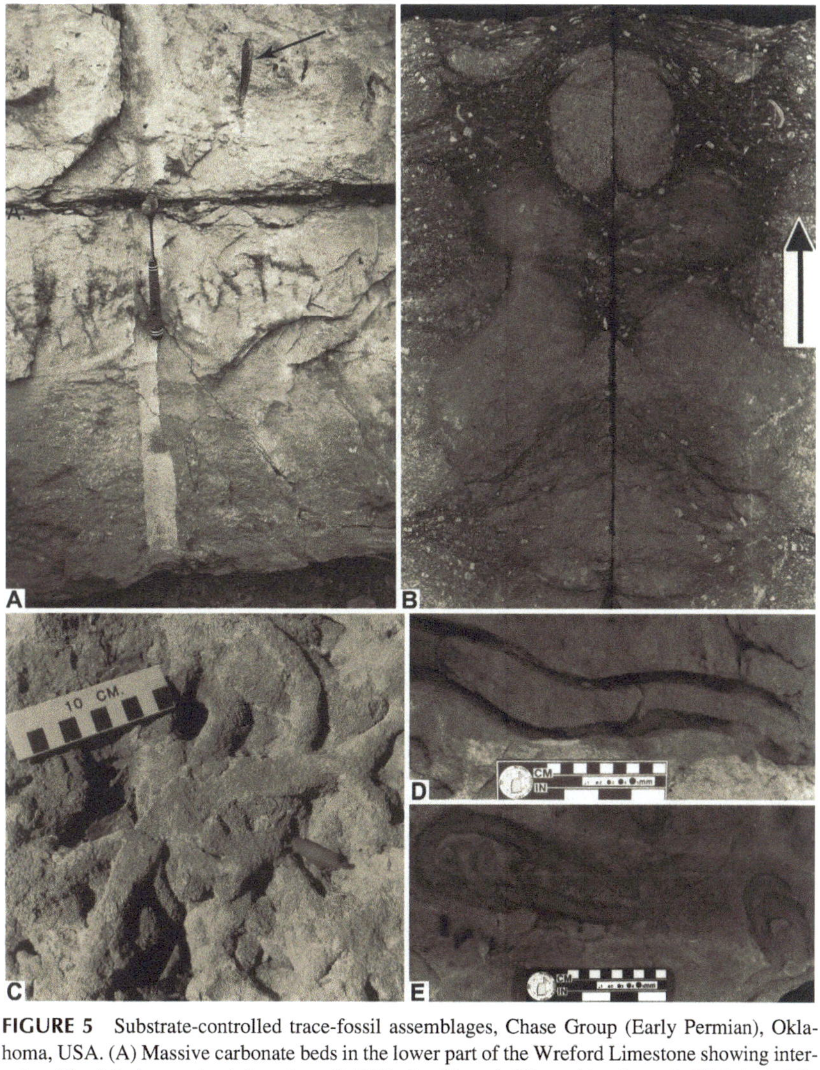

FIGURE 5 Substrate-controlled trace-fossil assemblages, Chase Group (Early Permian), Oklahoma, USA. (A) Massive carbonate beds in the lower part of the Wreford Limestone showing internal molds of the burrowing infaunal mytiloid bivalve *Pinna* in life position (arrow). (B) Paired slabs of a core showing oval to circular, Y-branching burrows of *Thalassinoides* isp. in the Barneston Limestone. Core hole 10, Kay County. Core slabs are 6 cm wide. *After Chaplin (1996).* (C) *Thalassinoides* isp. Distinctive Y-shaped burrow systems preserved as positive hyporeliefs on a sole of the Wreford Limestone, Kay County. *After Chaplin (1996).* (D) *Rhizocorallium* isp. Long, horizontal form interpreted as both a dwelling trace of a suspension-feeding animal and a feeding trace of a deposit feeder. *After Chaplin (1996).* (E) *Rhizocorallium* isp. Preserved as epirelief on a bedding-plane surface. Wreford Limestone, Kay County. *After Chaplin (1996).*

due to minor erosional events in comparison to the more major events indicated by the *Thalassinoides*-burrowed surfaces (Chaplin, 1996, 2010).

Glossifungites-demarcated discontinuity surfaces occur at numerous horizons in the Early Permian Chase Group (Chaplin, 1996, 2010). Despite the frequency with which they occur, they are invariably indicators of stratigraphically significant surfaces. The *Thalassinoides* suites typically penetrate claystone or shale substrates and are infilled with bioclastic wackestone, packstone, or grainstone. They typically have scratched walls and in several examples exhibit secondary ferruginization. The monotypic trace-fossil assemblages that characterize these transgressive surfaces are anomalous and likely reflect environmental or substrate preferences by the tracemaking organisms (likely arthropods). The upper few centimeters were likely removed by erosion, and shallow-tier trace-fossil communities are not represented in the preserved surface, thus lowering the apparent ichnodiversity (Gingras et al., 2001). Most likely, other deeply penetrating taxa were either incapable of penetrating the firm clay in which these traces occur or intolerant of the active wave-dominated setting in which this suite was emplaced.

The *Rhizocorallium* suite, which caps regressive surfaces and separates the underlying marine units from the overlying continental units, does not exhibit ornamentation although they are unlined and show detailed preservation of spreiten (Fig. 5E; Chaplin, 1996, 2010). Thus, it is believed that the substrate was not fully compacted, as the trace constructor was clearly able to adjust itself within it. Moreover, the higher diversity of this assemblage indicates environmental conditions more conducive to infaunal occupation than occurred at the transgressive boundaries.

Other successions, similar to the Chase Group, characterized by substrate-controlled trace-fossil concentrations at lithotype contacts are provided by Brett et al. (2010). These include the Cambrian Wheeler Shale, the Ordovician Kope Formation, and the Jurassic Blue Lias Formation.

3.2 Schei Point Group (Late Triassic), Sverdrup Basin

The Carnian–Norian Schei Point Group in the Sverdrup Basin, Nunavut, Northern Canada, is a dominantly siliciclastic succession that contains several carbonate intervals. These include the Gore Point Member of the Roche Point Formation and the Eden Bay Member of the Hoyle Bay Formation, as well as several horizons within the lowermost Barrow Formation (Embry, 1991; Fig. 6A). Major lithostratigraphic boundaries are commonly characterized by substrate-controlled trace-fossil assemblages. A low-diversity *Glossifungites* assemblage consisting of *Diplocraterion*, *Rhizocorallium*, and *Skolithos* characterizes the Hoyle Bay/Barrow formational contact (Fig. 6A and B). Coplanar *Glossifungites–Trypanites* assemblages characterize the top of the Gore Point Member (Fig. 6A and C). Formation of the *Trypanites* surface on the upper contact of the Gore Point Formation occurred first. Diminutive borings (*Trypanites* isp.) as well as large

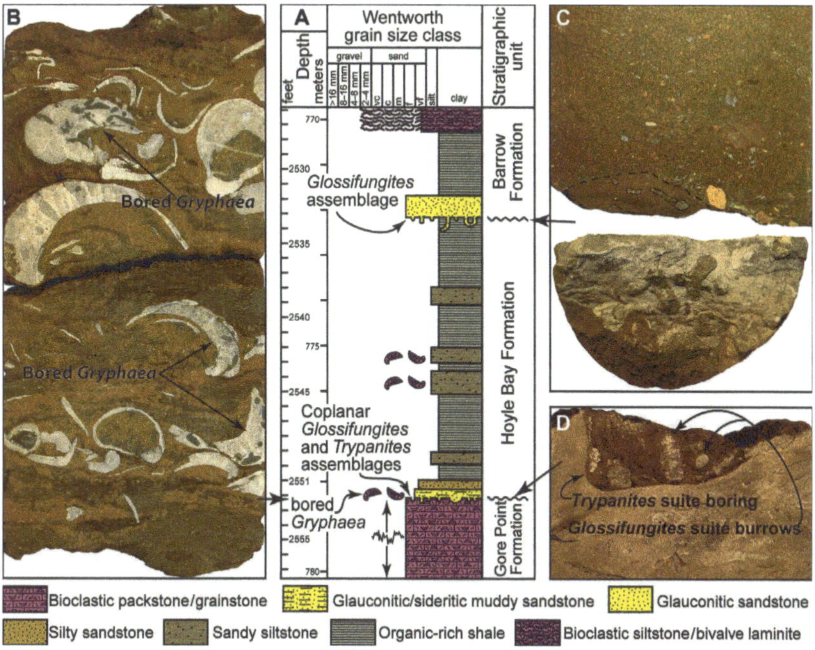

FIGURE 6 Trace fossils in the Hoyle Bay Formation (Middle Triassic), northwest Melville Island, Canada, Sandy Point L46 core 1, 2526–2557′ (769.3–780.2 m). Wavy lines denote erosional formation contacts. Wentworth size-classification acronyms include vf (very fine), f (fine), m (medium), c (coarse), and vc (very coarse). Note that all photographs are core slabs shown in vertical orientation except the bottom photograph in (C), which is shown in horizontal (bedding-plane) orientation. (A) Strip log showing the distribution of facies in the Gore Point, Hoyle Bay, and Barrow Formations. (B) *Gryphaea* sp. shells in a transgressive sandstone unit in the lowermost beds of the Hoyle Bay Formation are commonly bored and are clearly reworked from their original site of emplacement. (C) *Glossifungites*-demarcated discontinuity surface at the Hoyle Bay/Barrow contact. (D) Coplanar *Glossifungites* and *Trypanites* trace-fossil assemblages at the top Gore Point contact (=the Roche Point/Hoyle Bay contact).

pits (cf. *Gastrochaenolites* isp.) characterize this contact. The pits are infilled with silty to muddy sandstone. *Skolithos* burrows pipe clean sandstone from the overlying units into the glauconitic/sideritic sandstone. The sharp-walled nature of the burrows indicates that the sand was firm prior to penetration by the *Skolithos* constructors, supporting the assignment of this trace-fossil assemblage to the *Glossifungites* Ichnofacies. The co-occurrence of two distinct substrate-controlled ichnofacies at this contact underscores that this lithostratigraphic boundary represents a significant temporal break. Overlying this surface, within a transgressive silty to muddy sandstone unit, occur numerous bored *Gryphaea* (Fig. 6D). These ostreids occur primarily in convex-up position, indicating current mediation in their emplacement (Fig. 6D). It is believed that these bored bivalves may have been reworked either from the top Gore Point contact or from a unit that has been erosionally removed from the study area.

3.3 Baldonnel Formation (Late Triassic), British Columbia

The Baldonnel Formation (Middle to Late Carnian) in northeastern British Columbia, Canada, is a predominantly carbonate succession with a high proportion of quartz-dominated sand admixed into most facies and with subordinate clastic sandstone beds interfingering with bioclastic beds in many areas (Zonneveld and Orchard, 2002; Zonneveld et al., 2004). Parasequence boundaries (marine flooding surfaces) in the Baldonnel Formation are commonly characterized by abrupt bounding surfaces demarcated by substrate-controlled ichnofacies (Fig. 7; Zonneveld and Orchard, 2002; Zonneveld et al., 2004). These surfaces commonly exhibit a low diversity, although trace-fossil densities are typically quite high (Figs. 8 and 9). *Diplocraterion*, *Rhizocorallium*, *Thalassinoides*, and small, unlined tubes characterize *Glossifungites* surfaces in the Baldonnel Formation. *Trypanites*, flask-shaped traces (cf. *Gastrochaenolites*) and tube-shaped borings, characterize *Trypanites* assemblages (Fig. 8).

The Baldonnel Formation records an interval of regional marine transgression during the Middle and Late Carnian (Zonneveld and Orchard, 2002). The unit is highly diachronous, with western (basinward) occurrences older than eastern (landward) occurrences. These surfaces are invariably characterized by lags of coarse-clastic detritus including quartz sand, chert granules and pebbles, and phosphatic ooids and granules (Fig. 9B). Bioclastic detritus is also common at these surfaces. The presence of abundant phosphatic clasts concentrated at these horizons results in sharp increases in gamma radiation at these contacts.

Abundant substrate-controlled trace-fossil assemblages in the mixed siliciclastic/carbonate Baldonnel succession are consistent with erosional exhumation of previously buried strata prior to or during marine transgressions. Coplanar *Glossifungites–Trypanites* demarcated discontinuity surfaces record complex scenarios that must include initial deposition, burial, erosional exhumation, followed by burrowing by firmground-preferring organisms, then shallow burial and cementation, and finally boring by hardground-preferring organisms.

3.4 Green River and Wasatch Formations (Eocene), Wyoming

Glossifungites-demarcated discontinuity surfaces are common where fluvial systems incise into underlying layers within Eocene (Wasatchian–Bridgerian) strata in the Green River Basin of Wyoming, USA (Fig. 10). Although these surfaces are present in purely siliciclastic successions—where fluvial channel sandstone units occur incised into subjacent paleosols—they are particularly abundant in basin-margin settings where calcareous mudstone and bioclastic packstone/grainstone units of the lacustrine Green River Formation occur interbedded with clastic sandstone units of the fluvial Wasatch Formation (Fig. 10A–E; Zonneveld et al., 2003, 2006). The bioclastic grainstone and packstone lacustrine beds are dominated by gastropods and ostracods with subordinate unionid bivalves. Stromatolitic and oolitic horizons occur interspersed throughout the succession as well.

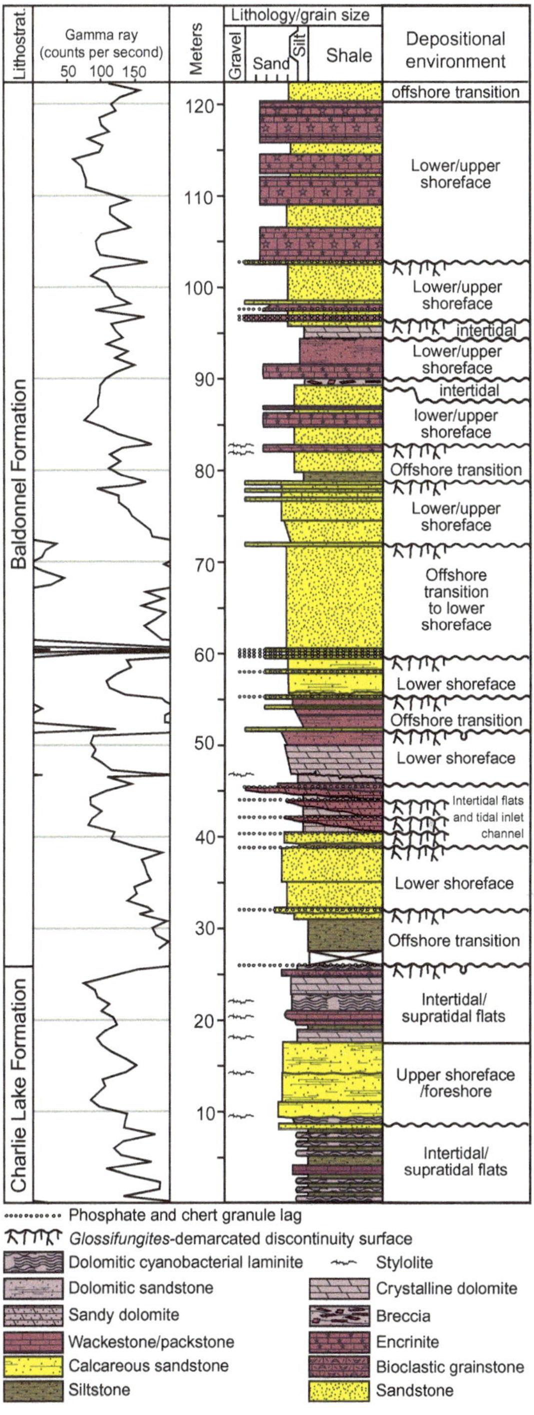

FIGURE 7 Outcrop section of the Upper Charlie Lake Formation and main body of the Baldonnel Formation (Late Triassic), West Schooler Creek, Williston Lake, British Columbia, Canada. The Charlie Lake/Baldonnel contact and numerous internal surfaces within the Baldonnel Formation are characterized by substrate-controlled trace-fossil assemblages.

FIGURE 8 Outcrop photographs of substrate-controlled trace-fossil assemblages in the medial Baldonnel Formation (Late Triassic) at Brown Hill (outcrop reference section for the Baldonnel Formation), Williston Lake, British Columbia. This *Glossifungites*-demarcated discontinuity surface is overwhelmingly dominated by *Rhizocorallium* isp., but other tube structures (cf. *Planolites*) also occur.

Where these bioclastic beds are overlain by siliciclastic claystone or silty to very fine-grained sandstone beds, the contact is typically sharp but appears conformable and trace fossils have not been observed crossing the facies boundary. Where the contact consists of medium-grained (or coarser) cross-stratified sandstone that incises into calcareous mudstone or bioclastic packstone/grainstone, low-diversity *Glossifungites* assemblages occur (Fig. 10C and D). Traces identified within these assemblages include *Camborygma*, *Lunulichnus*, *Planolites*, *Skolithos*, and cf. *Thalassinoides* (Zonneveld et al., 2003). Although these surfaces also occur subjacent to laterally restricted fluvial channels, they are particularly abundant subjacent to regionally correlatable erosional surfaces characterized by amalgamated fluvial-channel successions (Zonneveld et al., 2003). Although these strata were deposited within a continent-interior foreland basin at great distance (>1,000 km) from any possible marine influences, these substrate-controlled trace-fossil assemblages are similar to *Glossifungites* assemblages that occur within marginal and fully marine settings. It should be noted that important differences do occur between fluvial and marine/marginal-marine *Glossifungites* surfaces, including the nature of the traces that characterize the individual surfaces. Marine/marginal-marine *Thalassinoides* tend to be even in thickness, have smooth walls and branch regularly, whereas fluvial forms are less even, have rougher walls and branch irregularly. U-shaped traces (*Diplocraterion* and *Rhizocorallium*) are absent in fluvial successions but are common below marine-influenced *Glossifungites* surfaces.

FIGURE 9 Outcrop with substrate-controlled trace-fossil assemblages, Baldonnel Formation (Late Triassic), Williston Lake, British Columbia, Canada. (A) Outcrop section at 49–57 m, West Schooler Creek section, showing the location of the *Glossifungites* surfaces at a transgressive surface of erosion (downward-facing arrows) and at a regressive surface of erosion (upward-facing arrows). (B) A transgressive surface of erosion illustrating large flask-shaped burrows (cf. *Gastrochaenolites* isp.). This surface is the lower transgressive surface of erosion from (A). (C) Large *Thalassinoides* isp. at the regressive surface of erosion (upward facing arrows in A).

4. SHELL DEBRIS IN CLASTIC SUCCESSIONS

4.1 Influence by Infaunal Taxa

Infaunal organisms exhibit a well-documented influence on the nature and distribution of bioclastic material within the substrate. Bioturbators alter the sediment in which they dwell primarily as a result of domicile construction or while obtaining food (Brenchley, 1981, 1982; Murphey, 1985; Probert, 1984; Ricketts et al., 1985). Bioturbators have been subdivided into five categories based on the nature and direction of biogenic particle movement: downward-conveyors, upward-conveyors, biodiffusors, gallery diffusors, and regenerators (François et al., 2002; Gérino et al., 2003).

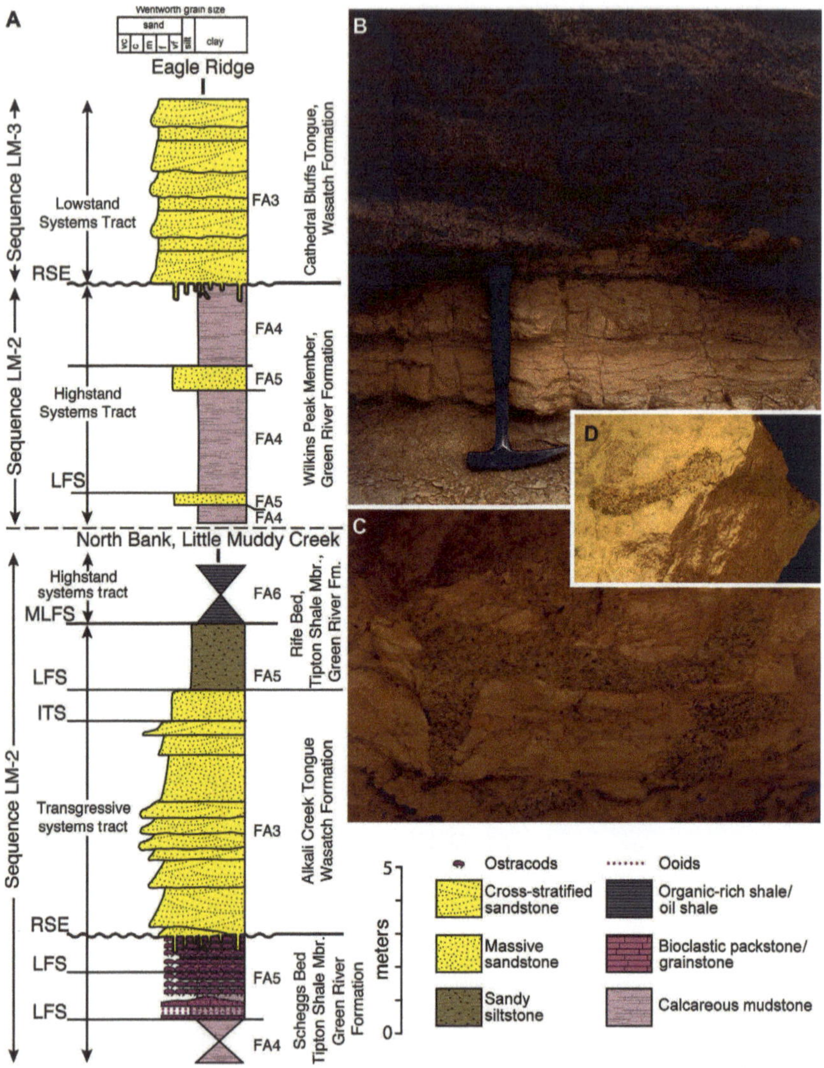

FIGURE 10 *Glossifungites* surfaces within the fluvial–lacustrine Green River-Wasatch section (Early–Middle Eocene), Little Muddy area, southwestern Wyoming, USA. *Adapted from Zonneveld et al. (2003).* FA1 through FA6 are facies associations discussed in depth in Zonneveld et al. (2003). Acronyms for the Wentworth Grain size scale at top are as follows: vf (very fine), f (fine), m (medium), c (coarse), and vc (very coarse). Sequence-stratigraphic divisions follow Zonneveld et al. (2003). Sequence-stratigraphic acronyms include LFS (lacustrine flooding surface), RSE (regressive surface of erosion), MLFS (maximum lacustrine flooding surface), and ITS (initial transgressive surface). (A) The two sections illustrated are separated by ~95 m of vertical section. (B) Cross-stratified sandstone erosionally overlying a carbonate mudstone, Tipton Shale/Wasatch Formation contact. (C) Close-up of the *Thalassinoides* network from (B). (D) Close-up of pseudo-*Diplocraterion* (serendipitous cut through a meandering *Planolites* isp. burrow) from (B).

Downward- and upward-conveyors actively transport particles through their guts resulting in both passive and advective transport of sediment particles (Bouchet et al., 2009). This may result in a transport distance of a few millimeters to several centimeters. Biodiffusors move sediment particles in a random manner over short distances (generally millimeters), resulting in diffusive transport that is most easily recognized as the partial to complete disruption of physical bedforms. Gallery diffusors dig open framework tube or tunnel systems that promote the transport of material from the sediment–water interface to deep within the burrow. Sediment transport distances are commonly in the order of decimeters. Regenerators are also gallery-digging species that promote both biodiffusive sediment admixtures, with large amounts of sediment transported out of the burrow and into the overlying water column, as well as passive downward transport of surface sediment to the bottom of the burrow after burrow abandonment.

With the exception of biodiffusors, all these behaviors can result in a stratigraphic concentration of certain particle types or grain sizes. In shell-rich intertidal and shoreface-sandstone systems, there is commonly a grain-size dichotomy between the two mineralogical end-members. In several examples from the North Sea, lugworms (*Arenicola marina*) were observed to play an active sediment unmixing role producing distinctly graded bedding (Bromley, 1996; Schäfer, 1962; Trewin and Welsh, 1976; Van Straaten, 1952, 1956). Grains of all sizes are drawn down the vertical shaft, but the larger fractions are too large to be eaten and accumulate at the head level of the worms (Bromley, 1996). Smaller grains adhere better to the mucus on the proboscis than do larger ones, and thus the fecal strings deposited on the sediment surface are generally finer grained. Lugworms rapidly turn over the upper few decimeters of the sediment (Cadée, 1976; Swinbanks, 1981) and thus can rapidly segregate coarse bioclastic detritus from finer-clastic detritus in mixed systems.

Under the right circumstances, excavation of deep, open-framework burrows may provide conduits for surface material to be transported from the sediment/water interface to deeper horizons. Mudbanks on the coast of Florida and on the Caicos platform are penetrated by numerous, deep *Callianassa* networks (Tedesco and Wanless, 1991). These banks are affected by numerous storms, and consequently the decapod burrow networks are frequently infilled with bioclastic detritus from the sea floor (Tedesco and Wanless, 1991; Wanless et al., 1988). Repetitive domicile excavation and subsequent storm infilling of burrow networks result in gradual replacement of the original mud substrate with bioclastic packstone and grainstone.

4.2 Influence on Infaunal Taxa

A major influence of mixed siliciclastic/carbonate sediment on trace-fossil communities is the function of grain size and sorting on infaunal distribution. Sufficient bioclastic material may promote colonization by shell-gravel dwellers (Kidwell, 1986) and result in trace-fossil assemblages distinct from those that occur in laterally adjacent sand-dominated successions. Some taxa, such as certain arthropods and bivalves, exhibit a strong preference for gravel substrates and, in the absence of significant clastic gravel in a succession, utilize bioclastic beds as an alternative. Coarse shell material is typically characterized by numerous voids, and thus many organisms live interstitially in shell hash beds. Commonly, these organisms modify their host medium minimally or not at all and thus their ichnological signature is absent, leading to the mistaken assumption of decreased biodiversity in ancient coarse-grained successions. Other organisms modify the shells for use as protective domiciles. Shells within shell-rich substrates are commonly modified by boring organisms, allowing for development of *Entobia*- or *Trypanites*-type assemblages in dominantly siliciclastic successions.

Abundant shell debris in sandstone successions may act as a baffle or barrier to infaunal organisms. In the case of shell debris disseminated throughout the sand, organisms may respond by burrowing around obstructions or, alternatively, moving the obstruction to allow continued passage (Goldring, 1995; Miller and Curran, 2001). If the shell debris is small enough to be easily manipulated, the organism may incorporate it into the burrow wall, either purposely, such as the annelid *Diopatra cuprea,* or accidentally, as the animal shifts its burrow upward in response to sedimentation (Fig. 11). Other organisms may be less adaptable and, in the presence of sufficient disseminated shells and shell debris, may be completely precluded from colonization. This issue is exacerbated where

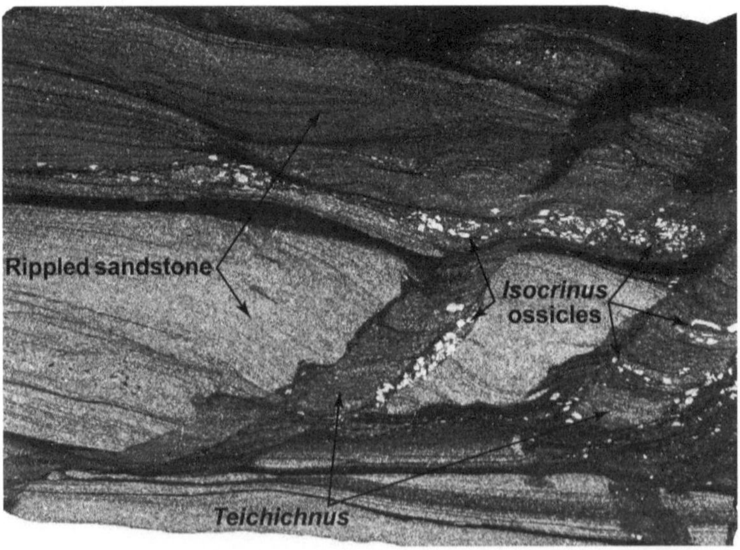

FIGURE 11 *Teichichnus* within the "A-Marker" Member of the lower Charlie Lake Formation (Late Triassic), British Columbia, Canada, at 1075.2 m in Well b-086-E/94-G-09. The main matrix is quartz-dominated fine-grained sand. The larger white grains are crinoid ossicles incorporated into the burrow wall and into burrow laminae as the organism shifted its burrow upward.

early diagenesis produces cemented horizons or patches in the substrate at shallow depths (i.e., intertidal and shallow subtidal beachrock formation).

Buried shell layers prevent some taxa from achieving optimum burrow depths (Frey et al., 1978; Goldring, 1995; Lunz, 1937). In areas with shell or consolidated mud layers buried by a sand veneer, the shrimp *Callichirus major* restricts its burrow to shallow depths (~30 cm), or it may be absent completely from these undesirable locations (Frey et al., 1978; Lunz, 1937). This phenomenon is exemplified in the rock record by the distribution of trace fossils proximal to the biostromes in the Middle Triassic Liard Formation, discussed above (Fig. 3), where distinct differences occur between quartzose sandstone beds with rare scattered shell debris and laterally equivalent shell-rich sandstone beds deposited adjacent to brachiopod-dominated biostromes. The presence of abundant shell debris has a clear effect on the composition of trace-fossil assemblages with fewer vertical dwelling and suspension-feeding structures occurring in biostrome-proximal settings.

4.3 Infaunal Exclusion by Living Taxa (with *Dendraster excentricus* as an Example)

The intertidal succession at Craig Bay, Vancouver Island, Canada, is characterized by a laterally extensive (~1100 m), long wave length (40–80 m), low amplitude (0.2–0.5 m), shore-parallel ridge, and runnel system (Hale and McCann, 1982). In the middle and lower intertidal zones, the runnels are characterized by dense populations of the Western Sand Dollar (*Dendraster excentricus*). These flattened echinoids live on the sediment surface and within the upper decimeter of the sand-dominated substrate. Preliminary field observations have indicated that intervals with exceptionally abundant sand dollars are devoid of burrowing bivalves, whereas the ridge areas have few sand dollars but numerous infaunal bivalves. Similar to Craig Bay, studies elsewhere have shown that the presence and activities of *Dendraster* has an exclusionary effect on the presence of other infaunal taxa (Brenchley, 1981).

Although the total bivalve faunas are moderately diverse, individual ridge systems were only observed to contain two to three taxa. The taxonomic composition changes gradually from the upper intertidal to the lower intertidal zone. Taxa observed include the Bent-nosed Macoma (*Macoma nasuta*), the Baltic Macoma (*M. balthica*), the White Sand Macoma (*M. secta*), the Fat Gaper (*Tresus capax*), and the Common Pacific Littleneck (*Protothaca staminea*). Lower intertidal successions are characterized by *T. capax*, which lives infaunally at considerable depth (15–70 cm below the surface). This large veneroid has an elongated fused siphon that, due to its size, cannot be fully retracted into its shell. Present within the upper ~15 cm of the substrate from the lower to middle intertidal zone is the smaller veneroid *P. staminea* (also with a fused siphon). The macomas all have broad environmental ranges, extending from the upper intertidal to the subtidal zone. *M. balthica* and *M. secta* live deep within the substrate (commonly >25 cm), whereas *M. nasuta* lives much shallower and, unlike other bivalve taxa, occurs with its shell emplaced horizontally in the substrate. All *Macoma* taxa have separated siphons.

Diminutive, opheliid polychaetes (*Euzonus mucronata*) occur in large numbers but dominantly within the ridge systems. Larger carnivorous nephtyid polychaetes (including *Nephtys* sp.) are less common but occur in both ridge and runnel systems. Other taxa occur as well. Among these, solitary anemones occur within all runnel systems occupied by sand dollars.

The lower to middle intertidal portion of the Craig Bay ridge and runnel system shows clear segregation of both extant biota and thus the biogenic sedimentary structures they created. Abundant *Dendraster*, to the exclusion of most other organisms, dominate the runnel systems. During intervals of high tide or falling tide, the *Dendraster* are commonly oriented vertically and are involved in suspension-feeding, catching mobile prey and fragments of algae with their spines, tube feet, and cilia (Ricketts et al., 1985; Timko, 1976). During intervals of low tide, the *Dendraster* typically either lay prone upon the surface or are shallowly buried and feed upon detritus in/on the sediment (Ricketts et al., 1985). At these times, the *Dendraster*, particularly those on the fringes of the runnels, may move about the surface leaving numerous horizontal meniscate structures. Of note, in areas of particularly high abundance, suspension-feeding becomes dominant and the proportion of deposit-feeding diminishes (Fodrie et al., 2007). Also present, albeit less common on the fringes of the runnels, are plug-shaped traces made by burrowing anemones (*Bergaueria* or *Conichnus*). The ridge systems are dominated by bivalve dwelling and adjustment traces (*Lockeia* and multiple forms of *Siphonichnus*). Diminutive horizontal (*Planolites*) and vertical (*Trichichnus*) feeding traces constructed by *Euzonus* occur dominantly within the ridge systems. Open-framework dwelling traces of carnivorous polychaetes (*Palaeophycus*) occur in both the ridges and the runnels.

Mixed siliciclastic/carbonate intertidal flats exhibit inherently complex distributions of fossils and trace fossils. In the Craig Bay example, the distribution of shell-dominated successions is controlled by the topography of the ridge and runnel system. *Dendraster* aggregates in high abundance only in the lows (i.e., in the areas where water remains the longest during low parts of the tidal cycle). The physical environmental control on the distribution of *Dendraster* results in biogenic control of the distribution of infaunal bioturbators. Thus, dense accumulations of shells may strongly influence the presence of infaunal bioturbators in portions of the intertidal flats.

5. SUMMARY

Mixed siliciclastic/carbonate successions are common in the rock record, arguably as common as the end-member successions. However, in many cases, the mineralogical mixing appears not to have affected trace-fossil assemblages. Indeed, with few exceptions, organisms do not care whether the sediment they burrow in is composed of carbonate grains, quartz grains, olivine grains, or any other grain type (Ekdale et al., 1984). However, some fundamental differences in the siliciclastic and carbonate components of the mixed systems discussed above have resulted in unique ichnological signatures.

One of these fundamental differences is the grain size. In many of the mixed systems discussed herein (e.g., Montney Formation, Moenkopi Formation, Liard Formation, modern Craig Bay intertidal), siliciclastic sand/sandstone successions occur interstratified with variably abundant bioclastic detritus. The grain sizes within the bioclastic component are highly variable but are generally significantly coarser than in the siliciclastic component. The presence of shell debris within a clastic succession introduces complexities into the infaunal habitat that are clearly reflected in trace-fossil assemblages. Some of these differences are preservational, with trace fossils being inherently more difficult to recognize in coarser bioclastic intervals than in fine-grained siliciclastic intervals. Other differences occur at the faunal level. Examples include those where some taxa are excluded by abundant admixed bioclastic detritus or by reduced ecospace availability at the sediment/water interface.

Moreover, diagenetic processes commonly operate at different scales and temporal intervals in siliciclastics and carbonates, leading to fundamental differences in preservation and in substrate consistency. In mixed systems, quintessential carbonate ichnotextures co-occur with prototypical siliciclastic ichnotextures. In numerous examples discussed in this chapter (i.e., the Chase Group, the Schei Point Group, and the Baldonnel Formation), mixed siliciclastic/carbonate deposition resulted in sedimentary successions characterized by numerous firmgrounds and co-occurring hardgrounds.

ACKNOWLEDGMENTS

Research on the ichnology of mixed siliciclastic/carbonate successions was supported by NSERC discovery Grant 371662 to the senior author. We gratefully acknowledge Richard Bromley and Dirk Knaust for inviting us to contribute to this volume. We also thank Carlton Brett, Francisco Rodríguez-Tovar, Richard Bromley, and Dirk Knaust for reviewing this chapter and greatly improving it with their many helpful suggestions.

REFERENCES

Blanchon, P., Perry, C.T., 2004. Taphonomic differentiation of *Acropora palmata* facies in cores from Campeche Bank Reefs, Gulf of Mexico. Sedimentology 51, 53–76.

Bouchet, V.M.P., Sauriau, P.-G., Debenay, J.-P., Mermillod-Blondin, F., Schmidt, S., Amiard, J.-C., Dupas, B., 2009. Influence of the mode of macrofauna-mediated bioturbation on the vertical distribution of living benthic foraminifera: first insight from axial tomodensitometry. J. Exp. Mar. Biol. Ecol. 371, 20–33.

Brenchley, G.A., 1981. Disturbance and community structure: an experimental study of bioturbation in marine soft-bottom environments. J. Mar. Res. 39, 767–790.

Brenchley, G.A., 1982. Mechanisms of spatial competition in marine soft-bottom communities. J. Exp. Mar. Biol. Ecol. 60, 17–33.

Brett, C.E., 1995. Sequence stratigraphy, biostratigraphy, and taphonomy in shallow marine environments. Palaios 10, 597–616.

Brett, C.E., Allison, P.A., Hendy, A.J.W., 2010. Comparative taphonomy and sedimentology of small-scale mixed carbonate/siliciclastic cycles: synopsis of Phanerozoic examples. In: Allison, P.A., Bottjer, D.J. (Eds.), Taphonomy: Process and Bias Through Time. Topics in Geobiology, vol. 32. Springer, New York, pp. 107–198.

Bromley, R.G., 1996. Trace Fossils: Biology, Taphonomy and Applications. Chapman & Hall, London, 361 pp.

Budd, D.A., Harris, P.M., 1990. Carbonate-siliciclastic mixtures. In: SEPM Reprint Series, vol. 14, 272 pp.

Cadée, G.C., 1976. Sediment reworking by *Arenicola marina* on tidal flats in the Dutch Wadden Sea. Neth. J. Sea Res. 10, 440–460.

Campbell, A.E., 2005. Shelf-geometry response to changes in relative sea level on a mixed carbonate–siliciclastic shelf in the Guyana Basin. Sediment. Geol. 175, 259–275.

Chaplin, J.R., 1996. Ichnology of transgressive-regressive surfaces in mixed carbonate-siliciclastic sequences, Early Permian Chase Group, Oklahoma. In: Witzke, B.J., Ludvigson, G.A., Day, J. (Eds.), Palaeozoic Sequence Stratigraphy: Views from the North American Craton. GSA Spec. Pap. 306, pp. 399–418.

Chaplin, J.R., 2010. Stratigraphic Analysis of the Permian Chase Group in Northern Oklahoma—Outcrop Analogs of Reservoir Rocks in the Hugo Embayment of Northwestern Oklahoma and Southwestern Kansas. Oklahoma Geol. Surv., Guidebook 36, 110 pp.

Ekdale, A.A., Bromley, R.G., Pemberton, S.G., 1984. Ichnology: The Use of Trace Fossils in Sedimentology and Stratigraphy. SEPM Short Course Notes 15, pp. 1–316.

Embry, A.F., 1991. Mesozoic history of the Arctic Islands. In: Trettin, H.P. (Ed.), Geology of the Innuitian Orogen and Arctic Platform of Canada and Greenland. Geology of Canada, Geol. Surv. Can., vol. 3, pp. 371–433.

Fodrie, J.F., Herzka, S.Z., Lucas, A.J., Francisco, V., 2007. Intraspecific density regulates positioning and feeding mode selection of the sand dollar *Dendraster excentricus*. J. Exp. Mar. Biol. Ecol. 340, 169–183.

François, F., Gérino, M., Stora, G., Durbec, J.-P., Poggiale, J.-C., 2002. Functional approach to sediment reworking by gallery-forming macrobenthic organisms: modeling and application with the polychaete *Nereis diversicolor*. Mar. Ecol. Progr. Ser. 229, 127–136.

Frey, R.W., Howard, J.D., Pryor, W.A., 1978. *Ophiomorpha*: its morphologic, taxonomic, and environmental significance. Palaeogeogr. Palaeoclimatol. Palaeoecol. 23, 199–229.

García-Hidalgo, J.F., Gil, J., Segura, M., Domínguez, C., 2007. Internal anatomy of a mixed siliciclastic-carbonate platform the Late Cenomanian-Mid Turonian at the southern margin of the Spanish Central System. Sedimentology 54, 1245–1271.

Gérino, M., Stora, G., François, F., Gilbert, J.C., Poggiale, J.-C., Mermillod-Blondin, F., Desrosiers, G., Vervier, P., 2003. Macro-invertebrate functional groups in freshwater and marine sediments: a common mechanistic classification. Vie Milieu 53, 222–231.

Gingras, M.K., Pemberton, S.G., Saunders, T.D.A., 2001. Bathymetry, sediment texture, and substrate cohesiveness; their impact on modern *Glossifungites* trace assemblages at Willapa Bay, Washington. Palaeogeogr. Palaeoclimatol. Palaeoecol. 169, 1–21.

Goldring, R., 1995. Organisms and substrate: response and effect. In: Bosence, D.W.J., Allison, P.A. (Eds.), Marine Palaeoenvironmental Analysis from Fossils. Geol. Soc. London, Spec. Publ., vol. 83, pp. 151–180.

Guertin, L.A., McNeill, D.F., Lidz, B.H., Cunningham, K.J., 1999. Biochronology and transgressive/regressive signature in the late Neogene siliciclastic foundation (Long Key Formation) of the Florida Keys. J. Sediment. Res. 69, 653–666.

Hale, P.B., McCann, S.B., 1982. Rhythmic topography in a mesotidal, low-wave-energy environment. J. Sediment. Petrol. 52, 415–429.

Kidwell, S.M., 1986. Taphonomic feedback in Miocene assemblages: testing the role of dead hardparts in benthic communities. Palaios 1, 239–255.

Kidwell, S.M., Jablonski, D., 1983. Taphonomic feedback: ecological consequences of shell accumulation. In: Tevesz, M.J.S., McCall, P.L. (Eds.), Biotic Interactions in Recent and Fossil Benthic Communities. Plenum Press, New York, pp. 195–248.

Knaust, D., 2007. Invertebrate trace fossils and ichnodiversity in shallow-marine carbonates of the German Middle Triassic (Muschelkalk). In: Bromley, R.G., Buatois, L.A., Mángano, M.G., Genise, J.F., Melchor, R.N. (Eds.), Sediment-Organism Interactions: A Multifaceted Ichnology, SEPM Spec. Publ., vol. 88, pp. 221–238.

Knaust, D., Curran, H.A., Dronov, A., 2012. Shallow-marine carbonates. In: Knaust, D., Bromley, R.G. (Eds.), Trace Fossils as Indicators of Sedimentary Environments. Developments in Sedimentology, vol. 64. Elsevier, Amsterdam, pp. 703–750.

Lomondo, A.J., Harris, P.M., 1991. Mixed carbonate-siliciclastic sequences. SEPM Core Workshop 15. 245–271.

Lunz Jr., G.R., 1937. Notes on *Callianassa major* Say. Charleston (South Carolina). Museum Leaflet 10, 3–15.

Martindale, R.C., Zonneveld, J.-P., Bottjer, D.J., 2010. Microbial framework in Upper Triassic (Carnian) patch reefs from Williston Lake, British Columbia, Canada. Palaeogeogr. Palaeoclimatol. Palaeoecol. 297, 609–620.

McLaughlin, P.I., Brett, C.E., Wilson, M.A., 2008. Hierarchy of sedimentary surfaces and condensed beds from the Middle Paleozoic of eastern North America: implications for cratonic sequence stratigraphy. In: Pratt, B.R., Holmden, C. (Eds.), Dynamics of Epeiric Seas. Geol. Ass. Can., Spec. Pap. 48, pp. 175–200.

McNeill, D.F., Cunningham, K.J., Guertin, L.A., Anselmetti, F.S., 2004. Depositional themes of mixed carbonate-siliciclastics in the south Florida Neogene: application to ancient deposits. AAPG Mem. 80, 23–43.

McNeill, D.F., Janson, X., Bergman, K.L., Eberli, G.P., 2010. Belize: a modern example of a mixed carbonate-siliciclastic shelf. In: Westphal, H., Reigl, B.M., Eberli, G.P. (Eds.), Carbonate Depositional Systems: Assessing Dimensions and Controlling Parameters. Springer, Berlin, pp. 81–143.

Miller, M.F., Curran, H.A., 2001. Behavioral plasticity of modern and Cenozoic thalassinidean shrimp. Palaeogeogr. Palaeoclimatol. Palaeoecol. 166, 219–236.

Moslow, T.F., Zonneveld, J.-P., 2005. Origin, architecture, sedimentary characteristics and reservoir properties of bioclastic accumulations (i.e., "coquina"), lower Middle Triassic of the WCSB. Can. Soc. Petrol. Geol. Reserv. 32, 12.

Mount, J.F., 1984. Mixing of siliciclastic and carbonate sediments in shallow shelf environments. Geology 12, 432–435.

Murphey, R.C., 1985. Factors affecting the distribution of the distribution of the introduced bivalve, *Mercenaria mercenaria*, in a California lagoon—the importance of bioturbation. J. Mar. Res. 43, 673–692.

Orchard, M.J., Zonneveld, J.-P., 2009. The Lower Triassic Sulphur Mountain Formation in the Wapiti Lake area: lithostratigraphy, conodont biostratigraphy, and a new biozonation for the lower Olenekian (Smithian). Can. J. Earth Sci. 46, 757–790.

Probert, P.K., 1984. Disturbance, sediment stability, and trophic structure of soft-bottom communities. J. Mar. Res. 43, 673–692.

Ricketts, E.F., Calvin, J., Hedgepeth, J.W., 1985. Between Pacific Tides, revised by D.W. Phillips, 5th ed. Stanford University Press, Stanford, 652 pp.

Rodríguez-Tovar, F.J., Perez-Valera, F., 2008. Trace fossil *Rhizocorallium* from the Middle Triassic of the Betic Cordillera, southern Spain: characterization and environmental implications. Palaios 23, 78–86.

Schäfer, W., 1962. Aktuo-Paläontologie nach Studien in der Nordsee. Kramer, Frankfurt am Main, 666 pp.

Schlager, W., 1991. Depositional bias and environmental change: important factors in sequence stratigraphy. Sediment. Geol. 70, 109–130.

Southgate, P.N., Kennard, J.M., Jackson, M.J., O'Brien, P.E., Sexton, M.J., 1993. Reciprocal lowstand clastic and highstand carbonate sedimentation, subsurface Devonian reef complex, Canning Basin, Western Australia. In: Loucks, R.G., Sarg, J.F. (Eds.), Carbonate Sequence Stratigraphy: Recent Developments and Applications. AAPG Mem. 57, pp. 157–179.

Swinbanks, D.D., 1981. Sediment reworking and the biogenic formation of clay laminae by *Abarenicola pacifica*. J. Sediment. Petrol. 51, 1137–1145.

Tapanila, L., Copper, P., Edinger, E., 2004. Environmental and substrate control on Palaeozoic bioerosion in corals and stromatoporoids, Anticosti Island, Eastern Canada. Palaios 19, 292–306.

Tedesco, L.P., Wanless, H.R., 1991. Generation of sedimentary fabrics and storm infilling of burrow networks, Holocene of South Florida and Caicos Platform, B.W.I. Palaios 6, 326–343.

Timko, P., 1976. Sand dollars as suspension feeders: a new description of feeding in *Dendraster excentricus*. Biol. Bull. Russ. Acad. Sci. 151, 247–259.

Trewin, N.H., Welsh, W., 1976. Formation and composition of a graded estuarine shell bed. Palaeogeogr. Palaeoclimatol. Palaeoecol. 19, 219–230.

Van Straaten, L.M.J.U., 1952. Biogenic textures and the formation of shell beds in the Dutch Wadden Sea. Kon. Netherl. Akad. Wetensch., Proc. B55, 500–516.

Van Straaten, L.M.J.U., 1956. Composition of shell beds formed in tidal flat environment in the Netherlands and in the Bay of Arcachon (France). Geol. Mijnbouw 18, 209–216.

Wanless, H.R., Tedesco, L.P., Tyrrell, K.M., 1988. Production of subtidal tubular and surficial tempestites by Hurricane Kate, Caicos Platform, British West Indies. J. Sediment. Petrol. 58, 739–750.

Webb, G.E., 1994. Palaeokarst, palaeosol, and rocky-shore deposits at the Mississippian-Pennsylvanian unconformity, northwestern Arkansas. GSA Bull. 106, 634–648.

Yang, B.C., Dalrymple, R.W., Gingras, M.K., Pemberton, S.G., 2009. Autogenic occurrence of *Glossifungites* ichnofacies; examples from wave-dominated, macrotidal flats, southwestern coast of Korea. Mar. Geol. 260, 1–5.

Zonneveld, J.-P., 2001. Middle Triassic biostromes from the Liard Formation, British Columbia, Canada: oldest examples from the Mesozoic of NM Pangea. Sediment. Geol. 145, 317–341.

Zonneveld, J.-P., Orchard, M.J., 2002. Stratal relationships of the Baldonnel Formation (Upper Triassic), Williston Lake, northeastern British Columbia. Geol. Surv. Can. Curr. Res. A8, 1–15.

Zonneveld, J.-P., Moslow, T.F., Henderson, C.M., 1997. Lithofacies associations and depositional environments in a mixed siliciclastic-carbonate coastal depositional system, upper Liard Formation, Triassic, northeastern British Columbia. In: Moslow, T.F., Wittenberg, J. (Eds.), Triassic of the Western Canada Sedimentary Basin. Bull. Can. Petrol. Geol., vol. 45, pp. 553–575.

Zonneveld, J.-P., Gingras, M.K., Pemberton, S.G., 2001. Trace fossil assemblages in a Middle Triassic mixed siliciclastic-carbonate marginal marine depositional system, British Columbia. Palaeogeogr. Palaeoclimatol. Palaeoecol. 166, 249–276.

Zonneveld, J.-P., Bartels, W.S., Clyde, W.C., 2003. Stratal architecture of an Early Eocene fluvial-lacustrine depositional system, Little Muddy Creek area, southwestern Green River Basin,

Wyoming. In: Raynolds, R.G., Flores, R.M. (Eds.), The Cenozoic of the Rocky Mountain Region. Rocky Mountain Section of the SEPM, pp. 253–288.

Zonneveld, J.-P., Carrelli, G.G., Riediger, C., 2004. Sedimentology of the Upper Triassic Charlie Lake, Baldonnel & Pardonet Formations from outcrop exposures in the southern Trutch region, northeastern British Columbia. In: Lane, L.S. (Ed.), Central Foreland NATMAP: Stratigraphic and Structural Evolution of the Cordilleran Foreland—Part 1. Bull. Can. Petrol. Geol., vol. 52, pp. 343–375.

Zonneveld, J.-P., Lavigne, J.M., Bartels, W.S., Gunnell, G.F., 2006. *Lunulichnus tuberosus* (ichnogen. and ichnosp. nov) from the Early Eocene Wasatch Formation, Fossil Butte National Monument, Wyoming: an arthropod-constructed trace fossil associated with alluvial firmgrounds. Ichnos 13, 87–94.

Zonneveld, J.-P., Henderson, C.M., Stanley Jr., G.D., Orchard, M.J., Gingras, M.K., 2007. Oldest scleractinian coral reefs on the North American Craton: Upper Triassic (Carnian), northeastern British Columbia. Palaeogeogr. Palaeoclimatol. Palaeoecol. 243, 421–450.

Part VI

Ichnology in Hydrocarbon-Reservoir and Aquifer Characterization

Chapter 27

Porosity and Permeability in Bioturbated Sediments

Murray K. Gingras,*[,1] Greg Baniak,* John Gordon,[†] Jussi Hovikoski,[‡] Kurt O. Konhauser,* Andrew La Croix,[§] Ryan Lemiski,[¶] Carl Mendoza,* S. George Pemberton,* Camilo Polo* and John-Paul Zonneveld*
*Department of Earth & Atmospheric Sciences, University of Alberta, Edmonton, Alberta, Canada, [†]Husky Energy, Calgary Alberta, Canada, [‡]Geological Survey of Denmark and Greenland (GEUS), Copenhagen, Denmark, [§]Department of Earth Sciences, Simon Fraser University, Burnaby, British Columbia, Canada, [¶]Talisman Energy Inc., Calgary, Alberta, Canada
[1]Corresponding author: e-mail: mgingras@ualberta.ca

1. INTRODUCTION

It is generally believed that bioturbation reduces the permeability of sedimentary strata. This is because biogenic churning of laminated sediment lowers the sorting of the sediment preserved within laminae. Also, bioturbation is commonly overlooked in intensely bioturbated media because of a lack of lithological definition and because of the complex fabrics that are sometimes present. However, several examples of bioturbation-enhanced bulk permeability have been reported in the literature (e.g., Cunningham et al., 2009, 2012; Dawson, 1978; Gingras et al., 1999, 2004a,b; Gordon et al., 2010; Gunatilaka et al., 1987; La Croix, 2010; Lemiski et al., 2011; McKinley et al., 2004; Mehrtens and Selleck, 2002; Sutton et al., 2004; Tonkin et al., 2010; Zenger, 1992). These range from biogenic modifications of the primary depositional fabric to diagenetic alterations (typically recrystallization) of the sedimentary matrix.

Permeability fabrics and the distribution of porosity result from physical heterogeneities in sedimentary rocks. Physical heterogeneity may result from the presence of laminae, the arrangement of grains or fossils, fabric-retentive and fabric-destructive diagenesis, and structural deformation. Bioturbation alters local grain distributions in the hosting sedimentary rock, and trace fossils commonly possess geochemical characteristics that differ from the surrounding sedimentary rock (e.g., Konhauser and Gingras, 2011; McIlroy et al., 2003; Over, 1990; Petrash et al., 2011). Thus, trace fossils can influence the distribution of porosity and permeability in sedimentary rocks by physically changing

Developments in Sedimentology, Vol. 64. http://dx.doi.org/10.1016/B978-0-444-53813-0.00027-7

the pore-throat distribution and by acting as *loci* of cementation and dissolution processes during early and late stages of diagenesis. These processes commonly act together, and the range of permeability fabrics generated by biogenic modification is broad.

Factors such as the morphology of the original burrow(s), the impact of bioturbation on grain-size distributions, the distribution of organics within incipient trace fossils, the mineralogy of fossils and grains, and a host of post-depositional chemical processes determine the distributions of porosity and permeability (summarized in Lemiski et al., 2011; Pemberton and Gingras, 2005; Tonkin et al., 2010). Burrowed fabrics are commonly three dimensional in their geometry and burrow shapes and sizes are variable. Thus, bioturbated sediment is more spatially complex than bedded sediment, and flow through it is difficult to model (see Knaust, 2012). Even very simple fabrics contain a range of burrow forms, and it is difficult to statistically characterize that range (cf. McIlroy, 2007). An important outcome of burrow-associated permeability is that burrows tend to improve the vertical permeability (k_v) over the horizontal permeability (k_h). In unfractured laminated media, k_h is almost always larger than k_v.

Many oil and gas reservoirs contain zones that are bioturbated, or are entirely composed of burrowed strata. In such cases, ichnological permeability and porosity likely contribute to storativity and deliverability of resource, and important relationships, such as k_v *versus* k_h are altered. The analysis of bioturbated strata in hydrocarbon reservoirs should constitute part of reservoir analyses. However, due to the aforementioned complexities, this seldom occurs.

This chapter provides a summary of the concepts associated with the applications of ichnology to reservoir fluid flow. We focus on the nature of permeability and porosity and scrutinize the ways in which burrows might influence those parameters. Physical and chemical modifications resulting from bioturbation are discussed and their relationships to a classification system for burrow-related permeability are considered.

2. THE NATURE OF ICHNOLOGICAL PERMEABILITY

Trace-fossil permeability is broadly comparable to fracture permeability in that trace fossils are discrete entities displaying predictable permeabilities that, due to their large surface areas, have significant flow interactions with the surrounding matrix. There are differences: trace fossils are tubular, trace-fossil permeabilities are lower than (open) fracture permeabilities (Gingras et al., 1999), and the volume comprising the biogenic fabric is commonly much higher than the volumes of fracture systems (10–100% *versus* 1–3%). Flow interactions between burrow fabrics and matrix are therefore more extensive than those between fractures and their hosting matrix. The burrow distribution can also be much extensive, whereas fractures present localized permeability streaks.

The permeability/porosity modification associated with burrow fabrics ranges between severe reductions of the bulk permeability and notable bulk-permeability enhancement. The nature of permeability enhancement can be described as either dual-porosity flow media or as dual-permeability flow media. Dual-porosity flow media include burrowed sediment wherein the matrix permeability is within two orders of magnitude of the burrow permeabilities (Fig. 1). In examples where the matrix permeability and the burrow permeability differ by three orders of magnitude or more, a dual-permeability flow network is established (Fig. 2).

In dual-porosity flow media, most of the rock volume is used to conduct flow. Although a larger fluid flux occurs through the higher permeability zones, flow interactions between burrows and the matrix are extensive (Fig. 1). This has several implications regarding reservoir behavior: (1) more pore volume contributes to fluid or gas production; (2) due to capillarity, multiphase fluid transport may lead to segregation of the different fluid phases (e.g., oil may be isolated in finer-grained parts of the reservoir); and (3) advection will be dominant over diffusion (in other words, resource flows through the stratum in response to a hydraulic gradient). Broadly speaking,

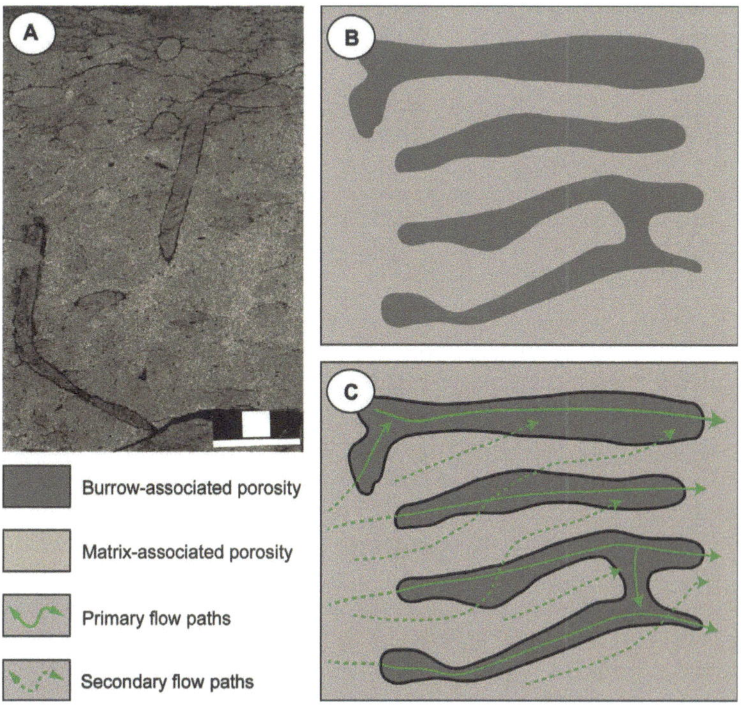

FIGURE 1 Fluid-flow behavior in dual-porosity biogenic fabrics. (A) Slabbed core image of sand-filled burrows encased in a sandstone matrix (*Ophiomorpha* and vague *Thalassinoides* are discernible). (B) Schematic sedimentary rock with bioturbation shown as darker gray. (C) Within dual-porosity flow media, fluid is advected through both the matrix and the trace-fossil fill. Where the trace-fossil permeability is higher, a greater flux exists in the burrows.

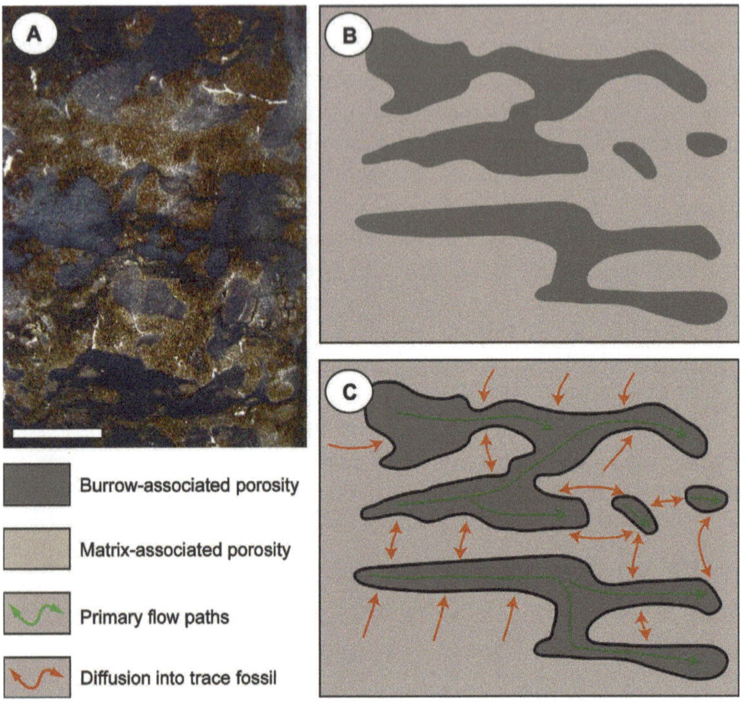

FIGURE 2 Fluid-flow behavior in dual-permeability biogenic fabrics. (A) Slabbed core image showing trace-fossil-associated dolomite in a limestone matrix. (B) Schematic sedimentary medium with bioturbation shown as darker gray. (C) Within dual-permeability flow media, fluid is advected through both the trace fossils and the matrix (dominantly by advection). The lower permeability matrix contributes to flow only via diffusion.

fluids transported within dual-porosity media are not likely to move uniformly. The flow follows preferentially higher permeability beds in which the flow paths are variably tortuous, depending on to the burrow morphology. Notably, flow within the lower permeability rock may move 10–100 times more slowly than in the higher permeability burrowed zones. With regard to cryptically bioturbated intervals, where the tracemaking organism is <1 mm in size, reduction of heterogeneities commonly occurs and k_v typically becomes equal to k_h.

In general, (bioturbate) dual porosity is either created by animals moving through the sediment, or is the result of sand-dwelling animals that selectively ingest and rework the sediment. Therefore, dual porosity is most commonly associated with cryptic bioturbation, the presence of *Macaronichnus*, and in bioturbated sandstone that contains *Thalassinoides*, *Ophiomorpha*, *Skolithos*, or *Arenicolites* (discussed below). Depositional settings that are associated with biogenic dual-porosity flow media thus include foreshore, sand-dominated intertidal flat, tidal bars, and lower shoreface deposits (Fig. 3).

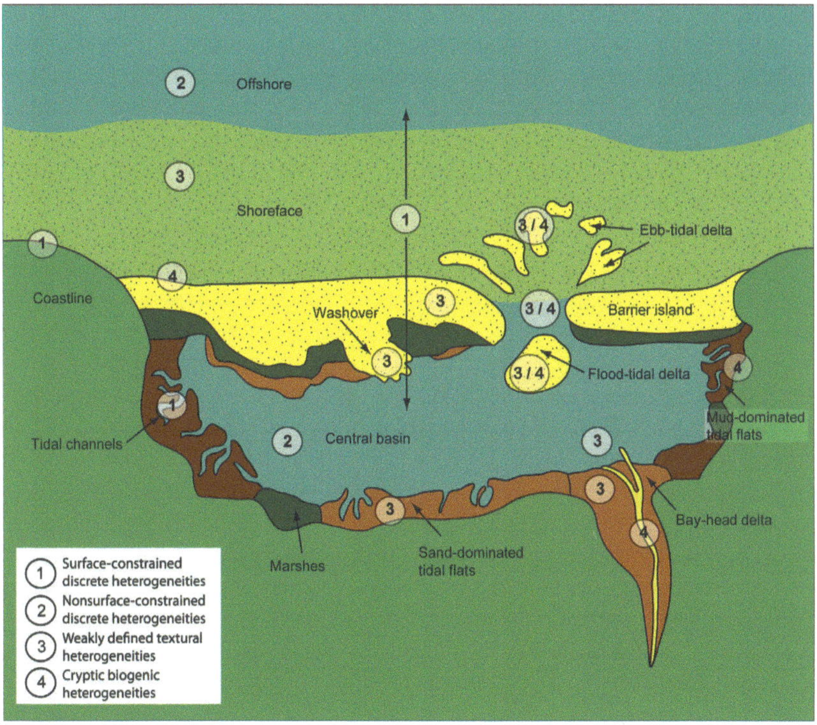

FIGURE 3 Biogenic flow-medium characteristics are strongly influenced by the environment of deposition. Biogenic dual-permeability flow media are associated with (1) surface-constrained discrete heterogeneities, and (2) non-constrained discrete heterogeneities. Dual porosity is strongly associated with (3) weakly defined textural heterogeneities, and (4) cryptic biogenic heterogeneities. Lagoonal and offshore settings are dominated by dual permeability and shoreface associations tend toward dual porosity.

For conventional oil production, dual-permeability flow media have comparably poorer resource characteristics (although they may provide good gas reservoirs). The impacts of dual-permeability systems in reservoirs and aquifers include that (1) only the higher permeability parts of the rock contribute to flow, (2) the matrix dominantly interacts with the flow conduits via diffusion, and (3) the highly contrasting permeability fields may lead to "watering out" of the reservoir and also mitigates the effectiveness of secondary recovery techniques (e.g., water flooding). On the other hand, gas production may not be impacted by the presence of dual-permeability flow media. This is because gas flows more readily through tortuous networks than liquids, and gas may be stored in the matrix and delivered by the permeability streaks.

Dual permeability is the result of open burrows passively filled in with coarser grains, burrows being actively backfilled with coarse material, or burrow-associated diagenesis in carbonate strata. Therefore, dual permeability is associated with elements of the *Cruziana*, *Zoophycos*, and *Glossifungites* ichnofacies, but dominantly *Zoophycos*, *Thalassinoides,* and *Planolites* (discussed in Section 4.3.2.). Sedimentary environments associated with biogenic dual porosity are offshore to lower shoreface, heterolithic tidal bars, and at the base of transgressive successions (Fig. 3). In the case of carbonate successions, burrow-associated diagenesis has been associated with carbonate accumulation in restricted platform and low-energy ramp settings (Beales, 1953; Corlett and Jones, 2012; Gingras et al., 2004b; Knaust, 2009).

Burrow-induced heterogeneity normally results in the presence of complex, two- or three-dimensional flow networks. Where the burrow permeability is markedly higher than the matrix, complex ichnofabrics are prone to flow-conduit cut-off and a volume of the transmissive rock is rendered ineffective. In these cases, the assessment of the ineffective flow volume is difficult to determine. Other major problems include fabric upscaling and determination of the three-dimensional fabric. The latter issue is solved in part through imaging techniques, such as Computed Tomography (CT) and Magnetic Resonance Imaging (MRI) (discussed below). Regarding the former, pervasive fabrics, including cryptic bioturbation and general biogenic mixing, are amenable to upscaling. However, the upscaling of biogenic fabrics composed of discrete trace fossils still remains a leap of faith.

Increased complexity of flow paths results in higher tortuosity, which leads to higher dispersivity, which is the rock property related to hydrodynamic spreading of a solute or phase. Low dispersivity means that a fluid can pass through the porous medium with minimal (primarily mechanical) mixing of fluids and gasses, and is a characteristic of homogeneous flow media. High dispersivity indicates more flow mixing, which is related to increased rock heterogeneity.

The impact of bioturbation on dispersivity was demonstrated by Gingras et al. (2004a). Using laboratory experiments and a burrowed (dual porosity) limestone medium, they showed that gas was transported rapidly through the higher-permeability burrows, whereas the matrix contributed minimally to bulk flux. The dispersivity of the burrowed limestone was higher than other flow media tested (sandstone and fractured limestone), an outcome of the high heterogeneity of the burrowed medium. This is important because dispersivity partly determines secondary recovery efficiency (Gingras et al., 1999, 2004a; Pemberton and Gingras, 2005; Fig. 4).

Simple ichnofabrics have been numerically modeled using flow-simulation software (e.g., Gingras et al., 1999), but complex ichnofabrics (i.e., more than three trace-fossil types and variably permeable burrow fills) have proven difficult to model.

3. FRAMEWORK FOR ASSESSING BURROW-ASSOCIATED PERMEABILITY

A number of factors strongly influence bulk-flow parameters. These are permeability contrast, bioturbation intensity, burrow connectivity, burrow-surface area, and burrow architecture.

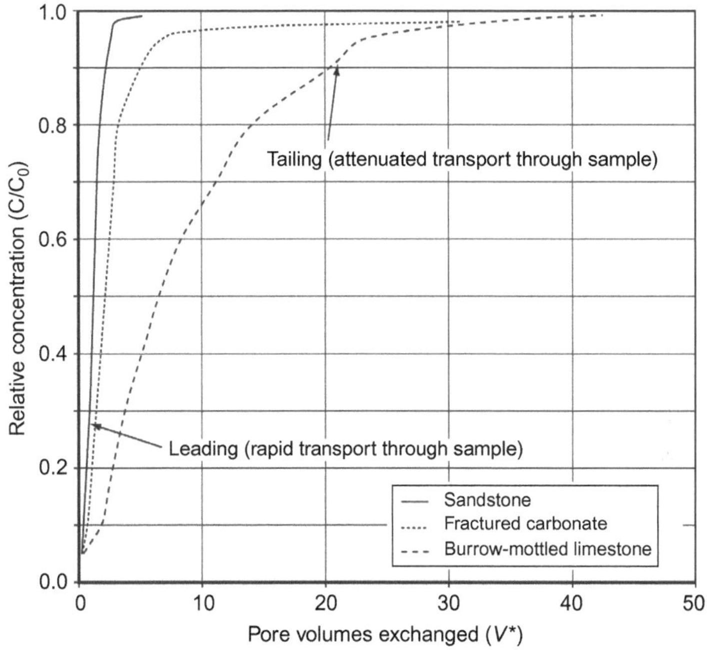

FIGURE 4 *Gingras et al. (2004a,b)* determined the relative dispersivities of three types of geological media: homogeneous sandstone, fractured limestone, and burrowed dolomitic limestone. Their results showed that the flow paths present in burrow-associated dolomite are tortuous, and that the interaction between the flow paths and the matrix is extensive. Results from their dispersivity experiments support this by indicating substantial leading of tracer gas through the sample (indicated). Leading is associated with permeability streaks in the rock, and tailing, which is associated with extensive matrix/flow path interactions.

3.1 Permeability Contrast

In stratified sedimentary rocks, fluid flow dominantly occurs either parallel or perpendicular to the layers. Assuming that each layer is homogeneous, the system as a whole can be characterized as a homogeneous, yet anisotropic medium (Freeze and Cherry, 1979). An estimate of the overall bulk permeability parallel to stratification is obtained from a simple volume-weighted arithmetic mean of the permeability of each layer:

$$K_{\text{arithmetic}} = \sum_{i=1}^{n} \frac{k_i d_i}{d},$$

where k_i is the permeability of each layer, d_i is the individual layer thickness, and d the total thickness.

For fluid flow occurring perpendicular to layered media, applying a volume-weighted harmonic mean provides an estimate of the bulk vertical permeability:

$$K_{\text{harmonic}} = \frac{1}{\sum_{i=1}^{n} \frac{d_i}{k_i d}}.$$

Finally, if the porous medium is not layered (i.e., it is homogeneous), the effective homogeneous permeability is isotropic and is best estimated by the volume-weighted geometric mean of the measured permeabilities (Warren et al., 1961):

$$\text{In}(k_{\text{geometric}}) = \sum_{i=1}^{n} \frac{\text{In}(K_i)d_i}{d}.$$

Gingras et al. (1999) explored the concept of applying arithmetic and harmonic means to estimate the bulk permeability of bioturbated media. By developing a set of empirical and numerically modeled solutions to bulk permeability, they showed that, under conditions of low burrow connectivity, bulk permeability could best be characterized using the harmonic mean. As burrow networks became increasingly interconnected, a modified arithmetic mean could be used, particularly for the estimation of k_v. The geometric mean is broadly applied in transitional situations where the permeable burrow structures were only locally connected (this is further explored in La Croix et al., 2012). For these studies, the above equations were modified such that the weighted volume (i.e., d_i/d) represents the proportional volume occupied by burrows, which is proportional to burrowing intensity.

3.2 Bioturbation Intensity and Connectivity

Higher intensities of bioturbation contribute to a higher effective permeability in two ways: (1) by increasing the amount of permeable flow medium present, and (2) by providing continuous flow paths through burrow interpenetration. Gingras et al. (1999) conducted a sensitivity analysis of flow responses in a numerical model with varying burrowing volumes (1%, 2%, 10%, 25%, and 50%). Because burrows possess at least a partial vertical component, changes in burrow abundance have an effect on both the bulk k_h and k_v. Generally, vertical permeability is semilogarithmically enhanced by increases in burrow density (Gingras et al., 1999; Weaver and Schultheiss, 1983).

As noted above, burrow interpenetrations increase in number with bioturbation intensity; exceptions to this generalization might include monospecific assemblages such as *Skolithos* piperock, wherein, due to the burrow morphology, interpenetrations are rare. Using a range of recognized burrow morphologies and populating a three-dimensional model using various Monte Carlo simulations, La Croix et al. (2012) produced randomly generated bioturbate flow media. They observed that between levels of 20% and 30% bioturbation (by volume), continuous flow paths were formed across the modeled flow medium. Above this degree of bioturbation, the geometric and arithmetic means of permeability can be used to characterize the bulk permeability of a flow medium.

3.3 Burrow-Surface Area and Trace Architecture

The presence of permeable burrows within a comparatively tight matrix increases the surface area of flow conduits within reservoirs. For instance, the surface area represented by 100 *Thalassinoides*/m^2, with 2 cm diameter tubes that descend 50 cm into the substrate, is a little over 3 m^2. This is three times more than the horizontal section and six times the vertical section. A simple fracture would possess a surface area roughly equivalent to twice the cross-sectional area of the block. It is easy to imagine extensive burrow-to-matrix interactions with higher burrow densities.

Storativity within burrow fabrics is simply related to the volumetric fraction the biogenic fabric occupies. Crustacean-generated traces, such as *Thalassinoides* and *Ophiomorpha*, can occupy more than 50% of the local sample volume. *Macaronichnus* may easily exceed 70% of the rock volume. Diagenetic fabrics around burrows (e.g., dolomitized burrows) range up to 80% of the volume before the matrix is no longer discernible. These examples illustrate that burrowed horizons not only change flow characteristics within a reservoir, but also have an impact on resource production and reserve calculations as well.

In addition to its size, a burrow's overall architecture can influence the degree of connectivity and tortuosity present in biogenically modified flow media (La Croix et al., 2012). Branching burrows with vertical elements provide the highest likelihood of developing an effective, isotropic network. A common example of this is *Thalassinoides*. Many trace fossils do not exhibit branches, and thus rely on chance interpenetration to connect the biogenic flow paths (e.g., *Zoophycos*). Cryptic bioturbation is so highly interconnected that the flow medium is essentially isotropic and permeability-dead zones likely are rare.

4. ANALYTICAL METHODS

The assessment of ichnological permeability revolves around measuring the permeability of matrix and trace fossils, and resolving the shape and distribution of trace fossils for upscaling. In both areas, analytical techniques have a range of limitations.

4.1 Permeability Assessment: Spot Permeametry

Permeability assessments most commonly are derived from spot permeametry, which provides small-scale measurements that are not particularly accurate. Plug, core, and bulk permeability measurements, which normally are some modifications of a Darcy experiment, are time-consuming and technically problematic as summarized in Gingras et al. (1999). Darcy experiments are not further discussed here, because they are well outlined in other works (e.g., Freeze and Cherry, 1979). The strategy for most ichnological spot-permeametry studies is outlined below.

In many instances, spot-permeametry testing is undertaken using a Pressure-Decay Profile Permeameter. Lab bench devices measure the permeability of gas in the range from 0.001 mD to >30 D through a probe tip (e.g., Knaust, 2009; Lemiski et al., 2011; Pemberton and Gingras, 2005; Tonkin et al., 2010). Permeability is normally measured using a grid pattern on slabbed and cleaned cores or rock faces. Good practice is to collect five measurements from each point on the grid, with the maximum and minimum values for each spot being discarded, and the remaining three values averaged (La Croix, 2010; Lemiski et al., 2011; Fig. 5). Anomalous measurements can result from poor probe-tip seals and should be removed from the dataset altogether. The averaged permeability values can then be plotted and contoured to produce plots of the permeability fields.

4.2 Trace-Fossil Architecture and Distributions

4.2.1 Outcrop and Core Data

Spatial data is commonly interpreted from two-dimensional data sets such as core and outcrop data. Observations that can be made pertaining to the bioturbate texture are limited by the size and quality of the rock exposure or sample available. The sample size is very important, as permeability fabrics must be upscaled. Unfortunately, outcrops seldom provide a three-dimensional understanding of their structure, but the spatial distribution of burrow fabrics may be displayed. Because it commonly samples the reservoir, core is a good source of permeability data; however, the representative elemental volume is small.

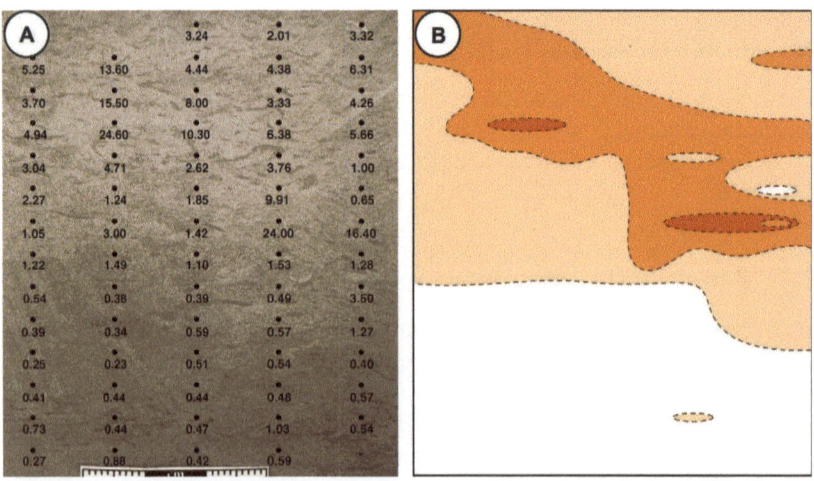

FIGURE 5 An example of contoured spot-permeameter data from the Alderson Member, Milk River Formation, Cretaceous of Alberta (Canada). This example is dominated by *Phycosiphon* and shows a range of permeability variance from <1 to >24 mD.

4.2.2 Non-Destructive Methods of Spatial Analysis

Non-destructive methods can be used to resolve the three-dimensional nature of sedimentary structures at the hand-sample scale. The most common of these are X-ray and X-ray CT. X-rays provide only two-dimensional data. CT scans are three dimensional. Neither X-ray nor CT scans are particularly sensitive to slight variations in density. Nevertheless, CT scans have the potential to yield high-resolution data that can be used to establish the three-dimensional nature of biogenic fabrics (Herringshaw et al., 2010; Pierret et al., 2002).

Microtomography (Micro-CT) is beginning to be used for ichnological applications (e.g., La Croix, 2010). It permits the resolution of micron- through to millimeter-scale density heterogeneities in geological media and is appropriate for the delineation of small to microscopic borings, burrows, and fecal pellets. Importantly, the technique can resolve the pore distribution in a small volume of sediment.

Unlike the aforementioned radiographic techniques, MRI does not measure the density of a rock directly. Rather, it is sensitive to fluids imbibed into the pore space of a rock. Therefore, MRI allows for three-dimensional mapping of a magnetic resonance signal, providing a tool that geologists can use to map the pore-space distribution in rocks (e.g., Gingras et al., 2002a,b). MRI can also be used to visualize the imbibition of porosity by gas or fluids, a true advantage in permeability studies (e.g., Chen et al., 2003).

4.3 Mechanisms and Styles of Burrow Permeability

In modern sediments, bioturbating animals, including earthworms, crustaceans, amphipods, polychaetes, and meiofauna, have been shown to enhance permeability and porosity primarily by creating macropores within the sediment (e.g., Aller, 1983, 1994; Katrak et al., 2003; Lavoie et al., 2000; Meadows and Tait, 1989; Pedley, 1992; Pierret et al., 2002). Although this type of flow-media enhancement can be a factor in the later developing of permeability fabrics, burrows do not remain open and unfilled in their passage to the rock record, and they are not preserved as macropores. In the rock record, the preserved patterns of permeability and porosity are strongly influenced by the character of burrow-filling sediment, the nature of burrow backfilling, compaction, and cementation. Recognizing this, Pemberton and Gingras (2005) classified biogenic flow media on the basis of their sedimentological, ichnological, diagenetic, and stratigraphical contexts. Pemberton and Gingras (2005) proposed that biogenic flow media could be separated into five classes (Fig. 6), including (1) surface-constrained discrete heterogeneities (SCD), (2) non-constrained discrete heterogeneities (NCD), (3) weakly defined heterogeneities, (4) cryptic heterogeneities, and (5) diagenetic heterogeneities. In this context, "surface-constrained" implies that the bioturbation descends from a single surface and "non-constrained" suggests that the burrows are distributed vertically and not associated with

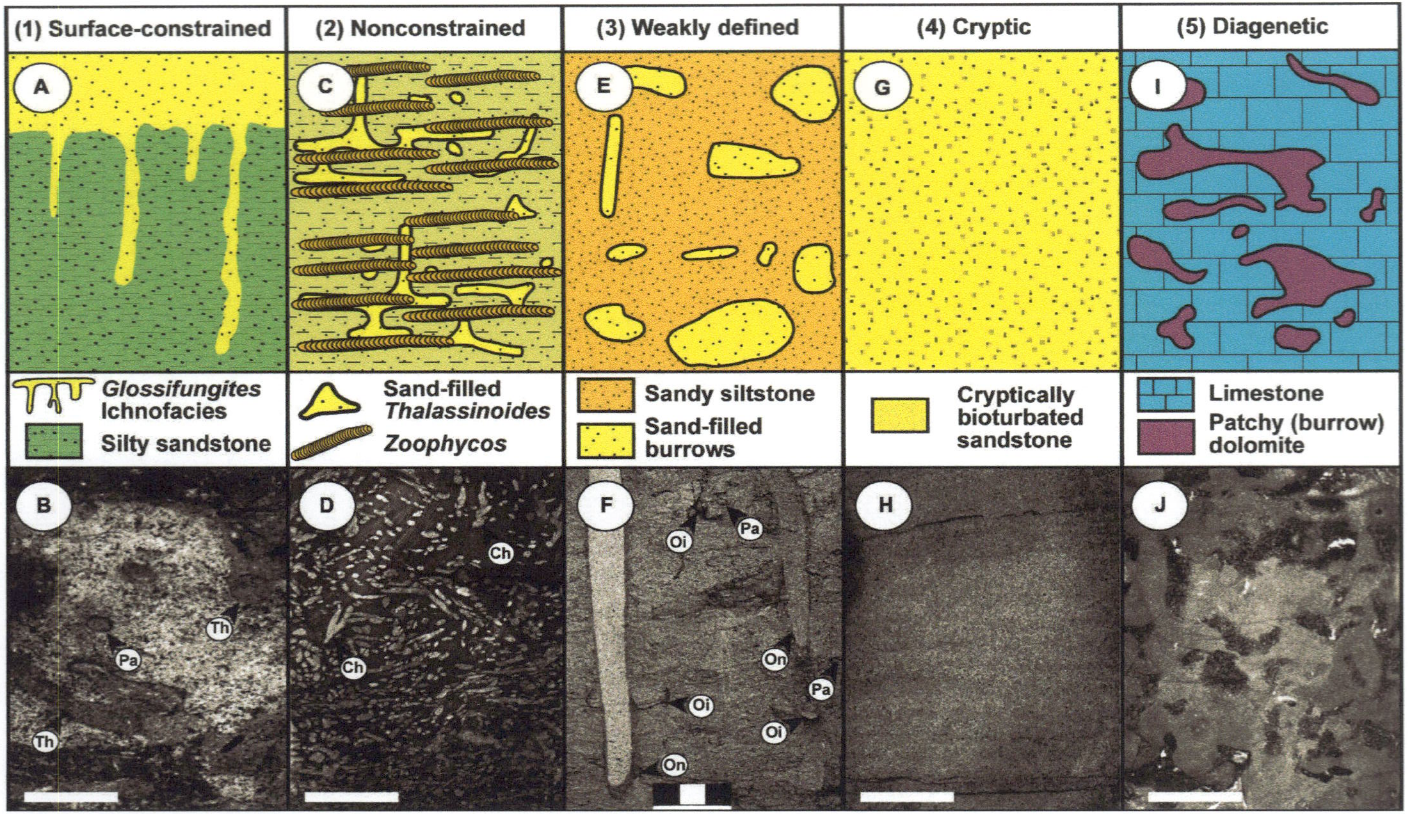

FIGURE 6 Examples of each class of biogenic flow media from drill cores, which is 10 cm in diameter. (A) Surface-constrained discrete heterogeneities (SCD). (B) Example from Jurassic strata of Germany with *Thalassinoides* (Th) and *Palaeophycus* (Pa) indicated. (C) Non-constrained discrete heterogeneities (NCD). (D) Example, with *Chondrites* (Ch) indicated, from the Debolt Formation, Mississippian, Alberta. (E) Weakly defined textural heterogeneities (WDH). (F) Example from Jurassic strata, North Sea. *Ophiomorpha irregulaire* (Oi), *O. nodosa* (On), and *Palaeophycus* (Pa) are indicated. (G) Cryptic biogenic heterogeneities (CBH). (H) Example from Jurassic strata, North Sea. (I) Diagenetic textural heterogeneities (DTH). (J) Example from Wabamun Formation, Devonian, Alberta (Canada).

a particular geological level. Discrete heterogeneities are sharply demarcated from the matrix by their texture, whereas cryptic heterogeneities are not. Diagenetic heterogeneities represent recrystallized cement accompanying burrow fabrics. Below, the permeability classifications are expanded upon case studies from the literature and are summarized with each example.

4.3.1 Surface-Constrained Discrete Heterogeneities

This class of biogenically enhanced flow media is characteristic of trace-fossil assemblages associated with the *Glossifungites* Ichnofacies. The *Glossifungites* Ichnofacies comprises unlined burrows that descend from a single surface. The burrows commonly exceed 10 cm in diameter and show little or no evidence of post-depositional compaction. These characteristics are associated with the development of *Glossifungites* assemblages, which dominantly result from animals burrowing into firm sediment (i.e., compacted substrates). The most common expression of the *Glossifungites* Ichnofacies is sand-filled burrows within a mud matrix (e.g., Gingras et al., 2001; MacEachern et al., 1992; Pemberton and Frey, 1985).

Trace-fossil assemblages that constitute the *Glossifungites* Ichnofacies typically penetrate as little as a few centimeters and as much as approximately 70 cm below their surface of origin. Some commonly associated ichnogenera include isolated burrows such as *Skolithos*, *Diplocraterion*, and clavate bivalve burrows (e.g., Gingras et al., 2001), and burrows that potentially interpenetrate one another such as *Thalassinoides* (e.g., Pemberton and Frey, 1985) and *Zoophycos* (MacEachern and Burton, 2000). Deeply emplaced burrows that interpenetrate, such as *Thalassinoides*-dominated *Glossifungites* assemblages, can constitute flow units that are, broadly speaking, isotropic in character. Owing to the presence of sand-filled burrows within the muddy matrix, SCD normally represent dual-permeability flow media (Fig. 5). The proportion of burrowing ranges from 10% to 80% of the volume. Notably, horizontal connectivity is limited until burrow volumes approach 20–30% (depending on the burrow architecture; La Croix et al., 2012).

Glossifungites-associated SCD are typically limited in their distribution. Due to their association with a single surface, the thickness of the permeability-enhanced zone is typically <1 m (e.g., Gingras et al., 2001; MacEachern et al., 1992). Additionally, the colonization of firmgrounds depends upon a number of factors, so their stratigraphic expression is discontinuous (cf. Gingras et al., 2001; Pemberton and Gingras, 2005); SCD are generally limited in scale to some hundreds of square meters, with a maximum areal extent of about a square kilometer (Pemberton and Gingras, 2005).

Surface-constrained permeability enhancement is probably not limited to *Glossifungites*-Ichnofacies-demarcated surfaces. For example, short-term colonization windows in deep-water environments (e.g., a turbidite fan) may promote the emplacement of sand-filled burrows along singular bedding planes. The porosity and permeability distributions in such occurrences are likely similar, but no case studies have as yet been conducted on such media.

4.3.1.1 Willapa Bay (Pleistocene), Washington, USA

The only detailed permeability study of SCD, by Gingras et al. (1999), investigated Pleistocene examples of *Glossifungites*-Ichnofacies-enhanced permeability using field slug tests, laboratory Darcy tests, and numerical models. They reported that three variables determine the flow characteristics of a *Glossifungites* surface: (1) the degree of contrast between burrow permeability (K_b) and matrix permeability (K_m), (2) the amount of burrow interpenetration, and (3) the burrow density. Their computer simulations indicated that the dominant influences are the permeability contrast (K_b/K_m) and the degree of burrow connectivity. Gingras et al. (1999) further showed that empirical and analytical formulae could be derived to determine the effective (bulk) permeability (Fig. 7). This was determined from K_b, K_m, and the degree of connectivity. The general formula can be applied in any geological medium as a simple, first-run approximation:

$$\log K_{\mathrm{bulk}} = \log[(1 - V_b)K_m + V_b K_b],$$

where V_b is the fractional volume of burrows within a block.

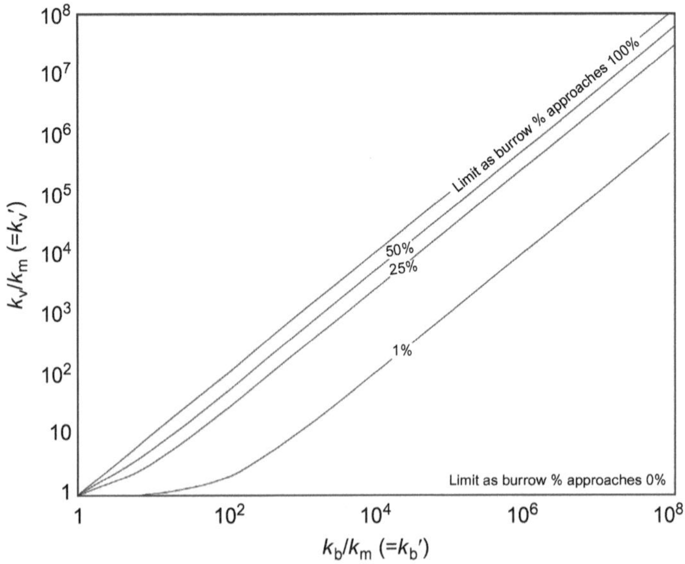

FIGURE 7 Curves showing the relationship $\log K_{\mathrm{bulk}} = \log[(1 - V_b)K_m + V_b K_b]$ for volumes of 1%, 25%, 50%, and 100% bioturbation. In this instance, enhancement of the vertical permeability is shown. The *x*-axis shows how the ratio of burrow (k_b) and matrix (k_m) permeabilities diverts the behavior from the predicted permeability. The *y*-axis indicates how much bulk permeability is enhanced of the matrix permeability (k_v/k_m), depending on (1) the burrow and matrix permeability ratio, and (2) the proportion of medium that is bioturbated. This relationship was derived empirically using Darcy experiments, slug tests, and numerical models. The formula is best applied to surface-constrained discrete heterogeneities and is useful for the estimation of the bulk permeability.

4.3.1.2 Hawiyah Portion of the Ghawar Field (Jurassic), Saudi Arabia

SCD have also been reported from carbonate strata. Pemberton and Gingras (2005) presented data from the Hawiyah portion of the Ghawar Field, where a distinct interval of stacked *Glossifungites*-Ichnofacies-demarcated surfaces occurs. The Ghawar example is rather extreme as it is characterized by *Thalassinoides* with 1–2 cm in burrow diameter that penetrates up to 2.1 m below the surface. The burrow-encapsulating matrix is composed of low-permeability micritic calcite. The burrow fills and the overlying strata consist of detrital sucrosic dolomite, providing a biogenic flow system for some of the Ghawar reservoirs. The permeability of the burrow fill is high enough to cause the biogenic flow medium to locally develop into high permeability (also known as "Super K") stratigraphic levels.

4.3.1.3 Baldonnel Formation (Triassic), British Columbia, Canada

Outcrops of the Baldonnel Formation display discontinuity surfaces delineated by granule/pebble lags, substrate-controlled trace-fossil assemblages, and pronounced diagenetic alteration of underlying strata in the Carnian Baldonnel Formation at Williston Lake, north-eastern British Columbia. Many of these surfaces are characterized by well-developed, low- to moderate-diversity trace-fossil assemblages attributed to the *Glossifungites* Ichnofacies that locally are overprinted by sparse, low-diversity trace-fossil assemblages characteristic of the *Trypanites* Ichnofacies, together comprising a polygenetic SCD (Fig. 8). Fabric-selective (interparticle, intraparticle, and biomoldic) and non-fabric-selective (vuggy and burrow) porosity are best developed in them, in association with bioclastic packstone and grainstone beds immediately beneath discontinuity surfaces, particularly those characterized by *Glossifungites* assemblages. Complex depositional/diagenetic relationships occurred where substrata characterized by moderately diverse *Skolithos* or proximal *Cruziana* trace-fossil assemblages were buried, exhumed, colonized by firmground-preferring infauna, lithified, colonized by hardground-preferring infauna, and reburied. Diagenetic alteration occurred via fluid penetration through burrows during, and shortly after, hardground formation.

4.3.2 Non-Surface-Constrained Discrete Heterogeneities

Non-surface-constrained discrete (textural) heterogeneities (NCD) comprise sharply defined burrows that are encased in a fine-grained matrix. Like surface-constrained heterogeneities, this architecture is common in clastic strata where sand-filled burrows descend into mudstone. However, NCD are laterally and vertically distributed (i.e., burrows originate from several depositional surfaces), and the bioturbation intensity must be high enough to afford connectivity in three dimensions. This type of permeability enhancement is most common in strata that accumulated in offshore settings, although lagoon and bay settings,

FIGURE 8 Intraformational unconformity within the Late Triassic Baldonnel Formation. This surface is characterized by low-diversity *Glossifungites* trace-fossil assemblages that consist of *Arenicolites*, *Rhizocorallium*, *Skolithos*, and *Thalassinoides*, as well as large, unnamed burrow networks. The lower lithofacies consists of a variably bioclastic fine-grained mudstone–wackestone (M–W) with low, mixed interparticle and intercrystalline porosity of ~2.5%. The upper lithofacies is a sandy bioclastic grainstone (Gr) with fine-to-medium-grained matrix and mixed interparticle, moldic and intercrystalline porosity of ~6–9%. These facies are separated by a discontinuous coarse-grained lag (Lag) that infills fractures and burrows that emanate downward from this surface. (A) Outcrop of the Baldonnel intraformation unconformity, East Carbon Creek, north-eastern British Columbia (Canada). Two burrows are identified, including a solitary *Skolithos* (*Sk*) and a large, approximately U-shaped burrow (U) that is believed to be part of a complex arthropod domichnion. (B) Polished hand sample illustrating the sharp boundary between the mudstone–wackestone (M–W) and the coarse-grained lag that infills burrows such as the obliquely-cut *Rhizocorallium* (*Rh*) illustrated here.

including deposits dominated by inclined heterolithic stratification, cannot be discounted.

NCD flow media are dominantly dual permeable in character (Figs. 2 and 6). Owing in part to the association of NCD with offshore settings, the bioturbation intensity can be quite high with burrows occupying between 30% and 90% of the sediment volume. In offshore and shelf settings, vertical and horizontal burrow architectures are common (e.g., *Cruziana* and *Zoophycos* ichnofacies); as such, NCD are likely to be highly interconnected (Fig. 9) and permeability should be isotropic. In increasingly distal settings, characterized by the *Nereites* Ichnofacies, bioturbation is dominantly restricted to bedding planes and the vertical permeability is not normally enhanced. The thicknesses of permeability enhanced zones range between 5 m and tens of meters (e.g., Pemberton and Gingras, 2005). Additionally, the permeability enhancement is stratigraphically

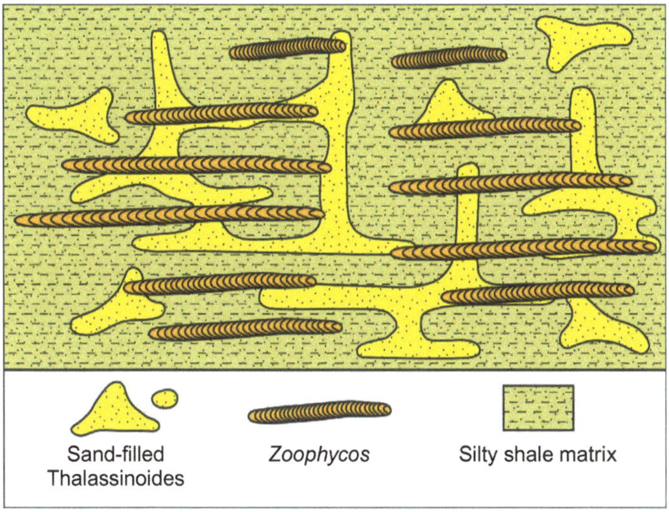

FIGURE 9 Foram-filled *Zoophycos* (Zo), encased in a lower permeability matrix, form an important part of the Sirasun Field in Indonesia. The burrows are further connected by *Thalassinoides* and *Chondrites*; thus, the biogenic reservoir is largely isotropic and is a prodigious natural gas producer.

continuous. Stratal units characterized by non-SCD can exceed several square kilometers in area (Pemberton and Gingras, 2005).

4.3.2.1 Lysing Formation (Cretaceous), Norwegian Sea

Within the Late Cretaceous Lysing Formation, biogenic fabrics appear to enhance the reservoir permeability significantly. The Lysing Formation is composed of interbedded very fine- to fine-grained sandstone, siltstone, and shale that contain a diverse trace-fossil suite. Sediment accumulation is the result of alternating fine- and coarse-grained sedimentation within a mid- to outer-shelf setting. The trace-fossil assemblage includes *Thalassinoides*, *Planolites*, *Scolicia*, *Chondrites*, *Palaeophycus,* and *Zoophycos* (i.e., distal *Cruziana* Ichnofacies). The trace fossils commonly contain coarser sediment (fine-grained sandstone) than the shale matrix.

Spot-permeability data taken from core datasets indicate that the burrow permeability can be up to two orders of magnitude higher than the matrix, thus proffering a dual-permeability flow medium. Importantly, bulk assessments of permeability based on core-plug data indicate that the harmonic mean of matrix *versus* burrow permeabilities provides the most accurate estimate in the Lysing example (Fig. 10). The harmonic relationship suggests that the flow is largely directed across the lower permeability matrix from burrow to burrow. This is, in fact, characteristic of NCD dominated by sparse

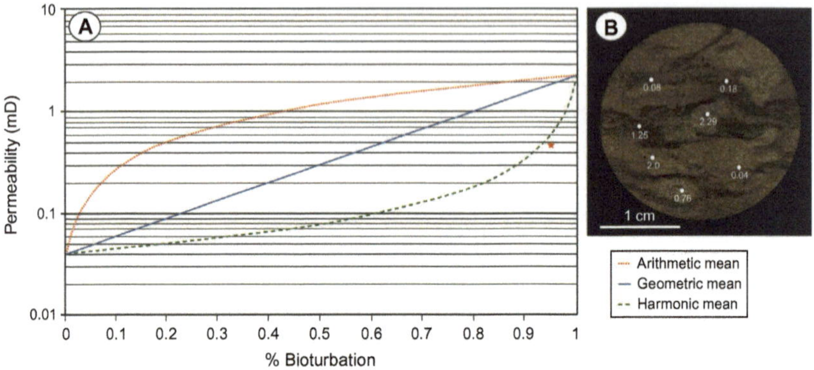

FIGURE 10 The predictable permeability relationships that may be present in bioturbated flow media (Lysing Formation, Norway). (A) Graph showing three possible solutions to estimating the bulk permeability (geometric, arithmetic, and harmonic means). The red star indicates the actual bulk permeability of a sampled plug (B), which is plotted according to its degree of bioturbation and measured permeability. Here, it appears that the burrows are conforming most closely to the harmonic mean, suggesting that, in spite of the high levels of bioturbation, some flow complexities are present.

bioturbation. As with most other biogenic flow media, the resulting flow network is an overall intricate, locally interconnected horizontal and vertical burrow system. In this example, bulk-permeability measurements also show that k_h is higher than k_v (Fig. 10) and that permeability distributions are anisotropic. This is likely the result of increased horizontal burrowing associated with shelfal trace-fossil assemblages or may be related to the small (2.54 cm) scale of the permeability measurement.

4.3.2.2 Terang-Sirasun (Miocene), East Java Sea

Pemberton and Gingras (2005) provided an example of non-SCD from the Terang-Sirasun gas field of the East Java Sea. Terang-Sirasun resides in the Kangean block approximately 140 km north of Bali (Matzke et al., 1992). The field comprises reserves residing in the Lower Paciran Sandstone Member (Late Miocene), the Upper Paciran Sandstone Member (Late Miocene), and the Paciran Limestone Member (Pliocene). Noble and Henk (1998) showed that the Paciran Limestone Member is a deep-marine pelagic carbonate deposit. The lower lithofacies are mainly grainstones, wackestones, and packstones, and the upper lithofacies predominately comprise marls. The upper marl lithofacies of the Paciran Limestone contains abundant, gas-charged *Zoophycos*. The *Zoophycos* are backfilled with hollow globeriginid tests (Fig. 9), and the trace-fossil permeability is substantially higher than the matrix. The ichnogenus is deeply penetrative; so, the burrows are interconnected and the permeability is isotropic.

4.3.3 Weakly Defined Textural Heterogeneities

Sedimentary media that exhibit less than two orders of magnitude between the trace-fossil and matrix permeabilities comprise weakly defined textural heterogeneities (WDH) (i.e., dual-porosity system). This class of flow media dominantly comprises sand-filled burrows within a sandstone matrix or very fine sand in siltstone matrices (Figs. 6 and 11). In examples of WDH, the sorting of the burrow infill tends to be better than the matrix, and the grain packing is in some cases likely to be looser—especially with passively infilled burrows. Biogenic permeability modification ascribed to WDT most commonly is associated with the presence of *Planolites*, *Thalassinoides*, *Skolithos,* and some expressions of *Macaronichnus*. In the light of their branching morphology, deeply penetrative nature and passive infill, *Thalassinoides* and *Ophiomorpha* are likely the most important permeability-enhancing trace fossils, although Tonkin et al. (2010) observed permeability reduction with *Ophiomorpha* as well. La Croix et al. (2012) and Lemiski et al. (2011) drew attention to another style of WDH comprising *Phycosiphon*-dominated intervals in siltstones. In these examples, the small trace fossils provided additional flow paths for natural gas to reach both the well bore and bedded reservoir units.

WDH are mostly associated with lower shoreface, sand-rich proximal offshore settings, and various bay-margin deposits (Pemberton and Gingras, 2005), with *Phycosiphon*-bearing (impoverished *Cruziana* Ichnofacies) examples largely identified within offshore siltstones. Sand-dominated intertidal-flat deposits should be included in WDH, but no studies have been conducted to demonstrate this. Due to the variable sedimentary conditions present in lower-shoreface and (particularly) tidal-flat deposits, the thicknesses and lateral extents of associated facies are generally limited. In strata hosting these fabrics, affected beds and bedsets reach thicknesses of up to 10 m (Pemberton and Gingras, 2005). The lateral extents of subtly burrowed facies range between hundreds of meters (tidal-flat settings) and several kilometers (offshore associations).

4.3.3.1 Ula Formation (Jurassic), Offshore Norway

Located in the north-eastern corner of the Norwegian Central Graben, the Late Jurassic (Kimmeridgian-Oxfordian) Ula Formation hosts significant hydrocarbon fields within the oil-rich Ula Trend (Spencer et al., 1986). Interpreted as a storm-influenced, fault-controlled shoreface deposit, hydrocarbon production from the Ula Formation occurs primarily from bioturbated units. Spot permeametry, Micro-CT, and Visual Modflow modeling of the Ula Formation demonstrated that the permeability enhancement within *Ophiomorpha*-dominated sandstone intervals represents a dual-porosity network (Fig. 11). The distal *Skolithos* Ichnofacies, that is gradational with proximal expressions of the *Cruziana* Ichnofacies, constitutes a good example of a dual-porosity rock fabric that can be categorized using the arithmetic mean. An example includes horizontally oriented networks of *Ophiomorpha* with high bioturbation intensities

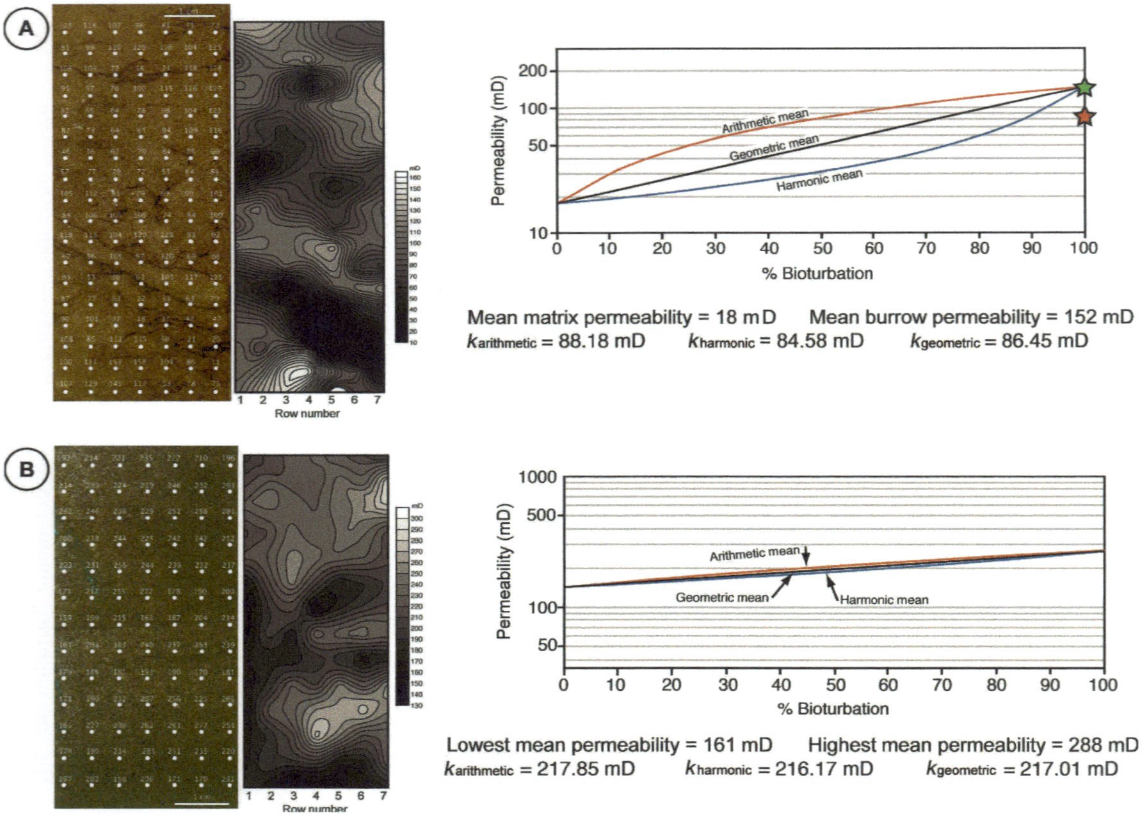

FIGURE 11 Biogenically modified flow media from the Late Jurassic Ula Formation, Norwegian North Sea. (A) Minipermeametry of a weakly defined textural heterogeneity for well 7/12-6, depth 3,439.00 m. From the contour map, it is apparent that the higher permeabilities are localized along well-defined burrow networks. This suggests that the arithmetic mean represents fluid flow along the most permeable conduits. Because of small-scale heterogeneities within the sample (i.e., burrow mud linings), the $k_{arithmetic}$ is only at 88.18 mD at 100% bioturbation (green star). This is opposed to the idealized 100% bioturbation intensity, which predicts 152 mD (red star). (B) Minipermeametry of a cryptic textural heterogeneity for well 7/12-6, depth 3,503.00 m. The highest permeabilities in this section are a result of higher levels of sediment reworking by microorganisms. Therefore, utilization of the geometric mean would be the most accurate for examples of cryptic bioturbation because it represents fluid movement in an unstructured medium (i.e., $k_v = k_h$).

(BI = 4–6). Conversely, the archetypal *Skolithos* Ichnofacies was found to be a good example of the harmonic mean in shoreface environments. Vertically oriented *Ophiomorpha* burrows with low bioturbation intensities (BI = 1–3) were found to be good examples.

4.3.3.2 Ben Nevis Formation (Cretaceous), Newfoundland, Canada

The Ben Nevis Formation largely represents a marine embayment deposit with several sand-rich levels (Pemberton et al., 2001; Spila et al., 2005). Tonkin et al. (2010) showed that *Thalassinoides* enhance the petrophysical properties in the net-pay intervals. Examples of *Ophiomorpha* also contain cleaner sediment surrounded by a muddy lining. Among Tonkin et al.'s (2010) key findings were the following: (1) the presence of *Thalassinoides* increases isotropy and enhanced porosity and permeability by up to 600%, (2) burrowing behavior (e.g., backfilling *versus* passive filling; tunneling *versus* sediment swimming) has a predictable effect on the permeability of sedimentary rock, and (3) characteristic bioturbate texture can be used to predict broad permeability classes.

4.3.3.3 Cusiana/Cupiagua Field (Eocene), Columbia

The Columbian Cusiana/Cupiagua Field is located in the Llanos Foothills of the Eastern Cordillera of the Andes. The field comprises three main sandstone units: quartz arenites of the Eocene Mirador Formation, quartz arenites of the Paleocene Barco Formation, and phosphatic litharenites and quartz arenites of the Early Cretaceous Guadalupe Group (Warren and Pulham, 2001). A suite of the *Glossifungites* Ichnofacies, associated with a basin-wide unconformity, occurs at the base of the Mirador Sandstone. The assemblage is characterized by *Thalassinoides*, which have a substantially higher porosity than the surrounding matrix (Fig. 12). In the case of the Mirador Sandstone, the highest rates of production are commonly associated with the *Thalassinoides*-dominated zone.

4.3.3.4 Medicine Hat Member, Niobrara Formation (Cretaceous), Alberta, Canada

The Santonian Medicine Hat Member was deposited in the broad, shallow Western Interior Seaway and is variably interpreted as offshore or delta-influenced offshore (La Croix, 2010; Pedersen and Nielsen, 2006). The Medicine Hat Member produces gas prolifically in the Medicine Hat Gas Field. La Croix et al. (2012) used spot-minipermeametry and Micro-CT to assess the effects of bioturbation on porosity and permeability within the Medicine Hat field (Fig. 13). The authors showed that the sandy burrows are two orders of magnitude more permeable than the mudstone matrix, resulting in a dual-porosity flow system. The authors further suggested that the bioturbate fabrics contribute to gas production within the field.

La Croix et al. (2012) modeled the permeability distributions and were able to establish that the permeability of the bioturbated intervals within the

FIGURE 12 Burrow-enhanced permeability in surface-constrained textural heterogeneities, Cusiana-Cupiaugua Field, Llanos Basin (Colombia). Core width is 8.9 cm. Core of the *Glossifungites* Ichnofacies at the base of the reservoir sandstone, Eocene Mirador Formation. Under ultraviolet light (right), it is showing that the burrow systems are filled with hydrocarbons and constitute the primary pathway for the migration of the oil (photos courtesy of A. Pulham).

Medicine Hat Member can be estimated using the arithmetic mean of the trace-fossil and matrix permeabilities. This led the authors to suggest that the burrows are intensely interpenetrating and, at the reservoir scale, fluid flow through the burrowed intervals is analogous to bedded sandstone.

4.3.3.5 Alderson Member, Lea Park Formation (Cretaceous), Alberta, Canada

The Alderson Member comprises bioturbated shale, mudstone, siltstone, and fine-grained sandstone. The gas fields contained within the Alderson Member produce from continuous and laterally extensive, thin-bedded, fine-grained sandstone within muddy units at depths <600 m (Hovikoski et al., 2008; Lemiski et al., 2011).

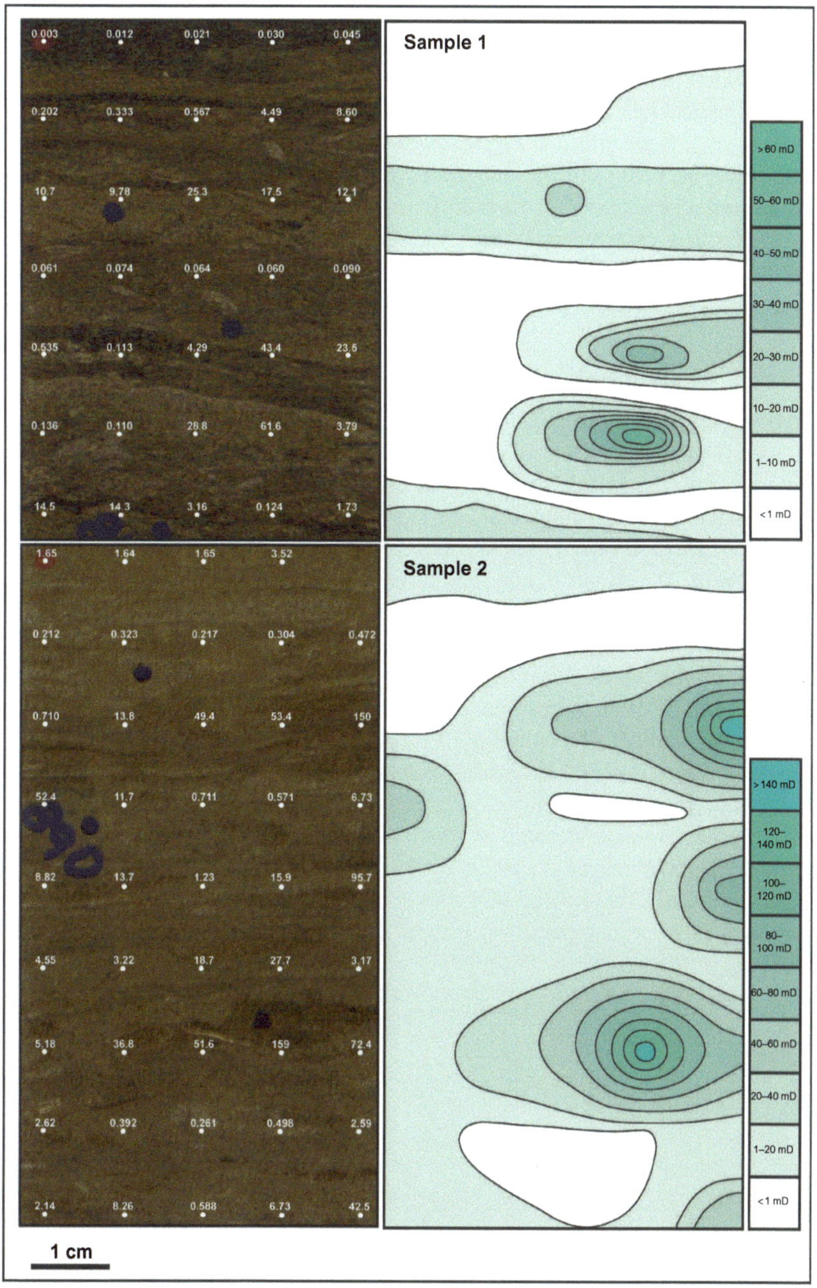

FIGURE 13 The results of the spot-minipermeametry testing on two examples (Sample 1 location Alberta, 06-31-14-01W4, 507.65 m depth; and Sample 2 location Alberta, 0606-31-14-01W4, 509.00 m depth). On the left, the core photographs are annotated with test locations and their corresponding permeability measurements in millidarcy (mD). On the right, contour plots highlight the distribution of permeability. The data points are arranged using 1-cm grid spacing. Gaps in the data set represent discarded measurements resulting from a poor probe-tip seal.

Lemiski et al. (2011) conducted spot-minipermeametry analyses of the Alderson Member. They demonstrated that, of the reservoir units, bioturbated heterolithic intervals make the most significant contribution to the overall storage and producibility of gas from the Alderson Member. The permeability is particularly enhanced in the *Phycosiphon*-dominated sandy mudstones (Fig. 5). Lemiski et al. (2011) further suggested that the vertical transmissivity of gas is substantially increased in the bioturbated levels. Their study concluded: "bioturbated zones may represent an important means of delivering gas from the tighter, encapsulating matrix. In other words, the burrows can act as biogenic 'fractures' (but) with higher surface area".

4.3.4 Cryptic Biogenic Heterogeneities

Cryptic biogenic heterogeneities (CBH) result from one of three processes: (1) meiofauna disrupting sedimentary lamination, which is referred to as cryptobioturbation; (2) some *Macaronichnus* behaviors, which preserve the vestigial sedimentary texture by implementing grain-selective deposit-feeding behaviors; and (3) eradication of sedimentary lamination by burrowing animals that churn the sediment. The effects of cryptobioturbation and *Macaronichnus*-like burrows on reservoir properties can be important (Gingras et al., 2002b; Gordon et al., 2010; Pemberton and Gingras, 2005; Tonkin et al., 2010). Although cryptobioturbated fabrics appear homogeneous (Fig. 14), Gingras et al. (2002b) and Gordon et al. (2010) demonstrate that the burrow fills relative to the unburrowed matrix commonly proffer a dual-porosity flow medium. Therefore, fluid production out of cryptobioturbated media presents challenges in terms of the presence of tortuous fluid-flow pathways, permeability streaks, diffusion in and out of the burrowed zones, and complex relationships between bulk porosity and bulk permeability (Gingras et al., 1999; Pemberton and Gingras, 2005). On the other hand, cryptically bioturbated and *Macaronichnus*-dominated sands have been shown to produce comparably isotropic flow units that may be amenable to

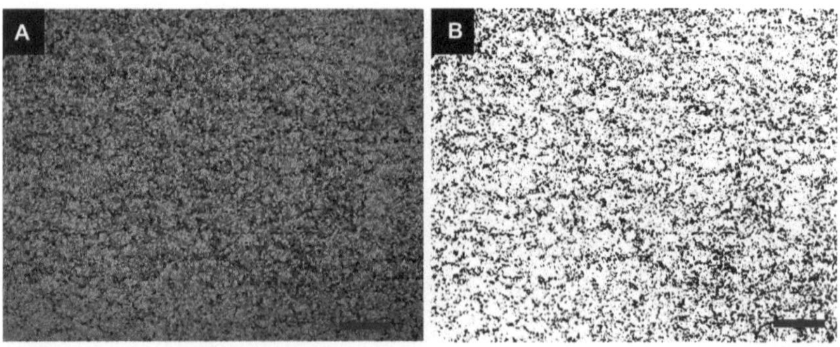

FIGURE 14 (A) *Macaronichnus*-burrowed sand. (B) The same field of view showing image-enhanced burrow structures. The burrow halo is defined by dark grains, and the white area is the burrow fill (scale bar = 0.5 cm).

hydrocarbon production (Gordon et al., 2010; Pemberton and Gingras, 2005; Tonkin et al., 2010). It is therefore important to interpret massive sandy beds in deep-water successions accurately, in order to create reliable depositional or reservoir models and to better anticipate a more complex porosity and permeability network in seemingly homogenous media.

Cryptic bioturbation (in sandstone) is common in clastic estuarine and distributary channels, as well as in upper shoreface deposits. More recently, Greene et al. (2012) identified cryptic bioturbation as a common feature of Neogene turbidite channels. The aforementioned depositional environments are dynamic, thus their thicknesses and lateral extents are limited. In general, this facies is 10 cm and 10 m thick. Amalgamated parasequences may generate stacked successions that are almost 100 m thick.

4.3.4.1 Bluesky Formation (Cretaceous), Alberta, Canada

Gordon et al. (2010) studied in detail the petrophysics of *Macaronichnus* and cryptically bioturbated sandstone. Their study focused on shoreface deposits of the Early Cretaceous Bluesky Formation in western Alberta, Canada. The studied reservoir comprises lithic sandstone and several levels that were heavily bioturbated with *Macaronichnus segregatus*. Petrographic techniques were used to assess the effects of *M. segregatus* on reservoir quality. Thin-section analysis and SEM (Scanning Electron Microscopy) have shown that the tracemaker of *M. segregatus* avoided iron-rich detrital fragments while exploiting the sediment for food (Fig. 14). Grain avoidance led to a re-sorting of the sand such that the dark-colored grains outline the burrows, and light-colored competent grains were dominant within the burrow fill. Notably, the light-colored burrow fills contain a higher chert/quartz ratio. Owing to the fact that pore-occluding quartz overgrowths do not form on chert fragments, the primary porosity and permeability were preferentially preserved within the burrow fills (Fig. 15). Gordon et al.'s (2010) study further showed that spot permeability in the *Macaronichnus*-burrowed zones was as much as four times higher than in laminated sandstone of similar grain size (Fig. 15).

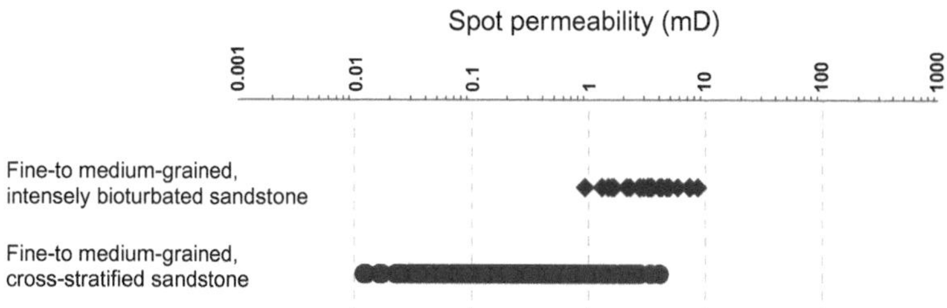

FIGURE 15 *Gordon et al. (2010)* showed that permeability was slightly enhanced in *Macaronichnus*-burrowed sand compared to laminated sand, and that the burrowed sediment had less variance. The above data are derived from a single-cored interval of the Bluesky Formation, Cretaceous, Alberta (Canada).

4.3.4.2 Ula Formation (Jurassic), Offshore Norway

Within the aforementioned Ula Formation, the influence of cryptically bioturbated sandstones on reservoir quality can be demonstrated. The studied reservoir units comprise upper fine- to lower medium-grained sandstone that contains rare sedimentary and biogenic structures. The massive appearance of the intervals is attributed to cryptobioturbation. A petrographic thin-section study showed that the porosity is enhanced slightly (by 5–10%) within the highly bioturbated sections. Spot permeametry revealed uniform permeability distributions (Fig. 14), which explain the correspondence of permeability to the geometric mean (Fig. 11).

4.3.5 Diagenetic Textural Heterogeneities

Diagenetic textural heterogeneities (DTH) occur typically as bioturbated, dolomite-mottled limestones or as nodular sandstones/siltstones. With carbonates, the dolomitic mottles result from dolomitization of calcareous sediment adjacent to body fossils, sedimentary lamination, and trace fossils (Fig. 6E; Abdel-Fattah et al., 2011; Corlett and Jones, 2012; Gingras et al., 2002a, 2004a,b; Kendall, 1977; Zenger, 1992, 1996). Another important diagenetic process is fabric-selective carbonate dissolution, which is discussed in detail by Cunningham et al. (2012).

The presence of fabric-selective dolomite in association with trace fossils is attributed to chemical and physical alteration of the substrate by burrowing organisms. Physical alteration of the substrate includes modification of the grain size, redistribution of grains, compaction, and sorting. Compositional heterogeneity is due to the incorporation of localized, concentrated organic material in the form of mucous or fecal material. It is most likely that physical and chemical factors together determine the resulting fabric in burrow-mottled limestones (Gingras et al., 2004b).

Burrow-associated cement precipitation and nodule formation have four main modes of occurrence: (1) cementation of a finer-grained burrow wall, such that siderite, dolomite, calcite, or pyrite occur as very fine-grained crystals in the burrow-wall material (preferred tube cementation); (2) cement precipitation or cement dissolution within a coarser-grained burrow fill, in which mosaic calcite is distributed heterogeneously throughout the burrowed medium (preferred burrow cementation); (3) cement—usually dolomite or calcite—precipitation or cement dissolution adjacent to a trace fossil, forming a diagenetic halo around discrete trace fossils (fabric-mimicking hypoburrow cementation); and (4) concretion formation, usually calcite and rarely dolomite, with a trace fossil as the nucleation point (nodular hypoburrow cementation). This range of diagenetic–ichnological associations shows that the impacts of diagenesis are extremely variable.

In most cases, burrow-induced diagenesis profoundly alters the rock's petrophysical characteristics. The permeabilities and porosities of burrow-affected and unburrowed zones are commonly different (e.g., Knaust, 2009), and diagenesis most commonly occurs in and around generations of interpenetrating trace fossils. Consequently, the heterogeneity in burrow-mottled carbonates is high. Although a dual-permeability network is generally produced, dual-porosity systems are also present.

Burrow-associated dolomite is commonly present in Paleozoic carbonates. Diagenetic heterogeneities are exemplified by the Ordovician Chazy Group (Mehrtens and Selleck, 2002), Devonian strata in western Canada (Beales, 1953; Corlett and Jones, 2012; Gingras et al., 2004b), the Late Ordovician Bighorn Dolomite in Wyoming (Zenger, 1992, 1996), and the Ordovician Yeoman Formation in the Williston Basin (Kendall, 1977; Pak, 2003) of North America. Younger examples are provided in Pedley (1992). The depositional environments range from platformal offshore (Palliser Formation) to restrict offshore. The successions result in significant volumes of altered/affected strata. Thicknesses typically approach 10 m, however, more or less continuous zones may be more than 100 m thick. Likewise, the areal extents of these ramp- and platform-associated strata are potentially extensive, generally falling between one and hundreds of square kilometers.

4.3.5.1 Red River Formation (Ordovician), Saskatchewan and Manitoba, Canada

The best-characterized example of burrow-associated diagenesis in North America is the Tyndall Stone, a dolomitic-mottled limestone quarried in Manitoba, Canada. The Tyndall Stone is assigned to the Ordovician Selkirk Member of the Red River Formation (Kendall, 1977). Similarly, mottled limestones comprise the subsurface equivalent Yeoman Formation in south-eastern Saskatchewan (Kendall, 1975). The Selkirk Member and the Yeoman Formation accumulated as an epicontinental carbonate platform in the Williston Basin. The most distinctive feature of the Tyndall Stone is the presence of pervasively distributed, dolomitized, branching trace fossils (Gingras et al., 2004b; Kendall, 1977). Spot-permeametry data from the Tyndall Stone show that the average permeability of the matrix is 1.65 mD, and the average burrow permeability is 19.2 mD (i.e., a dual-porosity system) (Gingras et al., 2004b; Fig. 4, inset). Oil production from the Yeoman Formation in Saskatchewan is primarily from the dolomitic phases that occur in layers between 1 and approximately 10 m thick.

4.3.5.2 Lonely Bay Formation (Devonian), Northwest Territories, Canada

Corlett and Jones (2012) show that the Lonely Bay Formation contains an intensely bioturbated facies, characterized by dolomite- and calcite-cemented burrows. The calcite-cemented burrows were formed proximal to the shoreline and the dolomite-cemented burrows in distal units. Within the calcite-cemented burrows, the primary sedimentary and ichnological fabric was preserved. The sedimentary fabric associated with the dolomite-cemented burrows appears massive. Recrystallization associated with dolomite formation generated intracrystalline porosity.

4.3.5.3 Palliser Formation (Devonian), Alberta, Canada

Another excellent example of burrow-selective dolomitization concerns the Devonian Palliser Formation (Wabamum Group in the subsurface) of the Western Canada Sedimentary Basin (Fig. 6). These rocks are composed of a series of highly bioturbated carbonate fabrics interpreted to represent a series of shallow-water ramp deposits. Spot permeametry of samples indicated that the burrows are preferentially higher in permeability than the matrix. In general, the samples have a low-permeability limestone matrix with negligible effective porosities. In turn, the limestone encompasses chaotically distributed, fabric-selective dolomite with an effective porosity as high as 5–6% (Gingras et al., 2004b). Analysis of SEM micrographs has indicated that preferential dissolution and precipitation of planar dolomite crystals within and adjacent to burrow fabrics and physical sedimentary structures occurred. This implies that bioturbation influenced the pore-water geochemistry, and in so doing encouraged dolomitization.

5. CONCLUSIONS

Biogenic fabrics in sedimentary rocks can strongly influence the permeability fabric and therefore fluid flow in sedimentary strata. Bioturbation may impose well-defined, highly contrasting permeability fields, referred to as dual permeability, or subtly contrasting permeability fields, referred to as dual porosity. Both types of flow media influence the reservoir quality of fluid-bearing rocks. Dual porosity leads to (1) the entire rock contributing to fluid or gas production; (2) in the presence of more than one fluid phase, flow is focused in higher porosity/permeability zones; and (3) moving of the fluid in the lower-permeability portions of the rock into higher-permeability media through diffusion and advection (mechanical movement).

Dual-permeability flow media have poorer resource characteristics, resulting in (1) burrows comprising the effective flow medium present, (2) fluid resources being more slowly produced by diffusion from the tighter rock, and (3) secondary recovery techniques possibly isolating large parts of the flow network. Other factors that influence the overall behavior of the flow-medium class are the burrow density, burrow connectivity, burrow/matrix permeability contrast, burrow-surface area, and burrow architecture. For the most part, these parameters can be assessed and their impact estimated.

Dual porosity and permeability fabrics are a component of the burrow-media classification presented in Pemberton and Gingras (2005): (1) SCD, (2) NCD, (3) non-constrained subtle heterogeneities, (4) CBH, and (5) diagenetic heterogeneities.

ACKNOWLEDGMENTS

M. K. G.'s research is supported by grants from NSERC (Grant number 238530) and ongoing support from Nexen Inc., ConocoPhillips Canada, Devon Energy Canada, BP Canada, Statoil and Shell. Thanks to Floyd Henk, Andrew Pulham, James MacEachern, Shahin Dashtgard, and Kerri Bann for thoughtful discussion and collaboration. S. G. P., C. M., and J.-P. Z. are also supported by NSERC through the Discovery Grant Program (Grant numbers A0816, 138826, and 371662, respectively).

REFERENCES

Abdel-Fattah, Z.A., Gingras, M.K., Pemberton, S.G., 2011. Significance of hypoburrow nodule formation associated with large biogenic sedimentary structures in open-marine bay siliciclastics of the upper Eocene Birket Qarun Formation, Wadi El-Hitan, Fayum, Egypt. Sediment. Geol. 233, 111–128.

Aller, R.C., 1983. The importance of the diffusive permeability of animal burrow linings in determining marine sediment chemistry. J. Mar. Res. 41, 299–322.

Aller, R.C., 1994. Bioturbation and remineralization of sedimentary organic matter: effects of redox oscillation: controls on carbon preservation. Chem. Geol. 114, 331–345.

Beales, F.W., 1953. Dolomitic mottling in Palliser (Devonian) limestone, Banff and Jasper National Parks, Alberta. AAPG Bull. 37, 2281–2293.

Chen, Q., Gingras, M.K., Balcom, B.J., 2003. A magnetic resonance study of pore filling processes during spontaneous imbibition in Berea sandstone. J. Chem. Phys. 119, 9609–9616.

Corlett, H.J., Jones, B.J., 2012. Petrographic and geochemical contrasts between calcite- and dolomite-filled burrows in the Middle Devonian Lonely Bay Formation, Northwest Territories, Canada. J. Sediment. Res., in press.

Cunningham, K.J., Sukop, M.C., Huang, H., Alvarez, P.F., Curran, H.A., Renken, R.A., Dixon, J.F., 2009. Prominence of ichnologically influenced macroporosity in the karst Biscayne aquifer: stratiform "super-K" zones. GSA Bull. 121, 164–186.

Cunningham, K.J., Sukop, M.C., Curran, H.A., 2012. Carbonate aquifers. In: Knaust, D., Bromley, R.G. (Eds.), Trace Fossils as Indicators of Sedimentary Environments. Developments in Sedimentology, vol. 64. Elsevier, Amsterdam, pp. 871–898.

Dawson, W.C., 1978. Improvement of sandstone porosity during bioturbation. AAPG Bull. 62, 508–509.

Freeze, R.A., Cherry, J.A., 1979. Groundwater. Prentice Hall, Englewood Cliffs, 604 pp.

Gingras, M.K., Pemberton, S.G., Mendoza, C.A., Henk, F., 1999. Assessing the anisotropic permeability of *Glossifungites* surfaces. Petrol. Geosci. 5, 349–357.

Gingras, M.K., Pemberton, S.G., Saunders, T., 2001. Bathymetry, sediment texture, and substrate cohesiveness; their impact on modern *Glossifungites* trace assemblages at Willapa Bay, Washington. Palaeogeogr. Palaeoclimatol. Palaeoecol. 169, 1–21.

Gingras, M.K., MacMillan, B., Balcom, B.J., 2002a. Visualizing the internal physical characteristics of carbonate sediments with magnetic resonance imaging and petrography. Bull. Can. Petrol. Geol. 50, 363–369.

Gingras, M.K., MacMillan, B., Balcom, B.J., Saunders, T., Pemberton, S.G., 2002b. Using magnetic resonance imaging and petrographic techniques to understand the textural attributes and porosity distribution in *Macaronichnus*-burrowed sandstone. J. Sediment. Res. 72, 552–558.

Gingras, M.K., Mendoza, C.A., Pemberton, S.G., 2004a. Fossilized worm burrows influence the resource quality of porous media. AAPG Bull. 88, 875–883.

Gingras, M.K., Pemberton, S.G., Muehlenbachs, K., Machel, H., 2004b. Conceptual models for burrow-related, selective dolomitization with textural and isotopic evidence from the Tyndall Stone, Canada. Geobiology 2, 21–30.

Gordon, J.B., Pemberton, S.G., Gingras, M.K., Konhauser, K.O., 2010. Biogenically enhanced permeability; a petrographic analysis of *Macaronichnus segregatus* in the Lower Cretaceous Bluesky Formation, Alberta, Canada. AAPG Bull. 94, 1779–1795.

Greene, T.J., Gingras, M.K., Gordon, G.S., McKeel, D.R., 2012. The significance of deep-water cryptic bioturbation in slope-channel massive sand deposits of the lower Rio Dell Formation, Eel River basin, California. Mar. Petrol. Geol. 29, 152–174.

Gunatilaka, A., Al-Zamel, A., Shearman, D.J., Reda, A., 1987. A spherulitic fabric in selectively dolomitized siliciclastic crustacean burrows, northern Kuwait. J. Sediment. Petrol. 57, 922–927.

Herringshaw, L.G., Sherwood, O.A., McIlroy, D., 2010. Ecosystem engineering by bioturbating polychaetes in event bed microcosms. Palaios 25, 46–58.

Hovikoski, J., Lemiski, R., Gingras, M., Pemberton, G., MacEachern, J.A., 2008. Ichnology and sedimentology of a mud-dominated deltaic coast: Upper Cretaceous Alderson Member (Lea Park Fm), Western Canada. J. Sediment. Res. 78, 803–824.

Katrak, G., Bird, F., Worthington, D., 2003. Comparative effects of the large bioturbators, *Trypaea australiensis* and *Heloecius cordiformis*, on intertidal sediments of Western Port, Victoria, Australia. Mar. Freshwat. Res. 54, 701–708.

Kendall, A.C., 1975. Anhydrite replacements of gypsum (satin-spar) veins in the Mississippian caprocks of southeastern Saskatchewan. Can. J. Earth Sci. 12, 1190–1195.

Kendall, A.C., 1977. Origin of dolomite mottling in Ordovician limestones from Saskatchewan and Manitoba. Bull. Can. Petrol. Geol. 25, 480–504.

Knaust, D., 2009. Ichnology as a tool in carbonate reservoir characterization: a case study from the Permian—Triassic Khuff Formation in the Middle East. GeoArabia 14, 17–38.

Knaust, D., 2012. Methodology and techniques. In: Knaust, D., Bromley, R.G. (Eds.), Trace Fossils as Indicators of Sedimentary Environments. Developments in Sedimentology, vol. 64. Elsevier, Amsterdam, pp. 245–271.

Konhauser, K.O., Gingras, M.K., 2011. Are animal burrows a major sedimentary sink for metals? Ichnos 18, 1–3.

La Croix, A.D., 2010. Ichnology, sedimentology, stratigraphy and trace-fossil permeability relationships in the Upper Cretaceous Medicine Hat Member, Medicine Hat Gas Field, Southeast Alberta, Canada. M.Sc. Thesis, University of Alberta, 166 pp.

La Croix, A.D., Gingras, M.K., Dashtgard, S.E., Pemberton, S.G., 2012. Computer modeling bioturbation: the creation of porous and permeable fluid-flow pathways. AAPG Bull. 96, 545–556.

Lavoie, D.L., Vaughan, W.C., Easley, D.H., 2000. Quantitative effects of bioturbation on permeability determined using a whole-tank permeameter. Eos 81, F650.

Lemiski, R.T., Hovikoski, J., Pemberton, S.G., Gingras, M., 2011. Sedimentological, ichnological and reservoir characteristics of the low-permeability, gas-charged Alderson Member (Hatton gas field, southwest Saskatchewan): implications for resource development. Bull. Can. Petrol. Geol. 59, 27–53.

MacEachern, J.A., Burton, J.A., 2000. Firmground *Zoophycos* in the Lower Cretaceous Viking Formation, Alberta: a distal expression of the *Glossifungites* Ichnofacies. Palaios 15, 387–398.

MacEachern, J.A., Raychaudhuri, I., Pemberton, S.G., 1992. Stratigraphic applications of the *Glossifungites* Ichnofacies: delineating discontinuities in the rock record. In: Pemberton, S.G. (Ed.), Applications of Ichnology to Petroleum Exploration: A Core Workshop. SEPM Core Workshop Notes 17, pp. 169–198.

Matzke, R.H., Smith, J.G., Foo, W.K., 1992. Iagifu/Hedinia field: first oil from the Papuan fold and thrust belt. In: Halbouty, M.T. (Ed.), Giant Oil and Gas Fields of the Decade 1978–1988. AAPG Mem. 54, pp. 471–482.

McIlroy, D., 2007. Lateral variability in shallow marine ichnofabrics: implications for the ichnofabric analysis method. J. Geol. Soc. Lond. 164, 359–369.

McIlroy, D., Worden, R.H., Needham, S.J., 2003. Faeces, clay minerals and reservoir potential. J. Geol. Soc. Lond. 160, 489–493.

McKinley, J.M., Lloyd, C.D., Ruffell, A.H., 2004. Use of variography in permeability characterization of visually homogeneous sandstone reservoirs with examples from outcrop studies. Math. Geol. 36, 761–779.

Meadows, P., Tait, J., 1989. Modification of sediment permeability and shear strength by two burrowing invertebrates. Mar. Biol. 101, 75–82.

Mehrtens, C., Selleck, B., 2002. Middle Ordovician section at Crown Point Peninsula. In: McLelland, J., Karabinos, P. (Eds.), Guidebook for Fieldtrips in New York and Vermont. University of Maine, Orono.

Noble, R.A., Henk Jr., F.H., 1998. Hydrocarbon charge of a bacterial gas field by prolonged methanogenesis: an example from the East Java Sea, Indonesia. In: Advances in Organic Geochemistry 1997. Proceedings of the 18th International Meeting on Organic Geochemistry, Part 1, Petroleum Geochemistry, Organic Geochemistry, vol. 29, pp. 301–314.

Over, D.J., 1990. Trace metals in burrow walls and sediments, Georgia Bight, USA. Ichnos 1, 31–41.

Pak, R., 2003. Implications on source rock accumulation from ichnology in the Ordovician Yeoman Formation, southeastern Saskatchewan. Reservoir 30, 17.

Pedersen, P., Nielsen, K., 2006. Cretaceous successions in the foothills of Alberta. Reservoir 33, 13.

Pedley, M., 1992. Bio-retexturing: early diagenetic fabric modifications in outer-ramp settings—a case study from the Oligo-Miocene of the Central Mediterranean. In: Sellwood, B.W. (Ed.), Ramps and Reefs. Sediment. Geol., vol. 79, pp. 173–188.

Pemberton, S.G., Frey, R.W., 1985. The *Glossifungites* ichnofacies: modern examples from the Georgia Coast, U.S.A. In: Curran, H.A. (Ed.), Biogenic Structures: Their Use in Interpreting Depositional Environments. SEPM Spec. Publ., vol. 35, pp. 237–259.

Pemberton, S.G., Gingras, M.K., 2005. Classification and characterizations of biogenically enhanced permeability. AAPG Bull. 89, 1493–1517.

Pemberton, S.G., Spila, M., Pulham, A.J., Saunders, T., MacEachern, J.A., Robbins, D., Sinclair, I.K., 2001. Ichnology & Sedimentology of Shallow to Marginal Marine Systems: Ben Nevis & Avalon Reservoirs, Jeanne d'Arc Basin. Geol. Ass. Can., Short Course Notes 15, 343 pp.

Petrash, D.A., Lalonde, S.V., Gingras, M.K., Konhauser, K.O., 2011. A surrogate approach to studying the chemical reactivity of burrow mucous linings in marine sediments. Palaios 26, 594–600.

Pierret, A., Capowiez, Y., Belzunces, L., Moran, C.J., 2002. 3D reconstruction and quantification of macropores using X-ray computed tomography and image analysis. Geoderma 106, 247–271.

Spencer, A.M., Home, P.C., Wiik, V., 1986. Habitat of hydrocarbons in the Jurassic Ula Trend, Central Graben, Norway. In: Spencer, A.M., Campbell, C.J., Hanslien, S.H., Nelson, P.H., Nysæther, E., Ormaasen, E.G. (Eds.), Habitat of Hydrocarbons on the Norwegian Continental Shelf. NPF/Graham and Trotman, London, pp. 111–127.

Spila, M.V., Pemberton, G., Sinclair, I.K., Hiscott, R., Pulham, A., 2005. Comparison of marine and brackish/stressed ichnological signatures in the Ben Nevis and Avalon Formations, Jeanne d'Arc Basin. In: Hiscott, R.N., Pulham, A.J. (Eds.), Petroleum Resources and Reservoirs of the Grand Banks, Eastern Canadian Margin. Geol. Ass. Can., Spec. Pap. 43, pp. 73–94.

Sutton, S.J., Ethridge, F.G., Almon, W.R., Dawson, W.C., Edwards, K.K., 2004. Textural and sequence-stratigraphic controls on sealing capacity of Lower and Upper Cretaceous shales, Denver Basin, Colorado. AAPG Bull. 88, 1185–1206.

Tonkin, N.S., McIlroy, D., Meyer, R., Moore-Turpin, A., 2010. Bioturbation influence on reservoir quality: a case study from the Cretaceous Ben Nevis Formation, Jeanne d'Arc Basin, offshore Newfoundland, Canada. AAPG Bull. 94, 1059–1078.

Warren, E.A., Pulham, A.J., 2001. Anomalous porosity and permeability preservation in deeply buried Tertiary and Mesozoic sandstones in the Cusiana Field, Llanos Foothills, Colombia. J. Sediment. Res. 71, 2–14.

Warren, J.E., Skiba, F.F., Price, H.S., 1961. An evaluation of the significance of permeability measurements. J. Petrol. Technol. 13, 739–744.

Weaver, P.P.E., Schultheiss, P.J., 1983. Vertical open burrows in deep-sea sediments 2 m in length. Nature 301, 329–331.

Zenger, D.H., 1992. Burrowing and dolomitization patterns in the Steamboat Point Member, Bighorn Dolomite (Upper Ordovician), Northwest Wyoming. Contr. Geol. 29, 133–142.

Zenger, D.H., 1996. Dolomitization patterns in widespread "Bighorn Facies" (Upper Ordovician), Western Craton, USA. Carb. Evap. 11, 219–225.

Chapter 28

Carbonate Aquifers

Kevin J. Cunningham,*[,1] Michael C. Sukop[†] and H. Allen Curran[‡]
*U.S. Geological Survey, Davie, Florida, USA, [†]Department of Earth and Environment, Florida International University, Miami, Florida, USA, [‡]Department of Geosciences, Smith College, Northampton, Massachusetts, USA
[1]Corresponding author: e-mail: kcunning@usgs.gov

1. INTRODUCTION

Recent research has demonstrated that trace fossils and their associated ichnofabrics can have an important impact on the fluid-flow properties of hydrocarbon reservoirs and groundwater aquifers. Ichnology, and specifically ichnofabric analysis, offers a powerful approach for reservoir and aquifer characterization when integrated with other data types and methodologies (see Ekdale et al., 2012). Most petroleum-industry-related studies have used ichnology to assist in definition of depositional environments and reservoir quality and have their focus on siliciclastic reservoir characterization (Bann and Fielding, 2004; Bockelie, 1991; Gerard and Bromley, 2008; Gingras et al., 1999, 2007, 2012; Knaust, 2009a; McIlroy, 2004; Spila et al., 2007; Tonkin et al., 2010). Ichnology has been used less as part of carbonate reservoir characterization, but its applications are becoming more prevalent in the scientific literature (Gingras et al., 2002, 2007; Knaust, 2009b; Pemberton and Gingras, 2005).

In the rocks of carbonate aquifers, zones with macroporous and highly permeable ichnofabrics have largely been overlooked by the hydrogeological community due to the lack of recognition of trace fossils and unawareness that high levels of bioturbation can influence groundwater flow. However, examples of burrow-associated macroporosity in carbonate aquifers have been reported by others in (1) the Cretaceous carbonate rocks of the Edwards–Trinity aquifer system in Texas (Fig. 1; Abbott, 1975; Barker and Ardis, 1996; Hovorka et al., 1996, 1998; Maclay and Small, 1986; Rose, 1972; Small et al., 1996), (2) the Cenozoic limestone of the Great Bahama Bank (Beach, 1995), and (3) the Pleistocene limestone of the Biscayne aquifer in southeastern Florida (Cunningham and Sukop, 2011; Cunningham et al., 2006a,b, 2009, 2010; Evans, 1984). Still, with the exception of Cunningham et al. (2009, 2010) and Cunningham and Sukop (2011), we have not identified any publications specifically focused on the link between carbonate ichnology and groundwater

Developments in Sedimentology, Vol. 64. http://dx.doi.org/10.1016/B978-0-444-53813-0.00028-9

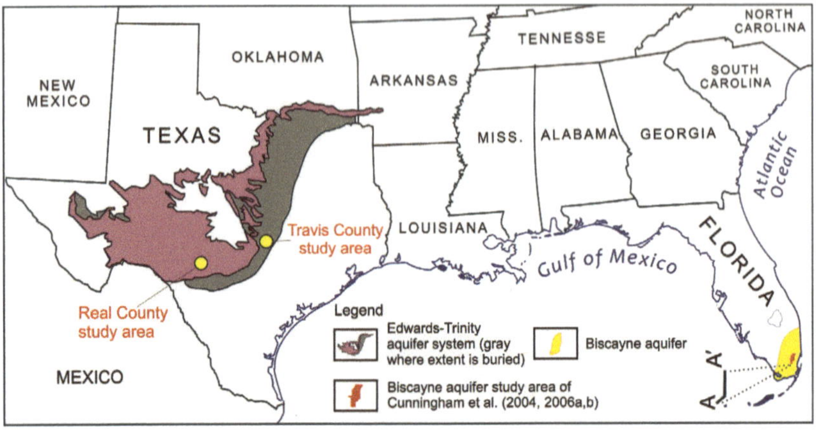

FIGURE 1 Map showing localities of (1) the Biscayne aquifer and the Edwards–Trinity aquifer system, (2) the two study areas in Texas, (3) the study area (solid red polygon) of Cunningham et al. (2004, 2006a,b) in southeastern Florida, and (4) the position of cross-section A–A′ shown in Fig. 2.

flow. Cunningham et al. (2009, 2010) and Cunningham and Sukop (2011) demonstrated well that groundwater flow, through parts of the mostly Pleistocene carbonate Biscayne aquifer in southeastern Florida (Fig. 1), can be concentrated within stratiform, very permeable macropore networks associated with *Ophiomorpha*-dominated ichnofabrics. Recent research by the U.S. Geological Survey, which includes the integration of ichnology, cyclostratigraphy, borehole geophysics, tracer tests, hydraulic tests, high-resolution X-ray computed tomography, Lattice Boltzmann simulations, a conduit flow process for MODFLOW-2005 simulations, and nuclear magnetic resonance imaging (Cunningham and Sukop, 2011; Cunningham et al., 2006a,b, 2009, 2010; Florea et al., 2009; Renken et al., 2005, 2008; Shapiro et al., 2008; Shoemaker et al., 2008), has produced a better understanding of the effect of carbonate macroporosity and permeability related to ichnofabrics on advective groundwater flow, and solute and colloidal transport in the Biscayne aquifer.

The purpose of this chapter is to show that (1) aspects of ichnology can offer a fruitful methodology for purposes of carbonate aquifer characterization, and (2) there can be a strong relationship between macroporosity and permeability related to ichnofabrics and groundwater flow (including paleogroundwater flow) in carbonate aquifers. This chapter is based primarily on ongoing carbonate aquifer characterization by Cunningham, chiefly on the rocks of the Biscayne aquifer in southeastern Florida and new research on outcrops of Cretaceous strata that comprise large parts of the Edwards–Trinity aquifer system in central Texas (Fig. 1). Sukop developed the methodologies for computing permeability of macroporous, vuggy carbonate rocks using Lattice Boltzmann simulations and calculations of macroporosity. Curran contributed to the ichnology of carbonate outcrops in the Biscayne aquifer's unsaturated zone with taxonomic and comparative analyses.

2. METHODS

This chapter presents existing data on the Biscayne aquifer (Cunningham and Sukop, 2011; Cunningham et al., 2004, 2006a,b, 2009, 2010) and summarizes new analyses and computations for carbonate rocks from the Biscayne aquifer and Edwards–Trinity aquifer system (Fig. 1). The new research includes (1) thin-section petrography; (2) high-resolution X-ray computed tomography scans of two carbonate rock samples analyzed at the high-resolution X-ray CT facility at the University of Texas; (3) Lattice Boltzmann calculation of the permeability; and (4) quantification of the macroporosity by pixel counting on computer renderings created from the CT scans. A description of the X-ray tomography acquisition, Lattice Boltzmann method (LBM; Sukop and Thorne, 2006), and pixel-counting technique can be found in Cunningham et al. (2009, 2010). The ichnofabric indices (II) scales of Droser and Bottjer (1986, 1989) were applied to measurements of the intensity of bioturbation.

3. APPLICATION OF ICHNOLOGY TO AQUIFER CHARACTERIZATION

Important applications of ichnology, especially the study of ichnofabrics, to carbonate aquifer characterization include its use in improving (1) delineation of cycles, (2) mapping of cycles based on correlation of biogenically altered surfaces, (3) mapping of zones of preferential groundwater flow and paleogroundwater flow, and (4) understanding of the origin of karst features where related to ichnogenic macroporosity and permeability (as defined by Cunningham et al., 2009). These applications are discussed individually for carbonate rocks of the Biscayne aquifer and Edwards–Trinity aquifer system (Fig. 1). The research results related to these two groundwater systems strongly reflect the practicality of integrating ichnology and hydrogeology for the purpose of carbonate aquifer characterization.

Three case studies are presented (Fig. 1), concentrating on the influence of thalassinidean or thalassinidean-like crustacean-produced ichnofabrics on carbonate aquifer characteristics. The first study focuses on interpretation of multidisciplinary data from the Biscayne aquifer and provides evidence for concentrated groundwater flow through stratiform, touching-vug, macroporous zones where the macroporosity is related to both, (1) primary intraburrow porosity, and (2) secondary fabric-selective intraburrow and interburrow macroporosity within *Ophiomorpha*-dominated ichnofabrics (see Cunningham et al., 2009 for definitions of porosity terms). These pore types occur as the principal groundwater-flow networks in broad areas of the aquifer. The other two studies of carbonate rocks that comprise parts of the Edwards–Trinity aquifer system show evidence for concentrated paleogroundwater flow through stratiform, touching-vug, macroporous flow zones linked to secondary fabric-selective intraburrow and interburrow macroporosity within *Thalassinoides*-dominated ichnofabrics.

3.1 Biscayne Aquifer in Southeastern Florida

The Biscayne aquifer is the principal aquifer in southeastern Florida and one of the most permeable in the world (Parker et al., 1955). The main lithostratigraphic units of the Biscayne aquifer are the Pleistocene Fort Thompson Formation and Miami Limestone (Fig. 2), both of which are characterized by eogenetic karst limestone (Cunningham et al., 2004, 2006a,b; Vacher and Mylroie, 2002). These carbonate units were deposited mostly in shallow, subtropical to tropical marine shelf or continental-marine transitional environments. The limestone has undergone only very shallow, near-surface burial, and multiple widespread subaerial discontinuities associated with relative sea-level lowstands are interspersed throughout the limestone system (Cunningham et al., 2004, 2006a,b; Multer et al., 2002; Perkins, 1977); thus, the limestone commonly has substantial primary interparticle and secondary moldic porosity. The Biscayne aquifer is characterized as a dual-porosity pore system (cf. Cunningham et al., 2004, 2006a,b, 2009, 2010), consisting of (1) matrix porosity (mostly interparticle pores and separate vugs), which in areas can provide much of the groundwater storage; and (2) touching-vug macroporosity. The touching-vug macroporosity can create stratiform, areally extensive, groundwater-flow pathways and less commonly observed bedding-plane and cavernous vugs, vertical solution pipes, and solution-enlarged fractures.

The trace fossil *Ophiomorpha* is almost everywhere the dominant ichnogenus in the shallow-marine, subtidal carbonate grainstone and grain-dominated packstone of the uppermost high-frequency cycle (HFC) of the Miami Limestone and is common in the HFCs composing the Fort Thompson Formation, Here its presence and paleoenvironmental setting are indicators of a *Skolithos* Ichnofacies (Curran, 2007; MacEachern et al., 2007a). These burrows are robust, with thick walls distinctly pelleted on exterior surfaces and smooth on the interior; their outside diameters typically are 1–3 cm or larger. The arrangement of burrows commonly forms multigenerational, three-dimensional, complex patterns. The burrows are filled or partly filled with sediment, or preserved completely open. In Pleistocene carbonates of the Bahamas and southeastern Florida, *Ophiomorpha* is attributed to the burrows of callianassid shrimp by analogy with modern callianassids that dominate deep-tier faunas of bank-top, shallow-subtidal environments throughout the Bahamas, Florida Keys, and other similar tropical carbonate settings (Curran, 2007; Knaust et al., 2012; Shinn, 1968).

Callianassids are powerful agents of bioturbation; their activities can completely transform original depositional fabrics to distinctive ichnofabrics,

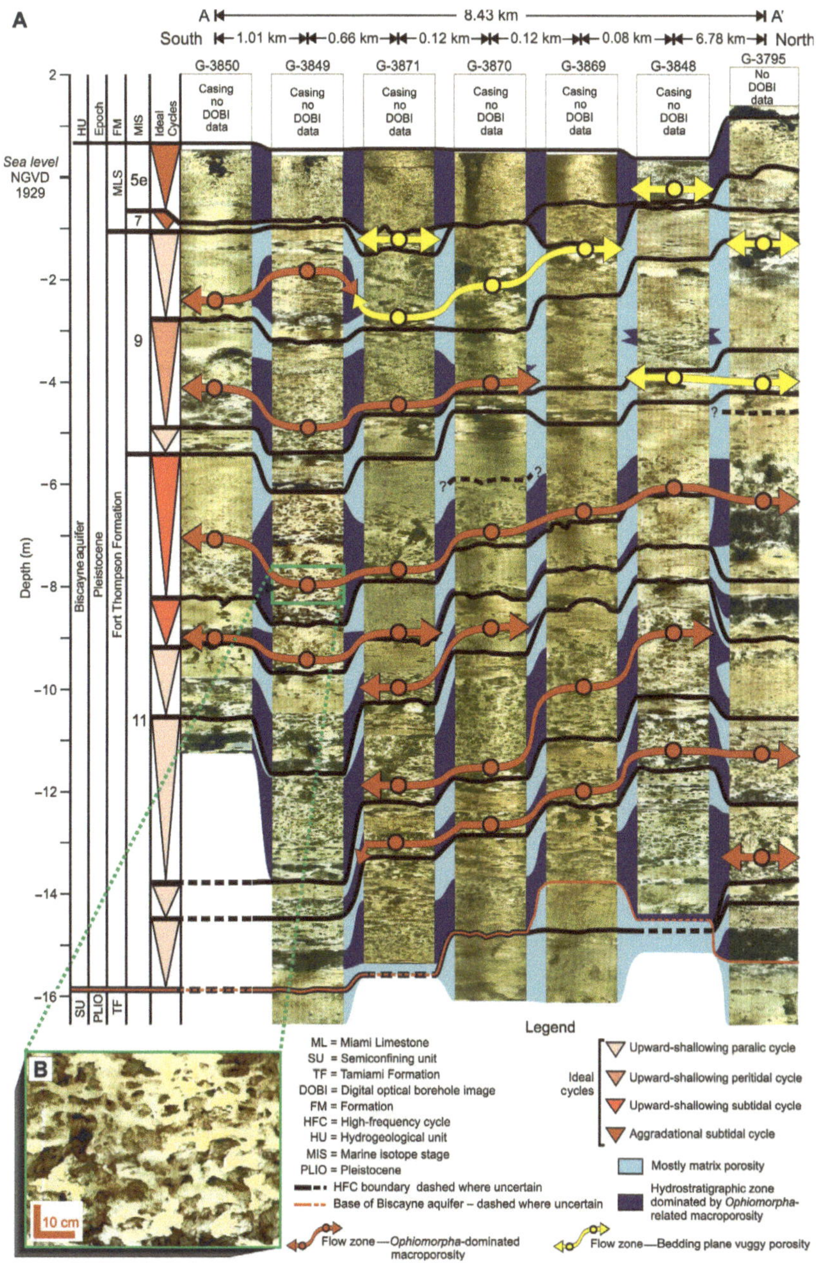

FIGURE 2 (A) North–south hydrogeological cross-section A–A′ (for location, see Fig. 1). A major point is that groundwater flow can be concentrated over a distance of many kilometers within cycle-bound, stratiform zones of a macroporous touching-vug, *Ophiomorpha*-dominated ichnofabric. A more detailed view of corehole G-3849 is displayed in Fig. 3. (B) Close-up view of macroporosity associated with an *Ophiomorpha*-dominated ichnofabric. *Modified from Cunningham and Sukop (2011).*

and their complex, pellet-lined burrows produce biologically engineered plumbing systems in sediments at shallow depths (up to 2 m or more) beneath the sea floor (Curran and Martin, 2003). *Ophiomorpha*-dominated ichnofabrics generated in tropical grainstone and grain-dominated packstone may be associated with substantial levels of touching-vug, intra- and interburrow macroporosity and permeability. The burrow tubes commonly remain open or are later washed clean or cleared of fill via fabric-selective dissolution, generating intraburrow macroporosity. Limestone dissolution can proceed beyond the periphery of a burrow fill and include the pellet-lined wall of the burrow, from the inside toward the outside, so that the pelletal lining can be entirely removed (Gameil and Sadek, 2010). Selective dissolution processes can also enhance the interburrow porosity within the matrix surrounding pellet-lined burrow walls of complex callianassid-burrow networks, resulting in large centimeter-scale interburrow voids. Commonly, intra- and interburrow macroporosities develop together, creating exceptionally permeable rock. The limestone of the Biscayne aquifer contains prime examples of stratiform ichnogenic macroporosity and permeability for comparison with other carbonate aquifers and reservoirs (Cunningham and Sukop, 2011; Cunningham et al., 2009, 2010).

3.1.1 Ichnology in the Delineation of Cycles

The vertical arrangement of ichnofabric types, ichnofabric index measurements, and assigned ichnofacies within cycles, provides criteria for use in the delineation and correlation of HFCs within any stack of cycles comprising the limestone of the Biscayne aquifer. In the context of ideal cycles (Kerans and Tinker, 1997), ichnofabric types and index measurements have a predictable vertical arrangement within HFCs and HFC sets (Cunningham et al., 2009). Cunningham et al. (2009) recognized four ideal HFC types: an upward-shallowing subtidal cycle, upward-shallowing paralic cycle, upward-shallowing peritidal cycle, and aggradational subtidal cycle. Zones with dense concentrations of *Ophiomorpha* (II = 5) are typical in the lower to middle parts in the upward-shallowing subtidal and paralic cycles that compose the Fort Thompson Formation (Figs. 2 and 3). The lateral continuity of the ichnofabric motifs within and between cycles helps in correlation of individual cycles from one borehole to another. The study area of Cunningham et al. (2004, 2006a,b) provides an example of the use of ichnofabric types and index measurements in the delineation and correlation of cycles. Here, a maximum *Thalassinoides* ichnofabric (II = 5) is very widespread in wackestone and mud-dominated packstone that compose an HFC assigned to marine isotope stage (MIS) 7 at the base of the Miami Limestone (Fig. 2). The occurrence of the *Thalassinoides*-dominated ichnofabric contrasts with an overlying HFC comprising peloidal grainstone and grain-dominated packstone with a maximum *Ophiomorpha* ichnofabric (II = 5). The recognition of this widespread vertical pattern in ichnofabrics and ichnofabric index measurements increases confidence in correlation of the two HFCs.

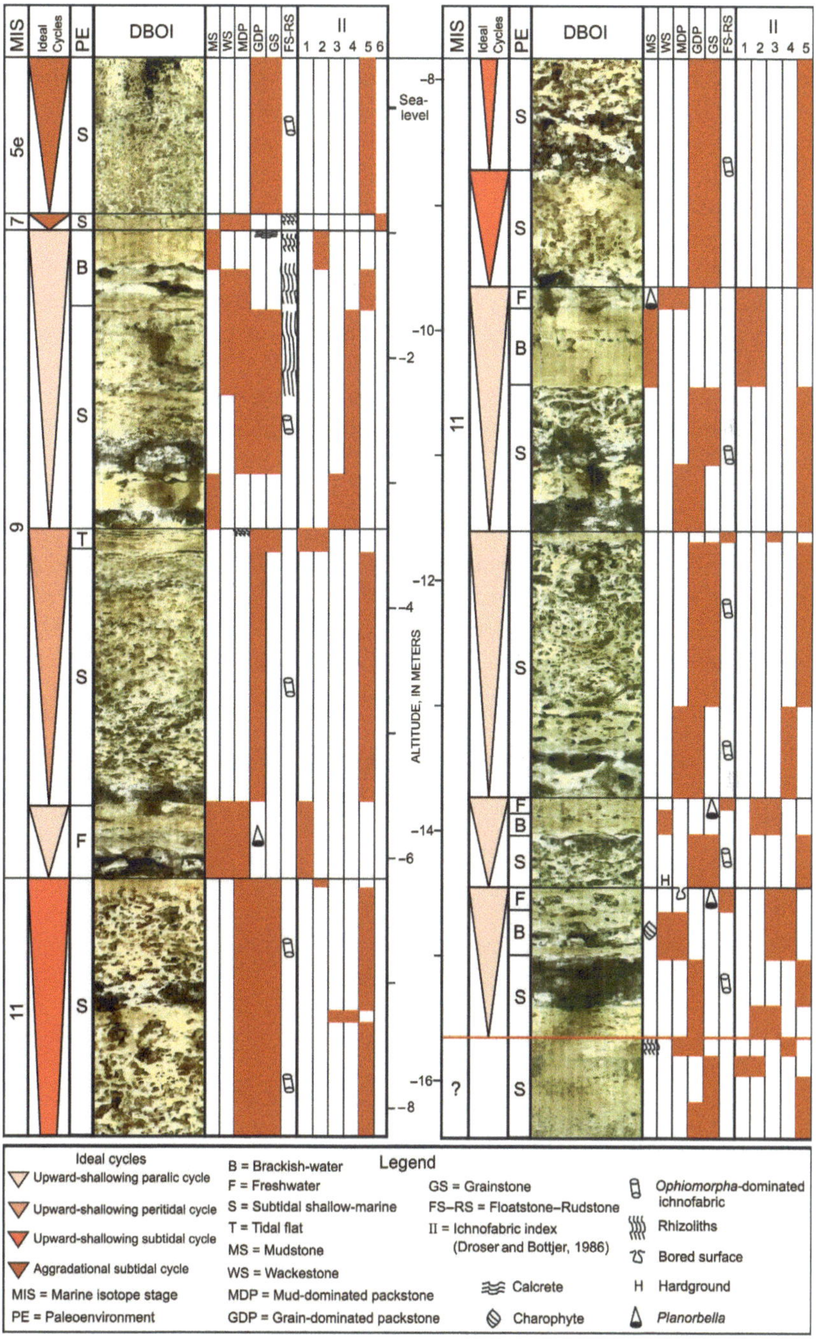

FIGURE 3 Stratigraphical section of corehole G-3849 (see also Figure 2) showing the prominence of cycle-bound, macroporous touching-vug, *Ophiomorpha*-dominated ichnofabrics.

Ichnofacies provides another means for characterization of HFCs composing the limestone of the Biscayne aquifer. In a very general sense, the dense accumulations of *Ophiomorpha* in the rocks of the Miami Limestone and Fort Thompson Formation correspond to moderate to high-energy, shallow-marine, inner to outer shelf paleoenvironments. These attributes are characteristic of the *Skolithos* Ichnofacies (cf. Curran, 2007; MacEachern et al., 2007a). In the study area of Cunningham et al. (2004, 2006a,b) shown in Fig. 1, peloidal and skeletal wackestones and packstones with a maximum *Thalassinoides*-dominated ichnofabric contribute most to the vertical thickness of the HFC assigned to MIS 7 at the base of the Miami Limestone. These depositional textures and ichnofabrics reflect a low- to moderate-energy, shallow-marine, inner shelf paleoenvironment that is attributed to a *Thalassinoides*-dominated *Cruziana* Ichnofacies (cf. MacEachern et al., 2007a,b). Consistent presence in wells of this ichnofacies within the HFC assigned to MIS 7 provides assurance in the certainty of correlation from well to well of this HFC.

3.1.2 Biogenically Altered Surfaces

Paleosols or exposure zones commonly cap some HFCs of the Fort Thompson Formation and Miami Limestone (Fig. 3). Rhizoliths (Klappa, 1980) are regularly a major element of the paleosols or subaerial exposure zones (Fig. 4), and thus the common physical characteristics of these paleosols at and near the tops

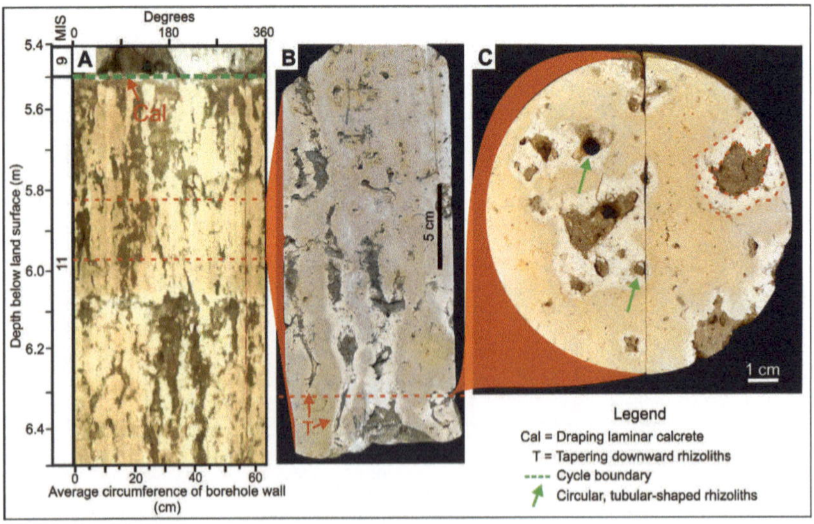

FIGURE 4 Rhizoliths in a digital optical borehole-wall image and core sample of the Fort Thompson Formation from test corehole G-3794 located in the Biscayne aquifer study area shown in Fig. 1. (A) Digital borehole image. (B) Slabbed core sample acquired from the depth interval shown by two dashed red lines in (A). (C) Cut core taken at dashed red line in (B). Halos of relatively white-colored, more porous, chalky-textured matrix (e.g., the area between the dashed red lines), perhaps due to dissolution enhanced by humic acid during decomposition of roots.

of each cycle can provide a means for long-distance correlation of upper cycle boundaries. Larger rhizoliths can have many shapes but are mostly circular to irregular in cuts perpendicular to the longitudinal dimension of core samples (Fig. 4C) and have cylindrical to irregularly tubular shapes in length (Fig. 4A and B). Longitudinally slabbed cores containing relatively large rhizoliths commonly show downward tapering and some downward-directed bifurcations (Fig. 4A and B). The distinctive exposure zone and accompanying rhizoliths shown in Fig. 4 are found over a widespread area (165 km^2) beneath the upper bounding surface of a stack of HFCs assigned to MIS 11 (Figs. 2 and 4). These rhizoliths can penetrate downward to about 1.5 m below the top of the uppermost HFC assigned to MIS 11. Perkins (1977) reported rhizoliths in some cases over 3 m in vertical length in carbonate rocks of South Florida and suggested an association with tap-root systems. Rhizoliths occurring below the MIS 11 upper surface are commonly filled with a gray to light-brown lime mudstone that can contain the freshwater gastropod *Planorbella* (Fig. 9), an occurrence that aids in the identification of the upper bounding surface of the MIS 11 limestone.

Firmgrounds and hardgrounds uncommonly bound the tops of HFCs within the carbonate rocks of the Biscayne aquifer. Structures that correspond to the ichnogenus *Thalassinoides* have a common association with the firmgrounds and as such are attributed to a *Glossifungites* Ichnofacies. The hardgrounds have borings with morphologies ascribed to excavation by bivalves (*Gastrochaenolites*) or sponges (*Entobia*) or both and correspond to the *Trypanites* Ichnofacies (cf. MacEachern et al., 2007a). Correlation of some of the firmgrounds may be over long distances (tens of kilometers). However, these hardgrounds have only been correlated at the kilometer scale and are possibly related to autocyclicity.

3.1.3 Ichnology in Defining Zones of Preferential Groundwater Flow

In the study area of Cunningham et al. (2004, 2006a,b), as shown in Fig. 1, stratiform zones of concentrated groundwater flow related to inter- and intra-burrow macroporosity are common within much of the limestone of the Biscayne aquifer. *Ophiomorpha* is the dominant ichnotaxon present within the ichnofabric of these macroporous and highly permeable zones. Cunningham et al. (2006a,b, 2009) and Cunningham and Sukop (2011) have produced reliable geologically based correlations of these *Ophiomorpha*-rich zones by combining ichnology, lithostratigraphy, cyclostratigraphy, paleontology, and borehole geophysics. Evidence for preferential groundwater flow within these zones has been corroborated by multiple methods, including geophysical borehole logs, computational methods, and hydraulic and chemical tracer tests (Cunningham and Sukop, 2011; Cunningham et al., 2006a,b, 2009, 2010, Renken et al., 2005, 2008; Shapiro et al., 2008).

Recent flowmeter logging in the Biscayne aquifer (Cunningham and Sukop, 2011) indicates that these stratiform, touching-vug, macroporous, bioturbated zones can stack to form an extremely transmissive aquifer over a wide-ranging area, with macroporous *Ophiomorpha*-dominated zones extending over a lateral distance of at least 8.4 km (Fig. 2). Macroporous zones with maximum *Ophiomorpha* ichnofabrics are even more widely distributed at the hundred-kilometer scale in the "bryozoan facies" of the Miami Limestone assigned to MIS 5e by Multer et al. (2002) and mapped by Hoffmeister et al. (1967) in southern Florida. But even these distances are conservative in comparison to the broad extent of modern, shallow-subtidal, carbonate-sand environments intensively burrowed by endobenthic callianassids (Curran and Martin, 2003; Shinn, 1968).

The range in intrinsic permeability values calculated from LBM for five Biscayne aquifer subsamples with macropore networks dominated by an *Ophiomorpha* ichnofabric is between 15×10^3 and 27×10^6 darcies (Cunningham and Sukop, 2011; Cunningham et al., 2009; Fig. 5 and Table 1). Intrinsic permeability is a fluid-free conductance parameter that depends only on the properties of a solid matrix, not on the properties of a fluid passing through it. The conductance parameter, hydraulic conductivity, which is commonly used by hydrologists, is related to both the properties of the matrix and the fluids moving through it (Freeze and Cherry, 1979). The important point is that the five intrinsic permeability values derived from LBM-calculated values (Cunningham and Sukop, 2011, Cunningham et al., 2009) are very high (Fig. 5 and Table 1).

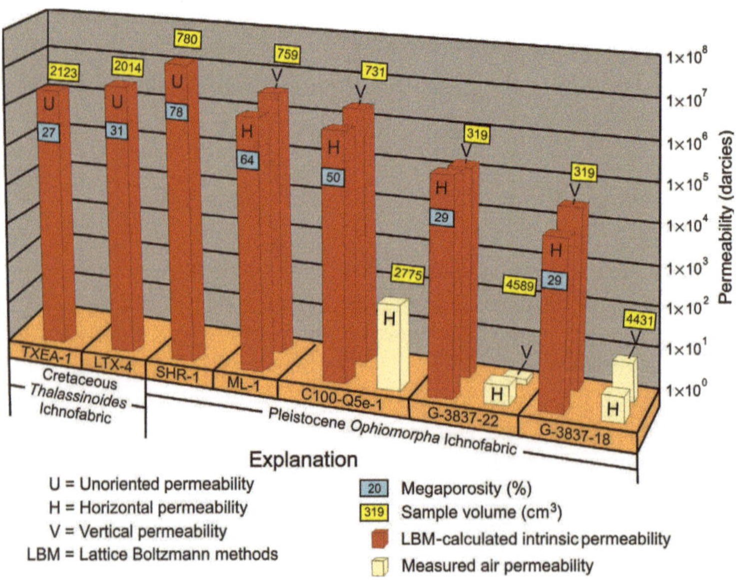

FIGURE 5 Lattice Boltzmann-calculated intrinsic permeability and laboratory air-permeability values derived from carbonate samples with networks of ichnofabric-related macroporosity *(modified from Cunningham and Sukop, 2011).*

TABLE 1 Ichnogenic Macroporosity and Intrinsic Permeability Determined by Computer Methods (See Also Fig. 5)

Sample number	Volume of rendering used in LBM (cm^3)	Macroporosity derived by pixel counting (%)	LBM-derived intrinsic permeability (darcies)	Dominant ichnotaxon	Ichnofacies	Age
SHR-1[a]	780	78	27×10^{6N}	*Ophiomorpha*	*Skolithos*	P
ML-1[b]	759	64	1.9×10^{6H} 3.7×10^{6V}	*Ophiomorpha*	*Skolithos*	P
C-100-Q5e-1[b]	731	50	1.3×10^{6H} 2.1×10^{6V}	*Ophiomorpha*	*Skolithos*	P
G-3837-18[b]	319	29	15×10^{3H} 31×10^{3V}	*Ophiomorpha*	*Skolithos*	P
G-3837-22[b]	319	29	220×10^{3H} 160×10^{3V}	*Ophiomorpha*	*Skolithos*	P
LTX-4	2014	31	5.6×10^{6N}	*Thalassinoides*	*Cruziana*	C
TXEA-1	2123	27	1.8×10^{6N}	*Thalassinoides*	*Cruziana*	C

H, horizontal permeability; V, vertical permeability; N, orientation not determined; P, Pleistocene; C, Cretaceous; LBM, Lattice Boltzmann method.
[a]*Cunningham and Sukop (2011).*
[b]*Cunningham et al. (2009, 2010).*

The LBM intrinsic permeability values (Fig. 5) are as much as two orders of magnitude higher than the intrinsic permeabilities reported for rocks and unconsolidated deposits by Freeze and Cherry (1979: table 2.2). This emphasizes the potential for exceptionally permeable macroporosities related to *Ophiomorpha*-dominated ichnofabrics in carbonate aquifers (Fig. 6A and 6B).

In some cases, tiering has impacted the development of macroporosity and high permeability within the carbonate rocks of the Biscayne aquifer, as in the oolitic barrier bar within the upper part of the Miami Limestone HFC assigned to MIS 5e (Figs. 7 and 8). Tiering is the vertical partitioning of endobenthic communities in the modern and fossil record (Bromley, 1996). In parts of the oolitic barrier bar, an upward decrease in the inner diameter of *Ophiomorpha* burrows and an upward decrease in the connectedness of *Ophiomorpha*-related macroporosity have been observed in outcrops and well bores. For the example shown in Fig. 7, the inner diameters of *Ophiomorpha* burrows average about 1.5 cm near the base of the HFC assigned to MIS 5e, whereas in the middle to upper part, the diameters average about 0.5 cm (Fig. 7). It is possible that larger and more energetic callianassids occupied the deep-tier sediment,

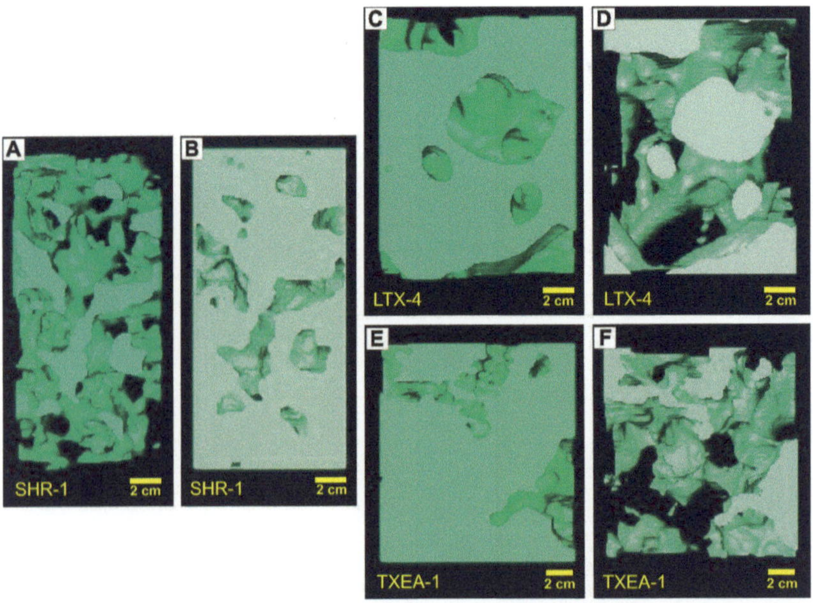

FIGURE 6 Cropped computer renderings of CT scans of carbonate samples with ichnogenic macroporosity and permeability. The light-blue areas in (A), (C), and (E) represent rock matrix and renderings used for LBM analyses (see Table 1). (A) Rendering of Pleistocene Miami Limestone outcrop sample SHR-1 (Cunningham and Sukop, 2011). (B) Complement of rendering in (A). Macropores are light-blue areas. (C) Rendering of Cretaceous sample LTX-4. Outcrop shown in Fig. 9. (D) Complement of rendering in (C). Macropores are light-blue areas. (E) Rendering of Cretaceous outcrop sample TXEA-1 (Fig. 10E). (F) Complement of rendering in (E). Macropores are light-blue areas.

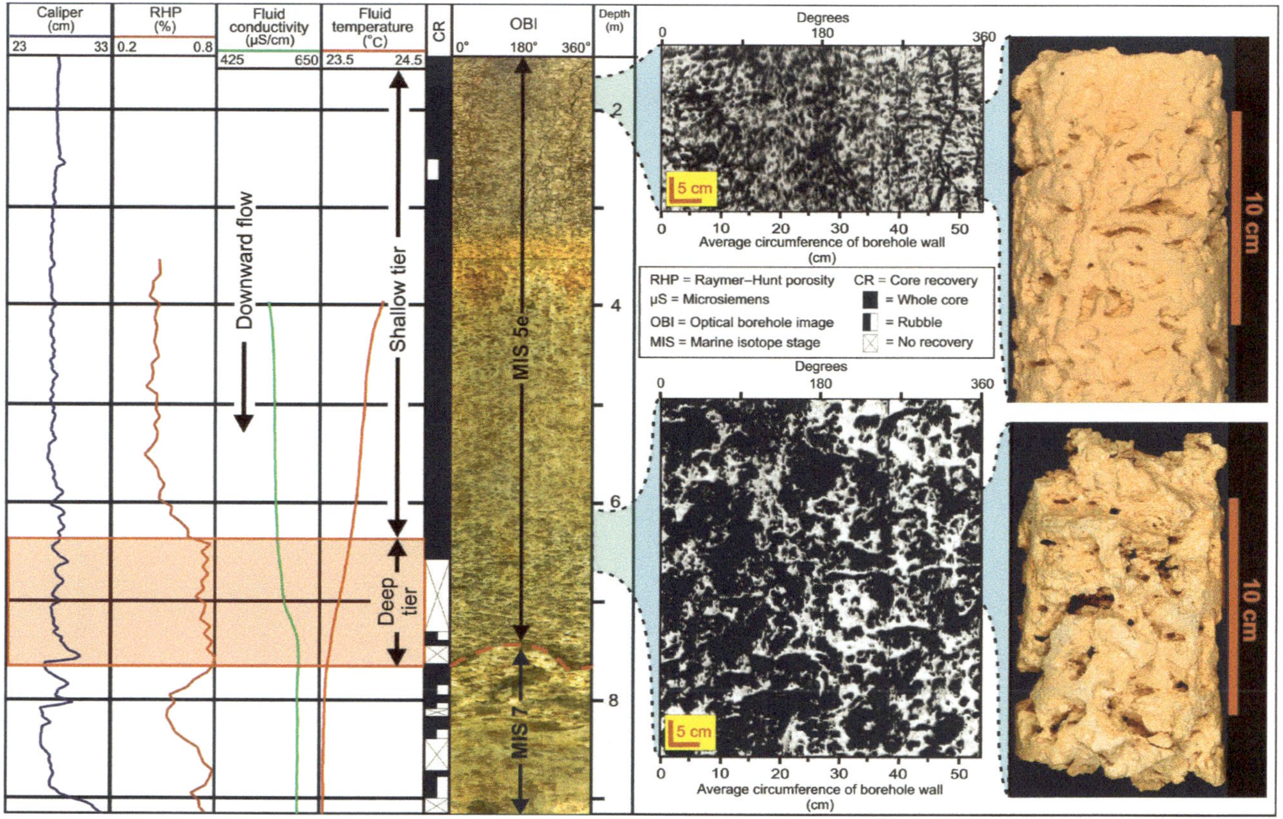

FIGURE 7 Comparison of geophysical logs, optical borehole image, gray-scale high-contrast OBIs, and core photos for test corehole G-3837 that penetrates the relict barrier bar of Halley et al. (1977) within the upper part of the Miami Limestone, which is assigned to marine isotope stage 5e. Data show evidence that the tiering within burrower paleocommunities is influencing development of ichnogenic macroporosity and permeability, permitting a concentrated flow of groundwater into or out of the borehole between the depths of about 6.1 and 7.7 m. This interval has a higher permeability that is linked to a deep-tiered, large-diameter, *Ophiomorpha*-dominated ichnofabric. An overlying shallow tier, small-diameter, *Ophiomorpha*-dominated ichnofabric is less permeable.

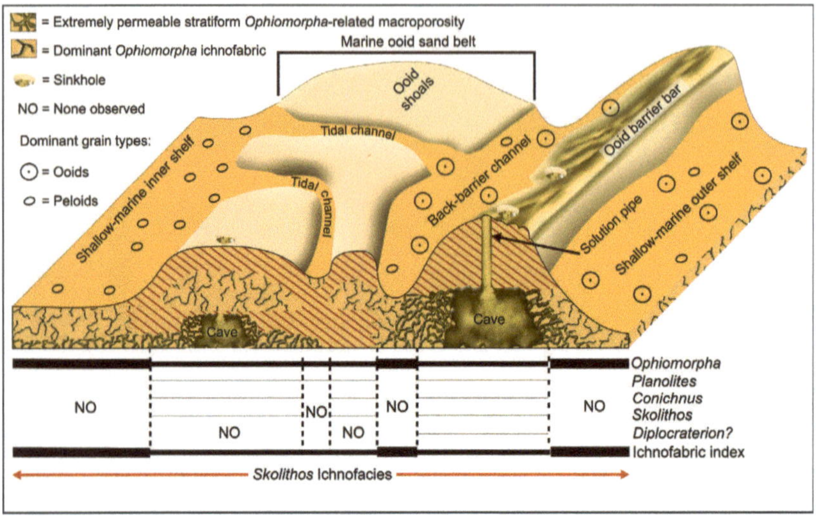

FIGURE 8 The uppermost high-frequency cycle of the Miami Limestone, assigned to marine isotope stage 5e (*modified from Halley et al., 1977*). Conceptual model showing relations between paleoenvironments, ichnologic parameters, and macroporosity related to *Ophiomorpha*-dominated ichnofabrics. In some areas, the *Ophiomorpha*-related macroporosity is linked to the development of karst features. The horizontal bars indicate the relative abundance of five ichnotaxa and the ichnofabric index scale with increasing line thickness corresponding to increasing abundance (few, moderate, and abundant).

while smaller and relatively less energetic callianassids dominated the shallow tier of the cycle. The shallow tier may, for instance, have been dominated by juveniles or maybe a smaller-bodied callianassid species. As shown in Figure 7, sonic-log porosity values calculated from the Raymer–Hunt equation are highest in the lower tiered zone of the *Ophiomorpha*-dominated ichnofabric at the base of the HFC assigned to MIS 5e. Also, across this zone, fluid-conductivity and fluid-temperature logs show marked changes in values, suggesting relatively larger fluid flow either into or out of the borehole over this interval and, thus, a relatively high permeability. Importantly, the geophysical logs show potential for concentrated flow of groundwater through deep-tiered zones of *Ophiomorpha*-dominated ichnofabric and a higher potential for diffuse flow in the shallower *Ophiomorpha*-dominated zone.

3.1.4 *Influence of Ichnology on Karst*

Stratiform, touching-vug, macroporous, *Ophiomorpha*-dominated zones can influence the development of relatively large-scale karst features (caves, sinkholes, and vertical solution pipes) in the upper HFC of the Miami Limestone assigned to MIS 5e. Examples of the stratiform, burrowed zones and related caves have been identified in outcrops of the Miami Limestone within the oolitic barrier bar and marine sand belt shown in Figure 8. The cave system at Deering Glade is the most striking. This system makes a transverse cut through the longitudinal barrier bar and was partly mapped by Cressler (1993); the interior of one of the caves is shown in Cunningham et al. (2008: figs. 3B and 4). Here, the highly burrowed zone is interpreted to have been a stabilized sand flat, later buried by southward progradation of the cross-bedded oolitic barrier bar (Fig. 8). The primary burrow porosity within the stabilized sand-flat sediment provided a template for stratiform fabric-selective dissolution by focusing aggressive groundwater dissolution, which produced caves, within the limits of the burrowed lithosome. The dissolution by groundwater bypassed the cross-bedded oolitic bar because it had a relatively lower primary porosity and permeability compared to the burrowed stabilized sand flat. The oolite barrier-bar forms the ceilings of many of these caves.

3.2 Edwards–Trinity Aquifer System in Real County, Central Texas

The study of outcrops in Real County focused on a freshly excavated road cut along Highway 83 about 0.7 km north of Bee Creek (Fig. 1). Regional observations indicate that cycle-bound *Thalassinoides*-dominated ichnofabrics, likely generated by thalassinidean or thalassinidean-like crustaceans (Goldring et al., 2007), are common in parts of the Early Cretaceous succession of carbonate rocks that form the Edwards Group in the eastern Edwards Plateau of central Texas (Rose, 1972). *Thalassinoides*-dominated ichnofabrics have a substantial volume of associated fabric-selective macroporosity in some carbonate outcrops atop saturated parts of the Edwards–Trinity aquifer system. Commonly, this macroporosity is either a result of fabric-selective leaching and removal of burrow fillings to form moldic, intraburrow macroporosity, previously referred to as "honeycomb porosity" (Abbott, 1975; Rose, 1972), or a result of dissolution of carbonate matrix surrounding burrow fills to form interburrow macroporosity. Honeycomb porosity is common in the Burrowed Member of the Fort Terrett Formation in the eastern Edwards Plateau, where it occurs in the basal parts of the Edwards–Trinity aquifer system (Rose, 1972). This pore type contributes to the Burrowed Member being the chief water-bearing zone of the Fort Terrett Formation in parts of central Texas (Rose, 1972). Barker and Ardis (1996) suggested that the Burrowed Member of the Fort Terrett Formation may be the most permeable part of the Edwards Group west of the intensely fractured and highly leached Balcones fault zone. Abbott (1975) described honeycomb porosity and permeability related to burrowing in the Kainer and Person Formations of the Edwards Group of the Edwards–Trinity aquifer system. Abbott (1975) noted that dissolution of burrow fills in these carbonate rocks can progress well beyond the burrow walls to form caverns. Similarly, Hovorka et al. (1998: 30) observed solution-collapse structures in outcropping carbonate rocks of the Edwards Group and suggested that the collapse could be related to advanced dissolution of burrows in a subtidal stratal unit.

3.2.1 Ichnology in the Delineation of Cycles

Hoffman (2000) recognized a hierarchy of cycles within the lowermost 46 m of the Fort Terrett Formation in the vicinity of Kerrville, Texas, composed of 15 meter-scale HFCs, 4 high-frequency sequences (HFSs) with thicknesses of about 7–18 m, and 1 or 2 composite sequences. The Burrowed Member of the Fort Terrett Formation is included in the 15 cycles of Hoffman (2000). Kerans (2005) placed the Fort Terrett Formation within a single Albian composite sequence.

Eight HFCs (Fig. 9) were delineated within the Burrowed Member of the Fort Terrett Formation at the Real County study area; they are of two types: upward-deepening and upward-shallowing cycles (Fig. 9A). Four upward-deepening HFCs (A–D, Fig. 9A) occur in the lowermost part of the Burrowed Member and are overlain by four upward-shallowing HFCs (E–H, Fig. 9A). The eight HFCs compose part of the transgressive and the highstand systems tract of one provisional HFS (HFS A, Fig. 9A) and part of the transgressive systems tracts of a second provisional HFS (HFS B, Fig. 9A). Combined, the eight HFCs indicate an overall landward shift of a paleoshoreline within the context of a transgressive systems tract included in a composite sequence (Fig. 9A).

The lower boundary of HFS A, which is exposed southwest of the road cut shown in Figure 9A, corresponds to the contact between the Glen Rose Formation and the Basal Nodular Member of the Fort Terrett Formation (Rose, 1972), a regional composite sequence boundary (Kerans, 2005). The maximum flooding surface of HFS A is at the upper bounding surface of HFC D (Fig. 9A). HFC G is mainly composed of micrite recrystallized to microspar and dolomite. The upper surface of HFC G forms the upper surface of HFS A. The top of HFC G in the West Frio section of Rose (1972: 181–182) is riddled with rhizoliths, which indicates that the upper boundary of HFS A is associated with a relative fall in sea level and subaerial exposure. A prominence of *Thalassinoides*, mudstone to grainstone depositional textures, peloids, miliolids, and bivalve fragments are found to characterize much of the Burrowed Member of the Fort Terrett Formation and are highly suggestive of a low-energy, restricted shallow-marine, shelf interior (cf. Rose, 1972). A monospecific ichnotaxon and paleoenvironment express a *Thalassinoides*-dominated *Cruziana* Ichnofacies for much of HFS A. An open shallow-marine depositional facies is characteristic of HFS B (Fig. 9A), as suggested by a substantial presence of green algae, and caprinid and toucasid rudists (Rose, 1972). Only a very few dense accumulations of *Thalassinoides* were observed in HFS B (Fig. 9A). A general upward, landward shift in depositional facies from HFS A to HFS B indicates a backstepping of the HFS B open shallow-marine shelf over the HFS A-restricted shallow-marine shelf interior. A more regional delineation of ichnofabrics and paleoenvironments should be useful for long-distance correlation of cycles within the Burrowed Member of the Fort Terrett Formation.

The stratiform *Thalassinoides*-dominated ichnofabric is mostly restricted to the micrite-rich parts of the HFCs of HFS A, where it is characterized by an

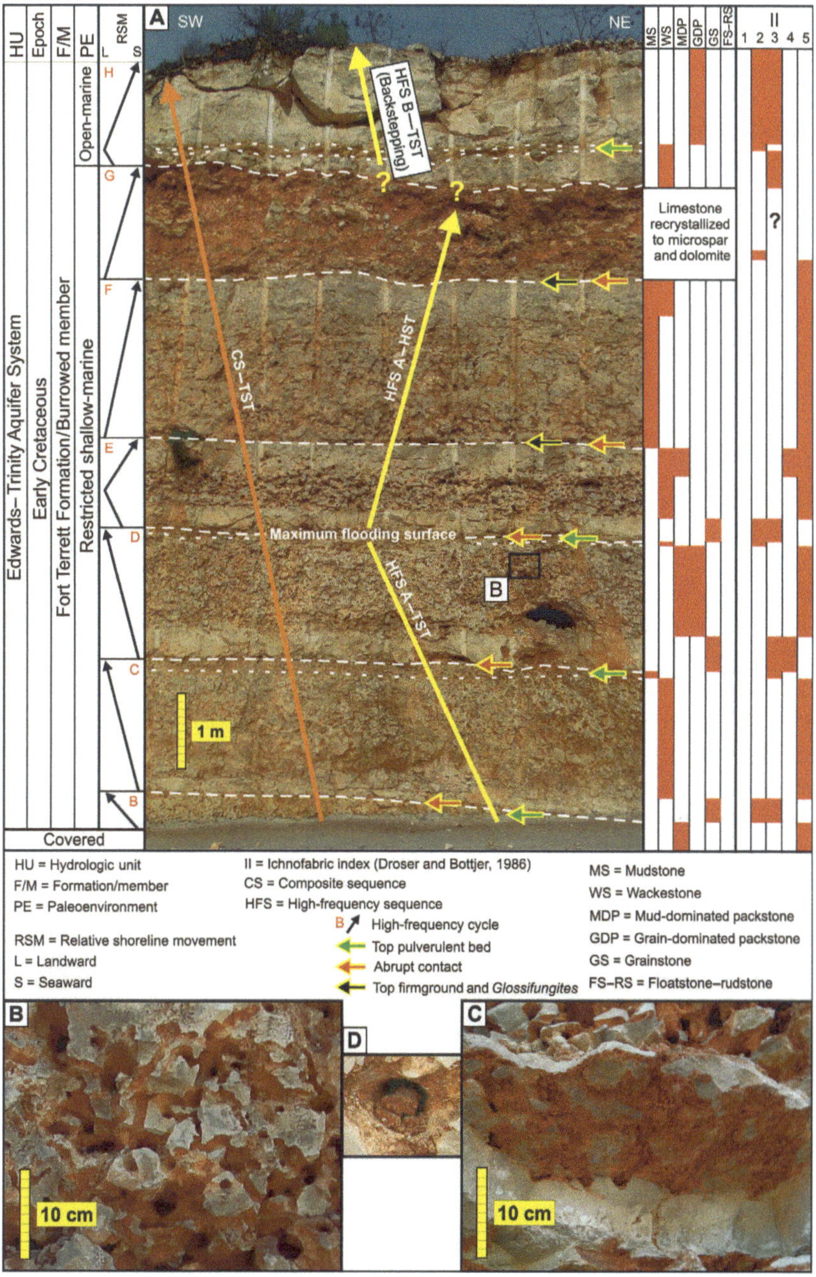

FIGURE 9 Subtidal high-frequency cycles (HFCs) within the Burrowed Member of the Fort Terrett Formation in Real County, Texas. (A) Hydrostratigraphy, lithostratigraphy, paleoenvironments, cycles, carbonate depositional textures, and ichnofabric indices. (B) Irregular boxwork of *Thalassinoides* tunnels and shafts that form intraburrow macroporosity within HFC D. See inset box on (A) for location. (C) Regular maze of *Thalassinoides* tunnels that form intraburrow macroporosity along a bedding plane within HFC E. (D) Interior of a *Thalassinoides* tunnel partly filled with terra rossa clay.

irregular boxwork configuration of open burrows (Fig. 9B). Within some cycles, *Thalassinoides* occurs parallel to the bedding plane of the rock, where burrows are repeatedly branched with a slight swelling at the junctions of the tubes and form a horizontal, regular maze (Fig. 9C). A high density of *Thalassinoides* in the middle to lower parts of HFCs E and F (Fig. 9A) may be due to a relatively concentrated abundance of deep-tiered burrow excavations (e.g., Bromley, 1996: fig. 4.25). The three-dimensional arrangement of the *Thalassinoides* burrows forms an irregular boxwork of well-connected, open tunnels and shafts with ichnofabric indices measuring as high as 5 (Fig. 9A). The open, inner diameters of burrows are cylindrical in shape and most commonly range from 0.5 to 3.5 cm in width.

3.2.2 Biogenically Altered Surfaces

Trace fossils have been shown to be useful for the genetic interpretation of discontinuity surfaces whether of sequence stratigraphic or autocyclic importance (Bromley, 1975; MacEachern et al., 2007a). A *Glossifungites* Ichnofacies assemblage and discontinuity surface occur at the tops of HFCs E and F (Fig. 9A). During the development of the discontinuities, firm substrates existed below the upper bounding surfaces of HFCs E and F as suggested by the presence of sharp-walled *Thalassinoides*, absence of borings penetrating the surface, and by the occurrence of post-burrow excavation infillings by transgressive sediments from superjacent cycles. Erosion of the surfaces is suggested by a sharp facies shift across them, and their irregular and uncommon overhanging paleotopography. Possible causative controls on the erosional surfaces include fair-weather waves, storm waves, or tidal currents in response to a seaward shift in paleoshoreline, followed by a period of transgression (Gerard and Bromley, 2008). The association of a *Glossifungites* Ichnofacies assemblage with both of the cycle-bounding surfaces could be a key to widespread correlation of HFC E and HFC F within the Burrowed Member of the Fort Terrett Formation.

3.2.3 Ichnology in Defining Zones of Preferential Groundwater Flow

The macroporosity associated with the interior of *Thalassinoides* tunnels and shafts is likely the result of fabric-selective dissolution of an internal burrow fill—a process proposed by Abbott (1975). Mapping of the burrow-related macroporosity shown in Figure 9 indicates that *Thalassinoides* are filled with grainstone and grain-dominated packstone toward the southwestern limit of the road cut outcrop as it approaches Bee Creek. From here, fabric-selective leaching and removal of burrow fill increase toward the northeastern limit of the road cut exposure, where most or all of the internal sediment that once filled the interiors of *Thalassinoides* has been leached and removed. The grain-rich fill in the *Thalassinoides* tubes has partly or entirely dissolved to produce a tangled, serpentine arrangement of highly permeable touching-vug porosity.

Lattice Boltzmann calculations of one whole-core-scale computer rendering of an outcrop sample (LTX-4, Fig. 6C and D) with a large volume of *Thalassinoides*-related intraburrow macroporosity produced an intrinsic permeability value of 5.6×10^6 darcies, a value within a range of *Ophiomorpha*-related LBM intrinsic permeabilities generated for Pleistocene limestone samples from the Biscayne aquifer (Fig. 5 and Table 1).

In the partly open or fully open burrows, the inner walls of *Thalassinoides* are stained light brown (Fig. 9B–D). The moldic interiors of *Thalassinoides* commonly are partly or completely filled (Fig. 9D) with a moderate brown terra rossa clay (cf. Young, 1986). The clay fill within *Thalassinoides* tubes provides evidence that the paleo-permeability was substantially enough within the paleo-macropore network to allow the transport of clay carried by groundwater through a transmissive paleo-pore network dominated by *Thalassinoides*. Through time the clays settled within the network of tube-like macroporosity to plug or partly plug it. Presumably, the clay was advected over distances of at least outcrop scale and perhaps much more. The clay is inferred to have entered a subsurface porosity network from an exposed paleosurface via sink-holes or vertically oriented solution-enhanced fractures that connected the *Thalassinoides*-related paleo-macroporosity to the paleosurface. Transport of suspended sediment through the Fort Terrett Formation limestone is suggestive of turbulent flow conditions (Mahler and Lynch, 1999). Located about 100 m north of the road cut shown in Figure 9, a sinkhole in the carbonate rocks of the Fort Terrett Formation is partly filled with terra rossa clay and may have acted as one source of clay that fills or partly fills molds of *Thalassinoides*. Alternatively, the clay was advected by groundwater flowing though a sinuous, solution-enlarged subterranean flow path converging on or diverging from a paleo-river valley (cf. Abbott, 1975).

3.2.4 Influence of Ichnology on Karst

Macroporosity associated with the *Thalassinoides*-dominated ichnofabric is present generally at a cm-scale but expands to m-scale at one paleo-cave exposed in the Highway 83 road cut. Like many *Thalassinoides* burrows (Fig. 9D), much of the paleo-cave is fully plugged with terra rossa clay. The paleo-cave has a stratiform geometry with its ceilings, floors, and walls mainly conforming to *Thalassinoides*-rich parts of HFCs B and C (Fig. 9A). The paleo-cave stair-steps downward between the two HFCs, conforming to the dissolved *Thalassinoides*-rich zones. Dissolution-produced dish-shaped indentations with prominent irregular polygonal boundaries demarcate a sculptured ceiling of the paleo-cave, and although similar in appearance to scallops, only some are asymmetric in shape typical of scallops (cf. Palmer, 2007). As scallops form during the flow of turbulent water across a bedrock surface (Palmer, 2007), their minor presence suggests cave formation when the paleo-cave was, at least at times, filled with groundwater. This speleogenetic scenario suggests cave formation in the phreatic groundwater zone and thus likely pre-Holocene. Speleogenesis probably began with fabric-selective dissolution of the internal sediments of *Thalassinoides* to produce an arrangement of molds that conforms to the stratiform distribution of a *Thalassinoides*-dominated ichnofabric confined within the upper and lower boundaries of upward-shallowing subtidal cycles (Fig. 9A). As dissolution progressed beyond the internal sediment of the burrows and into the carbonate matrix, cavern-sized voids were created. This process of burrow-mold formation and eventual development of caverns began during Cretaceous subaerial exposure of the Edwards Group (Abbott, 1975). Abbott (1975) proposed that the most aggressive dissolution within the ancestral Edwards aquifer occurred during or since the Miocene—an epoch when a through-flowing dynamic groundwater system was highly active. The filling of the macropore system at the Real County site by clay presumably occurred within a time span bracketing the Miocene to Holocene epochs.

3.3 Edwards–Trinity Aquifer System in Travis County, Central Texas

Burrow-related macroporosity was investigated in and around an abandoned quarry along Far West Boulevard in Travis County, where the dolomitic member of the Aptian–Albian Kainer Formation (Small et al., 1996) is exposed (Fig. 1). At this locality, these carbonate rocks compose an unsaturated part of the Edwards–Trinity aquifer system (Barker and Ardis, 1996).

3.3.1 Ichnology in the Delineation of Cycles

The sequence-stratigraphic framework of Kerans (2005) was used for the Travis County locality to infer that the lowest-order cycle shown partly in Fig. 10 is a tens-of-meter-scale upward-shallowing HFS. Figure 10 shows two stratiform burrowed lithological units (2–3) characterized by a *Thalassinoides*-dominated ichnofabric and peloidal mudstone to wackestone depositional textures that overlie a caprinid-rudisitid rudstone and floatstone. The very low-diversity ichnotaxa and peloidal micrite-rich depositional textures reflect a stressed, low-energy *Thalassinoides*-dominated *Cruziana* Ichnofacies (cf. MacEachern et al., 2007a,b). Above the burrowed zone, grainstone beds grade upward into beds composed of flat-pebble conglomerate and stromatolites that cap the upward-shallowing HFS. From bottom to top (Fig. 10A), the vertical lithofacies succession (lithological units 1–6) represents the seaward progradation of (1) an open shallow-marine, caprinid bioherm; (2) a restricted shallow-marine, shelf interior; (3) a shallow, high-energy, coastal foreshore; and (4) mud-rich and stromatolitic tidal to supratidal flats. The burrowed lithological units 2 and 3 (Fig. 10A) were mapped in discontinuous outcrops over an area of about 1.6 km^2.

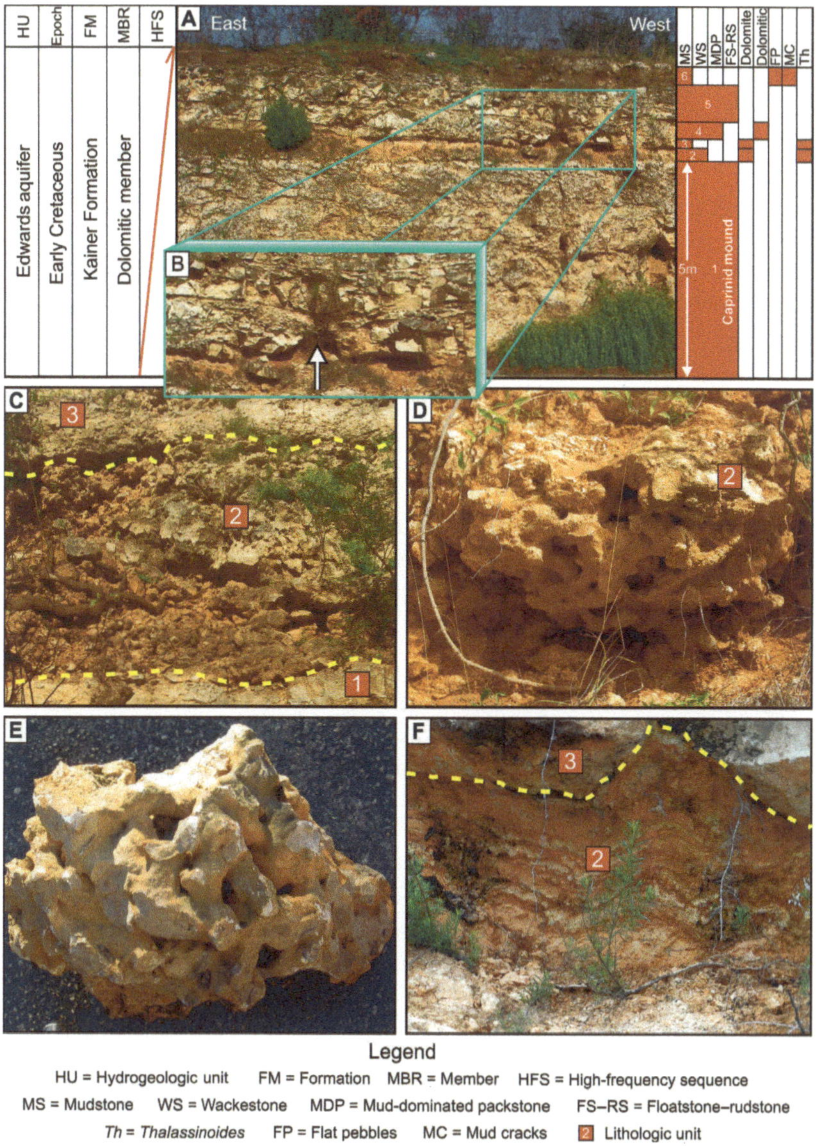

FIGURE 10 Ichnological and paleohydrogeological features of the Kainer Formation along the wall of an abandoned quarry in Travis County, Texas. (A) Hydrostratigraphy, lithostratigraphy, and carbonate textures of quarry-wall photomap. (B) Close-up view of inset (blue rectangle), showing beds that have collapsed into a stratiform zone of ichnogenic macroporosity related to a *Thalassinoides*-dominated ichnofabric developed in unit 2. (C) Serpentine, irregular boxwork of *Thalassinoides* tunnel and shaft casts, and associated interburrow macroporosity in unit 2. (D) Regular maze of casts of *Thalassinoides* tunnels near the base of unit 2. (E) Sample (TXEA-1) with a complexly arranged irregular boxwork of *Thalassinoides* casts from unit 2 (see Figs. 5 and 6; Table 1). (F) Small paleo-cave filled with terra rossa clay in unit 2.

3.3.2 Ichnology in Defining Zones of Preferential Groundwater Flow

The southern quarry wall shows a fabric-selective control on the development of interburrow macroporosity and karstification. The outer surfaces of *Thalassinoides* casts are commonly stained light brown, indicating the presence of iron oxide (Fig. 10C–E). The interburrow macroporosity is in places filled or partly filled with moderate brown-colored terra rossa clay. Lattice Boltzmann calculations of one whole-core-scale computer rendering (Fig. 6E and F) of an outcrop sample (TXEA-1; Fig. 10E) with a large volume of *Thalassinoides*-related interburrow macroporosity produced an intrinsic permeability value of 1.8×10^6 darcies (Fig. 5 and Table 1). This intrinsic permeability value is within the range of *Ophiomorpha*-related LBM permeabilities generated for limestone samples from the Biscayne aquifer (Fig. 5) and of the same magnitude as the intrinsic permeability calculated for the LTX-4 sample rendering (Fig. 6C and D) from the Fort Terrett Formation (Fig. 5 and Table 1). The extremely high value of interburrow permeability derived from LBM, the clay plugging of voids, and the widespread distribution of a high permeability throughout the burrowed lithological unit 2 (Fig. 10) indicate that this stratiform unit acted to concentrate paleo-groundwater flow.

3.3.3 Influence of Ichnology on Karst

At the Travis County study site, small clay-filled paleo-caves, open caves with speleothems, and a solution-collapse structure are associated with the burrowed lithological unit 2 (Fig. 10). The solution collapse resulted from overlying beds falling into an area of expansive interburrow macroporosity in burrowed unit 2 (Fig. 10A and B). The complete or partial filling of the burrow macroporosity and paleo-caves with terra rossa clay (Fig. 10F) is inferred to have resulted from transport by groundwater and deposition in the voids sometime within the Miocene-to-Holocene time span, based on Abbott's (1975) scenario for porosity development in the Edwards Group.

4. CONCLUSIONS

Three case studies, which include Pleistocene carbonate rocks of the Biscayne aquifer in southern Florida and Cretaceous carbonate strata of the Edwards–Trinity aquifer system in central Texas, demonstrate that (1) ichnology can offer a fruitful methodology for carbonate aquifer characterization, and (2) there can be a strong relation between ichnofabrics and groundwater flow in carbonate aquifers. Important applications of ichnology to carbonate aquifer characterization include its use in improving (1) the delineation of cycles, (2) the mapping of cycles based on correlation of biogenically altered surfaces, (3) the mapping of zones of preferential groundwater flow and paleogroundwater flow, and (4) the understanding of the origin of karst features where related to ichnofabrics.

Carbonate rocks with ichnofabrics dominated by *Ophiomorpha* in the Biscayne aquifer and *Thalassinoides* in the Edwards–Trinity aquifer system can possess exceptionally permeable stratiform networks of cm-scale, touching-vug ichnogenic macroporosity and permeability. The stratiform aquifer zones can evolve within two different scales of shallow-marine shelf cycles that include (1) multistoried stacks of HFCs delineated by predictable vertical successions of lithofacies, and (2) HFSs comprised in part by thick beds that act as aquifer zones. Burrowed zones can be laterally expansive, as in modern, flat-topped, tropical platforms; thus, these aquifer zones can have widespread continuity of up to the 100-km scale.

The stratiform highly permeable ichnofabric-related macroporous zones in the Biscayne aquifer compose a *Skolithos* Ichnofacies and in the Edwards–Trinity aquifer system a *Thalassinoides*-dominated *Cruziana* Ichnofacies. Other ichnofacies recognized in the carbonate rocks of the Biscayne aquifer are *Thalassinoides*-dominated *Cruziana*, *Glossifungites*, and *Trypanites*, but none have been recognized as part of the *Ophiomorpha*-related stratiform groundwater-flow zones. At the Real County site in central Texas, the burrows associated with the *Glossifungites* Ichnofacies are filled with sediment and would not have added any substantial input to paleogroundwater flow within the ancestral Edwards–Trinity aquifer system.

Multiple methodologies, such as geophysical logging and tracer tests, confirm preferential flow within many of the *Ophiomorpha*-dominated zones of the Biscayne aquifer. In parts of the Biscayne aquifer, the ichnologically associated networks of highly permeable macroporosity are related to the fabric-selective dissolution of *Ophiomorpha*-dominated limestone and primary burrow macroporosity. Five whole-core-scale-sized computer renderings of limestone samples from the Biscayne aquifer, which are totally bioturbated and characterized by *Ophiomorpha*-dominated ichnogenic macroporosity, have extremely high intrinsic permeabilities ranging between 15×10^3 and 27×10^6 darcies.

In the Edwards–Trinity aquifer system at two study sites in central Texas, examples of permeable *Thalassinoides*-dominated ichnofabrics occur within an unsaturated part of the aquifer system. At both sites, the macroporous zones within the Cretaceous carbonate rocks are representative of a restricted shallow-marine, shelf-interior paleoenvironment. Lattice Boltzmann-calculated intrinsic permeabilities for intra- and interburrow macroporosity represented in computer renderings of two whole-core-scale samples were 5.6×10^6 and 1.8×10^6 darcies, respectively. The Cretaceous macroporous, maximum *Thalassinoides* ichnofabrics have similar intrinsic permeability values as Pleistocene intra- and interburrow macroporosity related to *Ophiomorpha*-dominated ichnofabrics in limestone of the Biscayne aquifer. These highly permeable *Thalassinoides*-dominated zones could play an important role in groundwater movement, where present, in saturated parts of the Edwards–Trinity aquifer system.

In the Pleistocene and Cretaceous examples, fabric-selective dissolution of carbonate rocks with ichnofabrics related to an *Ophiomorpha*-dominated *Skolithos* Ichnofacies and a *Thalassinoides*-dominated *Cruziana* Ichnofacies, respectively, is shown to have a marked impact on karst development of (1) zones of highly permeable stratiform macroporosity; (2) caves and cavities, which commonly conform to upper and lower stratal boundaries in HFCs; and (3) solution-collapse structures. The stratiform networks of extremely permeable macroporosity are mappable and, as such, can be represented in groundwater flow and transport simulations. To improve the predictive conceptualization of aquifer properties of carbonate rocks with varying degrees of ichnofabric development at the field scale, future research should focus on better understanding the controls affecting the spatial distribution and hydraulic properties of porous and highly permeable bioturbated zones in carbonate aquifers.

ACKNOWLEDGMENTS

Priority Ecosystems Science (U.S. Geological Survey) and Critical Ecosystems Studies Initiative (Everglades National Park) programs provided major project funding. Mike Wacker contributed borehole geophysical data. Generous insight into the Cretaceous geology of central Texas came from Bill Ward (deceased), Rene Barker, Nico Hauwert, Scott Hiers, and Al Broun. Reviews by Rene Barker, Bill Ward, and Nico Hauwert improved an early draft. Edwards Aquifer Research & Data Center provided time for Rene Barker's review. Richard Westcott assisted with graphics. Final reviews were by Gene Shinn, Dirk Knaust, and Richard Bromley.

REFERENCES

Abbott, P.L., 1975. On the hydrology of the Edwards Limestone, South-central Texas. J. Hydrol. 24, 251–269.

Bann, K.L., Fielding, C.R., 2004. An integrated ichnological and sedimentological comparison of non-deltaic shoreface and subaqueous delta deposits in Permian reservoir units of Australia. In: McIlroy, D. (Ed.), The Application of Ichnology to Palaeoenvironmental and Stratigraphic Analysis. Geol. Soc., Spec. Publ., vol. 228, pp. 273–310.

Barker, R.A., Ardis, A.F., 1996. Hydrogeologic Framework of the Edwards-Trinity Aquifer System, West-Central Texas. U.S. Geol. Surv., Prof. Pap. 1421-B, 61 pp.

Beach, D.K., 1995. Controls and effects of subaerial exposure on cementation and development of secondary porosity in the subsurface of Great Bahama Bank. In: Budd, D.H., Saller, A.H., Harris, P.M. (Eds.), Unconformities and Porosity in Carbonate Strata. AAPG Mem. 63, pp. 1–33.

Bockelie, J.F., 1991. Ichnofabric mapping and interpretation of Jurassic reservoir rocks of the Norwegian North Sea. Palaios 6, 206–215.

Bromley, R.G., 1975. Trace fossils at omission surfaces. In: Frey, R.W. (Ed.), The Study of Trace Fossils. Springer, New York, pp. 399–428.

Bromley, R.G., 1996. Trace Fossils: Biology, Taphonomy and Applications. Chapman & Hall, London, 361 pp.

Cressler, A., 1993. The caves of Dade County, Florida. In: Fee, S. (Ed.), Speleo Digest. National Speleological Society, Huntsville, Alabama, pp. 59–63.

Cunningham, K.J., Sukop, M.C., 2011. Multiple technologies applied to characterization of porosity and permeability of the Biscayne aquifer. U.S. Geol. Surv., Open-File Report 2011–1037, 8 pp.

Cunningham, K.J., Carlson, J.L., Wingard, G.L., Robinson, E., Wacker, M.A., 2004. Characterization of aquifer heterogeneity using cyclostratigraphy and geophysical methods in the upper part of the karstic Biscayne aquifer, southeastern Florida. U.S. Geol. Surv., Water-Resources Investigations Report 03-4208, 46 pp.

Cunningham, K.J., Renken, R.A., Wacker, M.A., Zygnerski, M.R., Robinson, E., Shapiro, A.M., Wingard, G.L., 2006a. Application of carbonate cyclostratigraphy and borehole geophysics to delineate porosity and preferential flow in the karst limestone of the Biscayne aquifer, SE Florida. In: Harmon, R.S., Wicks, C. (Eds.), Perspectives on Karst Geomorphology, Hydrology, and Geochemistry—A Tribute Volume to Derek C. Ford and William B. White. GSA Spec. Pap. 404, pp. 191–208.

Cunningham, K.J., Wacker, M.A., Robinson, E., Dixon, J.F., Wingard, G.L., 2006b. A cyclostratigraphic and borehole geophysical approach to development of a three-dimensional conceptual hydrogeologic model of the karstic Biscayne aquifer, southeastern Florida. U.S. Geol. Surv., Scientific Investigations Report 2005–5235, 69 pp.

Cunningham, K.J., Sukop, M.C., Huang, H., Alvarez, P.F., Curran, H.A., Wacker, M.A., Florea, L.J., Renken, R.A., Dixon, J.F., 2008. Biogenic macroporosity and its Lattice Boltzmann method permeability in the karst Biscayne aquifer. In: Sasowsky, I.D., Feazel, C.T., Mylroie, J.E., Palmer, A.N., Palmer, M.V. (Eds.), Karst from Recent to Reservoirs, Karst Waters Institute, Spec. Publ., vol. 14, pp. 30–35.

Cunningham, K.J., Sukop, M.C., Huang, H., Alvarez, P.F., Curran, H.A., Renken, R.A., Dixon, J.F., 2009. Prominence of ichnologically influenced macroporosity in the karst Biscayne aquifer: stratiform "super-K" zones. GSA Bull. 121, 164–180.

Cunningham, K.J., Sukop, M.C., Huang, H., Alvarez, P.F., Curran, H.A., Renken, R.A., Dixon, J.F., 2010. Erratum for "Prominence of ichnologically influenced macroporosity in the karst Biscayne aquifer: stratiform 'super-K' zones" GSA Bull. 122, 2109.

Curran, H.A., 2007. Ichnofacies, ichnocoenoses, and ichnofabrics of Quaternary shallow-marine to dunal tropical carbonates: a model and implications. In: Miller III, W. (Ed.), Trace Fossils. Concepts, Problems, Prospects. Elsevier, Amsterdam, pp. 232–247.

Curran, H.A., Martin, A.J., 2003. Complex decapod burrows and ecological relationships in modern and Pleistocene intertidal carbonate environments, San Salvador Island, Bahamas. Palaeogeogr. Palaeoclimatol. Palaeoecol. 192, 229–245.

Droser, M.L., Bottjer, D.J., 1986. A semiquantitative field classification of ichnofabric. J. Sediment. Petrol. 56, 558–559.

Droser, M.L., Bottjer, D.J., 1989. Ichnofabric of sandstones deposited in high-energy nearshore environments: measurement and utilization. Palaios 4, 598–604.

Ekdale, A.A., Bromley, R.G., Knaust, D., 2012. The ichnofabric concept. In: Knaust, D., Bromley, R.G. (Eds.), Trace Fossils as Indicators of Sedimentary Environments. Developments in Sedimentology, vol. 64. Elsevier, Amsterdam, pp. 139–155.

Evans, C.C., 1984. Development of an ooid sand shoal complex: the importance of antecedent and syndepositional topography. In: Harris, P.M. (Ed.), Carbonate Sands—A Core Workshop. SEPM Core Workshop 5, pp. 392–428.

Florea, L.J., Cunningham, K.J., Altobelli, S., 2009. NMR imaging of fluid exchange between macropores and matrix in eogenetic karst. Ground Water 47, 382–390.

Freeze, R.A., Cherry, J.A., 1979. Groundwater. Prentice-Hall, Englewood Cliffs, 604 pp.

Gameil, M., Sadek, S., 2010. Unusual preservation of *Ophiomorpha* in middle Miocene rocks of Wadi Zablah, east Matruh area, northwestern Egypt. Acta Geol. Polon. 60, 39–44.

Gerard, J.R.F., Bromley, R.G., 2008. Ichnofabrics in Clastic Sediments—Application to Sedimentological Core Studies: A Practical Guide. Jean R.F. Gerard, Madrid, 97 pp.

Gingras, M.K., Pemberton, S.G., Mendoza, C.A., Henk, F., 1999. Assessing the anisotropic permeability of *Glossifungites* surfaces. Petrol. Geosci. 5, 349–357.

Gingras, M.K., MacMillan, B., Balcom, B.J., 2002. Visualizing the internal physical characteristics of carbonate sediments with magnetic resonance imaging and petrography. Bull. Can. Petrol. Geol. 50, 363–369.

Gingras, M.K., Pemberton, S.G., Henk, F., MacEachern, J.A., Mendoza, C., Rostron, B., O'Hare, R., Spila, M., Konhauser, K., 2007. Applications of ichnology to fluid and gas production in hydrocarbon reservoirs. In: MacEachern, J.A., Bann, K.L., Gingras, M.K., Pemberton, S.G. (Eds.), Applied Ichnology. SEPM Short Course Notes 52, pp. 131–145.

Gingras, M.K., Baniak, G., Gordon, J., Hovikoski, J., Konhauser, K.O., La Croix, A., Lemiski, R., Mendoza, C., Pemberton, S.G., Polo, C., Zonneveld, J.-P., 2012. Porosity and permeability in bioturbated sediments. In: Knaust, D., Bromley, R.G. (Eds.), Trace Fossils as Indicatorsof Sedimentary Environments. Developments in Sedimentology, vol. 64. Elsevier, Amsterdam, pp. 835–868.

Goldring, R., Cadée, G.C., Pollard, J.E., 2007. Climatic control of marine trace fossil distribution. In: Miller III, W. (Ed.), Trace Fossils. Concepts, Problems, Prospects. Elsevier, Amsterdam, pp. 159–171.

Halley, R.B., Shinn, E.A., Hudson, J.H., Lidz, B.H., 1977. Pleistocene barrier bar seaward of ooid shoal complex near Miami, Florida. AAPG Bull. 61, 519–526.

Hoffman, C.L., 2000. Evidence for Milankovitch orbital forcing in the Cretaceous Upper Glen Rose Formation of the East Texas Basin and the Fort Terrett Formation of the Central Texas Platform. Ph.D. Thesis, University of Texas, Austin, 324 pp.

Hoffmeister, J.E., Stockman, K.W., Multer, H.G., 1967. Miami Limestone of Florida and its Recent Bahamian counterpart. GSA Bull. 78, 175–190.

Hovorka, S.D., Dutton, A.R., Ruppel, S.C., Yeh, J.S., 1996. Edwards aquifer ground-water resources: geologic controls on porosity development in platform carbonates, South Texas. Bureau of Economic Geology, Austin, Report of Investigations 238, 75 pp.

Hovorka, S.D., Mace, R.E., Collins, E.W., 1998. Permeability structure of the Edwards aquifer, south Texas—implications for aquifer management. Bureau of Economic Geology, Austin, Report of Investigations 250, 55 pp.

Kerans, C., 2005. Field Trip Stop 4: Lower Glen Rose rudist bioherms, Pipe Creek, Bandera County. In: Filkorn, H.F., Johnson, C.C., Molineux, A., Scott, R.W. (Eds.), Abstracts and Post-Congress Field Guide: Seventh International Congress on Rudists. The University of Texas, Austin, pp. 156–166.

Kerans, C., Tinker, S.W., 1997. Sequence Stratigraphy and Characterization of Carbonate Reservoirs. SEPM Short Course Notes 40, 130 pp.

Klappa, C.F., 1980. Rhizoliths in terrestrial carbonates: classification, recognition, genesis and significance. Sedimentology 27, 613–629.

Knaust, D., 2009a. Characterization of a Campanian deep-sea fan system in the Norwegian Sea by means of ichnofacies. Mar. Petrol. Geol. 26, 1169–1211.

Knaust, D., 2009b. Ichnology as a tool in carbonate reservoir characterization: a case study from the Permian-Triassic Khuff Formation in the Middle East. GeoArabia 14, 17–38.

Knaust, D., Curran, H.A., Dronov, A.V., 2012. Shallow-marine carbonates. In: Knaust, D., Bromley, R.G. (Eds.), Trace Fossils as Indicators of Sedimentary Environments. In: Developments of Sedimentology, vol. 64. Elsevier, Amsterdam, pp. 707–752.

MacEachern, J.A., Bann, K.L., Pemberton, S.G., Gingras, M.K., 2007a. The ichnofacies paradigm: high-resolution paleoenvironmental interpretation of the rock record. In: MacEahern, J.A., Bann, K.L., Gingras, M.K., Pemberton, S.G. (Eds.), Applied Ichnology. SEPM Short Course Notes 52, pp. 27–64.

MacEachern, J.A., Pemberton, S.G., Bann, K.L., Gingras, M.K., 2007b. Departures from the archetypal ichnofacies: effective recognition of physico-chemical stresses in the rock record. In: MacEahern, J.A., Bann, K.L., Gingras, M.K., Pemberton, S.G. (Eds.), Applied Ichnology. SEPM Short Course Notes 52, pp. 65–93.

Maclay, R.W., Small, T.A., 1986. Carbonate geology and hydrology of the Edwards aquifer in the San Antonio area, Texas. Texas Water Development Board Report 296, 90 pp.

Mahler, B.J., Lynch, F.L., 1999. Muddy waters: temporal variation in sediment discharging from a karst spring. J. Hydrol. 214, 165–178.

McIlroy, D., 2004. Ichnofabrics and sedimentary facies of a tide-dominated delta: Jurassic Ile Formation of Kristin Field, Haltenbanken, offshore mid-Norway. In: McIlroy, D. (Ed.), The Application of Ichnology to Palaeoenvironmental and Stratigraphic Analysis. Geol. Soc., Spec. Publ., vol. 228, pp. 237–272.

Multer, H.G., Gischler, E., Lundberg, J., Simmons, K.R., Shinn, E.A., 2002. Key Largo Limestone revisited: Pleistocene shelf-edge facies, Florida Keys, USA. Facies 46, 229–272.

Palmer, A.N., 2007. Cave Geology. Cave Books, Dayton, 454 pp.

Parker, G.G., Ferguson, G.E., Love, S.K., 1955. Water resources of southeastern Florida. U.S. Geological Survey, Water-Supply Paper 1255, 965 pp.

Pemberton, S.G., Gingras, M.K., 2005. Classification and characterization of biogenically enhanced hydraulic conductivity. AAPG Bull. 89, 1493–1517.

Perkins, R.D., 1977. Depositional framework of Pleistocene rocks in south Florida. In: Enos, P., Perkins, R.D. (Eds.), Quaternary Sedimentation in South Florida. GSA Mem. 147, pp. 131–198.

Renken, R.A., Shapiro, A.M., Cunningham, K.J., Harvey, R.W., Metge, D.W., Zygnerski, M.R., Osborn, C.L., Wacker, M.A., Ryan, J.N., 2005. Assessing the vulnerability of a municipal well field to contamination in a karst aquifer. Environ. Eng. Geosci. 11, 341–354.

Renken, R.A., Cunningham, K.J., Shapiro, A.M., Harvey, R.W., Zygnerski, M.R., Metge, D.W., Wacker, M.A., 2008. Pathogen and chemical transport in the karst limestone of the Biscayne aquifer: 1. Revised conceptualization of groundwater flow. Water Resour. Res. 44, W08429. http://dx.doi.org/10.1029/2007WR006058.

Rose, P.R., 1972. Edwards Group, surface and subsurface, central Texas. Bureau of Economic Geology, Austin, Report of Investigations 74, 198 pp.

Shapiro, A.M., Renken, R.A., Harvey, R.W., Zygnerski, M., Metge, D.W., 2008. Pathogen and chemical transport in the karst limestone of the Biscayne aquifer: 2. Chemical retention from diffusion and slow advection. Water Resour. Res. 44, W08430. http://dx.doi.org/10.1029/2007WR00605.

Shinn, E.A., 1968. Burrowing in recent lime sediments of Florida and the Bahamas. J. Paleontol. 42, 879–894.

Shoemaker, W.B., Cunningham, K.J., Kuniansky, E.L., Dixon, J.F., 2008. Impacts of turbulence on hydraulic heads and parameter sensitivities in preferential groundwater flow layers. Water Resour. Res. 44, W03501. http://dx.doi.org/10.1029/2007WR006601.

Small, T.A., Hanson, J.A., Hauwert, N.M., 1996. Geologic framework and hydrogeologic characteristics of the Edwards aquifer outcrop (Barton Springs Segment), northeastern Hays and southwestern Travis Counties, Texas. U.S. Geol. Surv., Water-Resources Investigations Report 96-4306, 15 pp.

Spila, M.V., Pemberton, S.G., Rostron, B., Gingras, M.K., 2007. Biogenic textural heterogeneity, fluid flow, and hydrocarbon production: bioturbated facies Ben Nevis Formation, Hibernia Field, offshore Newfoundland. In: MacEachern, J.A., Bann, K.L., Gingras, M.K., Pemberton, S.G. (Eds.), Applied Ichnology. SEPM Short Course Notes 52, pp. 363–380.

Sukop, M.C., Thorne Jr., D.T., 2006. Lattice Boltzmann Modeling: An Introduction for Geoscientists and Engineers. Springer, New York, 172 pp.

Tonkin, N.S., McIlroy, D., Meyer, R., Moore-Turpin, A., 2010. Bioturbation influence on reservoir quality: a case study from the Cretaceous Ben Nevis Formation, Jeanne d'Arc Basin, offshore Newfoundland, Canada. AAPG Bull. 94, 1059–1078.

Vacher, H.L., Mylroie, J.E., 2002. Eogenetic karst from the perspective of an equivalent porous medium. Carb. Evap. 17, 182–196.

Young, K., 1986. The Pleistocene Terra Rossa of Central Texas. In: Abbott, P.L., Woodruff Jr., C.M. (Eds.), The Balcones Escarpment: Geology, Hydrology, Ecology, and Social Development in Central Texas. Field Trip Guidebook for the GSA Annual Meeting, San Antonio, pp. 63–73.

Epilogue

Trace fossils have attracted the interest of people since the Stone Age, although advanced models for the interpretation of trace fossils and sedimentary environments have been developed only during the second part of the twentieth century. Since then, much progress in the application of ichnological methods in sedimentology has been achieved, and ichnology as such has gained broad interest amongst geoscientists. Integrated sedimentological/ichnological analysis is a key driver in numerous case studies, but this requires a good understanding of existing classification schemes, concepts, and methods.

Today, an ethological classification is widely used for the interpretation of trace fossils, which is established on the basis of the inferred behavior of the tracemaker (such as resting, dwelling, locomotion, grazing, sediment-feeding, etc.). For many workers, this is the basis for an ichnological interpretation of paleoenvironments. However, a growing number of categories with gray transition zones have been realized, in addition to the fact that trace fossils can represent more than one mode of behavior. Therefore, the practicing sedimentologist/ichnologist may easily feel uncertain as to which ethology a particular trace fossil belongs to, and this, of course, directly impacts the ichnological value for sedimentological interpretations.

Therefore, a new approach is proposed, in which ichnogenera are classified according to their diagnostic features (e.g., orientation, branching, shape, fill, wall structure, etc.). Not only does this allow for an easy determination of the trace fossils found, but it further provides quick access to key reference data that can guide the interpretation. The enormous benefit of the knowledge of a particular tracemaker becomes obvious in the application of microbioerosion for paleoenvironmental interpretations. Future work will show if we succeed to extend this approach to other trace fossils and to interpret their producers at least at a higher taxonomic rank. Because this would allow for a better constrain of the trace-producing organism and the environments it is living in, this approach could be a big step forward in ichnological interpretations.

The ichnofacies and the ichnofabric concepts have received much attention among sedimentologists, even if their precise definition and applications often remain unclear. Nevertheless, ichnofacies assists in the identification of overall sedimentary environments, while ichnofabric analysis has a great value in revealing the interwoven nature of repeatedly bioturbated succession, both in well core and in outcrop data. Certain groups of trace fossils bear a great value for stratigraphical subdivisions. Following a carefully selected set of diagnostic features and well-defined ichnotaxa, in future work it may be possible to employ even more ichnotaxa in the solution of stratigraphic issues. Furthermore, sequence-stratigraphic surfaces are often delineated with particular trace-fossil suites, while sequences show significant trends in bioturbation, information which needs to be considered in applying interpretations.

It gives great pleasure to see the progress of implementing new methods and techniques in all fields of ichnology, which may be the biggest advantage for integrated studies. This is the case not only for the enhanced sampling and visualization methods but also for statistical analysis of large datasets and computer modeling. Likewise important are neoichnological studies and experiments, where a strong interdisciplinary link to biology is desired. Unfortunately, not all sedimentary environments received the same attention with this respect, and therefore, they are not equally well represented in this book.

Our journey through the ichnology of the major sedimentary environments aims to demonstrate the benefit of integrated ichnological studies in sedimentology. Each chapter has been carefully prepared by today's leading specialists in their field, and relevant case studies are intended to help illustrate significant characteristics. Different stages of research prohibit equal treatment of each environment, but hopefully, these overview chapters serve as the basis for further investigations, also by other research groups and with different subjects. Not only is this necessary for revealing the complexity in each environment, but also future innovative research will help in an improved definition of sedimentary environments, as well as the outlining of similarities and distinctions between them.

The facets of ichnology are many sided, and future work will come up with new ideas, concepts, and methods. This book aims for a comprehensive overview as to how trace fossils can be employed in interpreting sedimentary environments, whereas many other aspects are equally important but are only marginally or not dealt with herein. For instance, the ichnology of continental environments is not as equally developed as its marine counterpart (although ichnologists have called for "one ichnology") and partly requires different methods of investigation. This becomes obvious when dealing with vertebrate and arthropod trackways or in the study of paleosols including the complexity of insect and root traces. Coprolites are important for interpreting the diet, behavior, and environment of their producers, but more research is necessary to reveal such information from this fascinating group of trace fossils. Following initial discussions in this volume, burning topics include the marginal-marine/continental transition, brackish-water environments, marine *versus* lake carbonates, hyperpycnal-flow deposits, seep deposits and hybrid beds (turbidite/debrite couples), just to name a few. The discussion about the definition of trace fossils has not settled, and trace fossils in distinctive substrates (such as embedment structures or bioclaustration, plant/animal interactions, bones) certainly have their value in paleoecological interpretations. Finally, the evolutionary aspect of certain trace fossils has also attracted scientists for decades, and their refined definition will probably go hand in hand once ichnotaxa can be better assigned to their producers.

As part of many hydrocarbon reservoirs and aquifers, ichnofabrics more and more become the focus of reservoir characterization, although the limitations of core material call for an altered classification scheme of trace fossils. The effect of bioturbation on reservoir quality and connectivity has only recently started to be recognized, but more documentation, analysis, and modeling on the basis of various reservoir rocks is required for a better understanding of processes and products.

Dirk Knaust and Richard G. Bromley
May 13, 2012

Index

Note: Page numbers followed by "*f*" indicate figures, and "*t*" indicate tables.

B

C

D

E

F

G

H

I

J

K

M

N

O

P

S

Kolibri 360 GmbH
Ferdinand-Jühlke-Straße 7
99095 Erfurt, Deutschland
produktsicherheit@kolibri360.de